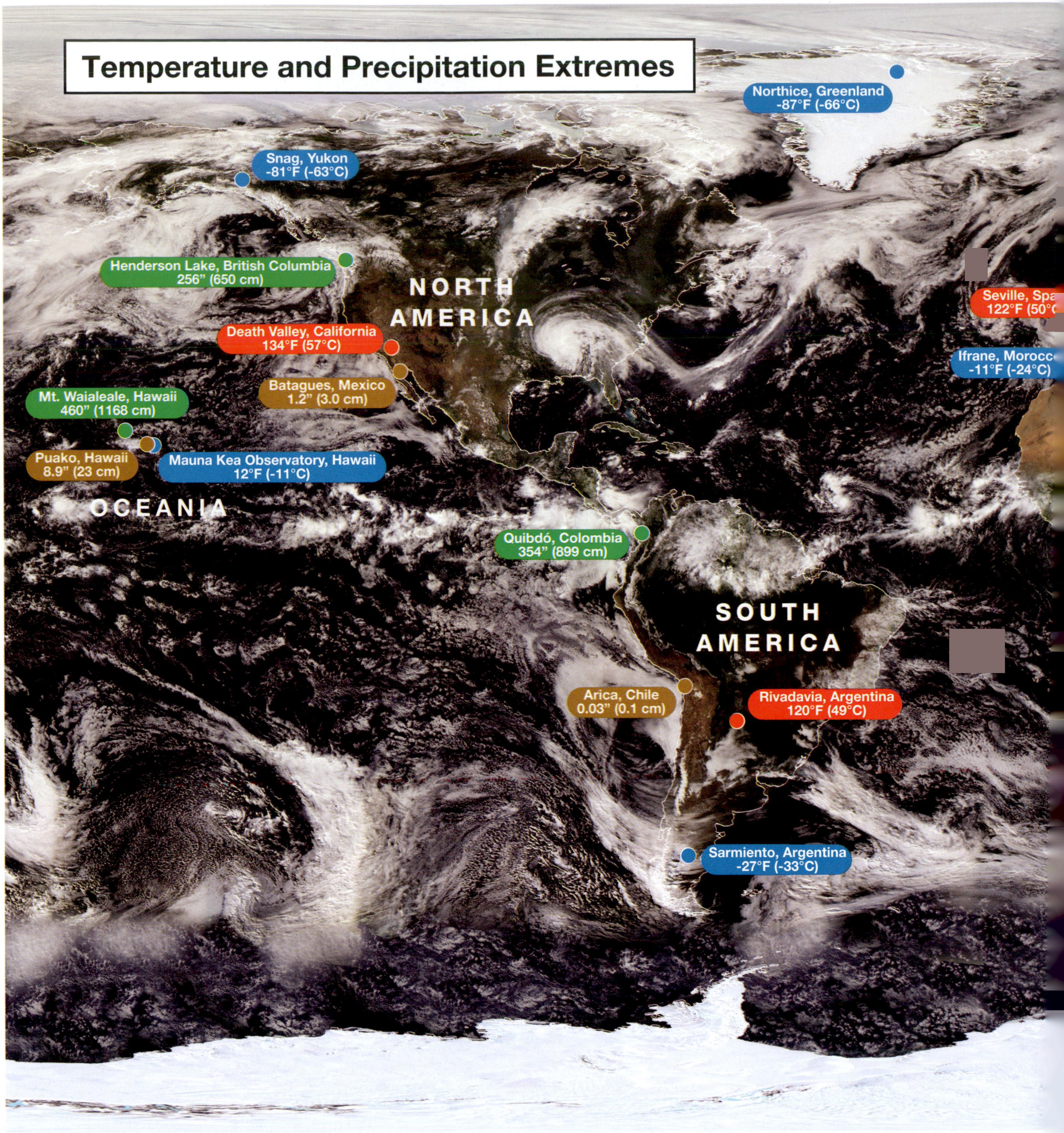


This photo-like view is based largely on observations from the Moderate Resolution Imaging Spectroradiometer (MODIS) on board NASAs *Terra* satellite.

Verkhoyansk, Russia
-90° (-68°C)
Ust'-Shchuger, Russia
-67°F (-55°C)
Oymyakon, Russia
-90°F (-68°C)
EUROPE
Astrakhan', Russia
6.4" (16.3 cm)
Crkvica, Bosnia & Herzegovina
183" (465 cm)
ASIA
Kebili, Tunisia
131°F (55°C)
Tirat Tsvi, Israel
129°F (54°C)
Mawsynram, India
467" (1187 cm)
Wadi Halfa, Sudan
<0.1" (<.03 cm)
Aden, Yemen
1.8" (4.6 cm)
Tuguegarao, Philippines
108°F (42°C)
AFRICA
OCEANIA
Debundscha, Cameroon
405" (1029 cm)
Bellenden Ker, Queensland
340" (864 cm)
Cloncurry, Queensland
128°F (53°C)
AUSTRALIA
Extremes
by Continent
Highest temperature
Lowest temperature
Highest precipitation
Lowest precipitation
Mulka, South Australia
4.05" (10.3 cm)
Charlotte Pass, New South Wales
-9°F (-23°C)
Vanda Station
59°F (15°C)
Vostok Station
-129°F (-89°C)
ANTARCTICA
ındsen-Scott South Pole Station
0.8" (2.0 cm)

# So Many Options for Your Meteorology Class!

Students today want options when it comes to their textbooks. *The Atmosphere* gives students the flexibility they desire, offering a wide range of formats for the book, and a large array of media and online learning resources. Find a version of the book that works best for YOU!

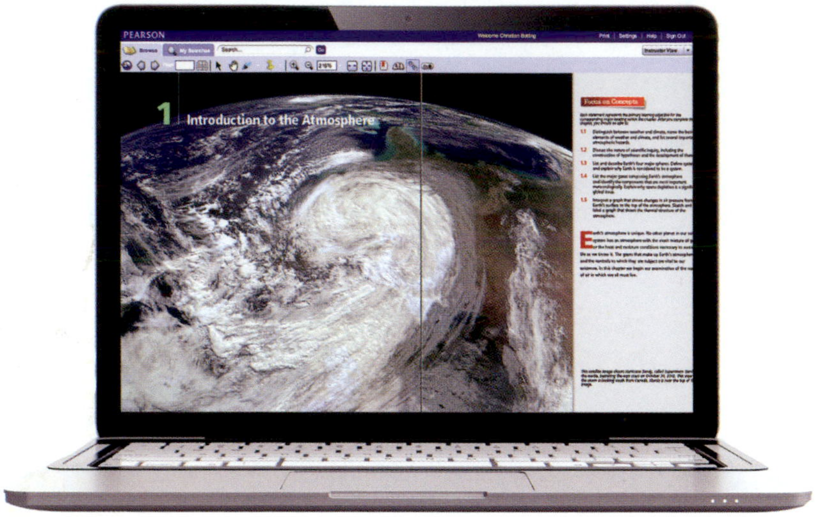


Whether it's on a laptop, tablet, smartphone, or other wired mobile devices, *The Atmosphere* lets students access media and other tools for learning meteorology.

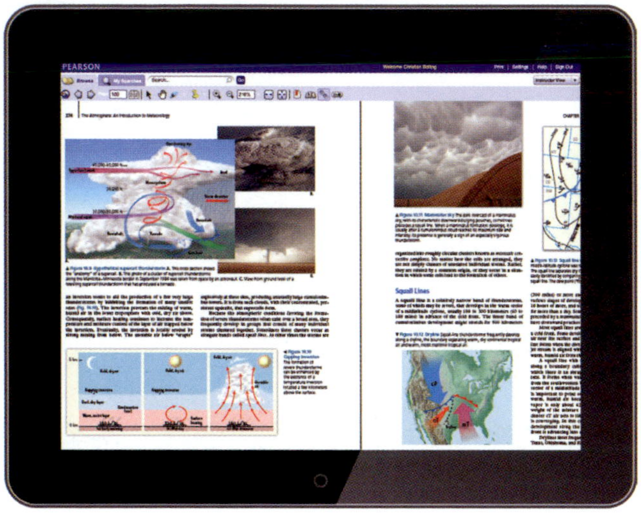


### *The Atmosphere* Plus MasteringMeteorology with eText
ISBN 0-321-98914-7 / 978-0-321-98914-7

Available at no additional charge with MasteringMeteorology, the Pearson eText version of *The Atmosphere*, 13th Edition, gives students access to the text whenever and wherever they are online. Features of Pearson eText:

- Now available on smartphones and tablets.
- Seamlessly integrated videos and other rich media.
- Fully accessible (screen-reader ready).
- Configurable reading settings, including resizable type and night reading mode.
- Instructor and student note-taking, highlighting, bookmarking, and search functionality.

### *The Atmosphere* CourseSmart eTextbook
ISBN 0-134-01601-7 / 978-0-134-01601-6

CourseSmart eTextbooks are an alternative to purchasing the print textbook, where students can subscribe to the same content online and save up the 40% off the suggested list price of the print text.

### *The Atmosphere* Books à La Carte
ISBN 0-321-98714-4 / 978-0-321-98714-3

Books à la Carte features the same exact content as *The Atmosphere* in a convenient, three-hole-punched, binder-ready, loose-leaf version. Books à la Carte offers a great value for students—this format costs 35% less than a new textbook package.

### Pearson Custom Library: You Create Your Perfect Text
http://www.pearsoncustomlibrary.com

*The Atmosphere* is available on the Pearson Custom Library, allowing instructors to create the perfect text for their courses. Select the chapters you need, in the sequence you want. Delete chapters you don't use: students pay only for the materials chosen.

### MasteringMeteorology Student Study Area
No matter the format, with each new copy of the text, students will receive full access to the Study Area in MasteringMeteorology™, providing a wealth of Animations, Videos, MapMaster™ Interactive Maps, GEODe Tutorials, *In the News* readings, Flashcards, practice quizzes, and much more.

# The Perfect Storm of Rich Media & Active Learning Tools

with

**The Atmosphere: An Introduction to Meteorology**

&

MasteringMeteorology™

# Real World Applications

The dynamic revision of *Atmosphere: An Introduction to Meteorology* incorporates the latest science, a new active-learning approach, integrated mobile media, and MasteringMeteorology™, the most complete, easy-to-use, engaging tutorial, media, and assessment platform available.

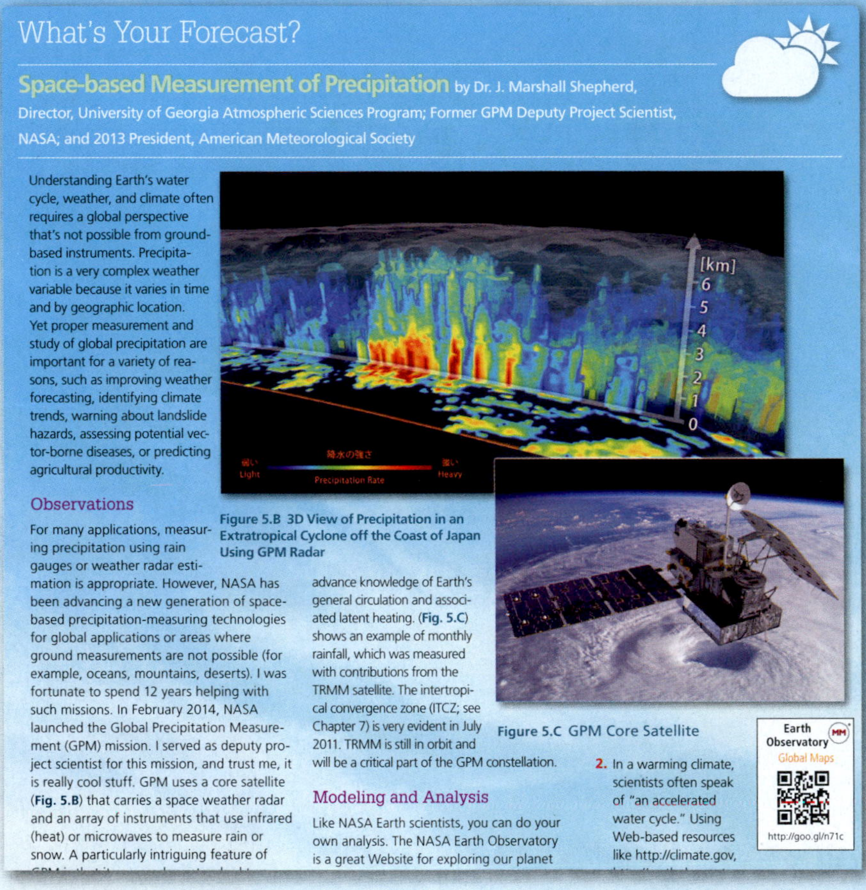
What's Your Forecast?

**Space-based Measurement of Precipitation** by Dr. J. Marshall Shepherd, Director, University of Georgia Atmospheric Sciences Program; Former GPM Deputy Project Scientist, NASA; and 2013 President, American Meteorological Society

Understanding Earth's water cycle, weather, and climate often requires a global perspective that's not possible from ground-based instruments. Precipitation is a very complex weather variable because it varies in time and by geographic location. Yet proper measurement and study of global precipitation are important for a variety of reasons, such as improving weather forecasting, identifying climate trends, warning about landslide hazards, assessing potential vector-borne diseases, or predicting agricultural productivity.



**Figure 5.B** 3D View of Precipitation in an Extratropical Cyclone off the Coast of Japan Using GPM Radar

Observations

For many applications, measuring precipitation using rain gauges or weather radar estimation is appropriate. However, NASA has been advancing a new generation of space-based precipitation-measuring technologies for global applications or areas where ground measurements are not possible (for example, oceans, mountains, deserts). I was fortunate to spend 12 years helping with such missions. In February 2014, NASA launched the Global Precipitation Measurement (GPM) mission. I served as deputy project scientist for this mission, and trust me, it is really cool stuff. GPM uses a core satellite (**Fig. 5.B**) that carries a space weather radar and an array of instruments that use infrared (heat) or microwaves to measure rain or snow. A particularly intriguing feature of

advance knowledge of Earth's general circulation and associated latent heating. (**Fig. 5.C**) shows an example of monthly rainfall, which was measured with contributions from the TRMM satellite. The intertropical convergence zone (ITCZ; see Chapter 7) is very evident in July 2011. TRMM is still in orbit and will be a critical part of the GPM constellation.

**Figure 5.C** GPM Core Satellite

Modeling and Analysis

Like NASA Earth scientists, you can do your own analysis. The NASA Earth Observatory is a great Website for exploring our planet

2. In a warming climate, scientists often speak of "an accelerated water cycle." Using Web-based resources like http://climate.gov,

Earth Observatory
Global Maps
http://goo.gl/n71c

## ◀ NEW! What's Your Forecast?

*What's Your Forecast?* features authored by expert contributors include active learning forecast activities tied to chapter content where students make predictions based on real-world scenarios and data.

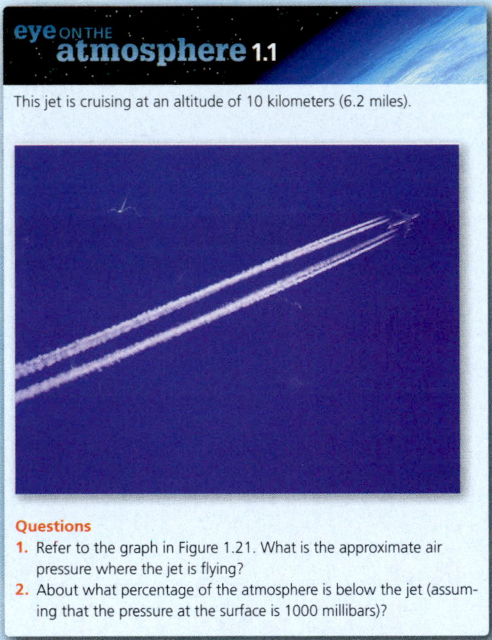
eye ON THE atmosphere 1.1

This jet is cruising at an altitude of 10 kilometers (6.2 miles).

Questions

1. Refer to the graph in Figure 1.21. What is the approximate air pressure where the jet is flying?
2. About what percentage of the atmosphere is below the jet (assuming that the pressure at the surface is 1000 millibars)?

eye ON THE atmosphere 4.1

Water is everywhere on Earth—in the oceans, glaciers, rivers, lakes, air, and living tissue. In addition, water can change from one state of matter to another at the temperatures and pressures experienced on Earth. Refer to this image, taken above the Grand Tetons, Wyoming, to answer the following questions.

Questions

1. What feature in this photo is composed of water in the liquid state?
2. Name the process by which ice changes directly from a solid to water vapor.
3. Identify where water vapor is found in this image.

## n the

ure boxes engage
sks, asking them to
nalysis, and make
ls of this course.

# Engaging & Empowering Students

## Severe and Hazardous Weather ▸

The text contains 15 *Severe and Hazardous Weather* features devoted to a broad variety of topics—heat waves, winter storms, floods, air pollution episodes, drought, wildfires, cold waves, and more.

severe& hazardous weather Box 5.1

**Worst Winter Weather**

Extremes, whether the *tallest* building or the *record low temperature* for a location, fascinate many humans. When it comes to weather, some places take pride in claiming to have the worst winters on record. In fact, both Fraser, Colorado, and International Falls, Minnesota, have proclaimed themselves the "ice box of the nation." Although Fraser recorded the lowest temperature for the 48 contiguous states 23 times in 1989, its neighbor Gunnison, Colorado, recorded the lowest temperature 62 times, far more than any other location.

Such facts do not impress the residents of Hibbing, Minnesota, where the temperature dropped to –38°C (–37°F) during the first week of March 1989. But this is mild stuff, say the old-timers in Parshall, North Dakota, where the temperature fell to –51°C (–60°F) on February 15, 1936. Not to be left out, Browning, Montana, holds the record for the most dramatic 24-hour temperature drop. Here the temperature plummeted 56°C (100°F), from a cool 7°C (44°F) to a frosty –49°C (–56°F) during a January evening in 1916.

Although impressive, the temperature extremes cited here represent only one aspect of winter weather.

Although impressive, the temperature extremes cited here represent only one aspect of winter weather. What about snowfall (**Fig. 5.A**)? Cooke City holds the seasonal snowfall record for Montana, with 1062 centimeters (418.1 inches) during the winter of 1977–1978. But what about cities like Sault Ste. Marie, Michigan, and Buffalo, New York? The winter snowfalls associated with the Great Lakes are legendary. Even larger snowfalls occur in many sparsely inhabited mountainous areas.

Try telling residents of the eastern United States that heavy snowfall alone makes for the worst weather. A blizzard in March 1993 produced heavy snowfall along with hurricane-force winds and

▲ **Figure 5.A** A winter snowstorm of historic proportions struck Chicago, Illinois, on February 2, 2011.

**Box 3.1** North America's Hottest and Coldest Places

Most people living in the United States have experienced temperatures of 38°C (100°F) or more. When statistics for the 50 states are examined for the past century or longer, we find that every state has a maximum temperature record of 38°C or higher. Even Alaska has recorded a temperature this high—set June 27, 1915, at Fort Yukon, a town along the Arctic Circle in the interior of the state.

**Maximum Temperature Records**

Surprisingly, the state that ties Alaska for the "lowest high" is Hawaii. Panala, on the south coast of the Big Island, recorded 38°C on April 27, 1931. Although humid tropical and subtropical places such as Hawaii are known for being warm throughout the year, they seldom experience maximum temperatures that surpass the low to mid-30s Celsius (90s Fahrenheit).

The highest accepted temperature record for the United States as well as the entire world is 57°C (134°F). This long-standing record was set at Death Valley, California, on July 10, 1913. Summer temperatures at Death Valley are consistently among the highest in the Western Hemisphere. During June, July, and August, temperatures exceeding 49°C (120°F) are to be expected. Fortunately, Death Valley has few human summertime residents (**Fig. 3.A**).

Why are summer temperatures at Death Valley so high? In addition to having the lowest elevation in the Western Hemisphere (53 meters [174 feet] below sea level), Death Valley is a desert. Although it is only about 300 kilometers (less than 200 miles) from the Pacific Ocean, mountains cut off the valley from the ocean's moderating influence and moisture. Clear skies allow a maximum of sunshine to strike the dry, barren surface. Because no energy is used to evaporate moisture, as occurs in humid regions, all the energy is available to heat the ground. In addition, subsiding air that warms by compression as it descends is also common to the region and contributes to its high maximum temperatures.

**Minimum Temperature Records**

The temperature controls that produce truly frigid temperatures are predictable and should come as no surprise. We expect extremely cold temperatures during winter in high-latitude places that lack the moderating influence of the ocean. Moreover, stations located on ice sheets and glaciers should be especially cold, as should stations positioned high in the mountains. All these criteria apply to Greenland's North Ice Station (elevation 2307 meters [7567 feet]). Here on January 9, 1954, the temperature plunged to –66°C (–87°F). If we exclude Greenland from consideration, Snag, in Canada's Yukon, holds the record for North America. This remote outpost experienced a temperature of –63°C (–81°F) on February 3, 1947. When only U.S. locations are considered, Prospect Creek, located north of the Arctic Circle in the Endicott Mountains of Alaska, came close to the North American record on January 23, 1971, when the temperature plunged to –62°C (–80°F). In the lower 48 states, the record of –57°C (–70°F) was set in the mountains at Rogers Pass, Montana, on January 20, 1954. Remember that many other places have no doubt experienced equally low or even lower temperatures; they just were not recorded.

FURNACE CREEK

129°F

190 feet (58 m) below sea level

▲ **Figure 3.A Almost a record!** On June 30, 2013, 100 years after Death Valley set the all-time high recorded temperature, it came close to equaling it. On that date, Death Valley's air temperature peaked at 54°C (129.2°F).

**Question**

1. Death Valley is not a great distance from the cool Pacific Ocean yet experiences very high temperatures. Why is there no moderating ocean influence?

## Students Sometimes Ask... ▾

*Students Sometimes Ask* features are integrated throughout the chapters, addressing high-interest topics and common student misconceptions.

students sometimes ask...

**If Earth's atmosphere had no greenhouse gases, what would surface-air temperatures be like?**

Cold! Earth's average surface temperature would be a chilly –18°C (–0.4°F) instead of the relatively comfortable 14.5°C (58°F) that it is today.

## Special Feature Box ▴

Special Feature Boxes throughout the chapters present compelling case studies or further illuminate interesting concepts discussed.

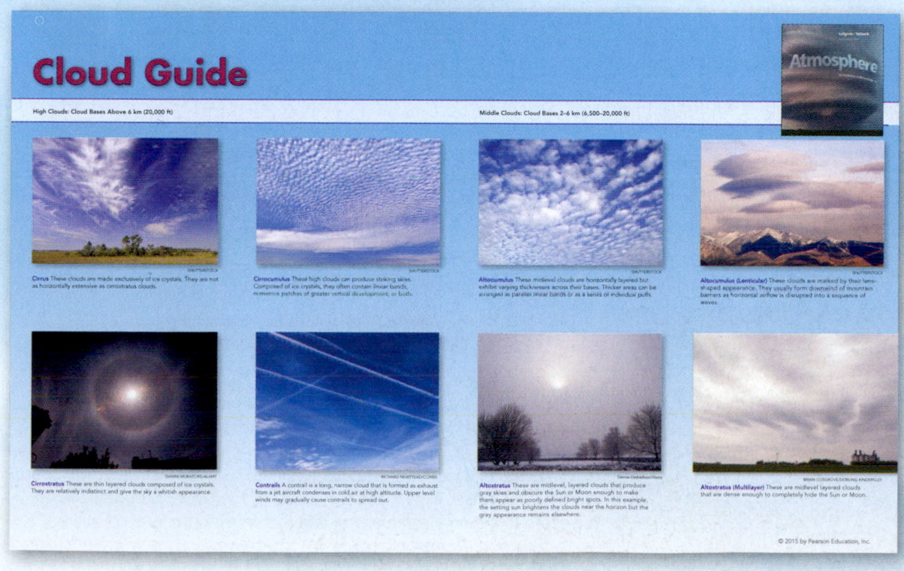


## ◂ Cloud Guide

A foldout cloud guide at the back of the book provides students with a tool and reference for real-world observation.

# Global Climate Change

This new edition features extended coverage of global climate change and includes the findings of the IPCC's 5th Assessment Report.

## ◂ Our Changing Climate

The latest data and applications related to global climate change are presented throughout the 13th edition, for the most current and comprehensive coverage.

### Thunderstorms and Climate Change

In the preceding discussion, you learned that the occurrence of thunderstorms varies seasonally and from place to place. Thunderstorm activity will likely increase in many areas in the years to come due to global climate change. Global temperatures have been warming for decades largely due to human activities that are altering the composition of the atmosphere. This trend is expected to continue for the foreseeable future. A thorough discussion of this phenomenon appears in Chapter 14.

**Video** (MM) Thundersleet and Thundersnow

http://goo.gl/GOm7IV

(MM) **MapMaster** ▸ North America Physical Environment ▸ Thunderstorm Occurrence Per Year

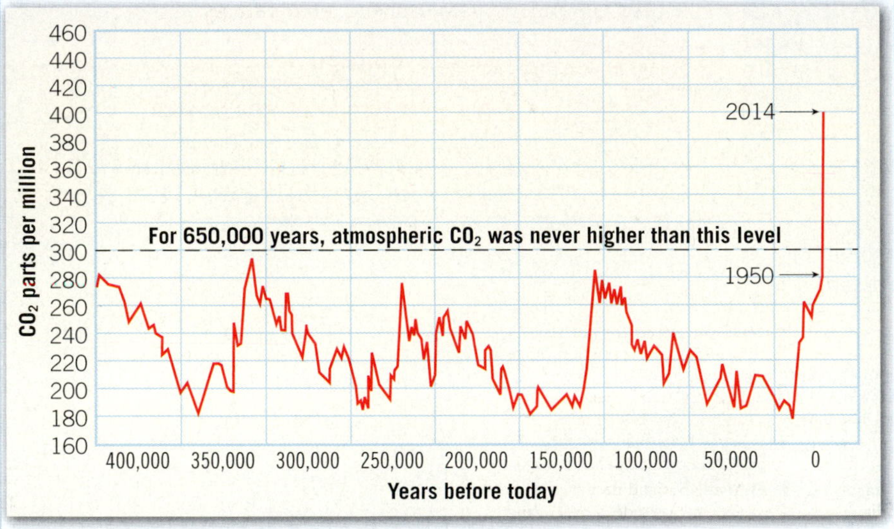


▲ **Figure 14.22 $CO_2$ concentrations over the past 400,000 years** Most of these data come from analyses of air bubbles trapped in ice cores. The record since 1958 comes from direct measurements at Mauna Loa Observatory, Hawaii. The rapid increase in $CO_2$ concentrations since the onset of the Industrial Revolution is obvious. (NOAA)

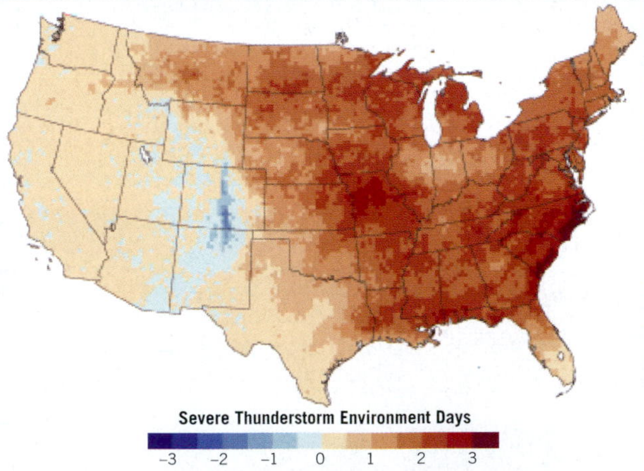


▲ **Figure 10.4 Future thunderstorm activity** This map shows changes in the number of days per year when the environmental conditions that promote severe thunderstorm activity occur. The map is based on a climate model comparing summer climate during 2072–2099 with a similar span during 1962–1989. Most of the area east of the Rocky Mountains is projected to experience an increase in these environmental conditions.

## Projected Impacts of Climate Change ▴

Integrated coverage of the findings and data of the 2013-2014 IPCC 5th Assessment Report are presented throughout the chapters, including discussion of possible future scenarios for Earth's climate.

# Structured Learning

The 13th Edition provides an active structured learning path to help guide students toward mastery of key meteorological concepts.

## Focus on Concepts

*Each statement represents the primary learning objective for the corresponding major heading within the chapter. After you complete the chapter, you should be able to:*

**3.1** Calculate five commonly used types of temperature data and interpret a map that depicts temperature data using isotherms.

**3.2** Name the principal controls of temperature and use examples to describe their effects.

**3.3** Interpret the patterns depicted on world temperature maps.

**3.4** Discuss the basic daily and annual cycles of air temperature.

**3.5** Explain how different types of thermometers work and why the placement of thermometers is an important factor in obtaining accurate readings. Distinguish among Fahrenheit, Celsius, and Kelvin temperature scales.

**3.6** Summarize several applications of temperature data.

## ◂ UPDATED! Focus on Concepts

Focus on Concepts learning goals are listed in the chapter-opening spreads and correlate to Concept Check and GIST questions to help students focus on and prioritize the learning goals for each chapter.

## ✔ Concept Checks 14.3

1. Why has the $CO_2$ level of the atmosphere been increasing over the past 200 years?
2. How has the atmosphere responded to the growing $CO_2$ levels? How are temperatures in the lower atmosphere likely to change as $CO_2$ levels continue to increase?
3. Aside from $CO_2$, what trace gases are contributing to global temperature change?
4. List the main sources of human-generated aerosols and describe their net effect on atmospheric temperatures.

## UPDATED! Give It Some Thought ▾

Give It Some Thought (GIST) questions are found at the end of each chapter and ask students to use higher-level thinking. They often involve chapter visuals, which help students apply and synthesize entire chapter concepts.

## UPDATED! Concept Checks ▴

Concept Check questions are integrated throughout each chapter. These serve as conceptual speed bumps, asking students to assess their understanding as they are reading.

## Give it Some Thought

1. If you were asked to identify the coldest city in the United States (or any other designated region), what statistics could you use? Can you list at least three different ways of selecting the coldest city?
2. The accompanying graph shows monthly high temperatures for Urbana, Illinois, and San Francisco, California. Although both cities are located at about the same latitude, the temperatures they experience are quite different. Which line on the graph represents Urbana, and which represents San Francisco? How did you figure this out?

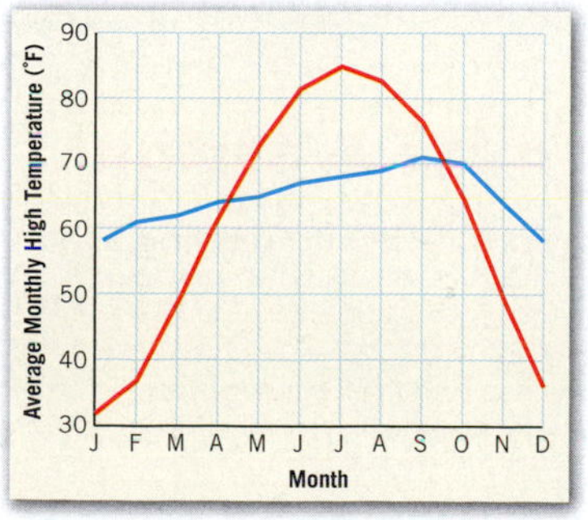


3. On which summer day would you expect the greatest temperature range? Which would have the smallest range in temperature? Explain your choices.
   - **a.** Cloudy skies during the day and clear skies at night
   - **b.** Clear skies during the day and cloudy skies at night
   - **c.** Clear skies during the day and clear skies at night
   - **d.** Cloudy skies during the day and cloudy skies at night
4. The accompanying scene shows an island near the equator in the Indian Ocean. Describe how latitude, altitude, and the differential heating of land and water influence the climate of this place.

# Continuous Learning Before, During & After Class with MasteringMeteorology

# BEFORE CLASS

## Mobile Media and Reading Assignments Ensure Students Come to Class Prepared

### NEW! Mobile-Enabled Videos and Animations (QR)▸

Mobile-Enabled Quick Response (QR) codes integrated throughout chapter sections empower students to use their mobile devices for learning as they read, providing instant access to over 130 SmartFigures, Videos, and Animations of real-world atmospheric phenomena and visualizations of key physical processes. These media can be assigned with quizzes in MasteringMeteorology.

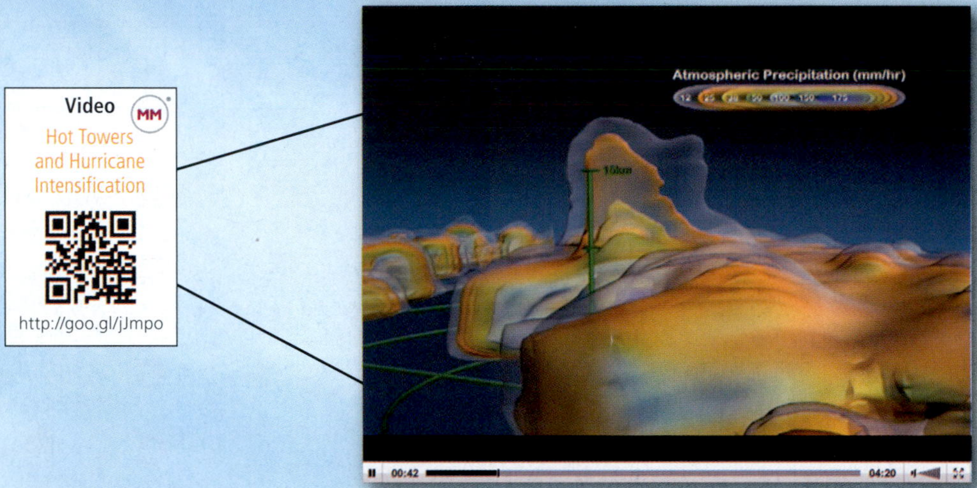


### NEW! Dynamic Study Modules▸

Dynamic Study Modules personalize each student's learning experience. Created to allow students to acquire knowledge on their own and be better prepared for class discussions and assessments, this mobile app is available for iOS and Android devices.

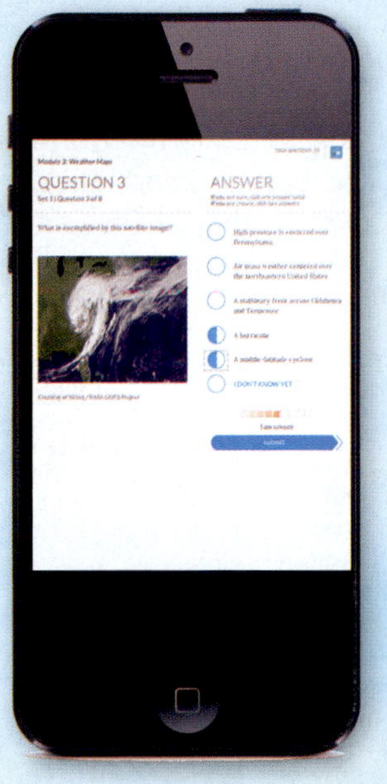


### eText▸

Pearson eText in MasteringMeteorology gives students access to the text whenever and wherever they can access the internet. Users can create notes, highlight text, create bookmarks, zoom and click hyperlinked words, phrases or media to view definitions, websites or view Pearson videos and animations.

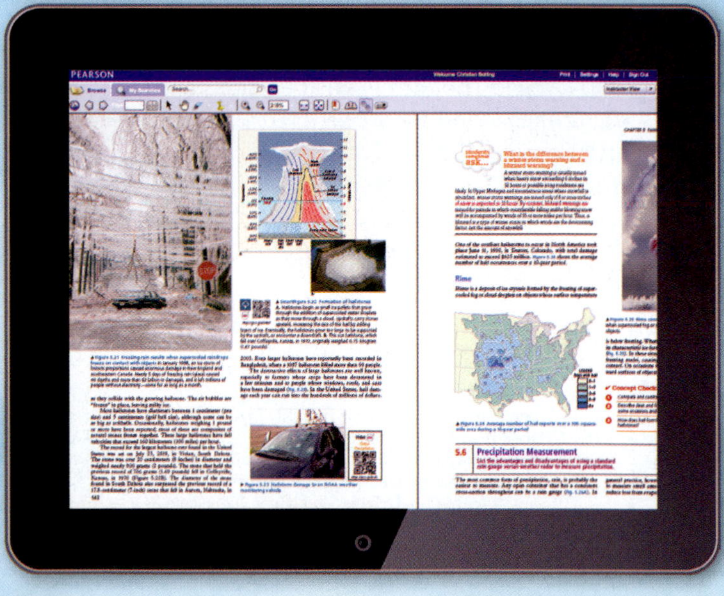


## Pre-Lecture Reading Quizzes are easy to customize & assign.

**NEW!** Reading Questions ensure that students complete the assigned reading before class and stay on track with reading assignments. Reading Questions are 100% mobile ready and can be completed by students on mobile devices.

# DURING CLASS

## Learning Catalytics

**"My students are so busy and engaged answering Learning Catalytics questions during lecture that they don't have time for Facebook"**
**Declan De Paor, Old Dominion University**

What has Professors and Students excited? Learning Cataltyics, a 'bring your own device' student engagement, assessment, and classroom intelligence system, allows students to use their smartphone, tablet, or laptop to respond to questions in class. With Learning Cataltyics, you can:

- Assess students in real-time using open ended question formats to uncover student misconceptions and adjust lecture accordingly.
- Automatically create groups for peer instruction based on student response patterns, to optimize discussion productivity.

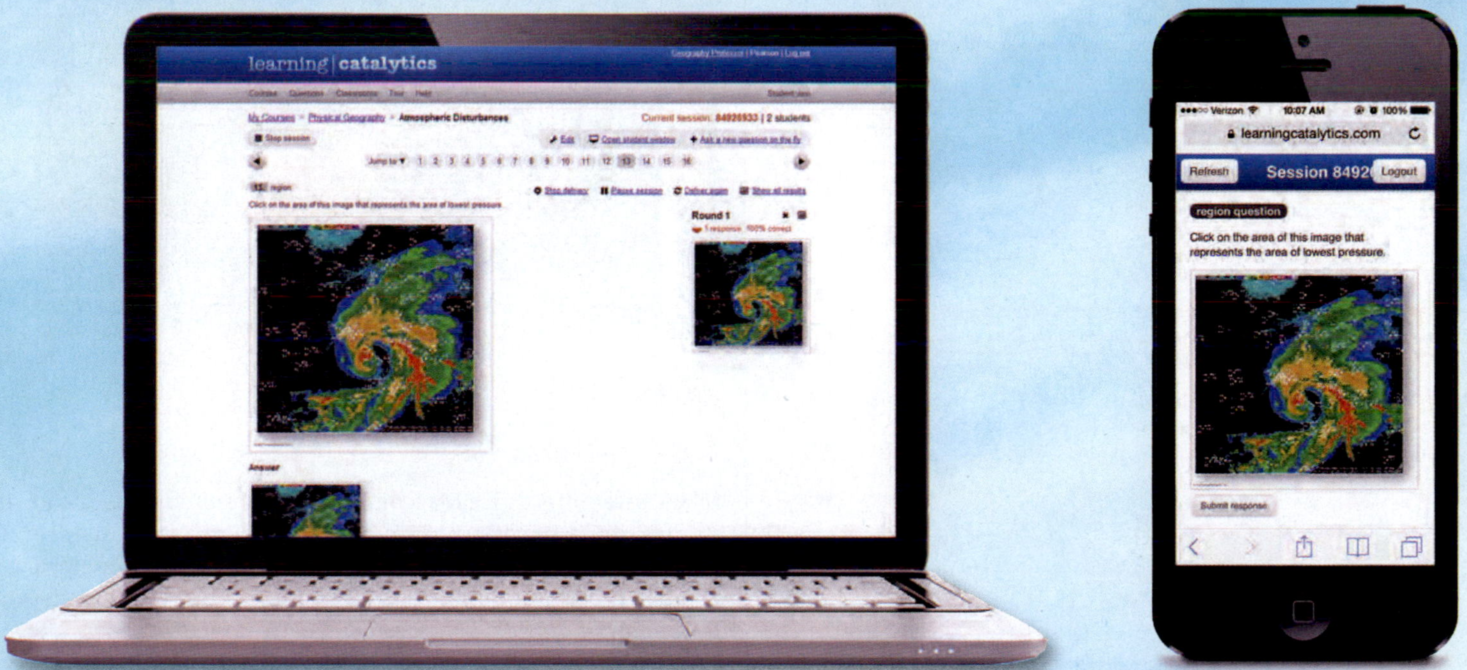


## Enrich Lecture with Dynamic Media

Teachers can incorporate dynamic media into lecture, such as Geoscience Animations, Videos, and MapMaster Interactive Maps.

# MasteringMeteorology™

# AFTER CLASS

## Easy to Assign, Customizable, Media-Rich, and Automatically Graded Assignments

The breadth and depth of media content available in MasteringMeteorology is unparalleled, allowing teachers to quickly and easily assign homework to reinforce key concepts. Most media activities are supported by automatically-graded multiple choice quizzes with hints and specific wrong answer feedback that helps coach students towards mastery of the concepts.

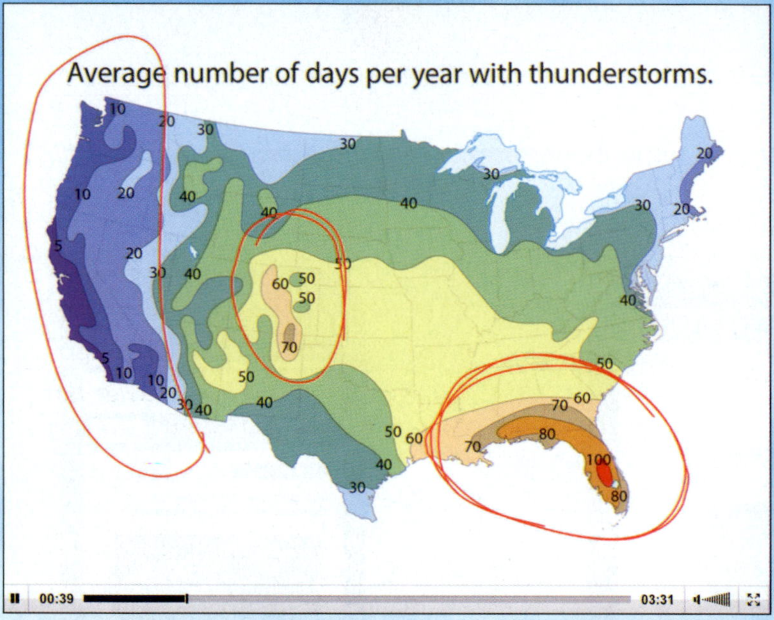


## Encounter Activities ▾

Encounter Activities provide rich, interactive Google Earth explorations of meteorology concepts to visualize and explore Earth's physical landscape and atmospheric processes.

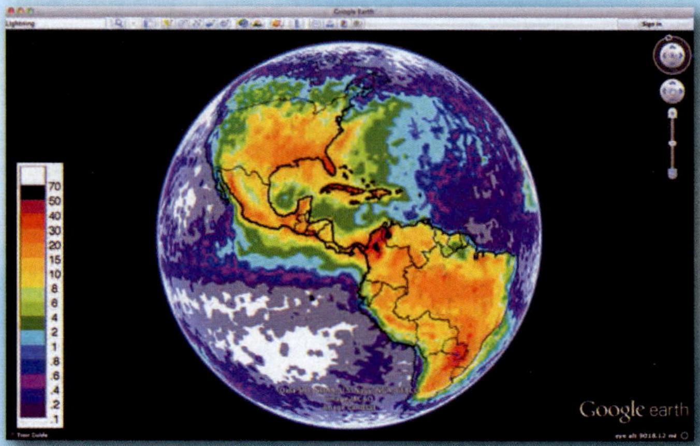


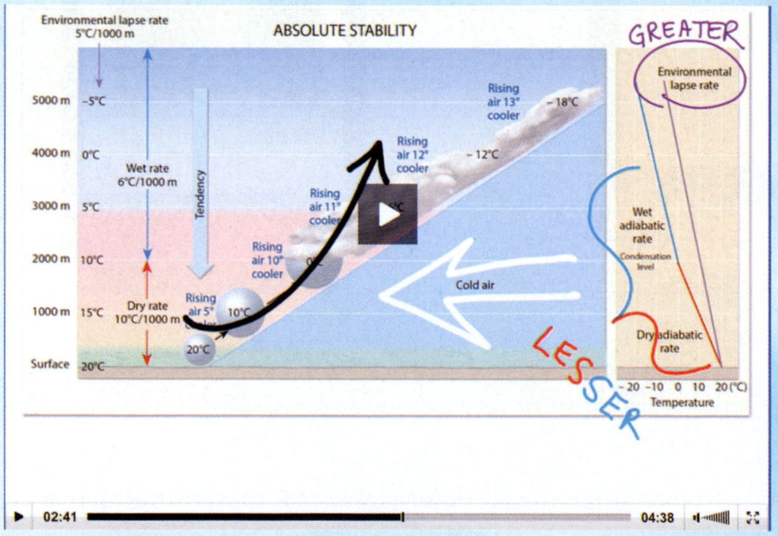


## SmartFigures ▴

SmartFigures are brief, narrated video lessons that examine and explain concepts illustrated by key figures within the text. Students access SmartFigures on their mobile devices by scanning Quick Response (QR) codes next to key figures. These media are also available in the Study Area of MasteringMeteorology and teachers can assign them with automatically-graded quizzes.

## MapMaster Interactive Map Activities ▾

MapMaster Interactive Map Activities are inspired by GIS, allowing students to layer various thematic maps to analyze spatial patterns and data at regional and global scales. This tool includes zoom and annotation functionality, with hundreds of map layers leveraging recent data from sources such as NOAA, NASA, USGS, United Nations, and the CIA.

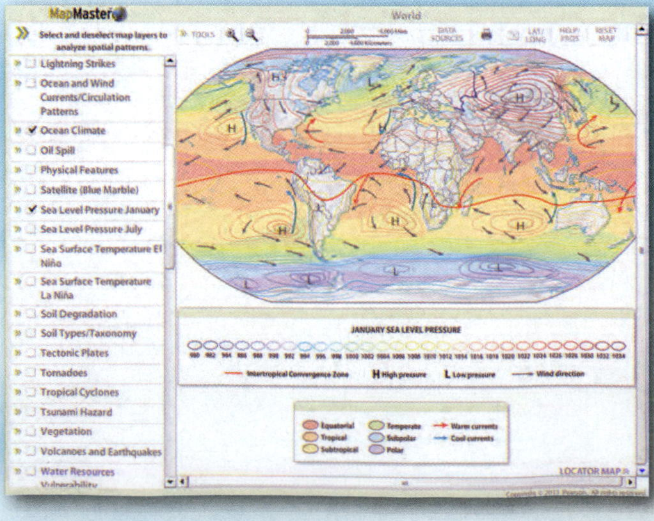

# MasteringMeteorology™

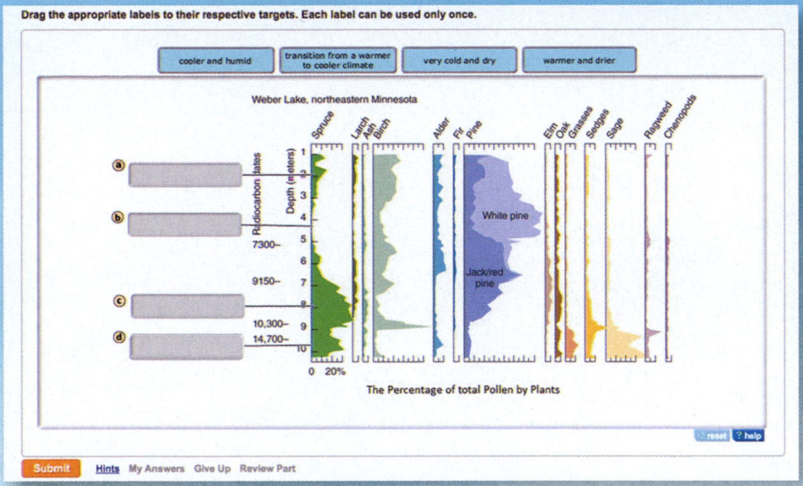


## NEW! GeoTutors▲

These coaching activities help students master the toughest physical geoscience concepts with highly visual, kinesthetic activities focused on critical thinking and application of core geoscience concepts.

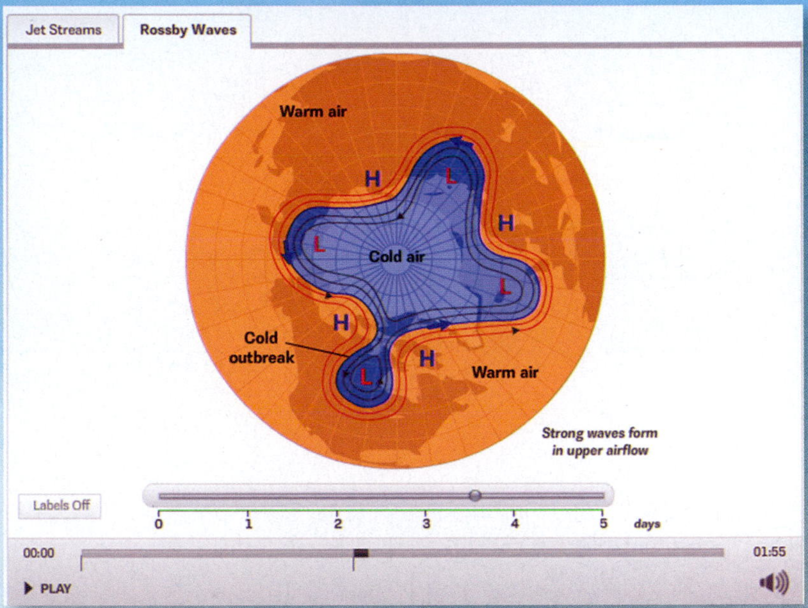


## Geoscience Animations Activities▲

Geoscience Animation Activities help students visualize the most challenging physical processes in the physical geosciences with schematic animations that include audio narration.

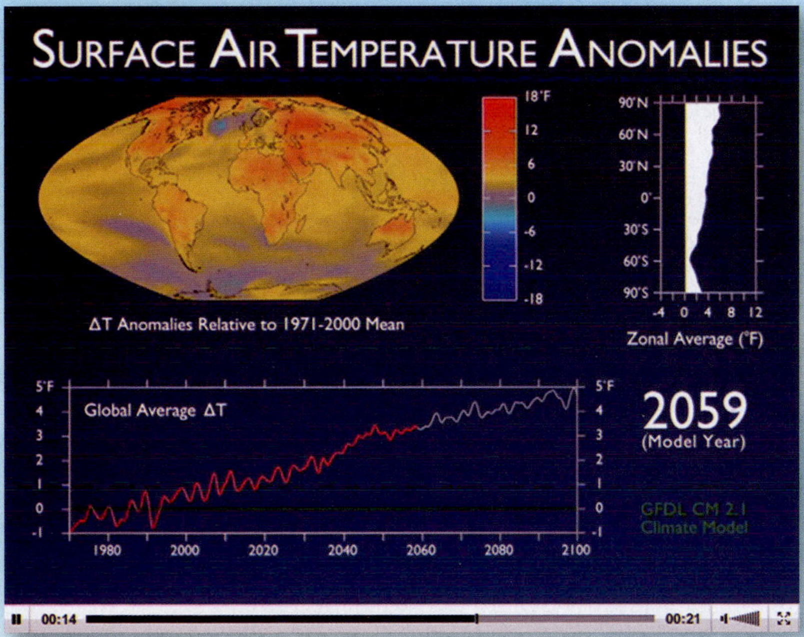


## NEW! Videos▲

Videos provide students with real-world case studies of atmospheric phenomena and engaging visualizations of critical data.

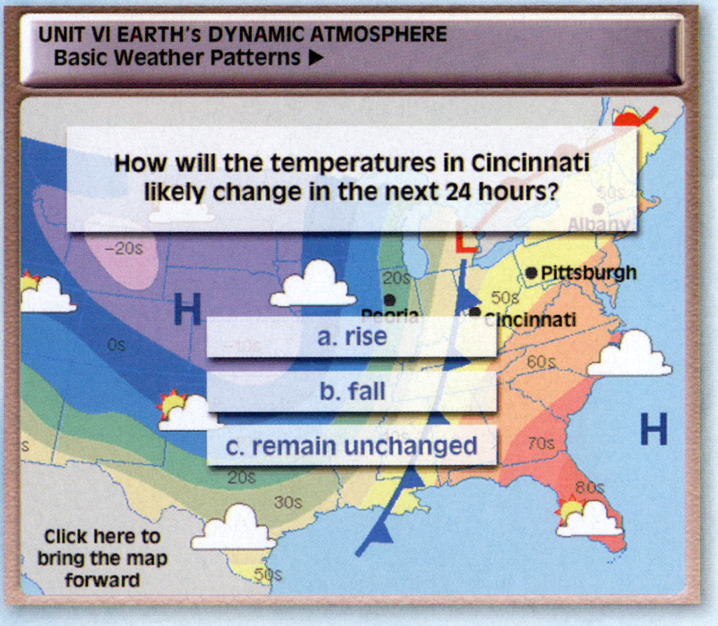


## GEODe: Atmosphere▲

GEODe: Atmosphere is a dynamic program that reinforces key meteorological concepts through animations, tutorials, interactive exercises, and review quizzes.

# THE Atmosphere 13e

## An Introduction to Meteorology

**Frederick K. Lutgens**

**Edward J. Tarbuck**

Illustrated by Dennis Tasa

PEARSON

Senior Meteorology Editor: Christian Botting
Executive Marketing Manager: Neena Bali
Program Manager: Anton Yakovlev
Project Manager: Crissy Dudonis
Editorial Assistant: Amy De Genaro
Director of Development: Jennifer Hart
Development Editor: Veronica Jurgena
Senior Project Manager, Text and Images: Rachel Youdelman
Text Permissions Specialist: Tom Wilcox, Lumina Datamatics
Program Management Team Lead: Kristen Flathman
Project Management Team Lead: David Zielonka
Production Management/Composition: Heidi Allgair/Cenveo® Publisher Services
Design Manager: Mark Ong
Interior and Cover Designer: Jeanne Calabrese
Photo and Illustration Support: International Mapping
Photo Researcher: Kristin Piljay
Operations Specialist: Maura Zaldivar-Garcia
Cover Photo Credit: Supercell moves across the country near West Point Nebraska. Mike Hollingshead/Corbis

Credits and acknowledgments for materials borrowed from other sources and reproduced, with permission, in this textbook appear on the appropriate page within the text or on p. C-1.

---



Library of Congress Cataloging-in-Publication Data
Lutgens, Frederick K.
The atmosphere : an introduction to meteorology / Frederick K. Lutgens, Edward J. Tarbuck ; illustrated by Dennis Tasa Pearson. — Thirteenth edition.
pages cm
Includes index.
ISBN 978-0-321-98462-3 — ISBN 0-321-98462-5
1. Atmosphere. 2. Meteorology. 3. Weather. I. Tarbuck, Edward J. II. Title.
QC861.2.L87 2014
551.5—dc23
2014031628

2 3 4 5 6 7 8 9 10—**V382**—16 15

Student edition ISBN 10: 0-321-98462-5; ISBN 13: 978-0-321-98462-3
Instructor's Review Copy ISBN 10: 0-321-98753-5; ISBN 13: 978-0-321-98753-2

PEARSON

**www.pearsonhighered.com**

# Brief Contents

# MasteringMeteorology™ Media

## 1 Introduction to the Atmosphere

**SmartFigures**
1.18 The Mauna Loa $CO_2$ Record
1.20 The Ozone Hole

**Geoscience Animations**
Ozone Depletion

**Videos**
Ozone Hole
Global Carbon Uptake by Plants
Global Changes in Carbon Dioxide Concentrations

**GEODe**
The Importance of Weather
Weather and Climate
Composition of the Atmosphere
Extent of the Atmosphere
Temperature Structure of the Atmosphere
In the Lab: Reading Weather Maps
The Influence of Volcanic Ash

## 2 Heating Earth's Surface and Atmosphere

**SmartFigures**
2.6 Solstices & Equinoxes
2.11 Three Mechanisms of Heat Transfer
2.15 Solar Radiation Paths
2.22 Three Planets, Three Climates

**Geoscience Animations**
Earth-Sun Relations
Atmospheric Energy Balance

**Videos**
Seasonal Changes in Global Snow Cover
July Global Movie
January Global Movie
Global Variations in Insolation Through the Year
Net Radiation at the Top of the Atmosphere
Solar Eclipse
Tour of the Electromagnetic Spectrum
The Sun in Ultraviolet
Solar Power

**GEODe**
Understanding Seasons, Part 1
Understanding Seasons, Part 2
Solar Radiation
What Happens to Incoming Solar Radiation
The Greenhouse Effect
In the Lab: The Influence of Color on Albedo

## 3 Temperature

**SmartFigures**
3.2 Isotherm Maps
3.5 Maritime Temperatures
3.14 Cloudy vs. Clear Days
3.18 January vs. July

**Geoscience Animations**
The Gulf Stream
Ocean Circulation Patterns

**Videos**
Global Sea-Surf Temperatures—Climatology
Seasonal Changes in Land Surface Temperatures
Heavy Convection over Florida
Urban Heat Islands
Temperatures and Agriculture

**GEODe**
Basic Temperature Data
Controls of Temperature

## 4 Moisture and Atmospheric Stability

**SmartFigures**
4.1 The Water Cycle
4.3 Changes of State in Water
4.13 Dew Point
4.20 Frontal Wedging and Convergence
4.24 Atmospheric Stability

**Geoscience Animations**
Earth's Water and Hydrologic Cycle
Water Phase Changes
Atmospheric Stability

**Videos**
Hydrologic Cycle
Global Evaporation Rates
January Water Vapor
July Water Vapor
Forecasting Relative Humidity
Gravity Wave Clouds

**GEODe**
Water's Changes of State
Humidity: Water Vapor in the Air
The Basics of Cloud Formation: Adiabatic Cooling
Processes That Lift Air
The Critical Weathermaker: Atmospheric Stability
In the Lab: Atmospheric Stability

## 5 Forms of Condensation and Precipitation

**SmartFigures**
5.1 Types of Clouds
5.12 Fog
5.22 Hail

**Videos**
Identifying Clouds in Satellite Imagery
Clouds Developing Over Florida
Is That a Cloud?
Clouds and Aviation
A Satellite View of Fog
The Importance of Wind Resistance
Global Precipitation
Record-Breaking Hailstorm as Seen by Radar
Earth Observatory-Global Maps

**MapMaster**
Days with Heavy Fog

**GEODe**
Classifying Clouds
Types of Fog
How Precipitation Forms
Forms of Precipitation

## 6 Air Pressure and Winds

**SmartFigures**
6.2 Air Pressure
6.14 The Coriolis Effect

**Geoscience Animations**
The Coriolis Effect
Wind Pattern Development
Cyclones and Anticyclones

**Videos**
The Coroliolis Effect on a Merry-Go-Round
Winds During a Drought
Hurricane Winds
The Growth of Wind Power in the U.S.
Forecasting Wind Patterns

**GEODe**
Measuring Air Pressure
Factors Affecting Wind
Highs and Lows

## 7 Circulation of the Atmosphere

**SmartFigures**
7.2 Local Winds

7.7 Global Circulation
7.21 Gyres

**Geoscience Animations**

Global Wind Patterns with Hadley Cell
Global Wind Patterns
Jet Stream and Rossby Waves
Ocean Circulation
Ekman Spiral Coastal Upwelling/Downwelling
El Niño and La Niña
Seasonal Pressure and Precipitation Patterns

**Videos**

Global Fire Patterns
Black Carbon Aerosols Trace Global Winds
El Niño
La Niña
Floods and Droughts

**MapMaster**

La Niña
El Niño

## 8 Air Masses

**Videos**

Radar Reflectivity and Air Masses
Effects of the 2011 Groundhog Day Blizzard
An Infrared View of the 2011 Groundhog Day Blizzard
Lake Effect Snow

**GEODe**

Air Masses

## 9 Midlatitude Cyclones

**SmartFigures**

9.3 Fronts

**Geoscience Animations**

Warm Fronts
Cold Fronts
Midlatitude Cyclones

**Videos**

A Midlatitude Cyclone's Dry Slot
A Midlatitude Cyclone's Effect on Society
Tornadoes Ahead of a Cold Front
Hurricanes and Air Masses
A Midlatitude Cyclone's Effect on Society
Water Vapor Transport by Midlatitude Cyclones
Short Waves and Long Waves
Winds During the Floods of 1993
A Midlatitude Cyclone's Dry Slot

**GEODe**

Fronts
Introducing Middle-Latitude Cyclones
In the Lab: Examining a Middle-Latitude Cyclone

## 10 Thunderstorms and Tornadoes

**SmartFigures**

10.05 Thunderstorms
10.22 Mesocyclones and Tornadoes

**Videos**

Forecasting Thunderstorms
Tornado Wind Patterns
The Deadliest Tornado Since Modern Recordkeeping Began
Identifying Tornadic Thunderstorms Using Radar Velocity Data
Identifying Tornadic Thunderstorms Using Radar Reflectivity Data

**MapMaster**

Lightning Strikes
Thunderstorm Occurrence Per Year
Tornado Incidence/Tornado Alley

## 11 Hurricanes

**Geoscience Animations**

Hurricane Wind Patterns

**Videos**

Hurricane Eye Wall
A Hurricane in the Middle Latitudes
Hurricane Katrina
Hot Towers and Hurricane Intensification
Improving Hurricane Predictions
The Making of a Super Storm

## 12 Weather Analysis and Forecasting

**Videos**

Uncertainty in Numerical Model
Modeling the Atmosphere on Your Desktop
Forecasting Thunderstorms
Forecasting Precipitation

## 13 Air Pollution

**SmartFigures**

13.7 U.S. Energy Consumption

**Videos**

Global Carbon Monoxide Concentrations
Smog Bloggers
Hello Crud

## 14 The Changing Climate

**SmartFigures**

14.9 The Climate Record in Glacial Ice
14.17 Orbital Forcing of Ice Ages
14.32 Shoreline Shift

**Geoscience Animations**

Orbital Variations and Climate Change
Global Warming

**Videos**

20,000 Years of Pine Pollen
Climate, Crops, and Bees
Taking Earth's Temperature
Climate Change Through Native Alaskan Eyes
Sea Level Rise
Retreat of Continental Ice Sheets

## 15 World Climates

**SmartFigures**

15.6 Tropical Climates
15.11 Deserts
15.25 Highland Climates

**Videos**

Diurnal Variability in Global Precipitation
Supercomputing the Climate
Lightning Seasonality
Studying Fires Using Multiple Sensors
Operation IceBridge in Greenland

**MapMaster**

Physical Environment
Climate: Tropical Wet and Dry
Climate: Humid Subtropical
Climate: Marine West Coast
Climate: Dry Summer Subtropical
Climate: Humid Continental
Climate: Ice Cap
Climate: Highland

# Contents

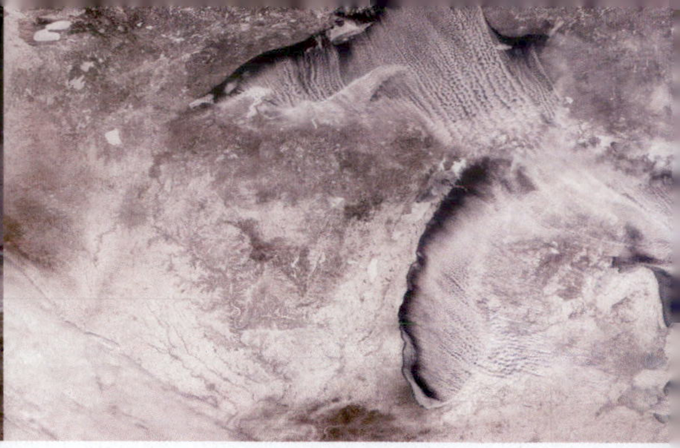

# Preface

The thirteenth edition of *The Atmosphere* is a college-level text for students taking their first and perhaps only course in meteorology. The text is intended to be a meaningful, nontechnical survey of atmospheric phenomena and weather forecasting to help students understand a field that impacts their daily lives. Our goal is for *The Atmosphere* to provide students with an informative, current, highly readable, text that is an engaging and usable tool for learning the basic principles and concepts of meteorology.

## New to This Edition

The 13th edition represents perhaps the *most extensive and thorough revision* in the long history of this textbook.

- **MasteringMeteorology™** delivers engaging, dynamic learning opportunities—focused on course objectives and responsive to each student's progress—that are proven to help students absorb course material and understand difficult concepts. Assignable activities in MasteringMeteorology include GIS-inspired *MapMaster Interactive Maps*, *Encounter Meteorology* Explorations using Google Earth™, *SmartFigure* activities, *GeoTutor*s on the most challenging topics in the geosciences, *Geoscience Animations*, *GEODe* tutorials, and more. MasteringMeteorology also includes all instructor resources and a robust Study Area with resources for students.
- **What's Your Forecast?** This new active-learning feature provides students with hands-on forecasting-themed activities. Prepared by experts in different areas of meteorology and climatology, each *What's Your Forecast?* feature highlights the relevance of meteorology in today's world by allowing students to make predictions based on real-world data. Examples of topics include using maps to identify precipitation patterns (Chapter 5), constructing and analyzing a surface weather map (Chapter 9), and predicting the probability of severe storm occurrences (Chapter 10). Critical thinking skills are reinforced as students apply concepts presented in the chapter.
- **SmartFigures** are brief, narrated video lessons that examine and explain concepts illustrated by key figures within the text. Students access SmartFigures on their mobile devices by scanning Quick Response (QR) codes next to key figures. These media are also available in the Study Area of MasteringMeteorology and teachers can assign them with automatically-graded quizzes.
- **Integrated Mobile Media.** QR links to mobile-enabled *Videos* and *Geoscience Animations* are integrated throughout the chapters, giving students just-in-time access to animations of key physical processes and videos of real-world case studies and data visualizations. These media are also available in the Study Area of MasteringMeteorology. Including SmartFigures, there are over 130 mobile media items linked to the 13th edition.
- **New and expanded active learning path.** *The Atmosphere*, 13th edition, is designed for learning. Every chapter begins with *Focus on Concepts*, which are numbered learning objectives that correspond to each major section in the chapter. The statements identify the knowledge and skills students should master by the end of the chapter, helping students prioritize key concepts. Within the chapter, each major section is also numbered and restates the relevant learning objective. Each section concludes with *Concept Checks* that allow students to check their understanding and comprehension of important ideas and terms before moving on to the next section. A new end-of-chapter feature, *Concepts in Review*, coordinates with the *Focus on Concepts* at the beginning of the chapter and with the numbered chapter sections. It is a readable, concise overview of key points and terms that often includes photos, diagrams, and questions to help students focus on important ideas and test their understanding of key concepts. Each chapter concludes with *Give It Some Thought*, a series of questions with illustrations that challenge learners with activities that require higher-order thinking skills, such as application, analysis, and synthesis of material in the chapter.
- **An unparalleled visual program.** In addition to the large number of new, high-quality photos and satellite images—many of which highlight recent weather events—dozens of figures are new or have been redrawn by renowned geoscience illustrator, Dennis Tasa. Maps and diagrams are frequently paired with photographs for greater effectiveness. Further, several new tables summarize key phenomena, and many new and revised figures have additional labels that narrate the process being illustrated to guide students as they examine the figures, resulting in a visual program that is clear and easy to understand.

- **Significant updating and revision of content.** A basic function of a college science textbook is to provide clear, understandable presentations that are accurate, engaging, and up-to-date. Our number-one goal is to keep *The Atmosphere* current, relevant, and highly readable for beginning students. Many discussions, case studies, and examples have been updated and revised, including:
  - An expanded discussion of greenhouse gases in Chapter 2
  - An updated discussion of El Niño/La Niña in Chapter 7
  - The extreme cold of the 2013/2014 U.S. Midwest winter (with new figure) integrated into the discussion of air mass movement in Chapter 8
  - The linkage between thunderstorms and climate change introduced in Chapter 10
  - Several new figures and a new *Severe and Hazardous Weather* box in Chapter 11 featuring 2013 Super Typhoon Haiyan
  - Chapter 12 has been expanded and rewritten to feature recent forecasting techniques and includes a new section, "The Role of the Forecaster," and a new box, "Thermodynamic Diagrams"
  - Chapter 14 presents key findings from the IPCC 5th Assessment report *Climate Change 2013: The Physical Science Basis*
  - Discussion of climate change and its possible impacts on weather and climate can be found throughout the book

# Distinguishing Features

## Readability

The language of this text is straightforward and *written to be understood*. Clear, readable discussions with a minimum of technical language are the rule. Frequent headings and subheadings help students follow discussions and identify the important ideas presented in each chapter. In the 13th edition, we have continued to improve readability by examining chapter organization and flow and by writing in a more personal style. Significant portions of several chapters were substantially rewritten in an effort to make the material easier to understand. For example, Chapter 1 was shortened and reorganized to improve flow and focus on key themes; Chapters 3, 4, and 6 have fewer main sections, but subsections were revised with more descriptive subheadings to help students understand linkages among phenomena. The order of main sections was changed for Chapters 9 and 12 to improve presentation of key concepts, and Chapter 12 was completely updated and rewritten to reflect new technologies in weather forecasting.

## Focus on Basic Principles and Instructor Flexibility

Although many topical issues are addressed in the 13th edition of *The Atmosphere*, it should be emphasized that the main focus of this new edition remains the same: to promote student understanding of basic principles. As much as possible, we have attempted to provide the reader with a sense of the observational techniques and reasoning processes that constitute the science of meteorology.

## Additional Learning Aids

In addition to the new and expanded learning path, the 13th edition continues to include these important learning aids:

- *Eye on the Atmosphere*, which feature real-world imagery paired with active-learning questions, giving students a chance to practice visual analysis tasks as they read. Instructors can discuss these in class or assign the questions to students from the book or MasteringMeteorology. Many of these have been updated with recent photos, and several chapters include new topics.
- Every chapter includes several *Students Sometimes Ask* features. Instructors and students continue to react favorably and indicate that the questions and answers that are sprinkled through each chapter add interest and relevance to discussions.
- The new edition continues to highlight severe and hazardous weather. Atmospheric hazards adversely affect millions of people worldwide every day. Severe weather events have a significance and fascination that go beyond ordinary weather phenomena. In addition to the two chapters (10, "Thunderstorms and Tornadoes," and 11, "Hurricanes") that focus entirely on such topics, the text contains 15 *Severe and Hazardous Weather* boxes devoted to a broad variety of topics—heat waves, winter storms, floods, air pollution episodes, drought, wildfires, cold waves, and more. Each box now includes one or two active-learning questions to help students test their understanding and link these events to critical chapter concepts.
- In many chapters, *Problems,* many with quantitative orientation, are included. Most problems require only basic math skills and allow students to enhance their understanding by applying concepts and principles explained in the chapter.

# Acknowledgments

Writing a college textbook requires the talents and cooperation of many people. It is truly a team effort, and the authors are fortunate to be part of an extraordinary team at Pearson Education. In addition to being great people to work with, all are committed to producing the best textbooks possible. Special thanks to our senior editor, Christian Botting, who invested a great deal of time, energy, and effort in this project. We appreciate his enthusiasm, hard work, and quest for excellence. We also appreciate our conscientious project manager, Crissy Dudonis, whose job it was to keep track of all that was going on—and a lot was going on. The 13th edition was certainly improved by the talents of our developmental editor, Veronica Jurgena. Many thanks. It was the job of the production team, led by Heidi Allgair at Cenveo® Publisher Services, to turn our manuscript into a finished product. The team also included copyeditor Kitty Wilson, compositor Annamarie Boley, and photo researcher Kristin Piljay. We think these talented people did great work. All are true professionals, with whom we are very fortunate to be associated.

Working with Dennis Tasa, who is responsible for all of the text's outstanding illustrations, is always special for us. He has been part of our team for more than 30 years. We not only value his artistic talents, hard work, patience, and imagination, but his friendship as well.

Many thanks also go to those colleagues who prepared in-depth reviews. Their critical comments and thoughtful input helped guide our work and clearly strengthened the text. Special thanks to:

Jason Allard, Valdosta State University
Mark Anderson, University of Nebraska–Lincoln
Deanna Bergondo, U.S. Coast Guard Academy
Anne Case Hanks, University of Louisiana–Monroe
William Conant, University of Arizona
Nathaniel Cunningham, Nebraska Wesleyan University
Kerry Doyle, Southern Illinois University–Edwardsville
Ron Dowey, Harrisburg Area Community College–Harrisburg
Douglas Gamble, University of North Carolina, Wilmington
Greg Gaston, University of North Alabama
Redina Herman, Western Illinois University
Mark Hildebrand, Southern Illinois University–Edwardsville
Helenmary Hotz, University of Massachusetts–Boston
Brennan Jordan, University of South Dakota
Timothy and Jennifer Klingler, Delta College
Mark Lemmon, Texas A&M University
Robert Mania, Kentucky State University
Jason Ortegren, University of West Florida
Robert Quinn, Eastern Washington University
Robert S. Rose, Tidewater Community College–Virginia Beach
Marshall Shepherd, University of Georgia
Roger D. Shew, University of North Carolina–Wilmington
Steve Simpson, Highland Community College
Eric Snodgrass, University of Illinois–Urbana-Champaign
Andrew Van Tuyl, Galivan College
Chuck Weidman, University of Arizona
Henry J. Zintambila, Illinois State University

We are very grateful to Neva Duncan-Tabb (St. Petersburg College), who authored the *Instructor Resource Manual,* to Jennifer Johnson (Ferris State University) for writing the *Test Bank,* and to the following expert contributors who have written the special *What's Your Forecast?* features throughout the book: Redina Herman (Western Illinois University), Marshall Shepherd (University of Georgia), Harold Brooks (NOAA National Severe Storms Laboratory), Brian McNoldy (University of Miami), Adam Clark (NOAA National Severe Storms Laboratory), and Donald Wuebbles (University of Illinois at Urbana-Champaign).

Last, but certainly not least, we gratefully acknowledge the support and encouragement of our wives, Nancy Lutgens and Joanne Bannon. Preparation of *The Atmosphere*, 13th edition, would have been far more difficult without their patience and understanding.

*Fred Lutgens*

*Ed Tarbuck*

# about our sustainability initiatives

# Digital & Print Resources

## For Students & Teachers

**MasteringMeteorology™ with Pearson eText.** The Mastering platform is the most widely used and effective online homework, tutorial, and assessment system for the sciences. It delivers self-paced tutorials that provide individualized coaching, that focus on course objectives, and that are responsive to each student's progress. The Mastering system helps teachers maximize class time with customizable, easy-to-assign, and automatically graded assessments that motivate students to learn outside of class and arrive prepared for lecture.

MasteringMeteorology offers:

- Assignable activities that include GIS-inspired MapMaster™ interactive maps, Encounter Meteorology Google Earth Explorations, Videos, Geoscience Animations, Map Projection Tutorials, GeoTutor coaching activities on the toughest topics in the geosciences, GEODe Tutorials, Dynamic Study Modules that provide each student with a customized learning experience, end-of-chapter questions and exercises, reading quizzes, Test Bank questions, and more.
- A student Study Area with GIS-inspired MapMaster™ interactive maps, Videos, Geoscience Animations, web links, glossary flashcards, *In the News* RSS feeds, chapter quizzes, an optional Pearson eText, and more.

Pearson eText gives students access to the text whenever and wherever they can access the Internet. Features of Pearson eText include:

- Now available on smartphones and tablets.
- Seamlessly integrated videos and other rich media.
- Fully accessible (screen-reader ready).
- Configurable reading settings, including resizable type and night reading mode.
- Instructor and student note-taking, highlighting, bookmarking, and search.

www.masteringmeteorology.com.

- ***Geoscience Animation Library on DVD,* 5th Edition [0321716841]** Geoscience Animations illuminate many difficult-to-visualize concepts. Animations include audio narration and text transcript, with assignable multiple-choice quizzes to select animations in MasteringMeteorology to help students master these core physical process concepts.
- ***Earth Report* Geography Videos on DVD [0321662989]** This three-DVD set is designed to help students visualize how human decisions and behavior have affected the environment and how individuals are taking steps toward recovery. With topics ranging from poor land management promoting the devastation of river systems in Central America to the struggles for electricity in China and Africa, these 13 videos from Television for the Environment's global *Earth Report* series recognize the efforts of individuals around the world to unite and protect the planet. Teachers can assign video clips with assessment in MasteringMeteorology.

## For Students

- ***Exercises for Weather & Climate,* 9th edition by Greg Carbone [0134041364]** This bestselling exercise manual's 17 exercises encourage students to review important ideas and concepts through problem solving, simulations, and guided thinking. The graphics program and computer-based simulations and tutorials help students grasp key concepts. Now with mobile-enabled Pre-Lab Videos and Pre- and Post-Lab quizzes in MasteringMeteorology, this manual is designed to complement any introductory meteorology or weather and climate course.
- ***Goode's World Atlas,* 23rd edition [0133864642]** *Goode's World Atlas* has been the world's premiere educational atlas since 1923—and for good reason. It features more than 250 pages of maps, from definitive physical and political maps to important thematic maps that illustrate the spatial aspects of many important topics. The 23rd edition includes 160 pages of digitally produced reference maps, as well as thematic maps on global climate change, sea-level rise, $CO_2$ emissions, polar ice fluctuations, deforestation, extreme weather events, infectious diseases, water resources, and energy production.
- ***Dire Predictions: Understanding Global Climate Change,* 2nd edition by Mike Mann and Lee Kump [0133909778]** Periodic reports from the Intergovernmental Panel on Climate Change (IPCC) evaluate the risk of climate change. But the sheer volume of scientific data remains inscrutable to the general public. In just over 200 pages, this practical text presents and expands on the essential findings of the IPCC's 5th Assessment Report in a visually stunning and undeniably powerful way to the lay reader. Scientific findings that provide validity to the implications of climate change are presented in clear-cut graphic elements, striking images, and understandable analogies.

- ***Encounter Physical Geography* by Jess C. Porter and Stephen O'Connell [0321672526]** Pearson's Encounter Series provides rich, interactive explorations of geoscience concepts through Google Earth activities, covering a range of topics in meteorology and physical geography. For those who do not use MasteringMeteorology, all chapter explorations are available in print workbooks, as well as in online quizzes at www.mygeoscienceplace.com, accommodating different classroom needs. Each exploration consists of a worksheet, a corresponding Google Earth KMZ file, and online quizzes whose results can be e-mailed to teachers.

## For Teachers

**Learning Catalytics** is a "bring your own device" student engagement, assessment, and classroom intelligence system. With Learning Catalytics, you can:

- Assess students in real time, using open-ended tasks to probe student understanding.
- Understand immediately where students are and adjust your lecture accordingly.
- Improve your students' critical thinking skills.
- Access rich analytics to understand student performance.
- Add your own questions to make Learning Catalytics fit your course exactly.
- Manage student interactions with intelligent grouping and timing.

Learning Catalytics is a technology that has grown out of 20 years of cutting-edge research, innovation, and implementation of interactive teaching and peer instruction. Available integrated with MasteringMeteorology.

- ***Instructor Resource Manual* (download only) by Neva Duncan-Tabb, St. Petersburg College [0321987640]** The *Instructor Resource Manual* is intended as a resource for both new and experienced instructors. It includes a variety of lecture outlines, teaching tips, advice about how to integrate visual supplements (including the MasteringMeteorology resources), answers to the textbook chapter questions, and various other ideas for the classroom.
  See www.pearsonhighered.com/irc.
- **TestGen® Computerized *Test Bank* (download only) by Jennifer Johnson, Ferris State University [0321987683]** TestGen® is a computerized test generator that lets instructors view and edit *Test Bank* questions, transfer questions to tests, and print tests in a variety of customized formats. This *Test Bank* includes more than 2000 multiple-choice, fill-in-the-blank, and short-answer/essay questions. Questions are correlated to the text's Learning Outcomes, Pearson's Global Science Outcomes, the section of each chapter, the revised U.S. National Geography Standards, and Bloom's taxonomy to help instructors better map the assessments against both broad and specific teaching and learning objectives. The *Test Bank* is also available in Microsoft Word and is importable into systems such as Blackboard.
  See www.pearsonhighered.com/irc.
- ***Instructor Resource DVD* [0321987659]** The Instructor Resource DVD provides a collection of resources to help teachers make efficient and effective use of their time. All digital resources can be found in one well-organized, easy-to-access place. The IRDVD includes:
  - All textbook images as JPEGs, PDFs, and PowerPoint™ presentations
  - Pre-authored Lecture Outline PowerPoint™ presentations, which outline the concepts of each chapter with embedded art and can be customized to fit teachers' lecture requirements
  - "Clicker" questions in PowerPoint™, which correlate to the text's Learning Outcomes, U.S. National Geography Standards, and Bloom's taxonomy
  - The TestGen software, *Test Bank* questions, and answers for both MACs and PCs
  - Electronic files of the *Instructor Resource Manual* and *Test Bank*

This Instructor Resource content is also available online via the Instructor Resources section of MasteringMeteorology and www.pearsonhighered.com/irc.

# 1 Introduction to the Atmosphere

## Focus on Concepts

*Each statement represents the primary learning objective for the corresponding major heading within the chapter. After you complete the chapter, you should be able to:*

**1.1** Distinguish between weather and climate, name the basic elements of weather and climate, and list several important atmospheric hazards.

**1.2** Discuss the nature of scientific inquiry, including the construction of hypotheses and the development of theories.

**1.3** List and describe Earth's four major spheres. Define *system* and explain why Earth is considered to be a system.

**1.4** List the major gases composing Earth's atmosphere and identify the components that are most important meteorologically. Explain why ozone depletion is a significant global issue.

**1.5** Interpret a graph that shows changes in air pressure from Earth's surface to the top of the atmosphere. Sketch and label a graph that shows the thermal structure of the atmosphere.

Earth's atmosphere is unique. No other planet in our solar system has an atmosphere with the exact mixture of gases or the heat and moisture conditions necessary to sustain life as we know it. The gases that make up Earth's atmosphere and the controls to which they are subject are vital to our existence. In this chapter we begin our examination of the ocean of air in which we all must live.

*This satellite image shows Hurricane Sandy, called Superstorm Sandy in the media, battering the east coast on October 30, 2012. This view of the storm is looking south from Canada. Florida is near the top of the image.*

# 1.1 Focus On the Atmosphere

**Distinguish between weather and climate, name the basic elements of weather and climate, and list several important atmospheric hazards.**

MM GEODe ▶ Introduction to the Atmosphere ▶ Weather and Climate

Weather influences our everyday activities, our jobs, and our health and comfort. Many of us pay little attention to the weather unless we are inconvenienced by it or when it adds to our enjoyment of outdoor activities. Nevertheless, there are few other aspects of our physical environment that affect our lives more than the phenomena we collectively call the weather.

## Weather in the United States

The United States occupies an area that stretches from the tropics to the Arctic Circle. It has thousands of miles of coastline and extensive regions that are far from the influence of the ocean. Some landscapes are mountainous, and others are dominated by plains. It is a place where Pacific storms strike the west coast, while the eastern states are sometimes influenced by events in the Atlantic and the Gulf of Mexico. For those in the center of the country, it is common to experience weather events triggered when frigid southward-bound Canadian air masses clash with northward-moving tropical air masses from the Gulf of Mexico.

Stories about weather are a routine part of the daily news. Articles and items about the effects of heat, cold, floods, drought, fog, snow, ice, and strong winds are commonplace (Fig. 1.1). Memorable weather events occur everywhere on our planet. The United States likely has the greatest variety of weather of any country in the world. Severe weather events, such as tornadoes, flash floods, and intense thunderstorms, as well as hurricanes and blizzards, are collectively more frequent and more damaging in the United States than in any other nation. Beyond its direct impact on the lives of individuals, the weather has a strong effect on the world economy, by influencing agriculture, energy use, water resources, transportation, and industry.

Weather influences our lives a great deal. Yet it is also important to realize that people influence the atmosphere and its behavior as well (Fig. 1.2). There are, and will continue to be, significant economic, political, and scientific decisions to

▼ **Figure 1.1 An extraordinary winter** The winter of 2013–2014 brought record-breaking cold and snow to much of the eastern half of the conterminous United States. Meanwhile, Alaska and much of the West were much warmer and drier than usual.

▲ **Figure 1.2 People influence the atmosphere** Smoke bellows from a coal-fired electricity generating plant in New Delhi, India, in June 2008. In addition to smoke, this power plant also emits gases such as sulfur dioxide and carbon dioxide that contribute to air pollution and global climate change.

make involving these impacts. Dealing with the effects of and controlling air pollution is one example. Another is the ongoing effort to assess and address global climate change. There is clearly a need for increased awareness and understanding of our atmosphere and its behavior.

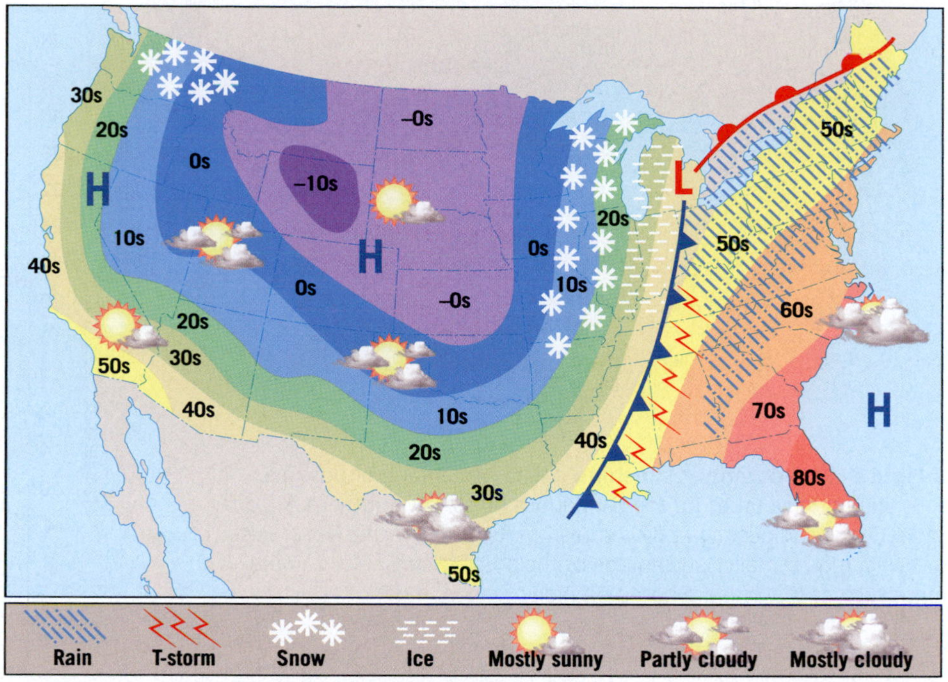


▲ **Figure 1.3 Newspaper weather map** A typical newspaper weather map for a day in late December. The color bands show predicted high temperatures for the day.

## Meteorology, Weather, and Climate

The subtitle of this book includes the word *meteorology*. **Meteorology** is the scientific study of the atmosphere and the phenomena that we usually refer to as *weather*. Along with geology, oceanography, and astronomy, meteorology is considered one of the *Earth sciences*—the sciences that seek to understand our planet. It is important to point out that there are not strict boundaries among the Earth sciences; in many situations, these sciences overlap. Moreover, all the Earth sciences involve an understanding and application of knowledge and principles from physics, chemistry, and biology. You will see many examples of this overlap in your study of meteorology.

Acted on by the combined effects of Earth's motions and energy from the Sun, our planet's formless and invisible envelope of air reacts by producing an infinite variety of weather, which in turn creates the basic pattern of global climates. Although not identical, weather and climate have much in common.

**Weather** is constantly changing, sometimes from hour to hour and at other times from day to day. It is a term that refers to the state of the atmosphere at a given time and place. Whereas changes in the weather are continuous and sometimes seemingly erratic, it is nevertheless possible to arrive at a generalization of these variations. Such a description of aggregate weather conditions is termed **climate**. It is based on observations that have been accumulated over many decades. Climate is often defined simply as "average weather," but this is an inadequate definition. In order to accurately portray the character of an area, variations and extremes must also be included, as well as the probabilities that such departures will take place. For example, it is necessary for farmers to know the average rainfall during the growing season, and it is also important to know the frequency of extremely wet and extremely dry years. Thus, climate is the sum of all statistical weather information that helps describe a place or region.

Maps similar to the one in Figure 1.3 are familiar to everyone who checks the weather report in the morning newspaper or on a television station. In addition to showing predicted high temperatures for the day, this type of map shows other basic weather information about cloud cover, precipitation, and fronts.

Suppose you were planning a vacation trip to an unfamiliar place. You would probably want to know what kind of weather to expect. Such information would help as you selected clothes to pack and could influence decisions regarding activities you might engage in during your stay. Unfortunately, weather forecasts that go beyond a few days are not very dependable. Thus, it may not be possible to get a reliable weather report about the conditions you are likely to encounter during your vacation.

Instead, you might ask someone who is familiar with the area about what kind of weather to expect. "Are thunderstorms common?" "Does it get cold at night?" "Are the afternoons sunny?" What you are seeking is information about the climate, the conditions that are typical for that place. Another useful source of such information is the great variety of climate tables, maps, and graphs that are available. For example, the map in Figure 1.4 shows the average percentage of possible sunshine in the United States for the month of November, and the graph in Figure 1.5 shows average daily high and low temperatures for each month, as well as extremes, for New York City.

**students sometimes ask...**

### Does meteorology have anything to do with meteors?

There is a connection. The word *meteor* refers to solid particles (meteoroids) that enter Earth's atmosphere from space and "burn up" due to friction ("shooting stars"). The term *meteorology* was coined in 340 B.C., when the Greek philosopher Aristotle wrote a book titled *Meteorlogica*, which described atmospheric and astronomical phenomena. In Aristotle's day *anything* that fell from or was seen in the sky was called a meteor. Today we distinguish between particles of ice or water in the atmosphere (*hydrometeors*) and extraterrestrial objects (meteoroids, or meteors).

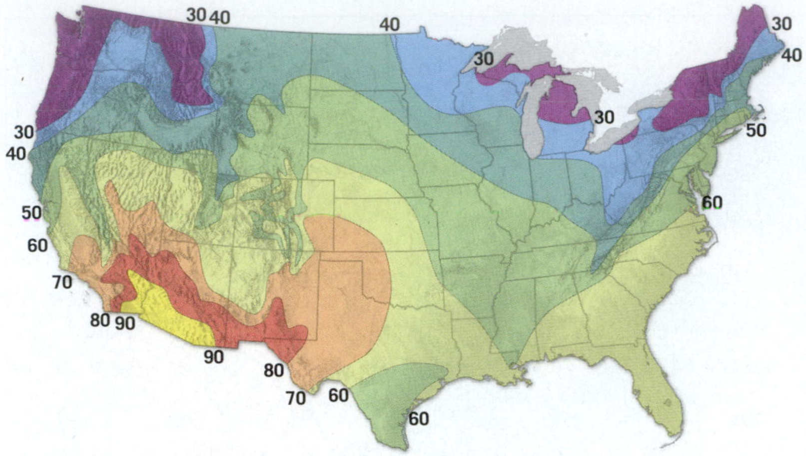


▲ **Figure 1.4 November sunshine** Mean percentage of possible sunshine for November for the contiguous 48 states. Southern Arizona is clearly the sunniest area. By contrast, parts of the Pacific Northwest receive a much smaller percentage of the possible sunshine. Climate maps such as this one are based on many years of data.

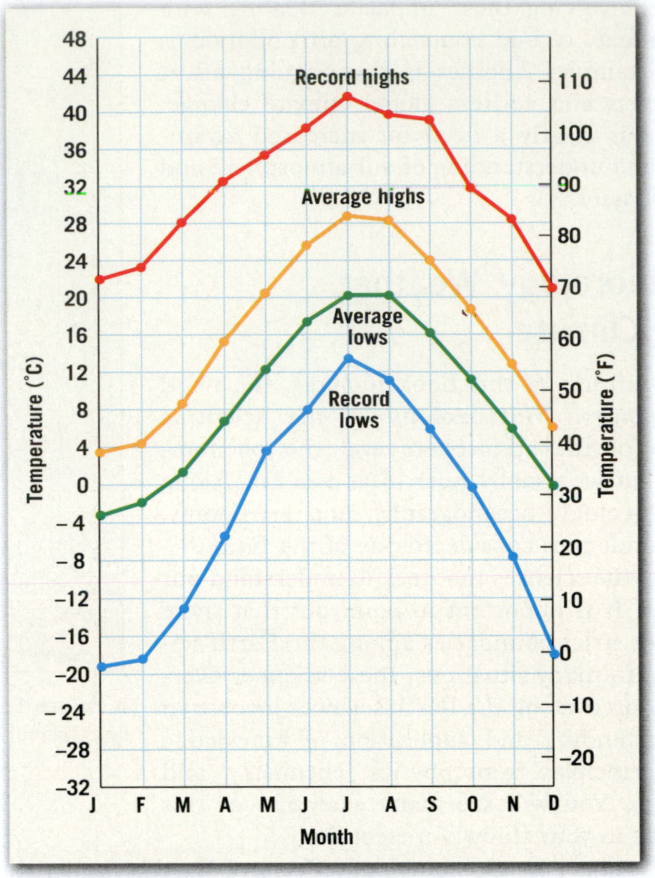


▲ **Figure 1.5 New York City temperatures** In addition to the average maximum and minimum temperatures for each month, extremes are also shown. The graph is based on data collected during a 30-year span. It shows that there can be significant departures from the average.

Such information could, no doubt, help as you planned your trip. But it is important to realize that *climate data cannot predict the weather.* Although the place may usually (climatically) be warm, sunny, and dry during the time of your planned vacation, you may actually experience cool, overcast, and rainy weather. There is a well-known saying that summarizes this idea: "Climate is what you expect, but weather is what you get."

The nature of both weather and climate is expressed in terms of the same basic **elements**—quantities or properties that are measured regularly. The most important are (1) the *temperature* of the air, (2) the *humidity* of the air, (3) the type and amount of *cloudiness,* (4) the type and amount of *precipitation,* (5) the *pressure* exerted by the air, and (6) the speed and direction of the *wind.* These elements constitute the variables by which weather patterns and climate types are depicted. Although you will study these elements separately at first, keep in mind that they are very much interrelated. A change in one of the elements often produces changes in the others.

▼ **Figure 1.6 Late season tornado** Although "tornado season" in central Illinois is spring and summer, a devastating tornado struck Washington, Illinois, on November 17, 2013. With maximum winds of 306 kilometers (190 miles) per hour, the storm caused complete destruction of well-built homes.

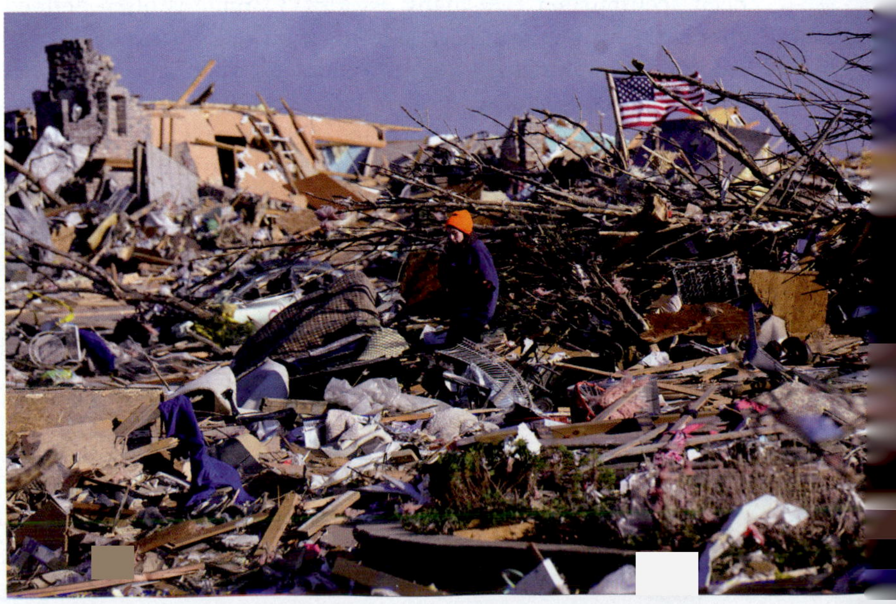

## Atmospheric Hazards: Assault by the Elements

Natural hazards are a part of living on Earth. Every day they adversely affect literally millions of people worldwide and are responsible for staggering damages. Some, such as earthquakes and volcanic eruptions, are geologic. Many others are related to the atmosphere.

For most people, occurrences of severe weather are far more fascinating than ordinary weather phenomena. A spectacular lightning display generated by a severe thunderstorm can elicit both awe and fear. Of course, hurricanes and tornadoes attract a great deal of much-deserved attention. A single tornado outbreak or hurricane can cause billions of dollars in property damage, much human suffering, and many deaths. The chapter-opening image of Hurricane Sandy and the tornado damage depicted in Figure 1.6 are good examples.

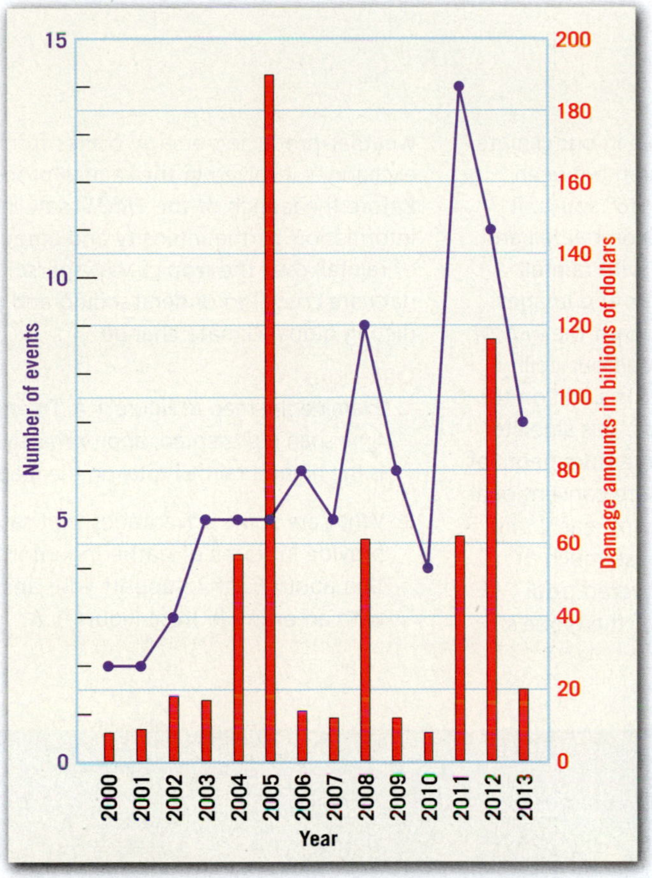


▲ **Figure 1.7 Billion-dollar weather events** Between 2000 and 2013, the United States experienced 84 weather-related disasters in which overall damages and costs reached or exceeded $1 billion. The line graph shows the number of events that occurred each year, and the bar graph shows damage amounts in billions of dollars (normalized to 2013 dollars). The total losses for the 84 events exceeded $600 billion! (Data from NOAA)

Of course, other atmospheric hazards adversely affect us. Some are storm related, such as blizzards, hail, and freezing rain. Others are not direct results of storms. Heat waves, cold waves, fog, wildfires, and drought are important examples. In some years the loss of human life due to excessive heat or bitter cold exceeds that caused by all other weather events combined. Moreover, although severe storms and floods usually generate more attention, droughts can be just as devastating and carry an even bigger price tag.

Between 2000 and 2013, the United States experienced 84 weather-related disasters in which overall damages and costs reached or exceeded $1 billion (**Fig. 1.7**). In addition to taking more than 4200 lives, the combined economic costs of these events exceeded $600 billion!

At appropriate places throughout this book, you will have an opportunity to learn about atmospheric hazards. Two entire chapters (Chapter 10 and Chapter 11) focus almost entirely on hazardous weather. In addition, a number of the book's special-interest boxes are devoted to a broad variety of severe and hazardous weather, including heat waves, winter storms, floods, dust storms, drought, mudflows, and lightning. Every day our planet experiences an incredible assault by the atmosphere, so it is important to develop an awareness and understanding of these significant weather events.

### ✓ Concept Checks 1.1

1. Define and distinguish among meteorology, weather, and climate.
2. List the basic elements of weather and climate.
3. List at least five storm-related atmospheric hazards and three atmospheric hazards that are not directly storm related.

## 1.2 The Nature of Scientific Inquiry

**Discuss the nature of scientific inquiry, including the construction of hypotheses and the development of theories.**

As members of a modern society, we are constantly reminded of the benefits derived from science. But what exactly is the nature of scientific inquiry? Science is a process of producing knowledge. The process depends both on making careful observations and on creating explanations that make sense of the observations. Developing an understanding of how science is done and how scientists work is an important theme in this book. You will explore the difficulties of gathering data and learn some of the ingenious methods that have been developed to overcome these difficulties. You will also see examples of how hypotheses are formulated and tested, as well as learn about the development of some significant scientific theories.

All science is based on the assumption that the natural world behaves in a consistent and predictable manner that is comprehensible through careful, systematic study. The overall goal of science is to discover the underlying patterns in nature and then to use the knowledge gained to make predictions about what should or should not be expected, given certain facts or circumstances. For example, by understanding the processes and conditions that produce certain cloud types, meteorologists are often able to predict the approximate time and place of their formation.

The development of new scientific knowledge involves some basic logical processes that are universally accepted. To determine what is occurring in the natural world, scientists collect scientific data through observation and measurement. The types of data that are collected often seek to answer a well-defined question about the natural world, such as "Why does fog frequently develop in this place?" or "What causes rain to form in this cloud type?" Because some error is inevitable, the accuracy of a particular measurement or observation is always open to question. Nevertheless, these data are essential to science and serve as a springboard for the development of scientific theories (**Box 1.1**).

## Box 1.1 Monitoring Earth from Space

Scientific data are gathered in many ways, including through laboratory experiments and field observations and measurements. Satellites provide another very important source of data. Satellite images give us perspectives that are difficult to gain from more traditional sources. The chapter-opening image of Hurricane Sandy is a good example. Moreover, the high-tech instruments aboard many satellites enable scientists to gather information from remote regions where data are otherwise scarce.

The image in **Figure 1.A** is from NASA's *Tropical Rainfall Measuring Mission* (*TRMM*). *TRMM* is a research satellite designed to expand our understanding of Earth's water (hydrologic) cycle and its role in our climate system. By covering the region between the latitudes 36° north and 36° south, it provides much-needed data on rainfall and the heat release associated with rainfall. Many types of measurements and images are possible. Instruments aboard the *TRMM* satellite have greatly expanded our ability to collect precipitation data. In addition to recording data for land areas, this satellite provides extremely precise measurements of rainfall over the oceans, where conventional land-based instruments cannot reach. This is especially important because much of Earth's rain falls in ocean-covered tropical areas, and a great deal of the globe's weather-producing energy comes from heat exchanges involved in the rainfall process. Before the launch of the *TRMM* satellite, information on the intensity and amount of rainfall over the tropics was sparse. Such data are crucial to understanding and predicting global climate change.

**Questions**

1. Examine the map in Figure 1.A. During the time span represented, approximately what is the highest rainfall total on the map?
2. What are some advantages that satellites provide in terms of gathering information about Earth? Support your answer with an example from Figure 1.A.

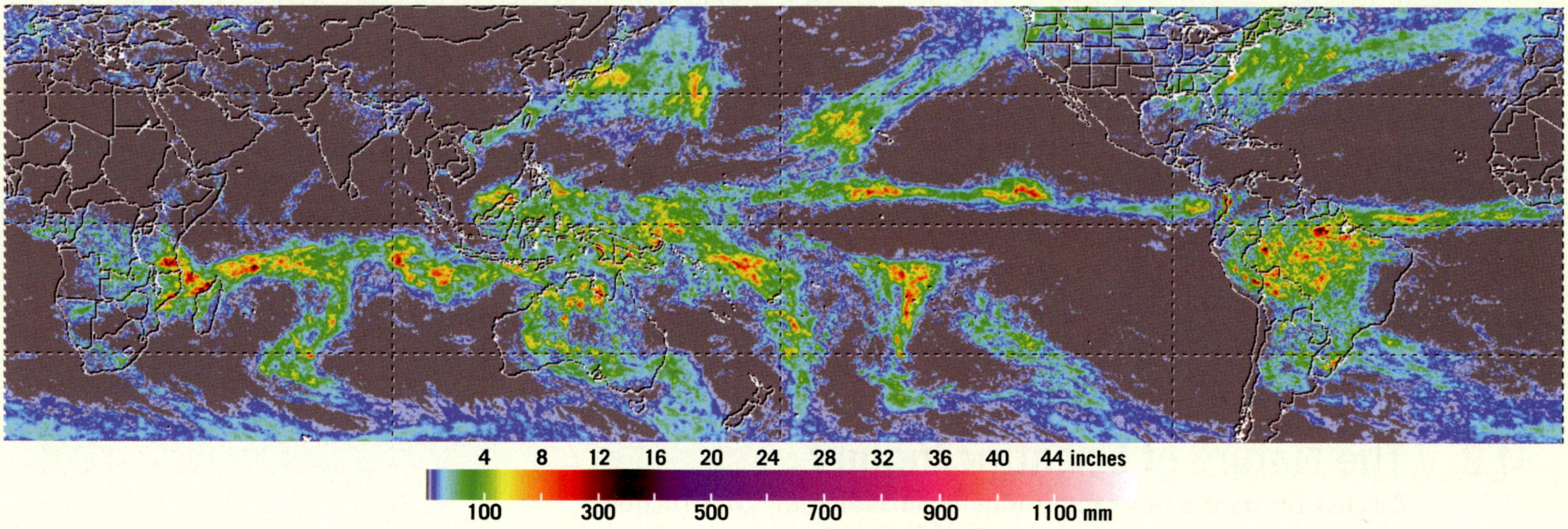


▲ **Figure 1.A Monitoring rainfall** This map shows rainfall for a 7-day period in February 2014. It was constructed using *TRMM* data.

## Hypothesis

Once data have been gathered and principles have been formulated to describe a natural phenomenon, investigators try to explain how or why things happen in the manner observed. They often do this by constructing a tentative (or untested) explanation, which is called a scientific **hypothesis**. It is best if an investigator can formulate more than one hypothesis to explain a given set of observations. If an individual scientist is unable to devise multiple hypotheses, others in the scientific community will almost always develop alternative explanations. A spirited debate frequently ensues. As a result, proponents of opposing hypotheses conduct extensive research, and scientific journals make the results available to the wider scientific community.

Before a hypothesis can become an accepted part of scientific knowledge, it must pass objective testing and analysis. If a hypothesis cannot be tested, it is not scientifically useful, no matter how interesting it may seem. The verification process requires that *predictions* be made based on the hypothesis being considered and that these predictions be tested by being compared against objective observations of nature. Put another

way, hypotheses must fit observations other than those used to formulate them in the first place. Hypotheses that fail rigorous testing are ultimately discarded. The history of science is littered with discarded hypotheses. One of the best known is the Earth-centered model of the universe—a proposal that was supported by the apparent daily motion of the Sun, Moon, and stars around Earth.

## Theory

When a hypothesis has survived extensive scrutiny and when competing hypotheses have been eliminated, it may be elevated to the status of a scientific **theory**. In everyday language, we may say that something is "only a theory." But a scientific theory is a well-tested and widely accepted view that the scientific community agrees best explains certain observable facts.

Some theories that are extensively documented and extremely well supported are comprehensive in scope. An example from the Earth sciences is the theory of plate tectonics, which provides the framework for understanding the origin of mountains, earthquakes, and volcanic activity. It also explains the evolution of continents and ocean basins through time. As you will see in Chapter 14, this theory also helps us understand some important aspects of climate change through long spans of geologic time.

## Scientific Methods

The processes just described, in which scientists gather data through observations and formulate scientific hypotheses and theories, is called the *scientific method*. Contrary to popular belief, the scientific method is not a standard recipe that scientists apply in a routine manner to unravel the secrets of our natural world. Rather, it is an endeavor that involves creativity and insight. Rutherford and Ahlgren put it this way: "Inventing hypotheses or theories to imagine how the world works and then figuring out how they can be put to the test of reality is as creative as writing poetry, composing music, or designing skyscrapers."*

There is no fixed path for scientists that leads unerringly to scientific knowledge. Nevertheless, many scientific investigations involve the following:

- A question is raised about the natural world.
- Scientific data that relate to the question are collected (**Fig. 1.8**).
- Questions that relate to the data are posed, and one or more working hypotheses are developed that may answer these questions.
- Observations, experiments, and models are developed to test the hypotheses.

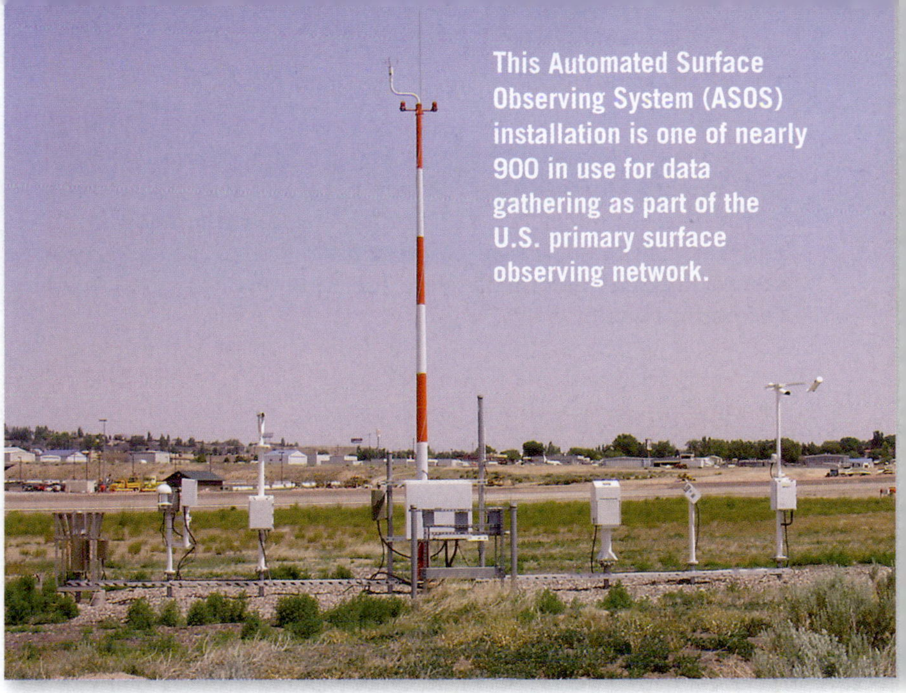
This Automated Surface Observing System (ASOS) installation is one of nearly 900 in use for data gathering as part of the U.S. primary surface observing network.

▲ **Figure 1.8 Observation and measurement** Gathering data and making careful observations are basic parts of scientific inquiry.

- The hypotheses are accepted, modified, or rejected, based on extensive testing.
- Data and results are shared with the scientific community for critical examination and further testing.

Some scientific discoveries may result from purely theoretical ideas that stand up to extensive examination. Some researchers use high-speed computers to simulate what is happening in the "real" world. These models are useful when dealing with natural processes that occur on very long time scales or take place in extreme or inaccessible locations. Still other scientific advancements have been made when a totally unexpected happening occurred during an experiment. These serendipitous discoveries are more than pure luck; as the nineteenth-century French scientist Louis Pasteur said, "In the field of observation, chance favors only the prepared mind."**

**students sometimes ask...**

### How do a hypothesis and a theory differ from a scientific law?

A *scientific law* is a basic principle that describes a particular behavior of nature that is generally narrow in scope and can be stated briefly—often as a simple mathematical equation. Because scientific laws have been shown time and time again to be consistent with observations and measurements, they are rarely discarded but may require modifications to fit new findings. For example, Newton's laws of motion are still useful for everyday applications (NASA uses them to calculate satellite trajectories), but they do not work at velocities approaching the speed of light. Einstein's theory of relativity is instead applied in these circumstances.

*F. James Rutherford and Andrew Ahlgren, *Science for All Americans* (New York: Oxford University Press, 1990), p. 7.

**Louis Pasteur quoted in "Science, History and Social Activism" by Everett Mendelsohn, Garland E. Allen, Roy M. MacLeod, Springer 2001.

Scientific knowledge is acquired through several avenues, so it might be best to describe the nature of scientific inquiry as the *methods* of science rather than *the* scientific method. In addition, it should always be remembered that even the most compelling scientific theories are still simplified explanations of the natural world.

### ✔ Concept Checks 1.2

1. How is a scientific hypothesis different from a scientific theory?
2. Summarize the basic steps followed in many scientific investigations.

**students sometimes ask...**

**Who provides all the data needed to prepare a weather forecast?**

Data from every part of the globe are needed to produce accurate weather forecasts. The World Meteorological Organization (WMO), consisting of 191 member states and territories, coordinates scientific activity related to weather and climate. Its World Weather Watch provides up-to-the-minute standardized observations through member-operated observation systems. This global system involves more than 15 satellites, 10,000 land-observation and 7300 ship stations, hundreds of automated data buoys, and thousands of aircraft.

## 1.3 Earth as a System

**List and describe Earth's four major spheres. Define *system* and explain why Earth is considered to be a system.**

Anyone who studies Earth soon learns that our planet is a dynamic body with many separate but highly interactive parts, or *spheres*. The atmosphere, hydrosphere, biosphere, and geosphere and all of their components can be studied separately. However, the parts are *not* isolated. Each is related in many ways to the others, producing a complex and continuously interacting whole that we call the *Earth system*.

### Earth's Spheres

The images in **Figure 1.9** are considered to be classics because they let humanity see Earth differently than ever before. These early views profoundly altered our conceptualizations of Earth and remain powerful images decades after they were first viewed. Seen from space, Earth is breathtaking in its beauty and startling in its solitude. The photos remind us that our home is, after all, a planet—small, self-contained, and in some ways even fragile. Bill Anders, the *Apollo 8* astronaut who took the "Earthrise" photo, expressed it this way: "We came all this way to explore the Moon, and the most important thing is that we discovered the Earth."

As we look closely at our planet from space, it becomes apparent that Earth is much more than rock and soil. In fact, the most conspicuous features in Figure 1.9 are not continents but swirling clouds suspended above the surface of the vast global ocean. These features emphasize the importance of water on our planet.

The closer view of Earth from space shown in Figure 1.9 helps us appreciate why the physical environment is traditionally divided into three major parts: the solid Earth, or *geosphere;* the water portion of our planet, the *hydrosphere;* and Earth's gaseous envelope, the *atmosphere.*

It should be emphasized that our environment is highly integrated and is not dominated by rock, water, or air alone.

View called "Earthrise" that greeted *Apollo 8* astronauts as their spacecraft emerged from behind the Moon in December 1968. This classic image let people see Earth differently than ever before.

This image taken from *Apollo 17* in December 1972 is perhaps the first to be called "The Blue Marble." The dark blue ocean and swirling cloud patterns remind us of the importance of the oceans and atmosphere.

▶ **Figure 1.9 Two classic views of Earth from space**

▲ **Figure 1.10 Interactions among Earth's spheres** The shoreline is one obvious example of an *interface*—a common boundary where different parts of a system interact. In this scene, ocean waves (*hydrosphere*) that were created by the force of moving air (*atmosphere*) break against a rocky shore (*geosphere*).

It is instead characterized by continuous interactions as air comes in contact with rock, rock with water, and water with air. Moreover, the *biosphere,* the totality of life-forms on our planet, extends into each of the three physical realms and is an equally integral part of the planet.

The interactions among Earth's spheres are incalculable. Figure 1.10 provides one easy-to-visualize example. The shoreline is an obvious meeting place for rock, water, and air. In this scene, ocean waves that were created by the drag of air moving across the water are breaking against the rocky shore. The force of the water can be powerful, and the erosional work that is accomplished can be great.

**The Geosphere** Beneath the atmosphere and the ocean is the solid Earth, or **geosphere**. The geosphere extends from the surface to the center of the planet, a depth of about 6400 kilometers (nearly 4000 miles), making it by far the largest of Earth's four spheres.

Based on compositional differences, the geosphere is divided into three principal regions: the dense inner sphere, called the *core;* the less dense *mantle;* and the *crust,* which is the light and very thin outer skin of Earth.

Soil, the thin veneer of material at Earth's surface that supports the growth of plants, may be thought of as part of all four spheres. The solid portion is a mixture of weathered rock debris (geosphere) and organic matter from decayed plant and animal life (biosphere). The decomposed and disintegrated rock debris is the product of weathering processes that require air (atmosphere) and water (hydrosphere). Air and water also occupy the open spaces between the solid particles.

**The Atmosphere** Earth is surrounded by a life-giving gaseous envelope called the **atmosphere** (Fig. 1.11). When we watch a high-flying jet plane cross the sky, it seems that the atmosphere extends upward for a great distance. However, when compared to the thickness (radius) of the solid Earth (about 6400 kilometers [4000 miles]), the atmosphere is a very shallow layer. More than 99 percent of the atmosphere is within 30 kilometers (20 miles) of Earth's surface. This thin blanket of air is nevertheless an integral part of the planet. It not only provides the air we breathe but also acts to protect us from the dangerous radiation emitted by the Sun. The energy exchanges that continually occur between the atmosphere and Earth's surface and between the atmosphere and space produce the effects we call *weather.* If, like the Moon, Earth had no atmosphere, our planet would not only be lifeless, but many of the processes and interactions that make the surface such a dynamic place could not operate.

**The Hydrosphere** Earth is sometimes called the *blue planet.* More than anything else, water makes Earth unique. The **hydrosphere** is a dynamic mass that is continually on the move, evaporating from the oceans to the atmosphere, precipitating to the land, and running back to the ocean again. The global ocean is certainly the most prominent feature of the hydrosphere, blanketing nearly 71 percent of Earth's surface to an average depth of about 3800 meters (12,500 feet). It accounts for about 97 percent of Earth's water (Fig. 1.12). However, the hydrosphere also includes the freshwater found in clouds, streams, lakes, and glaciers, as well as that found underground.

Although these latter sources constitute just a tiny fraction of the total, they are much more important than their meager percentage indicates. Clouds, of course, play a vital role

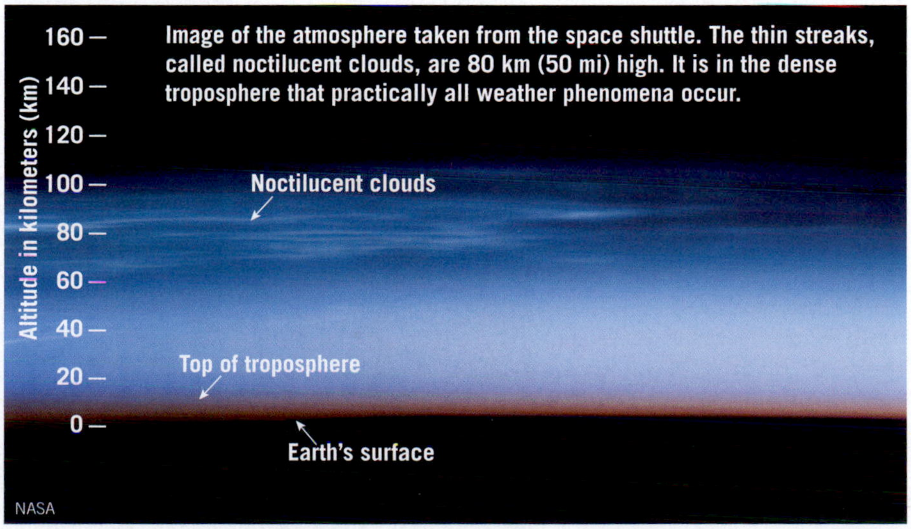
Image of the atmosphere taken from the space shuttle. The thin streaks, called noctilucent clouds, are 80 km (50 mi) high. It is in the dense troposphere that practically all weather phenomena occur.
Altitude in kilometers (km)
160
140
120
100
80
60
40
20
0
Noctilucent clouds
Top of troposphere
Earth's surface
NASA

▲ **Figure 1.11 A shallow layer** The atmosphere is an integral part of the planet.

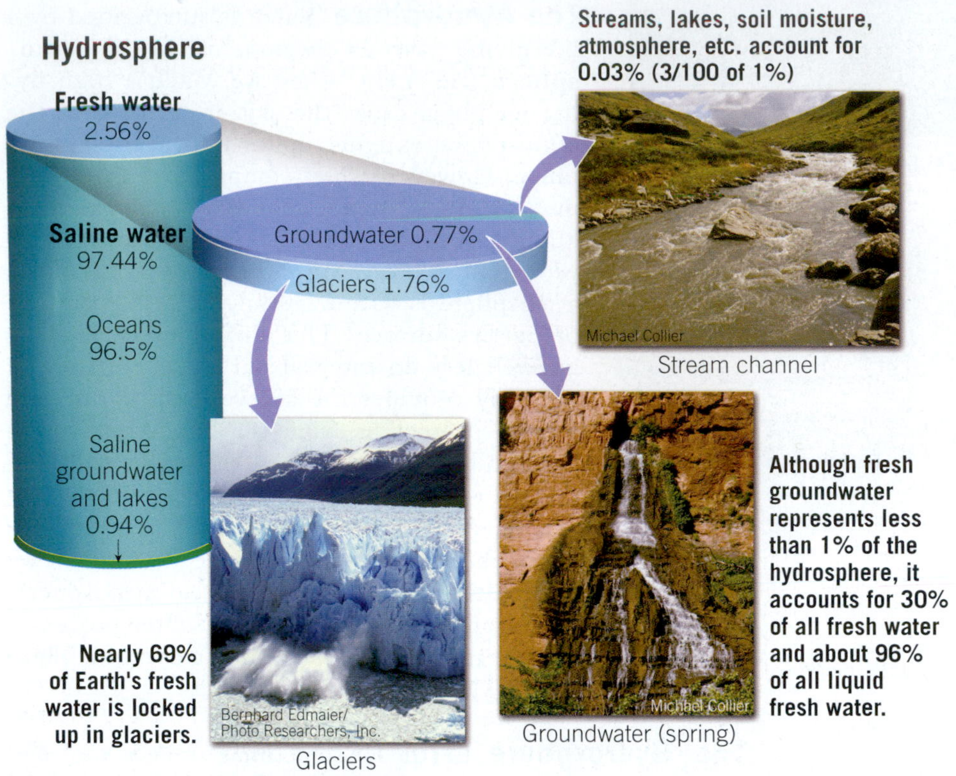


▲ **Figure 1.12 The water planet** Distribution of water in the hydrosphere.

in many weather and climate processes. In addition to providing the freshwater that is so vital to life on land, streams, glaciers, and groundwater are responsible for sculpting and creating many of our planet's varied landforms.

**The Biosphere** The **biosphere** includes all life on Earth (Fig. 1.13). Ocean life is concentrated in the sunlit surface waters of the sea. Most life on land is also concentrated near the surface, with tree roots and burrowing animals reaching a few meters underground and flying insects and birds reaching a kilometer or so above the surface. A surprising variety of life-forms are also adapted to extreme environments. For example, on the ocean floor, where pressures are extreme and no light penetrates, there are places where vents spew hot, mineral-rich fluids that support communities of exotic life-forms. On land, some bacteria thrive in rocks as deep as 4 kilometers (2.5 miles) and in boiling hot springs. Moreover, air currents can carry microorganisms many kilometers into the atmosphere. But even when we consider these extremes, life still must be thought of as being confined to a narrow band very near Earth's surface.

Plants and animals depend on the physical environment for the basics of life. However, organisms do more than just respond to their physical environment. Through countless interactions, life-forms help maintain and alter their physical environment. Without life, the makeup and nature of the geosphere, hydrosphere, and atmosphere would be very different.

## Earth System Science

A simple example of the interactions among different parts of the Earth system occurs most winters as moisture evaporates from the Pacific Ocean and subsequently falls as rain in the hills of southern California, sometimes triggering destructive debris flows (Fig. 1.14). The processes that move water from the hydrosphere to the atmosphere and then to the geosphere have a profound impact on the

The ocean contains a significant portion of Earth's biosphere. Modern coral reefs are unique and complex examples and are home to about 25% of all marine species. Because of this diversity, they are sometimes referred to as the ocean equivalent of a rain forest.
Tropical rain forests are characterized by hundreds of different species per square kilometer.

► **Figure 1.13 The biosphere** The biosphere, one of Earth's four spheres, includes all life.

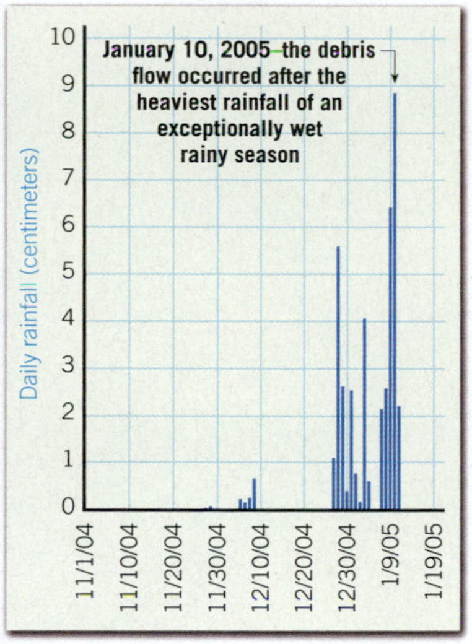


◄ **Figure 1.14 Heavy rains trigger debris flow** This image provides an example of interactions among different parts of the Earth system. On January 10, 2005, a massive debris flow (popularly called a mudslide) swept through the coastal community of La Conchita, California. The event occurred after a span of near-record amounts of rainfall.

Video (MM)
Global Carbon Uptake by Plants
http://goo.gl/hDHbF

physical environment and on the plants and animals (including humans) that inhabit the affected regions.

Scientists have recognized that in order to more fully understand our planet, they must learn how its individual components (land, water, air, and life-forms) are interconnected. This endeavor, called **Earth system science**, aims to study Earth as a *system* composed of numerous interacting parts, or *subsystems*. Using an interdisciplinary approach, those who practice Earth system science attempt to achieve the level of understanding necessary to comprehend and solve many of our global environmental problems.

A **system** is a group of interacting, or interdependent, parts that form a complex whole. Most of us hear and use the term *system* frequently. We may service our car's cooling *system*, make use of the city's transportation *system*, and participate in the political *system*. A news report might inform us of an approaching weather *system*. Further, we know that Earth is just a small part of a larger system known as the *solar system*, which in turn is a subsystem of an even larger *system* called the Milky Way Galaxy.

## The Earth System

The Earth system has a nearly endless array of subsystems in which matter is recycled over and over again. One familiar loop or subsystem is the *hydrologic cycle*. It represents the unending circulation of Earth's water among the hydrosphere, atmosphere, biosphere, and geosphere (**Fig. 1.15**). Water enters the atmosphere through evaporation from Earth's surface and transpiration from plants. Water vapor (water in the gaseous state) condenses in the atmosphere to form clouds, which in turn produce precipitation that falls back to Earth's surface. Some of the rain that falls onto the land infiltrates (soaks into the ground) and is later taken up by plants or becomes groundwater, and some flows across the surface toward the ocean.

The parts of the Earth system are linked so that a change in one part can produce changes in any or all of the other parts. For example, when a volcano erupts, lava from Earth's interior may flow out at the surface and block a nearby valley. This new obstruction influences the region's drainage system by creating a lake or causing streams to change course. The large quantities of volcanic ash and gases that can be emitted during an eruption might be blown high into the atmosphere and influence the amount of solar energy that can reach Earth's surface. The result could be a drop in air temperatures over the entire hemisphere.

Where the surface is covered by lava flows or a thick layer of volcanic ash, existing soils are buried. This causes the

▼ **Figure 1.15 The hydrologic cycle** Water readily changes state from liquid, to gas (vapor), to solid at the temperatures and pressures occurring on Earth. This cycle traces the movements of water among Earth's four spheres. It is one of many subsystems that collectively make up the Earth system.

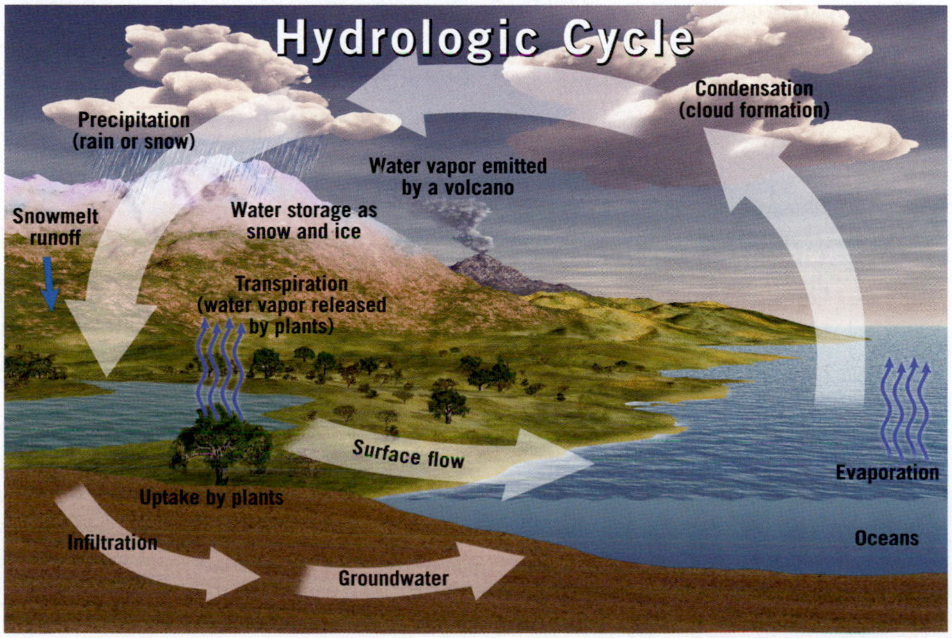

atmosphere, in the hydrosphere, and at Earth's surface. Weather and climate, ocean circulation, and erosional processes are driven by energy from the Sun. Earth's interior is the second source of energy. Heat remaining from when our planet formed and heat that is continuously generated by radioactive decay power the internal processes that produce volcanoes, earthquakes, and mountains.

Humans are *part of* the Earth system, a system in which the living and nonliving components are entwined and interconnected. Therefore, our actions produce changes in all the other parts. When we burn gasoline and coal, dispose of wastes, and clear the land, we cause other parts of the system to respond, often in unforeseen ways. Throughout this book, you will learn about some of Earth's subsystems, including the hydrologic system and the climate system. Remember that these components *and we humans* are all part of the complex interacting whole we call the Earth system.

▲ **Figure 1.16 Change is constant** When Mount St. Helens erupted in May 1980, the area shown here was buried by a volcanic mudflow. Now, plants are reestablished and new soil is forming.

soil-forming processes to begin anew and to transform the new surface material into soil (**Fig. 1.16**). The soil that eventually forms will reflect the interactions among many parts of the Earth system—the volcanic parent material, the climate, and the impact of biological activity. Of course, there would also be significant changes in the biosphere. Some organisms and their habitats would be eliminated by the lava and ash, whereas new settings for life, such as the lake, would be created. The potential climate change could also impact sensitive life-forms and geologic processes.

The Earth system is characterized by processes that vary on spatial scales from fractions of millimeters to thousands of kilometers. Time scales for Earth's processes range from milliseconds to billions of years. As we learn about Earth, it becomes increasingly clear that despite significant separations in distance or time, many processes are connected, and a change in one component can influence the entire system.

The Earth system is powered by energy from two sources. The Sun drives external processes that occur in the

### ✔ Concept Checks 1.3

1. List and briefly define the four spheres that constitute the Earth system.
2. Compare the height of the atmosphere to the thickness of the geosphere.
3. How much of Earth's surface do oceans cover? How much of the planet's total water supply do the oceans represent?
4. What is a system? List three examples.
5. What are the two sources of energy for the Earth system?

## 1.4 Composition of the Atmosphere

**List the major gases composing Earth's atmosphere and identify the components that are most important meteorologically. Explain why ozone depletion is a significant global issue.**

MM GEODe ▸ Introduction to the Atmosphere ▸ Composition of the Atmosphere

Sometimes the term *air* is used as if it were a specific gas, but it is not. Rather, **air** is a *mixture* of many discrete gases, each with its own physical properties, in which varying quantities of tiny solid and liquid particles are suspended (**Box 1.2**). The composition of air is not constant; it varies from time to time and from place to place. If the water vapor, dust, and other variable components were removed from the atmosphere, we would find that its makeup is very stable up to an altitude of about 80 kilometers (50 miles).

## Box 1.2 Origin and Evolution of Earth's Atmosphere

The air we breathe is a stable mixture of about 78 percent nitrogen, 21 percent oxygen, nearly 1 percent argon, and small amounts of gases such as carbon dioxide and water vapor. However, our planet's original atmosphere 4.6 billion years ago was substantially different.

### Earth's Primitive Atmosphere

Early in Earth's formation, its atmosphere likely consisted of gases most common in the early solar system: hydrogen, helium, methane, ammonia, carbon dioxide, and water vapor. The lightest of these gases, hydrogen and helium, escaped into space because Earth's gravity was too weak to hold them. Most of the remaining gases were probably scattered into space by strong *solar winds* (vast streams of particles) from a young active Sun. (All stars, including the Sun, apparently experience a highly active stage early in their evolution, during which solar winds are very intense.)

Earth's first enduring atmosphere was generated by a process called *outgassing,* through which gases trapped in the planet's interior are released. Outgassing from hundreds of active volcanoes still remains an important planetary function worldwide (**Fig. 1.B**). However, early in Earth's history, when massive heating and fluid-like motion occurred in the planet's interior, the gas output must have been immense. Our understanding of modern volcanic eruptions indicates that Earth's primitive atmosphere probably consisted of mostly water vapor, carbon dioxide, and sulfur dioxide, with minor amounts of other gases and minimal nitrogen. Most importantly, free oxygen was not present.

**Video** MM
The Influence of Volcanic Ash
http://goo.gl/oYEzbE

▼ **Figure 1.B Outgassing** Earth's first enduring atmosphere was formed by a process called *outgassing,* which continues today, from hundreds of active volcanoes worldwide.

### Oxygen in the Atmosphere

As Earth cooled, water vapor condensed and formed clouds, and torrential rains began to fill low-lying areas that became the oceans. In those oceans, nearly 3.5 billion years ago, photosynthesizing bacteria began to release oxygen into the water. During photosynthesis, organisms use the Sun's energy to produce organic material (energetic molecules of sugar containing hydrogen and carbon) from carbon dioxide ($CO_2$) and water ($H_2O$). The first bacteria probably used hydrogen sulfide ($H_2S$) as the source of hydrogen rather than water. One of the earliest bacteria, *cyanobacteria* (once called blue-green algae), began to produce oxygen as a by-product of photosynthesis.

Initially, the newly released oxygen was readily consumed by chemical reactions with other atoms and molecules (particularly iron) in the ocean (**Fig. 1.C**). Once the available iron satisfied its need for oxygen and as the number of oxygen-generating organisms increased, oxygen began to build up in the atmosphere. Chemical analyses of rocks suggest that a significant amount of oxygen appeared in the atmosphere as early as 2.2 billion years ago and increased steadily until it reached stable levels about 1.5 billion years ago. Obviously, the availability of free oxygen had a major impact on the development of life and vice versa. Earth's atmosphere evolved together with its life-forms from an oxygen-free envelope to an oxygen-rich environment.

▲ **Figure 1.C Atmospheric change recorded in the rocks** These ancient layered, iron-rich rocks, called *banded iron formations,* were deposited during a geologic span known as the Precambrian. Much of the oxygen generated as a byproduct of photosynthesis was readily consumed by chemical reactions with iron to produce these rocks.

Another significant benefit of the "oxygen explosion" is that oxygen molecules ($O_2$) readily absorb ultraviolet radiation and rearrange themselves to form *ozone* ($O_3$). Today, ozone is concentrated above the surface in a layer called the *stratosphere,* where it absorbs much of the Sun's ultraviolet radiation that strikes the upper atmosphere. For the first time, Earth's surface was protected from this type of solar radiation, which is particularly harmful to DNA. Marine organisms had always been shielded from ultraviolet radiation by the oceans, but the development of the atmosphere's protective ozone layer made the continents more hospitable.

### Questions

1. What was the source of the gases that composed Earth's first enduring atmosphere?
2. What was the source of the atmosphere's first free oxygen?

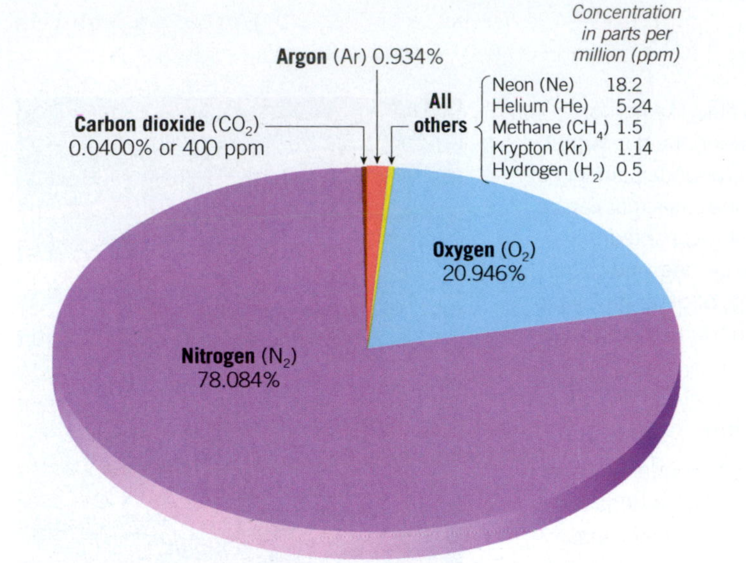


▲ **Figure 1.17 Composition of the atmosphere** Proportional volume of gases composing dry air. Nitrogen and oxygen obviously dominate.

As you can see in **Figure 1.17**, two gases—nitrogen and oxygen—make up about 99 percent of the volume of clean, dry air. Although these gases are the most plentiful components of the atmosphere and are of great significance to life on Earth, they are of little or no importance in affecting weather phenomena. The remaining 1 percent of dry air is mostly the inert gas argon (0.93 percent) plus tiny quantities of a number of other gases listed in Figure 1.17.

## Carbon Dioxide

Carbon dioxide, a gas present in only minute amounts (0.0400 percent, or 400 parts per million [ppm]), is nevertheless a meteorologically important constituent of air. Carbon dioxide is of great interest to meteorologists because it is an efficient absorber of energy emitted by Earth and thus influences the heating of the atmosphere. Although the proportion of carbon dioxide in the atmosphere is relatively uniform from place to place and at different heights in the atmosphere, its percentage has been rising steadily for more than a century. **Figure 1.18** is a graph that shows the growth in atmospheric $CO_2$ since 1958. Much of this rise is attributed to the burning of ever-increasing quantities of fossil fuels, such as coal and oil. Some of this additional carbon dioxide is absorbed by the waters of the ocean or is used by plants, but more than 40 percent remains in the air. Estimates project that by sometime in the second half of the twenty-first century, carbon dioxide levels will be twice as high as pre-industrial levels.

Most atmospheric scientists agree that increased carbon dioxide concentrations have contributed to a warming of Earth's atmosphere over the past several decades and will continue to do so in the decades to come. The magnitude of such temperature changes is uncertain and depends partly on the quantities of $CO_2$ contributed by human activities in the years ahead. The role of carbon dioxide in the atmosphere and its possible effects on climate are examined in more detail in Chapters 2 and 14.

**Video** (MM) Global Changes in Carbon Dioxide Concentrations
http://goo.gl/i7XlcL

## Variable Components

Air includes many gases and particles that vary significantly from time to time and place to place. Important examples include water vapor, aerosols, and ozone. Although usually present in small percentages, they can have significant effects on weather and climate.

**Water Vapor** You are probably familiar with the term *humidity* from watching weather reports on TV. Humidity is a reference to the amount of water vapor in the air. As you will learn in Chapter 4, there are several ways to express humidity. The amount of water vapor in the air varies considerably, from practically none at all up to about 4 percent by volume. Why is such a small fraction

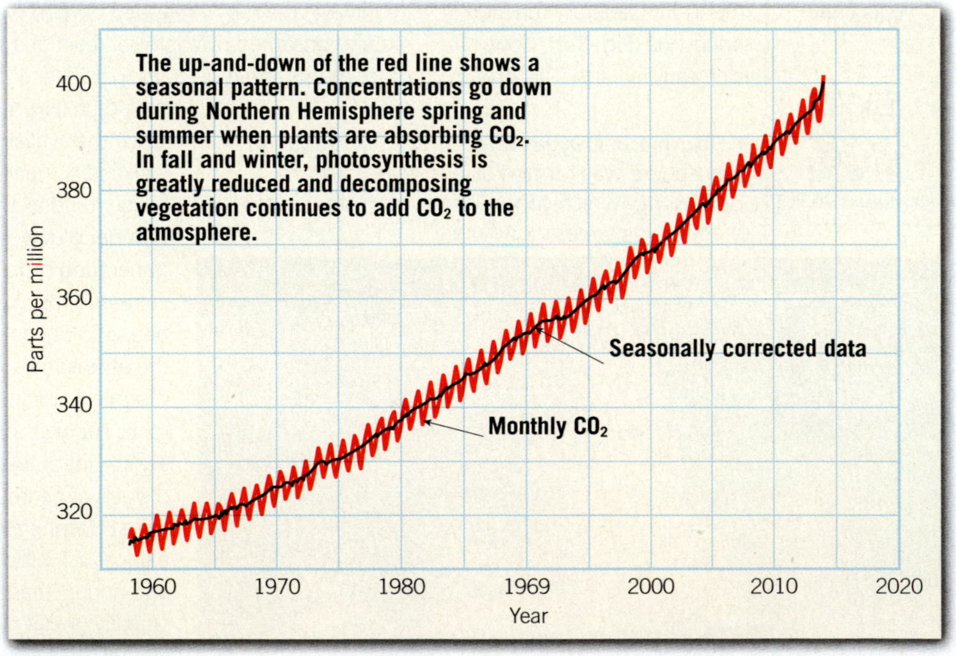


(MM) http://goo.gl/xcXbL

▲ **SmartFigure 1.18 Monthly $CO_2$ concentrations** Atmospheric $CO_2$ has been measured at Mauna Loa Observatory, Hawaii, since 1958. There has been a consistent increase since monitoring began. This graphic portrayal is known as the *Keeling Curve,* in honor of the scientist who originated the measurements.

of the atmosphere so significant? The fact that water vapor is the source of all clouds and precipitation would be enough to explain its importance. However, water vapor has other roles. Like carbon dioxide, water vapor absorbs heat given off by Earth as well as some solar energy. It is therefore important when we examine the heating of the atmosphere and the movement of energy on Earth.

When water changes from one state to another (see Figure 4.3, page 92), it absorbs or releases heat. This energy is termed *latent heat,* which means "hidden heat." As we shall see in later chapters, water vapor in the atmosphere transports this latent heat from one region to another, and it is the energy source that helps drive many storms.

**students sometimes ask...**

### Why does the graph in Figure 1.18 have so many ups and downs?

Photosynthesis, the process by which green plants convert sunlight into chemical energy, removes carbon dioxide from the air. In spring and summer, vigorous plant growth in the extensive land areas of the Northern Hemisphere removes atmospheric $CO_2$, so the graph takes a dip. As winter approaches, many plants die or shed leaves. The decay of organic matter returns $CO_2$ to the air, causing the graph to spike upward.

**Aerosols** The movements of the atmosphere are sufficient to keep a large quantity of solid and liquid particles suspended within it. Although visible dust sometimes obscures the sky, these relatively large particles are too heavy to stay in the air very long. Still, many particles are microscopic and remain suspended for considerable periods of time. They may originate from many sources, both natural and human made, and include sea salts from breaking waves, fine soil blown into the air, smoke and soot from fires, pollen and microorganisms lifted by the wind, ash and dust from volcanic eruptions, and more (Fig. 1.19). Collectively, these tiny solid and liquid particles are called **aerosols**.

Aerosols are most numerous in the lower atmosphere near their primary source, Earth's surface. Nevertheless, the upper atmosphere is not free of them because some dust is carried to great heights by rising currents of air, and other particles are contributed by meteoroids that disintegrate as they pass through the atmosphere.

From a meteorological standpoint, these tiny, often invisible particles can be significant. First, as you will see in Chapter 4, many act as surfaces on which water vapor may condense, a critical function in the formation of clouds and fog. Second, aerosols can absorb or reflect incoming solar radiation. Thus, when an air-pollution episode is occurring or when ash fills the sky following a volcanic eruption, the amount

▼ **Figure 1.19 Aerosols A.** The satellite image shows two examples of aerosols. First, a large dust storm is blowing across northeastern China toward the Korean Peninsula. Second, a dense haze toward the south (bottom center) is human-generated air pollution. **B.** As the photo on the right shows, dust in the air can cause sunsets to be especially colorful.

A.

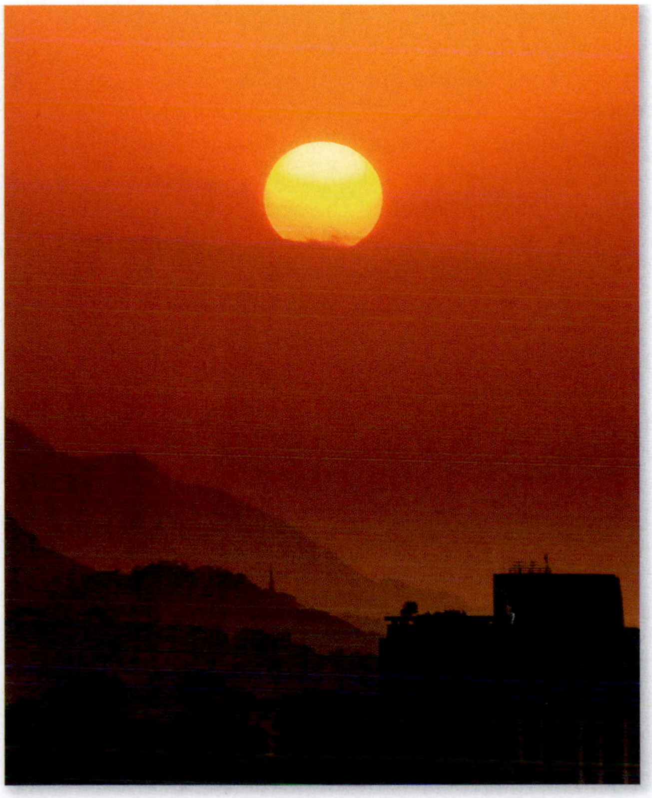

B.

of sunlight reaching Earth's surface can be measurably reduced. Finally, aerosols contribute to an optical phenomenon we have all observed—the varied hues of red and orange at sunrise and sunset. The photo on the right in Figure 1.19 illustrates this phenomenon.

**The ozone hole on September 16, 2013, reached 24 million km$^2$ (9.3 million mi$^2$)—an area about the size of North America.**

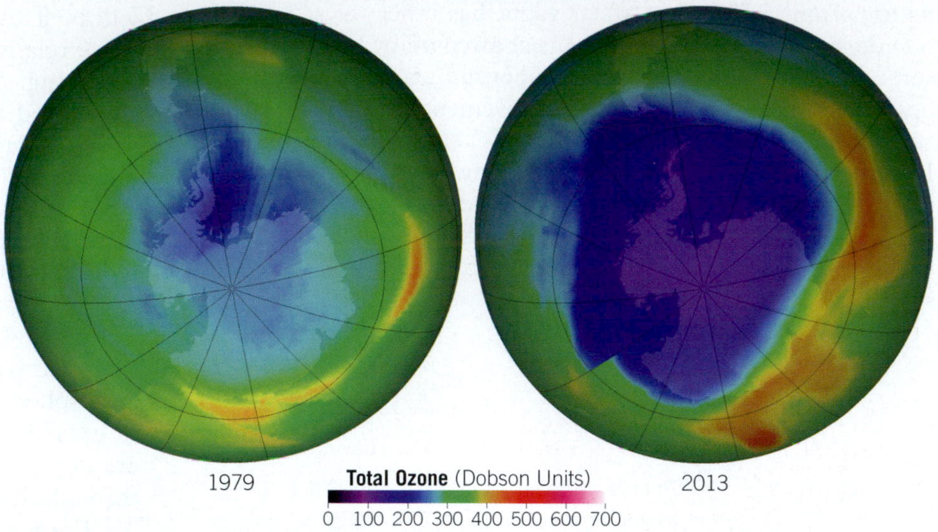


▲ **SmartFigure 1.20 Antarctic ozone hole** The two satellite images show ozone distribution in the Southern Hemisphere on the days in September 1979 and 2013 when the ozone hole was largest. The dark blue shades over Antarctica correspond to the region with the sparsest ozone. The ozone hole is not technically a "hole" where no ozone is present but a region of exceptionally depleted ozone in the stratosphere over the Antarctic that occurs in the spring.

MM http://goo.gl/qH9Hy

Animation MM Ozone Depletion http://goo.gl/llKhez

Animation MM Ozone Hole http://goo.gl/Z9LsBX

**Ozone** Another important component of the atmosphere is **ozone**. It is a form of oxygen that combines three oxygen atoms in each molecule ($O_3$). Ozone is not the same as the oxygen we breathe, which has two atoms per molecule ($O_2$). There is very little ozone in the atmosphere. Overall, it represents just 3 out of every 10 million molecules. Moreover, its distribution is not uniform. In the lowest portion of the atmosphere, ozone represents less than 1 part in 100 million. It is concentrated in a layer called the *stratosphere,* between 10 and 50 kilometers (6 and 31 miles) above the Earth's surface.

In this altitude range, oxygen molecules ($O_2$) are split into single atoms of oxygen (O) when they absorb ultraviolet radiation emitted by the Sun. Ozone is then created when a single atom of oxygen (O) and a molecule of oxygen ($O_2$) collide. This must happen in the presence of a third, neutral molecule that acts as a *catalyst* by allowing the reaction to take place without itself being consumed in the process. Ozone is concentrated in the 10- to 50-kilometer height range because a crucial balance exists there: The ultraviolet radiation from the Sun is sufficient to produce single atoms of oxygen, and there are enough gas molecules to bring about the required collisions.

The presence of the ozone layer in our atmosphere is essential to those of us who are land dwellers. The reason is that ozone absorbs the potentially harmful ultraviolet (UV) radiation from the Sun. If ozone did not filter a great deal of the ultraviolet radiation, and if the Sun's UV rays reached the surface of Earth undiminished, land areas on our planet would be uninhabitable for most life as we know it. Thus, anything that reduces the amount of ozone in the atmosphere could affect the well-being of life on Earth. Just such a problem is described next.

## Ozone Depletion: A Global Issue

Although stratospheric ozone is concentrated 10 to 50 kilometers (6 to 31 miles) above Earth's surface, it is vulnerable to human activities. Chemicals manufactured by people break up ozone molecules in the stratosphere, weakening our shield against UV rays. This loss of ozone is a serious global-scale environmental problem. Measurements over the past three decades confirm that ozone depletion is occurring worldwide and is especially pronounced above Earth's poles. You can see this effect over the South Pole in **Figure 1.20**.

Over the past 75 years, people have unintentionally placed the ozone layer in jeopardy by polluting the atmosphere. The most significant of the offending chemicals are known as *chlorofluorocarbons* (CFCs). Over the decades, CFCs were used as coolants for air-conditioning and refrigeration equipment, cleaning solvents for electronic components, and propellants for aerosol sprays, as well as in the production of certain plastic foams.

Because CFCs are practically inert (that is, not chemically active) in the lower atmosphere, some of these gases gradually make their way up to the ozone layer, where sunlight separates the chemicals into their constituent atoms. The chlorine atoms released this way break up some of the ozone molecules.

Because ozone filters out most of the UV radiation from the Sun, a decrease in its concentration permits more of these harmful wavelengths to reach Earth's surface. UV radiation's most serious threat to human health is an increased risk of skin cancer. An increase in damaging UV radiation also can impair the human immune system as well as promote cataracts, a clouding of the eye lens that reduces vision and may cause blindness if not treated.

In response to this problem, an international agreement known as the *Montreal Protocol* was developed in 1987, under the sponsorship of the United Nations, to eliminate the production and use of CFCs. More than 190 nations eventually ratified the treaty. Although relatively strong action has been taken, CFC levels in the atmosphere will not drop rapidly. Once CFC molecules are in the atmosphere, they can take many years to reach the ozone layer, and once there, they can remain active for decades. This does not promise a near-term reprieve for the ozone layer. Between 2060 and 2075, the abundance of ozone-depleting gases is projected to fall to values that existed before the ozone hole began to form in the 1980s.

## ✔ Concept Checks 1.4

1. Is *air* a specific gas? Explain.
2. What are the two major components of clean, dry air? What proportion does each represent?
3. Why is carbon dioxide an important component of Earth's atmosphere? Why are water vapor and aerosols important constituents?
4. What is ozone? Why is ozone important to life on Earth? What are CFCs, and what is their connection to ozone depletion?

**students sometimes ask...**

### Isn't ozone some sort of pollutant?

Although the naturally occurring ozone in the stratosphere is critical to life on Earth, it is considered a pollutant when produced at ground level because it can damage vegetation and harm human health. Ozone is a major component in a noxious mixture of gases and particles called *photochemical smog* formed from pollutants emitted by motor vehicles and industries. Chapter 13 provides more information about this.

## 1.5 Vertical Structure of the Atmosphere

**Interpret a graph that shows changes in air pressure from Earth's surface to the top of the atmosphere.**
**Sketch and label a graph that shows the thermal structure of the atmosphere.**

MM® **GEODe** ▶ Introduction to the Atmosphere ▶ Extent of the Atmosphere/Thermal Structure of the Atmosphere

To say that the atmosphere begins at Earth's surface and extends upward is obvious. However, where does the atmosphere end, and where does outer space begin? There is no sharp boundary; the atmosphere rapidly thins as you travel away from Earth, until there are too few gas molecules to detect.

### Pressure Changes

To understand the vertical extent of the atmosphere, let us examine the changes in atmospheric pressure with height. Atmospheric pressure is simply the weight of the air above. At sea level the average pressure is slightly more than 1000 millibars. This corresponds to a weight of slightly more than 1 kilogram per square centimeter (14.7 pounds per square inch). Obviously, the pressure at higher altitudes is less because there is less air above these altitudes (Fig. 1.21).

One-half of the atmosphere lies below an altitude of 5.6 kilometers (3.5 miles). At about 16 kilometers (10 miles), 90 percent of the atmosphere has been traversed, and above 100 kilometers (62 miles), only 0.00003 percent of all the gases composing the atmosphere remain.

At an altitude of 100 kilometers, the atmosphere is so thin that the density of air is less than could be found in the most perfect artificial vacuum at the surface. Nevertheless, the

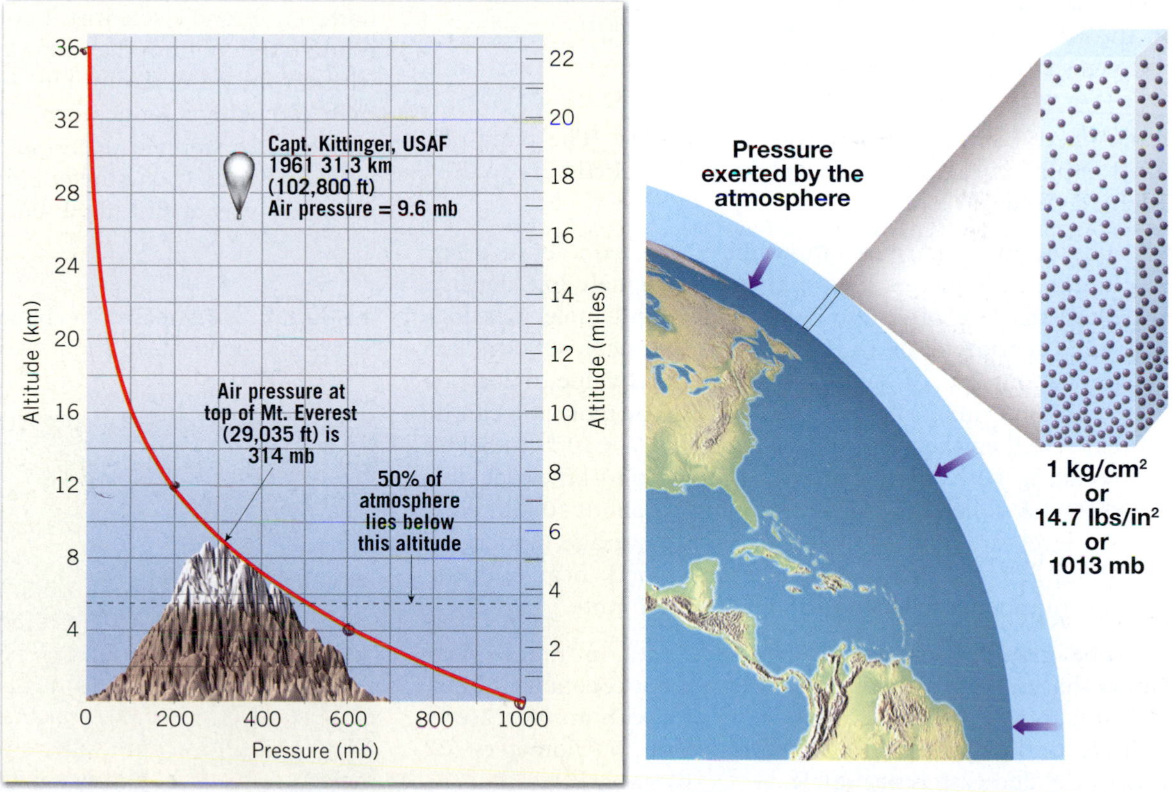


▲ **Figure 1.21 Air pressure changes with altitude** The rate of pressure decrease with an increase in altitude is not constant. Pressure decreases rapidly near Earth's surface and more gradually at greater heights. Put another way, the figure shows that the vast bulk of the gases making up the atmosphere is very near Earth's surface and that the gases gradually merge with the emptiness of space.

**eye ON THE atmosphere 1.1**

This jet is cruising at an altitude of 10 kilometers (6.2 miles).

**Questions**

1. Refer to the graph in Figure 1.21. What is the approximate air pressure where the jet is flying?
2. About what percentage of the atmosphere is below the jet (assuming that the pressure at the surface is 1000 millibars)?

atmosphere continues to even greater heights. The truly rarefied nature of the outer atmosphere is described very well by Richard Craig:

> The earth's outermost atmosphere, the part above a few hundred kilometers, is a region of extremely low density. Near sea level, the number of atoms and molecules in a cubic centimeter of air is about $2 \times 10^{19}$; near 600 km, it is only about $2 \times 10^{7}$, which is the sea-level value divided by a million million. At sea level, an atom or molecule can be expected, on the average, to move about $7 \times 10^{-6}$ cm before colliding with another particle; at the 600-km level, this distance, called the "mean free path," is about 10 km. Near sea level, an atom or molecule, on the average, undergoes about $7 \times 10^{9}$ such collisions each second; near 600 km, this number is reduced to about 1 each minute.*

The graphic portrayal of pressure data in Figure 1.21 shows that the rate of pressure decrease is not constant. Rather, air pressure decreases at a decreasing rate with an increase in altitude until, beyond an altitude of about 35 kilometers (22 miles), the decrease is negligible.

Put another way, data illustrate that air is highly compressible—that is, it expands with decreasing pressure and becomes compressed with increasing pressure. Consequently, traces of our atmosphere extend for thousands of kilometers beyond Earth's surface. Thus, to say where the atmosphere ends and outer space begins is arbitrary and, to a large extent, depends on what phenomenon one is studying. It is apparent that there is no sharp boundary.

In summary, data on vertical air pressure changes show that the vast bulk of the gases making up the atmosphere are very near Earth's surface and that the gases gradually merge with the emptiness of space. When compared with the size of the solid Earth, the envelope of air surrounding our planet is indeed very shallow.

## Temperature Changes

By the early twentieth century much had been learned about the lower atmosphere. The upper atmosphere was partly known from indirect methods. Data from balloons and kites had revealed that the air temperature dropped with increasing height above Earth's surface. This phenomenon is felt by anyone who has climbed a high mountain and is obvious in pictures of snow-capped mountaintops rising above snow-free lowlands (**Fig. 1.22**).

Although measurements had not been taken above a height of about 10 kilometers (6 miles), scientists believed that the temperature continued to decline with height to a value of absolute zero (–273°C) at the outer edge of the atmosphere. In 1902, however, French scientist Leon Philippe Teisserenc de Bort refuted the notion that temperature decreases continuously with an increase in altitude. In studying the results of more than 200 balloon launchings, Teisserenc de Bort found that the temperature stopped decreasing and leveled off at an altitude between 8 and 12 kilometers (5 and 7.5 miles). This surprising discovery was at first doubted, but subsequent data gathering

**▼ Figure 1.22 Temperature change in the troposphere**
Snow-capped mountains and snow-free lowlands are a reminder that temperatures decrease as we go higher in the troposphere.

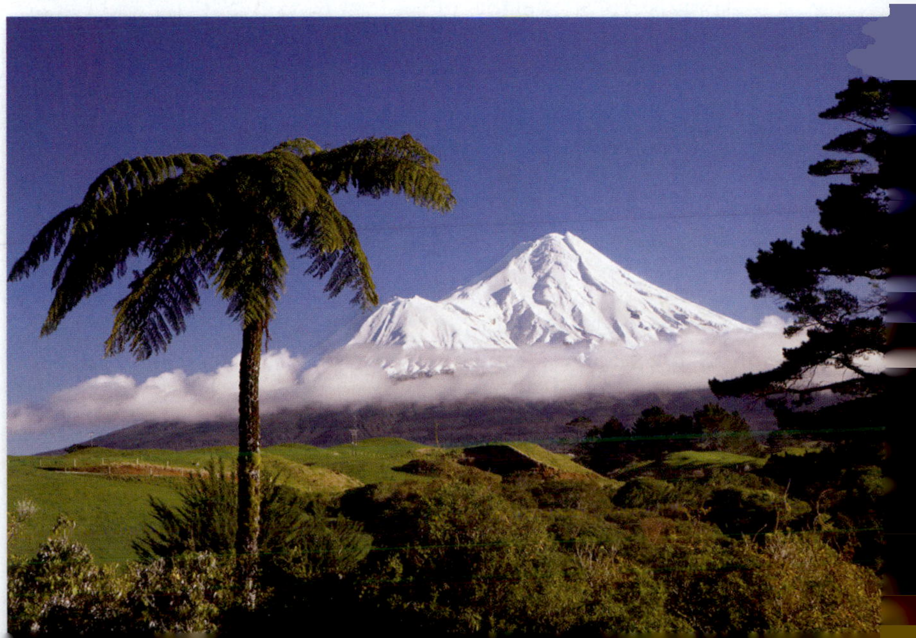

*Richard Craig, *The Edge of Space: Exploring the Upper Atmosphere* (New York: Doubleday & Company, Inc., 1968), p. 130.

confirmed his findings. Later, through the use of balloons and rocket-sounding techniques, the temperature structure of the atmosphere up to great heights became clear. Today the atmosphere is divided vertically into four layers, on the basis of temperature (Fig. 1.23).

**Troposphere** The bottom layer in which we live, where temperature decreases with an increase in altitude, is the **troposphere**. The term was coined in 1908 by Teisserenc de Bort and literally means the region where air "turns over," a reference to the appreciable vertical mixing of air in this lowermost zone.

The temperature decrease in the troposphere is called the **environmental lapse rate**. Its average value is 6.5°C per kilometer (3.5°F per 1000 feet), a figure known as the *normal lapse rate*. It should be emphasized, however, that the environmental lapse rate is not a constant but rather can be highly variable and must be regularly measured. To determine the actual environmental lapse rate as well as gather information about vertical changes in air pressure, wind, and humidity, radiosondes are used. A **radiosonde** is an instrument package that is attached to a balloon and transmits data by radio as it ascends through the atmosphere (Fig. 1.24). The environmental lapse rate can vary during the course of a day with

▼ **Figure 1.23 Thermal structure of the atmosphere** Earth's atmosphere is traditionally divided into four layers, based on temperature.

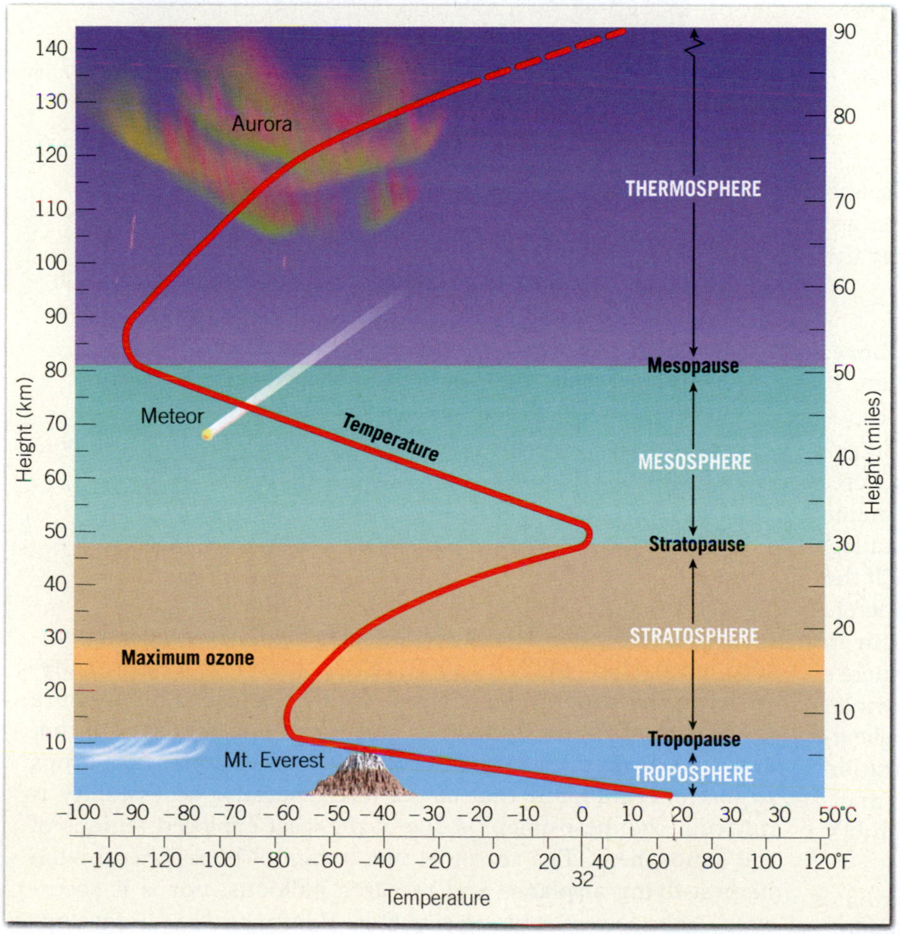

▲ **Figure 1.24 Radiosonde** This lightweight package of instruments is carried aloft by a small weather balloon. It transmits data on vertical changes in temperature, pressure, and humidity in the troposphere. The troposphere is where practically all weather phenomena occur, so it is very important to have frequent measurements.

fluctuations in weather, as well as seasonally and from place to place. Sometimes shallow layers where temperatures actually increase with height are observed in the troposphere. Such reversals are called *temperature inversions* and are described in greater detail in Chapter 13.

The temperature decrease continues to an average height of about 12 kilometers (7.5 miles). Yet the thickness of the troposphere is not the same everywhere. It reaches heights in excess of 16 kilometers (10 miles) in the tropics, but in polar regions it is lower, extending to 9 kilometers (5.5 miles) or less (Fig. 1.25). Warm surface temperatures and highly developed thermal mixing are responsible for the greater vertical extent of the troposphere near the equator. As a result, the environmental lapse rate extends to great heights; despite relatively high surface temperatures below, the lowest tropospheric temperatures are found aloft in the tropics and not at the poles.

The troposphere is the chief focus of meteorologists because it is in this layer that essentially all

68

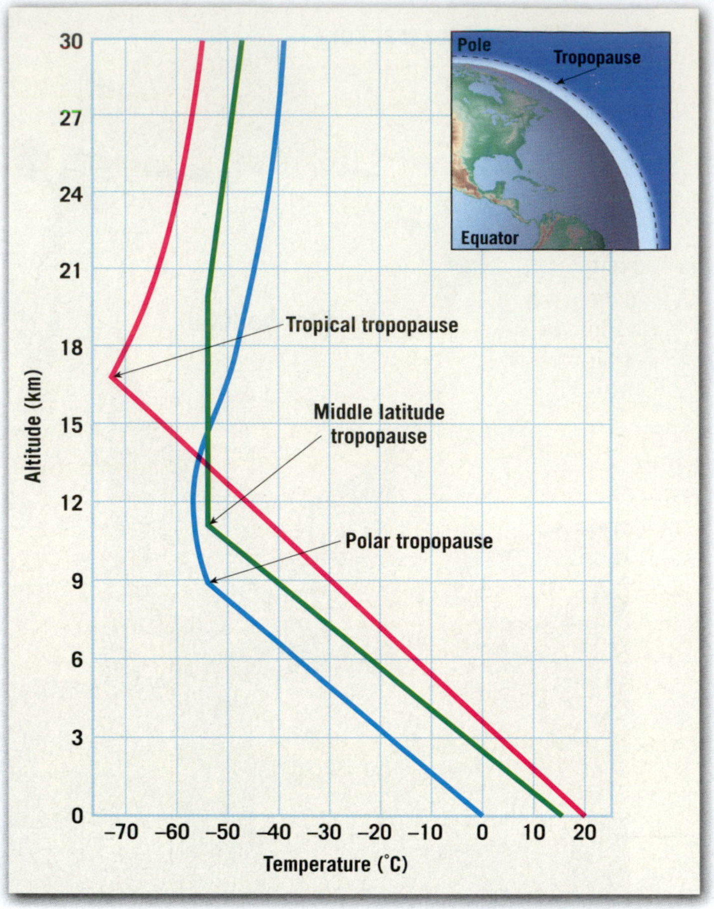


▲ **Figure 1.25 Differences in the height of the tropopause** The variation in the height of the tropopause, as shown on the small inset diagram, is greatly exaggerated.

## eye on the atmosphere 1.2

When this weather balloon was launched, the surface temperature was 17°C. It is now at an altitude of 1 kilometer.

**Questions**

1. What term is applied to the instrument package being carried aloft by the balloon?
2. In what layer of the atmosphere is the balloon?
3. If average conditions prevail, what air temperature is the instrument package recording? How did you figure this out?
4. How will the size of the balloon change, if at all, as it rises through the atmosphere? Explain.

important weather phenomena occur. Almost all clouds and certainly all precipitation, as well as all our violent storms, are born in this lowermost layer of the atmosphere. This is why the troposphere is often called the "weather sphere."

**Stratosphere** Beyond the troposphere lies the **stratosphere**; the boundary between the troposphere and the stratosphere is known as the **tropopause** (see Figure 1.23). Below the tropopause, atmospheric properties are readily transferred by large-scale turbulence and mixing, but above it, in the stratosphere, they are not. In the stratosphere, the temperature at first remains nearly constant to a height of about 20 kilometers (12 miles) before it begins a sharp increase that continues until the **stratopause** is encountered at a height of about 50 kilometers (30 miles) above Earth's surface. Higher temperatures occur in the stratosphere because it is in this layer that the atmosphere's ozone is concentrated. Recall that ozone absorbs ultraviolet radiation from the Sun. Consequently, the stratosphere is heated by the Sun. Although the maximum ozone concentration exists between 15 and 30 kilometers (9 and 19 miles), the smaller amounts of ozone above this height range absorb enough UV energy to cause the higher observed temperatures.

**Mesosphere** In the third layer, the **mesosphere**, temperatures again decrease with height until at the **mesopause**, some 80 kilometers (50 miles) above the surface, the average temperature approaches –90°C (–130°F). The coldest temperatures anywhere in the atmosphere occur at the mesopause. The pressure at the base of the mesosphere is only about one-thousandth that at sea level. At the mesopause, the atmospheric pressure drops to just one-millionth that at sea level. Because accessibility is difficult, the mesosphere is one of the least explored regions of the atmosphere. The reason is that it cannot be reached by the highest-flying airplanes and research balloons, nor is it accessible to the lowest-orbiting satellites. Recent technical developments are just beginning to fill this knowledge gap.

**Thermosphere** The fourth layer extends outward from the mesopause and has no well-defined upper limit. It is the **thermosphere**, a layer that contains only a tiny fraction of the atmosphere's mass. In the extremely rarified air of this outermost layer, temperatures again increase, due to the absorption of very shortwave, high-energy solar radiation by atoms of oxygen and nitrogen.

Temperatures rise to extremely high values of more than 1000°C (1800°F) in the thermosphere. But such temperatures are not comparable to those experienced near Earth's surface. Temperature is defined in terms of the average speed at which molecules move. Because the gases of the thermosphere are moving at very high speeds, the temperature is very high. But the gases are so sparse that collectively they possess only an insignificant quantity of heat. For this reason, the temperature of a satellite orbiting Earth in the thermosphere is determined chiefly by the amount of solar radiation it absorbs and not by the high temperature of the almost nonexistent surrounding air. If an astronaut inside were to expose his or her hand, the air in this layer would not feel hot.

## The Ionosphere

In addition to the layers defined by vertical variations in temperature, another layer is also recognized in the atmosphere. Located in the altitude range between 80 to 400 kilometers (50 to 250 miles), and thus coinciding with the lower portion of the thermosphere, is an electrically charged layer known as the **ionosphere**. Here molecules of nitrogen and atoms of oxygen are readily ionized as they absorb high-energy shortwave solar radiation. In this process, each affected molecule or atom loses one or more electrons and becomes a positively charged ion, and the electrons are set free to travel as electric currents.

Although ionization occurs at heights as great as 1000 kilometers (620 miles) and extends as low as perhaps 50 kilometers (30 miles), positively charged ions and negative electrons are most dense in the range of 80 to 400 kilometers (50 to 250 miles). The concentration of ions is not great below this zone because much of the short-wavelength radiation needed for ionization has already been depleted. In addition, low atmospheric density at this level results in a large percentage of free electrons being swiftly captured by positively charged ions. Beyond the 400-kilometer (250-mile) upward limit of the ionosphere, the concentration of ions is low because of the extremely low density of the air. Because so few molecules and atoms are present, relatively few ions and free electrons can be produced.

As best we can tell, the ionosphere has little impact on our daily weather. But this layer of the atmosphere is the site of one of nature's most interesting spectacles, the auroras (**Fig. 1.26**). The **aurora borealis** (northern lights) and its Southern Hemisphere counterpart, the **aurora australis** (southern lights), appear in a wide variety of forms. Sometimes the displays consist of vertical streamers in which there can be considerable movement. At other times the auroras appear as a series of luminous expanding arcs or as a quiet glow that has an almost foglike quality.

The occurrence of auroral displays is closely correlated in time with solar-flare activity and, in geographic location, with Earth's magnetic poles. Solar flares are massive magnetic storms on the Sun that emit enormous amounts of energy and great quantities of fast-moving atomic particles. As the clouds of protons and electrons from the solar storm approach Earth, they are captured by its magnetic field, which in turn guides them toward the magnetic poles. Then, as the ions impinge on the ionosphere, they energize the atoms of oxygen and molecules of nitrogen and cause them to emit light—the glow of the auroras. Because the occurrence of solar flares is closely correlated with sunspot activity, auroral displays increase conspicuously at times when sunspots are most numerous.

### ✔ Concept Checks 1.5

1. Does air pressure increase or decrease with an increase in altitude? Is the rate of change constant or variable? Explain.
2. Is the outer edge of the atmosphere clearly defined? Explain.
3. The atmosphere is divided vertically into four layers on the basis of temperature. List these layers in order from lowest to highest. In which layer does practically all of our weather occur?
4. Why does temperature increase in the stratosphere?
5. Why are temperatures in the thermosphere not strictly comparable to those experienced near Earth's surface?
6. What is the *ionosphere*? How is it related to the auroras?

► **Figure 1.26 The auroras** The aurora borealis (northern lights), as seen in Alaska. The same phenomenon occurs toward the South Pole, where it is called the aurora australis (southern lights).

# 1 Concepts in Review Introduction to the Atmosphere

## 1.1 Focus On the Atmosphere

► Distinguish between weather and climate, name the basic elements of weather and climate, and list several important atmospheric hazards.

**Key Terms:** meteorology, weather, climate, element (of weather and climate)

- Meteorology is the scientific study of the atmosphere. Weather refers to the state of the atmosphere at a given time and place. It is constantly changing, sometimes from hour to hour and other times from day to day. Climate is an aggregate of weather conditions, the sum of all statistical weather information that helps describe a place or region.
- The nature of both weather and climate is expressed in terms of the same basic elements, those quantities or properties measured regularly. The most important elements are (1) air temperature, (2) humidity, (3) type and amount of cloudiness, (4) type and amount of precipitation, (5) air pressure, and (6) the speed and direction of the wind.
- Some atmospheric hazards are storm related, such as lightning, blizzards, and hail. Others are not storm related, such as fog, heat waves, and drought.

**Q** This is a scene on a summer day in a portion of southern Utah called Sinbad Country. Hebes Mountain is in the center of the image. Write two brief statements about the place—one that relates to weather and another that might be part of a description of climate.

## 1.2 The Nature of Scientific Inquiry

► Discuss the nature of scientific inquiry, including the construction of hypotheses and the development of theories.

**Key Terms:** hypothesis, theory

- All science is based on the assumption that the natural world behaves in a consistent and predictable manner. Scientists make careful observations, construct tentative explanations for those observations (hypotheses), and then test those hypotheses with field investigations and laboratory work.
- In science, a theory is a well-tested and widely accepted explanation that the scientific community agrees best fits certain observable facts.
- As failed hypotheses are discarded, scientific knowledge moves closer to a correct understanding, but we can never be fully confident that we know all the answers. Scientists must always be open to new information that forces change in our model of the world.

## 1.3 Earth as a System

► List and describe Earth's four major spheres. Define *system* and explain why Earth is considered to be a system.

**Key Terms:** geosphere, atmosphere, hydrosphere, biosphere, Earth system science, system

- Earth's physical environment is traditionally divided into three major parts: the solid Earth, called the geosphere; the water portion of our planet, called the hydrosphere; and Earth's gaseous envelope, called the atmosphere.

- A fourth Earth sphere is the biosphere, the totality of life on Earth. It is concentrated in a relatively thin zone that extends a few kilometers into the hydrosphere and geosphere, and a few kilometers up into the atmosphere.
- Of all the water on Earth, more than 96 percent is in the oceans, which cover nearly 71 percent of the planet's surface.
- Although each of Earth's four spheres can be studied separately, they are all related in a complex and continuously interacting whole that is called the Earth system.
- Earth system science uses an interdisciplinary approach to integrate the knowledge of several academic fields in the study of our planet and its global environmental problems.
- The two sources of energy that power the Earth system are (1) the Sun, which drives the external processes that occur in the atmosphere, hydrosphere, and at Earth's surface, and (2) heat from Earth's interior that powers the internal processes that produce volcanoes, earthquakes, and mountains.

**Q** Is glacial ice part of the geosphere, or does it belong to the hydrosphere? Explain your answer.

## 1.4 Composition of the Atmosphere

▶ List the major gases composing Earth's atmosphere and identify the components that are most important meteorologically. Explain why ozone depletion is a significant global issue.

**Key Terms:** air, aerosols, ozone

- Air is a mixture of many discrete gases, and its composition varies from time to time and place to place. After water vapor, dust, and other variable components are removed, two gases, nitrogen and oxygen, make up 99 percent of the volume of the remaining clean, dry air. Carbon dioxide ($CO_2$), although present in only minute amounts (0.0400 percent, or 400 ppm), is an efficient absorber of energy emitted by Earth and thus influences the heating of the atmosphere.
- Among the variable components of air, water vapor is important because it is the source of all clouds and precipitation. Like carbon dioxide, water vapor can absorb heat emitted by Earth. When water changes from one state to another, it absorbs or releases heat. In the atmosphere, water vapor transports this latent ("hidden") heat from place to place and is the energy that helps to drive many storms.
- Aerosols are tiny solid and liquid particles that are important because they may act as surfaces onto which water vapor can condense, and they are also absorbers and reflectors of incoming solar radiation.
- Ozone, a form of oxygen that combines three oxygen atoms into each molecule ($O_3$), is a gas concentrated in the 10- to 50-kilometer (6- to 31-mile) height range in the atmosphere and is important to life because it can absorb potentially harmful ultraviolet radiation from the Sun. Over the past 75 years, people have placed Earth's ozone layer in jeopardy by polluting the atmosphere with chlorofluorocarbons (CFCs), which remove some of the gas. Ozone concentrations drop sharply over Antarctica during the Southern Hemisphere spring (September and October). The Montreal Protocol represents a positive international response to the ozone problem.

**Q** This graph shows changes in one atmospheric gas between 2011 and early 2014. Which gas is it? How did you figure it out? Why is the line so wavy?

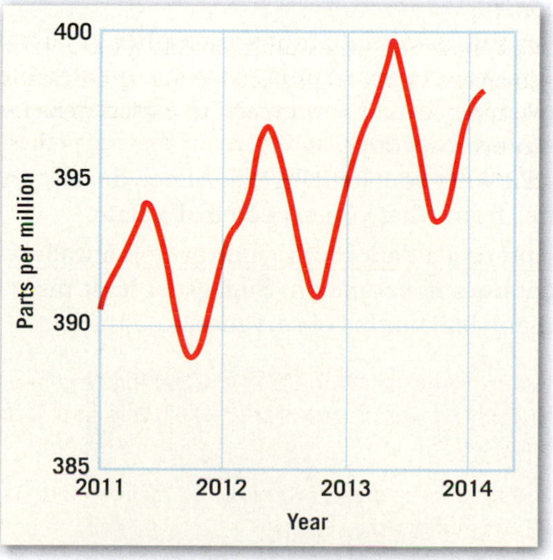

## 1.5 Vertical Structure of the Atmosphere

► Interpret a graph that shows changes in air pressure from Earth's surface to the top of the atmosphere. Sketch and label a graph that shows the thermal structure of the atmosphere.

**Key Terms:** troposphere, environmental lapse rate, radiosonde, stratosphere, tropopause, stratopause, mesosphere, mesopause, thermosphere, ionosphere, aurora borealis, aurora australis

- Because the atmosphere gradually thins with increasing altitude, it has no sharp upper boundary but simply blends into outer space.
- Based on temperature, the atmosphere is divided vertically into four layers. The troposphere is the lowermost layer. In the troposphere, temperature usually decreases with increasing altitude. This environmental lapse rate is variable but averages about 6.5°C per kilometer (3.5°F per 1000 feet). Essentially, all important weather phenomena occur in the troposphere.
- Above the troposphere is the stratosphere, which exhibits warming because of absorption of UV radiation by ozone. In the mesosphere, temperatures again decrease. Upward from the mesosphere is the thermosphere, a layer with only a tiny fraction of the atmosphere's mass and no well-defined upper limit.
- Between 80 and 400 kilometers (50 and 250 miles) above Earth's surface is an electrically charged layer of the atmosphere known as the ionosphere, where molecules of nitrogen and atoms of oxygen are readily ionized as they absorb high-energy, shortwave solar radiation. Auroras (the aurora borealis, northern lights, and its Southern Hemisphere counterpart, the aurora australis, southern lights) occur within the ionosphere. Auroras form as clouds of protons and electrons ejected from the Sun during solar-flare activity enter the atmosphere near Earth's magnetic poles and energize the atoms of oxygen and molecules of nitrogen, causing them to emit light.

## Give it Some Thought

1. Determine which statements refer to weather and which refer to climate. (*Note:* One statement includes aspects of both weather and climate.)
   **a.** The baseball game was rained out today.
   **b.** January is Omaha's coldest month.
   **c.** North Africa is a desert.
   **d.** The high this afternoon was 25°C.
   **e.** Last evening a tornado ripped through central Oklahoma.
   **f.** I am moving to southern Arizona because it is warm and sunny.
   **g.** Thursday's low of –20°C is the coldest temperature ever recorded for that city.
   **h.** It is partly cloudy.
2. Refer to the map in Figure 1.4 to answer the following questions:
   **a.** Does this map relate more to weather or climate?
   **b.** If you were to visit Yuma, Arizona, on a day in November, would you *expect* to experience a sunny day or an overcast day?
   **c.** Might what you actually experience during your visit differ from what you expected? Explain.
3. After entering a dark room, you turn on a wall switch, but the light does not come on. Suggest at least three hypotheses that might explain this observation.

4. Making accurate measurements and observations is a basic part of scientific inquiry. The accompanying radar image, showing the distribution and intensity of precipitation associated with a storm, provides one example. Identify three additional images in this chapter that illustrate ways in which scientific data are gathered. Suggest advantages that might be associated with each example.

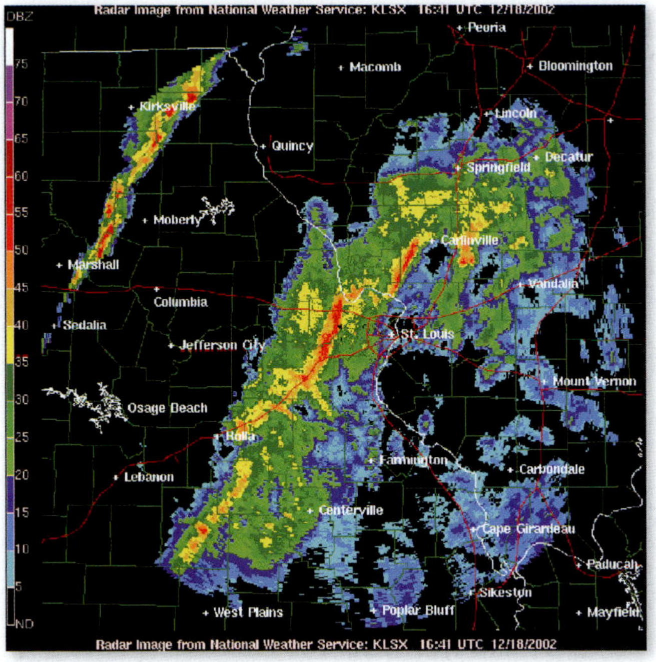


5. Refer to Figure 1.21 to answer the following questions.
   **a.** If you were to climb to the top of Mount Everest, how many breaths of air would you have to take at that altitude to equal one breath at sea level?
   **b.** If you are flying in a commercial jet at an altitude of 12 kilometers (7.5 miles), about what percentage of the atmosphere's mass is below you?

6. If you were ascending from the surface of Earth to the top of the atmosphere, which one of the following measurement devices would be most useful for determining the layer of the atmosphere you were in? Explain.
   **a.** Hygrometer (humidity)
   **b.** Barometer (air pressure)
   **c.** Thermometer (temperature)
7. Where would you expect the thickness of the troposphere (that is, the distance between Earth's surface and the tropopause) to be greater: over Hawaii or Alaska? Why? Do you think it is likely that the thickness of the troposphere over Alaska is different in January than in July? If so, why?
8. The accompanying photo provides an example of interactions among different parts of the Earth system. It is a view of a landslide triggered by extraordinary rains in March 2014. Which of Earth's four "spheres" were involved in this natural disaster that buried a 1-square-mile rural neighborhood near Oso, Washington, and caused more than 40 fatalities? Describe how each contributed to the mudflow.

## Problems

1. Refer to the newspaper-type weather map in Figure 1.3 to answer the following:
   **a.** Estimate the predicted high temperatures in central New York State and the northwestern corner of Arizona.
   **b.** Where is the coldest area on the weather map? Where is the warmest?
   **c.** On this weather map, H stands for the center of a region of high pressure. Does it appear as though high pressure is associated with precipitation or fair weather?
   **d.** Which is warmer—central Texas or central Maine? Would you normally expect this to be the case?
2. Refer to the graph in Figure 1.5 to answer the following questions about temperatures in New York City:
   **a.** What is the approximate average daily high temperature in January? In July?
   **b.** Approximately what are the highest and lowest temperatures ever recorded?
3. Refer to the graph in Figure 1.7. Which year had the greatest number of billion-dollar weather disasters? How many events occurred that year? In which year was the damage amount greatest?
4. Refer to the graph in Figure 1.21 to answer the following:
   **a.** Approximately how much does the air pressure drop (in millibars) between the surface and 4 kilometers (2.5 miles)? (Use a surface pressure of 1000 millibars.)
   **b.** How much does the pressure drop between 4 and 8 kilometers (2.5 and 5 miles)?
   **c.** Based on your answers to parts a and b, select the correct answer: With an increase in altitude, air pressure decreases at a(n) (constant, increasing, decreasing) rate.
5. If the temperature at sea level were 23°C, what would the air temperature be at a height of 2 kilometers, under average conditions?
6. Use the graph of the atmosphere's thermal structure (Figure 1.23) to answer the following:
   **a.** What are the approximate height and temperature of the stratopause?
   **b.** At what altitude is the temperature lowest? What is the temperature at that height?
7. Answer the following questions by examining the graph in Figure 1.25:
   **a.** In which one of the three regions (tropics, middle latitudes, poles) is the *surface* temperature lowest?
   **b.** In which region is the tropopause encountered at the lowest altitude? The highest? What are the altitudes and temperatures of the tropopause in those regions?
8. **a.** On a spring day, a middle-latitude city (about 40° north latitude) has a surface (sea-level) temperature of 10°C. If vertical soundings reveal a nearly constant environmental lapse rate of 6.5°C per kilometer and a temperature at the tropopause of –55°C, what is the height of the tropopause?
   **b.** On the same spring day as in part a, a station near the equator has a surface temperature of 25°C, which is 15°C higher than the middle-latitude city mentioned in part a. Vertical soundings reveal an environmental lapse rate of 6.5°C per kilometer and indicate that the tropopause is encountered at 16 kilometers. What is the air temperature at the tropopause?

## MasteringMeteorology™

Looking for additional review and test prep materials? Visit the Study Area in *MasteringMeteorology*™ to enhance your understanding of this chapter's content by accessing a variety of resources, including Videos, MapMaster™ interactive maps, Geoscience Animations, GEODe, *In the News* RSS feeds, flashcards, web links, self-study quizzes, and an eText version of *The Atmosphere*.

# 2 Heating Earth's Surface and Atmosphere

## Focus on Concepts

*Each statement represents the primary learning objective for the corresponding major heading within the chapter. After you complete the chapter, you should be able to:*

**2.1** Explain what causes the Sun angle and length of daylight to change throughout the year. Describe how these changes result in seasonal changes in temperature.

**2.2** Compare and contrast latent heat and sensible heat.

**2.3** List and describe the three mechanisms of heat transfer.

**2.4** Using the diagram in Figure 2.15 as a guide, describe what happens to incoming solar radiation.

**2.5** Explain what is meant by the statement "The atmosphere is heated from the ground up."

**2.6** Describe the major components of Earth's annual energy budget.

From our everyday experiences, we know that the Sun's rays feel warmer and paved roads become much hotter on clear, sunny days than on overcast days. Pictures of snowcapped mountains remind us that temperatures decrease with altitude. And we know that the fury of winter is always replaced by the newness of spring. What many don't know is that these are manifestations of the same phenomena that causes the blue color of the sky and the red color of a brilliant sunset. All are results of the interaction of solar radiation with Earth's atmosphere and its land–sea surface.

*Weather is the result of the interactions of solar radiation with Earth's atmosphere and its land-sea surface.*

# 2.1 Earth–Sun Relationships

**Explain what causes the Sun angle and length of daylight to change throughout the year. Describe how these changes result in seasonal changes in temperature.**

MM® **GEODe** ▶ Heating Earth's Surface and Atmosphere ▶ Understanding Seasons

The amount of solar energy received at any location varies with latitude, time of day, and season of the year. Contrasting images of polar bears and perpetual ice and palm trees along a tropical beach serve to illustrate the extremes. The unequal heating of Earth's land–sea surface creates winds and drives the ocean's currents, which in turn transport heat from the tropics toward the poles in an unending attempt to balance energy inequalities.

The consequences of these processes are the phenomena we call *weather*. If the Sun were "turned off," global winds and ocean currents would quickly cease. Yet as long as the Sun shines, winds *will* blow and weather *will* persist. So to understand how the dynamic weather machine works, we must know why different latitudes receive different quantities of solar energy and why the amount of solar energy received changes during the course of a year to produce the seasons (Fig. 2.1).

## Earth's Motions

Earth has two principal motions—rotation and revolution. **Rotation** is the spinning of Earth on its axis that takes 24 hours (1 day) and produces the cycle of day and night.

The other motion, **revolution**, refers to Earth's movement in a slightly elliptical orbit around the Sun that takes about 365¼ days (1 year). The distance between Earth and Sun averages about 150 million kilometers (93 million miles). Because Earth's orbit is not perfectly circular, however, the distance varies during the course of a year (Fig. 2.2). Each year, on about January 3, our planet is about 147 million kilometers (91.5 million miles) from the Sun, closer than at any other time—a position called **perihelion**. About 6 months later, on July 4, Earth is about 152 million kilometers (94.5 million miles) from the Sun, farther away than at any other time—a position called **aphelion**.

Although Earth is closest to the Sun and receives up to 7 percent more energy in January than in July, this difference plays only a minor role in producing seasonal temperature variations, as evidenced by the fact that Earth is closest to the Sun during the Northern Hemisphere winter.

## What Causes the Seasons?

If variations in the distance between the Sun and Earth do not cause seasonal temperature changes, what does? You have undoubtedly noticed that the *length of daylight* changes gradually throughout the year. This accounts for some of the differences in the temperatures experienced in summer versus winter: Longer daylight hours result in warmer days.

In addition, *the angle (altitude) of the Sun above the horizon affects the amount of solar energy that reaches Earth's surface*. When the Sun is directly overhead (at a 90° angle), the solar rays are most concentrated and thus most intense. At lower Sun angles, the

**Video** MM®
Season Changes in Global Snow Cover
http://goo.gl/A8D8Ay

▼ **Figure 2.1 An understanding of Earth–Sun relationships is basic to an understanding of the seasons** **A.** Cold winter scene in Chacago, Illinois. **B.** Clear, warm summer day in Chacago, Illinois.

A. B.

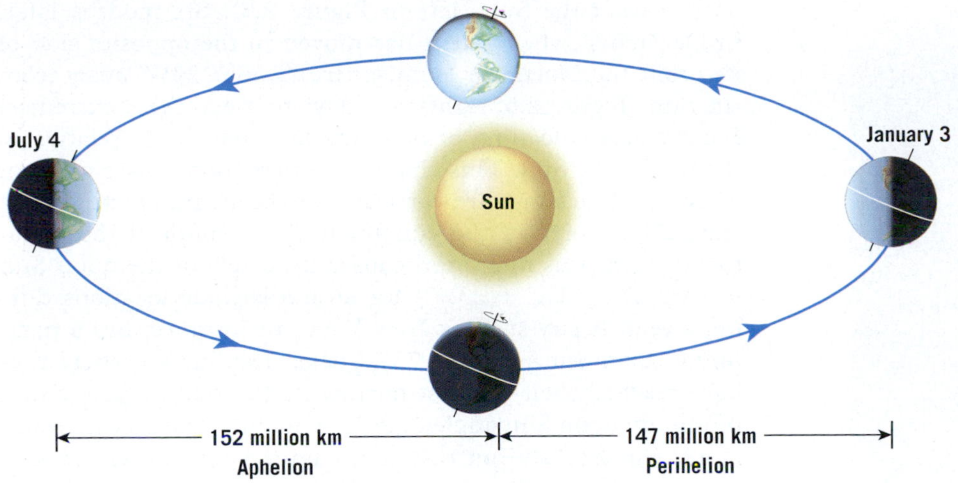


▲ **Figure 2.2 Earth's slightly elliptical orbit around the Sun** Notice that the Earth is farthest from the Sun on July 4 (aphelion) and closest to the Sun on January 3 (perihelion).

rays become more spread out and less intense. This explains why tropical areas, which experience consistently higher Sun angles throughout the year, are much warmer than polar regions, where the Sun angles are lower (Fig. 2.3A). You have probably experienced this when using a flashlight. If the flashlight beam strikes a surface at a 90° angle, a small intense spot is produced (Fig. 2.3B). By contrast, if the beam strikes at any other angle, the area illuminated is larger—but noticeably dimmer.

The angle at which the Sun strikes a particular location varies seasonally. For example, for someone living in Chicago, Illinois, the Sun is highest in the sky at noon on June 21–22. (When comparing seasonal changes in Sun angles, *noon solar time* is used because that is when the Sun in highest in the sky.) But as summer gives way to autumn, the noon Sun gradually appears lower in the sky from one day to the next, and sunset occurs earlier each evening. The lowest Sun angle and earliest sunset in Chicago occurs on December 21–22.

Although Chicago and other midlatitude cities in the Northern Hemisphere experience their shortest day and lowest Sun angle in late December, their lowest average temperatures are usually experienced a few weeks later, in January. The reason for this temperature lag will be discussed in Chapter 3.

Video MM
July Global Movie
http://goo.gl/wjjusN

Video MM
January Global Movie
http://goo.gl/zlDgr2

Sun angle also determines the path that solar rays take as they pass through the atmosphere (Fig. 2.4). When the Sun is directly overhead, the rays strike the atmosphere at a 90° angle and travel the shortest possible route to Earth's surface. Rays entering the atmosphere at a 30° angle must travel twice this distance before reaching the surface, while rays at a 5° angle travel a distance roughly equal to the thickness of 11 atmospheres. The longer the path the rays must travel, the greater the chance that sunlight will be dispersed or absorbed by Earth's atmosphere, which reduces the intensity of sunlight at the surface. These conditions account for the fact that we cannot look directly at the midday Sun, but we enjoy gazing at a sunset.

In summary, the most important reasons for variations in the amount of solar energy reaching a particular location are the seasonal changes in the angle at which the Sun's rays strike the surface and changes in the length of daylight.

## Earth's Orientation

What causes fluctuations in Sun angle and length of daylight during the course of a year? Variations occur because *Earth's orientation to the Sun continually changes*. Earth's axis (the imaginary line through the poles around which Earth rotates) is not perpendicular to the plane of its orbit around the Sun—called the **plane of the ecliptic**. Instead, it is tilted 23½° from the

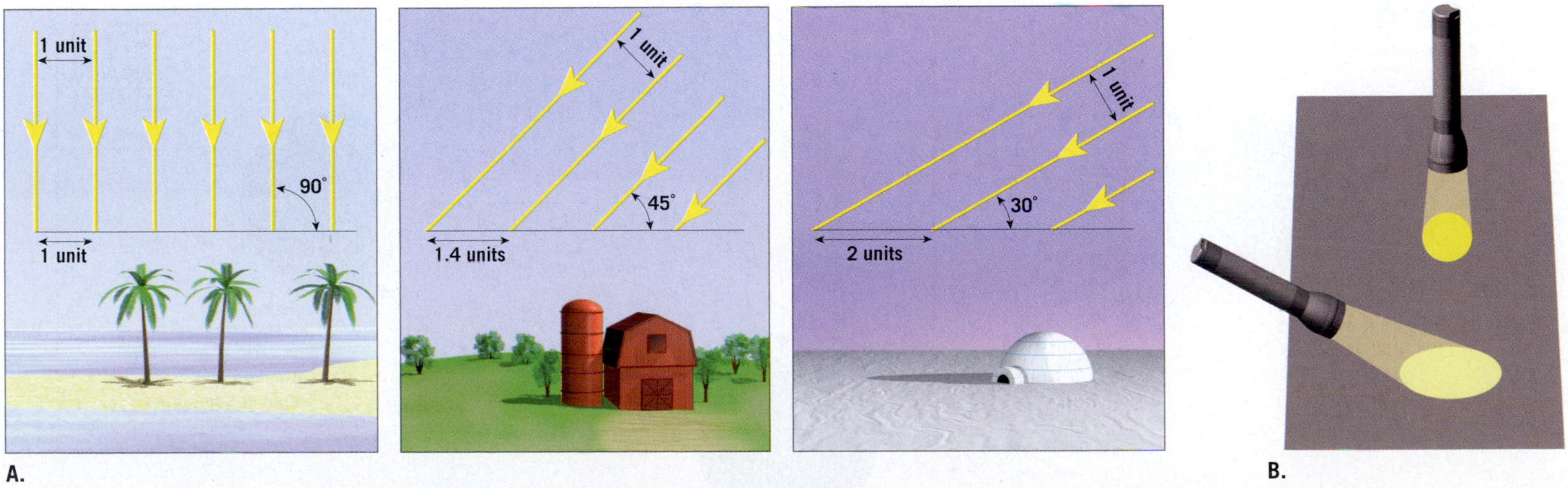


▲ **Figure 2.3 Changes in the Sun's angle cause variations in the amount of solar energy that reaches Earth's surface** **A.** The higher the angle, the more intense the solar radiation reaching the surface. **B.** If a flashlight beam strikes a surface at a 90° angle, a small intense spot is produced, however if it strikes at any other angle, the area illuminated is larger—but noticeably dimmer.

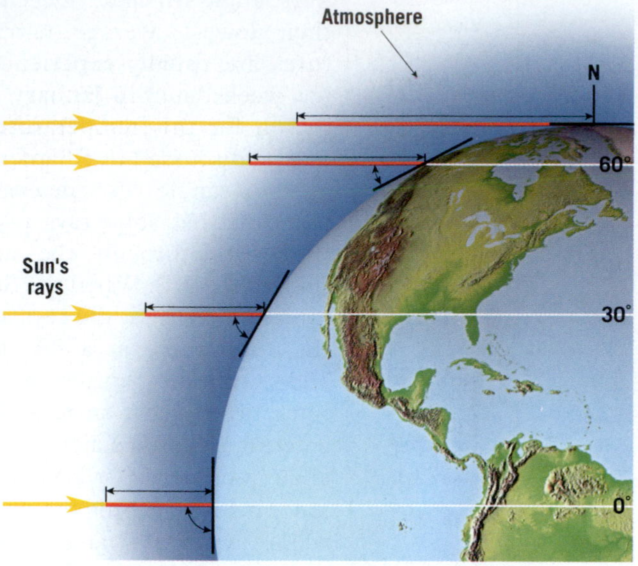


▲ **Figure 2.4 The amount of atmosphere sunlight must traverse before reaching the Earth's surface affects its intensity** Rays striking Earth at a low angle (near the poles) must traverse more of the atmosphere than rays striking at a high angle (around the equator) and thus are subject to greater depletion by reflection, scattering, and absorption.

plane of the ecliptic, called the **inclination of the axis**. If the axis were not inclined, Earth would lack seasons. Because the axis is always pointed in the same direction (toward the North Star), the orientation of Earth's axis to the Sun's rays is constantly changing (**Fig. 2.5**).

For example, on one day in June each year, Earth's position in orbit is such that the Northern Hemisphere is "leaning" 23½° *toward* the Sun (left in Figure 2.5). Six months later, in December, when Earth has moved to the opposite side of its orbit, the Northern Hemisphere "leans" 23½° *away* from the Sun (Figure 2.5, right). On days between these extremes, Earth's axis is leaning at amounts less than 23½° relative to the rays of the Sun. This change in orientation causes the spot where the Sun's rays are vertical to make an annual migration from 23½° north of the equator to 23½° south of the equator. In turn, this migration causes the angle of the noon Sun to vary 47° (23½° + 23½°) for all midlatitude locations during a year. A city such as New York, for instance, has a maximum noon Sun angle of 73½° when the Sun's vertical rays have reached their farthest northward location in June and a minimum noon Sun angle of 26½° 6 months later, a difference of 47° (**Fig. 2.6**). By contrast, a city on the equator will experience an annual migration of half that amount, 23½°. **Box 2.1** explains how the angle of the noon Sun can be calculated for a given latitude.

**students sometimes ask...**

### Is the Sun ever directly overhead anywhere in the United States?

Yes, but only in Hawaii. Honolulu, located on the island of Oahu at about 21° north latitude, experiences a 90° Sun angle twice each year—once at noon on about May 27 and again at noon on about July 20. All of the other states are located north of the Tropic of Cancer and therefore never experience the vertical rays of the Sun.

**Video** (MM)
Global Variations in Insolation Through the Year
http://goo.gl/2VT9ka

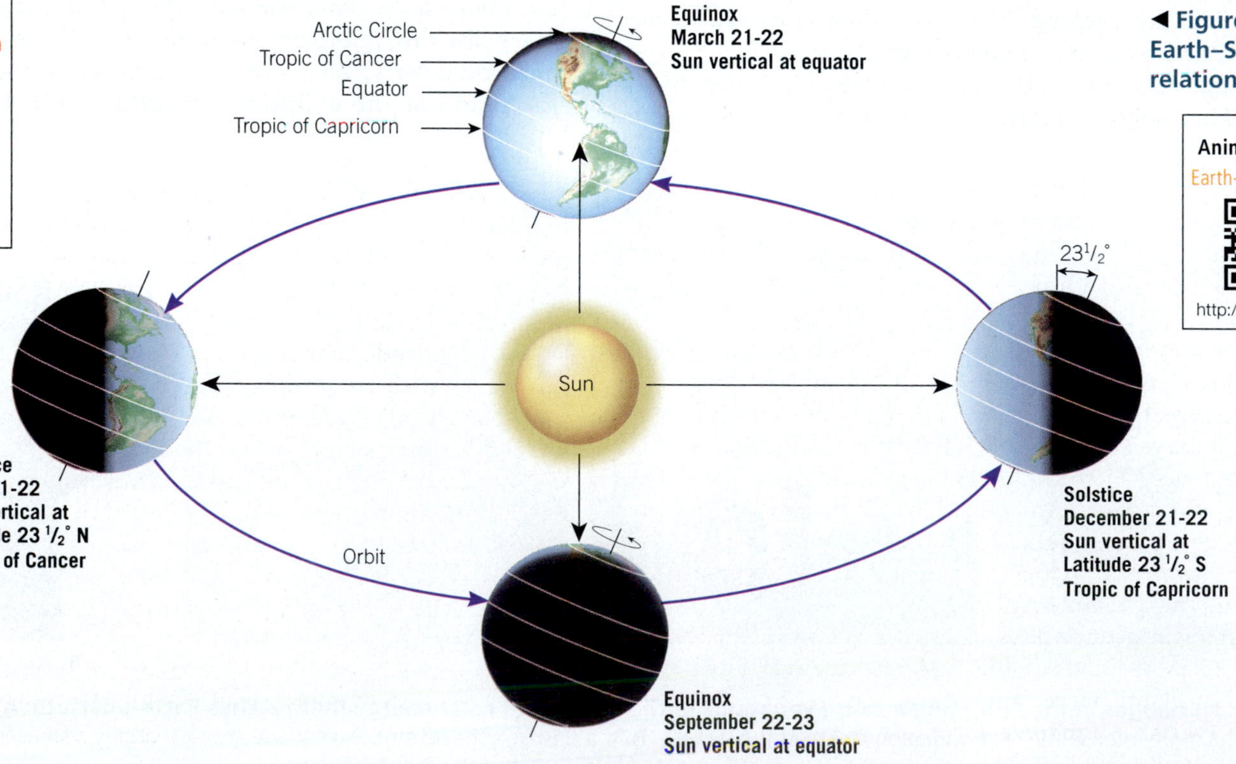


◄ **Figure 2.5 Earth–Sun relationships**

**Animation** (MM)
Earth-Sun Relations
http://goo.gl/HXefff

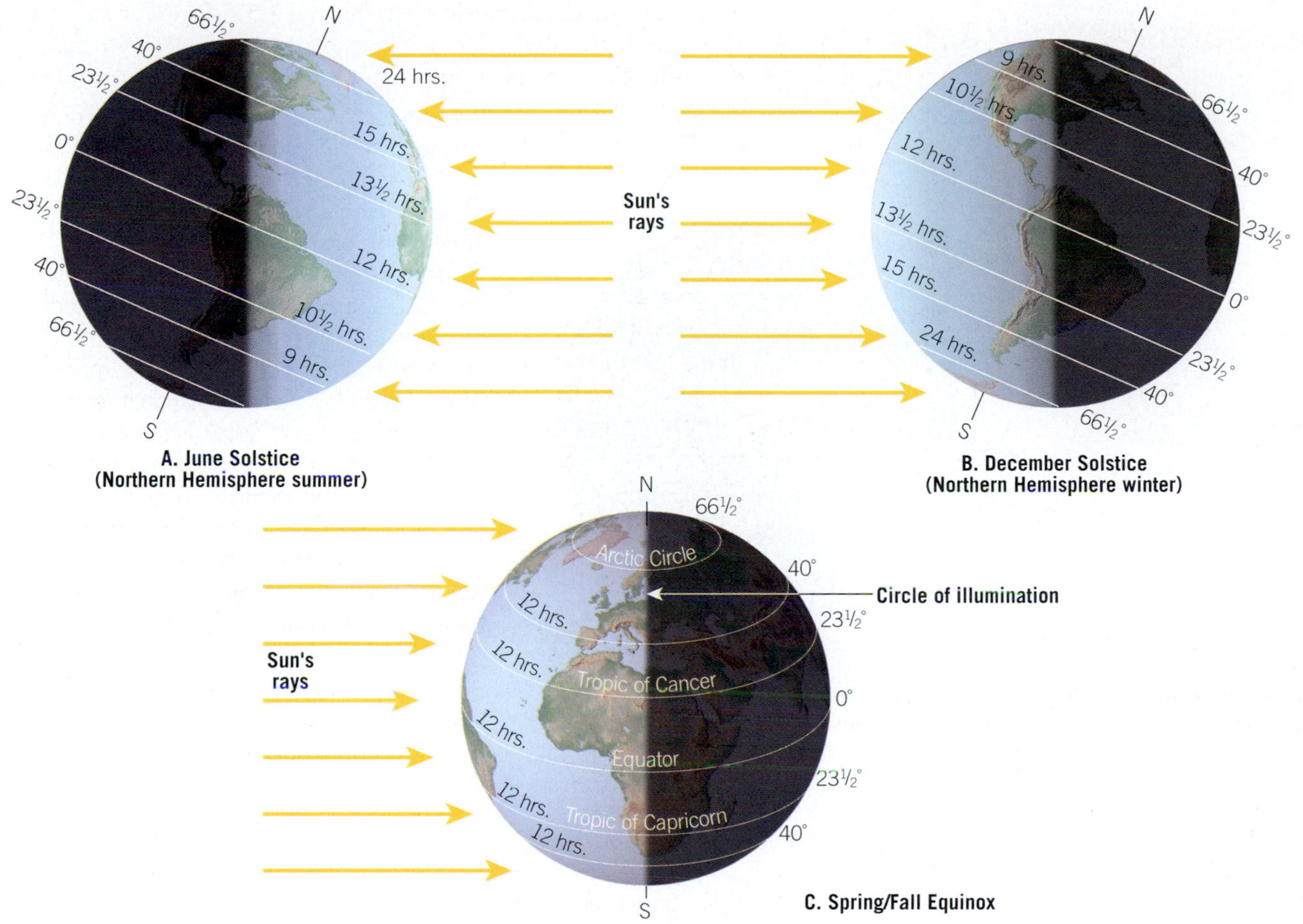


◀ **SmartFigure 2.6 Characteristics of the solstices and equinoxes**

http://goo.gl/tV8T5

## Solstices and Equinoxes

Based on the annual migration of the direct rays of the Sun, 4 days each year are especially significant. On June 21 or 22, the vertical rays of the Sun strike 23½° north latitude (23½° north of the equator), a line of latitude known as the **Tropic of Cancer** (Figure 2.5). For people living in the Northern Hemisphere, June 21 or 22 is known as the **summer solstice**, the first "official" day of summer (see Box 2.2).

Six months later, on December 21 or 22, Earth tilts (or leans) in the opposite direction, so the Sun's vertical rays strike at 23½° south latitude. (Recall that Earth's axis always points in the same direction; it is Earth's changing position relative to the Sun that causes the apparent change in tilt.) The line of latitude located at 23½° south latitude is known as the **Tropic of Capricorn**. For those in the Northern Hemisphere, December 21 or 22 is the **winter solstice**, the first day of winter. However, on this same day, people in the Southern Hemisphere are experiencing their summer solstice.

The *equinoxes* occur midway between the solstices. September 22 or 23 is the date of the **fall (autumnal) equinox** in the Northern Hemisphere, and March 21 or 22 is the date of the **spring (vernal) equinox**. On these dates, the vertical rays of the Sun strike the equator (0° latitude) because Earth's position is such that its axis is tilted neither toward nor away from the Sun.

The length of daylight versus darkness is also determined by the position of Earth relative to the Sun's rays. The length of daylight on the summer solstice in the Northern Hemisphere, June 21, is greater than the length of night. This fact can be established by examining Figure 2.6, which illustrates the **circle of illumination**—that is, the boundary separating the dark half of Earth from the lighted half. The length of daylight is established by comparing the fraction of a line of latitude that is on the "day" side of the circle of illumination with the fraction on the "night" side. Notice that on June 21, all locations in the Northern Hemisphere experience longer periods of daylight than darkness (Table 2.1). By contrast, during the Northern Hemisphere winter solstice in December, the length of darkness exceeds the length of daylight at all locations in the hemisphere. For example, consider that New York City (about 40° north latitude) has about 15 hours of daylight on June 21 and only 9 hours on December 21.

Also note from Table 2.1 that the farther north a location is from the equator on June 21, the longer the period of daylight. When you reach the Arctic Circle (66½° north latitude), the length of daylight is 24 hours. Places located on

## Box 2.1 Calculating the Noon Sun Angle

**Data:**

**Location: 40° N**

**Date: December 22**

**Location of 90° Sun: 23½° S**

**Calculations:**

**Step 1:**

**Distance in degrees between 23½° S and 40° N = 63½°**

**Step 2:**

$$\begin{array}{r} 90 \\ -63\frac{1}{2}^\circ \\ \hline 26\frac{1}{2}^\circ \end{array} = \text{Noon Sun angle at 40° N on December 22}$$

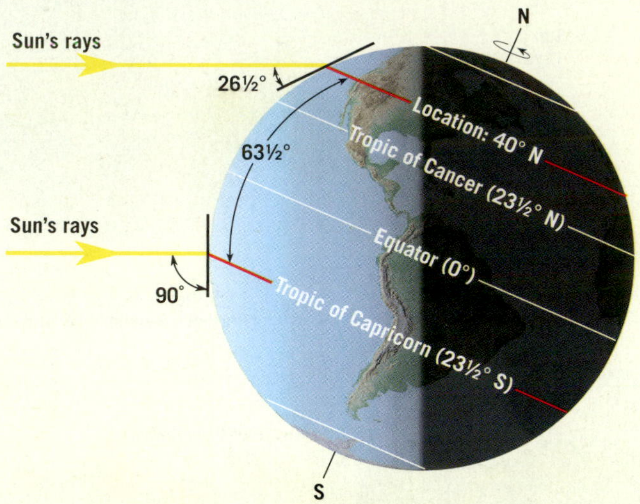


▲ **Figure 2.A Calculating the noon Sun angle** Recall that on any given day, only one latitude receives vertical (90°) rays of the Sun. A place located 1° away (either north or south) receives an 89° angle; a place 2° away, an 88° angle; and so forth. To calculate the noon Sun angle, simply find the number of degrees of latitude separating the location you want to know about from the latitude that is receiving the vertical rays of the Sun. Then subtract that value from 90°. The example in this figure illustrates how to calculate the noon Sun angle for a city located at 40° north latitude on December 22 (winter solstice).

Because Earth is roughly spherical, the only locations that receive vertical (90°) rays from the Sun are located along *one particular line of latitude* on any given day. As we move either north or south of this location, the Sun's rays strike at decreasing angles. Thus, the closer a place is situated to the latitude receiving the vertical rays of the Sun, the higher will be its noon Sun, and the more concentrated will be the radiation it receives.

A place located 1° away (either north or south) receives an 89° angle; a place 2° away, an 88° angle; and so forth. To calculate the noon Sun angle, simply find the number of degrees of latitude separating the location you want to know about from the latitude that is receiving the vertical rays of the Sun. Then subtract that value from 90°. The example in **Figure 2.A** illustrates how to calculate the noon Sun angle for a city located at 40° north latitude on December 22 (winter solstice).

**Question**

1. Calculate the noon Sun angle for your location on June 21 (summer solstice).

or north of the Arctic Circle experience the "midnight Sun," which means the Sun does not set for a period that ranges from 1 day at the Arctic Circle, to about 6 months at the poles (**Fig. 2.7**).

The seasonal changes in length of daylight and intensity of sunlight are the primary causes of the month-to-month variations in temperature observed at most locations outside the tropics. **Figure 2.8A,B,C** illustrate the daily paths of the Sun for a location at 40° north latitude for the seasons of the year. Notice that the length of daylight is longest and the Sun angle is highest in the sky during the summer solstice, while the opposite is true for the winter solstice. **Figure 2.8D** shows the path of the "midnight Sun" for a location located at 80° north latitude during the summer solstice.

**Figure 2.9** summarizes the characteristics of the solstices and equinoxes for a location in the Northern Hemisphere. An examination of Figure 2.9 should make it apparent why a mid-latitude location is warmest in the summer—when the days are longest and the angle of the Sun above the horizon is highest. The winter solstice facts are the reverse: The days are shortest, and the Sun angle is lowest. During an equinox (meaning "equal night"), the length of daylight is 12 hours everywhere on Earth because the circle of illumination passes directly through the poles, thus dividing the lines of latitude in half.

All locations situated at the same latitude have identical Sun angles and lengths of daylight. If the Earth–Sun relationships were the only controls of temperature, we would expect these places to have identical temperatures as well. Obviously, such is not the case. Although the angle of the Sun above the horizon and the length of daylight are the primary controls of temperature, other factors must be considered—a topic that will be addressed in Chapter 3.

**Table 2.1** | Length of Daylight

| Latitude | Summer Solstice | Winter Solstice | Equinoxes |
|---|---|---|---|
| 0° | 12 hr | 12 hr | 12 hr |
| 10° | 12 hr 35 min | 11 hr 25 min | 12 hr |
| 20° | 13 hr 12 min | 10 hr 48 min | 12 hr |
| 30° | 13 hr 56 min | 10 hr 04 min | 12 hr |
| 40° | 14 hr 52 min | 9 hr 08 min | 12 hr |
| 50° | 16 hr 18 min | 7 hr 42 min | 12 hr |
| 60° | 18 hr 27 min | 5 hr 33 min | 12 hr |
| 70° | 2 mo | 0 hr 00 min | 12 hr |
| 80° | 4 mo | 0 hr 00 min | 12 hr |
| 90° | 6 mo | 0 hr 00 min | 12 hr |

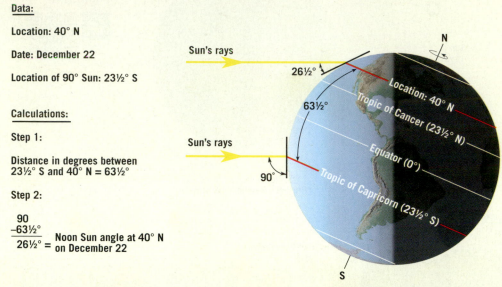

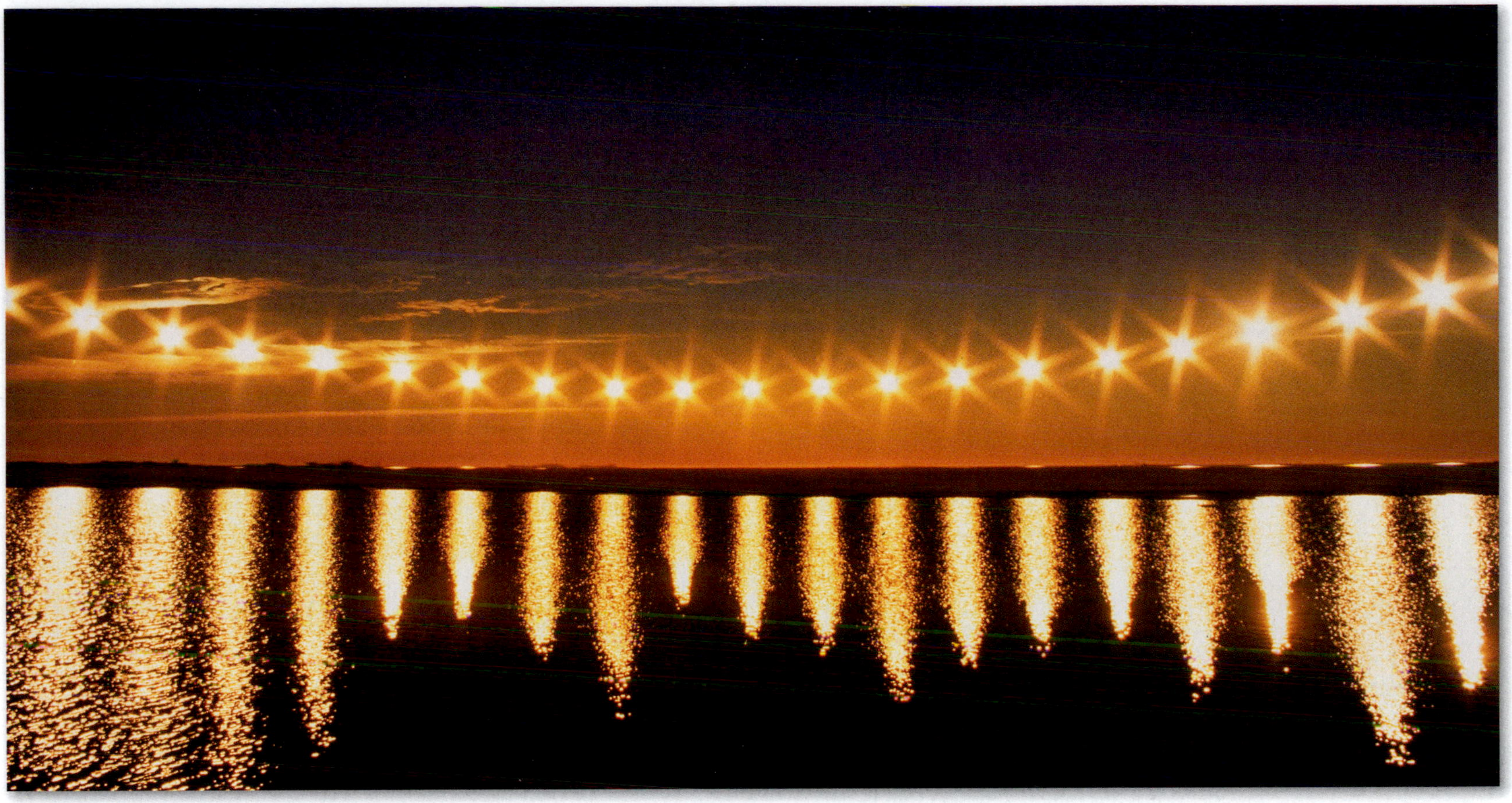

▲ **Figure 2.7 Midnight Sun** Multiple exposures, (taken on the same day) of the midnight Sun representative of midsummer in the high latitudes. This example shows the midnight Sun in Norway.

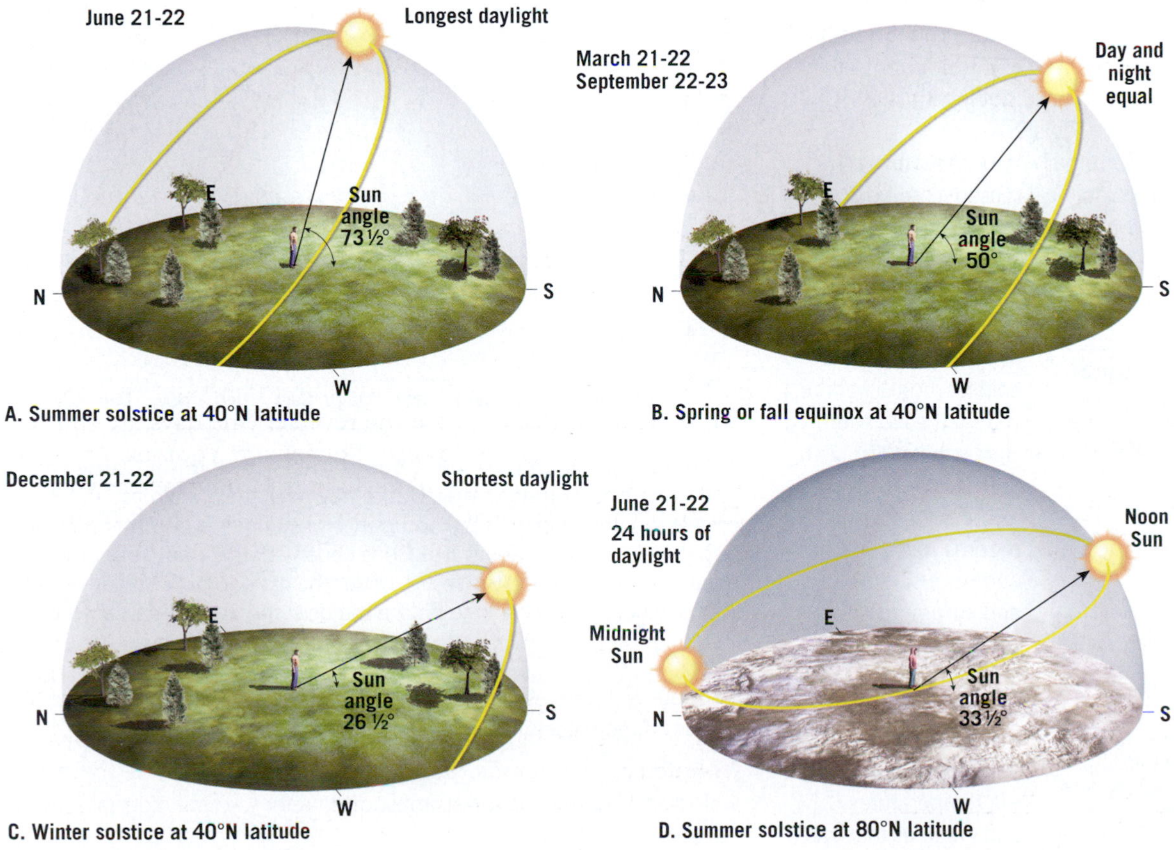


◀ **Figure 2.8 Daily paths of the Sun** Parts A through C illustrate the paths of the Sun for a place located at 40° north latitude at different times of the year. Part D illustrates the path of the Sun for a place located at 80° north latitude during the summer solstice.

Video MM
Net Radiation at the Top of the Atmosphere
http://goo.gl/E9I2G0

Video MM
Solar Eclipse
http://goo.gl/BfxU6Q

| Characteristics of the Solstices and Equinoxes for the Northern Hemisphere | | | |
|---|---|---|---|
| **Characteristics** | **Summer Solstice** | **Winter Solstice** | **Equinoxes** |
| Date of Occurrence | June 21-22 | December 21-22 | Spring: March 21-22<br>Fall: September 22-23 |
| Vertical Rays of the Sun | Tropic of Cancer (23½° N) | Tropic of Capricorn (23½° S) | Equator |
| Length of daylight | Longest Period of Daylight | Shortest Period of Daylight | Equal Days and Nights |
| Angle of Noon Sun | At its highest point above horizon | At its lowest point above horizon | At an intermediate position above horizon |

◀ **Figure 2.9 Characteristics of the solstices and equinoxes for the Northern Hemisphere**

In summary, seasonal fluctuations in the amount of solar energy reaching various places on Earth's surface are caused by the migrating vertical rays of the Sun and the resulting variations in Sun angle and length of daylight.

## ✔ Concept Checks 2.1

1. Do the annual variations in Earth–Sun distance adequately account for seasonal temperature changes? Explain.
2. Draw simple sketch to show why the intensity of solar radiation striking Earth's surface changes when the Sun angle changes.
3. Briefly explain the primary cause of the seasons.
4. What is the significance of the Tropic of Cancer and the Tropic of Capricorn?
5. After examining Table 2.1, write a general statement that relates the season, latitude, and the length of daylight.

## Box 2.2 When Are the Seasons?

Have you ever been caught in a snowstorm around Thanksgiving, even though winter does not begin until December 21? Or have you endured several consecutive days of 100° temperatures although summer has not "officially" started? The idea of dividing the year into four seasons originated from the Earth–Sun relationships discussed in this chapter (**Table 2.A**). This astronomical definition of the seasons defines winter in the Northern Hemisphere as the period from the winter solstice (December 21–22) to the spring equinox (March 21–22). This is also the definition used most widely by the news media, yet it is not unusual for portions of the United States and Canada to have significant snowfalls weeks before the "official" start of winter (Table 2.A).

Because the weather phenomena we normally associate with each season do not coincide well with the astronomical seasons, meteorologists prefer to divide the year into four 3-month periods based primarily on temperature. Thus, winter is defined as December, January, and February, the 3 coldest months of the year in the Northern Hemisphere (**Fig. 2.B**). Summer is defined as the 3 warmest months, June, July, and August. Spring and autumn are the transition periods between these two seasons. Inasmuch as these four 3-month periods better reflect the temperatures and weather that we associate with the respective climatological seasons, this definition of the seasons is more useful for meteorological discussions.

**Table 2.A** Occurrence of the Seasons in the Northern Hemisphere

| Season | Astronomical Season | Climatological Season |
|---|---|---|
| Spring | March 21 or 22 to June 21 or 22 | March, April, May |
| Summer | June 21 or 22 to September 22 or 23 | June, July, August |
| Autumn | September 22 or 23 to December 21 or 22 | September, October, November |
| Winter | December 21 or 22 to March 21 or 22 | December, January, February |

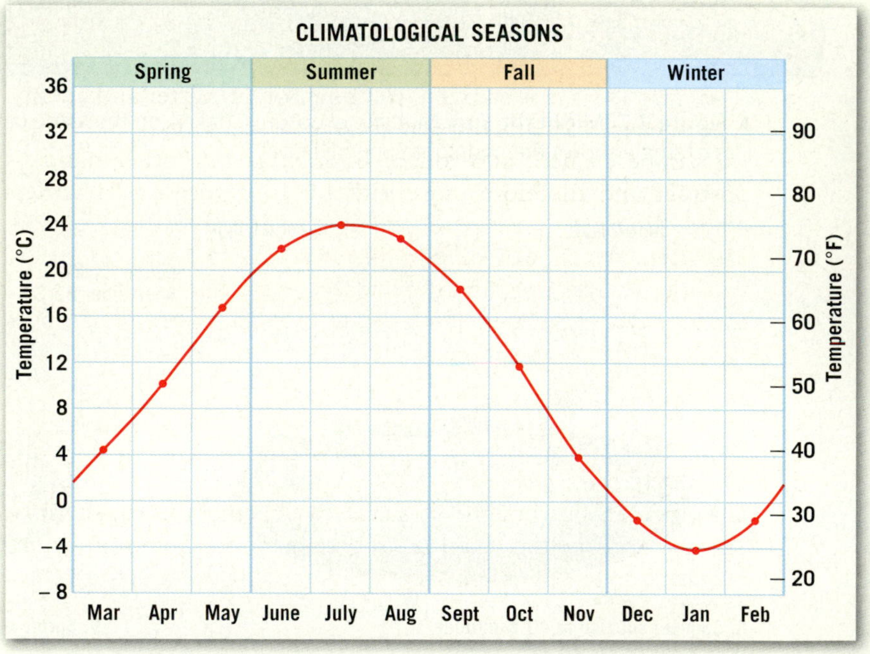


▲ **Figure 2.B Mean monthly temperatures for a midlatitude city in the central United States** Notice how well the 3 warmest and 3 coldest months align with the occurrence of summer and winter seasons, respectively. The astronomical seasons began about 21 days after the climatological seasons. As a result, under the astronomical seasons, winter-like conditions can occur long before the designated "first day of winter."

**Question**

1. Explain why meteorologists most often refer to climatological seasons rather than astronomical seasons.

## eye ON THE atmosphere 2.1

This image, which shows the *first sunrise* of 2008 at the South Pole, was taken at the U.S. Amundsen–Scott Station. At the moment the Sun cleared the horizon, the weathered American flag was seen whipping in the wind above a sign marking the location of the geographic South Pole.

**Questions**

1. What was the approximate date that this photograph was taken?
2. How long after this photo was taken did the Sun set at the South Pole?
3. Over the course of 1 year, what is the highest position the Sun can reach (measured in degrees) at the South Pole? On what date does this occur?

# 2.2 Energy, Temperature, and Heat

**Compare and contrast latent heat and sensible heat.**

The universe is made up of a combination of matter and energy. The concept of matter is easy to grasp because it is the "stuff" we can see, smell, and touch. Energy, on the other hand, is abstract and therefore more difficult to describe and understand. Energy comes to Earth from the Sun in the form of radiation, which we see as light and feel as heat. There are countless places and situations where energy is present—in the food we eat, the water flowing over a waterfall, and the waves that break along the shore.

## Forms of Energy

**Energy** can be thought of simply as having the capacity for doing work, such as making an object move. Common examples include the chemical energy from gasoline that powers automobiles, the heat energy from stoves that excites water molecules (boils water), and the gravitational energy that has the capacity to move snow down a mountain slope in the form of an avalanche. These examples illustrate that energy takes many forms and can also change from one form to another. For example, the chemical energy in gasoline is first converted to thermal energy (which we commonly refer to as heat) in the engine of an automobile, which is then converted to mechanical energy that moves the automobile along.

You are undoubtedly familiar with some of the common forms of energy, such as thermal energy, chemical, nuclear, radiant (light), and gravitational energy. Energy can also be placed into one of two major categories: *kinetic energy* and *potential energy.*

**Kinetic Energy** *Energy associated with an object by virtue of its motion* is described as **kinetic energy**. A simple example of kinetic energy is the motion of a hammer when driving a nail. Because of its motion, the hammer is able to move another object (do work). The faster the hammer is swung, the greater its kinetic energy (energy of motion). Similarly, a larger (more massive) hammer possesses more kinetic energy than a smaller one, provided that both are swung at the same velocity. Likewise, the winds associated with a hurricane possess much more kinetic energy than do light, localized breezes because they are both larger in scale (cover a larger area) and travel at higher velocities.

Kinetic energy is also significant at the atomic level. All matter is composed of atoms and molecules that are continually vibrating and, by virtue of this motion, have kinetic energy. For example, when a pan of water is heated on a stove, the water molecules begin to vibrate faster. Thus, when a solid, liquid, or gas is heated, its atoms or molecules move faster, and the material possess more kinetic energy.

**Potential Energy** As the term implies, **potential energy** has the capability to do work. For example, large hailstones suspended by an updraft in a towering cloud have gravitational potential energy. If the updraft subsides, these hailstones will be pulled to Earth and do destructive work on roofs and vehicles. Many substances, including wood, gasoline, and the food you eat, contain potential energy, which is capable of doing work, given the right circumstances.

## Temperature

In everyday use, temperature is used to describe how warm or cold an object is, using a standard measure. In the United States the Fahrenheit scale is used most often for the expressions of temperature. For example, the Weather Channel may forecast tomorrow's high temperature to be 88°F. However, scientists and the majority of other countries use the Celsius and Kelvin temperature scales. A discussion of all three scales is provided in Chapter 3.

**Temperature** can also be defined as a *measure of the average kinetic energy of the atoms or molecules in a substance.* When a substance is heated, its particles move faster, and its temperature rises. By contrast, when an object cools, the atoms and molecules vibrate more slowly, and its temperature drops.

It is important to note that temperature is *not* a measure of the total kinetic energy of an object. For example, a cup of boiling water has a much higher temperature than a bathtub of lukewarm water. However, the quantity of water in the cup is small, so it contains far less total kinetic energy than the water in the tub. Much more ice would melt in the tub of lukewarm water than in the cup of boiling water. The temperature of the water in the cup is higher because the atoms and molecules are vibrating faster, but the total amount of kinetic energy (also referred to as *heat,* or *thermal energy*) is much smaller because there are fewer particles.

## Heat

**Heat** is defined as *energy transferred into or out of an object because of temperature differences between that object and its surroundings.* If you hold a mug of hot coffee, your hand will begin to feel warm or even hot. By contrast, when you hold an ice cube, heat is transferred from your hand to the ice cube. Heat flows from a region of higher temperature to a region of lower temperature. Once the temperatures become equal, heat flow stops.

It is also common to use the word *heat* to describe *thermal energy,* which is the energy contained in a substance as a result of its temperature. A hot object is described as containing more thermal energy or heat than a cold object of equal mass and composition. Meteorologists subdivide heat into two categories: *latent heat* and *sensible heat.*

### Latent Heat

Heat is released or absorbed *when water changes from one state of matter to another,* a process often referred to as a *phase change.* For example, a phase change occurs when liquid water evaporates and becomes water vapor. During the process of evaporation, heat from the surroundings (environment) is absorbed by liquid water, causing its molecules to vibrate more rapidly. When the rate of vibration is great enough to overcome the strength of the hydrogen bonds holding together the water molecules, some of the molecules escape (evaporate) and become water vapor. Because the most energetic (fastest-moving) molecules escape, the average kinetic energy (temperature) of the remaining liquid water decreases. Therefore, *evaporation is considered a cooling process* because it removes heat from the environment. The cooling effect of evaporation is something you most certainly have experienced upon stepping out, dripping wet, from a swimming pool or shower.

During evaporation, the energy absorbed by the escaping water vapor molecules is called **latent heat** (*latent* = to lie hidden). The term *latent* is used to describe this phenomenon because the thermal energy (heat) required to evaporate the water is stored or "hidden" within the escaping water vapor. The latent heat stored in water vapor is eventually released, usually to the atmosphere during condensation—when water vapor returns to the liquid state during cloud formation. Therefore, *condensation, which is the opposite of evaporation, returns energy to the environment and is considered a warming process.*

Through the combined processes of evaporation and condensation, latent heat transports large amounts of energy from Earth's land–sea surface, mainly the oceans, to the atmosphere. The importance of latent heat in atmosphere processes will be explored in Chapter 4.

Keep in mind that latent heat is exchanged between the environment and water molecules anytime water undergoes a phase change. Figure 2.10 illustrates the phase changes water experiences; some involve the absorption and storage of latent heat, while the opposite processes release latent heat back to the environment. For example, when water vapor condenses and

▼ **Figure 2.10 Latent heat is either absorbed or released by each of these phase changes**

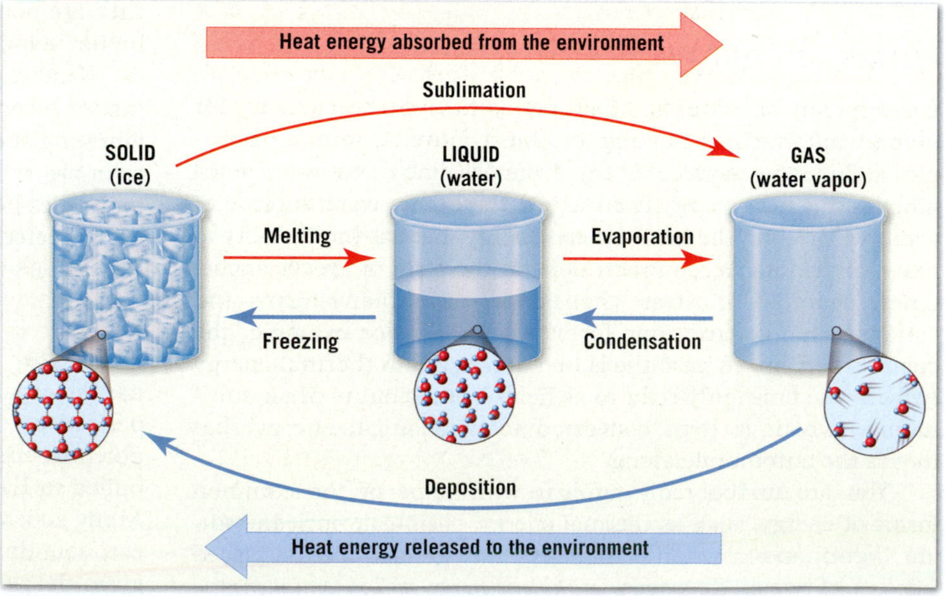

releases heat to the atmosphere, this is referred to as *latent heat of condensation*.

**Sensible Heat** In contrast to latent heat, **sensible heat** is the heat we can feel and measure with a thermometer but that does not involve a phase change. It is called *sensible* heat because it can be "sensed." On a clear summer day, the sunlight that is absorbed by the atmosphere, or your exposed skin, will cause an increase in temperature. Like latent heat, sensible heat can be transported from one location to another. Warm air that originates over the Gulf of Mexico and flows into the Great Plains in the winter is one example.

### ✔ Concept Checks 2.2

1. Define *kinetic energy*.
2. Briefly describe how latent heat is transferred from Earth's land–sea surface to the atmosphere.
3. Compare latent heat and sensible heat.

## 2.3 Mechanisms of Heat Transfer

**List and describe the three mechanisms of heat transfer.**

MM GEODe ▶ Heating Earth's Surface and Atmosphere ▶ Solar Radiation

The flow of energy can occur in three ways: *conduction, convection,* and *radiation* (Fig. 2.11). Although they are presented separately, all three mechanisms of heat transfer can operate simultaneously. In addition, working in tandum these processes may transfer heat between the Sun and Earth and between Earth's surface, the atmosphere, and outer space.

### Conduction

Anyone who attempts to pick up a metal spoon left in a boiling pot of soup realizes that heat is transmitted along the entire length of the spoon. The transfer of heat in this manner is called *conduction*. The hot soup causes the molecules at the bowl end of the spoon to vibrate more rapidly. These molecules and free electrons collide more vigorously with their neighbors and so on up the handle of the spoon. Thus, **conduction** is the transfer of heat through electron and molecular collisions from one molecule to another. The ability of substances to conduct heat varies considerably. Metals are good *conductors*, as those of us who have touched a hot spoon have quickly learned. Air, in contrast, is a very poor conductor of heat. Consequently, conduction is important only between Earth's surface and the air immediately in contact with the surface. As a means of heat transfer for the atmosphere as a whole, conduction is the least significant and can be disregarded when considering most meteorological phenomena.

Objects that are poor conductors, such as air, are called *insulators*. Most objects that are good insulators, such as cork, plastic foam, or goose down, contain many small air spaces. The poor conductivity of the trapped air gives these materials their insulating value. Snow is also a poor conductor, and like other good insulators, it contains numerous air spaces that impair the flow of heat. This is why wild animals may burrow into a snowbank to escape the "cold." The snow, like a down-filled comforter, does not supply heat; it simply retards the loss of the animal's own body heat.

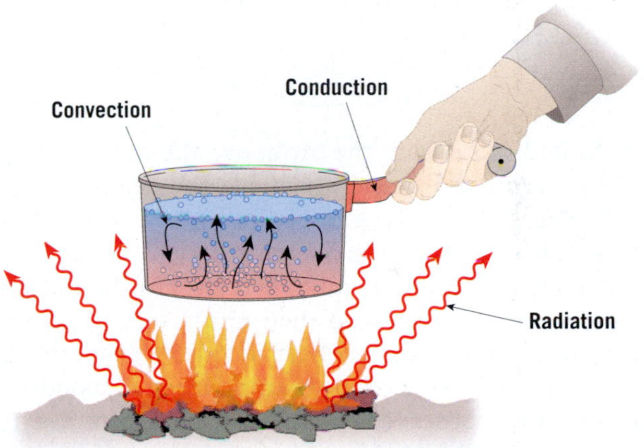


▲ SmartFigure 2.11 Three mechanisms of heat transfer: conduction, convection, and radiation

http://goo.gl/7sSdC

### Convection

Much of the heat transport in Earth's atmosphere and oceans occurs by convection. **Convection** is heat transfer that involves the actual movement or circulation of a substance. It takes place in fluids (liquids such as water and gases such as air) where the material is able to flow.

The pan of water being heated over a campfire in Figure 2.11 illustrates the nature of a simple convective circulation. The fire warms the bottom of the pan, which conducts heat to the water inside. Because water is a relatively poor conductor, only the water in close proximity to the bottom of the pan is heated by conduction. Heating causes water to expand and become less dense. Thus, the hot, buoyant water near the bottom of the pan rises, while the cooler, denser water above sinks. As long as the water is heated from the bottom and cools near the top, it will continue to "turn over," producing a *convective circulation*.

In a similar manner, some of the air in the lowest layer of the atmosphere that is heated by radiation and conduction is then transported by convection to higher layers of the atmosphere. For example, on a hot, sunny day the air above a plowed field

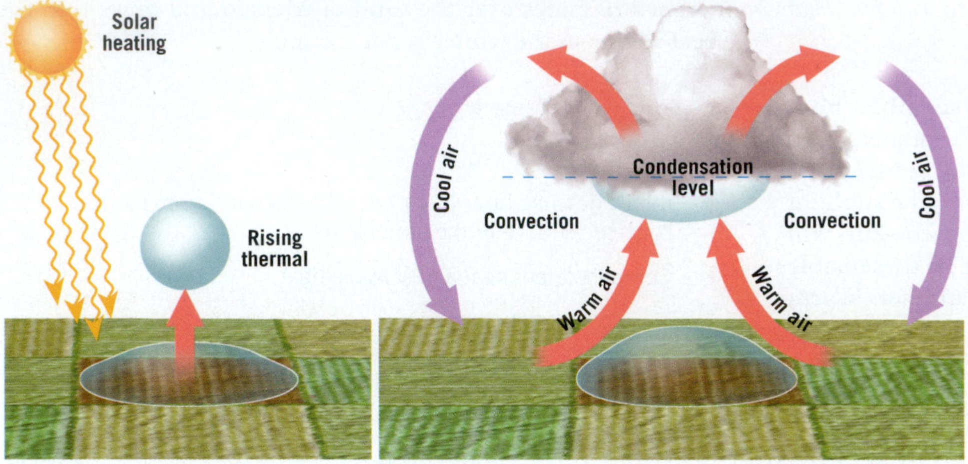


A. B.

▲ **Figure 2.12 Rising warmer air and descending cooler air are examples of convective circulation** **A.** Heating of Earth's surface produces thermals of rising air that transport heat and moisture aloft. **B.** The rising air cools, and if it reaches the condensation level, clouds form.

will be heated more than the air above the surrounding croplands. As warm, less-dense air above the plowed field buoys upward, it is replaced by the cooler air above the croplands (Fig. 2.12). In this way, a convective flow is established. The warm parcels of rising air are called **thermals** and are what hang-glider pilots use to keep their crafts soaring. Convection of this type not only transfers heat but also transports moisture (water vapor) aloft. The result is an increase in cloudiness that frequently can be observed on warm summer afternoons.

On a much larger scale is the global convective circulation of the atmosphere, which is driven by the unequal heating of Earth's surface. These complex movements are responsible for the redistribution of heat between hot equatorial regions and frigid polar latitudes and will be discussed in detail in Chapter 7.

Atmospheric circulation consists of vertical as well as horizontal components, so both vertical and horizontal energy transfer occurs. Meteorologists often use the term *convection* to describe the part of atmospheric circulation that involves *upward and downward* motion of air. By contrast, the term **advection** is used to denote the horizontal component of air flow. The common term for advection is *wind,* a phenomenon we will examine closely in later chapters. Residents of the midlatitudes often experience the effects of heat transfer by advection. For example, when frigid Canadian air invades the U.S. Midwest in January, it brings bitterly cold winter weather.

## Radiation

The third mechanism of heat transfer is radiation. Unlike conduction and convection, radiation is the only mechanism that can transfer thermal energy through the vacuum of space and thus is responsible for solar energy reaching Earth.

**Solar Radiation** The Sun is the ultimate source of energy that drives the weather machine. We know the Sun emits light of varying energy—including visible light, infrared radiation, and ultraviolet radiation. Although these forms of energy constitute a major portion of the total energy that radiates from the Sun, they are only a part of a large array of energy called **radiation**, or **electromagnetic radiation**. This array or spectrum of electromagnetic energy is shown in Figure 2.13.

All types of radiation, whether X-rays, radio waves, or heat waves, travel through the vacuum of space at 300,000 kilometers (186,000 miles) per second, a value known as the *speed of light.* To help visualize radiant energy, imagine ripples made in a calm pond when a pebble is tossed in. Like the waves produced in the pond, electromagnetic waves come in various sizes, or **wavelengths**—the distance from one crest to the next (Fig. 2.13). Radio waves have the longest wavelengths, up to thousands of meters in length. Gamma waves are the shortest, at less than one-billionth of a centimeter. Shortwave radiation is usually measured in **micrometers** (abbreviated μm), which are one-millionth of a meter.

Radiation is often identified by the effect that it produces when it interacts with an object. The retinas of our eyes, for instance, are sensitive to a range of wavelengths called **visible light**. We often refer to visible light as *white light* because it appears white in color. It is easy to show, however, that white light is really an array of colors, each color corresponding to a specific range of wavelengths. By using a prism, white light can be divided into the colors of the rainbow, from violet—with the shortest wavelength, 0.4 micrometer (μm)—to red—with the longest wavelength, 0.7 micrometer (Fig. 2.13).

Located adjacent to the color red, and having a longer wavelength, is **infrared radiation (IR)**, which cannot be seen by the human eye but is detected as heat. Only the infrared energy that is nearest the visible part of the spectrum is intense enough to be felt as heat and is referred to as *near infrared.* On the opposite side of the visible range, located next to violet,

**students sometimes ask...**

**On a cold morning, why does the bathroom's tile floor feel much colder than the bedroom carpet, even though both materials are the same temperature?**

The difference you feel is due mainly to the fact that the tile is a much better conductor of heat. Hence, energy is more rapidly conducted from your bare feet to the tile floor than from your feet to the carpet. Even at room temperature (20°C [68°F]), objects that are good conductors can feel chilly to the touch. (Remember that body temperature is about 37°C [98.6°F].)

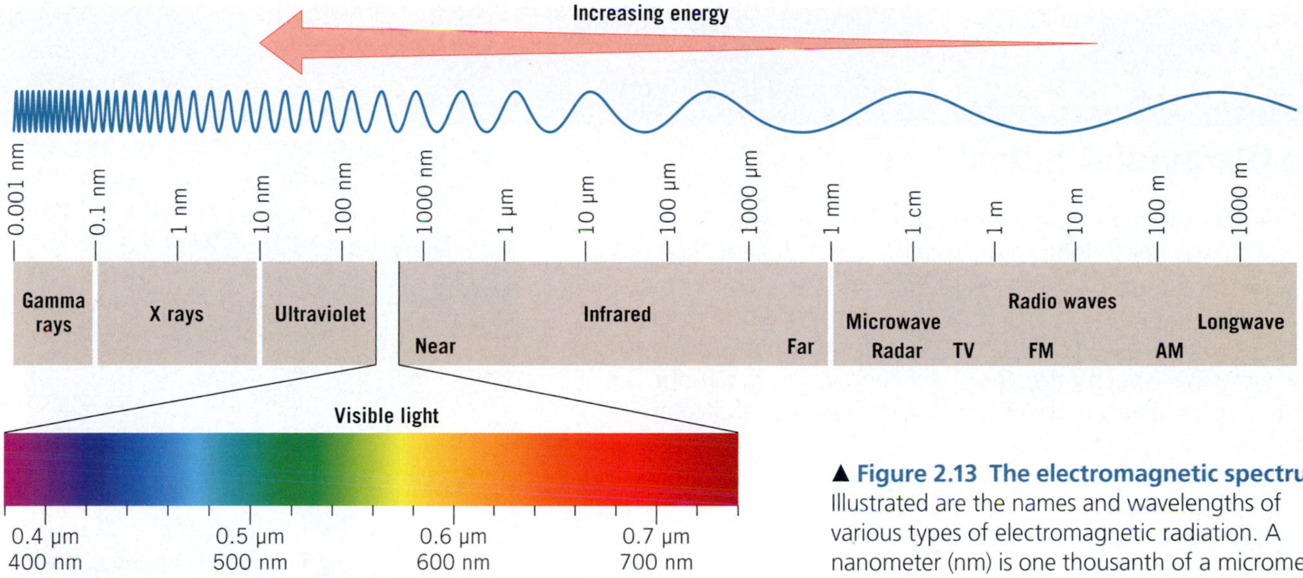


Video MM
Tour of the Electromagnetic Spectrum
http://goo.gl/q1q6Gq

▲ **Figure 2.13 The electromagnetic spectrum** Illustrated are the names and wavelengths of various types of electromagnetic radiation. A nanometer (nm) is one thousanth of a micrometer.

the energy emitted is called **ultraviolet (UV) radiation** and consists of shorter wavelengths that may cause skin to become sunburned.

Although we divide radiant energy into categories based on our ability to perceive them, all wavelengths of radiation behave similarly. When an object absorbs any form of electromagnetic energy, the waves excite subatomic particles (electrons). This results in an increase in molecular motion and a corresponding increase in temperature. Thus, electromagnetic waves from the Sun travel through space and, upon being absorbed, increase the molecular motion of other molecules—including those that make up the atmosphere, Earth's land–sea surfaces, and human bodies.

One important difference among the various wavelengths of radiant energy is that *shorter wavelengths are more energetic.* This accounts for the fact that exposure to relatively short (high-energy) ultraviolet waves can damage human tissue more readily than can similar exposure to longer-wavelength radiation. The damage can result in skin cancer and cataracts (see *Severe and Hazardous Weather* Box 2.3).

It is important to note that the Sun emits all forms of radiation, but in varying quantities. Over 95 percent of all solar radiation is emitted in a narrow band between 0.1 and 2.5 micrometers, much of which is concentrated in the visible and near-infrared parts of the electromagnetic spectrum (Fig. 2.14). Visible light, which lies between 0.4 and 0.7 micrometers, represents over 43

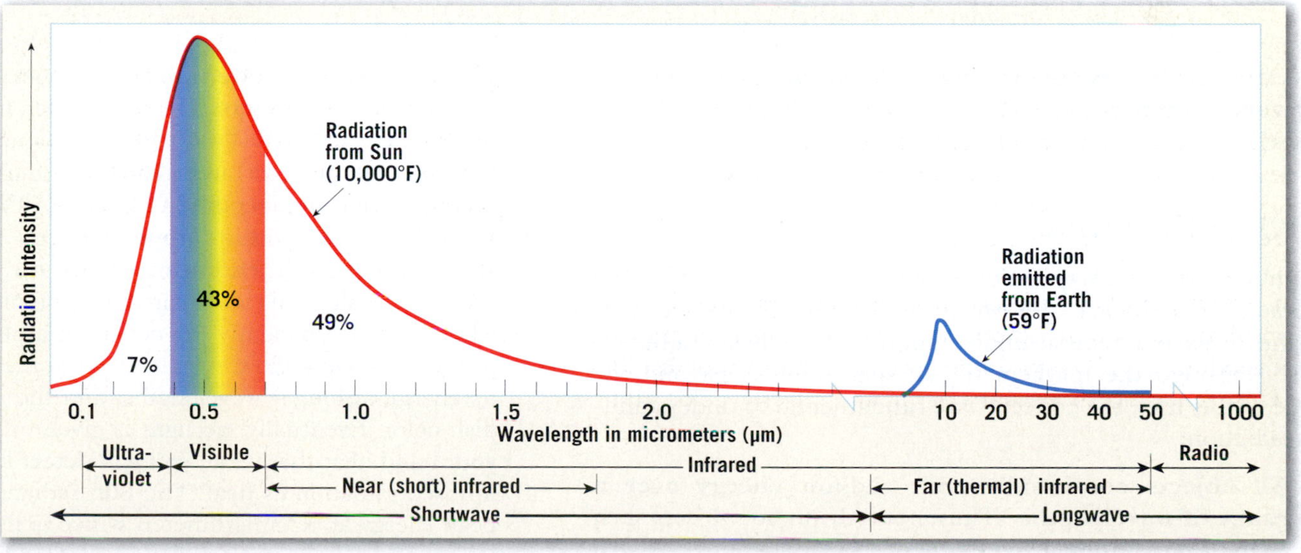


▲ **Figure 2.14 Comparison of the intensity of solar radiation and radiation emitted by Earth** Because of the Sun's high surface temperature, most of its energy is radiated at energetic wavelengths shorter than 2.5 micrometers (µm). The greatest intensity of solar radiation is in the visible range of the electromagnetic spectrum. Earth, in contrast, radiates most of its energy in wavelengths longer than 2.5 micrometers, primarily in the far end (less-energetic) of the infrared band. Thus, we call the Sun's radiation *shortwave* and Earth's radiation *longwave*.

severe & hazardous weather Box 2.3

## The Ultraviolet Index[*]

Most people welcome sunny weather. On warm days, when the sky is cloudless and bright, many spend a great deal of time outdoors, "soaking up" the sunshine (**Fig. 2.C**). For many, the goal is to develop a dark tan, one that sunbathers often describe as "healthy looking." Ironically, there is strong evidence that too much sunshine (specifically, too much ultraviolet radiation) can lead to serious health problems, mainly skin cancer and cataracts of the eyes.

The U.S. Environmental Protection Agency has established five exposure categories based on UV index values: Low, Moderate, High, Very High, and Extreme

Since June 1994 the National Weather Service (NWS) has issued the ultraviolet index (UVI) forecasts to warn the public of potential health risks of exposure to sunlight (**Fig. 2.D**). The UV index is determined by taking into account the predicted cloud cover and reflectivity of the surface, as well as the Sun angle and atmospheric depth for each forecast location. Because atmospheric ozone strongly absorbs ultraviolet radiation, the extent of the ozone layer is also considered. The UV index values lie on a scale from 0 to 11+, with larger values representing greatest risk.

The U.S. Environmental Protection Agency has established five exposure categories based on UV index values: Low, Moderate, High, Very High, and Extreme (**Table 2.B**). Precautionary measures have been developed for each category. The public is advised to minimize outdoor activities when the UV index is Very High or Extreme. Sunscreen with a sun-protection factor (SPF) of 30 or higher is recommended for all exposed skin (to be reapplied every 2 hours). This is especially important after swimming or while sunbathing, even on cloudy days with the UV index in the Low category. The public is advised to minimize outdoor activities when the UV index is Very High or Extreme.

[*]Based on material prepared by Professor Gong-Yuh Lin, a faculty member in the Department of Geography at California State University, Northridge.

▲ **Figure 2.C Exposing sensitive skin to too much solar ultraviolet radiation has potential health risks**

Video (MM) The Sun in Ultraviolet
http://goo.gl/ZdFhEG

percent of the total energy, infrared accounts for 49 percent and ultraviolet, 7 percent. Less than 1 percent of solar radiation is emitted as X-rays, gamma rays, and radio waves.

## Laws of Radiation

To obtain a better appreciation of how the Sun's radiant energy interacts with Earth's atmosphere and land–sea surface, it is helpful to have a general understanding of the basic radiation laws. Although the mathematics of these laws is beyond the scope of this text, the concepts are fundamental to understanding radiation:

1. **All objects continually emit radiant energy over a range of wavelengths.**[*] Thus, not only do hot objects such as the Sun continually emit energy, but Earth does as well (*terrestrial radiation*), even the polar ice caps.
2. **Hotter objects radiate more total energy per unit area than do colder objects.** The Sun, which has a surface temperature of 6000 K (10,000°F), emits about 160,000 times more energy per unit area than does Earth, which has an average surface temperature of 288 K (59°F).
3. **Hotter objects radiate more energy in the form of short-wavelength radiation than do cooler objects.** We can visualize this law by imagining a piece of metal that, when heated sufficiently (as occurs in a blacksmith's shop), produces a white glow. As it cools, the metal emits more of its energy in longer wavelengths, and the glow turns a reddish color. Eventually, no light is given off, but if you place your hand near the metal, you will detect longer-wavelength infrared radiation as heat. The Sun radiates the most, or its peak energy at 0.5 micrometer, which is in the visible range (Figure 2.14). The peak radiation emitted from Earth occurs at a wavelength of 10 micrometers, well within the infrared (heat) range. Because the peak terrestrial radiation is roughly 20 times longer than the peak solar radiation, it is often referred to as **longwave radiation**, whereas solar radiation is called **shortwave radiation**.

[*]The temperature of the object must be above a theoretical value called *absolute zero* (–273°C) in order to emit radiant energy. The letter K is used for values on the Kelvin temperature scale. For more explanation, see the subsection "Temperature Scales" in Chapter 3.

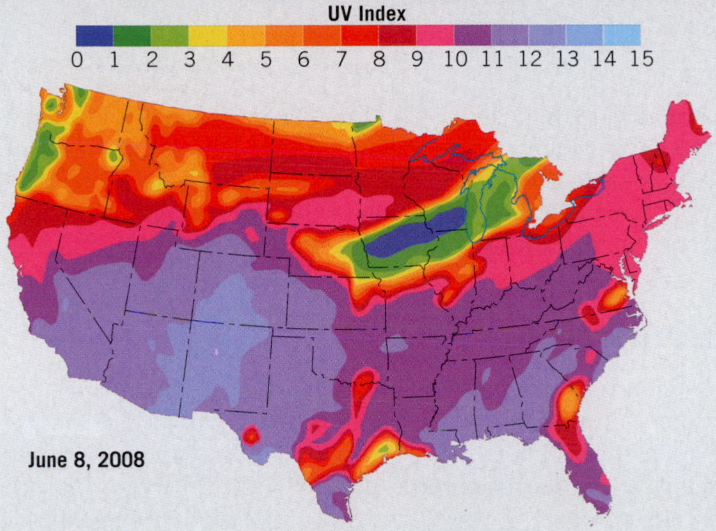


▲ **Figure 2.D UV index forecast for June 8, 2008** To view the current UV index forecast, go to www.epa.gov/sunwise/uvindex.html.

Table 2.B shows the range of minutes to burn for the most susceptible skin type (pale or milky white) for each exposure category. Note that the exposure that results in sunburn varies from over 60 minutes for the Low category to less than 10 minutes for the Extreme category. It takes approximately five times longer to cause sunburn of the least susceptible skin type, brown to dark. The most susceptible skin type develops red sunburn, painful swelling, and skin peeling when exposed to excessive sunlight. By contrast, the least susceptible skin type rarely burns and shows very rapid tanning response.

**Question**

1. List several factors that are used to determine the UV index for a location.

**Table 2.B** The UV Index: Minutes to Burn for the Most Susceptible Skin Type

| UV Index Value | Exposure Category | Description | Minutes to Burn |
|---|---|---|---|
| 0–2 | Low | Low danger from the Sun's UV rays for the average person. | > 60 |
| 3–5 | Moderate | Moderate risk from unprotected Sun exposure. Take precautions during the midday, when Sun is strongest. | 40–60 |
| 6–7 | High | Protection against sunburn is needed. Cover up, wear a hat and sunglasses, and use sunscreen. | 25–40 |
| 8–10 | Very High | Try to avoid the Sun between 11 A.M. and 4 P.M. Otherwise, cover up and use sunscreen. | 10–25 |
| 11 + | Extreme | Take all precautions. Unprotected skin will burn in minutes. Do not pursue outdoor activities if possible. If outdoors, apply sunscreen liberally every 2 hours. | < 10 |

4. **Objects that are good absorbers of radiation are also good emitters.** Earth's surface and the Sun are nearly perfect radiators because they absorb and radiate with nearly 100 percent efficiency. By contrast, the gases that compose our atmosphere are *selective* absorbers and emitters of radiation. For some wavelengths, the atmosphere is nearly transparent (that is, allowing most of the energy to pass through). For other wavelengths, however, it is nearly opaque (that is, absorbs most of the radiation that strikes it). Experience tells us that the atmosphere is quite transparent to visible light emitted by the Sun because these wavelengths readily reach Earth's surface.

To summarize, although the Sun is the ultimate source of radiant energy, all objects continually radiate energy over a range of wavelengths. Hot objects, such as the Sun, emit mostly shortwave (high-energy) radiation, but cooler objects, such as Earth, emit longwave (low-energy) radiation. Objects that are good absorbers of radiation, such as Earth's surface, are also good emitters. By contrast, most atmospheric gases are good absorbers (emitters) only in certain wavelengths but are poor absorbers (emitters) in other wavelengths.

## ✔ Concept Checks 2.3

1. Describe the three basic mechanisms of energy transfer. Which mechanism is least important meteorologically?
2. What is the difference between convection and advection?
3. Rank visible, infrared, and ultraviolet radiation, from longest to shortest wavelength and from most to least energetic.
4. At which wavelenths of the electromagnetic spectrum does the Sun radiate maximum energy? How does this compare to Earth?
5. Describe the relationship between the temperature of a radiating body and the wavelengths it emits.

# 2.4 What Happens to Incoming Solar Radiation?

**Using the diagram in Figure 2.15 as a guide, describe what happens to incoming solar radiation.**

MM® GEODe ▶ Heating Earth's Surface and Atmosphere ▶ What Happens to Incoming Solar Radiation?

When incoming solar radiation enters the atmosphere, three different things may occur simultaneously. First, air, which is transparent to certain wavelengths of radiation, may simply *transmit* energy—allowing it to pass through without redirecting or absorbing it. Second, some of the energy may be *absorbed.* Recall that when radiant energy is absorbed, the molecules begin to vibrate faster, which causes an increase in temperature. Third, some radiation may "bounce off" gas molecules or dust particles in the atmosphere, without being absorbed or transmitted.

What determines whether solar radiation will be transmitted to the Earth's land–sea surface, absorbed by the gases and particles in the atmosphere, or scattered or reflected by these gases and particles? As you will see, it depends greatly upon the *wavelength* of the radiation, as well as the *size and nature of the intervening material.*

## Transmission

**Transmission** is the process by which shortwave and longwave energy passes though the atmosphere (or any transparent media) without interacting with the gases or other particles in the atmosphere. About half of the incoming solar energy that reaches Earth's surface is transmitted by the atmosphere. The remainder is redirected by gas molecules and particles in the atmosphere and arrives as diffused light. Figure 2.15 illustrates what happens to incoming solar radiation, averaged for the entire globe. Notice that on average, about 55 percent of incoming solar energy reaches Earth's surface—about 50 percent is absorbed at the surface, and the remaining 5 percent is reflected back toward space.

## Absorption

The amount of energy absorbed by an object depends on the wavelength (intensity) of the radiation and the object's **absorptivity**. In the visible range, the degree of absorptivity is largely responsible for the brightness of an object. Surfaces that are good absorbers of all wavelengths of visible light appear black in color, whereas light-colored surfaces have a much lower absorptivity. That is why wearing light-colored clothing on a sunny summer day may help keep you cooler.

Although Earth's surface is a relatively good absorber (effectively absorbing most wavelengths of solar radiation), the atmosphere is not. As a result, only 20 percent of the solar radiation that reaches Earth is absorbed by gases in the atmosphere (Figure 2.15). The atmosphere is a less effective absorber because gases are selective absorbers (and emitters) of radiation.

Freshly fallen snow is another example of a selective absorber. Snow is a poor absorber of visible light (reflecting up to 90 percent) and, as a result, the temperature directly above a snow-covered

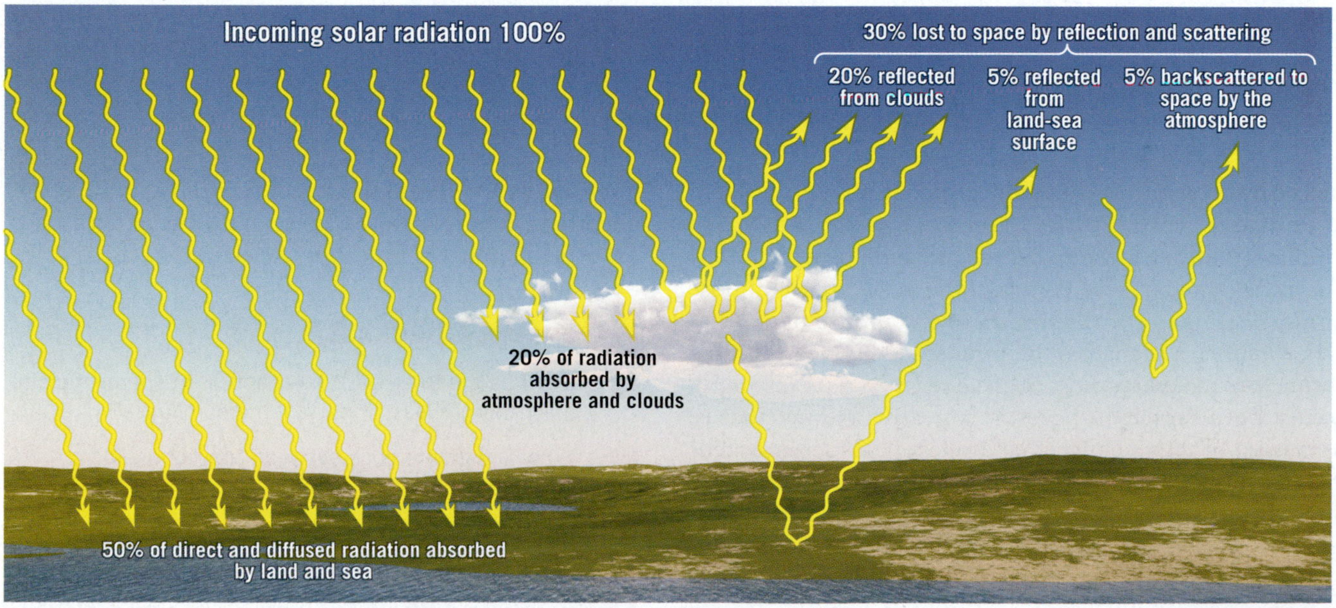


▲ **SmartFigure 2.15 Average distribution of incoming solar radiation** More solar energy is absorbed by Earth's surface than by the atmosphere.

http://goo.gl/NtE7Z

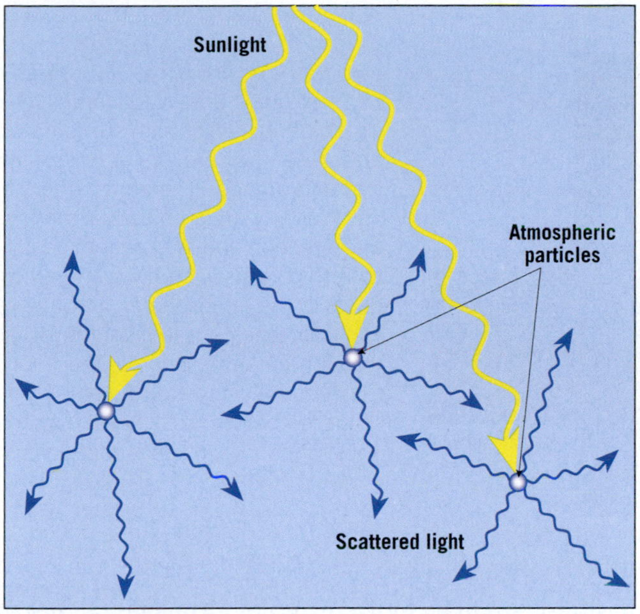


▲ **Figure 2.16 Scattering by atmospheric particles** When sunlight is scattered, it results in rays traveling in different directions. Usually more energy is scattered in the forward direction than is backscattered.

surface is colder than it would otherwise be because much of the incoming radiation is reflected away. By contrast, snow is a very good absorber (absorbing up to 95 percent) of the infrared (heat) radiation that is emitted from Earth's surface. As the ground radiates heat upward, the lowest layer of snow absorbs this energy and, in turn, radiates most of the energy downward. Thus, the depth at which a winter's frost can penetrate into the ground is much less when the ground is snow covered than in an equally cold region without snow—giving credence to the statement "The ground is blanketed with snow." Farmers who plant winter wheat desire a deep snow cover because it insulates their crops from bitter winter temperatures.

## Reflection and Scattering

**Reflection** is the process whereby light bounces back from an object at the same angle and intensity at which it was received. By contrast, **scattering** is a general process in which radiation is forced to deviate from a straight trajectory. When a beam of light strikes an atom, a molecule, or a tiny particle in the atmosphere, it can spread out in all directions (Fig. 2.16). Scattering disperses light both forward and backward, in a process called **backscattering**. Whether solar radiation is reflected or scattered depends largely on the size of the intervening particles and the wavelength of the light.

### Reflection and Earth's Albedo

The fraction of radiation that is reflected by an object is called its **albedo**. Figure 2.17 gives the albedos for various surfaces. Fresh snow and thick clouds have high albedos (that is, they're good reflectors). The high reflectivity of clouds can be observed when looking down on clouds during an airline flight. By contrast, dark soils and parking lots have low albedos and thus absorb much of the radiation they receive. In the case of a lake or the ocean, note that the angle at which the Sun's rays strike the water surface greatly affects its albedo.

Earth's total albedo, called *planetary albedo,* is 30 percent (see Figure 2.15). This energy is lost to Earth and does not play a role in heating the atmosphere or Earth's surface. The amount of light reflected from Earth's land–sea surface represents a small percentage of the total planetary albedo. Not surprisingly, thick clouds, which have high albedos, are largely responsible for most of Earth's "brightness," as seen from space.

The Moon, which has neither clouds nor an atmosphere, has an average albedo of only 7 percent (compared to Earth's 30 percent). Although a full Moon appears bright, the much brighter and larger Earth would, by comparison, provide far more light for an astronaut's "Earth-lit" stroll on the Moon.

### Scattering and Diffused Light

Although incoming solar radiation travels in a straight line, small dust particles and gas molecules in the atmosphere scatter some of this energy in different directions. The result, called **diffused light**, explains how light reaches the area under the limbs of a tree and how a room is lit in the absence of direct sunlight. In contrast, bodies like the Moon and Mercury, which are without atmospheres, have dark skies and "pitch-black" shadows, even during daylight hours. Overall, about one-half of the solar radiation that is absorbed at Earth's surface arrives as diffused (scattered) light.

▼ **Figure 2.17 Albedo (reflectivity) of various surfaces** In general, light-colored surfaces tend to be more reflective than dark-colored surfaces and thus have higher albedos.

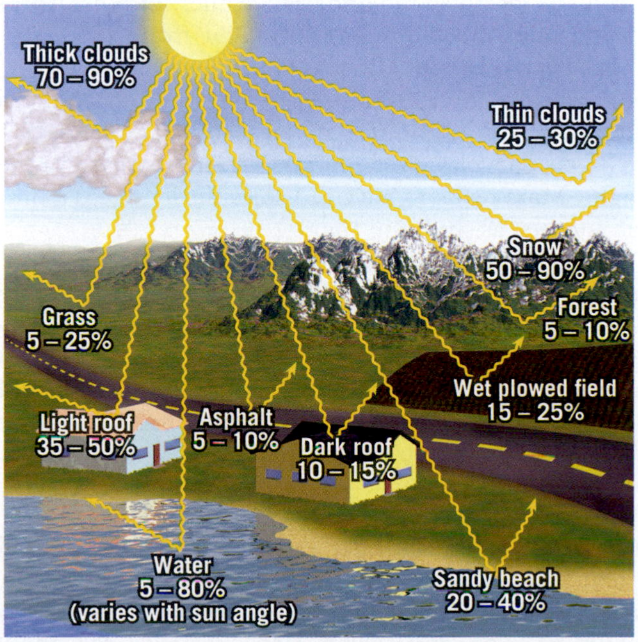

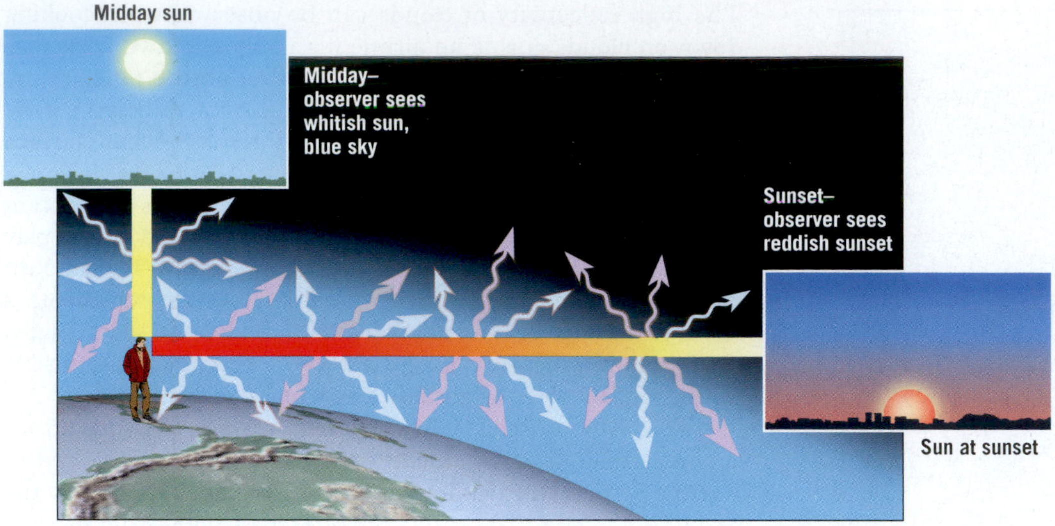


◀ **Figure 2.18 Preferential scattering by gas molecules in Earth's atmosphere produces blue skies and red sunsets** Short wavelengths (blue and violet) of visible light are scattered more effectively than are longer wavelengths (red and orange). Therefore, when the Sun is overhead, an observer can look in any direction and see predominantly blue light that was selectively scattered by the gases in the atmosphere. By contrast, at sunset, the path that light must take through the atmosphere is much longer. Consequently, most of the blue light is scattered before it reaches an observer. Thus, the Sun appears reddish in color.

**Blue Skies and Red Sunsets** The two factors that produce Earth's blue skies and red sunsets are the preferential scattering of solar radiation by atmospheric gases and the amount of atmosphere through which the Sun's rays travel before reaching Earth. Recall that sunlight appears white but is composed of all the colors of the rainbow. Atmospheric gases scatter shorter-wavelength (blue/violet) light more effectively than they scatter longer-wavelength (red/orange) light. Because shortwave radiation is preferentially scattered, when you look in any direction away from the direct Sun, you observe the short-wavelength (blue) light (Fig. 2.18).

The Sun appears reddish when viewed near Earth's horizon at sunrise or sunset because solar radiation must travel a greater distance through the atmosphere before it reaches your eyes. During its travel, shorter-wavelength blue and violet wavelengths are preferentially scattered, so the light that reaches your eyes consists mostly of red and orange hues. In other words, the sky and clouds are illuminated by light from which the blue color has been preferentially scattered away.

By contrast, Mars has a very thin atmosphere that often contains abundant amounts of large dust particles. Because these large particles preferentially scatter longer-wavelength red light, Mars tends to have red skies during the daytime and bluish sunsets, as shown in Figure 2.19.

On Earth, the most spectacular sunsets occur when large quantities of tiny dust or smoke particles penetrate the stratosphere. For 3 years after the great eruption of the Indonesian volcano Krakatau in 1883, brilliant sunsets occurred worldwide. In addition, the European summer that followed this colossal explosion was cooler than normal, which has been attributed to the loss of incoming solar radiation due to an increase in backscattering.

**Crepuscular Rays** Large particles associated with haze, fog, and smog scatter light more equally in all wavelengths. Because no color predominates over any other, the sky appears white or gray on days when large particles are abundant. Scattering of sunlight by haze, water droplets, or dust particles makes it possible for us to observe bands (or rays) of sunlight called *crepuscular rays*. These bright fan-shaped bands are most commonly seen when the Sun shines through a break in the clouds, as shown in Figure 2.20. Crepuscular rays

▼ **Figure 2.19 Bluish sunset on Mars** This image was transmitted by NASA's *Mars Exploration Rover Spirit.* The blue hue is produced by high-altitude dust particles, which preferentially scatter longer-wavelength red light. This is the opposite of what occurs in Earth's atmosphere where tiny gas molecules scatter out blue and violet light to produce a reddish sunset.

▲ **Figure 2.20 Crepuscular rays produced when haze scatters light** Crepuscular rays are most commonly seen when the Sun shines through a break in the clouds.

can also be observed around twilight, when towering clouds cause alternating lighter and darker bands (light rays and shadows) to streak across the sky.

In summary, the color of the sky gives an indication of the number of large or small particles present. Numerous small particles produce red sunsets, whereas large particles produce white (gray) skies. Thus, the bluer the sky, the less polluted, or dryer, the air.

## ✔ Concept Checks 2.4

1. Prepare and label a simple sketch that shows what happens to incoming solar radiation.
2. Considering the fact that solar radiation travels in a straight line, explain why you can see an apple on the ground directly under the apple tree.
3. Why does the daytime sky usually appear blue if the sky is clear?
4. Why may the sky have a red or orange hue near sunrise or sunset?

## eye on the atmosphere 2.2

This image was taken by astronauts aboard the International Space Station as it traveled over the west coast of South America. On average, these astronauts experience 16 sunrises and sunsets during a 24-hour orbital period. The separation between day and night is marked by a line called the *terminator*.

**Questions**

1. Locate the terminator in this image. Would you describe it as a sharp line? Explain.
2. Are the astronauts looking at a sunrise or a sunset? How do you know?

# 2.5 The Role of Gases in the Atmosphere

**Explain what is meant by the statement "The atmosphere is heated from the ground up."**

MM GEODe ▶ Heating Earth's Surface and Atmosphere ▶ The Greenhouse Effect

Understanding how the atmosphere is heated requires an understanding of how atmospheric gases interact with the short-wavelength *incoming* solar radiation and the long-wavelength *outgoing* radiation emitted by Earth. **Figure 2.21** shows that the majority of solar radiation is emitted in wavelengths shorter than 2.5 micrometers—shortwave radiation. By contrast, most radiation from Earth's surface is emitted at wavelengths between 2.5 and 30 micrometers, placing it in the far end (long-wavelength) of the infrared band of the electromagnetic spectrum.

At the bottom of Figure 2.21, you can see that for the atmosphere as a whole, none of the gases are effective absorbers of visible radiation with wavelengths between 0.3 and 0.7 micrometer. This visible light band constitutes about 43 percent of the energy radiated by the Sun. Because the atmosphere is a poor absorber of visible radiation, most of this energy is transmitted to Earth's surface. Thus, we say that

## Heating the Atmosphere

When a gas molecule absorbs radiation, the energy is transformed into internal molecular motion, which is detectable as a rise in temperature (sensible heat). For example, the absorption of UV energy by oxygen molecules in the stratosphere accounts for the high temperatures experienced there.

The lower part of Figure 2.21 gives the absorptivity of the principal atmospheric gases (see Figure 1.17, page 16). Note that nitrogen, the most abundant constituent in the atmosphere (about 78 percent), is a relatively poor absorber of incoming solar radiation. The only significant absorbers of incoming solar radiation are water vapor, oxygen, and ozone, which account for most of the solar energy absorbed directly by the atmosphere. Oxygen and ozone are efficient absorbers of high-energy, shortwave radiation. Oxygen removes most of the shorter-wavelength UV radiation high in the atmosphere, and ozone absorbs UV rays in the stratosphere between 10 and 50 kilometers (6 and 30 miles). Without the removal of most UV radiation, human life would not be possible because UV energy disrupts our genetic code.

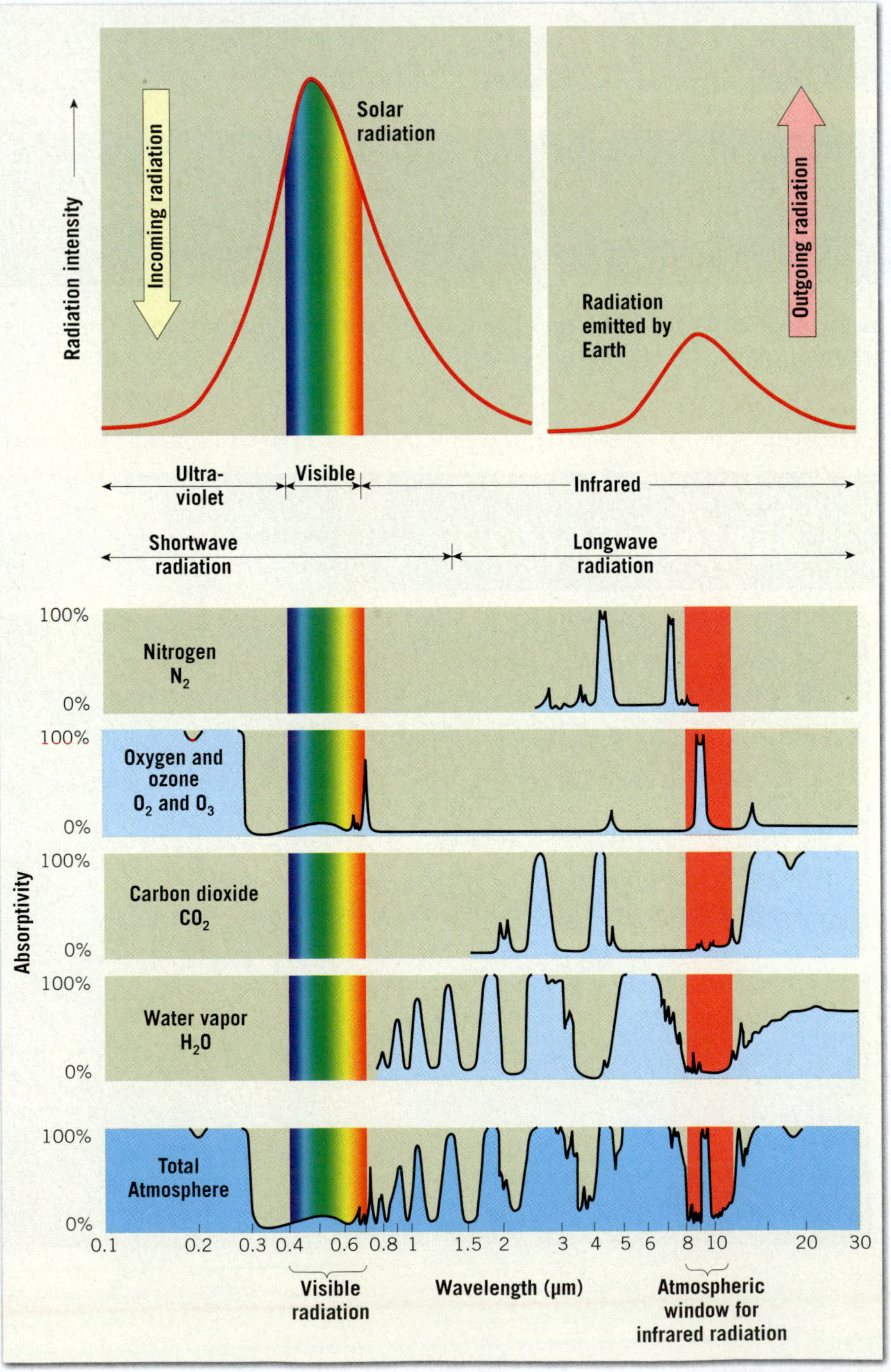


**▶ Figure 2.21 Absorption of solar and Earth's radiation by gases in the atmosphere** The effectiveness of selected gases of the atmosphere in absorbing incoming shortwave radiation (left side of graph) and outgoing long-wave terrestrial radiation (right side). The blue areas represent the percentage of radiation absorbed by the various gases. The atmosphere as a whole is quite transparent to solar radiation between 0.3 and 0.7 micrometer, which includes the band of visible light. Most solar radiation falls in this range, explaining why a large amount of solar radiation penetrates the atmosphere and reaches Earth's surface. Also, note that long-wave infrared radiation in the zone between 8 and 12 micrometers can escape the atmosphere most readily. This zone is called the atmospheric window.

students sometimes ask...

**Why do leaves on deciduous trees change color each fall?**

The leaves of all deciduous trees contain the pigment chlorophyll, which gives them a green color. The leaves of some trees also contain the pigment carotene, which is yellow, and still others produce a class of red pigments. During summer, leaves are factories that generate sugar from carbon dioxide and water by the action of light on chlorophyll. As the dominant pigment, chlorophyll causes the leaves of most trees to appear green. The shortening days and cool nights of autumn trigger changes in deciduous trees. With a drop in chlorophyll production, the green color of the leaves fades, allowing other pigments to be seen. Leaves containing carotene will change to bright yellow. The leaves of trees such as red maple and sumac display the brightest reds and purples in the autumn landscape.

*the atmosphere is nearly transparent to incoming solar radiation and that direct solar energy is not an effective "heater" of Earth's atmosphere.*

The atmosphere is generally a relatively efficient absorber of long-wave (infrared) radiation emitted by Earth (see the bottom right of Figure 2.21). Water vapor and carbon dioxide are the principal absorbing gases, with water vapor absorbing about 60 percent of the radiation emitted by Earth's surface. Therefore, water vapor, more than any other gas, accounts for the warm temperatures of the lower troposphere, where it is most highly concentrated.

Although the atmosphere is an effective absorber of radiation emitted by Earth's surface, it is nevertheless quite transparent to the band of radiation between 8 and 12 micrometers. Notice in Figure 2.21 (lower right) that the gases in the atmosphere ($N_2$, $CO_2$, $H_2O$) absorb minimal energy in these wavelengths. Because the atmosphere is transparent to radiation between 8 and 12 micrometers, much as window glass is transparent to visible light, this band is called the **atmospheric window**. Although other "atmospheric windows" exist, the one located between 8 and 12 micrometers is the most significant because it is located where Earth's radiation is most intense.

By contrast, clouds that are composed of tiny liquid droplets (not water vapor) are excellent absorbers of the energy in the atmospheric window. Clouds absorb outgoing longwave radiation and radiate much of this energy back to Earth's surface. Thus, clouds serve a purpose similar to window blinds because they effectively block the atmospheric window and lower the rate at which Earth's surface cools. This explains why nighttime temperatures remain higher on cloudy nights than on clear nights.

Because the atmosphere is largely transparent to solar (shortwave) radiation but more absorptive of the longwave radiation emitted by Earth, the atmosphere is heated from the ground up. This explains the general drop in temperature with increased altitude in the troposphere. The farther from the "radiator" (Earth's surface), the colder it gets. On average, the temperature drops 6.5°C for each kilometer (3.5°F per 1000 feet) increase in altitude, a figure known as the *normal lapse rate* (see Chapter 1). The fact that the atmosphere does not acquire the bulk of its energy directly from the Sun but is heated by Earth's surface is of utmost importance to the dynamics of the weather machine.

## The Greenhouse Effect

Research of "airless" planetary bodies such as the Moon have led scientists to determine that if Earth had no atmosphere, it would have an average surface temperature below freezing. Fortunately, Earth's atmosphere "traps" some of the outgoing radiation, which makes our planet habitable. The extremely important role that the atmosphere plays in heating Earth's land–sea surface is known as the **greenhouse effect**.

As discussed earlier, cloudless air is largely transparent to incoming shortwave solar radiation and, hence, transmits much of it to Earth's surface, where it can be absorbed and then eventually reradiated skyward. Two atmospheric gases, *water vapor* and *carbon dioxide*, are meteorologically important because they absorb a significant portion of the longwave radiation emitted by Earth's surface. As Earth's radiation heats these absorptive gases, the temperature of the atmosphere increases. The atmosphere, in turn, radiates some of this energy *out to space*, but more importantly, it radiates an equivalent amount *back toward Earth's surface*. Without this complicated game of "pass the hot potato," Earth's average temperature would be –18°C (0°F) rather than the current temperature of 15°C (59°F) (**Fig. 2.22**). As a result water vapor and carbon dioxide, which make Earth habitable for humans and other life-forms, are called **greenhouse gases**.

This natural phenomenon was named the greenhouse effect because greenhouses are heated in a similar manner. The glass in a greenhouse allows shortwave solar radiation to enter and be absorbed by the objects inside. These objects, in turn, radiate energy but at longer wavelengths, to which the glass is nearly opaque. The heat, therefore, is "trapped" in the greenhouse. Although this analogy is widely used, it has been shown that air inside greenhouses attains higher temperatures than outside air, in part due to the restricted exchange of warmer air inside and cooler air outside. Nevertheless, the term *greenhouse effect* is still used to describe atmospheric heating.

Media reports frequently and erroneously identify the greenhouse effect as the "villain" in the global warming problem. However, the greenhouse effect and global warming *are different concepts*. Without the greenhouse effect, Earth would be uninhabitable. Scientists have mounting evidence that human activities (particularly the release of carbon dioxide into the atmosphere) are responsible for a rise in global temperatures (see Chapter 14). Thus, humans are compounding the effects of an otherwise natural process (the greenhouse effect). It is incorrect to equate the greenhouse phenomenon, which makes life possible, with global warming—which involves undesirable changes to our atmosphere and is caused mainly by human activities.

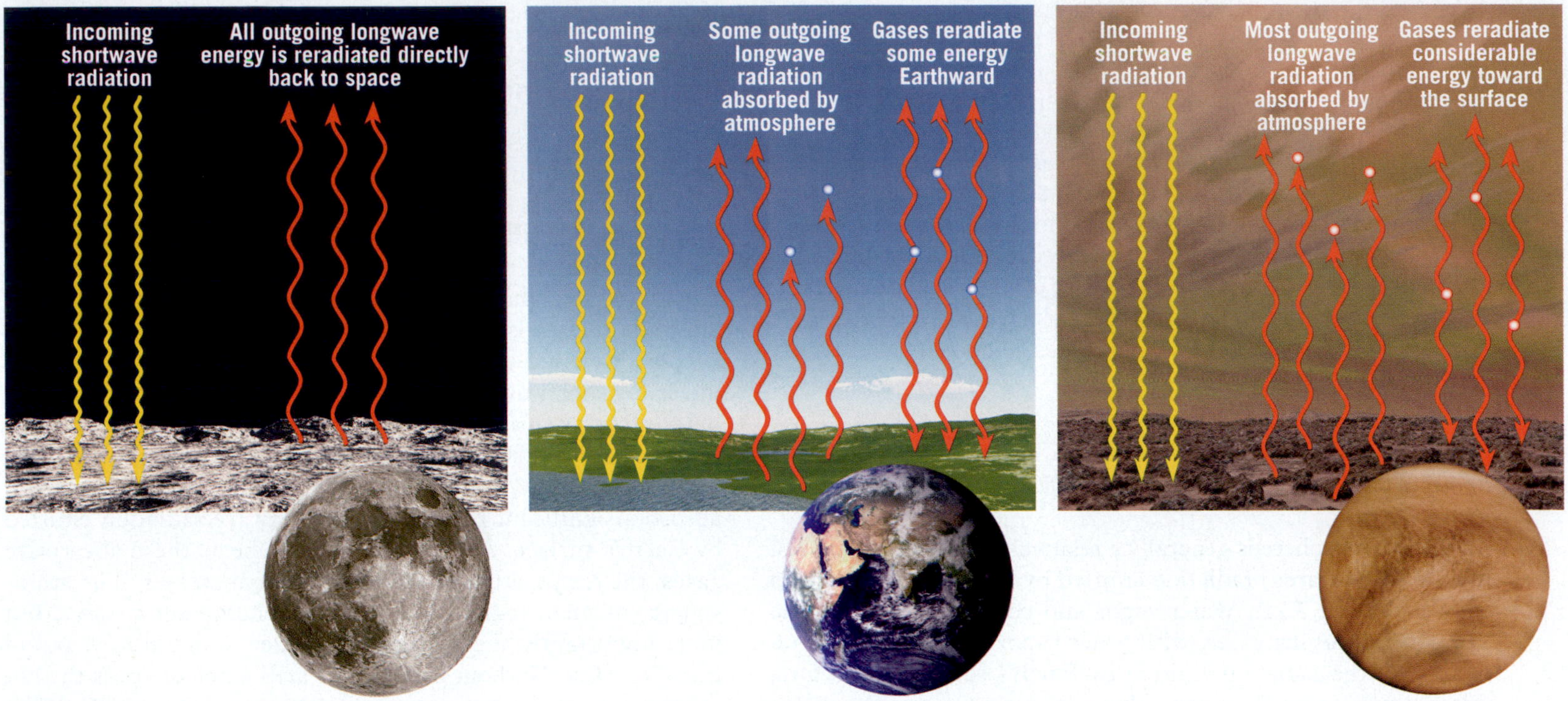
A. Airless bodies like the Moon All incoming solar radiation reaches the surface. Some is reflected back to space. The rest is absorbed by the surface and radiated directly back to space. As a result the lunar surface has a much lower average temperature than Earth.
Incoming shortwave radiation
All outgoing longwave energy is reradiated directly back to space
B. Bodies with modest amounts of greenhouse gases like Earth The atmosphere absorbs some of the longwave radiation emitted by the surface. A portion of this energy is radiated back to the surface and is responsible for keeping Earth's surface 33°C (59°F) warmer than it would otherwise be.
Incoming shortwave radiation
Some outgoing longwave radiation absorbed by atmosphere
Gases reradiate some energy Earthward
C. Bodies with abundant greenhouse gases like Venus Venus experiences extraordinary greenhouse warming, which is estimated to raise its surface temperature by 523°C (941°F).
Incoming shortwave radiation
Most outgoing longwave radiation absorbed by atmosphere
Gases reradiate considerable energy toward the surface

http://goo.gl/uC6OV

▲ **SmartFigure 2.22 The greenhouse effect** **A.** Airless bodies such as the Moon experience no greenhouse effect. **B.** On bodies with modest amounts of greenhouse gases, such as Earth, the greenhouse effect is responsible for keeping Earth's surface 33°C (59°F) warmer than it would be otherwise. **C.** Bodies with abundant greenhouse gases, such as Venus, experience extraordinary greenhouse warming, which is estimated to raise its surface temperature by 523°C (941°F).

### ✔ Concept Checks 2.5

1. Explain why the atmosphere is heated chiefly by radiation from Earth's surface rather than by direct solar radiation.
2. Which gases are the primary heat absorbers in the lower atmosphere?
3. What is the atmospheric window? How is it "closed"?
4. How does Earth's atmosphere act as a greenhouse?
5. What is the "villain" in the global warming problem?

## 2.6 Earth's Energy Budget

**Describe the major components of Earth's annual energy budget.**

Globally, Earth's average temperature remains relatively constant, despite seasonal cold spells and heat waves. This stability indicates that a balance exists between the amount of incoming solar radiation and the amount of radiation emitted back to space; otherwise, Earth would be getting progressively colder or warmer. In addition, the energy exchanged between the Earth's surface and the atmosphere must also remain stable. This surface-to-atmosphere equilibrium is accomplished through conduction, convection, and the transfer of latent heat as well as by the transmission of longwave radiation between the Earth's surface and the atmosphere. The annual balance of incoming and outgoing radiation, as well as the energy balance that exists between Earth's surface and its atmosphere, is generally referred to as Earth's **annual energy budget**. Further discussion of Earth's changing climate can be found in Chapter 14.

### Annual Energy Budget

**Figure 2.23** illustrates Earth's annual energy budget. For simplicity we will use 100 units to represent the solar radiation intercepted at the outer edge of the atmosphere. You have already seen in Figure 2.15 that, of the total radiation that

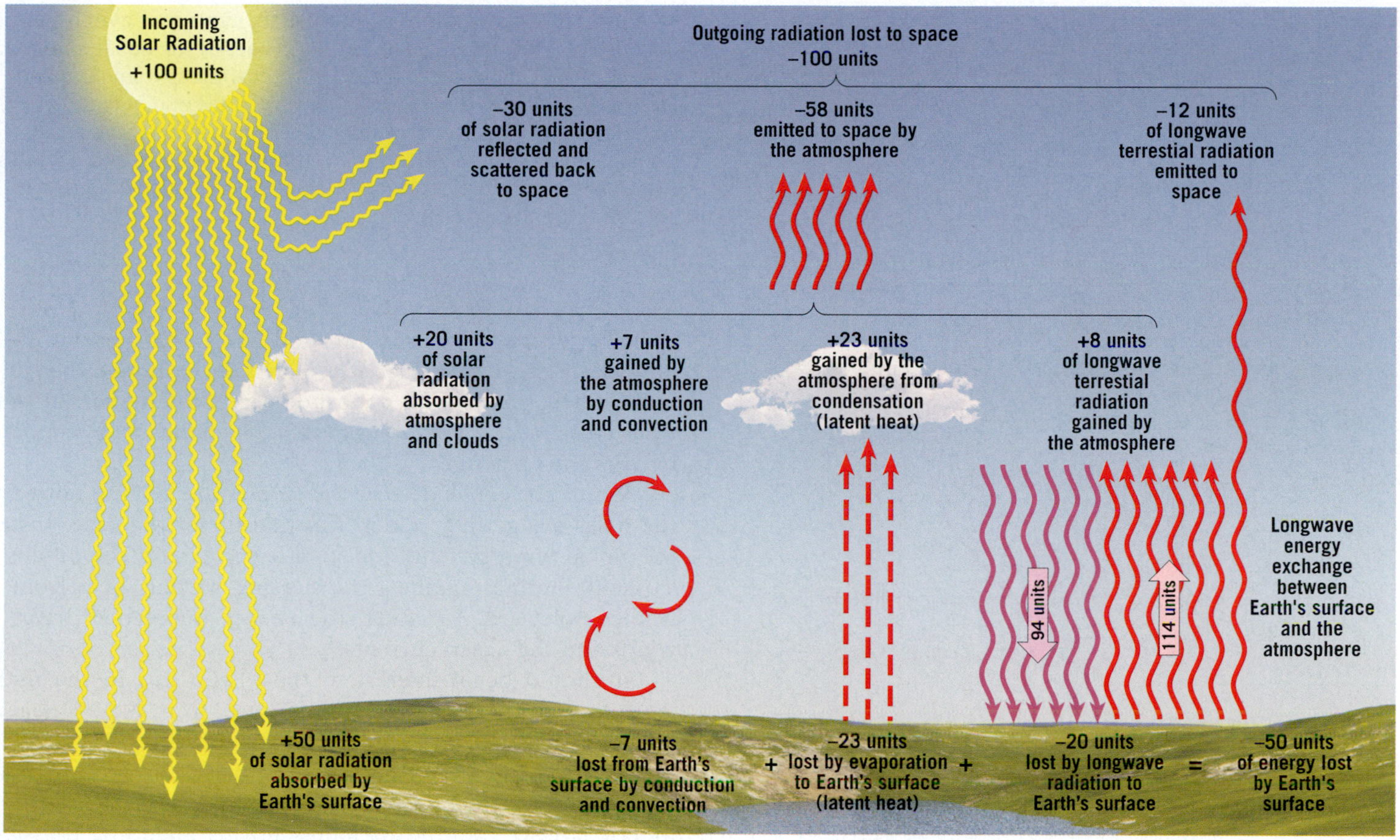


▲ **Figure 2.23 Earth's energy budget** These estimates of the average global energy budget come from satellite observations and radiation studies. As more data are accumulated, these numbers will be modified.

reaches Earth, roughly 30 units (30 percent) are reflected and scattered back to space. The remaining 70 units are absorbed: 20 units within the atmosphere and 50 units by Earth's land–sea surface. How does Earth transfer this energy back to space?

If all the energy absorbed by our planet were radiated directly and immediately back to space, Earth's heat budget would be simple: 100 units of radiation received and 100 units returned to space. In fact, this does happen *over time* (minus small quantities of energy that become locked up in biomass that may eventually become fossil fuel). What complicates the heat budget is the behavior of certain greenhouse gases, particularly water vapor and carbon dioxide. As you have learned, these greenhouse gases absorb a large share of outward-directed infrared radiation and radiate much of that energy back to Earth. This "recycled" energy significantly increases the radiation received by Earth's surface. In addition to the 50 units received directly from the Sun, Earth's surface receives longwave radiation emitted downward by the atmosphere (94 units).

A balance is maintained because all the energy absorbed by Earth's surface is returned to the atmosphere and eventually radiated back to space. Earth's surface loses energy through a variety of processes: the emission of longwave radiation; conduction and convection; and energy loss to Earth's surface through the process of evaporation—latent heat (Figure 2.23). Most of the long-wave radiation emitted skyward is re-absorbed by the atmosphere. Conduction results in the transfer of energy between Earth's surface to the air directly above, while convection carries the warm air located near the surface upward as thermals (7 units).

Video MM
Solar Power
http://goo.gl/9GvFmT

Earth's surface also loses a substantial amount of energy (23 units) through evaporation. This occurs because energy is required for liquid water molecules to leave the surface of a body of water and change to its gaseous form, water vapor. The energy lost by a water body is carried into the atmosphere by molecules of water vapor. Recall that the heat used to evaporate water does not produce a temperature change and is referred to as *latent heat* (hidden heat). If the water vapor condenses to form cloud droplets, the energy released by condensation will be detectable as *sensible heat* (heat we can feel and measure with a thermometer). Thus through the process of evaporation, water molecules in gas form carry latent heat into the atmosphere, where it is eventually released.

In summary, the quantity of incoming solar radiation is, over time, balanced by the quantity of longwave radiation that is radiated back to space.

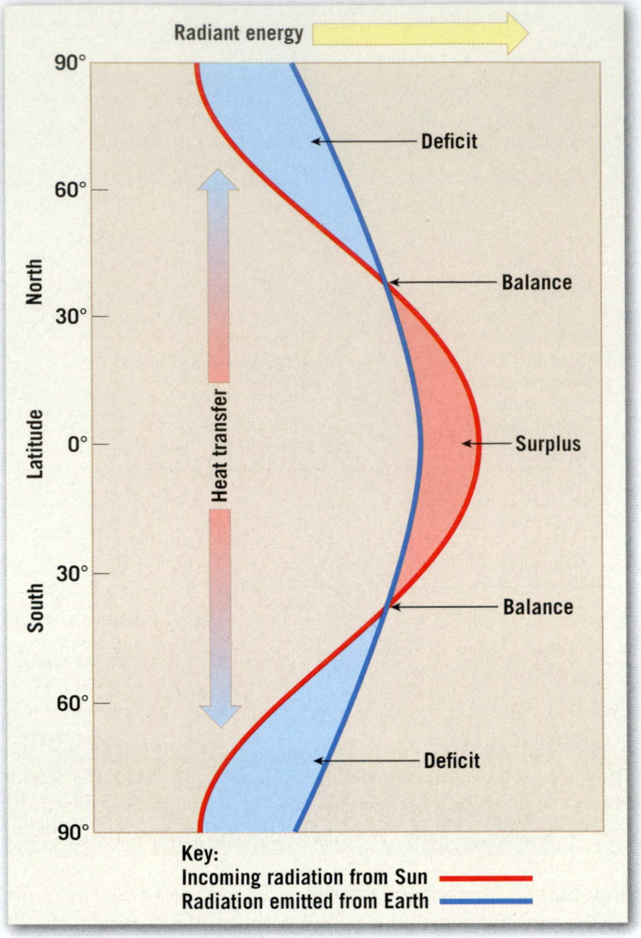


◀ **Figure 2.24 Latitudinal heat balance, averaged over an entire year** In the zone extending 38° on both sides of the equator, the amount of incoming solar radiation exceeds the loss from outgoing Earth radiation. The reverse is true for the middle and high polar latitudes, where losses from outgoing Earth radiation exceed gains from incoming solar radiation. The global wind systems and, to a lesser extent, the oceans act as giant thermal engines, transferring surplus heat from the tropics poleward.

## Latitudinal Energy Budget

Because incoming solar radiation is roughly equal to the amount of outgoing radiation, on average, worldwide temperatures remain nearly constant. However, the balance of incoming and outgoing radiation that is applicable for the entire planet is not maintained at each latitude. Averaged over the entire year, a zone around Earth that lies within 38° latitude of the equator *receives more solar radiation than is lost to space* (**Fig. 2.24**). The opposite is true for higher latitudes, where *more heat is lost through radiation emitted by Earth than is received from the Sun.*

We might conclude that the tropics are getting hotter and the poles are getting colder. But that is not the case. Instead, the global wind systems and, to a lesser extent, the oceans act as giant thermal engines, transferring surplus heat from the tropics poleward. In effect, the energy imbalance drives the winds and the ocean currents.

It should be of interest to those who live in the middle latitudes—in the Northern Hemisphere, from New Orleans at 30° north latitude to Winnipeg, Manitoba, at 50° north latitude—that most heat transfer takes place across this region. Consequently, much of the stormy weather experienced in the middle latitudes can be attributed to this unending transfer of heat from the tropics toward the poles. These processes are discussed in more detail in later chapters.

## ✔ Concept Checks 2.6

1. The tropics receive more solar radiation than is lost. Why then don't the tropics keep getting hotter?
2. What two phenomena result from the imbalance of heating that exists between the tropics and the poles?

# 2 Concepts in Review Heating Earth's Surface and Atmosphere

## 2.1 Earth–Sun Relationships

▶ Explain what causes the Sun angle and length of daylight to change throughout the year. Describe how these changes result in seasonal changes in temperature.

**Key Terms:** rotation, revolution, perihelion, aphelion, plane of the ecliptic, inclination of the axis, Tropic of Cancer, summer solstice, Tropic of Capricorn, winter solstice, autumnal (fall) equinox, spring (vernal) equinox, circle of illumination

- The seasons are caused by changes in the angle at which the Sun's rays strike the surface and the changes in the length of daylight at each latitude. These seasonal changes result from the tilt of Earth's axis as it revolves around the Sun.
- When the Sun is directly overhead (at a 90° angle to Earth's surface), the solar rays are most concentrated and thus most intense. At lower Sun angles, the rays become more spread out and less intense.

**Q** How would the seasons be affected if Earth's axis were not inclined to the plane of its orbit (see accompanying drawing) but were instead perpendicular?

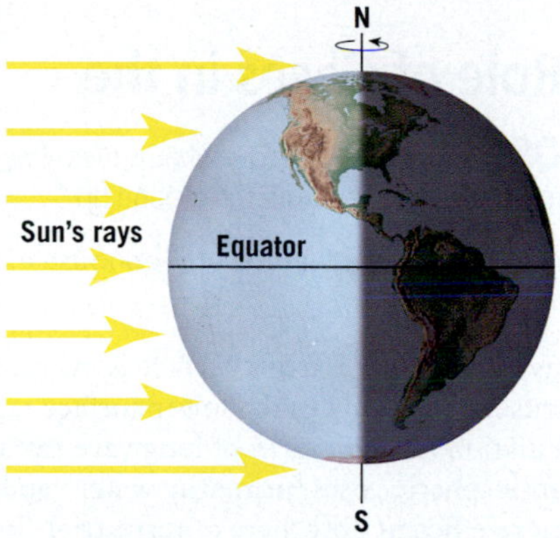


## 2.2 Energy, Temperature, and Heat

▶ Compare and contrast latent heat and sensible heat.

**Key Terms:** energy, kinetic energy, potential energy, temperature, heat, latent heat, sensible heat

- Energy is the ability to do work. The two major categories of energy are (1) kinetic energy, which can be thought of as energy of motion, and (2) potential energy, energy that has the capability to do work.
- Temperature is a measure of the average kinetic energy of the atoms or molecules in a substance.
- Heat is the transfer of energy into or out of an object because of temperature differences between that object and its surroundings. Heat flows from a region of higher temperature to a region of lower temperature.
- Latent heat is the energy involved when water changes from one state of matter to another. During evaporation, for example, energy is stored, or "hidden," within the escaping water vapor molecules, and this energy is eventually released when water vapor condenses to form water droplets in clouds.
- In contrast to latent heat, sensible heat is the heat we can feel and measure with a thermometer, and it does not involve a phase change. It is called sensible heat because it can be "sensed."

## 2.3 Mechanisms of Heat Transfer

▶ List and describe the three mechanisms of heat transfer.

**Key Terms:** conduction, convection, thermals, advection, radiation (electromagnetic radiation), wavelengths, micrometer, visible light, infrared radiation (IR), ultraviolet (UV) radiation, longwave radiation, shortwave radiation

- Conduction is the transfer of heat through matter by electron and molecular collisions between molecules. Because air is a poor conductor, conduction is significant only between Earth's surface and the air in immediate contact with the surface.
- Convection is heat transfer that involves the actual movement or circulation of a substance. Convection is an important mechanism of heat transfer in the atmosphere, where warm air rises and cooler air descends.
- Radiation or electromagnetic radiation consists of a large array of energy that includes X-rays, visible light, heat waves, and radio waves that travel as waves of various sizes. Shorter wavelengths of radiation have greater energy.
- These are four basic laws of radiation: (1) All objects emit radiant energy; (2) hotter objects radiate more total energy per unit area than colder objects; (3) the hotter the radiating body, the shorter is the wavelength of maximum radiation; and (4) objects that are good absorbers of radiation are also good emitters.

**Q** Describe how the three mechanisms of heat transfer are illustrated in this photo.

## 2.4 What Happens to Incoming Solar Radiation? ► Using the diagram in Figure 2.15 as a guide, describe what happens to incoming solar radiation.

**Key Terms:** transmission, absorptivity, reflection, scattering, backscattering, albedo, diffused light

- Approximately 50 percent of the solar radiation that strikes the atmosphere reaches Earth's surface. About 30 percent is reflected back to space. The fraction of radiation reflected by a surface is called its albedo. Clouds and the atmosphere's gases absorb the remaining 20 percent of the incoming solar energy.

## 2.5 The Role of Gases in the Atmosphere ► Explain what is meant by the statement "The atmosphere is heated from the ground up."

**Key Terms:** atmospheric window, greenhouse effect, greenhouse gases

- Radiant energy absorbed at Earth's surface is eventually radiated skyward. Because Earth has a much lower surface temperature than the Sun, its radiation is in the form of longwave infrared radiation. Because the atmospheric gases, primarily water vapor and carbon dioxide, are more efficient absorbers of terrestrial (longwave) radiation, the atmosphere is heated primarily from the ground up.
- The greenhouse effect refers to the selective absorption of Earth radiation by atmospheric gases, mainly water vapor and carbon dioxide, that results in Earth's average temperature being warmer than it would be otherwise.
- The greenhouse effect is a natural phenomenon that makes Earth habitable. The greenhouse effect is often, but inaccurately, portrayed as the "villain" of global warming, but human activities that release greenhouse gases (primarily carbon dioxide) are actually the villains.

## 2.6 Earth's Energy Budget ► Describe the major components of Earth's annual energy budget.

**Key Terms:** annual energy budget

- The annual balance of incoming and outgoing radiation, as well as the energy balance that exists between Earth's surface and the atmosphere, is generally referred to as Earth's annual energy budget.
- Averaged over an entire year, a zone around Earth between 38° north and 38° south latitudes receives more solar radiation than is lost to space. The opposite is true for more poleward locations, where more heat is lost through outgoing long-wave terrestrial radiation than is received. It is this energy imbalance between the low and high latitudes that drives the weather system and, in turn, transfers surplus heat from the tropics toward the poles.

## Give it Some Thought

1. Describe what the seasons would be like if Earth's axis were inclined 40° rather than 23½°, as is currently the situation. Where would the Tropics of Cancer and Capricorn be located? Where would the Arctic and Antarctic Circles be?

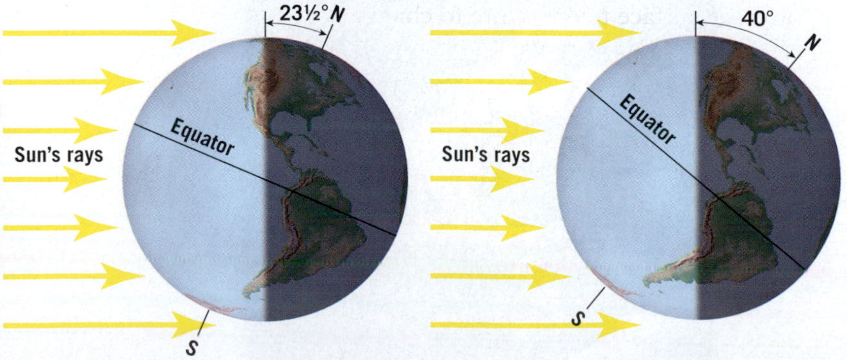


2. During a "shore lunch" on a fishing trip to a remote location, the fishing guide will sometimes place a pail of lake water next to the cooking fire, as shown in the accompanying illustration. When the water in the pail begins to boil, the guide will lift the pail from the fire with one hand and "impress" the guests by placing the other hand on the bottom. Use what you have learned about the three mechanisms of heat transfer to explain why the guide's hand isn't burned by touching the bottom of the pail.

3. On what date is Earth closest to the Sun? On that date, what season is it in the Northern Hemisphere? Explain this apparent contradiction.

4. Which of the three mechanisms of heat transfer is most significant in each of the following situations:
   a. Driving a car with the seat heater turned on
   b. Sitting in an outdoor hot tub
   c. Lying inside a tanning bed
   d. Driving a car with the air conditioning turned on
5. The accompanying four diagrams (labeled A–D) are intended to illustrate the Earth–Sun relationships that produce the seasons.
   a. Which one of these diagrams most accurately shows this relationship?
   b. Identify what is *inaccurately* shown in each of the other three diagrams.

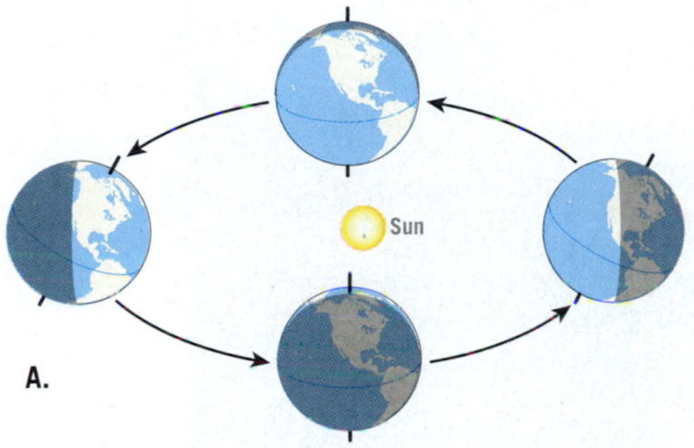


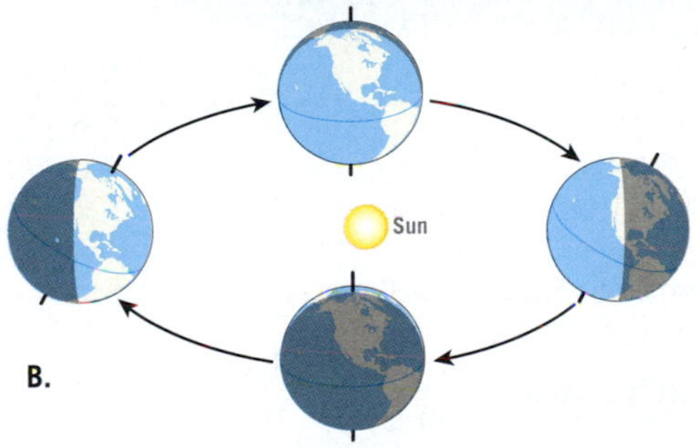


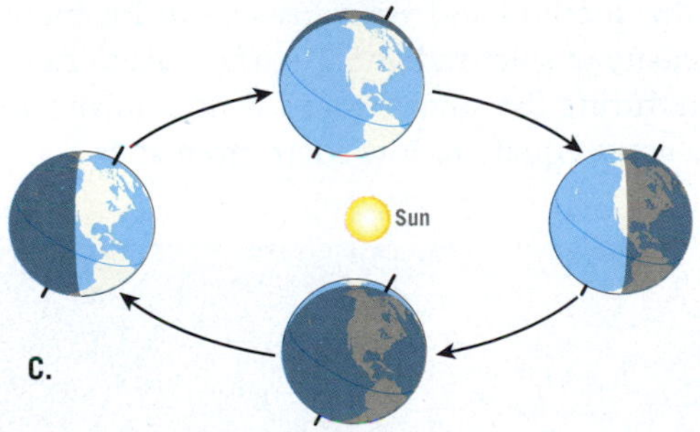


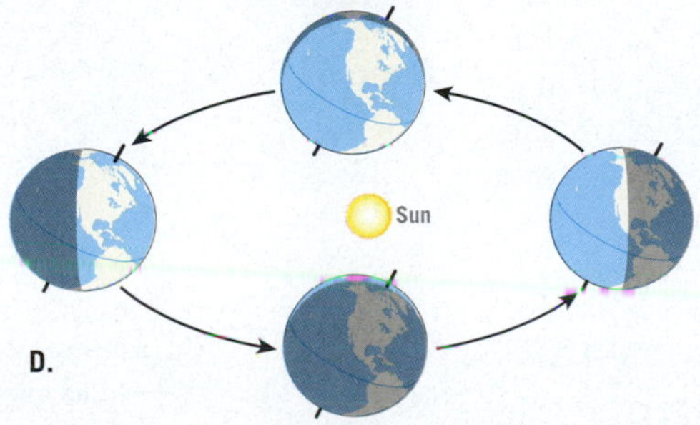


6. The Sun shines continually at the North Pole for 6 months, from the spring equinox until the fall equinox, yet temperatures never get very warm. Explain why this is the case.
7. The accompanying image shows an area of our galaxy where stars having surface temperatures much hotter than the Sun have recently formed. Imagine that an Earth-like planet formed around one of these stars, at a distance where it receives the same intensity of light as Earth. Use the laws of radiation to explain why this planet may not provide a habitable environment for humans.

8. Rank the following according to the wavelength of radiant energy each emits, from the shortest wavelength to the longest:
   a. A light bulb with a filament glowing at 4000°C
   b. A rock at room temperature
   c. A car engine at 140°C
9. Figure 2.15 shows that about 30 percent of the Sun's energy is reflected or scattered back to space. If Earth's albedo were to increase to 50 percent, how would you expect Earth's average surface temperature to change?
10. Explain why Earth's equatorial regions are not becoming warmer, despite the fact that they receive more incoming solar radiation than they radiate back to space.

11. The accompanying photo shows the explosive 1991 eruption of Mount Pinatubo in the Philippines. How would you expect global temperatures to respond to the ash and debris this volcano spewed high into the atmosphere?

# Problems

1. Refer to Figure 2.A in Box 2.1 (p. 34) and calculate the noon Sun angle on June 21 and December 21 at 50° north latitude, 0° latitude (the equator), and 20° south latitude. Which of these latitudes has the greatest variation in noon Sun angle between summer and winter?
2. For the latitudes listed in Problem 1, determine the length of daylight and darkness on June 21 and December 21 (refer to Table 2.1). Which of the latitudes listed has the largest seasonal variation in length of daylight? Which latitude has the smallest variation?
3. Calculate the noon Sun angle at your location for the equinoxes and solstices.
4. Using the accompanying figure and basic trigonometry, the intensity of solar radiation can be calculated. For simplicity, consider a solar beam of 1 unit width. The surface area over which the beam would be spread changes with Sun angle, such that:

$$\text{Surface area} = \frac{1 \text{ unit}}{\sin(\text{Sun angle})}$$

Therefore, if the Sun angle at solar noon is 56°:

$$\text{Surface area} = \frac{1 \text{ unit}}{\sin 56^\circ} = \frac{1 \text{ unit}}{0.829} = 1.260 \text{ units}$$

Using this method and your answers to Problem 3, calculate the intensity of solar radiation (surface area) for your location at noon during the summer and winter solstices. *Note:* Large surface areas equate to low solar intensities.

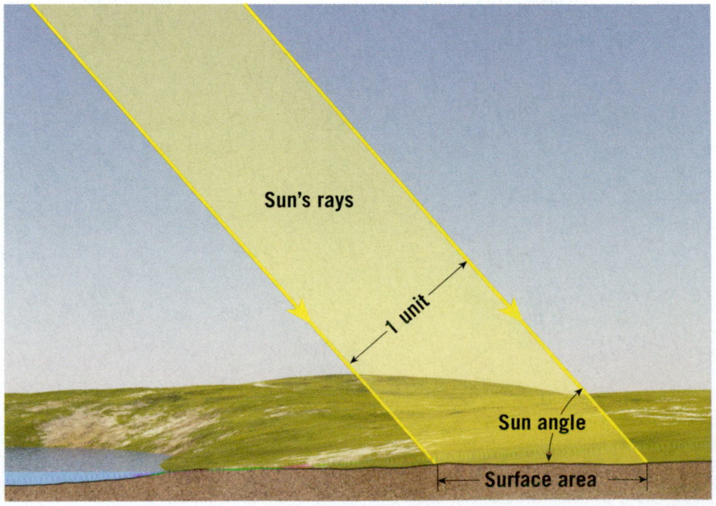

5. The accompanying analemma is a graph used to determine the latitude where the overhead noon Sun is located for any date. To determine the latitude of the overhead noon Sun from the analemma, find the desired date on the graph and read the coinciding latitude along the left axis. Determine the latitude of the overhead noon Sun for the following dates. Remember to indicate *north* (N) or *south* (S).
   **a.** March 21
   **b.** June 5
   **c.** December 10
6. Use Figure 2.A and the analemma used in Question 5 to calculate the noon Sun angle at your location (latitude) on the following dates:
   **a.** September 7
   **b.** July 5
   **c.** January 1

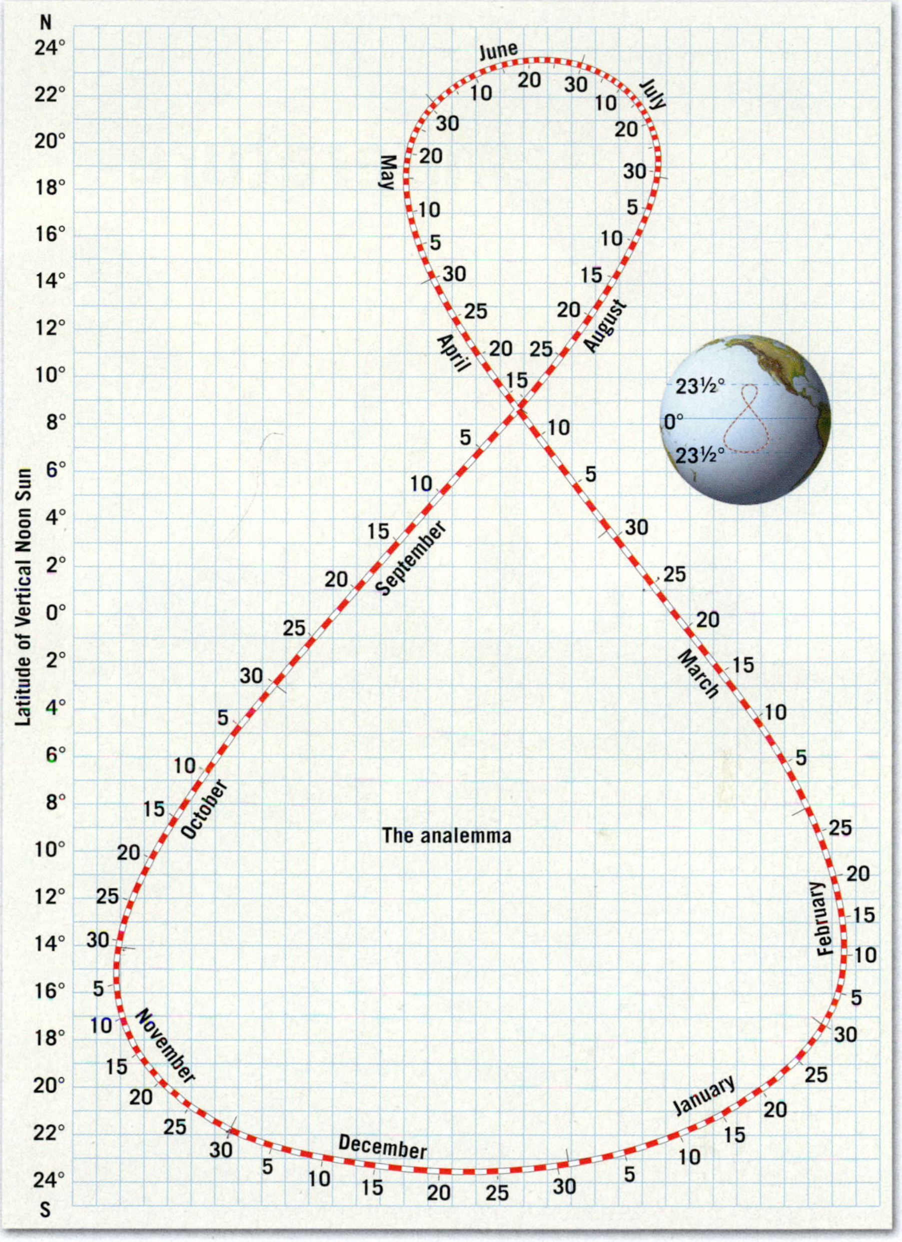


## MasteringMeteorology™

Looking for additional review and test prep materials? Visit the Study Area in *MasteringMeteorology*™ to enhance your understanding of this chapter's content by accessing a variety of resources, including Videos, **MapMaster**™ interactive maps, Geoscience Animations, GEODe, *In the News* RSS feeds, flashcards, web links, self-study quizzes, and an eText version of *The Atmosphere*.

# 3 Temperature

## Focus on Concepts

*Each statement represents the primary learning objective for the corresponding major heading within the chapter. After you complete the chapter, you should be able to:*

**3.1** Calculate five commonly used types of temperature data and interpret a map that depicts temperature data using isotherms.

**3.2** Name the principal controls of temperature and use examples to describe their effects.

**3.3** Interpret the patterns depicted on world temperature maps.

**3.4** Discuss the basic daily and annual cycles of air temperature.

**3.5** Explain how different types of thermometers work and why the placement of thermometers is an important factor in obtaining accurate readings. Distinguish among Fahrenheit, Celsius, and Kelvin temperature scales.

**3.6** Summarize several applications of temperature data.

Temperature is one of the basic elements of weather and climate. When someone asks what the weather is like outside, air temperature is often the first element we mention. From everyday experience, we know that temperatures vary on different time scales: seasonally, daily, and even hourly. Moreover, we all realize that substantial temperature differences exist from one place to another. In Chapter 2 you learned how air is heated and examined the role of Earth–Sun relationships in causing temperature variations from season to season and from latitude to latitude. In this chapter you will focus on several other aspects of this very important atmospheric property, including factors other than Earth–Sun relationships, that act as temperature controls. You will also look at how temperature is measured and expressed and see that temperature data can be of very practical value to us all. Applications include calculations that are useful in evaluating energy consumption, crop maturity, and human comfort.

*This snow-capped mountain reminds us that temperatures decrease with an increase in altitude in the troposphere. Altitude is one of several factors that influence temperature.*

# 3.1 For the Record: Air-Temperature Data

**Calculate five commonly used types of temperature data and interpret a map that depicts temperature data using isotherms.**

MM GEODe ▶ Temperature Data and the Controls of Temperature ▶ Basic Temperature Data

Temperatures recorded daily at thousands of weather stations worldwide provide much of the temperature data compiled by meteorologists and climatologists (Fig. 3.1). Hourly temperatures may be recorded by an observer or obtained from automated observing systems that continually monitor the atmosphere. At many locations, only the maximum and minimum temperatures are obtained (Box 3.1).

## Basic Calculations

The **daily mean temperature** is determined by averaging the 24 hourly readings or by adding the maximum and minimum temperatures for a 24-hour period and dividing by 2. From the maximum and minimum, the **daily temperature range** is computed by finding the difference between these figures. Other data involving longer periods are also compiled:

- The **monthly mean temperature** is calculated by adding together the daily means for each day of the month and dividing by the number of days in the month.
- The **annual mean temperature** is an average of the 12 monthly means.
- The **annual temperature range** is computed by finding the difference between the warmest and coldest monthly mean temperatures.

Mean temperatures are especially useful for making daily, monthly, and annual comparisons. It is common to hear a weather reporter state, "Last month was the warmest February on record" or "Today Omaha was 10° warmer than Chicago." Temperature ranges are also useful statistics because they give an indication of extremes, a necessary part of understanding the weather and climate of a place or an area.

## Isotherms

To examine the distribution of air temperatures over large areas, isotherms are commonly used. An **isotherm** is a line that connects points on a map that have the same temperature (*iso* = equal, *therm* = temperature). Therefore, all points through which an isotherm passes have identical temperatures for the time period indicated. Generally, isotherms

▼ **Figure 3.1 Temperatures vary widely in the midlatitudes** People living in the middle latitudes can experience a wide range of temperatures during a year. **A.** This pedestrian is navigating through a Chicago neighborhood during a February 2011 blizzard that dropped more than 50 centimeters (nearly 20 inches) of snow on the city. **B.** A few months later, people were beating the heat along Chicago's North Avenue beach on a hot summer day.

A.

B.

## Box 3.1 North America's Hottest and Coldest Places

Most people living in the United States have experienced temperatures of 38°C (100°F) or more. When statistics for the 50 states are examined for the past century or longer, we find that every state has a maximum temperature record of 38°C or higher. Even Alaska has recorded a temperature this high—set June 27, 1915, at Fort Yukon, a town along the Arctic Circle in the interior of the state.

### Maximum Temperature Records

Surprisingly, the state that ties Alaska for the "lowest high" is Hawaii. Panala, on the south coast of the Big Island, recorded 38°C on April 27, 1931. Although humid tropical and subtropical places such as Hawaii are known for being warm throughout the year, they seldom experience maximum temperatures that surpass the low to mid-30s Celsius (90s Fahrenheit).

The highest accepted temperature record for the United States as well as the entire world is 57°C (134°F). This long-standing record was set at Death Valley, California, on July 10, 1913. Summer temperatures at Death Valley are consistently among the highest in the Western Hemisphere. During June, July, and August, temperatures exceeding 49°C (120°F) are to be expected. Fortunately, Death Valley has few human summertime residents (**Fig. 3.A**).

Why are summer temperatures at Death Valley so high? In addition to having the lowest elevation in the Western Hemisphere (53 meters [174 feet] below sea level), Death Valley is a desert. Although it is only about 300 kilometers (less than 200 miles) from the Pacific Ocean, mountains cut off the valley from the ocean's moderating influence and moisture. Clear skies allow a maximum of sunshine to strike the dry, barren surface. Because no energy is used to evaporate moisture, as occurs in humid regions, all the energy is available to heat the ground. In addition, subsiding air that warms by compression as it descends is also common to the region and contributes to its high maximum temperatures.

▲ **Figure 3.A Almost a record!** On June 30, 2013, 100 years after Death Valley set the all-time high recorded temperature, it came close to equaling it. On that date, Death Valley's air temperature peaked at 54°C (129.2°F).

### Minimum Temperature Records

The temperature controls that produce truly frigid temperatures are predictable and should come as no surprise. We expect extremely cold temperatures during winter in high-latitude places that lack the moderating influence of the ocean. Moreover, stations located on ice sheets and glaciers should be especially cold, as should stations positioned high in the mountains. All these criteria apply to Greenland's North Ice Station (elevation 2307 meters [7567 feet]). Here on January 9, 1954, the temperature plunged to –66°C (–87°F). If we exclude Greenland from consideration, Snag, in Canada's Yukon, holds the record for North America. This remote outpost experienced a temperature of –63°C (–81°F) on February 3, 1947. When only U.S. locations are considered, Prospect Creek, located north of the Arctic Circle in the Endicott Mountains of Alaska, came close to the North American record on January 23, 1971, when the temperature plunged to –62°C (–80°F). In the lower 48 states, the record of –57°C (–70°F) was set in the mountains at Rogers Pass, Montana, on January 20, 1954. Remember that many other places have no doubt experienced equally low or even lower temperatures; they just were not recorded.

#### Question

1. Death Valley is not a great distance from the cool Pacific Ocean yet experiences very high temperatures. Why is there no moderating ocean influence?

representing 5° or 10° differences in temperature are used, but *any* interval may be chosen. **Figure. 3.2** illustrates how isotherms are drawn on a map. Notice that most isotherms do not pass directly through the observing stations because the station readings may not coincide with the values chosen for the isotherms. Only an occasional station temperature will be exactly the same as the value of the isotherm, so it is usually necessary to draw the lines by estimating the proper position between stations.

Isothermal maps are valuable tools because they make temperature distribution clearly visible at a glance. Areas of low and high temperatures are easy to pick out. In addition, the amount of temperature change per unit of distance, called the **temperature gradient**, is easy to visualize. Closely spaced isotherms indicate a rapid rate of temperature change, whereas more widely spaced lines indicate a more gradual rate of change. For example, notice in Figure 3.2 that the isotherms are more closely spaced in Colorado and Utah (steeper

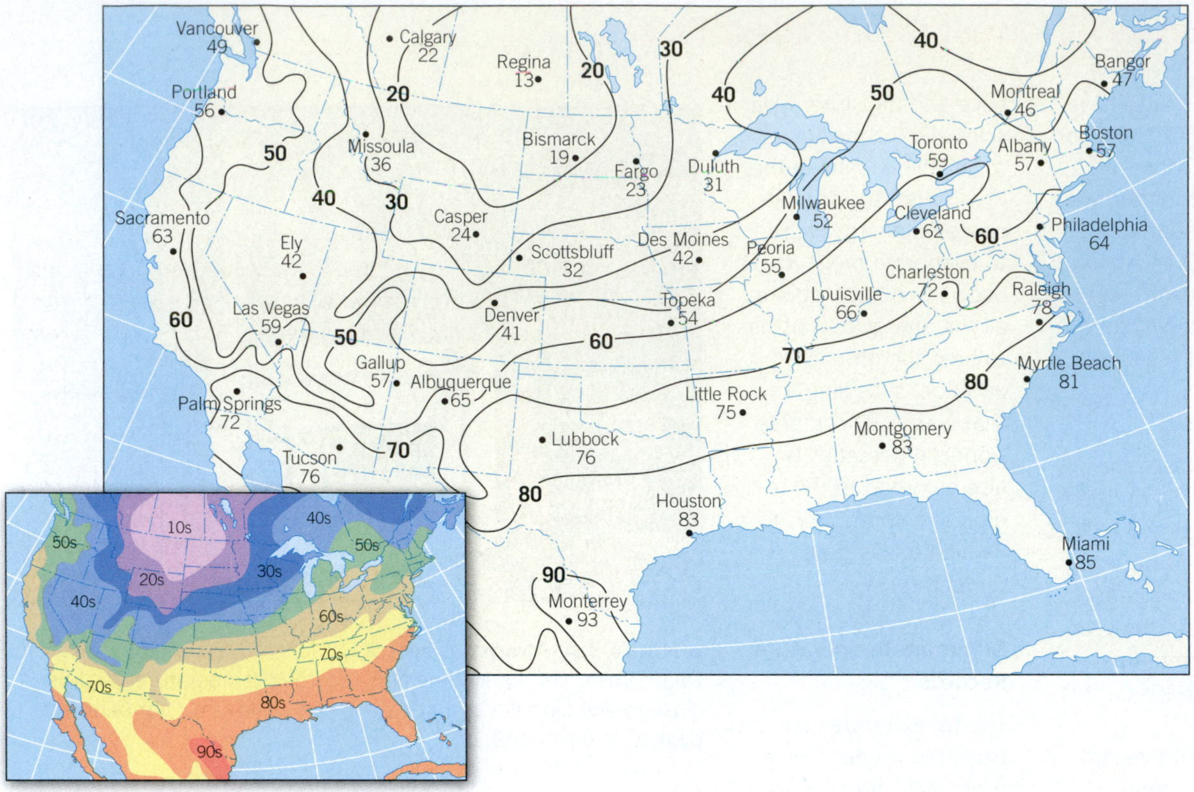


◀ **SmartFigure 3.2 Isotherms** The large map shows high temperatures for a spring day. Isotherms are lines that connect points of equal temperature. Showing temperature distribution in this way makes patterns easier to see. Notice that most isotherms do not pass directly through the observing stations. It is usually necessary to draw isotherms by estimating their proper position between stations. On television and in many newspapers, temperature maps are in color as shown in the inset map. Rather than label isotherms, these maps label the area *between* isotherms. For example, the zone between the 60° and 70° isotherms is labeled "60s."

http://goo.gl/a3dHr

temperature gradient), whereas the isotherms are spread farther apart in Texas (gentler temperature gradient). Without isotherms, a map would be covered with numbers representing temperatures at tens or hundreds of places, and that would make patterns difficult to see.

## ✔ Concept Checks 3.1

1. How are the following temperature data calculated: daily mean, daily range, monthly mean, annual mean, and annual range?
2. What are isotherms, and what is their purpose?

**students sometimes ask...**

### What's the hottest city in the United States?

It depends on how you define "hottest." If average annual temperature is used, then Key West, Florida, is the hottest, with an annual mean of 25.6°C (78°F) for the 30-year span 1981–2010. However, if we look at cities with the highest July maximums during the 1981–2010 span, then the desert community of Bullhead City, Arizona, has the distinction of being hottest. Its average daily high in July is a blistering 44.6°C (112.2°F)!

## 3.2 Why Temperatures Vary: The Controls of Temperature

**Name the principal controls of temperature and use examples to describe their effects.**

MM GEODe ▶ Temperature Data and the Controls of Temperature ▶ Controls of Temperature

A **temperature control** is any factor that causes temperatures to vary from place to place and from time to time. Chapter 2 examined the most important cause of temperature variation—differences in the receipt of solar radiation. Because variations in Sun angle and length of daylight depend on latitude, they are responsible for warm temperatures in the tropics and colder temperatures poleward. Of course, seasonal temperature changes at various latitudes occur as the Sun's vertical rays migrate toward and away from a place during the year. Figure 3.3 reminds us of the importance of latitude as a control of temperature.

But latitude is not the only control of temperature. If it were, we would expect all places along the same line of latitude to have identical temperatures. This is clearly not the case. For instance, Eureka, California, and New York City are both coastal cities at about the same latitude, and both places have an annual mean temperature of 11°C (51.8°F). Yet New York City is 9.4°C (16.9°F) warmer than Eureka in July and 9.4°C (16.9°F) colder than Eureka in January. In another example, two cities in Ecuador—Quito and Guayaquil—are relatively close to one another, but the mean annual temperatures at these two cities differ by 12.2°C (22°F). To explain these situations

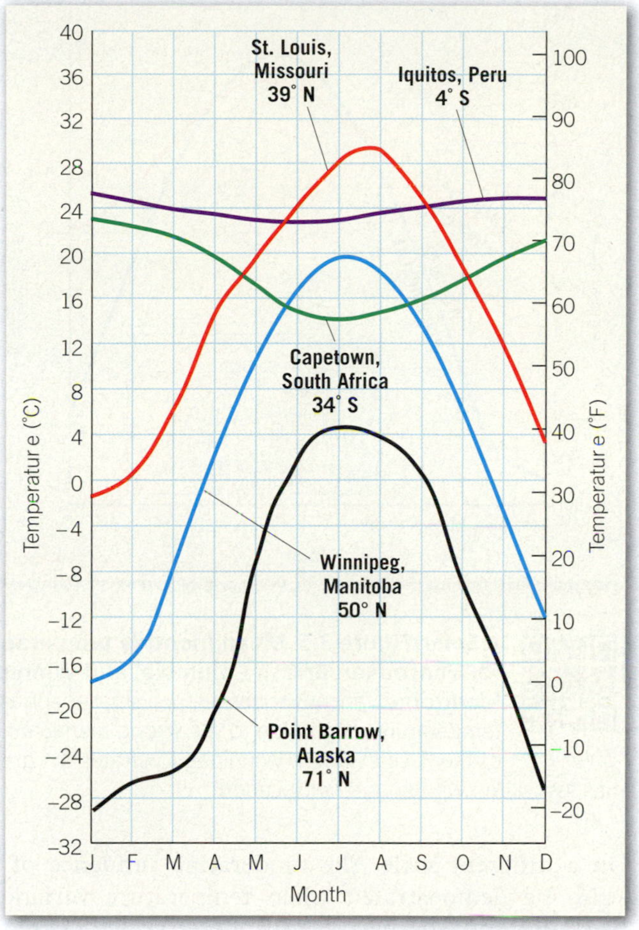


▲ **Figure 3.3 Latitude is a major control of temperature** The data for these five cities remind us that latitude (Earth–Sun relationships) is a significant factor influencing temperature.

and countless others, we must realize that factors other than latitude also strongly influence temperature. In the following subsections, we examine these other controls, which include:

- Differential heating of land and water
- Ocean currents
- Altitude
- Geographic position
- Albedo variations

## Differential Heating of Land and Water

In Chapter 2 you saw that the heating of Earth's surface controls the heating of the air above it. Therefore, to understand variations in air temperature, we must understand the variations in heating properties of the different surfaces that Earth presents to the Sun—soil, water, trees, ice, and so on. Different land surfaces reflect and absorb varying amounts of incoming solar energy, which in turn cause variations in the temperature of the air above. The greatest contrast, however, is not between different land surfaces but between land and water. **Figure 3.4** illustrates this idea nicely. This satellite image shows surface temperatures in portions of Nevada, California, and the adjacent Pacific Ocean on the afternoon of May 2, 2004, during a spring heat wave. Land-surface temperatures are clearly much higher than water-surface temperatures. The image shows the extremely high surface temperatures in southern California and Nevada in dark red. Surface temperatures in the Pacific Ocean are much lower. The peaks of the Sierra Nevada, still capped with snow, form a cool blue line down the eastern side of California.

For side-by-side bodies of land and water, such as those in Figure 3.4, *land heats more rapidly and to higher temperatures than water, and it cools more rapidly and to lower temperatures than water.* Variations in air temperatures, therefore, are much greater over land than over water. Why do land and water heat and cool differently? Several factors are responsible:

1. An important reason that the surface temperature of water rises and falls much more slowly than the surface temperature of land is that *water is highly mobile.* As water is heated, convection distributes the heat through a considerably larger mass. Daily temperature changes occur to depths of 6 meters (20 feet) or more below the surface, and yearly, oceans and deep lakes experience temperature variations through a layer between 200 and 600 meters (650 and 2000 feet) thick.

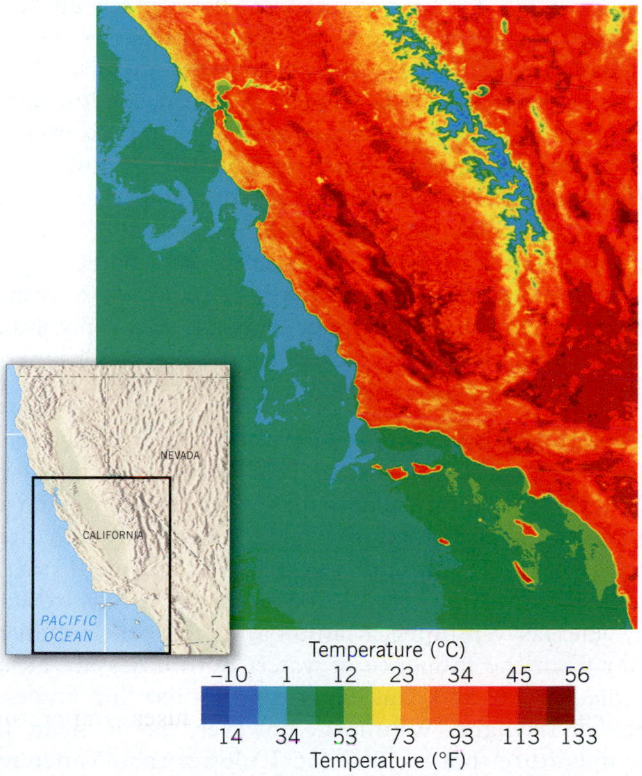


▲ **Figure 3.4 Differential heating of land and water** This satellite image shows land- and water-surface temperatures (not air temperatures) for the afternoon of May 2, 2004. Water-surface temperatures in the Pacific Ocean are much lower than land-surface temperatures in California and Nevada. The narrow band of cool temperatures in the center of the image is associated with snow-capped mountains (the Sierra Nevada). The cooler water temperatures immediately offshore are associated with the California Current and with the upwelling of deep cold water associated with the current. (see Figure 3.8).

In contrast, heat does not penetrate deeply into soil or rock; it remains near the surface. Obviously, no mixing can occur on land because it is not fluid. Instead, heat must be transferred by the slow process of conduction. Consequently, daily temperature changes are small below a depth of 10 centimeters (4 inches), although some change can occur to a depth of perhaps 1 meter (3 feet). Annual temperature variations usually reach depths of 15 meters (50 feet) or less. Thus, as a result of the mobility of water and the lack of mobility in the solid Earth, a relatively thick layer of water is heated to moderate temperatures during the summer. On land only a thin layer is heated, but to much higher temperatures.

During winter the shallow layer of rock and soil that was heated in summer cools rapidly. Water bodies, in contrast, cool slowly as they draw on the reserve of heat stored within. As the water surface cools, vertical motions are established. The chilled surface water, which is dense, sinks and is replaced from below by warmer water, which is less dense. Consequently, a larger mass of water must cool before the temperature at the surface will drop appreciably.

2. Because land surfaces are opaque, heat is absorbed only at the surface. This fact is easily demonstrated at a beach on a hot summer afternoon by comparing the surface temperature of the sand to the temperature just a few centimeters beneath the surface. Water, being more transparent, allows some solar radiation to penetrate to a depth of several meters.
3. The **specific heat** (the amount of heat needed to raise the temperature of 1 gram of a substance by 1°C) is more than three times greater for water than for land. Thus, considerably more heat is required to raise the temperature of water the same amount as an equal volume of land.
4. Evaporation (a cooling process) from water bodies is greater than from land surfaces. Energy is required to evaporate water. When energy is used for evaporation, it is not available for heating. (See "Water's Changes of State" in Chapter 4 for a detailed discussion of evaporation.)

All these factors collectively cause water to warm more slowly, store greater quantities of heat, and cool more slowly than land.

Monthly temperature data for two cities will demonstrate the moderating influence of a large water body and the extremes associated with land (**Fig. 3.5**). Vancouver, British Columbia, is a maritime city located along the windward Pacific coast, whereas Winnipeg, Manitoba, is in a continental position far from the influence of water. Both cities are at about the same latitude and thus experience similar Sun angles and lengths of daylight. Winnipeg, however, has a mean January temperature that is 20°C (36°F) lower than Vancouver's. Conversely, Winnipeg's July mean is 2.6°C (4.7°F) higher than Vancouver's. Although their latitudes are nearly the same, Winnipeg, which has no marine influence, experiences much greater temperature extremes than does Vancouver. The key to Vancouver's moderate year-round climate is the Pacific Ocean.

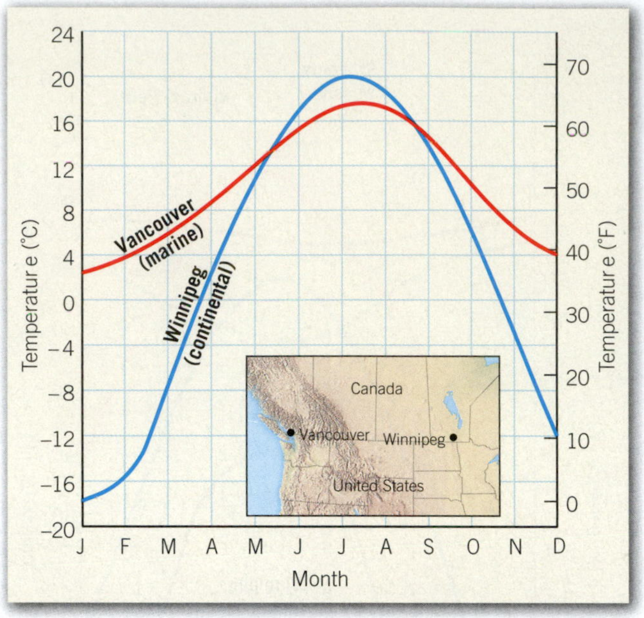


▲ **SmartFigure 3.5 Mean monthly temperatures for Vancouver, British Columbia, and Winnipeg, Manitoba** Vancouver has a much smaller annual temperature range, due to the strong marine influence of the Pacific Ocean. Winnipeg illustrates the greater extremes associated with an interior location.

http://goo.gl/JHy33

On a different scale, the moderating influence of water may also be demonstrated when temperature variations in the Northern and Southern Hemispheres are compared. The views of Earth in **Figure 3.6** show the uneven distribution of land and water over the globe. Water covers 61 percent of the Northern Hemisphere; land represents the remaining 39 percent. However, the figures for the Southern Hemisphere (81 percent water and 19 percent land) reveal why it is correctly called the *water hemisphere*. Between 45° north and 79° north latitude, there is actually more land than water, whereas between 40° south and 65° south latitude, there is almost no land to interrupt the oceanic and atmospheric circulation. **Table 3.1** portrays the considerably smaller annual temperature ranges in the water-dominated Southern Hemisphere compared with the Northern Hemisphere.

## Ocean Currents

You probably have heard of the Gulf Stream, an important surface current in the Atlantic Ocean that flows northward along the East Coast of the United States (**Fig. 3.7**). Surface currents like this one are set in motion by the wind. At the water surface, where the atmosphere and ocean meet, energy is passed from moving air to the water through friction. As a result, the drag exerted by winds blowing steadily across the ocean causes the surface layer of water to move. Thus, major horizontal movements of surface waters are closely related to the circulation of the atmosphere, which in turn is driven

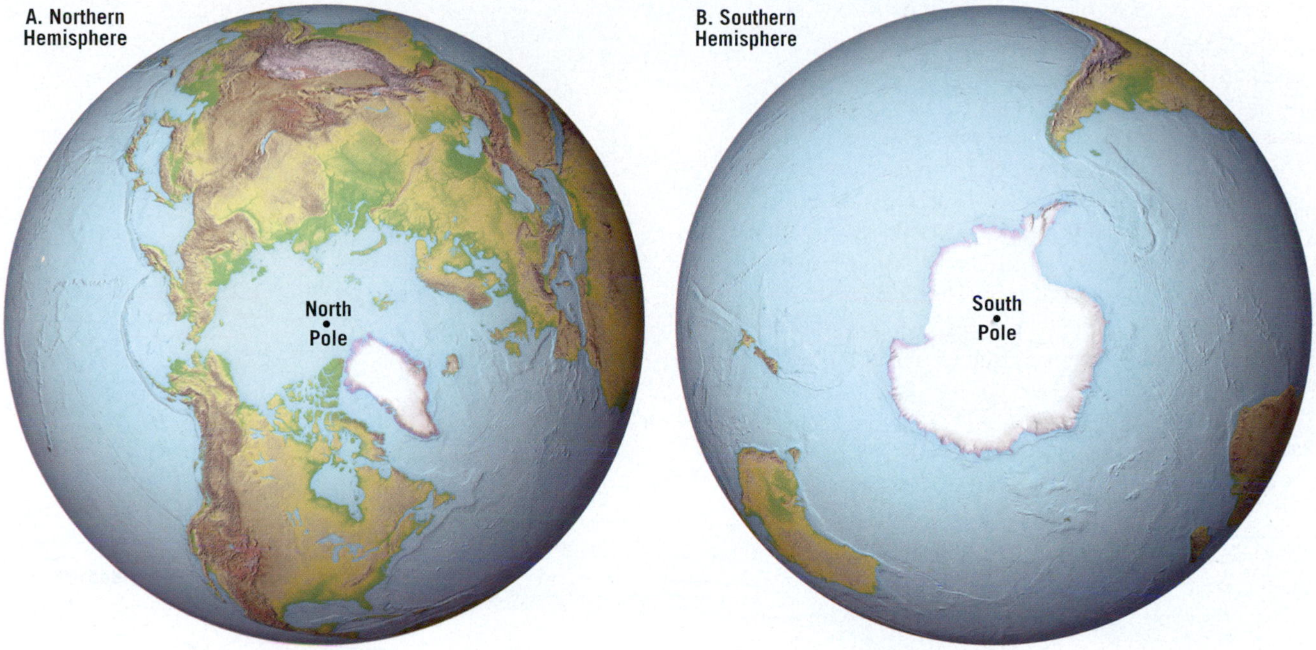


▲ **Figure 3.6 North versus south** These views of Earth show the uneven distribution of land and water between the **A.** Northern and **B.** Southern Hemispheres. Almost 81 percent of the Southern Hemisphere is covered by the oceans—20 percent more than the Northern Hemisphere.

by the unequal heating of Earth by the Sun (**Fig. 3.8**). The relationship between global winds and surface-ocean currents is examined in Chapter 7.

Surface-ocean currents have an important effect on climate. For Earth as a whole, the gains in solar energy equal the losses to space of heat radiated from the surface. When most latitudes are considered individually, however, this is not the case, as you learned in Chapter 2. There is a net gain of energy in lower latitudes, and there is a net loss at higher latitudes. Because the tropics are not becoming progressively warmer, nor are the polar regions becoming colder, there must be a large-scale transfer of heat from areas of excess to areas of deficit. This is indeed the case. *The transfer of heat by winds and ocean currents equalizes these latitudinal energy imbalances.* Ocean water movements account for about one-quarter of this total heat transport, and winds account for the remaining three-quarters.

The moderating effect of poleward-moving warm ocean currents is well known. The North Atlantic Drift, an extension

**Table 3.1** Variations in Annual Mean Temperature Range (°C) with Latitude

| Latitude | Northern Hemisphere | Southern Hemisphere |
|---|---|---|
| 0 | 0 | 0 |
| 15 | 3 | 4 |
| 30 | 13 | 7 |
| 45 | 23 | 6 |
| 60 | 30 | 11 |
| 75 | 32 | 26 |
| 90 | 40 | 31 |

Animation MM
The Gulf Stream
http://goo.gl/VxNztu

▼ **Figure 3.7 The Gulf Stream** In this satellite image off the east coast of the United States, red represents higher water temperatures and blue represents cooler water temperatures. The current transports heat from the tropics far into the North Atlantic.

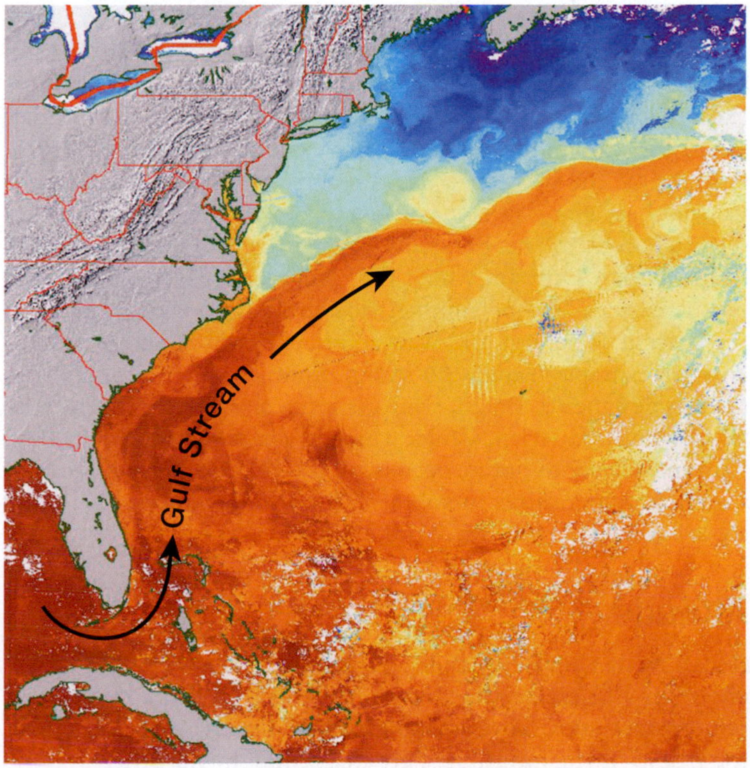

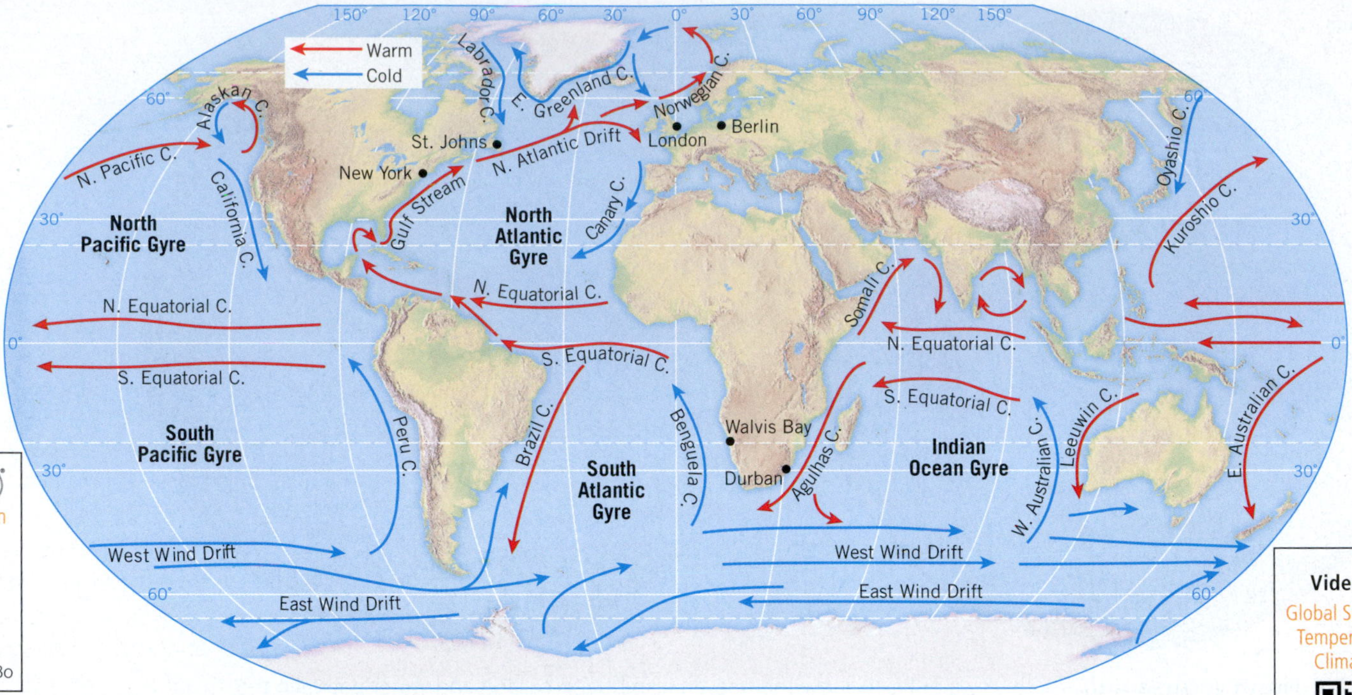


Animation MM
Ocean Circulation Patterns
http://goo.gl/QIYk8o

Video MM
Global Sea-Surface Temperatures—Climatology
http://goo.gl/VF3k7T

▲ **Figure 3.8 Major surface-ocean currents** Poleward-moving currents are warm, and equatorward-moving currents are cold. Surface-ocean currents are driven by global winds and play an important role in redistributing heat around the globe. Note that cities mentioned in the text discussion are shown on this map.

of the warm Gulf Stream, keeps wintertime temperatures in Great Britain and much of Western Europe warmer than would be expected for their latitudes. (London is farther north than St. John's, Newfoundland.) Because of the prevailing westerly winds, the moderating effects are carried far inland. For example, Berlin (52° north latitude) has a mean January temperature similar to that experienced in New York City, which lies 12° latitude farther south. The January mean in London (51° north latitude) is 4.5°C (8.1°F) higher than in New York City.

In contrast to warm ocean currents such as the Gulf Stream, the effects of which are felt most during the winter, cold currents exert their greatest influence in the tropics or during the summer months in the middle latitudes. For example, the cool Benguela Current off the western coast of southern Africa moderates the tropical heat along this coast. Walvis Bay (23° south latitude), a town adjacent to the Benguela Current, is 5°C (9°F) cooler in summer than Durban, which is 6° latitude farther poleward but on the eastern side of South Africa, away from the influence of the current. The east and west coasts of South America provide another example. **Figuge 3.9** shows monthly mean temperatures for Rio de Janeiro, Brazil, which is influenced by the warm Brazil Current and Arica, Chile, which is adjacent to the cold Peru Current. Closer to home, because of the cold California Current, summer temperatures in subtropical coastal southern California are lower by 6°C (10.8°F) or more compared to east coast stations.

## Altitude

Recall from Chapter 1 that temperatures decrease with an increase in altitude in the troposphere. As a result, some mountaintops are snow covered year round. This can even occur in the tropics if the mountains are high enough (**Fig. 3.10**).

The two cities in Ecuador mentioned earlier, Quito and Guayaquil, demonstrate the influence of altitude on mean temperature. Both cities are near the equator and relatively close to one another, but the annual mean temperature at Guayaquil is 25.5°C (77.9°F), compared with Quito's mean of 13.3°C (55.9°F). The difference may be understood when the cities'

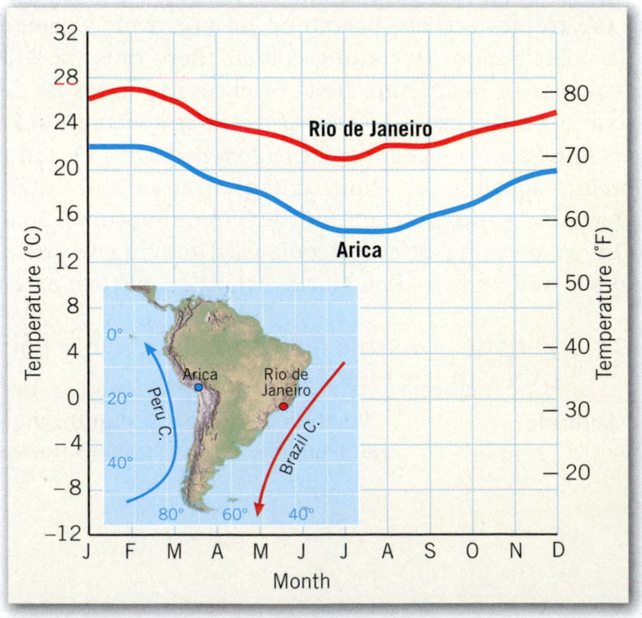


▲ **Figure 3.9 The chilling effect of a cold current** Monthly mean temperatures for Rio de Janeiro, Brazil, and Arica, Chile. Both are coastal cities near sea level. Even though Arica is closer to the equator than Rio de Janeiro, its temperatures are cooler. Arica is influenced by the cold Peru Current, whereas Rio de Janeiro is adjacent to the warm Brazil Current.

▲ **Figure 3.10 Cold mountaintop** Temperatures decrease with an increase in altitude in the troposphere. As a result, some mountaintops are snow covered all year. This can even occur in the tropics. Tanzania's Mt. Kilimanjaro (about 3° south latitude) actually has a small glacier at its summit.

elevations are noted. Guayaquil is only 12 meters (39 feet) above sea level, whereas Quito is high in the Andes Mountains, at 2800 meters (9200 feet). **Figure 3.11** provides another example.

In Chapter 1 you learned that temperatures drop an average of 6.5°C per kilometer (3.5°F per 1000 feet) in the troposphere. However, if this figure is applied, we would expect Quito to be about 18.2°C (32.7°F) cooler than Guayaquil, but the difference is only 12.2°C (22°F). The fact that high-altitude places, such as Quito, are warmer than the value calculated using the normal lapse rate results from the absorption and reradiation of solar energy by the ground surface.

In addition to the effect of altitude on mean temperatures, the daily temperature range also changes with variations in height. Not only do temperatures drop with an increase in altitude but atmospheric pressure and density also diminish. Because of the reduced density at high altitudes, the overlying atmosphere absorbs, reflects, and scatters a smaller portion of the incoming solar radiation. Consequently, with an increase in altitude, the intensity of solar radiation increases, resulting in relatively rapid and intense daytime heating. Conversely, rapid nighttime cooling is also the rule in high mountain locations. Therefore, stations located high in the mountains generally have a greater daily temperature range than do stations at lower elevations.

## Geographic Position

The geographic setting can greatly influence the temperatures experienced at a specific location. A coastal location where prevailing winds blow from the ocean onto the shore (a *windward* coast) experiences considerably different temperatures than does a coastal location where prevailing winds blow from the land toward the ocean (a *leeward* coast). In the first situation, the windward coast experiences the full moderating influence of the ocean—cool summers and mild winters—unlike an inland station at the same latitude.

A leeward coastal situation, however, has a more continental temperature regime because the winds do not carry the ocean's influence onshore. Eureka, California, and New York City, the two cities mentioned earlier, illustrate this aspect of geographic position (**Fig. 3.12**). The annual temperature range in New York City is 19°C (34°F) greater than Eureka's.

Seattle and Spokane, both in the state of Washington, illustrate a second aspect of geographic position: mountains acting as barriers. Although Spokane is only about 360 kilometers (225 miles) east of Seattle, the towering Cascade Range separates the cities. Consequently, Seattle's temperatures show a marked marine influence, whereas Spokane's are more typically continental (**Fig. 3.13**). Spokane is 7°C (12.6°F) cooler than Seattle in January and 4°C (7.2°F) warmer than Seattle in July. The annual range at Spokane is 11°C (nearly 20°F) greater than in Seattle. The Cascade Range effectively cuts off Spokane from the moderating influence of the Pacific Ocean.

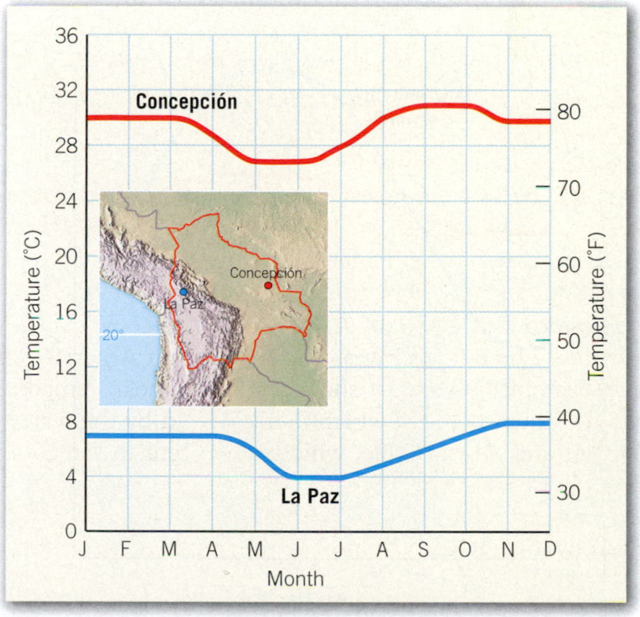


▲ **Figure 3.11 Monthly mean temperatures for Concepción and La Paz, Bolivia** Both cities have nearly the same latitude (about 16° south). However, because La Paz is high in the Andes, at 4103 meters (13,461 feet), it experiences much cooler temperatures than Concepción, which is at an elevation of 490 meters (1608 feet).

## eye ON THE atmosphere 3.1

Imagine being at this beach on a warm, sunny summer afternoon.

**Questions**

1. Describe the temperatures you would expect if you measured the surface of the beach and at a depth of 12 inches.
2. If you stood waist deep in the water and measured the water's surface temperature and the temperature at a depth of 12 inches, how would these measurements compare to those taken at the beach?

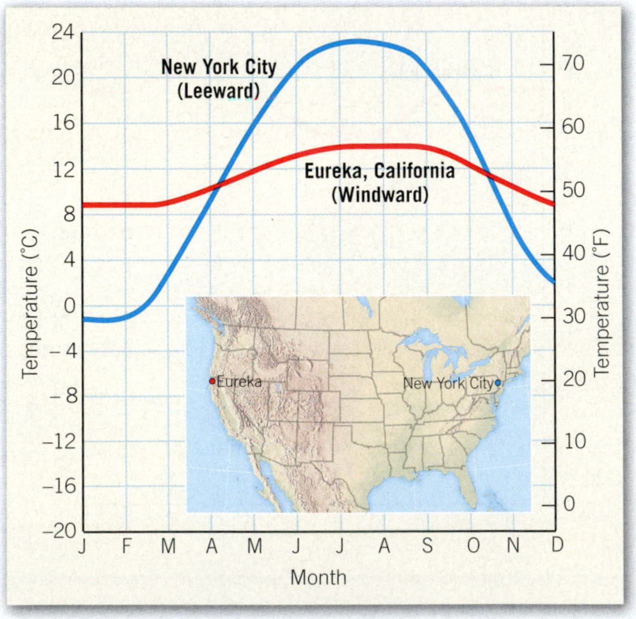


▲ **Figure 3.12 Monthly mean temperatures for Eureka, California, and New York City** Both cities are coastal and located at about the same latitude. Because Eureka is strongly influenced by prevailing winds from the ocean and New York City is not, the annual temperature range at Eureka is much smaller.

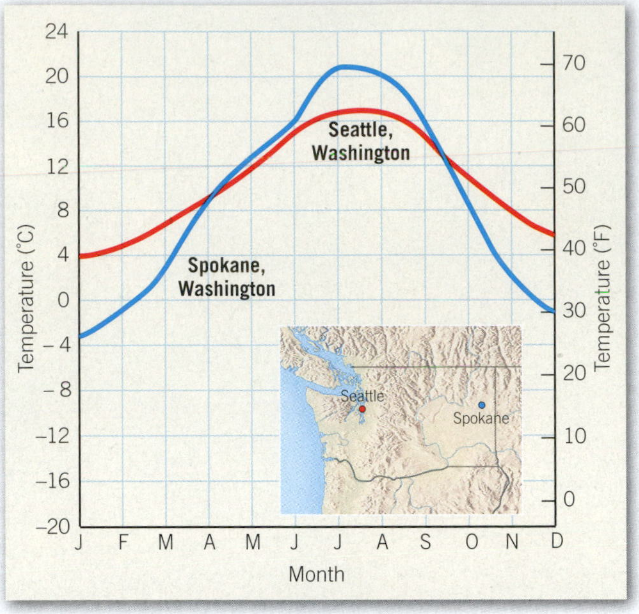


▲ **Figure 3.13 Monthly mean temperatures for Seattle and Spokane, Washington** Because the Cascade Mountains cut off Spokane from the moderating influence of the Pacific Ocean, its annual temperature range is greater than Seattle's.

## Albedo Variations

In Chapter 2 you learned that *albedo* refers to the fraction of radiation reflected by an object. Solar radiation reflected back to space is lost to Earth and does not play a role in heating Earth's surface and atmosphere. Any increase in albedo reduces the amount of energy available to heat the atmosphere. Conversely, a decrease in albedo means an increase in the quantity of energy absorbed by Earth's surface and available to heat the atmosphere.

**Cloud Cover Affects Daily Temperature Range** You may have noticed that clear days are often warmer than cloudy ones and that clear nights usually are cooler than cloudy ones. This demonstrates that cloud cover is another factor that influences temperature in the lower atmosphere. Studies using satellite images show that at any particular time, about half of our planet is covered by clouds. Cloud cover is important because many clouds have a high albedo and therefore reflect back to space a significant proportion of the sunlight that strikes them. By reducing the amount of incoming solar radiation, daytime temperatures will be lower than if the clouds were absent and the sky were clear (Fig. 3.14). As noted in Chapter 2, the albedo of clouds depends on the thickness of the cloud cover and can vary from 25 to 90 percent (see Figure 2.17, page 45).

At night, clouds have the opposite effect than during daylight. They absorb outgoing Earth radiation and emit a portion of it toward the surface. Consequently, some of the heat that otherwise would have been lost remains near the ground. Thus, nighttime air temperatures do not drop as low as they would on a clear night. The effect of cloud cover is to reduce the daily temperature range by lowering the daytime maximum and raising the nighttime minimum. This is illustrated nicely by the graph in Figure 3.14.

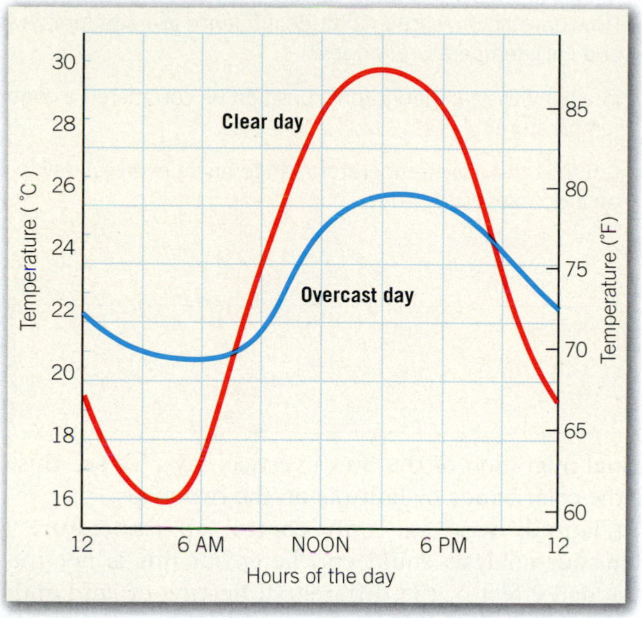


▲ **SmartFigure 3.14 The daily cycle of temperature at Peoria, Illinois, for two July days** Clouds reduce the daily temperature range. During daylight hours, clouds reflect solar radiation back to space. Therefore, the maximum temperature is lower than if the sky were clear. At night, the minimum temperature will not fall as low because clouds retard the loss of heat.

http://goo.gl/6CQrL

## Cloud Cover Affects Monthly Mean Temperatures

The effect of cloud cover on reducing maximum temperatures can also be detected when monthly mean temperatures are examined for some stations. For example, each year much of southern Asia experiences an extended dry period during the cooler low-Sun period; it is followed by heavy monsoon rains. This pattern is associated with the monsoon circulation and is discussed in Chapter 7. The graph for Yangon, Myanmar (also known as Rangoon, Burma), illustrates this pattern (**Fig. 3.15**). Notice that the highest monthly mean temperatures occur in April and May, before the summer solstice, rather than in July and August, as normally occurs at most stations in the Northern Hemisphere. Why? This is because during the summer months, when we would usually expect temperatures to climb, the extensive cloud cover increases the albedo of the region, which reduces incoming solar radiation at the surface. As a result, the highest monthly mean temperatures occur in late spring, when the skies are still relatively clear.

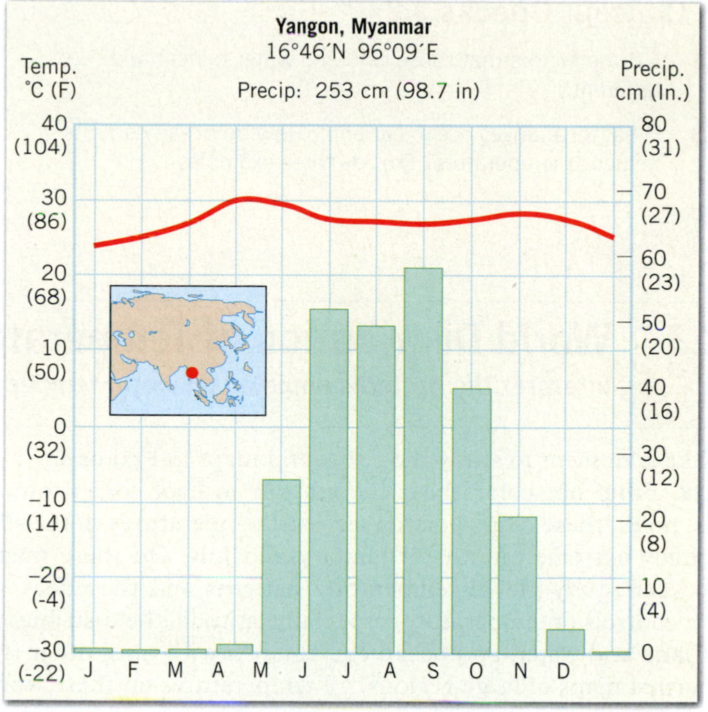


▲ **Figure 3.15 The influence of monsoon clouds** Monthly mean temperatures (line graph) and monthly mean precipitation (bar graph) for Yangon, Myanmar. The highest mean temperature occurs in April, just before the onset of heavy summer rains. The abundant cloud cover associated with the rainy period reflects back to space the solar energy that otherwise would strike the ground and raise summer temperatures.

**Influence of Snow and Ice** Cloudiness is not the only phenomenon that increases albedo and thereby reduces air temperature. We also recognize that snow- and ice-covered surfaces have high albedos. This is one reason why mountain glaciers do not melt away in the summer and why snow may still be present on a mild spring day. In addition, during the winter when snow covers the ground, daytime maximums on a sunny day are cooler than they otherwise would be because energy that the land would have absorbed and used to heat the air is reflected and lost.

**Shrinking Sea Ice** Large portions of the Arctic Ocean are covered by sea ice—frozen seawater that floats because it is less dense than liquid water. As you would expect, the area covered by sea ice changes with the seasons, expanding in winter and contracting in summer. Monitoring since the late 1970s has also shown that over the years, the area covered by sea ice is shrinking. Thus broad zones that were once covered by highly reflective ice are being replaced by the darker ocean surface that reflects less and absorbs more. This idea is illustrated in **Figure 3.16**. This lowering of albedo in the Arctic is contributing to rising temperatures in this region. You'll find more about this phenomenon in Chapter 14.

▼ **Figure 3.16 Contrasting albedos** Ice- and snow-covered surfaces have high albedos, thus keeping air temperatures lower than if the surface were not highly reflective. This view shows sea ice (frozen seawater) near Barrow, Alaska. When sea ice melts, as has occurred on the left side of the image, a bright reflective surface is replaced by a darker surface that absorbs a higher percentage of incoming solar radiation.

## ✔ Concept Checks 3.2

1. List the factors that cause land and water to heat and cool differently.
2. What force drives ocean currents? How do ocean currents influence temperature? Provide three examples.
3. How does an increase in altitude influence average temperatures and daily temperature ranges?
4. In what ways can geographic position be considered a control of temperature?
5. Contrast the daily temperature range on an overcast day with that on a cloudless sunny day.

# 3.3 World Distribution of Temperatures

**Interpret the patterns depicted on world temperature maps.**

Take a moment to study the two world maps in **Figures 3.17** and **3.18**. From hot colors near the equator to cool colors toward the poles, these maps portray sea-level temperatures in the seasonally extreme months of January and July. On these maps, you can study global temperature patterns and the effects of the controls of temperature, especially latitude, the distribution of land and water, and ocean currents. As with most other isothermal maps of large regions, all temperatures on these world maps have been reduced to sea level to eliminate the complications caused by differences in altitude.

On both maps the isotherms generally trend east and west and show a decrease in temperatures poleward from the tropics. They illustrate one of the most fundamental aspects of world temperature distribution: that the effectiveness of incoming solar radiation in heating Earth's surface and the atmosphere above it is largely a function of latitude. Moreover, there is a latitudinal shifting of temperatures caused by the seasonal migration of the Sun's vertical rays. To see this, compare the color bands by latitude on the two maps.

If latitude were the only control of temperature distribution, our analysis could end here, but this is not the case. The added effect of the differential heating of land and water is clearly reflected on the January and July temperature maps. The warmest and coldest temperatures are found over land—note the coldest area, a purple oval in Siberia, and the hottest areas, the deep orange ovals—all over land. Consequently, because temperatures do not fluctuate as much over water as over land, the north–south migration of isotherms is greater over the continents than over the oceans. In addition, it is clear that the isotherms in the Southern Hemisphere, where there is little land and where the oceans predominate, are much more regular than in the Northern Hemisphere, where they bend sharply northward in July and southward in January over the continents.

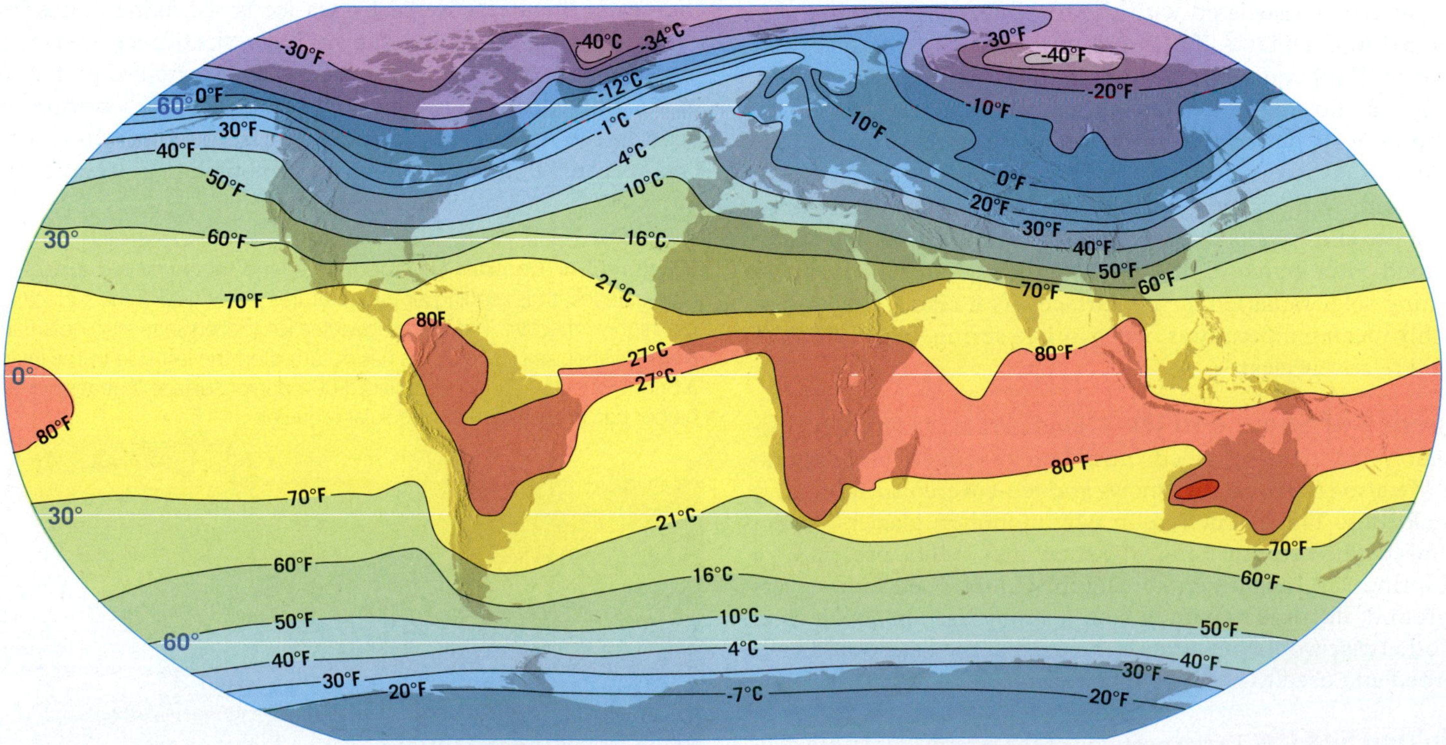


▲ **Figure 3.17 World mean sea-level temperatures in January, in Celsius (°C) and Fahrenheit (°F)**

Isotherms also reveal the presence of ocean currents. Warm currents cause isotherms to be deflected toward the poles, whereas cold currents cause an equatorward bending. The horizontal transport of water poleward warms the overlying air and results in air temperatures that are higher than would otherwise be expected for the latitude. Conversely, currents moving toward the equator produce cooler-than-expected air temperatures.

Figures 3.17 and 3.18 show the seasonal extremes of temperature, and comparing them enables us to see the annual range of temperature from place to place. Comparing the two maps shows that a station near the equator has a very small annual range because it experiences little variation in the length of daylight, and it always has a relatively high Sun angle. A station in the middle latitudes, however, experiences wide variations in Sun angle and length of daylight and hence large variations in temperature. Therefore, we can state that the annual temperature range increases with an increase in latitude (see **Box 3.2**).

Moreover, land and water also affect seasonal temperature variations, especially outside the tropics. A continental location must endure hotter summers and colder winters than a coastal location. Consequently, outside the tropics the annual range will increase with an increase in continentality.

**Figure 3.19**, which shows the global distribution of annual temperature ranges, serves to summarize the preceding two paragraphs. By examining this map, it is easy to see the influence of latitude and continentality on this temperature statistic. The tropics clearly experience small annual temperature variations. As expected, the highest values occur in the middle of large landmasses in the subpolar latitudes. It is also obvious that annual temperature ranges in the ocean-dominated Southern Hemisphere are much smaller than in the Northern Hemisphere, with its large continents.

## eye ON THE atmosphere 3.2

This image shows a snow-covered area in the middle latitudes on a sunny day in late winter. Assume that 1 week after this photo was taken, conditions were essentially identical except that the snow was gone.

**Questions**

1. Would you expect the air temperatures to be different on the two days? If so, which day would likely be warmer?
2. Suggest a reason for the different temperatures.

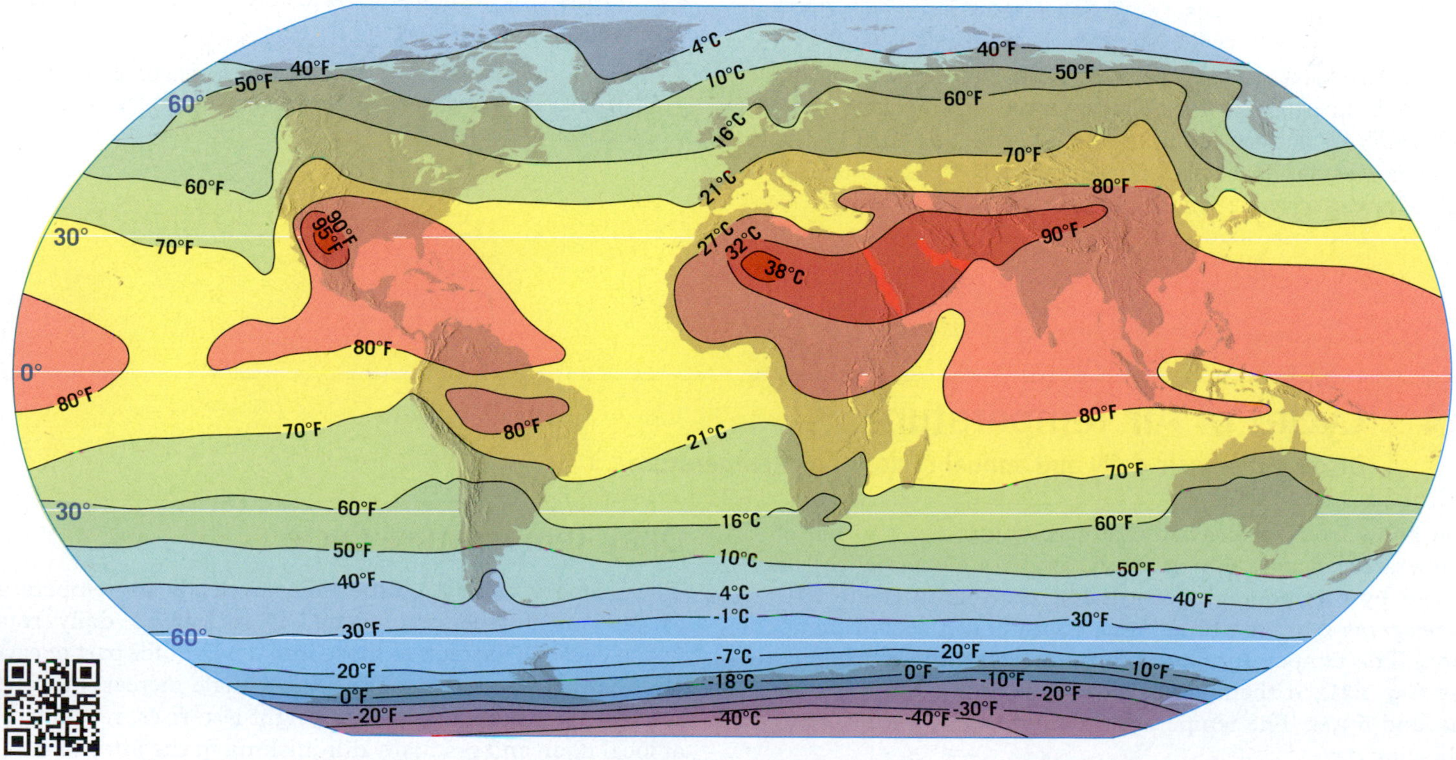


http://goo.gl/cRM01

▲ **SmartFigure 3.18 World mean sea-level temperatures in July, in Celsius (°C) and Fahrenheit (°F)**

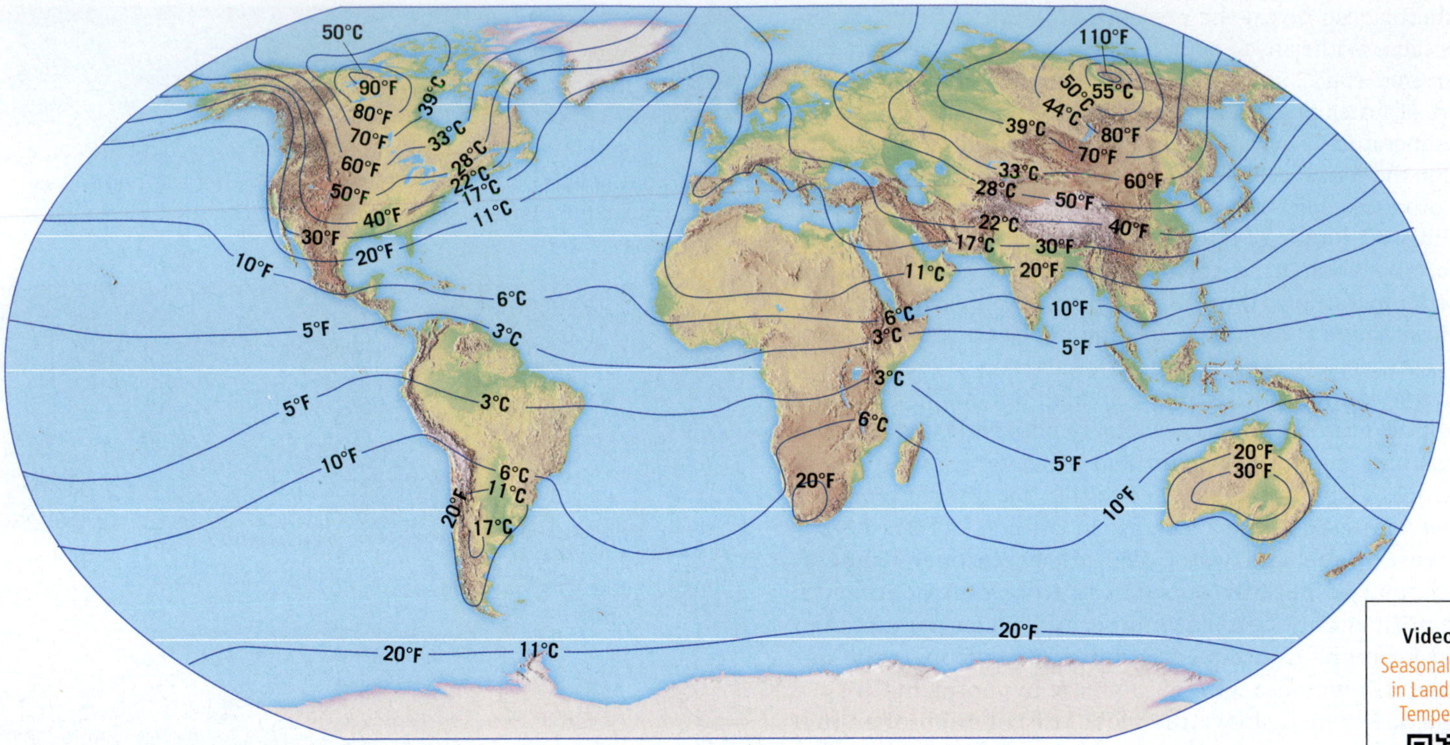


▲ **Figure 3.19 Global annual temperature ranges in Celsius (°C) and Fahrenheit (°F)** Annual ranges are small near the equator and increase toward the poles. Outside the tropics, annual temperature ranges increase as we move away from the ocean and toward the interior of large landmasses.

Video (MM) Seasonal Changes in Land Surface Temperatures

http://goo.gl/Lv1cl

students sometimes ask...

**What location experiences the greatest contrast between summer and winter temperatures?**

Among places for which records exist, it appears as though Yakutsk, a station in the heart of Siberia, is the best candidate. At 62° north latitude, Yakutsk is just a few degrees south of the Arctic Circle and far from the influence of water. The January mean at Yakutsk is a frigid −43°C (−45°F), whereas its July mean is a pleasant 20°C (68°F). The result is an average annual temperature range of 63°C (113°F), among the highest ranges anywhere on the globe.

## ✔ Concept Checks 3.3

1. Why do isotherms on the January and July temperature maps generally trend east–west?
2. Explain why the isotherms bend northward over North America on the July map (Figure 3.18).
3. Refer to Figure 3.19 to determine which area on Earth experiences the highest annual temperature range. Why is the annual range so high in this region?

# 3.4 Cycles of Air Temperature

**Discuss the basic daily and annual cycles of air temperature.**

You know from experience that a rhythmic rise and fall of air temperature occurs almost every day. Your experience is confirmed by thermograph records like the one in **Figure 3.20**. (A *thermograph* is an instrument that continuously records temperature.) The temperature curve reaches a minimum around sunrise (**Fig. 3.21**). It then climbs steadily to a maximum between 2 P.M. and 5 P.M. The temperature then declines until sunrise the following day.

## Daily Temperature Cycle

The primary control of the daily cycle of air temperature is as obvious as the cycle itself: It is Earth's daily rotation, which causes a location to move into daylight for part of each day and then into darkness. As the Sun's angle increases throughout the morning, the intensity of sunlight also rises, reaching a peak at local noon and gradually diminishing in the afternoon.

## Box 3.2 Latitude and Temperature Range

**Gregory J. Carbone**

**Professor Carbone is a faculty member in the Department of Geography at the University of South Carolina.*

Latitude, because of its influence on Sun angle, is the most important temperature control. Figures 3.17 and 3.18 clearly show higher temperatures in tropical locations and lower temperatures in polar regions. The maps also show that higher latitudes experience a greater range of temperatures during the year than do lower latitudes. Notice also that the temperature gradient between the subtropics and the poles is greatest during the winter season. A look at two cities—San Antonio, Texas, and Winnipeg, Manitoba—illustrates how seasonal differences in Sun angle and day length account for these temperature patterns. **Figure 3.B** shows the annual march of temperatures for the two cities, and **Figure 3.C** illustrates the noon Sun angles for the June and December solstices.

San Antonio and Winnipeg are a fixed distance apart (approximately 20.5° latitude), so the difference in Sun angles between the two cities is the same throughout the year. However, in December, when the Sun's rays are least direct, this difference more strongly affects the intensity of solar radiation received at Earth's surface. Therefore, we expect a greater difference in temperatures between the two stations in winter than in summer. Moreover, the seasonal difference in intensity (spreading out of the light beam) at Winnipeg is considerably greater than at San Antonio. This helps to explain the greater annual temperature range at the more northerly station. Table 2.1 (p. 34) shows that seasonal contrasts in day length also contribute to different temperature patterns at the two cities.

### Question

1. Use Figure 3.C to explain why the temperature contrast between San Antonio and Winnipeg is greater in winter than in summer.

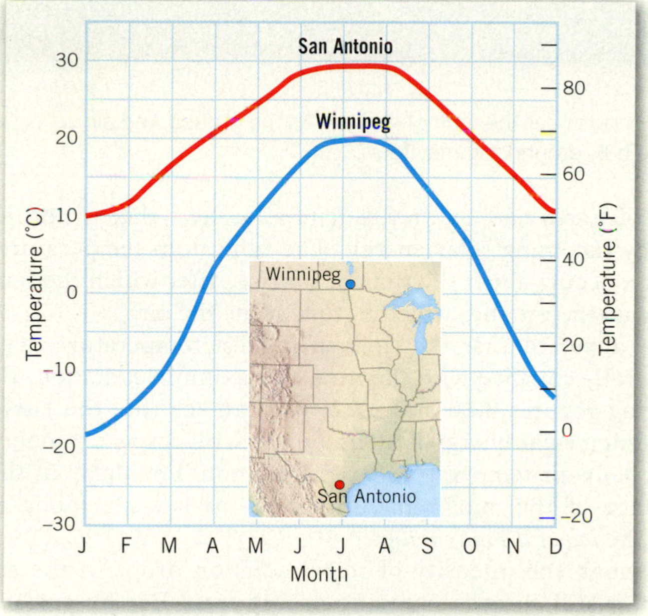


▲ **Figure 3.B Comparing temperatures** The annual temperature range at Winnipeg is much greater than at San Antonio.

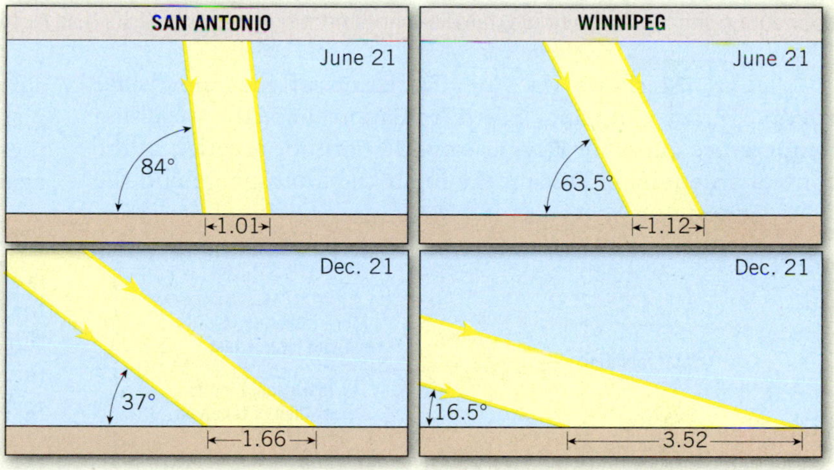


▲ **Figure 3.C Comparing noon sun angles** This diagram is a comparison of Sun angles (solar noon) for summer and winter solstices at San Antonio and Winnipeg. The space covered by a 90° Sun angle is 1.00.

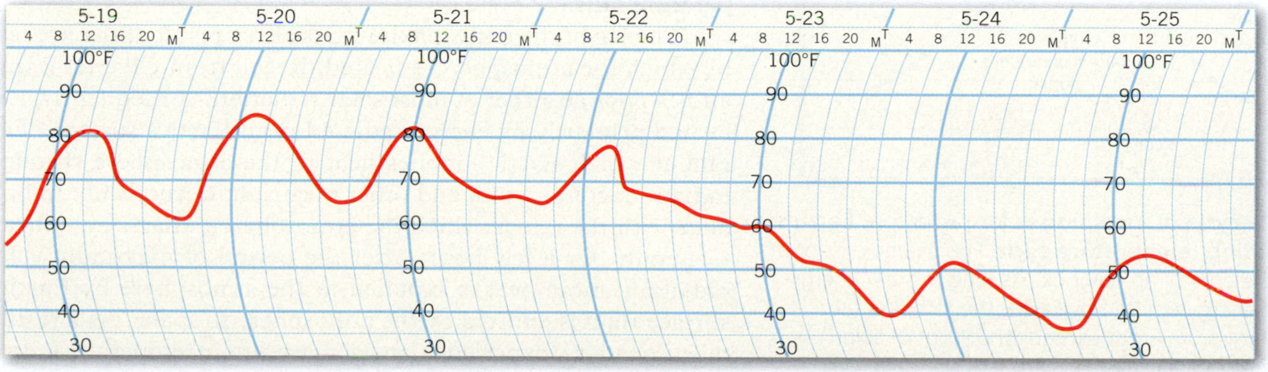


▲ **Figure 3.20 Thermograph record** Temperatures in Peoria, Illinois, during a 7-day span in May. The typical daily rhythm, with minimums around sunrise and maximums in mid- to late afternoon, occurred on most days. The obvious exception occurred on May 23, when the maximum was reached at midnight and temperatures dropped throughout the day.

A.

B.

▲ **Figure 3.21 Early morning phenomena** The minimum daily temperature usually occurs near the time of sunrise. As the ground and air cool during the nighttime hours, familiar early-morning phenomena such as **A.** frost and **B.** ground fog may form.

**Figure 3.22** shows the daily variation of incoming solar energy versus outgoing Earth radiation and the resulting temperature curve for a typical middle-latitude location at the time of an equinox. During the night the atmosphere and the surface of Earth cool as they radiate away heat that is not replaced by incoming solar energy. The minimum temperature, therefore, occurs about the time of sunrise, after which the Sun again heats the ground, which in turn heats the air.

It is apparent that the time of highest temperature does not generally coincide with the time of maximum radiation. By comparing Figures 3.20 and 3.22, you can see that the curve for incoming solar energy is symmetrical with respect to noon, but the daily air temperature curves are not. The delay in the occurrence of the maximum until mid- to late afternoon is termed the *lag of the maximum*.

Although the intensity of solar radiation drops in the afternoon, it still exceeds outgoing energy from Earth's surface for a period of time. This produces an energy surplus for up to several hours in the afternoon and contributes substantially to the lag of the maximum. In other words, as long as the solar energy gained exceeds the rate of Earth radiation lost, the air temperature continues to rise. When the input of solar energy no longer exceeds the rate of energy lost by Earth, the temperature falls.

The lag of the daily maximum is also a result of the process by which the atmosphere is heated. Recall from Chapter 2 that air is a poor absorber of most solar radiation; consequently, it is heated primarily by energy radiated from Earth's surface. The rate at which Earth supplies heat to the atmosphere through radiation, conduction, and other means, however, is not in balance with the rate at which the atmosphere radiates away heat. Generally, for a few hours after the period of maximum solar radiation, more heat is supplied to the atmosphere by Earth's surface than is emitted by the atmosphere to space. As a result, most locations experience an increase in air temperature during the afternoon.

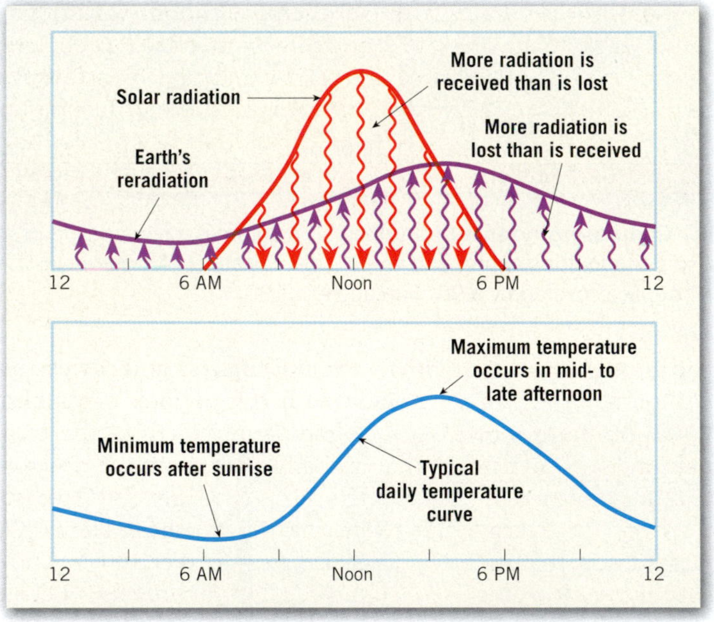


▲ **Figure 3.22 The daily cycle of incoming solar radiation, Earth's radiation, and the resulting temperature cycle** This example is for a midlatitude site around the time of an equinox. As long as solar energy gained exceeds outgoing energy emitted by Earth, the temperature rises. When outgoing energy from Earth exceeds the input of solar energy, temperature falls. Note that the daily temperature cycle *lags* behind the solar radiation input by a couple hours.

### eye ON THE atmosphere 3.3

For more than a decade, scientists have used the Moderate Resolution Imaging Spectroradiometer (MODIS, for short) aboard NASA's *Aqua* and *Terra* satellites to gather surface temperature data from around the globe. This image shows average land-surface temperatures for the month of February over a 10-year span (2001–2010).

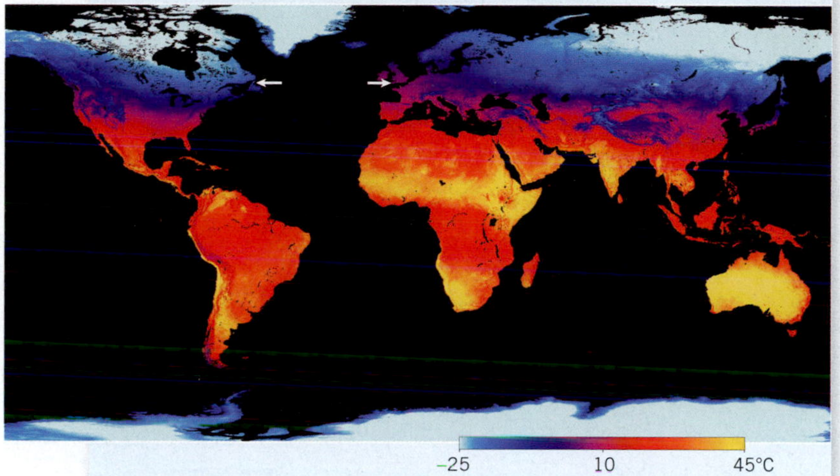


**Questions**

1. What are the approximate temperatures for southern Great Britain and northern Newfoundland (white arrows)?
2. Both are coastal areas at the same latitude, yet average February temperatures are quite different. Suggest a reason for this disparity.

In dry regions, particularly on cloud-free days, the amount of radiation absorbed by the surface is generally high. Therefore, the maximum temperature at these locales often occurs quite late in the afternoon. Humid locations, in contrast, frequently experience a shorter time lag in the occurrence of their temperature maximum.

## Variations in Daily Temperature Range

The magnitude of daily temperature changes is variable and may be influenced by locational factors or local weather conditions or both (see **Box 3.3**). Four common examples illustrate this point. The first two relate to location, and the second two pertain to local weather conditions:

Video (MM) Heavy Convection Over Florida http://goo.gl/PHJb7i

1. Variations in Sun angle are relatively great during the day in the middle and low latitudes. However, points near the poles experience a low Sun angle all day. Consequently, the daily temperature range experienced during a day in the high latitudes is small.
2. A windward coast is likely to experience only modest variations in the daily temperature cycle. During a typical 24-hour period, the ocean warms less than 1°C. As a result, the air above it shows a correspondingly slight change in temperature. For example, Eureka, California, a windward coastal station, consistently has a lower daily temperature range than Des Moines, Iowa, an inland city at about the same latitude. Annually the daily range at Des Moines averages 10.9°C (19.6°F) compared with 6.1°C (11°F) at Eureka, a difference of 4.8°C (8.6°F).
3. As mentioned earlier, an overcast day is responsible for a flattened daily temperature curve (see Figure 3.14). By day, clouds reflect incoming solar radiation and so reduce daytime heating. At night the clouds retard the loss of radiation by the ground and air. Therefore, nighttime temperatures are warmer than they otherwise would be.
4. The amount of water vapor in the air influences daily temperature range because water vapor is one of the atmosphere's important heat-absorbing gases. When the air is clear and dry, heat readily escapes at night, and the temperature falls rapidly. When the air is humid, absorption of outgoing longwave radiation by water vapor slows nighttime cooling, and the temperature does not fall to as low a value. Thus, dry conditions are associated with a higher daily temperature range because of greater nighttime cooling.

Although the rise and fall of daily temperatures usually reflects the general rise and fall of incoming solar radiation, such is not always the case. For example, a glance back at Figure 3.20 shows that on May 23, the maximum temperature occurred at midnight, after which temperatures fell throughout the day. If records for a station are examined for a period of several weeks, apparently random variations are seen. Obviously these are not Sun controlled. Such irregularities are caused primarily by the passage of atmospheric disturbances (weather systems) that are often accompanied by variable cloudiness and winds that bring air having contrasting temperatures. Under these circumstances, the maximum and minimum temperatures may occur at any time of the day or night.

## Annual Temperature Cycle

In most years the months with the highest and lowest mean temperatures do not coincide with the periods of maximum and minimum incoming solar radiation. North of the tropics, the greatest intensity of solar radiation occurs at the time of the summer solstice in June, yet the months of July and August are generally the warmest of the year in the Northern Hemisphere. Conversely, a minimum of solar energy is received in December at the time of the winter solstice, but January and February are usually colder.

The fact that the occurrence of annual maximum and minimum radiation does not coincide with the times of temperature maximums and minimums indicates that the amount of solar radiation received is not the only factor determining the temperature at a particular location. Recall from Chapter 2 that places equatorward of about 38° north and 38° south receive more solar radiation than is lost to space, and that the opposite is true of more poleward regions. Based on this

## Box 3.3 How Cities Influence Temperature: The Urban Heat Island

One of the most apparent human impacts on climate is the modification of the atmospheric environment by the building of cities. The construction of every factory, road, office building, and house destroys microclimates and creates new ones of great complexity. The most studied and well-documented urban climatic effect is the *urban heat island*. The term refers to the fact that temperatures within cities are generally higher than in rural areas.

**Figure 3.D**, which shows the distribution of average minimum temperatures in the Washington, DC, metropolitan area for the three-month winter period (December through February) over a 5-year span, illustrates a well-developed heat island. The warmest winter temperatures occurred in the heart of the city, whereas the suburbs and surrounding countryside experienced average minimum temperatures that were as much as 3.3°C (6°F) lower. Remember that these temperatures are averages. On many clear, calm nights, the temperature difference between the city center and the countryside was considerably greater, often 11°C (20°F) or more. Conversely, on many overcast or windy nights the temperature differential approached 0°.

Video MM Urban Heat Islands
http://goo.gl/M701Ba

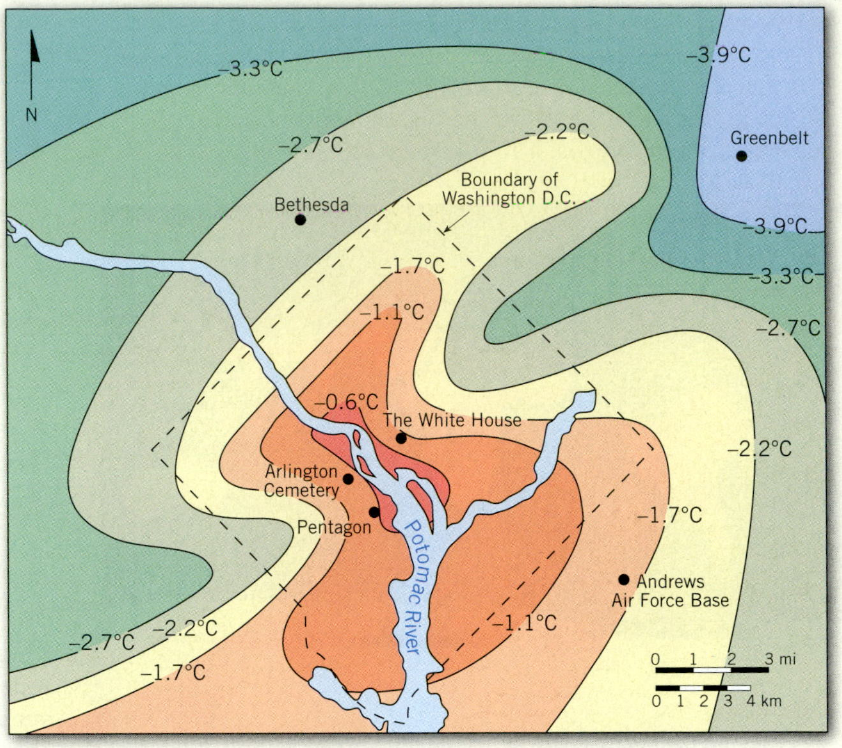


▲ **Figure 3.D Map of a heat island** The heat island of Washington, DC, as shown by the average minimum temperatures (°C) during the winter season (December through February). The city center had an average minimum that was more than 3°C higher than some outlying areas.

imbalance between incoming and outgoing radiation, any location in the southern United States, for example, should continue to get warmer late into autumn.

But this does not occur because more poleward locations begin experiencing a negative radiation balance shortly after the summer solstice. As the temperature contrasts become greater, the atmosphere and ocean currents "work harder" to transport heat from lower latitudes poleward.

### ✔ Concept Checks 3.4

1. Although the intensity of incoming solar radiation is greatest at local noon, the warmest part of the day is most often midafternoon. Why? Use Figure 3.22 to explain.
2. List at least three factors that might cause the daily temperature range to vary significantly from place to place and from time to time.

## 3.5 Temperature Measurement

**Explain how different types of thermometers work and why the placement of thermometers is an important factor in obtaining accurate readings. Distinguish among Fahrenheit, Celsius, and Kelvin temperature scales.**

MM GEODe ▶ Temperature Data and the Controls of Temperature ▶ Basic Temperature Data

During a typical day, many of us routinely check or encounter the current air temperature several times. Radio stations frequently quote the current temperature, and we may see a digital display of time and temperature outside a bank or another business. Many cars show the air temperature as part of the dashboard display. Are all these sources accurate?

### Thermometers

**Thermometers** are "meters of therms": They measure temperature (**Fig. 3.23**). Thermometers measure temperature either mechanically or electrically.

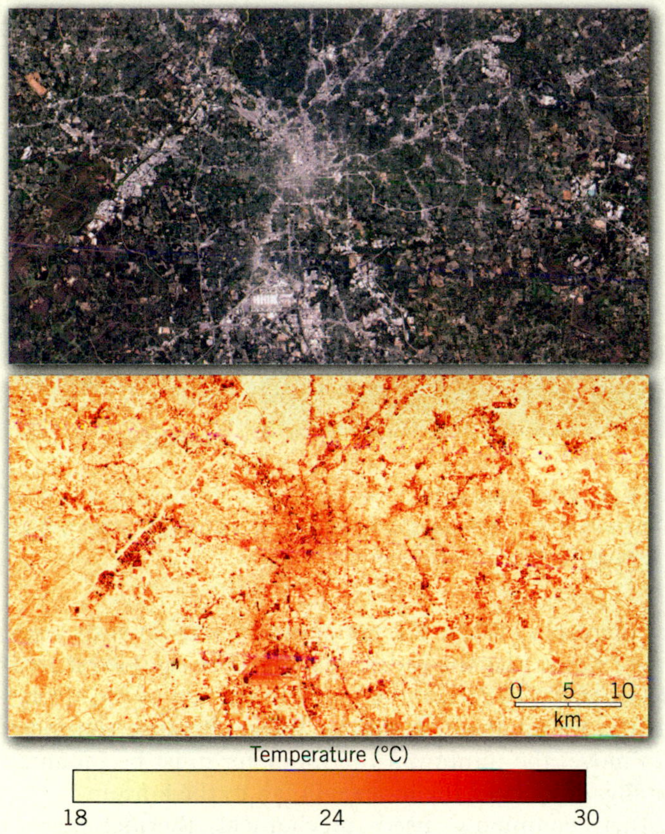


◄ **Figure 3.E Atlanta's heat island** This pair of satellite images provides two views of Atlanta, Georgia, on September 28, 2000. The urban core is in the center of the images. The top image is a photo-like view in which vegetation is green, roads and dense development appear gray, and bare ground is tan or brown. The bottom image is a land-surface temperature map in which cooler temperatures are yellow and hotter temperatures are red. It is clear that where development is densest, the land-surface temperatures are highest.

One effect of an urban heat island is to influence the biosphere by extending the plant cycle. In one study of 70 cities in eastern North America, researchers found that the growing cycle in cities was about 15 days longer than in surrounding rural areas. Plants began the growth cycle an average of about 7 days earlier in spring and continued growing an average of 8 days longer in fall.

Why are cities warmer than rural areas? The radical change in the surface that results when rural areas are transformed into cities is a significant cause of an urban heat island (**Fig. 3.E**). First, the tall buildings and the concrete and asphalt of the city absorb and store greater quantities of solar radiation than do the vegetation and soil typical of rural areas. In addition, the impermeable city surfaces cause rapid runoff of water following a rain, resulting in a significant reduction in the evaporation rate. Hence, heat that once would have been used to convert liquid water to a gas now goes to increase further the surface temperature. At night, as both the city and countryside cool by radiative losses, the stonelike surface of the city gradually releases the additional heat accumulated during the day, keeping the urban air warmer than that of the outlying areas.

A portion of the urban temperature rise is also attributable to waste heat from sources such as home heating and air conditioning, power generation, industry, and transportation. In addition, the "blanket" of pollutants over a city contributes to a heat island by absorbing a portion of the upward-directed longwave radiation emitted by the surface and re-emitting some of it back to the ground.

**Question**

1. List three factors that contribute to the development of urban heat islands.

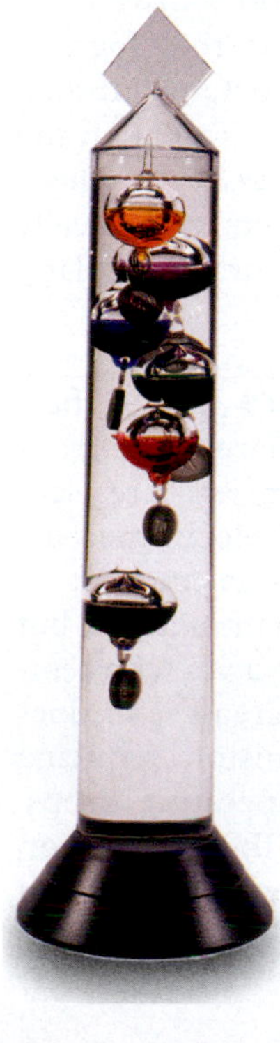

◄ **Figure 3.23 Galileo's thermoscope** The design of this thermometer is based on an instrument called a *thermoscope* that Galileo invented in the late 1500s. Today such devices, which are fairly accurate, are used mostly for decoration. The instrument is made of a sealed glass cylinder containing a clear liquid and a series of glass bulbs of different densities that can "float" up and down.

## Mechanical Thermometers

Most substances expand when heated and contract when cooled, and many common thermometers operate using this property. More precisely, they rely on the fact that different substances react to temperature changes differently.

The **liquid-in-glass thermometer** shown in Figure 3.24 is a simple instrument that provides relatively accurate readings over a wide temperature range. Its design has remained essentially unchanged since it was developed in the late 1600s. When temperature rises, the molecules of fluid grow more active and spread out (the fluid expands). Expansion of the fluid in the bulb is much greater than the expansion of the enclosing glass. As a consequence, a thin "thread" of fluid is forced up the capillary tube. Conversely, when temperature falls, the liquid contracts, and the thread of fluid moves back down the tube toward the bulb. The movement of the end of this thread (known as the *meniscus*) is calibrated against an established scale to indicate the temperature.

The highest and lowest temperatures that occur each day are of considerable importance and are often obtained by using specially designed liquid-in-glass thermometers. Mercury is the liquid used in the **maximum thermometer**, which has a narrowed passage called a *constriction* in the bore of the glass tube, just above the bulb (Fig. 3.25). As the temperature rises, the mercury expands and is forced through the constriction. When the temperature falls, the constriction prevents a return of mercury to the bulb. As a result, the top of the mercury column remains at the highest point (maximum temperature attained during the measurement period). The instrument is

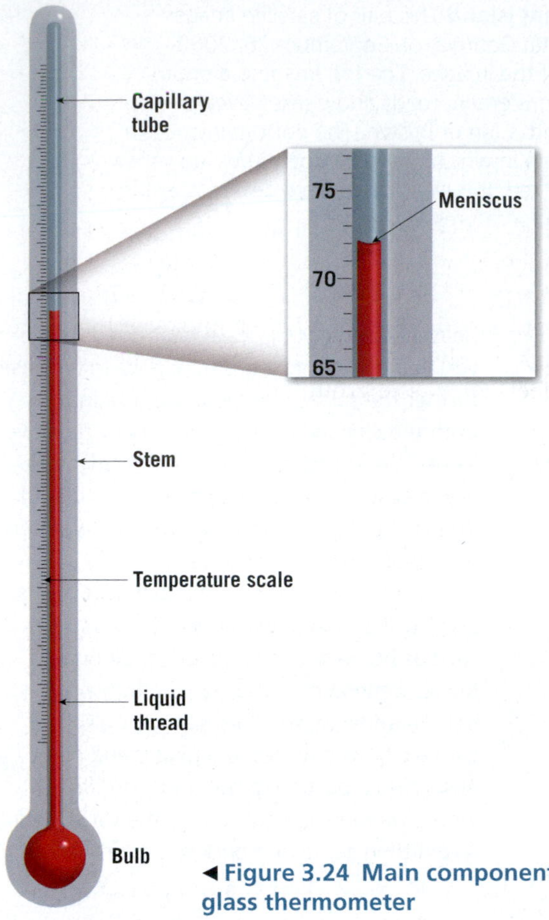


◄ **Figure 3.24 Main components of a liquid-in-glass thermometer**

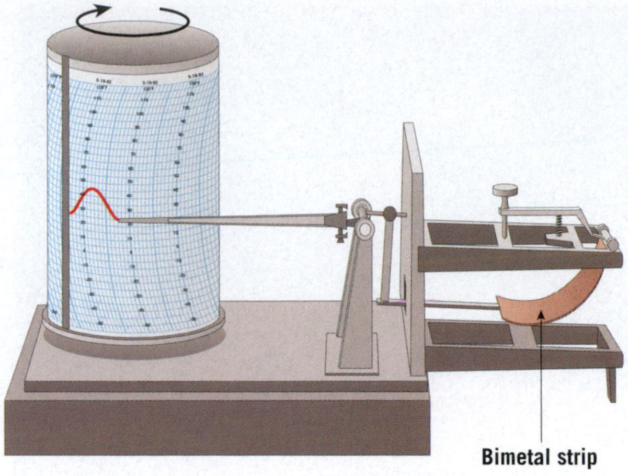


▲ **Figure 3.26 Thermograph** A common use of the bimetal strip is in the construction of a thermograph, an instrument that continuously records temperatures.

reset by shaking or by whirling it to force the mercury through the constriction back into the bulb. Once it is reset, the thermometer indicates the current air temperature.

In contrast to a maximum thermometer that contains mercury, a **minimum thermometer** contains a liquid of low density, such as alcohol. Within the alcohol, and resting at the top of the column, is a small dumbbell-shaped index (Fig. 3.25). As the air temperature drops, the column shortens, and the index is pulled toward the bulb by the effect of surface tension with the meniscus. When the temperature subsequently rises, the alcohol flows past the index, leaving it at the lowest temperature reached. To return the index to the top of the alcohol column, the thermometer is simply tilted. Because the index is free to move, the minimum thermometer must be mounted horizontally; otherwise, the index will fall to the bottom.

Another commonly used mechanical thermometer is the **bimetal strip**. As the name indicates, this thermometer consists of two thin strips of metal that are bonded together and have widely different expansion properties. When the temperature changes, both metals expand or contract, but they do so unequally, causing the strips to curl. This change corresponds to the change in temperature.

The primary meteorological use of the bimetal strip is in the construction of a **thermograph**, an instrument that continuously records temperature. The changes in the curvature of the strip can be used to move a pen arm that records the temperature on a calibrated chart that is attached to a clock-driven, rotating drum (Fig. 3.26). Although very convenient, thermograph records are generally less accurate than readings obtained from a mercury-in-glass thermometer. To obtain the most reliable values, it is necessary to check and correct the thermograph periodically by comparing it with an accurate, similarly exposed thermometer.

▼ **Figure 3.25 Maximum and minimum thermometers** Both examples are types of liquid-in-glass thermometers.

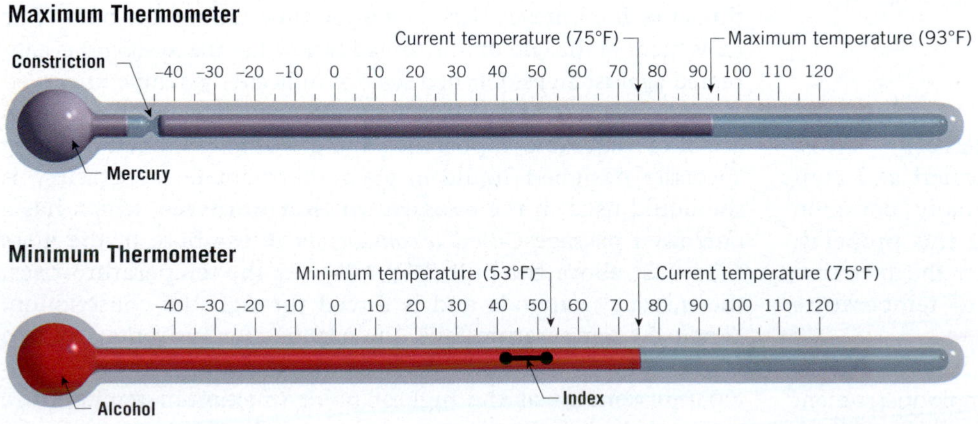


**Electrical Thermometers** Some thermometers do not rely on differential expansion but instead measure temperature electrically. A resistor is a small electronic part that resists the flow of electrical current. A **thermistor** (thermal resistor) is similar, but its resistance to current flow varies with temperature. As temperature increases, so does the resistance of the thermistor, reducing the flow of current. As temperature drops, so does the resistance of the thermistor, allowing more current to flow. The current

▲ **Figure 3.27 Measuring temperature electrically** This modern shelter contains an electrical thermometer called a *thermistor*.

operates a meter or digital display that is calibrated in degrees of temperature. The thermistor thus is used as a temperature sensor—an electrical thermometer.

Thermistors are rapid-response instruments that quickly register temperature changes. Therefore, they are commonly used in radiosondes where rapid temperature changes are often encountered. The National Weather Service also uses a thermistor system for ground-level readings. A sensor is mounted inside a shield made of louvered plastic rings, and a digital readout is placed indoors (**Fig. 3.27**).

**students sometimes ask...**

**What is lowest temperature ever recorded at Earth's surface?**

The lowest recorded temperature is –89°C (–129°F). This incredibly frigid temperature was recorded in Antarctica, at Russia's Vostok Station, on July 21, 1983.

## Instrument Shelters

How accurate are thermometer readings? Accuracy depends not only on the design and quality of the instruments but also on where they are placed. Placing a thermometer in direct sunlight will give a grossly excessive reading because the instrument itself absorbs solar energy much more efficiently than does the air. Placing a thermometer near a heat-radiating surface, such as a building or the ground, also yields inaccurate readings. Another way to assure false readings is to prevent air from moving freely around the thermometer.

Where should a thermometer be placed to read air temperature accurately? The ideal location is an instrument shelter (**Fig. 3.28**). An instrument shelter is a white box that has louvered sides to permit the free movement of air through it, while shielding the instruments from direct sunshine, heat from the ground, and precipitation. Furthermore, the shelter is placed over grass whenever possible and as far away from buildings as circumstances permit. Finally, the shelter must conform to a standardized height so that the thermometers will be mounted at 1.5 meters (5 feet) above the ground.

## Temperature Scales

In the United States, TV weather reporters give temperatures in degrees Fahrenheit. But scientists as well as most people outside the United States use degrees Celsius. Scientists sometimes also use the Kelvin, or absolute, scale. What are the differences among these three temperature scales? To make quantitative measurement of temperature possible, it was necessary to establish scales. Such temperature scales are based on the use of reference points, sometimes called **fixed points**. **Figure 3.29** compares three commonly used temperature scales.

**Fahrenheit Scale** In 1714, Gabriel Daniel Fahrenheit, a German physicist, devised the **Fahrenheit scale**.

▲ **Figure 3.28 Standard instrument shelter** This traditional shelter is white (for high albedo) and louvered (for ventilation). It protects instruments from direct sunlight and allows for the free flow of air.

He constructed a mercury-in-glass thermometer in which the zero point was the lowest temperature he could attain with a mixture of ice, water, and common salt. For his second fixed point, he chose human body temperature, which he arbitrarily set at 96°. On this scale, he determined that the melting point of ice (the **ice point**) was 32° and the boiling point of water (the **steam point**) was 212°. Because Fahrenheit's original reference points were difficult to reproduce accurately, his scale is now defined by using the ice point and the steam point. As thermometers improved, average human body temperature was later changed to 98.6°F.*

**Celsius Scale** In 1742, 28 years after Fahrenheit invented his scale, Anders Celsius, a Swedish astronomer, devised a decimal scale on which the melting point of ice was set at 0° and the boiling point of water at 100°.† For many years it was called the *centigrade scale*, but it is now known as the **Celsius scale**, after its inventor.

Because the interval between the melting point of ice and the boiling point of water is 100° on the Celsius scale and 180° on the Fahrenheit scale, a Celsius degree (°C) is larger than a Fahrenheit degree (°F), by a factor of 180/100, or 1.8. So, to convert from one system to the other, allowance must be made for this difference in the size of the degrees. Also, conversions must be adjusted because the ice point on the Celsius scale is at 0° rather than at 32°. This relationship is shown graphically in Figure 3.29.

The Celsius–Fahrenheit relationship also is shown by the following formulas:

$$°F = (1.8 \times °C) + 32$$

and:

$$°C = \frac{°F - 32}{1.8}$$

You can see that the formulas adjust for degree size with the 1.8 factor and adjust for the different 0° points with the ±32 factor.

**Kelvin Scale** For some scientific purposes, a third temperature scale is used: the **Kelvin**, or **absolute**, **scale**. On this scale, degrees Kelvin are called *Kelvins* (abbreviated K). This scale is similar to the Celsius scale because its divisions are exactly the same; there are 100° separating the melting point of ice and the boiling point of water. However, on the Kelvin scale, the ice point is set at 273 K, and the steam point is at 373 K (Fig. 3.29). This is because the zero point represents the temperature at which all molecular motion is presumed to cease (called **absolute zero**). Thus, unlike with the Celsius and Fahrenheit scales, it is not possible to have a negative value

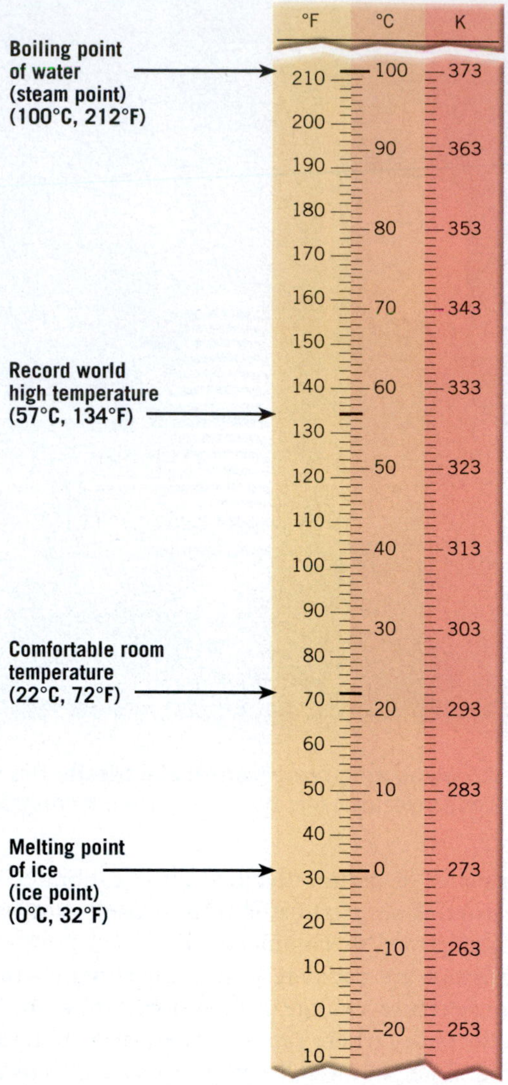


▲ **Figure 3.29 Three temperature scales compared**

when using the Kelvin scale, for there is no temperature lower than absolute zero. The relationship between the Kelvin and Celsius scales is easily written as follows:

$$°C = K - 273$$

or:

$$K = °C + 273$$

students sometimes ask...

**Which countries use the Fahrenheit scale?**

The United States and Belize (a small country in Central America) continue to use the Fahrenheit scale for everyday applications. The rest of the world has adopted the Celsius scale. The scientific community universally uses the Celsius and Kelvin scales.

*This traditional value for "normal body temperature" was established in 1868. A recent assessment places the value at 98.2°F, with a range of 4.8°F.

†The boiling point referred to in the Celsius and Fahrenheit scales pertains to pure water at standard sea-level pressure. It is necessary to remember this fact because the boiling point of water gradually decreases with altitude.

## ✔ Concept Checks 3.5

1. Describe how each of the following thermometers works: liquid-in-glass, maximum, minimum, bimetal strip, and thermistor.
2. What is a thermograph? Which type of mechanical thermometer is commonly used in the construction of a thermograph?
3. In addition to having an accurate thermometer, which other factors must be considered to obtain a meaningful air temperature reading?
4. What is meant by the terms *steam point* and *ice point*? What values are given these points on each of the three temperature scales presented here?

# 3.6 Applying Temperature Data

**Summarize several applications of temperature data.**

There are many useful and practical applications for temperature data. In this section you will learn about some simple but important indices that relate to energy use, agriculture, and human comfort.

## Degree Days

Three indices have the term *degree days* as part of their name: heating degree days, cooling degree days, and growing degree days. The first two are relative measures that allow us to evaluate the weather-produced needs and costs of heating and cooling. The third is a basic index used by farmers to estimate the maturity of crops.

**Heating Degree Days** A commonly used method for evaluating energy demand and consumption is **heating degree days**. This index starts from the assumption that heating is not required in a building when the daily mean temperature outdoors is 65°F (18.3°C) or higher. Because the National Weather Service and the U.S. news media still compute and report degree-day information in Fahrenheit, we will use the Fahrenheit scale throughout this discussion. Simply, each degree of temperature below 65°F is counted as 1 heating degree day. Therefore, heating degree days are determined each day by subtracting the daily mean below 65°F from 65°F. Thus, a day with a mean temperature of 50°F has 15 heating degree days ($65 - 50 = 15$), and one with an average temperature of 65°F or higher has none.

The amount of heat required to maintain a certain temperature in a building is proportional to the total heating degree days. This linear relationship means that doubling the heating degree days usually doubles the fuel consumption. Consequently, a fuel bill is generally twice as high for a month with 1000 heating degree days as for a month with 500. When seasonal totals are compared for different places, we can estimate differences in seasonal fuel consumption (Table 3.2). For example, more than five times as much fuel is required to heat a building in Chicago (nearly 6500 total heating degree days) than to heat a similar building in Los Angeles (almost 1300 heating degree days). This statement is true, however, only if we assume that building construction and living habits in these areas are the same.

Each day, the previous day's accumulation is reported along with the total thus far in the season. For reporting purposes, the heating season is defined as the period from July 1 through June 30. These reports often include a comparison with the total up to this date last year or with the long-term average for this date or both, and so it is a relatively simple matter to judge whether the season thus far is above, below, or near normal.

**Cooling Degree Days** Just as fuel needs for heating can be estimated and compared by using heating degree days, the amount of power required to cool a building can be estimated by using a similar index called **cooling degree days**. Because the 65°F base temperature is also used in calculating this index, cooling degree days are determined each day by subtracting 65°F from the daily mean. Thus, if the mean temperature for a given day is 80°F, 15 cooling degree days would be accumulated. Mean annual totals of cooling degree days for selected cities are shown in Table 3.2. By comparing the totals for Baltimore and Miami, we can see that the fuel requirements for

**Table 3.2** Average Annual Heating and Cooling Degree Days for Selected Cities

| City | Heating Degree Days | Cooling Degree Days |
|---|---|---|
| Anchorage, AK | 10,470 | 3 |
| Baltimore, MD | 3807 | 1774 |
| Boston, MA | 5630 | 777 |
| Chicago, IL | 6498 | 830 |
| Denver, CO | 6128 | 695 |
| Detroit, MI | 6422 | 736 |
| Great Falls, MT | 7828 | 288 |
| International Falls, MN | 10,269 | 233 |
| Las Vegas, NV | 2239 | 3214 |
| Los Angeles, CA | 1274 | 679 |
| Miami, FL | 149 | 4361 |
| New York City, NY | 4754 | 1151 |
| Phoenix. AZ | 1125 | 4189 |
| San Antonio, TX | 1573 | 3038 |
| Seattle, WA | 4797 | 173 |

Source: NOAA, National Climatic Data Center.

severe& hazardous weather Box 3.4

## Heat Waves

A *heat wave* is a prolonged period of abnormally hot and usually humid weather that typically lasts from a few days to several weeks. The impact of heat waves on individuals varies greatly. Elderly people are the most vulnerable because heat puts more stress on weak hearts and bodies. Poor people, who often cannot afford air conditioning, also suffer disproportionately. Studies also show that the temperature at which death rates increase varies from city to city. In Dallas, Texas, a temperature of 39°C (103°F) is required before the death rate climbs. In San Francisco, the key temperature is just 29°C (84°F).

Heat waves do not elicit the same sense of fear or urgency as tornadoes, hurricanes, and flash floods. One reason is that it may take many days of oppressive temperatures for a heat wave to exact its toll rather than just a few minutes or a few hours. Another reason is that they cause far less property damage than other extreme weather events. Nevertheless, heat waves can be deadly and costly.

### Summer 1936

The 1936 North American heat wave is likely the continent's most severe in modern history. It took place during the economic hard times of the Great Depression and coincided with a significant drought in many parts of the Great Plains and Midwest. The prolonged heat wave began in late June and did not end until early September. The estimated death toll exceeded 5000, and agricultural losses were catastrophic in many areas. Many records from this summer of exceptional temperatures still stand. In fact, current record high temperatures for 13 states are from July and August 1936 (**Table 3.A**). There are other extraordinary records as well. For example, Mt. Vernon, Illinois, experienced 18 straight days (August 12–29) when temperatures surpassed 38°C (100°F).

The severity of heat waves is usually greatest in cities.

### Deadly Impacts

A more recent example occurred in the summer of 2003, when much of Europe experienced perhaps its worst heat wave in more than a century. The image in **Figure 3.F** relates to this deadly event. An estimate based on government records puts the death toll at between 20,000 and 35,000. The majority died during the first two weeks of August—the hottest span. France suffered the greatest number of heat-related fatalities—about 14,000.

The dangerous impact of summer heat is also reinforced when examining **Figure 3.G**, which shows average annual weather-related deaths in the United States for the 10-year period 2004–2013. A comparison of values reveals that the number of heat deaths is among the highest.

The severity of heat waves is usually greatest in cities because of the *urban heat island* (see Box 3.3). Large cities do not cool off as much at night during heat waves as rural areas do, and this can be a critical difference in the amount of heat stress experienced in inner cities. In addition, the stagnant atmospheric conditions usually associated with heat waves trap pollutants in urban areas and add the stresses of severe air pollution to the already dangerous stresses caused by the high temperatures.

In July 1995, a brief but intense heat wave developed in the central United States. A total of

**Table 3.A** State Temperature Records Remaining from 1936

| State | Temperature (°F) | Date |
|---|---|---|
| Arkansas | 120 | August 10 |
| Indiana | 116 | July 14 |
| Kansas | 121 | July 24 |
| Louisiana | 114 | August 10 |
| Maryland | 109 | July 10 |
| Michigan | 112 | July 13 |
| Minnesota | 114 | July 13 |
| Nebraska | 118 | July 24 |
| New Jersey | 110 | July 10 |
| North Dakota | 121 | July 6 |
| Pennsylvania | 111 | July 10 |
| West Virginia | 112 | July 10 |
| Wisconsin | 114 | July 13 |

cooling a building in Miami are almost 2½ times as great as for a similar building in Baltimore. The "cooling season" is conventionally measured from January 1 through December 31. Therefore, when cooling degree-day totals are reported, the number represents the accumulation since January 1 of that year.

Although indices that are more sophisticated than heating and cooling degree days have been proposed to take into account the effects of wind speed, solar radiation, and humidity, degree days continue to be widely used.

**Growing Degree Days** Another practical application of temperature data is used in agriculture to determine the approximate date when crops will be ready for harvest. This simple index is called **growing degree days**. The number of growing degree days for a particular crop on any day is the difference between the daily mean temperature and the base temperature of the crop, which is the minimum temperature required for it to grow. For example, the base temperature for sweet corn is 50°F, and for peas it is 40°F. Thus, on a day when the mean temperature is 75°F, the number of growing degree days for sweet corn is 25 and the number for peas is 35.

Starting with the onset of the growth season, the daily growing degree-day values are added. Thus, if 2000 growing degree days are needed for a crop to mature, it should be ready to harvest when the accumulation reaches 2000. Although many factors important to plant growth are not included in the index, such as moisture conditions and sunlight, this system nevertheless serves as a simple and widely used tool in determining approximate dates of crop maturity.

Video MM
Temperatures and Agriculture
http://goo.gl/H0y00

## Indices of Human Discomfort

Summertime weather reports often try to make people aware of potential harmful effects of high humidity coupled with high temperatures (see **Box 3.4**). In winter we are reminded about the effect of strong winds combined with low temperatures.

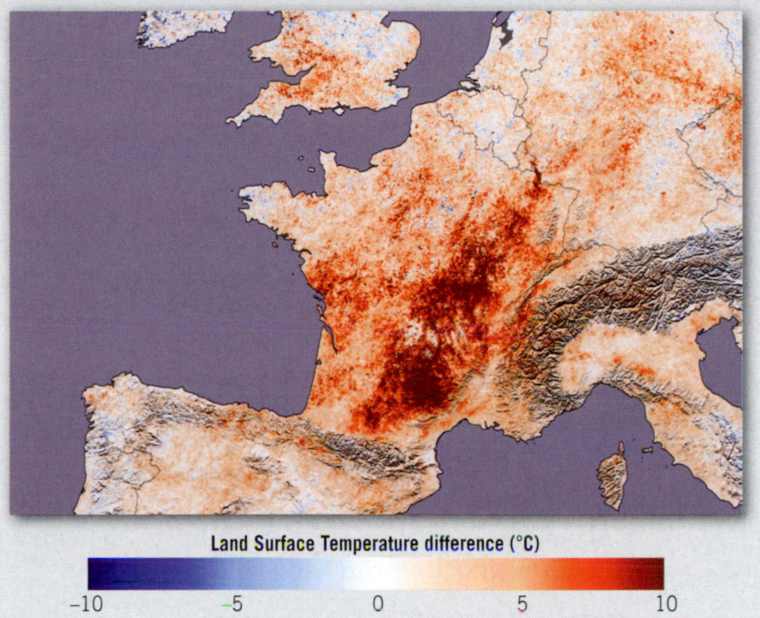


**Figure 3.F French heat wave** This image is derived from satellite data and shows the difference in daytime land surface temperatures during the 2003 European heat wave (July 20–August 20) as compared to the years 2000, 2001, 2002, and 2004. The zone of deep red shows where temperatures were 10°C (18°F) hotter than in the other years. France was hardest hit.

830 deaths were attributed to this severe 5-day event, the worst in 50 years in the northern Midwest. The greatest loss of life occurred in Chicago, where there were 525 fatalities. This event provided a sobering lesson by focusing attention on the need for more effective warning and response plans, especially in major urban areas where heat stress is greatest.

### Heat Waves and Climate Change

Among the possible consequences of global warming related to human activities is an increase in the frequency and severity of heat waves. A 2013 report by the Intergovernmental Panel on Climate Change (IPCC) states that it is very likely that human activities have contributed to observed changes in temperature extremes since the mid-twentieth century. The term *very likely* indicates a probability between 90 and 99 percent. The report goes on to state that as we advance through the twenty-first century, warmer and/or more frequent hot days and nights over most land areas is virtually certain, and that it is *very likely* that heat waves will occur with a higher frequency and have longer durations.* There is much more about the IPCC and the possible consequences of global climate change in Chapter 14.

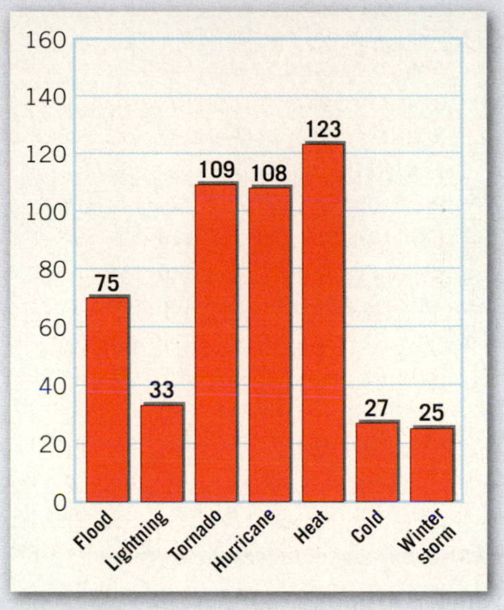


**Figure 3.G** Average annual weather-related fatalities for the 10-year period 2004–2013. The figure for the hurricane category was dramatically affected by Hurricane Katrina in 2005. Katrina was responsible for 1016 of the 1079 hurricane-related fatalities that occurred during this 10-year span.

**Question**

1. How do the number of fatalities related to heat waves compare to the number of deaths caused by other weather phenomena?

*IPCC, 2013–2014: Summary for Policymakers. In *Climate Change 2013: The Physical Science Basis. Contribution of Working Group I to the fifth Assessment Report of the Intergovernmental Panel on Climate Change*. Cambridge and New York: Cambridge University Press.

In the first instance, we are cautioned about heat stress and the possibility of heat stroke, and in the second case we are warned about windchill and the potential dangers of frostbite. These indices are expressions of **apparent temperature**—the temperature that a person perceives. Heat stress and windchill are based on the fact that our sensation of temperature is often different from the actual air temperature recorded by a thermometer.

The human body is a heat generator that continually releases energy. Anything that influences the rate of heat loss from the body also influences our sensation of temperature, thereby affecting our feeling of comfort. Several factors control the thermal comfort of the human body, and certainly air temperature is a major one. Other environmental conditions are also significant, such as relative humidity, wind, and sunshine.

**Heat Stress** High humidity contributes significantly to the discomfort people feel during a heat wave. Why are hot, muggy days so uncomfortable? Humans, like other mammals, are warm-blooded creatures who maintain a constant body temperature, regardless of the temperature of the environment. One of the ways the body prevents overheating is by perspiring. However, this process does little to cool the body unless the perspiration can evaporate. It is the cooling created by the evaporation of perspiration that reduces body temperature. Because high humidities retard evaporation, people are more uncomfortable on a hot and humid day than on a hot and dry day.

Generally, temperature and humidity are the most important elements influencing summertime human comfort. One index widely used by the National Weather Service that combines these factors to establish the degree of comfort or discomfort is called the **heat stress index**, or simply the heat index. In Figure 3.30, you will see that as relative humidity increases, the apparent temperature, and heat stress, increases as well. Further, when the relative humidity is low, the apparent temperature can have a value that is less than the actual air temperature.

**Heat Index**

Relative Humidity (%)

| Air Temperature (°F) | 40 | 45 | 50 | 55 | 60 | 65 | 70 | 75 | 80 | 85 | 90 | 95 | 100 |
|---|---|---|---|---|---|---|---|---|---|---|---|---|---|
| 110 | 136 | | | | | | | | | | | | |
| 108 | 130 | 137 | | | | | | | | | | | |
| 106 | 124 | 130 | 137 | | | | | | | | | | |
| 104 | 119 | 124 | 131 | 137 | | | | | | | | | |
| 102 | 114 | 119 | 124 | 130 | 137 | | | | | | | | |
| 100 | 109 | 114 | 118 | 124 | 129 | 136 | | | | | | | |
| 98 | 105 | 109 | 113 | 117 | 123 | 128 | 134 | | | | | | |
| 96 | 101 | 104 | 108 | 112 | 116 | 121 | 126 | 132 | | | | | |
| 94 | 97 | 100 | 102 | 106 | 110 | 114 | 119 | 124 | 129 | 135 | | | |
| 92 | 94 | 96 | 99 | 101 | 105 | 108 | 112 | 116 | 121 | 126 | 131 | | |
| 90 | 91 | 93 | 95 | 97 | 100 | 103 | 106 | 109 | 113 | 117 | 122 | 127 | 132 |
| 88 | 88 | 89 | 91 | 93 | 95 | 98 | 100 | 103 | 106 | 110 | 113 | 117 | 121 |
| 86 | 85 | 87 | 88 | 89 | 91 | 93 | 95 | 97 | 100 | 102 | 105 | 108 | 112 |
| 84 | 83 | 84 | 85 | 86 | 88 | 89 | 90 | 92 | 94 | 96 | 98 | 100 | 103 |
| 82 | 81 | 82 | 83 | 84 | 84 | 85 | 86 | 88 | 89 | 90 | 91 | 93 | 95 |
| 80 | 80 | 80 | 81 | 81 | 82 | 82 | 83 | 84 | 84 | 85 | 86 | 86 | 87 |

With prolonged exposure and/or physical activity

**Extreme danger** Heat stroke or sunstroke highly likely

**Danger** Sunstroke, muscle cramps, and/or heat exhaustion likely

**Extreme caution** Sunstroke, muscle cramps, and/or heat exhaustion possible

**Caution** Fatigue possible

▲ **Figure 3.30 Heat index expresses apparent temperature** As relative humidity increases, apparent temperature increases as well. For example, if the air temperature is 90°F and the relative humidity is 65 percent, it would "feel like" 103°F.

It is important to note that factors such as the length of exposure to direct sunlight, the wind speed, and the general health of the individual greatly affect the amount of stress a person will experience. In addition, while a period of hot, humid weather in New Orleans might be reasonably well tolerated by its residents, a similar event in a northern city such as Minneapolis would not be well tolerated. This occurs because hot and humid weather is more taxing on people who live where these conditions are relatively rare than it is on people who live where prolonged periods of heat and high humidity are the rule.

**Windchill.** Most everyone is familiar with the wintertime cooling power of moving air. When the wind blows on a cold day, we realize that comfort would improve if the wind would stop. A stiff breeze penetrates ordinary clothing and reduces its capacity to retain body heat while causing exposed parts of the body to chill rapidly. Not only is cooling by evaporation heightened in this situation, but the wind is also acting to carry heat away from the body by constantly replacing warmer air next to the body with colder air.

The U.S. National Weather Service and the Meteorological Services of Canada use the **Windchill Temperature (WCT)** index, which is designed to calculate how the wind and cold feel on human skin (Fig. 3.31). The index accounts for wind effects at face level and takes into account body heat-loss estimates. It was tested on human subjects in a chilled wind tunnel. The results of those trials were used to validate and improve the accuracy of the formula. The windchill chart includes a frostbite indicator that shows where temperature, wind speed, and exposure time produce frostbite (Fig. 3.31).

It is worth pointing out that in contrast to a cold and windy day, a calm and sunny day in winter feels warmer than the thermometer reading. In this situation, the warm feeling is caused by the absorption of direct solar radiation by the body. The index does not take into account any offsetting effect of solar radiation on windchill. Such a factor may be added in the future.

It is important to remember that the windchill temperature is only an estimate of human discomfort. The degree of discomfort felt by different people varies because it is influenced by many factors. Even if clothing is assumed to be the same, individuals vary widely in their responses because of such factors as age, physical condition, state of health, and level of activity. Nevertheless, as a relative measure, the WCT index is useful because it allows people to make more informed judgments regarding the potential harmful effects of wind and cold.

## ✓ Concept Checks 3.6

1. Distinguish among the three temperature applications that have *degree days* in their name.
2. What is apparent temperature?
3. Why does high humidity contribute to summertime discomfort?
4. Why do strong winds make apparent temperatures in winter feel lower than the thermometer reading?

▼ **Figure 3.31 Windchill chart** Fahrenheit temperatures are used because this is how the National Weather Service and the news media in the United States commonly report windchill information. The shaded areas on the chart indicate frostbite danger. Each shaded zone shows how long a person can be exposed before frostbite develops. For example, a temperature of 0°F and a wind speed of 15 miles per hour will produce a windchill temperature of –19°F. Under these conditions, exposed skin can freeze in 30 minutes.

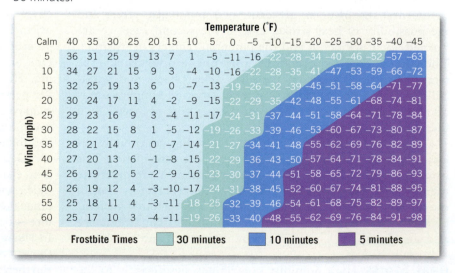

Temperature (°F)

| Wind (mph) / Calm | 40 | 35 | 30 | 25 | 20 | 15 | 10 | 5 | 0 | -5 | -10 | -15 | -20 | -25 | -30 | -35 | -40 | -45 |
|---|---|---|---|---|---|---|---|---|---|---|---|---|---|---|---|---|---|---|
| 5 | 36 | 31 | 25 | 19 | 13 | 7 | 1 | -5 | -11 | -16 | -22 | -28 | -34 | -40 | -46 | -52 | -57 | -63 |
| 10 | 34 | 27 | 21 | 15 | 9 | 3 | -4 | -10 | -16 | -22 | -28 | -35 | -41 | -47 | -53 | -59 | -66 | -72 |
| 15 | 32 | 25 | 19 | 13 | 6 | 0 | -7 | -13 | -19 | -26 | -32 | -39 | -45 | -51 | -58 | -64 | -71 | -77 |
| 20 | 30 | 24 | 17 | 11 | 4 | -2 | -9 | -15 | -22 | -29 | -35 | -42 | -48 | -55 | -61 | -68 | -74 | -81 |
| 25 | 29 | 23 | 16 | 9 | 3 | -4 | -11 | -17 | -24 | -31 | -37 | -44 | -51 | -58 | -64 | -71 | -78 | -84 |
| 30 | 28 | 22 | 15 | 8 | 1 | -5 | -12 | -19 | -26 | -33 | -39 | -46 | -53 | -60 | -67 | -73 | -80 | -87 |
| 35 | 28 | 21 | 14 | 7 | 0 | -7 | -14 | -21 | -27 | -34 | -41 | -48 | -55 | -62 | -69 | -76 | -82 | -89 |
| 40 | 27 | 20 | 13 | 6 | -1 | -8 | -15 | -22 | -29 | -36 | -43 | -50 | -57 | -64 | -71 | -78 | -84 | -91 |
| 45 | 26 | 19 | 12 | 5 | -2 | -9 | -16 | -23 | -30 | -37 | -44 | -51 | -58 | -65 | -72 | -79 | -86 | -93 |
| 50 | 26 | 19 | 12 | 4 | -3 | -10 | -17 | -24 | -31 | -38 | -45 | -52 | -60 | -67 | -74 | -81 | -88 | -95 |
| 55 | 25 | 18 | 11 | 4 | -3 | -11 | -18 | -25 | -32 | -39 | -46 | -54 | -61 | -68 | -75 | -82 | -89 | -97 |
| 60 | 25 | 17 | 10 | 3 | -4 | -11 | -19 | -26 | -33 | -40 | -48 | -55 | -62 | -69 | -76 | -84 | -91 | -98 |

Frostbite Times: 30 minutes, 10 minutes, 5 minutes

# 3 Concepts in Review Temperature

## 3.1 For the Record: Air-Temperature Data

▶ Calculate five commonly used types of temperature data and interpret a map that depicts temperature data using isotherms.

**Key Term:** daily mean temperature, daily temperature range, monthly mean temperature, annual mean temperature, annual temperature range, isotherm, temperature gradient

- Daily mean temperature is an average of the daily maximum and daily minimum temperatures, whereas the daily range is the difference between the daily maximum and daily minimum temperatures. The monthly mean is determined by averaging the daily means for a particular month. The annual mean is an average of the 12 monthly means, whereas the annual temperature range is the difference between the highest and lowest monthly means.
- Temperature distribution is shown on a map by using isotherms, which are lines of equal temperature. Temperature gradient is the amount of temperature change per unit of distance. Closely spaced isotherms indicate a rapid rate of change.

## 3.2 Why Temperatures Vary: The Controls of Temperature

▶ Name the principal controls of temperature and use examples to describe their effects.

**Key Terms:** temperature control, specific heat

- Controls of temperature are factors that cause temperature to vary from place to place and from time to time. Latitude (Earth–Sun relationships) is an important example that was a focus of Chapter 2.
- Unequal heating of land and water is a temperature control. Because land and water heat and cool differently, land areas experience greater temperature extremes than do water-dominated areas.
- Poleward-moving warm ocean currents moderate winter temperatures in the middle latitudes. Cold currents exert their greatest influence during summer in the middle latitudes and year-round in the tropics.
- Altitude is an easy-to-visualize control: The higher up you go, the colder it gets; therefore, mountains are cooler than adjacent lowlands.
- Geographic position as a temperature control involves such factors as mountains acting as barriers to marine influence or whether a place is on a windward or a leeward coast.

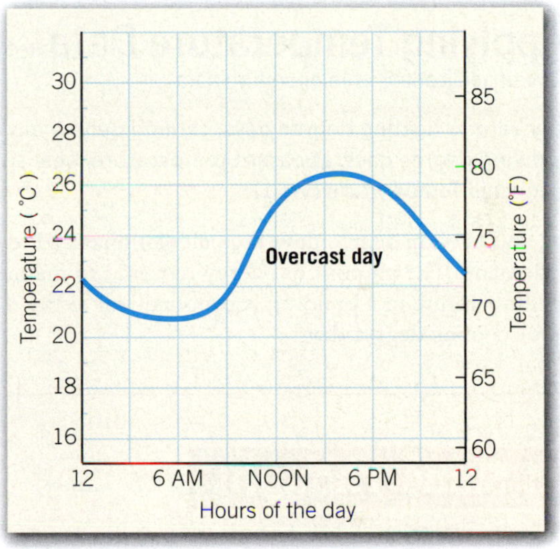


**Q** This graph is a plot of hourly temperatures for an overcast day. How would the graph be different if it had not been overcast? Explain.

## 3.3 World Distribution of Temperature

▶ Interpret the patterns depicted on world temperature maps.

- On world maps showing January and July mean temperatures, isotherms generally trend east–west and show a decrease in temperature moving poleward from the tropics. When the two maps are compared, a latitudinal shifting of temperatures is easily seen. Bending isotherms reveal the locations of ocean currents.
- Annual temperature range is small near the equator and increases with an increase in latitude. Outside the tropics, annual temperature range also increases as marine influence diminishes.

**Q** Refer to Figure 3.17. What causes the isotherms to bend in the North Atlantic?

## 3.4 Cycles of Air Temperature

▶ Discuss the basic daily and annual cycles of air temperature.

- The main control of the daily cycle of air temperature is Earth's rotation, which causes a place to move into and out of daylight. The time of highest temperature usually does not coincide with the time of maximum intensity of solar radiation. This delay is termed the lag of the maximum.
- Daily temperature range is influenced by locational factors such as latitude and whether a place is coastal. It is also affected by cloud cover and humidity.
- As a result of the mechanism by which Earth's atmosphere is heated, the highest and lowest monthly means usually do not coincide with the periods of maximum and minimum incoming solar radiation.

## 3.5 Temperature Measurement ▸ Explain how different types of thermometers work and why the placement of thermometers is an important factor in obtaining accurate readings. Distinguish among Fahrenheit, Celsius, and Kelvin temperature scales.

**Key Terms:** thermometer, liquid-in-glass thermometer, maximum thermometer, minimum thermometer, bimetal strip, thermograph, thermistor, fixed points, Fahrenheit scale, ice point, steam point, Celsius scale, Kelvin (absolute) scale, absolute zero

- Thermometers measure temperature either mechanically or electrically. Most mechanical thermometers are based on the ability of a material to expand when heated and contract when cooled. Electrical thermometers use a thermistor (a thermal resistor) to measure temperature.
- To obtain an accurate air temperature, thermometer placement is important. The best location is a properly situated instrument shelter.
- Temperature scales are established using reference points called fixed points. Three common scales are the Fahrenheit scale, Celsius scale, and Kelvin, or absolute, scale.

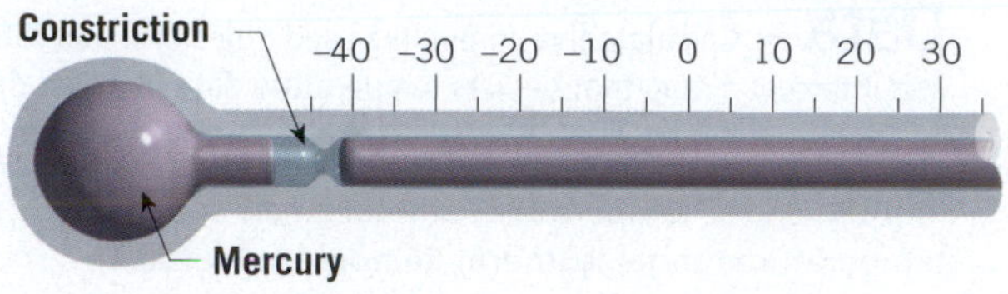


**Q** This sketch shows a portion of a thermometer that has a specific function. What is that function? How did you figure it out?

## 3.6 Applying Temperature Data ▸ Summarize several applications of temperature data.

**Key Terms:** heating degree days, cooling degree days, growing degree days, apparent temperature, heat stress index, Windchill Temperature (WCT)

- Heating and cooling degree days are calculations used to evaluate energy demand. The amount of energy required to maintain a certain temperature in a building is proportional to the total heating or cooling degree days.
- Growing degree days is a practical application used to determine the approximate date when crops will be ready to harvest.
- Heat stress and windchill are two familiar uses of temperature data that relate to apparent temperature—the temperature people perceive.

## Give it Some Thought

1. If you were asked to identify the coldest city in the United States (or any other designated region), what statistics could you use? Can you list at least three different ways of selecting the coldest city?
2. The accompanying graph shows monthly high temperatures for Urbana, Illinois, and San Francisco, California. Although both cities are located at about the same latitude, the temperatures they experience are quite different. Which line on the graph represents Urbana, and which represents San Francisco? How did you figure this out?

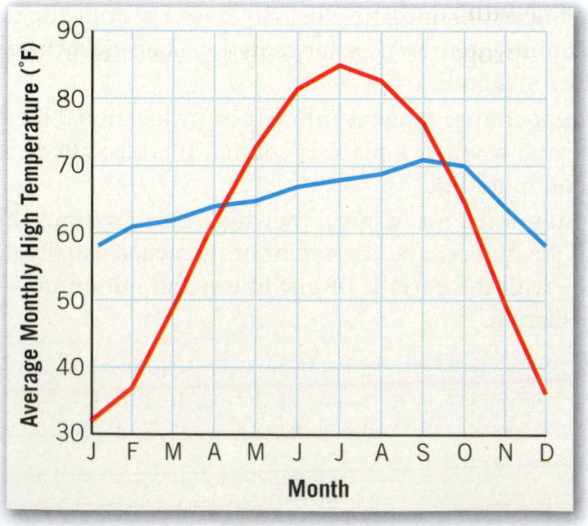


3. On which summer day would you expect the greatest temperature range? Which would have the smallest range in temperature? Explain your choices.
   a. Cloudy skies during the day and clear skies at night
   b. Clear skies during the day and cloudy skies at night
   c. Clear skies during the day and clear skies at night
   d. Cloudy skies during the day and cloudy skies at night
4. The accompanying scene shows an island near the equator in the Indian Ocean. Describe how latitude, altitude, and the differential heating of land and water influence the climate of this place.

5. The accompanying sketch map represents a hypothetical continent in the Northern Hemisphere. One isotherm has been placed on the map.

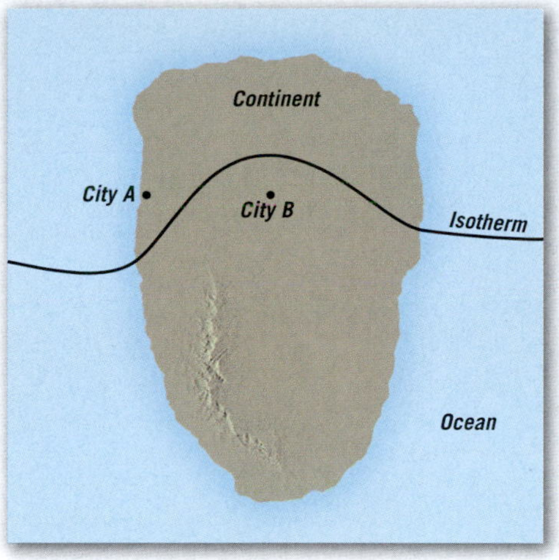


a. Is the temperature higher at city A or city B? Explain.
b. Is the season winter or summer? How are you able to determine this?
c. Describe (or sketch) the position of this isotherm 6 months later.

6. The data below are mean monthly temperatures in °C for an inland location that lacks any significant ocean influence. *Based on the annual temperature range*, what is the approximate latitude of this place? Are these temperatures what you would normally expect for this latitude? If not, what control would explain these temperatures?

| J | F | M | A | M | J | J | A | S | O | N | D |
|---|---|---|---|---|---|---|---|---|---|---|---|
| 6.1 | 6.6 | 6.6 | 6.6 | 6.6 | 6.1 | 6.1 | 6.1 | 6.1 | 6.1 | 6.6 | 6.6 |

7. Palm trees in Scotland? Yes, in the 1850s and 1860s, amateur gardeners planted palm trees on the western shore of Scotland. The latitude pictured is 57° north, about the same as the northern portion of Labrador, across the Atlantic in Canada. Surprisingly, these exotic plants flourished. Suggest a possible explanation for how these palms can survive at such a high latitude.

## Problems

1. Examine the number of hurricane-related deaths in Figure 3.G (p. 83). Be sure to read the entire caption. If Hurricane Katrina had *not* occurred, what number would have been plotted on the graph for hurricane fatalities?
2. Refer to the thermograph record in Figure 3.20. Determine the maximum and minimum temperatures for each day of the week. Use these data to calculate the daily mean and daily range for each day.
3. By referring to the world maps of temperature distribution for January and July (Figures 3.17 and 3.18), determine the approximate January mean, July mean, and annual temperature range for a place located at 60° north latitude, 80° east longitude and a place located at 60° south latitude, 80° east longitude.
4. Calculate the annual temperature range for three cities in Appendix G. Try to choose cities with different ranges and explain these differences in terms of the controls of temperature.
5. Referring to Figure 3.31, determine windchill temperatures under the following circumstances:
   **a.** Temperature = 5°F, wind speed = 15 mph.
   **b.** Temperature = 5°F, wind speed = 30 mph.
6. The mean temperature is 55°F on a particular day. The following day the mean drops to 45°F. Calculate the number of heating degree days for each day. How much more fuel would be needed to heat a building on the second day compared with the first day?
7. Use the appropriate formulas to convert the following temperatures:

   20°C = ____°F
   −25°C = ____K
   59°F = ____°C

## MasteringMeteorology™

Looking for additional review and test prep materials? Visit the Study Area in *MasteringMeteorology*™ to enhance your understanding of this chapter's content by accessing a variety of resources, including **MapMaster**™ interactive maps, Geoscience Animations, GEODe, *In the News* RSS feeds, flashcards, web links, self-study quizzes, and an eText version of *The Atmosphere*.

# 4 Moisture and Atmospheric Stability

## Focus on Concepts

*Each statement represents the primary learning objective for the corresponding major heading within the chapter. After you complete the chapter, you should be able to:*

**4.1** Describe the movement of water through the hydrologic cycle. List and describe water's unique properties.

**4.2** Summarize the six processes by which water changes from one state of matter to another. For each, indicate whether energy is absorbed or released.

**4.3** Write a generalization relating air temperature and the amount of water vapor needed to saturate air.

**4.4** List and describe the ways relative humidity changes in nature. Compare relative humidity to dew-point temperature.

**4.5** Describe adiabatic temperature changes and explain why the wet adiabatic rate of cooling is less than the dry adiabatic rate.

**4.6** List and describe four mechanisms that cause air to rise.

**4.7** Write a statement relating the environmental lapse rate to stability.

**4.8** List the primary factors that influence the stability of air.

Water vapor is an odorless, colorless gas that mixes freely with the other gases of the atmosphere. Unlike oxygen and nitrogen—the two most abundant components of the atmosphere—water can change from one state of matter to another (solid, liquid, or gas) at the temperatures and pressures experienced on Earth. (By comparison, nitrogen in the atmosphere condenses to a liquid at a temperature of –196°C [–371°F].) Because of this unique property, water leaves the oceans as a gas and returns to the oceans as a liquid.

*Rain showers near the Mitten Buttes, Monument Valley, Arizona.*

# 4.1 Water on Earth

**Describe the movement of water through the hydrologic cycle. List and describe water's unique properties.**

Water is found everywhere on Earth—in the oceans, glaciers, rivers, lakes, air, soil, and living tissue. The vast majority of the water on or close to Earth's surface (over 97 percent) is saltwater found in the oceans. Much of the remaining 3 percent is stored in the ice sheets of Antarctica and Greenland. Only a meager 0.001 percent is found in the atmosphere, and most of this is in the form of water vapor.

## Movement of Water Through the Atmosphere

The continuous exchange of water among the oceans, the atmosphere, and the continents is called the **hydrologic cycle** (Fig. 4.1). Water from the oceans and, to a lesser extent, from land areas evaporates into the atmosphere. Winds transport this moisture-laden air, often over great distances, until the process of cloud formation causes the water vapor to condense into tiny liquid cloud droplets.

The process of cloud formation may result in precipitation. The precipitation that falls into the ocean has ended its cycle and is ready to begin another. The remaining precipitation falls over the land. A portion of the water that falls on land soaks into the ground (*infiltration*), some of it moving downward and then laterally, eventually seeping into lakes and streams. The remainder, which flows along Earth's surface, is called *runoff*.

Much of the groundwater and runoff eventually returns to the atmosphere through evaporation. A smaller amount of groundwater is taken up by plants, which release it into the atmosphere through a process called *evapotransporation*.

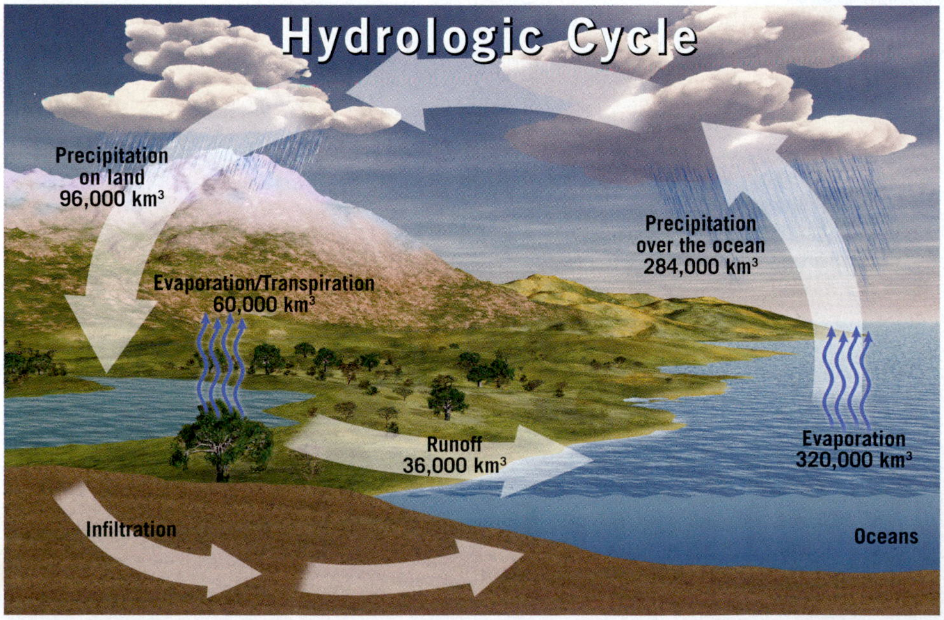


▲ **SmartFigure 4.1 Earth's hydrologic cycle**

http://goo.gl/F4cdm

The total amount of water vapor in the atmosphere remains fairly constant. Therefore, the average annual precipitation over Earth must be roughly equal to the quantity of water lost through evaporation. However, over the continents, precipitation exceeds evaporation. Evidence for the roughly balanced hydrologic cycle is found in the fact that the level of the world's oceans is not dropping.

This continuous movement of water through the hydrologic cycle holds the key to the distribution of moisture over the surface of our planet and is intricately related to all atmospheric phenomena.

## Water: A Unique Substance

Water has unique properties that set it apart from most other substances. For instance, (1) water is the only *liquid* found at Earth's surface in large quantities; (2) water is readily converted from one state of matter to another (solid, liquid, gas); (3) water's solid phase, ice, is less dense than liquid water; and (4) water has a high heat capacity—meaning changing its temperature requires considerable energy. All these properties influence Earth's weather and climate and are favorable to life as we know it.

These unique properties are largely a consequence of water's ability to form hydrogen bonds. **Hydrogen bonds** are the attractive forces that exist between hydrogen atoms in one water molecule and oxygen atoms of any other water molecule. To better grasp the nature of hydrogen bonds, let's examine a water molecule. Water molecules ($H_2O$) consist of two hydrogen atoms that are strongly bonded to an oxygen atom (Fig. 4.2A). Because oxygen atoms have a greater affinity for the bonding electrons (negatively charged subatomic particles) than do hydrogen atoms, the oxygen end of a water molecule acquires a partial negative charge. For the same reason, the hydrogen atoms of a water molecule acquire a partial positive charge. Because oppositely charged particles attract, a hydrogen atom on one water molecule is attracted to an oxygen atom on another water molecule (Fig. 4.2B).

Hydrogen bonds hold water molecules together to form the solid we call *ice*. In ice, hydrogen bonds produce a rigid hexagonal network (Fig. 4.2C). The resulting molecular configuration is very open (lots of empty spaces). When ice is heated sufficiently, it melts. Melting causes some, but not all, of the hydrogen bonds to break. As a result, the water molecules in liquid water display a more compact arrangement (Fig. 4.2D). This explains why water in its liquid phase is denser than it is in the solid phase.

Because ice is less dense than the liquid water beneath it, a water body freezes from the top

Video MM
Hydrologic Cycle
http://goo.gl/h7pTt6

Animation MM
Earth's Water and Hydrologic Cycle
http://goo.gl/a10BWC

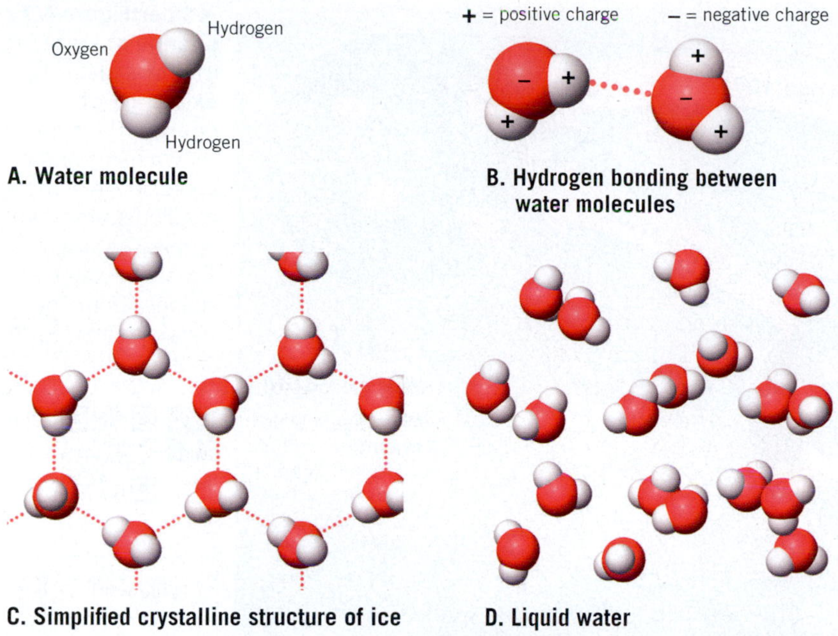


▲ **Figure 4.2 Hydrogen bonding in water** **A.** A water molecule consists of one oxygen atom and two hydrogen atoms. **B.** Water molecules are joined together by hydrogen bonds that loosely bond a hydrogen atom on one water molecule with an oxygen atom on another. **C.** Crystalline structure of ice. For simplicity, the illustration is two-dimensional rather than the actual three-dimensional structure. **D.** Water in the liquid state consists of clusters of water molecules joined together by hydrogen bonds. In liquid water, hydrogen bonds continually break apart and are replaced by new ones, which give liquid water its fluid nature.

**Animation** MM
Water Phase Changes
http://goo.gl/ZChN5n

down. This has far-reaching effects, both for our daily weather and aquatic life. When ice forms on a water body, it insulates the underlying liquid and slows the rate of freezing at depth. If a water body froze from the bottom, imagine the consequences. Many lakes would freeze solid during the winter, killing the aquatic life. In addition, deep bodies of water, such as the Arctic Ocean, would never be ice covered. This would alter Earth's energy budget, which in turn would modify global atmospheric and oceanic circulations.

Water's heat capacity is also related to hydrogen bonding. When water is heated, some of the energy is used to break hydrogen bonds rather than to increase molecular motion. (Recall that an increase in average molecular motion corresponds to an increase in temperature.) Thus, under similar conditions, water heats up and cools down more slowly than most other common substances. As a result, large water bodies tend to moderate temperatures by remaining warmer than adjacent landmasses in winter and cooler in summer.

### ✔ Concept Checks 4.1

1. What serves as a reservoir (storage facility) for most of Earth's water?
2. Briefly describe the hydrologic cycle.
3. Water's solid phase, ice, is less dense than liquid water. Why is this unique property of water important?
4. Explain what happens as ice melts to become liquid water.
5. What property of water causes large water bodies to remain warmer than adjacent landmasses in winter but cooler in summer?

## 4.2 Water's Changes of State

**Summarize the six processes by which water changes from one state of matter to another. For each, indicate whether energy is absorbed or released.**

MM **GEODe** ▶ Moisture and Cloud Formation ▶ Water's Changes of State

Water is the only substance that naturally exists on Earth as a solid (ice), liquid, and gas (water vapor). Because all forms of water are composed of hydrogen and oxygen atoms that are bonded together to form water molecules ($H_2O$), the primary difference among water's three phases is the arrangement of these water molecules.

### Ice, Liquid Water, and Water Vapor

Ice is composed of water molecules that have low kinetic energies (motion) and are held together by mutual molecular attractions (hydrogen bonds). These molecules form a tight, orderly network (see Figure 4.2C) and are not free to move relative to each other; rather, they vibrate about fixed sites. When ice is heated, the molecules oscillate more rapidly. When the rate of molecular movement increases sufficiently, the hydrogen bonds between some of the water molecules are broken, resulting in melting.

In the liquid state, water molecules are still tightly packed but move fast enough to be able to easily slide past one another. As a result, liquid water is fluid and will take the shape of the container that holds it.

When liquid water gains heat from the environment, some of the molecules acquire enough energy to break their hydrogen bonds and escape from the surface, becoming water vapor. Water vapor molecules are widely spaced compared to liquid water and exhibit very energetic and random motion.

These processes are reversed when water vapor returns to its liquid state and when water freezes. When water changes state, hydrogen bonds either form or are broken.

### Latent Heat

Whenever water changes state, energy is exchanged between water and its surroundings (Fig. 4.3). For example, heat is required to evaporate water. The heat involved when water changes

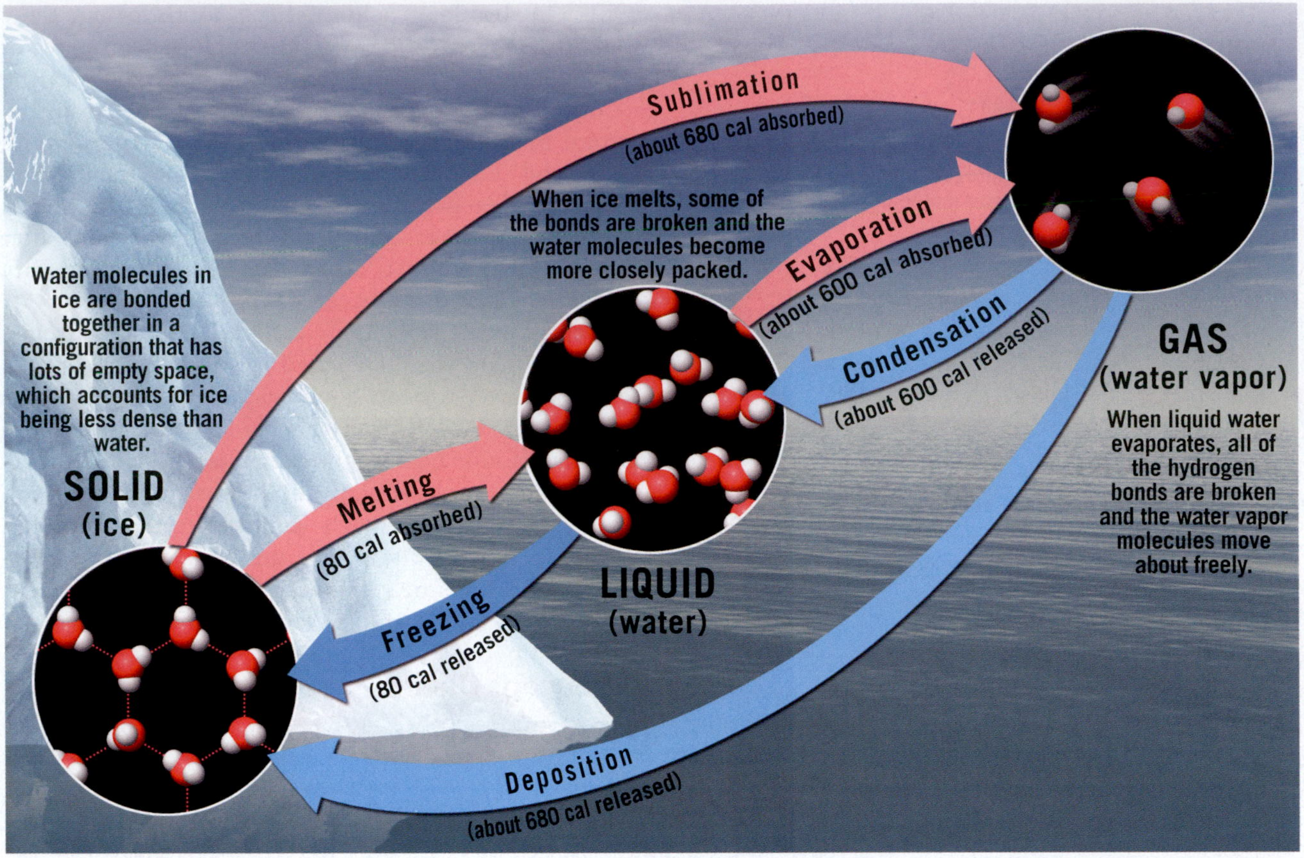


◀ **SmartFigure 4.3 Change of state involves an exchange of energy** The amounts shown here are the approximate numbers of calories absorbed or released when 1 gram of water changes from one state of matter to another.

http://goo.gl/eCqeY

Video
Global Evaporation Rates

http://goo.gl/bBKAbl

state is measured in units called *calories*. One **calorie** is the amount of energy required to raise the temperature of 1 gram of water 1°C (1.8°F). Thus, when 10 calories of heat are absorbed by 1 gram of water, the molecules vibrate faster, and a 10°C (18°F) temperature increase occurs. (In the International System of Units [SI], joules [J] are used to denote energy; 1 calorie = 4.184 J.)

Under certain conditions, energy may be added to a substance without an accompanying rise in temperature. For example, when the ice in a glass of ice water melts, the temperature of the mixture remains a constant 0°C (32°F) until all the ice has melted. If the added energy does not raise the temperature of ice water, where does this energy go? In this case, the added energy breaks the hydrogen bonds that once bound the water molecules into a crystalline structure.

Because the energy used to melt ice does not produce a temperature change, it is referred to as **latent heat**. (Latent means *hidden*, like the fingerprints hidden at a crime scene.) The energy, stored in liquid water, is released to its surroundings when the water freezes.

Melting 1 gram of ice requires 80 calories (334 J), an amount referred to as the *latent heat of melting*. The reverse process, *freezing*, releases these 80 calories per gram to the environment as the *latent heat of fusion*. We will consider the importance of latent heat of fusion in Chapter 5, in the section on frost prevention.

**Evaporation and Condensation** Latent heat is also involved in **evaporation**, the process of converting a liquid to a gas (vapor). The energy absorbed by water molecules during evaporation is used to give them the motion needed to escape the surface of the liquid and become a gas. This energy is referred to as the *latent heat of vaporization* and varies from about 600 calories (2500 J) per gram for water at 0°C to 540 calories (2260 J) per gram at 100°C. (Notice from Figure 4.3 that it takes much more energy to evaporate 1 gram of water than it does to melt the same amount of ice.) During the evaporation process, the faster-moving molecules escape the surface. As a result, the average molecular motion (temperature) of the remaining water is lowered—hence the expression "evaporation is a cooling process." You have undoubtedly experienced this cooling effect when you step, dripping wet, out of a swimming pool or shower. In this situation, the energy used to evaporate water comes from your skin—hence, you feel cool.

**Condensation**, the reverse process, occurs when water vapor changes to the liquid state. During condensation, water vapor molecules release energy (*latent heat of condensation*) in an amount equivalent to what was absorbed during evaporation. When condensation occurs in the atmosphere, it results in the formation of fog or clouds (**Fig. 4.4A**).

Latent heat plays an important role in many atmospheric processes. In particular, when water vapor condenses to form cloud droplets, latent heat is released, which warms the surrounding air and gives it buoyancy. When the moisture content of air is high, this process can spur the growth of towering storm clouds. In addition, the evaporation of water over the tropical oceans and the subsequent condensation at higher latitudes results in significant energy transfer from equatorial to more poleward locations.

Condensation of water vapor generates phenomena such as dew, clouds, and fog.
A.
Frost on a windowpane, an example of deposition.
B.


▲ **Figure 4.4 Examples of condensation and deposition**

**Sublimation and Deposition** You are probably least familiar with the last two processes illustrated in Figure 4.3—sublimation and deposition. **Sublimation** is the conversion of a solid directly to a gas, without passing through the liquid state. Examples you may have observed include the gradual shrinkage of unused ice cubes in a freezer and the rapid conversion of dry ice (frozen carbon dioxide) to wispy clouds that quickly disappear.

**Deposition** is the reverse process: the conversion of a vapor directly to a solid. An example is water vapor deposited as ice on solid objects such as grass or a window pane (Fig. 4.4B). These deposits are called *white frost* or simply *frost*. As shown in Figure 4.3, the process of deposition returns the combined energy released by condensation and freezing to the environment.

## ✔ Concept Checks 4.2

1. Summarize the six processes by which water changes from one state to another. Indicate whether heat is absorbed or released in each case.
2. Why is evaporation called a cooling process?
3. Define *latent heat* and explain the role that latent heat of condensation plays in the growth of towering clouds.
4. Compared to the other phase changes, why does the process of sublimation (or deposition) involve the exchange of more latent heat?

**students sometimes ask...**

### What is "freezer burn"?

Poorly wrapped food stored in a frost-free freezer for an extended time can become "freezer burned." Because modern refrigerators are designed to remove moisture from the freezer compartment, the air inside them is relatively dry. As a result, the moisture in food sublimates—turns from ice to water vapor—and escapes. Thus, the food is not actually burned; it is simply dried out.

# 4.3 Humidity: Water Vapor in the Air

**Write a generalization relating air temperature and the amount of water vapor needed to saturate air.**

MM GEODe ▶ Moisture and Cloud Formation ▶ Humidity: Water Vapor in the Air

**Humidity** is the general term used to describe the amount of water vapor in the air (Fig. 4.5). Water vapor constitutes only a small fraction of the atmosphere, varying from as little as 0.1 percent up to about 4 percent by volume. But the importance of water in the air is far greater than these small percentages indicate. In fact, *water vapor* is the most important gas in the atmosphere when it comes to atmospheric processes.

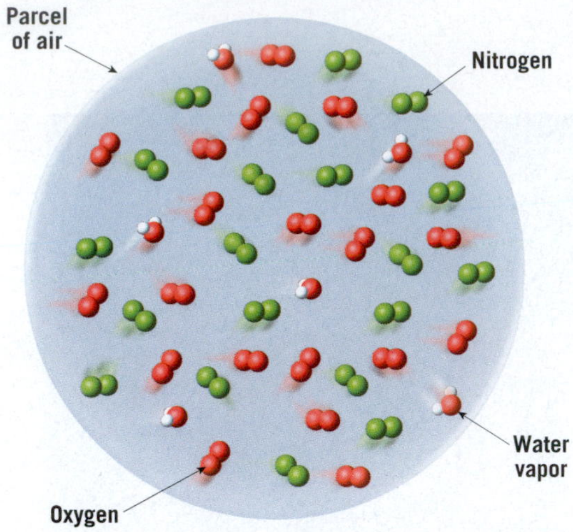


▲ **Figure 4.5 Meteorologists use several methods to express the water-vapor content of air**

## How Is Humidity Expressed?

Meteorologists employ several methods to express the water vapor content of the air, including (1) absolute humidity, (2) mixing ratio, (3) vapor pressure, (4) relative humidity, and (5) dew point. Two of these methods, *absolute humidity* and *mixing ratio*, are similar in that both are expressed as the quantity of water vapor contained in a specific amount of air.

**Absolute Humidity** *The mass of water vapor in a given volume of air* (usually as grams per cubic meter) is known as **absolute humidity**:

$$\text{Absolute humidity} = \frac{\text{Mass of water vapor (grams)}}{\text{Volume of air (cubic meters)}}$$

As air moves from one place to another, changes in pressure and temperature cause changes in its volume. When such volume changes occur, the absolute humidity also changes, even if no water vapor is added or removed. Consequently, it is difficult to monitor the water vapor content of a moving mass of air when using the absolute humidity index. Therefore, meteorologists generally prefer to use mixing ratio to express the water vapor content of air.

**Mixing Ratio** The **mixing ratio** is *the mass of water vapor in a unit of air compared to the remaining mass of dry air*:

$$\text{Mixing ratio} = \frac{\text{Mass of water vapor (grams)}}{\text{Mass of dry air (kilograms)}}$$

Because it is measured in units of mass (usually grams per kilogram), the mixing ratio is not affected by changes in pressure or temperature (Fig. 4.6).*

Absolute humidity and mixing ratio are difficult to measure. Therefore, weather reports commonly use relative humidity and dew point to express the moisture content of the air; these terms are discussed in the next section.

## Vapor Pressure and Saturation

We can determine the moisture content of the air from the pressure exerted by water vapor. To understand how water vapor exerts pressure, imagine a closed flask containing pure water and overlain by dry air, as shown in Figure 4.7A. Almost immediately some of the water molecules begin to leave the water surface and evaporate into the dry air above. The addition of water vapor into the air can be detected by a small increase in pressure (Fig. 4.7B). This increase in pressure is a result of the motion of the water vapor molecules that were added to the air through evaporation. In the atmosphere, this pressure is called **vapor pressure** and is defined as *the part of the total atmospheric pressure attributable to its water vapor content.*

Initially, many more molecules will leave the water surface (evaporate) than will return (condense). However, as more and more molecules evaporate from the water surface, the steadily increasing vapor pressure in the air above forces more and more water molecules to return to the liquid. Eventually a balance is reached in which the number of water molecules returning to the surface equals the number leaving. At that point, the air is said to have reached an equilibrium called **saturation** (Fig. 4.7C). When air is saturated, the pressure exerted by the motion of the water vapor molecules is called the **saturation vapor pressure**.

If the water in the closed container is heated, the equilibrium between evaporation and condensation will be disrupted,

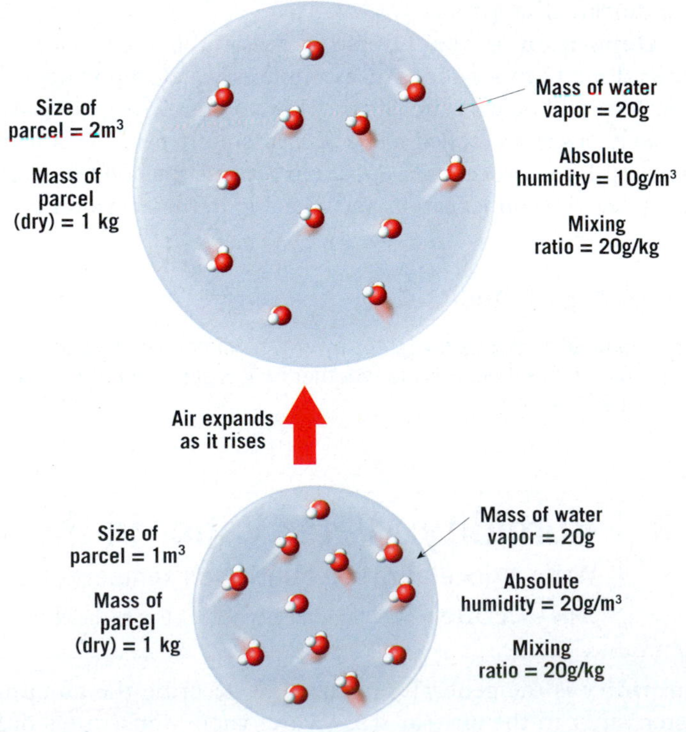


▲ **Figure 4.6 Comparison of absolute humidity and mixing ratio for a rising parcel of air** Notice that the mixing ratio is not affected by changes in pressure as the parcel of air rises and expands.

*Another commonly used expression is *specific humidity,* which is the mass of water vapor in a unit mass of air, including the water vapor. Because water vapor constitutes just a few percent of the total mass, specific humidity is equivalent to the mixing ratio for all practical purposes.

as illustrated in **Figure 4.7D**. The added energy increases the rate of evaporation, which causes the vapor pressure in the air above to increase, until a new equilibrium is reached. Thus, we conclude that the saturation vapor pressure is temperature dependent, such that at higher temperatures it takes more water vapor to saturate air (**Fig. 4.8**).

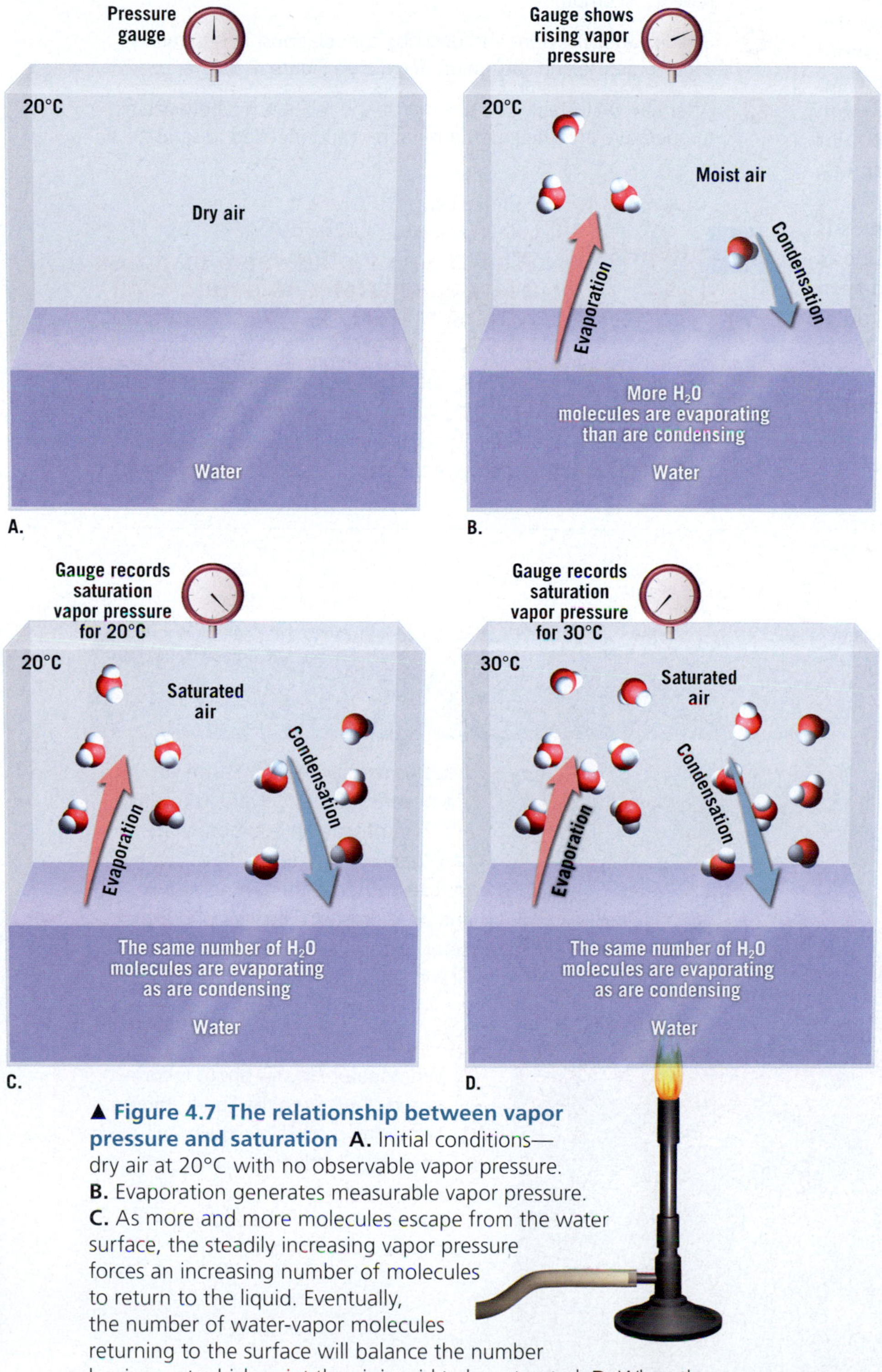


▲ **Figure 4.7 The relationship between vapor pressure and saturation** **A.** Initial conditions—dry air at 20°C with no observable vapor pressure. **B.** Evaporation generates measurable vapor pressure. **C.** As more and more molecules escape from the water surface, the steadily increasing vapor pressure forces an increasing number of molecules to return to the liquid. Eventually, the number of water-vapor molecules returning to the surface will balance the number leaving—at which point the air is said to be saturated. **D.** When the container is heated from 20°C to 30°C, the rate of evaporation increases, causing the vapor pressure to increase until a new balance is reached.

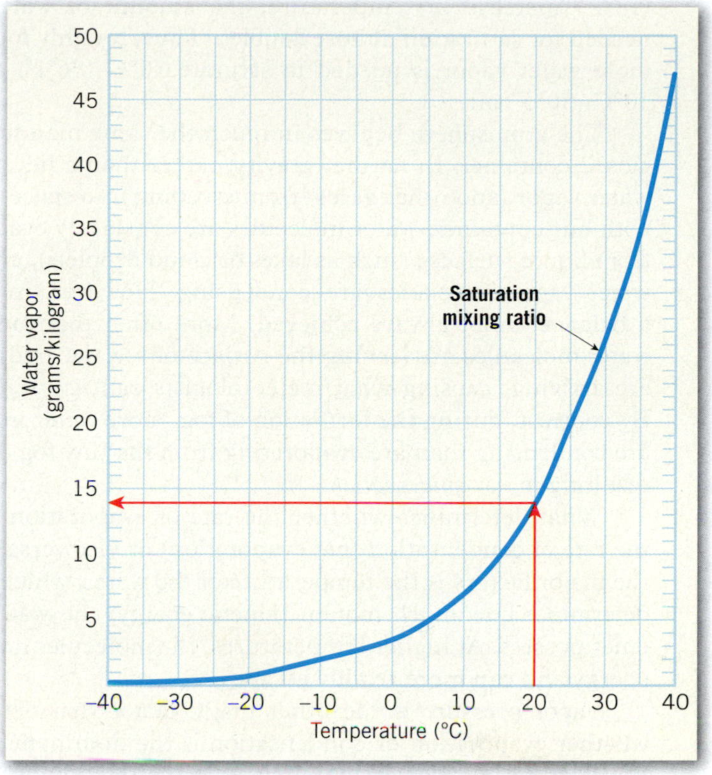


▲ **Figure 4.8 Graph illustrating the amount of water vapor required to saturate 1 kilogram of dry air at various temperatures** For example, the red arrows show that saturated air at 20°C contains 14 grams of water vapor per kilogram of dry air.

The amount of water vapor required to saturate 1 kilogram (2.2 pounds) of dry air at various temperatures is shown in **Table 4.1**. Note that for every 10°C

**Table 4.1** | Saturation Mixing Ratio (at Sea-Level Pressure)

| Temperature, °C (°F) | Saturation Mixing Ratio, g/kg |
|---|---|
| –40 (–40) | 0.1 |
| –30 (–22) | 0.3 |
| –20 (–4) | 0.75 |
| –10 (14) | 2 |
| 0 (32) | 3.5 |
| 5 (41) | 5 |
| 10 (50) | 7 |
| 15 (59) | 10 |
| 20 (68) | 14 |
| 25 (77) | 20 |
| 30 (86) | 26.5 |
| 35 (95) | 35 |
| 40 (104) | 47 |

(18°F) increase in temperature, the amount of water vapor needed for saturation almost doubles. Thus, roughly four times more water vapor is needed to saturate 30°C (86°F) air than 10°C (50°F) air.

The atmosphere behaves in much the same manner as our closed container. In nature, gravity, rather than a lid, prevents water vapor (and other gases) from escaping into space. Also as with our container, water molecules are constantly evaporating from liquid surfaces (such as lakes or cloud droplets), and other water vapor molecules are condensing. However, in nature, a balance is not always achieved. More often than not, more water molecules are leaving the surface of a water puddle than are arriving, causing what meteorologists call *net evaporation*. By contrast, during the formation of fog, more water molecules are condensing than are evaporating from the tiny fog droplets, resulting in *net condensation*.

What determines whether the rate of evaporation exceeds the rate of condensation (net evaporation) or vice versa? One of the major factors is the temperature of the water, which in turn determines how much motion (kinetic energy) the water molecules possess. At higher temperatures, the molecules have more energy and can more readily escape.

Vapor pressure is the other major factor that determines whether evaporation or condensation is the dominant process. Recall from our closed container example that vapor pressure influences the rate at which the water molecules leave (evaporate) and also the rate at which they return to the surface (condense). When the air is dry (low vapor pressure), the rate at which water molecules escape from a liquid surface is high. As the vapor pressure increases, the rate at which water vapor returns to the liquid phase increases as well.

## ✔ Concept Checks 4.3

1. How do absolute humidity and mixing ratio differ? What do they have in common?
2. Define *vapor pressure* and describe the relationship between vapor pressure and saturation. (*Hint:* See Figure 4.7.)
3. After reviewing Table 4.1, summarize the relationship between air temperature and the amount of water vapor needed to saturate air.

**students sometimes ask...**

**Why do snow piles seem to shrink a few days after a snowfall, even when the temperatures remain below freezing?**

On clear, cold days following a snowfall, the air can be very dry. This fact, plus solar heating, causes the ice crystals to sublimate—turn from a solid to a gas. Thus, even without any appreciable melting, these accumulations of snow gradually get smaller.

## eye ON THE atmosphere 4.1

Water is everywhere on Earth—in the oceans, glaciers, rivers, lakes, air, and living tissue. In addition, water can change from one state of matter to another at the temperatures and pressures experienced on Earth. Refer to this image, taken above the Grand Tetons, Wyoming, to answer the following questions.

**Questions**

1. What feature in this photo is composed of water in the liquid state?
2. Name the process by which ice changes directly from a solid to water vapor.
3. Identify where water vapor is found in this image.

# 4.4 Relative Humidity and Dew-Point Temperature

**List and describe the ways relative humidity changes in nature. Compare relative humidity to dew-point temperature.**

MM GEODe ▶ Moisture and Cloud Formation ▶ Humidity: Water Vapor in the Air

The most familiar and, unfortunately, the most misunderstood term used to describe the moisture content of air is relative humidity. **Relative humidity** *is a ratio of the air's actual water vapor content compared with the amount of water vapor required for saturation at that temperature (and pressure).* Thus, relative humidity indicates how near the air is to saturation rather than the actual quantity of water vapor in the air (see Box 4.1). Relative humidity can be determined as follows:

$$\text{Relative humidity (RH)} = \frac{\text{Mixing ratio (amount of water vapor in the air, g/kg)}}{\text{Saturation mixing ratio (g/kg)}} \times 100$$

To illustrate, we see from Table 4.1 that at 25°C, air is saturated when it contains 20 grams of water vapor per kilogram of air (*saturation mixing ratio*). Thus, if the air contains 10 grams per kilogram (*mixing ratio*) on a 25°C day, the relative humidity is expressed as 10/20, or 50 percent. Further, if air with a temperature of 25°C has a water-vapor content of 20 grams per kilogram, the relative humidity would be expressed as 20/20, or 100 percent. On occasions when the relative humidity reaches 100 percent, the air is said to be *saturated*.

## How Relative Humidity Changes

Because relative humidity is based on the air's water-vapor content, as well as the amount of moisture required for saturation, it can change in one of two ways. First, relative humidity changes when water vapor is added to or removed from the atmosphere. Second, because the amount of moisture required for saturation is a function of air temperature, relative humidity varies with temperature.

**How Changes in Moisture Affect Relative Humidity** Notice in Figure 4.9 that when water vapor is added to air through evaporation, the relative humidity of the air increases until saturation occurs (100 percent relative humidity). What if even more moisture is added to this parcel of saturated air? Does the relative humidity exceed 100 percent? Normally, this situation does not occur. Instead, the excess water vapor condenses to form liquid water.

You may have experienced such a situation while taking a hot shower. The water is composed of very energetic (hot) molecules, which means that the rate of evaporation is high. As long as you run the shower, the process of evaporation continually adds water vapor to the unsaturated air in the bathroom. If you stay in a hot shower long enough, the air eventually becomes saturated, and the excess water vapor begins to condense on the mirror, window, tile, and other surfaces in the room.

In nature, moisture is added to the air mainly via evaporation from the oceans. However, plants, soil, and smaller bodies of water also make substantial contributions. Unlike with your shower, however, the natural processes that add water vapor to the air generally do not operate at rates fast enough to cause saturation to occur directly. One exception is when you exhale on a cold winter day and "see your breath": The warm, moist air from your lungs mixes with the cold outside air. Your breath has enough moisture to saturate a small quantity of cold outside air, producing a miniature "cloud." Almost as fast as the "cloud" forms, it mixes with the surrounding dry air and evaporates.

### How Relative Humidity Changes with Temperature

The second condition that affects relative humidity is air temperature (see Box 4.2). Examine Figure 4.10A carefully and note that when air at 20°C contains 7 grams of water vapor per kilogram, it has a relative humidity of 50 percent. When the flask

▼ **Figure 4.9 At a constant temperature (in this example, it is 25°C), the relative humidity will increase as water vapor is added to the air** The saturation mixing ratio for air at 25°C is 20 g/kg (see Table 4.1). As the water-vapor content in the flask increases, the relative humidity rises from 25 percent in **A** to 100 percent in **C**.

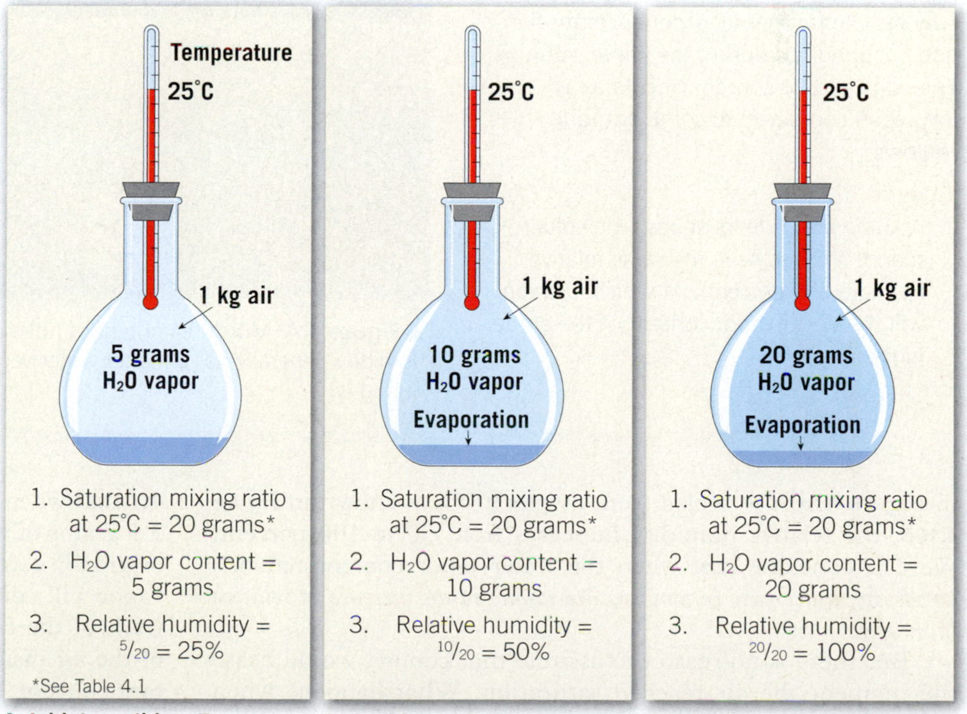


**A. Initial condition: 5 grams of water vapor**
**B. Addition of 5 grams of water vapor = 10 grams**
**C. Addition of 10 grams of water vapor = 20 grams**

## Box 4.1 Dry Air at 100 Percent Relative Humidity?

A common misconception is the notion that air with a higher relative humidity has greater water-vapor content than air with a lower relative humidity. This is not always the case (**Fig. 4.A**). Let us compare a typical January day in Chicago, Illinois, to one in the desert of Death Valley, California. On this hypothetical day, the temperature in Chicago is a cold –10°C (14°F), and the relative humidity is 100 percent. By referring to Table 4.1, we can see that saturated –10°C air has a water-vapor content (mixing ratio) of 2 grams per kilogram (g/kg). By contrast, the desert air at Death Valley on this January day is a warm 25°C (77°F), and the relative humidity is just 20 percent. A look at Table 4.1 reveals that 25°C air has a saturation mixing ratio of 20 g/kg. Therefore, with a relative humidity of 20 percent, the air at Death Valley has a water-vapor content of 4 g/kg (20 grams × 20 percent). Consequently, the "dry" air in Death Valley actually contains twice the water vapor as the air in Chicago, with a relative humidity of 100 percent.

This example illustrates that places that are very cold are also very dry. The low water-vapor content of frigid air (even when saturated) helps explain why many arctic areas receive only meager amounts of precipitation and are referred to as "polar deserts." This also helps us understand why people frequently experience dry skin and chapped lips during the winter months. The water-vapor content of cold air is low, even compared to some hot, arid regions.

**Question**

1. Assume that two locations, one colder than the other, have the same relative humidity (50 percent). At which location will the water-vapor content of the air be higher?

▲ **Figure 4.A** Moisture content of hot air versus frigid air. **A.** Hot desert air with a low relative humidity generally has a higher water-vapor content than **B.** frigid air with a high relative humidity.

in Figure 4.10A is cooled from 20° to 10°C, as shown in **Figure 4.10B**, the relative humidity increases from 50 to 100 percent. We can conclude that when the water-vapor content remains constant, *a decrease in temperature results in an increase in relative humidity*.

But there is no reason to assume that cooling would cease the moment the air reached saturation. What happens when the air is cooled below the temperature at which saturation occurs? **Figure 4.10C** illustrates this situation. Notice from Table 4.1 that when the flask is cooled to 0°C, the air is saturated, at 3.5 grams of water vapor per kilogram of air. Because this flask originally contained 7 grams of water vapor, 3.5 grams of water vapor will condense to form liquid droplets that collect on the walls of the container. In the meantime, the relative humidity of the air inside remains at 100 percent. This raises an important concept: When air aloft is cooled below its saturation level, some of the water vapor condenses to form clouds. Since clouds

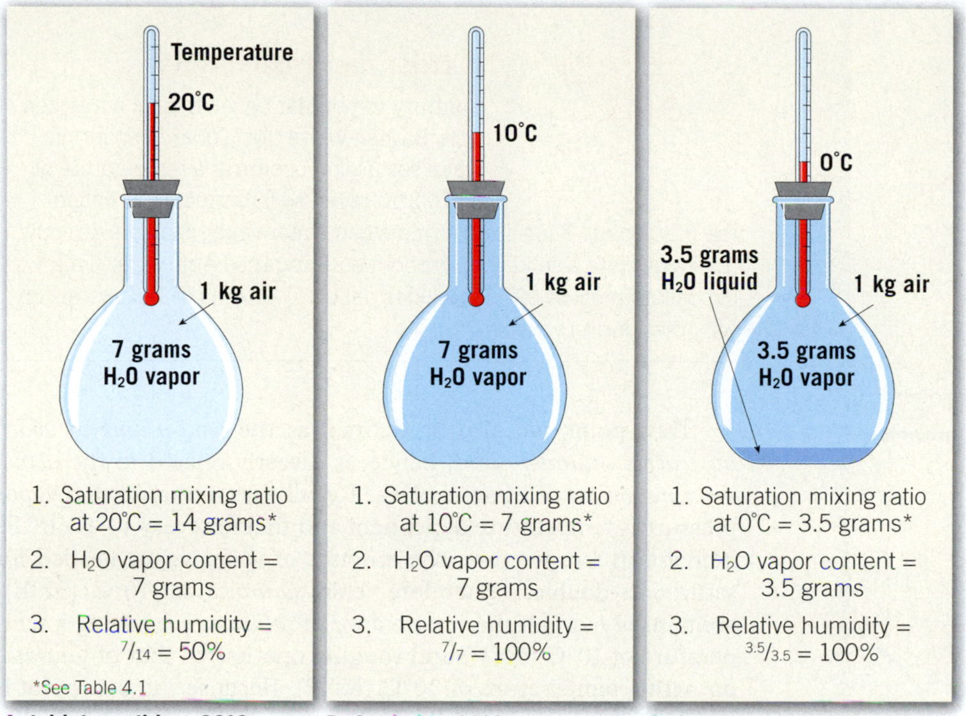


**A. Initial condition: 20°C** **B. Cooled to 10°C** **C. Cooled to 0°C**

▲ **Figure 4.10 Relative humidity varies with temperature** When the water-vapor content (mixing ratio) remains constant, the relative humidity will change when the air temperature either increases or decreases. In this example, when the temperature of the air in the flask was lowered from 20°C in **A** to 10°C in **B**, the relative humidity increased from 50 to 100 percent. Further cooling from 10°C in **B** to 0°C in **C** causes one-half of the water vapor to condense. In nature, when saturated air cools, it causes condensation in the form of clouds, dew, or fog.

are made of liquid droplets (or ice crystals), this moisture is no longer part of the water-vapor content of the air.

Conversely, *an increase in temperature results in a decrease in relative humidity*. For example, assume that the flask in Figure 4.10A containing 7 grams of water vapor is heated from 20°C to 35°C. Table 4.1 indicates that at 35°C, saturation occurs at 35 grams of water vapor per kilogram of air. Consequently, by heating the air from 20° to 35°C, the relative humidity will drop from 7/14 (50 percent) to 7/35 (20 percent).

## Natural Changes in Relative Humidity

In nature there are three major ways that air temperatures change (over relatively short time spans) to cause corresponding changes in relative humidity:

- Daily changes in temperatures (daylight versus nighttime temperatures)
- Temperature changes that result when air moves horizontally from one location to another
- Temperature changes caused when air moves vertically in the atmosphere

The effect of the first of these three processes (daily changes) is shown in Figure 4.11. Notice that during midafternoon, relative humidity reaches its lowest level, whereas the cooler evening hours are associated with higher relative humidity. In this example, the actual water-vapor content (mixing ratio) of the air remains unchanged; only the relative humidity varies. We will consider the other two processes in more detail in later chapters.

In summary, relative humidity indicates how near the air is to being saturated, whereas the air's mixing ratio denotes the actual quantity of water vapor contained in that air. When

## Box 4.2 Humidifiers and Dehumidifiers

In summer, stores sell *dehumidifiers*. As winter rolls around, these same retailers feature *humidifiers*. Why are so many homes equipped with both appliances? The answer lies in the relationship between temperature and relative humidity. Recall that if the water-vapor content of air remains at a constant level, an increase in temperature *lowers the relative humidity* and a decrease in temperature *increases relative humidity.*

During the summer months, warm, moist air frequently dominates the weather of the central and eastern United States. When hot, humid air enters a home, some of it circulates into the cool basement, which causes an increase in relative humidity of the air. The result is a damp, musty-smelling basement. The homeowner may install a dehumidifier to alleviate the problem. As air is drawn over the cold coils of the dehumidifier, water vapor condenses and collects in a bucket or flows down a drain. This process reduces the moisture content of the air and makes for a drier, more comfortable basement.

By contrast, during the winter months, outside air is cool and dry. When this air is drawn into the home, it is heated to room temperature. This process causes the relative humidity to plunge, often to uncomfortable levels of 25 percent or lower. Living with dry air can mean static electrical shocks, dry skin, sinus headaches, and even nosebleeds. Consequently, the homeowner may install a humidifier, which adds water vapor to the air and increases the relative humidity to a more comfortable level.

**Question**

1. During what season of the year do homeowners usually run dehumidifiers? Explain.

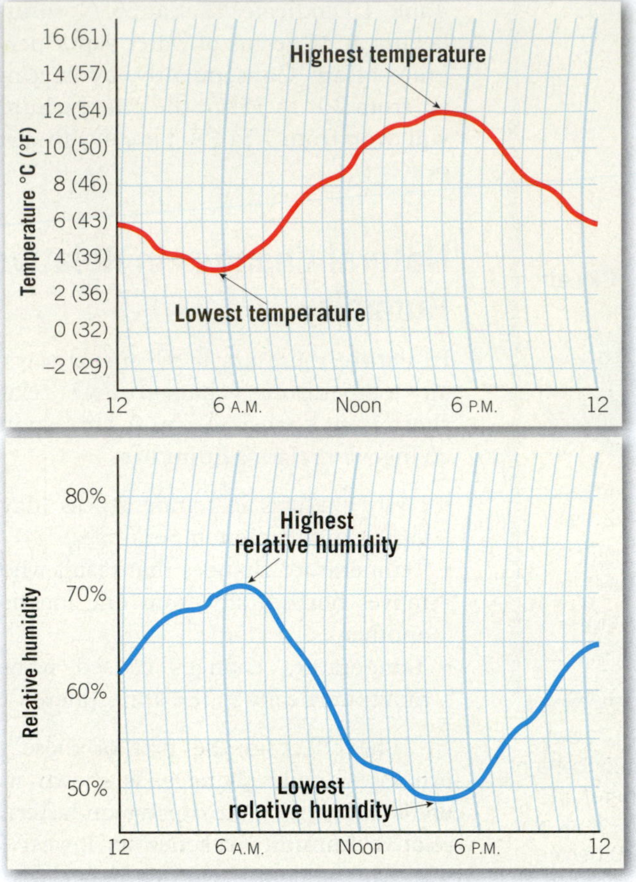


▲ **Figure 4.11 Typical daily variation in temperature and relative humidity during a spring day in Washington, DC**

water-vapor content remains constant, relative humidity drops with an increase in air temperature and, conversely, rises with a decrease in air temperature.

## Dew-Point Temperature

The **dew-point temperature,** or simply the **dew point,** is *the temperature at which water vapor begins to condense*. The term *dew point* stems from the fact that during nighttime hours, objects near the ground often cool below the dew-point temperature and become coated with dew. You have undoubtedly seen "dew" form on an ice-cold drink on a humid summer day (**Fig. 4.12**). In nature, cooling air below its dew-point temperature typically generates dew, fog, or clouds when the dew point is above freezing and frost when it is below freezing (0°C, 32°F).

students sometimes ask...

### Why do my lips get chapped in the winter?

During the winter months, outside air is comparatively cool and dry. When this air is drawn into a home, it is heated, which causes the relative humidity to plunge. Unless your home is equipped with a humidifier, you are likely to experience chapped lips and dry skin at that time of year.

students sometimes ask...

### Is frost just frozen dew?

Contrary to popular belief, frost is not frozen dew. Rather, white frost (*hoar frost*) forms when saturation occurs at a temperature of 0°C (32°F) or below (a temperature called the *frost point*). Thus, frost forms when water vapor changes directly from a gas into a solid (ice), without entering the liquid state. This deposition process often produces delicate patterns of ice crystals on windows during winter.

Dew point can also be defined as the *temperature at which air reaches saturation* and, hence, is directly related to the *actual moisture content* of a parcel of air. Recall that the saturation vapor pressure is temperature dependent and that for every 10°C (18°F) increase in temperature, the amount of water vapor needed for saturation doubles. Therefore, cold *saturated air* (0°C [32°F]) contains about half the water vapor of *saturated air* having a temperature of 10°C (50°F) and roughly one-fourth that of *saturated air* with a temperature of 20°C (68°F). Because the dew point is the temperature at which saturation occurs, we can conclude that high dew-point temperatures equate to moist air and, conversely, low dew-point temperatures indicate dry air (**Table 4.2**). More precisely, based on what we have learned about vapor pressure and saturation, we can state that for every 10°C (18°F) increase in the dew-point temperature, air contains about twice as much water vapor. Therefore, we know that when the dew-point temperature is 25°C (77°F), air contains about twice the water vapor as when the dew point is 15°C (59°F) and four times that of air with a dew point of 5°C (41°F).

Because the dew-point temperature is a good measure of the amount of water vapor in the air, it commonly appears on weather maps. When the dew point exceeds 65°F (18°C), most people

▼ **Figure 4.12 Condensation and dew-point temperature** Condensation, or "dew," occurs when a cold drinking glass chills the surrounding layer of air below the dew-point temperature.

**Table 4.2** Dew-Point Thresholds

| Dew-Point Temperature | Threshold |
|---|---|
| ≤ 10°F | Significant snowfall is inhibited |
| ≥ 55°F | Minimum for severe thunderstorms to form |
| ≥ 65°F | Considered humid by most people |
| ≥ 70°F | Typical of the rainy tropics |
| ≥ 75°F | Considered oppressive by most |

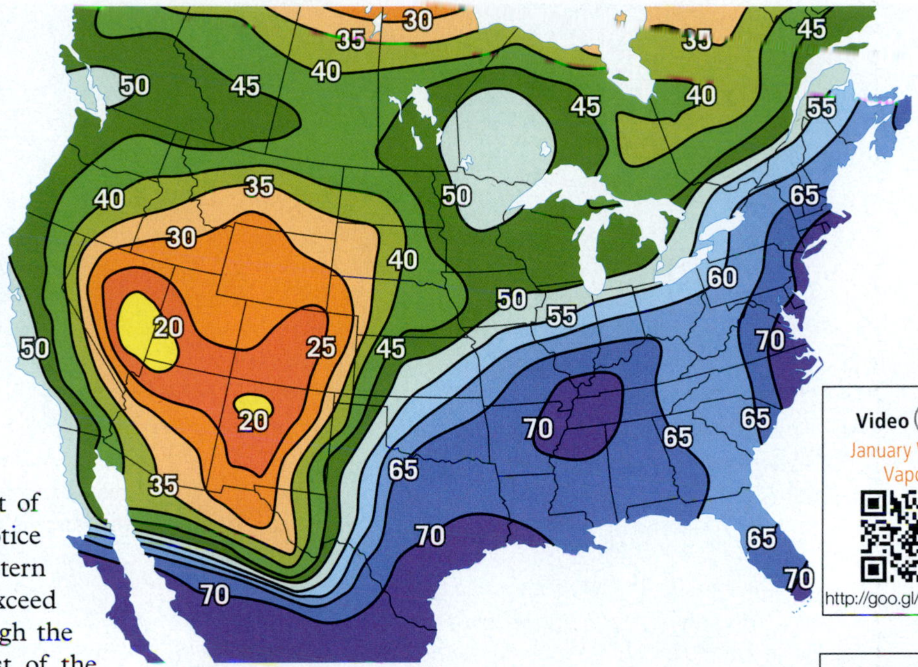


▲ **SmartFigure 4.13 Surface map showing dew-point temperatures for typical September day** Dew-point temperatures above 60°F dominate the southeastern United States, indicating that this region is blanketed with humid air.

http://goo.gl/fNjH8

Video MM January Water Vapor http://goo.gl/QzzGWg

Video MM July Water Vapor http://goo.gl/gUHFiO

consider the air to feel humid; air with a dew point of 75°F (24°C) or higher is considered oppressive. Notice on the map in **Figure 4.13** that much of the southeastern United States has dew-point temperatures that exceed 65°F (18°C). Also notice in Figure 4.13 that although the Southeast is dominated by humid conditions, most of the remainder of the country is experiencing comparatively drier air.

## How Is Humidity Measured?

MapMaster MM North America Physical Environment: Average Distribution of Dew Point January and July

Instruments called **hygrometers** are used to measure the moisture content of the air. In addition to being used in meteorology, hygrometers are used in greenhouses, humidors, museums, and numerous industrial settings that are sensitive to humidity, such as paint booths where protective coatings are applied to products. Because it is difficult to measure absolute humidity and the mixing ratio directly, most hygrometers measure either relative humidity or dew-point temperature. Once either of these is known, it is relatively easy to convert to any of the other humidity measurements.

**Psychrometers** One of the simplest hygrometers, a **psychrometer** (called a *sling psychrometer* when connected to a handle and spun) consists of two identical thermometers mounted side by side (**Fig. 4.14A**). One thermometer, called the *dry bulb*, measures air temperature, and the other, called the *wet bulb*, has a thin cloth wick tied at the bottom. This cloth wick is saturated with water, and a continuous current of air is passed over the wick, either by swinging the psychrometer or by using an electric fan to move air past the instrument

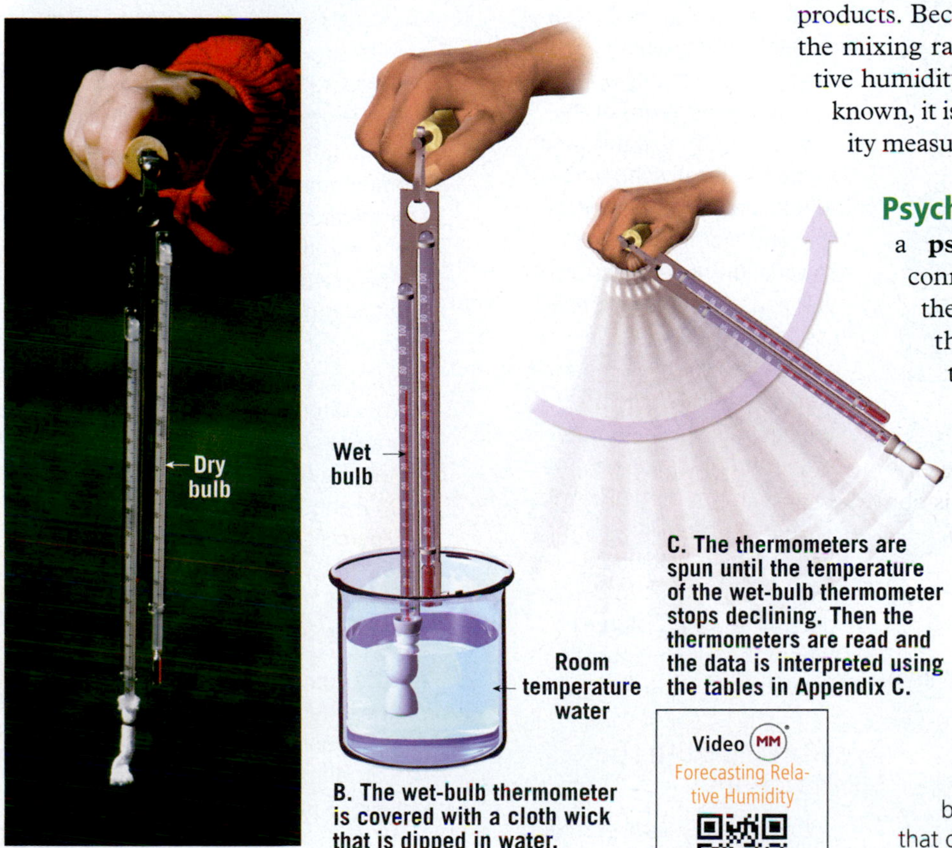


**A. The dry-bulb thermometer gives the current air temperature.**

**B. The wet-bulb thermometer is covered with a cloth wick that is dipped in water.**

**C. The thermometers are spun until the temperature of the wet-bulb thermometer stops declining. Then the thermometers are read and the data is interpreted using the tables in Appendix C.**

Video MM Forecasting Relative Humidity http://goo.gl/0zww9F

◀ **Figure 4.14 Sling psychrometer A.** A sling psychrometer consists of one dry-bulb and one wet-bulb thermometers. **B.** The dry-bulb thermometer measures the current air temperature. The wet-bulb thermometer is covered with a cloth wick dipped in water. **C.** As the instrument is spun, evaporation cooling causes the temperature of the wet-bulb thermometer to decrease. The amount of cooling that occurs is directly proportional to the dryness of the air. The temperature difference between the dry- and wet-bulb thermometers is used in conjunction with the tables in Appendix C to determine relative humidity and dew-point temperature.

# What's Your Forecast?

## Humidity and Human Comfort

by Dr. Redina L. Herman, Western Illinois University

If the air conditioning in your car, house, or apartment has ever malfunctioned, you know the result is an uncomfortable environment. The air conditioner makes your surroundings more comfortable because it cools the air *and* reduces humidity. Humidity, the amount of moisture in the air, is important to meteorologists and nonmeteorologists alike. To a meteorologist, humidity is a controlling factor for the weather we experience. Humidity in the atmosphere influences when it is likely to rain and whether the atmosphere will become unstable enough to produce a thunderstorm. For all of us, humidity determines our level of comfort and how healthy our environment is (**Fig 4.B**). A humid day in the middle of summer makes you want to stay in your air-conditioned house. High humidity can also lead to the growth of mold and dust mites, which impact human health. On the other hand, if the humidity is too low, your skin will feel uncomfortably dry and may possibly crack. In cold climates, people often add moisture to the air in their homes during the winter to feel more comfortable (see Box 4.2).

▲ **Figure 4.B** Working outdoors in the summer months can be humid and uncomfortable, because warm air can hold more water vapor. Heat stress can occur when temperature and humidity levels increase.

### Measuring Humidity

Knowing the amount of moisture in the air is very important, but it can be difficult to measure accurately. Humidity is measured with reference to another quantity—like mass, pressure, temperature, or electrical resistance—so you have to know at least two independent bits of information to determine moisture. In fact, most hygrometers sense relative humidity (how close the air is to saturation) instead of absolute humidity or mixing ratio (expressions of the actual amount of water vapor in the air). Relative humidity is a function of both temperature and moisture content, so variations in temperature will lead to variations in relative humidity. For example, a chilled-mirror hygrometer works by chilling a mirror to the dew-point temperature of the surrounding air so that fog forms on the mirror. Therefore, the chilled-mirror hygrometer is actually sensing relative humidity. In order to determine the moisture content of the air, you must also know its temperature.

### Working with the Numbers

A common way to determine relative humidity and dew-point temperature is to use a sling psychrometer (Fig. 4.14). The temperature of the wet bulb is compared to the temperature of the dry bulb to determine moisture. Referring to Appendix C, if the dry-bulb (air) temperature is 70°F and the wet-bulb temperature is 57°F, then the depression of wet-bulb temperature is 13°F. According to Table C.3 (page A-12), the relative humidity is 44 percent. According to Table C.4 (page A-12), the dew-point temperature is 47°F. The heat index chart (Fig. 3.30) does not include temperatures as low as 68°F, which indicates no ill effects due to relative humidity. In fact, this would be a pretty comfortable day!

**Questions**

1. Use the tables in Appendix C to determine the approximate value for the missing pieces of information.
   a. If the dry-bulb (air) temperature is 85°F and the wet-bulb temperature is 81°F, find the depression of wet-bulb temperature, the dew-point temperature, and the relative humidity.
   b. If the dry-bulb (air) temperature is 75°F and the depression of wet-bulb temperature is 20°F, find the wet-bulb temperature, the dew point temperature, and the relative humidity.
   c. If the dry-bulb (air) temperature is 90°F and the relative humidity is 74 percent, find the wet-bulb temperature, the depression of wet-bulb temperature, and the dew point temperature.
   d. If the dry-bulb (air) temperature is 100°F and the relative humidity is 65 percent, find the wet-bulb temperature, the depression of wet-bulb temperature, and the dew point temperature.
2. Based on your answers in Question 1, rank the letters in terms of how comfortable you would be on a day with the given conditions. Refer to the heat index chart (Fig. 3.30) in Chapter 3 to help you decide.

(Fig. 4.14B,C). As a result, water evaporates from the wick, absorbing heat energy from the wet-bulb thermometer, which causes its temperature to drop. The amount of cooling that takes place is directly proportional to the dryness of the air: The drier the air, the greater the cooling. Therefore, the larger the difference between the wet- and dry-bulb temperatures, the lower the relative humidity. By contrast, if the air is saturated, no evaporation will occur, and the two thermometers will have identical readings. By using a psychrometer and the tables provided in Appendix C, the relative humidity and the dew-point temperature can be easily determined.

**Hair Hygrometers** One of the oldest instruments used for measuring relative humidity, called a *hair hygrometer*, operates on the principle that hair changes length in proportion to changes in relative humidity. Hair lengthens as relative humidity increases and shrinks as relative humidity drops. People with naturally curly hair experience this phenomenon: In humid weather their hair lengthens and hence becomes curlier. A hair hygrometer uses a bundle of hairs linked mechanically to an indicator that is calibrated between 0 and 100 percent. However, these instruments have become largely obsolete due to the development of more accurate tools.

**Electric Hygrometers** Today, a variety of *electric hygrometers* are widely used to measure humidity. One type of electric hygrometer uses a chilled mirror and a mechanism that detects the temperature at which condensation begins to form on the mirror. Thus, a *chilled mirror hydrometer* measures the dew-point temperature of the air.

The Automated Weather Observing System (AWOS) operated by the National Weather Service (NWS) employs an electric hygrometer that works on the principle of *capacitance*—a material's ability to store an electrical charge. The sensor consists of a thin hygroscopic (water-absorbent) film that is connected to an electric current. As the film absorbs or releases water the capacitance of the sensor changes at rate proportional to the relative humidity of the surrounding air. Thus, relative humidity can be measured by monitoring the change in the film's capacitance. Higher capacitance equates to higher relative humidity.

### ✔ Concept Checks 4.4

1. How is relative humidity different from absolute humidity and the mixing ratio?
2. Refer to Figure 4.11 to answer the following questions:
   - **a.** When is relative humidity highest during a typical day? When is it lowest?
   - **b.** At what time of day would dew most likely form?
   - **c.** Write a generalization relating changes in air temperature to changes in relative humidity.
3. If the temperature remains unchanged, and if the mixing ratio decreases, how will relative humidity change?
4. Define *dew-point temperature*.
5. Which measure of humidity, relative humidity or dew point, best describes the actual quantity of water vapor in a mass of air?
6. Briefly describe the principle of a psychrometer.

## 4.5 Adiabatic Temperature Changes and Cloud Formation

**Describe adiabatic temperature changes and explain why the wet adiabatic rate of cooling is less than the dry adiabatic rate.**

MM® **GEODe** ▶ Moisture and Cloud Formation ▶ The Basics of Cloud Formation: Adiabatic Cooling

Recall that condensation occurs when sufficient water vapor is added to the air or, more commonly, when the air is cooled to its dew-point temperature. Condensation may produce dew, fog, or clouds. Heat near Earth's surface is readily exchanged between the ground and the air directly above. As the ground loses heat in the evening (radiation cooling), dew may condense on the grass, while fog may form slightly above Earth's surface. Thus, surface cooling that occurs after sunset produces some condensation. Cloud formation, however, often takes place during the warmest part of the day—an indication that another mechanism must operate aloft that cools air sufficiently to generate clouds.

The process that generates most clouds is easily visualized. Have you ever pumped up a bicycle tire with a hand pump and noticed that the pump barrel became very warm? When you applied energy to *compress* the air, the motion of the gas molecules increased, and the temperature of the air rose. Conversely, if you allow air to escape from a bicycle tire, the air *expands*; the gas molecules move less rapidly, and the air cools. You have probably felt the cooling effect of the propellant gas expanding as you applied hair spray or spray deodorant. The temperature changes just described, in which heat energy was neither added nor subtracted, are called **adiabatic temperature changes**. Instead, changes in pressure result in temperature changes. When air is compressed, it warms, and when air is allowed to expand, it cools.

### Adiabatic Cooling and Condensation

To simplify the discussion of adiabatic cooling, imagine a volume of air enclosed in a thin balloon-like bubble. Meteorologists call this imaginary volume of air a **parcel**. Typically, we consider a parcel to be a few hundred cubic meters in volume, and we assume that it acts independently of the surrounding air. It is also assumed that no heat is transferred into or out of the parcel. Although this image is highly idealized, over short time spans, a parcel of air behaves much like an actual volume of air moving up or down in the atmosphere. In nature, sometimes

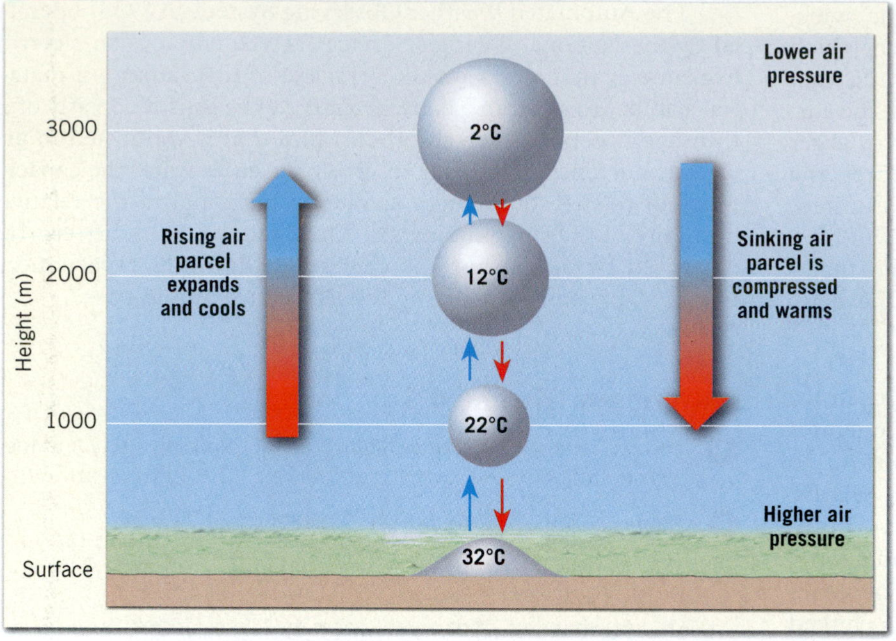


▲ **Figure 4.15 Dry adiabatic rate of cooling and heating** Whenever an unsaturated parcel of air is lifted, it expands and cools at the *dry adiabatic rate* of 10°C per 1000 meters. Conversely, when air sinks, it is compressed and heats at the same rate.

the surrounding air infiltrates a rising or descending column of air, a process called **entrainment**. For the following discussion, however, we assume that no mixing of this type occurs.

Recall from Chapter 1 that atmospheric pressure decreased with height. Any time a parcel of air moves upward, it passes through regions of successively lower pressure. As a result, ascending air expands and cools adiabatically. Unsaturated air cools at a constant rate of 10°C for every 1000 meters of ascent (5.5°F per 1000 feet). Conversely, descending air comes under increasing pressure and is compressed and heated 10°C for every 1000 meters of descent (**Fig. 4.15**). This rate of cooling or heating applies only to *unsaturated air* and is known as the **dry adiabatic rate** ("dry" because the air is unsaturated).

If an air parcel rises high enough, it will eventually cool to its dew point and trigger the process of condensation. The altitude at which a parcel reaches saturation and cloud formation begins is called the **lifting condensation level**, or simply **condensation level**. At the lifting condensation level, an important change occurs: The *latent heat* that was absorbed by the water vapor when it evaporated is released as *sensible heat*—energy that can be measured with a thermometer. Although the parcel will continue to cool adiabatically, the release of latent heat slows the rate of cooling. In other words, when a parcel of air ascends above the lifting condensation level, the rate at which it cools is reduced. This slower rate of cooling is called the **wet adiabatic rate** (also commonly refered to as the moist or saturated adiabatic rate).

Because the amount of latent heat released depends on the quantity of moisture present in the air (generally between 0 and 4 percent), the wet adiabatic rate varies from 5°C per 1000 meters for air with a high moisture content to 9°C per 1000 meters for air with a low moisture content. **Figure 4.16** illustrates the role of adiabatic cooling in the formation of clouds.

To summarize, rising air cools at the dry adiabatic rate from the surface up to the lifting condensation level, at which point it cools at the wet adiabatic rate.

## ✔ Concept Checks 4.5

1. What name is given to the processes whereby the temperature of air changes without the addition or subtraction of energy?
2. Why does air expand as it moves upward through the atmosphere?
3. At what rate does unsaturated air cool when it rises through the atmosphere?
4. Why does the adiabatic rate of cooling change when condensation begins?
5. Why is the wet adiabatic rate not a constant figure?

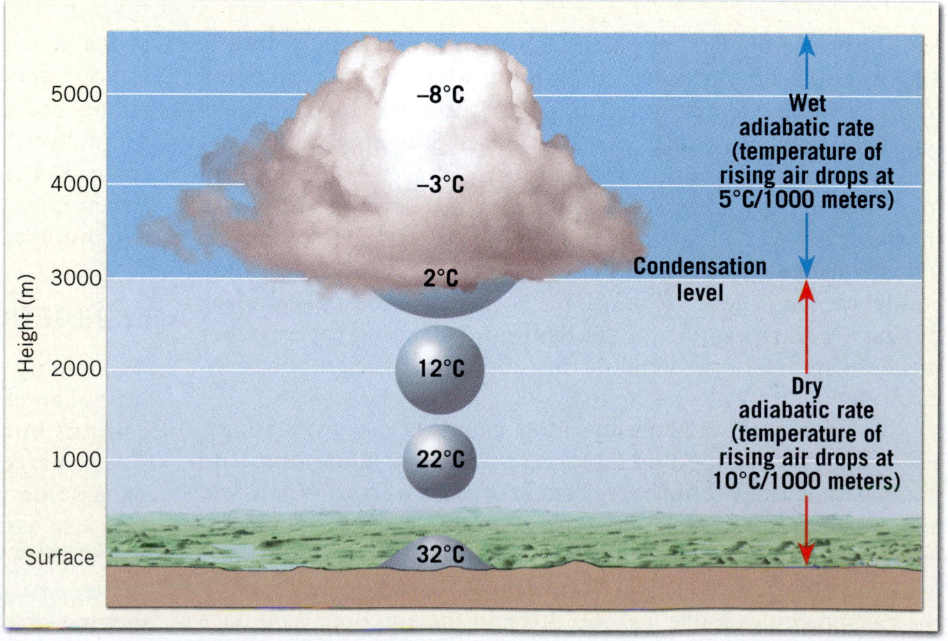


► **Figure 4.16 Lifting condensation level and the wet adiabatic rate** Rising air expands and cools at the dry adiabatic rate of 10°C per 1000 meters until the air reaches the dew point and condensation (cloud formation) begins. As air continues to rise, the latent heat released by condensation reduces the rate of cooling. The wet adiabatic rate is therefore always less than the dry adiabatic rate.

# 4.6 Processes That Lift Air

**List and describe four mechanisms that cause air to rise.**

MM GEODe ▶ Moisture and Cloud Formation ▶ Processes That Lift Air

Why does air rise on some occasions to produce clouds, but not on others? Generally, the tendency is for air to resist vertical movement; air near the surface "wants" to stay near the surface, and air aloft tends to remain aloft. However, the following four processes cause air to rise, thereby generating clouds:

1. *Orographic lifting*, in which air is forced to rise over a mountainous barrier
2. *Frontal lifting*, in which warmer, less dense air is forced over cooler, denser air
3. *Convergence*, which is a pileup of horizontal airflow that results in upward movement
4. *Localized convective lifting*, in which unequal surface heating causes localized pockets of air to rise because of their buoyancy

## Orographic Lifting

**Orographic lifting** occurs when elevated terrains, such as mountains, act as barriers to the flow of air (Fig. 4.17). As air ascends a mountain slope, adiabatic cooling often generates clouds and copious precipitation. In fact, many of the rainiest places in the world are located on windward mountain slopes (Box 4.3).

By the time air reaches the leeward side of a mountain, much of its moisture has been lost. If the air descends, it warms adiabatically, making condensation and precipitation even less likely. As shown in Figure 4.17, the result can be a **rain shadow desert** (Box 4.4, page 114). The Great Basin Desert of the western United States lies only a few hundred kilometers from the Pacific Ocean, but it is effectively cut off from the ocean's moisture by the imposing Sierra Nevada (Figure 4.17). The Gobi Desert of Mongolia, the Takla Makan of China, and the Patagonia Desert of Argentina are other examples of deserts that exist because they are on the leeward sides of large mountain systems.

## Frontal Lifting

If orographic lifting were the only mechanism that forced air aloft, the relatively flat central portion of North America would be an expansive desert rather than the area known as "the nation's breadbasket." Fortunately, this is not the case.

In central North America, warm and cold air masses often collide, producing boundaries called **fronts**. Rather than mixing, the cooler, denser air acts as a barrier over which the warmer, less dense air rises. This process, called **frontal lifting**, also referred to as **frontal wedging**, is illustrated in Figure 4.18.

It should be noted that weather-producing fronts are associated with storm systems called *midlatitude cyclones*. Because these storms are responsible for producing a high percentage of the precipitation in the middle latitudes, we will examine them in detail in Chapter 9.

## Convergence

When the wind pattern near Earth's surface is such that more air is entering an area than is leaving—a phenomenon called **convergence**—lifting occurs

**A. Orographic lifting leads to precipitation on windward slopes.**

**B. By the time air reaches the leeward side of the mountains, much of the moisture has been lost, resulting in a *rain shadow desert*.**

◀ **Figure 4.17 Orographic lifting and precipitation** **A.** Orographic lifting leads to precipitation on windward slopes of a topographic barrier, such as a mountain. **B.** By the time air reaches the leeward side of the mountains, much of the moisture has been lost. The Great Basin desert is a rain shadow desert that covers nearly all of Nevada and portions of adjacent states.

## Box 4.3 Precipitation Records and Mountainous Terrain

Many of the rainiest places in the world are located on windward mountain slopes. Typically, these areas are rainy because the mountains act as barriers to Earth's natural circulation. The prevailing winds are forced to ascend the sloping terrain, thereby generating clouds and often abundant precipitation. Mount Waialeale, Hawaii, for example, records the highest average annual rainfall in the world, some 1234 centimeters (486 inches). The station is located on the windward (northeastern) coast of the island of Kauai, at an elevation of 1569 meters (5148 feet). By contrast, only 31 kilometers (19 miles) away lies sunny Barking Sands, with annual precipitation averaging less than 50 centimeters (20 inches).

The largest recorded rainfall for a 12-month period occurred at Cherrapunji, India, where an astounding 2647 centimeters (1042 inches), over 86 feet, fell from August 1860 through July 1861. Most of this rainfall occurred in the summer; that July, a record 930 centimeters (366 inches) fell. By comparison, 10 times more rain fell in a *month* at Cherrapunji, India, than falls in Chicago in an average *year*. Cherrapunji's location just north of the Bay of Bengal and its elevation of 1293 meters (4309 feet) makes it an ideal location to receive the full effect of India's wet summer monsoons.

Because mountains receive abundant precipitation, they are typically very important sources of water, especially for many dry locations in the western United States. The snow pack that accumulates high in the mountains during the winter is a major source of water for the summer season, when precipitation is light and demand for water is great (**Fig. 4.C**). The record for greatest annual snowfall in the United States goes to the Mount Baker ski area north of Seattle, Washington, where 2896 centimeters (1140 inches) of snow fell during the winter of 1998–1999.

▲ **Figure 4.C** This heavy snowpack is at Gotthard Pass in the Swiss Alps.

In addition to providing lift, mountains remove additional moisture in other ways. By slowing the horizontal flow of air, they cause convergence and impede the passage of storm systems. Moreover, the irregular topography of mountains enhances the differential heating that causes some localized convective lifting. These combined effects account for the generally higher precipitation associated with mountainous regions compared to surrounding lowlands.

### Questions

1. What location has the record for the highest average annual rainfall in the world?
2. Why do mountainous areas usually receive abundant precipitation?

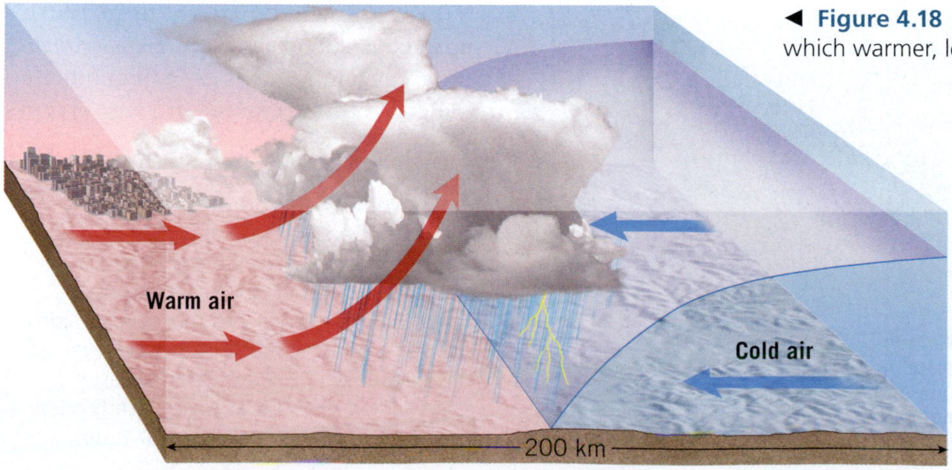


◀ **Figure 4.18 Frontal lifting** Colder, denser air acts as a barrier over which warmer, less dense air rises.

(**Fig. 4.19**). Convergence as a mechanism of lifting is most often associated with large centers of *low pressure*, mainly midlatitude cyclones and hurricanes. The inward flow of air at the surface of these systems is balanced by rising air, cloud formation, and usually precipitation.

Convergence can also occur when an obstacle slows or restricts horizontal airflow (wind). For example, when air moves from a relatively smooth surface, such as the ocean, onto an irregular landscape, increased friction

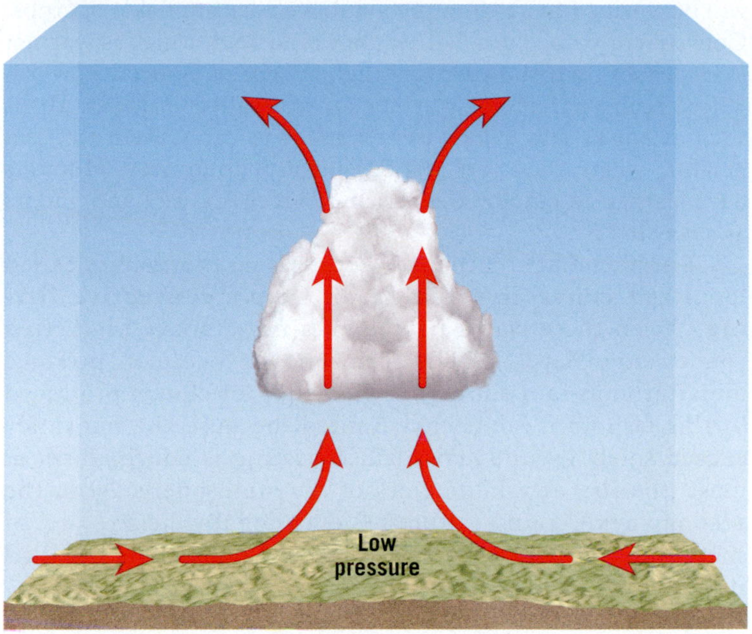


▲ **Figure 4.19 Convergence at the surface causes air to rise** When the wind pattern near Earth's surface is such that more air is entering an area than is leaving—a phenomenon called convergence—lifting occurs.

reduces its speed. The result is a pileup of air (convergence). When air converges, there is an upward flow of air molecules rather than a simple squeezing together of molecules (as happens with people are entering a crowded building).

The Florida peninsula provides an excellent example of the role that convergence can play in initiating cloud development and precipitation. On warm days, the airflow is from the ocean to the land along both coasts of Florida. This leads to a pileup of air along the coasts and general convergence over the peninsula. This pattern of convergence and uplift is aided by intense solar heating of the land. As a result, Florida's peninsula experiences the greatest frequency of midafternoon thunderstorms in the United States (**Fig. 4.20**).

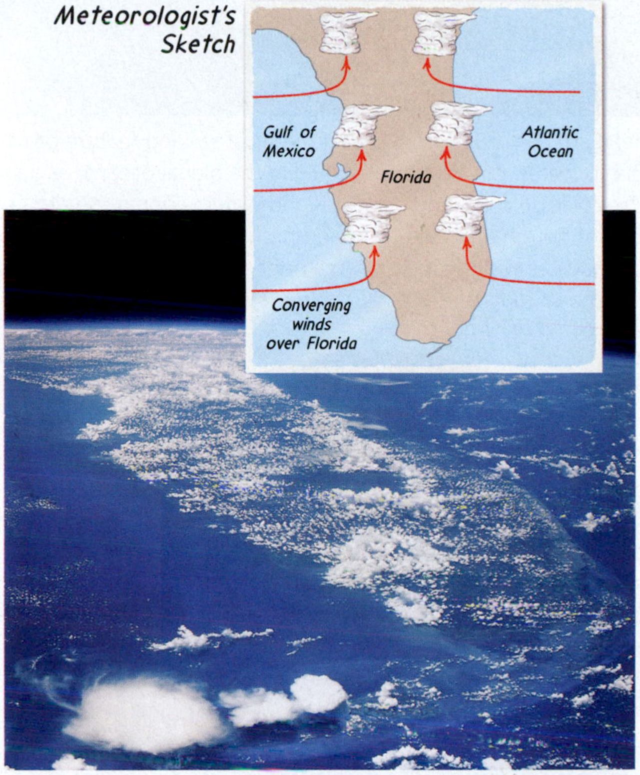


MM http://goo.gl/Mm2y6

▲ **SmartFigure 4.20 Convergence over the Florida peninsula** When surface air converges, the column of air increases in height to allow for the decreased area it occupies. Florida provides a good example. On warm days, airflow from the Atlantic Ocean and Gulf of Mexico onto the Florida peninsula generates many midafternoon thunderstorms.

## Localized Convective Lifting

On warm summer days, unequal heating of Earth's surface may cause pockets of air to be warmed more than the surrounding air (**Fig. 4.21**). For instance, air above a plowed field

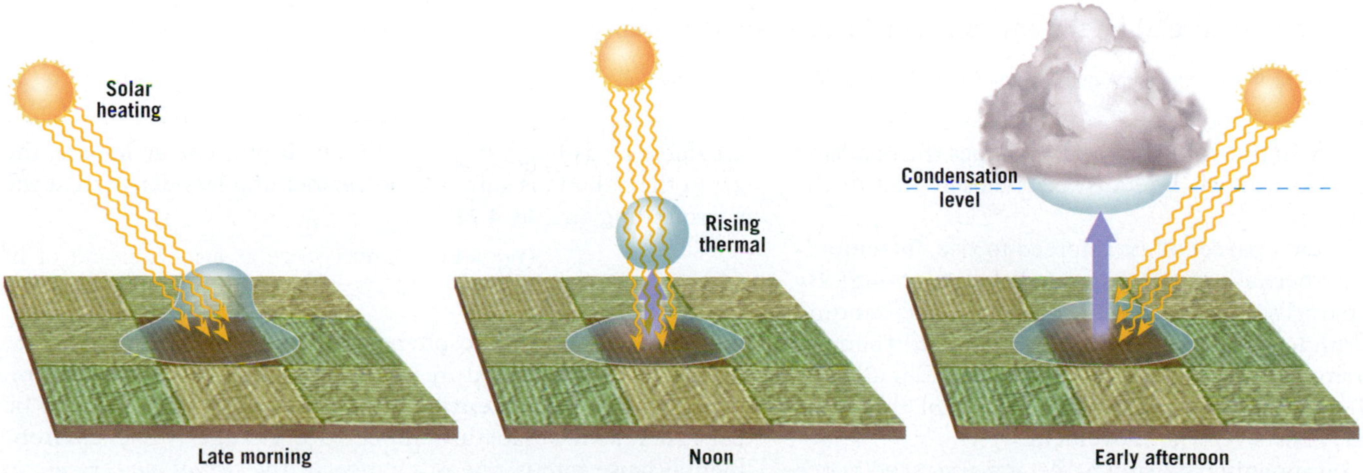


▲ **Figure 4.21 Localized convective lifting** Unequal heating of Earth's surface causes pockets of air to be warmed more than the surrounding air. These buoyant parcels of hot air rise, producing thermals, and if they reach the condensation level, clouds form.

## eye ON THE atmosphere 4.2

When Earth is viewed from space, the most striking feature of the planet is *water*. It is found as a liquid in the global oceans, as a solid in the polar ice caps, and as clouds and water vapor in the atmosphere. Although only one-thousandth of 1 percent of the water on Earth exists as water vapor, it has a huge influence on our planet's weather and climate.

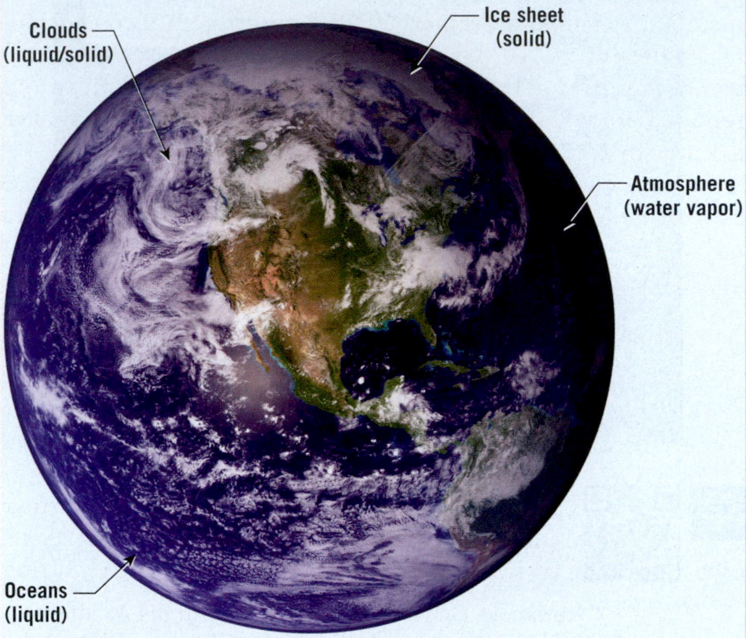


**Question**

1. What role does water vapor play in heating Earth's surface?
2. How does water vapor act to transfer heat from Earth's land–sea surface to the atmosphere?

will be warmed more than the air above adjacent fields of crops. Consequently, the parcel of air above the field, which is warmer (less dense) than the surrounding air, will be buoyed upward. These rising parcels of warmer air are called *thermals*. Birds such as hawks and eagles use thermals to carry them to great heights, where they can identify unsuspecting prey. Humans have learned to employ these rising parcels to use hang gliders as a way to "fly."

The phenomenon that produces rising thermals is called **localized convective lifting**, or simply **convective lifting**. When these warm parcels of air rise above the lifting condensation level, clouds form and on occasion produce midafternoon rain showers. The height of clouds produced in this fashion is somewhat limited, because the buoyancy caused solely by unequal surface heating is confined to, at most, the first few kilometers of the atmosphere. Also, the accompanying rains, although occasionally heavy, are of short duration and widely scattered, a phenomenon called *sun showers*.

### ✔ Concept Checks 4.6

1. Explain why the Great Basin of the western United States is dry. What term is applied to these locations?
2. How does frontal lifting cause air to rise?
3. Define *convergence*. Identify two weather systems associated with convergence in the lower atmosphere.
4. Why does Florida have abundant midafternoon thunderstorms?
5. Describe convective lifting.

Video MM
Gravity Wave Clouds
http://goo.gl/ykZrY5

## 4.7 The Critical Weathermaker: Atmospheric Stability

**Write a statement relating the environmental lapse rate to stability.**

MM GEODe ▸ Moisture and Cloud Formation ▸ The Critical Weathermaker: Atmospheric Stability

Why do clouds vary so much in size, and why does the resulting precipitation vary so widely? The answer is closely tied to the *stability* of the air.

Recall that when a parcel of air is forced to rise, its temperature will decrease because of expansion (adiabatic cooling). By comparing the parcel's temperature to that of the surrounding air, we can determine its stability. If the parcel is *cooler* than the surrounding environment, it will be more dense; and if allowed to do so, it will sink to its original position. Air of this type, called **stable air**, resists vertical movement.

If, however, our imaginary rising parcel is *warmer* and hence less dense than the surrounding air, it will continue to rise until it reaches an altitude where its temperature equals that of its surroundings. This type of air is classified as **unstable air**. Unstable air is like a hot-air balloon: It will rise as long as the air in the balloon is sufficiently warmer and less dense than the surrounding air (Fig. 4.22).

### Types of Stability

The stability of the atmosphere is determined by regularly measuring the air temperature at various heights. This measure, called the **environmental lapse rate (ELR)**, should not be confused with adiabatic temperature changes. The environmental lapse rate is rate of change in the *actual temperature* of the atmosphere with height, as determined from observations made by radiosondes and aircraft. (Recall that a radiosonde is an instrument package that is attached to a balloon and

▲ **Figure 4.22 Hot air rises** As long as air is warmer than its surroundings, it will rise. Hot-air balloons rise up through the atmosphere for this reason.

transmits data by radio as it ascends though the atmosphere.) Adiabatic temperature changes, on the other hand, are the *changes in temperature* that a parcel of air experiences as it moves vertically through the atmosphere.

**Figure 4.23** illustrates how the stability of the atmosphere is determined, where the air at 1000 meters above the surface is 5°C cooler than the air at the surface, whereas the air at 2000 meters is 10°C cooler, and so forth. Thus, the prevailing environmental lapse rate is 5°C per 1000 meters. The air at the surface appears to be less dense than the air at 1000 meters because it is 5°C warmer. However, if the surface air were forced to rise to 1000 meters, it would expand and cool at the dry adiabatic rate of 10°C per 1000 meters. Therefore, on reaching 1000 meters, the temperature of the rising parcel would have dropped from 25°C to 15°C. Because the rising air would be 5°C cooler than its environment, it would be more dense and, if allowed to do so, it would sink to its original position. Thus, we say that the air near the surface is *potentially cooler* than the air aloft, and therefore, it will not rise unless forced to do so. The air just described is referred to as *stable* and resists vertical movement.

We will now look at three fundamental conditions of the atmosphere: absolute stability, absolute instability, and conditional instability.

**Absolute Stability** Stated quantitatively, **absolute stability** prevails when *the environmental lapse rate is less than the wet adiabatic rate*. **Figure 4.24** depicts this situation by using an environmental lapse rate of 5°C per 1000 meters (and a wet adiabatic rate of 6°C per 1000 meters). Note that at 1000 meters, the temperature of the rising parcel is 5°C cooler than its environment, which makes it denser. Even if this stable air were forced above the lifting condensation level, it would remain cooler and denser than its environment and would have a tendency to return to the surface.

Despite its tendency to remain near Earth's surface, stable air can be forced aloft, most commonly by frontal lifting, and to a lesser extent by orographic lifting. If stable air is forced above the condensation level, relatively thin widespread clouds may be generated. Precipitation, if any, will be light to moderate, depending on the moisture content of the air.

**Absolute Instability** At the other extreme, a layer of air is said to exhibit **absolute instability** when *the environmental lapse rate is greater than the dry adiabatic rate*. As shown in **Figure 4.25**, the ascending parcel of air is always warmer than its environment and will continue to rise because of its own buoyancy. Absolute instability occurs most often during

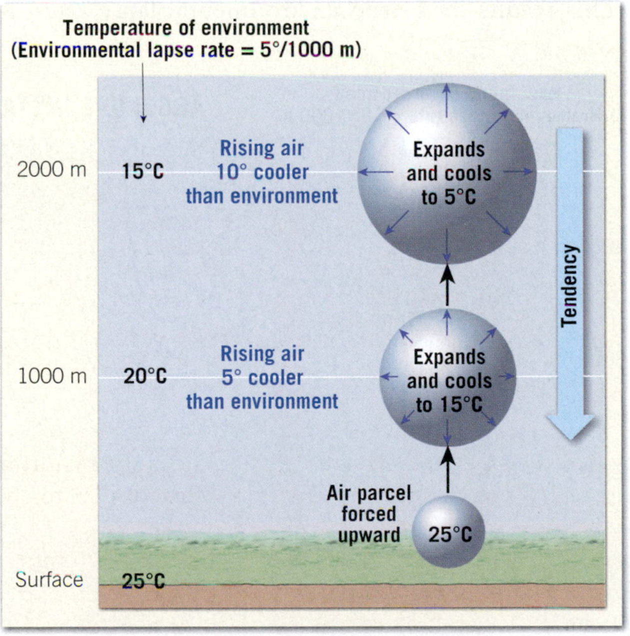


▲ **Figure 4.23 How the stability of the air is determined** When an unsaturated parcel of air is lifted, it expands and cools at the dry adiabatic rate of 10°C per 1000 meters. In this example the temperature of the rising parcel of air is lower than that of the surrounding environment; therefore, it will be heavier and, if allowed to do so, will sink to its original position.

Animation (MM) Atmospheric Stability
http://goo.gl/bqxtGo

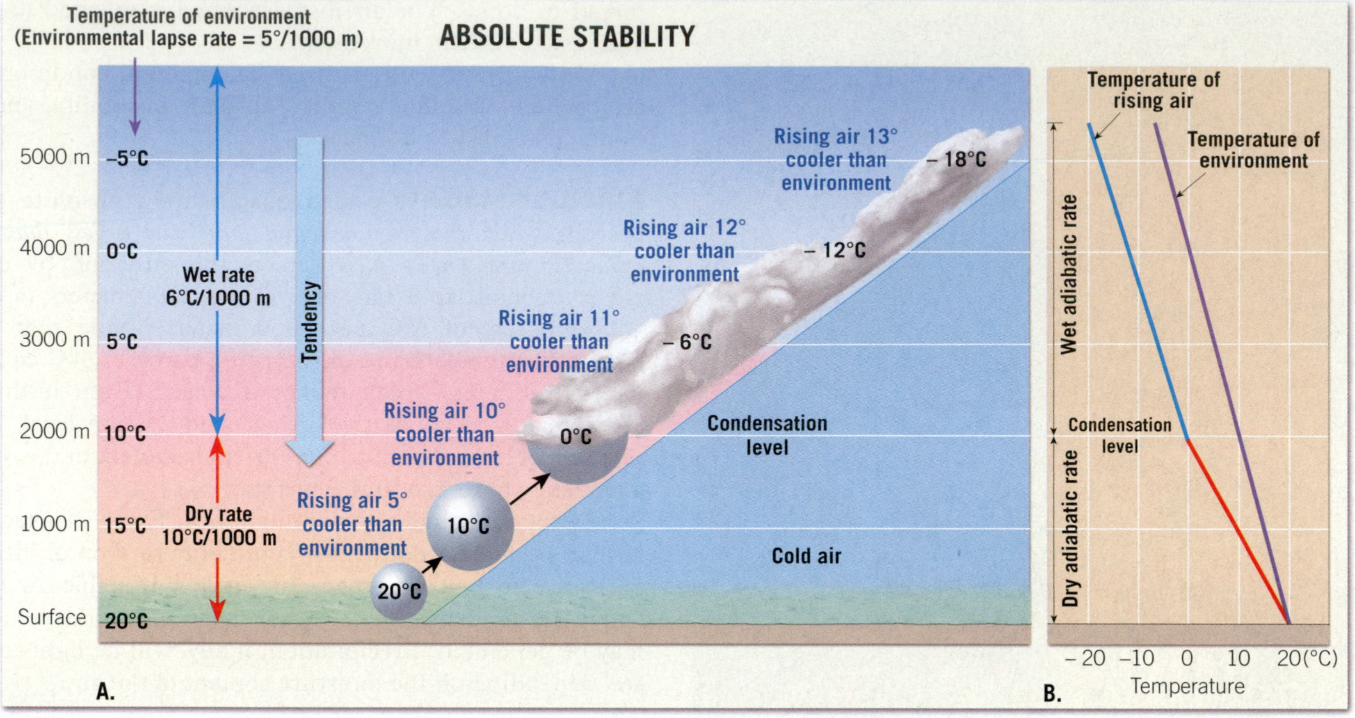


▲ **SmartFigure 4.24 Atmospheric conditions that result in absolute stability** Absolute stability prevails when the environmental lapse rate is less than the wet adiabatic rate. **A.** The rising parcel of air is always cooler and heavier than the surrounding air, producing stability. **B.** Graphical representation of the conditions shown in part **A**.

http://goo.gl/MbKFE

the warmest months and on clear days, when solar heating is intense. Under these conditions the lowermost layer of the atmosphere is heated to a much higher temperature than the air aloft. This results in a steep environmental lapse rate—in other words, air temperature rapidly decreases with height—and an unstable atmosphere. Convective lifting of the air near Earth's surface generates towering clouds and the potential for midafternoon thunderstorms that tend to dissipate after sunset.

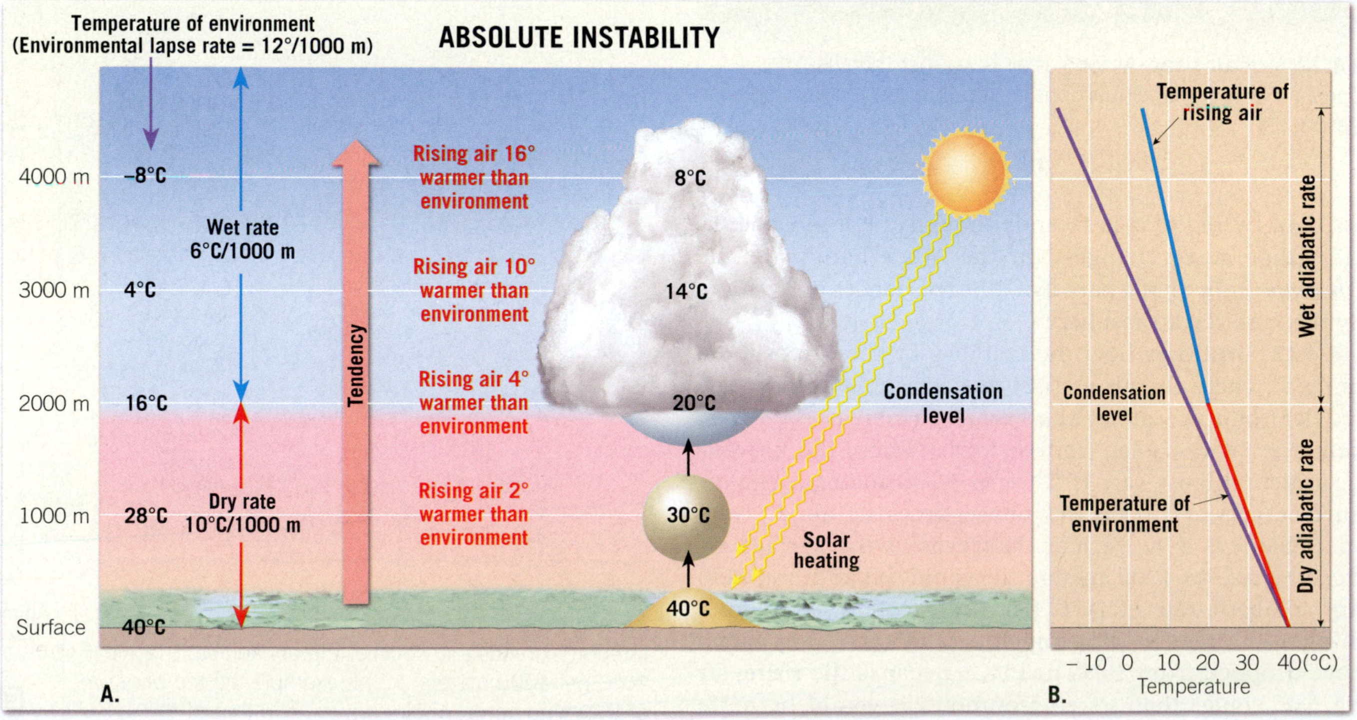


▲ **Figure 4.25 Atmospheric conditions that result in absolute instability** *Absolute instability* can develop when solar heating causes the lowermost layer of the atmosphere to be warmed to a much higher temperature than the air aloft. The result is a steep environmental lapse rate that renders the atmosphere unstable. **B.** Graphical representation of the conditions shown in part **A**.

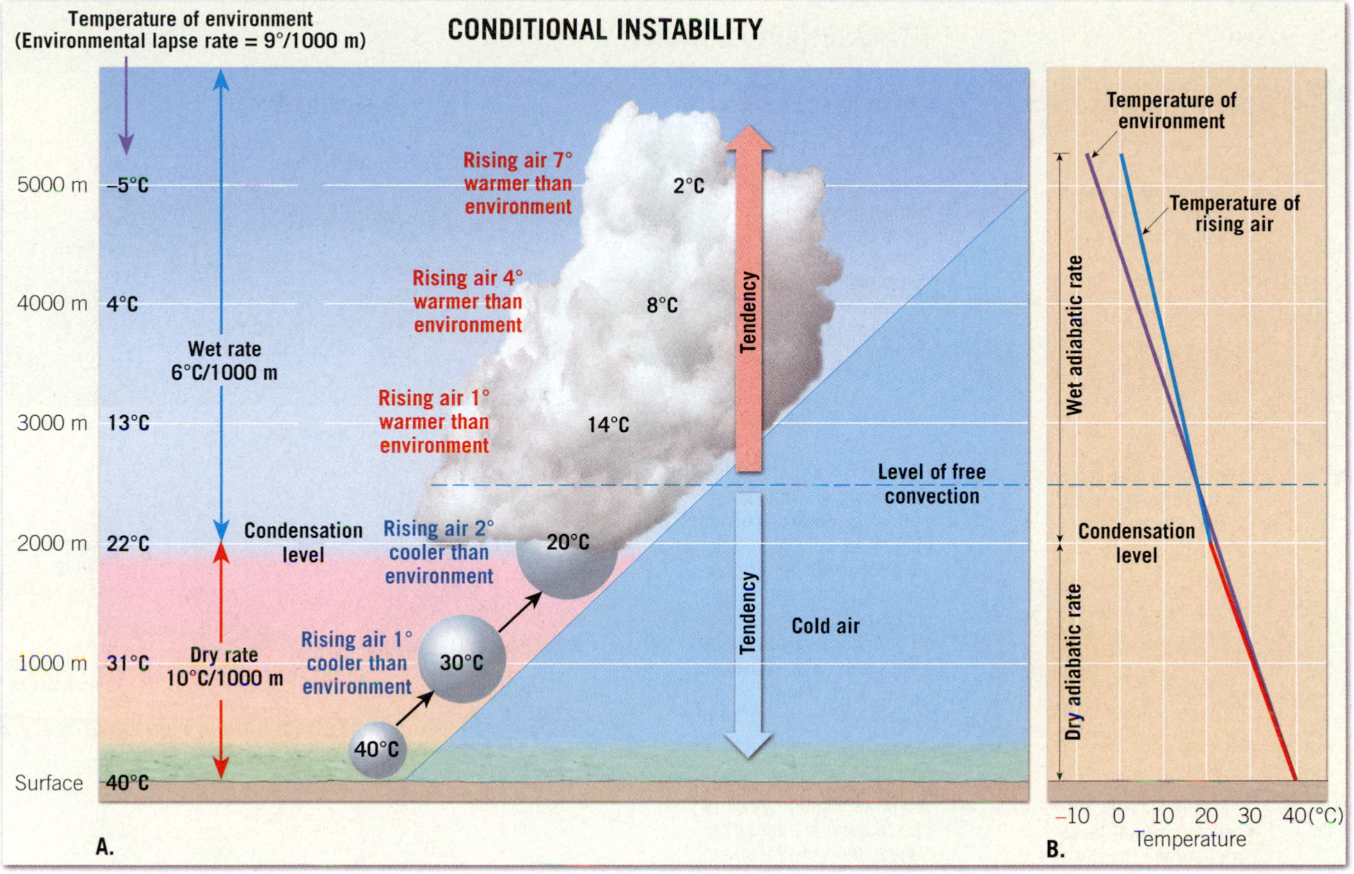


▲ **Figure 4.26 Atmospheric conditions that result in conditional instability** Conditional instability may result when warm air is forced to rise along a frontal boundary. Note that the environmental lapse rate of 9°C per 1000 meters lies between the dry and wet adiabatic rates. **A.** The parcel of air is cooler than the surrounding air up to nearly 3000 meters, where its tendency is to sink toward the surface (stable). Above this level, however, the parcel is warmer than its environment and will rise because of its own buoyancy (unstable). Thus, when conditionally unstable air is forced to rise, the result can be towering cumulus clouds. **B.** Graphical representation of the conditions shown in part **A**.

**Conditional Instability** A common type of atmospheric instability is called **conditional instability**. This situation prevails when *moist air has an environmental lapse rate between the dry and wet adiabatic rates* (between about 5° and 10°C per 1000 meters). Simply stated, the atmosphere is said to be conditionally unstable when it is *stable* with respect to an *unsaturated* parcel of air but *unstable* with respect to a *saturated* parcel of air.

Notice in Figure 4.26 that the rising parcel of air is cooler than the surrounding air for nearly about 2500 meters. However, because of the release of latent heat that occurs above the condensation level, the parcel eventually becomes warmer than the surrounding air. From this point along its ascent, the parcel will continue to rise without an outside force. The word *conditional* is used because the air must be forced upward before it reaches the level where it becomes unstable and rises on its own. The altitude at which air rises because of its own buoyancy is called the **level of free convection (LFC)**.

Conditional instability is usually a summertime phenomenon associated with warm, humid air. When conditionally unstable air is lifted (usually along a front) above the condensation level, the resulting weather often consists of thunderstorms and occasionally tornadoes.

Figure 4.27 summarizes the three types of atmospheric stability. The stability of air is determined by comparing the temperature of the atmosphere at various heights (environmental lapse rate) with the wet and dry adiabatic rates.

## ✔ Concept Checks 4.7

1. How does stable air differ from unstable air?
2. Explain the difference between the environmental lapse rate and adiabatic cooling.
3. How is the stability of air determined?
4. Write a statement relating the environmental lapse rate to stability.
5. Describe conditional instability.

**ATMOSPHERIC STABILITY**

| TYPES | ATMOSPHERIC CONDITIONS | ASSOCIATED WEATHER | ILLUSTRATION |
| --- | --- | --- | --- |
| Absolute Stability | Environmental lapse rate is less than wet adiabatic rate | Surface air does not rise unless forcefully lifted. If air is forced aloft such as along a warm front, the clouds that form will be thin and widespread. Precipitation, if any, will be light to moderate. | 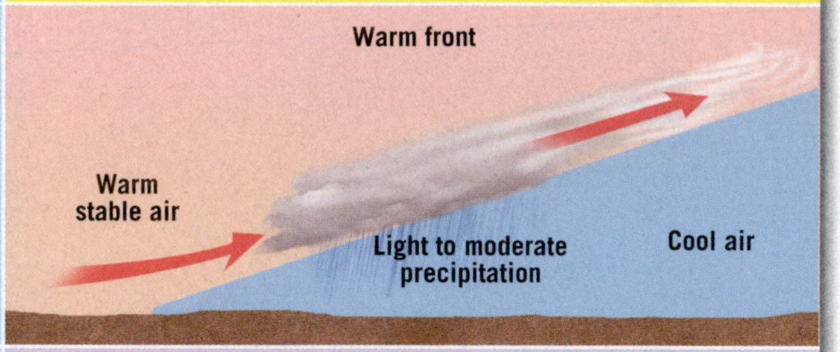 |
| Absolute Instability | Environmental lapse rate is greater than dry adiabatic rate | Occurs on warm, humid summer days when solar heating is intense, causing the lower atmosphere to become much hotter than air aloft. Bouyant lifting may produce midafternoon thunderstorms that usually dissipate after sunset. | 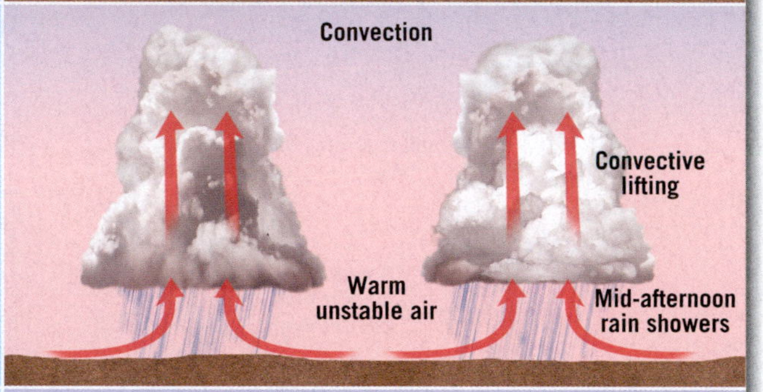 |
| Conditional Instability | Moist air with an environmental lapse rate between the dry and wet adiabatic rates | A type of instability where the release of latent heat above the condensation level is suffiicent to render the air unstable. Usually associated with frontal lifting of humid air to produce strong thunderstorms and occasionally tornadoes. | 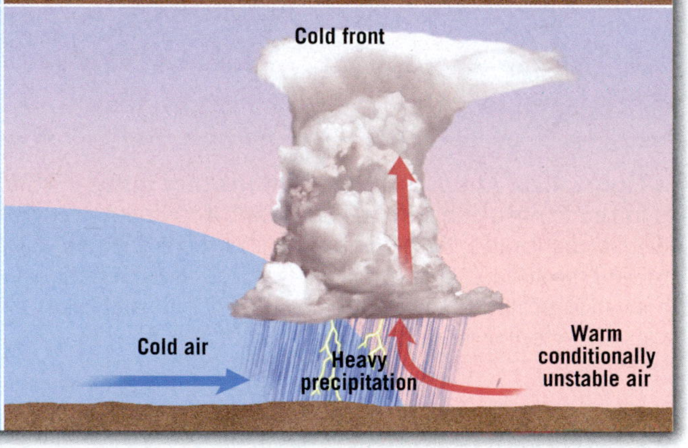 |

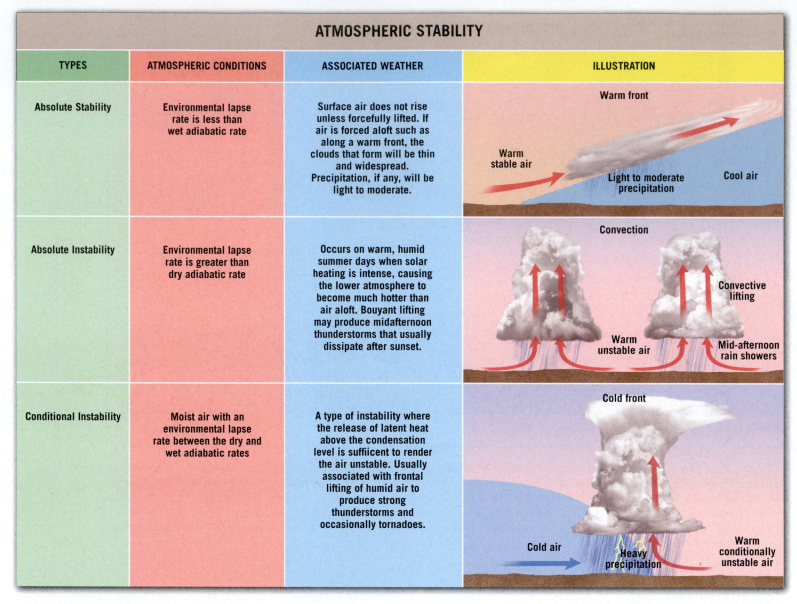

▲ **Figure 4.27 Comparison of the three types of atmospheric stability**

# 4.8 Stability and Daily Weather

**List the primary factors that influence the stability of air.**

How does air stability manifest itself in our daily weather? When stable air is forced aloft, relatively thin widespread clouds typically form, and any precipitation that results is light to moderate. In contrast, as unstable air rises, towering clouds are generated that are usually accompanied by heavy precipitation (**Fig. 4.28**).

## How Stability Changes

Any factor that causes air near the surface to become warmed relative to the air aloft makes air more unstable (increases instability), whereas any factor that causes the surface air to be chilled increases stability. Stated another way, any factor that increases the environmental lapse rate renders the air less stable, whereas any factor that reduces the environmental lapse rate increases the air's stability.

*Instability* is enhanced by the following:

1. Warming of the lowermost layer of the atmosphere by solar radiation during daylight hours
2. Heating of a cold air mass from below as it passes over a warm surface
3. Upward movement of air caused by processes such as orographic lifting, frontal lifting, and convergence
4. Radiation cooling of cloud tops

*Stability* is enhanced by the following:

1. Radiation cooling of Earth's surface after sunset
2. The cooling of an air mass from below as it traverses a cold surface
3. Subsidence within an air column

Note that most processes that alter stability result from temperature changes caused by horizontal or vertical air movement, although daily temperature changes are important as well.

**Solar Heating and Stability** On clear summer days, when there is abundant surface heating, the lower atmosphere may become warmed sufficiently to cause parcels of air to rise—localized convection. After the Sun sets, surface cooling generally renders the atmosphere stable again.

**Horizontal Air Movement and Stability** Changes in stability may occur as air moves horizontally over surfaces that have markedly different temperatures. For example, in the winter, when warm air from the Gulf of Mexico moves northward over the cold, snow-covered Midwest, the air is cooled from below. This *increases the stability* of the air and often produces widespread fog but no cloud development.

On the other hand, *instability* can be enhanced in the winter when frigid polar air moves southward over the open waters of the Great Lakes. Although the Great Lakes are cold in the winter, they are as much as 25°C warmer than a subfreezing polar air mass as it pushes southward across the lakes. During its journey, moisture and heat is added to the frigid polar air from the comparatively warm water below, rendering the air humid and unstable. The result can be heavy snowfalls on the downwind shores of the Great Lakes—called "lake-effect snows"—a topic covered in more detail in Chapter 8.

## eye on the atmosphere 4.3

The Navajo Generating Station, located on the Navajo Indian Reservation near Page, Arizona, has three 236-meter (560-feet) stacks.

**Questions**

1. Why do power-generating facilities such as this have such tall stacks?
2. Can you explain why the "smoke" changes color from bright white to pale yellow when it reaches a height of about 200 feet above the stacks?

**Vertical Air Movement and Stability** When there is a general downward flow, called **subsidence**, the upper part of the column of the subsiding layer is heated by compression, more so than the air below. (Usually the air just above the surface is not involved in subsidence, so its temperature remains unchanged.) Because the air aloft is warmed more than the air near the surface, subsidence tends to stabilize the atmosphere. The warming effect of a few hundred meters of subsidence is enough to

▶ **Figure 4.28 Towering clouds provide evidence of unstable conditions in the atmosphere**

## Box 4.4 Orographic Effects: Windward Precipitation and Leeward Rain Shadows

Orographic lifting is a significant factor in the development of windward precipitation and leeward rain shadows. A simplified hypothetical situation, illustrated in **Figure 4.D**, shows prevailing winds forcing warm moist air over a nearly 3000-meter-high mountain range. As the unsaturated air ascends the windward side of the range, it cools at a rate of 10°C per 1000 meters (dry adiabatic rate) until it reaches the dew-point temperature of 20°C. Because the dew-point temperature is reached at 1000 meters, we can say that this height represents the lifting condensation level and the height of the cloud base. Notice that above the lifting condensation level, latent heat is released, which results in a slower rate of cooling, the wet adiabatic rate.

From the cloud base to the top of the mountain, water vapor within the rising air condenses to form more and more cloud droplets. As a result, the windward side of the mountain range experiences abundant precipitation.

For simplicity, we will assume that the air that was forced to the top of the mountain is cooler than the surrounding air and hence begins to flow down the leeward slope of the mountain. As the air descends, it is compressed and *heated* at the dry adiabatic rate. Upon reaching the base of the mountain range, the temperature of the descending air has risen to 40°C, or 10°C warmer than the temperature at the base of the mountain on the windward side. The higher temperature on the leeward side is a result of the latent heat that was released during condensation as the air ascended the windward slope of the mountain range.

Two factors account for the rain shadow commonly observed on leeward mountain slopes. First, water is extracted from air in the form of precipitation on the windward side. Second, the air on the leeward side is warmer than the air on the windward side. (Recall that an increase in temperature results in a drop in relative humidity.)

A classic example of windward precipitation and leeward rain shadows is found in western Washington State. As moist Pacific air flows inland over the Olympic and Cascade Mountains, orographic precipitation is abundant (**Fig. 4.E**). By contrast, precipitation data for Sequim and Yakima indicate the presence of rain shadows on the leeward sides of these highlands.

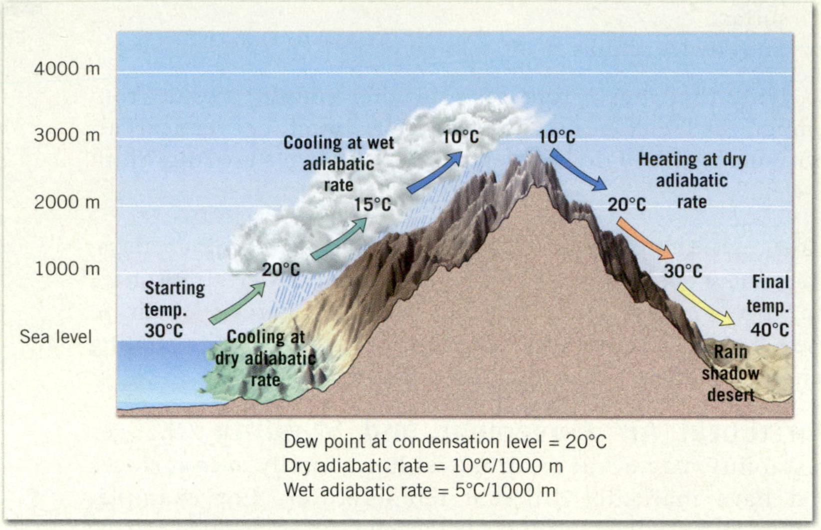


▲ **Figure 4.D** Orographic lifting and the formation of rain shadow deserts.

### Question

1. Give two reasons why the air on the windward side of a topographic barrier has a higher relative humidity than when it arrives on the leeward side.

evaporate clouds. Thus, one sign of subsiding air is a deep blue, cloudless sky.

Upward movement of air generally enhances *instability* and is particularly significant in generating towering clouds and thunderstorms during the warm summer months. Recall that when *conditionally unstable air* is forcefully lifted, it can become *unstable* and continue to rise because of its buoyancy (see Figure 4.26).

**Radiation Cooling from Clouds** The loss of heat by radiation emitted from cloud tops during evening hours adds to their instability and growth. Unlike air, which is a poor radiator of heat, cloud droplets emit considerable energy to space. Towering clouds that owe their growth to surface heating lose that source of energy at sunset. After sunset, however, radiation cooling at their tops steepens the lapse rate near the tops of these clouds and can lead to additional upward flow of warmer air below. This process is responsible for producing nocturnal thunderstorms from clouds whose growth temporarily ceased around sunset.

## Temperature Inversions and Stability

The most stable atmospheric conditions are associated with temperature inversions. A **temperature inversion** is a layer of the atmosphere in which temperature increases with altitude, rather

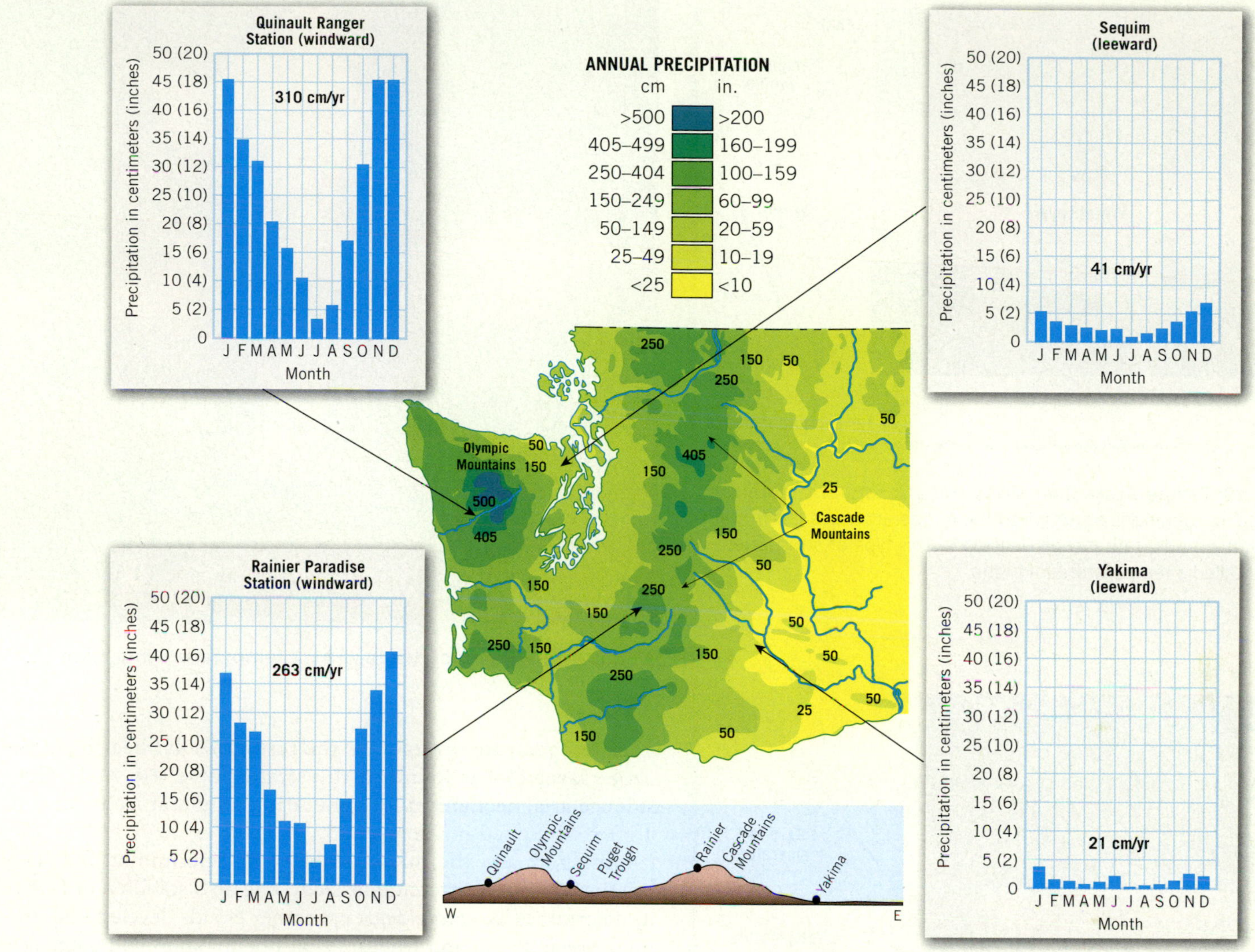


▲ **Figure 4.E** Distribution of precipitation in western Washington State. Data from four stations provide examples of wetter windward locations and drier leeward rain shadows.

than the more common condition of decreasing with altitude. Temperature inversions act like a lid, keeping convective motions in the atmosphere from penetrating the inversion. There are two main types of inversions—those that occur near the surface and those that develop aloft.

Many processes can generate a temperature inversion, such as radiation cooling of Earth's surface on a clear night (**Fig. 4.29**). After sunset, Earth's surface loses energy quickly and, through conduction, cools the air near the surface. However, because air is a poor conductor of heat, the air aloft remains comparatively warm.

When the air near the surface is cooler and heavier than a layer of air aloft, minimal vertical mixing occurs between the two layers. Because pollutants are generally added to the atmosphere from below, a temperature inversion confines the pollutants to the lowermost layer, where their concentration will continue to increase until the temperature inversion dissipates (**Fig. 4.30**).

Widespread fog can also be enhanced by a temperature inversion. Fog often forms after sunset because of radiation cooling. If an inversion develops, mixing between the moist, fog-laden air near the surface and dryer air aloft is inhibited—thus preventing the fog from dissipating.

Temperature inversions that occur aloft can cause convective clouds to spread out and take on a flattened appearance. One example is the flattened tops of towering storm clouds that reach the top of the troposphere, called the *tropopause*.

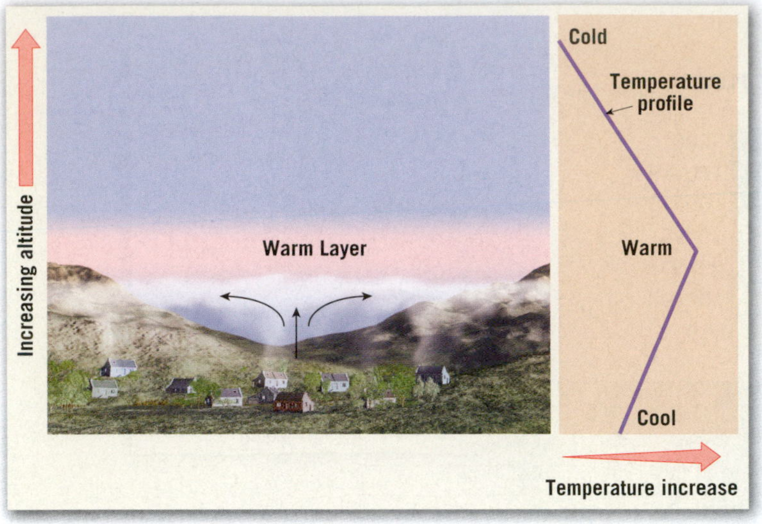


▲ **Figure 4.29 Temperature inversions and atmospheric stability** The most stable conditions occur when the temperature in a layer of air increases with altitude, rather than decreases. This type of atmosphere condition is called a *temperature inversion*.

▲ **Figure 4.30 Pollution trapped by a temperature inversion**

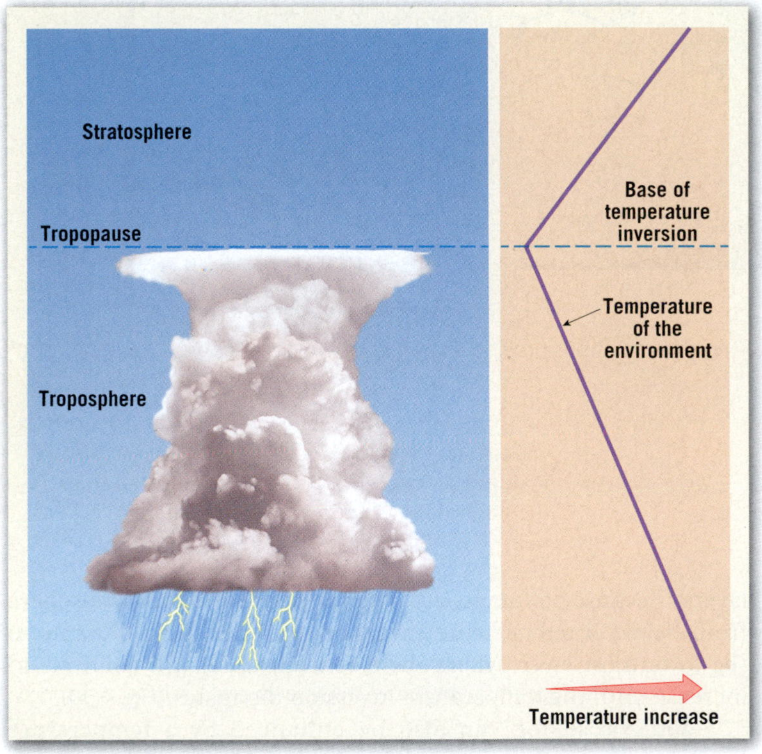


▲ **Figure 4.31 Temperature inversions aloft tend to inhibit cloud growth** In this example, The stratosphere forms a warm inversion layer (caused by solar heating of ozone) and therefore serves as a lid to stop the growth of convective clouds.

This temperature inversion is the result of solar heating of the ozone layer that is found in the stratosphere (**Fig. 4.31**). Subsidence as a mechanism for generating temperature inversions aloft is discussed in Chapter 13.

In summary, the role of stability in determining our daily weather cannot be overemphasized. The air's stability, or lack of it, determines to a large degree whether clouds develop and produce precipitation and whether that precipitation will come as a gentle shower or a violent downpour. When stable air is forced aloft, the associated clouds generally have little vertical thickness and precipitation, if any, is light. In contrast, unstable air can result in towering clouds frequently accompanied by thunderstorms and heavy precipitation. The most stable conditions occur during a temperature inversion, when the air temperature increases with height and inhibit vertical air movement.

## ✔ Concept Checks 4.8

1. What weather condition would lead you to believe that air is unstable?
2. What weather condition would lead you to believe that air is stable?
3. List four ways instability can be enhanced.
4. List three ways stability can be enhanced.

# 4 Concepts in Review Moisture and Atmospheric Stability

## 4.1 Water on Earth

► Describe the movement of water through the hydrologic cycle. List and describe water's unique properties.

**Key Terms:** hydrologic cycle, hydrogen bonds

- The unending circulation of Earth's water supply is called the hydrologic cycle. The cycle illustrates the continuous movement of water from the oceans to the atmosphere, from the atmosphere to the land, and from the land back to the sea.
- Water has unique properties: (1) It is the only *liquid* found at Earth's surface in large quantities; (2) it is readily converted from one state of matter to another (solid, liquid, gas); (3) its solid phase, ice, is less dense than liquid water; and (4) it has a high heat capacity—meaning it requires considerable energy to change its temperature.

## 4.2 Water's Changes of State

► Summarize the six processes by which water changes from one state of matter to another. For each, indicate whether energy is absorbed or released.

**Key Terms:** calorie, latent heat, evaporation, condensation, sublimation, deposition

- Water vapor, an odorless, colorless gas, can change from one state of matter (solid, liquid, or gas) to another at the temperatures and pressures experienced on Earth.
- The processes involved in changes of state include evaporation (liquid to gas), condensation (gas to liquid), melting (solid to liquid), freezing (liquid to solid), sublimation (solid to gas), and deposition (gas to solid). During each change, latent (hidden, or stored) heat is either absorbed or released.

**Q** Label the accompanying diagram with the appropriate terms for the changes in state that are shown.

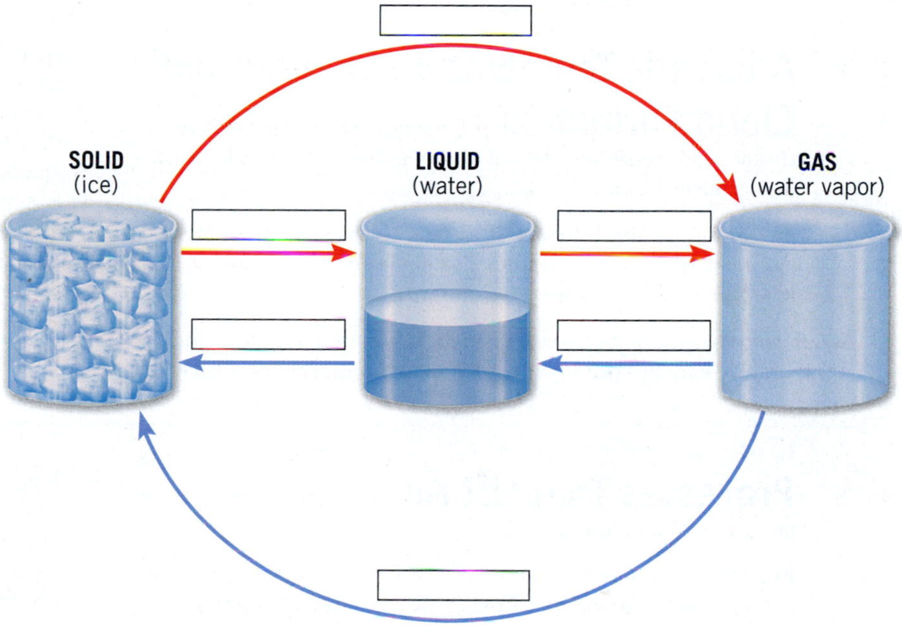


## 4.3 Humidity: Water Vapor in the Air

► Write a generalization relating air temperature and the amount of water vapor needed to saturate air.

**Key Terms:** humidity, absolute humidity, mixing ratio, vapor pressure, saturation, saturation vapor pressure

- Humidity is the general term used to describe the amount of water vapor in the air. The methods used to quantitatively express humidity include (1) absolute humidity, the mass of water vapor in a given volume of air; (2) mixing ratio, the mass of water vapor in a unit of air compared to the remaining mass of dry air; (3) vapor pressure, that part of the total atmospheric pressure attributable to its water-vapor content; (4) relative humidity, the ratio of the air's actual water-vapor content compared with the amount of water vapor required for saturation at that temperature; and (5) dew point, the temperature at which saturation occurs.
- When air is saturated, the pressure exerted by the water vapor, called the saturation vapor pressure, produces a balance between the number of water molecules leaving the surface of the water and the number returning. Because the saturation vapor pressure is temperature dependent, at higher temperatures more water vapor is required for saturation to occur.

## 4.4 Relative Humidity and Dew-Point Temperature

► List and describe the ways relative humidity changes in nature. Compare relative humidity to dew-point temperature.

**Key Terms:** relative humidity, dew-point temperature (dew point), hygrometer, psychrometer

- Relative humidity can change in two ways: (1) when the amount of moisture in the air increases or decreases and (2) through temperature change. When the water-vapor content of air remains at a constant level, a decrease in air temperature results in an increase in relative humidity, and an increase in temperature causes a decrease in relative humidity.

- The dew-point temperature (or simply dew point) is the temperature to which a parcel of air must be cooled to reach saturation. Unlike relative humidity, dew-point temperature is a measure of the air's actual moisture content. High dew-point temperatures equate to moist air, and low dew-point temperatures indicate dry air, regardless of the air's relative humidity.
- A variety of instruments called hygrometers are used to measure relative humidity.

**Q** Refer to the accompanying photo and explain how the relative humidity inside the house compares to the relative humidity outside the house on this particular day.

## 4.5 Adiabatic Temperature Changes and Cloud Formation

▶ Describe adiabatic temperature changes and explain why the wet adiabatic rate of cooling is less than the dry adiabatic rate.

**Key Terms:** adiabatic temperature change, parcel, entrainment, dry adiabatic rate, lifting condensation level (condensation level), wet adiabatic rate

- When air expands, it cools, and when air is compressed, it warms. Temperature changes produced in this manner, in which thermal energy is neither added nor subtracted, are called adiabatic temperature changes. The rate of cooling or warming of vertically moving unsaturated ("dry") air is 10°C for every 1000 meters (5.5°F per 1000 feet), the dry adiabatic rate. At the lifting condensation level latent heat is released, and the rate of cooling is reduced. The slower rate of cooling, called the wet adiabatic rate of cooling ("wet" because the air is saturated) varies from 5°C per 1000 meters to 9°C per 1000 meters.
- When air rises, it expands and cools adiabatically. If air is lifted sufficiently, it will cool to its dew-point temperature, and clouds will develop.

## 4.6 Processes That Lift Air

▶ List and describe four mechanisms that cause air to rise.

**Key Terms:** orographic lifting, rain shadow desert, front, frontal lifting (frontal wedging), convergence, localized convective lifting (convective lifting)

- Four mechanisms that cause air to rise are (1) orographic lifting, where air is forced to rise over a mountainous barrier; (2) frontal lifting, where warmer, less dense air is forced over cooler, denser air along a front; (3) convergence, a pile-up of horizontal airflow resulting in an upward flow; and (4) localized convective lifting, where unequal surface heating causes localized pockets of air to rise because of their buoyancy.

## 4.7 The Critical Weathermaker: Atmospheric Stability

▶ Write a statement relating the environmental lapse rate to stability.

**Key Terms:** stable air, unstable air, environmental lapse rate (ELR), absolute stability, absolute instability, conditional instability, level of free convection (LFC)

- Stable air resists vertical movement, whereas unstable air rises because of its buoyancy. The stability of air is determined from the environmental lapse rate, the temperature of the atmosphere at various heights. The three fundamental conditions of the atmosphere are (1) absolute stability, when the environmental lapse rate is less than the wet adiabatic rate; (2) absolute instability, when the environmental lapse rate is greater than the dry adiabatic rate; and (3) conditional instability, when moist air has an environmental lapse rate between the dry and wet adiabatic rates.
- In general, when stable air is forced aloft, the associated clouds have little vertical thickness, and precipitation, if any, is light. In contrast, clouds associated with unstable air are towering and frequently accompanied by heavy rain.

**Q** Describe the atmospheric conditions that were likely associated with the development of the towering clouds shown in the accompanying photo.

## 4.8 Stability and Daily Weather ▸ List the primary factors that influence the stability of air.

**Key Terms:** subsidence, temperature inversion

- Any factor that causes air near the surface to become warmed in relation to the air aloft increases the air's instability. The opposite is also true: Any factor that causes the surface air to be chilled compared to air aloft results in the air becoming more stable.
- The most stable atmospheric conditions are associated with temperature inversions. A temperature inversion is a layer of the atmosphere in which temperature increases with altitude, rather than the more common condition of decreasing with altitude. Temperature inversions act like a lid, keeping convective motions in the atmosphere from penetrating the inversion.

## Give it Some Thought

1. Refer to Figure 4.3 to complete the following.
   **a.** In which state of matter is water the most dense?
   **b.** In which state of matter are water molecules most energetic?
   **c.** In which state of matter is water compressible?
2. The accompanying photo shows a cup of hot coffee. In what state of matter is the "steam" rising from the liquid? (*Hint:* Can you see water vapor?)

3. The primary mechanism by which the human body cools itself is perspiration.
   **a.** Explain how perspiring cools the skin.
   **b.** Referring to the data for Phoenix, Arizona, and Tampa, Florida (Table A), in which city would it be easier to stay cool by perspiring? Explain your choice.

**TABLE A**

| City | Temperature | Dew point temperature |
|---|---|---|
| Phoenix, AZ | 101°F | 47°F |
| Tampa, FL | 101°F | 77°F |

4. As shown in the accompanying photo, during hot summer weather, many people put "koozies" around their beverages to keep drinks cold. Describe at least two ways the koozies help keep beverages cold.

5. Refer to Table 4.1 to answer this question. How much more water is contained in saturated air at a tropical location with a temperature of 40°C compared to a polar location with a temperature of –10°C?
6. Refer to the data for Phoenix, Arizona, and Bismarck, North Dakota (Table B), to complete the following:
   **a.** Which city has a higher relative humidity?
   **b.** Which city has the greatest quantity of water vapor in the air?
   **c.** In which city is the air closest to its saturation point with respect to water vapor?
   **d.** In which city does the air have the greatest holding capacity for water vapor?

**TABLE B**

| City | Temperature | Dew point temperature |
|---|---|---|
| Phoenix, AZ | 101°F | 47°F |
| Bismark, ND | 39°F | 38°F |

7. The accompanying graph shows how air temperature and relative humidity change on a typical summer day in the Midwest.
   **a.** Assuming that the dew-point temperature remained constant, what would be the best time of day to water a lawn to minimize evaporation of the water sprayed on the grass?
   **b.** Use this graph to explain why dew almost always forms early in the morning.

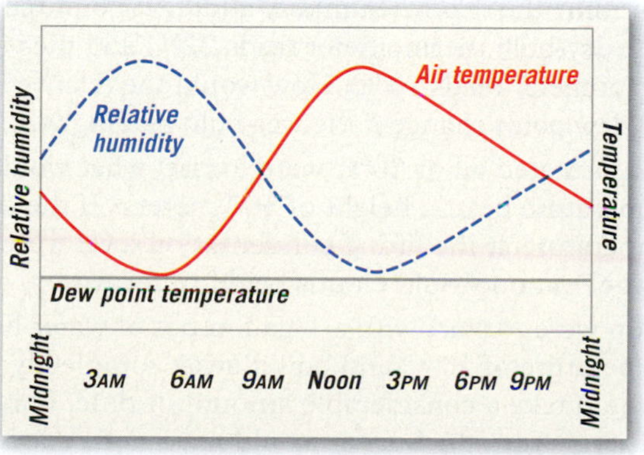

8. Are the atmospheric conditions illustrated in Figure 4.21 an example of absolute stability, absolute instability, or conditional instability?
9. This chapter examines four processes that cause air to rise. Describe how convective lifting is different from the other three processes.
10. Imagine a parcel of air that has a temperature of 40°C at the surface and a dew-point temperature of 20°C at the lifting condensation level. Assume that the environmental lapse rate is 8°C per 1000 meters, the dry adiabatic lapse rate is 10°C per 1000 meters, and the wet adiabatic lapse rate is 6°C/1000 meters. In Table C, record the environmental temperature, the parcel temperature, and the temperature difference (parcel temperature minus the environmental temperature) and then determine whether the atmosphere is *stable* or *unstable* at each height.
    **a.** What is the height of the lifting condensation level?
    **b.** Does this example describe absolute stability, absolute instability, or conditional instability?
    **c.** Would you forecast thunderstorms under these conditions?

**TABLE C**

| Height (meters) | Parcel Temperature °C | Environmental Temperature °C | Temperature difference (Parcel-environment | Stable or unstable |
|---|---|---|---|---|
| 7000 | | | | |
| 6000 | | | | |
| 5000 | | | | |
| 4000 | | | | |
| 3000 | | | | |
| 2000 | | | | |
| 1000 | | | | |
| Surface | 40°C | 40°C | 0°C | Stable |

11. Calculate the lifting condensation level (LCL) for the two following examples. (Assume that the dew point does not change until after the LCL is reached.)

| | Surface Temperature | Surface Dew Point | LCL |
|---|---|---|---|
| Example A | 35°C | 20°C | _____ |
| Example B | 35°C | 14°C | _____ |

What do these calculations tell you about the relationship between surface dew-point temperature and the height at which clouds develop?

## Problems

1. Using Table 4.1, answer the following:
    **a.** If a parcel of air at 25°C contains 10 grams of water vapor per kilogram of air, what is its relative humidity?
    **b.** If a parcel of air at 35°C contains 5 grams of water vapor per kilogram of air, what is its relative humidity?
    **c.** If a parcel of air at 15°C contains 5 grams of water vapor per kilogram of air, what is its relative humidity?
    **d.** If the temperature of the parcel of air in part c dropped to 5°C, how would its relative humidity change?
    **e.** If 20°C air contains 7 grams of water vapor per kilogram of air, what is its dew point?
2. Using the standard tables (Appendix C Tables C-1 and C-2), determine the relative humidity and dew-point temperature if the dry-bulb thermometer reads 22°C and the wet-bulb thermometer reads 16°C. How would the relative humidity and dew point change if the wet-bulb reading were 19°C?
3. If unsaturated air at 20°C were to rise, what would its temperature be at a height of 500 meters? If the dew-point temperature at the lifting condensation level were 11°C, at what elevation would clouds begin to form?
4. If you were to start with a 1-gallon pot of water having a temperature of 10°C and boil it away completely on a stove, it would take a considerable amount of time. Large amounts of energy from the burner would have to be conducted into the pot of water to bring it up to its boiling temperature (100°C), and even more energy would be required to convert it to a gas. If you could take all of the vaporized water that you boiled away (which now occupies part of the air in your kitchen) and instantly condensed it back into the pot, it would release enough energy to blow your house off its foundation. Do the calculation to prove that this statement is true by comparing the amount of energy that would be released to that contained in a stick of dynamite.

    *Important information:*
    1 gallon of water = 3785 grams
    J = Joule, a unit of energy used in the SI system
    4.186 J/g = amount of energy required to raise 1 gram of water 1°C
    2260 J/g = energy required to vaporize 1 gram of liquid water when the water's temperature is 100°C
    10°C = starting temperature of water
    $2.1 \times 10^6$ J = amount of energy contained in one stick of dynamite

    How much energy (measured in sticks of dynamite) is needed to completely boil away 1 gallon of water? This is the same amount of energy that would be released if the water vapor condensed back into the pot.
5. Fill out the accompanying diagram to answer the following. (*Hint:* Read Box 4.4.)
    **a.** What is the elevation of the cloud base?
    **b.** What is the temperature of the ascending air when it reaches the top of the mountain?
    **c.** What is the dew-point temperature of the rising air at the top of the mountain? (Assume 100 percent relative humidity.)

**d.** Estimate the amount of water vapor that must have condensed (in grams per kilogram) as the air moved from the cloud base to the top of the mountain.

**e.** What will the temperature of the air be if it descends to point G? (Assume that the moisture that condensed fell as precipitation on the windward side of the mountain.)

**f.** What is the approximate capacity of the air to hold water vapor at point G?

**g.** Assuming that no moisture was added or subtracted from the air as it traveled downslope, estimate the relative humidity at point G.

**h.** What is the *approximate* relative humidity at point A? (Use the dew-point temperature at the lifting condensation level for the surface dew point.)

**i.** Give two reasons for the difference in relative humidity between points A and G.

**j.** Needles, California, is situated on the dry leeward side of a mountain range similar to the position of point G. What term describes this situation?

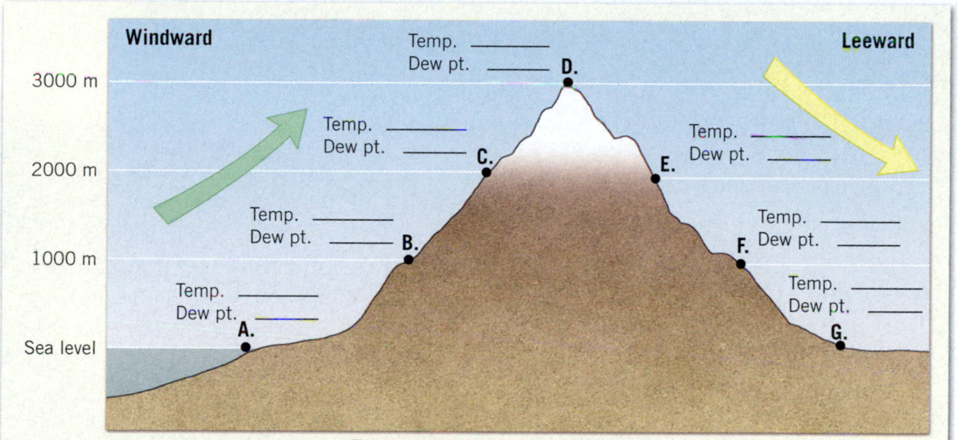


**6.** Figure 4.8 shows the nonlinear relationship between air temperature and the saturation mixing ratio. This relationship makes it possible for two unsaturated air parcels to mix and form a saturated parcel. For example, consider the following two parcels of air:

| | A | B |
|---|---|---|
| Temperature | 10°C | 40°C |
| Relative humidity | 75% | 85% |

**a.** Use Table 4.1 to find the *saturation mixing ratios* of parcels A and B.

**b.** What are the actual mixing ratios of Parcels A and B?

Answer c–g, assuming that the two air masses mix together and the resulting temperature is halfway between 10°C and 40°C, and the actual mixing ratio is halfway between the values you found for the two parcels in part b.

**c.** What is the temperature of the combined parcel? _____ °C

**d.** What is the saturation mixing ratio of the combined parcel? ____ g/kg

**e.** What is the actual mixing ratio of the combined parcel? _____ g/kg

**f.** How much does the actual mixing ratio differ from the saturation mixing ratio? _____ g/kg

**g.** Because relative humidity normally does not exceed 100 percent, what must happen to the excess water vapor in the combined parcel?

**h.** Describe a situation in which the mixing of two unsaturated parcels of air produces a parcel of saturated air.

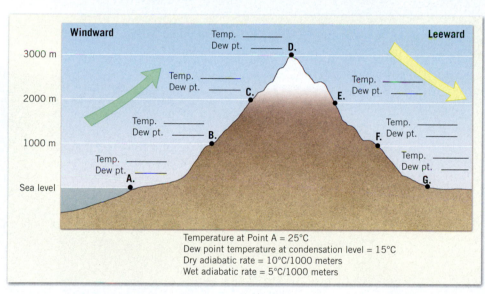

## MasteringMeteorology™

Looking for additional review and test prep materials? Visit the Study Area in *MasteringMeteorology*™ to enhance your understanding of this chapter's content by accessing a variety of resources, including **MapMaster**™ interactive maps, Geoscience Animations, GEODe, *In the News* RSS feeds, flashcards, web links, self-study quizzes, and an eText version of *The Atmosphere*.

# 5 Forms of Condensation and Precipitation

## Focus on Concepts

*Each statement represents the primary learning objective for the corresponding major heading within the chapter. After you complete the chapter, you should be able to:*

**5.1** Explain the roles of adiabatic cooling and cloud condensation nuclei in cloud formation.

**5.2** Name and describe the 10 basic cloud types, based on form and height. Contrast nimbostratus and cumulonimbus clouds and their associated weather.

**5.3** Identify the basic types of fog and describe how each forms.

**5.4** Describe the Bergeron process and explain how it differs from the collision–coalescence process.

**5.5** Describe the atmospheric conditions that produce sleet, freezing rain (glaze), and hail.

**5.6** List the advantages and disadvantages of using a standard rain gauge versus weather radar to measure precipitation.

**5.7** Discuss several ways that humans attempt to modify the weather.

Clouds, fog, and the various forms of precipitation are among the most observable weather phenomena. The primary focus of this chapter is to provide a basic understanding of each. In addition to learning how clouds are classified and named, you will learn that the formation of an average raindrop involves complex processes requiring water from roughly a million tiny water droplets.

*Scenic clouds forming above the Eastern Sierra Nevada, California.*

# 5.1 Cloud Formation

**Explain the roles of adiabatic cooling and cloud condensation nuclei in cloud formation.**

**Clouds** consist of billions of minute water droplets and/or ice crystals that are suspended above Earth's surface. In addition to being prominent and sometimes spectacular features in the sky, clouds are of continual interest to meteorologists because they provide a visual indication of atmospheric conditions. For condensation to generate clouds, the air must reach saturation, and there must be a surface on which the water vapor can condense to form liquid droplets.

## How Does Air Reach Saturation?

Saturation occurs in air aloft in one of two ways. First, cooling air to its dew-point temperature causes saturation, which results in condensation and cloud formation. Recall from Chapter 4 that clouds most often form when air rises and cools to its dew-point temperature by the process of *adiabatic cooling*. When a parcel of air ascends, it passes through regions of successively lower air pressure causing the parcel to expand and cool adiabatically. At a height called the *lifting condensation level*, the ascending parcel will have cooled to its dew-point temperature, and saturation is reached.

Saturation also occurs when cool, unsaturated air passes over a warm water body and sufficient water vapor is added from below. This process is mainly responsible for the formation of low clouds, particularly those that form over the subtropical oceans.

## The Role of Condensation Nuclei

Another requirement for condensation is that there must be a *surface* on which water vapor can condense. Objects at or near the ground, such as blades of grass, are such surfaces. When condensation occurs aloft, tiny particles known as **cloud condensation nuclei** serve this purpose. Without condensation nuclei, a relative humidity well in excess of 100 percent is necessary to produce cloud droplets. (At very low temperatures—low kinetic energies—water molecules will "stick together" in tiny clusters even in the absence of condensation nuclei.)

Dust storms, volcanic eruptions, and pollen from plants are major sources of cloud condensation nuclei. In addition, condensation nuclei are introduced into the atmosphere as byproducts of combustion (burning) from such sources as forest fires, automobiles, and coal-burning furnaces.

The most effective particles for condensation aloft are **hygroscopic (water-seeking) nuclei**. Common food items such as crackers and cereals are hygroscopic: When exposed to humid air, they absorb moisture and quickly become stale. Over the ocean, salt particles are released into the atmosphere when sea spray evaporates. Because salt is hygroscopic, water droplets begin to form around sea salt particles at relative humidities less than 100 percent. As a result, the cloud droplets that form on hygroscopic particles such as sea salt are generally much larger than those that grow on **hydrophobic (water-repelling) nuclei**. Although hydrophobic particles are not efficient condensation nuclei, cloud droplets will form on them when the relative humidity reaches 100 percent.

Because cloud condensation nuclei have a wide range of affinities for water, cloud droplets of various sizes often coexist in the same cloud—an important factor for the formation of precipitation.

## Growth of Cloud Droplets

Initially, the growth of cloud droplets occurs rapidly. However, the rate of growth slows as the available water vapor is consumed by the large number of competing droplets. The result is the formation of a cloud consisting of billions of tiny water droplets—usually having radii of 20 micrometers (μm) or less. These cloud droplets are so minute that they remain suspended in air by the smallest updraft.

Even in very moist air, the growth of cloud droplets by additional condensation is quite slow. Furthermore, the vast size difference between cloud droplets and raindrops (it takes about one million cloud droplets to form a single raindrop) indicates that condensation is not responsible for the formation of raindrops (or ice crystals) large enough to fall to the ground without evaporating. We will investigate the processes that generate precipitation later in this chapter.

### ✔ Concept Checks 5.1

1. Describe the process of cloud formation.
2. What role do cloud condensation nuclei play in the formation of clouds?
3. Define *hygroscopic nuclei*.
4. Why isn't growth by condensation able to generate droplets large enough to fall as rain?

## 5.2 Cloud Classification

**Name and describe the 10 basic cloud types, based on form and height.**
**Contrast nimbostratus and cumulonimbus clouds and their associated weather.**

MM GEODe ▶ Forms of Condensation and Precipitation ▶ Classifying Clouds

In 1803, English naturalist Luke Howard published a cloud classification scheme that serves as the basis of our present-day system. According to Howard's system, clouds are classified on the basis of two criteria: *form* and *height* (**Fig. 5.1**). We will look at the basic cloud forms or shapes first and then examine cloud height.

### Cloud Forms

Clouds are classified based on how they appear when viewed from Earth's surface. The basic forms or shapes are:

- *Cirrus* (*cirriform*) clouds are high, white, and thin. They form delicate veil-like patches or wisplike strands and often have a feathery appearance. (*Cirrus* is Latin for "curl" or "filament.")
- *Cumulus* (*cumuliform*) clouds consist of globular cloud masses that are often described as cottonlike in appearance. Normally cumulus clouds exhibit a flat base and appear as rising domes or towers. (*Cumulus* means "heap" or "pile" in Latin.) Cumulus clouds form within a layer of the atmosphere where there is some convection and rising air.
- *Stratus* (*stratiform*) clouds consist of sheets or layers (*strata*) that cover much or all of the sky.

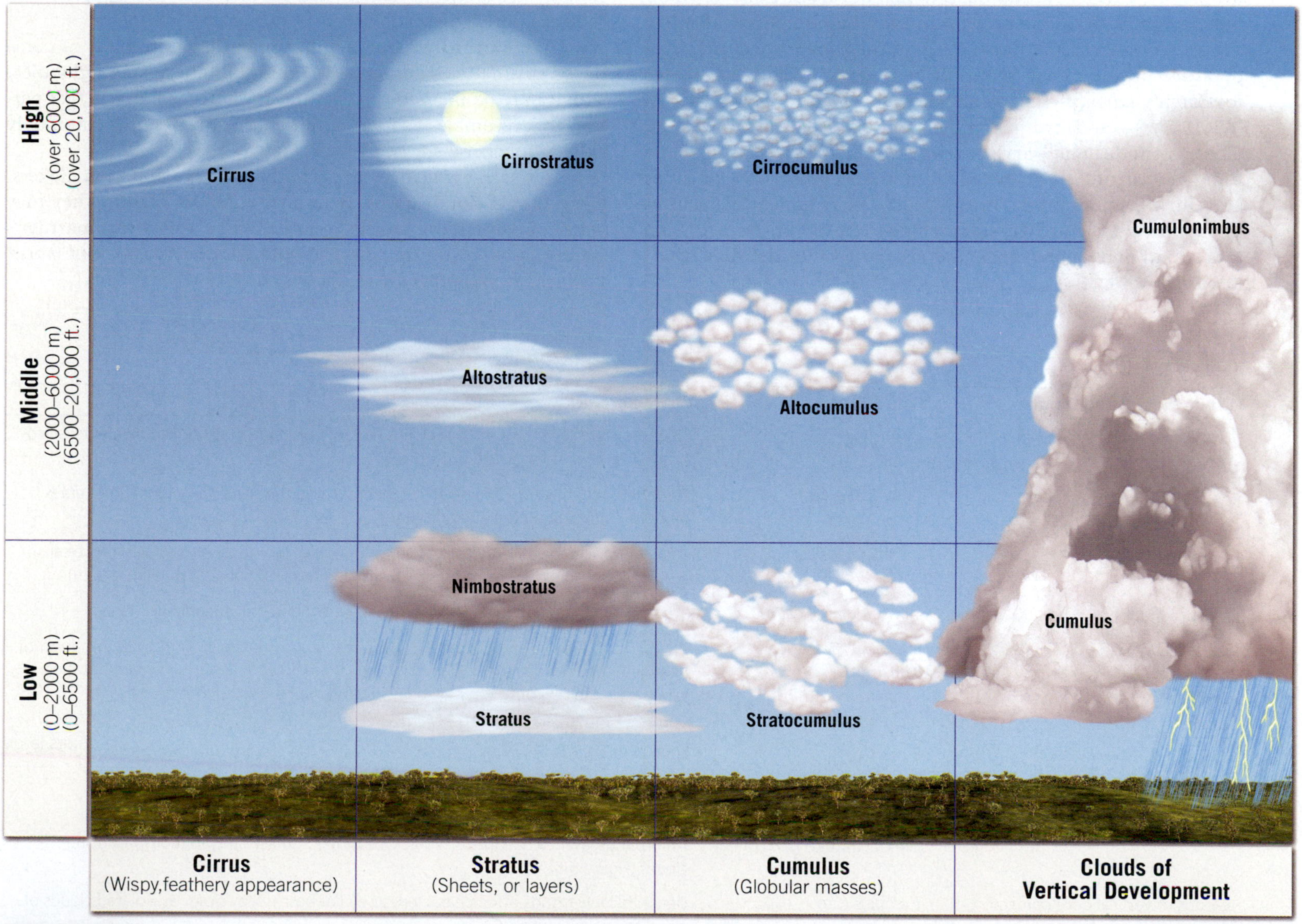


▲ **SmartFigure 5.1 Classification of clouds based on form and height**

http://goo.gl/15sC8

Although there may be minor breaks, there are no distinct individual cloud units. All clouds have at least one of these three basic forms, and some are a combination of two of them; for example, stratocumulus clouds are mostly sheetlike structures composed of long parallel rolls or broken globular patches. In addition, the term **nimbus** (Latin for "violent rain") is used in the name of a cloud that is a major producer of precipitation. Thus, *nimbostratus* denotes a flat-lying rain cloud.

## Cloud Height

The second aspect of cloud classification—height—recognizes three levels: high, middle, and low. **High clouds** form in the highest and coldest region of the troposphere and normally have bases above 6000 meters (20,000 feet). Temperatures at these altitudes are usually below freezing, so the high clouds are generally composed of ice crystals or supercooled water droplets. **Middle clouds** occupy heights from 2000 to 6000 meters (6500 to 20,000 feet) and may be composed of water droplets or ice crystals depending on the time of year and temperature profile of the atmosphere. **Low clouds** form nearer to Earth's surface—up to an altitude of about 2000 meters (6500 feet)—and are generally composed of water droplets. These altitudes may vary somewhat according to season of the year and latitude. For example, at high (poleward) latitudes and during cold winter months, high clouds generally occur at lower altitudes. Further, some clouds extend upward to span more than one height range and are called **clouds of vertical development**.

The 10 internationally recognized cloud types are summarized in Table 5.1 and described in the sections that follow.

## High Clouds

The family of high clouds (above 6000 meters [20,000 feet]) include *cirrus*, *cirrostratus*, and *cirrocumulus*. Low temperatures and small quantities of water vapor present at high altitudes result in high clouds that are thin, white, and made up primarily of ice crystals.

**Cirrus** (Ci) clouds are composed of delicate, icy filaments. Winds aloft often cause these fibrous ice trails to bend or curl. Cirrus clouds with hooked filaments are called "mares' tails" (Fig. 5.2A).

**Cirrostratus** (Cs) are transparent, whitish cloud veils with a fibrous or sometimes smooth appearance that may cover much or all of the sky. These clouds are easily recognized when they produce a halo around the Sun or Moon (Fig. 5.2C). Occasionally, cirrostratus clouds are so thin and transparent that they are barely discernible.

**Cirrocumulus** (Cc) clouds appear as white patches composed of small cells or ripples (Fig. 5.2B). These small globules, which may be merged or separate, are often arranged in a pattern that resembles fish scales. When this occurs, it is commonly called "mackerel sky."

Although high clouds are generally not precipitation makers, when cirrus clouds give way to cirrocumulus clouds, they may warn of impending stormy weather. The following mariner's phrase is based on this observation: *Mackerel scales and mares' tails make lofty ships carry low sails.*

**Table 5.1** | Basic Cloud Types

| Cloud Family and Height | Cloud Type | Characteristics |
|---|---|---|
| High clouds—above 6000 m (20,000 ft) | Cirrus (Ci) | Thin, delicate, fibrous, ice-crystal clouds. Sometimes appear as hooked filaments called "mares' tails" or cirrus uncinus (Figure 5.2A). |
| | Cirrostratus (Cs) | Thin sheet of white, ice-crystal clouds that may give the sky a milky look. Sometimes produce halos around the Sun and Moon (Figure 5.2C). |
| | Cirrocumulus (Cc) | Thin, white, ice-crystal clouds. In the form of ripples or waves, or globular masses all in a row. May produce a "mackerel sky." Least common of high clouds (Figure 5.2B). |
| Middle clouds—2000–6000 m (6500–20,000 ft) | Altocumulus (Ac) | White to gray clouds, often made up of separate globules; "sheepback" clouds (Figure 5.3A). |
| | Altostratus (As) | Stratified veil of clouds that is generally thin and may produce very light precipitation. When thin, the Sun or Moon may be visible as a "bright spot," but no halos are produced (Figure 5.3B). |
| Low clouds—below 2000 m (6500 ft) | Stratus (St) | Low uniform layer resembling fog but not resting on the ground. May produce drizzle. |
| | Stratocumulus (Sc) | Soft, gray clouds in globular patches or rolls. Rolls may join together to make a continuous cloud (Figure 5.4). |
| | Nimbostratus (Ns) | Amorphous layer of dark gray clouds. One of the primary precipitation-producing clouds (Figure 5.5). |
| Clouds of vertical development | Cumulus (Cu) | Dense, billowy clouds often characterized by flat bases. May occur as isolated clouds or closely packed (Figure 5.6). |
| | Cumulonimbus (Cb) | Towering cloud, sometimes spreading out on top to form an "anvil head." Associated with heavy rainfall, thunder, lightning, hail, and tornadoes (Figure 5.7). |

A.

B.

C.

▲ **Figure 5.2 Three basic cloud types make up the family of high clouds** **A.** Cirrus **B.** Cirrocumulus **C.** Cirrostratus.

## Middle Clouds

Clouds that form in the middle altitude range (2000–6000 meters [6500–20,000 feet]) are described with the prefix *alto* (meaning "middle") and include two types: *altocumulus* and *altostratus*.

**Altocumulus** (Ac) tend to form in large patches composed of rounded masses or rolls that may or may not merge (**Fig. 5.3A**). Because they are generally composed of water droplets rather than ice crystals, the individual cells usually have a more distinct outline. Altocumulus are sometimes confused with cirrocumulus (which are smaller and less dense) and stratocumulus (which are thicker).

**Altostratus** (As) is the name given to a formless layer of grayish clouds that cover all or large portions of the sky. Generally, the Sun is visible through altostratus clouds as a bright spot but with the edge of its disc not discernible (**Fig. 5.3B**). However, unlike cirrostratus clouds, altostratus do not produce halos. Infrequent precipitation in the form of light snow or drizzle may accompany these clouds. Altostratus clouds, commonly associated with approaching warm fronts, thicken into a dark gray layer of nimbostratus clouds capable of producing copious rainfall.

## Low Clouds

There are three members of the family of low clouds (below 2000 meters [6500 feet]): *stratus*, *stratocumulus*, and *nimbostratus*.

**Stratus** (St) clouds form in low, horizontal layers that on occasion may produce light drizzle or mist. White to light-gray in color, stratus clouds have very uniform bases and appear to blanket the entire sky.

Stratus-like clouds that develop a scalloped bottom that appear as long parallel rolls or broken globular patches are called **stratocumulus** (Sc) (**Fig. 5.4**). Although stratocumulus clouds are similar in appearance to altocumulus, they are located lower in the sky and consist of broken patches that are generally much larger than those of altostratus. A simple way to distinguish between these is to point your hand in the direction of an individual cloud mass, and if the cloud is about the size of your thumbnail, it is an altocumulus; if it is the size of your fist, it is a stratocumulus cloud.

Stratocumulus clouds often cover vast stretches of the subtropical oceans, which provide a ready supply of surface moisture. Because stratocumulus clouds cover such large areas, they are extremely important for Earth's energy balance, primarily because they reflect considerable amounts of incoming solar radiation.

A.

B.

▲ **Figure 5.3 Clouds found in the middle-altitude range A.** Altocumulus tend to form in patches composed of rolls or rounded masses. **B.** Altostratus occur as grayish sheets covering a large portion of the sky. When visible, the Sun appears as a bright spot through these clouds.

**Nimbostratus** (Ns) clouds derive their name from the Latin *nimbus*, "rain cloud," and *stratus*, "to cover with a layer" (**Fig. 5.5**). Nimbostratus clouds tend to produce constant precipitation and low visibility. These clouds normally form under stable conditions when air is forced to rise, as along a front (discussed in Chapter 9). Such forced ascent of stable air leads to the formation of a stratified cloud deck that is widespread and that may grow into the middle level of the troposphere. Precipitation associated with nimbostratus clouds is generally light to moderate (but can be heavy) and is usually of long duration and covering a large area.

▼ **Figure 5.4 Stratocumulus clouds commonly form over midlatitude oceans** Satellite view of a large bank of stratocumulus clouds over the Pacific Ocean just south of San Diego, California.

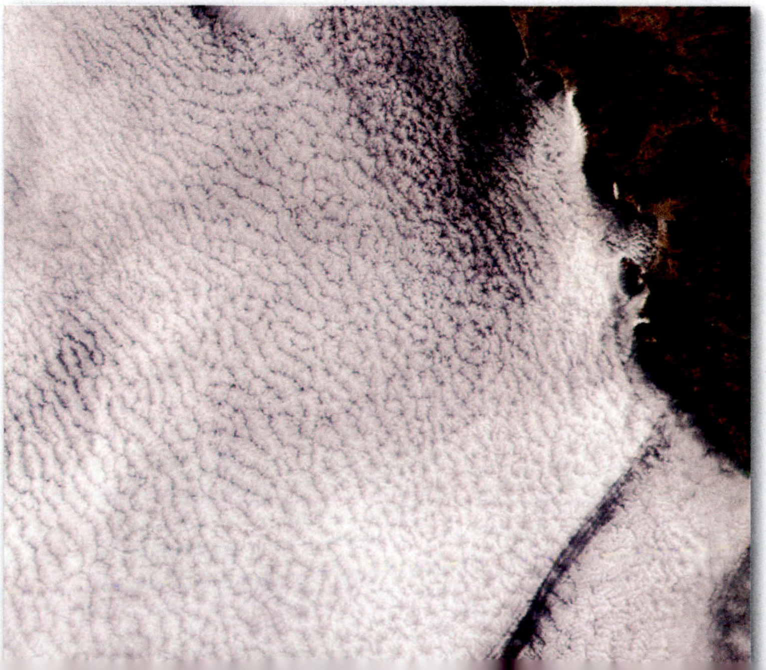

## Clouds of Vertical Development

Clouds that do not fit into any of the three height categories but instead have their bases in the low height range and extend upward into the middle or high altitudes are referred to as **clouds of vertical development.**

The most familiar type, **cumulus** (Cu) clouds, are individual masses that develop into vertical domes or towers having tops that resemble a head of cauliflower. Cumulus clouds most often form on clear days when unequal surface heating causes parcels of air to rise convectively above the lifting condensation level (**Fig. 5.6**).

When cumulus clouds are present early in the day, we can expect an increase in cloudiness in the afternoon as solar heating intensifies. Furthermore, because small cumulus clouds (*cumulus humilis*) form on "sunny" days and rarely produce appreciable precipitation, they are often called "fair-weather clouds." However, when the air is unstable, cumulus clouds can grow dramatically in height. As such a cloud grows, its top enters the middle height range, and it is called a *cumulus congestus*. Finally, if the cloud continues to grow and rain begins to fall, it is called a *cumulonimbus*.

▲ **Figure 5.5 Nimbostratus clouds are significant precipitation producers** These dark gray layers often exhibit a ragged-appearing base.

**Cumulonimbus** (Cb) are large, dense, billowy clouds of considerable vertical extent in the form of huge towers (**Fig. 5.7**). In late stages of development, the upper part of a cumulonimbus turns to ice, appears fibrous, and frequently spreads out in the shape of an anvil. Cumulonimbus towers extend from a few hundred meters above the surface upward to 12 kilometers (7 miles) or, on rare occasions, 20 kilometers (12 miles). These huge towers are capable of producing heavy precipitation with accompanying lightning, hail, and occasionally tornadoes. We consider the development of these important weather producers in Chapter 10.

▼ **Figure 5.6 Cumulus clouds, often called "fair-weather clouds"** These small, white, billowy clouds generally form on sunny days.

## Cloud Varieties

The 10 basic cloud types can be further subdivided into varieties that are named using adjectives that describe particular cloud characteristics. For example, the term *uncinus*, meaning "hook shaped," is applied to streaks of cirrus clouds that are shaped like a comma resting on its side. Named *cirrus uncinus*, these clouds are often precursors of bad weather.

When stratus or cumulus clouds appear broken (or fractured), the adjective *fractus* is used in their description. In addition, some clouds have rounded protuberances on their

▼ **Figure 5.7 Cumulonimbus clouds** These dense, billowy clouds have great vertical extent and can produce heavy precipitation and violent thunderstorms.

A.

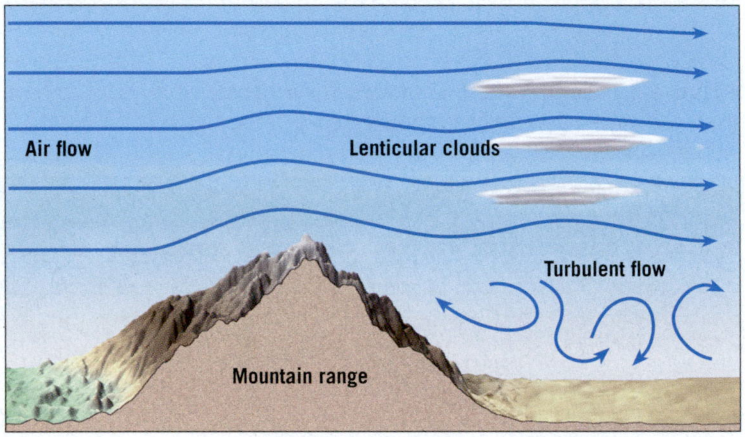


B.

▲ **Figure 5.8 Lenticular clouds A.** These lens-shaped clouds are relatively common in mountainous areas. **B.** This diagram depicts the formation of lenticular clouds in the turbulent flow that develops in the lee of a mountain range.

## eye ON THE atmosphere 5.1

The cap cloud perched above this mountain may remain in place for hours. Cap clouds belong to a group referred to as *orographic clouds*.

**Questions**

1. Describe how these clouds form, based on the fact that they are referred to as orographic clouds.
2. Why does this cloud have a relatively flat bottom?
3. Cap clouds are related to another cloud type that has a very similar shape. Can you name the cloud type?

**Video** MM
Identifying Clouds in Satellite Imagery
http://goo.gl/94G0U4

bottom surface, similar to a cow udder. When these structures are present, the term *mammatus* is applied. This configuration is usually associated with stormy weather and cumulonimbus clouds.

Stationary lens-shaped clouds, referred to as **lenticular clouds** (formal name *altocumulus lenticularis*), are common in rugged or mountainous topographies (**Fig. 5.8A**). Although lenticular clouds can develop whenever the airflow develops a wavy pattern, they most frequently form on the leeward side of mountains. As moist stable air passes over mountainous terrain, a series of standing waves form on the downwind side, as shown in **Figure 5.8B**. As the air ascends the wave crest, it cools adiabatically. If the air reaches its dew point temperature, moisture in the air will condense to form a lenticular cloud. As the moist air moves down into the trough of the wave, the cloud droplets evaporate, leaving areas with descending air cloud-free.

**Video** MM
Clouds Developing Over Florida
http://goo.gl/sVW2cl

## ✔ Concept Checks 5.2

1. What are the two criteria by which clouds are classified?
2. Why are high clouds always thin in comparison to low and middle clouds?
3. List the 10 basic cloud types and describe each based on its form (shape) and height (altitude).
4. Explain how lenticular clouds form.

**Video** MM
Is That a Cloud?
http://goo.gl/W80Jyw

# 5.3 Types of Fog

**Identify the basic types of fog and describe how each forms.**

MM GEODe ▶ Forms of Condensation and Precipitation ▶ Types of Fog

**Fog** is defined as *a cloud with its base at or very near the ground.* Physically, there are no differences between fog and a cloud; their appearances and structures are the same. The essential difference is the method and place of formation. While clouds result when air rises and cools adiabatically, fog results from cooling or when air becomes saturated through the addition of water vapor (evaporation fog).

Although fog is not inherently dangerous, it is generally considered an atmospheric hazard (Fig. 5.9). During daylight hours, fog reduces visibility to 2 or 3 kilometers (1 or 2 miles). When the fog is particularly dense, visibility may be cut to a few dozen meters or less, making travel by any mode difficult and dangerous. Official weather stations report fog only when it is thick enough to reduce visibility to 1 kilometer (0.6 mile) or less. Table 5.2 summarizes the basic fog types.

## Fogs Formed by Cooling

When the temperature of a layer of air in contact with the ground falls below its dew point, condensation produces fog. Depending on the prevailing conditions, fogs formed by cooling are called either *radiation fog*, *advection fog*, or *upslope fog*.

**Radiation Fog** As the name implies, **radiation fog** results from radiation cooling of the ground and adjacent air. It is a nighttime phenomenon that requires clear skies and a high relative humidity. Under clear skies, the ground and the air immediately above cool rapidly. Because of the high relative humidity, a small amount of cooling lowers the temperature to the dew point. If the air is calm, the fog is usually patchy and less than 1 meter (3 feet) deep. For radiation fog to be more extensive vertically, a light breeze of 3 to 5 kilometers (2 to 3 miles) per hour is necessary, to create enough turbulence to carry the fog upward 10 to 30 meters (30 to 100 feet) without dispersing it. High winds, on the other hand, mix the air with drier air above and disperse the fog.

Because the air containing the fog is relatively cold and dense, it flows downslope in hilly terrain. As a result, radiation fog is thickest in valleys, whereas the surrounding hills may remain clear (Fig. 5.9A). Normally, radiation fog dissipates within one to three hours after sunrise—and is often said to "lift." However, the fog does not actually "lift." Instead, as the Sun warms the ground, the lowest layer of air is heated first, and the fog evaporates from the bottom up. The last vestiges of radiation fog may appear as a low layer of stratus clouds.

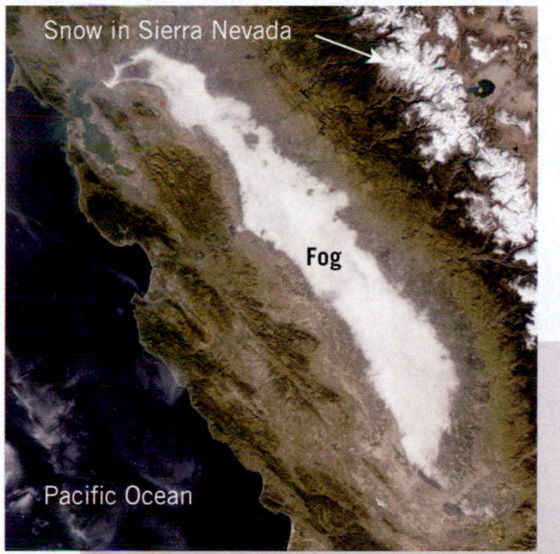


A.

B.

◀ **Figure 5.9 Radiation fog is generated by radiation cooling of Earth's surface A.** Satellite image of dense fog in California's San Joaquin Valley on November 20, 2002. This early-morning radiation fog was responsible for several car accidents in the region, including a 14-car pileup. The white areas to the east of the fog are the snow-capped Sierra Nevadas. **B.** Radiation fog can make a morning commute quite hazardous.

**Advection Fog** When warm, moist air blows over a cold surface, it becomes chilled by contact with the cold surface below. If cooling is sufficient, the result will be a blanket of fog called **advection fog**. (The term *advection* refers to air moving horizontally.) A classic example is the frequent advection fog around San Francisco's Golden Gate Bridge (Fig. 5.10). The fog experienced in San Francisco, California, as well as many other west coast locations, is produced when warm, moist air from the Pacific Ocean moves over the cold California Current.

A certain amount of turbulence is needed for proper development of advection fog; typically wind between 10 and 30 kilometers (6 and 18 miles) per hour are required. Not only does the turbulence facilitate cooling through a thicker layer of air, but it also carries the fog to greater heights.

Video MM
Clouds and Aviation
http://goo.gl/R5aC2p

**Table 5.2** Basic Fog Types

| Fog Groups | Fog Types | Mode of Formation and Characteristics |
|---|---|---|
| Fogs formed by cooling | Radiation fog | A nighttime phenomenon generated by radiation cooling of the ground and adjacent air. Usually forms in valleys with the surrounding hills fog free. |
| | Advection fog | Forms when warm, moist air flows over a cold surface and is chilled from below. Often a wintertime phenomenon in the Midwest, forming when warm Gulf air flows inland or when moist air flows over a cool ocean current. |
| | Upslope fog | When air flows up a mountain slope, or sometimes a gradually sloping landform, it expands and cools adiabatically. |
| Fogs formed by the addition of water vapor | Steam fog | When cool air moves over warm water, enough moisture may evaporate from the water surface to saturate the air above. Common on autumn mornings when the air is cool and the water in lakes or streams is still relatively warm. |
| | Frontal (precipitation) fog | Forms when raindrops falling from warm air above a frontal surface evaporate into the cooler air below and causes saturation. A wintertime phenomenon that produces cool damp days. |

Thus, advection fogs often extend 300 to 600 meters (1000 to 2000 feet) above the surface and persist longer than radiation fogs. An example of such fog can be found at Cape Disappointment, Washington—the foggiest location in the United States. The name is indeed appropriate because the station averages about 2552 hours of fog each year—equivalent to 106 days.

Advection fog is also a common wintertime phenomenon in the Southeast and Midwest when relatively warm, moist air from the Gulf of Mexico and Atlantic moves over cold and occasionally snow-covered surfaces to produce widespread foggy conditions. This type of advection fog tends to be thick and produce hazardous driving conditions.

**Upslope Fog** As its name implies, **upslope fog** is created when relatively humid air moves up a gradually sloping landform or, in some cases, up the steep slopes of a mountain. Because of the upward movement, air expands and cools adiabatically. If the dew point is reached, an extensive layer of fog will form.

It is easy to visualize how upslope fog might form in mountainous terrain. However, in the United States, upslope fog also occurs in the Great Plains, when humid air moves from the Gulf of Mexico toward the Rocky Mountains. (Recall that Denver, Colorado, is called the "mile-high city," and the Gulf of Mexico is at sea level.) Air flowing "up" the Great Plains expands and cools adiabatically by as much as 12°C (22°F), which can result in extensive upslope fog in the western plains.

## Evaporation Fogs

When saturation occurs primarily because of the addition of water vapor, the resulting fogs are called *evaporation fogs*. Two types of evaporation fogs are recognized: *steam fog* and *frontal (precipitation) fog*.

Meteorologist's Sketch

▶ **Figure 5.10 Advection fog forms when warm, moist air moves over a cool surface** This fog bank, rolling into San Francisco Bay, was generated as moist air passed over the cold California Current.

▲ **Figure 5.11 Steam fog occurs in the fall when cool air flows over a comparatively warm water body** This image shows steam fog rising from Sierra Blanca Lake, Arizona.

**Steam Fog** When cool, unsaturated air moves over a warm water body, enough moisture may evaporate to saturate the air directly above, generating a layer of fog. The added moisture and energy often makes the saturated air buoyant enough to cause it to rise. Because the foggy air looks like the "steam" that forms above a hot cup of coffee, the phenomenon is called **steam fog** (Fig. 5.11). Steam fog is a fairly common occurrence over lakes and rivers on clear, crisp mornings in the autumn when the water is still relatively warm but the air is comparatively cold. Steam fog usually forms a shallow foggy layer because as it rises, the water droplets mix with the unsaturated air above and evaporate.

In a few settings, steam fogs can be dense—especially during the winter, as cold arctic air pours off the continents and ice shelves over the comparatively warm open ocean. The temperature contrast between the warm ocean surface and overlying cold air mass has been known to exceed 30°C (54°F). The result is thick steam fog produced as the rising water vapor saturates a large volume of air. Because of its source and appearance, this type of dense steam fog is given the name *arctic sea smoke*.

**students sometimes ask...**

**Why do I see my breath on cold mornings?**

On cold days when you "see your breath," you are actually creating steam fog. The moist air that you exhale saturates a small volume of cold air, causing tiny droplets to form. As with steam fogs, the droplets quickly evaporate as the "fog" mixes with the unsaturated air around it.

**Frontal (Precipitation) Fog** Frontal boundaries where a warm, moist air mass is forced to rise over cooler, dryer air below generates **frontal (precipitation) fog**. The foggy conditions result because the raindrops falling from relatively warm air above the frontal surface evaporate in the cooler air below, causing it to become saturated. Frontal fog, which can be quite thick, is most common on cool days during extended periods of light rainfall.

The frequency of dense fog varies considerably from place to place (Fig. 5.12). As might be expected, fog incidence is highest in coastal areas, especially where cold currents prevail, as along the Pacific and New England coasts. Relatively high frequencies are also found in the Great Lakes region and in the humid Appalachian Mountains of the Eastern United States. In contrast, fogs are rare in the interior of the continent, especially in the arid and semiarid areas of the West (the yellow areas in Figure 5.12).

## ✔ Concept Checks 5.3

1. Distinguish between clouds and fog.
2. List five types of fog and discuss how they form.
3. What actually happens when a radiation fog "lifts"?
4. Why is there a relatively high frequency of dense fog along the Pacific coast?

**Video** (MM)
A Satellite View of Fog
http://goo.gl/ECLZLM

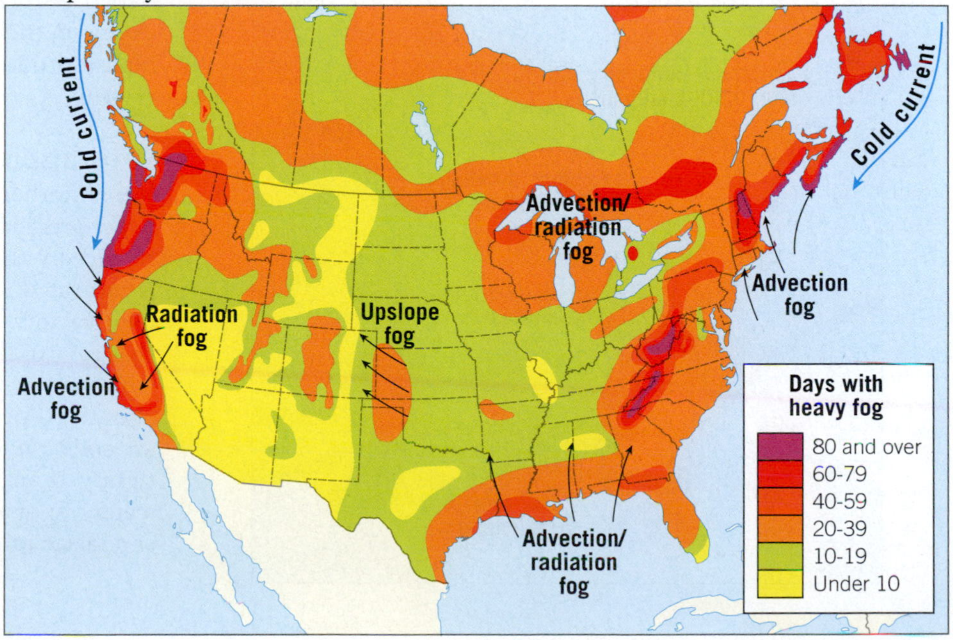


http://goo.gl/FJVFO

▶ **SmartFigure 5.12 Map showing average numbers of days per year with heavy fog** Coastal areas where cold currents prevail, particularly the Pacific Northwest and New England, have high occurrences of dense fog.

**MapMaster** ▶ North America ▶ Physical Environment ▶ Days with Heavy Fog

# 5.4 How Precipitation Forms

**Describe the Bergeron process and explain how it differs from the collision–coalescence process.**

MM GEODe ▶ Forms of Condensation and Precipitation ▶ How Precipitation Forms

If all clouds contain water, why do some produce precipitation while others drift placidly overhead? This seemingly simple question perplexed meteorologists for many years.

Typical cloud droplets are miniscule—20 micrometers (0.02 millimeter) in diameter (Fig. 5.13). In comparison, a human hair is about 75 micrometers in diameter. Because of their small size, cloud droplets fall in still air incredibly slowly. An average cloud droplet falling from a cloud base at 1000 meters (3280 feet) would require several hours to reach the ground. However, it would never complete its journey. Instead, the cloud droplet would evaporate before it fell a few meters from the cloud base into the unsaturated air below.

How large must a cloud droplet grow in order to fall as precipitation? A typical raindrop has a diameter of about 2 millimeters, or 100 times that of the average cloud droplet (Figure 5.13). However, the *volume* of a typical raindrop is 1 million times that of a cloud droplet. Thus, for precipitation to form, cloud droplets must grow in volume by roughly 1 million times. You might suspect that additional condensation creates drops large enough to survive the descent to the surface. However, clouds consist of many billions of tiny cloud droplets that all compete for the available water. Thus, condensation provides an inefficient means of raindrop formation.

Two processes are responsible for the formation of precipitation: the *Bergeron process* and the *collision–coalescence process*.

▼ **Figure 5.13 Diameters of particles involved in condensation and precipitation processes**

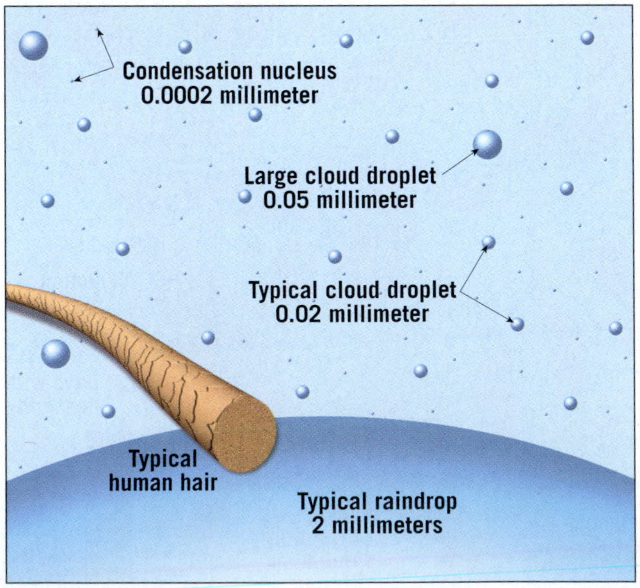


## Precipitation from Cold Clouds: The Bergeron Process

The **Bergeron process**, which generates much of the precipitation in the middle and high latitudes, is named for its discoverer, the highly respected Swedish meteorologist Tor Bergeron. To understand how this mechanism operates, we must first examine two important properties of water.

**Supercooled Water** The Bergeron process operates in *cold clouds*, at temperatures below 0°C (32°F), where liquid cloud droplets and ice crystals coexist. Contrary to what you might expect, *cloud droplets do not usually freeze at 0°C (32°F)*. In fact, pure water suspended in air will not freeze until it reaches a temperature of about −40°C (−40°F). Water in the liquid state below 0°C (32°F) is referred to as **supercooled water**. Supercooled water readily freezes if it impacts an object, which explains why airplanes collect ice when they pass through a cold cloud made up of supercooled droplets. Supercooled water droplets also cause *freezing rain*, which falls as a liquid, but then turns to a sheet of ice when it strikes the pavement, tree branches, and car windshields.

In the atmosphere, supercooled droplets freeze on contact with solid particles that have a shape that closely resembles that of ice (silver iodide, for example). These materials, called **freezing nuclei**, are sparse in the atmosphere and do not generally become active until the air temperature is about −15°C (5°F) or colder. Thus, at temperatures between 0 and −15°C, most clouds consist only of supercooled water droplets (Fig. 5.14). Between −15 and −40°C, most clouds consist of supercooled droplets that coexist with ice crystals, and at temperatures colder than −40°C (−40°F), clouds are composed entirely of ice crystals. For example, the tops of towering cumulonimbus clouds and wispy high-altitude cirrus clouds are usually composed entirely of ice crystals.

**Saturation Vapor Pressure over Water Versus over Ice** Another important property of water is that *the saturation vapor pressure above ice crystals is lower than above water droplets*. Stated another way, when the air surrounding a water droplet is saturated (100 percent relative humidity), it is supersaturated relative to a nearby ice crystal. For example, Table 5.3 shows that at −10°C (14°F), when the relative humidity is 100 percent with respect to *water*, the relative humidity with respect to *ice* is about 110 percent. This difference results because ice crystals are solid, so the individual ice molecules are held together more tightly than those of a liquid droplet. For the same reason, water vapor molecules escape (evaporate) from water droplets at a faster rate than from ice crystals, at a given temperature.

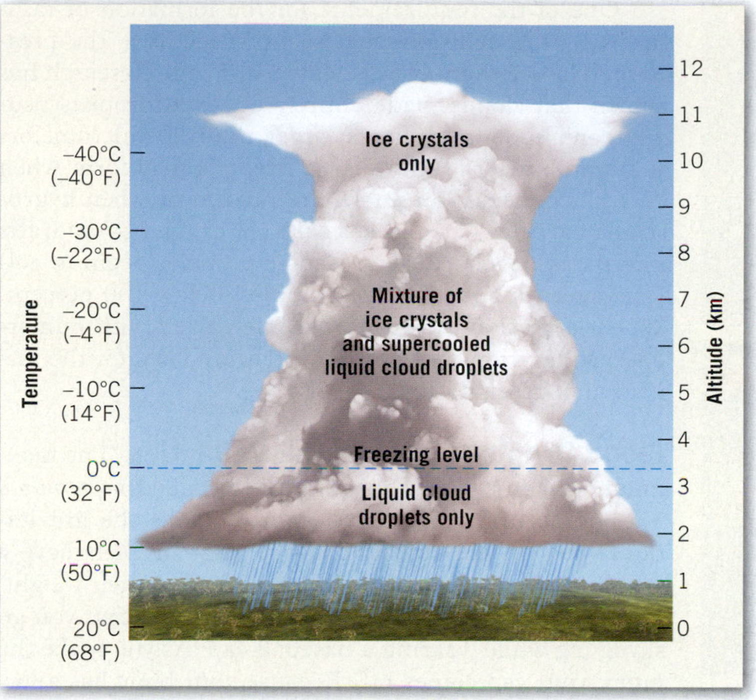


▲ **Figure 5.14 Nature of the particles that are found in towering cumulonimbus clouds**

## How the Bergeron Process Generates Precipitation

When ice crystals and supercooled water droplets coexist, the conditions are ideal for creating precipitation. Because freezing nuclei are sparse, cold clouds consist of relatively few ice crystals (snow crystals) surrounded by numerous liquid droplets (**Fig. 5.15**). Since the air is supersaturated with respect to the comparatively few ice crystals, the ice crystals will begin to collect water molecules by the process of deposition. This in turn lowers the overall relative humidity of the air. In response, the surrounding water droplets will begin to evaporate to replenish the lost water vapor. Thus the growth of ice crystals is fed by the continued evaporation and shrinkage of the liquid droplets.

When ice crystals become sufficiently large, they begin to fall. Air movement will sometimes break up these delicate crystals, and the fragments will serve as freezing nuclei for other liquid droplets. A chain reaction ensues and produces many snow crystals, which may join together into larger masses called *snowflakes*.

The Bergeron process can produce precipitation throughout the year in the middle latitudes, provided that at least a portion of a cloud is cold enough, about –15°C (5°F), to

**Table 5.3** Relative Humidity with Respect to Ice When Relative Humidity with Respect to Water Is 100 Percent

| Temperature (°C) | Relative Humidity with Respect To: | |
|---|---|---|
| | Water | Ice |
| 0 | 100% | 100% |
| –5 | 100% | 105% |
| –10 | 100% | 110% |
| –15 | 100% | 115% |
| –20 | 100% | 121% |

**eye ON THE atmosphere 5.2**

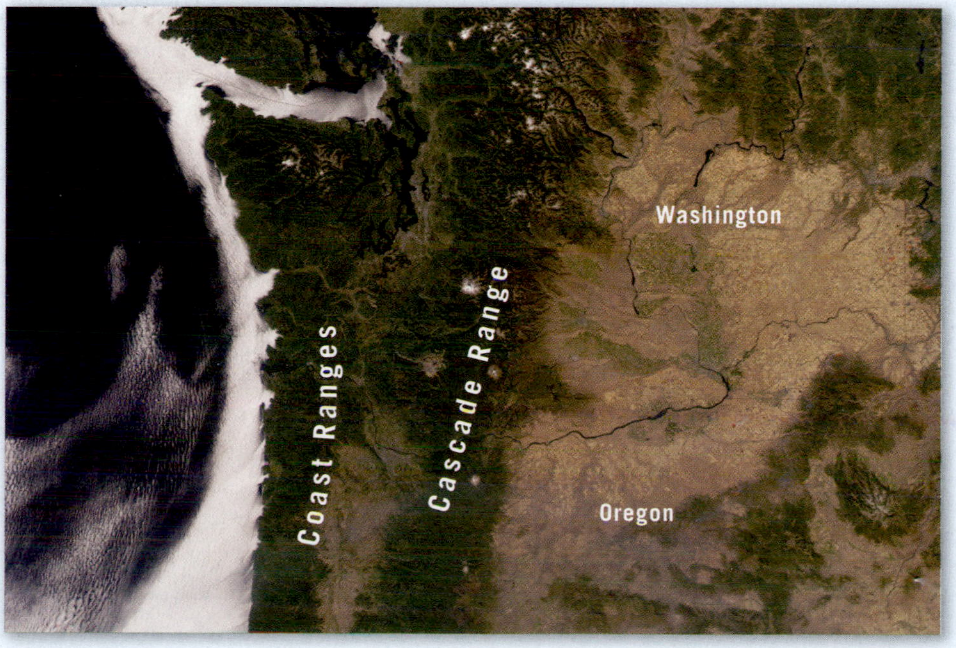


This satellite image shows a layer of low-lying clouds moving eastward into the bays and waterways of the northwestern United States. Lining the coast are the forested Coast Ranges and further inland is the Cascade Range, which includes several large volcanoes.

**Questions**

1. Why do you think the clouds formed over the ocean, while the land ares are cloud free?
2. Why are the mountainous areas of western Washington State covered with lush vegetation while much of the eastern half of the state appears semiarid?

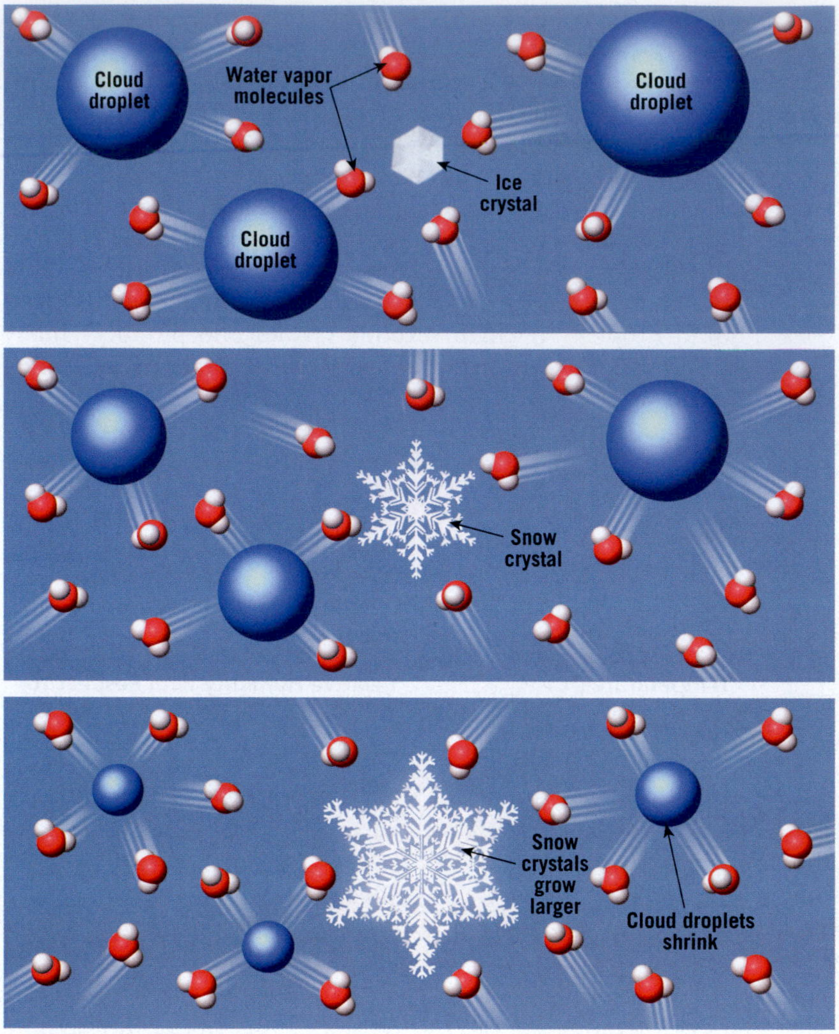


▲ **Figure 5.15 The Bergeron process** Ice crystals grow at the expense of cloud droplets until they are large enough to fall. The size of these particles has been greatly exaggerated.

generate ice crystals. The type of precipitation (snow, sleet, rain, or freezing rain) that reaches the ground depends on the temperature profile in the lower few kilometers of the atmosphere. When the surface temperature is above 4°C (39°F), snowflakes usually melt before they reach the ground and continue their descent as rain. Even on a hot summer day, a heavy rainfall may have begun as a snowstorm high in the clouds overhead. During a middle-latitude winter, even low clouds are cold enough to trigger precipitation via the Bergeron process.

## Precipitation from Warm Clouds: The Collision–Coalescence Process

The **collision–coalescence process** is the dominant process for generating precipitation in *warm clouds*—clouds with tops warmer than –15°C (5°F). Simply, the collision–coalescence process involves multiple collisions of tiny cloud droplets that stick together (coalesce) to form raindrops large enough to reach the ground before evaporating.

One of the requirements for the formation of raindrops by the collision–coalescence process is the presence of larger-than-average cloud droplets. Research has shown that clouds made entirely of liquid droplets usually contain some droplets larger than 20 micrometers (0.02 millimeter). These large droplets may form when "giant" condensation nuclei are present or when hygroscopic particles (such as sea salt) are carried by updrafts into the atmosphere. Hygroscopic particles begin to collect water vapor at relative humidity below 100 percent. When large cloud droplets are intermixed with numerous smaller droplets, the conditions are ideal for the formation of precipitation.

### Cloud Droplet Size and Fall Velocities

The maximum speed at which an object falls, called its *terminal velocity*, occurs when air resistance equals the gravitational pull on the object. Because large droplets have a smaller ratio of surface area as compared to their weight, they fall faster than small droplets. Imagine that you go skydiving while wearing a baseball cap. As you make the jump, your cap comes off. Because your body has a low ratio of surface area compared to your weight, you will have a much higher terminal velocity than your baseball cap. Table 5.4 summarizes how this principle applies to cloud droplets and their fall velocities.

As the larger droplets fall through a cloud, they collide with smaller, slower droplets and coalesce. They become larger in the process and fall even more rapidly (or, in an updraft, they rise more slowly), which increases their chances of collision and rate of growth (Fig. 5.16A). After a million or so cloud droplets coalesce, they form a raindrop that is large enough to fall to the surface without evaporating.

Because of the huge number of collisions required for growth to raindrop size, clouds that have great vertical thickness and contain large cloud droplets have the best chance of producing precipitation. Updrafts associated with unstable air also aid this process because the droplets can traverse the cloud repeatedly, which results in more collisions.

As raindrops grow in size, their fall velocity increases. This in turn increases the frictional resistance of the air, which causes the drop's "bottom" to flatten out (Fig. 5.16B). As a drop approaches 4 millimeters in diameter, it develops a depression,

**Table 5.4** | Fall Velocity of Water Drops

| Types | Diameter (millimeters) | Fall Velocity (km/hr) | (mi/hr) |
|---|---|---|---|
| Small cloud droplets | 0.01 | 0.01 | 0.006 |
| Typical cloud droplets | 0.02 | 0.04 | 0.03 |
| Large cloud droplets | 0.05 | 0.3 | 0.2 |
| Drizzle droplets | 0.5 | 7 | 4 |
| Typical rain drops | 2.0 | 23 | 14 |
| Large rain drops | 5.0 | 33 | 20 |

Data from Smithsonian Meteorological Tables.

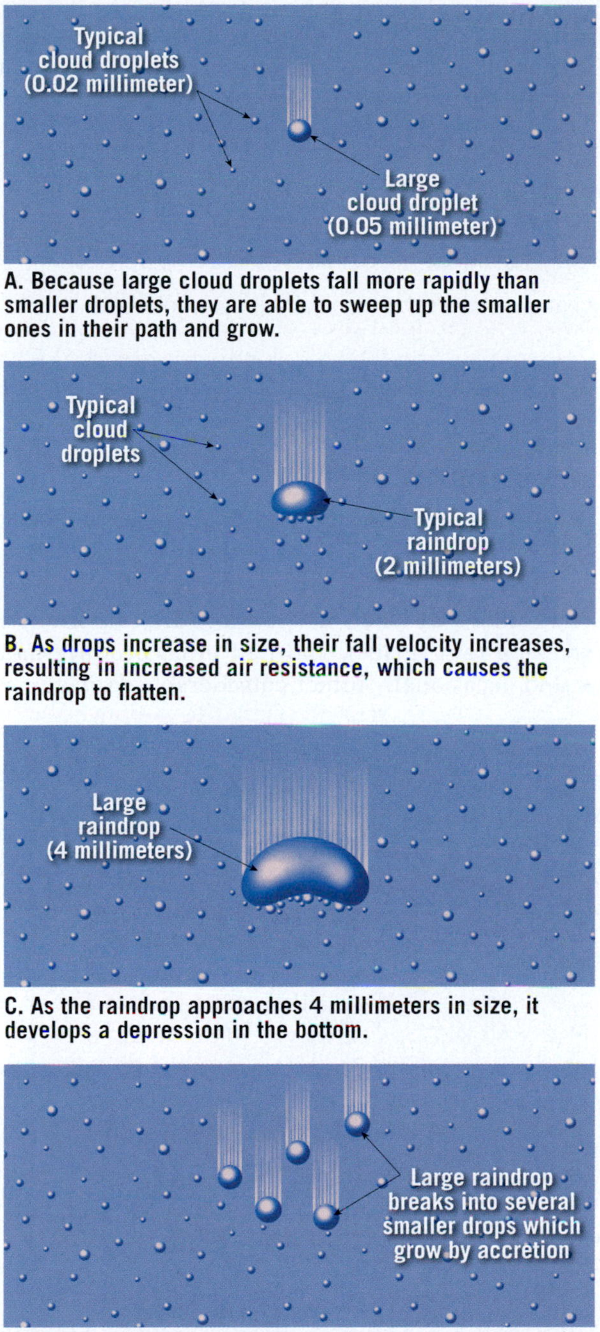


**A. Because large cloud droplets fall more rapidly than smaller droplets, they are able to sweep up the smaller ones in their path and grow.**

**B. As drops increase in size, their fall velocity increases, resulting in increased air resistance, which causes the raindrop to flatten.**

**C. As the raindrop approaches 4 millimeters in size, it develops a depression in the bottom.**

**D. Finally, when the diameter exceeds about 5 millimeters, the depression grows upward almost explosively, forming a donut–like ring of water that immediately breaks into smaller drops.**

▲ **Figure 5.16 The collision–coalescence process** The collision–coalescence process involves multiple collisions of tiny cloud droplets that stick together (coalesce) to form raindrops large enough to reach the ground before evaporating.

as shown in **Figure 5.16C**. Raindrops can grow to a maximum of 5 millimeters when they fall at the rate of 33 kilometers (20 miles) per hour. At this size, the water's surface tension, which holds the drop together, is surpassed by the frictional drag of the air. The depression grows almost explosively, forming a donutlike ring that immediately breaks apart. The resulting breakup of a large raindrop produces numerous smaller drops that begin anew the task of sweeping up cloud droplets (**Fig. 5.16D**).

**How Do Cloud Droplets Coalesce?** The collision–coalescence process is not as simple as it may seem. First, as the larger droplets descend, they produce an airstream around them similar to that produced by an automobile when driven rapidly down the highway. The airstream repels objects, especially the smallest cloud droplets. Imagine driving on a summer night along a country road. The bugs in the air are like cloud droplets: Most are pushed aside, but larger bugs (cloud droplets) have an increased chance of colliding with the car (giant droplet).

Further, collision does not guarantee coalescence. Experimentation has indicated that the presence of atmospheric electricity may be the key to what holds these droplets together once they collide. If a droplet with a negative charge collides with a positively charged droplet, their electrical attraction may bind them together.

The air over the tropical oceans is an ideal setting for the development of precipitation by the collision–coalescence process because the relatively clean air contains fewer condensation nuclei compared to the air over populated urban regions. With fewer condensation nuclei to compete for available water vapor (which is plentiful), condensation is fast-paced and produces comparatively few large cloud droplets. Within developing cumulus clouds, the largest drops quickly gather smaller droplets to generate the warm afternoon showers associated with tropical climates.

In the middle latitudes, the collision–coalescence process may contribute to the precipitation from a large cumulonimbus cloud by working in tandem with the Bergeron process—particularly during the hot, humid summer months. High in these towers, the Bergeron process generates snow that melts as it passes below the freezing level. When snowflakes melt, they generate relatively large drops with fast fall velocities. As these large drops descend, they overtake and coalesce with the slower and smaller cloud droplets that comprise much of the lower regions of the cloud. The result can be a heavy downpour.

**students sometimes ask...**

**Why does it often seem like the roads are slippery when it rains after a long dry period?**

Studies suggest that a buildup of oily residue emitted by automobiles during extended periods of dry weather is the cause of those slippery road conditions after a rainfall. One traffic study indicates that rain results in no increase in risk of fatal crashes if it also rained the previous day. However, if 2 days have lapsed since the last rain, then the risk for a deadly accident increases by 3.7 percent. If it has been 21 days since the last rain, the risk increases by 9.2 percent.

In summary, two mechanisms are known to generate precipitation: the Bergeron process and the collision–coalescence process. The Bergeron process dominates in the middle and high latitudes, where cold clouds (or cold cloud tops) are the rule. In the tropics, abundant water vapor and comparatively few condensation nuclei are more typical. This leads to the formation of fewer, larger drops with fast fall velocities that grow by collision and coalescence.

Video MM
The Importance of Wind Resistance
http://goo.gl/WcRqp7

## ✔ Concept Checks 5.4

1. Describe the temperature conditions in clouds that are required to form precipitation by the Bergeron process.
2. What is supercooled water?
3. Explain how snow that formed high in a towering cloud might produce rain.
4. Briefly summarize the collision–coalescence process.
5. What determines the terminal velocity of objects?

# 5.5 Forms of Precipitation

**Describe the atmospheric conditions that produce sleet, freezing rain (glaze), and hail.**

MM GEODe ▶ Forms of Condensation and Precipitation ▶ Forms of Precipitation

Atmospheric conditions vary greatly both geographically and seasonally, resulting in several different types of precipitation (Fig. 5.17). Rain and snow are the most common and familiar forms, but others, as listed in Table 5.5, are important as well. Sleet, freezing rain (glaze), and hail often produce hazardous weather and occasionally inflict considerable damage.

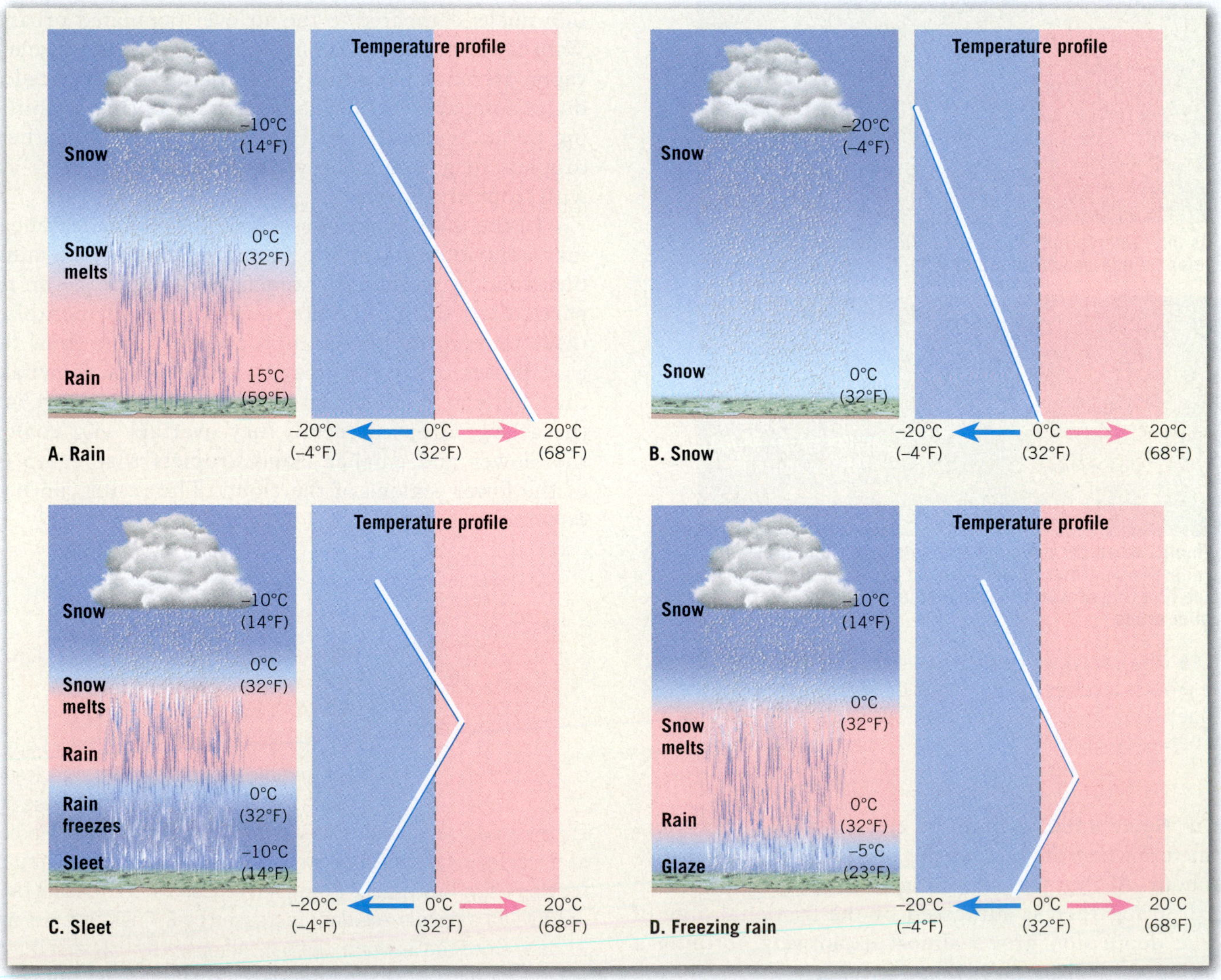


▲ **Figure 5.17 Four precipitation types and their temperature profiles**

**Table 5.5** | Types of Precipitation

| Type | Approximate Size | State of Water | Description |
|---|---|---|---|
| Mist | 0.005–0.05 mm | Liquid | Droplets large enough to be felt on the face when air is moving 1 meter/second. Associated with stratus clouds. |
| Drizzle | 0.05–0.5 mm | Liquid | Small, uniform droplets that fall from stratus clouds, generally for several hours. |
| Rain | 0.5–5 mm | Liquid | Generally produced by nimbostratus or cumulonimbus clouds. When heavy, size can be highly variable from one place to another. |
| Sleet | 0.5–5 mm | Solid | Small, spherical to lumpy ice particles that form when raindrops freeze while falling through a layer of subfreezing air. Because the ice particles are small, any damage is generally minor. Sleet can make travel hazardous. |
| Freezing Rain (glaze) | Layers 1 mm–2 cm thick | Solid | Produced when supercooled raindrops freeze on contact with solid objects. Freezing rain can form a thick coating of ice that has sufficient weight to seriously damage trees and power lines. |
| Rime | Variable accumulations | Solid | Deposits usually consisting of ice feathers that point into the wind. These delicate frostlike accumulations form as supercooled cloud or fog droplets encounter objects and freeze on contact. |
| Snow | 1 mm–2 cm | Solid | The crystalline nature of snow allows it to assume many shapes, including six-sided crystals, plates, and needles. Produced in supercooled clouds where water vapor is deposited as ice crystals that remain frozen during their descent. |
| Hail | 5–10 cm or larger | Solid | Precipitation in the form of hard, rounded pellets or irregular lumps of ice. Produced in large convective cumulonimbus clouds, where frozen ice particles and supercooled water coexist. |
| Graupel | 2–5 mm | Solid | "Soft hail" that forms as rime collects on snow crystals to produce irregular masses of "soft" ice. Because these particles are softer than hailstones, they normally flatten out upon impact. |

## Rain, Drizzle, and Mist

In meteorology, the term **rain** is restricted to drops of water that fall from a cloud and have a diameter of at least 0.5 millimeter. Most rain originates in either nimbostratus clouds or in towering cumulonimbus clouds that are capable of producing unusually heavy rainfalls known as *cloudbursts*.

As rain enters the unsaturated air below the cloud, it begins to evaporate. Depending on the humidity of the air and the size of the drops, rain may completely evaporate before reaching the ground. This phenomenon produces **virga**, which appear as streaks of precipitation falling from a cloud that extend toward Earth's surface without reaching it (**Fig. 5.18**). Similar to virga, ice crystals may sublimate (see Chapter 4) when they enter the dry air below. These wisps of ice particles are called **fallstreaks**.

Fine, uniform droplets of water with diameters less than 0.5 millimeters are called **drizzle**. Drizzle and small raindrops generally are produced in stratus or nimbostratus clouds, from which precipitation may be continuous for several hours or, on rare occasions, for days.

Precipitation containing the very smallest droplets able to reach the ground is called **mist**. Mist can be so fine that the tiny droplets appear to float, and their impact is almost

▶ **Figure 5.18 Virga is precipitation that evaporates before reaching the surface** Virga, Latin for "streak," is common in the arid Southwest.

imperceptible. Mist closely resembles fog. Meteorologists use the word *fog* when the visibility is less than 1 kilometer (0.6 miles) and mist when the visibility is greater than 1 kilometer.

## Snow and Graupel

**Snow** is a form of winter precipitation in the form of ice crystals, or aggregates of ice crystals. The size, shape, and concentration of snowflakes depend to a great extent on the temperature profile of the atmosphere.

Recall that at very low temperatures, the moisture content of air is low. The result is the generation of very light and fluffy snow made up of individual six-sided ice crystals (**Fig. 5.19**). This is the "powder" that downhill skiers covet. By contrast, at temperatures warmer than about −5°C (23°F), the ice crystals may join together into larger clumps consisting of tangled aggregates of crystals. Snowfalls consisting of these composite snowflakes are generally heavy and have a high moisture content, which makes them ideal for making snowballs.

Snow can reach Earth's surface when the temperature of a thin layer of air near the ground is above freezing. In this situation, the snow does not have enough time to melt before reaching the ground. This type of snow is usually quite wet.

Under certain atmospheric conditions, falling snow crystals grow as they intercept tiny supercooled cloud droplets that freeze on them. The resulting snowflakes are described as being *rimed*. If riming continues and makes the shape of the original six-sided snow crystal no longer identifiable, the soft ice pellet is called **graupel**. Usually oblong in shape and fragile enough that it will fall apart when touched, graupel is also known as *soft hail* or *snow pellets*.

▼ **Figure 5.19 Snow crystals** Snow crystals are usually six-sided, but they come in an infinite variety of forms. The snowflakes that reach the ground often consist of multiple ice crystals stuck together.

**students sometimes ask...**

**What is the snowiest city in the United States?**

According to National Weather Service records, Rochester, New York, is the snowiest U.S. city, averaging nearly 239 centimeters (94 inches) of snow annually. However, Buffalo, New York, is a close runner-up.

## Sleet and Freezing Rain or Glaze

**Sleet**, a wintertime phenomenon, consists of clear to translucent ice pellets. Depending on intensity and duration, sleet can cover the ground much like a thin blanket of snow. **Freezing rain**, or **glaze**, on the other hand, falls as supercooled raindrops that freeze on contact with roads, power lines, and other structures.

As shown in **Figure 5.20**, both sleet and freezing rain occur in the winter, and they most often form along a warm front where a mass of relatively warm air is forced over a layer of subfreezing air near the ground. Both begin as snow, which melts to form raindrops as it falls though the layer of warm air below.

However, the conditions present beyond this point determine whether the drops become sleet or freezing rain. When the newly formed raindrops encounter a thick cold layer of air below the frontal boundary, sleet results. In this setting, as raindrops fall through the subfreezing air, they refreeze and reach the ground as small pellets of ice roughly the size of the raindrops from which they formed. If, however, the layer of cold air near the ground is not thick enough to cause the raindrops to refreeze, they instead become supercooled—that is, they remain liquid at temperatures below freezing (Figure 5.20). Upon striking subfreezing objects on Earth's surface, these supercooled raindrops instantly turn to ice. The result is thick coating of freezing rain that has sufficient weight to break tree limbs, down power lines, and make walking and driving extremely hazardous.

In January 1998 an ice storm of historic proportions caused enormous damage in New England and southeastern Canada. Five days of freezing rain deposited a heavy layer of ice on exposed surfaces from eastern Ontario to the Atlantic coast. The 8 centimeters (3 inches) of precipitation caused trees, power lines, and high-voltage towers to collapse, leaving over 1 million households without power—many for nearly a month following the

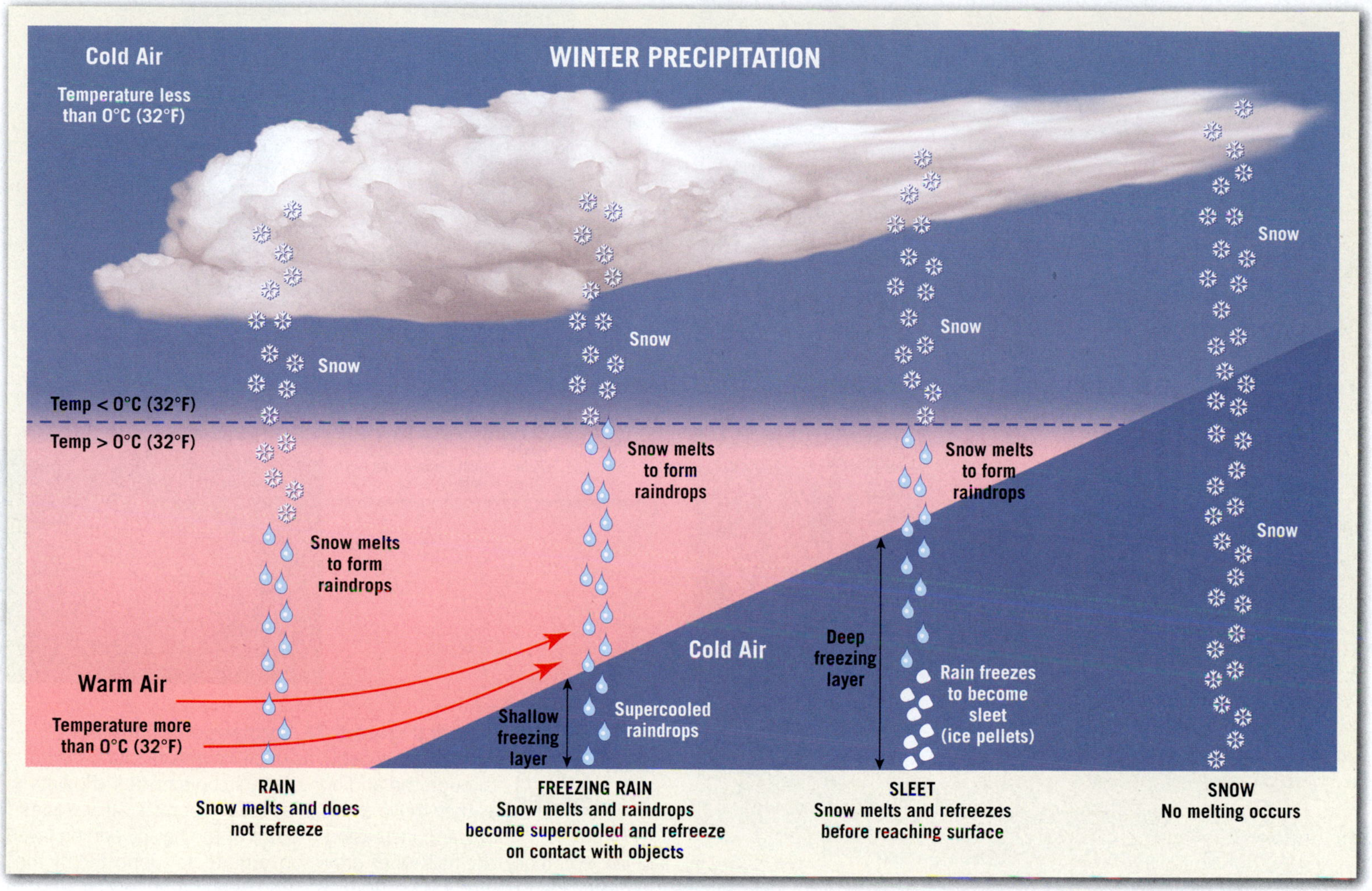


▲ **Figure 5.20 Formation of sleet and freezing rain** When rain passes through a cold layer of air and freezes, the resulting ice pellets are called sleet. Freezing rain forms under similar conditions, except the cold layer of air is not deep enough to refreeze the raindrops. These forms of precipitation occur often in the winter, when warm air (along a warm front) is forced over a layer of subfreezing air.

storm (Fig. 5.21). At least 40 deaths were blamed on the storm, which caused damages in excess of $3 billion. Much of the damage was to the electrical grid, which one Canadian climatologist summed up this way: "What it took human beings a half-century to construct, took nature a matter of hours to knock down."

## Hail

**Hail** is precipitation in the form of hard, rounded pellets or irregular lumps of ice with diameters of 5 millimeters (0.20 inches) or more. Hail is produced in the middle to upper reaches of tall cumulonimbus clouds where updrafts can sometimes exceed speeds of 160 kilometers (100 miles) per hour and where the air temperature is below freezing. Hailstones begin as small embryonic ice pellets or graupel that coexist with supercooled droplets. The ice pellets grow by collecting supercooled water droplets, and sometimes other small pieces of hail, as they are lifted by updrafts within the cloud.

Cumulonimbus clouds that produce hail have a complex system of updrafts and downdrafts. As shown in Figure 5.22A, a region of intense updrafts suspends rain and hail aloft, producing a rain-free region surrounded by an area of downdrafts and heavy precipitation. The largest hailstones are generated around the core of the most intense zone of updraft where they rise slowly enough to collect appreciable amounts of supercooled water. The process continues until the hailstone grows too heavy to be supported by the updraft or encounters a downdraft and falls to the surface.

There are two methods by which large hailstones grow, *wet growth* and *dry growth*, which tends to produce a layered look that alternates between clear and milky ice (Fig. 5.22B). Clear ice is produced by wet growth in the lower and warmer regions of the clouds, where colliding droplets wet the surface of the hailstones. As these droplets slowly freeze, any air bubbles in the water escape—producing relatively bubble-free clear ice. By contrast, high in the clouds where the temperatures are well below freezing, the supercooled droplets immediately freeze

▲ **Figure 5.21 Freezing rain results when supercooled raindrops freeze on contact with objects** In January 1998, an ice storm of historic proportions caused enormous damage in New England and southeastern Canada. Nearly 5 days of freezing rain (glaze) caused 40 deaths and more than $3 billion in damages, and it left millions of people without electricity—some for as long as a month.

as they collide with the growing hailstone. The air bubbles are "frozen" in place, leaving milky ice.

Most hailstones have diameters between 1 centimeter (pea size) and 5 centimeters (golf ball size), although some can be as big as softballs. Occasionally, hailstones weighing 1 pound or more have been reported; most of these are composites of several stones frozen together. These large hailstones have fall velocities that exceed 160 kilometers (100 miles) per hour.

The record for the largest hailstone ever found in the United States was set on July 23, 2010, in Vivian, South Dakota. The stone was over 20 centimeters (8 inches) in diameter and weighed nearly 900 grams (2 pounds). The stone that held the previous record of 766 grams (1.69 pounds) fell in Coffeyville, Kansas, in 1970 (Figure 5.22B). The diameter of the stone found in South Dakota also surpassed the previous record of a 17.8-centimeter (7-inch) stone that fell in Aurora, Nebraska, in 2003. Even larger hailstones have reportedly been recorded in Bangladesh, where a 1987 hailstorm killed more than 90 people.

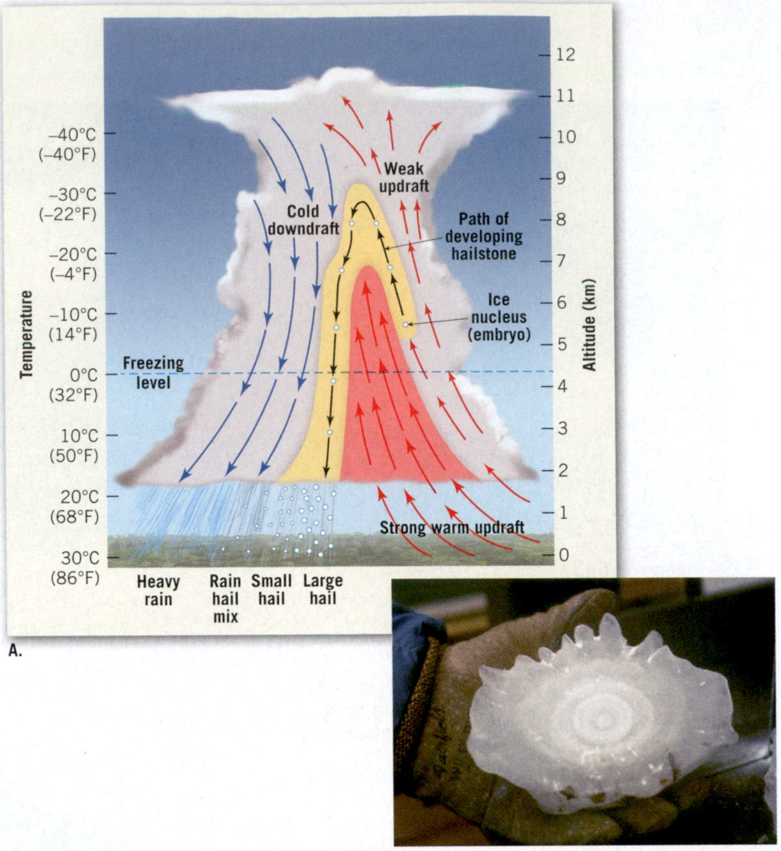


▲ **SmartFigure 5.22 Formation of hailstones** **A.** Hailstones begin as small ice pellets that grow through the addition of supercooled water droplets as they move through a cloud. Updrafts carry stones upward, increasing the size of the hail by adding layers of ice. Eventually, the hailstones grow too large to be supported by the updraft, or encounter a downdraft. **B.** This cut hailstone, which fell over Coffeyville, Kansas, in 1970, originally weighed 0.75 kilogram (1.67 pounds).

MM http://goo.gl/a34bP

The destructive effects of large hailstones are well known, especially to farmers whose crops have been devastated in a few minutes and to people whose windows, roofs, and cars have been damaged (Fig. 5.23). In the United States, hail damage each year can run into the hundreds of millions of dollars.

► **Figure 5.23 Hailstorm damage to an NOAA weather monitoring vehicle**

Video MM
Global Precipitation
http://goo.gl/6rvTi

**students sometimes ask...**

### What is the difference between a winter storm warning and a blizzard warning?

A winter storm warning is usually issued when heavy snow exceeding 6 inches in 12 hours or possible icing conditions are likely. In Upper Michigan and mountainous areas where snowfall is abundant, winter storm warnings are issued only if 8 or more inches of snow is expected in 12 hours. By contrast, blizzard warnings are issued for periods in which considerable falling and/or blowing snow will be accompanied by winds of 35 or more miles per hour. Thus, a blizzard is a type of winter storm in which winds are the determining factor, not the amount of snowfall.

One of the costliest hailstorms to occur in North America took place June 11, 1990, in Denver, Colorado, with total damage estimated to exceed $625 million. Figure 5.24 shows the average number of hail occurrences over a 10-year period.

## Rime

**Rime** is a deposit of ice crystals formed by the freezing of supercooled fog or cloud droplets on objects whose surface temperature is below freezing. When rime forms on trees, it adorns them with its characteristic ice feathers, which can be spectacular to observe (Fig. 5.25). In these situations, objects such as pine needles act as freezing nuclei, causing the supercooled droplets to freeze on contact. On occasions when the wind is blowing, only the windward surfaces of objects will accumulate the layer of rime.

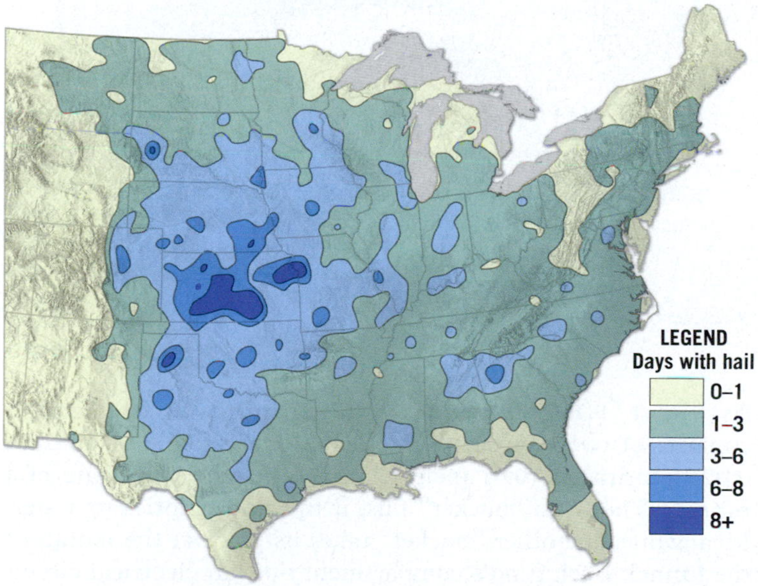


▲ **Figure 5.24 Average number of hail reports over a 100-square-mile area during a 10-year period**

▲ **Figure 5.25 Rime consists of delicate ice crystals** Rime forms when supercooled fog or cloud droplets freeze on contact with objects.

## ✔ Concept Checks 5.5

1. Compare and contrast rain, drizzle, and mist.
2. Describe sleet and freezing rain. Why does freezing rain result on some occasions and sleet on others?
3. How does hail form? What factors govern the ultimate size of hailstones?

# 5.6 Precipitation Measurement

**List the advantages and disadvantages of using a standard rain gauge versus weather radar to measure precipitation.**

The most common form of precipitation, rain, is probably the easiest to measure. Any open container that has a consistent cross-section throughout can be a rain gauge (Fig. 5.26A). In general practice, however, more sophisticated devices are used to measure small amounts of rainfall more accurately and to reduce loss from evaporation.

severe& hazardous weather Box 5.1

## Worst Winter Weather

Extremes, whether the *tallest* building or the *record low temperature* for a location, fascinate many humans. When it comes to weather, some places take pride in claiming to have the worst winters on record. In fact, both Fraser, Colorado, and International Falls, Minnesota, have proclaimed themselves the "ice box of the nation." Although Fraser recorded the lowest temperature for the 48 contiguous states 23 times in 1989, its neighbor Gunnison, Colorado, recorded the lowest temperature 62 times, far more than any other location.

Such facts do not impress the residents of Hibbing, Minnesota, where the temperature dropped to –38°C (–37°F) during the first week of March 1989. But this is mild stuff, say the old-timers in Parshall, North Dakota, where the temperature fell to –51°C (–60°F) on February 15, 1936. Not to be left out, Browning, Montana, holds the record for the most dramatic 24-hour temperature drop. Here the temperature plummeted 56°C (100°F), from a cool 7°C (44°F) to a frosty –49°C (–56°F) during a January evening in 1916.

Although impressive, the temperature extremes cited here represent only one aspect of winter weather.

Although impressive, the temperature extremes cited here represent only one aspect of winter weather. What about snowfall (Fig. 5.A)? Cooke City holds the seasonal snowfall record for Montana, with 1062 centimeters (418.1 inches) during the winter of 1977–1978. But what about cities like Sault Ste. Marie, Michigan, and Buffalo, New York? The winter snowfalls associated with the Great Lakes are legendary. Even larger snowfalls occur in many sparsely inhabited mountainous areas.

Try telling residents of the eastern United States that heavy snowfall alone makes for the worst weather. A blizzard in March 1993 produced heavy snowfall along with hurricane-force winds and

▲ **Figure 5.A** A winter snowstorm of historic proportions struck Chicago, Illinois, on February 2, 2011.

## Standard Instruments

The **standard rain gauge** (Fig. 5.26B) has a diameter of about 20 centimeters (8 inches) at the top. Once the water is caught, a funnel conducts the rain through a narrow opening into a cylindrical measuring tube that has a cross-sectional area only one-tenth as large as the receiver. Consequently, rainfall depth is magnified 10 times, which allows for accurate measurements to the nearest 0.025 centimeter (0.01 inch). When the amount of rain is less than 0.025 centimeter (0.01 inch), it is generally reported as being a **trace of precipitation**.

In addition to the standard rain gauge, several types of recording gauges are routinely used. These instruments not only record the amount of rain but also its time of occurrence and intensity (amount per unit of time). Two of the most common gauges are the tipping-bucket gauge and the weighing gauge.

As Figure 5.26C illustrates, the **tipping-bucket gauge** consists of two compartments, each one capable of holding 0.025 centimeter (0.01 inch) of rain, situated at the base of a funnel. When one "bucket" fills, it tips and empties its water. Meanwhile, the other "bucket" takes its place at the mouth of the funnel. Each time a compartment tips, an electrical circuit is closed, and 0.025 centimeter (0.01 inch) of precipitation is automatically recorded on a graph.

A **weighing gauge** collects precipitation in a cylinder that rests on a spring balance. As the cylinder fills, the movement is transmitted to a pen that records the data.

All rain gauges are susceptible to inaccuracies. The tipping-bucket rain gauge is known to underestimate heavy rainfall by perhaps 25 percent because of the rainwater that is not collected during the tipping movement of the bucket. Also, wind can lead to measurement errors, by either causing too much or too little precipitation to enter the collecting record low temperatures that immobilized much of the region from Alabama to the Maritime Provinces of eastern Canada. This event quickly earned the well-deserved title Storm of the Century.

So, determining which location has the worst winter weather depends on how it is measured. Most snowfall in a season? Longest cold spell? Coldest temperature? Most disruptive storm?

### Winter Weather Terms

Here are the meanings of some common terms that the National Weather Service uses for winter weather events.

#### Snow flurries

Snow falling for short durations at intermittent periods and resulting in generally little or no accumulation.

#### Blowing snow

Snow lifted from the surface by the wind and blown about to such a degree that horizontal visibility is reduced.

#### Drifting snow

Significant accumulations of falling or loose snow caused by strong wind.

#### Blizzard

A winter storm characterized by winds of at least 56 kilometers (35 miles) per hour for at least 3 hours. The storm must also be accompanied by low temperatures and considerable falling and/or blowing snow that reduces visibility to 0.25 mile or less.

#### Severe blizzard

A storm with winds of at least 72 kilometers (45 miles) per hour, a great amount of falling or drifting snow, and temperatures –12°C (10°F) or lower.

#### Heavy snow warning

A snowfall in which at least 4 inches (10 centimeters) in 12 hours or 6 inches (15 centimeters) in 24 hours is expected.

#### Freezing rain

Rain falling through a shallow subfreezing layer of air near the ground. The rain (or drizzle) freezes on impact with the ground or other objects, resulting in a clear coating of ice also known as *glaze*.

#### Sleet

Formed when raindrops or melted snowflakes freeze as they pass through a subfreezing layer of air near Earth's surface. Also called *ice pellets*, sleet does not stick to trees and wires, and it usually bounces when it hits the ground. An accumulation of sleet sometimes has the consistency of dry sand.

#### Travel advisory

An alert issued to inform the public of hazardous driving conditions caused by snow, sleet, freezing precipitation, fog, wind, or dust.

#### Cold wave

A rapid fall of temperature in a 24-hour period, usually signifying the beginning of a spell of very cold weather.

#### Wind chill

A measure of apparent temperature on the human body caused by the cooling power of wind. It is an approximation only for the human body and has no significance for cars, buildings, or other objects.

**Question**

1. Explain why it is difficult to determine a record weather event.

container. In addition, because rain gauges provide data for a specific location, the significant variations in the amount of precipitation that reaches the ground over a region cannot be accurately estimated.

## Measuring Snowfall

When snow records are kept, two measurements are normally used: depth and water equivalent. One way to measure the depth of snow is by using a calibrated stick. The actual measurement is not difficult, but choosing a representative spot can be. Even when winds are light or moderate, snow drifts freely. As a rule, it is best to take several measurements in an open place, away from trees and obstructions, and then average them. To obtain the water equivalent, samples may be melted and then weighed or measured as rain.

The quantity of water in a given volume of snow is not constant. You may have heard media weathercasters say, "Every 10 inches of snow equals 1 inch of rain." But the actual water content of snow may deviate widely from this figure. It may take as much as 30 inches of light and fluffy dry snow (30:1) or as little as 4 inches of wet snow (4:1) to produce 1 inch of water.

To measure the mountain snowpack, which produces 75 percent of the water supply for the western United States, automated weather stations employ **snow pillows** that have been installed at more than 600 sites. A snow pillow typically consists of two, three, or four large panels that measure the pressure exerted by the weight of snow that collects on them. Because snow pillows have large surface areas and measure the water equivalent of snow, they provide a good estimate of precipitation amounts.

## eye ON THE atmosphere 5.3

This image shows a phenomenon called a *hole punch cloud* that was produced when a jet aircraft ascended through a cloud deck composed of supercooled water droplets. As the aircraft passed through the clouds, tiny particles in the jet engine exhaust interacted with some of the supercooled water droplets—which froze instantly. The dark area in the center of the image is actually white and consists of large ice crystals that formed within the area of the cloud occupied by the punch hole and began to fall as precipitation.

**Questions**

1. What is the name of the process whereby some cloud droplets freeze and grow in size at the expense of the remaining liquid cloud droplets?
2. Explain why only clouds that are composed of supercooled droplets can develop "punch holes."
3. Describe the role that a jet aircraft plays in the formation of punch hole clouds.

# Precipitation Measurement by Weather Radar

Using **weather radar**, the National Weather Service (NWS) produces maps like the one in **Figure 5.27** in which colors illustrate precipitation intensity. The development of weather radar has given meteorologists an important tool to track storm systems and the precipitation patterns they produce, even when the storms are as far as a few hundred kilometers away.

Radar units have transmitters that send out short pulses of radio waves. The specific wavelengths selected depend on the objects being detected. When radar is used to monitor precipitation, wavelengths between 3 and 10 centimeters are employed. Radio waves, at these wavelengths, are able to penetrate clouds composed of small droplets, but are reflected by larger raindrops, ice crystals, and hailstones. The reflected signal, called an *echo*, is received and displayed on a TV monitor. Because the echo is "brighter" when the precipitation is more intense, modern weather radar is able to depict both the regional extent and rate of precipitation. Also, because the measurements are in real time, they are particularly useful in short-term forecasting.

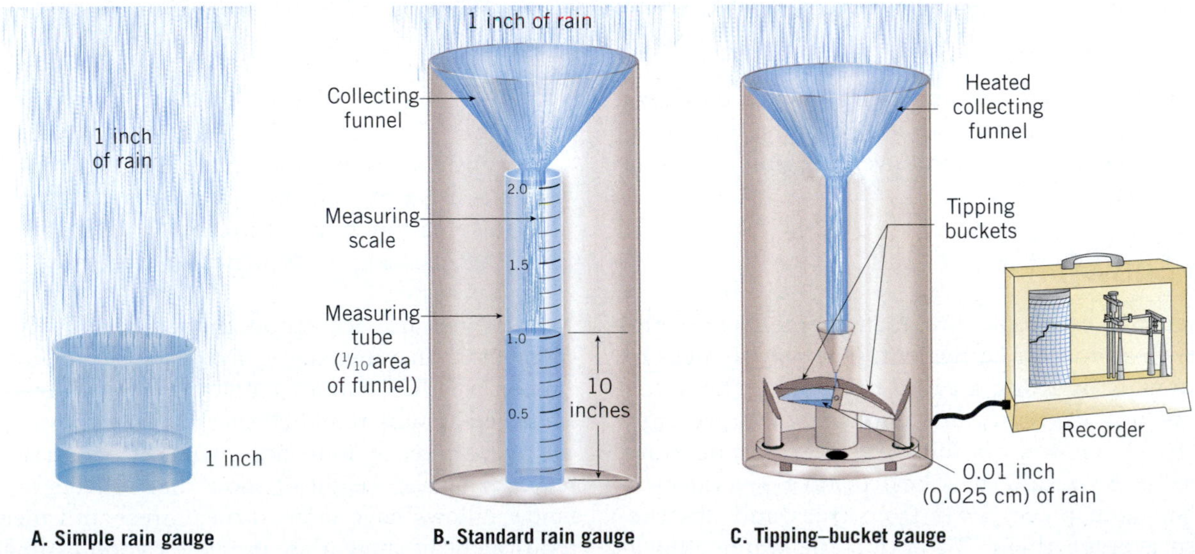


▲ **Figure 5.26 Precipitation measurement A.** The simplest gauge is any container left in the rain. **B.** The standard rain gauge increases the height of water collected by a factor of 10, allowing for accurate rainfall measurement to the nearest 0.025 centimeter (0.01 inch). Because the cross-sectional area of the measuring tube is only one-tenth as large as the collector, rainfall is magnified 10 times. **C.** The tipping-bucket rain gauge contains two "buckets," each holding the equivalent of 0.025 centimeter (0.01 inch) of liquid precipitation. When one bucket fills, it tips and the other bucket takes its place. Each event is recorded as 0.01 inch of rainfall.

# What's Your Forecast?

## Space-based Measurement of Precipitation

by Dr. J. Marshall Shepherd, Director, University of Georgia Atmospheric Sciences Program; Former GPM Deputy Project Scientist, NASA; and 2013 President, American Meteorological Society

Understanding Earth's water cycle, weather, and climate often requires a global perspective that's not possible from ground-based instruments. Precipitation is a very complex weather variable because it varies in time and by geographic location. Yet proper measurement and study of global precipitation are important for a variety of reasons, such as improving weather forecasting, identifying climate trends, warning about landslide hazards, assessing potential vector-borne diseases, or predicting agricultural productivity.

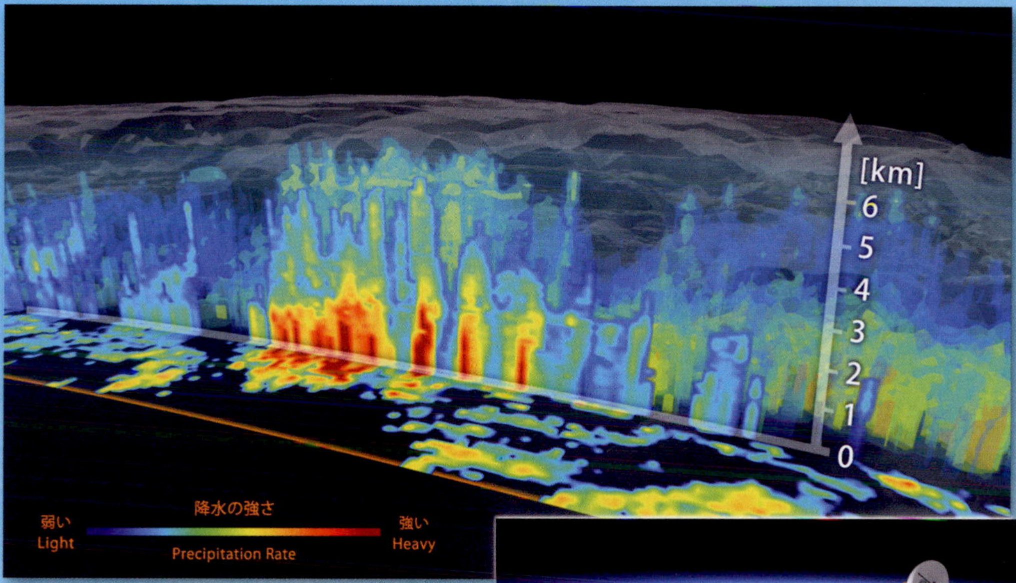

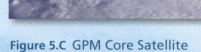

**Figure 5.B 3D View of Precipitation in an Extratropical Cyclone off the Coast of Japan Using GPM Radar**

### Observations

For many applications, measuring precipitation using rain gauges or weather radar estimation is appropriate. However, NASA has been advancing a new generation of space-based precipitation-measuring technologies for global applications or areas where ground measurements are not possible (for example, oceans, mountains, deserts). I was fortunate to spend 12 years helping with such missions. In February 2014, NASA launched the Global Precipitation Measurement (GPM) mission. I served as deputy project scientist for this mission, and trust me, it is really cool stuff. GPM uses a core satellite (**Fig. 5.B**) that carries a space weather radar and an array of instruments that use infrared (heat) or microwaves to measure rain or snow. A particularly intriguing feature of GPM is that it can produce standard two-dimensional maps or three-dimensional CAT scan–type images of weather systems, like the one shown in Figure 5.B.

To produce nearly global maps of precipitation, the measurements combine data from an international fleet of existing and future satellites. GPM evolved from the success of the Tropical Rainfall Measuring Mission (TRMM) satellite, which was launched in 1997 to measure rainfall in the tropical regions of our planet and to advance knowledge of Earth's general circulation and associated latent heating. (**Fig. 5.C**) shows an example of monthly rainfall, which was measured with contributions from the TRMM satellite. The intertropical convergence zone (ITCZ; see Chapter 7) is very evident in July 2011. TRMM is still in orbit and will be a critical part of the GPM constellation.

**Figure 5.C GPM Core Satellite**

### Modeling and Analysis

Like NASA Earth scientists, you can do your own analysis. The NASA Earth Observatory is a great Website for exploring our planet using real satellite data.

- Go to http://earthobservatory.nasa.gov/GlobalMaps/ and scroll down to Total Rainfall.
- Click the play button under the globe to see global rainfall totals over several years.

**Questions**

1. Using the map that you generated, what patterns can you identify in the rainfall? Why might such patterns exist (for example, mountains, ITCZ, frequent hurricanes)?
2. In a warming climate, scientists often speak of "an accelerated water cycle." Using Web-based resources like http://climate.gov, http://earthobservatory.nasa.gov, or http://ipcc.ch, look for useful information on the concept of accelerated water cycle and what that means. How would you describe it in simple terms to a friend?

Earth Observatory MM
Global Maps
http://goo.gl/n71c

For more information on the GPM or TRMM missions, visit the NASA Precipitation Measurement Missions page at http://pmm.nasa.gov.

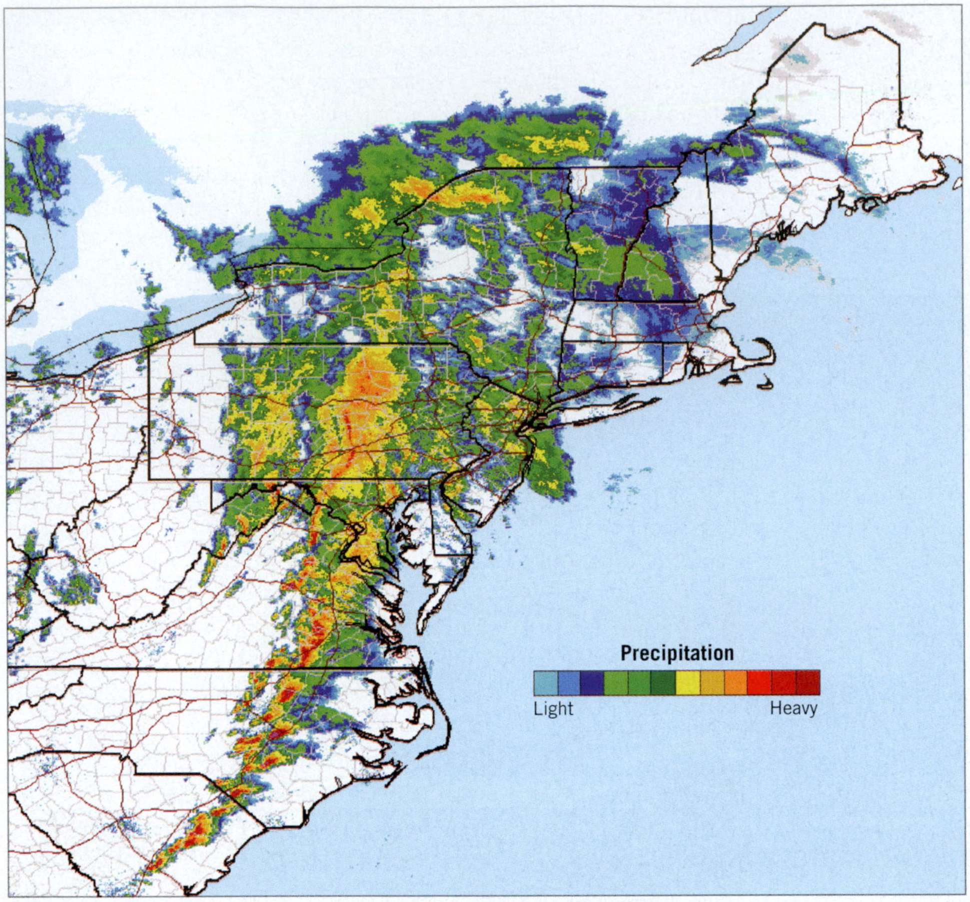


▲ **Figure 5.27 Doppler radar display produced by the National Weather Service** Colors indicate different intensities of precipitation. Note the band of heavy precipitation along the eastern seaboard.

Despite its usefulness, weather radar does not always show what is occurring at ground level. For example, radar may detect precipitation high in the atmosphere that doesn't reach the ground (virga). Also, weather radar detects "ground clutter" such as trees and buildings that complicate the images received. Occasionally, radar receives echoes produced by dense swarms of insects or birds. In addition, conventional radar systems cannot distinguish rain (liquid water) from solid forms of precipitation. Fortunately, the National Weather Service has upgraded most of its local forecast centers to *dual polarization radar*, which produces two-dimensional images. These images will help forecasters distinguish rain, hail, snow, or sleet from other flying objects, including tornado debris.

**Video** MM
Record-Breaking Hailstorm as Seen by Radar
http://goo.gl/sE6wy9

## ✔ Concept Checks 5.6

1. Although any open container can serve as a rain gauge, what advantages does a standard rain gauge provide?
2. What measurement inaccuracies are associated with the standard rain gauge?
3. What advantage does weather radar have over a standard rain gauge?

# 5.7 Planned and Inadvertent Weather Modification

**Discuss several ways that humans attempt to modify the weather.**

People modify the weather deliberately as well as unintentionally. Cloud seeding efforts to enhance precipitation at ski resorts and to dissipate fog at some airports are two examples of planned intervention, while an example of inadvertent weather modification is the increase in cloudiness from condensation trails produce by jet aircraft along major transportation corridors.

## Planned Weather Modification

*Planned weather modification* is deliberate human intervention to influence atmospheric processes that constitute the weather—that is, to alter the weather for human purposes. The desire to change or enhance certain weather phenomena dates back to ancient history, when people used prayer, wizardry, dances, and even black magic in attempts to alter the weather.

**Snow and Rain Making** The first breakthrough in weather modification came in 1946, when Vincent J. Schaefer discovered that dry ice, dropped into a supercooled cloud, spurred the growth of ice crystals. Recall that once ice crystals form in a supercooled cloud, they grow larger (at the expense of the remaining liquid cloud droplets) and, upon reaching a sufficient size, fall as precipitation.

Scientists later learned that silver iodide crystals could also be used for **cloud seeding**. Unlike dry ice, which simply chills the air, silver iodide crystals act as freezing nuclei. A critical atmospheric condition required for silver iodide to trigger precipitation is that the cloud must contain supercooled droplets—that is, liquid droplets with temperatures below 0°C (32°F). Because silver iodide can be easily delivered to clouds from burners on the ground or from aircraft, it is a more cost-effective alternative than dry ice (**Fig. 5.28**).

Seeding of winter clouds that form along mountain barriers (orographic clouds) has been attempted on numerous occasions. Since 1977, Colorado's Vail and Beaver Creek ski areas have used this method to increase winter snows. An additional benefit of cloud seeding is that the increased precipitation, which melts and runs off during spring and summer months, can be collected in reservoirs for irrigation and hydroelectric power generation.

▲ **Figure 5.28 Cloud seeding** Cessna aircraft equipped with silver iodide flares to supply freezing nuclei to supercooled clouds in order to trigger precipitation.

In recent years, the seeding of warm convective clouds with hygroscopic (water-seeking) particles has received renewed attention. The interest in this technique arose when it was discovered that a pollution-belching paper mill near Nelspruit, South Africa, seemed to be triggering precipitation. Research aircraft flying through clouds near the paper mill collected samples of the particulate matter emitted from the mill. It turned out that the mill was emitting tiny salt crystals (potassium chloride and sodium chloride), which rose into the clouds. Because these salts attract moisture, they quickly form large cloud droplets, which grow into raindrops through the collision–coalescence process. Thus, seeding of warm clouds using hygroscopic particles seems to show promise for accelerating the precipitation process.

In the United States, numerous research projects are currently under way. Researchers estimate a 10 percent increase in rainfall from clouds seeded with silver iodide compared to those left unseeded. Because cloud seeding has shown some promising results and is relatively inexpensive, it has been a primary focus of modern weather-modification technology.

**Fog and Stratus Cloud Dispersal** One of the most successful applications of cloud seeding involves spreading dry ice (solid carbon dioxide) into layers of supercooled fog or stratus clouds to disperse them and thereby improve visibility. Airports, harbors, and foggy stretches of interstate highway are obvious candidates. Such applications trigger a transformation in cloud composition from supercooled water droplets to ice crystals. The ice crystals then settle out, leaving an opening in the cloud or fog. The U.S. Air Force has practiced this technology for many years at air bases, and commercial airlines have used this method at selected foggy airports in the western United States.

Unfortunately, most fog does not consist of supercooled water droplets. The more common "warm fogs" are more expensive to combat because seeding will not diminish them. Successful attempts at dispersing warm fogs have involved mixing drier air from above into the fog. When the layer of fog is very shallow, helicopters have been used. By flying just above the fog, the helicopter creates a strong downdraft that forces drier air toward the surface, where it mixes with the saturated foggy air. Thermal techniques are generally too expensive to implement at most municipal airports.

**Hail Suppression** Each year hailstorms inflict on average $500 million in property damage and crop loss in the United States (Fig. 5.29). Occasionally a single severe hailstorm can produce damages that exceed that amount. As a result, some of

▲ **Figure 5.29 Hail damage to a soybean field north of Sioux Falls, South Dakota**

A.

B.

▲ **Figure 5.30 Two common frost-prevention methods**
**A.** Sprinklers distribute water, which releases latent heat as it freezes on citrus. **B.** Wind machines mix warmer air aloft with cooler surface air.

history's most interesting efforts at weather modification have focused on hail suppression.

Farmers desperate to find ways to save their crops have long believed that strong noises—explosions, cannon shots, or ringing church bells—can help reduce the amount of hail that is produced during a thunderstorm. In Europe it was common practice for village priests to ring church bells to shield nearby farms from hail. Although this practice was banned in 1780 due to bell ringers being killed by lightning, it is still employed in a few locations.

Modern attempts at hail suppression have employed various methods of cloud seeding using silver iodide crystals to disrupt the growth of hailstones. In an effort to verify the effectiveness of cloud seeding as a mechanism of hail suppression, the U.S. government established the National Hail Research Experiment in northeastern Colorado. This effort included several randomized cloud-seeding experiments. An analysis of the data collected after 3 years revealed no statistically significant difference in the occurrence of hail between the seeded and non-seeded clouds, so the planned 5-year experiment was abandoned. Nevertheless, cloud seeding is still employed today to combat hail damage, and research on hail suppression continues.

**Frost Prevention** **Frost or freeze hazards** are strictly temperature-dependent phenomena that occur when the air temperature falls to 0°C (32°F) or below. The word *frost* is commonly used for ice crystals that form on surfaces near the ground during the night. According to the World Meteorological Organization, the correct term for the deposits of ice crystals are *hoar frost* or *white frost*, which form only when air becomes saturated at subfreezing temperatures.

A frost or freeze hazard can be generated in two ways: when a cold air mass moves into a region or when sufficient radiation cooling occurs on a clear night. Frost associated with an invasion of cold air, which can produce widespread crop damage, is characterized by low daytime temperatures and long periods of freezing conditions. By contrast, frost induced by radiation cooling is strictly a nighttime phenomenon that tends to be confined to low-lying areas. Obviously, the latter phenomenon is much easier to combat.

Several methods of frost prevention are being used with varying success. They either conserve heat (reduce heat loss at night) or add heat to warm the lowermost layer of air.

Heat-conservation methods include covering plants with insulating material, such as paper or cloth, and generating particles that, when suspended in air, reduce the rate of radiation cooling. Warming methods employ *water sprinklers*, *air-mixing techniques*, and/or *orchard heaters*. Water sprinklers (**Fig. 5.30A**) add heat in two ways: first, from the warmth of the water and, more importantly, from the latent heat of fusion released when the water freezes. As long as an ice–water mixture remains on the plants, the latent heat released will keep the temperature from dropping below 0°C (32°F).

Air mixing works best when the air temperature 15 meters (50 feet) above the ground is at least 5°C (9°F) warmer than the surface temperature. By using a wind machine, the warmer air aloft is mixed with the colder surface air (**Fig. 5.30B**).

Orchard heaters probably produce the most successful results. However, because as many as 30 to 40 heaters are required per acre, fuel costs and pollution emissions are significant.

## Inadvertent Weather Modification

The effects of *inadvertent* or *unintentional weather modification* are numerous, and scientists are beginning to understand them better. The focus of this section is to examine unintentional weather modifications that affect cloud formation and precipitation patterns. Changes in air quality, visibility, and the formation of acid rain caused by human activity are covered in Chapter 13, while human impacts on global climate is a major topic in Chapter 14. In addition, human effects on temperature, referred to as "the urban heat island," are discussed in Chapter 3.

**Effects of Large Transportation Corridors** You have undoubtedly seen *contrails* (from *condensation trail*) in the wake of an aircraft flying on a clear day (**Fig. 5.31**). Contrails, which are simply human-made clouds, form because jet aircraft engines expel large quantities of hot, moist air. Upon mixing with the frigid air aloft, the moist air cools sufficiently to reach saturation.

Contrails typically form above 9 kilometers (6 miles), where air temperatures are a frigid –50°C (–58°F) or colder. Thus, it is not surprising that contrails are composed of minute ice crystals. Most contrails have a very short life span. Once formed, these streamlined clouds mix with surrounding cold, dry air and ultimately sublimate. However, if the air aloft is near saturation, contrails may survive for long periods. Under these conditions, the upper airflow usually spreads these narrow clouds into broad bands of cirrus or cirrostratus clouds. Similar increases in cloud cover have also been detected recently along major shipping routes.

With the increase in air traffic over the past few decades, an overall increase in cloudiness has been recorded, particularly near major transportation hubs (**Fig. 5.32**). It appears these human-made clouds reduce the amount of incoming solar radiation that reach the ground and can lead to lower daytime maximum temperatures. Because cloud cover also reduces the amount of radiation cooling that occurs at night, additional research is still needed to accurately determine the impact of contrails on climate change.

▲ **Figure 5.31 Aircraft contrail** Condensation trails produced by jet aircraft often spread out to form broad bands of cirrus clouds.

▲ **Figure 5.32 Inadvertent increase in cloud cover** This photograph, taken through the window of the International Space Station, shows contrails over the Rhone Valley in eastern France. It is estimated that these "artificial clouds" cover 0.1 percent of the planet's surface, and the percentages are far higher in places such as southern California and some parts of Europe, as illustrated here.

**Urban Effects** Major population centers enhance the formation of warm clouds and increase precipitation in these urban areas by as much as 10 to 20 percent. This effect does not appear to alter winter precipitation patterns.

Increased industrialization associated with urbanization also influences weather conditions downwind of major cities. The added particulates and aerosols that reduce air quality and visibility also contribute to increased cloudiness and precipitation. In addition, large cooling lakes adjacent to industrial facilities—used to dispose of hot water that was heated during manufacturing—have been shown to cause localized fog, low clouds, and even icing in the winter.

### ✔ Concept Checks 5.7

1. What is meant by the phrase *intentional weather modification*?
2. What atmospheric condition must exist in order for cloud seeding to work?
3. How can "warm fog" be dispersed to improve visibility?
4. Describe how orchard heaters, sprinkling, and air mixing are used in frost prevention.
5. Explain how jet aircraft can increase cloudiness along major transportation corridors.

# 5 Concepts in Review Forms of Condensation and Precipitation

## 5.1 Cloud Formation ► Explain the roles of adiabatic cooling and cloud condensation nuclei in cloud formation.

**Key Terms:** cloud, cloud condensation nucleus, hygroscopic (water-seeking) nucleus, hydrophobic (water-repelling) nucleus

- Condensation occurs when water vapor changes to a liquid. For condensation to occur aloft, the air must reach saturation, and there must be a surface on which the water vapor can condense to form liquid droplets. Condensation produces tiny cloud droplets that are held aloft by the slightest updrafts.
- Clouds, visible aggregates of minute droplets of water and/or tiny crystals of ice, are one form of condensation.

## 5.2 Cloud Classification ► Name and describe the 10 basic cloud types, based on form and height. Contrast nimbostratus and cumulonimbus clouds and their associated weather.

**Key Terms:** nimbus, high cloud, middle cloud, low cloud, clouds of vertical development, cirrus, cirrostratus, cirrocumulus, altocumulus, altostratus, stratus, stratocumulus, nimbostratus, clouds of vertical development, cumulus, cumulonimbus, lenticular clouds

- Clouds are classified on the basis of two criteria: form and height. The three basic cloud forms are *cirrus* (high, white, and thin), *cumulus* (globular, individual cloud masses), and *stratus* (sheets or layers).
- Cloud heights can be *high*, with bases above 6000 meters (20,000 feet); *middle*, from 2000 to 6000 meters; or *low*, below 2000 meters (6500 feet). *Clouds of vertical development* have bases in the low height range and extend upward into the middle or high range.
- Considering both form and height, 10 basic cloud types can be classified.

**Q** Which of the three basic cloud forms (cirrus, cumulus, or stratus) is illustrated by the accompanying images, A–C?

A.

B.

C.

## 5.3 Types of Fog ► Identify the basic types of fog and describe how each forms.

**Key Terms:** fog, radiation fog, advection fog, upslope fog, steam fog, frontal (precipitation) fog

- Fog, generally considered an atmospheric hazard, is a cloud with its base at or very near the ground. Fogs formed by cooling include radiation fog, advection fog, and upslope fog. Fogs formed by the addition of water vapor are steam fog and frontal fog.

**Q** The accompanying three images (A–C) depict three types of fog. Identify the fog type shown and describe the mechanism that generated it.

A.

B.

C.

## 5.4 How Precipitation Forms ► Describe the Bergeron process and explain how it differs from the collision–coalescence process.

**Key Terms:** Bergeron process, supercooled water, freezing nucleus, collision–coalescence process

- For precipitation to form, millions of cloud droplets must coalesce into drops large enough to sustain themselves during their descent.
- The two mechanisms that have been proposed to explain this phenomenon are the Bergeron process, which produces precipitation from cold clouds primarily in the middle and high latitudes, and the warm-cloud process, most associated with the tropics, called the collision–coalescence process.

## 5.5 Forms of Precipitation ▶ Describe the atmospheric conditions that produce sleet, freezing rain (glaze), and hail.

**Key Terms:** rain, virga, fallstreaks, drizzle, mist, snow, graupel, sleet, freezing rain (glaze), hail, rime

- The two most common and familiar forms of precipitation are rain and snow. Rain can form in either warm or cold clouds. When it falls from cold clouds, it begins as snow that melts before reaching the ground.
- Sleet consists of spherical to lumpy ice particles that form when raindrops freeze while falling through a thick layer of subfreezing air. Freezing rain results when supercooled raindrops freeze upon contact with cold objects. Rime consists of delicate frostlike accumulations that form as supercooled fog droplets encounter objects and freeze on contact. Hail consists of hard, rounded pellets or irregular lumps of ice produced in towering, cumulonimbus clouds, where frozen ice particles and supercooled water coexist.

## 5.6 Precipitation Measurement ▶ List the advantages and disadvantages of using a standard rain gauge versus weather radar to measure precipitation.

**Key Terms:** standard rain gauge, trace of precipitation, tipping-bucket gauge, weighing gauge, snow pillow, weather radar

- The instruments most commonly used to measure rain are the standard rain gauge, which is read directly, and the tipping-bucket gauge and the weighing gauge, both of which record the amount and intensity of rain. The two most common measurements of snow are depth and water equivalent. Although the quantity of water in a given volume of snow is not constant, a general ratio of 10 units of snow to 1 unit of water is often used when exact information is not available.
- Modern weather radar has given meteorologists an important tool to track storm systems and precipitation patterns they produce, even when the storms are as far as a few hundred kilometers away.

## 5.7 Planned and Inadvertent Weather Modification ▶ Discuss several ways that humans attempt to modify the weather.

**Key Terms:** cloud seeding, frost or freeze hazard

- Planned weather modification is deliberate human intervention to influence atmospheric processes that constitute the weather. Both planned and inadvertent weather modification use energy to forcefully alter the weather; modify land and water surfaces to change their natural interaction with the lower atmosphere; and trigger, intensify, or redirect atmospheric processes.
- The effects of inadvertent or unintentional weather modification include increased cloudiness, which may lower surface temperatures, and an increase in precipitation.

## Give it Some Thought

1. Clouds are classified into four major height categories: low, middle, high, and clouds of vertical development. Explain why low clouds (or clouds of vertical development) are much more likely to produce precipitation than middle or high clouds.
2. Which cloud types are associated with the following characteristics:
   **a.** Halos
   **b.** Light to moderate precipitation
   **c.** Hail
   **d.** Mackerel sky
   **e.** Mares' tails
3. Use the accompanying diagram, which shows a towering cumulonimbus cloud and the temperature profile of the atmosphere, to complete the following:
   **a.** What is the temperature at the base of this cloud?
   **b.** At point A, would the consist of liquid droplets only, ice crystals only, or both liquid droplets and ice crystals?
   **c.** At point B, would the consist of liquid droplets only, ice crystals only, or both liquid droplets and ice crystals?
   **d.** At point C, would the consist of liquid droplets only, ice crystals only, or both liquid droplets and ice crystals?

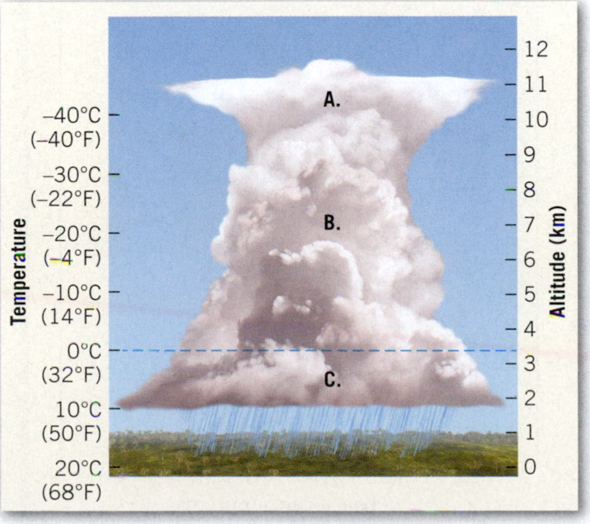

4. Fog can be defined as a cloud with its base at or very near the ground, yet fogs and clouds form by different processes. Describe the similarities and differences in the formation of fogs and clouds.
5. Why does radiation fog form mainly on clear nights as opposed to cloudy nights?
6. Assume that it is a midwinter day in central Illinois. Mild conditions prevail because of a steady wind from the south. As the day progresses, fog forms across a broad area. Identify the likely type of fog.
7. Imagine that you are driving in a hilly area early in the morning and encounter fog as you descend into the valleys, but as you drive out of the valleys, the conditions clear. Identify the likely type of fog.
8. Describe or sketch the vertical temperature profile that would cause precipitation that began as snow to reach the ground as each of the following:
   **a.** Snow
   **b.** Rain
   **c.** Freezing rain
9. Use the accompanying diagram, which shows changes in temperature and the dew-point temperature with altitudes, to complete the following. (*Hint:* The dew point is the temperature at which the air reaches saturation.)
   **a.** At what altitude would clouds be found: 0–4 kilometers, 4–8 kilometers, or 8–12 kilometers?
   **b.** What would this cloud consist of: liquid droplets only, ice crystals only, or both liquid droplets and ice crystals?
   **c.** If this cloud produced precipitation, what type would likely reach Earth's surface: rain, snow, sleet, or freezing rain?

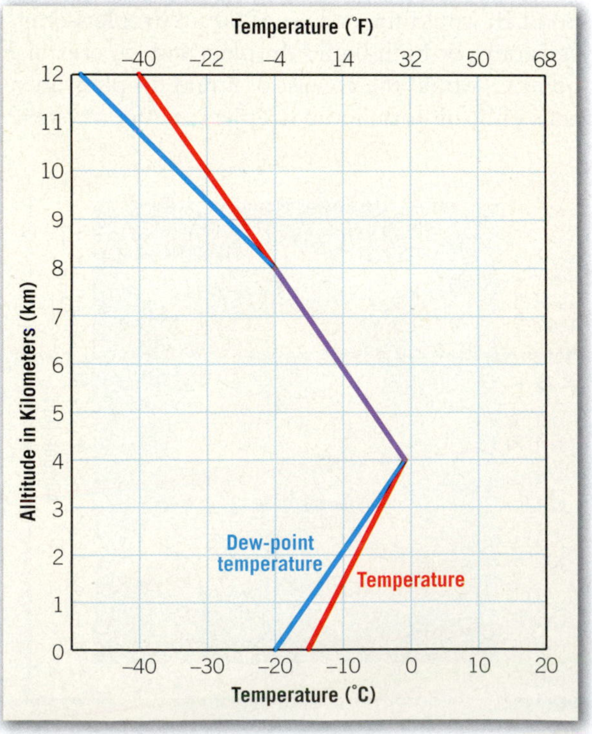


10. Why is it unlikely to have virga and fog occur simultaneously?
11. The accompanying images show two different types of precipitation after they have been deposited.

A. B.

   **a.** Name the types of precipitation shown.
   **b.** Describe the nature (form) of each type of precipitation before it was deposited.
   **c.** In what way are these types of precipitation similar to one another?
   **d.** In what way are they different?
12. Weather radar provides information on the intensity as well as the total amount of precipitation. Table A shows the relationship between radar reflectivity values and rainfall rates. If radar measured a reflectivity value of 47 dBZ for 2½ hours at a particular location, how much rain has fallen there?

***Table A***
***Conversion of radar reflectivity to rainfall rate***

| Radar Reflectivity (dBZ) | Rainfall Rate (inches/hr) |
|---|---|
| 65 | 16+ |
| 60 | 8.0 |
| 55 | 4.0 |
| 52 | 2.5 |
| 47 | 1.3 |
| 41 | 0.5 |
| 36 | 0.3 |
| 30 | 0.1 |
| 20 | trace |

## Problems

1. Suppose that the air temperature is 20°C (68°F) and the relative humidity is 50 percent at 6:00 P.M. and that during the evening, the air temperature drops, but its water vapor content does not change. If the air temperature drops 1°C (1.8°F) every 2 hours, will fog occur by sunrise (6:00 A.M.) the next morning? Explain your answer. (*Hint:* The data you need are found in Table 4.1, page 96.)
2. By using the same conditions as in Problem 1, will fog occur if the air temperature drops 1°C (1.8°F) every hour? If so, when will it first appear? What name is given to fog of this type that forms because of surface cooling during the night?
3. Assuming that the air is still, how long would it take a large raindrop (5 mm) to reach the ground if it fell from a cloud base at 3000 meters? (See Table 5.4, page 136.) How long would a typical raindrop (2 mm) take to fall to the ground from the same cloud? How long if it were a drizzle drop (0.5 mm)?
4. Assuming that the air is still, how long would it take for a typical cloud droplet (0.02 mm) to reach the ground if it fell from a cloud base at 1000 meters? (See Table 5.4.) It is very unlikely that a cloud droplet would ever reach the ground, even if the air were perfectly still. Explain why.
5. The record for the largest hailstone in the United States was set on July 23, 2010, in Vivian, South Dakota. The stone, pictured in the upper right, was 8 inches in diameter and 18.62 inches in circumference, and it weighed nearly 2 pounds. The maximum fall speed of a spherical hailstone of diameter $d$ can be calculated using $V = k2d$, where $k = 20$ if $d$ is in centimeters and $V$ is in meters per second. Estimate the strength (speed) of the updrafts in this storm that were necessary to support the stone just before it began to fall. Convert your answer from meters per second to feet per second and then to miles per hour.

(NOAA)

6. The dimensions of the large cumulonimbus cloud pictured below are roughly 12 kilometers high, 8 kilometers wide, and 8 kilometers long. Assume that the droplets in every 1 cubic meter of the cloud total 0.5 cubic centimeter of water. How much liquid water does the cloud contain? How many gallons does it contain (3785 $cm^3$ = 1 gallon)?

## MasteringMeteorology™

Looking for additional review and test prep materials? Visit the Study Area in *MasteringMeteorology*™ to enhance your understanding of this chapter's content by accessing a variety of resources, including **MapMaster**™ interactive maps, Geoscience Animations, GEODe, *In the News* RSS feeds, flashcards, web links, self-study quizzes, and an eText version of *The Atmosphere*.

# 6 Air Pressure and Winds

## Focus on Concepts

*Each statement represents the primary learning objective for the corresponding major heading within the chapter. After you complete the chapter, you should be able to:*

**6.1** Define *atmospheric pressure* and explain how it is displayed on a weather map.

**6.2** Name and describe the factors affecting air pressure at Earth's surface as well as aloft.

**6.3** List and describe the three forces that act on the atmosphere to either create or alter winds.

**6.4** Explain why winds aloft flow roughly parallel to the isobars, while surface winds travel at an angle across the isobars.

**6.5** Describe the airflow around a low-pressure center (cyclone) and a high-pressure center (anticyclone) and the weather associated with each.

**6.6** Define *prevailing wind* and explain how wind direction is described.

Of the various elements of weather and climate, changes in air pressure are the least perceptible to humans; however, they are very important in producing changes in our weather because variations in air pressure generate winds that trigger changes in temperature and humidity. In addition, air pressure is a significant factor in weather forecasting and is closely tied to the other elements of weather (temperature, moisture, and wind) in cause-and-effect relationships. For example, horizontal differences in air pressure create the winds propelling the sailboats in the chapter-opening photo.

*Horizontal differences in air pressure created the winds that propel these sailboats.*

# 6.1 Atmospheric Pressure and Wind

**Define *atmospheric pressure* and explain how it is displayed on a weather map.**

MM GEODe ▶ Air Pressure and Winds ▶ Measuring Air Pressure

We know that air moves *vertically* if it is forced over a barrier or if it is warmer and thus more buoyant than surrounding air. But what causes air to move *horizontally*—the phenomenon we call **wind**? Simply stated, *wind is the result of horizontal differences in atmospheric pressure*. Air flows from areas of *higher pressure* to areas of *lower pressure*. You may have experienced this phenomenon when opening a can of warm carbonated soda. When you open the can, the gaseous carbon dioxide that was dissolved in the liquid rushes from the area of higher pressure inside the can to the area of lower pressure outside. Wind is nature's attempt to balance inequalities in air pressure.

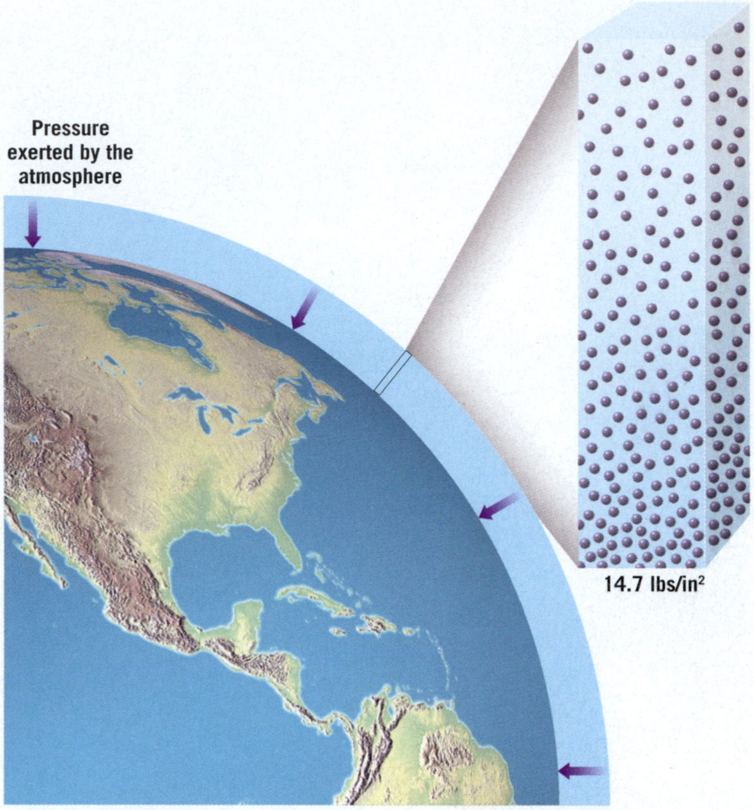


▲ **Figure 6.1 Average air pressure at sea level is about 14.7 pounds per square inch**

## What Is Atmospheric Pressure?

We live at the bottom of the atmosphere, and just as the creatures living at the bottom of the ocean are subjected to pressure exerted by water, humans are subjected to the pressure exerted by the weight of the atmosphere above. Although we do not generally notice the pressure exerted by the ocean of air around us (except when rapidly ascending or descending in an elevator or airplane), it is nonetheless substantial.

We define **atmospheric pressure**, or simply **air pressure**, as the force per unit area on a surface exerted by the weight of the air above. Average atmospheric pressure at sea level is about 14.7 pounds per square inch. (In the metric system, average atmospheric pressure at sea level is equivalent to the weight of 1 kg/cm$^2$.) Specifically, a column of air 1 square inch in cross section, measured from sea level to the top of the atmosphere, would weigh about 14.7 pounds (Fig. 6.1). This is roughly the same pressure produced by a 1-square-inch column of water 10 meters (33 feet) in height.

The pressure that air exerts at Earth's surface is much greater than most people realize. For example, the air pressure exerted on the top of a small (50-centimeter-by-100-centimeter) school desk exceeds 5000 kilograms (11,000 pounds), or about the weight of a 50-passenger school bus. Why doesn't the desk collapse under the weight of the air above? Simply, air pressure is exerted in all directions—down, up, and sideways. Thus, the air pressure around all sides of the desk is exactly balanced.

## Measuring Atmospheric Pressure

To describe atmospheric pressure, the National Weather Service (NWS) uses a unit of force from physics called the **newton**.* Under average conditions (at sea level), the atmosphere exerts a force of 101,325 newtons per square meter. To simplify this large number, the NWS adopted the **millibar (mb)**, which equals 100 newtons per square meter. Thus, average sea-level pressure is stated as 1013.25 millibars (Fig. 6.2).**

You may have heard the expression "inches of mercury," which is also used to describe atmospheric pressure. This expression dates from 1643, when Torricelli, a student of the famous Italian scientist Galileo, invented the **mercury barometer**. Torricelli correctly described the atmosphere as a vast ocean of air that exerts pressure on Earth's surface. To measure this force, he closed one end of a glass tube and filled it with mercury, then inverted the tube into a dish of mercury (Fig. 6.3A). Torricelli found that the mercury flowed out of the tube until the weight of the mercury column was balanced by the pressure exerted on the surface of the mercury by the air above. In other words, the weight of the mercury in the column equaled the weight of a similar-diameter column of air that extended from the ground to the top of the atmosphere.

*A newton is the force needed to accelerate a 1 kilogram mass 1 meter per second squared.

**The standard unit of pressure in the SI system is the pascal, which is 1 newton per square meter (N/m$^2$). In this notation, 1 standard atmosphere has a value of 101,325 pascals, or 101.325 kilopascals. If the National Weather Service officially converts to the metric system, it will probably adopt this unit.

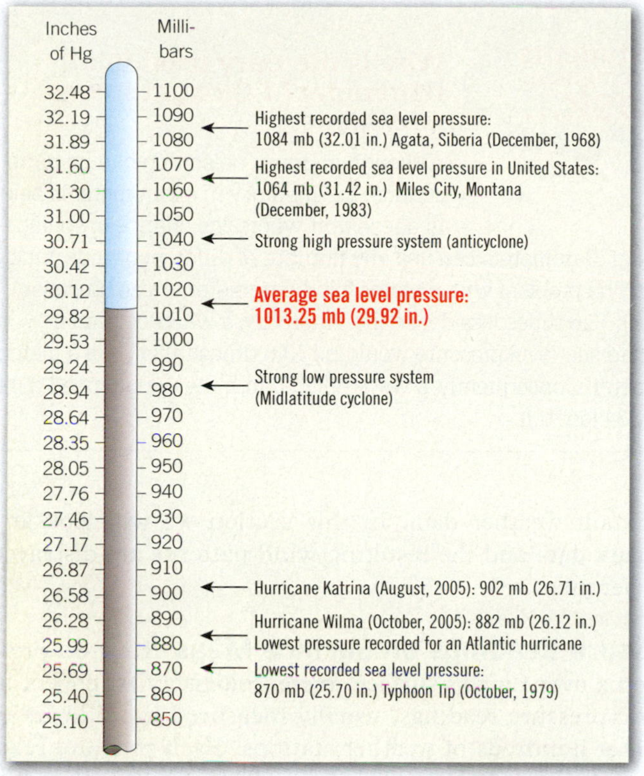


▲ **SmartFigure 6.2 A comparison of atmospheric pressure in inches of mercury and in millibars**

http://goo.gl/DH6pb

Torricelli noted that when air pressure increases, the mercury in the tube rises; conversely, when air pressure decreases, so does the height of the column of mercury. The height of the column of mercury, therefore, became the measure of the air pressure—"inches of mercury." With some refinements, Torricelli's mercury barometer remains the standard pressure-measuring instrument. Because air pressure is measured with a barometer, it is also commonly referred to as **barometric pressure**.

Standard atmospheric pressure at sea level equals 29.92 inches (760 millimeters) of mercury. In the United States, the National Weather Service uses millibars on weather maps and charts but reports surface pressure to the public in inches of mercury.

The need for a smaller, more portable instrument for measuring atmospheric pressure led to the development of the **aneroid barometer** (*aneroid* means "without liquid" **Fig. 6.4A**). Rather than use a mercury column held up by air pressure, the aneroid barometer uses a partially evacuated metal chamber (**Fig. 6.4B**). The chamber, being very sensitive to pressure variations, changes shape, compressing as pressure increases and expanding as pressure decreases.

As shown in Figure 6.4A, the face of an aneroid barometer intended for home use includes the words *fair*, *change*, *rain*, and *stormy*. Notice that "fair weather" corresponds with high-pressure readings, whereas "rain" is associated with low pressures. To "predict" weather in a local area, the change in air

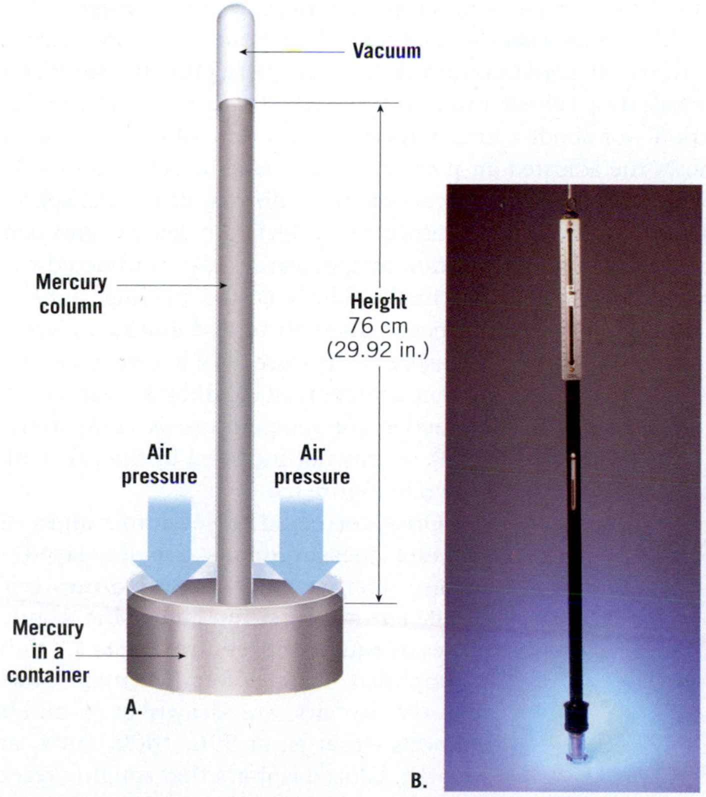


▲ **Figure 6.3 Mercury barometer A.** The weight of the column of mercury is balanced by the pressure exerted on the dish of mercury by the air above. If the pressure decreases, the column of mercury falls; if the pressure increases, the column rises. **B.** Image of a mercury barometer.

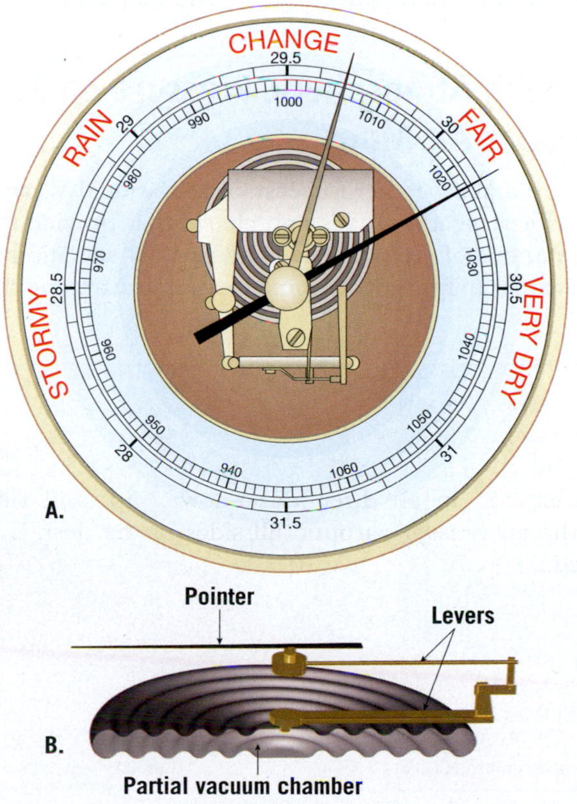


▲ **Figure 6.4 Aneroid barometer A.** Illustration of an aneroid barometer. **B.** The aneroid barometer has a partially evacuated chamber that changes shape, compressing as atmospheric pressure increases and expanding as pressure decreases.

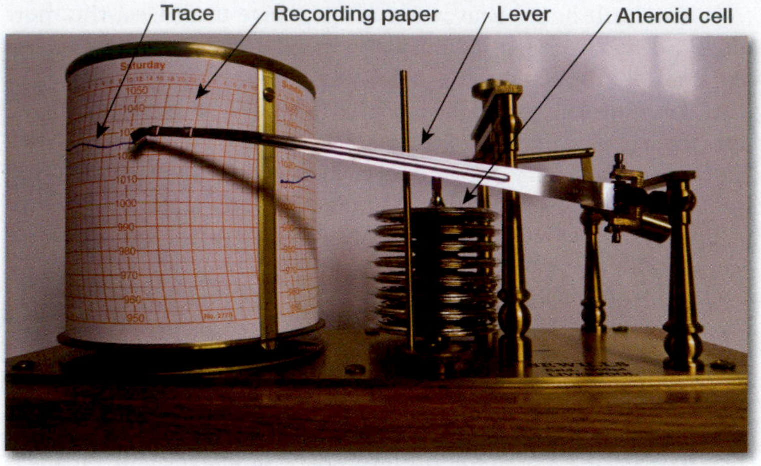


▲ **Figure 6.5 Aneroid barograph** This instrument makes a continuous record of air pressure changes.

pressure over the past few hours is more important than the current pressure reading. Falling pressure is often associated with increasing cloudiness and the possibility of precipitation, whereas rising air pressure generally indicates clearing conditions.

Another advantage of the aneroid barometer is that it can easily be connected to a recording mechanism. This recording instrument, called a **barograph**, provides a continuous record of pressure changes over time (Fig. 6.5). Another important adaptation of the aneroid barometer is its use to indicate altitude for aircraft, mountain climbers, and mapmakers.

## Displaying Atmospheric Pressure on Surface and Upper-Air Maps

In order to make weather forecasts, the National Weather Service produces weather maps and charts that provide symbolic representations of atmospheric conditions at specific points in time. Weather maps show barometric pressure among the other important weather data. In this section we will look at how pressure data and the resulting wind patterns are displayed on weather maps.

**students sometimes ask...**

**Why is mercury used in a barometer? I thought it was poisonous.**

Although mercury poisoning can be quite serious, the mercury in a barometer is held in a reservoir where the chance of spillage is minimal. Barometers can use any number of different liquids, including water. The problem with a water-filled barometer is size. Because water is 13.6 times less dense than mercury, a water column at standard sea-level pressure would be 13.6 times taller than a mercury barometer. Consequently, a water-filled barometer would need to be nearly 34 feet tall.

**Pressure Readings on Surface Maps** To show pressure patterns over Earth's surface, meteorologists first collect barometric pressure readings, usually measured in millibars (mb), taken at hundreds of weather stations. Each pressure reading, called **station pressure**, is then adjusted to account for the *elevation* at the station where it was collected. (Recall that air pressure drops with increased elevation.) Without these adjustments, all regions located at high elevations would be mapped as having low pressure, and locations below sea level like Death Valley, California, would report high pressure readings.

To compensate for altitude, all pressure measurements are converted to sea-level equivalents. Generally, pressure near Earth's surface drops about 1 millibar for each 10 meters of increased elevation—or about 1 inch of mercury for every 1000 feet. Figure 6.6 shows the adjusted air pressure values, usually referred to as *corrected values*, for three locations in California, using this approximation. Because temperature also affects air density, and hence the weight of an air column, temperature is also considered when altitude corrections are made. The adjusted reading, considering both temperature and altitude, gives the atmospheric pressure, as if it were taken at sea level under identical conditions. Normally the correction for temperature is comparatively small, so it is not included in the calculation shown in Figure 6.6.

Once corrected to sea-level values, the pressure measurements are displayed on surface weather maps using **isobars** (*iso* = equal, *bar* = pressure), lines that connect places of equal air pressure. Figure 6.7 shows a simplified surface weather map. Notice that the isobars are drawn at 4-millibar intervals—that is, at 996, 1000, 1004, and so forth. Closed isobars that roughly resemble circles identify *highs* and *lows*. The pressure in a *high* is greater than that of the surrounding air and is labeled with a large blue

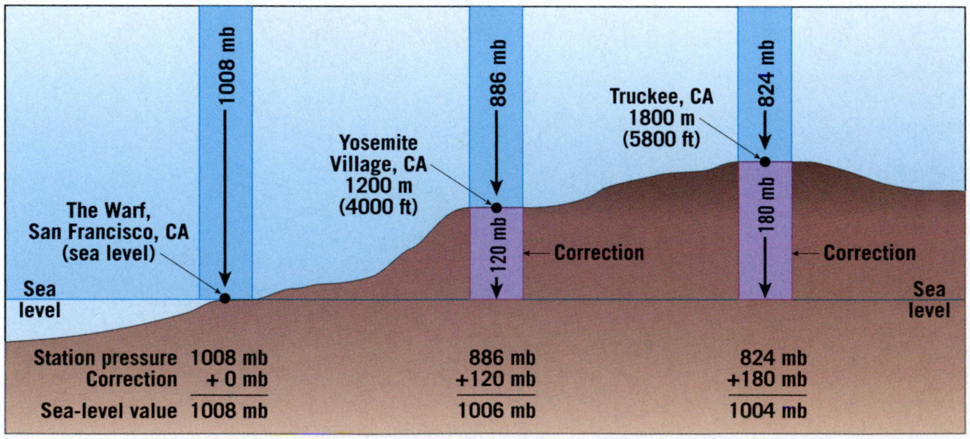


▲ **Figure 6.6 Correcting pressure readings to sea-level equivalence** This is done by adding the pressure that would be exerted by an imaginary column of air to the station's pressure reading. The higher the recording station's elevation, the greater the correction.

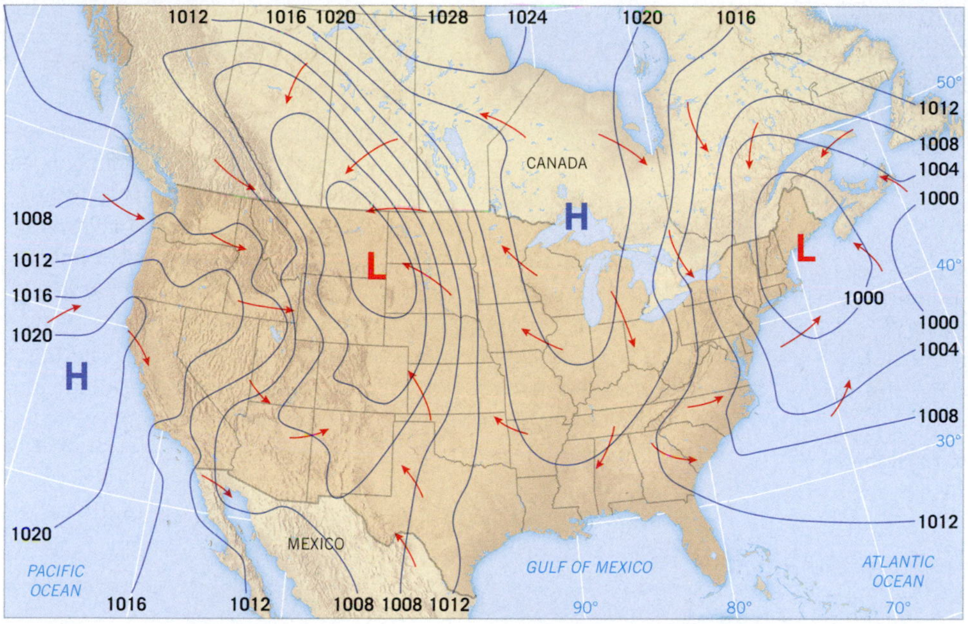


▲ **Figure 6.7 Surface weather map** This simplified surface weather map uses isobars to show high-pressure (anticyclones) and low-pressure (cyclones) systems. The idealized wind patterns associated with these pressure systems are shown with red arrows.

*H*, whereas the pressure in a *low* is lower than the surrounding air and is labeled with a large red *L*.

In general, *high-pressure systems*, also called **anticyclones**, are associated with dry conditions. *Low-pressure systems* that occur in the middle latitudes are called **cyclones**, or **midlatitude cyclones**, to differentiate them from *tropical cyclones*. (Tropical cyclones are also called *hurricanes* or *typhoons*, depending on their intensity and location.) In contrast to anticyclones, midlatitude cyclones tend to produce stormy weather.

In addition to isobars, the weather maps in Figure 6.7 show the idealized wind patterns with red arrows. Notice that surface winds generally blow at an angle across the isobars *away* from the centers of high pressure and *toward* the centers of low pressure.

### Upper-Level Weather Charts

In addition to surface maps, the National Weather Service produces weather charts that show atmospheric conditions at 850-, 700-, 500-, 300-, and 200-millibar levels, twice daily. Recall from Chapter 1 that upper-air data comes mainly from *radiosondes*, instrument packages carried aloft by weather balloons. Of particular interest is the 500-millibar chart showing the circulation of the atmosphere at roughly 5600 meters (18,000 feet) (**Fig. 6.8**). Because the 500-millibar pressure level is in the middle of the atmosphere in terms of mass, about half of the air is below that level, and half is above. The 500-millibar chart is important because major weather systems tend to travel in the direction of the winds at that level.

Like other upper-air charts, the 500-millibar chart in Figure 6.8 shows variations in the height (in meters) at which 500-millibar level is found. Rather than show variations in pressure at fixed heights, upper-level charts are similar to topographic maps. That is, the contour pattern reveals the "hills" and "valleys" of a constant-pressure surface, in this case the 500-millibar surface. There is a simple relationship between height contours and isobars; places that experience 500-millibar pressure at higher altitudes are subject to higher pressures than places where the height contours indicate lower altitudes. In other words, *higher-elevation* contours indicate *higher* pressures, and *lower-elevation* contours indicate *lower* pressures.

Notice that the contour lines in Figure 6.8 mainly consist of large sweeping curves labeled ridges and troughs. **Ridges** are elongated high-pressure areas that extend toward the poles and are associated with warm, dry air. **Troughs**, by contrast, are elongated areas of low pressure that extend equatorward, and they are usually associated with cool, wet weather.

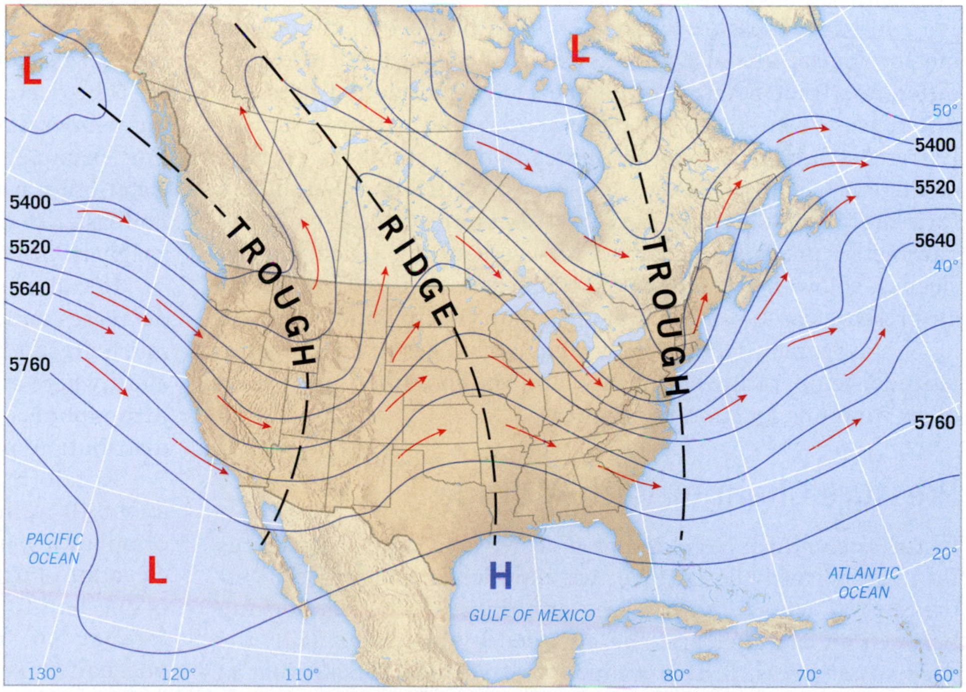


▲ **Figure 6.8 Upper-air weather chart** This upper-air weather chart shows the height contours at the 500-millibar level. Rather than show variations in pressure at fixed heights, upper-level charts are similar to topographic maps in that the contour pattern reveals the "hills" (ridges) and "valleys" (troughs) of a constant-pressure surface. Therefore, *higher-elevation* contours indicate *higher* pressures, and *lower-elevation* contours indicate *lower* pressures. The idealized wind patterns associated with these pressure systems are shown with red arrows.

The importance of upper-level charts in weather forecasting is discussed in Chapter 12. However, introducing weather maps at this point provides a tool to explore how winds are generated and how they are related to pressure patterns.

### ✔ Concept Checks 6.1

1. What is wind, and what generates it?
2. What is average (standard) sea-level pressure, measured in pounds per square inch, millibars, and inches of mercury?
3. Describe atmospheric pressure in your own words.
4. What is an isobar?
5. Describe the atmospheric pressure associated with a cyclone compared to an anticyclone.

**students sometimes ask...**

**Why do my ears sometimes feel painful when I fly?**

When airplanes take off and land, some people experience ear pain because of slight changes in cabin pressure. Normally, the air pressure in one's middle ear is the same as the pressure of the surrounding atmosphere because the eustachian tube connects the ear to the throat. But the eustachian tubes of a passenger with a cold may become blocked, preventing the flow of air either into or out of the middle ear. When the plane ascends or descends, the resulting pressure change can cause discomfort or, less often, excruciating pain, which subsides when the ears "pop" to equalize the pressure.

## 6.2 Why Does Air Pressure Vary?

**Name and describe the factors affecting air pressure at Earth's surface as well as aloft.**

What is the importance of atmospheric pressure and its daily variations? Recall that variations in air pressure cause the wind to blow, which in turn causes changes in temperature and humidity. In short, differences in air pressure create global winds that become organized into the systems that bring us our weather. Therefore, the National Weather Service closely monitors daily changes in air pressure.

Although pressure changes with altitude are important, meteorologists are also interested in the horizontal pressure differences that occur around the globe. Horizontal pressure differences from place to place tend to be relatively small. Extreme readings are rarely more than 30 millibars (1 inch of mercury) above average sea-level pressure or 60 millibars (2 inches) below average sea-level pressure. Occasionally the barometric pressure measured in severe storms, such as hurricanes, is lower (see Figure 6.2). Yet these small differences in air pressure can be sufficient to generate violent winds.

Four primary factors contribute to changes in air pressure: altitude, temperature, humidity, and the movement of a mass of air from one location to another.

### Pressure Changes with Altitude

Just as scuba divers experience a decrease in water pressure as they rise toward the surface, we experience a decrease in air pressure as we ascend into the atmosphere. The relationship between air pressure and the air's density largely explains the drop in air pressure that occurs with altitude. Recall that at sea level, a column of air weighs 14.7 pounds per square inch and therefore exerts that amount of pressure. As we ascend through the atmosphere, we find that the air becomes less dense because of the continual decrease in the amount (weight) of air above. Therefore, as would be expected, there is a corresponding *decrease in pressure with an increase in altitude.*

Because density decreases with altitude, the term *thin air* is normally associated with mountainous regions. Except for the Sherpas (indigenous peoples of Nepal), most of the climbers who reach the summit of Mount Everest use supplementary oxygen for the final leg of the journey. Even with the extra oxygen, many of these climbers experience periods of disorientation because of an inadequate supply of oxygen to their brains.

The decrease in pressure with altitude also affects the boiling temperature of water, which at sea level is 100°C (212°F). For example, in Denver, Colorado—the Mile High City—water boils at about 95°C (203°F). Although water comes to a boil faster in Denver than in San Diego, it takes longer to cook spaghetti in Denver because of its lower boiling temperature.

Recall from Chapter 1 that the rate at which pressure decreases with altitude is not constant. The rate of decrease is much greater near Earth's surface, where pressure is high, than aloft, where air pressure is low. A model of the **U.S. standard atmosphere**, shown in Figure 6.9, depicts the idealized vertical distribution of atmospheric pressure at various altitudes.

Recall that near Earth's surface, air pressure decreases by about 10 millibars for every 100-meter increase in elevation, or about a 1-inch drop of mercury for every 1000-foot rise in elevation. Further, atmospheric pressure falls by approximately one-half for each 5.6-kilometer increase in altitude. Therefore, at 5.6 kilometers, the pressure is 500 millibars, about one-half its sea-level value; at 11.2 kilometers, it is about 250 millibars, or one-fourth its sea-level value, and so forth. Thus, at the altitude at which commercial jets fly (about 10–12 kilometers), the air exerts a pressure equal to only one-fourth that at sea level.

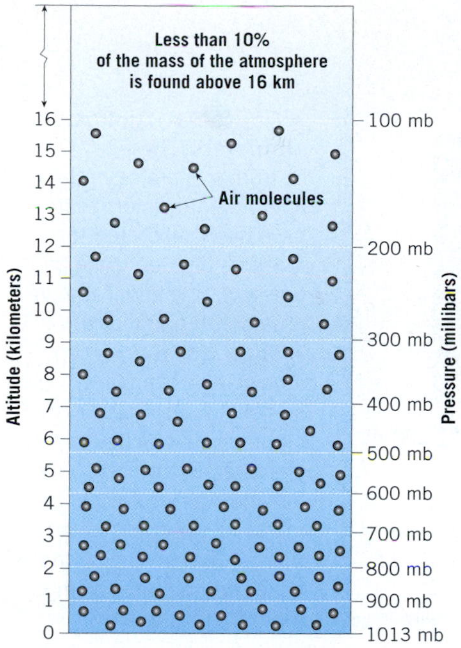


◄ **Figure 6.9 Illustration of the U.S. standard atmosphere** Each layer contains about 10 percent of the atmosphere. Notice that at roughly 5.6 kilometers (18,000 feet), the pressure is about one-half of its sea-level value.

## Pressure Changes with Temperature

How do horizontal pressure differences arise? Picture northern Canada in midwinter. Here the snow-covered surface is continually radiating heat to space, while receiving minimal incoming solar radiation. The frigid ground cools the air above so that daily lows of –34°C (–30°F) are common.

Recall that temperature is a measure of the average molecular motion (kinetic energy) of a substance. As a result, cold Canadian air is composed of comparatively slow-moving gas molecules that are packed closely together. These cold, dense air masses are associated with high surface pressures and are labeled *highs* (H) on weather maps. By contrast, during the summer, large areas of the American Southwest experience extremely high temperatures accompanied by low surface pressures labeled *lows* (L) on weather maps. Thus, all else being equal, a cold air column is associated with high surface pressure, and a warm air column is associated with low surface pressure. These pressure differences, in turn, generate a force called the *pressure gradient force* that causes the air to flow. Air flows from areas of high pressure to areas of low pressure.

Another important difference between cold and warm air is that *air pressure drops more rapidly with altitude in a column of cold (dense) air than in a column of warm (less dense) air.* This concept is illustrated in Figure 6.10, where we assume that both columns of air exert the same surface pressure, and (although greatly exaggerated) differences in the spacing of air molecules represent differences in density. As shown, air pressure decreases more rapidly with altitude in a cold air column because the molecules are closely packed (denser). Therefore, if we could move up this imaginary column of cold air, we would quickly pass through the dense molecules near the surface. But in a warmer, less dense column, the pressure drops more slowly because we would pass through fewer air molecules in the same vertical distance. Because of this difference, we can conclude that at any given altitude above the surface, there are more air molecules (equating to higher pressure) in a warm air column than in a cold air column.

If we look at the line drawn halfway up the two columns shown in Figure 6.10, we see that there are more air molecules above this line in the warm air column than there are in the cold column. As a result, above the surface, *warm air tends to exhibit a higher pressure than cold air.* Note that this is just the opposite of what occurs at the surface, where cold air usually exerts a higher pressure than warm air. This important phenomenon also has implications in aviation (Box 6.1).

## Pressure Changes with Moisture Content

Although less important than temperature, the amount of water vapor contained in a volume of air influences its density. Contrary to popular perception, water vapor *reduces* the density of air. The air may feel "heavy" on hot, humid days, but it is not. You can easily verify this fact for yourself by examining a periodic table of the elements and noting that the molecular weights of nitrogen ($N_2$) and oxygen ($O_2$) are greater than that of water vapor ($H_2O$). (These gases have relative mass of 28, 38, and 18 grams, respectively.)

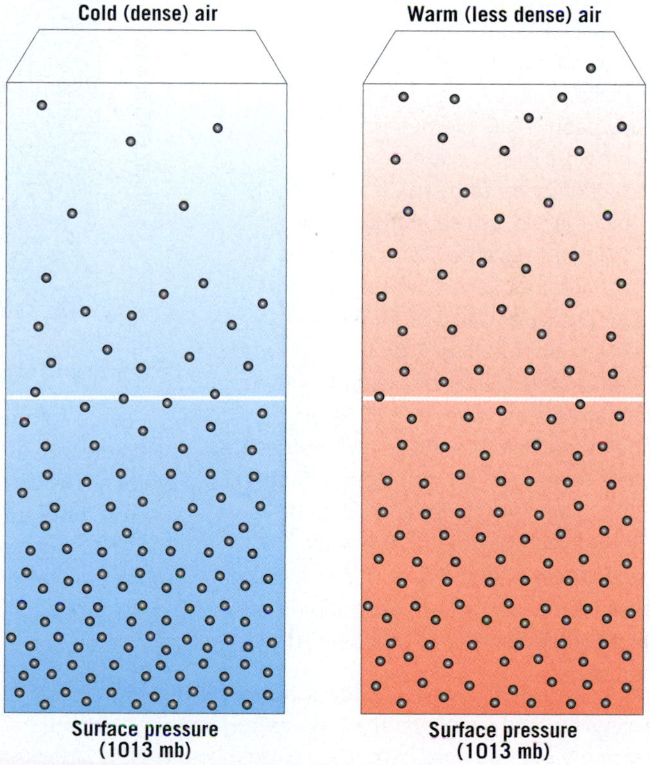


▲ **Figure 6.10 A comparison of the densities of a column of cold air and a column of warm air** Air pressure decreases more rapidly with altitude in a cold air column because the molecules are closely packed (more dense). Looking at the line drawn halfway up the two columns, notice that there are more air molecules above this line in the warm air column than there are in the cold column. As a result, *warm air tends to exhibit a higher pressure than cold air* above Earth's surface.

## Box 6.1 Air Pressure and Aviation

The cockpit of nearly every aircraft contains a *pressure altimeter*, an instrument that allows a pilot to determine the plane's altitude. A pressure altimeter is essentially an aneroid barometer marked in meters instead of millibars and, as such, responds to changes in air pressure. For example, Figure 6.9 shows that the standard atmospheric pressure of about 800 millibars "normally" occurs at a height of 2 kilometers. Therefore, when the pressure reaches 800 millibars, the altimeter will indicate an altitude of 2 kilometers.

Because of temperature variations and moving pressure systems, actual conditions usually differ from that shown by an aircraft's altimeter. When the air is warmer than predicted by the standard atmosphere, the plane will fly higher than the height indicated by the altimeter. By contrast, in cold air, the plane will fly lower than indicated. This could be especially dangerous if the pilot is flying a small plane through mountainous terrain with poor visibility (**Fig. 6.A**). To avoid dangerous situations, pilots make altimeter corrections before takeoffs and landings, and they sometimes make corrections en route as well.

Above about 5.6 kilometers (18,000 feet), pressure changes are more gradual, so corrections cannot be made as precisely as at lower levels. Consequently, commercial aircraft have their altimeters set at the standard atmosphere and fly paths of constant pressure, called *flight levels*, instead of constant altitude (**Fig. 6.B**). In other words, when an aircraft flies at a constant altimeter setting, a pressure variation will result in a change in the plane's altitude. When pressure increases (warm air column) along a flight path, the plane will climb, and when pressure decreases (cold air column), the plane will descend. There is little risk of midair collisions because nearby aircraft are assigned different flight levels in order to maintain sufficient separation.

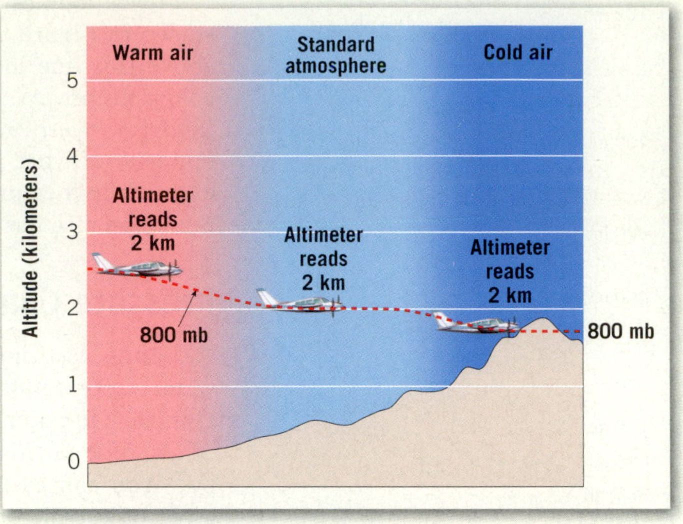


**Figure 6.A Aircraft altimeters** These instruments are essentially aneroid barometers that have been calibrated to measure altitude using air pressure. The lower the air pressure, the higher the altitude. However, when a plane enters a warm air column, its altitude is higher than indicated by the altimeter. By contrast, when a plane flies into a cold air mass, its altitude is lower than shown on the altimeter, which can be a problem in mountainous terrains.

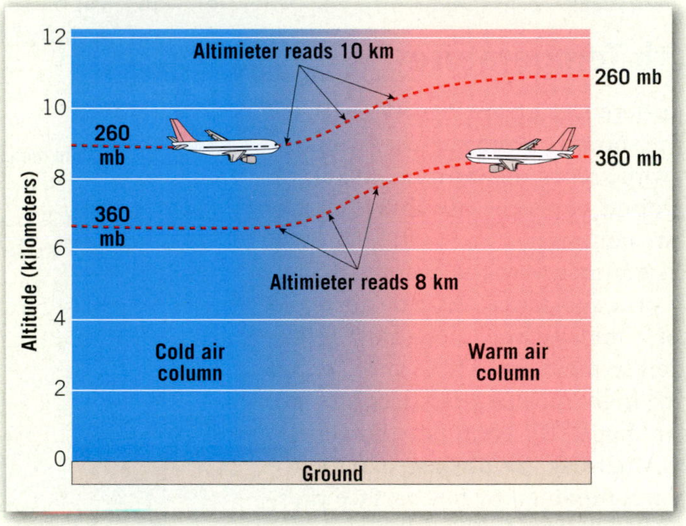


**Figure 6.B Flying a path of constant pressure** Commercial aircraft above 5.6 kilometers (18,000 feet) generally fly paths of constant pressure instead of constant altitude.

Large commercial aircraft also use radar altimeters to measure heights above the terrain. The time required for a radio signal to reach the surface and return is used to accurately determine the height of the plane above the ground. This system is not without drawbacks; a radar altimeter provides the elevation above the ground rather than above sea level, so knowledge of the underlying terrain is required. However, radar altimeters are useful when measuring the height above ground level during landing.

### Questions

1. Compare and contrast a pressure altimeter and an aneroid barometer.
2. When are radar altimeters used?

In a volume of air, the molecules of these gases are intermixed, and each takes up roughly the same amount of space. As the water content of an air mass increases, lighter water vapor molecules displace heavier nitrogen and oxygen molecules. Therefore, humid air is lighter (less dense) than dry air. Nevertheless, even very humid air is only about 2 percent less dense than dry air at the same temperature.

In summary, temperature differences and to a lesser extent moisture content cause horizontal pressure differences, producing a force that causes air to flow from areas of high pressure to areas of low pressure. In general, cold, dry air produces higher surface pressures than warm, humid air. Further, a warm, dry air mass exhibits higher surface pressure than an equally warm, but humid, air mass. Aloft, the situation is reversed: Warm air aloft air tends to exhibit a higher air pressure than cold air, at the same height above the surface.

## Pressure Changes Caused by Airflow Aloft

The movement of air causes changes in air pressure. For example, in situations where there is a net flow of air into a region, air accumulates. This phenomenon, called **convergence**, causes air to be squeezed into a smaller space, which results in a more massive air column exerting more pressure at the surface. By contrast, in regions where there is a net outflow of air, a situation referred to as **divergence**, the surface pressure drops. We will return to the important mechanisms of convergence and divergence, which generate areas of high and low pressure, later in this chapter.

### ✔ Concept Checks 6.2

1. Explain why air pressure decreases with an increase in altitude.
2. Describe how winds blow in relation to areas of high pressure and low pressure.
3. In a column of cold air, does the pressure aloft tend to be higher or lower than in a column of warm air? Explain.
4. If all other factors are equal, does a *dry* or *moist* air mass exert more air pressure? Explain.
5. Explain why a cold, dry air mass produces a higher surface pressure than a warm, humid air mass.
6. Explain how horizontal *convergence* aloft affects surface pressure.

# 6.3 Factors Affecting Wind

**List and describe the three forces that act on the atmosphere to either create or alter winds.**

MM GEODe ► Air Pressure and Winds ► Factors Affecting Wind

If Earth did not rotate and if there were no friction, air would flow directly from areas of higher pressure to areas of lower pressure. However, because both factors exist, wind is controlled by a combination of forces, including:

1. Pressure gradient force
2. Coriolis force
3. Friction

## Pressure Gradient Force

If an object experiences an unbalanced force in one direction, it will accelerate (experience a change in velocity). The force that generates winds results from horizontal pressure differences. When air is subjected to greater pressure on one side than on another, the imbalance produces a force that is directed from the region of higher pressure toward the area of lower pressure. Thus, pressure differences cause the wind to blow, and the greater these differences, the greater the wind speed.

Recall that isobars are used on surface maps to show horizontal pressure patterns. The *spacing* of the isobars indicates the amount of pressure change occurring over a given distance and is called the **pressure gradient force (PGF)**. Pressure gradient is analogous to gravity acting on a ball rolling down a hill. A steep (or *strong*) pressure gradient, like a steep hill, causes greater acceleration of a parcel of air than does a weak pressure gradient (a gentle hill). Thus, the relationship between wind speed and the pressure gradient is straightforward: *Closely spaced isobars indicate a steep pressure gradient and strong winds; widely spaced isobars indicate a weak pressure gradient and light winds*. Figure 6.11A illustrates the relationship between the spacing of isobars and wind speed. Note also that the pressure gradient force is always directed at *right angles* to the isobars. When the isobars are curved, the pressure gradient force radiates out from the area of high pressure toward areas of low pressure, as illustrated in Figure 6.11B.

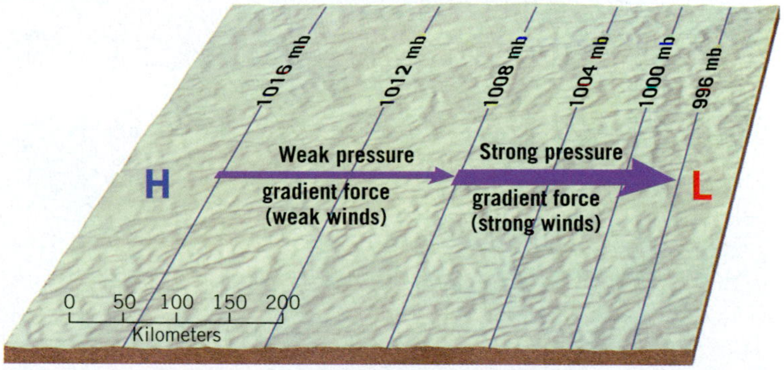


**A. Pressure gradient force when the isobars are nearly straight.**

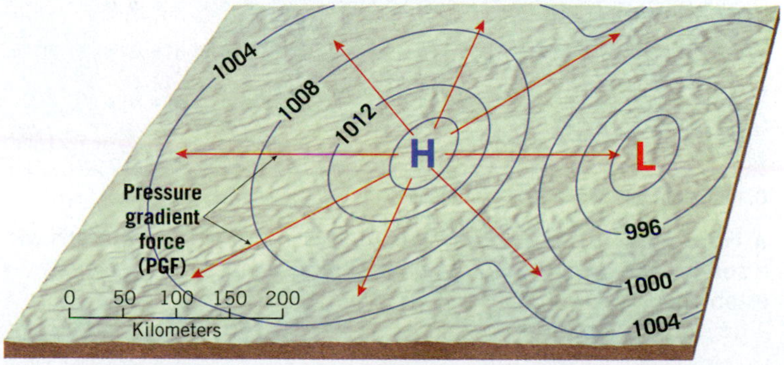


**B. The pressure gradient force when the isobars are curved or form nearly concentric circles.**

► **Figure 6.11 Isobars are lines connecting places of equal atmospheric pressure** The spacing of isobars indicates the amount of pressure change occurring over a given distance—called the *pressure gradient force*. Closely spaced isobars indicate a strong pressure gradient and high wind speeds, whereas widely spaced isobars indicate a weak pressure gradient and low wind speeds.

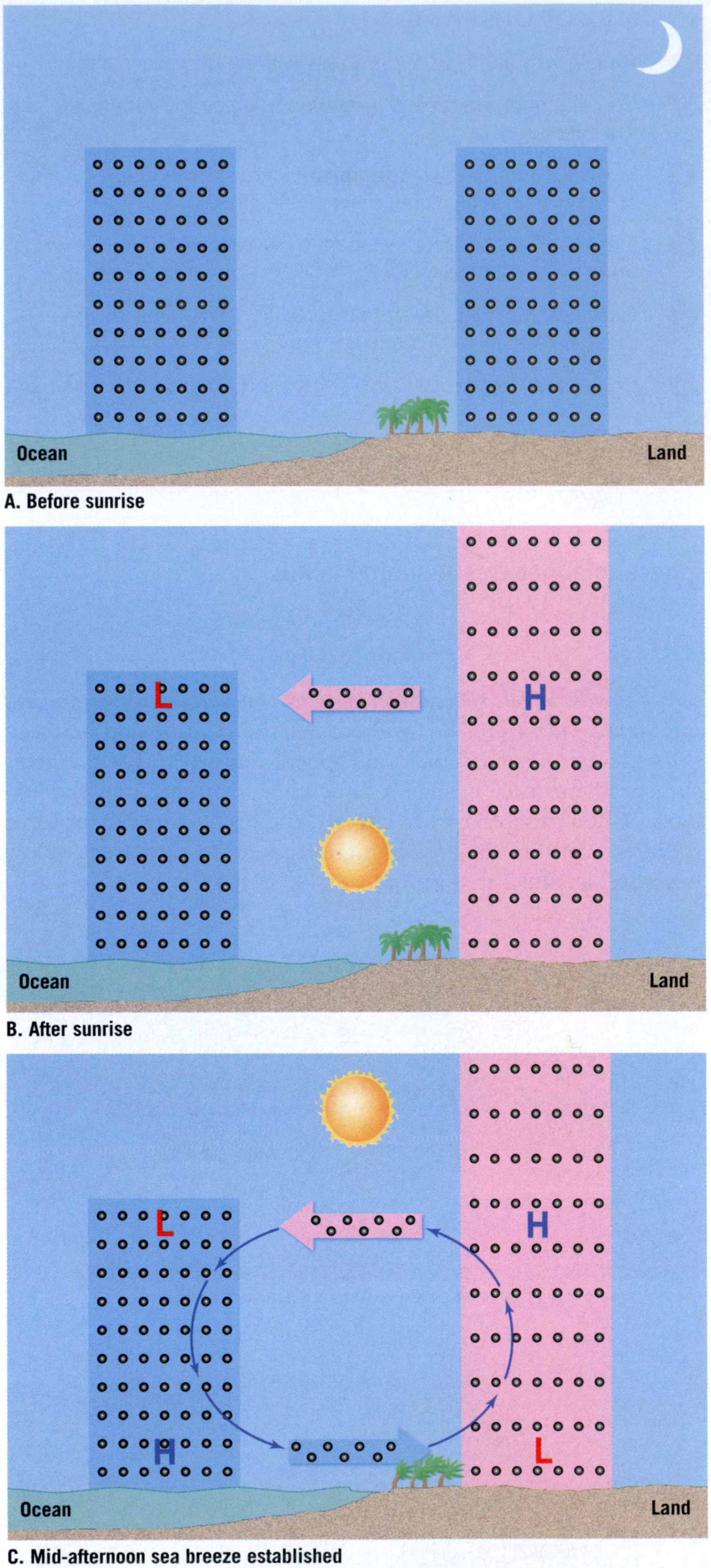


▲ **Figure 6.12 Cross-sectional view illustrating the formation of a sea breeze** **A.** Just before sunrise; **B.** after sunrise; **C.** sea breeze established.

**How the Pressure Gradient Force Generates Wind: The Sea Breeze** To illustrate how temperature differences can generate a horizontal pressure gradient and winds, consider a common example, the *sea breeze*. **Figure 6.12A** shows a hypothetical cross section of a coastal location before sunrise. The dots in the columns represent air molecules. At this time of day, we will assume that temperatures and pressures over land and water are the same, so Figure 6.12A shows two columns of air with identical masses (same number of air molecules) that exert the same surface pressure. Because there is no horizontal variation in pressure (zero pressure gradient) either aloft or at the surface, there is no wind.

After sunrise, the temperature of the atmosphere above Earth's land surface begins to increase, while the temperature over the ocean remains nearly constant. (Recall that land heats up more rapidly than a large body of water.) As the air over land warms, it expands and forms a taller, less dense air column. Although the air pressure at the surface remains essentially the same, this is not the case at higher altitudes. Notice in **Figure 6.12B** that there are more air molecules over the *H* (high pressure) in the column of air over land than over the *L* (low pressure) located at the same elevation over water. Consequently, the pressure above land is higher than the pressure over water at the same altitude—which results in the air aloft flowing away from the land (area of high pressure) toward the water (area of low pressure).

The mass transfer of air aloft, in turn, creates a high-pressure area over the ocean, where the air collects, and a surface low over land. The *surface* circulation that develops from this redistribution of mass is from the sea toward the land (sea breeze), as shown in **Figure 6.12C**. Thus, a simple thermal circulation develops, with a seaward flow aloft and a landward flow at the surface. Note that *vertical* movement is required to make the circulation complete.

An important relationship exists between air pressure and temperature, as you saw in the preceding discussion. Temperature variations create pressure differences and ultimately wind. Daily temperature differences caused by unequal heating as in the sea breeze example tend to be confined to a zone only a few kilometers thick. On a global scale, however, variations in the amount of solar radiation received in the polar versus the equatorial latitudes generate the much larger pressure systems that in turn produce the planetary atmospheric circulation. Therefore, the underlying cause of global pressure differences and, by extension, wind is mainly the *unequal heating of Earth's land–sea surface.*

In summary, the horizontal pressure gradient is the driving force of wind. The magnitude of the pressure gradient force is shown by the *spacing* of isobars, whereas the direction of force is always from areas of higher pressure toward areas of lower pressure and at right angles to the isobars.

## Coriolis Force

The weather map in **Figure 6.13** shows the typical airflow associated with surface high- and low-pressure systems. As expected, the air moves out of the regions of higher pressure and into the

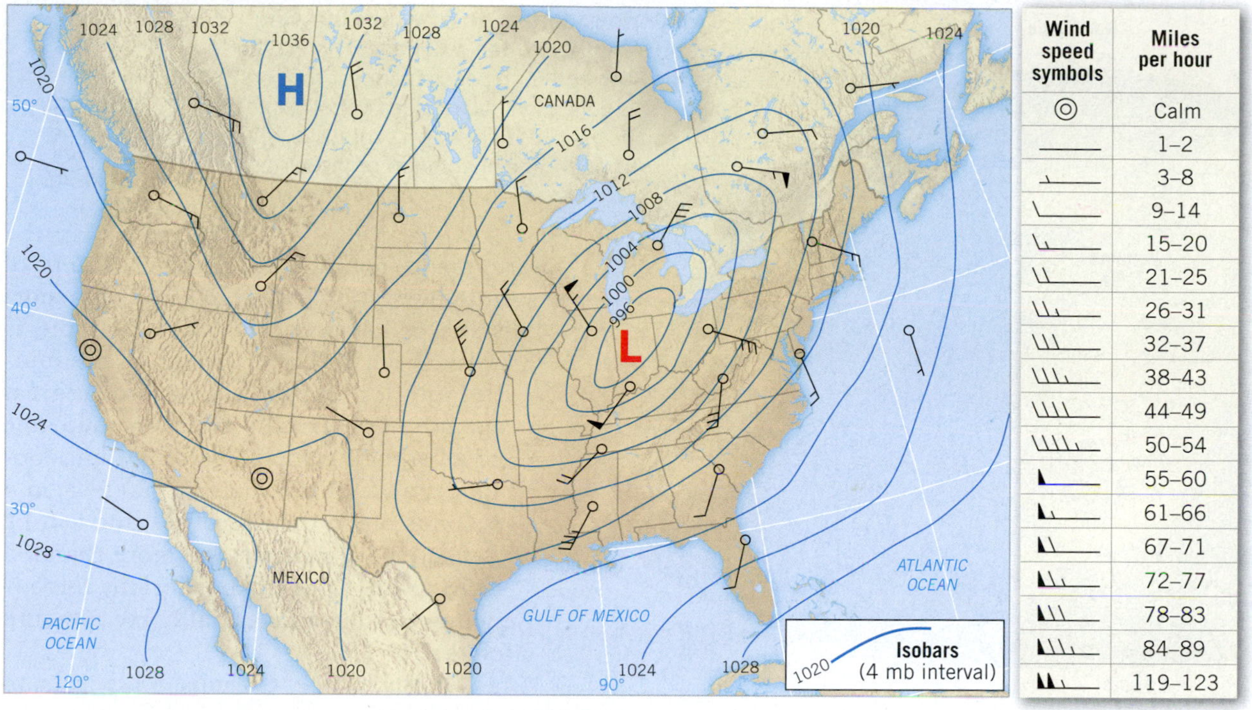


◀ **Figure 6.13 Isobars show the distribution of pressure on surface weather maps** Isobars are seldom straight but usually form broad curves. Concentric isobars indicate cells of high and low pressure. The "wind flags" indicate the expected airflow surrounding pressure cells and are plotted as "flying" with the wind (that is, the wind blows toward the station circle). Notice on this map that the isobars are more closely spaced and the wind speed is faster around the low-pressure center than around the high.

regions of lower pressure. However, the wind rarely crosses the isobars at right angles, as the pressure gradient force directs. This deviation is a result of Earth's rotation and has been named the **Coriolis force**, after the French scientist Gaspard-Gustave Coriolis, who first expressed its magnitude quantitatively. It is important to note that the Coriolis force does not generate wind; rather, *it modifies the direction of airflow*.

The Coriolis force causes all free-moving objects, including wind, to be deflected to the *right* of their path of motion in the Northern Hemisphere and to the *left* in the Southern Hemisphere. The reason for this deflection can be illustrated by imagining the path of a rocket launched from the North Pole toward a target on the equator (**Fig. 6.14**). If the rocket travels 1 hour toward its target, Earth would have rotated 15° to the east during its flight. To someone watching the rocket's path from the location of the intended target, it would look as if the rocket veered off its path and hit Earth 15° west of its target. The true path of the rocket was straight and would appear as such to someone in space looking toward Earth. Earth's rotation under the rocket produced the *apparent* deflection.

Although it is usually easy to visualize Coriolis deflection when the motion is from north to south, as in our rocket example, it is not as easy to see how a west-to-east flow would

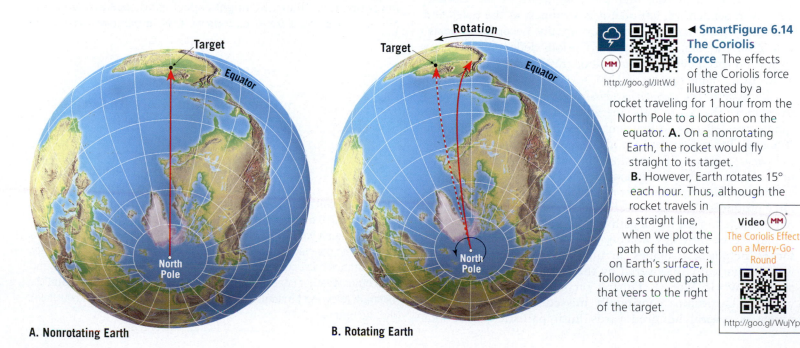

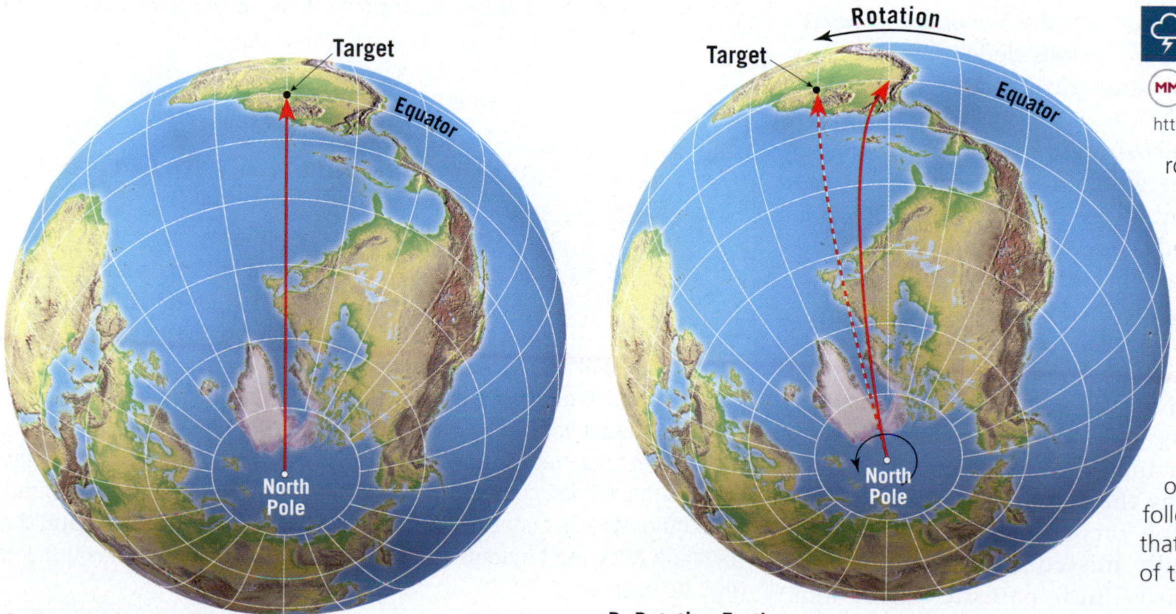


**A. Nonrotating Earth**

**B. Rotating Earth**

http://goo.gl/JltWd

◀ **SmartFigure 6.14 The Coriolis force** The effects of the Coriolis force illustrated by a rocket traveling for 1 hour from the North Pole to a location on the equator. **A.** On a nonrotating Earth, the rocket would fly straight to its target. **B.** However, Earth rotates 15° each hour. Thus, although the rocket travels in a straight line, when we plot the path of the rocket on Earth's surface, it follows a curved path that veers to the right of the target.

**Video** The Coriolis Effect on a Merry-Go-Round

http://goo.gl/WujYp

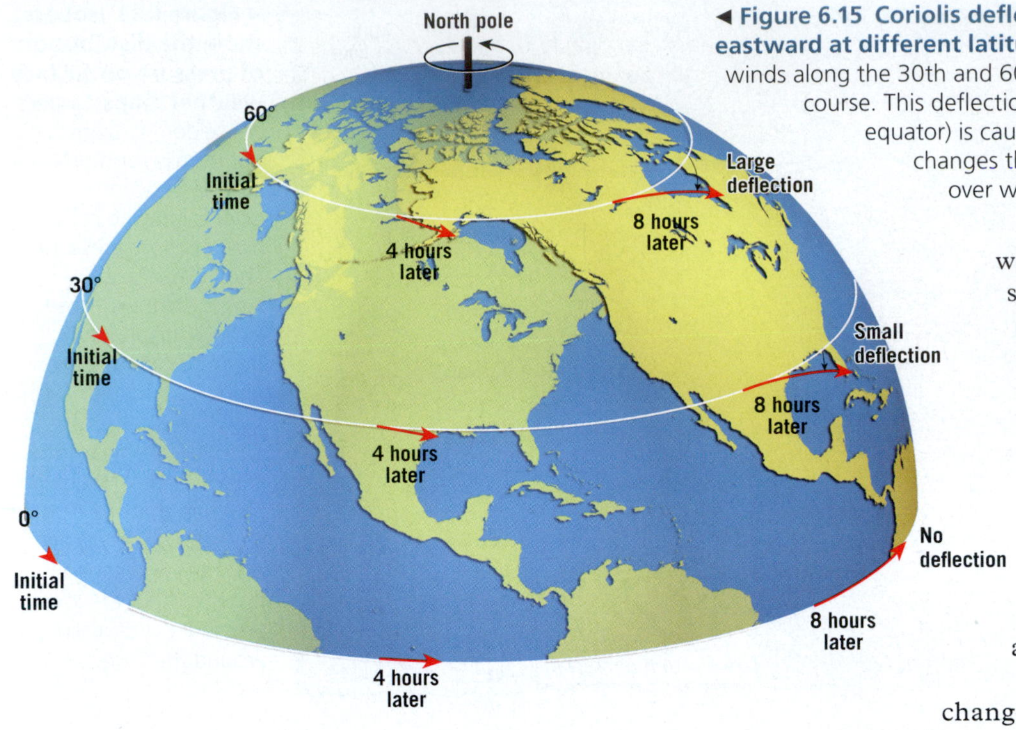


◄ **Figure 6.15 Coriolis deflection of winds blowing eastward at different latitudes** After a few hours, the winds along the 30th and 60th parallels appear to veer off course. This deflection (which does not occur at the equator) is caused by Earth's rotation, which changes the orientation of the surface over which the winds are moving.

Animation MM
The Coriolis Effect
http://goo.gl/slLc3a

be deflected. **Figure 6.15** illustrates this situation, using winds blowing eastward at three different latitudes (0°, 30°, and 60°). Notice that after a few hours, the winds along the 30th and 60th parallels appear to be veering off course. However, when viewed from space, it is apparent that these winds have maintained their original direction. It is the "change" of orientation of North America as Earth rotates on its axis that produces the deflection we observe in Figure 6.15.

We can also see in Figure 6.15 that the amount of deflection is greater at 60° latitude than at 30° latitude. Furthermore, there is no deflection observed for the airflow along the equator. We conclude, therefore, that the magnitude of the Coriolis force is dependent on latitude; it is strongest at the poles and weakens equatorward, where it eventually becomes nonexistent. Also note that the amount of Coriolis deflection increases with wind speed because faster winds travel farther than slower winds in the same time period.

Because of the *counterclockwise* rotation of Earth as viewed from above the North Pole, the winds in the Northern Hemisphere are deflected to the right of their initial path of motion (**Fig. 6.16**). In the Southern Hemisphere, however, Earth exhibits a *clockwise* rotation. (To visualize this, look down on a globe that has a counterclockwise rotation and then look at it from below—a Southern Hemisphere point of view.) As a result, the Coriolis force produces a similar deflection in the Southern Hemisphere, but to the left of the initial path of motion (Figure 6.16).

All "free-moving" objects are affected by the Coriolis force, including airplanes, artillery, and rockets. This fact was dramatically discovered by the U.S. Navy at the beginning of World War II. During target practice, long-range guns on battleships continually missed their targets by as much as several hundred yards until ballistic corrections were made for the changing position of seemingly stationary targets. Across short distances, however, the Coriolis force is relatively small. Nevertheless, in the middle latitudes, this deflecting force is great enough to potentially affect the outcome of a baseball game. A ball hit a horizontal distance of 100 meters (330 feet) in 4 seconds down the right field line will be deflected 1.5 centimeters (more than 0.5 inch) to the right by the Coriolis force—enough to turn a potential home run into a foul ball!

In summary, the Coriolis force acts to change the direction of a moving body to the right in the Northern Hemisphere and to the left in the Southern Hemisphere. This deflecting force (1) is always directed at right angles to the direction of airflow; (2) affects only wind direction, not speed; (3) is proportional to wind speed, so that the stronger the wind, the greater the deflecting force; and (4) is strongest at the poles and weakens equatorward, becoming nonexistent at the equator.

## Friction

Earlier we stated that the pressure gradient force is the primary driving force of wind. As an unbalanced force, it causes air to accelerate from regions of higher pressure toward regions of lower pressure. Thus, we might expect wind speeds to continually increase (accelerate) as long as this imbalance exists. But

**students sometimes ask...**

### Do baseballs really fly farther at Denver's Coors Field?

Coors Field has become known as the "homerun hitter's ballpark" because it led all Major League ballparks in total home runs during its first decade. In theory, a well-struck baseball should travel roughly 10 percent farther in Denver (elevation 5280 feet) because of lower air density (and therefore less friction) than it would in a ballpark at sea level. However, a group of researchers concluded that the assumed elevation enhancement of fly ball distance has been greatly overestimated; these researchers found that other weather conditions, mainly the prevailing winds around the ballpark (blowing predominantly from home plate toward the outfield), aid the hitters.

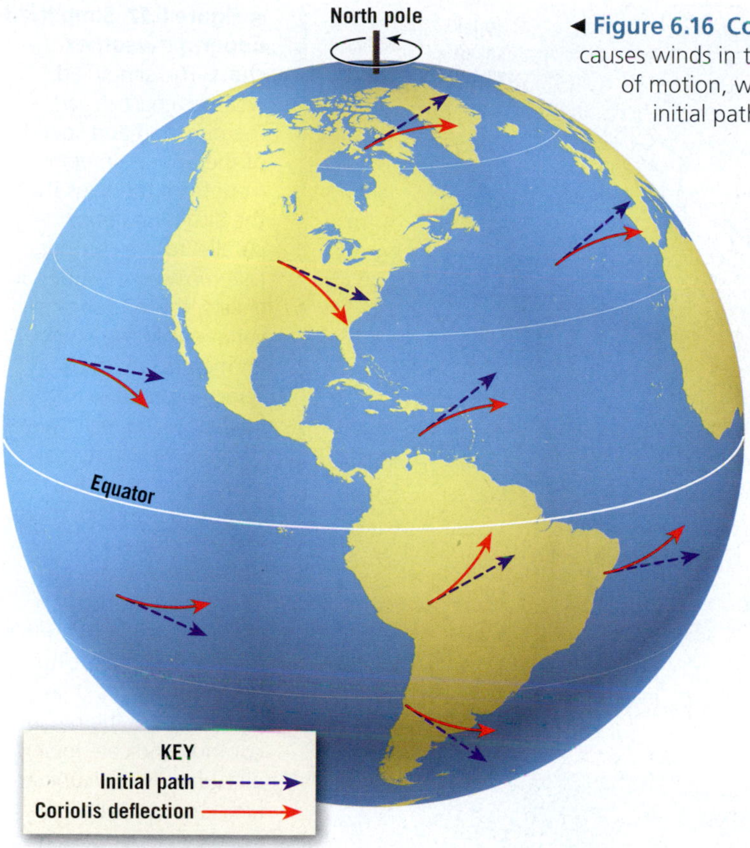


◀ **Figure 6.16 Coriolis deflection of winds in both hemispheres** The Coriolis force causes winds in the Northern Hemisphere to be deflected to the right of their initial path of motion, whereas in the Southern Hemisphere they are deflected to the left of their initial path.

we know from personal experience that winds do not become faster indefinitely. Rather, **friction** acts to slow moving objects and, thus, affects wind speed.

By slowing airflow, friction also reduces the Coriolis force. Thus not only is airflow near the surface slower than winds aloft, the direction of flow is also impacted. As we will see in the next section, winds aloft flow nearly parallel to the isobars, whereas wind near the surface flows across the isobars at an angle, dependent on the region's topography.

Friction significantly influences airflow only in the first 1.5 kilometers (1 mile) of Earth's atmosphere, often referred to as the **boundary layer**. The effect of friction is negligible above that height. As a result, upper-level flow is less complicated than surface winds. We analyze both types of flow in the next section.

## ✔ Concept Checks 6.3

1. List three forces that combine to direct wind (horizontal airflow). Which of these forces is responsible for *generating* wind?
2. Write a generalization relating the spacing of isobars to wind speed.
3. Briefly describe how the Coriolis force modifies air movement.
4. Identify the two factors that influence the magnitude of the Coriolis force.

**students sometimes ask...**

**I've heard that water goes down a sink in one direction in the Northern Hemisphere and in the opposite direction in the Southern Hemisphere. Is that true?**

This is a myth that comes from applying a scientific principle to a situation where it does not fit. Coriolis deflection causes cyclonic systems to rotate counterclockwise in the Northern Hemisphere and clockwise in the Southern Hemisphere, so it was inevitable that someone would suggest (without checking) that a sink should drain in a similar manner. However, a cyclone is more than 1000 kilometers in diameter and may exist for several days, while a typical sink is less than 1 meter in diameter and drains in a matter of seconds. On this scale, the Coriolis force is minuscule. The shape of the sink and how level it is has more to do with the direction of water flow than the Coriolis force.

# 6.4 Winds Aloft Versus Surface Winds

**Explain why winds aloft flow roughly parallel to the isobars, while surface winds travel at an angle across the isobars.**

In this section we address airflow aloft that occurs at least 1.5 kilometers above Earth's surface, where the effects of friction are not significant enough to influence the flow of winds, and then we analyze winds at the surface, where friction significantly influences airflow.

## Straight-line Flow and Geostrophic Winds

As shown in **Figure 6.17**, the winds aloft tend to flow parallel to the isobars. In regions where the isobars are relatively straight and evenly spaced, the resulting winds flow in

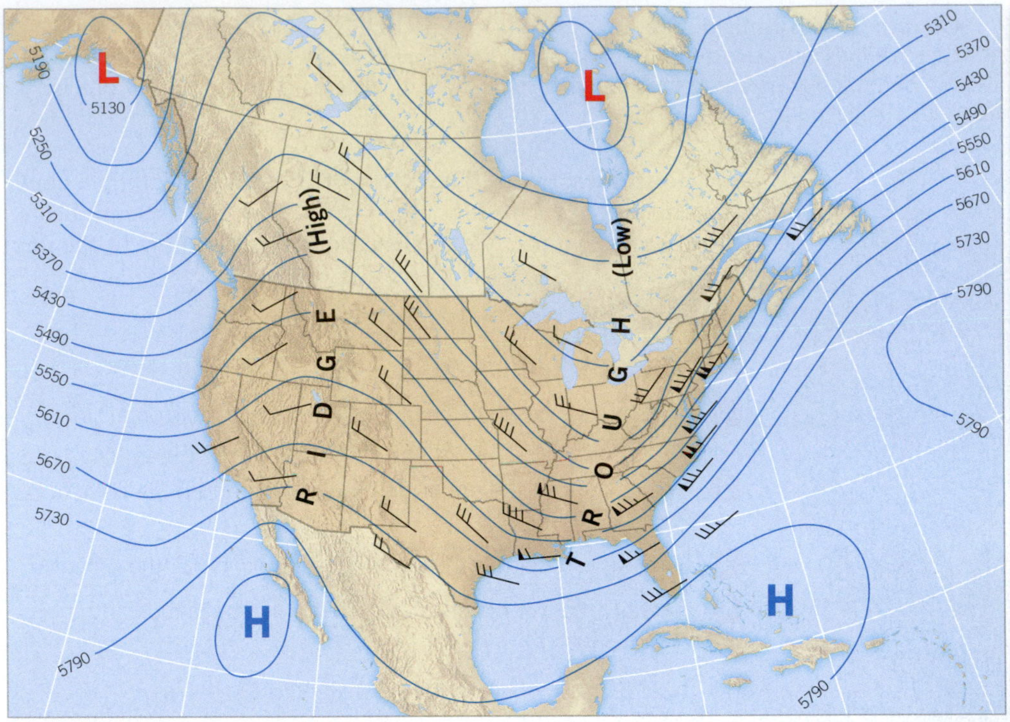

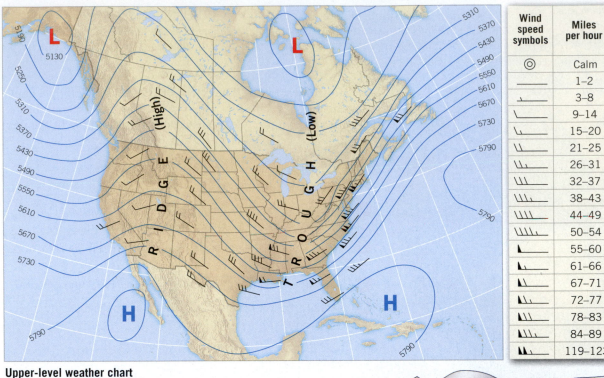

Upper-level weather chart

| Wind speed symbols | Miles per hour |
|---|---|
| ◎ | Calm |
| | 1–2 |
| | 3–8 |
| | 9–14 |
| | 15–20 |
| | 21–25 |
| | 26–31 |
| | 32–37 |
| | 38–43 |
| | 44–49 |
| | 50–54 |
| | 55–60 |
| | 61–66 |
| | 67–71 |
| | 72–77 |
| | 78–83 |
| | 84–89 |
| | 119–123 |

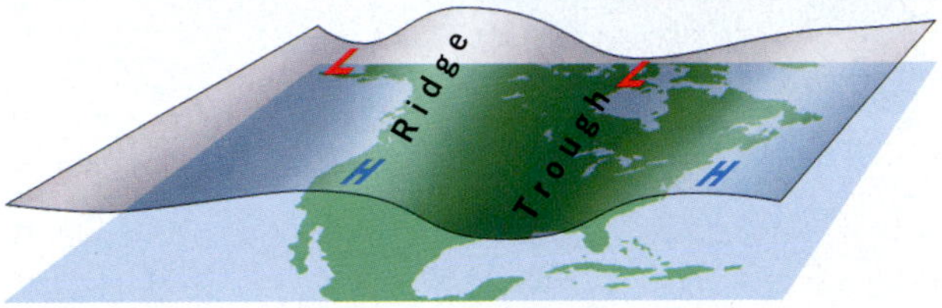


◄ **Figure 6.17 Simplified upper-air weather chart** This simplified weather chart shows the direction and speed of the upper-air winds. Note from the flags that the airflow is almost parallel to the contours. Like most other upper-air charts, this one shows variations in the height (in meters) at which a selected pressure (500 millibars) is found instead of showing variations in pressure at a fixed height, like surface maps. Places experiencing 500-millibar pressure at higher altitudes (toward the south on this map) are experiencing higher pressures than places where the height contours indicate lower altitudes. Thus, *higher-elevation* contours indicate *higher* pressures, and *lower-elevation* contours indicate *lower* pressures.

roughly a straight line and parallel to the isobars. This phenomenon, called **geostrophic wind**, is generated when a balance is reached between the Coriolis force (CF) and the pressure gradient force (PGF).

Figure 6.18 illustrates the formation of geostrophic winds. We begin with a nonmoving parcel of air directed by the pressure gradient force from an area of high pressure near the bottom of the diagram to the low-pressure area at the top. Initially our parcel of air has no motion, and the Coriolis force is nonexistent. Under the influence of the pressure gradient force, the parcel begins to accelerate from the area of high pressure to the area of low pressure. As soon as the parcel begins to move, the Coriolis force commences and causes a deflection to the right for winds in the Northern Hemisphere. As the parcel accelerates, the strength of the Coriolis force increases. Recall that the magnitude of the Coriolis force is proportional to wind speed: The greater the wind speed, the greater the deflection.

Eventually, the wind turns so that it is flowing parallel to the isobars, at which point the pressure gradient force is exactly balanced by the opposing Coriolis force (Figure 6.18). As long as these forces remain balanced, the resulting wind continues to flow parallel to the isobars at a constant speed. Stated another way, the wind is coasting (not accelerating or decelerating) along a pathway defined by the isobars.

Under these idealized conditions, when the Coriolis force is exactly equal in strength to the pressure gradient force,

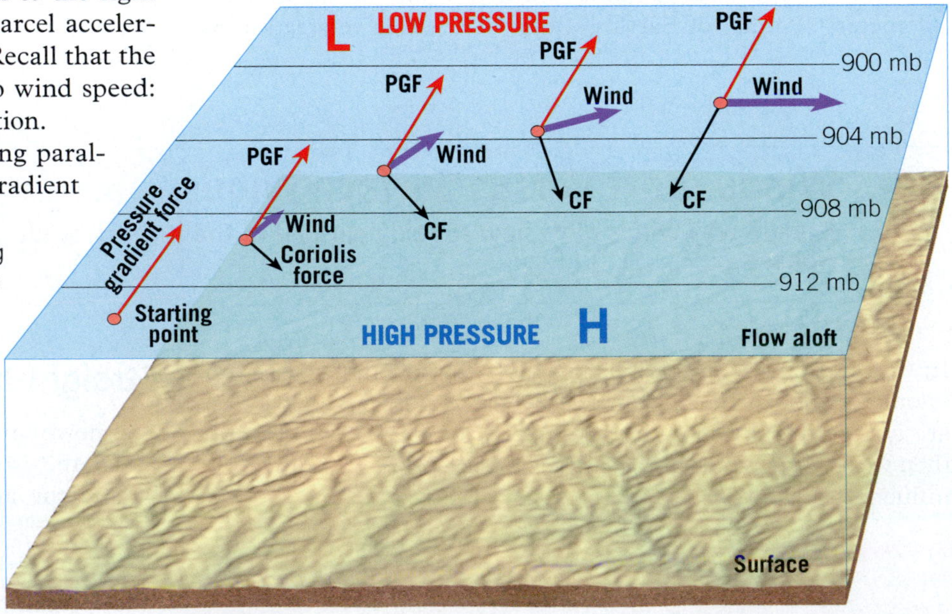


► **Figure 6.18 Geostrophic wind** The only force acting on a stationary parcel of air is the pressure gradient force (PGF). Once the air begins to accelerate, the Coriolis force (CF) deflects it to the right in the Northern Hemisphere. Greater wind speeds result in a stronger Coriolis force (deflection) until the flow is parallel to the isobars. At this point the pressure gradient force and Coriolis force are in balance, and the flow is called a *geostrophic wind*. It is important to note that in the "real" atmosphere, airflow continually adjusts for variations in the pressure field. As a result, the adjustment to geostrophic equilibrium is much more irregular than shown.

but acting in the opposite direction, the airflow is said to be in *geostrophic balance*. Geostrophic winds flow in relatively straight paths, parallel to the isobars, with velocities proportional to the pressure gradient force. A strong pressure gradient (contour lines closer together) creates strong winds, and a weak pressure gradient (contour lines farther apart) produces light winds.

It is important to note that the geostrophic wind is an idealized model that only approximates the actual behavior of airflow aloft. Under normal conditions, the isobars are rarely straight and uniformly spaced, so winds are never purely geostrophic. Nonetheless, the geostrophic model offers a useful estimation of the actual winds aloft. By measuring the pressure field (orientation and spacing of isobars) that exists aloft, meteorologists can determine both wind direction and speed (see Figure 6.17).

**Buys Ballot's Law** As we have seen, wind direction is directly linked to the prevailing pressure pattern. Therefore, if we know the wind direction, we can also establish a rough approximation of the pressure distribution. This rather straightforward relationship between wind direction and pressure distribution was first formulated by the Dutch meteorologist Buys Ballot in 1857. Essentially, **Buys Ballot's law** states that in the Northern Hemisphere, *if you stand with your back to the wind, low pressure will be found to your left and high pressure to your right*. In the Southern Hemisphere, the situation is reversed.

Although Buys Ballot's law holds for airflow aloft, it must be used with caution when applied to surface winds. At the surface, friction and topography interfere with the idealized circulation. At the surface, if you stand with your back to the wind and then turn clockwise about 30°, low pressure will be to your left and high pressure to your right.

In summary, winds above a few kilometers can be considered geostrophic—that is, flowing in a straight path parallel to the isobars at speeds that can be calculated from the pressure gradient. The major discrepancy from true geostrophic winds involves the flow along highly curved paths, a topic considered next.

## Curved Flow and Gradient Winds

A casual glance at the upper-level weather chart in Figure 6.17 shows that the isobars (contours) are not generally straight; instead, they make broad, sweeping curves. Occasionally the isobars connect to form roughly circular cells of either high or low pressure. Thus, unlike geostrophic winds that flow along relatively straight paths, winds around bends and cells of high or low pressure follow highly curved paths. Winds of this nature, which blow at a constant speed parallel to curved isobars, are called **gradient winds**.

Let us examine how the pressure gradient force and Coriolis force combine to produce gradient winds. **Figure 6.19A** shows the gradient flow around a center of low pressure. As soon as the flow begins, the Coriolis force causes the air to be deflected. In the Northern Hemisphere, where the Coriolis force deflects the flow to the right, the resulting wind blows counterclockwise about a low. Conversely, around a high-pressure cell, the outward-directed pressure gradient force is opposed by the inward-directed Coriolis force, and a clockwise flow results. **Figure 6.19B** illustrates this idea.

Because the Coriolis force deflects the winds to the left in the Southern Hemisphere, the flow is reversed—clockwise around low-pressure centers and counterclockwise around high-pressure centers.

Recall that meteorologists call centers of low pressure *cyclones* and the flow around them *cyclonic*. There are several types and scales of cyclones. Large low-pressure systems that are major weather makers in the United States are called *midlatitude cyclones*. Other examples include *tropical cyclones* (hurricanes), which are generally smaller than midlatitude cyclones, and *tornadoes*, which are tiny and extremely intense cyclonic storms. **Cyclonic flow** has the same direction of rotation as Earth: counterclockwise in the Northern Hemisphere and clockwise in the Southern Hemisphere. Centers of high pressure are called *anticyclones* and exhibit **anticyclonic flow** (opposite that of Earth's rotation).

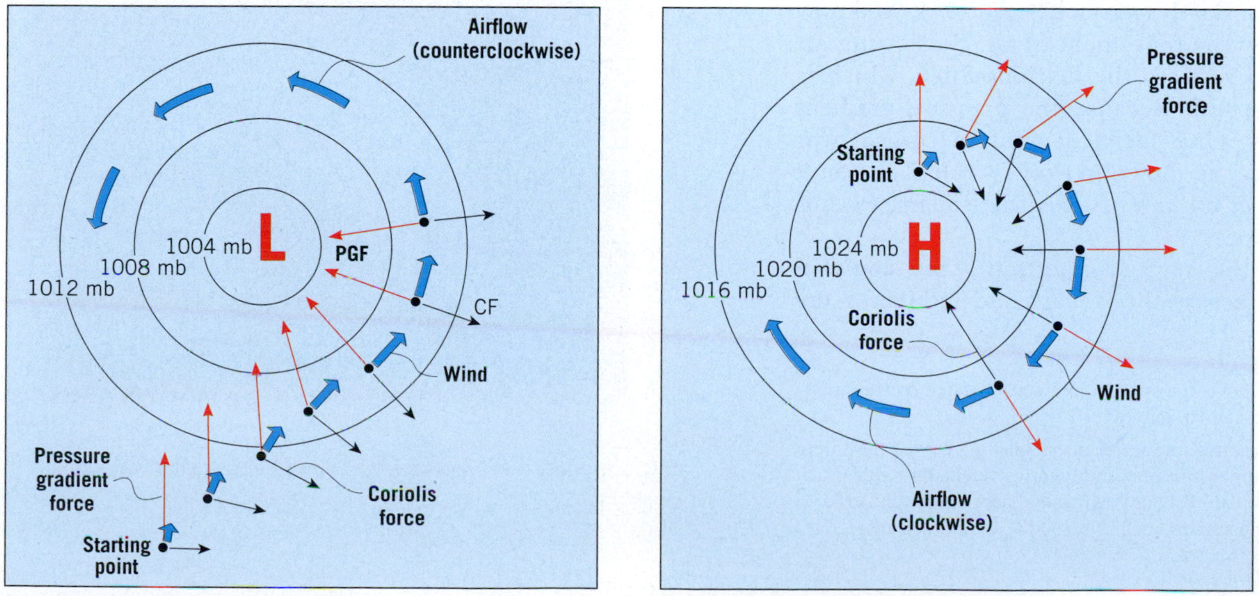


A. Cyclonic flow (Northern Hemisphere)

B. Anticyclonic flow (Northern Hemisphere)

◀ **Figure 6.19 Gradient wind A.** Idealized airflow aloft around a low-pressure center (cyclone). **B.** Idealized airflow aloft around a high-pressure center (anticyclone).

Animation MM
Wind Pattern Development
http://goo.gl/MTT9bU

Recall that whenever isobars curve to form elongated regions of low and high pressure, these areas are called *troughs* and *ridges*, respectively (see Figure 6.17). The flow around a trough is cyclonic; conversely, the flow around a ridge is anticyclonic.

Now let us consider the forces that produce the gradient flow associated with cyclonic and anticyclonic circulations. Wherever the flow is curved, a force has deflected the air (changed its direction), even when no change in speed results. This is a consequence of Newton's first law of motion, which states that a moving object will continue to move in a straight line* unless acted upon by an unbalanced force. You have undoubtedly experienced the effect of Newton's law when the automobile in which you were riding turned sharply and your body tried to continue moving straight ahead (see Appendix E).

Figure 6.19A shows us that in a low-pressure center, the inward-directed pressure gradient force is opposed by the outward-directed Coriolis force. But to keep the path curved (parallel to the isobars), the inward pull of the pressure gradient force must be strong enough to balance the Coriolis force as well as to turn (accelerate) the air inward. The inward turning of the air is called *centripetal acceleration*. Stated another way, the pressure gradient force must exceed the Coriolis force to overcome the air's tendency to continue moving in a straight line.**

The opposite situation exists in anticyclonic flows, where the inward-directed Coriolis force must balance the pressure gradient force as well as provide the inward acceleration needed to turn the air. Notice in Figure 6.19 that the pressure gradient force and Coriolis force are not balanced (the arrow lengths are different), as they are in geostrophic flow. This imbalance provides a change in direction (centripetal acceleration) that generates curved flow.

## Surface Winds

Friction as a factor affecting wind is important only within the first 1.5 kilometers of Earth's surface (**Fig. 6.20A**). We know that friction acts to slow the movement of air. By slowing air movement, friction also reduces the Coriolis force, which is proportional to wind speed. Because the pressure gradient force is not affected by wind speed, it wins the tug of war against the Coriolis force, as shown in **Figure 6.20B**. The result is the movement of air *at an angle* across the isobars, toward the area of lower pressure.

The roughness of the surface determines the angle at which the air will flow across the isobars and influences the speed at which it will move. Over relatively smooth surfaces, where friction is low, air moves at an angle of 10° to 20° to the isobars and at speeds roughly two-thirds of geostrophic flow (Figure 6.20B). Over rugged terrain where friction is high, the angle can be as great as 45° from the isobars, with wind speeds reduced by as much as 50 percent (**Fig. 6.20C**). We have learned that above the friction layer in the Northern Hemisphere, winds blow counterclockwise around a cyclone and clockwise around an anticyclone, with winds nearly parallel to the isobars. Combined with the effect of friction, we find that surface airflow crosses the isobars at varying angles, depending on the terrain, but always from higher to lower pressure (see Figure 6.7). In a cyclone, in which pressure decreases inward, friction causes a net flow *toward* its center. In an anticyclone, the opposite is true: Pressure decreases outward, and friction causes a net flow *away* from the center. Therefore, the resultant winds blow into and counterclockwise about a cyclone (**Fig. 6.21A**) and outward and clockwise about an anticyclone. Of course, in the Southern Hemisphere the Coriolis force deflects the winds

*The tendency of a particle to move in a straight line when rotated creates an imaginary outward force called centrifugal force.

**Note that centripetal acceleration is important in establishing curved flow aloft, but near the surface, friction comes into play and greatly overshadows this much weaker force. Consequently, we can discount centripetal acceleration in surface flow except with rapidly rotating storms such as tornadoes and hurricanes.

### eye ON THE atmosphere 6.1

In May and June 2011, the Wallow Fire burned more than 538 thousand acres (840 square miles) in southeastern Arizona. It was the largest wildfire in Arizona history.

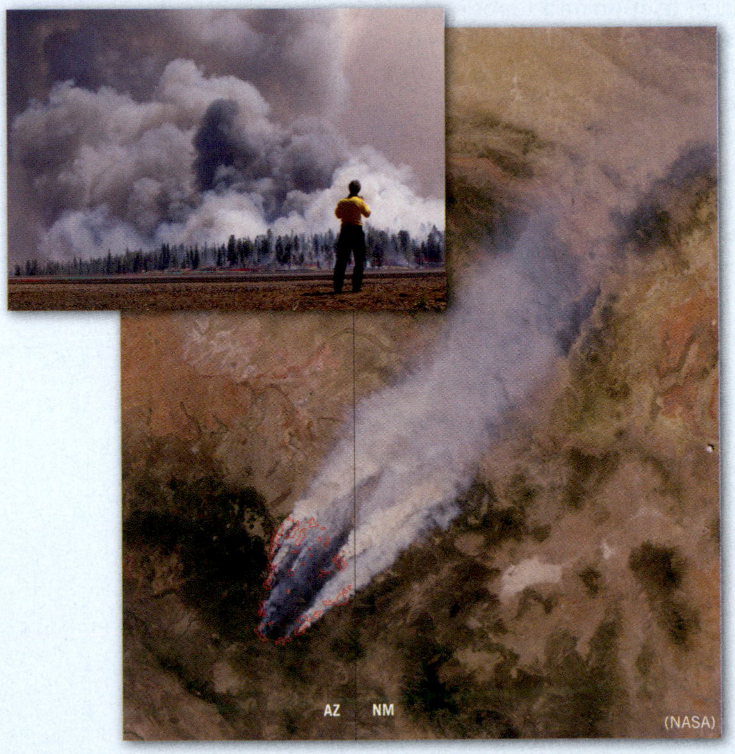


**Question**

1. Can you suggest two ways in which wind helps sustain wildfires such as this?

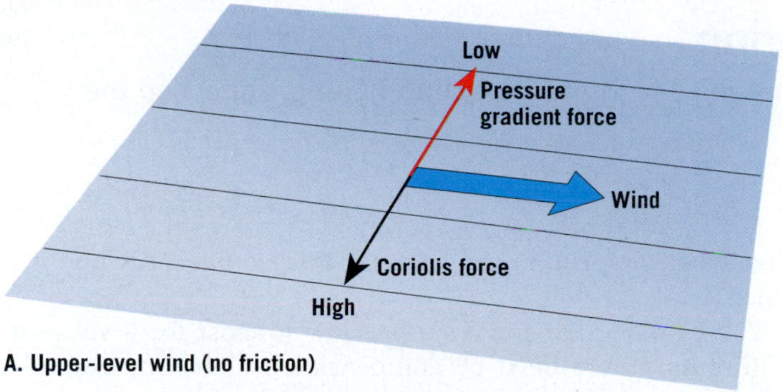


A. Upper-level wind (no friction)

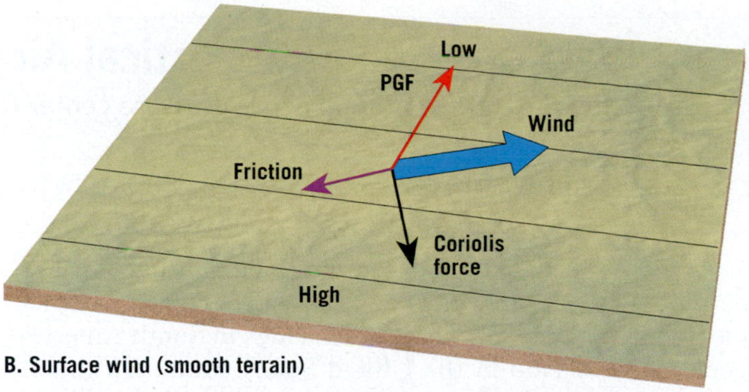


B. Surface wind (smooth terrain)

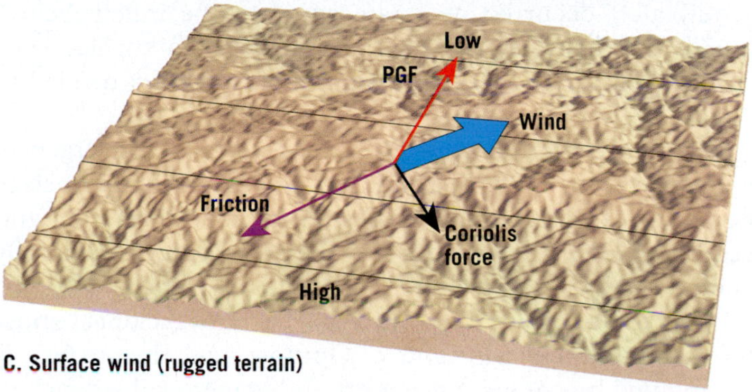


C. Surface wind (rugged terrain)

▲ **Figure 6.20 Winds aloft versus surface winds A.,B.** Comparison between upper-level winds and surface winds, showing the effects of friction on airflow. Friction slows surface wind speed, which weakens the Coriolis force, causing the winds to cross the isobars. **C.** Notice that over rugged terrain, wind speed is slower and directed at a steeper angle across the isobars than winds over smooth terrain.

to the left and reverses the direction of flow (**Fig. 6.21B**). Regardless of hemisphere, however, friction causes a net inflow (*convergence*) around a cyclone and a net outflow (*divergence*) around an anticyclone.

Video MM
Winds During a Drought
http://goo.gl/Br203E

## ✔ Concept Checks 6.4

1. Explain the formation of geostrophic wind and describe how geostrophic winds blow relative to isobars.
2. Describe the direction of cyclonic and anticyclonic flow in both the Northern and Southern Hemispheres.
3. Unlike winds aloft, which blow nearly parallel to the isobars, surface winds generally cross the isobars. Explain what causes this difference.
4. Prepare a diagram with isobars and wind arrows that shows the winds associated with surface cyclones and anticyclones in both the Northern and Southern Hemispheres.

▼ **Figure 6.21 Cyclonic circulation in the Northern and Southern Hemispheres** The cloud patterns in these images allow us to see the circulation pattern in the lower atmosphere.

**A. This satellite image shows a large low-pressure center in the Gulf of Alaska. The cloud pattern clearly shows an inward and counterclockwise spiral.**

**B. This satellite image shows a strong cyclonic storm in the Southern Hemisphere. The cloud pattern shows an inward and clockwise circulation.**

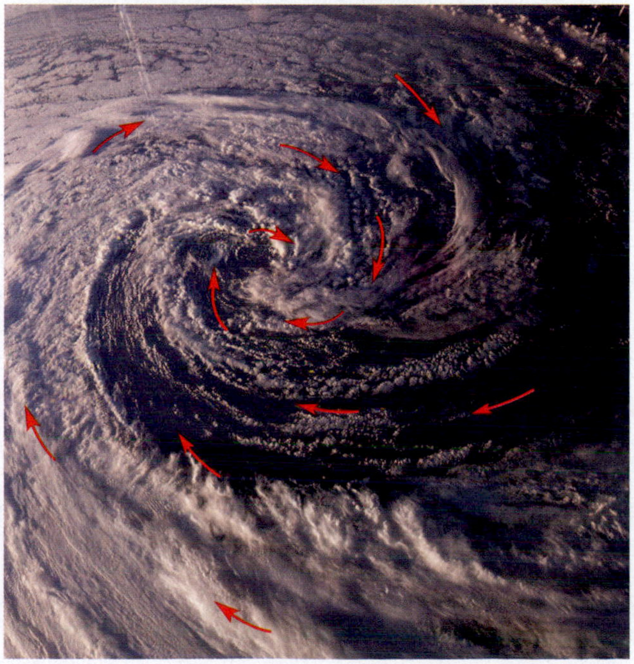

Video MM
Hurricane Winds
http://goo.gl/Pp6GnB

# 6.5 How Winds Generate Vertical Air Motion

**Describe the airflow around a low-pressure center (cyclone) and a high-pressure center (anticyclone) and the weather associated with each.**

MM GEODe ▶ Air Pressure and Winds ▶ Highs and Lows

So far we have discussed wind without regard to how airflow in one region might affect airflow elsewhere. One researcher commented that a butterfly flapping its wings in South America can generate a tornado in the United States. Although this is an exaggeration, it suggests that airflow in one region might cause a change in weather at some later time at a different location.

Just how horizontal airflow (wind) relates to vertical flow is an important issue. Although the rate at which air ascends or descends (except in violent storms) is slow compared to horizontal flow, it is very important as a weather maker. You learned in Chapter 4 that rising air is associated with cloudy conditions and often precipitation, whereas subsidence produces adiabatic heating and clearing conditions. In this section we examine how the movement of air can create pressure changes and hence generate winds.

## Vertical Airflow Associated with Cyclones and Anticyclones

Let us first examine a surface low-pressure system (cyclone) in which the air is spiraling inward. The net inward transport of air causes a shrinking of the area it occupies, a process called *convergence* (Fig. 6.22). Whenever air converges horizontally, it piles up—that is, it increases in mass to allow for the decreased area it occupies. This process generates a "denser" column. We have encountered a paradox: Low-pressure centers cause a net accumulation of air, which increases their pressure. Consequently, a surface cyclone should quickly eradicate itself, not unlike what happens to the vacuum in a coffee can when it is opened.

You can see that for a surface low to exist for a substantial duration, there must be compensation aloft. For example, surface convergence could be maintained if *divergence* (spreading out) aloft occurred at a rate equal to the inflow below. Figure 6.22 illustrates the relationship between surface convergence (inflow) and the divergence aloft (outflow) needed to maintain a low-pressure center.

Divergence aloft may even exceed surface convergence, thereby accelerating vertical motion and intensifying surface inflow. Because rising air often results in cloud formation and precipitation, the passage of a low-pressure center is generally accompanied by "bad weather."

Like their cyclonic counterparts, anticyclones must also be maintained from above. Outflow near the surface is accompanied by convergence aloft and general subsidence of the air column (Figure 6.22). Because descending air is compressed and warmed, cloud formation and precipitation are less likely in an anticyclone. Thus, fair weather can usually be expected with the approach of a high-pressure system.

For these reasons, it is common to see "stormy" at the low-pressure end of household barometers and "fair" at the high end. By noting the pressure trend—rising, falling, or steady—we have a good indication of forthcoming weather. Such a determination, called the **pressure tendency**, or **barometric tendency**, is useful in short-range weather prediction. The generalizations relating cyclones and anticyclones to weather conditions (Fig. 6.23) are stated nicely in this verse (where "glass" refers to the barometer):

When the glass falls low,
Prepare for a blow;
When it rises high,
Let all your kites fly.

In conclusion, it should be obvious why local television weather broadcasters emphasize the positions and projected paths of cyclones and anticyclones. The "villain" on these weather programs is always the low-pressure system, which produces "foul" weather in any season. Lows move in roughly a west-to-east direction across the United States and require a few days to more than a week

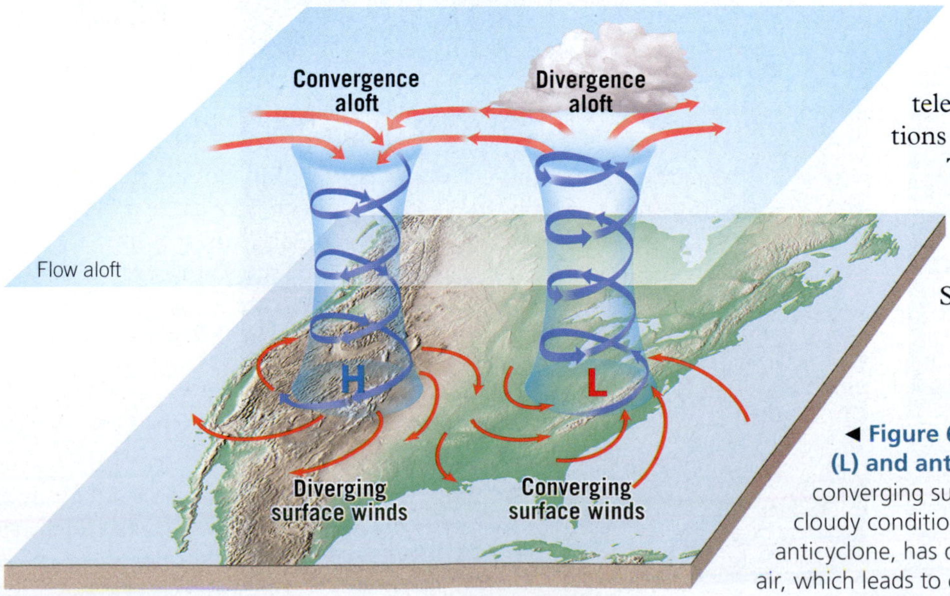


◀ **Figure 6.22 Airflow associated with cyclones (L) and anticyclones (H)** A low, or cyclone, has converging surface winds and rising air, resulting in cloudy conditions and often precipitation. A high, or anticyclone, has diverging surface winds and descending air, which leads to clear skies and fair weather.

Animation MM
Cyclones and Anticyclones
http://goo.gl/0HarBu

for the journey. Because their paths can be erratic, accurate prediction of their migration is difficult, yet it is essential for short-range forecasting. Meteorologists must also determine whether the flow aloft will intensify an embryo storm or suppress its development.

## eye ON THE atmosphere 6.2

Cold winds blowing from Greenland encountered moist air over the Greenland Sea, and their convergence generated parallel rows of clouds called *cloud streets* in the skies around Jan Mayen Island. The island acts as an obstacle that causes the wind to form spiraling eddies (called *van Karman vortices*) on the south (leeward) side of the island.

**Questions**

1. Based on the orientation of the cloud streets, what direction is the wind blowing over the Greenland Sea?
2. Can you think of an analogy for the spiraling eddies that formed downwind of Jan Mayen Island?
3. Eddies are also generated by bicycles, cars, and airplanes. Bicycle and car racers use these eddies, which reduce wind resistance, by traveling directly behind another competitor. What term is used to describe this competitive advantage?

A.

B.

▲ **Figure 6.23 Basic weather generalizations associated with pressure centers** **A.** A rainy day in London. Low-pressure systems are frequently associated with cloudy conditions and precipitation. **B.** By contrast, clear skies and "fair" weather may be expected when an area is under the influence of high pressure.

## Other Factors Promoting Vertical Airflow

Because of the close tie between vertical motion in the atmosphere and our daily weather, we will consider some other factors that contribute to surface convergence and surface divergence.

Friction can cause both convergence and divergence. When air moves from the relatively smooth ocean surface to land, for instance, the increased friction causes an abrupt drop in wind speed. This reduction of wind speed downstream results in a pileup of air upstream. Thus, converging winds and ascending air accompany flow from the ocean to the land. This effect contributes to the cloudy conditions over the humid coastal regions of Florida. Conversely, when air moves from the land to the ocean, general divergence and subsidence accompany the seaward flow of air because of lower friction and increasing wind speed over the water. The result is often subsidence and clearing conditions.

## eye ON THE atmosphere 6.3

These satellite images show four different tropical cyclones (hurricanes) that occurred on different dates in different parts of the world.

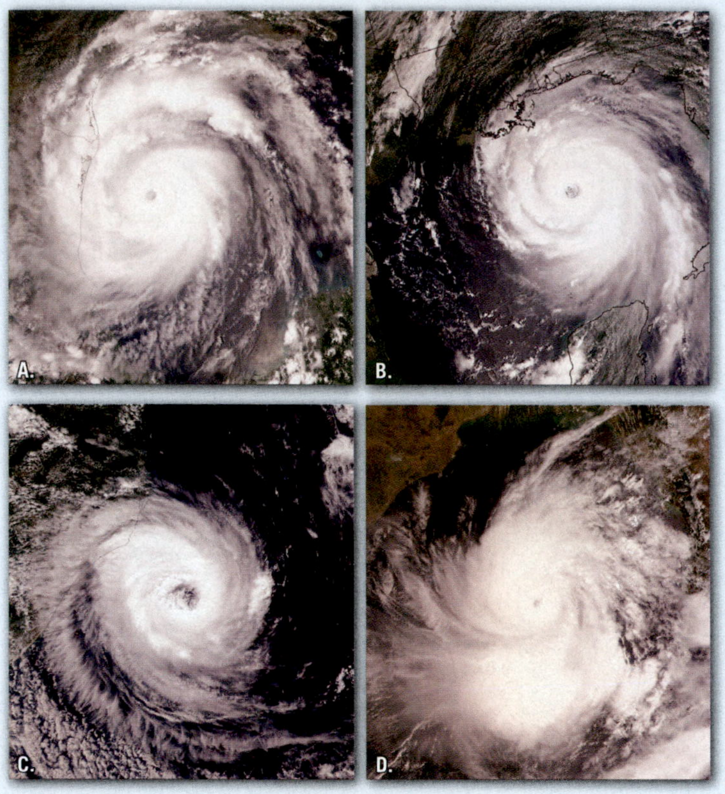


**Questions**

1. For each storm, examine the cloud pattern and determine whether the flow is clockwise or counterclockwise.
2. In which hemisphere is each storm located, Northern or Southern?

Mountains also hinder the flow of air and cause divergence and convergence. As air passes over a mountain range, it is compressed vertically, which produces horizontal spreading (divergence) aloft. Air reaching the leeward side of the mountain experiences vertical expansion, which causes convergence aloft. This effect greatly influences the weather in the United States east of the Rocky Mountains, as we shall examine later.

The connections between surface conditions and those aloft have led to significant research on understanding atmospheric circulation, especially in the midlatitudes. After we examine global atmospheric circulation in Chapter 7, we will again consider the relationships between horizontal airflow (wind) and vertical motions (rising and descending air currents).

**students sometimes ask...**

### What causes "mountain sickness"?

When visitors hike at elevations above 3000 meters (10,000 feet), they typically become tired and short of breath. These symptoms are caused by breathing air with roughly 30 percent less oxygen than at sea level. At these altitudes, our bodies try to compensate for the air's oxygen deficiency by breathing more deeply and increasing the heart rate, thereby pumping more blood to the body's tissues. The additional blood is thought to cause brain tissues to swell, resulting in headaches, insomnia, and nausea—the main symptoms of *acute mountain sickness*. Mountain sickness usually can be alleviated with a night's rest at a lower altitude, but some people suffer from *high-altitude pulmonary edema*, a buildup of fluid in the lungs that requires prompt medical attention.

### ✔ Concept Checks 6.5

1. For surface low pressure to exist for an extended period, what condition must exist aloft?
2. What general weather conditions can we expect when surface pressure is rising? When the pressure tendency is falling?
3. Converging winds and ascending air are often associated with the flow of air from the oceans onto land. Conversely, divergence and subsidence often accompany the flow of air from land to sea. What causes this convergence over land and divergence over the ocean?

# 6.6 Wind Measurement

**Define *prevailing wind* and explain how wind direction is described.**

Two basic wind measurements—direction and speed—are important to weather observers.

## Measuring Wind Direction

Winds are always labeled by the direction *from* which they blow. A north wind blows from the north toward the south; an east wind blows from the east toward the west. One instrument commonly used to determine wind direction, the **wind vane**, is often seen on the tops of buildings (Fig. 6.24A). Sometimes the wind direction is shown on a dial connected to the wind vane. The dial indicates the direction of the wind either by points of the compass—that is, N, NE, E, SE, and so on—or by a scale of 0° to 360°. On the latter scale, 0° (or 360°) is north, 90° is east, 180° is south, and 270° is west.

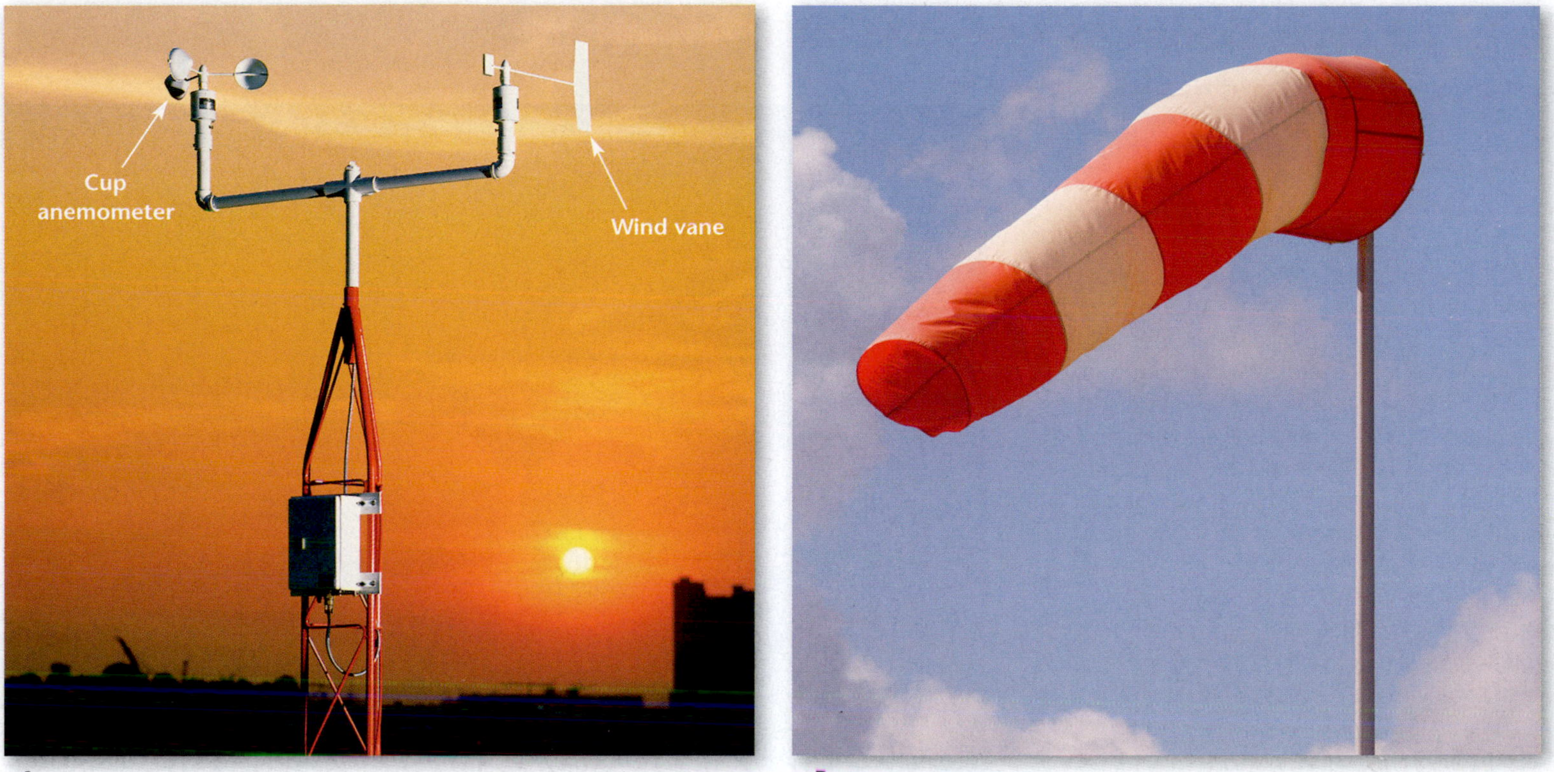


A.

B.

▲ **Figure 6.24 Wind measurement** **A.** Wind vane (right) and cup anemometer (left). The wind vane shows wind direction, and the anemometer measures wind speed. **B.** A wind sock is a device for determining wind direction and estimating wind speed. Wind socks are common sights at small airports and landing strips.

When the wind consistently blows more often from one direction than from any other, it is called a **prevailing wind**. You may be familiar with the prevailing westerlies that dominate midlatitude circulation. In the United States, for example, these winds consistently move the "weather" from west to east across the continent. Embedded within this general eastward flow are cells of high and low pressure, with their characteristic clockwise and counterclockwise flows. As a result, the winds associated with the westerlies, as measured at the surface, often vary considerably from day to day and from place to place. A *wind rose* provides a way to represent prevailing winds by indicating the percentage of time the wind blows from various directions (Fig. 6.25). The length of the lines on the wind rose indicates the percentage of time the wind blew from that direction. As seen in Figure 6.25B, the direction of airflow associated with the belt of trade winds is much more consistent than the westerlies (Fig. 6.25A.)

Knowledge of the wind patterns for a particular area can be useful. For example, when constructing an airport, the runways are aligned with the prevailing wind to assist in takeoffs and landings. Furthermore, prevailing winds greatly affect a region's weather and climate. Mountain ranges that trend north–south, such as the Cascade Range of the Pacific

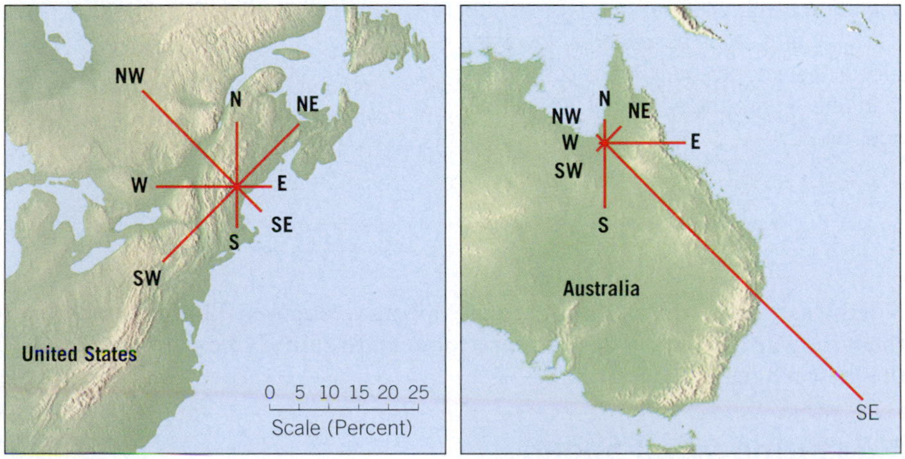


**A. Wind frequency for winter in the northeastern United States.**

**B. Wind frequency for winter in northeastern Australia. Note the reliability of the southeast trade winds in Australia as compared to the westerlies in the northeastern United States.**

► **Figure 6.25 Wind roses** The percentage of directional windflow is charted on a *wind rose* and may represent daily, weekly, monthly, seasonal, or annual totals.

## Box 6.2 Wind Energy: An Alternative with Potential

Air has mass, and when it moves it has energy of motion—kinetic energy. A portion of that energy can be converted into mechanical energy or electricity, both of which power our modern society.

Mechanical energy from wind was commonly used for pumping water and grinding wheat and corn until the advent of readily available electrical energy, and the farm windmill is still a familiar sight in many rural areas. By contrast, modern wind-powered electric turbines generate electricity for homes, businesses, and manufacturing (**Fig. 6.C**). Global wind-generating capacity has been doubling every 3 years (**Fig. 6.D**). According to the Global Wind Energy Council, at the end of 2013, China had the world's greatest installed wind-generation capacity (28.7 percent), followed by the United States (19.2 percent), Germany (10.8 percent), Spain (7.2 percent), and India (6.3 percent).

Wind speed is crucial in determining whether a place is a suitable site for a wind-energy facility. Generally, a minimum average wind speed of about 6 meters per second (13 miles per hour) is necessary for a large-scale wind-power facility to be profitable. A small difference in wind speed results in a large difference in energy production and, therefore, a large difference in the cost of the electricity generated. For example, a turbine operating on a site with an average wind speed of 13 miles per hour generates roughly 30 percent more electricity than one operating at 12 miles per hour. Further, there is little energy to be harvested at low wind speeds: 6-mile-per-hour winds contain less than one-eighth the energy of 12-mile-per-hour winds.

▲ **Figure 6.C Modern wind farm** These wind turbines are operating near Tehachapi Pass, Kern County, California. California was the first state to develop significant wind power, although it has now been surpassed by Texas.

Although the modern U.S. wind industry began in California, many states have greater wind potential. **Figure 6.E** shows the estimated average wind speeds at a height of 80 meters (260 feet) above the surface—the level that most commercial wind turbines operate.

Northwest, cause the ascent of the prevailing westerlies. Thus, the windward (west) slopes of these ranges are rainy, whereas the leeward (east) sides are dry.

## Measuring Wind Speed

Wind speed is often measured with a **cup anemometer**, which has a dial much like the speedometer of an automobile (Figure 6.24A). Sometimes an **aerovane** is used instead of a wind vane and cup anemometer. As illustrated in **Figure 6.26**, this instrument resembles a wind vane with a propeller at one end. The fin keeps the propeller facing into the wind, allowing the blades to rotate at a rate proportional to the wind speed. This instrument is commonly attached to a recorder that produces a continuous record of wind speed and direction. This information is valuable for determining locations where winds are steady

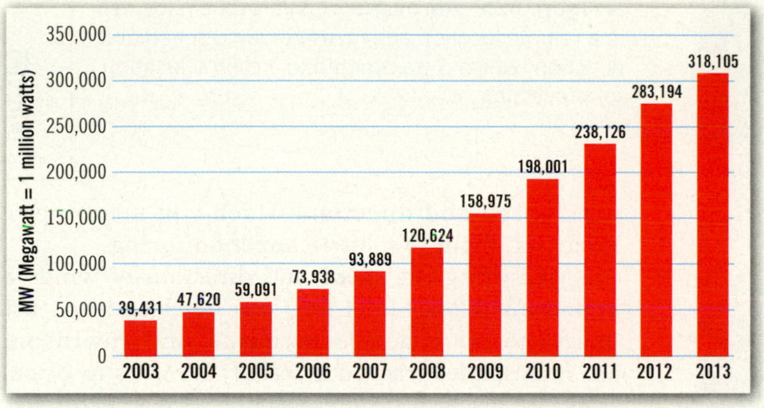


▲ **Figure 6.D Global cumulative installed wind capacity, 2003–2013**

Areas with average wind speeds greater than about 6 meters per second (13 miles per hour) are considered to have potential for development. At the end of 2013, Texas (12,355 MW) had the most installed wind capacity, followed by California (5,830 MW), Iowa (5,178 MW), Illinois (3,568 MW), and Oregon (3,153 MW). Some of the largest wind farms include the Alta Wind Energy Center, California; the Shepherds Flat Wind Farm, Oregon; and the Roscoe Wind Farm, Texas.

Compared to burning fossil fuels to generate electricity, wind power produces little impact in terms of pollution. However, there are other problems related to wind energy, most of which are local. One of the most critical is bird kills, because birds occasionally collide with wind turbines. Land erosion caused by improper installation, noise, and visual impact are also cited as potential environmental issues.

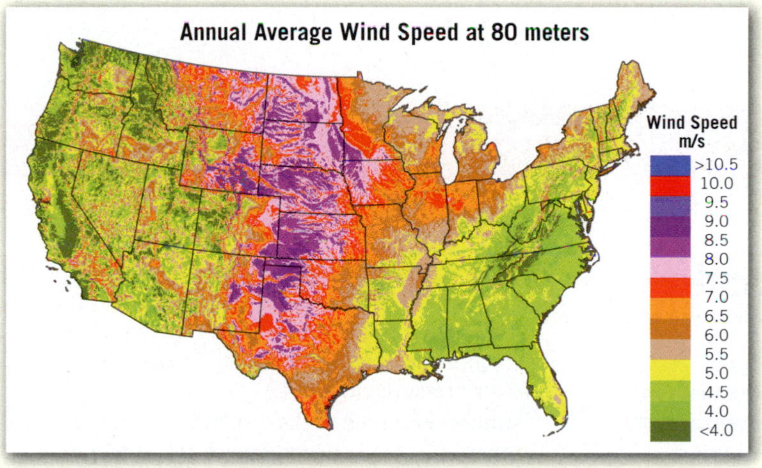


▲ **Figure 6.E The wind energy potential for the United States** Large wind systems require average wind speeds of about 6 meters per second (13 miles per hour).

Video MM
The Growth of Wind Power in the U.S.
http://goo.gl/FhzBgs

The U.S. Department of Energy currently provides funds to develop offshore wind technologies. Offshore winds are abundant, stronger, and blow more consistently than winds over land. Data suggest that more than 4,000,000 megawatts (MW)* of capacity is available on public land along the coasts off the United States and the Great Lakes. This is four times more than the total electrical generating capacity of the United States. Unfortunately, development of many of these sites is unlikely due to our reluctance to dot scenic coastal land with wind turbines.

Although only a small fraction of U.S. electricity currently generated comes from wind energy, the Department of Energy's goal is to obtain 20 percent of U.S. electricity from wind by 2030, with 4 percent of that amount anticipated from offshore wind farms. This target seems consistent with the current growth rate of wind energy nationwide. Thus, wind-generated electricity appears to be shifting from being an alternative to being a mainstream energy source.

### Questions

1. What country has the largest wind-generating capacity?
2. List the potential environmental issues associated with the production of wind energy.

*One megawatt (MW) is enough electricity to supply 240–400 American households.

and speeds are relatively high—potential sites for tapping wind energy (Box 6.2).

At small airstrips, *wind socks* are frequently used (Figure 6.24B). A wind sock consists of a cone-shaped bag that is open at both ends and free to change position with shifts in wind direction. The degree to which the sock is inflated indicates the strength of the wind.

Recall that 70 percent of Earth's surface is covered by water, making conventional methods of measuring wind speed difficult. Weather buoys and ships at sea provide limited coverage, but the availability of satellite-derived wind data has dramatically improved weather forecasts. One example is an instrument that NASA plans to attach to the International Space Station that can measure ocean surface

◀ **Figure 6.26 Aerovane** This aerovane is located in a remote location and measures wind speed and direction, which is transmitted to a central location for processing.

Video
Forecasting Wind Patterns
http://goo.gl/pliWTF

wind speed and direction, which will help improve weather forecasts, including hurricane monitoring.

Measuring the speed and direction of winds aloft is also important. Upper-level flow can be established using satellite images to track cloud movements, and **rawinsondes**, which are radiosondes tracked by radar, allow us to calculate airflow at several levels of the atmosphere.

## ✔ Concept Checks 6.6

1. A southwest wind blows from the ________ (direction) toward the ________ (direction).
2. When the wind direction is 315°, from what compass direction is it blowing?
3. What is the name of the prevailing winds that affect the contiguous United States?

# 6 Concepts in Review Air Pressure and Winds

## 6.1 Atmospheric Pressure and Wind

▶ Define *atmospheric pressure* and explain how it is displayed on a weather map.

**Key Terms:** wind, atmospheric pressure (air pressure), newton, millibar (mb), mercury barometer, barometric pressure, aneroid barometer, barograph, station pressure, isobar, anticyclone, cyclone (midladitude cyclone), ridge, trough

- Wind, the horizontal movement of air, is a result of horizontal differences in air pressure.
- Air pressure is the force exerted by the weight of air above. Average air pressure at sea level is about 14.7 pounds per square inch, or 1013.25 millibars, or 29.92 inches of mercury.
- Two instruments are used to measure atmospheric pressure: the mercury barometer, where the height of a mercury column provides a measure of air pressure; and the aneroid barometer, which uses a partially evacuated metal chamber that changes shape as air pressure changes.
- Atmospheric pressure is displayed on surface weather maps using isobars—lines connecting places of equal pressure. On upper-level weather charts, height contours are used. There is a simple relationship between height contours and isobars: Higher-elevation contours indicate higher pressures, and lower-elevation contours indicate lower pressures.

## 6.2 Why Does Air Pressure Vary?

▶ Name and describe the factors affecting air pressure at Earth's surface as well as aloft.

**Key Terms:** U.S. standard atmosphere, convergence, divergence

- The pressure at any given altitude is equal to the weight of the air above that point. The rate at which pressure decreases with an increase in altitude is much greater near Earth's surface than aloft.
- Two factors that largely determine the air pressure exerted at the surface by an air mass are temperature and humidity. A cold, dry air mass will produce higher surface pressures than a warm, humid air mass.
- Temperature differences cause horizontal pressure differences, which produces a force that causes air to flow from areas of high pressure to areas of low pressure. In general, cold air is associated with higher surface pressures, and warm air is associated with lower surface pressures. Aloft, the situation is reversed: Warm air aloft air tends to exhibit a higher pressure than cold air.

## 6.3 Factors Affecting Wind

▶ List and describe the three forces that act on the atmosphere to either create or alter winds.

**Key Terms:** pressure gradient force (PGF), Coriolis force, friction, boundary layer

- Wind is controlled by a combination of (1) the pressure gradient force, (2) the Coriolis force, and (3) friction. The pressure gradient force is the primary driving force of wind that results from pressure differences that are depicted by the spacing of isobars on a map. Closely spaced isobars indicate a steep pressure gradient and strong winds; widely spaced isobars indicate a weak pressure gradient and light winds.
- The Coriolis force produces a deviation in the path of wind due to Earth's rotation (to the right in the Northern Hemisphere and to the left in the Southern Hemisphere).
- Friction significantly influences airflow near Earth's surface but is negligible above a height of a few kilometers.

## 6.4 Winds Aloft Versus Surface Winds

▶ Explain why winds aloft flow roughly parallel to the isobars, while surface winds travel at an angle across the isobars.

**Key Terms:** geostrophic wind, Buys Ballot's law, gradient wind, cyclonic flow, anticyclonic flow

- Above a height of a few kilometers, geostrophic winds are generated when a balance is reached between the pressure gradient force and the opposing Coriolis force. Geostrophic winds flow in a nearly straight path, parallel to the isobars, with velocities proportional to the pressure gradient force.
- Winds that blow at a constant speed parallel to curved isobars are termed gradient winds. In low-pressure systems, such as midlatitude cyclones, the circulation of air, referred to as cyclonic flow, is counterclockwise in the Northern Hemisphere and clockwise in the Southern Hemisphere.
- Centers of high pressure, called anticyclones, exhibit anticyclonic flow, which is clockwise in the Northern Hemisphere and counterclockwise in the Southern Hemisphere.
- Near the surface, friction plays a major role in determining the direction of airflow. The result is a movement of air at an angle across the isobars, toward the area of lower pressure. Therefore, the resultant winds blow into and counterclockwise about a Northern Hemisphere surface cyclone. In a Northern Hemisphere surface anticyclone, winds blow outward and clockwise.

**Q** Use the accompanying drawing to answer the following.

**a.** Does the arrow labeled A indicate the pressure gradient force, the Coriolis force, or friction?

**b.** Does the arrow labeled B indicate the pressure gradient force, the Coriolis force, or friction?

**c.** Is the wind shown in this drawing located near Earth's surface or aloft?

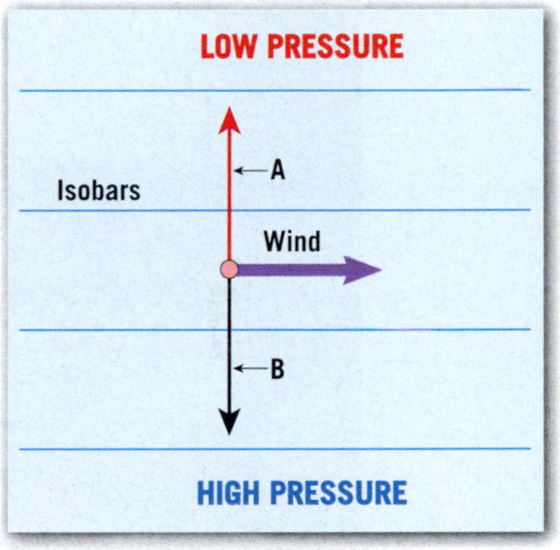


## 6.5 How Winds Generate Vertical Air Motion

▶ Describe the airflow around a low-pressure center (cyclone) and a high-pressure center (anticyclone) and the weather associated with each.

**Key Terms:** pressure tendency (barometric tendency)

- A surface low-pressure system (or cyclone) with its associated horizontal convergence is maintained or intensified by divergence (spreading out) aloft. Surface convergence in a cyclone accompanied by divergence aloft causes a net upward movement of air. Therefore, the passage of a low-pressure center is often associated with stormy weather.
- Fair weather can usually be expected with the approach of a high-pressure system or anticyclone.

**Q** Use the accompanying simplified cross sections of two pressure systems that show possible surface flow, vertical flow, flow aloft, and cloud cover, to answer the following.

**a.** Which of these cross sections (A or B) illustrates the idealized flow of a cyclone?

**b.** Is the surface flow around a cyclone inward (converging) or outward (diverging)?

**c.** Is the vertical airflow within an anticyclone downward or upward?

**d.** Would you expect to experience cloudy or clear conditions to be associated with a cyclone?

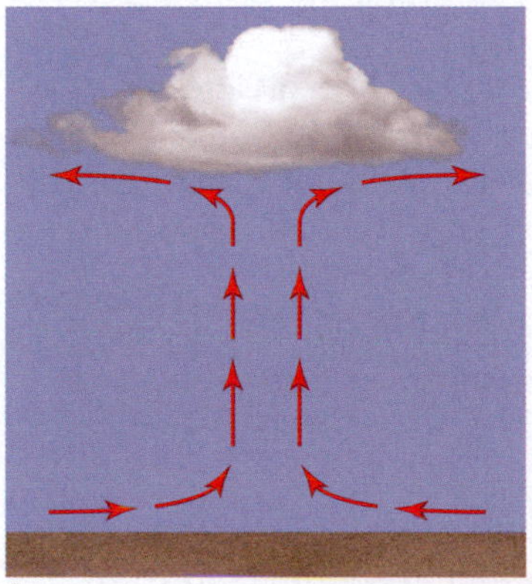

A.

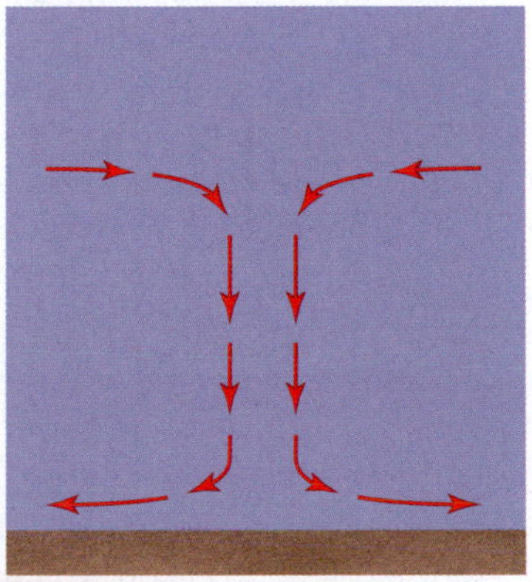

B.

## 6.6 Wind Measurement

▶ Define *prevailing wind* and explain how wind direction is described.

**Key Terms:** wind vane, prevailing wind, cup anemometer, aerovane, rawinsonde

- Two basic wind measurements—direction and speed—are measured and recorded as part of daily weather maps. Winds are always labeled by the direction from which they blow.
- Anemometers measure wind speed; wind vanes measure wind direction; aerovanes and satellites measure both wind speed and direction.

## Give it Some Thought

1. The accompanying satellite images show the cloud patterns associated with two cyclones (low-pressure systems).
   a. Determine the airflow around each of these pressure cells (clockwise or counterclockwise).
   b. Which of these is located in the Northern Hemisphere?
   c. Can you identify which one of these storm systems is a tropical cyclone (hurricane)? (*Hint:* Compare these images to Figures 11.9, page 306.)

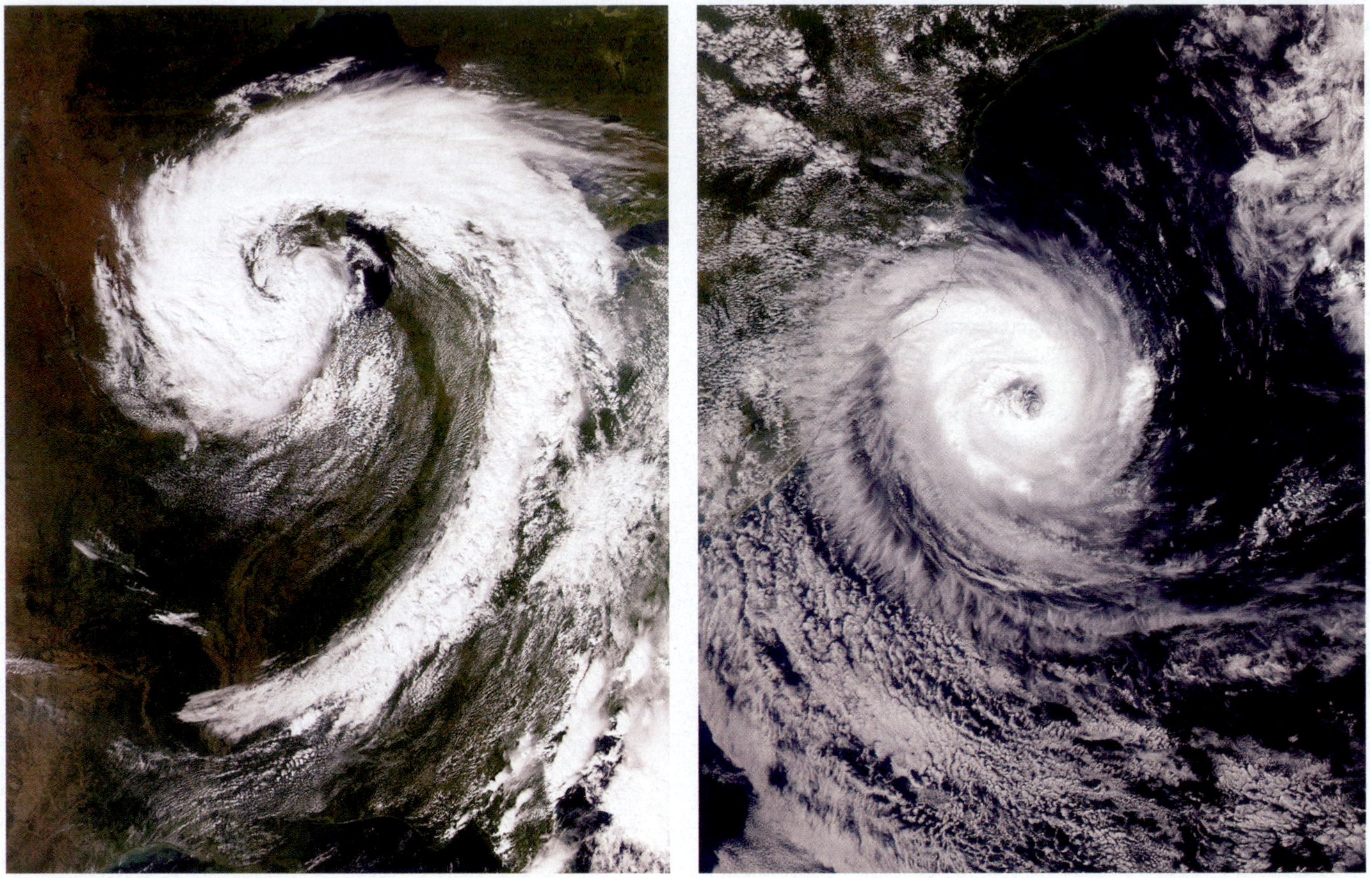

A. B.

2. Given the following descriptions, identify the direction (for example, east, west, northwest) in which the Coriolis force is acting on the moving object.
   a. A commercial jet flying from New York to Chicago
   b. A baseball thrown from south to north in South Dakota
   c. A blimp floating from St. Louis northeast toward Detroit
   d. A boomerang thrown from west to east in Australia
   e. A football thrown along the equator
3. If a weather system with strong winds approached Lake Michigan from the west, how might the speed of the winds change as the system traversed the lake? Explain your answer.
4. The accompanying map is a simplified surface weather map for April 2, 2011, on which the centers of three pressure cells are numbered.
   a. Which of the pressure cells are anticyclones (highs), and which are cyclones (lows)?
   b. Which pressure system has the steepest pressure gradient and hence exhibits the strongest winds?
   c. Refer to Figure 6.2 to determine whether pressure system 3 should be considered strong or weak.

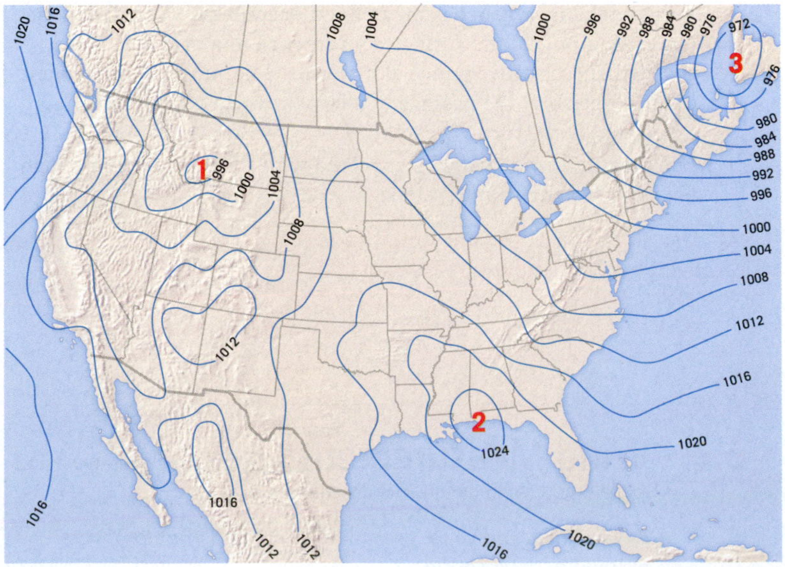

5. If you live in the Northern Hemisphere and are directly west of the center of a midlatitude cyclone, what is the probable wind direction?
6. Given the following time line of barometric pressure readings, interpret the *likely* weather conditions, with regard to cloudiness and precipitation, for each of the following days:
   Day 1: Pressure steady at 1025 millibars
   Day 2: Pressure at 1010 millibars and falling
   Day 3: Pressure reaches a 4-day minimum of 992 millibars
   Day 4: Pressure at 1008 millibars and rising
7. If you wanted to erect wind turbines to generate electricity, would you search for a location that typically experiences a strong pressure gradient or a weak pressure gradient? Explain.
8. When designing an airport, it is important to have the runways positioned so that planes take off into the wind. Refer to the accompanying wind rose and discuss the orientation of the runway and the direction the planes would travel when they take off.

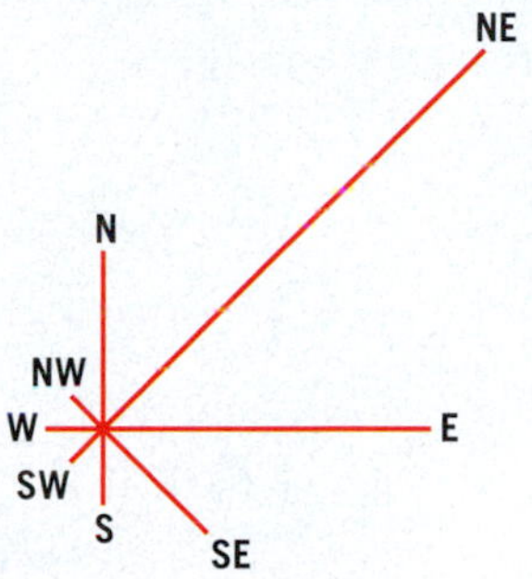


9. You and a friend are watching TV on a rainy day when the weather reporter states, "The barometric pressure is 28.8 inches and rising." Hearing this, you say, "It looks like fair weather is on its way." How would you respond if your friend asked the following questions?
   **a.** "I thought air pressure had something to do with the weight of air. How does 'inches' relate to weight?"
   **b.** "Why do you think the weather is going to improve?"
10. Use the accompanying diagram to show the wind directions associated with high- and low-pressure cells in both hemispheres. The diagram for a low-pressure system in the Northern Hemisphere is already completed. Add arrows to show the wind in the other pressure cells.

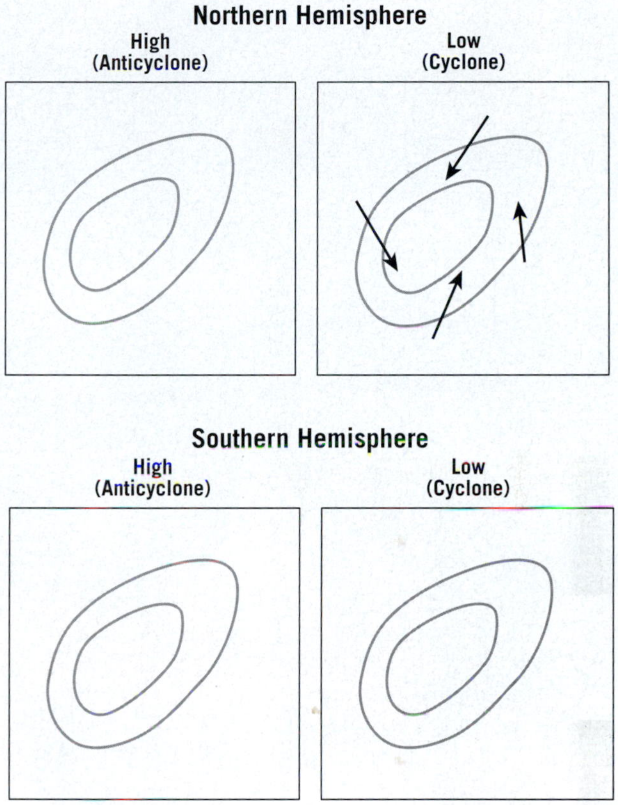


## Problems

1. Figure 6.3 illustrates a simple mercury barometer. When a glass tube is completely evacuated of air and placed into the dish of mercury, the mercury rises to a height such that the force of the air pushing on the open dish matches the gravity pulling the mercury back down the tube. The density of mercury is 13,534 kilograms per cubic meter, which is very dense compared to water (1000 kilograms per cubic meter). In the mercury barometer, the mercury will rise to 29.92 inches under standard sea-level pressure. How tall would the barometer need to be if you used water instead of mercury?
2. Average air pressure at sea level is about 14.7 pounds per square inch. Using this information, how much does the entire atmosphere weigh? (*Hint:* The radius of Earth is 3963 miles.)

## MasteringMeteorology™

Looking for additional review and test prep materials? Visit the Study Area in *MasteringMeteorology*™ to enhance your understanding of this chapter's content by accessing a variety of resources, including MapMaster™ interactive maps, Geoscience Animations, GEODe, *In the News* RSS feeds, flashcards, web links, self-study quizzes, and an eText version of *The Atmosphere*.

# 7 Circulation of the Atmosphere

## Focus on Concepts

*Each statement represents the primary learning objective for the corresponding major heading within the chapter. After you complete the chapter, you should be able to:*

**7.1** Distinguish between microscale, mesoscale, and macroscale winds and give an example of each.

**7.2** List four types of local winds and describe their formation.

**7.3** Describe or sketch the three-cell model of global circulation.

**7.4** Summarize Earth's idealized zonal pressure belts. Explain how continents and seasonal temperature changes complicate the idealized pattern.

**7.5** Describe the seasonal changes in global circulation that produce the Asian monsoon.

**7.6** Explain why the airflow aloft in the middle latitudes has a strong west-to-east component.

**7.7** Explain the origin of the polar jet stream and its relationship to midlatitude cyclonic storms.

**7.8** Sketch and label the major ocean currents on a world map.

**7.9** Describe the Southern Oscillation and its relationship to El Niño and La Niña. List the climate impacts of El Niño and La Niña on North America.

**7.10** Discuss the major factors that influence the global distribution of precipitation.

Pressure differences caused by unequal heating of Earth's surface generate the global wind system. These winds, which come in various scales, blow in an unending attempt to balance these surface temperature differences. Because the zone of maximum solar heating migrates with the seasons—moving northward during the Northern Hemisphere summer and southward as winter approaches—the wind patterns that make up the general circulation also migrate latitudinally. This chapter focuses on models describing the distribution of Earth's pressure zones that, in turn, generate the global wind system. This global wind system drives ocean circulation and creates global precipitation patterns.

*Winds are the result of pressure differences. These wind driven waves are pounding the coastal village of Sainte-Luce, Quebec.*

# 7.1 Scales of Atmospheric Motion

**Distinguish between microscale, mesoscale, and macroscale winds and give an example of each.**

Earth's highly integrated wind system can be thought of as a series of deep rivers of air encircling the planet. Embedded in the main currents are vortices of various sizes, including hurricanes, tornadoes, and midlatitude cyclones. Like eddies in a stream, these rotating wind systems develop and die out with somewhat predictable regularity.

Residents of the United States and Canada are familiar with the term *westerlies,* which describes winds that predominantly blow across the midlatitudes from west to east. However, over short time periods, the winds may blow from any direction. You may recall being in a storm when shifts in wind direction and speed came in rapid succession. With such variations, how can we describe our winds as westerly? The answer lies in our attempt to simplify descriptions of the atmospheric circulation by sorting out events according to *size* and the *time frame* in which the wind systems occur. On the scale of a weather map, for instance, where observing stations are spaced about 150 kilometers (nearly 100 miles) apart, small whirlwinds that carry dust skyward are far too small to be identified. Instead, weather maps reveal larger-scale wind patterns, such as those associated with traveling cyclones and anticyclones.

In general, large weather patterns persist longer than their smaller counterparts. For example, dust devils usually last a few minutes, but midlatitude cyclones typically take a few days to cross the United States and sometimes dominate the weather for a week or longer.

Winds are divided into the three categories of atmospheric circulation: microscale, mesoscale, and macroscale.

## Microscale Winds

We call the smallest scale of air motion **microscale winds**. These small, often chaotic winds normally last for seconds or at most minutes. Examples include simple gusts that hurl debris into the air (Fig. 7.1A) and small, well-developed vortices such as dust devils. Although dust devils resemble tornadoes, they are much smaller and less intense than tornadoes (see Box 7.1).

## Mesoscale Winds

**Mesoscale winds** generally last for several minutes and occasionally exist for hours. These middle-sized phenomena are usually less than 100 kilometers (60 miles) across and include strong updrafts and downdrafts, tornadoes, as well as a number of unique wind systems referred to as *local winds* (Fig. 7.1B). Some mesoscale winds—for example, the updrafts and downdrafts within thunderstorms—have a strong vertical component. Strong downdrafts with speeds exceeding 100 kilometers per hour can produce considerable damage at Earth's surface and may be accompanied by heavy rain and hail. Tornadoes, the most destructive mesoscale winds, will be considered in Chapter 10.

Land and sea breezes, chinooks, and katabatic winds are other examples of mesoscale winds. These will be discussed in the next section, along with other local winds.

## Macroscale Winds

The largest wind patterns, called **macroscale winds**, are divided into two categories: *planetary-scale* and *synoptic-scale*. **Planetary-scale winds** are exemplified by the westerlies and trade winds that carried sailing vessels back and forth across the Atlantic during the opening of the New World. These large-scale flow patterns extend around the entire globe and can remain essentially unchanged for weeks at a time.

The somewhat smaller macroscale circulation, called **synoptic-scale winds**—also referred to as *weather-map scale*—are about 1000 kilometers (600 miles) in diameter and are easily identified on weather maps. Two well-known synoptic-scale systems are the traveling *midlatitude cyclones* and *anticyclones* that appear on weather maps as areas of low and high pressure, respectively. These weather producers are confined largely to the middle latitudes.

The smallest macroscale weather systems include *tropical storms* and *hurricanes* that develop in late summer and early fall over warm tropical oceans (Fig. 7.1C). Like the larger midlatitude cyclones, airflow in these systems is inward and upward, but the winds associated with hurricanes are much stronger than those of their more poleward cousins.

## Wind Patterns on All Scales

Although it is common practice to divide atmospheric motions according to size, remember that global winds are a composite of motion on all scales—much like a meandering river that contains large eddies composed of smaller eddies containing still smaller eddies. As an example, we will examine winds associated with hurricanes that form over the North Atlantic. When we view one of these tropical cyclones on a satellite image, the storm appears as a large whirling cloud migrating slowly across the ocean (Fig. 7.1C). From this perspective, which is at the weather-map (synoptic) scale, the general counterclockwise rotation of the storm is easily seen.

In addition to their rotating motions, hurricanes often move from east to west or northwest. (Once hurricanes move into the belt of the westerlies, they tend to change course and move in a northeasterly direction.) This motion demonstrates that these large eddies are embedded in a still larger flow (planetary scale) that is moving westward across the tropical portion of the North Atlantic.

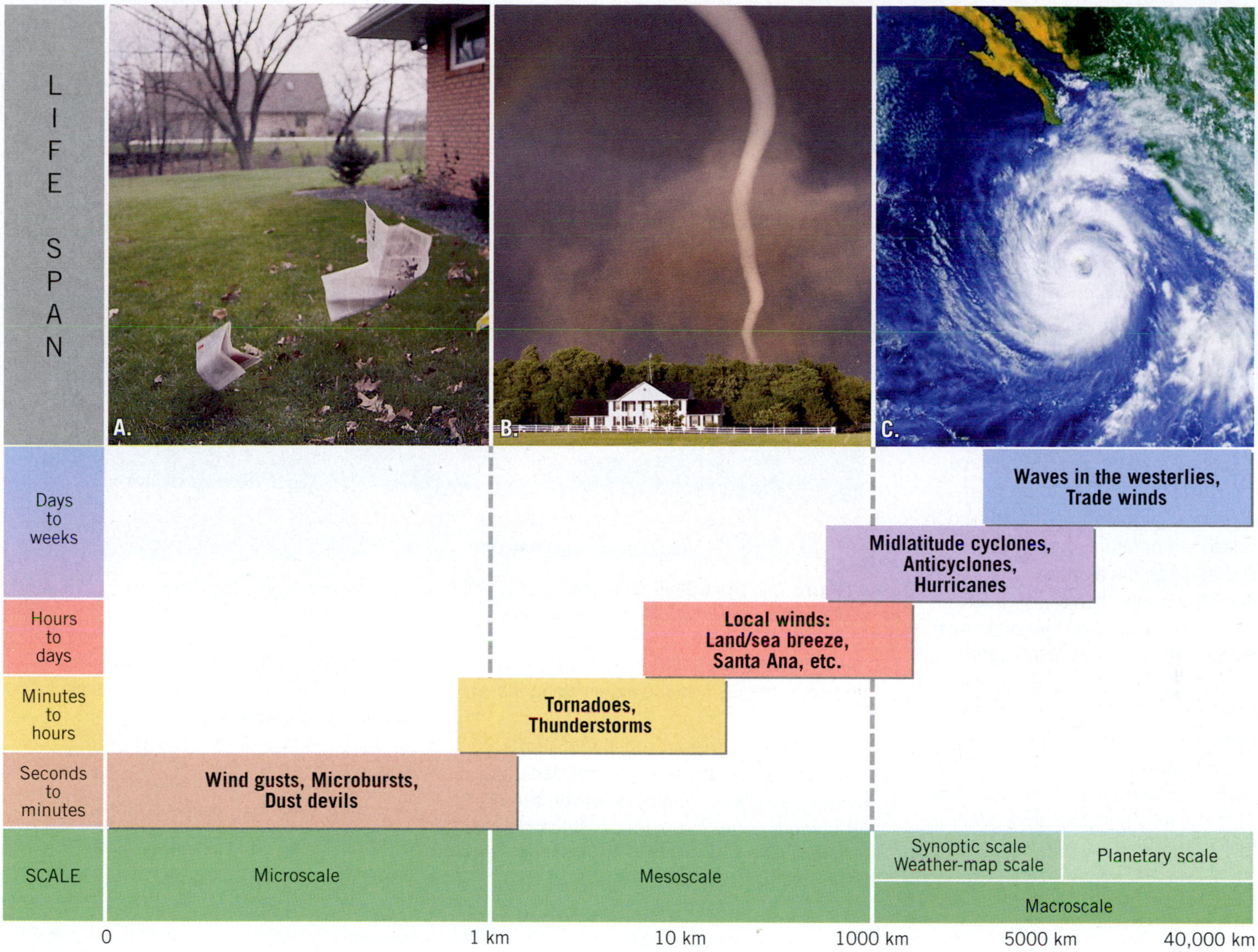


▲ **Figure 7.1 Three scales of atmospheric motion** **A.** Gusts illustrate microscale winds. **B.** Tornado-producing supercell thunderstorms exemplify mesoscale wind systems. **C.** Satellite image of a hurricane, an example of macroscale circulation.

When we examine a hurricane more closely by flying an airplane through it, some of the small-scale aspects of the storm become noticeable. As the plane approaches the outer edge of the system, it is evident that the large rotating cloud that we see in the satellite image consists of many individual cumulonimbus towers (thunderstorms). Each of these mesoscale phenomena lasts for a few hours, and they must be continually replaced by new ones for the hurricane to persist. During the flight, we also realize that the individual thunderstorms are made up of even smaller-scale turbulences. The small thermals of rising and descending air in these clouds make for a rough flight.

In summary, the global wind system is divided into three groups, based on size and the time frame in which the wind systems occur. Small microscale winds, such as dust devils, may last only a few minutes, whereas larger mesoscale winds that include strong updrafts and downdrafts, tornadoes, and local winds, may last for hours. The largest, macroscale winds, include the westerlies and the trade winds and may persist for weeks. Somewhat smaller macroscale systems include midlatitude cyclones, tropical storms, and hurricanes.

## ✔ Concept Checks 7.1

1. List the three major categories of atmospheric circulation and give at least one example of each.
2. Describe how the size of a wind system is related to its duration (life span).
3. What scale of atmospheric circulation includes midlatitude cyclones, anticyclones, and tropical cyclones (hurricanes)?
4. Explain in your own words what is meant by the statement "Global winds are a composite of winds of all scales."

## Box 7.1 Dust Devils

Common phenomena in arid regions of the world are the whirling vortices called *dust devils* (**Fig. 7.A**). Although they resemble tornadoes, dust devils are generally much smaller and less intense than their destructive cousins. Most dust devils are only a few meters in diameter and reach heights no greater than about 100 meters (300 feet). Further, these whirlwinds are usually short-lived phenomena that die out within minutes.

Unlike tornadoes, which are associated with convective clouds, dust devils form on days when clear skies dominate and develop from the ground upward. Because surface heating is critical to their formation, dust devils occur most frequently in the afternoon, when surface temperatures are highest.

Recall that when the air near the surface is considerably warmer than the air a few dozen meters overhead, the layer of air near Earth's surface becomes unstable. In this situation, warm surface air begins to rise, causing air near the ground to be drawn into the developing whirlwind. The rotating winds associated with dust devils are produced by the same phenomenon that causes ice skaters to spin faster as they pull their arms closer to their body. As the inwardly spiraling air rises, it carries sand, dust, and other loose debris dozens of meters into the air—making a dust devil visible.

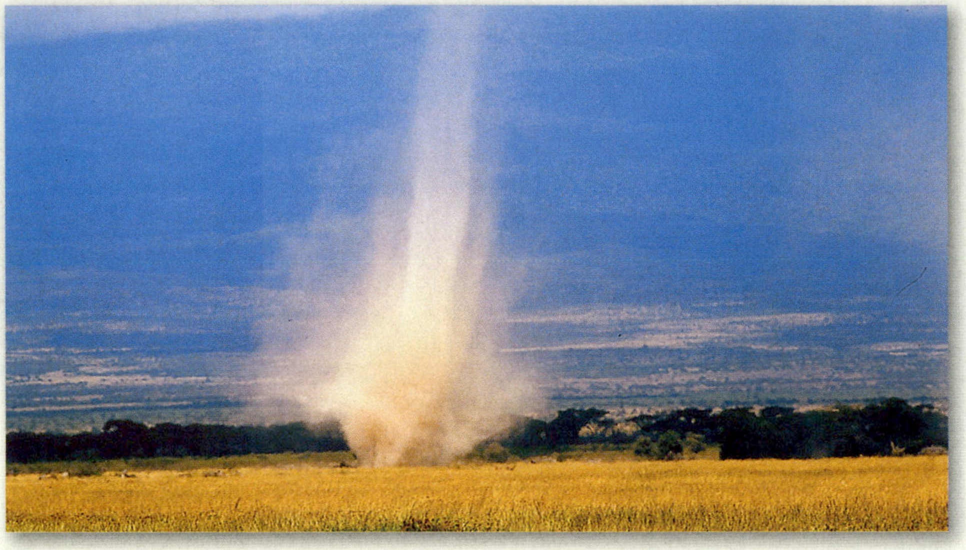

▲ **Figure 7.A Dust devil** Although these whirling vortices resemble tornadoes, they have a different origin and are much smaller and less intense.

Most dust devils are small and short-lived; consequently, they are not generally destructive. Occasionally, however, these whirlwinds grow to be 100 meters or more in diameter and over a kilometer high. With wind speeds that may reach 100 kilometers (60 miles) per hour, large dust devils can do considerable damage.

**Questions**

1. Where do most dust devils form?
2. In what ways are dust devils different from tornadoes?

# 7.2 Local Winds

**List four types of local winds and describe their formation.**

Local winds are examples of mesoscale winds (time frame of minutes to hours and size of 1 to 1000 kilometers). Remember that most winds have the same cause: pressure differences that arise because of temperature differences caused by unequal heating of Earth's surface. Most local winds are linked to temperature and pressure differences that result from variations in topography or in local surface conditions.

Recall that winds are named for the direction *from which they blow*. This holds true for local winds. Thus, a sea breeze originates over water and blows toward the land, whereas a valley breeze blows upslope, away from its source.

## Land and Sea Breezes

The daily temperature differences that develop between the sea and adjacent land areas, and the resulting pressure pattern that creates a sea breeze, were discussed in Chapter 6 (see Figure 6.12). Because land is heated more intensely during daylight hours than is an adjacent body of water, the air above the land heats and expands, creating an area of high pressure aloft. This in turn causes the air above the land to move seaward. This mass transfer of air aloft creates a surface low over the land. A **sea breeze** develops as cooler air over the water moves landward

**students sometimes ask...**

**What is the highest wind speed ever recorded in the United States?**

The highest wind speed recorded at a surface station is 372 kilometers (231 miles) per hour, measured April 12, 1934, at Mount Washington, New Hampshire. Wind speed at the observatory atop Mount Washington, 1879 meters (6262 feet) above sea level, averages 56 kilometers (35 miles) per hour. Faster wind speeds have undoubtedly occurred on mountain peaks, but no instruments were in place to record them.

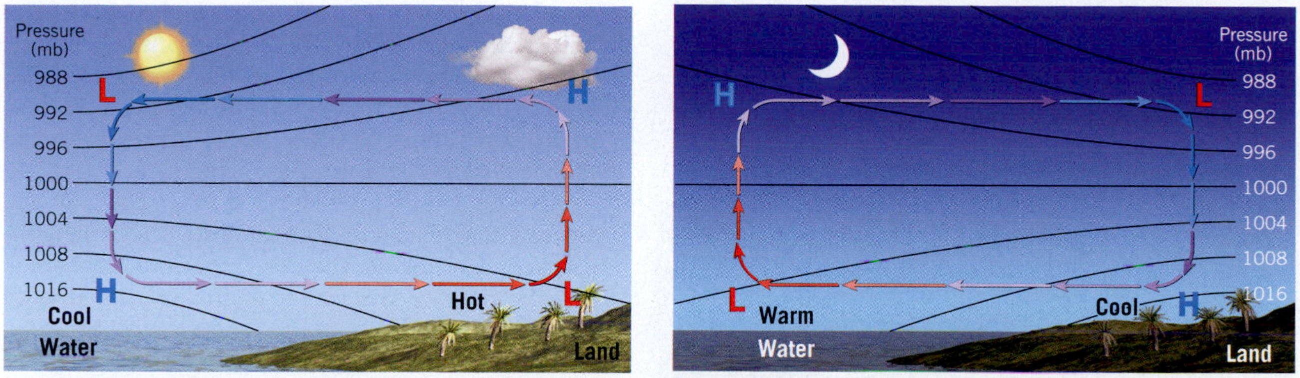


A. During daylight hours, cooler and denser air over the water moves onto the land, generating a sea breeze.

B. At night the land cools more rapidly than the sea, generating an offshore flow called a land breeze.

◀ SmartFigure 7.2 Sea breeze and land breeze

http://goo.gl/SquWt

toward the area of lower pressure (Fig. 7.2A). At night, the land cools more rapidly than the sea, and a **land breeze** may develop, with airflow off the land (Fig. 7.2B).

A sea breeze has a significant moderating influence in coastal areas. Shortly after a sea breeze begins, air temperatures over the land may drop by as much as 5° to 10°C. However, the cooling effect of these breezes is generally noticeable for only 100 kilometers (60 miles) inland in the tropics and often less than half that distance in the middle latitudes. These cool sea breezes generally begin shortly before noon and reach their greatest intensity—about 10 to 20 kilometers per hour—by midafternoon.

Smaller-scale sea breezes can also develop along the shores of large lakes. Cities near the Great Lakes, such as Chicago, benefit from the "lake effect" during the summer, when residents typically enjoy cooler temperatures near the lake compared to warmer inland areas. In many places sea breezes also affect the amount of cloud cover and rainfall. The Florida peninsula, for example, experiences increased summer precipitation partly due to convergence associated with sea breezes from both the Atlantic and Gulf coasts (see Figure 4.20).

The intensity and extent of land and sea breezes vary by location and season. Tropical areas where intense solar heating is continuous throughout the year experience more frequent and stronger sea breezes than midlatitude locations. The most intense sea breezes develop along tropical coastlines adjacent to cool ocean currents. In the middle latitudes, sea breezes are best developed during the warmest months, but land breezes are often missing because at night the land does not always cool below the temperature of the ocean surface.

## Mountain and Valley Breezes

A daily wind similar to land and sea breezes occurs in mountainous regions. During the day, air along mountain slopes is heated more intensely than air at the same elevation over the valley floor (Fig. 7.3A). This warmer air glides up the mountain slope and generates a **valley breeze**. Valley breezes can often be identified by the cumulus clouds that develop over adjacent mountain peaks and may account for late afternoon thundershowers that occur on warm summer days (Fig. 7.4).

After sunset the pattern is reversed. Rapid heat loss along the mountain slopes cools the air, which drains into the valley and causes a **mountain breeze** (Fig. 7.3B). Similar cool air drainage can occur in hilly regions with modest slopes. The result is that the coldest pockets of air are usually found in the lowest spots.

Like many other winds, mountain and valley breezes vary by season. Valley breezes are most common during warm seasons, when solar heating is most intense, whereas mountain breezes tend to occur more frequently during cold seasons.

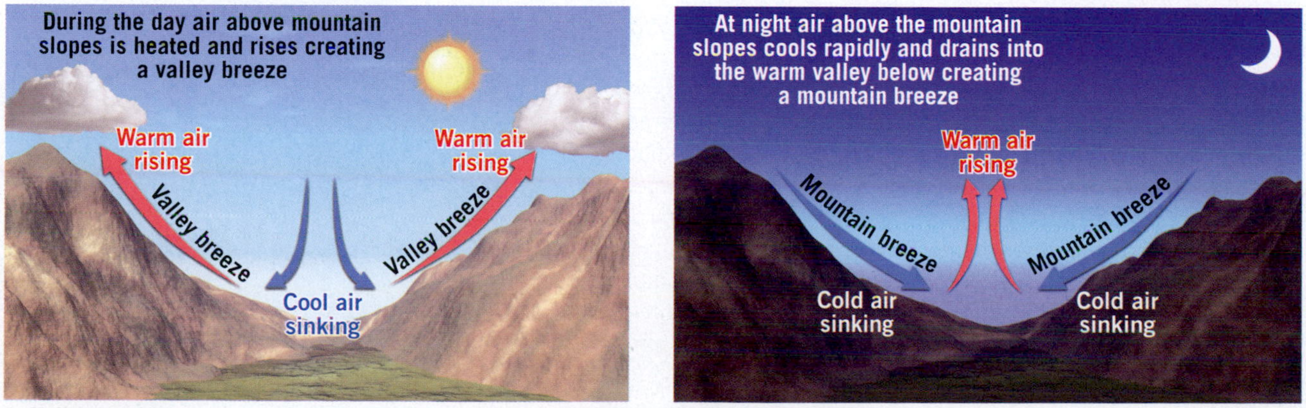


▶ Figure 7.3 Valley and mountain breeze

A. Valley breeze

B. Mountain breeze

▲ **Figure 7.4 The occurrence of a daytime upslope (valley) breeze is identified by cloud development on mountain peaks** Sometimes this cloud development can produce a midafternoon thunderstorm.

## Chinook (Foehn) Winds

Warm, dry winds called **chinooks** sometimes move down the slopes of mountains in the United States. Similar winds in the Alps are called **foehns**. Such winds are usually created when a strong pressure gradient develops in a mountainous region. As the air descends the leeward slopes of the mountains, it is heated adiabatically (by compression; see Chapter 4). Because condensation may have occurred as the air ascended the windward side, releasing latent heat, the air descending the leeward side will be warmer and drier than at a similar elevation on the windward side.

Chinooks commonly flow down the east slopes of the Colorado Rockies in the winter and spring, when the affected area may be experiencing subfreezing temperatures. Thus, these dry, warm winds often bring drastic change. Within minutes of a chinook's arrival, the temperature may climb 20°C (36°F). These winds can rapidly melt snow cover, which explains why they are named *chinook*, a Native American word for "snow-eater." Chinook winds have been known to melt more than a foot of snow in a single day. A chinook that moved through Granville, North Dakota, on February 21, 1918, caused the temperature to rise from –33°F to 50°F—an increase of 83°F!

**students sometimes ask...**

### What is a haboob?

A haboob (from the Arabic word *habb*, meaning "wind") is a type of local wind that occurs in arid regions. The name was originally applied to strong dust storms in the African Sudan, where one city experiences an average of 24 haboobs per year. Haboobs generally occur when downdrafts from large thunderstorms reach the surface and swiftly spread out across the desert. Tons of silt, sand, and dust are lifted, forming a whirling wall of debris hundreds of meters high. These dense, dark "clouds" can completely engulf desert towns and deposit enormous quantities of sediment. The deserts of the southwestern United States occasionally experience these dust storms.

Chinooks are sometimes viewed as beneficial to ranchers east of the Rockies because they keep the grasslands clear of snow during much of the winter. However, this benefit is offset by the loss of moisture that the snow would bequeath to the land if it remained until the spring melt.

Another chinook-like wind that occurs in the United States is the **Santa Ana**. Occurring in southern California, the hot, dry Santa Ana winds greatly increase the threat of fire in this already dry area (see Box 7.2).

## Katabatic (Fall) Winds

In the winter, areas adjacent to highlands may experience a **katabatic wind**, or **fall wind**. These local winds originate when cold, dense air situated over a highland area such as the ice sheets of Greenland and Antarctica, begins to move (Fig. 7.5). Under the influence of gravity, the cold air cascades over the rim of a highland like a waterfall. Although the air is heated adiabatically, the initial temperatures are so low that the wind arrives in the lowlands still colder and more dense than the air it displaces. As this frigid air descends, it occasionally is channeled into narrow valleys, where it acquires velocities capable of great destruction.

A few of the better-known katabatic winds have local names. Most famous is the **mistral**, which blows from the French Alps toward the Mediterranean Sea. Another is the **bora**, which originates in the mountains of the Balkan Peninsula and blows to the Adriatic Sea.

## Country Breezes

One mesoscale wind, the **country breeze**, is associated with large urban areas. As the name implies, this circulation pattern is characterized by a light wind blowing into the city from the surrounding countryside. In cities, massive buildings composed of rocklike materials tend to retain the heat accumulated during the day more than the open landscape of outlying areas (see Box 3.3, page 76, on the urban heat island). The result is that the warm, less-dense air over cities rises, which in turn initiates the country-to-city flow. A country breeze is most likely to develop on a relatively

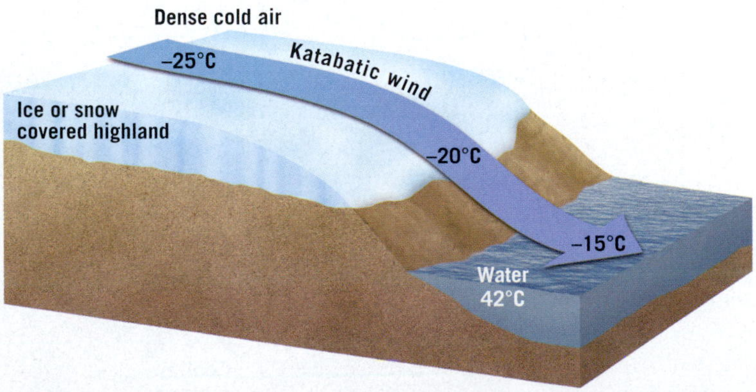


▲ **Figure 7.5 Katabatic winds** These winds, also referred to as *fall winds*, travel from ice- or snow-covered highlands driven mainly by the force of gravity.

clear, calm night. One unfortunate consequence of the country breeze is that pollutants emitted near the urban perimeter tend to drift in and concentrate near the city's center.

## ✔ Concept Checks 7.2

1. The most intense sea breezes develop along tropical coasts adjacent to cool ocean currents. Explain.
2. In what way are land and sea breezes similar to mountain and valley breezes?
3. What are chinook winds? Name two areas where they are common.
4. In what way are katabatic (fall) winds different from most other types of local winds?
5. Explain how cities create their own local winds.

# 7.3 Global Circulation

**Describe or sketch the three-cell model of global circulation.**

Our knowledge of global winds comes from two sources: the patterns of pressure and winds observed worldwide and theoretical studies of fluid motion. We will first consider the classical model of global circulation that was developed largely from average worldwide pressure distribution. We will then modify this idealized model by adding more recently discovered aspects of the atmosphere's complex motions.

## Single-Cell Circulation Model

One of the first contributions to the classical model of global circulation came from George Hadley in 1735. Well aware that solar energy drives the winds, Hadley proposed that the large temperature contrast between the poles and the equator creates a large *convection cell* in both the Northern and Southern Hemispheres (Fig. 7.6).

In Hadley's model, warm equatorial air rises until it reaches the tropopause, where it spreads toward the poles. Eventually, this upper-level flow reaches the poles, where cooling causes it to sink and spread out at the surface as equatorward-moving winds.

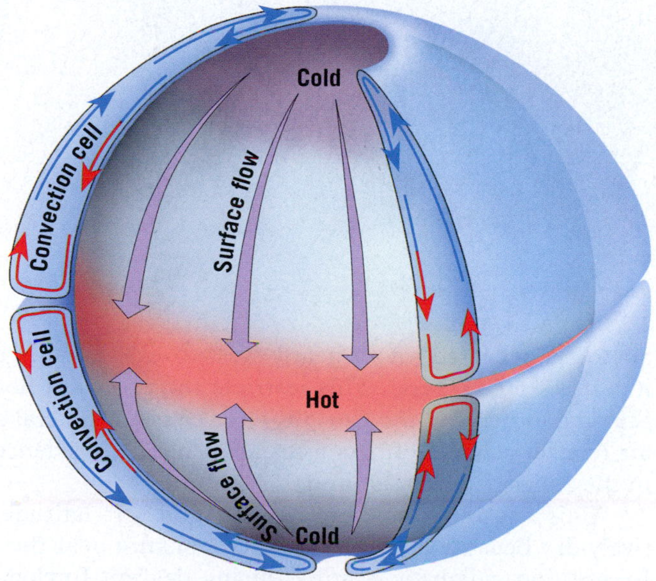


▲ **Figure 7.6 Global circulation on a nonrotating Earth** A simple convection system is produced by unequal heating of the atmosphere on a nonrotating Earth.

As this cold polar air approaches the equator, it is reheated and rises again. Thus, the circulation proposed by Hadley has upper-level air flowing poleward and surface air moving equatorward. Although correct in principle, Hadley's model does not take into account Earth's rotation.

## Three-Cell Circulation Model

In the 1920s a three-cell circulation model, which considers Earth's rotation, was proposed. Although this model has been modified to fit upper-air observations, it remains a useful tool for examining global circulation. Figure 7.7 illustrates the idealized three-cell model and the surface winds that result.

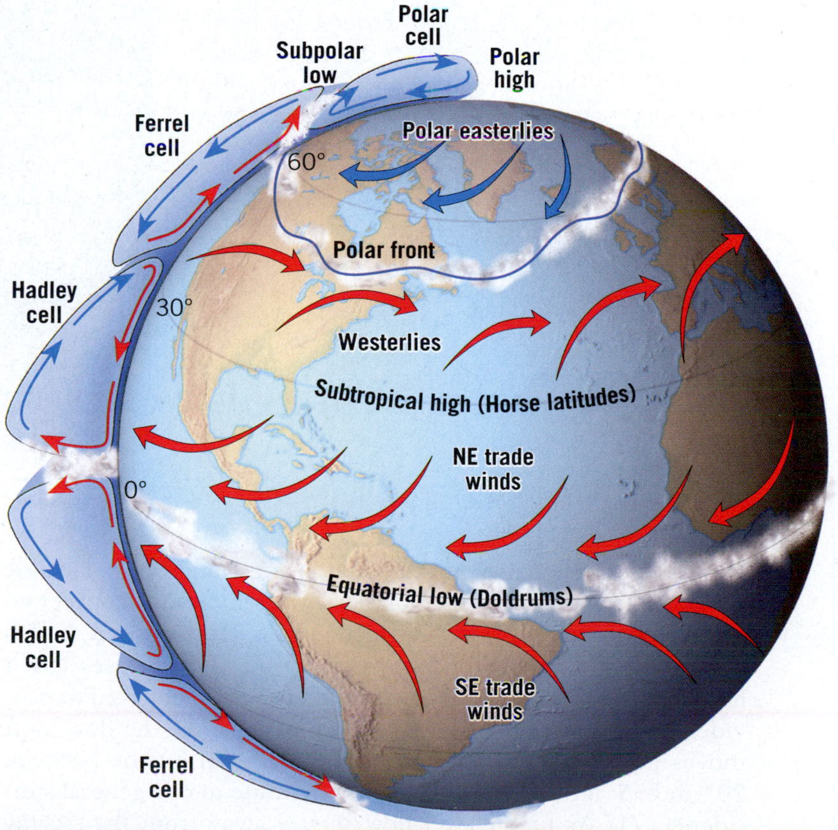


▲ **SmartFigure 7.7 Idealized global circulation for the three-cell circulation model on a rotating Earth**

http://goo.gl/In1LI

severe& hazardous weather Box 7.2

## Santa Ana Winds and Wildfires

Santa Ana is the local name for Chinook-like winds that characteristically sweep through southern California and northwestern Mexico in the fall and winter. These hot, dry winds are infamous for fanning regional wildfires.

Santa Ana winds are driven by strong high-pressure systems with subsiding air that tend to develop in the fall over the Great Basin. The clockwise flow from the anticyclone directs desert air from Arizona and Nevada westward toward the Pacific (**Fig. 7.B inset**). The wind gains speed as it is funneled through the canyons of the Coast Ranges, in particular the Santa Ana Canyon, from which the winds derive their name. Adiabatic heating of this already warm, dry air as it descends mountain slopes further accentuates the already parched conditions. Vegetation, seared by the summer heat, is dried even further by these hot, dry winds.

Although Santa Ana winds occur every year, they were particularly hazardous in the fall of 2003, and to a lesser extent in 2007, when hundreds of thousands of acres were scorched. In late October 2003, Santa Anas began blowing toward the coast of southern California at speeds that sometimes exceeded 100 kilometers (60 miles) per hour. Much of this area is covered by brush known as chaparral and related shrubs. It didn't take much—a careless camper or motorist, a lightning strike, or an arsonist—to ignite fires. Soon

Video (MM) Global Fire Patterns
http://goo.gl/gne63L

▶ **Figure 7.B Wildfires driven by Santa Ana winds** Ten large wildfires rage across southern California in this image taken on October 27, 2003, by NASA's *Aqua* satellite. Inset shows an idealized high-pressure area composed of cool, dry air that drives Santa Ana winds. Adiabatic heating causes the air temperature to increase and the relative humidity to decrease.

In the zones between the equator and roughly 30° latitude north and south, the circulation closely resembles the convection model proposed by Hadley—called the **Hadley cell** in his honor. Near the equator, warm rising air that releases latent heat during the formation of cumulus towers is believed to provide the energy that drives the Hadley cells. As the flow aloft moves poleward, the air begins to subside in a zone between 20° and 35° latitude. Two factors contribute to this general subsidence: (1) As upper-level flow moves away from the stormy equatorial region, radiation cooling becomes the dominant process. As a result, the air cools, becomes more dense, and sinks. (2) The Coriolis force becomes stronger with increasing distance from the equator, causing the poleward-moving upper air to be deflected into a nearly west-to-east flow by the time it reaches 30° latitude. This restricts the poleward flow of air. Stated another way, the Coriolis force causes a general pileup of air (convergence) aloft. As a result, general subsidence occurs in the zones between 20° and 35° latitude.

This subsiding air between 20° and 35° latitude is relatively dry because it has released its moisture near the equator. In addition, adiabatic heating during descent further reduces the air's relative humidity. Consequently, this subtropical zone of subsidence is the site of many of the world's great deserts, such as the Sahara of North Africa and the Great Australian Desert. Further, because surface winds are sometimes weak between 20° and 35° latitude, this belt was named the **horse**

▲ **Figure 7.C Flames from a wildfire move toward a home south of Valley Center, California** Image taken on October 27, 2003.

a number of small fires occurred in portions of Los Angeles, San Bernardino, Riverside, and San Diego Counties (**Fig. 7.B**). Several quickly developed into wildfires that moved almost as fast as the ferocious Santa Ana winds sweeping through the canyons.

Within a few days more than 13,000 firefighters were on fire lines extending from north of Los Angeles to the Mexican border. Nearly 2 months later, when all the fires were officially extinguished, more than 742,000 acres had been scorched, more than 3000 homes destroyed, and 26 people killed (**Fig. 7.C**). The Federal Emergency Management Agency put the dollar losses at over $2.5 billion. The 2003 southern California wildfires became the worst fire disaster in the state's history.

Strong Santa Ana winds, coupled with dry summers, have produced wildfires in southern California for millennia.

Strong Santa Ana winds, coupled with dry summers, have produced wildfires in southern California for millennia. These fires are nature's way of burning out chaparral thickets and sage scrub to prepare the land for new growth. When people began building homes and crowding into the fire-prone area between Santa Barbara and San Diego, the otherwise natural problems were compounded. Landscaped properties consisting of highly flammable eucalyptus and pine trees have further increased the hazard risk. In addition, fire prevention efforts have resulted in the accumulation of large quantities of flammable material over time, ultimately producing fewer but larger and more destructive fires. Clearly, wildfires will remain a major threat in these areas into the foreseeable future.

**Questions**

1. To what class of local winds do the Santa Ana winds belong?
2. During what time of year do Santa Ana winds occur?
3. Why are Santa Ana winds a threat for coastal areas of southern California?

**latitudes** (Fig. 7.7) because early Spanish sailing ships crossing the Atlantic were sometimes becalmed and stalled for long periods of time in these waters. If food and water supplies for the horses on board became depleted, the Spanish sailors were forced to throw the horses overboard.

From the center of the horse latitudes, the surface flow splits into two branches—one flowing poleward and one flowing toward the equator. The equatorward flow is deflected by the Coriolis force to form the reliable **trade winds**, so called because they enabled early sailing ships to move goods between Europe and North America. In the Northern Hemisphere, the trades blow from the northeast, while in the Southern Hemisphere, the trades are from the southeast. The trade winds from both hemispheres meet near the equator, in a region that has a weak pressure gradient. This zone is called the **doldrums**. Here light winds and humid conditions provide the monotonous weather that is the basis for the expression "the doldrums."

In the three-cell model, the circulation between 30° and 60° latitude (north and south), called the **Ferrel cell**, was proposed by William Ferrel to account for the westerly surface winds in the middle latitudes (see Fig. 7.7). These **prevailing westerlies** were known to Benjamin Franklin, perhaps the first American weather forecaster, who noted that storms migrated from west to east across the colonies. Franklin also observed that the westerlies were much more sporadic and therefore less

### eye ON THE atmosphere 7.1

This mountain area was cloud free as this summer day began. By afternoon these clouds had formed.

**Questions**

1. With which local wind are the clouds in this photo most likely associated?
2. Describe the process that created the local wind that was associated with the formation of these clouds.
3. Would you expect clouds such as these to form at night?

reliable than the trade winds for sail power. We now know that it is the migration of cyclones and anticyclones across the midlatitudes that disrupts the general westerly flow at the surface. Because of the significance of the midlatitude circulation in producing our daily weather, we will consider the westerlies in more detail later in this chapter.

The circulation in a **polar cell** is driven by subsidence near the poles that produces a surface flow that moves equatorward; this is called the **polar easterlies** in both hemispheres. As these cold polar winds move equatorward, they eventually encounter the warmer westerly flow of the midlatitudes. The region where the flow of cold air clashes with warm air has been named the **polar front**. The significance of this region will be considered later.

In summary, global circulation in driven mainly by temperature differences between the equator and the poles. However, due to Earth's rotation, rather than being a single convection cell that transfers energy between the equator and the poles, the global winds crudely resemble a three-cell system. The convection cell from the equator to roughly 30° latitude is called the Hadley cell; from 30° to about 60° latitude is the Ferrel cell; and from 60° to the poles is the polar cell.

### ✔ Concept Checks 7.3

1. Briefly describe the idealized global circulation proposed by George Hadley. What are the shortcomings of the Hadley model?
2. Name two factors that cause air to subside between 20° and 35° latitude.
3. In the idealized three-cell model of atmospheric circulation, most of the United States is situated in which belt of prevailing winds?
4. Which winds are found between the equator and 30° latitude?

## 7.4 Pressure Zones Drive the Wind

**Summarize Earth's idealized zonal pressure belts. Describe how continents and seasonal temperature changes complicate the idealized pattern.**

The idealized three-cell model provides a foundation for Earth's global wind patterns, but actual wind patterns are derived from a distinct and complex distribution of surface air pressure. To simplify this discussion, we will first examine the idealized pressure distribution that would be expected if Earth's surface were uniform—that is, composed entirely of water or smooth land. We will then turn to real-world pressure systems, and in subsequent sections we will look at seasonal and prevailing wind patterns produced by these systems.

### Idealized Zonal Pressure Belts

If Earth's surface were uniform, each hemisphere would have two east–west oriented belts of high pressure and two of low pressure (Fig. 7.8A). Near the equator, the warm rising branch of the Hadley cells is associated with the low-pressure zone known as the **equatorial low**. This region of ascending moist, hot air is marked by abundant precipitation. Because this region of low pressure is where the trade winds converge, it is also referred to as the **intertropical convergence zone (ITCZ)**. In Figure 7.9, the ITCZ is visible as a band of clouds near the equator.

About 20° to 35° on either side of the equator, where the westerlies and trade winds originate and go their separate ways, are the high-pressure zones known as the **subtropical highs**. In these zones a subsiding air column produces weather that is normally warm and dry.

Another low-pressure region is situated at about 50° to 60° latitude, in a position corresponding to the polar front. Here

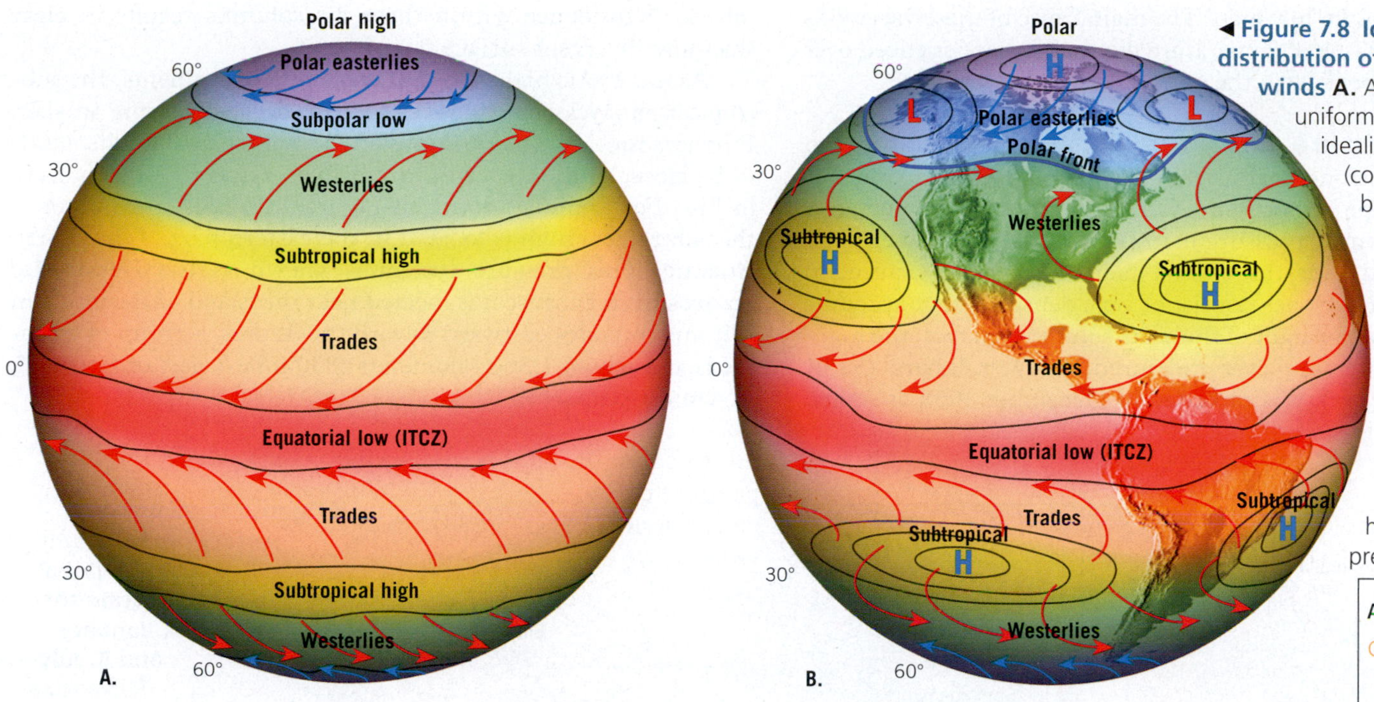

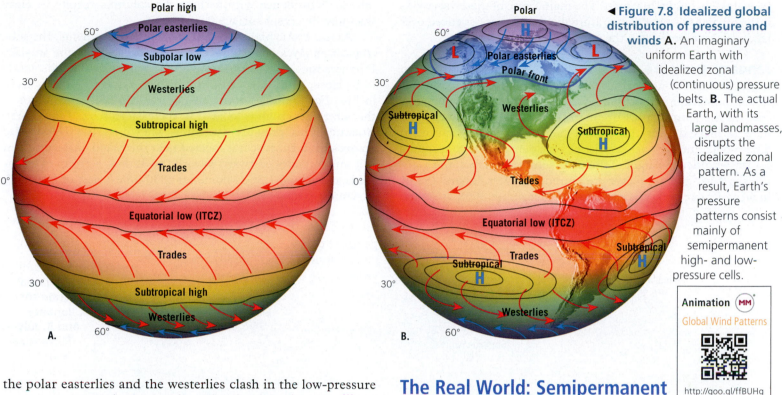

◀ **Figure 7.8 Idealized global distribution of pressure and winds A.** An imaginary uniform Earth with idealized zonal (continuous) pressure belts. **B.** The actual Earth, with its large landmasses, disrupts the idealized zonal pattern. As a result, Earth's pressure patterns consist mainly of semipermanent high- and low-pressure cells.

Animation MM
Global Wind Patterns
http://goo.gl/ffBUHg

the polar easterlies and the westerlies clash in the low-pressure convergence zone known as the **subpolar low**. As you will see later, this zone is responsible for much of the stormy weather in the middle latitudes, particularly in the winter.

Finally, near Earth's poles are the **polar highs**, from which the polar easterlies originate (see Fig. 7.8A). These polar highs result from surface cooling. Because air near the poles is cold and dense, it exerts higher-than-average surface pressure.

▼ **Figure 7.9 The intertropical convergence zone (ITCZ)** This zone of low pressure and convergence is seen as a band of clouds that extends east–west slightly north of the equator.

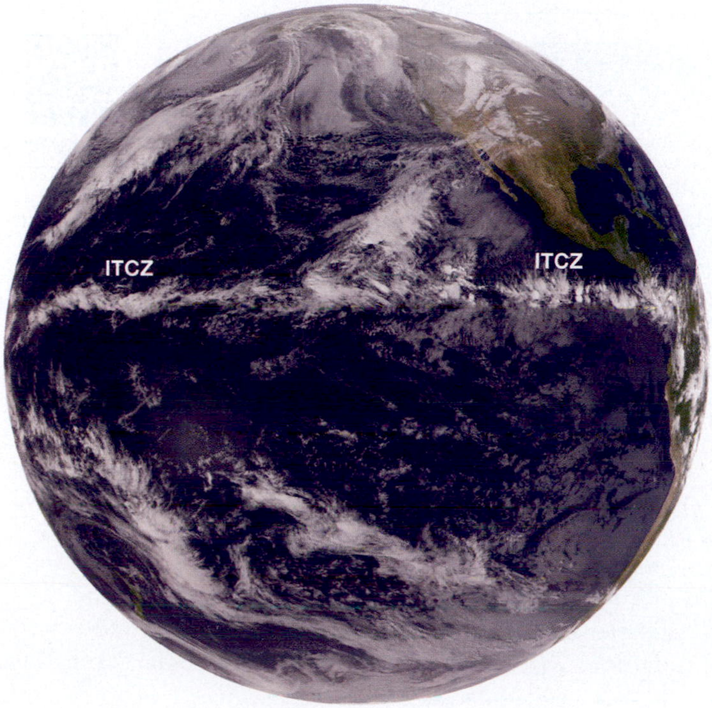


## The Real World: Semipermanent Pressure Systems

Up to this point, we have considered the global pressure systems as if they were continuous belts around Earth. However, because Earth's surface is not uniform, the only true zonal distribution of pressure exists along the subpolar low in the Southern Hemisphere, where the ocean is continuous. To a lesser extent, the equatorial low is also continuous. At other latitudes, particularly in the Northern Hemisphere, where there is a higher proportion of land compared to ocean, the zonal pattern is replaced by semipermanent cells of high and low pressure.

The idealized pattern of pressure and winds for the "real" Earth is illustrated in Figure 7.8B. This pattern is always in a state of flux because of seasonal temperature changes, which serve to either strengthen or weaken these pressure cells. In addition, the position of these pressure systems moves either poleward or equatorward with the seasonal migration of the zone of maximum solar heating. As a result of these factors, Earth's pressure patterns vary in strength and location during the course of the year.

Average global pressure patterns and resulting winds for the months of January and July are shown in Figure 7.10. Notice on these maps that the observed pressure patterns are circular (or elongated) instead of zonal (east–west bands). The most prominent features on both maps are the subtropical highs. These systems are centered between 20° and 35° latitude over the subtropical oceans.

When we compare Figures 7.10A (January) and 7.10B (July), we see that some pressure cells are year-round features—the subtropical highs, for example. Others, however, are seasonal. For example, the low-pressure cell over northern Mexico and the southwestern United States is a summer phenomenon and appears only on the July map. The main cause of these variations is the greater seasonal temperature fluctuations experienced over landmasses, especially in the middle and higher latitudes.

**January Pressure and Wind Patterns** The **Siberian high**, a very strong high-pressure center positioned over the frozen landscape of northern Asia, is the most prominent feature on the January pressure map (Fig. 7.10A). A weaker polar high is located over the chilled North American continent. These cold anticyclones consist of very dense air that accounts for the significant weight of these air columns. In fact, the highest sea-level pressure ever measured, 1084 millibars (32.01 inches of mercury), was recorded in December 1968 at Agata, Siberia. Subsidence within these air columns results in clear skies and divergent surface flow.

As the Arctic highs strengthen over the continents, the subtropical anticyclones situated over the oceans become weaker. Further, the average position of the subtropical highs tends to be closer to the eastern shore of the oceans in January than in July. For example, notice in Figure 7.10A that the center of the subtropical high is located in the eastern part of the North Atlantic. This pressure system is known as the **Bermuda/Azores high** because it is located near the island of Bermuda in July and migrates eastward toward the Azores, a group of volcanic islands about 1,360 kilometers (850 miles) west of Portugal, as winter approaches.

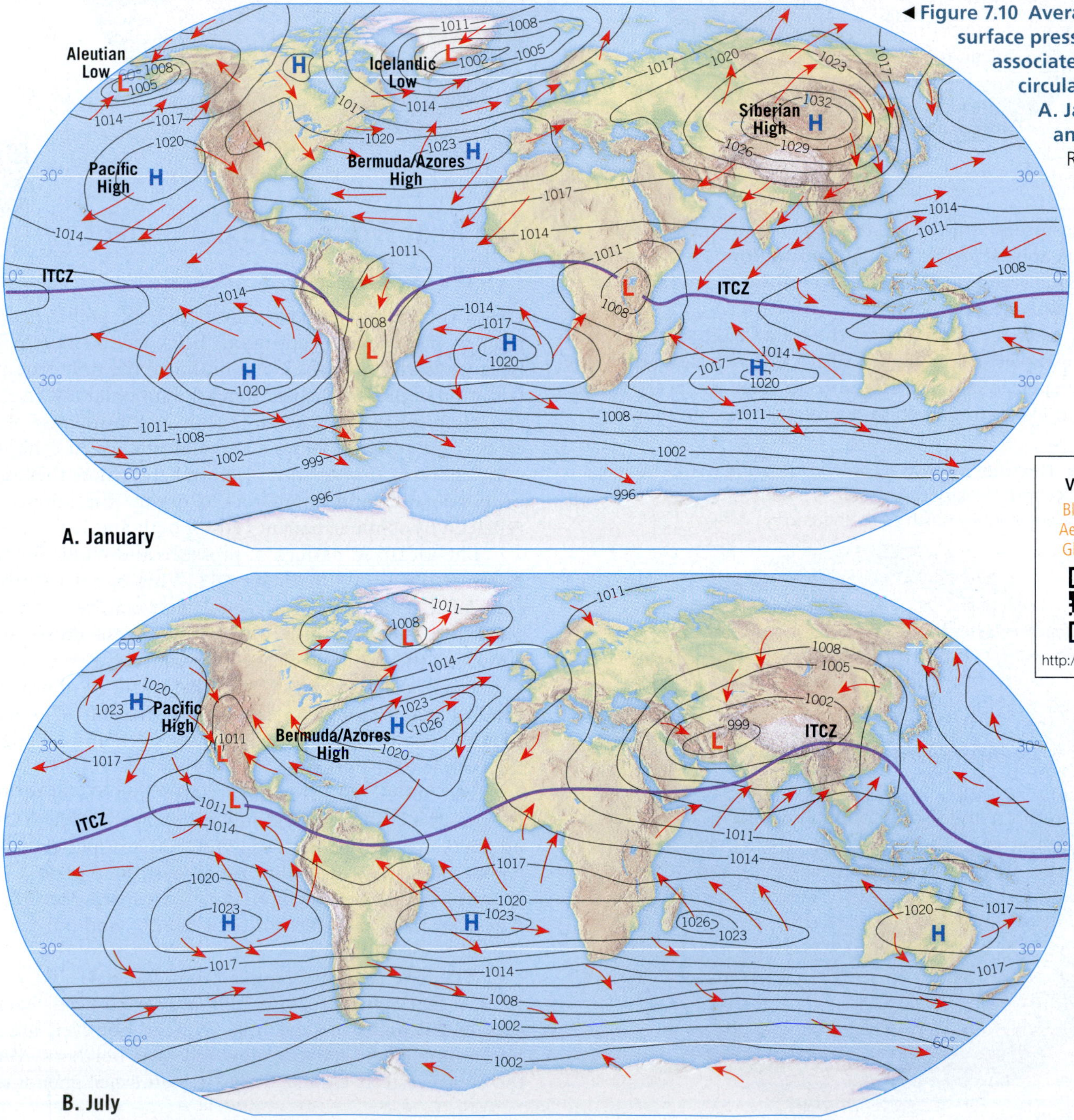


◄ **Figure 7.10 Average surface pressure and associated global circulation for A. January and B. July** Red arrows indicate surface airflow.

Video MM
Black Carbon Aerosols Trace Global Winds
http://goo.gl/Pqfv5u

Also shown on the January map but absent in July are two intense semipermanent low-pressure centers. Named the **Aleutian low** and the **Icelandic low**, these cyclonic cells are situated over the North Pacific and North Atlantic, respectively. They are not stationary cells but rather composites of numerous cyclonic storms that traverse these regions. In other words, so many midlatitude cyclones occur during the winter that these regions almost always experience low pressure, hence the term *semipermanent*. As a result, the areas affected by the Aleutian and Icelandic lows are frequently cloudy and receive abundant winter precipitation.

Because a large number of cyclonic storms form over the North Pacific and travel eastward, the southern Alaska coast receives abundant precipitation. This fact is exemplified by Sitka, Alaska, a coastal town that receives 215 centimeters (85 inches) of precipitation each year, more than five times that received in Churchill, Manitoba, Canada. Although both towns are situated at roughly the same latitude, Churchill is located in the continental interior, far removed from the influence of the cyclonic storms associated with the Aleutian low.

**July Pressure and Wind Patterns** The pressure pattern over the Northern Hemisphere changes dramatically in summer (see Fig. 7.10B). High surface temperatures over the continents generate lows that replace wintertime highs. These thermal lows consist of warm ascending air that induces inward-directed surface flow. The strongest of these low-pressure centers develops over southern Asia, while a weaker thermal low is found in the southwestern United States.

Notice in Figure 7.10 that during the summer months, the subtropical highs in the Northern Hemisphere migrate westward and become stronger than during the winter months. These strong high-pressure centers dominate the summer circulation over the oceans and pump warm moist air onto the continents that lie to the west of these highs. This results in an increase in precipitation over parts of eastern North America and Southeast Asia.

### ✔ Concept Checks 7.4

1. What is the intertropical convergence zone (ITCZ)?
2. If Earth had a uniform surface, east–west belts of high and low pressures would exist. Name these zones and the approximate latitude in which each would be found.
3. During what season is the Siberian high strongest, and why?
4. During what season is the Bermuda/Azores high strongest?

## 7.5 Monsoons

**Describe the seasonal changes in Earth's global circulation that produce the Asian monsoon.**

Large *seasonal* changes in Earth's global circulation are called monsoons. Contrary to popular belief, **monsoon** does not mean "rainy season"; rather, it refers to a particular wind system that reverses its direction twice each year. In general, winter is associated with winds that blow predominantly off the continents, called the *winter monsoon*. In contrast, in summer, warm moisture-laden air blows from the sea toward the land. Thus, the *summer monsoon* is usually associated with abundant precipitation over affected land areas, and this is the source of the misconception.

### The Asian Monsoon

The best-known and best-developed monsoon circulation occurs in southern and southeastern Asia—affecting India and the surrounding areas as well as parts of China, Korea, and Japan. As with most other winds, the Asian monsoon is driven by pressure differences that are generated by unequal heating of Earth's surface.

As winter approaches, long nights and low Sun angles result in the accumulation of frigid air over the vast landscape of northern Russia. This generates the cold Siberian high, which ultimately dominates Asia's winter circulation. The subsiding dry air of the Siberian high produces surface flow that moves across southern Asia, producing predominantly offshore winds (**Fig. 7.11A**). By the time this flow reaches India, it has warmed considerably but remains extremely dry. For example, Kolkata, India, receives less than 2 percent of its annual precipitation in the cooler 6 months. The remainder comes in the warmer 6 months, with the vast majority falling from June through September.

In contrast, summertime temperatures in the interior of southern Asia often exceed 40°C (104°F). This intense solar heating generates a low-pressure area over the region similar to that associated with a sea breeze, but on a much larger scale. This in turn generates outflow aloft that encourages inward flow at the surface. With the development of the low-pressure center over southeastern Asia, moisture-laden air from the Indian and Pacific Oceans flows landward, thereby generating a pattern of precipitation typical of the summer monsoon.

One of the world's rainiest regions is found on the slopes of the Himalayas, where orographic lifting of incoming moist air from the Indian Ocean produces copious precipitation. Cherrapunji, India, once recorded an annual rainfall of 25 meters (82.5 feet), most of which fell during the 4 months of the summer monsoon (**Fig. 7.11B**).

The Asian monsoon is complex and strongly influenced by the seasonal change in solar heating received by the vast Asian continent. However, another factor, related to the annual migration of the Sun's vertical rays, also contributes to the monsoon circulation of southern Asia. As shown in Figure 7.11, the Asian monsoon is associated with a large seasonal migration of the ITCZ. With the onset of summer, the ITCZ moves

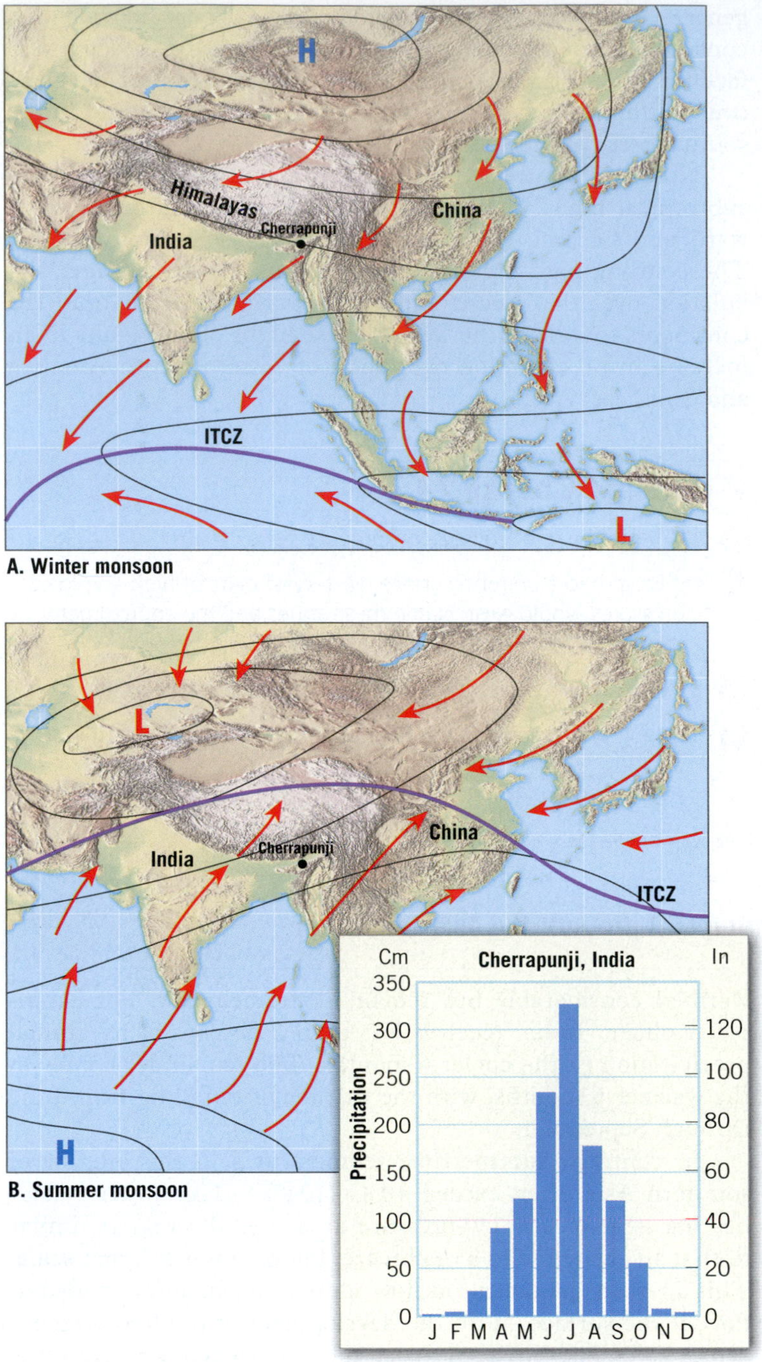


▲ **Figure 7.11 Asia's monsoon circulation** This circulation pattern occurs in conjunction with the seasonal shift of the intertropical convergence zone (ITCZ). **A.** In January a strong high pressure develops over Asia. The resulting flow of cool air off the continent generates the dry winter monsoon. **B.** With the onset of summer, the ITCZ migrates northward and draws warm, moist air onto the continent.

northward over the continent and is accompanied by peak rainfall. The opposite occurs in the Asian winter, as the ITCZ moves south of the equator.

Nearly half the world's population inhabits regions affected by the Asian monsoons. Many of these people depend on subsistence agriculture for their survival. The timely arrival of monsoon rains often means the difference between adequate nutrition and widespread malnutrition.

## The North American Monsoon

Other regions experience seasonal wind shifts like those associated with the Asian monsoon. For example, a seasonal wind shift influences a portion of North America. Sometimes called the *North American monsoon,* this circulation pattern produces a dry spring followed by a relatively rainy summer that impacts large areas of the southwestern United States and northwestern Mexico.* This is illustrated by precipitation patterns observed in Tucson, Arizona, which typically receives nearly 10 times more precipitation in August than in May. As shown in **Figure 7.12**, summer rains typically last into September before drier conditions are reestablished.

Summer daytime temperatures in the American Southwest, particularly in the low deserts, can be extremely high. This intense surface heating generates a low-pressure center over Arizona. The resulting circulation pattern brings warm, moist air from the Gulf of California and to a lesser extent the Gulf of Mexico (**Fig. 7.13**). The supply of atmospheric moisture from nearby marine sources coupled with the convergence and upward flow of the thermal low generates the precipitation this

*This event is also called the *Arizona monsoon* and the *Southwest monsoon* because it has been extensively studied in this part of the United States.

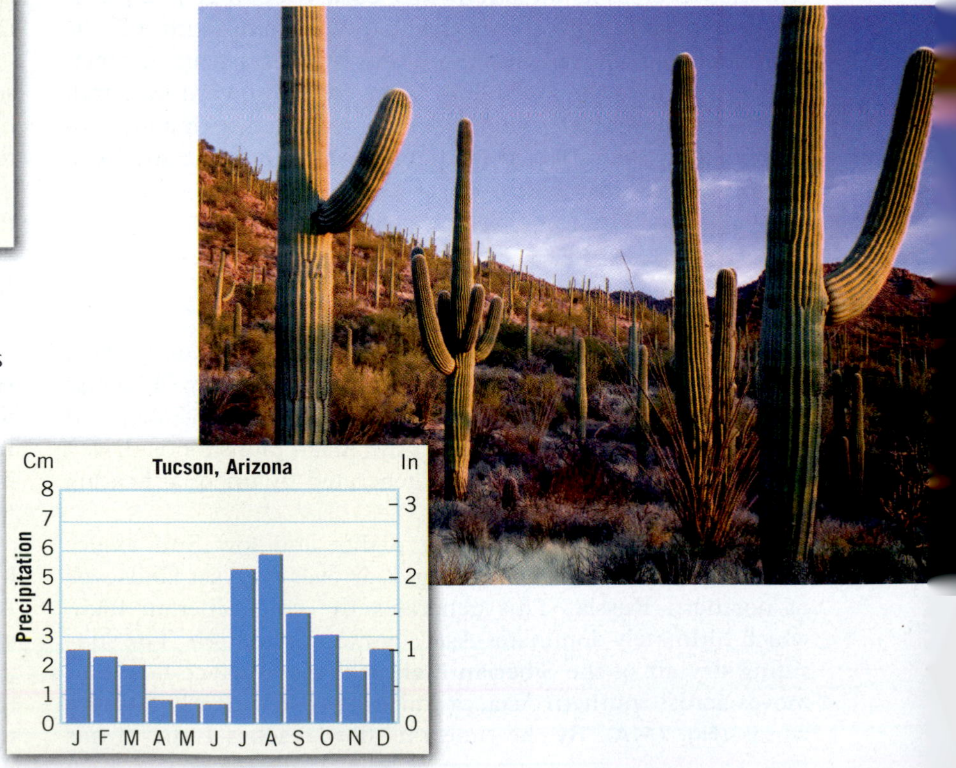


► **Figure 7.12 The North American monsoon** This seasonal circulation is illustrated by the precipitation pattern for Tucson, Arizona.

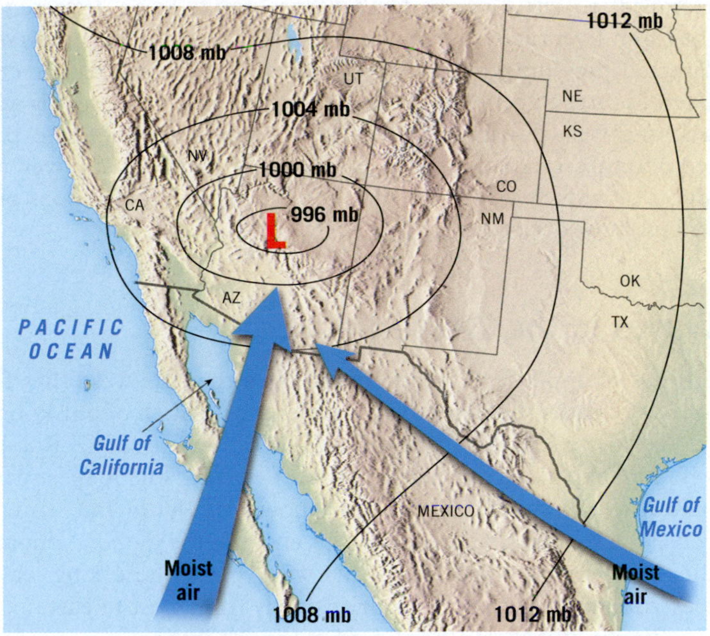


**◀ Figure 7.13 High summer temperatures over the southwestern United States generate the North American monsoon** High temperatures create a thermal low that draws moist air from the Gulf of California and the Gulf of Mexico. This summer monsoon produces an increase in precipitation, which often comes in the form of thunderstorms, over the southwestern United States and northwestern Mexico.

region experiences during the hottest months. Although often associated with the state of Arizona, this monsoon is actually strongest in northwestern Mexico and is also quite pronounced in New Mexico.

## ✔ Concept Checks 7.5

1. Define *monsoon.*
2. Explain the cause of the Asian monsoon. Which season (summer or winter) is the rainy season?
3. What areas of North America experience a pronounced monsoon circulation?

# 7.6 The Westerlies

**Explain why the airflow aloft in the middle latitudes has a strong west-to-east component.**

Prior to World War II, upper-air observations were scarce. Since then, aircraft and radiosondes have provided a great deal of data about the upper troposphere. Among the most important discoveries was that airflow aloft in the middle latitudes has a strong west-to-east component, thus the name *westerlies.*

## Why Westerlies?

Let us consider the reason for the predominance of westerly flow aloft. In the case of the westerlies, the temperature contrast between the poles and equator drives these winds. **Figure 7.14** illustrates the pressure distribution with height over the cold polar region as compared to the much warmer tropics. Recall that because cold air is more dense (compact) than warm air, air pressure decreases more rapidly in a column of cold air than in a column of warm air (see Fig. 6.10). The pressure surfaces (planes) in Figure 7.14 represent a simplified view of the pressure distribution we would expect to observe from pole to equator.

Over the equator, where temperatures are higher, air pressure decreases more gradually than over the cold polar regions. Consequently, at the same altitude above Earth's surface, higher pressure exists over the tropics, and lower pressure is the norm above the poles. Thus, the pressure gradient aloft is directed from the equator (area of higher pressure) toward the poles (area of lower pressure).

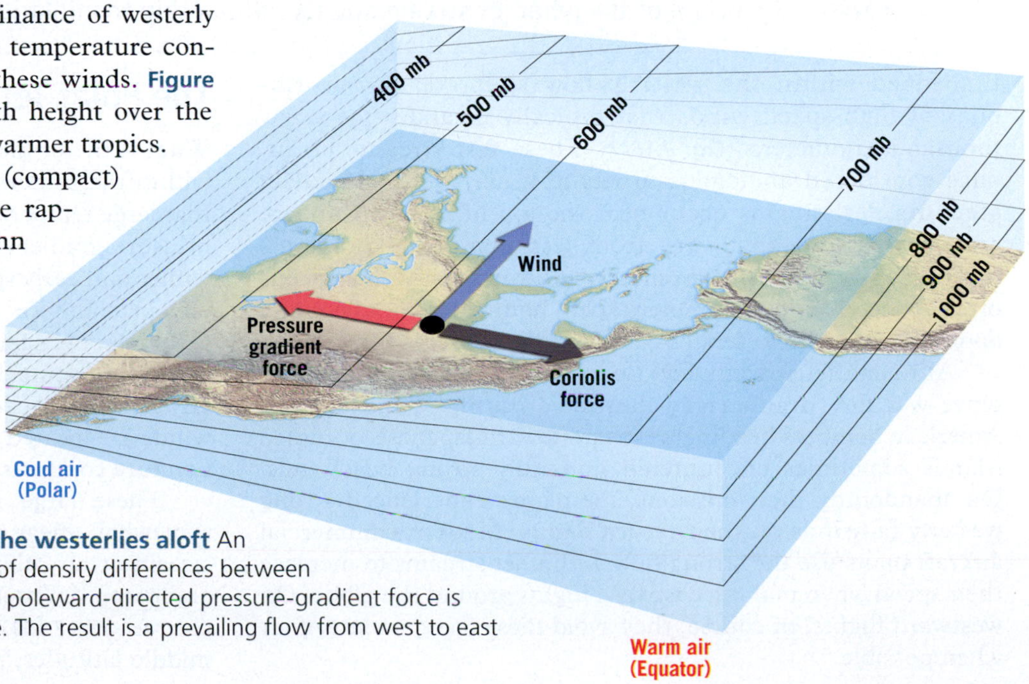


**▶ Figure 7.14 Pressure pattern that produces the westerlies aloft** An idealized pressure gradient develops aloft because of density differences between cold polar air and warm tropical air. Notice that the poleward-directed pressure-gradient force is balanced by an equatorward-directed Coriolis force. The result is a prevailing flow from west to east called the *westerlies.*

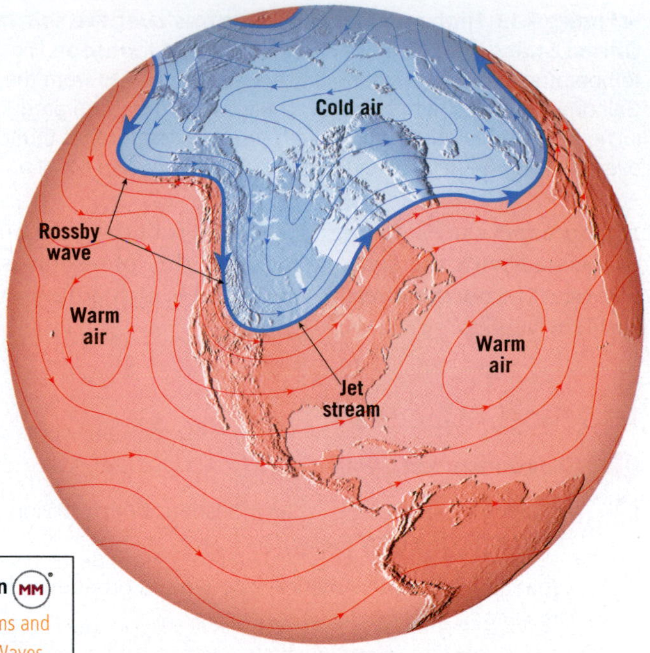


▲ **Figure 7.15 Idealized airflow of the westerlies aloft** The five long-wavelength undulations, called *Rossby waves*, compose this flow. The jet stream is the fast core of this wavy flow.

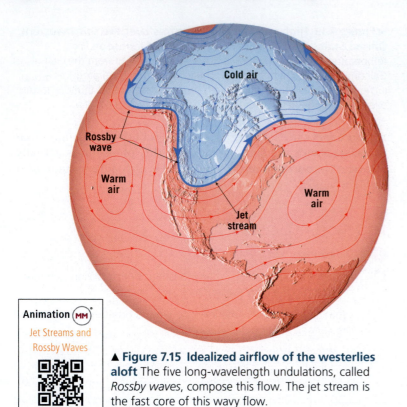

Animation MM
Jet Streams and Rossby Waves
http://goo.gl/QUMn5i

When the air aloft that originated in the tropics begins to advance poleward in response to this pressure-gradient force (red arrow in Fig. 7.14), the Coriolis force causes a change in the direction of airflow. Recall from Chapter 6 that in the Northern Hemisphere, the Coriolis force causes winds to be deflected to the right. Eventually, a balance is reached between the poleward-directed pressure-gradient force and the equatorward-directed Coriolis force to generate *geostrophic* winds, which are winds with a strong west-to-east component. Because the equator-to-pole temperature gradient shown in Figure 7.14 is typical over the globe, a westerly flow aloft should be expected, and it does prevail on most occasions.

### Waves in the Westerlies

Studies of upper-level wind charts show that the westerlies follow wavy paths that have long wavelengths. Much of our knowledge of these large-scale motions is attributed to C. G. Rossby, who first explained the nature of these waves. The longest wave patterns, called **Rossby waves**, shown in Figure 7.15, usually consist of four to six meanders that encircle the globe. Although the air flows eastward along this wavy path, these long waves tend to remain stationary or drift slowly from west to east.

Rossby waves can have a tremendous impact on our daily weather, especially when they meander widely from north to south. We will consider this role in the following section and again in Chapter 9.

#### ✔ Concept Checks 7.6

1. Why is the flow aloft in the midlatitudes predominantly westerly?
2. What name is given to the long-wavelength flow that is apparent on upper-level charts?

## 7.7 Jet Streams

**Explain the origin of the polar jet stream and its relationship to midlatitude cyclonic storms.**

Embedded within the westerly flow aloft are narrow ribbons of high-speed winds that typically meander for a few thousand kilometers (Fig. 7.16A). These fast streams of air, once considered analogous to jets of water, were named **jet streams**. Jet streams occur near the top of the troposphere and have widths that vary from less than 100 kilometers (60 miles) to over 500 kilometers (300 miles). Wind speeds often exceed 100 kilometers per hour and occasionally approach 400 kilometers (240 miles) per hour.

Although jet streams had been detected earlier, their existence was first dramatically illustrated during World War II. American bombers heading westward toward Japanese-occupied islands sometimes encountered unusually strong headwinds. On abandoning their missions, the planes experienced strong westerly tailwinds on their return flights. Modern commercial aircraft pilots use the strong flow within jet streams to increase their speed when making eastward flights around the globe. On westward flights, of course, they avoid these fast currents of air when possible.

### The Polar Jet Stream

What is the origin of the distinctive energetic winds that exist within the somewhat slower general westerly flow? The key is that large temperature differences at the surface produce steep pressure gradients aloft and hence faster upper-air winds. In winter and early spring, it is not unusual to have a warm balmy day in southern Florida and near-freezing temperatures in Georgia, only a few hundred kilometers to the north. Such large wintertime temperature contrasts lead us to expect faster westerly flow at that time of year. In general, the fastest upper-air winds are located above regions of the globe having large temperature contrasts across very narrow zones.

These large temperature contrasts occur along narrow, somewhat linear zones called *fronts*. The most prevalent jet stream occurs along a major frontal zone called the *polar front* and is appropriately named the **polar jet stream**, or simply the *polar jet* (Fig. 7.16B). Because this jet stream is often found in the middle latitudes, particularly in the winter, it is also known as

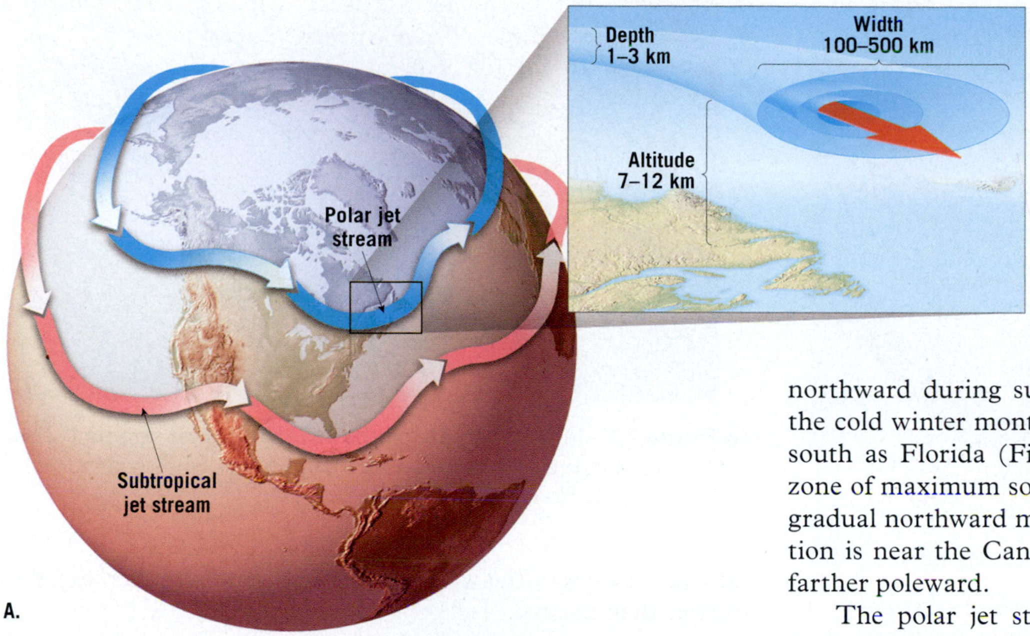


A.

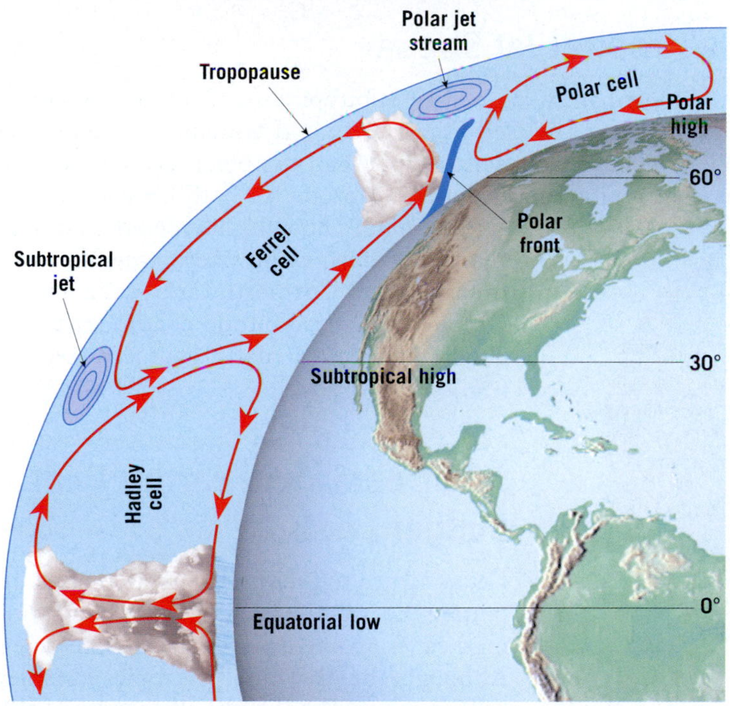


B.

▲ **Figure 7.16 Jet streams A.** Approximate positions of the polar and subtropical jet streams. **B.** A cross-sectional view of the polar and subtropical jets in relation to the three-cell model of global circulation.

the *midlatitude jet stream* (**Fig. 7.17**). Recall that the polar front is situated between the cool winds of the polar easterlies and the relatively warm westerlies. Instead of flowing nearly straight west to east, the polar jet stream usually has a meandering path. Occasionally, it flows almost due north–south. Sometimes it splits into two jets that may or may not rejoin. Like the polar front, this zone of high-velocity airflow is not continuous around the globe.

On average, the polar jet travels at 125 kilometers (75 miles) per hour in the winter and roughly half that speed in the summer (**Fig. 7.18**). This seasonal difference is due to the much stronger temperature gradient that exists in the middle latitudes during the winter.

Because the location of the polar jet roughly coincides with that of the polar front, its latitudinal position migrates with the seasons. Thus, like the zone of maximum solar heating, the jet moves northward during summer and southward in winter. During the cold winter months, the polar jet stream may extend as far south as Florida (Fig. 7.18). With the coming of spring, the zone of maximum solar heating, and therefore the jet, begins a gradual northward migration. By midsummer, its *average* position is near the Canadian border, but it can be located much farther poleward.

The polar jet stream plays a very important role in the weather of the midlatitudes. In addition to supplying energy to help drive the rotational motion of surface storms, it also directs the paths of these storms. Consequently, determining changes in the location and flow pattern of the polar jet is an important part of modern weather forecasting. As the polar jet shifts northward, there is a corresponding change in the region where outbreaks of severe thunderstorms and tornadoes occur. In February, most thunderstorms and tornadoes occur in the states bordering the Gulf of Mexico, but by midsummer the center of this activity shifts to the Northern Plains and Great Lakes states.

The location of the polar jet stream also affects other surface conditions, particularly temperature and humidity. When it is situated substantially equatorward of your location, the weather will be colder and drier than normal. Conversely, when the polar jet moves poleward of your location, warmer and more humid conditions will prevail. Thus, depending on the jet

▼ **Figure 7.17 Simplified 200-millibar height-contour for January**
The position of the jet stream core (or streak) is shown in dark pink.

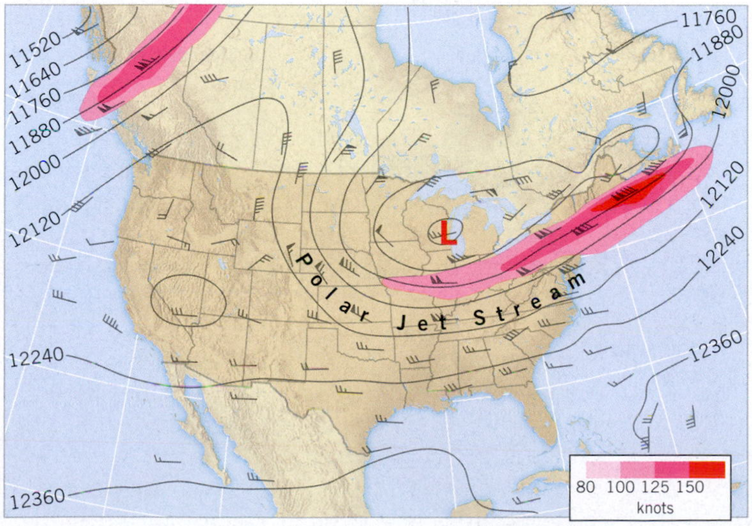

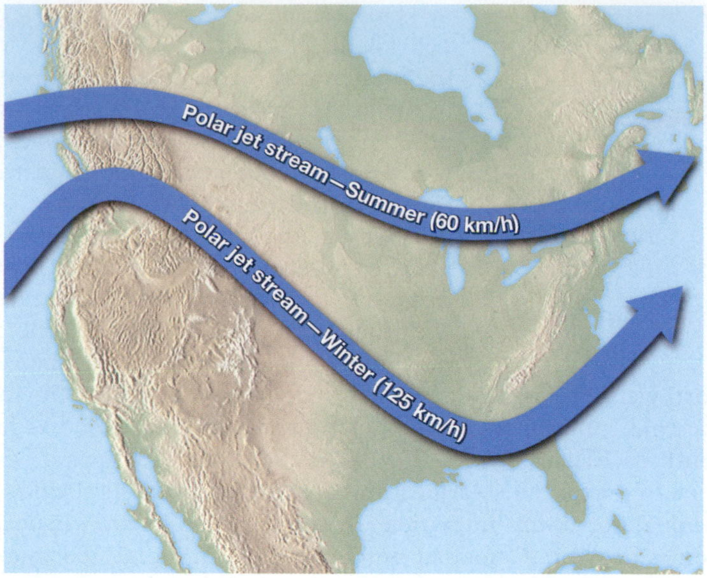


▲ **Figure 7.18 The position and speed of the polar jet stream changes with the seasons** The polar jet stream migrates freely between about 30° and 70° latitude. Shown are flow patterns that are common for summer and winter.

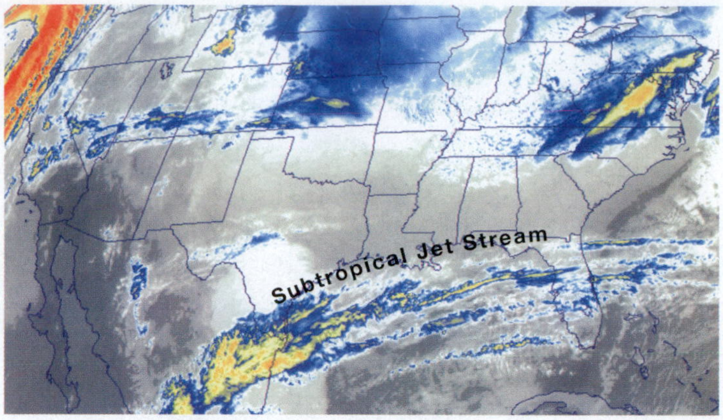


▲ **Figure 7.19 Infrared image of the subtropical jet stream** In this image, the subtropical jet appears as a band of cloudiness extending from Mexico to Florida.

stream's position, the weather could be hotter, colder, drier, or wetter than normal.

▼ **Figure 7.20 Cyclic changes occur in the upper-level airflow of the westerlies** The flow, which has the jet stream as its axis, starts out nearly straight and then develops meanders and cyclonic activity that dominate the weather.

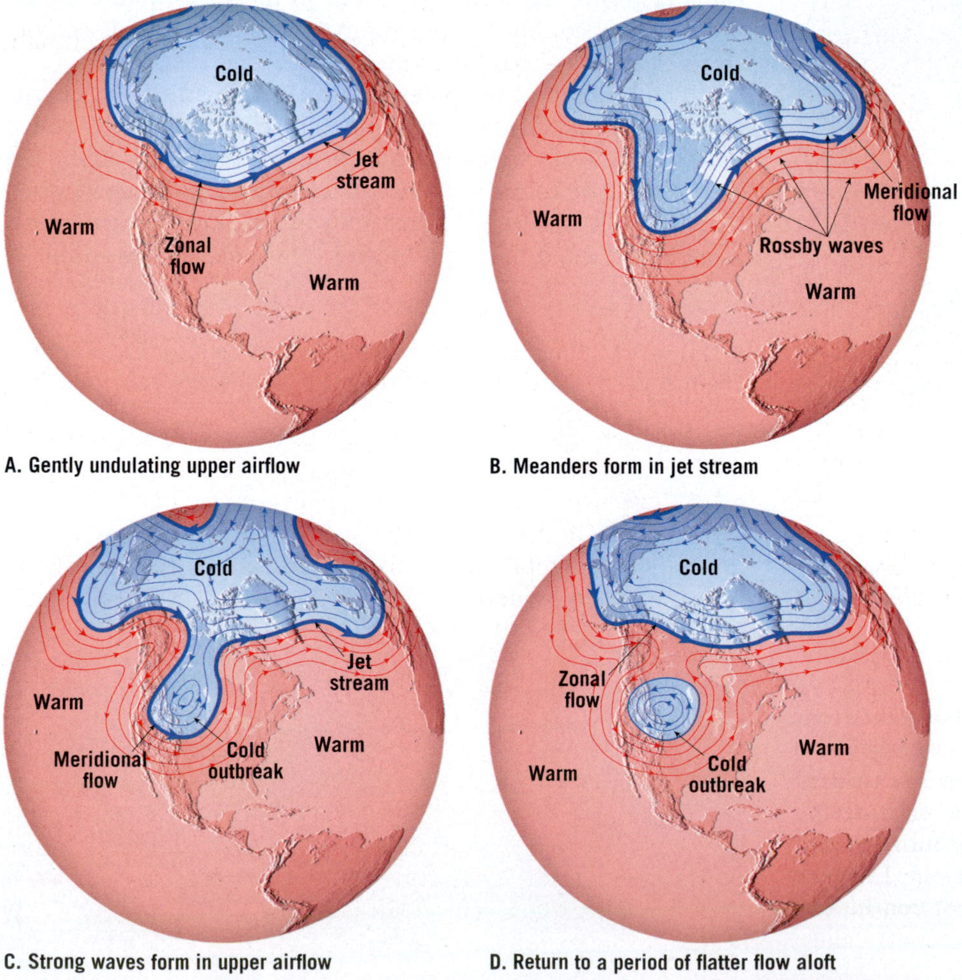


## Subtropical Jet Stream

A semipermanent jet over the subtropics is called the **subtropical jet stream** (see Fig. 7.16B), and is mainly a wintertime phenomenon. Somewhat slower than the polar jet, this west-to-east current is centered at 25° latitude (in both hemispheres) at an altitude of about 13 kilometers (8 miles). In the Northern Hemisphere, when the subtropical jet sweeps northward in the winter, it may bring warm humid conditions to the Gulf states, particularly southern Florida (Fig. 7.19).

## Jet Streams and Earth's Heat Budget

Let us return to the wind's function of maintaining Earth's heat budget by transporting heat from the equator toward the poles. Although the flow in the tropics (Hadley cell) is somewhat *meridional* (north–south), at most other latitudes the flow is *zonal* (west–east). The reason for the zonal flow, as we have seen, is the Coriolis force. The question we now consider is: *How can wind with a west-to-east flow transfer heat from south to north?*

The important function of heat transfer is accomplished by the wavy flow (Rossby waves) of the westerlies centered on the polar jet stream. There may be periods of a week or more when the flow in the polar jet is essentially west to east, as illustrated in Figure 7.20A. When this condition prevails, relatively mild temperatures occur south of the jet stream, and cooler temperatures prevail to the north. Then, with minimal warning, the flow aloft may begin to meander and produce large-amplitude waves that exhibit a more pronounced north-to-south flow, as shown in Figure 7.20B and Figure 7.20C. Such a change causes cold air to advance equatorward and warm air to flow poleward. In addition, a cold air mass may become detached and produce an outbreak of cold air, as shown in Figure 7.20D.

This redistribution of energy eventually results in a weakened temperature gradient and a return to a flatter flow aloft (Fig. 7.20D). Therefore, the wavy flow of the westerlies centered on the polar jet plays an important role in Earth's heat budget.

**students sometimes ask...**

**Why do pilots of commercial aircraft always remind passengers to keep their seat belts fastened, even in ideal flying conditions?**

The reason for this request is *clear air turbulence*, a phenomenon that occurs when airflow in two adjacent layers moves at different velocities. This can happen when the air at one level travels in a different direction than the air above or below it. More often, however, it occurs when air at one level travels faster than air in an adjacent layer. Such movements create eddies (turbulence) that can cause the plane to move suddenly up or down.

## ✔ Concept Checks 7.7

1. How are jet streams generated?
2. At what time of year should we expect the fastest polar jet streams? Explain.
3. Why is the polar jet sometimes referred to as the midlatitude jet stream?
4. Describe the expected winter temperatures in the north-central states when the polar jet stream is located over central Florida.
5. Explain how the wavy flow centered on the jet stream helps balance Earth's heat budget.

# 7.8 Global Winds and Ocean Currents

**Sketch and label the major ocean currents on a world map.**

Energy is passed from moving air to the surface of the ocean through friction. As a consequence, winds blowing steadily across the ocean drag the water along with them. Because winds are the primary driving force of surface ocean currents, a relationship exists between atmospheric circulation and oceanic circulation—as illustrated by a comparison of Figures 7.21 and 7.10.

As shown in Figure 7.21, positioned north and south of the equator are two westward-moving currents, the North and South Equatorial Currents. These currents derive their energy principally from the trade winds that blow from the northeast and southeast, respectively, toward the equator. The Coriolis force deflects surface currents poleward to form clockwise spirals in the Northern Hemisphere and counterclockwise spirals in the Southern Hemisphere. These nearly circular ocean currents, called **gyres**, are centered over the five major subtropical high-pressure systems—located in both the southern and northern Atlantic and Pacific Oceans and in the Indian Ocean (Fig. 7.21).

In the North Atlantic, the equatorial current is deflected northward through the Caribbean, where it becomes the *Gulf Stream*. As the Gulf Stream moves along the eastern coast of the United States, it is strengthened by the prevailing westerly winds. As it continues northeastward beyond the Grand Banks, it gradually widens and slows until it becomes a vast, slowly moving current known as the North Atlantic Drift. The North Atlantic Drift splits as it approaches Western Europe. Part moves northward past Great Britain toward Norway, while the other portion is deflected southward as the cool Canary Current. As the Canary Current moves south, it eventually merges with the North Equatorial Current.

## Ocean Currents Influence Climate

Surface-ocean currents have an important effect on climate. We know that for Earth as a whole, the gains in solar energy equal the losses to space of heat radiated from the surface. When most latitudes are considered individually, however, this is not the case. There is a net gain of energy in lower latitudes and a net loss at higher latitudes. Because the tropics are not becoming progressively warmer, nor are the polar regions becoming colder, there must be a large-scale transfer of heat from areas of excess to areas of deficit. This is indeed the case. *The transfer of heat by winds and ocean currents equalizes these latitudinal energy imbalances*. Ocean water movements account for about one-quarter of this total heat transport, and winds account for the remaining three-quarters.

**The Effect of Warm Currents** The moderating effect of poleward-moving, warm ocean currents is well known. The North Atlantic Drift, an extension of the warm Gulf Stream, keeps wintertime temperatures in Great Britain and much of Western Europe warmer than would be expected for their latitudes. London is farther north than St. John's, Newfoundland, yet is not nearly so frigid in winter. Because of the prevailing westerly winds, the moderating effects are carried far inland. For example, Berlin (52° north latitude) has a mean January temperature similar to that experienced at New York City, which

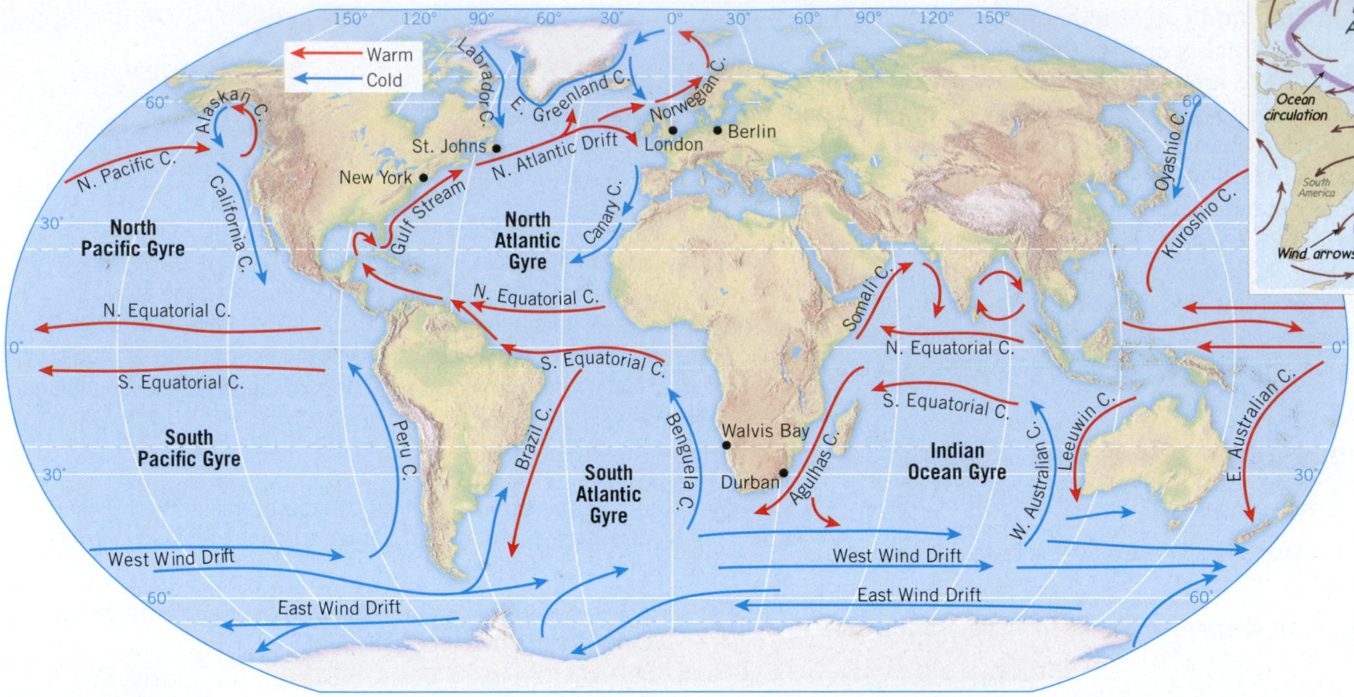


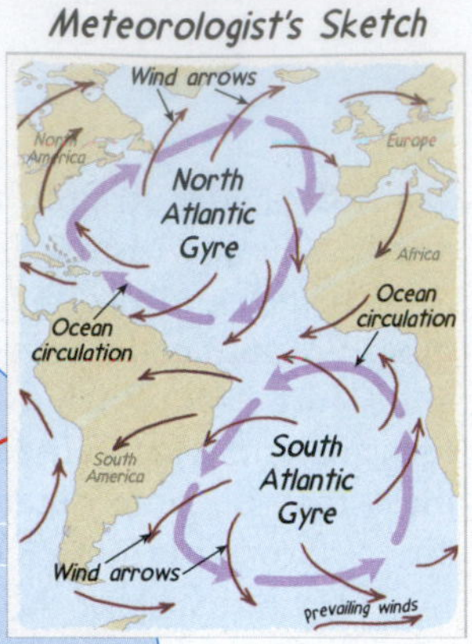


Animation
Ocean Currents
http://goo.gl/QIYk8o

http://goo.gl/I6q8q

▲ **SmartFigure 7.21 Major surface-ocean currents** The ocean's surface circulation is organized into five major gyres. Poleward-moving currents are warm, and equatorward-moving currents are cold. Ocean currents play an important role in redistributing heat around the globe. Note that cities mentioned in the text discussion are shown on this map. In the smaller inset map, broad arrows show the idealized surface circulation for the Atlantic, and the thin arrows show prevailing winds. Winds provide the energy that drives the ocean's surface circulation.

lies 12° latitude farther south. The January mean at London (51° north latitude) is 4.5°C (8.1°F) higher than at New York City.

**Cold Currents Chill the Air** In contrast to warm ocean currents like the Gulf Stream, the effects of which are felt most during the winter, cold currents exert their greatest influence in the tropics and during the summer months in the middle latitudes. For example, the cool Benguela Current off the western coast of southern Africa moderates the tropical heat along this coast. Walvis Bay (23° south latitude), a town adjacent to the Benguela Current, is 5°C (9°F) cooler in summer than Durban, which is 6° latitude farther poleward but on the eastern side of South Africa, away from the influence of the cold current. Closer to home, the cold California Current lowers summer temperatures in subtropical coastal southern California by 6°C (10.8°F) or more compared to east coast stations.

**Cold Currents Increase Aridity** In addition to influencing temperatures of adjacent land areas, cold currents have other climatic influences. For example, where tropical deserts exist along the west coasts of continents, cold ocean currents have a dramatic impact. The principal west coast deserts are the Atacama in Peru and Chile and the Namib in southwestern Africa (**Fig. 7.22**). The aridity along these coasts is intensified because the lower atmosphere is chilled by cold offshore waters. When this occurs, the air becomes very stable and resists the upward movement necessary to create precipitation-producing clouds. In addition, the presence of cold currents causes temperatures to approach and often reach the dew point, the temperature at which water vapor condenses. As a result, these areas are characterized by high relative humidity and fog. Thus, not all subtropical deserts are hot with low humidity and clear skies. Rather, the presence of cold currents transforms some subtropical deserts into relatively cool, damp places that are often shrouded in fog.

► **Figure 7.22 Chile's Atacama Desert** This is the driest desert on Earth. Average rainfall at the wettest locations is not more than 3 millimeters (0.12 inch) per year. Stretching nearly 1000 kilometers (600 miles), the Atacama is situated between the Pacific Ocean and the towering Andes Mountains. The cold Peru Current makes this slender zone cooler and drier than it would otherwise be.

## eye ON THE atmosphere 7.2

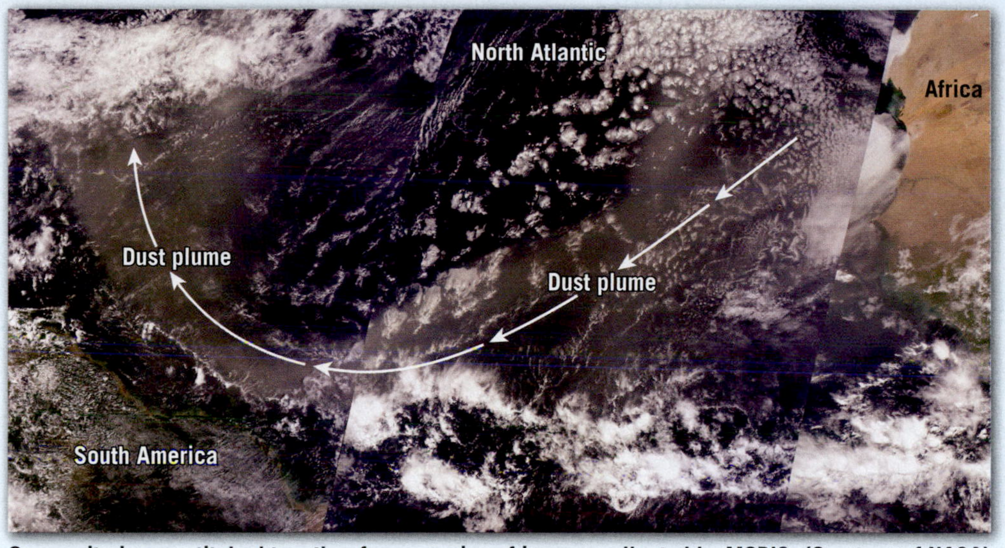


**Composite image stitched together from a series of images collected by MODIS. (Courtesy of NASA)**

This image shows a January dust storm in Africa that produced a dust plume that reached all the way to the northeast coast of South America. Plumes such as this transport an estimated 40 million tons of dust from the Sahara Desert to the Amazon basin each year. The minerals carried by these dust plumes help to replenish nutrients that are continually being washed out of rain forest soils by heavy tropical rains.

**Questions**

1. Notice that the dust plume follows a curved path. Does the atmospheric circulation carrying this dust plume exhibit a clockwise or counterclockwise rotation?
2. What global pressure system was responsible for transporting this dust from Africa to South America?

## Ocean Currents and Upwelling

**Upwelling**, the rising of cold water from deeper layers to replace warmer surface water, is a common wind-induced vertical movement. It is most characteristic along the eastern shores of the global oceans, most notably along the coasts of California, Peru, and West Africa.

Upwelling occurs in areas where winds blow parallel to the coast toward the equator. Because of the Coriolis force, the surface water is directed (deflected) away from the shore. As the surface layer moves away from the coast, it is replaced by nutrient-filled water that "upwells" from below the surface. This slow upward flow from depths of 50 to 300 meters (165 to 1000 feet) brings water that is cooler than the water it replaces and creates a characteristic coastal zone of relatively cool water favorable to plant growth and marine life.

Swimmers accustomed to the waters along the mid-Atlantic states of the United States might find a dip in the Pacific on central California's coast a chilling surprise. In August, when water temperatures along the Atlantic shore usually exceed 21°C (70°F), the surf along the coast of California is only about 15°C (60°F).

### ✔ Concept Checks 7.8

1. What is the primary driving force of surface-ocean currents?
2. How does the Coriolis force influence ocean currents?
3. Name the five subtropical gyres.
4. How do ocean currents influence climate? Provide at least three examples.
5. Describe the process of upwelling. Why is an abundance of marine life associated with these areas?

**Animation** (MM)
Ekman Spiral
Coastal Upwelling/
Downwelling.
http://goo.gl/aJMMST

## 7.9 El Niño, La Niña, and the Southern Oscillation

**Describe the Southern Oscillation and its relationship to El Niño and La Niña.**
**List the climate impacts of El Niño and La Niña on North America.**

**El Niño** was first recognized by fishermen from Ecuador and Peru, who noted a gradual warming of waters in the eastern Pacific in December or January. Because the warming usually occurred near the Christmas season, the event was named El Niño—"little boy," or "Christ child," in Spanish. These periods of abnormal sea-surface warming occur at irregular intervals of 2 to 7 years and usually persist for spans of 9 months to 2 years.

El Niño is noted for its potentially catastrophic impact on the weather and economies of Peru, Chile, and Australia, among other countries. As shown in **Figure 7.23A**, during an El Niño, strong equatorial countercurrents amass large quantities of *warmer-than-normal* water along the west coast of South America. The unusually warm water and associated low pressure cause some normally arid areas of Peru and Chile to receive above-average rainfall. The result can be major

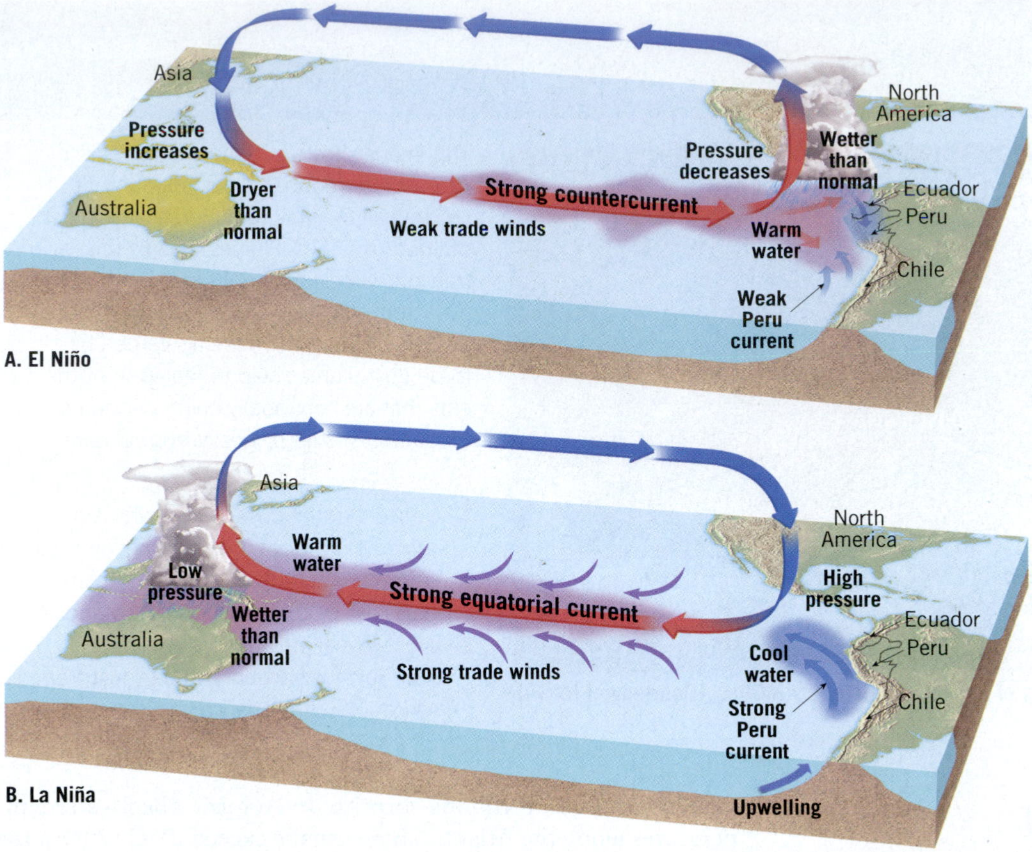


◄ **Figure 7.23 The relationship between El Niño, La Niña, and the Southern Oscillation** **A.** During an El Niño event the pressure over the eastern Pacific drops, while the pressure over the western Pacific rises. This causes the trade winds to diminish, leading to an eastward movement of warm water along the equator. This strengthens the equatorial countercurrent, causing surface waters of the central and eastern Pacific to warm, with far-reaching consequences for weather patterns. **B.** During a La Niña event, strong trade winds drive the equatorial currents toward the west. At the same time, the strong Peru Current causes upwelling of cold water along the west coast of South America.

**Video** MM
El Niño
http://goo.gl/lzs5E

flooding in the affected areas. In addition, the warm surface water blocks the upwelling of colder, nutrient-filled water—the primary food source for millions of small feeder fish (mainly anchovies)—that typically rises from below.

Within a year or two, the circulation associated with El Niño is usually replaced by a La Niña event (**Fig. 7.23B**). **La Niña,** which means "little girl," is essentially the opposite of El Niño and refers to *colder-than-normal sea-surface temperatures* in the central and eastern Pacific (**Fig. 7.24**). As Figure 7.23B illustrates, the atmospheric circulation in the equatorial Pacific during La Niña is dominated by strong trade winds. These wind systems, in turn, generate a strong equatorial current that flows westward from South America toward Australia and Indonesia. This circulation pattern can result in flooding in northeastern Australia and Indonesia, whereas dry conditions occur along the western coastal areas of South America.

In addition, during La Niña the cold ocean current that flows equatorward along the coast of Chile and Peru intensifies. This surface current, called the Peru Current, encourages upwelling. Thus, La Niña has become known as the "gift giver" because the nutrients carried upward are a boon to marine life, making fishing particularly good during these periods of strong upwelling.

## Global Impact of El Niño

The climatic fluctuations associated with El Niño and La Niña have been known for years but were considered a local phenomenon. Scientists now recognize that El Niño and La Niña are part of the global atmospheric circulation pattern that affects the weather at great distances from the equatorial regions of the Pacific. Although the effects of these circulation patterns are somewhat variable, some locales appear to be affected more consistently.

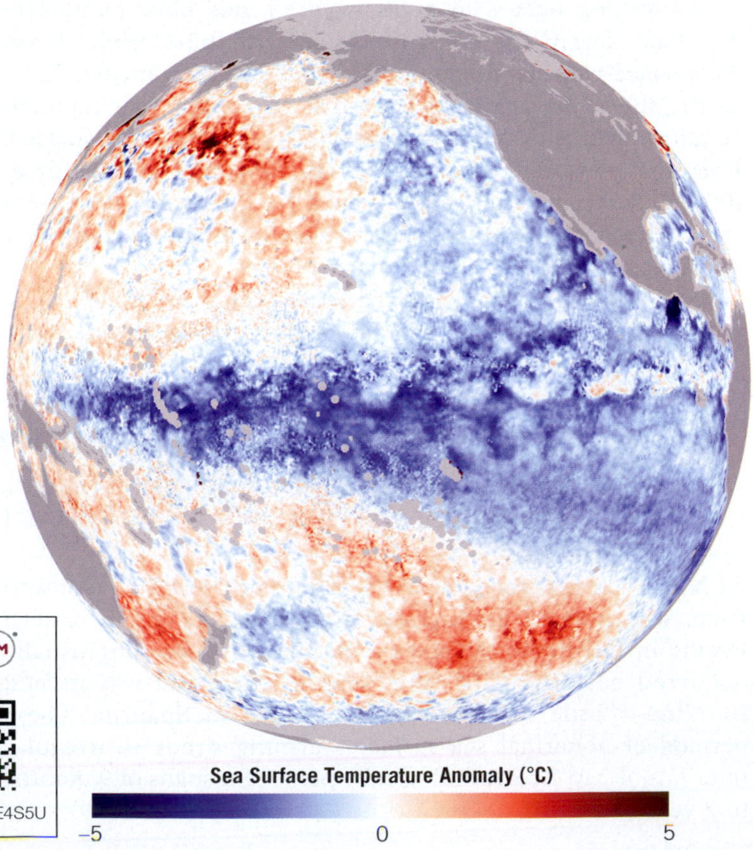


► **Figure 7.24 Sea-surface temperature anomaly associated with La Niña** This image, acquired December 15, 2010, shows that the sea-surface temperatures were much colder than average across the central and eastern equatorial Pacific Ocean. The 2010–2011 La Niña was one of the strongest in the past half-century and was accompanied by heavy flooding in Queensland, Australia, shown in Figure 7.26.

**Video** MM
La Niña
http://goo.gl/E4S5U

During an El Niño event, winter temperatures are warmer than normal in the north-central United States and parts of Canada (**Fig. 7.25A**). In addition, significantly wetter winters are experienced in the southwestern United States and northwestern Mexico, while the southeastern United States experiences wetter and cooler conditions. One major benefit of El Niño is a lower-than-average number of Atlantic hurricanes. El Niño is credited with suppressing hurricanes during the 2009 hurricane season, the least active in 12 years.

One of the most severe El Niño events on record occurred in 1997–1998 and was responsible for a variety of weather extremes in many parts of the world. During that El Niño episode, ferocious winter storms struck the California coast, causing unprecedented beach erosion, landslides, and floods. In the southern United States, heavy rains also brought floods to Texas and the Gulf states.

As this chapter was being prepared in July 2014, an El Niño event was beginning to develop. It is too early to predict whether this event will materialize.

## Global Impact of La Niña

La Niña was once thought to be the normal conditions that occur between two El Niño events, but meteorologists now consider La Niña an important atmospheric phenomenon in its own right. Researchers have come to recognize that colder-than-average surface temperatures in the central and eastern Pacific can trigger a La Niña event that exhibits a distinctive set of weather patterns.

Typical La Niña winter weather includes cooler and wetter conditions over the northwestern United States and especially cold winter temperatures in the Northern Plains states, while unusually warm conditions occur in the Southwest and Southeast (**Fig. 7.25C**). In the western Pacific, La Niña events are associated with wetter-than-normal

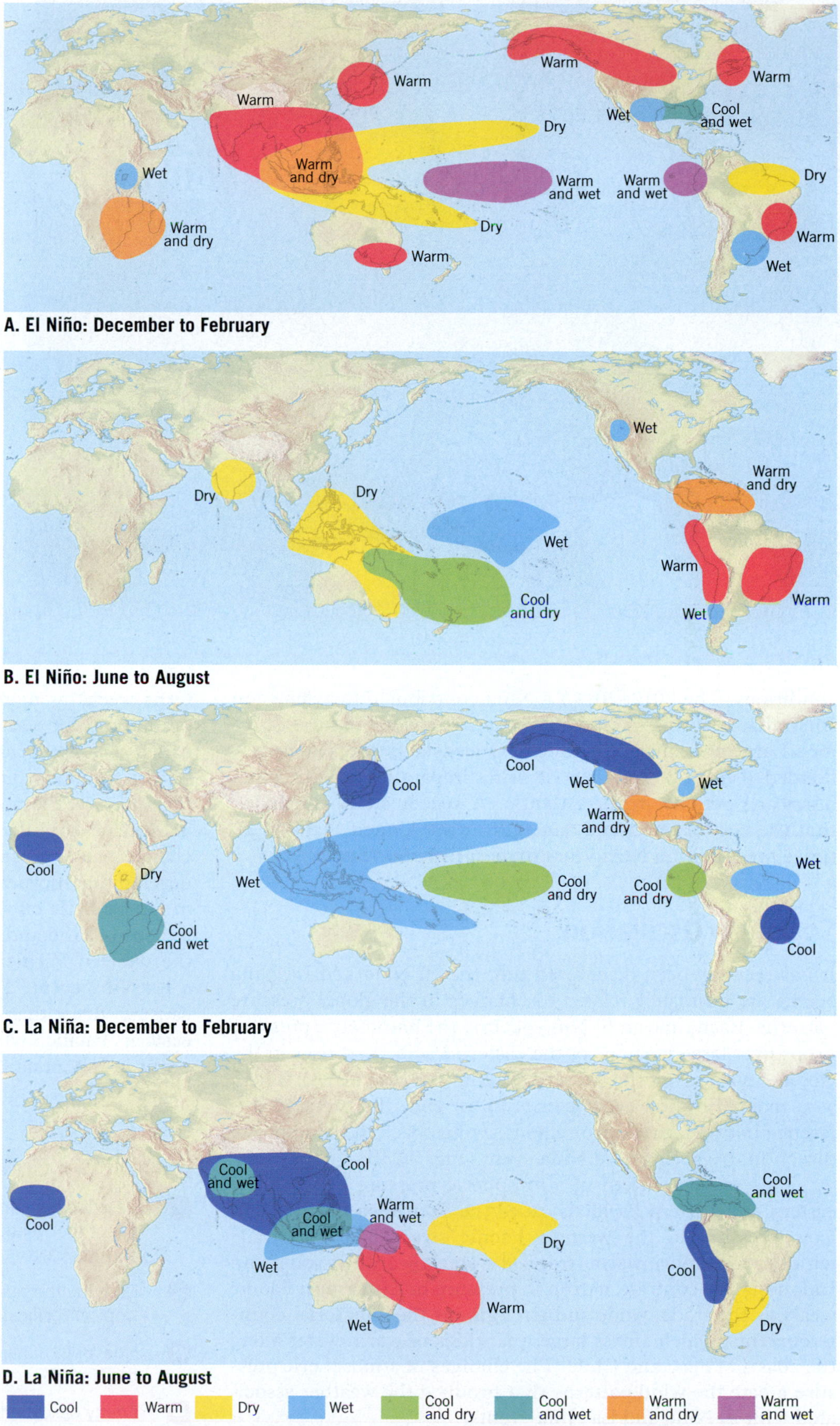

A. El Niño: December to February

B. El Niño: June to August

C. La Niña: December to February

D. La Niña: June to August

► **Figure 7.25 Climatic impacts of El Niño and La Niña in various locations** El Niño has the most significant impact on the climate of North America during the winter. In addition, El Niño affects the areas around the tropical Pacific in both winter and summer. La Niña has its most significant impact on North America in the winter but affects other areas during all seasons.

Animation MM
El Niño and La Niña
http://goo.gl/pm1Wb6

◀ **Figure 7.26 Queensland, Australia, floods, January 2011** The Queensland floods of 2010–2011 have been attributed to one of the strongest La Niña events as far back as records have been kept. Unusually warm sea-surface temperatures around Australia contributed to the heavy rains. In other parts of Australia, this strong La Niña event brought relief from a decade-long drought.

conditions. The 2010–2011 La Niña contributed to a deluge in Australia, resulting in one of the country's worst natural disasters: Large portions of the state of Queensland were extensively flooded (**Fig. 7.26**). Another La Niña impact is more frequent hurricane activity in the Atlantic. A recent study concluded that the cost of hurricane damages in the United States is 20 times greater in La Niña years than in El Niño years.

## Southern Oscillation

It has recently been discovered that the El Niño and La Niña events are intimately related to changes in the global pressure patterns. Each time an El Niño occurs, the barometric pressure drops over large portions of the eastern Pacific and rises in the tropical portions of the western Pacific (see Fig. 7.23A). Then, as a major El Niño event comes to an end, the pressure difference between these two regions swings back in the opposite direction, triggering a La Niña event (Fig. 7.23B).

This seesaw pattern of atmospheric pressure between the eastern and western Pacific is called the **Southern Oscillation**. Low pressure in the western Pacific strengthens the trade winds, which move warm, tropical water toward Australia and Indonesia. By contrast, a drop in pressure in the eastern Pacific weakens the trade winds and strengthens the equatorial countercurrents, which amass large quantities of warm water along the coasts of Peru and Chile. The changes in atmospheric pressure create the wind patterns that produce the weather associated with El Niño and La Niña events.

The link between the weather occurring in widely separated regions of the globe is called a **teleconnection**. The sea-surface temperature anomaly associated with an El Niño event, where warming of the eastern Pacific is associated with winter flooding in southern California, is an example. In addition, there appears to be a teleconnection between a strong El Niño and a weak Asian monsoon.

Because sea-surface temperatures are often predictable months in advance, understanding teleconnection patterns allows meteorologists to make climatic predictions for distant locations. For instance, predicting the start of an El Niño enables prediction of North American precipitation and temperature patterns weeks or months in advance. National Weather Service predictions for periods from 1 to 13 months into the future are called *climate outlooks*.

In summary, an El Niño event begins with a decrease in pressure over the coastal areas of Peru and Chile and an increase in surface pressure over the western Pacific. This pressure change causes the trade winds to weaken and countercurrents to develop that move warm water eastward. The resulting changes in atmospheric circulation increase rainfall in the normally dry regions of Peru and Chile, while a drought commonly occurs in the western Pacific. The opposite circulation develops during a La Niña event: The trade winds strengthen, causing dryer-than-normal conditions in the eastern Pacific, while extreme flooding may occur in Indonesia and northeastern Australia.

**Video** MM
Floods and Droughts
http://goo.gl/ECRmJ9

### ✔ Concept Checks 7.9

1. Describe how a major El Niño event tends to affect the weather in Peru and Chile as compared to Indonesia and Australia.
2. Describe the sea-surface temperatures on both sides of the tropical Pacific during a La Niña event.
3. How does a major La Niña event influence the hurricane season in the North Atlantic?
4. Briefly describe the Southern Oscillation and how it is related to El Niño and La Niña.
5. Describe how an El Niño event might affect the climate in North America during the winter. Describe the same for a La Niña event.

# 7.10 Global Distribution of Precipitation

**Discuss the major factors that influence the global distribution of precipitation.**

Figure 7.27 shows the distribution pattern for average annual precipitation around the globe. Although this map may appear complicated, the general features of the pattern can be explained using knowledge of global winds and pressure systems. In general, regions influenced by high pressure, with its associated subsidence and divergent winds, experience dry conditions. Conversely, regions under the influence of low pressure, with its converging winds and ascending air, receive ample precipitation. However, if the wind-pressure regimes were the only control of precipitation, the pattern shown in Figure 7.27 would be much simpler.

Air temperature is also important in determining precipitation potential. Because cold air has a lower capacity for moisture than does warm air, we would expect a latitudinal variation in precipitation, with low latitudes (warm regions) receiving the greatest amounts of precipitation and high latitudes (cold regions) receiving the least.

In addition to latitudinal variations in precipitation, the distribution of land and water complicates the precipitation pattern. Large landmasses in the middle latitudes commonly experience decreased precipitation toward their interiors. For example, North Platte, Nebraska, receives less than half the precipitation that falls on the coastal community of Bridgeport, Connecticut, despite being located at the same latitude. Furthermore, mountain barriers alter precipitation patterns. Windward mountain slopes receive abundant precipitation, whereas leeward slopes and adjacent lowlands are usually deficient in moisture.

## Zonal Distribution of Precipitation

We will first examine the zonal distribution of precipitation that we would expect on a uniform Earth covered entirely in water, and then we will add the variations caused by the distribution of land and water. Recall from our earlier discussion that on a uniform Earth, four major pressure zones emerge in each hemisphere (see Fig. 7.8A): the equatorial low (ITCZ), the subtropical high, the subpolar low, and the polar high. Also recall that these pressure belts show a marked seasonal shift toward the hemisphere that is experiencing summer.

The idealized precipitation regimes expected from these pressure systems are shown in Figure 7.28. Near the equator the trade winds converge (ITCZ), resulting in heavy precipitation in all seasons. Poleward of the equatorial low in each hemisphere lie the belts of subtropical high pressure. In these regions, subsidence contributes to dry conditions throughout the year. Between the wet equatorial regime and the dry subtropical regime lies a zone that is influenced by both high- and low-pressure systems.

**Animation** MM
Seasonal Pressure and Precipitation Patterns
http://goo.gl/GGgL4q

▼ **Figure 7.27 Global distribution of average annual precipitation**

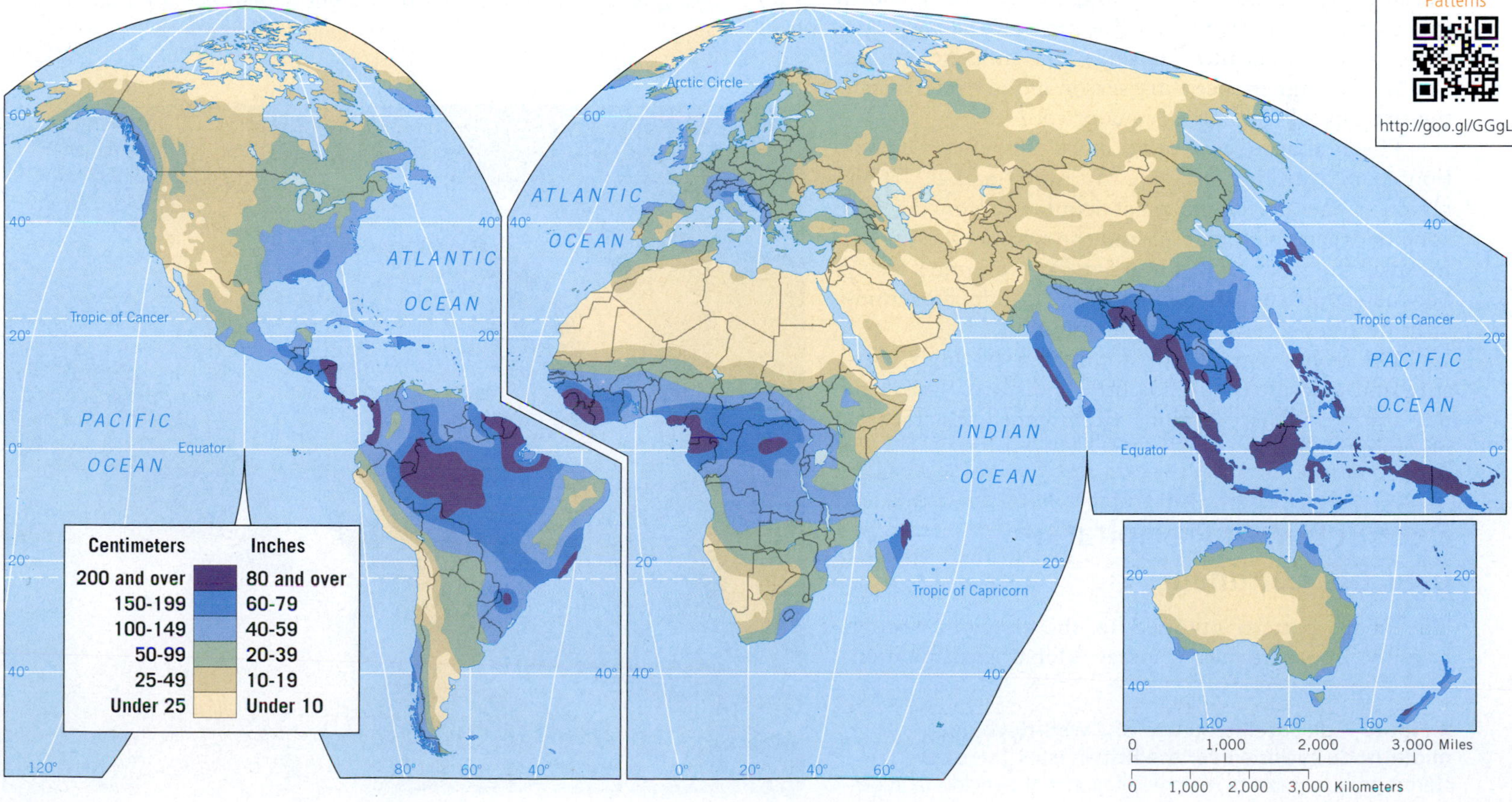

► **Figure 7.28 Zonal precipitation patterns**

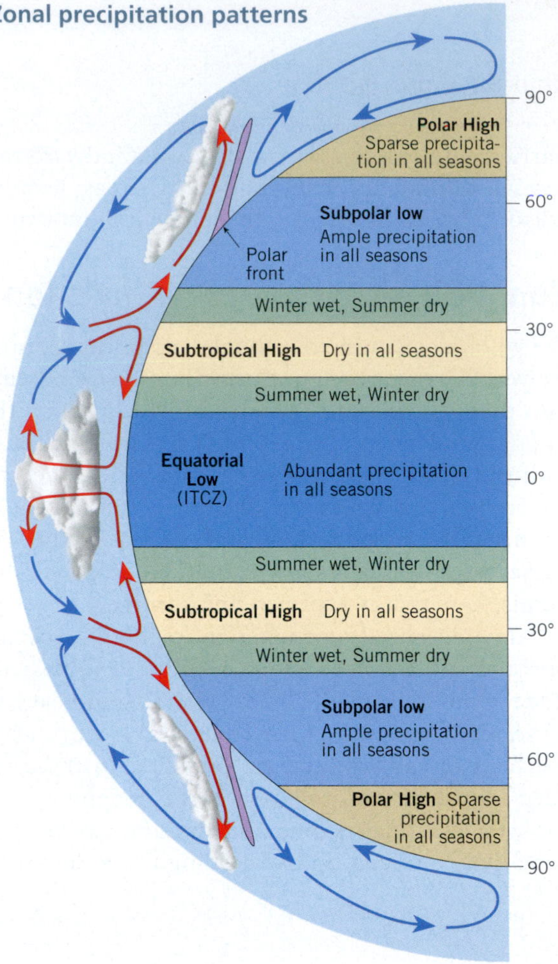


Because the pressure systems migrate seasonally with the Sun, these transitional regions receive most of their precipitation in the summer, when they are under the influence of the ITCZ. They experience a dry season in the winter, when the subtropical high moves equatorward.

The midlatitudes receive most of their precipitation from traveling cyclonic storms (**Fig. 7.29**). This region is the site of the polar front, the convergence zone between cold polar air and the warmer westerlies. Because the position of the polar front migrates freely between approximately 30° and 70° latitude, most midlatitude areas receive ample precipitation.

The polar regions are dominated by high pressure and cold air that holds little moisture. Throughout the year, these regions experience only meager precipitation.

## Distribution of Precipitation over the Continents

The zonal pattern outlined in the previous section roughly approximates general global precipitation. Abundant precipitation occurs in the equatorial and midlatitude regions, whereas substantial portions of the subtropical and polar realms are relatively dry.

Numerous exceptions to this idealized zonal pattern are obvious in Figure 7.27. For example, several arid areas are found in the midlatitudes. The desert region of southern South America, known as Patagonia, is one example. Midlatitude deserts such as Patagonia exist mostly on the leeward (rain shadow) side of a mountain barrier or in the interior of a continent, cut off from a source of moisture.

The most notable anomaly in the zonal distribution of precipitation occurs in the subtropics. Here we find not only many of the world's great deserts but also regions of abundant rainfall (Fig. 7.27). This pattern results because the subtropical high-pressure centers that dominate the circulation in these latitudes have different characteristics on their eastern and western sides (**Fig. 7.30**). Subsidence is most pronounced on the eastern side, which results in stable atmospheric conditions. Because these anticyclones tend to crowd the eastern side of an ocean, particularly in the winter, the western portions of the continents adjacent to subtropical highs are arid (Fig. 7.30). Centered at approximately 25° north or south latitude on the west side of their respective continents, we find the Sahara Desert of North Africa, the Namib of southwest Africa, the Atacama of South America, the deserts of northwestern Mexico, and Australia.

On the western flanks of these highs, however, subsidence is less pronounced. In addition, the surface air that flows out of these highs often traverses large expanses of warm water. As a result, this air acquires moisture through evaporation that acts to enhance its instability. Consequently, landmasses located west of a subtropical high generally receive ample precipitation throughout the year. Southern Florida is a good example (see Fig. 7.27).

► **Figure 7.29 Satellite image of a well-developed midlatitude cyclone over the British Isles** These traveling storms produce most of the precipitation in the middle latitudes.

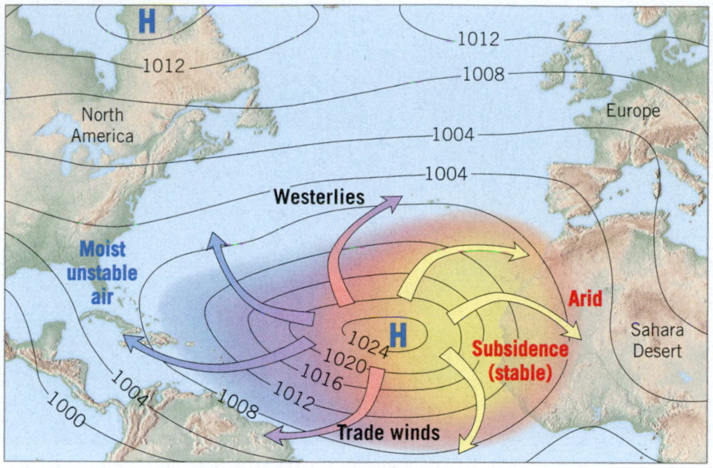


▲ **Figure 7.30 Characteristics of subtropical high-pressure systems** Subsidence on the east side of these systems produces stable conditions and aridity. Surface air that flows out of the western flanks of these highs and traverse large expanses of warm water acquire moisture and may become unstable.

## ✔ Concept Checks 7.10

1. Are regions that are dry throughout the year dominated by high or low pressure?
2. Describe the precipitation pattern for locations near the equator and near the poles.
3. Earth's major subtropical deserts are located between about 20° and 35° latitude.
   a. Name five subtropical deserts.
   b. On what side (western or eastern) of the continent are subtropical deserts located?
   c. What is the general name of the pressure systems responsible for subtropical deserts?
4. List two reasons that explain why polar regions experience meager precipitation.
5. What factors, in addition to global wind and pressure systems, influence the global distribution of precipitation?

## eye ON THE atmosphere 7.3

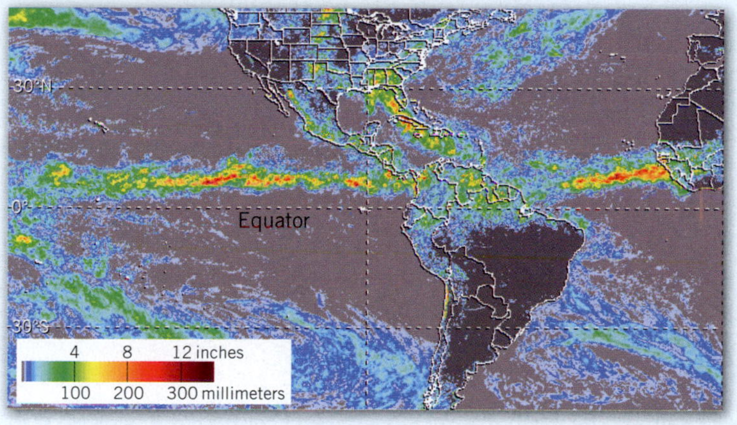


This satellite image was produced with data from the *Tropical Rainfall Measuring Mission* (*TRMM*). Notice the band of heavy rainfall shown in reds and yellows that extends east–west across the image.

**Questions**

1. With which pressure zone is this band of rainy weather associated?
2. Is it more likely that this image was acquired in July or January? Explain.

# 7 Concepts in Review Circulation of the Atmosphere

## 7.1 Scales of Atmospheric Motion

► Distinguish between microscale, mesoscale, and macroscale winds and give an example of each.

**Key Terms:** microscale wind, mesoscale wind, macroscale wind, planetary-scale wind, synoptic-scale wind

- The smallest scale of air motion is the microscale, which includes gusts and dust devils that normally last for seconds or at most minutes.
- Mesoscale winds, such as thunderstorms, tornadoes, and land-sea breezes, are usually less than 100 kilometers (60miles) across and often exhibit intense vertical flow.

- The largest wind patterns, called macroscale winds, are divided into two categories. Planetary-scale winds are exemplified by the westerlies and trade winds. The somewhat smaller macroscale circulation, called synoptic scale, are about 1000 kilometers (600 miles) in diameter and are easily identified on weather maps. Well-known synoptic-scale wind systems are the traveling midlatitude cyclones and anticyclones that appear on weather maps as areas of low and high pressure, respectively, and somewhat smaller tropical storms and hurricanes that form over the tropical oceans.

**Q** This satellite image shows a large rotating wind system, shown with red arrows, moving toward North America. What scale of atmospheric motion do these rotating clouds exhibit?

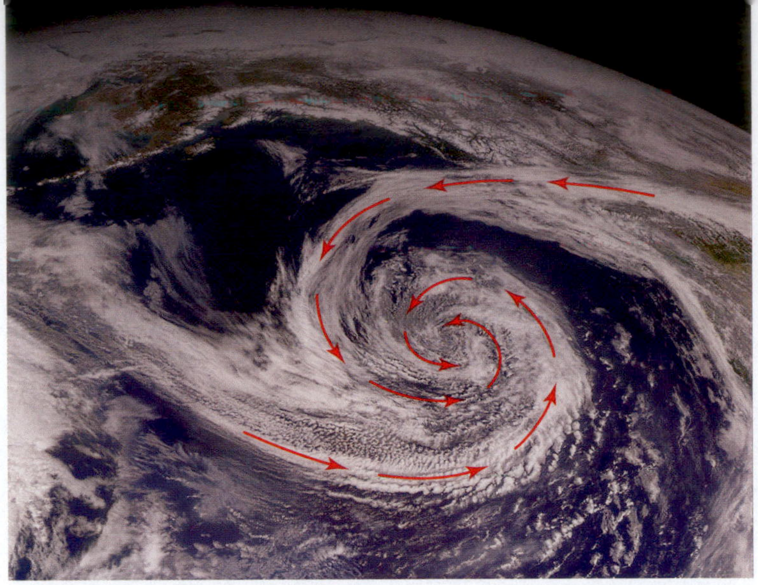

## 7.2 Local Winds ► List four types of local winds and describe their formation.

**Key Terms:** sea breeze, land breeze, valley breeze, mountain breeze, chinook, foehn, Santa Ana, katabatic (fall) wind, mistral, bora, country breeze

- Most winds are caused by pressure differences that arise because of temperature differences due to unequal heating of Earth's surface. Most local winds are linked to temperature and pressure differences that result from variations in topography or in local surface conditions.
- Sea and land breezes form along coasts and are brought about by temperature contrasts between land and water. Valley and mountain breezes occur in mountainous areas where the air along slopes heats differently than does the air at the same elevation over the valley floor. Chinook and Santa Ana winds are warm, dry winds created when air descends the leeward side of a mountain and warms by compression.

**Q** It is late afternoon on a warm summer day at a beach. Until the last hour or two, winds were calm. Then a breeze began to develop. Is it more likely a cool breeze from the water or a warm breeze from the adjacent land area? Explain.

## 7.3 Global Circulation ► Describe or sketch the three-cell model of global circulation.

**Key Terms:** Hadley cell, horse latitudes, trade winds, doldrums, Ferrel cell, prevailing westerlies, polar cell, polar easterlies, polar front

- The three-cell circulation model for each hemisphere provides a simplified view of global circulation. According to this model, atmospheric circulation cells are located between the equator and 30° latitude, 30° and 60° latitude, and 60° latitude and the poles. The areas of general subsidence in the zone between 20° and 35° are called the horse latitudes. In each hemisphere, the equatorward flow from the horse latitudes forms the reliable trade winds.
- The circulation between 30° and 60° latitude (north and south) results in the prevailing westerlies.
- Air that moves equatorward from the poles produces the polar easterlies in both hemispheres.

## 7.4 Pressure Zones Drive the Wind ► Summarize Earth's idealized zonal pressure belts. Describe how continents and seasonal temperature changes complicate the idealized pattern.

**Key Terms:** equatorial low, intertropical convergence zone (ITCZ), subtropical high, subpolar low, polar high, Siberian high, Bermuda/Azores high, Aleutian low, Icelandic low

- If Earth's surface were uniform, four latitudinally oriented belts of pressure would exist in each hemisphere—two high and two low. Beginning at the equator, the four belts would be the (1) equatorial low, also referred to as the intertropical convergence zone (ITCZ), (2) subtropical high, at about 20° to 35° on either side of the equator, (3) subpolar low, situated at about 50° to 60° latitude, and (4) polar high, near Earth's poles.
- Because Earth's surface is not uniform, the only true zonal distribution of pressure exists along the subpolar low in the Southern Hemisphere, where the ocean is continuous. At other latitudes, particularly in the Northern Hemisphere, with its higher proportion of land compared to ocean, the zonal pattern is replaced by semipermanent cells of high and low pressure.

## 7.5 Monsoons ► Describe the seasonal changes in Earth's global circulation that produce the Asian monsoon.

**Key Term:** monsoon

- The greatest seasonal change in Earth's global circulation is the development of monsoons, wind systems that exhibit a pronounced seasonal reversal in direction. A commonly known and pronounced monsoonal circulation is the Asian monsoon.
- The North American monsoon, a relatively small seasonal wind shift, produces a dry spring followed by a comparatively rainy summer that impacts large areas of the southwestern United States and northwestern Mexico.

## 7.6 The Westerlies

► Explain why the airflow aloft in the middle latitudes has a strong west-to-east component.

**Key Term:** Rossby wave

- The airflow aloft in the middle latitudes is predominately from west to east, called the westerlies. The westerlies are driven by the pressure gradient aloft caused by the temperature contrast between the poles and equator.
- The westerlies follow wavy paths that have long wavelengths, called Rossby waves, which usually consist of four to six meanders that encircle the globe.

## 7.7 Jet Streams

► Explain the origin of the polar jet stream and its relationship to midlatitude cyclonic storms.

**Key Terms:** jet stream, polar jet stream, subtropical jet stream

- The temperature contrast between the poles and the equator drives the westerlies in the middle latitudes. Embedded within the flow aloft are narrow ribbons of high-speed winds, called jet streams, that meander for thousands of kilometers. Polar jet streams result from large temperature contrasts at Earth's surface.

## 7.8 Global Winds and Ocean Currents

► Sketch and label the major ocean currents on a world map.

**Key Terms:** gyre, upwelling

- The global wind system is the primary driving force of ocean currents. These nearly circular ocean currents, called gyres, are centered over the five major subtropical high-pressure systems, located in both the southern and northern Atlantic and Pacific Oceans and the Indian Ocean.
- Poleward-moving, warm ocean currents have a moderating effect on high-latitude locations in the winter months. By contrast, equatorward-moving cold ocean currents exert their greatest influence in the tropics and during the summer months in the middle latitudes.
- Cold ocean currents impact the aridity of west coast deserts, mainly the Atacama in Peru and Chile and the Namib in southwestern Africa. The aridity along these coasts is intensified because the lower atmosphere is chilled by cold offshore waters, causing the air to become stable. This unusually stable air resists the upward movement necessary to create precipitation-producing clouds.

**Q** Which of the globes shown here most closely illustrates the circulation you would expect for ocean currents on an Earth-like planet with hypothetical continents?

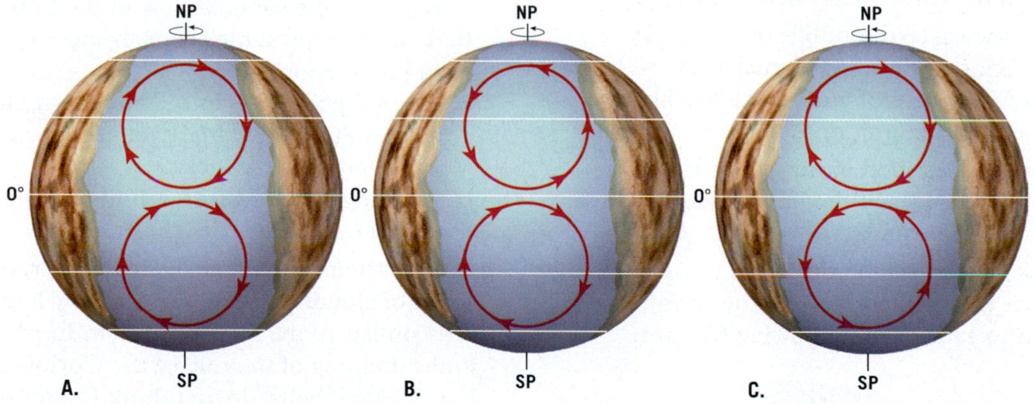


## 7.9 El Niño, La Niña, and the Southern Oscillation

► Describe the Southern Oscillation and its relationship to El Niño and La Niña. List the climate impacts of El Niño and La Niña on North America.

**Key Terms:** El Niño, La Niña, Southern Oscillation, teleconnection

- El Niño refers to episodes of ocean warming in the eastern Pacific along the coasts of Ecuador and Peru. It is associated with weak trade winds, a strong eastward-moving equatorial countercurrent, a weakened Peru Current, and diminished upwelling along the western margin of South America.
- A La Niña event is associated with colder-than-average surface temperatures in the eastern Pacific. La Niña is linked to strong trade winds, a strong westward-moving equatorial current, and a strong Peru Current with significant coastal upwelling.
- El Niño and La Niña events are part of the global circulation and are related to a seesaw pattern of atmospheric pressure between the eastern and western Pacific called the Southern Oscillation. El Niño and La Niña events influence weather on both sides of the tropical Pacific Ocean as well as weather in the United States.

**Q** This image shows floods occurring in eastern Australia. Is this region more likely being influenced by El Niño or La Niña?

## 7.10 Global Distribution of Precipitation

▶ Discuss the major factors that influence the global distribution of precipitation.

- The general features of the global distribution of precipitation can be explained by global winds and pressure systems. In general, regions influenced by high pressure, with its associated subsidence and divergent surface winds, experience dry conditions. Regions under the influence of low pressure and its converging surface winds and ascending air receive ample precipitation.
- On a uniform Earth throughout most of the year, heavy precipitation would occur in the equatorial region, the midlatitudes would receive most of their precipitation from traveling cyclonic storms, and polar regions would be dominated by cold air that holds little moisture.
- Air temperature, the distribution of continents and oceans, and the location of mountains also influence the global distribution of precipitation.

**Q** Examine this classic image of Africa from space and pick out the region dominated by the equatorial low and the areas influenced by the subtropical highs in each hemisphere. What clue(s) did you use?

## Give it Some Thought

1. It is a warm summer day, and you are shopping in downtown Chicago, just a few blocks from Lake Michigan. All morning the winds have been calm, suggesting that no major weather systems are nearby. By midafternoon, should you expect a cool breeze from Lake Michigan or a warm breeze originating from the rural areas outside the city?
2. Boulder, Colorado, in the eastern foothills of the Rocky Mountains, is experiencing a warm, dry January day with strong westerly winds. What type of local winds are likely responsible for these weather conditions?
3. Which of the local winds described in this chapter is not heavily dependent on differences in the rates at which various ground surfaces are heated?
4. The accompanying sketch shows the three-cell circulation model in the Northern Hemisphere. Match the appropriate number on the sketch to each of the following features:

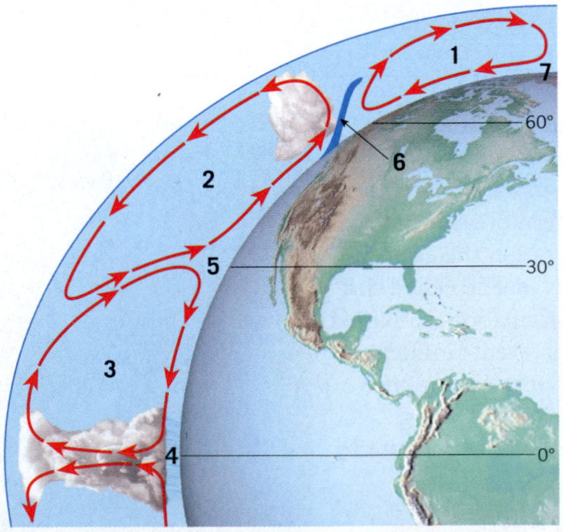


   **a.** Hadley cell
   **b.** Equatorial low
   **c.** Polar front
   **d.** Ferrel cell
   **e.** Subtropical high
   **f.** Polar cell
   **g.** Polar high

5. If Earth did not rotate on its axis and if its surface were completely covered with water, what direction would a boat drift if it started its journey in the middle latitudes of the Northern Hemisphere? (*Hint:* What would the global circulation pattern be like for a nonrotating Earth?)
6. Briefly explain each of the following statements that relate to the global distribution of surface pressure:
   **a.** The only true zonal distribution of pressure exists in the region of the subpolar low in the Southern Hemisphere.
   **b.** The subtropical highs are stronger in the North Atlantic in July than in January.
   **c.** The subpolar lows in the Northern Hemisphere are the result of individual cyclonic storms that are more common in the winter.
   **d.** A strong high-pressure cell develops in the winter over northern Asia.
7. On the accompanying image of Jupiter, notice the many zones of clouds that are produced by large convective cells similar to the Hadley cells on Earth. Given your understanding of the role of the Coriolis force in producing Earth's wind belts, do you think Jupiter rotates faster or slower on its axis than Earth? Explain.

**Jupiter** (NASA)

8. Refer to Figure 7.27, page 209 and Figure 7.10, page 196, to determine what aspect of global circulation (polar highs, equatorial low, etc.) is responsible for each of the following:
   **a.** Dry conditions over North Africa.
   **b.** The wet summer monsoon over Southeast Asia.
   **c.** Dry conditions over west-central Australia.
   **d.** Wet conditions over northeastern South America.
9. Explain why the west coasts of continents generally experience cold ocean currents. How do these cold currents contribute to desert conditions in some coastal areas like the Atacama region of Peru?
10. The accompanying map shows sea-surface temperature anomalies (difference from normal) over the equatorial Pacific Ocean. Based on this map, answer the following questions:

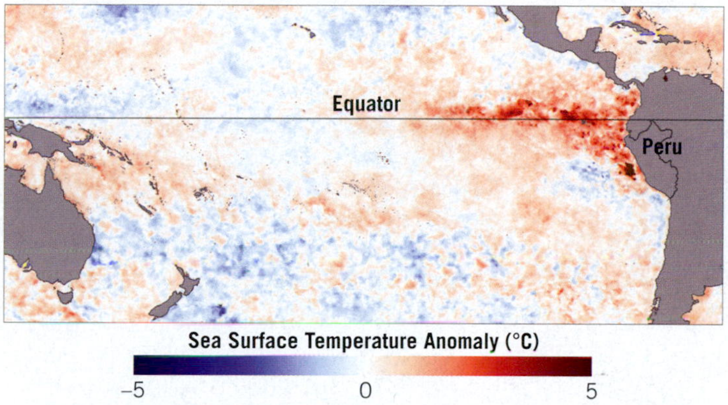


   **a.** In what phase was the Southern Oscillation (El Niño or La Niña) when this image was made?
   **b.** Would the trade winds be strong or weak at this time?
   **c.** If you lived in Australia during this event, what weather conditions would you expect?
   **d.** If you were attending college in the southeastern United States during winter months, what type of weather conditions would you expect? (*Hint:* See Fig. 7.25, page 207.)
11. Refer to Figure 7.27 on page 209 to determine the number of inches of precipitation received annually in the area in which you reside.
12. The accompanying maps of Africa show the distribution of precipitation for July and January. Which map represents July, and which represents January? How did you determine your answer?

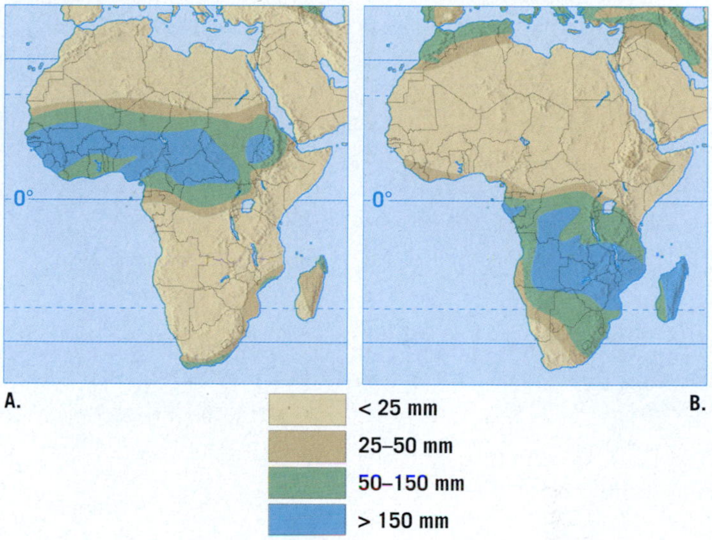


## MasteringMeteorology™

Looking for additional review and test prep materials? Visit the Study Area in *MasteringMeteorology*™ to enhance your understanding of this chapter's content by accessing a variety of resources, including **MapMaster**™ interactive maps, Geoscience Animations, GEODe, *In the News* RSS feeds, flashcards, web links, self-study quizzes, and an eText version of *The Atmosphere*.

# 8 Air Masses

## Focus on Concepts

*Each statement represents the primary learning objective for the corresponding major heading within the chapter. After you complete the chapter, you should be able to:*

**8.1** Define *air mass* and *air-mass weather.*

**8.2** List the basic criteria for an air-mass source region and identify the source regions that influence North America.

**8.3** Describe the processes by which traveling air masses are modified and discuss two examples.

**8.4** Summarize the weather conditions associated with each of the air masses that influence North America during the summer and the winter.

Most people living in the middle latitudes have experienced hot, "sticky" summer heat waves and frigid winter cold spells. In the case of a heat wave, several days of sultry weather may come to an abrupt end that is marked by thundershowers, followed by a few days of relatively cool relief. In the case of a cold spell, thick stratus clouds and snow may replace the clear skies that had prevailed, and temperatures may climb to values that seem mild compared with what preceded them. Both examples feature a period of generally uniform weather conditions, followed by a relatively short period of change and the subsequent reestablishment of a new set of weather conditions that remain for perhaps several days before changing again.

*This satellite view of Lake Superior and Lake Michigan illustrates what happens when a frigid, cloudless air mass from Canada moves across the lakes. Notice the narrow white bands of snow along the downwind shores of the lakes. There is more about this in the discussion of lake-effect snow.*

## 8.1 What Is an Air Mass?

**Define *air mass* and *air-mass weather*.**

MM GEODe ▶ Basic Weather Patterns ▶ Air Masses

The weather patterns just described result from the movements of large bodies of air, called air masses. An **air mass**, as the term implies, is an immense body of air, usually 1600 kilometers (1000 miles) or more across and perhaps several kilometers thick, that is characterized by homogeneous physical properties (in particular, temperature and moisture content) at any given altitude. When this air moves out of its region of origin, it carries these temperatures and moisture conditions elsewhere, eventually affecting a large portion of a continent (Fig. 8.1).

An excellent example of the influence of an air mass is illustrated in Figure 8.2. Here a cold, dry mass from northern Canada moves southward. With a beginning temperature of −46°C (−51°F), the air mass warms 13°C (24°F), to −33°C (−27°F), by the time it reaches Winnipeg. It continues to warm as it moves southward through the Great Plains and into Mexico. Throughout its southward journey, the air mass becomes warmer, but it also brings some of the coldest weather of the winter to the places in its path. Thus, the air mass is modified, but it also modifies the weather in the areas over which it moves.

The horizontal uniformity of an air mass is not complete because it may extend through 20° or more of latitude and cover hundreds of thousands to millions of square kilometers. Consequently, small differences in temperature and humidity from one point to another at the same level are to be expected. Still, the differences observed within an air mass are small in comparison to the rapid rates of change experienced across air-mass boundaries.

Because it may take several days for an air mass to traverse an area, the region under its influence is likely to experience generally constant weather conditions, a situation called **air-mass weather**. Certainly, some day-to-day variations may exist, but the events are very unlike those in an adjacent air mass having different temperature and moisture characteristics.

The air-mass concept is an important one because it is closely related to the study of atmospheric disturbances. Many significant middle-latitude disturbances and related weather events originate along the boundary zones that separate different air masses.

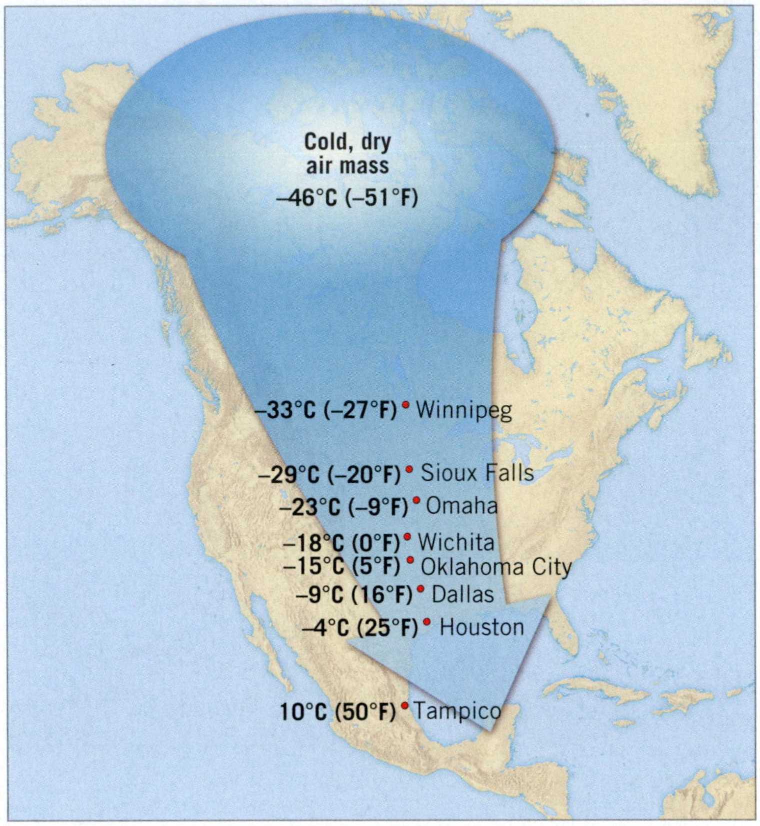


▲ **Figure 8.2 Frigid Canadian air mass** As this air mass moved southward, it brought some of the coldest weather of the winter to the areas in its path. As it advanced out of Canada, the air mass slowly got warmer. Thus, the air mass was gradually modified at the same time that it modified the weather in the areas over which it moved.

▼ **Figure 8.1 North Africa's Sahara Desert** Air masses that form over land areas in the subtropics are hot and dry. The Sahara Desert is an extensive region where such air masses form.

### ✔ Concept Checks 8.1

1. Define *air mass*.
2. What is *air-mass weather?*

# 8.2 Classifying Air Masses

**List the basic criteria for an air-mass source region and identify the source regions that influence North America.**

MM GEODe ▶ Basic Weather Patterns ▶ Air Masses

Where do air masses form? What factors determine the nature and degree of uniformity of an air mass? How are air masses classified? These basic questions are closely related because the area over which an air mass forms vitally affects the properties that characterize it and its resulting classification.

## Source Regions

The large areas in which air masses originate are called **source regions**. Because the atmosphere is heated chiefly from below and gains its moisture by evaporation from Earth's surface, the nature of the source region largely determines the initial characteristics of an air mass. An ideal source region must meet two essential criteria. First, it must be an extensive and physically uniform area. A region having highly irregular topography or one that has a surface consisting of both water and land is not satisfactory.

The second criterion is that the area must be characterized by a general stagnation of atmospheric circulation so that air stays over the region long enough to come to some measure of equilibrium with the surface. In general, it means regions dominated by stationary or slow-moving anticyclones (see Chapter 6), with their extensive areas of calm or light winds.

Regions under the influence of cyclones are not likely to produce air masses because such systems are characterized by converging surface winds. The winds in these lows are constantly bringing air with unlike temperature and humidity properties into the area. Because the air isn't still long enough to eliminate these differences, steep temperature gradients result, and air-mass formation cannot take place.

Figure 8.3 shows the source regions that produce the air masses that most often influence North America. The waters of the Gulf of Mexico and Caribbean Sea and similar regions in the Pacific Ocean west of Mexico yield warm air masses, as does the land area that encompasses the southwestern United States and northern Mexico. In contrast, the North Pacific, the North Atlantic, and the snow- and ice-covered areas comprising northern North America and the adjacent Arctic Ocean are major source regions for cold air masses. It is also clear that the size of the source regions and the intensity of temperatures change seasonally.

Notice in Figure 8.3 that major source regions are not found in the middle latitudes but instead are confined to subtropical and subpolar locations. The middle latitudes are in fact where cold and warm air masses clash, often because the converging winds of a traveling cyclone draw them together. This means that this zone lacks the stagnation of atmospheric circulation necessary for a source region. Instead, this latitude belt is one of the stormiest on the planet.

## Air-Mass Classification

The classification of an air mass depends on the latitude of the source region and the nature of the surface in the area of origin—ocean or continent. The latitude of the source region

▼ **Figure 8.3 Air-mass source regions for North America** Source regions are largely confined to subtropical and subpolar locations. Arrows show the common paths that air masses follow as they move out of their source regions. The fact that the middle latitudes are the site where cold and warm air masses clash, often because the converging winds of a traveling cyclone draw them together, means that this zone lacks the conditions necessary to be a source region. The differences between polar and arctic are relatively small and serve to indicate the degree of coldness of the respective air masses. By comparing winter (**A**) and summer (**B**) maps, it is clear that the extent and temperature characteristics of source regions fluctuate.

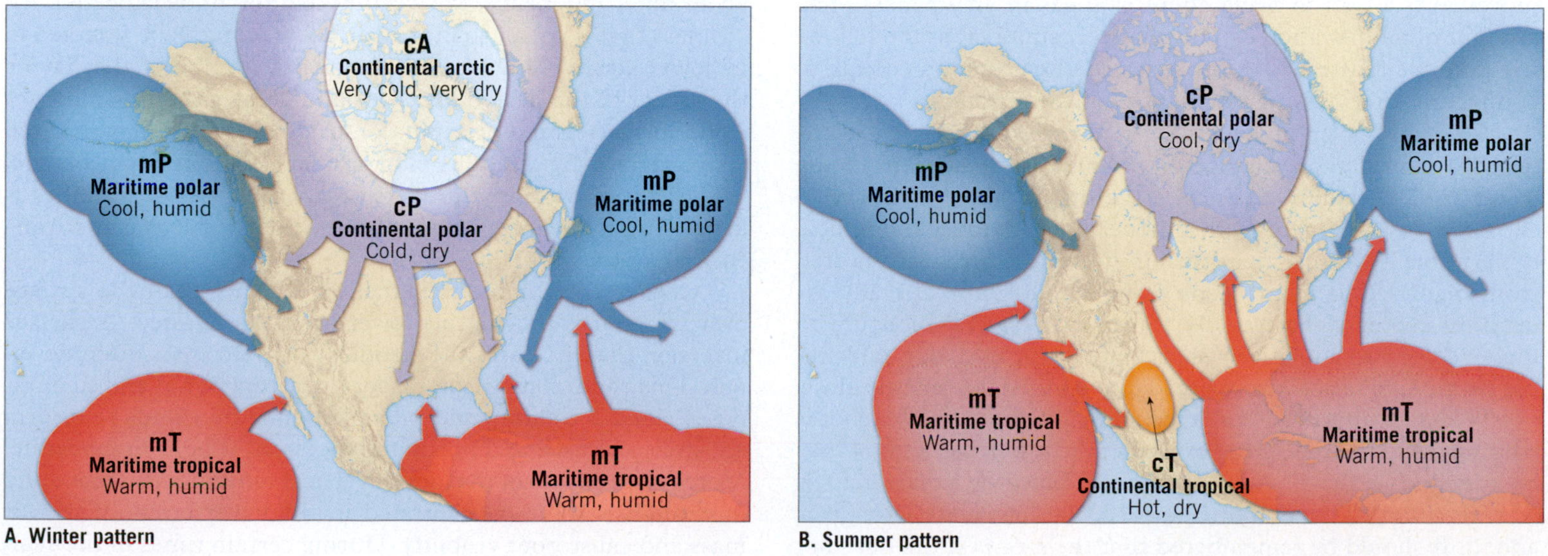

indicates the temperature conditions within the air mass, and the nature of the surface below strongly influences the moisture content of the air.

Air masses are identified by two-letter codes. With reference to latitude (temperature), air masses are placed into one of three categories: **polar (P)**, **arctic (A)**, or **tropical (T)**. The differences between polar and arctic are usually small and simply serve to indicate the degree of coldness of the respective air masses.

The lowercase letter **m** (for **maritime**) or the lowercase letter **c** (for **continental**) is used to designate the nature of the surface in the source region and, hence, the humidity characteristics of the air mass. Because maritime air masses form over oceans, they have a relatively high water-vapor content compared to continental air masses that originate over landmasses.

When this classification scheme is applied, the following air masses can be identified:

| | | |
|---|---|---|
| cA | continental | arctic |
| cP | continental | polar |
| cT | continental | tropical |
| mT | maritime | tropical |
| mP | maritime | polar |

Notice that the list does not include mA (maritime arctic). These air masses are not listed because they seldom, if ever, form. Although arctic air masses do form over the Arctic Ocean, this water body is largely ice covered throughout the year. Consequently, the air masses that originate here consistently have the moisture characteristics associated with a continental source region.

## ✔ Concept Checks 8.2

1. What two criteria must be met for an area to be an air-mass source region?
2. Why are regions that have a cyclonic circulation generally not conducive to air-mass formation?
3. On what basis are air masses classified?
4. Compare the temperature and moisture characteristics of the following air masses: cA, cP, mP, mT, and cT.

Video MM
Radar Reflectivity and Air Masses
http://goo.gl/MUX5nR

# 8.3 Air-Mass Modification

**Describe the processes by which traveling air masses are modified and discuss two examples.**

After an air mass forms, changes in atmospheric circulation eventually cause it to migrate from the area where it acquired its distinctive properties to a region with different surface characteristics. Once the air mass moves from its source region, it not only modifies the weather of the area it is traversing, but it is also gradually modified by the surface over which it is moving. This idea is clearly shown in Figure 8.2. Warming or cooling from below, the addition or loss of moisture, and vertical movements all act to bring about changes in an air mass. The amount of modification can be relatively small or, as the following example illustrates, the changes can be profound enough to completely alter the original identity of the air mass.

When cA or cP air moves over the ocean in winter, it undergoes considerable change (Fig. 8.4). Evaporation from the water surface rapidly transfers large quantities of moisture to the once-dry continental air. Furthermore, because the underlying water is warmer than the air above, the air is also heated from below. This factor leads to instability and vertically ascending currents that rapidly transport heat and moisture to higher levels. In a relatively short time span, cold, dry, and stable continental air is transformed into an unstable mP air mass.

When an air mass is colder than the surface over which it is passing, as in the preceding example, the lowercase letter *k* may be added after the air-mass symbol. If, however, an air mass is warmer than the underlying surface, the lowercase letter *w* is added. It should be remembered that the *k* or *w* suffix does not mean that the air mass itself is cold or warm. It means only that the air is *relatively* cold or warm in comparison with the underlying surface over which it is traveling. For example, an mT air mass from the Gulf of Mexico is usually classified as mTk as it moves over the southeastern states in summer. Although the air mass is warm, it is still cooler than the highly heated landmass over which it is passing.

The *k* or *w* designation gives an indication of the stability of an air mass and, hence, the weather that might be expected. An air mass that is colder than the surface is going to be warmed in its lower layers. This fact causes greater instability that favors the ascent of the heated lower air and creates the possibility of cloud formation and precipitation. Indeed, a *k* air mass is often characterized by cumulus clouds, and if precipitation occurs, it is of the shower or thunderstorm variety. Also, visibility is generally good (except in rain) because of the stirring and overturning of the air.

Conversely, when an air mass is warmer than the surface over which it is moving, its lower layers are chilled. A surface inversion that increases the stability of the air mass often develops. This condition does not favor the ascent of air, and so it opposes cloud formation and precipitation. Any clouds that form are stratus clouds, and precipitation, if any, is light to moderate. Moreover, because of the lack of vertical movements, smoke and dust often become concentrated in the lower layers of the air mass and cause poor visibility. During certain times of the year,

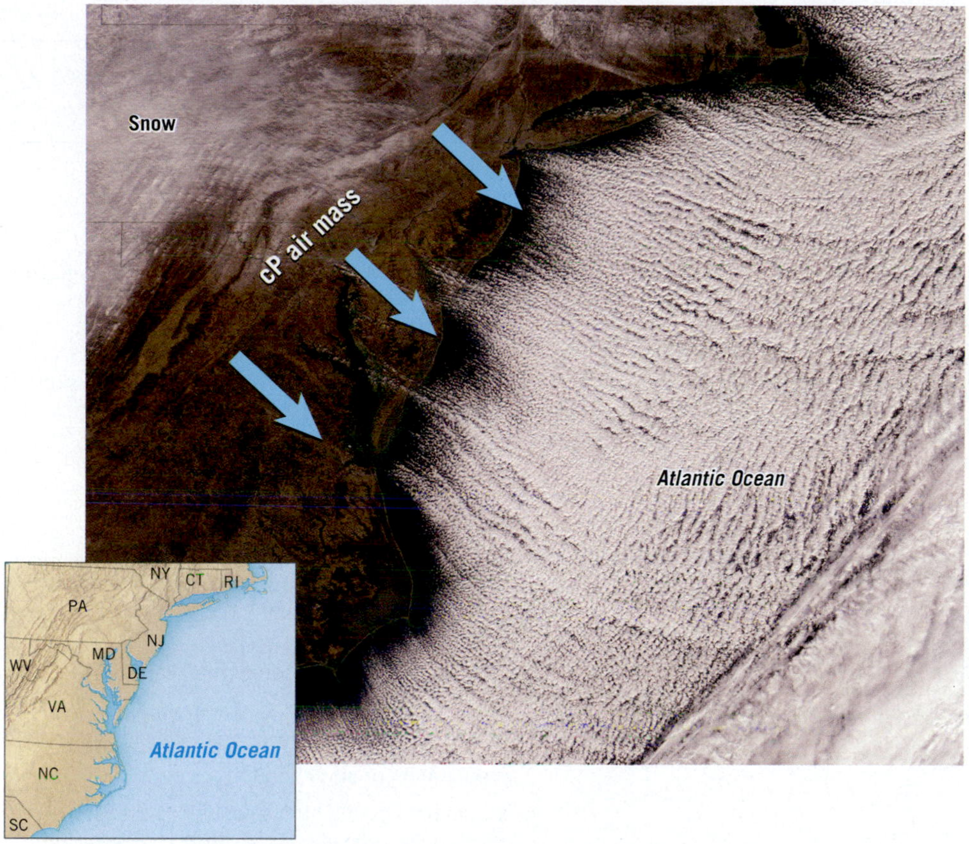


▲ **Figure 8.4 Air-mass modification** This satellite image from January 7, 2014, shows the modification of cold, dry, and cloud-free cP air that produced a cold snap in the eastern United States. As the air mass moved over the Atlantic, the addition of heat and water vapor from the relatively warm water quickly modified the air mass and created instability, as evidenced by the development of clouds.

fogs, especially the advection type (see Chapter 5), may also be common in some regions.

In addition to modifications resulting from temperature differences between an air mass and the surface below, upward and downward movements induced by cyclones and anticyclones or topography can also affect the stability of an air mass. Such modifications are often called *mechanical* or *dynamic*, and they are usually independent of the changes caused by surface cooling or heating. For example, significant modification can result when an air mass is drawn into a surface low. Here convergence and lifting dominate, and the air mass is rendered more unstable. Conversely, the subsidence associated with anticyclones acts to stabilize an air mass. Similar alterations in stability occur when an air mass is lifted over highlands or descends the leeward side of a mountain barrier. In the first case, the air's stability is reduced; in the second case, the air becomes more stable.

### ✔ Concept Checks 8.3

1. What do the lowercase letters *k* and *w* indicate about an air mass? List the general weather conditions associated with *k* and *w* air masses.
2. How might vertical movements induced by a pressure system or topography act to modify an air mass?

## 8.4 Properties of North American Air Masses

**Summarize the weather conditions associated with each of the air masses that influence North America during the summer and the winter.**

Air masses frequently pass over us, which means that the day-to-day weather we experience often depends on the temperature, stability, and moisture content of these large bodies of air. In this section, we briefly examine the properties of the principal North American air masses. In addition, **Table 8.1** serves as a useful summary.

### Continental Polar (cP) and Continental Arctic (cA) Air Masses

Continental polar and continental arctic air masses are, as their classification implies, cold and dry. Continental polar air originates over the often snow-covered interior regions of Canada and Alaska, poleward of the 50th parallel. Continental arctic air forms even farther north, over the Arctic basin and the Greenland ice cap (see Figure 8.3).

Continental arctic air is distinguished from cP air by its lower temperatures, although at times the differences may be slight. In fact, some meteorologists do not differentiate between cP and cA.

**Winter Characteristics** During the winter, both cP and cA air masses are bitterly cold and very dry. Winter nights are long, and the daytime Sun is short-lived and low in the sky. Consequently, as winter advances, Earth's surface and atmosphere lose heat that, for the most part, is not replenished by incoming solar energy. Therefore, the surface reaches very low temperatures, and the air near the ground is gradually chilled to heights of 1 kilometer (0.6 mile) or more. The result is a strong and persistent temperature inversion, with the coldest temperatures near the ground. Marked stability is, therefore, the rule. Because the air is very cold and the surface below is frozen, the mixing ratio (water vapor content; see Chapter 4)

**Table 8.1** | Weather Characteristics of North American Air Masses

| Air Mass | Source Region | Temperature and Moisture Characteristics in Source Region | Stability in Source Region | Associated Weather |
|---|---|---|---|---|
| cA | Arctic basin and Greenland ice cap (winter only) | Bitterly cold and very dry in winter | Stable | Cold spells in winter |
| cP | Interior Canada and Alaska | Very cold and dry in winter | Stable entire year | a. Cold spells in winter<br>b. Modified to cPk in winter over Great Lakes, bringing lake-effect snow to leeward shores |
| mP | North Pacific | Mild (cool) and humid entire year | Unstable in winter<br>Stable in summer | a. Low clouds and showers in winter<br>b. Heavy orographic precipitation on windward side of western mountains in winter<br>c. Low stratus and fog along coast in summer; modified to cP inland |
| mP | Northwestern Atlantic | Cold and humid in winter<br>Cool and humid in summer | Unstable in winter<br>Stable in summer | a. Occasional nor'easter in winter<br>b. Occasional periods of clear, cool weather in summer |
| cT | Northern interior Mexico and southwestern U.S. (summer only) | Hot and dry | Unstable | a. Hot, dry, and cloudless; rarely influencing areas outside source region<br>b. Occasional drought to southern Great Plains |
| mT | Gulf of Mexico, Caribbean Sea, western Atlantic | Warm and humid entire year | Unstable entire year | a. In winter it usually becomes mTw, moving northward and bringing occasional widespread precipitation or advection fog<br>b. In summer hot and humid conditions, frequent cumulus development, and showers or thunderstorms |
| mT | Eastern subtropical Pacific | Warm and humid entire year | Stable entire year | a. In winter it brings fog, drizzle, and occasional moderate precipitation to northwestern Mexico and the southwestern U.S.<br>b. In summer occasionally reaches the western U.S. and is a source of moisture for infrequent convectional thunderstorms |

of these air masses is necessarily low, ranging from perhaps 0.1 gram per kilogram in cA up to 1.5 grams per kilogram in some cP air.

As wintertime cP or cA air moves outward from its source region, it carries its cold, dry weather to the United States, normally entering between the Great Lakes and the Rockies. Because there are no major barriers between the high-latitude source regions and the Gulf of Mexico, cP and cA air masses can sweep rapidly and with relative ease far southward into the United States. The winter cold spells experienced in much of the central and eastern United States are closely associated with such polar outbreaks. One such cold spell is described in **Box 8.1**.

The winter of 2013–2014 provides another example of the chilling effects of arctic and polar air masses (**Fig. 8.5**). Directed by the meandering polar jet stream (see Figure 7.18, page 202), frigid cA and cP air poured into the central and eastern United States. Seven Midwestern states experienced one of their top-10 coldest winters on record. A December cold spell was followed by a period of extreme cold in January over much of the eastern half of the country. Many cities experienced their coldest February in decades, and during the first week of March, 18 states set all-time records for cold for that time of year. Notice in Figure 8.5, however, that even though millions of people experienced an especially cold winter, many others were enjoying warmer-than-average conditions.

**Video** MM
Effects of the 2011 Groundhog Day Blizzard
http://goo.gl/5OODwD

students sometimes ask...

## When a cold air mass moves south from Canada into the United States, how rapidly can temperatures change?

When a fast-moving frigid air mass advances into the northern Great Plains, temperatures can plunge 20° to 30°C (40° to 50°F) in just a few hours. One notable example is a drop of 55.5°C (100°F), from 6.7°C to −48.8°C (44° to −56°F), in 24 hours at Browning, Montana, on January 23–24, 1916. Another remarkable event occurred on Christmas Eve, 1924, when the temperature at Fairfield, Montana, dropped from 17°C (63°F) at noon to −29°C (−21°F) at midnight—an amazing 46°C (83°F) change in just 12 hours.

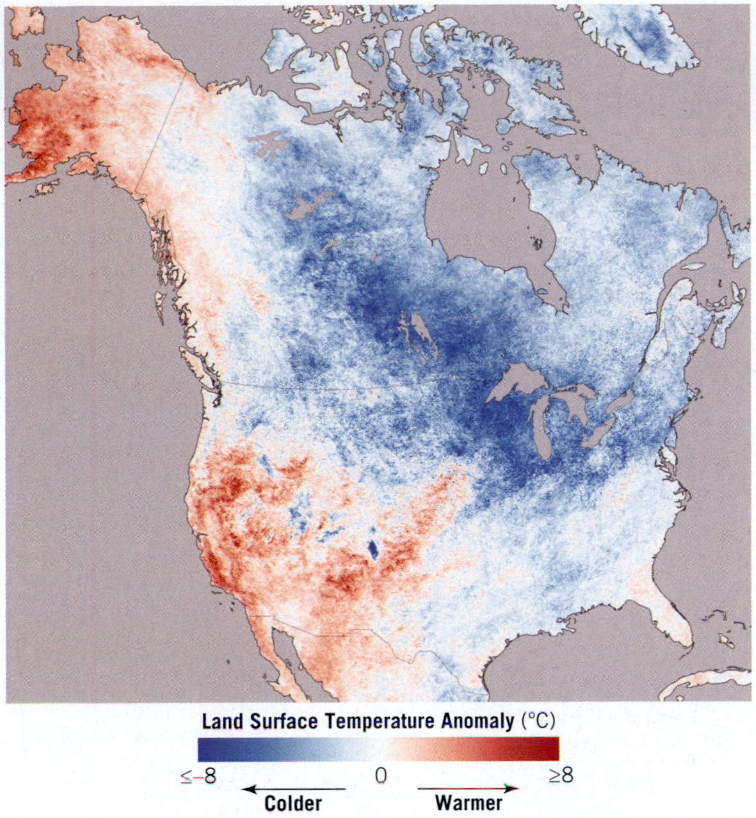


◀ **Figure 8.5 A cold winter in the Midwest and East** This satellite-generated map shows land-surface temperature anomalies for North America for December 2013 through February 2014. *Anomaly* refers to the deviation or departure from what is expected. In this case, the map shows departures from the December through February average for 2000–2013. Land-surface temperatures are not the same as air temperatures, but they are a reasonable proxy for how warm or cold each region was. Notice that some parts of the continent were considerably warmer than average.

**Summer Characteristics** Because cA air is present principally in the winter, only cP air has any influence on our summer weather, and this effect is considerably reduced compared to winter. During summer months the properties of the source region for cP air are very different from those during winter. Instead of being chilled by the ground, the air is warmed from below as the long days and higher Sun angle warm the snow-free land surface. Although summer cP air is warmer and has a higher moisture content than its wintertime counterpart, the air is still cool and relatively dry compared with air in areas farther south. Summer heat waves in the northern portions of the eastern and central United States are often ended by the southward advance of cP air, which brings cooling relief and bright, pleasant weather for a day or two.

## Lake-Effect Snow: Cold Air over Warm Water

A glance at the chapter-opening image provides a perspective from space of the process discussed in this section. The skies over Lake Superior and Lake Michigan exhibit long rows of dense, white, snow-producing clouds. They formed as cold, dry cP air moved from the land surface across the open water. Continental polar air masses are not, as a rule, associated with dense clouds and heavy precipitation. Yet during late autumn and winter, a unique and interesting weather phenomenon takes place along the downwind shores of the Great Lakes. Periodically, heavy snow showers issue from dark clouds that move onshore from the lakes (see **Box 8.2**). Seldom do these storms move more than about 80 kilometers (50 miles) from the lakeshore before the snows come to an end. These highly localized storms produce what is known as **lake-effect snow** and are associated primarily with the Great Lakes, although other large lakes can experience this phenomenon.

Lake-effect storms account for a high percentage of the snowfall in many areas adjacent to the lakes. The strips of land that are most frequently affected, called *snowbelts*, are shown in **Figure 8.6**. A comparison of average snowfall totals at Thunder Bay, Ontario, on the northern shore of Lake Superior, and Marquette, Michigan, along the southern shore, provides another excellent example. Because Marquette is situated on the leeward shore of the lake, it receives substantial lake-effect snow and therefore has a much higher snowfall total than does Thunder Bay (**Table 8.2**).

Video MM
An Infrared View of the 2011 Groundhog Day Blizzard
http://goo.gl/hSLiub

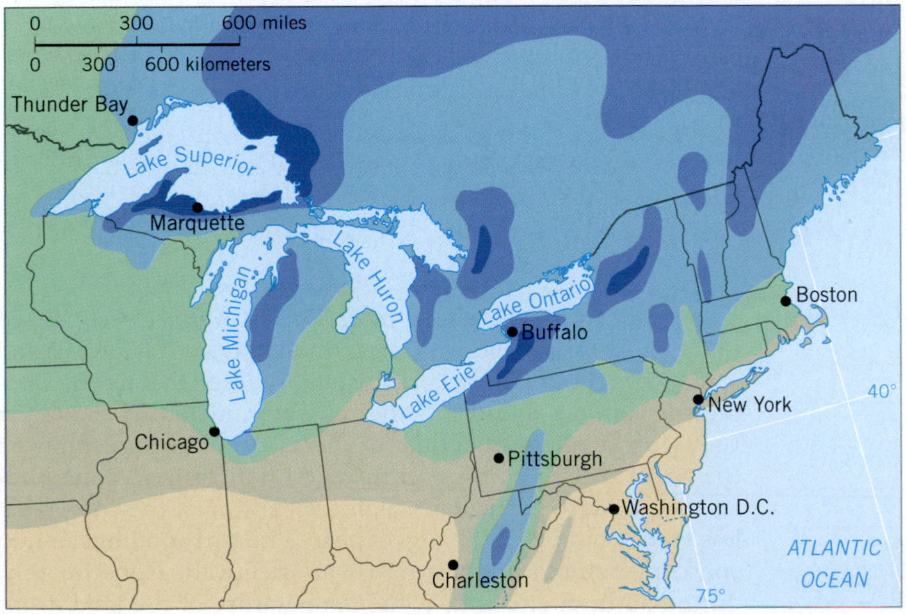


**AVERAGE ANNUAL SNOWFALL**

| cm | in. |
|---|---|
| >330 | >130 |
| 250–329 | 100–129 |
| 150–249 | 60–99 |
| 90–149 | 36–59 |
| 60–89 | 24–35 |
| <60 | <24 |

◀ **Figure 8.6 Average annual snowfall** The snowbelts of the Great Lakes region are easy to pick out on this snowfall map.

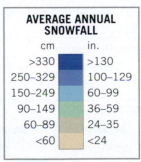

severe& hazardous weather Box 8.1

## The Siberian Express

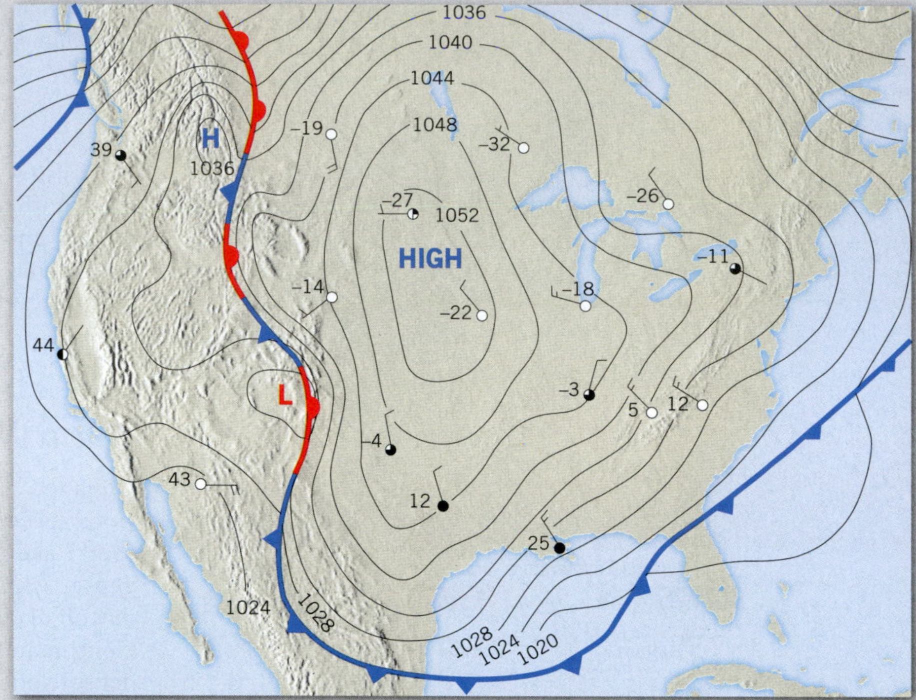


▲ **Figure 8.A Arctic air invasion** A surface weather map for 7 A.M. EST, December 22, 1989. This simplified National Weather Service (NWS) map shows an intense winter cold spell caused by an outbreak of frigid continental arctic air. This event brought subfreezing temperatures as far south as the Gulf of Mexico. Temperatures on NWS maps are in degrees Fahrenheit.

The surface weather map for December 22, 1989, shows a very large high-pressure center covering the eastern two-thirds of the United States and a substantial portion of Canada (**Fig. 8.A**). As is usually the case in winter, a large anticyclone such as this is associated with a huge mass of dense and bitterly cold arctic air. After such an air mass forms over the frozen expanses near the Arctic Circle, the winds aloft sometimes direct it toward the south and east. When an outbreak takes place, it is popularly called the "Siberian Express" by the news media, even though the air mass did not originate in Siberia.

November 1989 was unusually mild for late autumn. In fact, across the United States, more than 200 daily high-temperature records were set. December, however, was different. East of the Rockies, the month's weather was dominated by two arctic outbreaks. The second brought record-breaking cold.

Between December 21 and 25, as the frigid dome of high pressure advanced southward and eastward, more than 370 record low temperatures were reported. On December 21, Havre, Montana, had an overnight low of −42.2°C (−44°F), breaking a record set in 1884. Meanwhile, Topeka's −32.2°C (−26°F) was that city's lowest temperature for any date since recordkeeping had begun 102 years earlier.

The 3 days that followed saw the arctic air migrate toward the south and east. By December 24, Tallahassee, Florida, had a low temperature of −10°C (14°F). In the center of the state, the daily minimum at Orlando was −5.6°C (22°F). It was actually warmer in North Dakota on Christmas Eve than in central and northern Florida!

It was actually warmer in North Dakota on Christmas Eve than in central and northern Florida!

As we would expect, utility companies in many states reported record demand. When the arctic air advanced into Texas and Florida, agriculture was especially hard hit. Some Florida citrus growers lost 40 percent of their crop, and many vegetable crops were wiped out completely.

After Christmas the circulation pattern that had brought this record-breaking Siberian Express from the arctic deep into the United States changed. As a result, temperatures for much of the country during January and February 1990 were well above normal. In fact, January 1990 was the second-warmest January in 96 years. Thus, despite frigid December temperatures, the winter of 1989–1990 "averaged out" to be a relatively warm one.

**Questions**

1. What air mass (classification) was most likely associated with this event?
2. Did the air mass originate in Siberia?

**Table 8.2** Monthly Snowfall at Thunder Bay, Ontario, and Marquette, Michigan

| | | | |
|---|---|---|---|
| **Thunder Bay, Ontario** | | | |
| **October** | **November** | **December** | **January** |
| 3.0 cm (1.2 in.) | 14.9 cm (5.8 in.) | 19.0 cm (7.4 in.) | 22.6 cm (8.8 in.) |
| **Marquette, Michigan** | | | |
| **October** | **November** | **December** | **January** |
| 5.3 cm (2.1 in.) | 37.6 cm (14.7 in.) | 56.4 cm (22.0 in.) | 53.1 cm (20.7 in.) |

What causes lake-effect snow? The answer is closely linked to the differential heating of water and land (see Chapter 3) and to the concept of atmospheric instability (see Chapter 4). During the summer months, bodies of water, including the Great Lakes, absorb huge quantities of energy from the Sun and from the warm air that passes over them. Although these water bodies do not reach particularly high temperatures, they nevertheless represent huge reservoirs of heat. The surrounding land, in contrast, cannot store heat nearly as effectively. Consequently, during autumn and winter, the temperature of the land drops

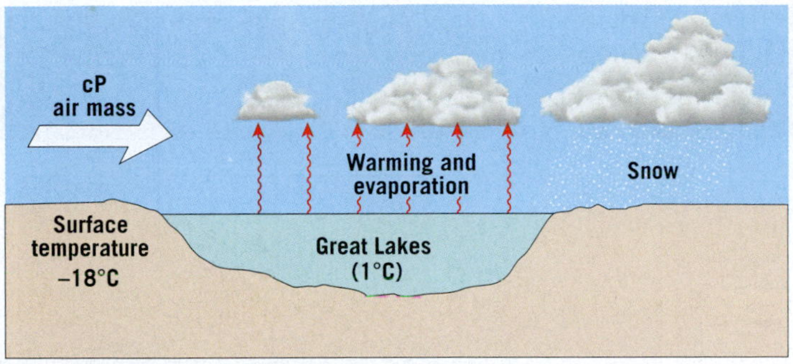


▲ **Figure 8.7 Lake-effect snow** As continental polar air crosses the Great Lakes in winter, it acquires moisture and becomes unstable because of warming from below. Lake-effect snow showers on the downwind side of the lakes often result from this air-mass modification.

quickly, whereas water bodies lose their heat more gradually and cool slowly.

From late November through late January, the contrasts in average temperatures between water and land range from about 8°C in the southern Great Lakes to 17°C farther north. However, the temperature differences can be much greater (perhaps 25°C) when a very cold cP or cA air mass pushes southward across the lakes. When such a dramatic temperature contrast exists, the lakes interact with the air to produce major lake-effect storms. Figure 8.7 depicts the movement of a cP air mass across one of the Great Lakes. During its journey, the air acquires large quantities of heat and moisture from the relatively warm lake surface. By the time it reaches the opposite shore, this cPk air is humid and unstable, and heavy snow showers are likely.

## Maritime Polar (mP) Air Masses

Maritime polar air masses form over oceans at high latitudes. As the classification indicates, mP air is cool to cold and humid, but compared with cP and cA air masses in winter, mP air is relatively mild because of the higher temperatures of the ocean surface compared to the colder continents.

Two regions are important sources for mP air that influences North America: the North Pacific and the northwestern Atlantic from Newfoundland to Cape Cod (see Figure 8.3). Because of the general west-to-east atmospheric circulation in the middle latitudes, mP air masses from the North Pacific source region usually influence North American weather more than mP air masses generated in the northwest Atlantic. Whereas air masses that form in the Atlantic generally move eastward toward Europe, mP air from the North Pacific has a strong influence on the weather along the western coast of North America, especially in the winter.

**Pacific mP Air Masses** During the winter, mP air masses from the Pacific usually begin as cP air in Siberia (Fig. 8.8). Although air rarely stagnates over this area, the source region is extensive enough to allow the air moving across it to acquire its characteristic properties. As the air advances eastward over the relatively warm water, active evaporation and heating occur in the lower levels. Consequently, what was once a very cold, dry, and stable air mass is changed into one that is mild and humid near the surface and relatively unstable. As this mP air arrives at the western coast of North America, it is often accompanied by low clouds and shower activity. When the mP air advances inland against the western mountains, orographic uplift can produce heavy rain or snow on the windward slopes of the mountains.

Summer brings a change in the characteristics of mP air masses from the North Pacific. During the warm season, the ocean is cooler than the surrounding continents. In addition, the Pacific high-pressure cell lies off the western coast of the United States (see Figure 7.10B). Consequently, there is almost continuous southward flow of moderate-temperature air. Although the air near the surface may often be conditionally unstable, the presence of the Pacific high means that there is subsidence and stability aloft. Thus, low stratus clouds and summer fogs characterize much of the western coast. Once summer mP air from the Pacific moves inland, it is heated at the surface over the hot and dry interior. The heating and resulting turbulence act to reduce the relative humidity in the lower layers, and the clouds dissipate.

**Maritime Polar Air from the North Atlantic** As is the case for mP air from the Pacific, air masses forming in the northwestern Atlantic source region were originally cP air

**students sometimes ask...**

**I know that Buffalo, New York, is famous for its lake-effect snows. Just how bad can it get?**

Located along Lake Erie's eastern shore, Buffalo does indeed receive a great deal of lake-effect snow (see Figure 8.6). Between December 24, 2001, and January 1, 2002, Buffalo's longest-lasting lake-effect event buried the city under 207.3 centimeters (81.6 inches) of snow. Prior to this storm, the record for the *entire month* of December had been 173.7 centimeters (68.4 inches)! The eastern shore of Lake Ontario was also hard hit, with one station recording more than 317 centimeters (125 inches) of snow.

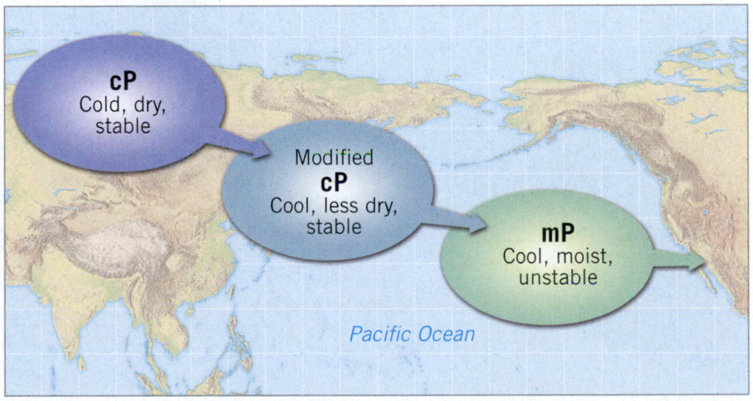


▲ **Figure 8.8 Formation of a Pacific mP air mass** During winter, maritime polar (mP) air masses in the North Pacific usually begin as continental polar (cP) air masses in Siberia. The cP air is modified to mP as it slowly crosses the ocean.

severe& hazardous weather Box 8.2

## An Extraordinary Lake-Effect Snowstorm

Northeastern Ohio is part of the Lake Erie snowbelt, a zone that extends eastward into northwestern Pennsylvania and western New York (see Figure 8.6). Here snowfall is enhanced when cold winds blow from the west or northwest across the relatively warm, unfrozen waters of Lake Erie. Average annual snowfall in northeastern Ohio is between 200 and 280 centimeters (80 to 110 inches) and increases to 450 centimeters (175 inches) in western New York.

Video MM
Lake Effect Snow
http://goo.gl/3hPLBF

Although residents here are accustomed to abundant snowstorms, even they were surprised in November 1996 by an especially early and strong storm. During a 6-day span, from November 9 to 14, northeastern Ohio experienced record-breaking lake-effect snows. Rather than raking fall leaves, people were shoveling sidewalks and clearing snow from overloaded roofs (**Fig. 8.B**).

Rather than raking fall leaves, people were shoveling sidewalks and clearing snow from overloaded roofs

Persistent snow squalls (narrow bands of heavy snow) frequently produced accumulation rates of 5 centimeters (2 inches) per hour during the 6-day siege. The snow resulted from especially deep and prolonged lower atmospheric instability created by the movement of cold air across a relatively warm Lake Erie. The surface temperature of the lake was 12°C (54°F), several degrees above normal. The air temperature at a height of 1.5 kilometers (nearly 5000 feet) was −5°C (23°F).

The flow of cold air across Lake Erie and the 17°C lapse rate generated lake-effect rain and snow almost immediately.* During some periods, loud thunder accompanied the snow squalls!

▲ **Figure 8.B State record** A 6-day lake-effect snowstorm in November 1996 dropped 175 centimeters (nearly 69 inches) of snow on Chardon, Ohio, setting a new state record.

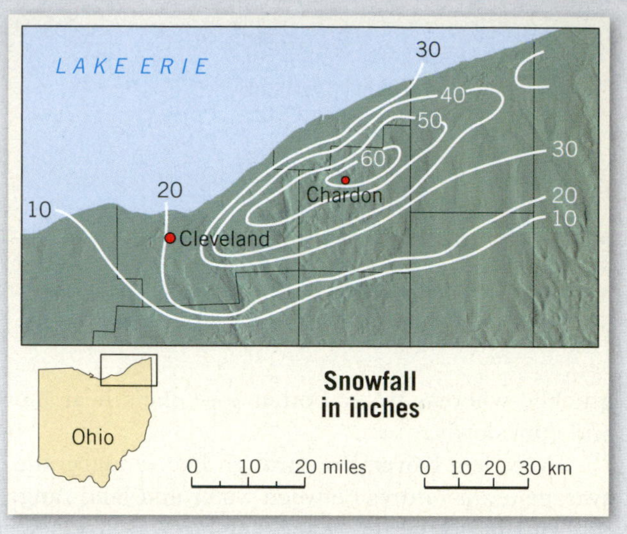


▲ **Figure 8.C Snow depths** Snowfall totals (in inches) for northeastern Ohio for the November 9–14, 1996, lake-effect storm. The deepest accumulation, 175 centimeters (nearly 69 inches), occurred near Chardon. (After National Weather Service)

Data from the Cleveland National Weather Service Snow Spotters Network showed 100 to 125 centimeters (39 to 49 inches) of snow through the core of the Ohio snowbelt (**Fig. 8.C**). The deepest accumulation was measured near Chardon, where the 6-day total of 175 centimeters (68.9 inches) far exceeded the previous Ohio record of 107 centimeters (42 inches) set in 1901. In addition, the November 1996 snowfall total of 194.8 centimeters (76.7 inches) at the same site set a new monthly snowfall record for Ohio. The previous 1-month record had been 176.5 centimeters (69.5 inches).

The impact of the storm was considerable. Ohio's governor declared a state of emergency on November 12, and National Guard troops were dispatched to assist in snow removal and the rescue of snowbound residents. There was widespread damage to trees and shrubs across the region because the snow was particularly wet and dense and readily clung to objects. Press reports indicated that about 168,000 homes were without electricity, some for several days. Roofs of numerous buildings collapsed under the excessive snow loads. Although residents of the northeastern Ohio snowbelt are winter storm veterans, the 6-day storm in November 1996 will be long remembered as an extraordinary event.

### Questions

1. From what direction was the wind blowing during the storm?
2. How much more snow fell at Chardon than at Cleveland? Check Figure 8.C.

*Thomas W. Schmidlin and James Kasarik, "A Record Ohio Snowfall During 9–14 November 1996," *Bulletin of the American Meteorological Society*, Vol. 80, No. 6, June 1999, p. 1109.

## eye ON THE atmosphere 8.1

An intense cold snap in early January 2014 brought thick and widespread ice to the Great Lakes. This satellite image from January 9 shows Lake Erie 90 percent ice covered.

**Questions**

1. When compared to an ice-free condition, would the situation depicted here promote more lake-effect snow storms, reduce the chances of lake-effect snow, or likely have no effect?
2. Explain the reasoning you used to answer Question 1.

masses that moved from the continent and were transformed over the ocean. However, unlike air masses from the North Pacific, mP air from the Atlantic only occasionally affects the weather of North America. Nevertheless, this air mass does have an effect when the northeastern United States is on the northern or northwestern edge of a passing low-pressure center. In winter, strong cyclonic winds can draw mP air into the region. Its influence is generally confined to the area east of the Appalachians and north of Cape Hatteras, North Carolina. The weather associated with a wintertime invasion of mP air from the Atlantic is known locally as a **nor'easter**. Strong northeast winds, freezing or near-freezing temperatures, high relative humidity, and the likelihood of precipitation make this weather phenomenon an unwelcome event.

A classic example is shown in **Figure 8.9**. This nor'easter moved up the east coast on January 12, 2011, dumping heavy snow on the New England states for the third time in 3 weeks. The storm had begun developing to the south a day earlier. As it moved northward along the coast, it merged with another system crossing from the Midwest. The satellite image in Figure 8.9 shows that the storm had a distinctive comma shape—which forms from the counterclockwise circulation around a low-pressure center. Cold, humid mP air from the North Atlantic was drawn toward the storm center, producing dense clouds, especially on the north and west sides of the storm. In parts of New England, snow fell as fast as 7.6 centimeters (3 inches) per hour. More than 61 centimeters (24 inches) fell in many areas by the evening of January 12. Blizzard conditions—with visibility cut to less than 0.4 kilometer (0.25 mile) and gale-force winds for more than 3 hours—developed in parts of Connecticut and Massachusetts. The storm left more than 100,000 people without electricity and shut down portions of Interstate 95 and the northeastern railroad service.

Whereas mP air masses from the Atlantic may produce an occasional unwelcome nor'easter during the winter, summertime incursions of this air mass may bring pleasant weather. Like the Pacific source region, the northwestern Atlantic is dominated by high pressure during the summer (see Figure 7.10B). Thus, the upper air is stable because of subsidence, and the lower air is essentially stable because of the chilling effect of the relatively cool water. As the circulation on the southern side of the anticyclone carries this stable mP air into New England, and

► **Figure 8.9 Classic nor'easter** The satellite image shows a strong nor'easter along the coast of New England on January 12, 2011. In winter, a nor'easter exhibits a weather pattern in which strong northeast winds carry cold, humid mP air from the North Atlantic into New England and the middle Atlantic states. The ground-level view of the storm in Boston shows that the combination of ample moisture and strong convergence can result in heavy snow.

**eye ON THE atmosphere 8.2**

This satellite image from December 27, 2010, shows a strong winter storm off the east coast.

**Questions**

1. Can you identify the very center of the storm?
2. What air mass is being drawn into the storm to produce the dense clouds in the upper right?
3. What name is applied to a storm such as this?
4. Farther south, the cold air mass over the southeastern states is cloud free. What is its likely classification? Explain how it is being modified as it moves over the Atlantic.

occasionally as far south as Virginia, the region enjoys clear, cool weather and good visibility.

## Maritime Tropical (mT) Air Masses

Maritime tropical air masses affecting North America most often originate over the warm waters of the Gulf of Mexico, the Caribbean Sea, or the adjacent western Atlantic Ocean (see Figure 8.3). The tropical Pacific is also a source region for mT air. However, the land area affected by this latter source is small compared with the size of the region influenced by air masses produced in the Gulf of Mexico and adjacent waters.

As expected, mT air masses are warm to hot, and they are humid. In addition, they are often unstable. It is through invasions of mT air masses that the subtropics export much heat and moisture to the cooler and drier areas to the north. Consequently, these air masses make important contributions to a region's weather because they are capable of generating significant precipitation.

**North Atlantic mT Air** Maritime tropical air masses from the Gulf–Caribbean–Atlantic source region greatly affect the weather of the United States east of the Rocky Mountains. Although the source region is dominated by the North Atlantic subtropical high, the air masses produced are not stable but are neutral or unstable because the source region is located on the weak western edge of the anticyclone, where pronounced subsidence is absent.

During winter, when cP air dominates the central and eastern United States, mT air only occasionally enters this part of the country. When an invasion does occur, the lower portions of the air mass are chilled and stabilized as it moves northward over the cold land. Its classification is changed to mTw. As a result, the formation of convective showers is unlikely. Widespread precipitation does occur, however, when a northward-moving mT air mass is pulled into a traveling cyclone and forced to ascend. In fact, much of the wintertime precipitation over the eastern and central states results when mT air from the Gulf of Mexico is lifted along fronts in traveling cyclones.

Another weather phenomenon associated with a northward-moving wintertime mT air mass is advection fog. Dense fogs can develop as the warm, humid air is chilled as it moves over the cold land surface.

During the summer, mT air masses from the Gulf, Caribbean, and adjacent Atlantic affect a much wider area of North America and are present for a greater percentage of the time than during the winter. As a result, they exert a strong and often dominating influence over the summer weather of the United States east of the Rocky Mountains. This influence is due to the general sea-to-land airflow over the eastern portion of North America during the warm months, which brings more frequent incursions of mT air that penetrate much deeper into the continent than during the winter months. Consequently, these air masses are largely responsible for the hot and humid conditions that prevail over the eastern and central United States.

Initially, summertime mT air from the Gulf is unstable. As it moves inland over the warmer land, it becomes an mTk air mass as daytime heating of the surface layers further increases the air's instability. Because the relative humidity is high, only modest lifting is necessary to bring about active convection, cumulus development, and thunderstorm or shower activity (Fig. 8.10). This is, indeed, a common warm-weather phenomenon associated with mT air.

We should also note here that air masses from the Gulf–Caribbean–Atlantic region are the primary source of much, if not most, of the precipitation received in the eastern two-thirds of the United States. Pacific air masses contribute little to the water supply east of the Rockies because the western mountains effectively "wring dry" the moisture from the air through numerous episodes of orographic uplift.

Figure 8.11, showing the distribution of average annual precipitation for the eastern two-thirds of the United States by using **isohyets** (lines connecting places having equal rainfall), illustrates this situation nicely. The pattern of isohyets shows the greatest rainfall in the Gulf region and a decrease in precipitation with increasing distance from the mT source region.

▲ **Figure 8.10 Gulf air moving north in summer** As mT air from the Gulf of Mexico moves over the heated land in summer, cumulus development and afternoon showers frequently result.

**North Pacific mT Air** Compared to mT air from the Gulf of Mexico, mT air masses from the Pacific source region have much less of an impact on North American weather. In winter, only northwestern Mexico and the extreme southwestern United States are influenced by air from the tropical Pacific. Because the source region lies along the eastern side of the Pacific anticyclone, subsidence aloft produces upper-level stability. When the air mass moves northward, cooling at the surface also causes the lower layers to become more stable, often resulting in fog or drizzle. If the air mass is lifted along a front or forced over mountains, moderate precipitation results.

There are times, however, when mT air from the subtropical North Pacific is involved in weather phenomena known as **atmospheric rivers**, narrow zones in the atmosphere that can transport huge amounts of water vapor to regions outside the tropics. One well-known example is popularly called the *Pineapple Express*. Unlike the *Siberian Express* described earlier in the chapter, which delivers bitter cold spells to the nation's midsection, this atmospheric river is a narrow band of enhanced water vapor transport that can bring extraordinary rains to southern California and other west coast locations.

Most precipitation along the west coast results from wintertime storms that pass across the Gulf of Alaska. These storms are dominated by humid, cool mP air. However, in some years, a strong southern branch of the polar jet stream acts as a conduit that transports humid, warm mT air from the tropics near the Hawaiian Islands northeastward to the west coast (**Fig. 8.12**). This mT air feeds into storm systems that can bring torrential rains to lower elevations and heavy snows to the Sierra Nevada range. Mudflows (popularly called mudslides) can result when the soaking rains saturate hillsides that have lost their anchoring vegetation to recent wildfires.

For many years, the summertime influence of air masses from the tropical Pacific source region on the weather of the southwestern United States and northern Mexico was believed to be minimal. It was thought that the moisture for the region's infrequent summer thunderstorms came from occasional westward thrusts of mT air from the Gulf of Mexico. However, the Gulf of Mexico is no longer believed to be the primary supplier of moisture for the area west of the Continental Divide. Rather, it has been demonstrated that the tropical North Pacific west of central Mexico is a more important source of moisture for this area.

In summer the mT air moves northward from its Pacific source region up the Gulf of California and into the interior of the western United States (see Figure 7.13, page 199). This movement, which is confined largely to July and August, is essentially monsoonal in character. That is, the inflow of moist air is a response to thermally produced low pressure that develops over the heated landmass. The July–August rainfall maximum for Tucson, Arizona, is a response to this incursion of Pacific mT air (**Fig. 8.13**).

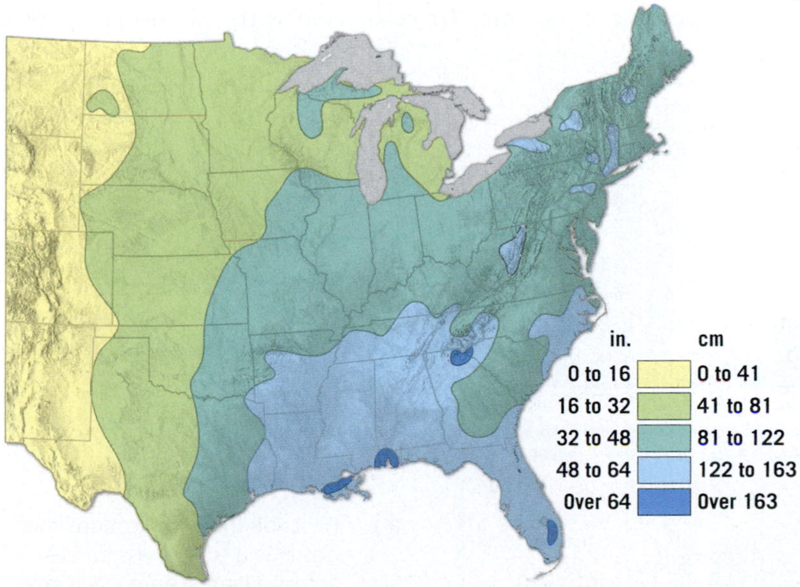


▲ **Figure 8.11 Average annual precipitation for the eastern two-thirds of the United States** Note the general decrease in yearly precipitation totals with increasing distance from the Gulf of Mexico, the source region for mT air masses. Isohyets are labeled in inches.

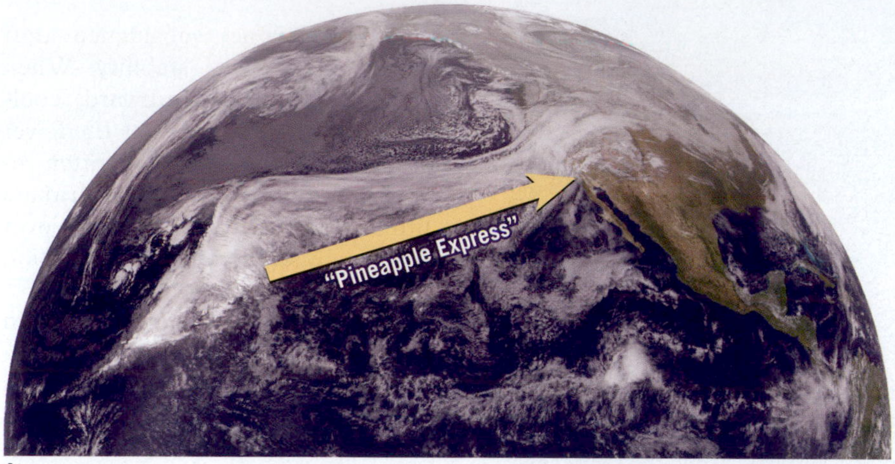


A.

B.

▲ **Figure 8.12 Atmospheric river A.** This satellite image of clouds over the Pacific Ocean on December 19, 2010, illustrates the "Pineapple Express," a phenomenon in which a strong jet stream carries mT air from the vicinity of Hawaii to California. **B.** The Pineapple Express battered much of California December 17–22, 2010, bringing as much as 50 centimeters (20 inches) of rain to the San Gabriel Mountains and more than 1.5 meters (5 feet) of snow to the Sierra Nevada. Southern California bore the brunt of the storms, as coastal and hillside areas experienced mudflows and floods.

## Continental Tropical (cT) Air Masses

North America narrows as it extends southward through Mexico; therefore, the continent has no extensive source region for continental tropical air masses. By checking the maps in Figure 8.3, you can see that only in summer do northern interior Mexico and adjacent parts of the arid southwestern United States produce hot, dry cT air. Because of the intense daytime heating at the surface, both a steep environmental lapse rate and turbulence extending to considerable heights are found. Nevertheless, although the air is unstable, it generally remains nearly cloudless because of extremely low humidity. Consequently, the prevailing weather is hot, with an almost complete lack of rainfall. Large daily temperature ranges are the rule. Although cT air masses are usually confined to their source region, occasionally they move into the southern Great Plains. When this occurs, the cT air may be associated with the formation of a *dryline*, a narrow zone of stormy weather. There is more about drylines in chapters 9 and 10 (see Figures 9.8, 10.12 and 10.13).

▼ **Figure 8.13 Summer monsoon in the Southwest** The photo shows cumulonimbus clouds developing over the Sonoran Desert in southern Arizona on a July afternoon. The source of moisture for these summer storms is maritime tropical air from the eastern North Pacific.

### ✔ Concept Checks 8.4

1. Which two air masses have the greatest influence on weather east of the Rocky Mountains? Explain your choice.
2. What air mass influences the weather of the Pacific Coast more than any other?
3. Describe the modifications that occur as a cP air mass passes across a large ice-free lake in winter.
4. What air mass and source region provide the greatest amount of moisture to the eastern and central United States?

# 8 Concepts in Review Air Masses

## 8.1 What Is an Air Mass?

▶ Define *air mass* and *air-mass weather.*

**Key Terms:** air mass, air-mass weather

- An air mass is a large body of air, usually 1600 kilometers (1000 miles) or more across, that is characterized by a sameness of temperature and moisture at any given height. When this air moves out of its region of origin, it carries these temperatures and moisture conditions elsewhere, perhaps eventually affecting a large portion of a continent.
- The day-to-day weather we experience depends on the temperature, stability, and moisture content of the air mass affecting our location. A region under the influence of an air mass usually exhibits relatively uniform weather conditions perhaps for several days, a situation referred to as air-mass weather.

## 8.2 Classifying Air Masses

▶ List the basic criteria for an air-mass source region and identify the source regions that influence North America.

**Key Terms:** source region, polar (P) air mass, arctic (A) air mass, tropical (T) air mass, maritime (m) air mass, continental (c) air mass

- Areas in which air masses originate, called source regions, must be extensive and physically uniform areas and must be characterized by a general stagnation of atmospheric circulation.
- The classification of an air mass depends on the latitude of the source region and the nature of the surface in the area of origin—ocean or continent. Air masses are identified by two-letter codes.
- Continental (c) designates an air mass of land origin, with the air likely to be dry, whereas a maritime (m) air mass originates over water and therefore is humid. Polar (P) and arctic (A) air masses originate in high latitudes and are cold. Tropical (T) air masses form in low latitudes and are warm. According to this classification scheme, the basic types of air masses are continental polar (cP), continental arctic (cA), continental tropical (cT), maritime polar (mP), and maritime tropical (mT).

**Q** Identify the source region associated with each letter on this map. One of the letters on the map is not associated with a source region. Which one is it? Explain.

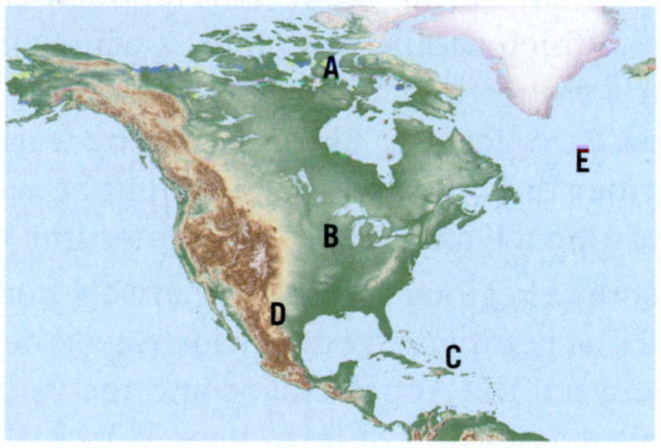


## 8.3 Air-Mass Modification

▶ Describe the processes by which traveling air masses are modified and discuss two examples.

- Changes to the stability of an air mass can result from temperature differences between an air mass and the surface and/or vertical movements induced by cyclones, anticyclones, or topography.
- An air mass that is colder than the surface beneath it tends to become unstable. An air mass that is chilled from below tends to become stable.
- Convergence and lifting cause an air mass to become more unstable, whereas subsidence acts to stabilize an air mass.

**Q** What is the correct designation for the air mass pictured in Figure 8.10: mTk or mTw? Explain.

## 8.4 Properties of North American Air Masses

▶ Summarize the weather conditions associated with each of the air masses that influence North America during the summer and the winter.

**Key Terms:** lake-effect snow, nor'easter, isohyet, atmospheric river

- Continental polar (cP) and maritime tropical (mT) air masses influence the weather of North America most, especially east of the Rocky Mountains, because the convergence associated with traveling cyclones draws these contrasting air masses together.
- Lake-effect snows occur on the downwind sides of large lakes and are associated with cP air masses that move across relatively warm lakes, where they are supplied with moisture and become unstable.
- Maritime polar (mP) air masses that influence the Pacific coast of North America tend to be unstable in winter and stable in summer. Storms called nor'easters result when mP air from the North Atlantic is drawn into a low-pressure center along the east coast.
- Maritime tropical (mT) air masses from the Gulf of Mexico and adjacent Atlantic Ocean are a major source of precipitation in the eastern two-thirds of the United States.

- Pacific mT air masses affect North America much less than mT air masses from the Gulf of Mexico and the adjacent North Atlantic. Sometimes mT air from the subtropical Pacific is part of an atmospheric river, a narrow corridor of concentrated moisture that can produce heavy rains along the Pacific coast.

**Q** This image shows the effects of a major December snowstorm at Rochester, New York, a city on the southern shore of Lake Ontario. Places not far from Rochester received only modest amounts of snow or none at all. Name and describe the process that likely produced this scene.

## Give it Some Thought

1. Air masses can be classified as cold or warm, but there is variation within this designation. In each case that follows, provide a brief explanation for your answer.
   **a.** We know that, during the winter, all polar (P) air masses are cold. Which should be colder: a wintertime mP air mass or a wintertime cP air mass?
   **b.** We expect tropical (T) air masses to be warm, but some are warmer than others. Which should be warmer: a summertime cT air mass or a summertime mT air mass?
2. Air-mass source regions are large, relatively homogenous areas. As the accompanying map illustrates, the broad expanse between the Appalachians and the Rockies is such a zone, yet air masses do not form here. Why is this area not a source region?

3. What is the proper classification for an air mass that forms over the Arctic Ocean in winter: cA or mA? Explain your choice.

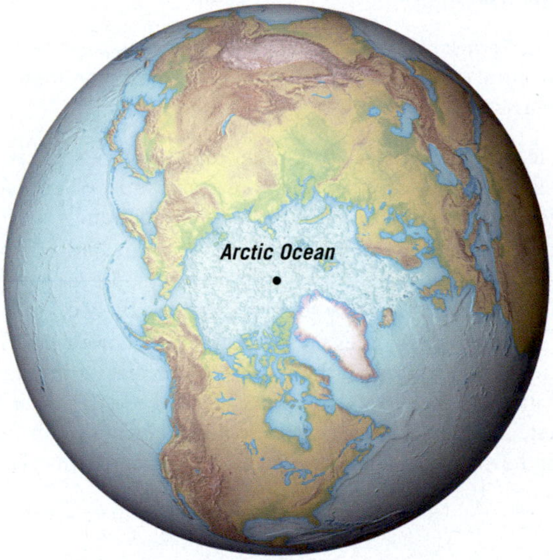


4. The Great Lakes are not the only water bodies associated with lake-effect snow. For example, large lakes in Canada also experience this phenomenon. Shown below are snowfall data for Fort Resolution, a settlement on the southeastern shore of Great Slave Lake.

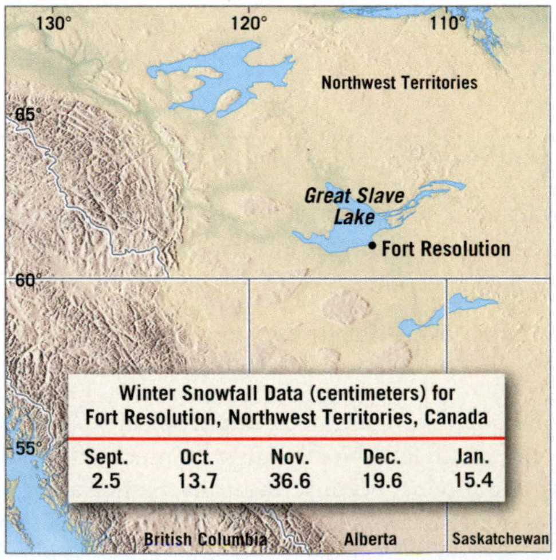


Winter Snowfall Data (centimeters) for Fort Resolution, Northwest Territories, Canada

| Sept. | Oct. | Nov. | Dec. | Jan. |
|---|---|---|---|---|
| 2.5 | 13.7 | 36.6 | 19.6 | 15.4 |

During what month is snowfall greatest? Suggest an explanation as to why the maximum occurs when it does.

5. In each of the situations described here, indicate whether the air mass is becoming more stable or more unstable. Briefly explain each choice.
   **a.** An mT air mass moving northward from the Gulf of Mexico over the southeastern states in winter
   **b.** A cP air mass moving southward across Lake Superior in late November
   **c.** An mP air mass in the North Atlantic drawn into a low-pressure center off the coast of New England in January
   **d.** A wintertime cP air mass from Siberia moving eastward from Asia across the North Pacific

## Problems

1. The accompanying map shows the distribution of air temperatures (top number) and dew-point temperatures (lower number) for a December morning. Two well-developed air masses are influencing North America at this time. The air masses are separated by a broad zone that is not affected by either air mass. Draw lines on the map to show the boundaries of each air mass. Label each air mass with the proper classification.

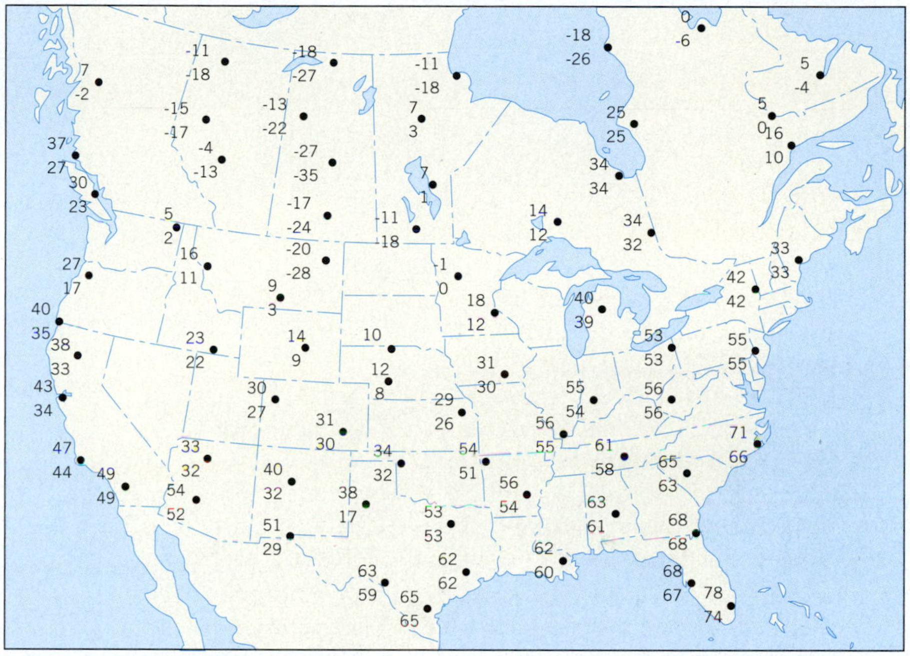


Air temperatures and dew-point temperatures in Fahrenheit for weather stations on a December morning.

2. Refer to Figure 8.6. Notice the narrow, north–south-oriented zone of relatively heavy snowfall east of Pittsburgh and Charleston. This region is too far from the Great Lakes to receive lake-effect snows. Speculate on a likely reason for the higher snowfalls here. Does your explanation explain the shape of this snowy zone?
3. Albuquerque, New Mexico, is situated in the desert Southwest. Its annual precipitation is just 21.2 centimeters (8.3 inches). Month-by-month data (in centimeters) are as follows:

| Jan. | Feb. | Mar. | Apr. | May | Jun. | Jul. | Aug. | Sep. | Oct. | Nov. | Dec. |
|---|---|---|---|---|---|---|---|---|---|---|---|
| 1.0 | 1.0 | 1.3 | 1.3 | 1.3 | 1.3 | 3.3 | 3.8 | 2.3 | 2.3 | 1.0 | 1.3 |

What are the two rainiest months for Albuquerque? The pattern here is similar to the pattern in other southwestern cities, including Tucson, Arizona. Briefly explain why the rainiest months occur when they do.

# 9 Midlatitude Cyclones

## Focus on Concepts

*Each statement represents the primary learning objective for the corresponding major heading within the chapter. After you complete the chapter, you should be able to:*

**9.1** Compare and contrast typical weather associated with a warm front and a cold front.

**9.2** Outline the stages in the life cycle of a typical midlatitude cyclone.

**9.3** Describe the general weather conditions associated with the passage of a mature midlatitude cyclone.

**9.4** Explain why divergence in the flow aloft is a necessary condition for the development and intensification of a midlatitude cyclone.

**9.5** List the primary sites for the development of midlatitude cyclones that affect North America.

**9.6** Explain the conveyor belt model of a midlatitude cyclone and sketch the three interacting air streams (conveyor belts) on which it is based.

**9.7** Describe blocking high-pressure systems and how they influence weather over the midlatitudes.

**9.8** Summarize the weather associated with a midlatitude cyclone over the north-central United States in winter.

The winter of 1992–1993 came to a stormy conclusion in eastern North America one weekend in March. Daffodils were blooming across the South, and people were thinking about spring when the blizzard of '93 struck on March 13 and 14. The huge storm brought record-low temperatures and barometric-pressure readings, accompanied by record-high snowfalls from Alabama to the Maritime Provinces of eastern Canada. The monster storm, with its driving winds and heavy snow, combined the attributes of a hurricane and a blizzard as it moved up the spine of the Appalachians, lashing and burying a huge swath of territory. Although the atmospheric pressure at the storm's center was lower than the pressures at the centers of some hurricanes and the winds were frequently as strong as those in hurricanes, this was definitely not a tropical storm—but instead a classic winter cyclone.

*Supercell thunderstorm that formed along a cold front over Kansas.*

# 9.1 Frontal Weather

**Compare and contrast typical weather associated with a warm front and a cold front.**

MM GEODe ▶ Basic Weather Patterns ▶ Fronts

In previous chapters we examined the basic elements of weather as well as the dynamics of atmospheric motions. Our knowledge of these diverse phenomena applies directly to an understanding of day-to-day weather patterns in the middle latitudes. For our purposes, *middle latitudes* refers to the area of North America roughly between southern Alaska and Florida—essentially the area of the prevailing westerlies, where the primary weather producer is the **midlatitude**, or **middle-latitude, cyclone**. Midlatitude cyclones go by a number of different names, including wave cyclones, frontal cyclones, extratropical cyclones, low-pressure systems, and, simply, lows.

Because *weather fronts* are the major weather producers imbedded within midlatitude cyclones, we will begin our discussion with these basic structures.

## What Is a Front?

One prominent feature of middle-latitude weather is how suddenly and dramatically it can change (see the chapter-opening photo). Most of these sudden changes are associated with the passage of weather fronts. **Fronts** are boundary surfaces that separate air masses of different densities—one of which is usually warmer and contains more moisture than the other. However, fronts can form between any two contrasting air masses. When the vast sizes of air masses are considered, the zones (fronts) that separate them are relatively narrow and are shown as lines on weather maps (Fig. 9.1).

Generally, the air mass located on one side of a front moves faster than the air mass on the other side. Thus, one air mass actively advances into the region occupied by another and collides with it. During World War I, Norwegian meteorologists visualized these zones of air-mass interactions as analogous to battle lines and tagged them "fronts," as in battlefronts. It is along these zones of "conflict" that midlatitude cyclones develop and produce much of the precipitation and severe weather in the belt of the westerlies.

As one air mass moves into a region occupied by another, minimal mixing occurs along the frontal surface. Instead, the air masses retain their identity as one is displaced upward over the other. No matter which air mass is advancing, it is always the warmer, less-dense air that is forced aloft, whereas the cooler, denser air acts as a wedge on which lifting occurs. The process of warm air gliding up and over a cold air mass is termed **overrunning**.

There are five basic types of fronts—*warm fronts*, *cold fronts*, *stationary fronts*, *occluded fronts*, and *drylines*. Each type of front separates air masses of different densities, which results because of the temperature or humidity differences that exist on opposite sides of the front.

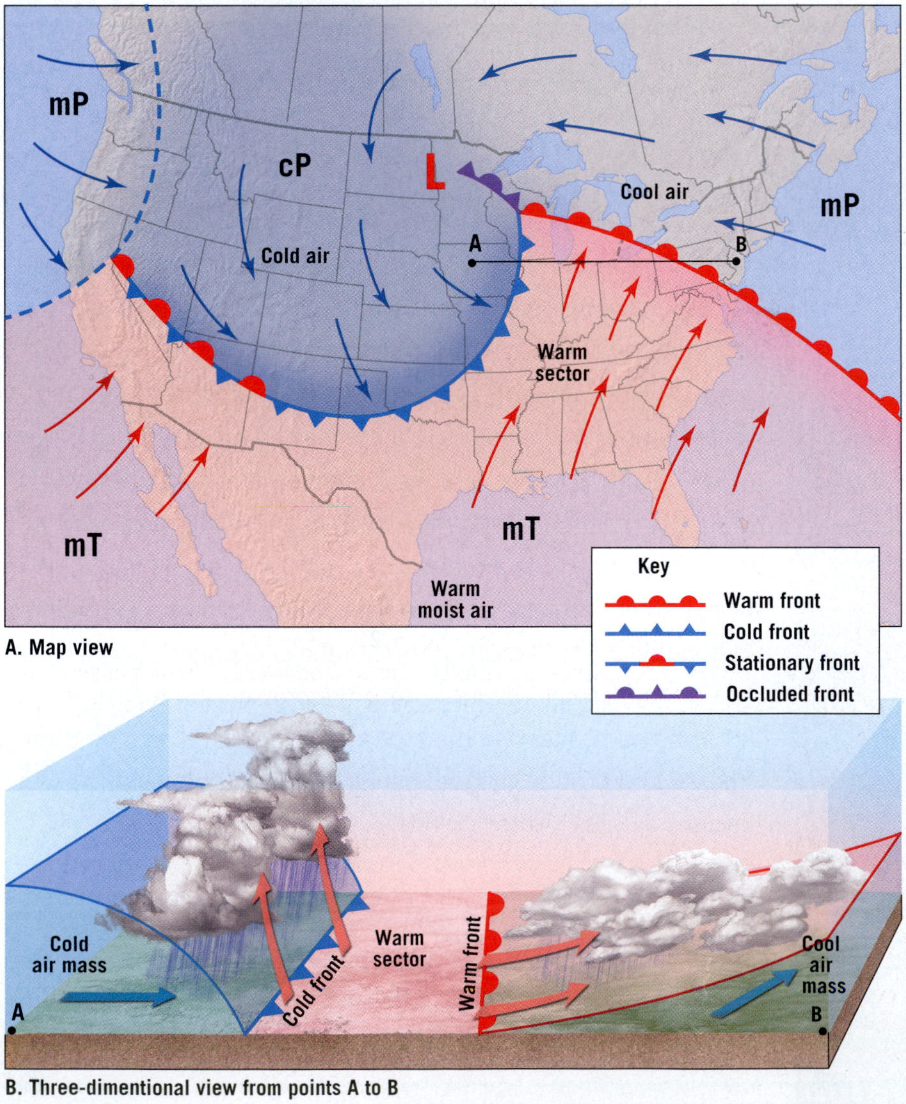


◀ **Figure 9.1 Idealized structure of a midlatitude cyclone** **A.** Map view showing fronts, air masses, and surface winds. **B.** Three-dimensional view of the warm and cold fronts along a line from point A to point B.

## Warm Fronts

When the surface position of a front moves so that warmer air invades territory formerly occupied by cooler air, it is called a **warm front** (Fig. 9.2A). On a weather map, a warm front is shown as a red line with red semicircles protruding into the area of

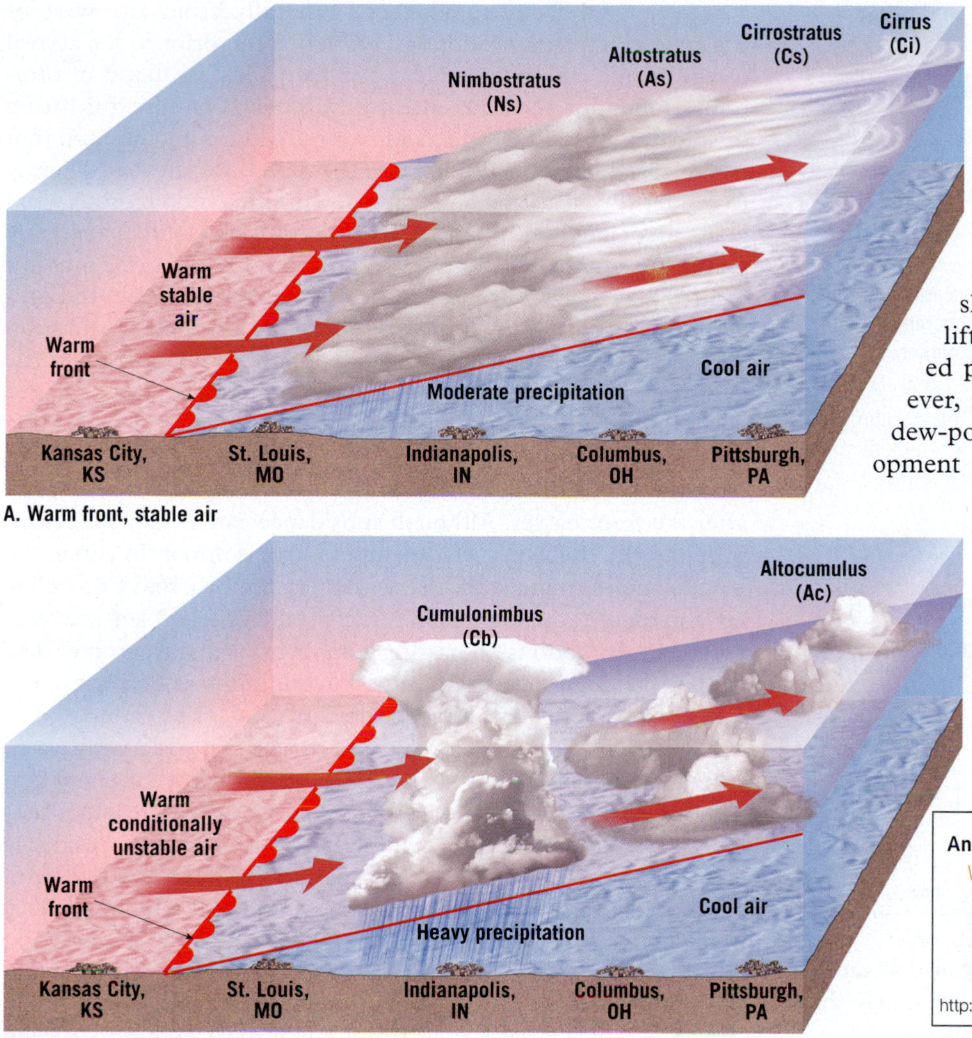


◀ **Figure 9.2 Warm fronts A.** Idealized clouds and weather associated with a warm front. During most of the year, warm fronts produce light to moderate precipitation over a wide area. **B.** During the warm season, when conditionally unstable air is forced aloft, cumulonimbus clouds and thunderstorms often arise.

cooler air. East of the Rockies, warm fronts are usually associated with maritime tropical (mT) air that enters the United States from the Gulf of Mexico and "glides" over cooler air positioned over land. The boundaries separating these air masses have very gradual slopes that average about 1:200 (height compared to horizontal distance). This means that if you traveled 200 kilometers (120 miles) ahead of the surface location of a warm front, the frontal surface would be 1 kilometer (0.6 mile) overhead.

As warm air ascends over the retreating wedge of cold air, it expands and cools adiabatically. As a result, moisture in the ascending air condenses to generate clouds that may produce precipitation. The cloud sequence in **Figure 9.2A** typically *precedes* the approach of a warm front. The first sign of the approaching warm front is cirrus clouds that form 1000 kilometers (600 miles) or more ahead of the surface front. Aircraft contrails provide another clue that a warm front is approaching. On a clear day, when condensation trails persist for several hours, you can be fairly certain that comparatively warm, moist air is ascending overhead.

As the front nears, cirrus clouds grade into cirrostratus that gradually blend into denser sheets of altostratus. About 300 kilometers (180 miles) ahead of the front, thicker stratus and nimbostratus clouds begin to form and may generate precipitation.

Because warm fronts have relatively gentle slopes, the cloud deck that results from frontal lifting covers a large area and produces an extended period of light-to-moderate precipitation. However, if the overriding air mass is relatively dry (low dew-point temperatures), there is minimal cloud development and no precipitation. By contrast, during the hot summer months when moist, conditionally unstable air is often forced aloft, towering cumulonimbus clouds and thunderstorms may occur (**Fig. 9.2B**).

As you can see from Figure 9.2A, the precipitation associated with a warm front occurs ahead of its surface position. Therefore, any precipitation that forms must fall through the cool layer below. During extended periods of light rainfall, enough of these raindrops may evaporate for saturation to occur, resulting in the development of a low-level stratus cloud deck. These clouds occasionally grow rapidly downward, causing problems for pilots of small aircraft that require visual landings. In these situations, aircraft pilots may have adequate visibility one minute but the next may be in a cloud mass, called *frontal fog*, that has the landing strip "socked in."

**Animation** MM
Warm Fronts
http://goo.gl/hdT4BM

In winter, snow may replace rain as the dominant form of precipitation associated with a warm front. In addition, when a relatively warm air mass is forced over a body of subfreezing air, hazardous driving conditions may occur ahead of a warm front. The raindrops may become supercooled as they fall through the subfreezing air and freeze on contact with road surfaces to produce an icy layer called *freezing rain*. Alternatively, the raindrops freeze as they pass though the layer of cold air and fall as ice pellets called *sleet* (see Fig. 5.20, page 141).

Temperatures gradually rise as the warm front passes. As you would expect, the increase is most apparent when a large contrast exists between adjacent air masses. Moreover, a wind shift from the east or southeast to the south or southwest is generally noticeable. (The shift in winds that accompanies frontal passage will be explained later.) The moisture content and stability of the encroaching warm air mass largely determine the time required for clear skies to return. During the summer, cumulus and occasionally cumulonimbus clouds are embedded in the warm, unstable air mass that follows the front. These clouds may produce precipitation, which can be heavy but is usually scattered and short in duration. **Table 9.1** shows the typical weather conditions that can be expected with the passage of a warm front in the Northern Hemisphere.

**Table 9.1** Weather Typically Associated with a Warm Front (North America)

| Weather Element | Before Passage | During Passage | After Passage |
|---|---|---|---|
| Temperature | Cool or cold | Rising | Warmer |
| Winds | East or southeast | Variable | South or southwest |
| Precipitation | Light to moderate rain, snow, or freezing rain in winter; heavy rain possible in summer | None or light rain | None, occasionally showers in summer |
| Clouds | Cirrus, cirrostratus, stratus, nimbostratus when air is stable; cumulonimbus when air is conditionally unstable | None, stratus, or fog | Clearing, cumulus, or cumulonimbus in summer |
| Pressure | Falling | Falling or steady | Falling then rising |
| Humidity | Moderate to high | Rising | High, particularly in summer |

Animation MM
Cold Fronts
http://goo.gl/XIlinH

## Cold Fronts

When cold air actively advances into a region occupied by warmer air, the zone of discontinuity is called a **cold front** (Fig. 9.3). On a weather map, a cold front is shown as a blue line with blue triangles protruding into the area of warmer air. Because of surface friction, the cold air near the ground advances more slowly than the air aloft. As a result, cold fronts steepen as they move.

Cold fronts are on average twice as steep as warm fronts, having slopes of perhaps 1:100. In addition, cold fronts advance at speeds up to 80 kilometers (50 miles) per hour, about 50 percent faster than warm fronts. These two differences—steepness of slope and rate of movement—largely account for the more violent nature of cold-front weather compared to the weather generally accompanying a warm front.

As a cold front approaches, generally from the west or northwest, towering clouds can often be seen prior to the arrival of the frontal boundary. Near the front, a dark band of ominous clouds foretells the ensuing weather. The forceful lifting of warm, moist air along a cold front is often rapid enough that the released latent heat increases the air's buoyancy sufficiently to render the air unstable. Heavy downpours and vigorous wind gusts associated with mature cumulonimbus clouds frequently result. Because a cold front produces roughly the same amount of lifting as a warm front, but over a shorter distance, the precipitation is generally more intense but of shorter duration (Fig. 9.4). A marked temperature drop and wind shift from the southwest to the northwest usually accompany frontal passage.

The weather behind a cold front is dominated by subsiding air within a continental polar (cP) air mass. Thus, the drop in temperature is usually accompanied by clearing that begins soon after the front passes. Although subsidence causes adiabatic heating aloft, the effect on surface temperatures is minor. In winter the long, cloudless nights that follow the passage of a cold front allow for abundant radiation cooling that produces frigid surface temperatures. By contrast, a passing cold front during a summer heat wave produces a welcome change in conditions as hot, hazy, and sometimes polluted mT air is replaced by cool, clear cP air.

When the air behind a cold front moves over a relatively warm surface, radiation emitted from Earth can heat the air enough to produce shallow convection. This in turn may generate low cumulus or stratocumulus clouds behind the front. However, subsidence aloft keeps these air masses relatively stable. Any clouds that form will not develop great vertical thickness and will seldom produce precipitation. One exception is the lake-effect snow discussed in Chapter 8, in which the cold air behind a front acquires heat and moisture as it traverses a comparatively warm body of water.

In North America, cold fronts form most commonly when a continental polar air mass clashes with maritime tropical air. However, wintertime cold fronts can form when even colder, dryer continental arctic (cA) air invades a continental polar or maritime polar air mass. Over land, arctic cold fronts tend to produce very light snow because the continental polar air masses they invade are quite dry. By contrast, an arctic cold front passing over a relatively warm water body may yield heavy snowfall and gusty winds. Table 9.2 shows the typical weather conditions associated with the passage of a cold front in North America.

A type of cold front termed a **backdoor cold front** sometimes affects the eastern seaboard of North America. Most cold fronts arrive from the west or northwest, whereas backdoor cold fronts come in from the east or northeast, hence their name. Driven by clockwise circulation from a strong high-pressure

Video MM
Tornados Ahead of a Cold Front
http://goo.gl/UITBgL

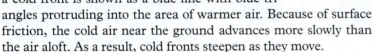

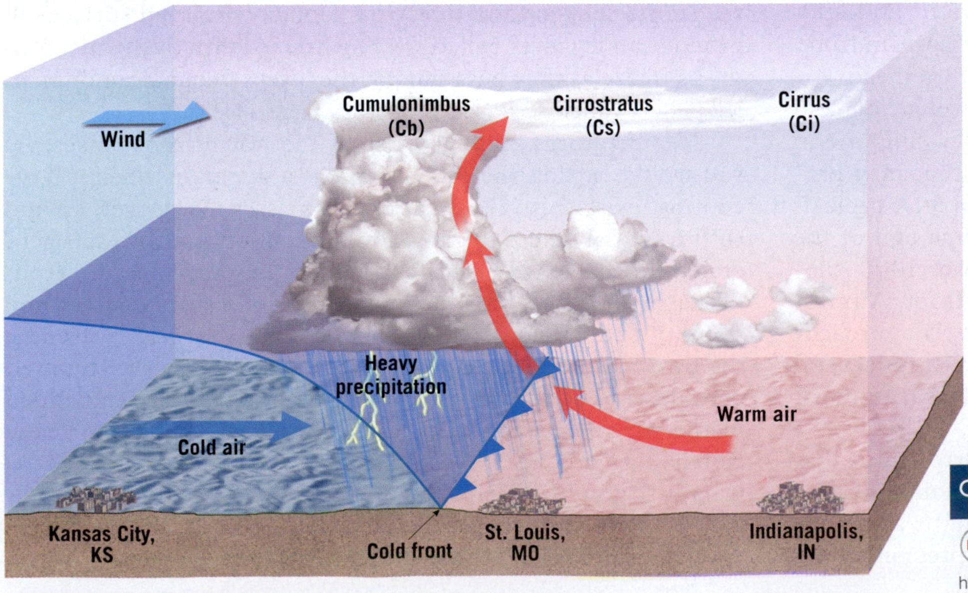


◀ **SmartFigure 9.3 Fast-moving cold front with cumulonimbus clouds** Thunderstorms often occur if the warm air is unstable.

http://goo.gl/YwZkL

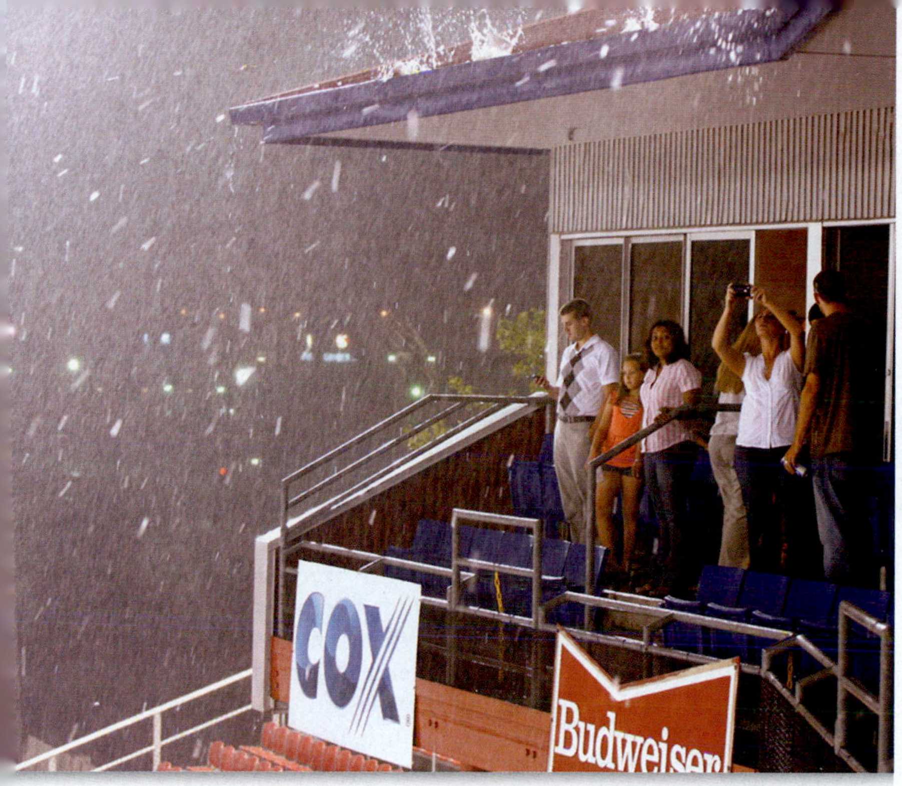


▲ **Figure 9.4 A cold front produced hail and heavy rain at this ballpark in Wichita, Kansas** Dozens of cars in the parking lot were damaged.

center over northeastern Canada, colder, denser maritime polar (mP) air from the North Atlantic displaces warmer, lighter air over the continent, as shown in Figure 9.5. Backdoor fronts are primarily springtime events that tend to bring cold temperatures, low clouds, and drizzle, although thunderstorms occur occasionally. Backdoor cold fronts are less frequent in summer, but when they occur, the cool air can provide welcome relief from midsummer heat waves in the northeastern United States.

## Stationary Fronts

Occasionally, airflow on both sides of a front is neither toward the cold air mass nor toward the warm air mass. Rather, it is almost parallel to the line of the front. Consequently, the surface position of the front does not move, or it moves sluggishly. This condition is called a **stationary front**. On a weather map, a stationary front is shown with blue triangles pointing into the warm air and red semicircles pointing into the cold air (see Fig. 9.1A). Because some overrunning usually occurs along stationary fronts, gentle to moderate precipitation is likely. Stationary fronts may remain over an area for several days, in which case flooding is possible. When stationary fronts begin to move, they become cold or warm fronts, depending on which air mass advances.

**Table 9.2** Weather Typically Associated with a Cold Front (North America)

| Weather Element | Before Passage | During Passage | After Passage |
|---|---|---|---|
| Temperature | Warm | Sharp drop | Colder |
| Winds | South or southwest | Variable and gusty | West or northwest |
| Precipitation | None or showers | Thunderstorms in summer, rain or snow in winter | Clearing |
| Clouds | None, cumulus, or cumulonimbus | Cumulonimbus | None or cumulus in summer |
| Pressure | Falling then rising | Rising | Rising |
| Humidity | High, particularly in summer | Dropping | Low, particularly in winter |

## Occluded Fronts

The fourth major front type is an **occluded front**, which most often forms when a rapidly moving cold front overtakes a warm

▼ **Figure 9.5 Weather associated with a backdoor cold front in the Northeast** In early spring, a warm, sunny day can become chilly and damp as cool, moist maritime polar (mP) air moves inland from the North Atlantic.

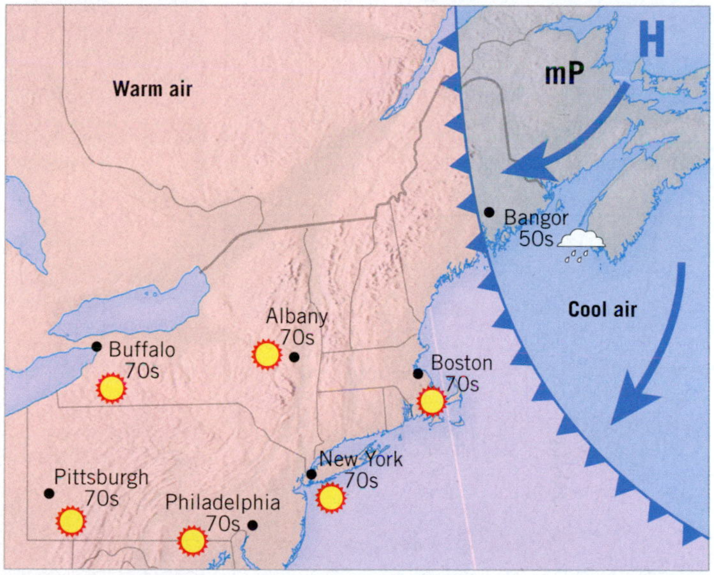


A.

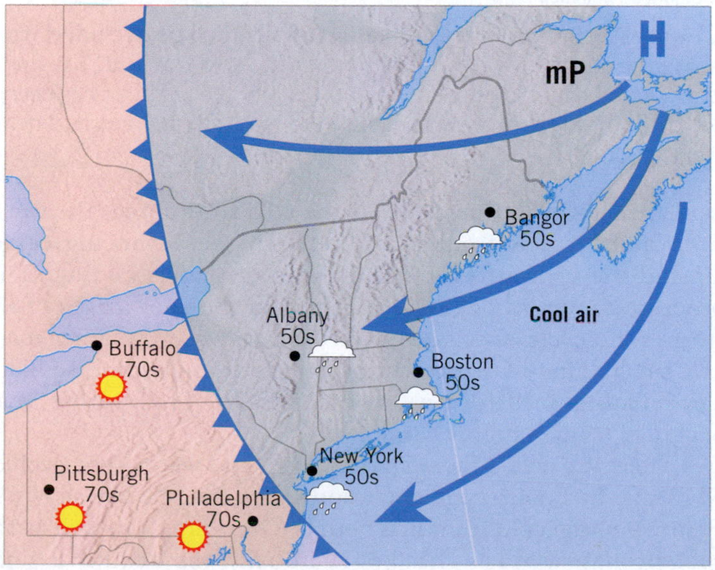


B.

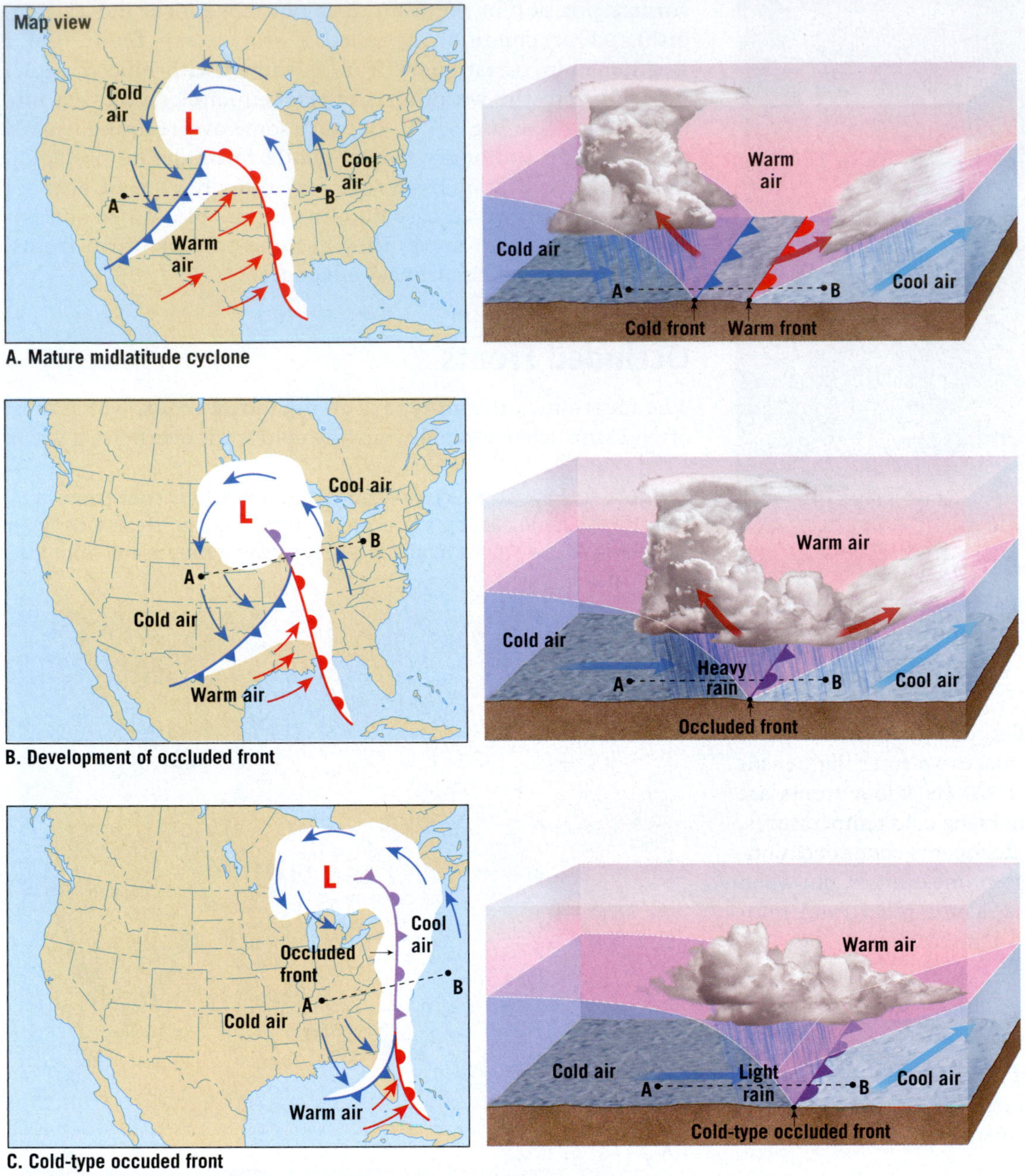


▲ **Figure 9.6 Stages in the formation of a cold-type occluded front**
**A.** Mature midlatitude cyclone with warm and cold fronts. **B.** The cold front overtakes the warm front to produce a cold-type occluded front. **C.** After the warm air has been forced aloft, the system begins to dissipate. The areas shown in white on the map indicate regions where clouds and precipitation are most likely to occur.

front, as shown in Figure 9.6. As the cold front forces the warm front aloft, a new front forms between the advancing colder air and the cool air over which the warm air is gliding. This process, known as **occlusion**, occurs in the later stages of the storm's life cycle. Occluded fronts are drawn on weather maps as a purple line with alternating purple triangles and semicircles pointing in the direction of movement. The weather of an occluded front is highly variable.

There are cold-type occluded fronts and warm-type occluded fronts. In a **cold-type occluded front**, like the one shown in Figure 9.6, the cold front lifts both the warm front as well as the cool air mass that lies ahead of it. Initially, the weather is similar to that associated with a warm front. However, as the occlusion develops and the warm air is lifted increasingly higher, thunderstorms may develop. Thus, fully developed cold-type occluded fronts may resemble cold fronts in the type of weather generated. Cold-type occlusions are more common than warm-type occlusions.

A **warm-type occluded front** develops when the air behind an advancing front is warmer than the cold air it is overtaking. This type of occluded front most often forms along the west coasts of continents, where milder maritime polar air invades frigid polar air that had its origin over the adjacent continent (Fig. 9.7). In this situation, the cool air is warmer and lighter than the cold air ahead of the front. Consequently, the cool air ascends and moves over the denser cold air ahead of the newly developed occluded front. The weather associated with a warm-type occlusion is usually similar to that of warm front: gentle to moderate precipitation. However, if the warm air that is lifted is conditionally unstable, thunderstorms can develop.

## Drylines

Although most fronts separate air masses of different temperatures, frontal boundaries can also separate air masses having different humidity. All other factors being equal, dry air is more dense than humid air. Therefore, when a dry warm air mass advances into a region occupied by an equally warm but more humid air mass, a type of frontal boundary called a **dryline** develops.

Drylines most often develop over the southern Great Plains. This occurs when dry, continental tropical (cT) air originating in the Southwest meets moist, maritime tropical (mT) air from the Gulf of Mexico. Drylines are spring and summer phenomena, most often generating a band of severe thunderstorms along a line extending from Texas to Nebraska that moves eastward across the Great Plains. A dryline is easily identified by comparing the dew-point temperatures of the cT air west of the boundary with the dew points of the mT air mass to the east (Fig. 9.8).

## eye ON THE atmosphere 9.1

The accompanying five images show clouds that commonly form along frontal boundaries. Four of these cloud types are produced when stable air ascends a warm frontal boundary, and the other tends to form along a cold front.

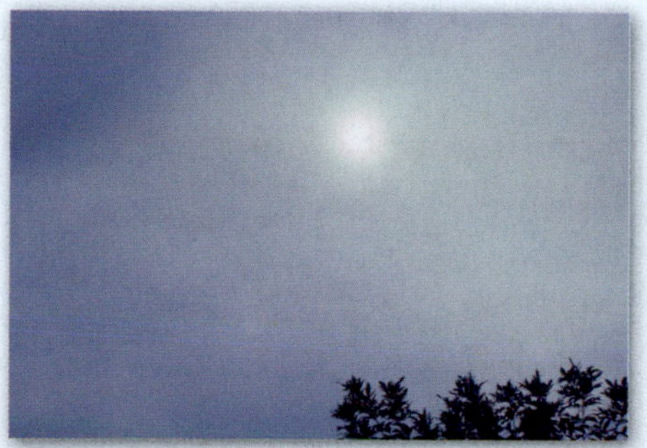

A.

B.

C.

D.

E.

### Questions

1. Which one of these five cloud types (A–E) tends to be generated along a cold front?
2. Assuming that a warm front is approaching your location, list the names and appropriate letters of the other four clouds in the order in which they would pass overhead.

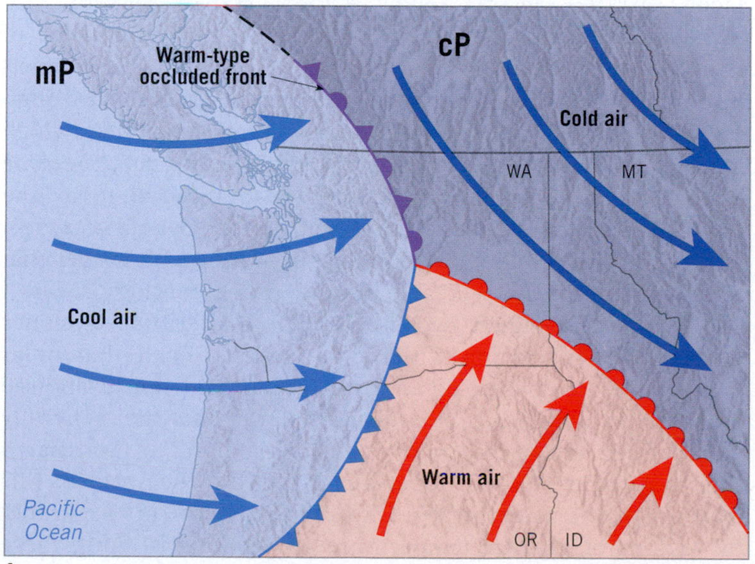


A.

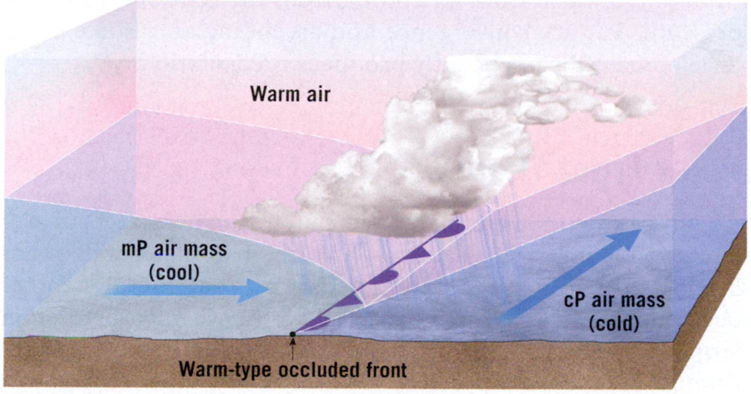


B.

▲ **Figure 9.7 Warm-type occluded front A.** Map showing the location of a warm-type occluded front in relationship to maritime polar (mP) and continental polar (cP) air masses. **B.** Block diagram of a warm-type occluded front.

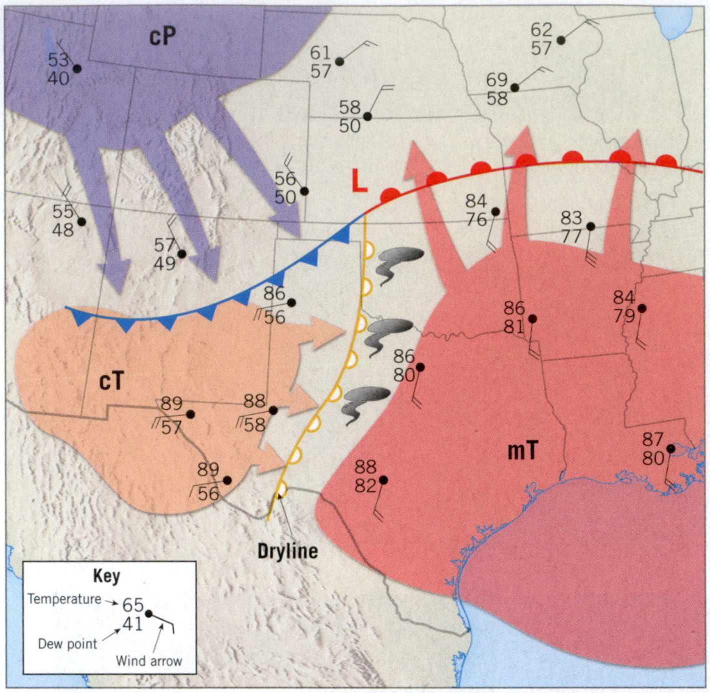


◀ **Figure 9.8 Dryline** This dryline developed over Texas and Oklahoma and generated thunderstorms and tornadoes. Notice that the dry (low–dew point) cT air is pushing eastward and displacing warm, moist mT air. The result is weather that resembles a rapidly moving cold front.

Video MM
Hurricanes and Air Masses
http://goo.gl/evwO0

The role of drylines in generating a line of severe thunderstorms called a *squall line* is discussed in Chapter 10 (see Fig. 10.13, page 275).

## ✔ Concept Checks 9.1

1. Compare the weather of a typical warm front with that of a typical cold front.
2. List two reasons why cold-front weather is usually more severe than warm-front weather.
3. What is a backdoor cold front?
4. How does a stationary front produce precipitation when its position does not change or changes very slowly?
5. In what way are drylines different from warm and cold fronts?

# 9.2 Midlatitude Cyclones and the Polar-Front Theory

**Outline the stages in the life cycle of a typical midlatitude cyclone.**

MM GEODe ▶ Basic Weather Patterns ▶ Introducing Middle-Latitude Cyclones

*Midlatitude cyclones* are synoptic-scale, low-pressure systems with diameters that often exceed 1000 kilometers (600 miles) and travel from west to east across the middle latitudes in both hemispheres (see Fig. 9.1). Lasting from a few days to more than a week, a midlatitude cyclone in the Northern Hemisphere has a counterclockwise circulation pattern, with airflow directed inward toward its center. Most midlatitude cyclones have a cold front and a warm front extending from the central area of low pressure. Surface convergence and ascending air initiate cloud development that frequently produces precipitation.

## Polar-Front Theory

As early as the 1800s, cyclones were known to be bearers of precipitation and severe weather. Thus, the barometer was established as the primary tool in "forecasting" day-to-day weather changes. However, this early method of weather prediction largely ignored the role of air-mass interactions in the formation of these weather systems. Consequently, it was impossible to pinpoint the conditions favorable to cyclone development.

The first comprehensive model of cyclone development and intensification was formulated by a group of Norwegian scientists during World War I. German restrictions on international communications to Norway—including critical weather reports pertaining to conditions in the Atlantic Ocean—led to the establishment of a closely spaced network of weather stations throughout Norway. Using this network, Norwegian-trained meteorologists greatly advanced our understanding of the weather and, in particular, weather associated with a midlatitude cyclone. Included in this group were Vilhelm Bjerknes (pronounced "Bee-yurk-ness"), his son Jacob Bjerknes, Jacob's fellow student Halvor Solberg, and Swedish meteorologist Tor Bergeron. In 1921 the work of these scientists resulted in a publication outlining a compelling model of how midlatitude cyclones progress through stages of birth, growth, and decay. These insights, which marked a turning point in atmospheric science, became known as the **polar-front theory**—also referred to as the **Norwegian cyclone model**. Even without the benefit of upper-air charts, these skilled meteorologists presented a model that remains remarkably applicable in modern meteorology.

In the Norwegian cyclone model, midlatitude cyclones develop in conjunction with polar fronts. Recall that polar fronts separate cold polar air from warm subtropical air (see Chapter 7). During cool months polar fronts are generally well defined and form a nearly continuous band around Earth that is recognizable on upper-air charts. At the surface, this frontal zone is often broken into distinct segments separated by regions of gradual temperature changes. It is along these

**students sometimes ask...**

**What is an extratropical cyclone?** Meaning "outside the tropics," *extratropical* is simply another name for a midlatitude cyclone. The term *cyclone* refers to the circulation around any low-pressure center, regardless of its size or intensity. Hence, midlatitude cyclones and hurricanes are two types of cyclones; *extratropical cyclone* is another name for a midlatitude cyclone, and *tropical cyclone* is often used to describe a hurricane.

frontal zones that cold, equatorward-moving air collides with warm, poleward-moving air to produce most midlatitude cyclones.

## Life Cycle of a Midlatitude Cyclone

According to the Norwegian model, cyclones form along fronts and proceed through a generally predictable life cycle. The development and strengthening of these storm systems is referred to as **cyclogenesis**—a process that can last from a few days to more than a week, depending on atmospheric conditions. One study found that more than 200 midlatitude cyclones form in the Northern Hemisphere annually. The stages in the life of a typical midlatitude cyclone are illustrated in Figure 9.9.

### Formation: Two Air Masses Clash

A midlatitude cyclone is born when two air masses of different densities (temperatures) move roughly parallel to a front but in opposite directions (Fig. 9.9A). In the classic polar-front model, this would be continental polar air associated with the *polar easterlies* on the north side of the front and maritime tropical air driven by the *westerlies* on the south side of the front.

**A Wave Develops** Under suitable conditions, the frontal surface that separates these two contrasting air masses takes the shape of a wave that is usually several hundred kilometers long (Fig. 9.9B). These waves are similar to, but much larger than, the swells that form on water. Some waves tend to dampen or die out, whereas others grow in amplitude. As these storms intensify, or "deepen," the waves change shape, much like a gentle ocean swell does as it moves into shallow water and becomes a tall, breaking wave.

Two factors cause a cyclonic storm to develop further: (1) diverging airflow aloft that compensates for the inflow at the surface and (2) lifting of warm, moist air along a warm front that leads to the release of latent heat as water vapor condenses to form clouds. This release of latent heat, in turn, enhances instability (convection).

**Cyclonic Flow** As a wave evolves, warm air advances poleward to form a warm front, while cold air moves equatorward to form a cold front (Fig. 9.9C). This change in the direction

Animation Midlatitude Cyclones
http://goo.gl/mah9eC

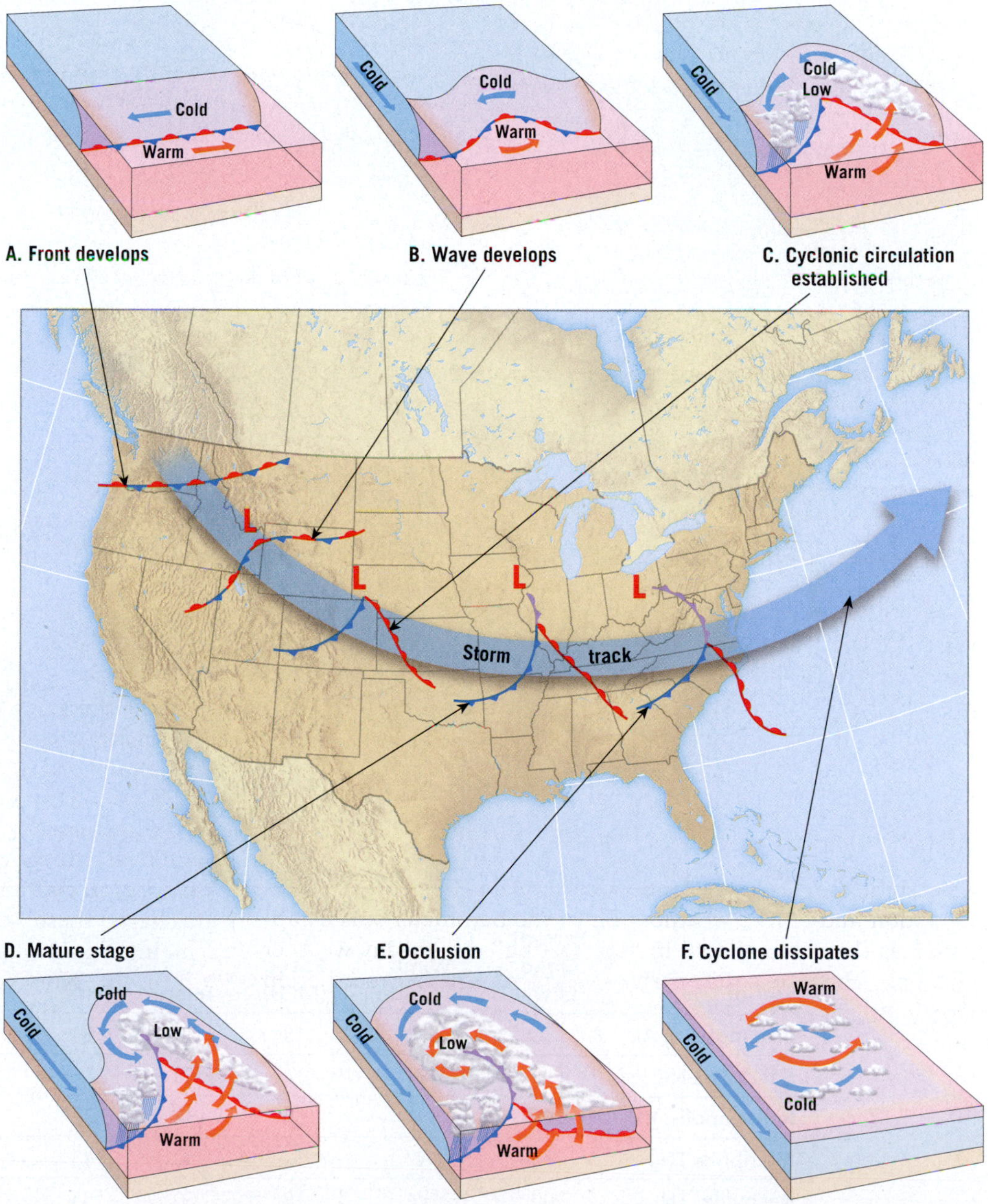


► **Figure 9.9 Stages in the life cycle of a midlatitude cyclone**

# What's Your Forecast?

## Constructing and Analyzing a Surface Weather Map and Forecast

Understanding day-to-day weather in the continental United States and many parts of Canada requires familiarity with the weather patterns associated with *midlatitude cyclones*. These large centers of low pressure generally travel from west to east and last from a few days to more than a week. A midlatitude cyclone typically contains a cold front and a warm front, and when they reach maturity, together they may develop an occluded front that extends from the central area of low pressure. One way to become familiar with these weather systems is to prepare and analyze a surface weather map.

### Preparing a Weather Map

To manage the large quantity of observational data that needs to be plotted on surface weather maps, meteorologists have developed a system for coding weather data. Appendix B (pages A-5–A-10) details that system and the symbols used to display data for a weather station. **Table 9.A** contains weather data for five central and eastern U.S. cities on a December day. Data for several cities have been plotted on the map in **Figure 9.A**. To complete the weather map in Figure 9.A, perform the following:

- Plot the data in Table 9.A for the appropriate stations on a copy of the map. Refer to Figure 12.7 or Appendix B, if needed. (*Note*: Plot the barometric pressure readings as written.)
- Beginning with the 996 isobar, draw and label the appropriate isobars as accurately as possible at 4-millibar intervals (996 mb, 1000 mb, 1004 mb, etc.). The 992 isobar has been completed, as have partial isobars for 1008 and 1012 millibars. You will have to estimate pressures between cities to determine the locations of isobars. Also, it is a good idea to first lightly sketch the isobars in pencil.

### Analyzing a Weather Map

Once it is prepared, a weather map must be analyzed. You can analyze your weather map by completing the following:

- Locate one center of either high or low pressure on the map and label it with the appropriate symbol (H or L).
- Use the weather data plotted on the map to determine the location of one cold front and one warm front. Draw and label these fronts on the map using the proper symbols.
- Label the air masses that are likely located to the northwest of the cold front, to the southeast of the cold front, and to the northeast of the warm front. (*Hint:* See Chapter 8.)
- Indicate areas of precipitation by lightly shading the map with a pencil.

Now that you have completed your analysis, you have the tools needed to complete a forecast.

### Making a Forecast

Assume that the midlatitude cyclone shown on the map is moving northeastward, as shown by the red arrow. Using your knowledge of the weather associated with a midlatitude cyclone, provide a forecast that includes changes in temperature, wind direction, probability of precipitation, cloud cover, and pressure tendency for the next 12–24 hours at the following locations: Chattanooga, Tennessee; Little Rock, Arkansas; Jackson, Mississippi; and Roanoke, Virginia.

### Investigating Further

Go to www.hpc.ncep.noaa.gov/html/sfc2.shtml. About halfway down the page is a scrollbar labeled North America; use it to scroll to United States (CONUS) and click on Get Image. Carefully examine that map, which is an analysis of current weather, to answer the following questions:

1. Locate the centers of high and low pressures as well as any frontal boundaries.
2. Find your location and determine whether temperature, cloud cover, and other data shown on the map match what you are currently experiencing.
3. Assuming that any low-pressure systems and frontal boundaries shown on the map are moving from west to east, describe how your weather might change in the next day or two.

**Table 9.A** December Surface Weather Data for Selected Cities in the Central and Eastern United States

| Station | Percent Cloud Cover | Wind Direction | Wind Speed (MPH) | Temp. | Dew Pt. Temp. | Pressure (MB) | Pressure Last 3 Hours (MB) | Precip. |
|---|---|---|---|---|---|---|---|---|
| **Birmingham, AL** | 80 | SW | 15 | 70 | 64 | 1004 | –1.4 | |
| **Charlotte, NC** | 70 | SW | 14 | 60 | 54 | 1002 | –4.4 | |
| **Indianapolis, IN** | 100 | NE | 30 | 34 | 32 | 996 | –5.6 | Snow |
| **Memphis, TN** | 80 | NW | 12 | 50 | 45 | 1103 | +5.8 | |
| **Nashville, TN** | 100 | SW | 18 | 56 | 55 | 996 | –0.1 | Rain |

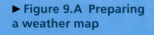

► **Figure 9.A Preparing a weather map**

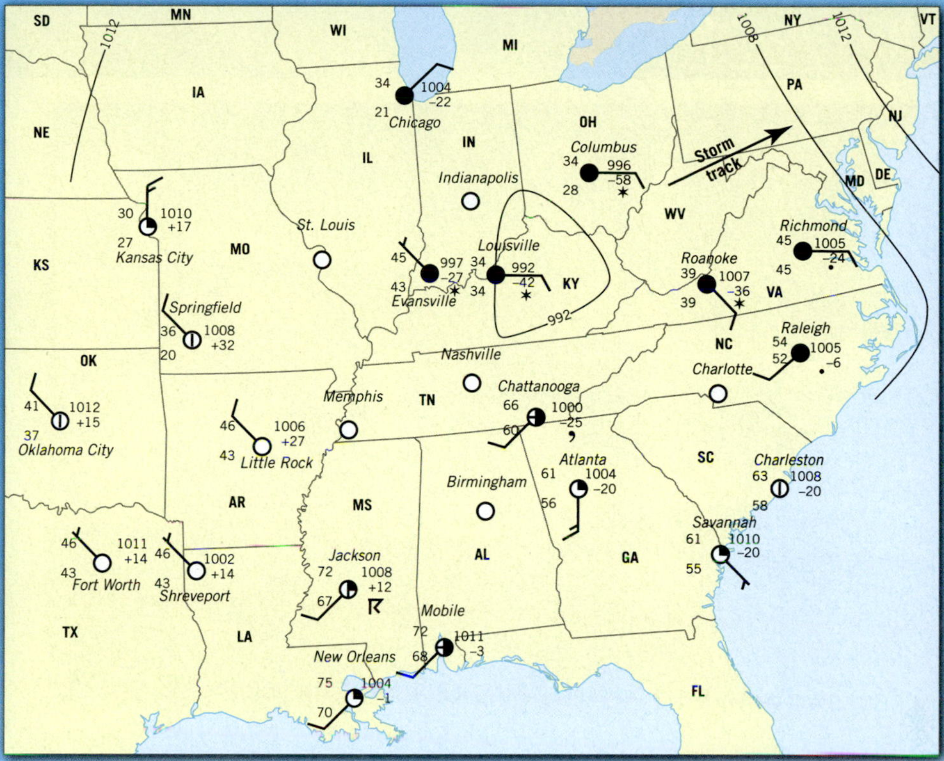


of the surface flow is accompanied by a readjustment in the pressure pattern and results in somewhat circular isobars, with the lowest pressure located at the crest of the wave. The resulting flow is an inward-directed, counterclockwise circulation that can be seen clearly on the weather map shown in Figure 9.10. Once the cyclonic circulation develops, convergence results in forceful lifting, especially where warm air is overrunning colder air. Figure 9.10 illustrates that the air in the *warm sector* (over the southern states) is flowing northeastward, toward cooler air that is moving toward the northwest. Because the warm air is moving perpendicular to the front, we can conclude that the warm air is invading a region formerly occupied by cold air. Therefore, this must be a warm front. Similar reasoning indicates that to the left (west) of the wave front, cold air from the northwest is displacing the air of the warm sector and generating a cold front.

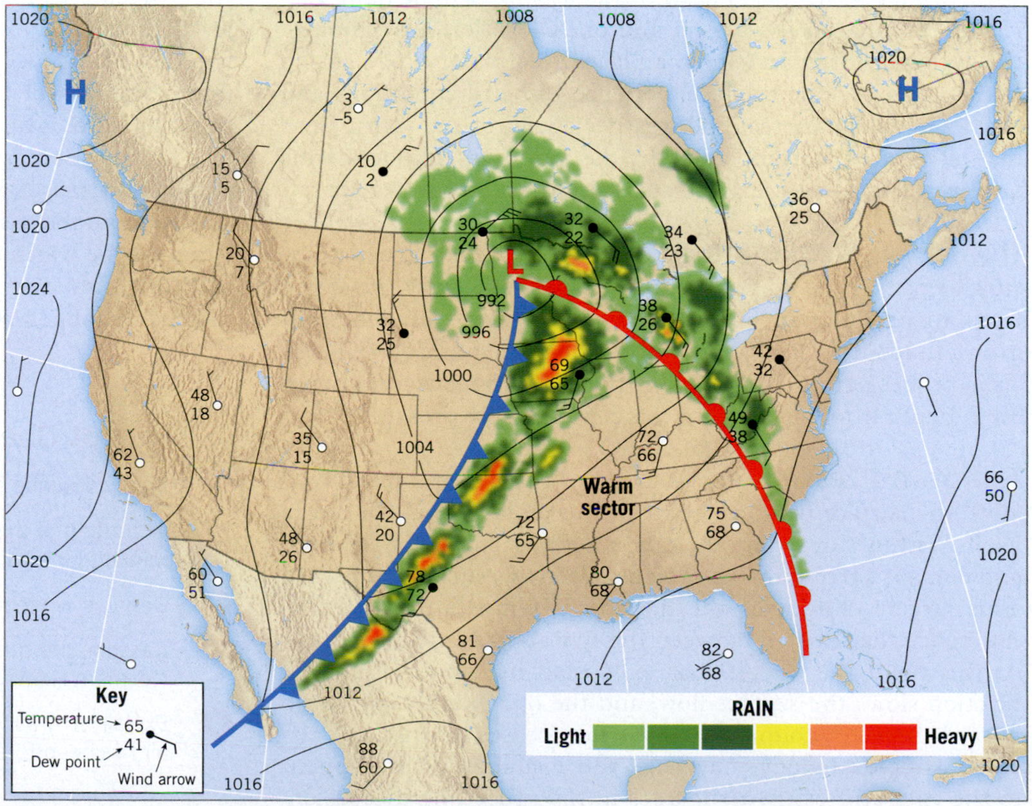


▲ **Figure 9.10 Simplified surface weather map showing the circulation of a midlatitude cyclone** The colored areas indicate the intensity of precipitation.

**Mature Stage** During the *mature stage* of a midlatitude cyclone, the pressure

## eye ON THE atmosphere 9.2

This image of a line of clouds was taken by astronauts aboard the International Space Station. The dashed line on the image shows the approximate surface position of the front responsible for the cloud development. Assume that this front is located over the central United States as you answer the following questions.

**Questions**

1. What is the cloud type of the tallest clouds in this image?
2. Is the cloud pattern shown typical of a cold front or a warm front?
3. Is this front moving toward the southeast or toward the northwest?
4. Is the air mass located to the southeast of the front a continental polar (cP) or a maritime tropical (mT) air mass?

surrounding the center low reaches its lowest level, and the *occluded front* begins to form (**Fig. 9.9D**). Although this mature stage produces the most hazardous weather, it can be quite varied, depending on the season and one's location in relation to the cyclonic storm. In winter, a place directly north of a strong midlatitude cyclone might experience blizzard conditions and heavy snowfall, whereas the weather ahead of the warm front may consist of freezing rain. In the summer, thunderstorms and even tornadoes are associated with these storm systems.

**Occlusion: Beginning of the End** Recall that occlusion begins as a cold front starts to overtake (lift) the warm front to produce an occluded front. As the occluded front grows in length, the warm sector is displaced aloft (**Fig. 9.9E**). This process gradually weakens the storm because the warm air is being lifted away from the center of low pressure.

**The Storm Dissipates** As more of the warm air is forced aloft, the surface pressure gradient weakens, as does the storm itself. Within 1 or 2 days the entire warm sector is forced aloft, and cold air surrounds the cyclone at Earth's surface (**Fig. 9.9F**). This largely eliminates the horizontal temperature (density) difference that existed between the two contrasting air masses. At this point the cyclone has exhausted its source of energy. Friction slows the surface flow, and the once highly organized inward-directed, counterclockwise flow ceases to exist.

A simple analogy may help you visualize what happens to the cold and warm air masses in the preceding discussion. Imagine a large water tank with a vertical divider separating it into two equal parts. Half of the tank is filled with hot water dyed red, and the other half is filled with blue-colored icy cold water. Now imagine what happens when the divider is removed. The cold, dense water will flow under the less-dense warm water, displacing it upward. This rush of water will come to a halt as soon as all of the warm water is displaced toward the top of the tank. In a similar manner, a midlatitude cyclone dies when all the warm air is displaced aloft and the horizontal discontinuity between the air masses no longer exists.

**Video** MM
A Midlatitude Cyclone's Effects on Society
http://goo.gl/hO5bz5

## ✔ Concept Checks 9.2

1. Briefly describe four characteristics of a midlatitude cyclone.
2. According to the Norwegian cyclone model, where do midlatitude cyclones form?
3. Define *cyclogenesis*.
4. Describe the surface circulation of a midlatitude cyclone in the Northern Hemisphere.
5. Explain why a cyclone begins to dissipate when most of the warm air has been forced aloft late in the process of occlusion.

## 9.3 Idealized Weather of a Midlatitude Cyclone

**Describe the general weather conditions associated with the passage of a mature midlatitude cyclone.**

MM® **GEODe** ▶ Basic Weather Patterns ▶ Introducing Middle-Latitude Cyclones

The Norwegian model is a valuable tool for interpreting midlatitude weather patterns, so keeping it in mind may help you understand, and possibly anticipate, changes in daily weather.

Guided by the westerlies aloft, cyclones generally move eastward across the United States. Therefore, we can expect the first signs of a cyclone's arrival to appear in the western sky. Upon reaching the Mississippi Valley, however, cyclones often begin a more northeasterly trajectory and occasionally move directly northward. Typically, a midlatitude cyclone requires 2 or more days to pass completely over a region. During that span, abrupt changes in atmospheric conditions may occur, particularly in late winter and spring, when the greatest temperature contrasts occur across the middle latitudes.

**Figure 9.11** illustrates a mature midlatitude cyclone; note the distribution of clouds and thus the regions of possible precipitation. Compare this map to the satellite image of a cyclone in **Figure 9.12**. It is easy to see why we often describe the cloud pattern of a cyclone as having a "comma" shape.

Depending on your location, two vastly different types of weather can be expected relative to the low pressure located near the center of the storm. Persons located south of the storm's center will encounter clouds and precipitation patterns shown by the profile provided below the map that corresponds to line *A–E*. Those located north of the storm's center will encounter clouds and precipitation patterns shown by the profile provided above the map that corresponds to line *F–G*. Because these storms move from west to east, the right side of the cyclone shown in Figure 9.11 will be the first to pass over a particular region.

First, imagine the change in weather as you move from right to left along profile *A–E*. At point *A* the sighting of cirrus clouds is the first sign of the approaching cyclone. These high clouds can precede the surface front by 1000 kilometers (600 miles) or more and are normally accompanied by falling pressure. As the warm front advances, lowering and thickening of the cloud deck occurs. Within 12 to 24 hours after the first sighting of cirrus clouds, light precipitation usually commences (point *B*). As the front nears, the rate of precipitation increases, the temperature rises, and winds begin to change from an easterly to a southwesterly flow. Summer precipitation comes in the form of rain, which at times can be heavy. In the winter, sleet and freezing rain can occur, making driving hazardous.

With the passage of the warm front, the area behind (west of) the front, called the *warm sector*, is under the influence of a maritime tropical air mass (point *C*). Depending on the season, the region

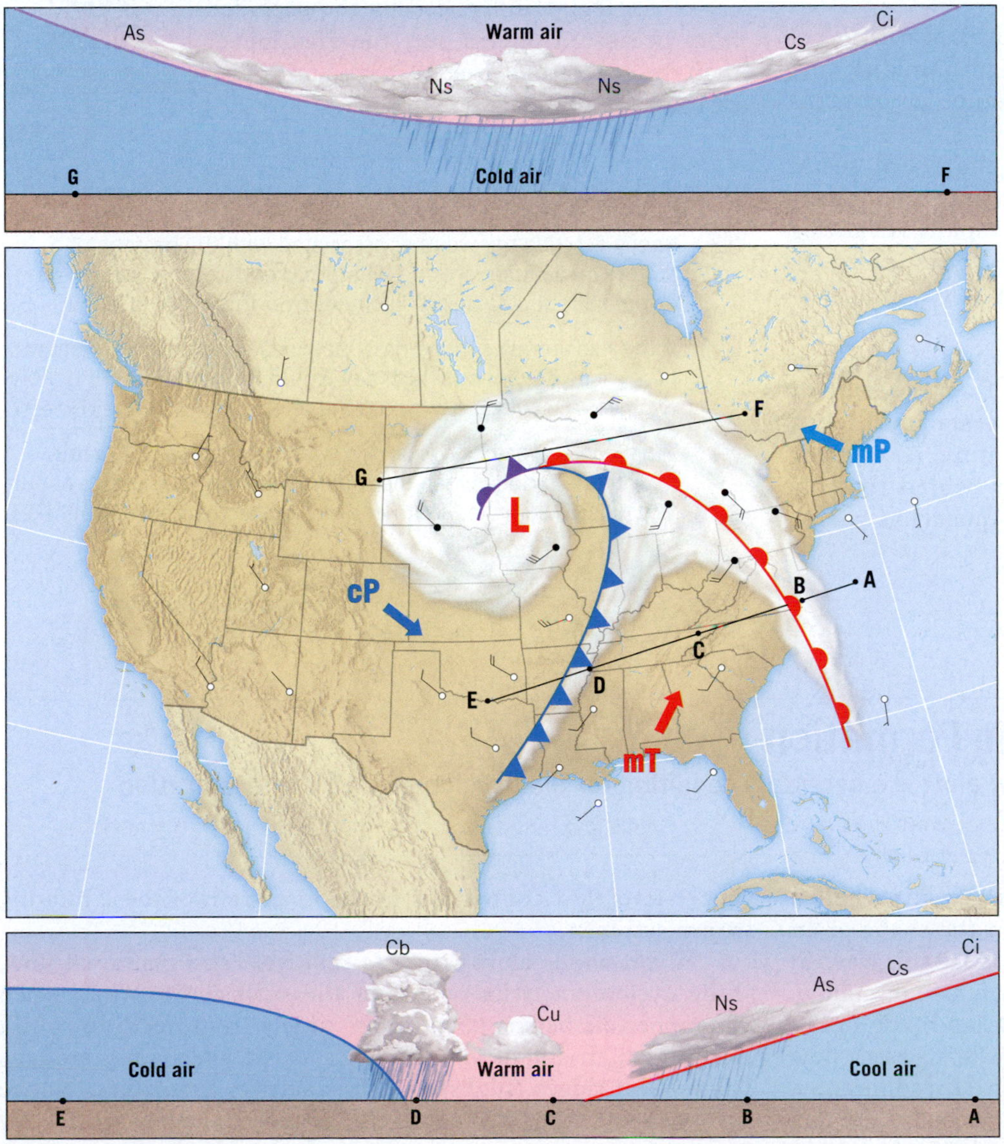


◀ **Figure 9.11 Cloud patterns typically associated with a mature midlatitude cyclone** Above the map is a vertical cross section along line *F–G*. The middle section is a map view with lines showing the surface positions of the cross sections. Below the map is a cross section along *A–E*. For cloud abbreviations, refer to Figures 9.2 and 9.3.

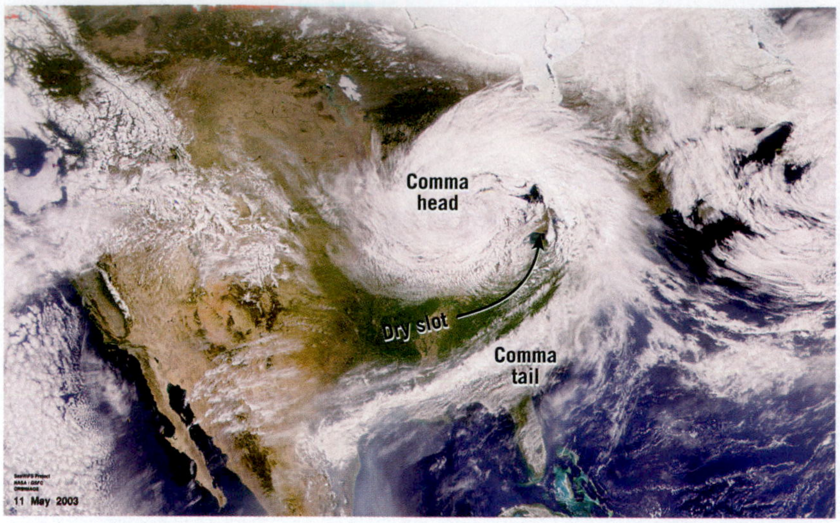


▲ **Figure 9.12 Satellite view of a mature midlatitude cyclone over the eastern half of the United States** It is easy to see why we often refer to the cloud pattern of a cyclone as having a "comma" shape.

affected by this part of the cyclone experiences warm to hot temperatures, southwesterly winds, fairly high humidity, and clear to partly cloudy skies containing cumulus or cumulus congestus clouds.

The warm conditions associated with the warm sector are quickly replaced by gusty winds and precipitation generated along the cold front. The approach of a rapidly advancing cold front is marked by a wall of rolling black clouds (point *D*). Severe weather accompanied by heavy precipitation, and occasionally hail or tornadoes, can be expected.

The passage of the cold front is easily detected by a dramatic shift in wind direction. The warm flow from the south or southwest is replaced by cold winds from the west to northwest, resulting in a pronounced decrease in temperature. Also, rising pressure hints of the subsiding cool, dry air behind the cold front. Once the front passes, the skies clear quickly as cooler, drier air invades the region (point *E*). A day or two of almost cloudless blue skies is often experienced, unless another cyclone is edging into the region.

A very different set of weather conditions prevails in the portion of the cyclone located north of the center of low pressure—profile *F*–*G* at the top of Figure 9.11. In this part of the storm, temperatures remain cool. The first hints of the approaching low-pressure center are a continual drop in air pressure and increasingly overcast conditions that bring varying amounts of precipitation. This section of the cyclone most often generates heavy snowfall during the winter months.

Once the process of occlusion begins, the character of the storm changes. Because occluded fronts tend to move more slowly than other fronts, the entire wishbone-shaped frontal structure shown in Figure 9.11 rotates counterclockwise. As a result, the occluded front appears to "bend over backward." This effect adds to the misery of the region influenced by the occluded front because it lingers over the area longer than the other fronts (see "What's Your Forecast?", page 244).

**Video** (MM)
Water Vapor Transport by Midlatitude Cyclones
http://goo.gl/aRLhHB

## ✔ Concept Checks 9.3

1. Briefly describe the weather associated with the passage of a mature midlatitude cyclone when the center of low pressure is located about 200 to 300 kilometers north of your location.
2. If the midlatitude cyclone described in Question 1 took 3 days to pass your location, on which day would the temperatures be the warmest? On which day would the temperatures be the coldest?
3. What winter weather is expected with the passage of a mature midlatitude cyclone when the center of low pressure passes about 100 to 200 kilometers south of a city located near the Great Lakes?

# 9.4 Flow Aloft and Cyclone Formation

**Explain why divergence in the flow aloft is a necessary condition for the development and intensification of a midlatitude cyclone.**

The polar-front model shows that cyclogenesis occurs where a frontal surface is distorted so that it takes on the shape of an ocean wave. Several factors are thought to produce this wave in a frontal zone. Topographic irregularities (such as mountains), temperature contrasts (as between sea and land), or ocean current influences can disrupt the zonal (west-to-east) flow sufficiently to produce a wave along a front. In addition, a strong jet stream in the flow aloft frequently precedes the formation of a surface cyclone. This fact strongly suggests that upper-level flow contributes to the formation of these rotating storm systems.

When winds aloft exhibit a relatively straight zonal flow, little cyclonic activity occurs at the surface. However, when the upper air begins to meander widely from north to south, forming high-amplitude rising waves of alternating troughs (lows) and ridges (highs), cyclonic activity intensifies. Notice in **Figure 9.13** that when surface cyclones form, they are usually

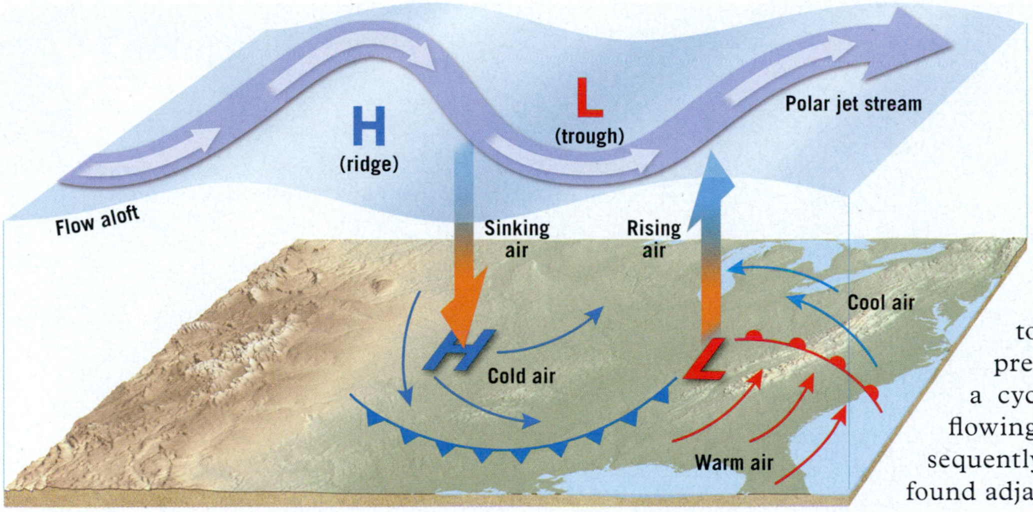


▲ **Figure 9.13 Relationship between the meandering flow in the jet stream aloft and cyclone development at the surface** Midlatitude cyclones tend to form downstream of an upper-level low (trough).

Video MM
Short Waves and Long Waves
http://goo.gl/WLIc81

centered below the jet stream core and downwind (east) of an upper-level low (trough).

## Cyclonic and Anticyclonic Circulation

Before discussing how surface cyclones are generated and supported by the flow aloft, let us review the nature of cyclonic and anticyclonic winds. Recall that in the Northern Hemisphere airflow around a surface low is counterclockwise and inward, which leads to mass convergence (coming together). Because the accumulation of air is accompanied by a corresponding increase in surface pressure, we would expect a surface low-pressure center to "fill" rapidly and be eliminated, just as the vacuum in a coffee can is quickly equalized when opened. When this occurs in a cyclonic storm, the surface pressure rises and the storm weakens, a process called *filling*.

However, cyclones often exist for a week or longer. For this to occur, surface convergence must be offset by rising air in the column and divergence aloft (**Fig. 9.14**). As long as outflow aloft is greater (more air is removed) than the amount of air that arrives at the surface (convergence), the low pressure at the surface will intensify. This process is referred to as *deepening*.

Surface anticyclones, where the airflow is clockwise and outward, are also supported by the flow aloft. In an anticyclone, divergence at the surface must be exceeded by convergence aloft and general subsidence (sinking) of the air column for the system to intensify (Fig. 9.14).

Because cyclones bring stormy weather, they have received far more attention than their counterparts, anticyclones. Yet the close relationship between them makes it difficult to totally separate a discussion of these two pressure systems. The surface air that feeds a cyclone generally originates as surface air flowing out of an anticyclone (Fig. 9.14). Consequently, cyclones and anticyclones are typically found adjacent to one another.

## Divergence and Convergence Aloft

Because divergence aloft is essential to cyclogenesis, a basic understanding of its role is important. Divergence aloft does not produce outward flow in all directions. Instead, the winds aloft generally flow from west to east along sweeping curves. How does zonal flow aloft cause upper-level divergence?

One mechanism responsible for divergence aloft is a phenomenon known as *speed divergence*. Wind speeds can change dramatically in the vicinity of the jet stream. On entering a zone of high wind speed, air accelerates and stretches out (divergence). In contrast, when air enters a zone of slower wind speed, an air pileup (convergence) results. Analogous situations occur every day on a toll highway. When exiting a toll booth and entering the zone of maximum speed, we find automobiles diverging (increasing the number of car lengths between them). As automobiles slow to pay the toll, they experience convergence (coming together).

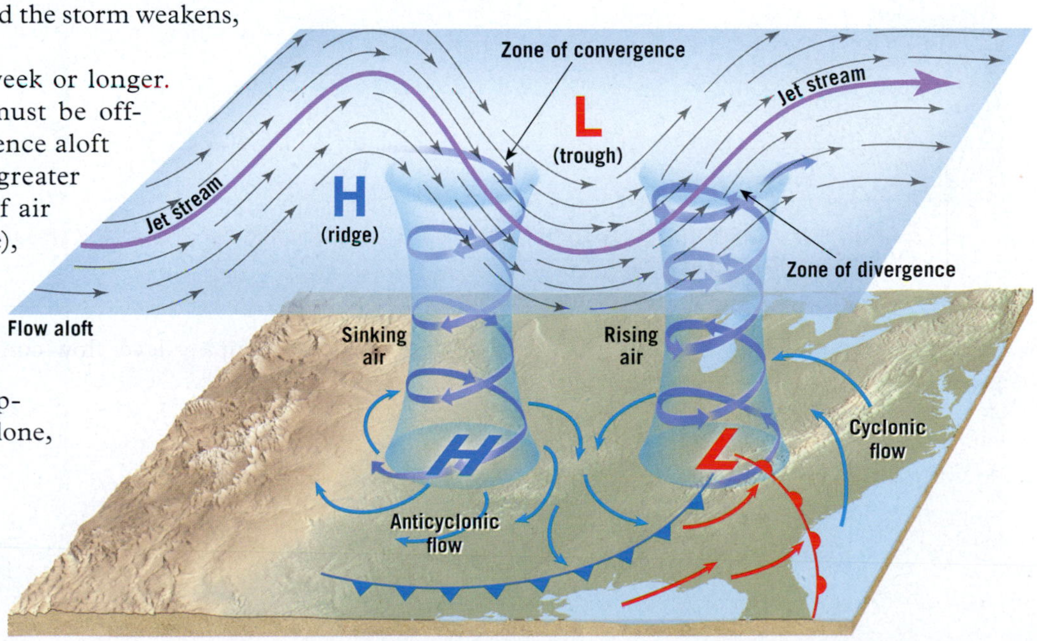


► **Figure 9.14 Idealized view of divergence and convergence aloft that supports cyclonic and anticyclonic circulation at the surface**

# What's Your Forecast?

## Winds as a Forecasting Tool

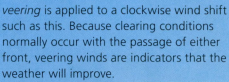

**"Every wind has its weather."**
**Francis Bacon, English philosoper and scientist (1561–1626)**

People living in the middle latitudes know well that during the winter, north winds can chill a person to the bone (**Fig. 9.B**). Conversely, a sudden change to a southerly flow can bring welcome relief from these frigid conditions. Especially observant people (or those who have completed a meteorology course!) might notice that when the wind direction switches from southeast to south, foul weather often follows. By contrast, a change in wind direction from the southwest to the northwest is usually accompanied by clearing conditions. Just how are the winds related to the weather to come?

Modern weather forecasts require the processing capabilities of high-speed computers as well as the expertise of people with considerable professional training. Nevertheless, careful observation yields some reasonable insights into the impending weather. The two most significant weather elements for this purpose are barometric pressure and wind direction. Recall that anticyclones (high-pressure cells) are associated with clear skies and that cyclones (low-pressure cells) frequently bring clouds and precipitation. Thus, we can roughly predict the forthcoming weather by noting whether the barometer is rising, falling, or steady. For example, rising pressure indicates the approach of a high-pressure system and generally fair weather.

The use of winds in weather forecasting is also straightforward. Because cyclones are the "villains" in the weather game, we are most concerned with the circulation around these storm centers. In particular, changes in wind direction that occur with the passage of warm and cold fronts are useful in predicting impending weather. Notice in Figure 9.11 that with the passage of both the warm and cold fronts, the wind arrows change positions in a clockwise manner. For example, with the passage of a cold front, the wind shifts from southwest to northwest. From nautical terminology, the word *veering* is applied to a clockwise wind shift such as this. Because clearing conditions normally occur with the passage of either front, veering winds are indicators that the weather will improve.

In contrast, Figure 9.11 shows that the area in the northern portion of the cyclone will experience winds that shift in a counterclockwise direction. Winds that shift in this manner are said to be *backing*. With the approach of a midlatitude cyclone, backing winds indicate cool temperatures and continued foul weather.

A summary of the relationship among barometer readings, winds, and the impending weather is provided in **Table 9.B**. Although this information is applicable in a very general way to much of the United States and Canada, local influences must be taken into account. For example, a rising barometer and a change in wind direction from southwest to northwest is usually associated with the passage of a cold front, and clearing conditions should follow. However, in the winter, residents of the southeast shore of one of the Great Lakes may not be so lucky. As cold, dry northwest winds cross large expanses of open water, they acquire heat and moisture from the relatively warm lake surface. By the time this air reaches the leeward shore, it is often humid and unstable enough to produce heavy lake-effect snow (see Chapter 8).

### Questions

1. You just experienced a veering wind shift. Make a forecast for your area.
2. Assume it is winter and you notice that the winds were out of the east a few hours ago, but are now blowing out of the northeast. Use this information plus the fact that the barometric pressure is falling rapidly to make a forecast for your area.

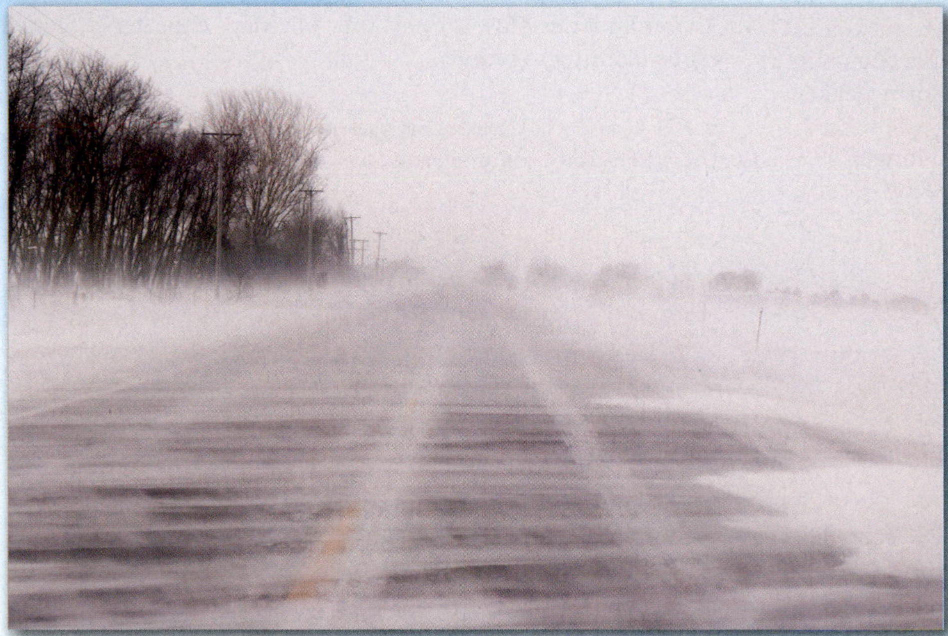

◄ **Figure 9.B Blizzard conditions make for hazardous driving.** This blizzard occurred along Highway 30 in western Iowa.

**Table 9.B** Wind, Barometric Pressure, and Impending Weather

| Changes in Wind Direction | Barometric Pressure | Pressure Tendency | Impending Weather |
|---|---|---|---|
| Any direction | 1023 mb and above (30.20 in.) | Steady or rising | Continued fair with no temperature change |
| SW to NW | 1013 mb and below (29.92 in.) | Rising rapidly | Clearing within 12 to 24 hours and colder |
| S to SW | 1013 mb and below (29.92 in.) | Rising slowly | Clearing within a few hours and fair for several days |
| SE to SW | 1013 mb and below (29.92 in.) | Steady or slowly falling | Clearing and warmer, followed by possible precipitation |
| E to NE | 1019 mb and above (30.10 in.) | Falling slowly | In summer, with light wind, rain may not fall for several days; in winter, rain within 24 hours |
| E to NE | 1019 mb and above (30.10 in.) | Falling rapidly | In summer, rain probable within 12 to 24 hours; in winter, rain or snow with strong winds likely |
| SE to NE | 1013 mb and below (29.92 in.) | Falling slowly | Rain will continue for 1 to 2 days |
| SE to NE | 1013 mb and below (29.92 in.) | Falling rapidly | Stormy conditions followed within 36 hours by clearing and, in winter, colder temperatures |

Source: Adapted from the National Weather Service.

In addition to speed divergence, other factors contribute to divergence (or convergence) aloft. These include *directional divergence* and *directional convergence*, which result from changes in wind direction. For example, directional convergence (also called *confluence*) is the result of air being funneled into a restricted area. On an upper-air chart, convergence occurs in regions where height contours move progressively closer together. Returning to our interstate highway analogy, convergence occurs where a crowded three-lane highway is reduced to two lanes due to construction. Directional divergence, the opposite phenomenon, is analogous to the location where the two lanes change back to three.

The combined effect of these and other factors is that an area of upper-air divergence and corresponding surface cyclonic circulation generally develop downstream from an upper-level trough, as illustrated in Figure 9.14. Consequently, in the United States midlatitude cyclones generally form downstream (east) of an upper-level trough. As long as divergence aloft exceeds convergence at ground level, surface pressures will fall, and the cyclonic storm will intensify.

Conversely, the zone in the jet stream that experiences upper-air convergence and corresponding anticyclonic rotation is located downstream from a ridge, or an upper-level high (Fig. 9.14). The accumulation of air in this region of the jet stream leads to subsidence (sinking) and increased surface pressure. Hence, it is a favorable site for the development of a surface anticyclone.

Because of the significant role that upper-level flow has on cyclogenesis, it should be evident that any attempt at weather prediction must account for the airflow aloft. This is why TV weather reporters frequently illustrate the flow within the jet stream.

In summary, the flow aloft contributes to the formation and intensification of surface low- and high-pressure systems. Changes in wind speeds and/or directions cause air either to pile up (convergence) or spread out (divergence). Upper-level convergence is favored downstream (east) from a ridge, whereas divergence occurs downstream (east) from an upper-level trough. At the surface, below regions of upper-level convergence are areas of high pressure (anticyclones), whereas upper-level divergence supports the formation and development of surface cyclonic systems (lows).

## Flow Aloft and Cyclone Migration

The wavy airflow aloft not only influences the development and evolution of a surface cyclonic storm; it also affects cyclone intensity and movement. Middle- and upper-troposphere flow strongly influences the rate at which these pressure systems

students sometimes ask...

### Why are winter storms now being named?

Beginning in the 2012–2013 winter season, The Weather Channel began naming noteworthy winter storms. Europe has been naming storms for some time, and the United States has been naming hurricanes and tropical storms since the 1940s. In the United States, a few major winter storms have been given informal names, including the "President's Day Storm" of 2003. The lack of a formal naming system may stem from the fact that we have no national center for winter storms, like the National Hurricane Center, which monitors tropical cyclones.

advance and the direction they follow. Generally, surface cyclones move in the same direction as winds aloft at the 500-millibar level, but at about one-quarter to one-half the speed. Normally these systems travel at 25 to 50 kilometers (15 to 30 miles) per hour, advancing roughly 600 to 1200 kilometers (400 to 800 miles) daily. The faster speeds occur during the coldest months, when temperature gradients are greatest.

One of the most difficult tasks in weather forecasting is predicting the paths of cyclonic storms. We have seen that the flow aloft tends to steer developing pressure systems. Let us examine an example of this steering effect by seeing how changes in the upper-level flow correspond to changes in the path taken by a cyclone.

Figure 9.15A illustrates the changing position of a midlatitude cyclone over a 4-day period. Notice in Figure 9.15B that on March 21, the 500-millibar contours are relatively flat. Also notice that for the following 2 days, the cyclone moves in a southeasterly direction. By March 23 the 500-millibar contours make a sharp bend northward on the eastern side of a trough situated over Wyoming (Fig. 9.15C). Likewise, the next day the path of the cyclone makes a similar northward migration.

Our example shows the influence of upper airflow on cyclonic movement after the fact. Useful predictions of cyclone movements require accurate appraisals of changes in the westerly flow aloft. For this reason, predicting the behavior of the wavy flow in the middle and upper troposphere is an important part of modern weather forecasting.

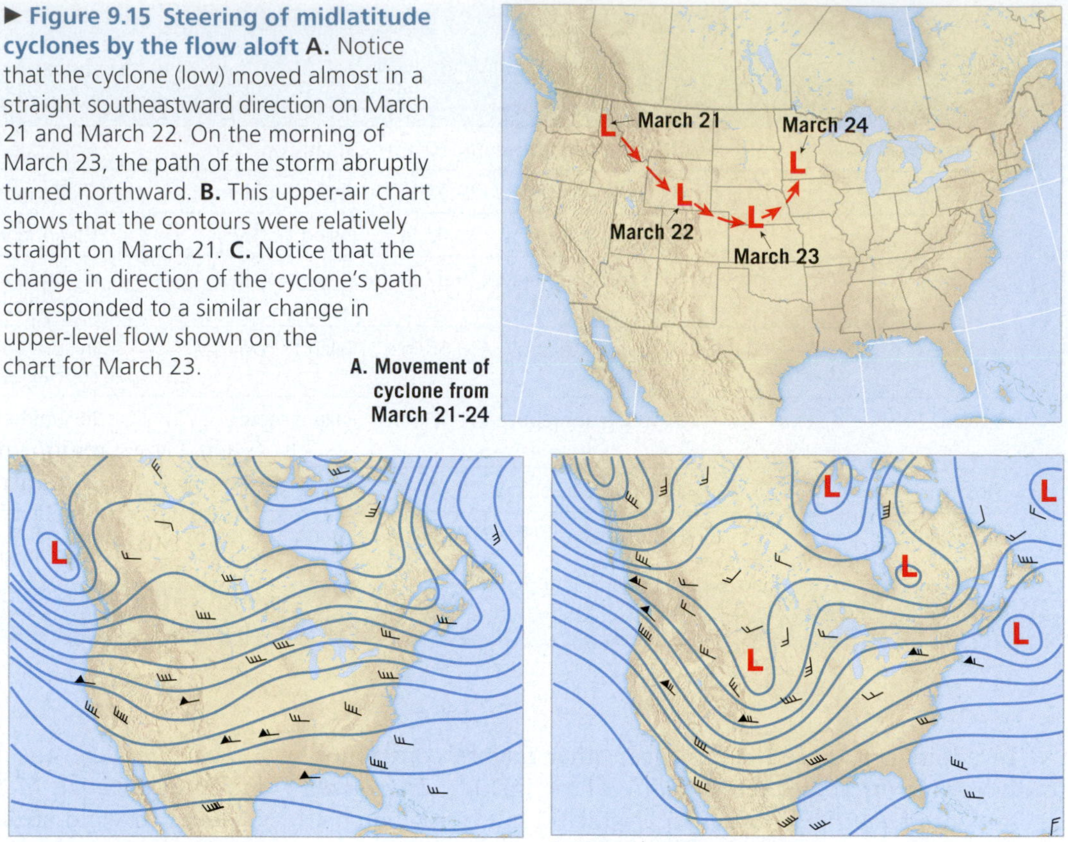


▶ **Figure 9.15 Steering of midlatitude cyclones by the flow aloft** **A.** Notice that the cyclone (low) moved almost in a straight southeastward direction on March 21 and March 22. On the morning of March 23, the path of the storm abruptly turned northward. **B.** This upper-air chart shows that the contours were relatively straight on March 21. **C.** Notice that the change in direction of the cyclone's path corresponded to a similar change in upper-level flow shown on the chart for March 23.

**A. Movement of cyclone from March 21-24**

**B. 500-mb chart for March 21**

**C. 500-mb chart for March 23**

## ✔ Concept Checks 9.4

1. Briefly explain how the flow aloft maintains cyclones at the surface.
2. What is speed divergence? Speed convergence?
3. Given an upper-air chart, where do forecasters usually look to find favorable sites for cyclogenesis? Where do anticyclones usually form in relation to the wavy flow aloft?
4. Describe the motion of a midlatitude cyclone in relation to the flow at the 500-millibar level.

# 9.5 Where Do Midlatitude Cyclones Form?

**List the primary sites for the development of midlatitude cyclones that affect North America.**

The development of midlatitude cyclones does not occur uniformly over Earth's surface but tends to favor certain locations, such as the leeward (eastern) sides of mountains and along coastal areas. In general, midlatitude cyclones form in areas where significant temperature contrasts occur in the lower troposphere.

## Sites of Midlatitude Cyclone Formation that Affect North America

Figure 9.16 shows the main areas of cyclone development over North America and adjacent oceans. Notice that a prime site for cyclone formation is along the east side of the Rocky Mountains,

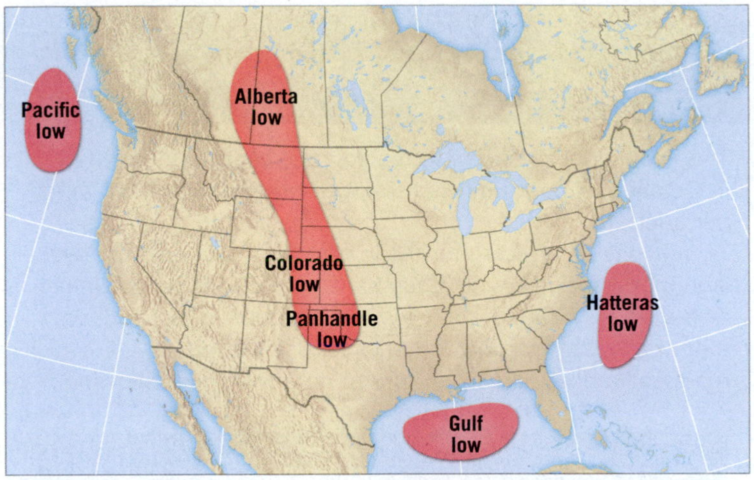


▲ **Figure 9.16 Sites of cyclone formation**

near the Colorado low. As the westerly flow meets the Rockies, the low-level flow is blocked by mountainous terrain while the air aloft continues on its eastward path. When the upper-level flow reaches the eastern slopes of the Rockies, it expands vertically. This produces a trough in the flow aloft, which enhances the development of a surface low. The storms that form near the Colorado low are fed at lower levels by warm Gulf coast air, whereas the air flowing into the low from the north is relatively cold. These temperature differences can produce intense low-pressure systems that migrate eastward across the United States. In winter, sleet and freezing rain occur along the warm front, strong thunderstorms are associated with the cold front, and heavy snowfall usually occurs north of the storm's center.

Other important sites of development are in the North Pacific, along the North Atlantic coastal areas, and in the Gulf of Mexico. These are areas of temperature contrasts between relatively warm ocean waters and a cool landmass. For example, midlatitude cyclones that develop over the Hatteras low form along the boundary between the warm waters of the Gulf Stream and cold Atlantic coast. These storms, with a vast source of moisture from the warm ocean waters, can develop quickly, creating pressure drops of over 24 millibars in a single day. As a result, these systems are capable of producing flooding or heavy snowfall along the eastern seaboard. When these storms move northeastward, they are called *nor'easters*.

## Patterns of Movement

Once formed, most midlatitude cyclones tend to travel in an easterly direction across North America and then follow a more northeasterly path into the North Atlantic (**Fig. 9.17**). However, numerous exceptions occur.

Midlatitude cyclones that influence western North America originate over the North Pacific. Many of these systems migrate from the Gulf of Alaska and affect Alaska and western Canada. However, during the winter months, these storms travel farther southward and often reach the west coast of the contiguous 48 states, occasionally traveling as far south as southern California. These low-pressure systems provide the winter rainy season that affects much of the west coast.

Most Pacific storms diminish in strength as they cross the Rockies, but they often redevelop on the eastern side of these mountains. A common area for redevelopment is Colorado, but other sites exist as far south as Texas and as far north as Alberta, Canada. Cyclones that redevelop over the Great Plains generally migrate eastward until they reach the central United States, where they follow a northeastward or even northward trajectory. Many of these cyclones traverse the Great Lakes region, making it one of the most storm-ridden portions of the country. In addition, some storms develop off the coast of the Carolinas and tend to move northward with the warm Gulf Stream, bringing stormy conditions to the entire Northeast.

### Panhandle Hook

One well-known storm pattern anomaly is known as a *Panhandle hook*. The "hook" describes the curved path these storms follow (Fig. 9.17). Developing in southern Colorado near the Texas and Oklahoma panhandles, these cyclones first travel toward the southeast and then bend and travel sharply northward across Wisconsin and into Canada.

### Alberta Clipper

An *Alberta Clipper* is a cold, windy cyclonic storm that forms on the eastern side of the Canadian Rockies in the province of Alberta (Fig. 9.17). Noted for their speed, they are called "clippers" because in colonial times, the fastest vehicles were small ships by the same name. Alberta Clippers dive southeastward into Montana or the Dakotas and then track across the Great Lakes, bringing dramatically lower temperatures. Winds associated with clippers frequently exceed 50 kilometers (30 miles) per hour. Because clippers

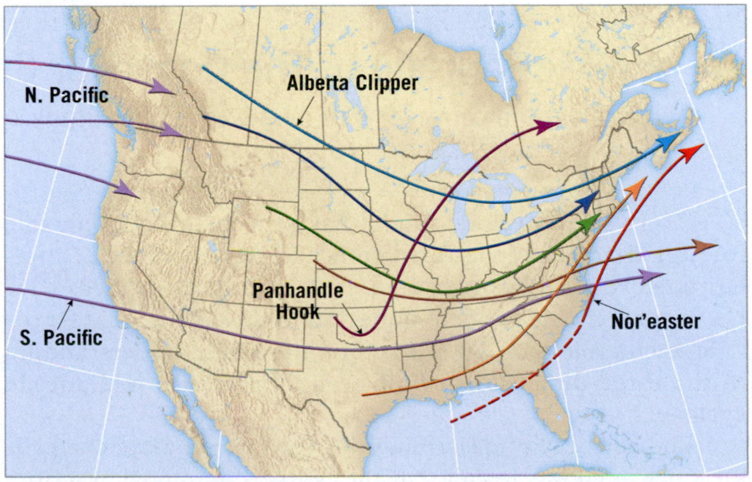


▲ **Figure 9.17 Typical paths of cyclonic storms that affect the lower 48 states**

move rapidly and remain great distances from the mild waters of the Gulf of Mexico, they tend to be moisture deprived and do not drop large amounts of snow. Instead, they may leave a few inches in a narrow band from the Dakotas to New York over a span as short as 2 days. However, because these winter storms are relatively frequent occurrences, they make a significant contribution to the total winter snowfall in the northern tier of states.

**Nor'easter** From the Mid-Atlantic coast to New England, the classic storm is called a *nor'easter* (Fig. 9.17) because the winds preceding such a storm in coastal areas are from the northeast. These storms are most frequent and violent between September and April, when cold air pouring south from Canada meets relatively warm, humid air from the Atlantic. Once formed, a nor'easter follows the coast and often brings rain, sleet, and heavy snowfall to the Northeast. The circulation produces strong onshore winds, and these storms can cause considerable coastal erosion, flooding, and property damage. A classic nor'easter is described in Chapter 8, page 227.

### ✔ Concept Checks 9.5

1. List four locations where midlatitude cyclones that affect North America tend to form.
2. Where do the midlatitude cyclones that affect the Pacific coast of the United States originate?
3. How did the *Alberta Clipper* get its name?
4. What portion of the United States is most affected by nor'easters?

Video MM
Winds During the Floods of 1993
http://goo.gl/AJBtvB

### eye ON THE atmosphere 9.3

This midlatitude cyclone swept across the central United States in October 2010. It produced strong wind gusts (up to 78 miles per hour), rain, hail, and snow, and it spawned 61 tornadoes. The cyclone set a record for the lowest measured pressure over land (not associated with a hurricane) in the continental United States: 28.21 inches of mercury. This pressure corresponds to that of a typical category 3 hurricane. Use the letters A–E, which represent various parts of the storm, to answer the following.

**Questions**

1. What part of the storm is dominated by a maritime tropical (mT) air mass?
2. Where is the area dominated by a continental polar (cP) air mass located?
3. Where did the strongest thunderstorms occur when this image was produced?
4. What area of the storm experienced the heaviest snowfall?
5. Where is the location of lowest pressure?

## 9.6 A Modern View: The Conveyor Belt Model

**Explain the conveyor belt model of a midlatitude cyclone and sketch the three interacting air streams (conveyor belts) on which it is based.**

The Norwegian cyclone model has proven to be a valuable tool for describing the formation and development of midlatitude cyclones through time. Although modern ideas have not replaced this model, upper-air and satellite data provide meteorologists with a more thorough understanding of the three-dimensional flow around a mature midlatitude cyclone.

The new way of describing the flow within a cyclone calls for a new analogy. Recall that the Norwegian model describes cyclone development in terms of the interactions of air masses along frontal boundaries, similar to armies clashing along battlefronts. The new model employs a modern example from industry—conveyor belts. Just as conveyor belts transport goods or people from one location to another, atmospheric conveyor belts transport air with distinct characteristics from one location to another.

This modern view of cyclogenesis, called the **conveyor belt model**, provides a good picture of the airflow within a cyclonic system at a particular point in time. It consists of three interacting air streams: two that originate near the surface and ascend, and a third that originates in the uppermost troposphere.

### Warm Conveyor Belt

A schematic representation of these air streams is shown in Figure 9.18. The **warm conveyor belt** (shown in red) is the

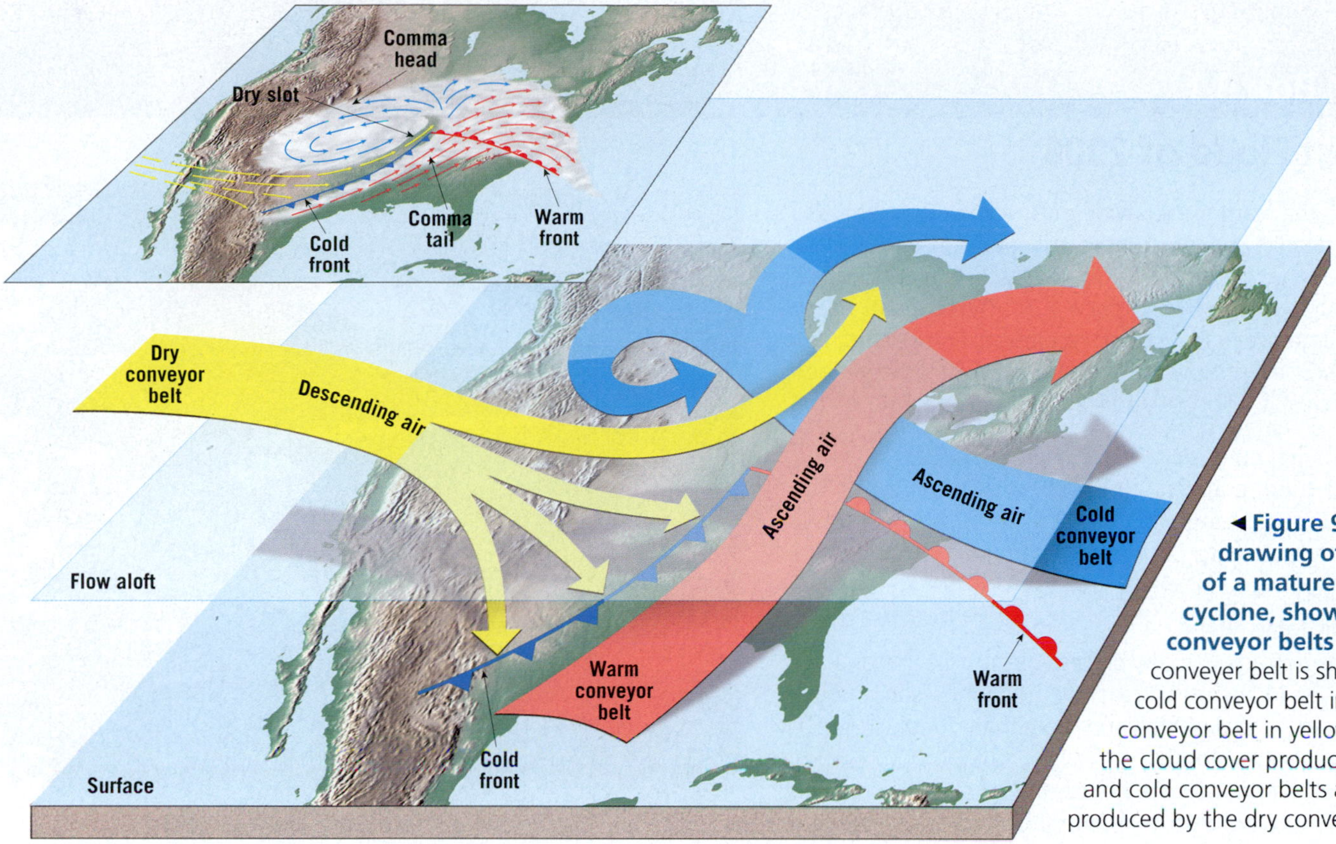


Video MM
A Midlatitude Cyclone's Dry Slot
http://goo.gl/LbiVlX

◀ **Figure 9.18 Schematic drawing of the circulation of a mature midlatitude cyclone, showing the conveyor belts** The warm conveyer belt is shown in red, the cold conveyor belt in blue, and the dry conveyor belt in yellow. The inset shows the cloud cover produced by the warm and cold conveyor belts and the dry slot produced by the dry conveyor belt.

main source of warm, moist air feeding the cyclone. In North America, the Gulf of Mexico is the major source of the warm, moist air that flows into the warm sector of the developing midlatitude cyclone (Fig. 9.18). This air stream flows northward, roughly parallel to the cold front. As it approaches the warm front, it slowly begins to rise. When it reaches the sloping boundary of the warm front, it rises even more rapidly over the cold air that lies beyond (poleward of) the front.

During its ascent, the warm, humid air cools adiabatically and produces a wide band of clouds and precipitation. Depending on atmospheric conditions, rain, drizzle, freezing rain, and snow are possible. When this air stream reaches the middle troposphere, it begins to turn right (eastward) and eventually joins the general west-to-east flow aloft. (In the later stages of its life cycle, a portion of the warm conveyor belt may be deflected westward, wrapping around the center of low pressure.)

The warm conveyor belt is the main precipitation-producing air stream in a midlatitude cyclone. When a well-organized midlatitude cyclone has a good source of moisture, abundant precipitation results.

## Cold Conveyor Belt

The **cold conveyor belt** (blue arrow) is airflow that starts at the surface ahead (poleward) of the warm front and flows westward (counterclockwise) toward the center of the cyclone (Fig. 9.18). Flowing beneath the warm conveyor belt, this air is moistened by the evaporation of raindrops falling through it. (Near the Atlantic Ocean, this conveyor belt has a marine origin and feeds significant moisture into the storm.) Convergence causes this air stream to rise as it nears the cyclone's center. During its ascent, this air cools adiabatically, becomes saturated, and contributes to the cyclone's precipitation.

As the cold conveyor belt reaches the middle troposphere, some of the flow rotates cyclonically around the low to produce the distinctive "comma head" of the mature storm system (see Fig. 9.12). The remaining flow turns right (clockwise) and becomes incorporated into the general westerly flow. Here it parallels the flow of the warm conveyor belt and may generate precipitation.

## Dry Conveyor Belt

The third air stream, called the **dry conveyor belt**, is shown as a yellow arrow in Figure 9.18. Whereas both the warm and cold conveyor belts begin at the surface, the dry air stream originates in the uppermost troposphere. As part of the upper-level westerly flow, the dry conveyor belt is relatively cold and dry.

As this air stream enters the cyclone, it splits. One branch descends behind the cold front, resulting in the clear, cool conditions normally associated with the passage of a cold front. In addition, this flow maintains the strong temperature contrast observed across the cold front. The other branch of the dry conveyor belt maintains its westerly flow aloft and forms the *dry*

severe& hazardous weather Box 9.1

## The Midwest Flood of 2008

Floods are part of the natural behavior of streams. They are also among the most deadly and destructive of natural hazards. Flash floods in small valleys are frequently triggered by torrential rains that last just a few hours (see Box 10.1, page 276). By contrast, major regional floods in large river valleys often result from an extraordinary series of precipitation events over a broad region for an extended time span. The Midwest flood of June 2008 is an example of the latter situation.

> Rainfall totals were more than twice the average in many places.

June was a very wet month across a significant part of this region, with numerous heavy rain events resulting in record flooding in Iowa, Wisconsin, Indiana, and Illinois. Rainfall totals were more than twice the average in many places. Martinsville, Indiana, for example, reported 20.11 inches of rain for the month—an amount equal to about half its annual rainfall and more than double its previous record high for a single month.

During the 2 months prior to the floods, the jet stream had been regularly dipping southward into the central United States, placing the favored storm track for rainy low-pressure centers over the Midwest. As a result, at least 65 locations in the Midwest set new June rainfall records, while more than 100 other stations had rainfall totals

▲ **Figure 9.C The flooding Cedar River covers a large portion of Cedar Rapids, Iowa** This image was taken on June 14, 2008.

*slot* (cloudless area) that separates the head and tail of a comma cloud pattern (Fig. 9.18).

In summary, the conveyor belt model provides a three-dimensional picture of the major circulation of a midlatitude cyclone. It also accounts for the distribution of precipitation and the comma-shaped cloud pattern characteristic of mature cyclonic storms.

### ✔ Concept Checks 9.6

1. Briefly describe the conveyor belt model of midlatitude cyclones.
2. What is the source of air that is carried by the warm conveyor belt?
3. In what way is the dry conveyor belt different from both the warm and cold conveyor belts?

## 9.7 Anticyclonic Weather and Atmospheric Blocking

**Describe blocking high-pressure systems and how they influence weather over the middle latitudes.**

The gradual subsidence within anticyclones generally produce clear skies and calm conditions. Because these high-pressure systems are not associated with stormy weather, both their development and movement have not been studied as extensively as those of midlatitude cyclones.

However, anticyclones do not always bring desirable weather. Large anticyclones often develop over the Arctic during the winter. These cold high-pressure centers are known to migrate as far south as the Gulf coast, where they can impact the weather over more than two-thirds of the United States (**Fig. 9.19**). This dense, frigid air often brings record-breaking cold temperatures.

Occasionally, large anticyclones persist over a region for several days or even weeks. Once in place, these stagnant anticyclones block or redirect the migration of midlatitude cyclones. Thus, they are sometimes called **blocking highs**. During these

▲ **Figure 9.D Water rushes through a break in an artificial levee** During the record-breaking floods of 2008, many levees could not withstand the force of the floodwaters. Sections of many weakened structures were overtopped or simply collapsed.

that ranked second to fifth on record. Prior to the heavy June rains, many of these stations had an exceptionally wet winter and spring. Soils that were already waterlogged provided no place for additional rain to go, forcing the flow across the surface and causing streams to rise.

The June 2008 floods represented the costliest weather disaster in Indiana history.

The June 2008 floods represented the costliest weather disaster in Indiana history, and Iowa also suffered widespread losses, with 83 of its 99 counties declared disaster areas. Nine of Iowa's rivers were at or above previous record flood levels, and millions of acres of productive farmland (an estimated 16 percent of the state's total) were submerged. Residential and commercial areas were evacuated, mostly in Cedar Rapids, where more than 400 city blocks were underwater (**Fig. 9.C**). The list of affected towns and rural areas throughout the Midwest was astonishing. Many riverside communities were flooded when swollen streams breached levees that had been built to protect the towns (**Fig. 9.D**).

**Question**

1. What is the major difference between weather systems that cause flash floods in small watersheds and weather systems that cause regional floods in large watersheds?

events, one section of the nation is kept dry for a week or more while another region remains continually under the influence of cyclonic storms. Such a situation prevailed during the summer of 1993, when a strong high-pressure system became anchored over the southeastern United States and caused migrating storms to stall over the Midwest. The result was severe drought in the southeast and devastating flooding for the central and upper Mississippi Valley. A similar situation occurred in June 2008 (see Box 9.1).

Strong blocking highs are also responsible for long periods of drought. One example is the devastating droughts that occurred in the Sahel from 1968 through 1974 and again in 2010. The Sahel includes the areas located south of the Sahara Desert of Africa that usually receive meager precipitation in summer, when the intertropical convergence zone (ITCZ) moves northward.

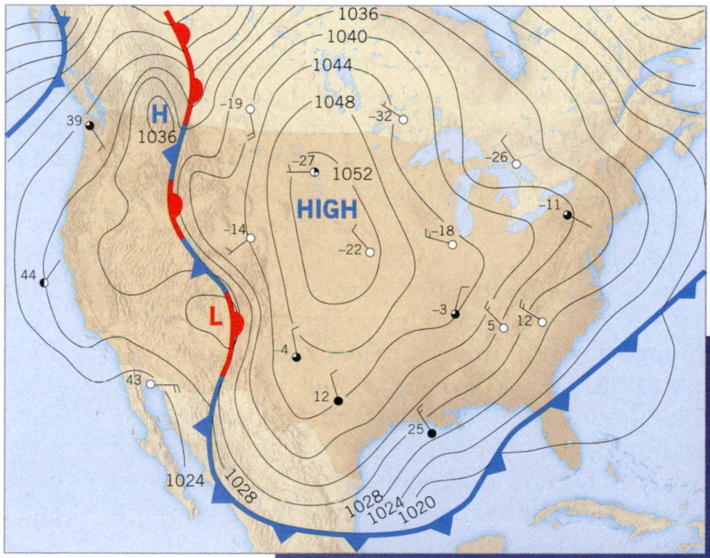


A.

B.

► **Figure 9.19 Anticyclonic weather A.** A cold anticyclone associated with an outbreak of frigid arctic air impacts the eastern two-thirds of North America. Temperatures are shown in degrees Fahrenheit. **B.** An outbreak of arctic air invades New England, bringing subzero temperatures and mostly clear skies.

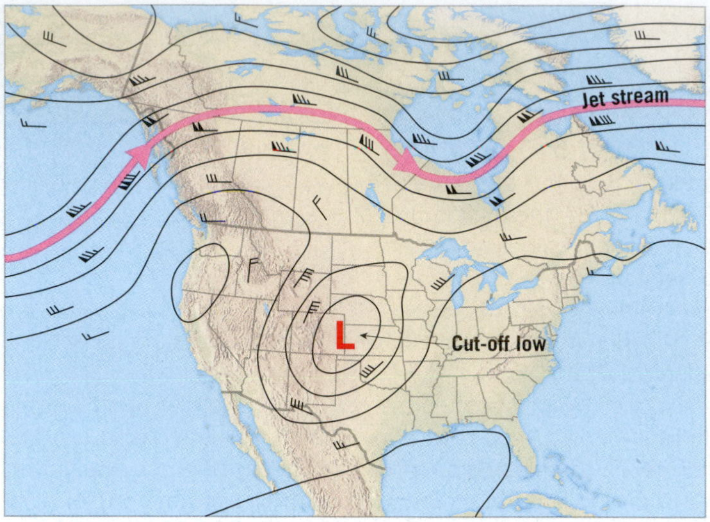


▲ **Figure 9.20 Cut-off low pressure systems** This cut-off low pressure system is literally cut off from the west-to-east flow in the jet stream. As a result, these systems can spin for days over the same area and are capable of producing large quantities of precipitation.

However, blocking highs during these periods prevented the ITCZ from moving sufficiently far poleward to bring much-needed precipitation.

Low-pressure systems can also generate a blocking pattern. These lows, called **cut-off lows**, are literally cut off from the west-to-east flow in the jet stream (Fig. 9.20). Without a connection to the prevailing flow aloft, these lows remain over the same area for days, often producing dreary weather and large quantities of precipitation.

### ✔ Concept Checks 9.7

1. Describe the weather associated with a strong anticyclone that penetrates the southern tier of the United States in winter.
2. What is a blocking high-pressure system?
3. What type of weather does a cut-off low-pressure system typically bring to an area?

## 9.8 Case Study of a Midlatitude Cyclone

**Summarize the weather associated with a midlatitude cyclone over the north-central United States in winter.**

MM GEODe ▶ Basic Weather Patterns ▶ In the Lab: Examining a Mature Middle-Latitude Cyclone

To visualize the weather one might expect from a strong late-winter midlatitude cyclone, we will look at the evolution of a storm migrating across the United States in late March. This cyclone reached the west coast on March 21, a few hundred kilometers northwest of Seattle, Washington. Like many other Pacific storms, this one rejuvenated east of the Rockies and moved eastward into the Plains. By the morning of March 23, it was centered near the Kansas–Nebraska border (Fig. 9.21). At that time, the central pressure had reached 985 millibars, and the storm's well-developed cyclonic circulation exhibited a warm front and a cold front.

During the next 24 hours, the forward motion of the storm's center became sluggish. The storm curved slowly northward through Iowa, and the pressure deepened to 982 millibars (Fig. 9.22). Although the storm center advanced slowly, the associated fronts moved vigorously toward the east and somewhat northward. The northern sector of the cold front overtook the warm front and generated an occluded front (shown in purple), which by the morning of March 24 was oriented nearly east–west (Fig. 9.22).

At that point the storm produced one of the worst blizzards ever to hit the north-central states. While the winter storm was brewing in the north, the cold front marched from western Texas (on March 23) to the Atlantic Ocean (March 25). During its 2-day trek, this violent cold front generated numerous severe thunderstorms and 19 tornadoes.

By March 25 the low pressure had diminished in intensity (to 1000 millibars) and had split into two centers (Fig. 9.23). Although the remnant low from this system, which was situated over the Great Lakes, generated some snow for the remainder of March 25, it had completely dissipated by the following day.

Now that you have read an overview of this storm, let us revisit this cyclone's passage in detail, using the weather charts for March 23 through March 25 (see Figs. 9.21 to 9.23). Fort Worth, Texas, in the system's warm sector, as shown on the March 23 weather map (Fig. 9.21), is under the influence of a warm, humid air mass having a temperature of 70°F and a dew-point temperature of 64°F. Notice that winds in the warm sector are from the south and are overrunning cooler air situated north of the warm front. In contrast, the air behind the cold front is 20° to 40°F cooler than the air of the warm sector and is flowing from the northwest.

On the morning of March 23, little activity was occurring along the fronts. As the day progressed, however, the storm intensified and changed dramatically. The map for March 24 illustrates the highly developed midlatitude cyclone. The extent and spacing of the isobars indicate a strong system that affected the circulation of the eastern two-thirds of the United States. A glance at the winds reveals a robust counterclockwise flow converging on the low.

The cold sector of the storm, located just north of the occluded front, produced one of the worst March blizzards ever to affect the north-central United States. In the Duluth–Superior area of Minnesota and Wisconsin, winds up to 81 miles per hour were measured and speeds exceeding 100 miles per hour were estimated for the bridge connecting these cities. Winds blew 12 inches of snow into 10- to 15-foot drifts, and some roads were closed for 3 days (**Fig. 9.24**). One restaurant in Superior was destroyed by fire when high winds and snowdrifts prevented firefighters and necessary equipment from reaching the blaze.

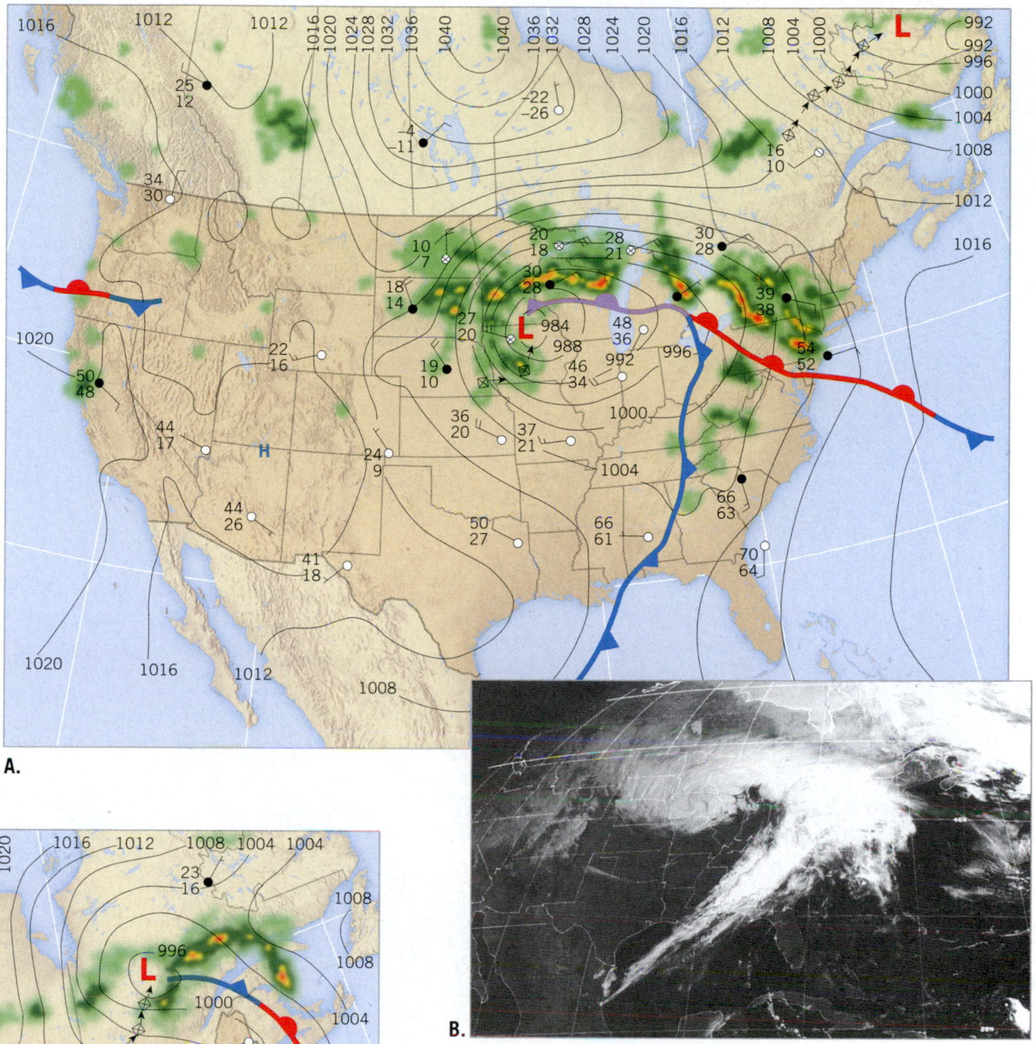


▲ **Figure 9.22 Weather on March 24** **A.** Surface weather map. **B.** Satellite image showing the cloud patterns.

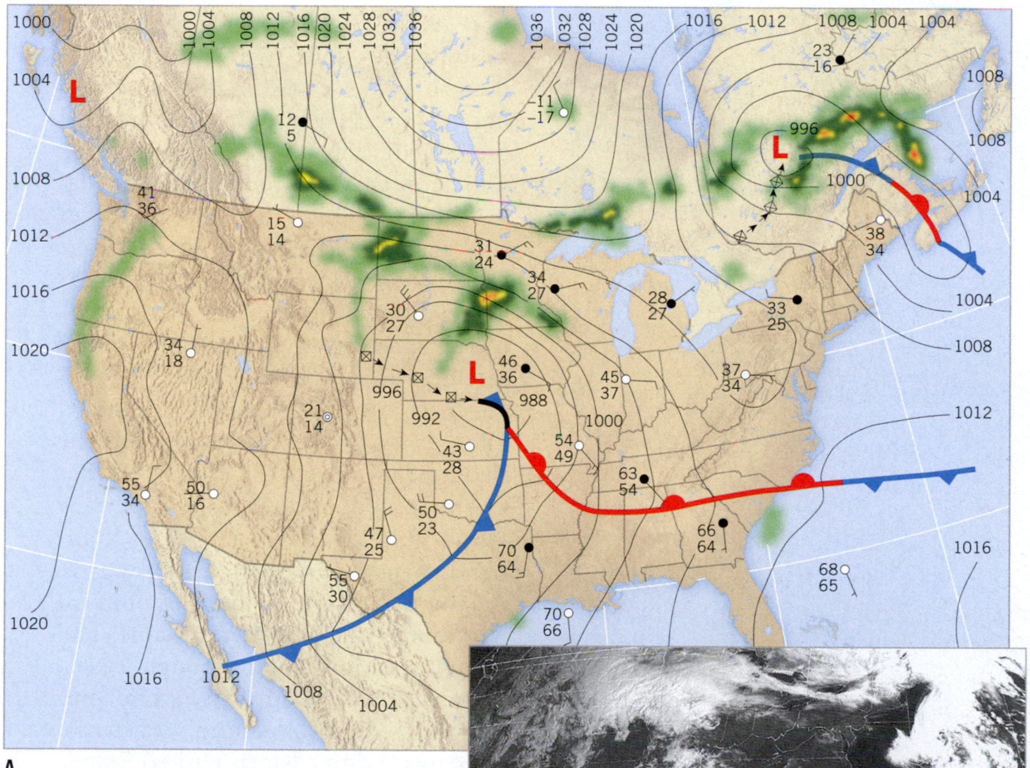

▲ **Figure 9.21 Weather on March 23** **A.** Surface weather map. **B.** Satellite image showing the cloud patterns.

Meanwhile, the cold front had generated a hailstorm in parts of eastern Texas by late afternoon on March 23. As the cold front moved eastward, it affected the entire southeastern United States except southernmost Florida. Throughout this region, the storm spawned numerous thunderstorms. High winds, hail, and lightning caused extensive damage, but the 19 tornadoes generated by the storm caused even greater death and destruction.

The path of the front can be easily traced from the reports of storm damage. By the evening of March 23, hail and wind damage were reported as far east as Mississippi and Tennessee. Early on the morning of March 24, golf ball–size hail was reported in Selma, Alabama. About 6:30 A.M. that day, the "Governor's Tornado" struck Atlanta, Georgia, where the storm

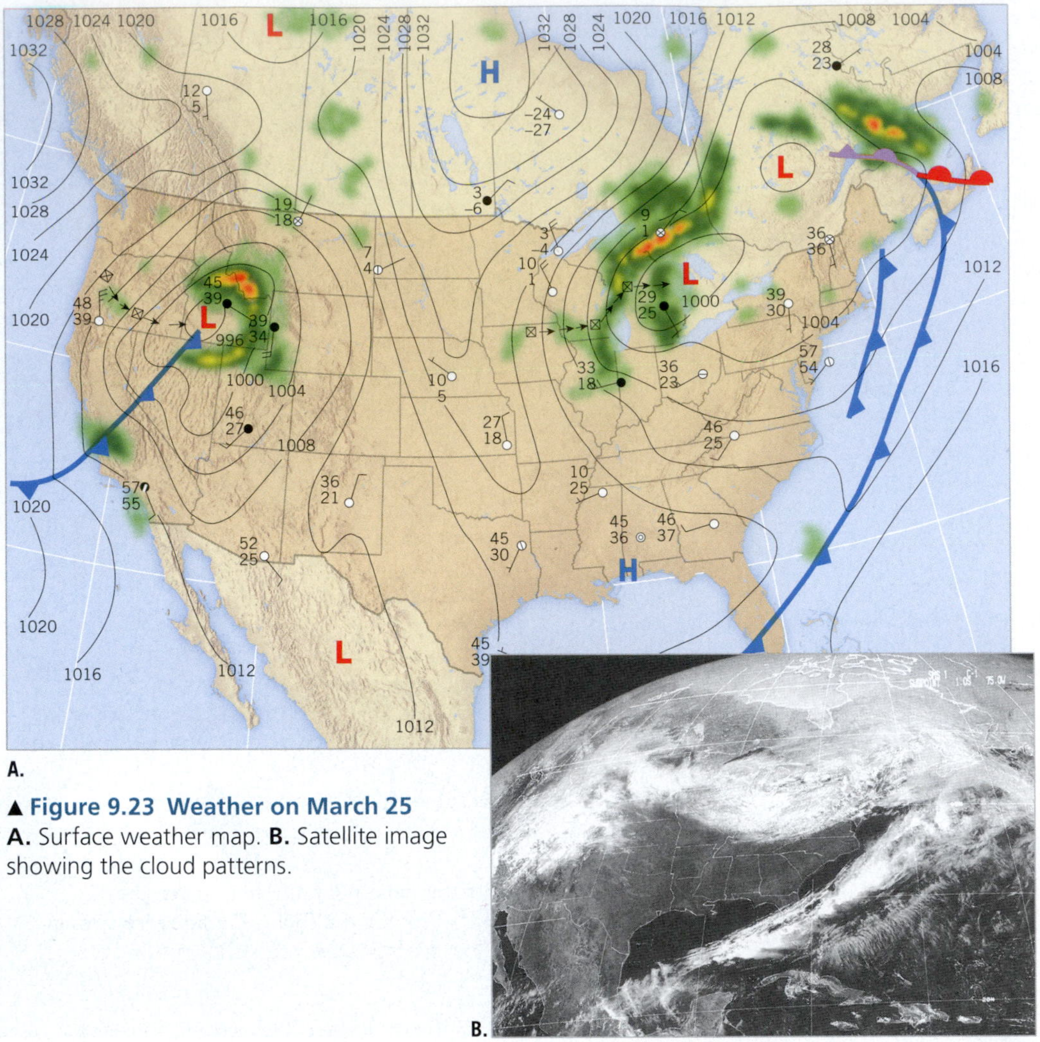


**▲ Figure 9.23 Weather on March 25**
**A.** Surface weather map. **B.** Satellite image showing the cloud patterns.

produced its worst destruction. Damage was estimated at over $50 million; 3 people died, and another 152 were injured. The 12-mile path of the Governor's Tornado cut through an affluent residential area of Atlanta that included the governor's mansion (hence the name of the storm). The final damage (hail and a small tornado) along the cold front was reported at 4:00 A.M. on March 25 in northeastern Florida. Thus, a day and a half and some 1200 kilometers after the cold front became active in Texas, the front left the United States to vent its energy over the Atlantic Ocean.

By the morning of March 24, cold polar air had penetrated deep into the United States behind the cold front (see Fig. 9.22). Fort Worth, which just the day before had been in the warm sector, experienced cool northwest winds. Subfreezing temperatures moved as far as northern Oklahoma. Notice, however, that by March 25, Fort Worth was again experiencing a southerly flow. We can conclude that this was a result of the decaying cyclone that no longer dominated the circulation in the region. Also notice on the map for March 25 (Fig. 9.23) that a high was situated over the Alabama-Mississippi border. The clear skies and weak surface winds associated with the center of a subsiding air mass are well illustrated on the satellite image.

You may have already noticed another cyclone moving in from the Pacific on March 25, as the earlier storm exited to the east. This new storm developed in a similar manner but was centered farther north. As you might guess, another blizzard hammered the Northern Plains, and a few tornadoes spun through Texas, Arkansas, and Kentucky, while precipitation dominated the weather in the central and eastern United States.

This example demonstrates the effect of a spring cyclone on middle-latitude weather. Within 3 days, temperatures in Fort Worth, Texas, changed from warm to unseasonably

**◄ Figure 9.24 Paralyzing blizzard strikes the north-central United States**

cold. Thunderstorms with hail were followed by cold, clear skies. The cold sector of the storm, located just north of the occluded front, produced one of the worst March blizzards ever to affect the north-central United States.

## ✔ Concept Check 9.8

1. Briefly describe the weather at Fort Worth, Texas, from March 23 through March 25, during the passage of this well-developed spring storm.
2. How did the weather in the Duluth–Superior area of Minnesota and Wisconsin compare to that in the Fort Worth area during March 23–25?

**students sometimes ask...**

### Sometimes a major flood is described as a 100-year flood. What does that mean?

The phrase "100-year flood" is misleading because it implies that such an event happens only once every 100 years. In fact, an uncommonly big flood can happen any year. The phrase is really a statistical designation indicating that there is a 1-in-100 chance that a flood of a certain size will happen during any year. Perhaps a better term would be the "1-in-100-chances flood." Many flood designations are reevaluated and changed over time as more data are collected or when a river basin is altered in a way—for example, through dam construction or urban development—that affects the flow of water.

# 9 Concepts in Review Midlatitude Cyclones

## 9.1 Frontal Weather

▸ Compare and contrast typical weather associated with a warm front and a cold front.

**Key Terms:** midlatitude (middle-latitude) cyclone, front, overrunning, warm front, cold front, backdoor cold front, stationary front, occluded front, occlusion, cold-type occluded front, warm-type occluded front, dryline

- Fronts are boundary surfaces that separate air masses of different densities, one usually warmer and more humid than the other. As one air mass moves into another, the warmer, less dense air mass is forced aloft in a process called overrunning.
- Warm fronts form when warmer air invades territory formerly occupied by cooler air. By contrast, cold fronts develop when cold air actively advances into a region occupied by warmer air. Warm fronts tend to generate light to moderate precipitation, whereas cold fronts are associated with stormy weather.
- Stationary fronts form when the airflow on both sides of a front is neither toward the cold air mass nor toward the warm air mass. In addition, they tend to produce precipitation patterns similar to those of a warm front.
- Occluded fronts most often form when a rapidly moving cold front overtakes a warm front. Occluded fronts produce a variety of weather.
- Drylines most often develop when dry, continental tropical (cT) air meets moist, maritime tropical (mT) air. Drylines are spring and summer phenomena that often generate a band of severe thunderstorms.

**Q** The accompanying diagrams are cross-sections of three types of fronts. Select the name for each of the fronts shown from the following: warm front, cold front, warm-type occluded front, cold-type occluded front, or stationary front.

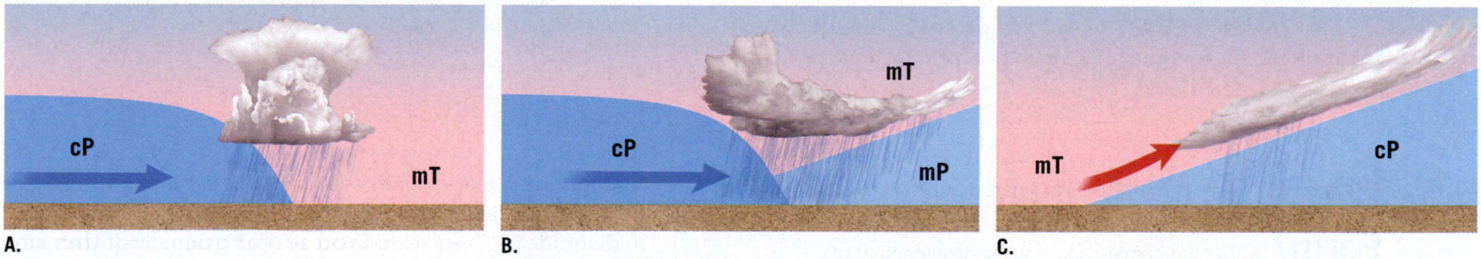


## 9.2 Midlatitude Cyclones and the Polar-Front Theory

▸ Outline the stages in the life cycle of a typical midlatitude cyclone.

**Key Terms:** polar-front theory (Norwegian cyclone model), cyclogenesis

- The primary weather producer in the middle latitudes is the midlatitude, or middle-latitude, cyclone. Midlatitude cyclones are large low-pressure systems with diameters often exceeding 1000 kilometers (600 miles) that generally travel from west to east. They last a few days to more than a week, have a counterclockwise circulation pattern in the Northern Hemisphere, with a flow inward toward their centers, and have a cold front and a warm front extending from the central area of low pressure.
- According to the polar-front theory, midlatitude cyclones form along fronts and proceed through a generally predictable life cycle. Along the polar front, where two air masses of different densities are moving parallel to the front and in opposite directions, cyclogenesis (cyclone formation) occurs, and the frontal surface takes on a wave shape that is usually several hundred kilometers long.

- When a wave forms, warm air advances poleward, invading the area formerly occupied by cooler air. This change in the direction of the surface flow causes a readjustment in the pressure pattern that results in somewhat circular isobars, with the low pressure centered at the apex of the wave. Usually, the cold front advances faster than the warm front and gradually closes the warm sector and lifts the warm front in a process known as occlusion. Eventually, the warm sector is forced aloft, and cold air surrounds the cyclone at low levels. At this point, the cyclone has exhausted its source of energy, and the once highly organized counterclockwise flow ceases to exist.

## 9.3 Idealized Weather of a Midlatitude Cyclone ▶ Describe the general weather conditions associated with the passage of a mature midlatitude cyclone.

- The Norwegian model is a valuable tool for interpreting the weather conditions associated with the passage of a midlatitude cyclone. Different types of weather can be expected, depending on your location relative to the storm's center of low pressure. South of the storm's center, you will encounter clouds and precipitation patterns associated the passage of a warm front, followed by a cold front. North of the storm's center, you will encounter clouds and precipitation patterns associated with an occluded front.

**Q** The accompanying illustration shows the fronts associated with a midlatitude cyclone. Match the numbered regions with the type of precipitation that is most typically associated with each region during *winter*: thunderstorms, light-to-moderate rain, sleet or freezing rain, or heavy snow.

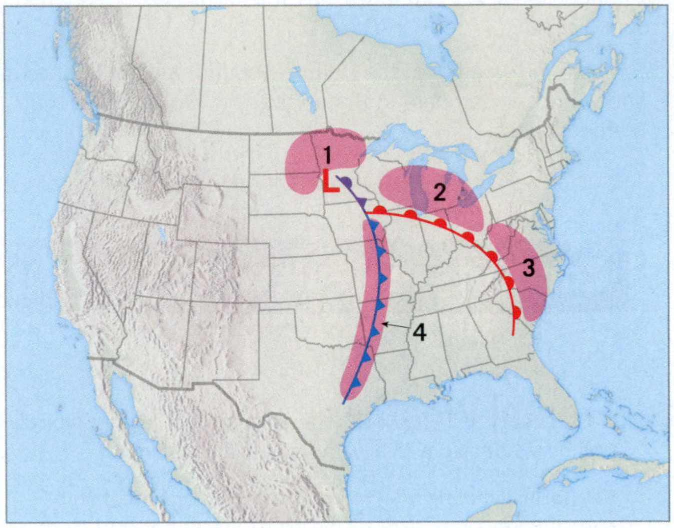


## 9.4 Flow Aloft and Cyclone Formation ▶ Explain why divergence in the flow aloft is a necessary condition for the development and intensification of a midlatitude cyclone.

- Guided by the westerlies aloft, cyclones generally move eastward across the United States. Airflow aloft (divergence and convergence) plays an important role in maintaining cyclonic and anticyclonic circulation. In cyclones, divergence aloft supports the inward flow at the surface.
- During the colder months when temperature gradients are steepest, cyclonic storms move at their fastest rate. Furthermore, the westerly airflow aloft tends to steer these developing pressure systems in a general west-to-east direction.

**Q** The accompanying illustration shows the position of the polar jet stream on a particular day. Below which location, A or B, would a surface midlatitude cyclone most likely form?

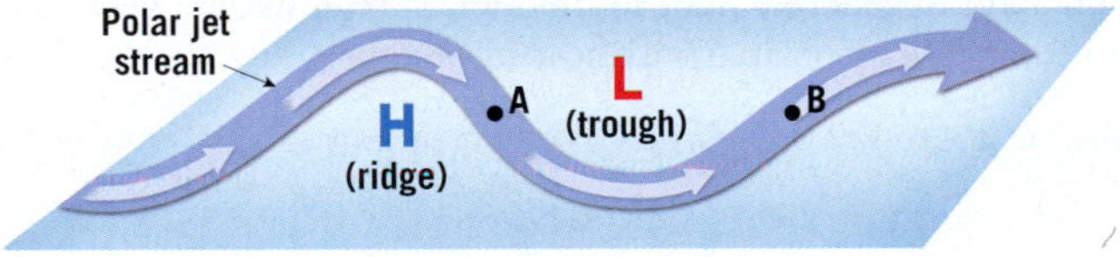


## 9.5 Where Do Midlatitude Cyclones Form? ▶ List the primary sites for the development of midlatitude cyclones that affect North America.

- Midlatitude cyclones tends to develop in certain locations, such as the leeward sides of mountains and along coastal areas. In general, midlatitude cyclones form in areas where significant temperature contrasts occur in the lower troposphere.

## 9.6 A Modern View: The Conveyor Belt Model ▶ Explain the conveyor belt model of a midlatitude cyclone and sketch the three interacting air streams (conveyor belts) on which it is based.

**Key Terms:** conveyor belt model, warm conveyor belt, cold conveyor belt, dry conveyor belt

- The modern view of cyclogenesis, called the conveyor belt model, consists of three interacting air streams: two that originate at the surface and ascend, and a third that originates aloft. The conveyor belt model provides a three-dimensional view of the airflow within a cyclonic storm.

## 9.7 Anticyclonic Weather and Atmospheric Blocking

▶ Describe blocking high-pressure systems and how they influence weather over the middle latitudes.

**Key Terms:** blocking high, cut-off low

- Due to the gradual subsidence within them, anticyclones tend to produce clear skies and generally calm conditions.
- Anticyclones can sometimes remain in nearly the same location for weeks at a time and are known as blocking highs. Blocking highs can slow the movement of storm systems, causing some areas to experience extreme amounts of precipitation while other areas may incur drought.

## 9.8 Case Study of a Midlatitude Cyclone

▶ Summarize the weather associated with a midlatitude cyclone over the north-central United States in winter.

- This section describes the effects of a spring cyclone on middle-latitude weather. Within 3 days, temperatures in Fort Worth, Texas, changed from warm to unseasonably cold. Thunderstorms with hail were followed by clear skies and cold temperatures. The activity in the storm's cold sector, located just north of the occluded front, produced one of the worst March blizzards ever to affect the north-central United States.

## Give it Some Thought

1. Sketch a vertical cross-section (side view) of a cold front and a warm front and include the following elements:
   **a.** Shape and slope of each type of front
   **b.** Air mass on both sides of each front
   **c.** Type of clouds typically associated with each front
   **d.** Type of rainfall associated with each front
   **e.** Characteristics of temperature and humidity found on both sides of each front
2. The accompanying diagrams show surface temperatures (in degrees Fahrenheit) for noon and 6 P.M. on January 29, 2008. On this day, an incredibly powerful weather front moved through Missouri and Illinois.
   **a.** What type of front passed through the Midwest?
   **b.** Describe how the temperature changed in St. Louis, Missouri, over the 6-hour period that began at noon and ended at 6 P.M.
   **c.** Describe the likely shift in wind direction in St. Louis during this time period.
3. The following questions refer to midlatitude cyclones and the frontal weather associated with them.
   **a.** Describe the weather several hours after the passage of a cold front. With what type of pressure system are these weather conditions associated?
   **b.** What is the source region for the mT air mass that produces most of the clouds and precipitation in the eastern two-thirds of the United States?
   **c.** What type of severe weather would most concern a meteorologist when a stationary front is present over a region? (*Hint:* This severe weather type claims the most lives, on average, each year in the United States.)
4. Refer to the accompanying weather map and answer the following questions:
   **a.** What is the probable wind direction at each of the lettered cities?
   **b.** What air mass is likely to be influencing each city?
   **c.** Identify the cold front, warm front, and occluded front.
   **d.** What is the barometric tendency at city A and at city C?
   **e.** Which of the three cities is coldest? Which is warmest?

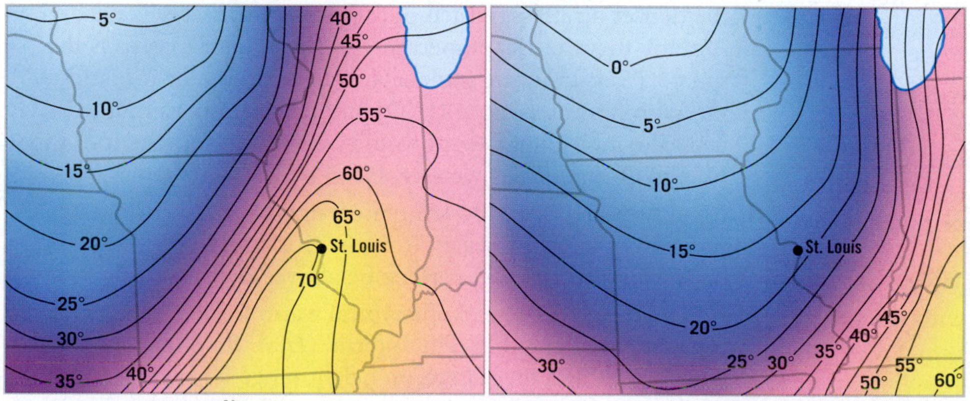


Noon 6 P.M.

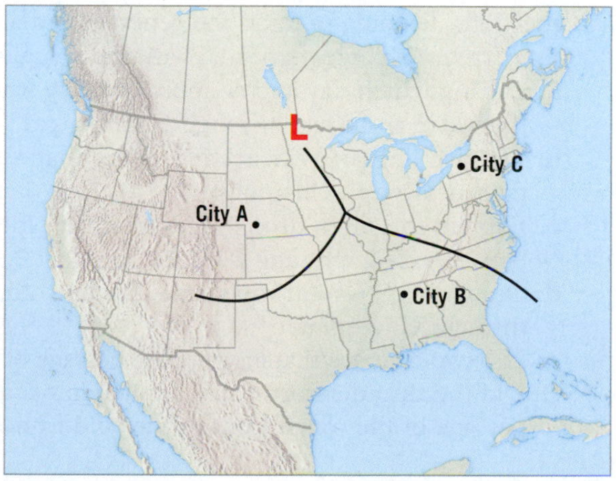

5. Explain the basis for the following weather proverb:
   Rain long foretold, long last;
   Short notice, soon past.
6. The accompanying satellite image shows the comma-shaped cloud pattern of a well-developed midlatitude cyclone over the United States in late winter.
   **a.** Describe the weather in northern Minnesota (A), located under the comma head of the winter storm.
   **b.** What is the probable weather in eastern Alabama (B), which is located under the comma tail?

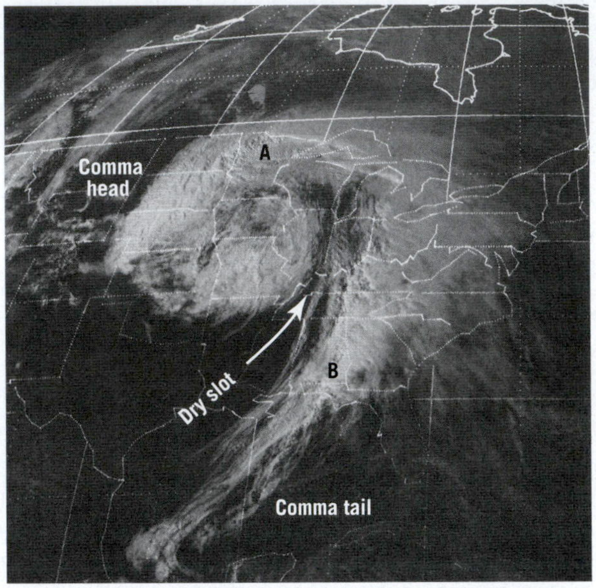


7. Why are occurrences of midlatitude cyclones less common over the United States in late summer than in winter and spring?
8. Write a statement explaining the relationship between the circulation around a surface low-pressure system and the flow aloft, using terms such as *ascending air*, *descending air*, *divergence*, and *convergence*.
9. The accompanying map shows the path aloft of the jet stream over the United States.
   **a.** Where is the upper-air low (trough) located?
   **b.** Where is the upper-air high (ridge) located?
   **c.** In what region of the country or states would you expect the center of a low pressure system to form?
   **d.** In what region of the country or states would you expect a high-pressure system to be located?

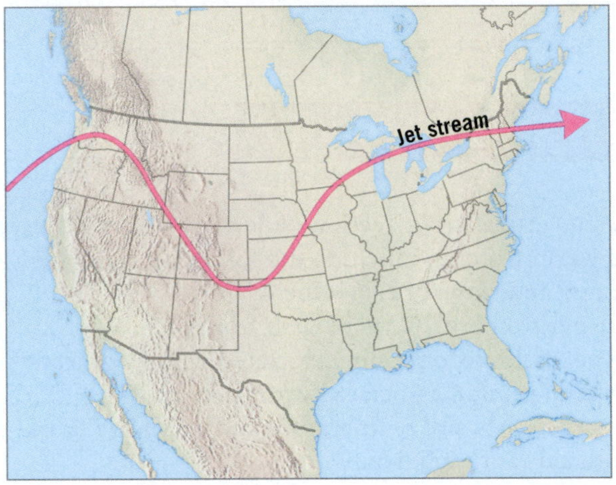


## Problems

1. Refer to Table A, which provides weather observations for Champaign, Illinois, over a 3-day period as a midlatitude cyclone passed through the area. Answer the following questions, keeping in mind the general wind and temperature changes expected with the passage of fronts.
   **a.** During which day and at approximately what time did the warm front pass through Champaign?
   **b.** List two lines of evidence indicating that a warm front passed through Champaign.
   **c.** Explain the slight drop in temperature experienced between midnight and 6:00 A.M. on Day 2.
   **d.** On what day and at what time did the cold front pass through Champaign?
   **e.** List two changes that indicate the passage of the cold front.
   **f.** Did the thunderstorms in Champaign occur with the passage of the warm front or the cold front?
2. In the spring and summer, a warm moist (mT) air mass from the Gulf of Mexico occasionally collides with a warm dry (cT) air mass from the desert Southwest. These air masses meet over Texas, Oklahoma, and Kansas, producing what meteorologists call a dryline. Thunderstorms that erupt along drylines can produce some of the most severe storms on the planet. When these two air masses meet, stormy weather is triggered when the less-dense air mass overruns the denser air mass. Which of these two air masses is denser? Assume that the air temperature of the two air masses is the same. (*Hint:* Find the molecular weight of dry air [$N_2$ and $O_2$] with no water vapor [$H_2O$] and then compare that to what the molecular weight would be if about 4 percent of the $N_2$ and $O_2$ molecules were replaced by $H_2O$ molecules.)
3. If you were located 400 kilometers ahead of the surface position of a typical warm front (with slope 1:200), how high would the frontal surface be above you?
4. Calculate how far ahead of a typical warm front you are when the first cirrus clouds begin to form. (*Hint:* Refer to Fig. 5.1, page 125 to find the minimum height range for high clouds.)

**Table A** Weather Data for Champaign, Illinois

| | Temperature (°F) | Wind Direction | Weather and Precipitation |
|---|---|---|---|
| **Day 1** | | | |
| 00:00 | 46 | E | Partly cloudy |
| 3:00 A.M. | 46 | ENE | Partly cloudy |
| 6:00 A.M. | 48 | E | Overcast |
| 9:00 A.M. | 49 | ESE | Drizzle |
| 12:00 P.M. | 52 | ESE | Light rain |
| 3:00 P.M. | 53 | SE | Rain |
| 6:00 P.M. | 68 | SSW | Partly cloudy |
| 9:00 P.M. | 67 | SW | Partly cloudy |

| | Temperature (°F) | Wind Direction | Weather and Precipitation |
|---|---|---|---|
| **Day 2** | | | |
| 00:00 | 66 | SW | Partly cloudy |
| 3:00 A.M. | 64 | SW | Mostly sunny |
| 6:00 A.M. | 63 | SSW | Mostly sunny |
| 9:00 A.M. | 69 | SW | Mostly sunny |
| 12:00 P.M. | 72 | SSW | Mostly sunny |
| 3:00 P.M. | 76 | SW | Mostly sunny |
| 6:00 P.M. | 74 | SW | Cloudy |
| 9:00 P.M. | 64 | W | Thunderstorm, gusty winds |

| | Temperature (°F) | Wind Direction | Weather and Precipitation |
|---|---|---|---|
| **Day 3** | | | |
| 00:00 | 52 | WNW | Isolated thunderstorms |
| 3:00 A.M. | 48 | WNW | Cloudy |
| 6:00 A.M. | 42 | NW | Partly cloudy |
| 9:00 A.M. | 39 | NW | Mostly sunny |
| 12:00 P.M. | 38 | NW | Sunny |
| 3:00 P.M. | 40 | NW | Sunny |
| 6:00 P.M. | 42 | NW | Sunny |
| 9:00 P.M. | 40 | NW | Sunny |

**MasteringMeteorology™**

Looking for additional review and test prep materials? Visit the Study Area in *MasteringMeteorology*™ to enhance your understanding of this chapter's content by accessing a variety of resources, including **MapMaster**™ interactive maps, Geoscience Animations, GEODe, *In the News* RSS feeds, flashcards, web links, self-study quizzes, and an eText version of *The Atmosphere*.

# 10 Thunderstorms and Tornadoes

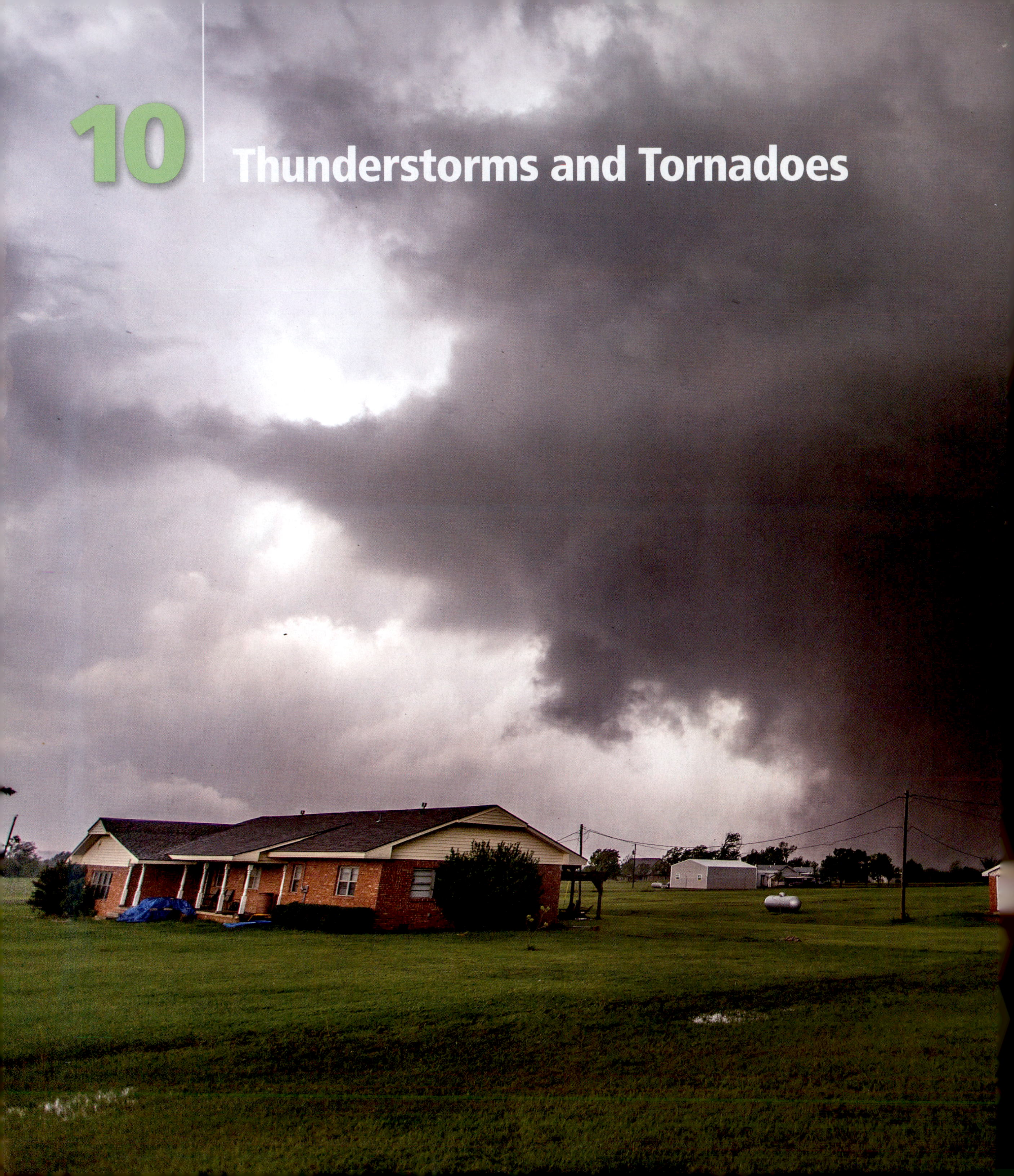

## Focus on Concepts

*Each statement represents the primary learning objective for the corresponding major heading within the chapter. After you complete the chapter, you should be able to:*

**10.1** Distinguish among three types of storm-producing cyclones.

**10.2** List the basic requirements for thunderstorm formation and locate places on a map that exhibit frequent thunderstorm activity.

**10.3** Sketch and explain simple diagrams that illustrate the three stages of an air-mass thunderstorm.

**10.4** List the characteristics of a severe thunderstorm and distinguish among different circumstances associated with their formation.

**10.5** Explain what causes lightning and thunder.

**10.6** Describe the structure and basic characteristics of tornadoes.

**10.7** Summarize the atmospheric conditions and locations that are favorable to the formation of tornadoes.

**10.8** Describe tornado intensity. Distinguish between a tornado watch and a tornado warning and discuss the role of Doppler radar in the warning process.

This chapter and Chapter 11 focus on severe and hazardous weather. In this chapter we examine the severe local weather produced in association with cumulonimbus clouds—namely, thunderstorms and tornadoes. In Chapter 11 we will turn to the large tropical cyclones we call hurricanes.

Occurrences of severe weather are more fascinating than ordinary weather phenomena. The lightning display generated by a thunderstorm can be a spectacular event that elicits both awe and fear. Of course, tornadoes and hurricanes also attract a great deal of much-deserved attention. A single tornado outbreak or hurricane can cause billions of dollars in property damage as well as many deaths.

*On May 20, 2013, a deadly EF5 tornado, struck Moore, Oklahoma. This most extreme tornado to strike the United States in 2013 was responsible for at least 25 deaths.*

## 10.1 What's in a Name?

**Distinguish among three types of storm-producing cyclones.**

In Chapter 9 we examined the midlatitude cyclones that play an important role in causing day-to-day weather changes, but the use of the term *cyclone* is often confusing. For many people, the term implies only an intense storm, such as a tornado or a hurricane. When a hurricane unleashes its fury on India, Bangladesh, or Myanmar, for example, it is usually reported in the media as a cyclone (the term denoting a hurricane in that part of the world).

Similarly, tornadoes are called cyclones in some places. This custom is particularly common in portions of the Great Plains of the United States. Recall that in the *Wizard of Oz*, Dorothy's house was carried from her Kansas farm to the land of Oz by a cyclone. Indeed, the nickname for the athletic teams at Iowa State University is the *Cyclones* (Fig. 10.1). Although hurricanes and tornadoes are, in fact, cyclones, the vast majority of cyclones are not hurricanes or tornadoes. The term *cyclone* simply refers to the circulation around any low-pressure center, no matter how large or intense it is.

Tornadoes and hurricanes are both smaller and more violent than midlatitude cyclones. A midlatitude cyclone may have a diameter of 1600 kilometers (1000 miles) or more. By contrast, hurricanes average only 600 kilometers (375 miles) across, and tornadoes, with a diameter of just 0.25 kilometer (0.16 mile), are much too small to show up on a weather map.

Thunderstorms, much more familiar weather events, hardly need to be distinguished from tornadoes, hurricanes, and midlatitude cyclones. Unlike the flow of air about these storms, the circulation associated with thunderstorms is characterized by strong up-and-down movements. Winds in the vicinity of a thunderstorm do not follow the inward spiral of a cyclone, but they are typically variable and gusty.

Although thunderstorms form "on their own," away from cyclonic storms, they also form in conjunction with cyclones. For instance, thunderstorms are frequently spawned along the cold front of a midlatitude cyclone, and on rare occasions, a tornado may descend from a thunderstorm's cumulonimbus tower. Hurricanes also generate widespread thunderstorm activity. Thus, thunderstorms are related in some manner to all three types of cyclones mentioned here.

In southern Asia and Australia, the term *cyclone* is applied to storms that are called *hurricanes* in the United States. This image shows Cyclone Yasi, which struck eastern Australia in February 2011.

**In parts of the Great Plains, *cyclone* is a synonym for *tornado*. The nickname for the athletic teams at Iowa State University is the *Cyclones*.***

* Iowa State University is the only Division I school to use cyclones as its team name. The pictured logo was created to better communicate the school's image by combining the mascot, a cardinal bird named Cy, andthe Cyclone team name.

◄ **Figure 10.1 The term *cyclone*** Sometimes the use of the term *cyclone* can be confusing.

### ✔ Concept Checks 10.1

1. List and briefly describe three different ways in which the term *cyclone* is used.
2. Compare the wind speeds and sizes of middle-latitude cyclones, tornadoes, and hurricanes.

## 10.2 Thunderstorms

**List the basic requirements for thunderstorm formation and locate places on a map that exhibit frequent thunderstorm activity.**

Almost everyone has observed various small-scale phenomena that result from the vertical movements of relatively warm, unstable air. Perhaps you have seen a dust devil over an open field on a hot day, whirling its dusty load to great heights, or have watched a bird glide effortlessly skyward on an invisible thermal. These examples illustrate the dynamic thermal instability that occurs during the development of a *thunderstorm*. A **thunderstorm** is a storm that generates lightning and thunder. It frequently produces gusty winds, heavy rain, and hail. A thunderstorm may be produced by a single cumulonimbus

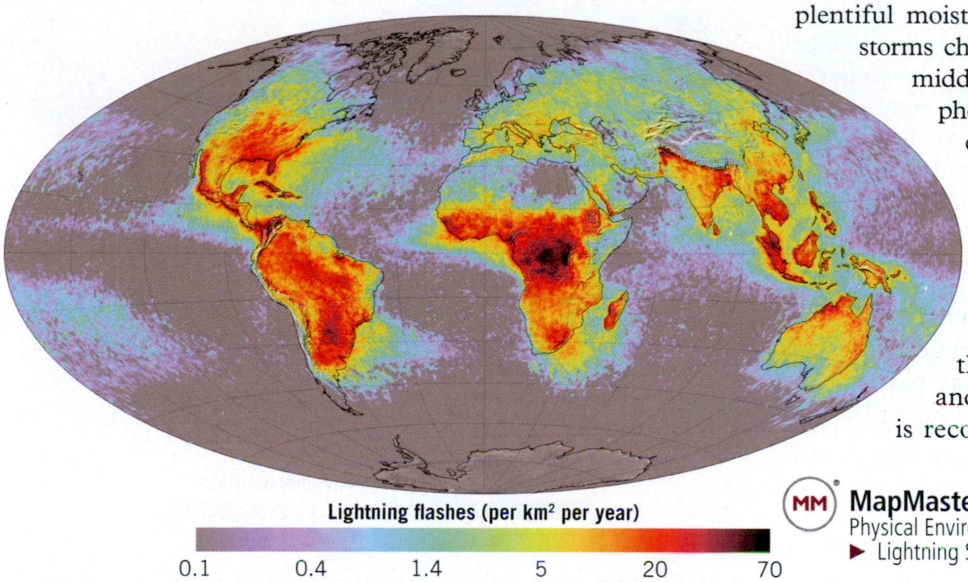


▲ **Figure 10.2 World distribution of lightning** Data from space-based optical sensors show the worldwide distribution of lightning, with color variations indicating the average annual number of lightning flashes per square kilometer.

cloud and influence only a small area, or it may be associated with clusters of cumulonimbus clouds covering a large area.

Thunderstorms form when warm, humid air rises in an unstable environment. Various mechanisms can trigger the upward air movement needed to create thunderstorm-producing cumulonimbus clouds. One mechanism, the unequal heating of Earth's surface, significantly contributes to the formation of *air-mass thunderstorms*. These storms are associated with the scattered puffy cumulonimbus clouds that commonly form *within* maritime tropical air masses and produce scattered thunderstorms on summer days. Such storms are usually short-lived and seldom produce strong winds or hail.

In contrast, thunderstorms in a second category not only benefit from uneven surface heating but are associated with the lifting of warm air, as occurs along a front or a mountain slope. Moreover, diverging winds aloft frequently contribute to the formation of these storms because they tend to draw air from lower levels upward beneath them. Some of the thunderstorms in this second category may produce high winds, damaging hail, flash floods, and tornadoes. Such storms are described as *severe*.

## Distribution and Frequency

At any given time, an estimated 2000 thunderstorms are in progress. As we would expect, the greatest proportion of these storms occur where warmth, plentiful moisture, and instability are present. Thus, thunderstorms characterize many tropical areas year-round. In the middle latitudes, these storms are largely warm-season phenomena. About 45,000 thunderstorms take place each day, and more than 16 million occur annually around the world. The lightning from these storms strikes Earth 100 times each second (Fig. 10.2).

Annually the United States experiences about 100,000 thunderstorms and millions of lightning strikes. A glance at Figure 10.3 shows that thunderstorms are most frequent in Florida and the eastern Gulf Coast region, where activity is recorded between 70 and 100 days each year. The region on the east side of the Rockies in Colorado and New Mexico is next, with thunderstorms occurring 60 to 70 days annually. Most of the rest of the nation experiences thunderstorms 30 to 50 days a year. Clearly, the western margin of the United States has little thunderstorm activity. The same is true for the northern tier of states and for Canada, where warm, moist, unstable maritime tropical (mT) air seldom penetrates.

MM MapMaster ▶ World Physical Environment ▶ Lightning Strikes

## Thunderstorms and Climate Change

In the preceding discussion, you learned that the occurrence of thunderstorms varies seasonally and from place to place. Thunderstorm activity will likely increase in many areas in the years to come due to global climate change. Global temperatures have been warming for decades largely due to human activities that are altering the composition of the atmosphere. This trend is expected to continue for the foreseeable future. A thorough discussion of this phenomenon appears in Chapter 14.

Video MM
Thundersleet and Thundersnow
http://goo.gl/GOm7IV

MM MapMaster ▶ North America Physical Environment ▶ Thunderstorm Occurrence Per Year

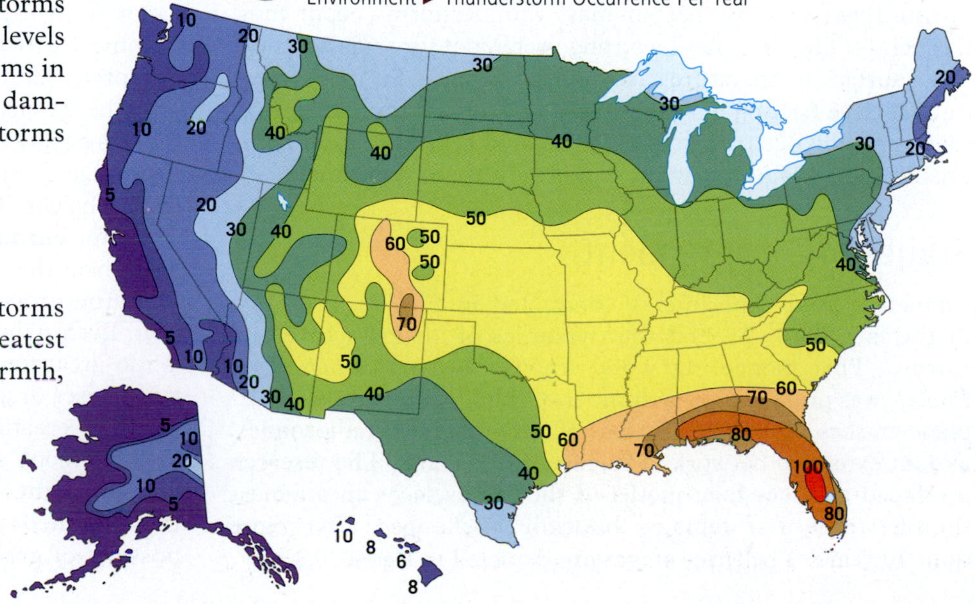


▶ **Figure 10.3 Average number of days per year with thunderstorms** The humid subtropical climate that dominates the southeastern United States receives much of its precipitation in the form of thunderstorms. Most of the Southeast averages 50 or more days each year with thunderstorms.

Recent studies using sophisticated climate model simulations indicate that continued global warming will enhance the atmospheric conditions that promote severe thunderstorm formation. Springtime severe thunderstorms in some areas east of the Rockies could increase by as much as 40 percent by 2100. As **Figure 10.4** illustrates, the number of days per year that severe thunderstorms occur may increase significantly in the eastern and southern United States. Cities such as Atlanta and New York City could see a doubling of the number of days per year when conditions are conducive to the development of severe thunderstorms.

Each year, significant weather-related damages and fatalities are associated with the lightning, strong winds, hail, tornadoes, and floods that accompany severe thunderstorms. This impact is likely to increase in the decades to come due to the influence of human-induced climate change.

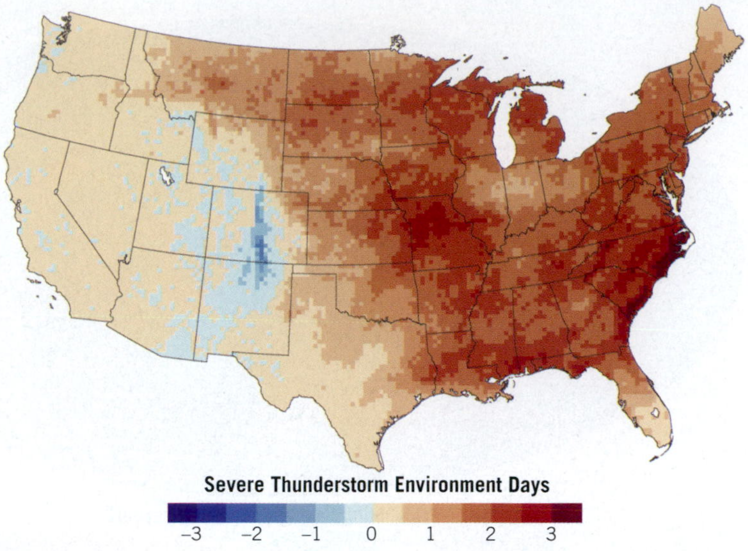


▲ **Figure 10.4 Future thunderstorm activity** This map shows changes in the number of days per year when the environmental conditions that promote severe thunderstorm activity occur. The map is based on a climate model comparing summer climate during 2072–2099 with a similar span during 1962–1989. Most of the area east of the Rocky Mountains is projected to experience an increase in these environmental conditions.

### ✔ Concept Checks 10.2

1. What are the primary requirements for the formation of thunderstorms?
2. Where would you expect thunderstorms to be most common on Earth? In the United States?
3. How might severe thunderstorm activity east of the Rockies change in the years to come?

## 10.3 Air-Mass Thunderstorms

**Sketch and explain simple diagrams that illustrate the three stages of an air-mass thunderstorm.**

In the United States, **air-mass thunderstorms** frequently occur in maritime tropical (mT) air that moves northward from the Gulf of Mexico. These warm, humid air masses contain abundant moisture in their lower levels and can be rendered unstable when heated from below or lifted along a front. Because mT air most often becomes unstable in spring and summer, when it is warmed from below by the heated land surface, it is during these seasons that air-mass thunderstorms occur most frequently. They also have a strong preference for midafternoon, when surface temperatures are highest. Because local differences in surface heating aid the growth of air-mass thunderstorms, they generally occur as scattered, isolated cells instead of being organized in relatively narrow bands or other configurations.

### Stages of Development

Important field experiments conducted in Florida and Ohio in the late 1940s probed the dynamics of air-mass thunderstorms. This pioneering work, known as the *Thunderstorm Project*, was prompted by a number of thunderstorm-related airplane crashes. It involved the use of radar, aircraft, radiosondes, and an extensive network of surface instruments. The research produced a three-stage model of the life cycle of an air-mass thunderstorm that remains basically unchanged after more than 70 years. The three stages are depicted in **Figure 10.5**.

**Cumulus Stage** Recall that an air-mass thunderstorm is largely a product of the uneven heating of the surface, which leads to rising currents of air that ultimately produce a cumulonimbus cloud. At first the buoyant thermals produce fair-weather cumulus clouds that may exist for just minutes before evaporating into the drier surrounding air (**Fig. 10.6A**). This initial cumulus development is important because it moves water vapor from the surface to greater heights. Ultimately, the air becomes sufficiently humid that newly forming clouds do not evaporate but instead continue to grow vertically.

The development of a cumulonimbus tower requires a continuous supply of moist air. The release of latent heat allows each new surge of warm air to rise higher than the last, adding to the cloud's height. This phase in the development of a thunderstorm, called the **cumulus stage**, is dominated by updrafts.

Once the cloud passes beyond the freezing level, the Bergeron process (see Chapter 5) begins producing precipitation. Eventually, the accumulation of precipitation in the cloud is too great for the updrafts to support. The falling precipitation causes drag on the air and initiates a downdraft.

The creation of the downdraft is further aided by the influx of cool, dry air surrounding the cloud, a process termed **entrainment**. This process intensifies the downdraft because the air added during entrainment is cool and therefore heavy; possibly of greater importance, it is dry. It thus causes some

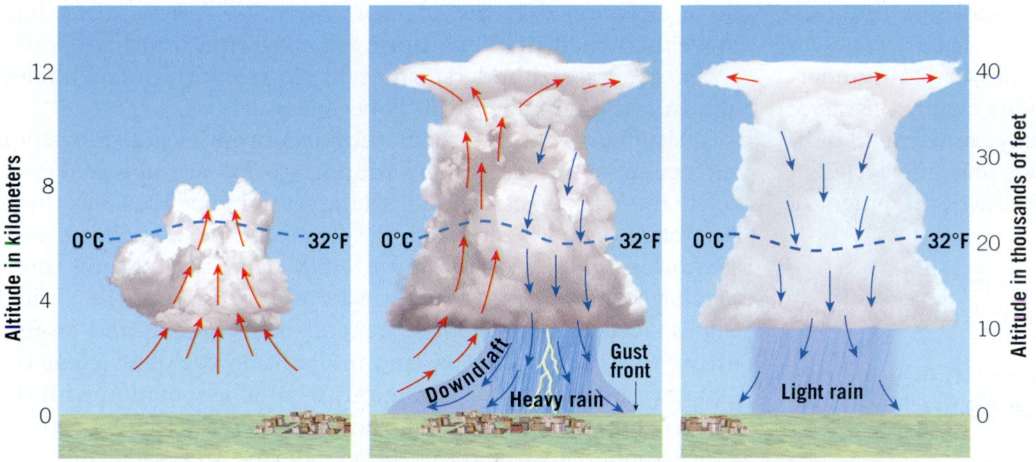


During the cumulus stage, strong updrafts provide moisture that condenses and builds the cloud.

The mature stage is marked by heavy precipitation. Updrafts exist side by side with downdrafts and continue to enlarge the cloud.

When updrafts disappear, precipitation becomes light and then stops. Without a supply of moisture from updrafts, the cloud evaporates.

◀ **SmartFigure 10.5 Stages in the development of an air-mass thunderstorm** http://goo.gl/S03JC Once a cloud passes beyond the freezing level, the Bergeron process begins producing precipitation. Eventually, the accumulation of precipitation in the cloud is too great for updrafts to support. The falling precipitation causes drag on the air and initiates a downdraft. Once downdrafts dominate, rainfall diminishes, and the cloud starts to dissipate.

of the falling precipitation to evaporate (a cooling process), thereby cooling the air within the downdraft.

**Mature Stage** As a downdraft leaves the base of a cloud, precipitation is released, marking the beginning of the cloud's **mature stage**. At the surface, the cool downdrafts spread laterally and can be felt before the precipitation reaches the ground. The sharp, cool gusts at the surface are indicative of the downdrafts aloft. During the mature stage, updrafts exist side by side with downdrafts and continue to enlarge the cloud. When the cloud grows to the top of the unstable region, often located at the base of the stratosphere, the updrafts spread laterally and produce the cloud's characteristic anvil top (Figure 10.5 center).

Generally, ice-laden cirrus clouds make up the top and are spread downwind by strong winds aloft. The mature stage is the most active period of a thunderstorm. Gusty winds, lightning, heavy precipitation, and sometimes small hail are common.

**Dissipating Stage** Once downdrafts begin, the vacating air and precipitation encourage more entrainment of the cool, dry air surrounding the cell. Eventually, the downdrafts dominate throughout the cloud and initiate the **dissipating stage** (Figure 10.5 right). The cooling effect of falling precipitation and the influx of colder air aloft mark the end of thunderstorm activity. Without a supply of moisture from updrafts, the cloud will soon evaporate. An interesting fact is that only a modest portion—on the order of 20 percent—of the moisture that condenses in an air-mass thunderstorm actually leaves the cloud as precipitation. The remaining 80 percent evaporates back into the atmosphere.

▼ **Figure 10.6 Cumulus development** **A.** Buoyant thermals produce fair-weather cumulus clouds that soon evaporate into the surrounding air, making the air more humid. As this process of cumulus development and evaporation continues, the air eventually becomes sufficiently humid so that newly forming clouds do not evaporate but continue to grow. **B.** This developing cumulonimbus cloud became a towering August thunderstorm over central Illinois.

A.

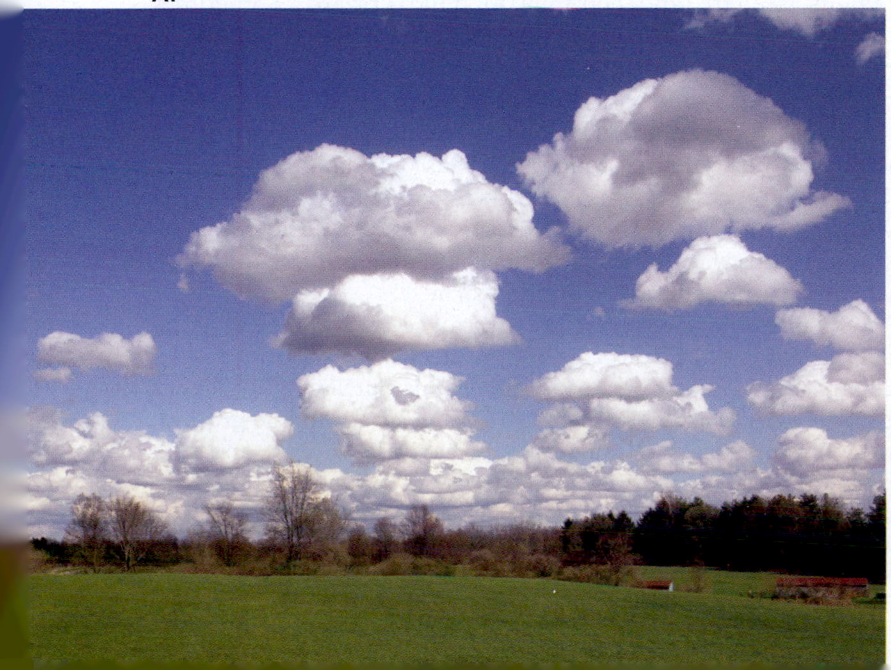

B.

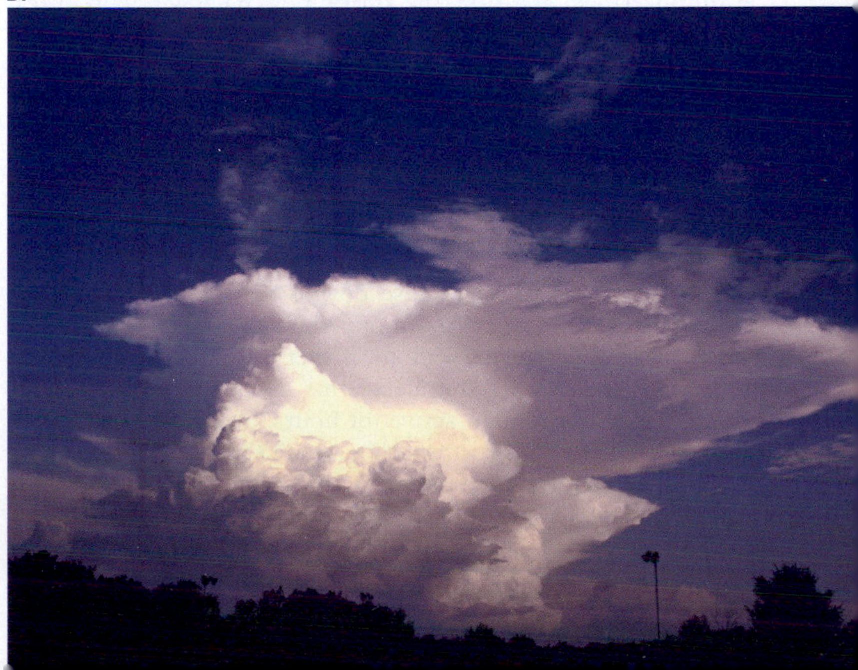

It should be noted that within a single air-mass thunderstorm, there may be several individual *cells*—that is, zones of adjacent updrafts and downdrafts. When you view a thunderstorm, you may notice that the cumulonimbus cloud consists of several towers (Figure 10.6B). Each tower may represent an individual cell that is in a somewhat different part of its life cycle.

To summarize, there are three stages in the development of an air-mass thunderstorm:

1. The *cumulus stage*, in which updrafts dominate throughout the cloud and growth from a cumulus to a cumulonimbus cloud occurs
2. The *mature stage*, the most intense phase, with heavy rain and possibly small hail, in which downdrafts are found side by side with updrafts
3. The *dissipating stage*, dominated by downdrafts and entrainment, which causes evaporation of the structure

## Occurrence

Mountainous regions, such as the Rockies in the West and the Appalachians in the East, experience a greater number of air-mass thunderstorms than do the Plains states. The air near the mountain slope is heated more intensely than air at the same elevation over the adjacent lowlands. A general upslope movement then develops during the daytime that can sometimes generate thunderstorm cells. These cells may remain almost stationary above the slopes below.

Although the growth of thunderstorms is aided by high surface temperatures, many thunderstorms are not generated solely by surface heating. For example, many of Florida's thunderstorms are triggered by the convergence associated with sea-to-land airflow (see Figure 4.20, page 107). Many thunderstorms that form over the eastern two-thirds of the United States occur as part of the general convergence and frontal lifting that accompany passing midlatitude cyclones. Near the equator, thunderstorms commonly form in association with the convergence along the equatorial low (the intertropical convergence zone). Most of these thunderstorms are not severe, and their life cycles are similar to the three-stage model described for air-mass thunderstorms.

### ✔ Concept Checks 10.3

1. During what seasons and at what time of day is air-mass thunderstorm activity greatest? Why?
2. Why does entrainment intensify thunderstorm downdrafts?
3. Summarize the three stages of an air-mass thunderstorm.

# 10.4 Severe Thunderstorms

**List the characteristics of a severe thunderstorm and distinguish among different circumstances associated with their formation.**

**Severe thunderstorms** are capable of producing heavy downpours and flash floods as well as strong, gusty, straight-line winds; large hail; frequent lightning; and perhaps tornadoes. Section 10.5 and **Boxes 10.1** and **10.2** explore three of the dangerous weather phenomena associated with severe thunderstorms. For the National Weather Service to officially classify a thunderstorm as *severe*, the storm must have winds in excess of 93 kilometers (58 miles) per hour (50 knots) or produce hailstones larger than 2.5 centimeters (1 inch) in diameter or generate a tornado. Of the estimated 100,000 thunderstorms that occur annually in the United States, about 10 percent (10,000 storms) reach severe status.

As you learned in the preceding section, air-mass thunderstorms are localized, relatively short-lived phenomena that dissipate after a brief, well-defined life cycle. They actually extinguish themselves because downdrafts cut off the supply of moisture necessary to maintain the storm. For this reason, air-mass thunderstorms seldom, if ever, produce severe weather. By contrast, other thunderstorms do not quickly dissipate but instead may remain active for hours. Some of these larger, longer-lived thunderstorms attain severe status.

Why do some thunderstorms persist for many hours? A key factor is the existence of strong vertical **wind shear**—that is, changes in wind direction and/or speed at different heights. When such conditions prevail, the updrafts that provide the storm with moisture do not remain vertical but become tilted. Because of this, the precipitation that forms high in the cloud falls into the downdraft rather than into the updraft, as occurs in air-mass thunderstorms. This allows the updraft to maintain its strength and continue to build upward. Sometimes the updrafts are sufficiently strong that the cloud top is able to push its way into the stable lower stratosphere, a situation called an *overshooting top*. (**Fig. 10.7**).

Beneath the cumulonimbus tower, where downdrafts reach the surface, the denser cool air spreads out along the ground. The leading edge of this outflowing downdraft acts like a wedge, forcing warm, moist surface air up into the thunderstorm. In this way, the downdrafts act to maintain the updrafts, which in turn sustain the thunderstorm.

By examining Figure 10.7, you can see that the outflowing cool air of the downdraft acts as a "mini cold front" as it advances into the warmer surrounding air. This outflow boundary is called a **gust front**. As the gust front moves across the ground, the very turbulent air sometimes picks up loose dust and soil, making the advancing boundary visible. Frequently a *roll cloud* or *shelf cloud* may form as warm air is lifted along the leading

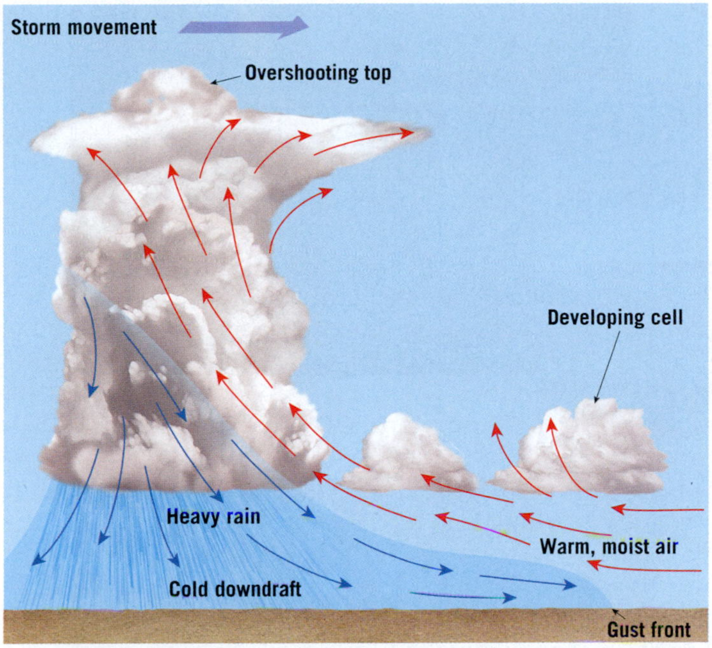


▲ **Figure 10.7 Diagram of a well-developed cumulonimbus tower** The cloud is characterized by updrafts, downdrafts, and an overshooting top. Precipitation forming in the tilted updraft falls into the downdraft. Beneath the cloud, the denser cool air of the downdraft spreads out along the ground. The leading edge of the outflowing downdraft acts to wedge moist surface air up into the cloud. Eventually the outflow boundary may become a gust front that initiates new cumulonimbus development

Video MM
Forecasting Thunderstorms
http://goo.gl/T26p2r

edge of the gust front (Fig. 10.8). The gust front's advance can provide the lifting needed for the formation of new thunderstorms many kilometers away from the initial cumulonimbus clouds.

## Supercell Thunderstorms

Some of our most dangerous weather is caused by a type of thunderstorm called a **supercell**. Few other weather phenomena are as awesome (Fig. 10.9). An estimated 2000 to 3000 supercell thunderstorms occur annually in the United States. They represent just a small fraction of all thunderstorms, but they are responsible for a disproportionate share of the deaths, injuries, and property damage associated with severe weather. Fewer than half of all supercells produce tornadoes, yet virtually all of the strongest and most violent tornadoes are spawned by supercells.

A supercell consists of a single, very powerful cell that may extend to a height of 20 kilometers (65,000 feet) and persist for many hours. These massive clouds have diameters ranging between about 20 and 50 kilometers (12 and 30 miles).

Despite their single-cell structure, supercells are remarkably complex. The vertical wind profile may cause the updraft to rotate. For example, if the surface airflow is from the south or southeast and the winds aloft increase in speed and become more westerly with height, the updraft of a thunderstorm developing in such a wind environment is made to rotate. It is within this column of cyclonically rotating air, called the **mesocyclone**, that tornadoes often form (see Figure 10.9C). More about mesocyclones can be found in the section "Tornado Development," later in the chapter.

The huge quantities of latent heat needed to sustain a supercell require special conditions that keep the lower troposphere warm and moisture rich. Studies suggest that the existence of a capping inversion layer a few kilometers above the surface helps provide this basic requirement. Recall from Chapter 4 that temperature inversions represent very stable atmospheric conditions that restrict vertical air motions. The presence of

### students sometimes ask... What would it be like inside a towering cumulonimbus cloud?

The words *violent* and *dangerous* come to mind! A German paragliding champion was pulled into an erupting thunderstorm near Tamworth, New South Wales, Australia, and blacked out.* Updrafts lofted her to 32,600 feet, covered her with ice, and pelted her with hailstones. She finally regained consciousness around 22,600 feet. Her GPS equipment and a computer tracked her movements as the storm carried her away. Shaking vigorously and with lightning all around, she managed to slowly descend and eventually land 40 miles from where she started.

*Reported in the *Bulletin of the American Meteorological Society*, Vol. 88, No. 4 (April 2007): 490.

▼ **Figure 10.8 Roll cloud** This roll cloud (shelf cloud) at Miles City, Montana, was produced along a gust front in an eddy between the inflow and the downdraft.

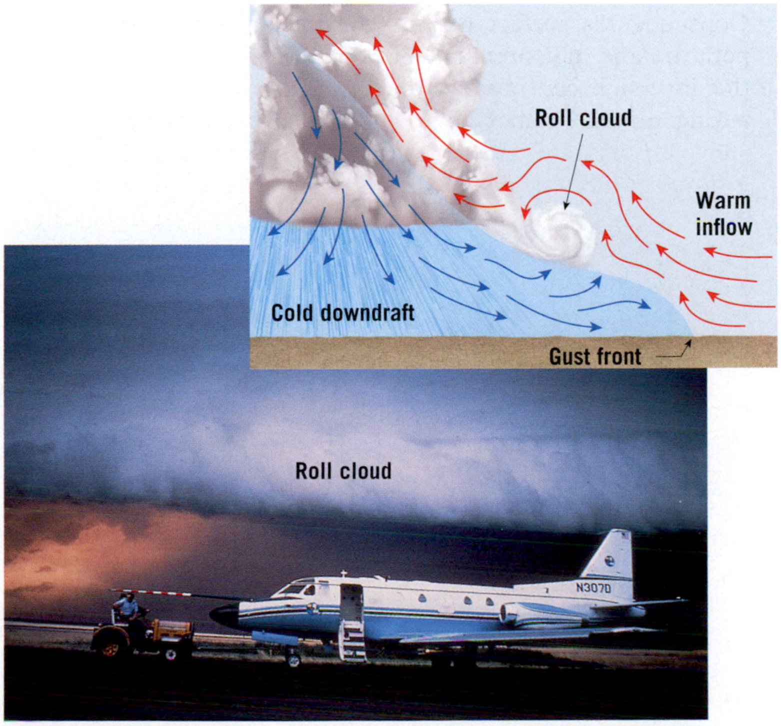

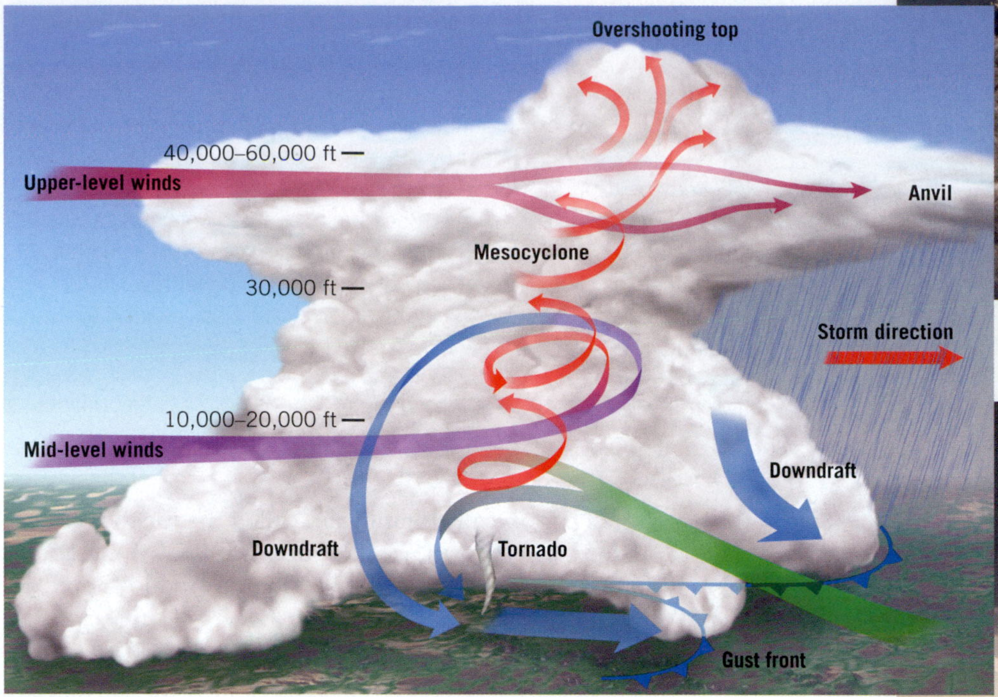


A.

B.

C.

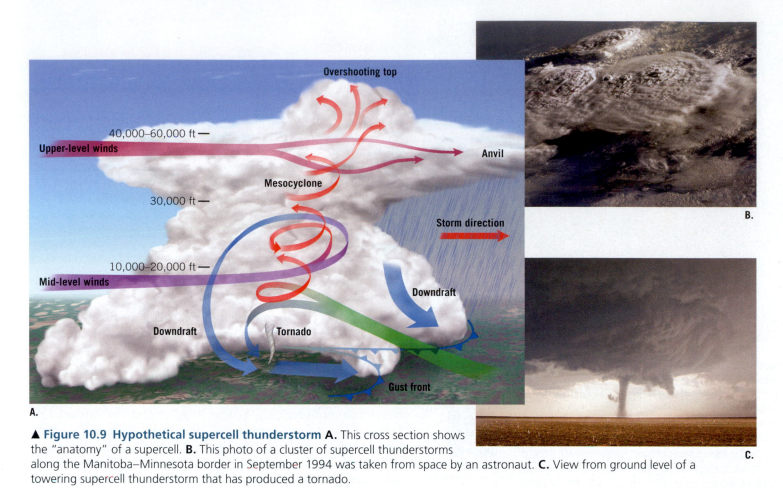

▲ **Figure 10.9 Hypothetical supercell thunderstorm A.** This cross section shows the "anatomy" of a supercell. **B.** This photo of a cluster of supercell thunderstorms along the Manitoba–Minnesota border in September 1994 was taken from space by an astronaut. **C.** View from ground level of a towering supercell thunderstorm that has produced a tornado.

an inversion seems to aid the production of a few very large thunderstorms by inhibiting the formation of many smaller ones (**Fig. 10.10**). The inversion prevents the mixing of warm, humid air in the lower troposphere with cold, dry air above. Consequently, surface heating continues to increase the temperature and moisture content of the layer of air trapped below the inversion. Eventually, the inversion is locally eroded by strong mixing from below. The unstable air below "erupts" explosively at these sites, producing unusually large cumulonimbus towers. It is from such clouds, with their concentrated, persistent updrafts, that supercells form.

Because the atmospheric conditions favoring the formation of severe thunderstorms often exist over a broad area, they frequently develop in groups that consist of many individual storms clustered together. Sometimes these clusters occur as elongate bands called *squall lines*. At other times the storms are

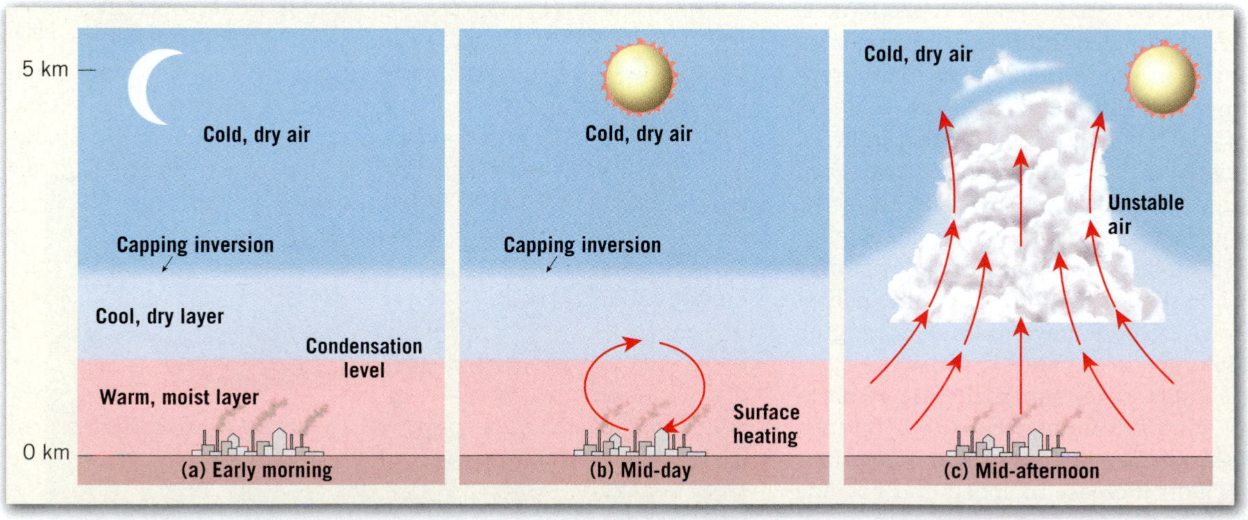


◀ **Figure 10.10 Capping inversion** The formation of severe thunderstorms can be enhanced by the existence of a temperature inversion located a few kilometers above the surface.

▲ **Figure 10.11 Mammatus sky** The dark overcast of a mammatus sky, with its characteristic downward-bulging pouches, sometimes precedes a squall line. When a mammatus formation develops, it is usually after a cumulonimbus cloud reaches its maximum size and intensity. Its presence is generally a sign of an especially vigorous thunderstorm.

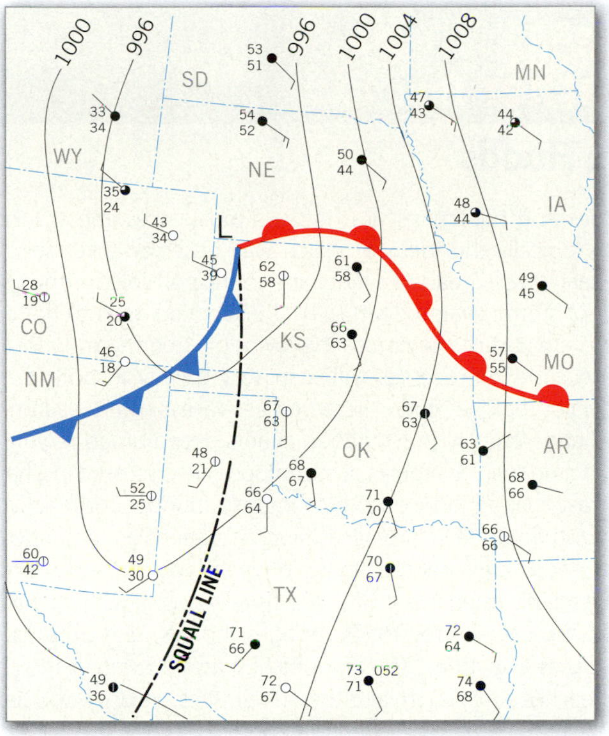


▲ **Figure 10.13 Squall line on a weather map** The squall line of this middle-latitude cyclone was responsible for a major outbreak of tornadoes. The squall line separates dry cT air and very humid mT air. The dryline is easily identified by comparing dew-point temperatures on either side of the squall line. The dew point (°F) is the lower number at each station.

organized into roughly circular clusters known as *mesoscale convective complexes*. No matter how the cells are arranged, they are not simply clusters of unrelated individual storms. Rather, they are related by a common origin, or they occur in a situation in which some cells lead to the formation of others.

## Squall Lines

A **squall line** is a relatively narrow band of thunderstorms, some of which may be severe, that develops in the warm sector of a midlatitude cyclone, usually 100 to 300 kilometers (60 to 180 miles) in advance of the cold front. The linear band of cumulonimbus development might stretch for 500 kilometers (300 miles) or more and consist of many individual cells in various stages of development. An average squall line can last 10 hours or more, and some have been known to remain active for more than a day. Sometimes the approach of a squall line is preceded by a *mammatus sky*, consisting of dark cloud rolls that have downward pouches (Fig. 10.11).

▼ **Figure 10.12 Dryline** Squall-line thunderstorms frequently develop along a dryline, the boundary separating warm, dry continental tropical air and warm, moist maritime tropical air.

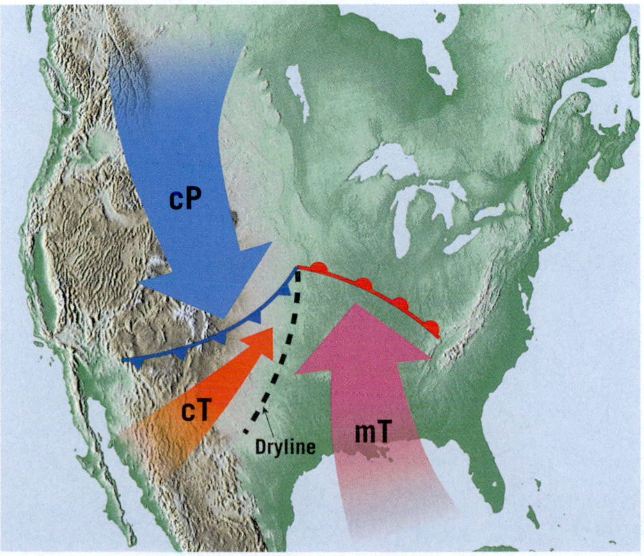


Most squall lines are not produced by forceful lifting along a cold front. Some develop from a combination of warm, moist air near the surface and an active jet stream aloft. The squall line forms when the divergence and resulting lift created by the jet stream is aligned with a strong, persistent low-level flow of warm, humid air from the south.

A squall line with severe thunderstorms can also form along a boundary called a **dryline**, a narrow zone along which there is an abrupt change in the air's moisture content. It forms when warm, dry continental tropical (cT) air from the southwestern United States is pulled into the warm sector of a midlatitude cyclone, as shown in Figure 10.12. It is important to point out that warm, dry air is denser than warm, humid air because the molecular weight of water vapor is only about 62 percent as great as the molecular weight of the mixture of gases that make up dry air. The denser cT air acts to lift the less dense mT air with which it is converging. In this case, both cloud formation and storm development along the cold front are minimal because the front is advancing into dry cT air.

Drylines most frequently develop in the western portions of Texas, Oklahoma, and Kansas. Such a situation is illustrated by

severe& hazardous weather Box 10.1

## Flash Floods

Tornadoes and hurricanes are nature's most awesome storms, and they are logically the focus of much well-deserved attention. Yet in many years, these dreaded events are not responsible for the greatest number of storm-related deaths. That distinction goes to floods.

Floods are part of the *natural* behavior of streams, and most begin with atmospheric processes that can vary greatly in both time and space. Major regional floods in large river valleys often result from an extraordinary series of precipitation events over a broad region for an extended time span. Examples of such floods are discussed in Chapter 9. By contrast, just an hour or two of intense thunderstorm activity can trigger flash floods in small valleys. Such situations are described here.

*Flash floods* are localized floods of great volume and short duration. The rapidly rising surge of water usually occurs with little advance warning and can destroy roads, bridges, homes, and other substantial structures (**Fig. 10.A**). The amount of water flowing in the channel quickly reaches a maximum and then diminishes rapidly. Flood flows often contain large quantities of sediment and debris as they sweep channels clean.

Just an hour or two of intense thunderstorm activity can trigger flash floods in small valleys.

Several factors influence flash flooding. Among them are rainfall intensity and duration, surface conditions, and topography. Urban areas are susceptible to flash floods because a high percentage of the surface area is composed of impervious roofs, streets, and parking lots, where runoff is very rapid (**Fig. 10.B**). In fact, a recent study indicated that the area of impervious surfaces in the conterminous United States amounts to more than 112,600 square kilometers (nearly 44,000 square miles), which is slightly smaller than the area of the state of Ohio.*

▲ **Figure 10.A Colorado flash floods** Extraordinary rains falling on saturated ground in an area with steep slopes led to record-breaking flash floods along the Front Range and foothills of northern Colorado in September 2013.

Frequently, flash floods result from the torrential rains associated with a slow-moving severe thunderstorm or take place when a series of thunderstorms repeatedly pass over the same location. Sometimes they are triggered by heavy rains from hurricanes and tropical storms. Occasionally, floating debris or ice can accumulate at a natural or

*C. C. Elvidge, et al., "U.S. Constructed Area Approaches the Size of Ohio," in *EOS, Transactions, American Geophysical Union*, Vol. 85, No. 24 (15 June 2004): 233.

**Figure 10.13**. The dryline is easily identified by comparing the dew-point temperatures on either side of the squall line. The dew points in the mT air to the east are 30° to 45°F higher than those in the cT air to the west. Much severe weather was generated as this extraordinary squall line moved eastward, including 55 tornadoes over a six-state region.

## Mesoscale Convective Complexes

A **mesoscale convective complex (MCC)** consists of many individual thunderstorms organized into a large oval to circular cluster. A typical MCC is large, covering an area of at least 100,000 square kilometers (39,000 square miles). The

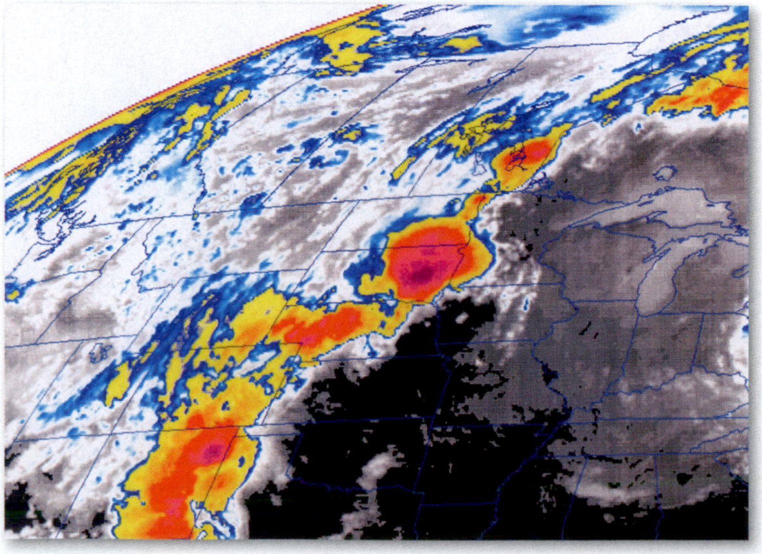

► **Figure 10.14 Mesoscale convective complex** This satellite image shows a mesoscale convective complex (MCC) over the eastern Dakotas.

artificial obstruction and restrict the flow of a stream. When such temporary dams fail, torrents of water can be released as flash floods.

Flash floods can take place in almost any area of the country. They are particularly common in mountainous terrain, where steep slopes can quickly channel runoff into narrow valleys. The hazard is most acute when the soil is already nearly saturated from earlier rains or consists of impermeable materials.

Why do so many people perish in flash floods? Aside from the surprise factor (many are caught sleeping), people do not appreciate the power of moving water. Just 15 centimeters (6 inches) of fast-moving flood water can knock a person down. Most automobiles will float and be swept away in only 0.6 meter (2 feet) of water. *More than half of all U.S. flash-flood fatalities are auto related!* Clearly, people should never attempt to drive over a flooded road (**Fig. 10.C**).

▼ **Figure 10.B Effect of urban development on flooding** Urban areas are susceptible to flash floods because runoff following heavy rains is rapid due to the high percentage of the surface that is impervious. Streamflow in Mercer Creek, an urban stream in western Washington, increases more quickly, reaches a higher peak flow, and has a larger volume during a 1-day storm on February 1, 2000, than streamflow in Newaukum Creek, a nearby rural stream. Streamflow during the following week, however, was greater in Newaukum Creek.

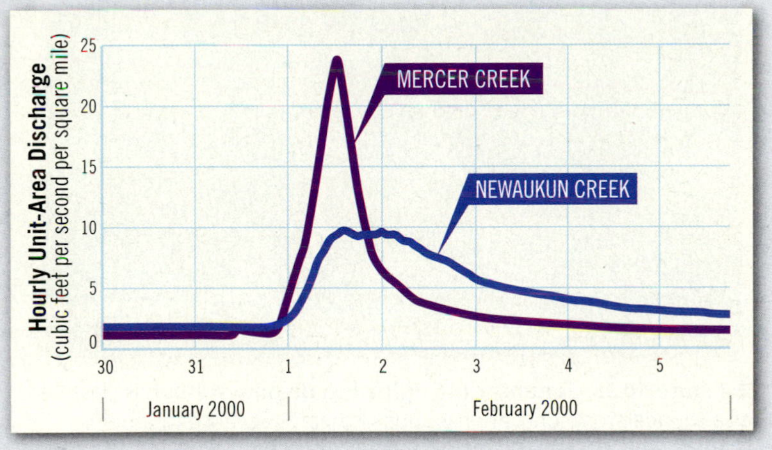

▲ **Figure 10.C Vulnerable cars** Most people do not appreciate the power of moving water. Many automobiles will float and be swept away in a strong current that is only 2 feet deep. More than half of all U.S. flash-flood fatalities are auto related.

**Questions**

1. List at least three factors that influence flash flooding.
2. When rural areas become urbanized, the chances of flash floods increases. Explain.

Video MM
Rain and Flooding
http://goo.gl/KokICh

usually slow-moving complex may persist for 12 hours or more (**Fig. 10.14**).

MCCs tend to form most frequently in the Great Plains. When conditions are favorable, an MCC develops from a group of afternoon air-mass thunderstorms. In the evening, as the local storms decay, the MCC starts developing. The transformation of afternoon air-mass thunderstorms into an MCC requires a strong low-level flow of very warm and moist air. This flow enhances instability, which in turn spurs convection and cloud development. As long as favorable conditions prevail, MCCs remain self-propagating as gust fronts from existing cells lead to the formation of new powerful cells nearby. New thunderstorms tend to develop near the side of the complex that faces the incoming low-level flow of warm, moist air.

Although mesoscale convective complexes sometimes produce severe weather, they are also beneficial because they provide a significant portion of the growing-season rainfall to the agricultural regions of the central United States.

## ✔ Concept Checks 10.4

1. How does a severe thunderstorm differ from an air-mass thunderstorm?
2. Describe how downdrafts in a severe thunderstorm act to maintain updrafts.
3. What is a gust front?
4. What is a dryline? Briefly describe the formation of a squall line along a dryline.
5. What circumstances favor the development of mesoscale convective complexes?

## 10.5 Lightning and Thunder

**Explain what causes lightning and thunder.**

In most years, **lightning** ranks second only to floods in the number of storm-related deaths in the United States. Although an average of 60 lightning deaths are reported each year, it is estimated that as many as 100 people are killed and more than 1000 injured by lightning every year in the United States.

Watches, warnings, and forecasts are routinely issued for floods, tornadoes, and hurricanes, but not for lightning. Why? Lightning strikes the ground tens of millions of times each year. Because lightning is so widespread and strikes the ground with such great frequency, it is not possible to warn each person of every flash. For this reason, lightning is the most dangerous weather hazard many people encounter during the year (Table 10.1).

A storm is classified as a thunderstorm only after thunder is heard. Because thunder is produced by lightning, lightning must also be present (Fig. 10.15). Lightning is similar to the electrical shock you may have experienced when touching a metal object on a very dry day. However, the intensity is dramatically different.

During the formation of a large cumulonimbus cloud, a separation of charge occurs; this means that part of the cloud develops an excess negative charge, whereas another part acquires an excess positive charge. The objective of lightning is to equalize these electrical differences by producing a negative flow of current from the region of excess negative charge to the region with excess positive charge or vice versa. Because air is a poor conductor of electricity (good insulator), the electrical potential (charge difference) must be very high before lightning will occur.

The most common type of lightning occurs between oppositely charged zones *within* a cloud or between clouds. About 80 percent of all lightning is of this type. It is often called **sheet lightning** because it produces a bright but diffuse illumination of the parts of the cloud in which the flash occurred. Sheet lightning is not a unique form; rather, it is ordinary lightning in which the flash is obscured by the clouds. The second type of lightning, in which the electrical discharge occurs between the cloud and Earth's surface, is often more dramatic. This **cloud-to-ground lightning** represents about 20 percent of lightning strokes and is the most damaging and dangerous form.

▲ **Figure 10.15 Summertime lightning display** A storm is classified as a thunderstorm only after thunder is heard. Because thunder is produced by lightning, lightning must also occur.

**Table 10.1 Lightning Casualties in the United States, by Location or Activity**

| Rank | Location/Activity | Relative Frequency |
|---|---|---|
| 1 | Open areas (including sports fields) | 45% |
| 2 | Going under trees to keep dry | 23% |
| 3 | Water-related activities (swimming, boating, and fishing) | 14% |
| 4 | Golfing (while in the open) | 6% |
| 5 | Farm and construction vehicles (with open exposed cockpits) | 5% |
| 6 | Corded telephone (number-one indoor source of lightning casualties) | 4% |
| 7 | Golfing (while mistakenly seeking "shelter" under trees) | 2% |
| 8 | Using radios and radio equipment | 1% |

Source: National Weather Service.

### What Causes Lightning?

The origin of charge separation in clouds, although not fully understood, must hinge on rapid vertical movements within clouds because lightning occurs primarily in the violent mature stage of a cumulonimbus cloud. In the midlatitudes, the formation of these towering clouds is chiefly a summertime phenomenon, which explains why lightning is seldom observed there severe & hazardous weather Box 10.2

# Downbursts

*Downbursts* are strong localized zones of sinking air that originate in the lower part of some cumulus and cumulonimbus clouds. Downbursts are different from typical thunderstorm downdrafts in that they are more intense and concentrated over smaller areas (**Fig. 10.D**). Their small horizontal dimension, usually less than 4 kilometers (2.5 miles), is the reason the alternate term *microburst* is often used. When a downburst reaches the ground, air spreads out in all directions, similar to a jet of water from a faucet splashing in a sink. Within minutes the downburst dissipates, while the outflow of air at the ground continues to expand. The straight-line winds from a downburst can exceed 160 kilometers (100 miles) per hour and can cause damages equivalent to that of a weak tornado.

The acceleration of air in a downburst occurs when evaporating raindrops cool the air. Remember that the colder the air, the denser it is, and the denser the air, the faster it will "fall." A second mechanism that contributes to downbursts is the drag of falling precipitation. Although a single raindrop does not exert much drag, the force of millions of falling drops can be substantial.

Downbursts are extremely hazardous to aircraft, especially during takeoff and landing

The violent winds of a downburst can be dangerous and destructive. For example, in July 1993, millions of trees were uprooted by a downburst near Pakwash, Ontario. In July 1984, 11 people drowned when a downburst caused a 28-meter- (90-foot-) long sternwheeler boat to capsize in the Tennessee River. Downbursts are extremely hazardous to aircraft, especially during takeoff and landing, when planes are nearest the ground. In just a matter of minutes, a cloud can change from being dominated by updrafts to having a downburst. Imagine an aircraft attempting to land and being confronted by a downburst, as in **Fig. 10.E**. As the airplane flies into the downburst, it initially encounters a strong headwind, which tends to carry it upward. To reduce the lift, the pilot points the nose of the aircraft downward. Then, seconds later, a tailwind is encountered. Now, because the wind is moving with the airplane, the amount of air flowing over the wings and providing lift is dramatically reduced, causing the craft to suddenly lose altitude and possibly crash. This serious aviation hazard has been reduced significantly because systems to detect the wind shifts associated with downbursts have been developed and deployed at major airports. Moreover, pilots receive training on how to handle downbursts during takeoff and landing.

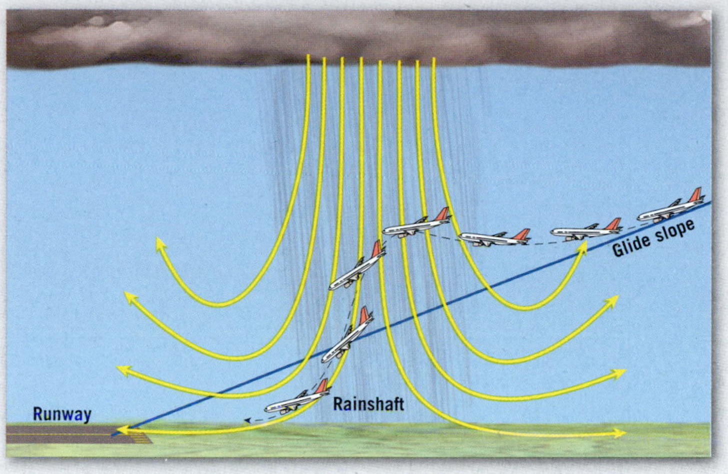


▲ **Figure 10.E Airport hazard** The arrows in this sketch represent the downward and outward movement of air in a downburst. An airplane passing through a downburst as it attempts to land initially experiences a strong headwind and lift. That is followed by an abrupt descent, caused by the downward motion of air, and a rapid loss of air speed.

▼ **Figure 10.D Dangerous downburst** The rain shaft extending downward from a cumulonimbus cloud identifies a powerful downburst near Denver's Stapleton Airport.

**Questions**

1. What is another name for a downburst?
2. List two factors that contribute to the formation of a downburst.

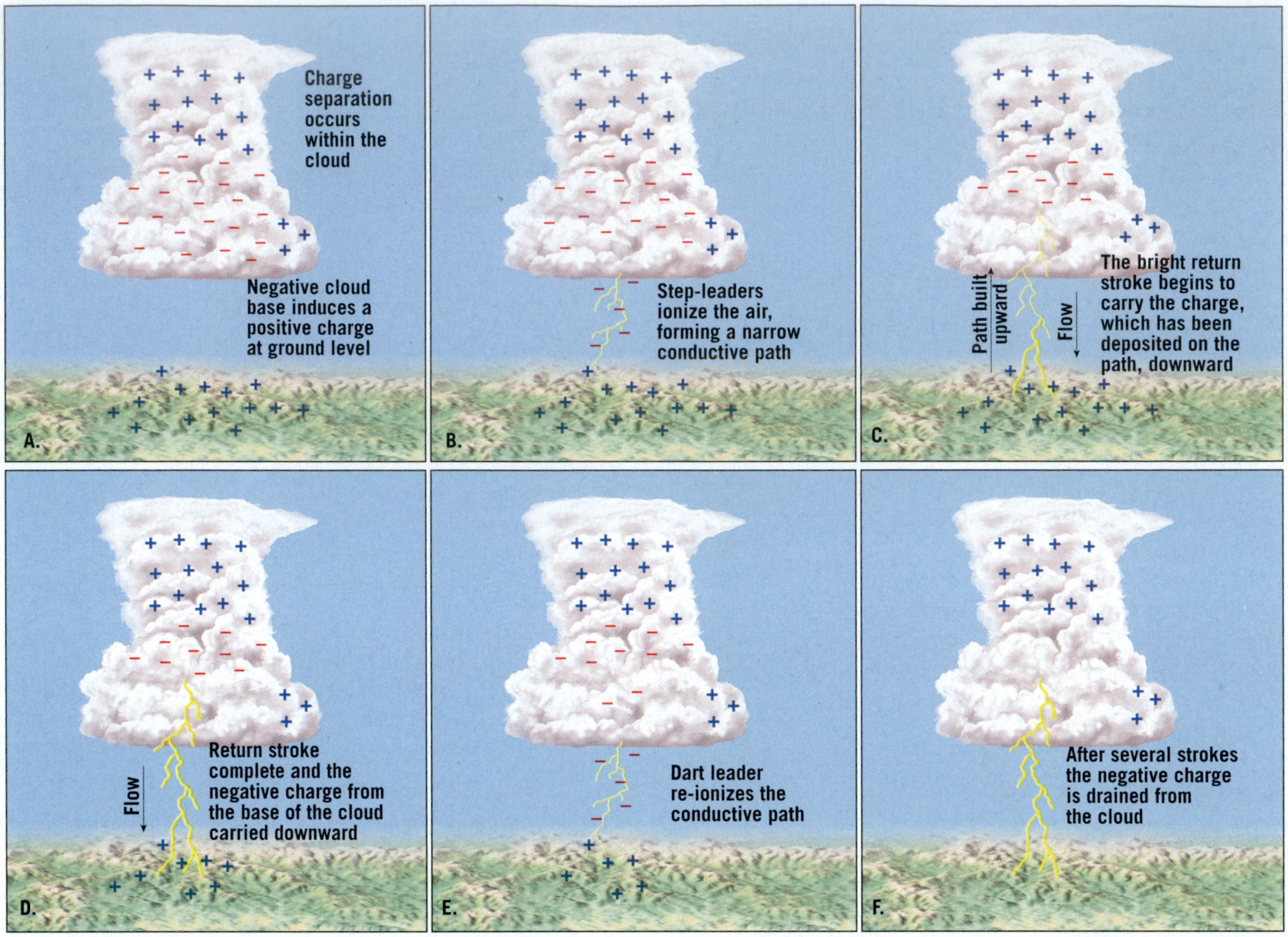


▲ **Figure 10.16 Discharge of a cloud via cloud-to-ground lightning** Examine this drawing carefully while reading the text.

in the winter. Furthermore, lightning rarely occurs before the growing cloud penetrates the 5-kilometer (16,000-foot) level, where sufficient cooling begins to generate ice crystals.

Some cloud physicists believe that charge separation occurs during the formation of ice pellets. Experimentation shows that as droplets begin to freeze, positively charged ions are concentrated in the colder regions of the droplets, whereas negatively charged ions are concentrated in the warmer regions. Thus, as a droplet freezes from the outside in, it develops a positively charged ice shell and a negatively charged interior. As the interior begins to freeze, it expands and shatters the outside shell.

**students sometimes ask...**

### How frequently does lightning strike the ground in the United States?

Since 1989, the National Lightning Detection Network has recorded an average of about 25 million cloud-to-ground flashes per year in the contiguous 48 states. About half of all flashes have more than one ground-strike point, so more than 40 million points on the ground are struck in an average year. In addition to cloud-to-ground flashes, there are roughly 5 to 10 times as many flashes within clouds.

The positively charged ice fragments are carried upward by turbulence, while the relatively heavy droplets eventually carry their negative charge toward the cloud base. As a result, the upper part of the cloud is left with a positive charge, and the lower portion of the cloud maintains an overall negative charge, with small positively charged pockets (**Fig. 10.16**).

As the cloud moves, the negatively charged cloud base alters the charge at the surface directly below by repelling negatively charged particles. Thus, the surface beneath the cloud acquires a net positive charge. These charge differences build to millions and even hundreds of millions of volts before a lightning stroke acts to discharge the negative region of the cloud by striking the positive area of the ground below, or, more frequently, the positively charged portion of that cloud or a nearby cloud.

## Lightning Strokes

Cloud-to-ground strokes are of most interest and have been studied in detail. Moving-film cameras have greatly aided in these studies (**Fig. 10.17**). They show that the lightning we see as a single flash is really several very rapid strokes between the cloud and the ground. We call the total discharge—which lasts only a few tenths of a second and appears as a bright streak—the **flash**. Individual components that make up each flash are termed **strokes**. Each stroke is separated by roughly 50 milliseconds, and there are usually three to four strokes per flash.* When a lightning flash appears to flicker, it is because your eyes discern the individual strokes that make up this discharge. Moreover, each stroke consists of a downward propagating leader that is immediately followed by a luminous return stroke.

Each stroke is believed to begin when the electrical field near the cloud base frees electrons in the air immediately below, thereby ionizing the air (Figure 10.16). Once ionized, the air becomes a conductive path with a radius of roughly 10 centimeters and a length of 50 meters. This path is called a **leader**. During this electrical breakdown, the mobile electrons in the cloud base begin to flow down this channel. This flow increases the electrical potential at the head of the leader, which causes a further extension of the conductive path through further ionization. Because this initial path extends itself toward Earth in short, nearly invisible bursts, it is called a **step leader**. Once this channel nears the ground, the electrical field at the surface ionizes the remaining section of the path. With the path completed, the electrons that were deposited along the channel begin to flow downward. The initial flow begins near the ground.

As the electrons at the lower end of the conductive path move toward Earth, electrons positioned successively higher up the channel begin to migrate downward. Because the path of electron flow is continually being extended upward, the accompanying electrical discharge has been appropriately named a **return stroke**. As the wave front of the return stroke moves upward, the negative charge that was deposited on the channel is effectively lowered to the ground. This intense return stroke illuminates the conductive path and discharges the lowest kilometer or so of the cloud. During this phase, tens of coulombs of negative charge are lowered to the ground.*

The first stroke is usually followed by additional strokes that apparently drain charges from higher areas within the cloud. Each subsequent stroke begins with a **dart leader** that once again ionizes the channel and carries the cloud potential toward the ground. The dart leader is continuous and less branched than the step leader. When the current between strokes has ceased for a period greater than 0.1 second, further strokes will be preceded by a step leader whose path is different from that of the initial stroke. The total time of each flash consisting of three or four strokes is about 0.2 second.

**eye ON THE atmosphere 10.1**

This satellite image is a composite showing 3 months' worth of data from NASA's Lightning Imaging Sensor, which records all lightning strikes between 35° north and 35° south latitude.

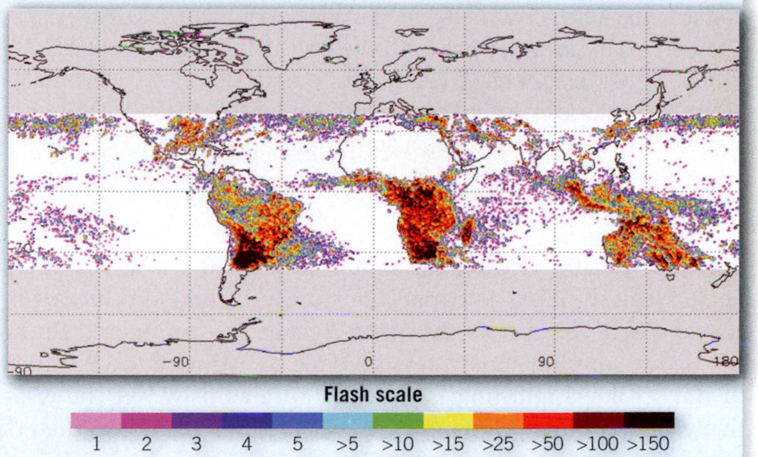


**Questions**

1. What 3-month span does this image represent: June–August or December–February? How did you figure this out?
2. Why is there so much more lightning over land areas than over the oceans?

▲ **Figure 10.17 Multiple lightning stroke of a single flash as recorded by a moving-film camera**

## Thunder

The electrical discharge of lightning superheats the air immediately around the lightning channel. In less than a second,

*One millisecond equals one one-thousandth ($^{1}/_{1000}$) of a second.

*A coulomb is a unit of electrical charge equal to the quantity of charge transferred in 1 second by a steady current of 1 ampere.

**students sometimes ask...**

### What are the chances of being struck by lightning?

Most people greatly underestimate the probability of being involved in a lightning strike. According to the National Weather Service, the chance of an individual in the United States being killed or injured during a given year is 1 in 240,000. Assuming a life span of 80 years, a person's lifetime odds become 1 in 3000. The average person has 10 family members and others with whom he or she is close, so the chances are 1 in 300 that a lightning strike will closely affect a person during his or her lifetime.

the temperature rises by as much as 33,000°C. When air is heated this quickly, it expands explosively and produces the sound waves we hear as **thunder**. Because lightning and thunder occur simultaneously, it is possible to estimate the distance to the stroke. Lightning is seen instantaneously, but the relatively slow sound waves, which travel approximately 330 meters (1000 feet) per second, reach us a little later. If thunder is heard 5 seconds after the lightning is seen, the lightning occurred about 1650 meters away (approximately 1 mile).

The thunder that we hear as a rumble is produced along a long lightning path located at some distance from the observer. The sound that originates along the path nearest the observer arrives before the sound that originated farthest away. This factor lengthens the duration of the thunder. Reflection of the sound waves by mountains or buildings further delays their arrival and adds to this effect. When lightning occurs more than 20 kilometers (12 miles) away, thunder is rarely heard. This type of lightning, popularly called *heat lightning*, is no different from the lightning that we associate with thunder.

### ✔ Concept Checks 10.5

1. Which is more common: sheet lightning or cloud-to-ground lightning?
2. How is thunder produced?
3. What is heat lightning?

**students sometimes ask...**

### About how many people who are struck by lightning are actually killed?

According to the National Weather Service, about 10 percent of lightning strike victims are killed; 90 percent survive. However, survivors are not unaffected. Many suffer severe, lifelong injury and disability.

## 10.6 Tornadoes

**Describe the structure and basic characteristics of tornadoes.**

Tornadoes are local storms of short duration that must be ranked high among nature's most destructive forces (**Fig. 10.18**). Their sporadic occurrence and violent winds cause many deaths each year. The nearly total destruction in some stricken areas has led many to liken their passage to bombing raids during war.

Such was the case during a very stormy period in late May 2013 in central Oklahoma. On May 20, an EF-5 tornado, the most severe category (see Table 10.3, page 289), struck the city of Moore. Peak winds were estimated at 340 kilometers (210 miles) per hour. The storm took 25 lives and injured more than 350 people. Entire neighborhoods were destroyed; at least 13,000 structures were destroyed or damaged. Estimated damages exceeded $2 billion (**Fig. 10.19**). The tornado was one of many that occurred across the Great Plains over a 2-day span, including five that struck central Oklahoma on May 19. This was not the first time Moore had experienced such a storm. Thirteen years earlier, in May 1999, an even stronger and deadlier tornado hit this community.

**Tornadoes**, sometimes called *twisters* or *cyclones*, are violent windstorms that take the form of a rotating column of air, or *vortex*, that extends downward from a cumulonimbus cloud. Pressures within some tornadoes have been estimated to be as much as 10 percent lower than immediately outside the storm. Drawn by the much lower pressure in the center of the vortex, air near the ground rushes into the tornado from all directions. As the air streams inward, it is spiraled upward around the core until it eventually merges with the airflow of the parent thunderstorm deep in the cumulonimbus tower.

Because of this rapid drop in pressure, air sucked into the storm expands and cools adiabatically. If the air cools below its dew point, the resulting condensation creates a pale and ominous-appearing cloud that may darken as it moves across the ground, picking up dust and debris. Occasionally, when the inward spiraling air is relatively dry, no condensation funnel forms because the drop in pressure is not sufficient to cause the necessary adiabatic cooling. In such cases, the vortex is made visible only by the material that it vacuums from the surface and carries aloft.

A tornado may consist of a single vortex, but within many stronger tornadoes are smaller intense whirls called *suction vortices* that orbit the center of the larger tornado circulation

◀ **Figure 10.18 Condensation and debris make tornadoes visible** A tornado is a violently rotating column of air in contact with the ground. The air column is visible when it contains condensation or when it contains dust and debris. Often the appearance is a result of both. When the column of air is aloft and does not produce damage, the visible portion is properly called a *funnel cloud.*

(**Fig. 10.20**). The tornadoes in this latter category are called **multiple-vortex tornadoes**. Suction vortices have diameters of only about 10 meters (30 feet) and usually form and die out in less than a minute. They can occur in all sorts of tornado sizes, from huge "wedges" to narrow "ropes." Suction vortices are responsible for most of the narrow, short swaths of extreme damage that sometimes occur along tornado tracks. It is now believed that most reports of several tornadoes at once—from news accounts and early-twentieth-century tornado tales—actually were multiple-vortex tornadoes.

Because of the tremendous pressure gradient associated with a strong tornado, maximum winds can sometimes exceed 480 kilometers (300 miles) per hour. Reliable wind-speed measurements using traditional anemometers are lacking, but using Doppler radar observations, scientists measured wind speeds of 486 kilometers (302 miles) per hour in a devastating tornado that struck the Oklahoma City area in May 1999.

Most records of changes in atmospheric pressure associated with the passage of a tornado are estimates based on a few storms that happened to pass a nearby weather station or were studied by storm-chasing meteorologists with mobile equipment. Many attempts have been made to deploy instruments in the path of a tornado, but only a few have met with success. One successful measurement occurred in Manchester, South Dakota, on June 24, 2003. Meteorologists chasing a supercell thunderstorm deployed a specially designed probe directly in the path of a violent tornado. The tornado passed directly over the instrument

▶ **Figure 10.19 Tornado destruction at Moore, Oklahoma** On May 20, 2013, central Oklahoma was devastated by an EF-5 tornado, the most severe category. At its peak, the tornado was 2.1 kilometers (1.3 miles) wide and had winds of 340 kilometers (210 miles) per hour.

Video (MM)
Radar Research at NOAA's National Severe Storms Lab
http://goo.gl/bQHPEs

package, which measured a sharp drop of 100 millibars over a span of about 40 seconds (Fig. 10.21). Gathering such data is challenging because tornadoes are short-lived, highly localized, and dangerous. The development of Doppler radar, however, has improved our ability to study tornado-producing thunderstorms. As you will see, this technology is allowing meteorologists to expand our understanding from a safe distance.

▼ **Figure 10.20 Multiple-vortex tornado** Some tornadoes have multiple suction vortices. These small and very intense vortices are roughly 10 meters (30 feet) across and move in a counterclockwise path around the tornado center. Because of this multiple-vortex structure, one building might be heavily damaged and another one, just 10 meters away, might suffer little damage.

Animation MM
Tornado Wind Patterns
http://goo.gl/IuOPiY

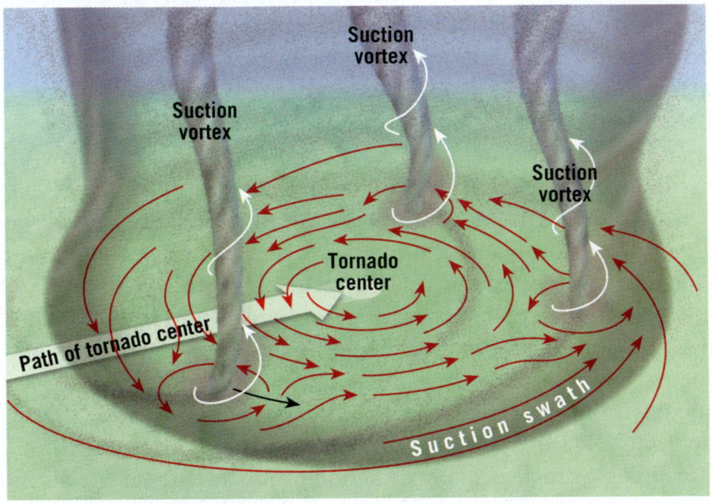


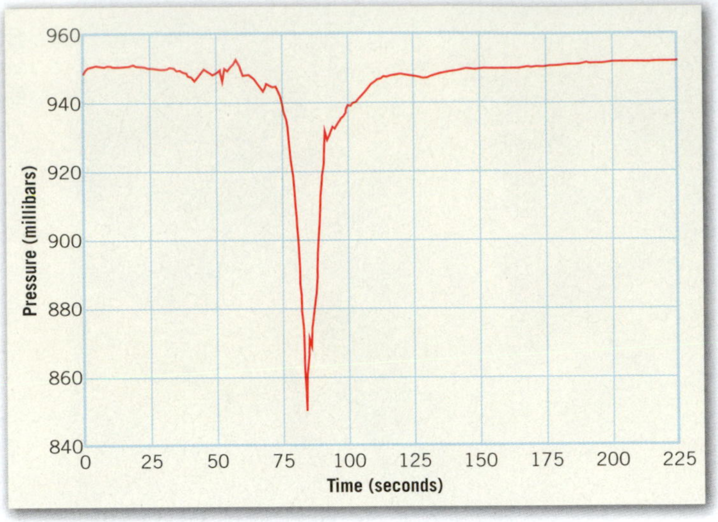


▲ **Figure 10.21 Pressure change with the passage of a tornado** This graph shows the dramatic pressure change of 100 millibars in just 40 seconds that occurred as a violent tornado passed directly over a specially designed probe that had been placed in the storm's path just moments earlier. Manchester, South Dakota, June 24, 2003.

Video MM
NSSL in the Field
http://goo.gl/fttN0K

## ✔ Concept Checks 10.6

1. Why do tornadoes have such high wind speeds?
2. What makes the rotating air column of a tornado visible?
3. What are multiple-vortex tornadoes? Are all tornadoes of this type?

# 10.7 Development and Occurrence of Tornadoes

**Summarize the atmospheric conditions and locations that are favorable to the formation of tornadoes.**

Tornadoes form in association with severe thunderstorms that produce high winds, heavy (sometimes torrential) rainfall, and often damaging hail. Although hail may or may not precede a tornado, the portion of the thunderstorm adjacent to large hail is often the area where strong tornadoes are most likely to occur.

Tornadoes are the product of the interaction between strong updrafts in a thunderstorm and the winds in the troposphere. Fortunately, fewer than 1 percent of all thunderstorms produce tornadoes. Nevertheless, a much higher number must be monitored as potential tornado producers.

## Tornado Development

Tornadoes can form in any situation that produces severe weather, including cold fronts, squall lines, and tropical cyclones (hurricanes). Usually the most intense tornadoes are those that form in association with supercells. An important precondition linked to tornado formation in severe thunderstorms is the development of a mesocyclone. A *mesocyclone* is a vertical cylinder of rotating air, typically about 3 to 10 kilometers (2 to 6 miles) across, that develops in the updraft of a severe thunderstorm. The formation of this large vortex often precedes tornado formation by 30 minutes or so.

Mesocyclone formation depends on the presence of vertical wind shear. Moving upward from the surface, winds change direction from southerly to westerly, and the wind speed increases. The speed wind shear (that is, stronger winds aloft and weaker winds near the surface) produces a rolling motion about a horizontal axis, as shown in Figure 10.22A. If conditions are right, strong updrafts in the storm tilt the horizontally rotating air to a nearly vertical alignment (see Figure 10.22B). This produces the initial rotation within the cloud interior.

At first, the mesocyclone is wider, shorter, and rotating more slowly than will be the case in later stages. Subsequently, the mesocyclone is stretched vertically and narrowed horizontally, causing wind speeds to accelerate in an inward vortex (just as spinning ice skaters accelerate by pulling arms in, or

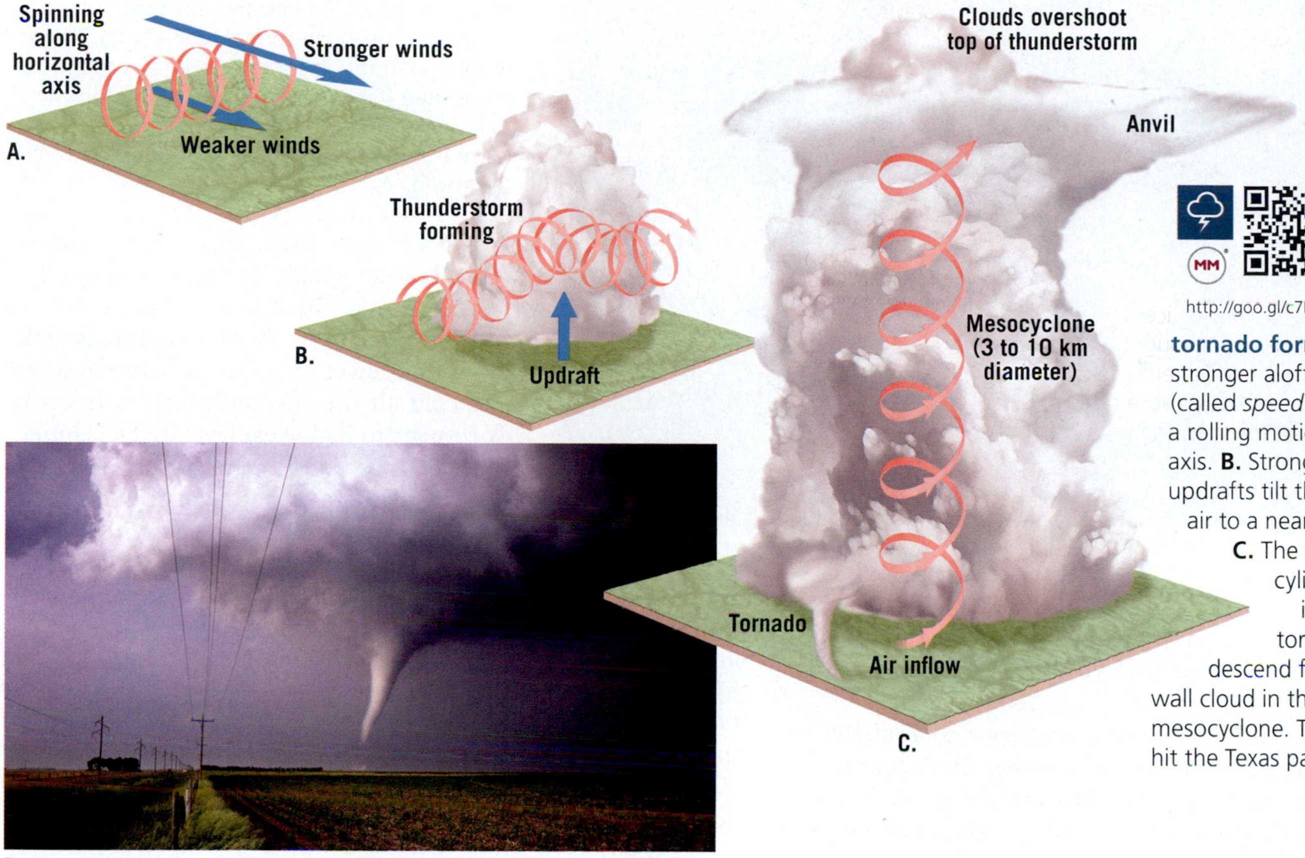

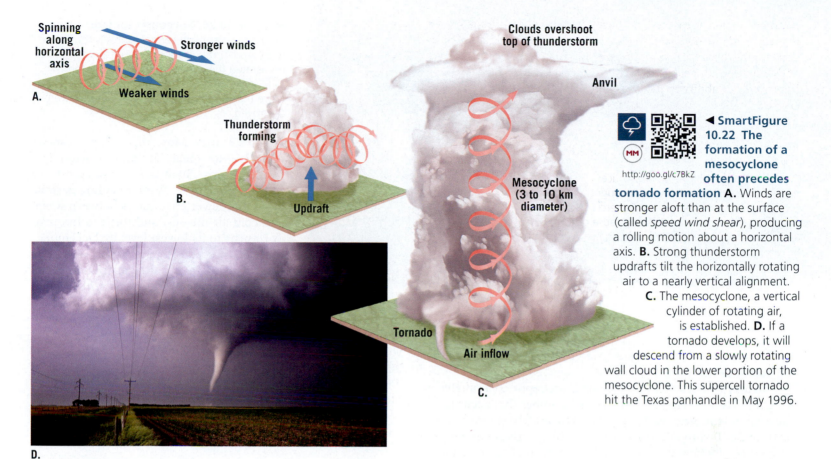

◀ **SmartFigure 10.22 The formation of a mesocyclone often precedes tornado formation** **A.** Winds are stronger aloft than at the surface (called *speed wind shear*), producing a rolling motion about a horizontal axis. **B.** Strong thunderstorm updrafts tilt the horizontally rotating air to a nearly vertical alignment. **C.** The mesocyclone, a vertical cylinder of rotating air, is established. **D.** If a tornado develops, it will descend from a slowly rotating wall cloud in the lower portion of the mesocyclone. This supercell tornado hit the Texas panhandle in May 1996.

http://goo.gl/c7BkZ

as a sink full of water accelerates as it spirals down a drain; see Box 11.1, page 302). Next, the narrowing column of rotating air stretches downward until a portion of the cloud protrudes below the cloud base to produce a very dark, slowly rotating *wall cloud*. Finally, a slender and rapidly spinning vortex emerges from the base of the wall cloud to form a *funnel cloud*. If the funnel cloud makes contact with the surface, it is then classified as a *tornado*.

The formation of a mesocyclone does not necessarily mean that tornado formation will follow. Only about half of all mesocyclones produce tornadoes. The reason for this is not yet understood, and therefore forecasters cannot determine in advance which mesocyclones will spawn tornadoes.

## Tornado Climatology

Recall that severe thunderstorms—and hence tornadoes—are most often spawned along the cold front or squall line of a midlatitude cyclone or in association with supercell thunderstorms. Throughout the spring, air masses associated with midlatitude cyclones are most likely to have greatly contrasting conditions. Continental polar air from Canada may still be very cold and dry, whereas maritime tropical air from the Gulf of Mexico is warm, humid, and unstable. The greater the contrast, the more intense the storm tends to be.

These two contrasting air masses are most likely to meet in the central United States because there is no significant natural barrier separating the center of the country from the arctic or the Gulf of Mexico. Consequently, this region generates more tornadoes than any other part of the country or, in fact, the world. **Figure 10.23**, which depicts the average annual tornado

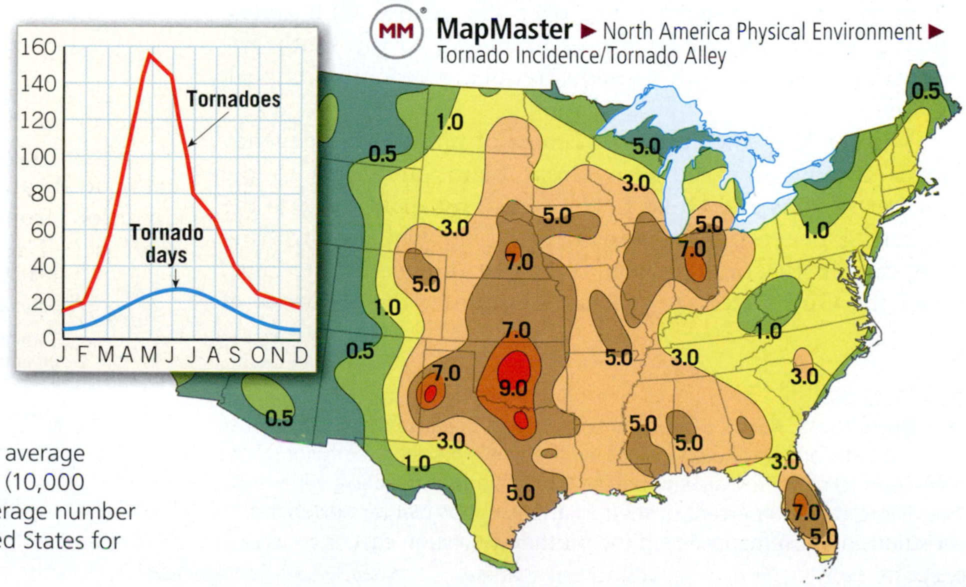

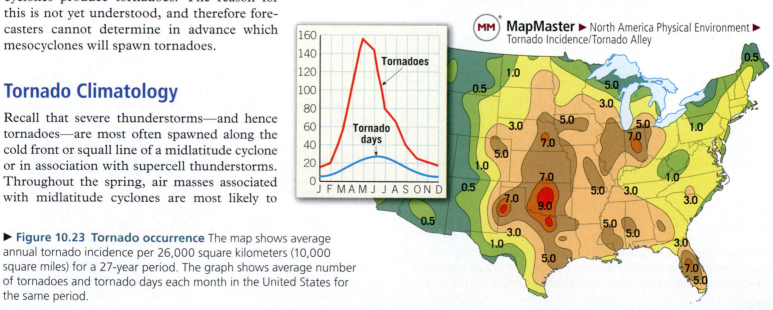

▶ **Figure 10.23 Tornado occurrence** The map shows average annual tornado incidence per 26,000 square kilometers (10,000 square miles) for a 27-year period. The graph shows average number of tornadoes and tornado days each month in the United States for the same period.

◀ **Figure 10.24 November tornado** Although "tornado season" in central Illinois is considered to be April through June, a major tornado devastated the community of Washington, Illinois, on Sunday morning, November 17, 2013. More than 630 homes and 2500 vehicles were destroyed.

incidence in the United States over a 27-year period, readily substantiates this fact.

On average, nearly 1300 tornadoes are reported annually in the United States. However, the actual number that occur from one year to the next varies greatly. During the 14-year period from 2000 through 2013, for example, yearly totals ranged from a low of 935 in 2002 to a high of 1894 in 2011. Tornadoes occur during every month of the year. April through June is the period of greatest tornado frequency in the United States, and December and January are the months of lowest activity. Of the nearly 40,522 confirmed tornadoes reported over the contiguous 48 states during the 50-year period 1950–1999, an average of almost 6 per day occurred during May. At the other extreme, a tornado was reported only about every other day in December and January.

More than 40 percent of all tornadoes take place during the spring. Fall and winter, by contrast, together account for only 19 percent (Figure 10.23). In late January and February, when the incidence of tornadoes begins to increase, the center of maximum frequency lies over the central Gulf states. During March this center moves eastward, to the southeastern Atlantic states, with tornado frequency reaching its peak in April.

During May and June, the center of maximum frequency moves through the southern Great Plains and then to the Northern Plains and Great Lakes area. This drift is due to the increasing penetration of warm, moist air while contrasting cool, dry air still surges in from the north and northwest. Thus, when the Gulf states are substantially under the influence of warm air after May, there is no cold-air intrusion to speak of, and tornado frequency drops. Such is the case across the country after June. Winter cooling permits fewer and fewer encounters between warm and cold air masses, and tornado frequency returns to its lowest level by December.

The preceding paragraphs described tornado climatology. Recall from the discussion of weather and climate in Chapter 1 that climate provides a statistical perspective of atmospheric behavior; that discussion included the well-known saying, "Climate is what you expect, but weather is what you get." The tornado that struck Moore, Oklahoma, in May 2013 met climatological expectations: It occurred in the heart of "Tornado Alley" during a time considered to be "tornado season." However, as the saying warns, weather events do not always occur when statistical probabilities suggest. Figure 10.24 provides an example from November 2013.

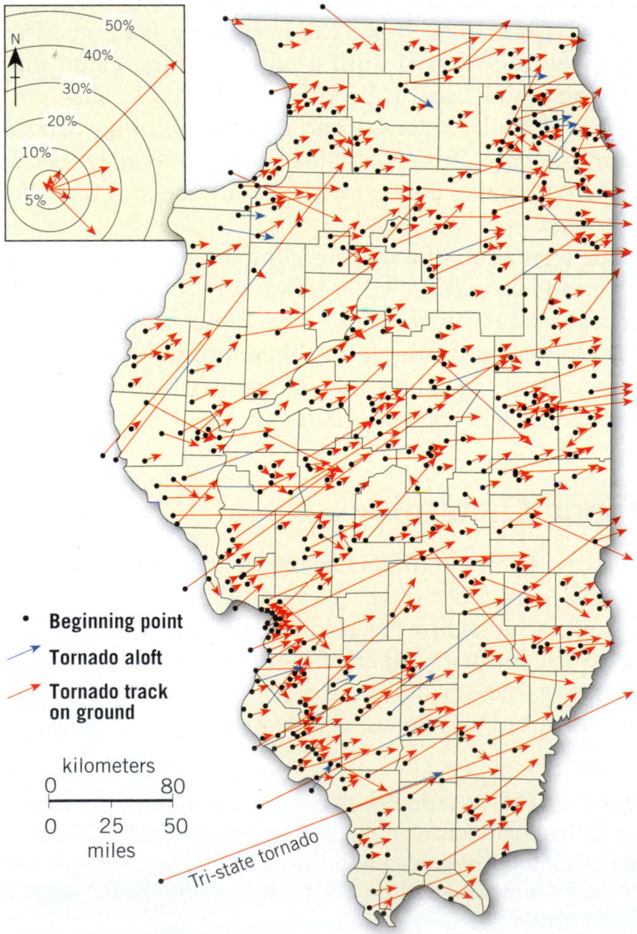


▶ **Figure 10.25 Paths of Illinois tornadoes** Because most tornadoes occur slightly ahead of a cold front, in the zone of southwest winds, they tend to move toward the northeast. Tornadoes in Illinois verify this. More than 80 percent of the tornadoes in this diagram exhibited directions of movement toward the northeast through east.

**Table 10.2** | Tornado Extremes

| Tornado Characteristic | Value | Date | Location |
|---|---|---|---|
| World's deadliest single tornado | 1300 dead, 12,000 injured | April 26, 1989 | Salturia and Manikganj, Bangladesh |
| U.S. deadliest single tornado | 695 dead | March 18, 1925 | MO–IL–IN |
| U.S. deadliest tornado outbreak | 747 dead | March 18, 1925 | MO–IL–IN (includes the Tri-State Tornado deaths) |
| Biggest 24-hour total tornado outbreak | 147 tornadoes | April 3–4, 1974 | 13 central U.S. states |
| Calendar month with greatest number of tornadoes | 753 tornadoes | April 2011 | United States |
| Widest tornado (diameter*) | Nearly 4000 m (2.5 miles) in width | May 22, 2004 | Hallam, NE; EF-4 tornado |
| Highest recorded tornadic wind speed† | 135 m/s (302 mph) | May 8, 1999 | Bridge Creek, OK |
| Highest-elevation tornado | 3650 m (12,000 ft) | Jul 7, 2004 | Sequoia National Park, CA |
| Longest tornado transport | A personal check carried 359 km (223 miles) | April 11, 1991 | Stockton, KS, to Winnetoon, NE |

Source: National Weather Service, Storm Prediction Center.

*The National Weather Service now defines "widest" as the maximum width of tornado damage.

† By necessity, this value is restricted to the small number of tornadoes sampled by mobile Doppler radars.

**Video** MM
The Deadliest Tornado Since Modern Record-keeping Began
http://goo.gl/EA3ry7

## Profile of a Tornado

The average tornado has a diameter between 150 and 600 meters (500 to 2000 feet), travels across the landscape at approximately 45 kilometers (30 miles) per hour, and cuts a path about 26 kilometers (16 miles) long. Because many tornadoes occur slightly ahead of a cold front, in the zone of southwest winds, most move toward the northeast. The Illinois example shown in **Figure 10.25** demonstrates this fact well. The figure also shows that many tornadoes do not fit the description of the "average" tornado.

Of the hundreds of tornadoes reported in the United States each year, more than half are comparatively weak and short-lived. Most of these small tornadoes have lifetimes of 3 minutes or less and paths that seldom exceed 1 kilometer (0.6 mile) in length and 100 meters (330 feet) in width. Typical wind speeds are on the order of 150 kilometers (90 miles) per hour or less. On the other end of the tornado spectrum are the infrequent and often long-lived violent tornadoes (**Table 10.2**).

Although large tornadoes constitute only a small percentage of the total reported, their effects are often devastating. Such tornadoes may exist for periods in excess of 3 hours and produce an essentially continuous damage path more than 150 kilometers (90 miles) long and perhaps 1 kilometer (0.6 mile) or more wide. Maximum winds range beyond 480 kilometers (300 miles) per hour (**Fig. 10.26**).

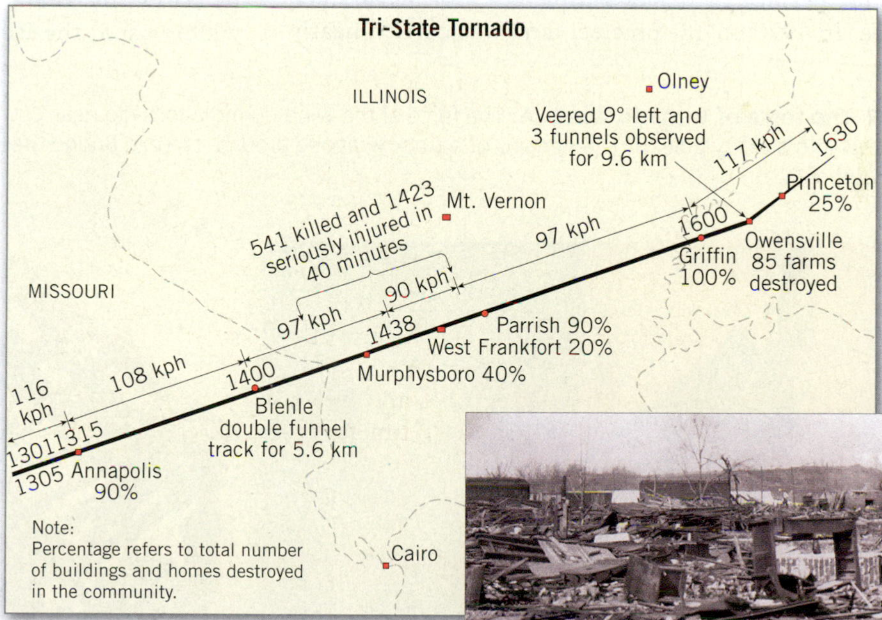

A.

B.

▶ **Figure 10.26 Tri-State Tornado**
**A.** This tornado, which occurred on March 18, 1925, is the deadliest U.S. tornado on record. The tornado remained on the ground for more than 350 kilometers (219 miles), killing 695 and injuring 2027 people. The numbers immediately above the track line are a reference to time of day (24-hour clock). **B.** The historic black-and-white image shows Murphysboro, Illinois, on March 21, 1925, 3 days after the Tri-State Tornado devastated the community.

students sometimes ask...

### What is "Tornado Alley"?

Tornado Alley is a nickname the popular media and others use to refer to the broad swath of high tornado occurrence in the central United States (see Figure 10.23). The heart of Tornado Alley stretches from the Texas panhandle through Oklahoma and Kansas to Nebraska. It's important to remember that violent (killer) tornadoes occur outside Tornado Alley every year. Tornadoes can occur almost anywhere in the United States.

### ✔ Concept Checks 10.7

1. What general atmospheric conditions are most conducive to tornado formation?
2. When is "tornado season"? Why does it occur at this time?
3. Why does the area of greatest tornado frequency migrate?
4. In what direction do the majority of tornadoes move? Explain.

## 10.8 Tornado Destruction and Tornado Forecasting

**Describe tornado intensity. Distinguish between a tornado watch and a tornado warning and discuss the role of Doppler radar in the warning process.**

Because tornadoes generate the strongest winds in nature, they have accomplished many seemingly impossible tasks, such as driving a piece of straw through a thick wooden plank and uprooting huge trees (Fig. 10.27). Although it may seem impossible for winds to cause some of the fantastic damage attributed to tornadoes, tests in engineering facilities have repeatedly demonstrated that winds in excess of 320 kilometers (200 miles) per hour are capable of incredible feats.

There is a long list of documented examples. In 1931 a tornado actually carried an 83-ton railroad coach and its 117 passengers 24 meters (80 feet) through the air and dropped them in a ditch. A year later, near Sioux Falls, South Dakota, a steel beam 15 centimeters (6 inches) thick and 4 meters (13 feet) long was ripped from a bridge, flew more than 300 meters (nearly 1000 feet), and perforated a 35-centimeter- (14-inch-) thick hardwood tree. In 1970 an 18-ton steel tank was carried nearly 1 kilometer (0.6 mile) at Lubbock, Texas. Fortunately, the winds associated with most tornadoes are not this strong.

### Tornado Intensity

Most tornado losses are associated with a few storms that strike urban areas or devastate entire small communities. The destruction wrought by such storms depends to a significant degree (but not completely) on the strength of the winds. A wide spectrum of tornado strengths, sizes, and lifetimes are observed. The commonly used guide to tornado intensity is the **Enhanced Fujita Scale**, or the **EF Scale** for short (Table 10.3). Because tornado winds cannot be measured directly, a rating on the EF Scale is determined by assessing the damage produced by a storm. Although widely used, the EF Scale is not perfect. Estimating tornado

▼ **Figure 10.27 The force of tornado winds A.** The force of the wind during a tornado near Wichita, Kansas, in April 1991 was enough to drive this piece of metal into a utility pole. **B.** The remains of a truck wrapped around a tree in Bridge Creek, Oklahoma, on May 4, 1999, following a major tornado outbreak.

A.

B.

**Table 10.3** | Enhanced Fujita Scale*

| Scale | Wind Speed km/hr | Wind Speed mi/hr | Damage |
|---|---|---|---|
| EF-0 | 105–137 | 65–85 | *Light.* Some damage to siding and shingles. |
| EF-1 | 138–177 | 86–110 | *Moderate.* Considerable roof damage. Winds can uproot trees and overturn single-wide mobile homes. Flagpoles bend. |
| EF-2 | 178–217 | 111–135 | *Considerable.* Most single-wide mobile homes destroyed. Permanent homes can shift off foundations. Flagpoles collapse. Softwood trees debarked. |
| EF-3 | 218–265 | 136–165 | *Severe.* Hardwood trees debarked. All but small portions of houses destroyed. |
| EF-4 | 266–322 | 166–200 | *Devastating.* Complete destruction of well-built residences, large sections of school buildings. |
| EF-5 | >322 | >200 | *Incredible.* Significant structural deformation of mid- and high-rise buildings. |

*The original Fujita scale was developed by T. Theodore Fujita in 1971 and put into use in 1973. The Enhanced Fujita Scale is a revision that was adopted in February 2007. Wind speeds are estimates (not measurements) based on damage and represent three-second gusts at the point of damage. More information about the criteria used to evaluate tornado intensity can be found at www.spc.noaa.gov/efscale/.

intensity based on damage alone does not take into account the structural integrity of the objects hit by a tornado. A well-constructed building can withstand very high winds, whereas a poorly built structure can suffer devastating damage from the same or even weaker winds.

The drop in atmospheric pressure associated with the passage of a tornado plays a minor role in the damage process. Most structures have sufficient venting to allow for the sudden drop in pressure. Opening a window, once thought to be a way to minimize damage by allowing inside and outside atmospheric pressure to equalize, is no longer recommended. In fact, if a tornado gets close enough to a structure for the pressure drop to be experienced, the strong winds probably will have already caused significant damage.

Although the greatest part of tornado damage is caused by violent winds, most tornado injuries and deaths result from flying debris. On average, tornadoes cause more deaths each year than any other weather events except lightning and flash floods. For the United States, the average annual death toll from tornadoes is about 60 people. However, the actual number of deaths each year can depart significantly from the average. On April 3–4, 1974, for example, an outbreak of 147 tornadoes brought death and destruction to a 13-state region east of the Mississippi River. More than 300 people died, and nearly 5500 people were injured (**Fig. 10.28**).

## Loss of Life

The proportion of tornadoes that result in loss of life is small. In 2013, there were 943 tornadoes reported in the United States. Of this total, just 14 were "killer tornadoes." In most years, slightly less than 2 percent of all reported tornadoes in the United States are "killers." Although the percentage of tornadoes that result in death is small, every tornado is potentially lethal. **Figure 10.29** compares tornado fatalities with storm intensities, and the results are quite interesting. It is clear from this graph that the majority (63 percent) of all tornadoes are weak and that the number of storms decreases as tornado intensity increases. The distribution of tornado fatalities, however, is just the opposite. Although only 2 percent of tornadoes are classified as violent, they account for nearly 70 percent of the deaths.

If there is some question about the causes of tornadoes, there certainly is none about the destructive effects of these violent storms. A severe tornado leaves the affected area stunned and disorganized and may require a response of the magnitude demanded in war.

## Tornado Forecasting

Because tornadoes are small and short-lived phenomena, they are among the most difficult weather features to forecast precisely.

▼ **Figure 10.28 Major tornado outbreak** On April 3–4, 1974, in the span of just 16 hours, 147 tornadoes hit 13 states. Many of the tornado paths are shown here. It was the largest and costliest outbreak on record. The storms took 315 lives and injured 5500 more people. Forty-eight tornadoes were killers, with 7 rated F-5 on the Fujita Scale and 23 rated F-4. This event took place when tornado intensities were described using the Fujita Scale (F Scale), the predecessor of the EF Scale. Wind speeds (mph) for the F Scale are F-0 (<72), F-1 (72–112), F-2 (113–157), F-3 (158–206), F-4 (207–260), F-5 (>260).

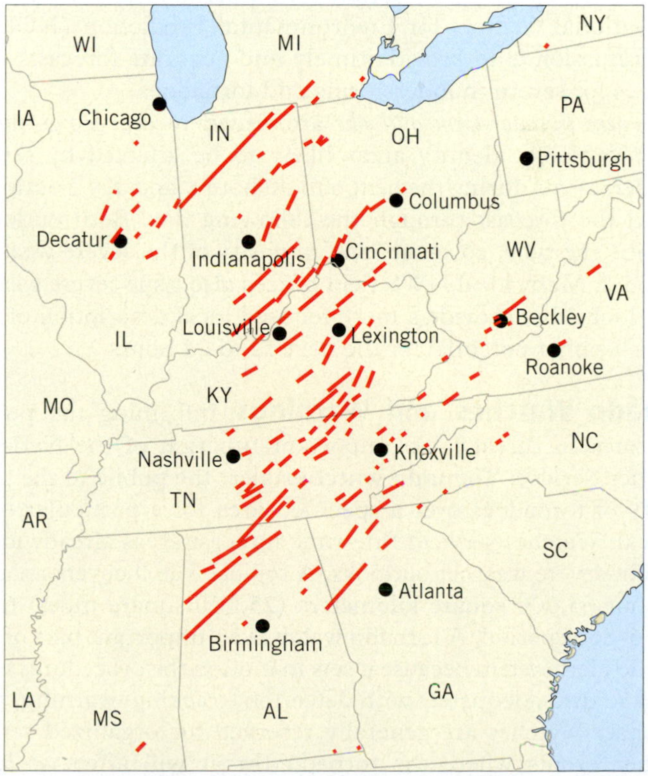

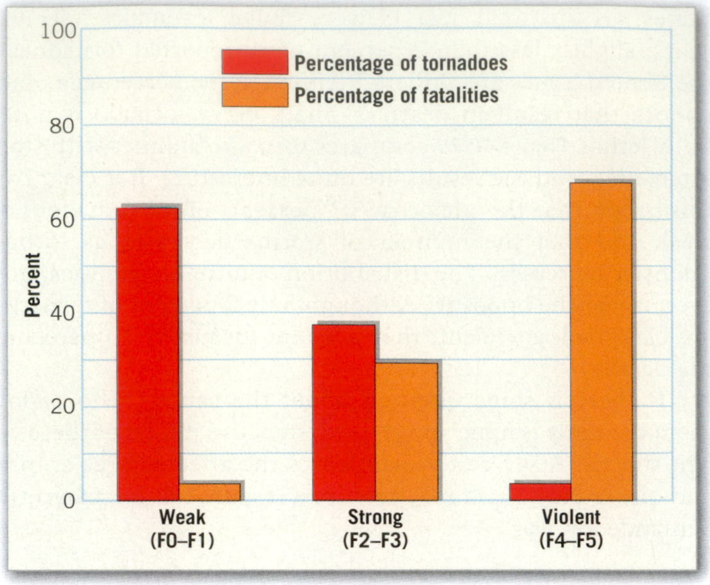


▲ **Figure 10.29 Tornado intensity and fatalities** The bar graph compares the percentage of tornadoes in each intensity category with the percentage of fatalities associated with that category. Because this study was completed prior to the adoption of the EF Scale, storm intensities are expressed using the F Scale. Wind speeds (mph) for this scale are F-0 (<72), F-1 (72–112), F-2 (113–157), F-3 (158–206), F-4 (207–260), F-5 (>260).

Nevertheless, the prediction, detection, and monitoring of such storms are among the most important services provided by professional meteorologists. The timely issuance and dissemination of watches and warnings are both critical to the protection of life and property (see Box 10.3).

The Storm Prediction Center (SPC) located in Norman, Oklahoma, is part of the National Weather Service (NWS) and the National Centers for Environmental Prediction (NCEP). SPC's mission is to provide timely and accurate forecasts and watches for severe thunderstorms and tornadoes.

*Severe thunderstorm outlooks* are issued several times daily. *Day 1* outlooks identify areas likely to be affected by severe thunderstorms during the next 6 to 30 hours, and *day 2* outlooks extend the forecast through the following day. Both outlooks describe the type, coverage, and intensity of the severe weather expected. Many local NWS field offices also issue severe weather outlooks that provide a more focused local description of the severe weather potential for the next 12 to 24 hours.

## Tornado Watches and Warnings

Informing the public of a tornado threat is an important function of the National Weather Service. **Tornado watches** alert the public to the possibility of tornadoes over a specified area for a particular time interval. Watches serve to fine-tune forecast areas already identified in severe weather outlooks. A typical watch covers an area of about 65,000 square kilometers (25,000 square miles) for a 4- to 6-hour period. A tornado watch is an important part of the tornado alert system because it sets in motion the procedures necessary to deal adequately with detection, tracking, warning, and response. Watches are generally reserved for organized severe weather events where the tornado threat will affect at least 26,000 square kilometers (10,000 square miles) and/or persist for at least 3 hours. Watches typically are not issued when the threat is thought to be isolated and/or short-lived.

Whereas a tornado watch is designed to alert people to the possibility of tornadoes, a **tornado warning** is issued by local offices of the National Weather Service when a tornado has actually been sighted in an area or is indicated by weather radar. It warns of a high probability of imminent danger. Warnings are issued for much smaller areas than watches, usually covering portions of a county or counties. In addition, they are in effect for much shorter periods, typically 30 to 60 minutes. Because a tornado warning may be based on an actual sighting, warnings are occasionally issued after a tornado has already developed. However, most warnings are issued prior to tornado formation, sometimes by several tens of minutes, based on Doppler radar data and/or spotter reports of funnel clouds or cloud-base rotation.

If the direction and the approximate speed of the storm are known, an estimate of its most probable path can be made. Because tornadoes often move erratically, the warning area is fan shaped downwind from the point where the tornado has been spotted. Improved forecasts and advances in technology have contributed to a significant decline in tornado deaths over the past 50 years. **Figure 10.30** illustrates this trend. During a span when the U.S. population grew rapidly, tornado deaths trended downward.

### eye on the atmosphere 10.2

This satellite image shows a portion of the diagonal path left by a tornado as it moved across northern Wisconsin in 2007.

**Questions**

1. Toward what direction did the storm advance: the northeast or the southwest?
2. Did the tornado more likely occur ahead of or behind a cold front? Explain.
3. Is it more probable that the storm took place in March or June? Why is the month you selected more likely?

**severe & hazardous weather Box 10.3**

## Surviving a Violent Tornado

About 11 A.M. on July 13, 2004, much of northern and central Illinois was put on a tornado watch. A large supercell had developed in the northwestern part of the state and was moving southeast, into a very unstable environment (**Fig. 10.F**). A few hours later, as the supercell entered Woodford County, rain began to fall, and the storm showed signs of becoming severe. The National Weather Service issued a *severe thunderstorm warning* at 2:29 P.M. Minutes afterward, a tornado developed. Twenty-three minutes later, the quarter-mile-wide twister had carved a 9.6-mile-long path across the rural Illinois countryside.

What, if anything, made this storm special or unique? After all, it was just one of a record-high 1819 tornadoes reported in the United States in 2004. For one, the National Weather Service estimated that maximum winds reached 240 miles per hour. This tornado attained EF-4 status for a portion of its life; fewer than 1 percent of tornadoes reach this level of severity. But what was most remarkable is that no one was killed or injured when the Parsons Manufacturing facility west of the small town of Roanoke took a direct hit while the storm was most intense. At the time, 150 people were in three buildings comprising the plant. The 250,000-square-foot facility was flattened, cars were twisted into gnarled masses, and debris was strewn for miles (**Fig. 10.G**).

How did 150 people escape death or injury?

How did 150 people escape death or injury? Foresight and planning saved them. More than 30 years earlier, company owner Bob Parsons was inside his first factory when a small tornado passed close enough to blow out windows. Later, when he built a new plant, he made sure that the restrooms were constructed to double as tornado shelters, with steel-reinforced concrete walls and 8-inch-thick concrete ceilings. In addition, the company developed a severe weather plan. When the severe thunderstorm warning was issued at 2:29 P.M. that day, the emergency response team leader at the Parsons plant was immediately notified. A few moments later, he went outside and observed a rotating wall cloud with a developing funnel cloud. He radioed back to the office to institute the company's severe weather plan. Employees were told to immediately go to their designated storm shelters. Everyone knew where to go and what to do because the plant conducted semi-annual tornado drills. All 150 people reached shelters in less than 4 minutes. The emergency response team leader was the last person to reach a shelter, less than 2 minutes before the tornado destroyed the plant at 2:41 P.M.

The total number of U.S. tornado deaths in 2004 was just 36. The toll could have been much higher. The building of tornado shelters and the development of an effective severe storm plan made the difference between life and death for 150 people at Parsons Manufacturing.

▲ **Figure 10.F Violent central Illinois tornado** On July 13, 2004, an EF-4 tornado cut a 23-mile path through the rural countryside near the Woodford County town of Roanoke, Illinois. The Parsons Manufacturing plant was just west of town.

**Questions**

1. About what percentage of tornadoes attain EF-4 status?
2. List two factors that allowed employees of Parsons Manufacturing to avoid injury or death from the tornado.

▲ **Figure 10.G Aftermath of the storm** The quarter-mile-wide tornado had wind speeds reaching 240 miles per hour. The destruction at Parsons Manufacturing was devastating.

# What's Your Forecast?

## Putting the Ingredients Together to Forecast Severe Thunderstorms

By Harold Brooks, Senior Research Scientist, Forecast Research and Development Division, NOAA/National Severe Storms Laboratory

Tornadoes and severe thunderstorms are frightening, not only because of their destruction, but also because pinpointing their location and potential impacts hours in advance is challenging (**Fig. 10.H**). To step up to this challenge, we can look at the ingredients necessary to make a tornado or severe thunderstorm. Understanding those ingredients can tell us a lot about where and when tornadoes are likely to occur.

Storms need warm, moist surface air and relatively dry, cold air above that. A simple model of these elements helps us estimate the amount of energy available in the atmosphere to fuel a storm's updraft. A great way to express that energy is as the strongest updraft that could be formed. A typical thunderstorm may have updrafts of 10 meters per second (m/s), whereas the strongest have updrafts approaching 100 m/s. Severe storms usually form when the atmosphere can support an updraft of about 20 m/s. An updraft of a little more than 40 m/s can support baseball-size hail, which would fall out of the storm at about the same speed as a Major League fastball.

But a severe storm requires more than just energy. It also needs to be organized and rotating. Rotation results from the increase in wind speed that occurs as you ascend from the ground to a height of 6 km. The larger the change, the stronger the rotation will be. Most severe storms need a wind speed difference of at least 20 m/s between the surface and a height of 6 km.

Meteorologists have collected a great deal of data on the energy needed to fuel strong updrafts and the wind differences associated with severe storm development. This knowledge allows them to forecast when severe thunderstorms will occur and when they won't. By looking at millions of data points, we can estimate the probability that a severe thunderstorm will occur given a particular combination of available energy and wind difference (**Fig. 10.I**). The highest probability occurs when a great deal of energy is available and a significant wind difference exists between the ground and an altitude of 6 kilometers. Fortunately those conditions don't happen very often, but when they do, damage and loss of life are serious concerns.

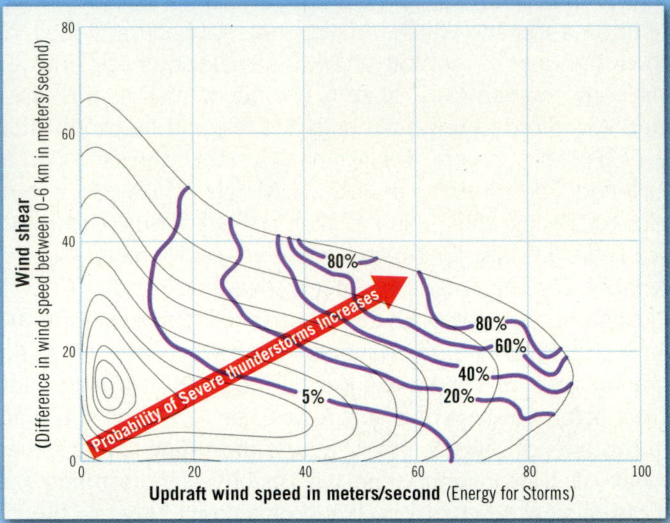


▲ **Figure 10.I Predicting the Probability of a Severe Thunderstorm** Probability in percent (shown in blue) of a thunderstorm in the United States producing at least golf-ball size hail, hurricane force winds, or an EF2 or stronger tornado, given a combination of energy available for a storm expressed as updraft speed and wind shear, expressed as the difference in the wind speeds between 0 and 6 kilometers. The gray contours indicate how frequently the combination is observed in the United States. The bull's-eye in the lower left indicates the most likely values for the combination of updraft wind speed and wind shear. Most storms have updrafts and wind speed differences of 10 m/s or less. More energetic storms (going to the right on the graph) or storms with greater wind shear (going upward on the graph) are less likely to occur, but produce more severe weather.

▲ **Figure 10.H Dimmitt, Texas, Tornado on June 2, 1995** The photo was taken during the VORTEX field project. Because tornadoes are small and short-lived phenomena, they are difficult to forecast precisely.

### Questions

1. In the winter months in the U.S. Great Plains, there's very little moisture in the air so the energy available for storms is low, but the difference in surface and upper-air wind speeds associated with the jet stream tends to be large. In the summer, there's a lot of moisture and needed energy for storms, but the jet stream moves far to the north, so the vertical wind difference is small. In the spring, moisture from the Gulf of Mexico moves slowly north, and the jet stream also moves north and gets weaker. Explain how this combination of factors affects when and where we expect to see severe thunderstorms and tornadoes.
2. Our current understanding of future climate change indicates that a likely scenario is that the energy needed for thunderstorms will increase as the atmosphere warms up, but the wind difference will decrease as the poles warm more than the equator. In this scenario, how might the distribution of severe thunderstorms change in the United States?

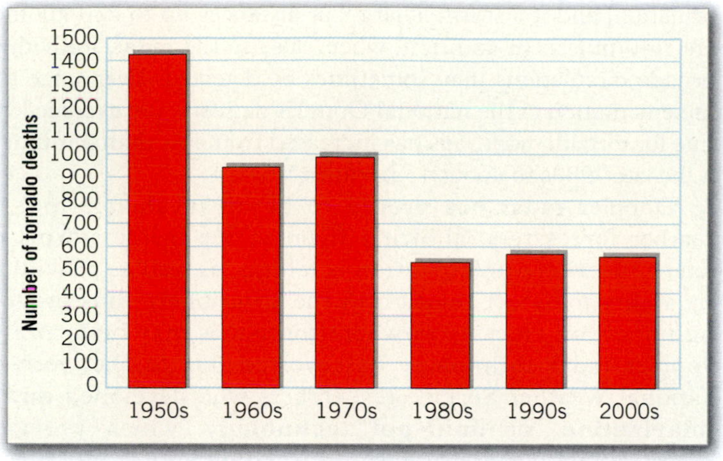


▲ **Figure 10.30 Number of tornado deaths in the United States by decade, 1950–2009** Even though the population has risen sharply since 1950, there has been a general downward trend in tornado deaths.

As noted earlier, the probability of one place being struck by a tornado, even in the area of greatest frequency, is slight. Nevertheless, although the probabilities may be small, tornadoes have provided many mathematical exceptions. For example, the small town of Codell, Kansas, was hit 3 years in a row—1916, 1917, and 1918—and each time on the same date, May 20! Needless to say, tornado watches and warnings should never be taken lightly.

**Doppler Radar** The installation of **Doppler radar** across the United States significantly improved our ability to track thunderstorms and issue warnings based on their potential to produce tornadoes (Fig. 10.31). Conventional weather radar works by transmitting short pulses of electromagnetic energy. A small fraction of the waves that are sent out is scattered by a storm and returned to the radar, creating an "echo." The strength of the returning signal indicates rainfall intensity, and the time difference between the transmission and return of the signal indicates the distance to the storm.

However, to identify tornadoes and severe thunderstorms, we must be able to detect the characteristic circulation patterns associated with them. Conventional radar cannot do so except occasionally, when spiral rain bands occur in association with a tornado and give rise to a hook-shaped echo.

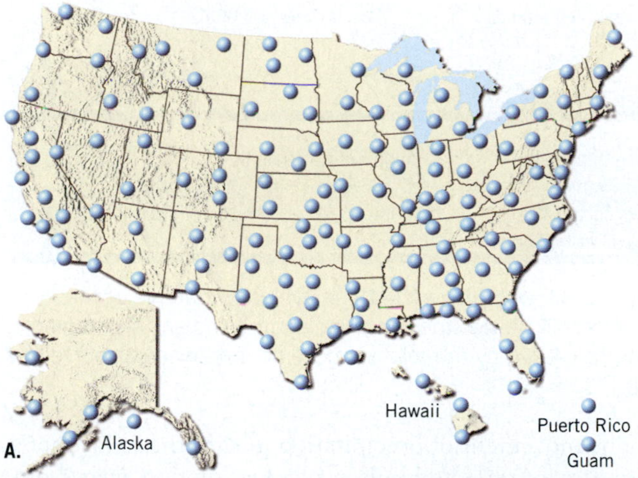


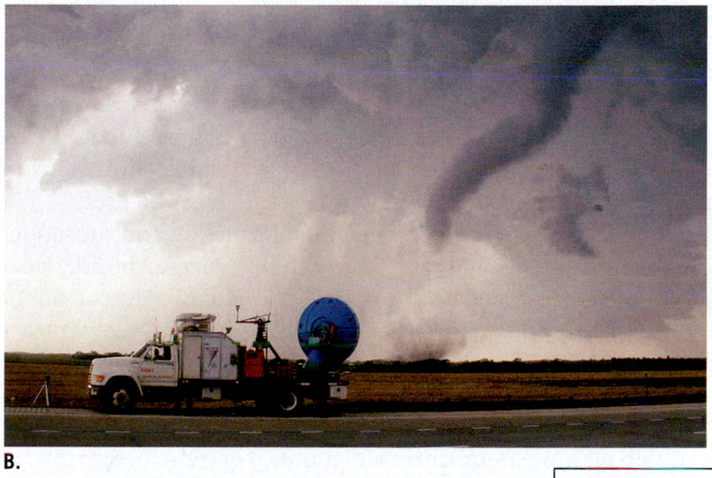


▲ **Figure 10.31 Doppler radar A.** Doppler radar sites in the United States. If you go to http://radar.weather.gov, you will see a similar map. You can click on any site to see the current National Weather Service Doppler radar display. **B.** Doppler on Wheels is a portable unit that researchers use in field studies of severe weather events.

**Video** MM
Identifying Tornadic Thunderstorms Using Radar Velocity Data
http://goo.gl/szdHJQ

Doppler radar not only performs the same tasks as conventional radar but also has the ability to detect motion directly (Fig. 10.32). The principle involved is known as the *Doppler effect* (Fig. 10.33). Air movement in clouds is determined by comparing the frequency of the reflected signal to that of the original

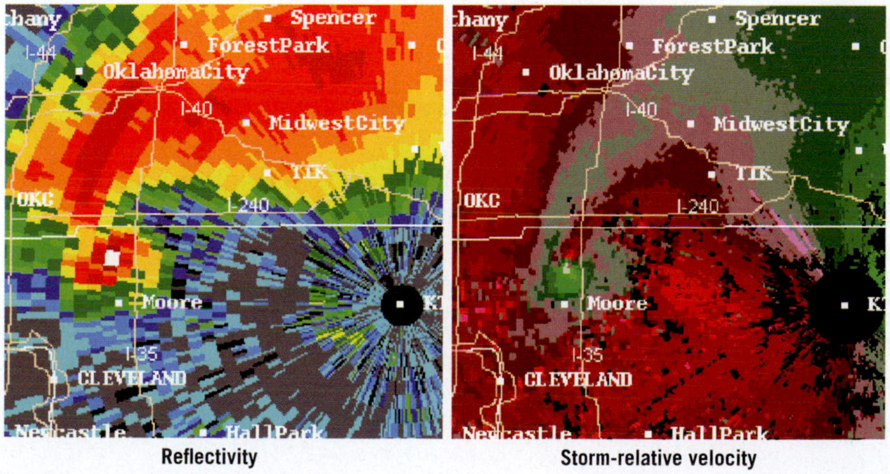


◀ **Figure 10.32 Doppler radar images** This is a dual Doppler radar image of an EF-5 tornado near Moore, Oklahoma, on May 3, 1999. The left image (reflectivity) shows precipitation in the supercell thunderstorm. The right image shows motion of the precipitation along the radar beam—that is, how fast rain or hail is moving toward or away from the radar. In this example, the radar was unusually close to the tornado—close enough to make out the signature of the tornado itself. (Most of the time only the weaker and larger mesocyclone is detected.)

**Video** MM
Identifying Tornadic Thunderstorms Using Radar Reflectivity Data
http://goo.gl/SbbmOj

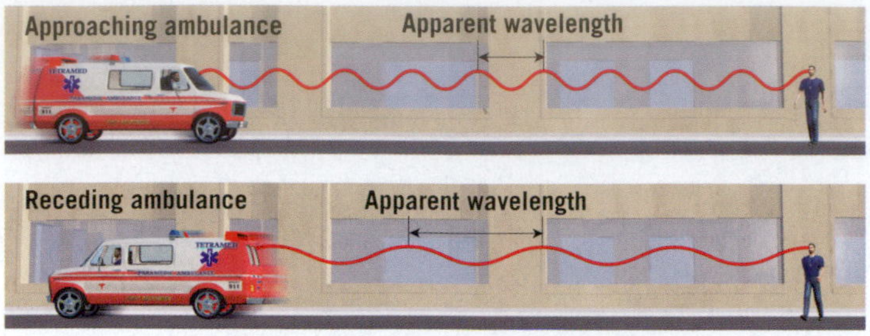


▲ **Figure 10.33 Doppler effect** This everyday example of the Doppler effect illustrates the apparent lengthening and shortening of wavelengths caused by the relative movement between a source and an observer.

pulse. The movement of precipitation toward the radar increases the frequency of reflected pulses, whereas motion away from the radar decreases the frequency. These frequency changes are then interpreted in terms of speed toward or away from the Doppler radar unit. This same principle allows police radar to determine the speed of moving cars. Unfortunately, a single Doppler radar unit cannot detect air movements that occur parallel to it. Therefore, when a more complete picture of the winds within a cloud mass is desired, two or more Doppler units must be used.

Doppler radar can detect the initial formation and subsequent development of the mesocyclone within a severe thunderstorm that frequently precedes tornado development. Almost all (96 percent) mesocyclones produce damaging hail, severe winds, or tornadoes. Those that produce tornadoes (about 50 percent) can sometimes be distinguished by their stronger wind speeds and sharper wind speed gradients. Mesocyclones can sometimes be identified within parent storms 30 minutes or more before tornado formation, and if a storm is large, at distances up to 230 kilometers (140 miles). In addition, when close to the radar, individual tornado circulations may sometimes be detected. Ever since the implementation of the national Doppler network, the average lead time for tornado warnings has increased from less than 5 minutes in the late 1980s to about 13 minutes today.

Doppler radar has proved to be an invaluable tool to weather forecasters, allowing them to see inside storms to determine where precipitation is falling, as well as providing the ability to detect rotation in the atmosphere thus providing important clues as to where tornadoes may be forming. Doppler radar technology has evolved due to the work of National Weather Service researchers who developed **dual-polarization**, or **dual-pol technology**, which provides forecasters with more precise information about the type of precipitation and its intensity, size, and location.

By April 2013, the dual-pol upgrade had been installed at every National Weather Service Doppler radar in the United States. It involved the development of new software and a hardware attachment to the radar dish that sends and receives both horizontal and vertical pulses of energy, providing a much more informative picture of what is happening in the atmosphere. It helps forecasters clearly identify and distinguish among rain, hail, snow or ice pellets, and other flying objects. Another important benefit is that dual-pol more clearly detects airborne debris. This allows forecasters to confirm a tornado is on the ground and causing damage so they can more confidently warn communities in its path. This is especially helpful at night when ground spotters are unable to see the tornado.

Video MM
The Benefits of Doppler Radar
http://goo.gl/wNyeKp

students sometimes ask...

**How dangerous is it to be in a mobile home during a tornado?**

Mobile homes represent a relatively small fraction of all residences in the United States. Yet, according to the National Weather Service, during the span 2000–2010, 52 percent of all tornado fatalities (314 of 604) occurred in mobile homes.

### ✔ Concept Checks 10.8

1. Name the scale commonly used to rate tornado intensity. How is a rating on this scale determined?
2. In an average year, about what percentage of tornadoes in the United States are "killer tornadoes"?
3. Distinguish between a tornado watch and a tornado warning.
4. What advantages does Doppler radar have over conventional radar?

# 10 Concepts in Review Thunderstorms and Tornadoes

## 10.1 What's in a Name?

▸ Distinguish among three types of storm-producing cyclones.

- Although tornadoes and hurricanes are cyclones, the vast majority of cyclones are not hurricanes or tornadoes. The term *cyclone* refers to the circulation around any low-pressure center, no matter how large or intense it is.

**Q** Relate thunderstorms to the three types of cyclones discussed in this section.

## 10.2 Occurrence of Thunderstorms ▸ List the basic requirements for thunderstorm formation and locate places on a map that exhibit frequent thunderstorm activity.

**Key Terms:** thunderstorm

- Dynamic thermal instability occurs during the development of thunderstorms, which form when warm, humid air rises in an unstable environment.
- A number of mechanisms, such as unequal heating of Earth's surface or lifting of warm air along a front or mountain slope, can trigger the upward air movement needed to create thunderstorm-producing cumulonimbus clouds.
- In the coming decades, global warming will likely enhance the conditions that promote thunderstorm development. In the United States, it appears that such change will be most pronounced in eastern and southern states.

**Q** Explain why the western margin of North America has fewer days per year with thunderstorms than the southeastern states.

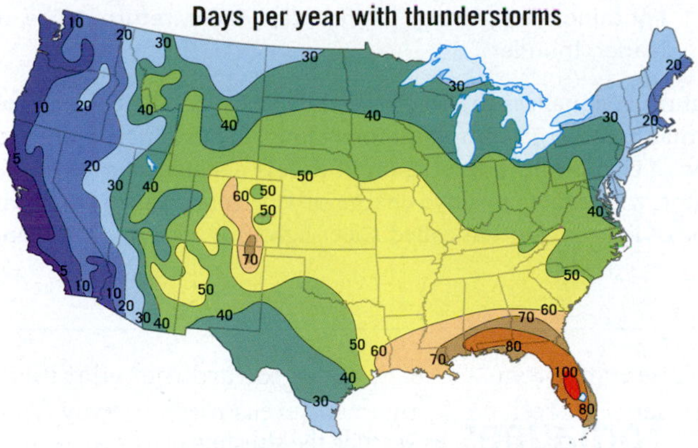


## 10.3 Air-Mass Thunderstorms ▸ Sketch and explain simple diagrams that illustrate the three stages of an air-mass thunderstorm.

**Key Terms:** air-mass thunderstorm, cumulus stage, entrainment, mature stage, dissipating stage

- Air-mass thunderstorms frequently occur in maritime tropical (mT) air during the spring and summer. Generally, three stages are involved in the development of these storms: the cumulus stage, mature stage, and dissipating stage.
- Mountainous regions, such as the Rockies and the Appalachians, experience a greater number of air-mass thunderstorms than do the Plains states. Many thunderstorms that form over the eastern two-thirds of the United States occur as part of the general convergence and frontal wedging that accompany passing midlatitude cyclones.

**Q** What stage in the development of a thunderstorm is shown in this sketch? Describe what is occurring. Is there a stage that follows this one? If so, describe what occurs during that stage.

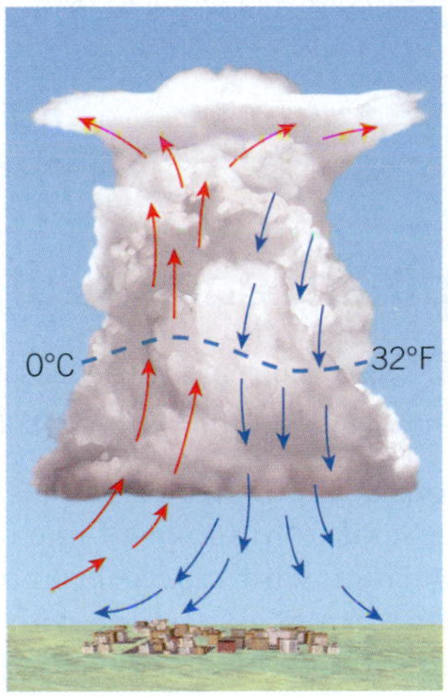


## 10.4 Severe Thunderstorms ▸ List the characteristics of a severe thunderstorm and distinguish among different circumstances associated with their formation.

**Key Terms:** severe thunderstorm, wind shear, gust front, supercell, mesocyclone, squall line, dryline, mesoscale convective complex (MCC)

- Severe thunderstorms are capable of producing heavy downpours and flash flooding as well as strong, gusty, straight-line winds. They are influenced by strong vertical wind shear—that is, changes in wind direction and/or speed between different heights, causing updrafts to become tilted and continue to build upward.
- Downdrafts from the thunderstorm cells reach the surface and spread out to produce an advancing wedge of cold air, called a gust front.
- Some of the most dangerous weather is produced by a type of thunderstorm called a supercell, a single, very powerful thunderstorm cell that at times may extend to heights of 20 kilometers (65,000 feet) and persist for many hours. These cells may produce a mesocyclone, a column of cyclonically rotating air, within which tornadoes sometimes form.
- Squall lines are relatively narrow, elongated bands of thunderstorms that develop in the warm sector of a midlatitude cyclone, usually in advance of a cold front. A mesoscale convective complex (MCC) consists of many individual thunderstorms that are organized into a large oval to circular cluster. They form most frequently in the Great Plains from groups of afternoon air-mass thunderstorms.
- Some squall lines are associated with drylines, narrow boundaries along which less dense mT air rises over denser cT air.

## 10.5 Lightning and Thunder ▸ Explain what causes lightning and thunder.

**Key Terms:** lightning, sheet lightning, cloud-to-ground lightning, flash, stroke, leader, step leader, return stroke, dart leader, thunder

- Lightning equalizes the electrical difference associated with the formation of large cumulonimbus clouds by producing a negative flow of current from the region of excess negative charge to the region with excess positive charge or vice versa. The most common type of lightning, often called sheet lightning, occurs within and between clouds. The less common but more dangerous type of lightning is cloud-to-ground lightning.
- The origin of charge separation in clouds, although not fully understood, hinges on rapid vertical movements within a cloud. The lightning we see as single flashes is really several very rapid strokes between the cloud and the ground. When air is heated by the electrical discharge of lightning, it expands explosively and produces the sound waves we hear as thunder.

**Q** Is it possible to experience a thunderstorm that has no lightning? Explain.

## 10.6 Tornadoes ▸ Describe the structure and basic characteristics of tornadoes.

**Key Terms:** tornado, multiple-vortex tornado

- A tornado is a violent windstorm that takes the form of a rotating column of air, or vortex, that extends downward from a cumulonimbus cloud. Some tornadoes consist of a single vortex. Within many stronger tornadoes, called multiple-vortex tornadoes, are smaller intense whirls called suction vortices that rotate within the main vortex.
- Pressures within some tornadoes have been estimated to be as much as 10 percent lower than immediately outside the storm. Because of the tremendous pressure gradient associated with a strong tornado, maximum winds approach 480 kilometers (300 miles) per hour.

## 10.7 Development and Occurrence of Tornadoes ▸ Summarize the atmospheric conditions and locations that are favorable to the formation of tornadoes.

- Severe thunderstorms, and hence tornadoes, are most often spawned along the cold front or squall line of a midlatitude cyclone or in association with supercell thunderstorms. Tornadoes can also form in association with tropical cyclones (hurricanes). April through June is the period of greatest tornado activity, but tornadoes occur during every month of the year.
- The average tornado has a diameter between 150 and 600 meters (500 to 2000 feet), travels across the landscape toward the northeast at approximately 45 kilometers (30 miles) per hour, and cuts a path about 26 kilometers (16 miles) long.

**Q** Is a tornado in the central United States more likely to move from southwest to northeast or from southeast to northwest? Explain.

## 10.8 Tornado Destruction and Tornado Forecasting ▸ Describe tornado intensity. Distinguish between a tornado watch and a tornado warning and discuss the role of Doppler radar in the warning process.

**Key Terms:** Enhanced Fujita Scale (EF Scale), tornado watch, tornado warning, Doppler radar, dual-polarization (dual-pol) technology

- Most tornado damage is caused by tremendously strong winds. One commonly used guide to tornado intensity is the Enhanced Fujita Scale (EF Scale). A rating on the EF Scale is determined by assessing damages produced by a tornado.
- Because severe thunderstorms and tornadoes are small and short-lived phenomena, they are among the most difficult weather features to forecast precisely.
- When weather conditions favor the formation of tornadoes, a tornado watch is issued to alert the public to the possibility of tornadoes over a specified area for a particular time interval. A tornado warning is issued by local offices of the National Weather Service when a tornado has been sighted in an area or is indicated by weather radar.
- With its ability to detect the movement of precipitation within a cloud, Doppler radar technology has greatly advanced the accuracy of tornado warnings.

## Give it Some Thought

1. If you are a resident of central Ohio and hear that a cyclone is approaching, should you immediately seek shelter? What if you live in western Iowa?
2. Which one of the locations shown on the accompanying map is more likely to have dryline thunderstorms? Why is this the case?

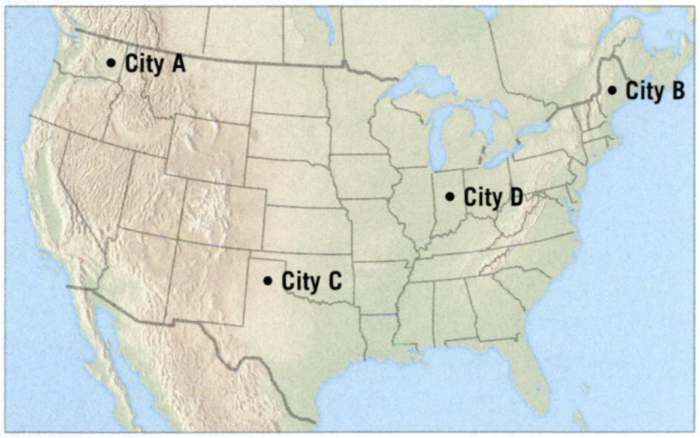


3. Sinking air warms by compression (adiabatically), yet thunderstorm downdrafts are usually cold. Explain this apparent contradiction.
4. Studies have linked the formation of supercell thunderstorms to the presence of temperature inversions. However, cumulonimbus clouds form in an unstable environment, whereas temperature inversions are associated with very stable atmospheric conditions. Explain the connection between these two phenomena.
5. The table below lists the number of tornadoes reported in the United States by decade. Propose a reason that might explain why the total for the 2000s is so much higher than for the 1950s.

| Number of U.S. Tornadoes Reported, by Decade | |
|---|---|
| **Decade** | **Number of Tornadoes Reported** |
| 1950–1959 | 4796 |
| 1960–1969 | 6613 |
| 1970–1979 | 8579 |
| 1980–1989 | 8196 |
| 1990–1999 | 12,138 |
| 2000–2009 | 12,914 |

6. Figure 10.30 shows that the number of tornado deaths in the United States in the 2000s fell to less than 40 percent of 1950s fatalities, even though there was a significant rise in the population during that span. What explains this decline in the death toll?
7. As you will learn in Chapter 11, hurricane intensity is monitored and reported as the storm approaches. However, the intensity of a tornado is not determined and reported until *after* the storm passes. Why is this the case?

## Problems

1. If thunder is heard 15 seconds after lightning is seen, about how far away was the lightning stroke?
2. Examine the upper-left portion of Figure 10.25 and determine the percentage of tornadoes that exhibited directions of movement toward the E through NNE.
3. These maps represent two common ways that U.S. tornado statistics are graphically presented to the public. Which four states experience the greatest number of tornadoes? Are these the states with the greatest tornado threat? Which map is most useful for depicting the tornado hazard? Does the map in Figure 10.23 have an advantage over either or both of these maps?

**Annual average number of tornadoes by state for a 45-year period**

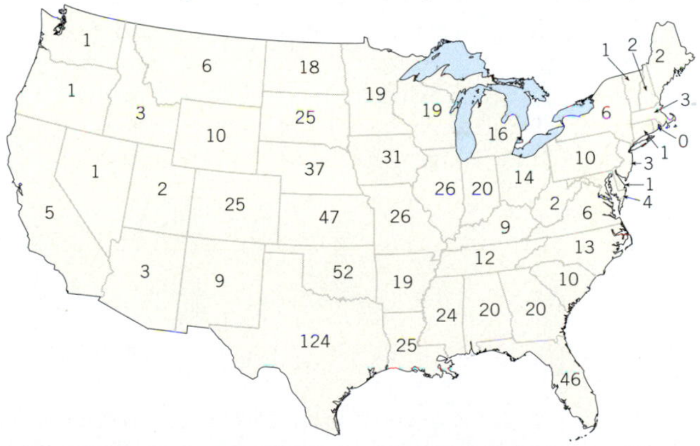

**Annual average number of tornadoes per 10,000 square miles by state for a 45-year period**

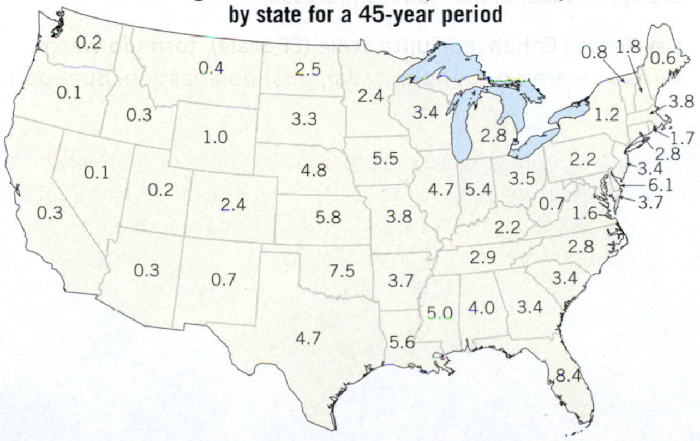

# 11 Hurricanes

## Focus on Concepts

*Each statement represents the primary learning objective for the corresponding major heading within the chapter. After you complete the chapter, you should be able to:*

**11.1** Define *hurricane* and describe the basic structure and characteristics of this storm.

**11.2** Discuss the conditions that promote hurricane formation and list the factors that cause hurricanes to dissipate.

**11.3** Explain how hurricane intensity is determined and summarize the three broad categories of hurricane destruction.

**11.4** Compare two methods used to determine the intensity of a hurricane and explain why storm intensity may change in the decades to come.

**11.5** List four tools that provide data used to track hurricanes and develop forecasts. Contrast *hurricane watch* and *hurricane warning*.

The whirling tropical cyclones that occasionally have wind speeds exceeding 300 kilometers (185 miles) per hour are known in the United States as *hurricanes*—the greatest storms on Earth. Hurricanes are among the most destructive of natural disasters. When a hurricane reaches land, it is capable of annihilating low-lying coastal areas and killing thousands of people. On the positive side, hurricanes provide essential rainfall over many areas they cross. Consequently, a resort owner along the Florida coast may dread the coming of hurricane season, whereas a farmer in Japan may welcome its arrival.

*Super Typhoon Haiyan, among the strongest storms on record, devastated portions of the central Philippines in November 2013. This storm is discussed in detail in Box 11.3.*

# 11.1 Profile of a Hurricane

**Define *hurricane* and describe the basic structure and characteristics of this storm.**

Many view the weather in the tropics with favor—and rightfully so. Places such as islands in the South Pacific and the Caribbean are known for their lack of significant day-to-day variations. Warm breezes, steady temperatures, and rains that come as heavy but brief tropical showers are expected. It is ironic that these relatively tranquil regions occasionally produce some of the most violent storms on Earth. Once formed, these storms can carry severe conditions far from the tropics (Fig. 11.1).

**Hurricanes** are intense centers of low pressure that form over tropical or subtropical oceans and are characterized by intense convective (thunderstorm) activity and strong cyclonic circulation. Sustained winds must equal or exceed 119 kilometers (74 miles) per hour. Unlike midlatitude cyclones, hurricanes lack contrasting air masses and fronts. Rather, the source of energy that produces and maintains hurricane-force winds is the huge quantity of latent heat liberated during the formation of the storm's cumulonimbus towers.

▲ **Figure 11.1 Destruction by Hurricane Sandy** In late October 2012, this very large and powerful storm, unofficially called "Superstorm Sandy" by the media, made landfall just south of New York City along the New Jersey coastline. The fact that it struck the most populated metropolitan region in the United States clearly contributed to the storm's great financial impact.

Video MM
The Making of a Superstorm
http://goo.gl/4dm6PN

These intense tropical storms are known in various parts of the world by different names. In the northwestern Pacific, they are called *typhoons*, and in the southwestern Pacific and the Indian Ocean, they are called *cyclones*. In the following discussion, we will refer to these storms as hurricanes. The term *hurricane* is derived from *Huracan*, a Carib god of evil.

Most hurricanes form between the latitudes of 5° and 30° over all the tropical oceans, except only rarely in the South Atlantic and the eastern South Pacific (Fig. 11.2). The western North Pacific has the greatest number of storms, averaging 20 per year. Fortunately for those living in the coastal regions of the southern and eastern United States, only about 5 hurricanes, on average, develop each year in the warm sector of the North Atlantic.

Although many tropical disturbances develop each year, only a few reach hurricane status. By international agreement, a hurricane has sustained wind speeds of at least 119 kilometers (74 miles) per hour and a rotary circulation. **Sustained winds** are averaged over a 1-minute interval. Mature hur-

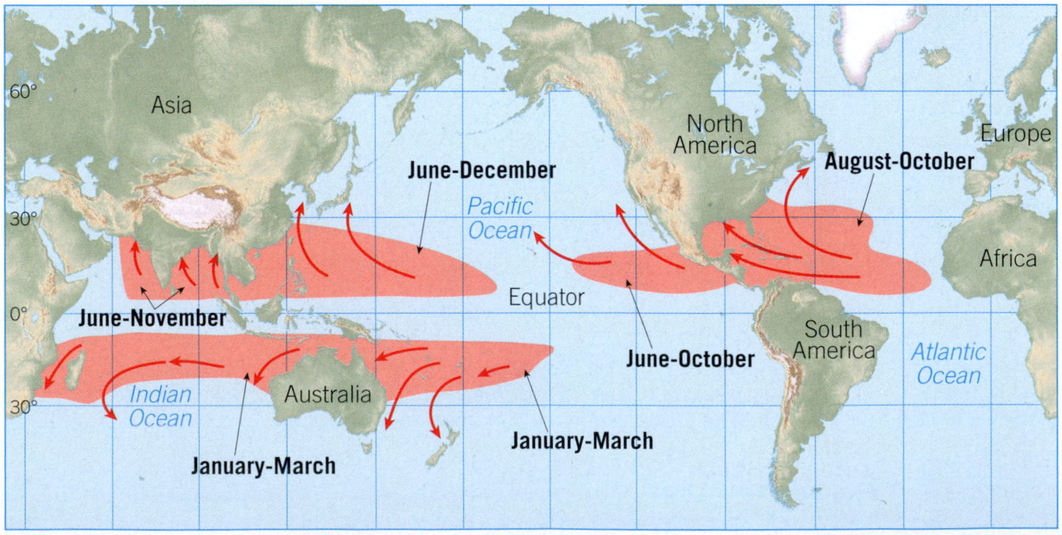


◀ **Figure 11.2 Regions of hurricane formation** This world map shows the regions where most hurricanes form as well as their principal months of occurrence and the most common tracks they follow. Hurricanes do not develop within about 5° of the equator because the Coriolis force is too weak in that region. Because warm surface ocean temperatures are necessary for hurricane formation, hurricanes seldom form poleward of 30° latitude nor over the cool waters of the South Atlantic and the eastern South Pacific.

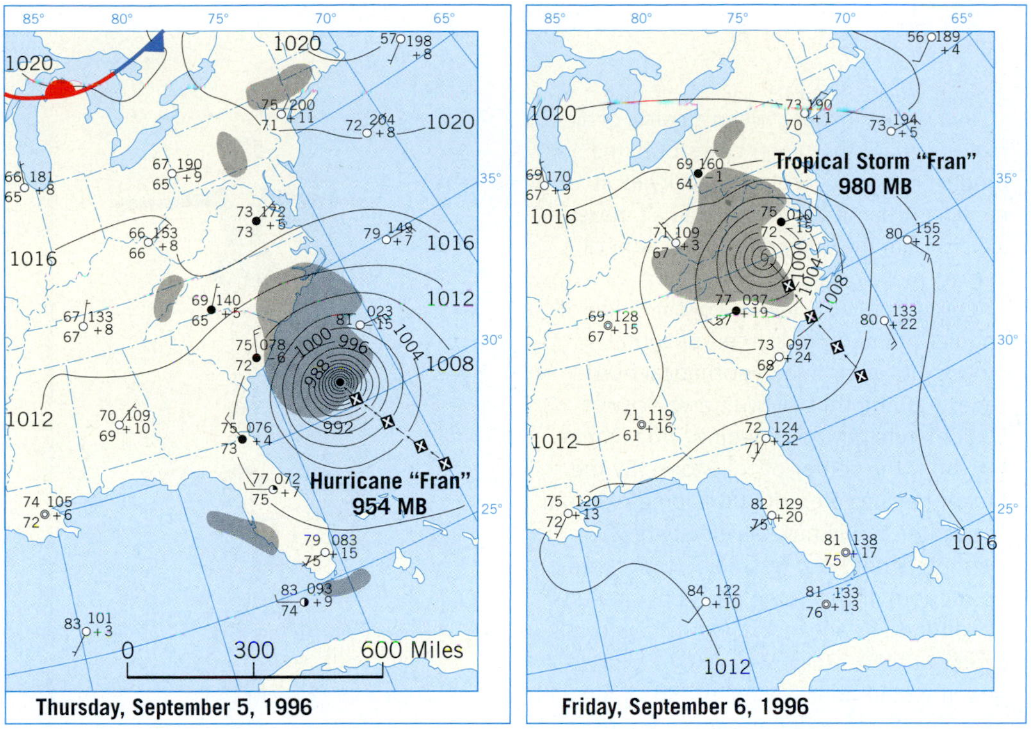

Tropical Storm "Fran"
980 MB
Hurricane "Fran"
954 MB
0
300
600 Miles
Thursday, September 5, 1996
Friday, September 6, 1996

◀ **Figure 11.3 Hurricane Fran** These weather maps show Hurricane Fran at 7 A.M. on two successive days, September 5 and 6, 1996. On September 5, winds exceeded 190 kilometers per hour. As the storm moved inland, heavy rains caused flash floods, killed 30 people, and caused more than $3 billion in damages. The station information plotted off the Gulf and Atlantic coasts is from data buoys, which are remote floating instrument packages. The small boxes extending southeast from the storm's center show the position of the eye at 6-hour intervals.

**Animation** MM
Hurricane Wind Patterns
http://goo.gl/gqBLv2

▼ **Figure 11.4 Anatomy of a hurricane**
**A.** Cross-section of a hypothetical hurricane. Note that the vertical dimension is greatly exaggerated. **B.** Changes in wind and pressure with the passage of a hurricane.

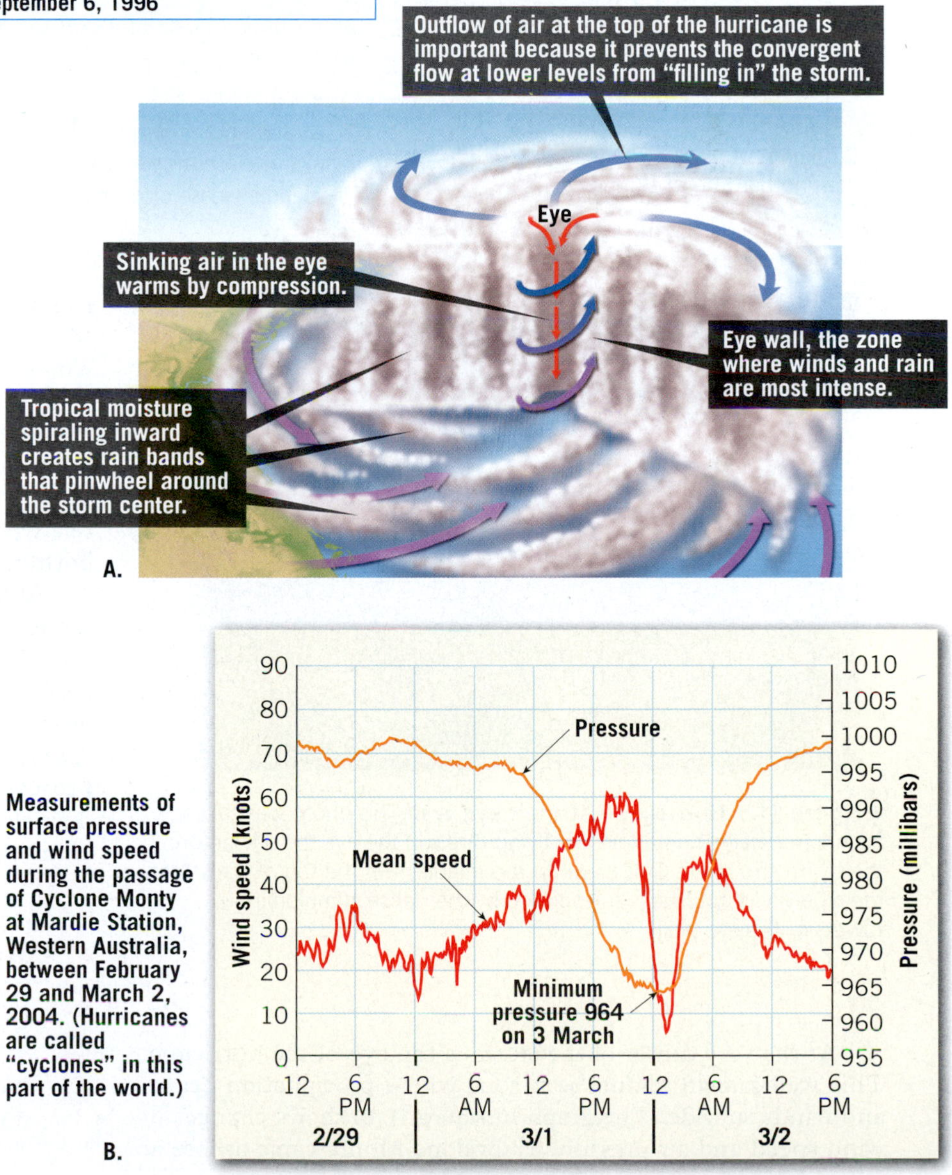

Outflow of air at the top of the hurricane is important because it prevents the convergent flow at lower levels from "filling in" the storm.
Eye
Sinking air in the eye warms by compression.
Eye wall, the zone where winds and rain are most intense.
Tropical moisture spiraling inward creates rain bands that pinwheel around the storm center.
A.
Measurements of surface pressure and wind speed during the passage of Cyclone Monty at Mardie Station, Western Australia, between February 29 and March 2, 2004. (Hurricanes are called "cyclones" in this part of the world.)
Pressure
Mean speed
Minimum pressure 964 on 3 March
Wind speed (knots)
Pressure (millibars)
12
6 PM
12
6 AM
12
6 PM
12
6 AM
12
6 PM
2/29
3/1
3/2
B.

ricanes average about 600 kilometers (375 miles) across, although they can range in diameter from 100 kilometers (60 miles) up to about 1500 kilometers (930 miles). From the outer edge of a hurricane to the center, the barometric pressure has at times dropped 60 millibars, from 1010 to 950 millibars.

A steep pressure gradient like that shown in **Figure 11.3** generates the rapid, inward spiraling winds of a hurricane. As the air moves closer to the center of the storm, its velocity increases. This acceleration is explained by the law of conservation of angular momentum (**Box 11.1**). It is important to understand that hurricane-strength winds of 119 kilometers (74 miles) per hour or greater occur primarily in the interior region of a hurricane, a zone about 160 kilometers (100 miles) across in an average storm. Outer areas of the storm have gale-force winds that exceed 50 kilometers (30 miles) per hour. Thus, the winds associated with much of a hurricane are actually less than hurricane strength.

As the inward rush of warm, moist surface air approaches the core of the storm, it turns upward and ascends in a ring of cumulonimbus towers (**Fig. 11.4A**). This doughnut-shaped wall of intense convective activity surrounding the center of the storm is called the **eye wall**. It is here that the greatest wind speeds and heaviest rainfall occur. Surrounding the eye wall are curved bands of thunderstorm-producing cumulonimbus clouds, called **rain bands**, that trail away in a spiral fashion. Near the top of the hurricane, the air flows outward, carrying the rising air away from the storm center, thereby providing room for more inward flow at the surface.

## Box 11.1 Conservation of Angular Momentum

Why do winds blowing around a storm move faster near the center and more slowly near the edge? To understand this phenomenon, we must examine the *law of conservation of angular momentum*. This law states that the product of the velocity of an object around a center of rotation (axis) and the distance of the object from the axis is constant.

Picture an object on the end of a string being swung in a circle. If the string is pulled inward, the distance of the object from the axis of rotation decreases and the speed of the spinning object increases. The change in radius of the rotating mass is balanced by a change in its rotational speed.

Another common example of the conservation of angular momentum occurs when a figure skater starts whirling on the ice with both arms extended (**Fig. 11.A**). Her arms are traveling in a circular path about an axis (her body). When the skater pulls her arms inward, she decreases the radius of the circular path of her arms. As a result, her arms go faster, and the rest of her body must follow, thereby increasing her rate of spinning.

In a similar manner, when a parcel of air moves toward the center of a storm, the product of its distance and velocity must remain unchanged. Therefore, as air moves inward from the outer edge, its rotational velocity must increase.

Let us apply the law of conservation of angular momentum to the horizontal movement of air in a hypothetical hurricane. Assume that air with a velocity of 5 kilometers per hour begins 500 kilometers from the center of the storm. By the time it reaches a point 100 kilometers from the center, it will have a velocity of 25 kilometers per hour (assuming that there is no friction). If this same parcel of air were to continue to advance toward the storm's center until its radius was just 10 kilometers, it would be traveling at 250 kilometers per hour. Friction reduces these values somewhat.

**Question**

1. As air in the outer portion of a hurricane moves closer to the storm's center, does its speed increase or decrease?

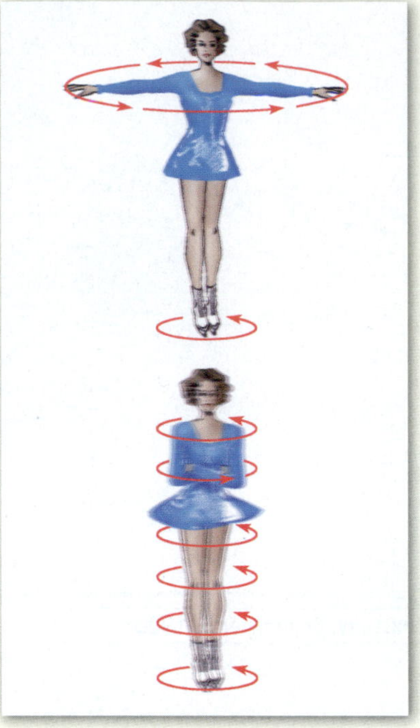

▲ **Figure 11.A Conservation of angular momentum** When the skater's arms are extended, she spins slowly. When her arms are pulled in, she spins much faster.

▲ **Figure 11.5 Hurricane Katrina's eye wall** This photo was taken from a hurricane-hunter aircraft flying through the eye the day before the storm struck the Gulf coast. In this image from the top of the storm, the blue sky of the eye is surrounded by the dense cumulonimbus towers of the eye wall.

At the very center of the storm is the **eye** of the hurricane. This well-known feature is a zone where precipitation ceases and winds subside. The graph in **Figure 11.4B** shows changes in wind speed and air pressure as Cyclone Monty came ashore at Mardie Station in Western Australia between February 29 and March 2, 2004. The very steep pressure gradient and strong winds associated with the eye wall are evident, as is the relative calm of the eye. The diameter of the eye varies from about 5 to more than 60 kilometers (3 to more than 35 miles). In most hurricanes, the eye is smaller at the surface and widens going upward. Near the top of the storm, it may eventually form a relatively large, nearly cloud-free zone surrounded by the upper portions of the eye-wall clouds (**Fig. 11.5**).

At the surface, the eye offers a brief but deceptive break from the extreme weather that is occurring in the enormous curving wall clouds that surround it. The air within the eye gradually descends and heats by compression, making it the warmest part of the storm. Many people believe that the entire eye is a zone characterized by clear blue skies, but this is usually not the case because the subsidence in the eye is seldom strong enough to produce cloudless conditions. Although the sky appears much brighter in this region, scattered clouds at various levels are common.

## ✔ Concept Checks 11.1

1. Define *hurricane*. What other names are used for this storm?
2. In what latitude zone do hurricanes develop?
3. Distinguish between the eye and the eye wall of a hurricane. How do conditions differ in these zones?

# 11.2 Hurricane Formation and Decay

**Discuss the conditions that promote hurricane formation and list the factors that cause hurricanes to dissipate.**

A hurricane is a heat engine fueled by the latent heat liberated when huge quantities of water vapor condense. The amount of energy produced by a typical hurricane in just a single day is truly immense. The release of latent heat warms the air and provides buoyancy for its upward flight. The result is to reduce the pressure near the surface, which in turn encourages more rapid inflow of air. A large quantity of warm, moist air is required to get this engine started, and a continuous supply is needed to keep it going.

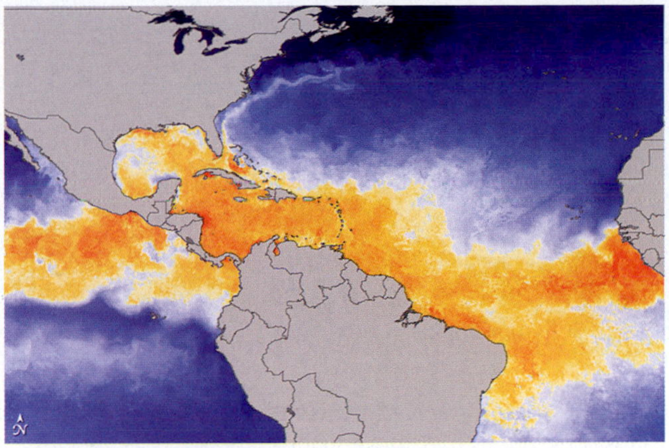


▲ **Figure 11.7 Sea-surface temperatures** Among the necessary ingredients for a hurricane is warm ocean temperatures above 27°C (80°F). This color-coded satellite image from June 1, 2010, shows sea-surface temperatures at the beginning of hurricane season.

## Hurricane Formation

As the graph in **Figure 11.6** illustrates, hurricanes most often form in late summer and early fall. It is during this span that sea-surface temperatures reach 27°C (80°F) or higher and are thus able to provide the necessary heat and moisture to the air (**Fig. 11.7**). This ocean-water temperature requirement explains why hurricane formation over the relatively cool waters of the South Atlantic and the eastern South Pacific is extremely rare. For the same reason, few hurricanes form poleward of 30° latitude (see Figure 11.2). Hurricanes do not form within 5° of the equator, even though water temperatures are sufficiently high, because the Coriolis force is too weak in that region to initiate the necessary rotary motion.

Many tropical storms achieve hurricane status in the western parts of oceans, but their origins often lie far to the east. In such locations, disorganized arrays of clouds and thunderstorms, called **tropical disturbances**, sometimes develop and exhibit weak pressure gradients and little or no rotation. Most of the time these zones of convective activity die out. However, tropical disturbances occasionally grow larger and develop strong cyclonic rotation.

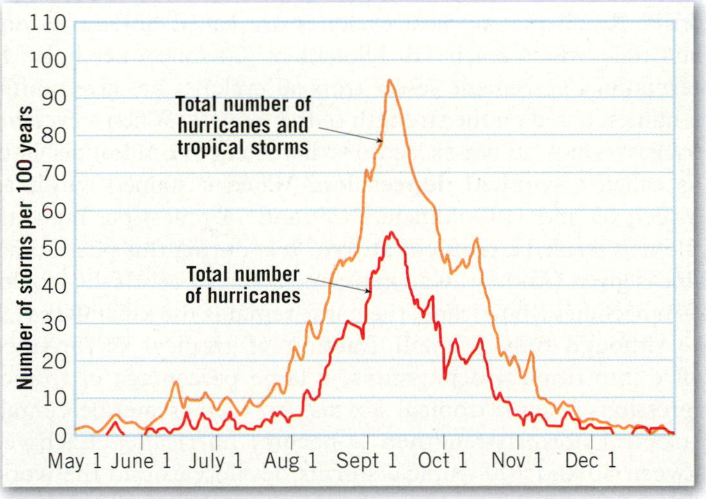


▲ **Figure 11.6 Frequency of tropical storms and hurricanes in the Atlantic basin** The graph shows the number of storms between May 1 and December 31 to be expected over a span of 100 years. The period from late August through October is clearly the most active.

Several different situations can trigger tropical disturbances. They are sometimes initiated by the powerful convergence and lifting associated with the intertropical convergence zone (ITCZ). Others form when a trough from the middle latitudes intrudes into the tropics.

**Easterly Waves** Tropical disturbances that produce many of the strongest hurricanes that enter the western North Atlantic and threaten North America often begin as large undulations or ripples in the trade winds known as **easterly waves**, so named because they gradually move from east to west.

**Figure 11.8** illustrates an easterly wave. The lines on this simple map are not isobars. Rather, they are **streamlines**, lines drawn parallel to the wind direction used to depict surface airflow. When middle-latitude weather is analyzed, isobars are usually drawn on the weather map. By contrast, in the tropics the differences in sea-level pressure are quite small, so isobars are not always useful. Streamlines are helpful because they show where surface winds converge and diverge.

To the east of the wave axis, the streamlines move poleward and get progressively closer together, indicating that the surface flow is convergent. Recall from Chapter 4 that convergence encourages air to rise and form clouds. Therefore, the tropical disturbance is located on the east side of the wave. To the west of the wave axis, surface flow diverges as it turns toward the equator. Consequently, clear skies are the rule here.

Easterly waves frequently originate as disturbances in Africa. As these storms head westward with the prevailing

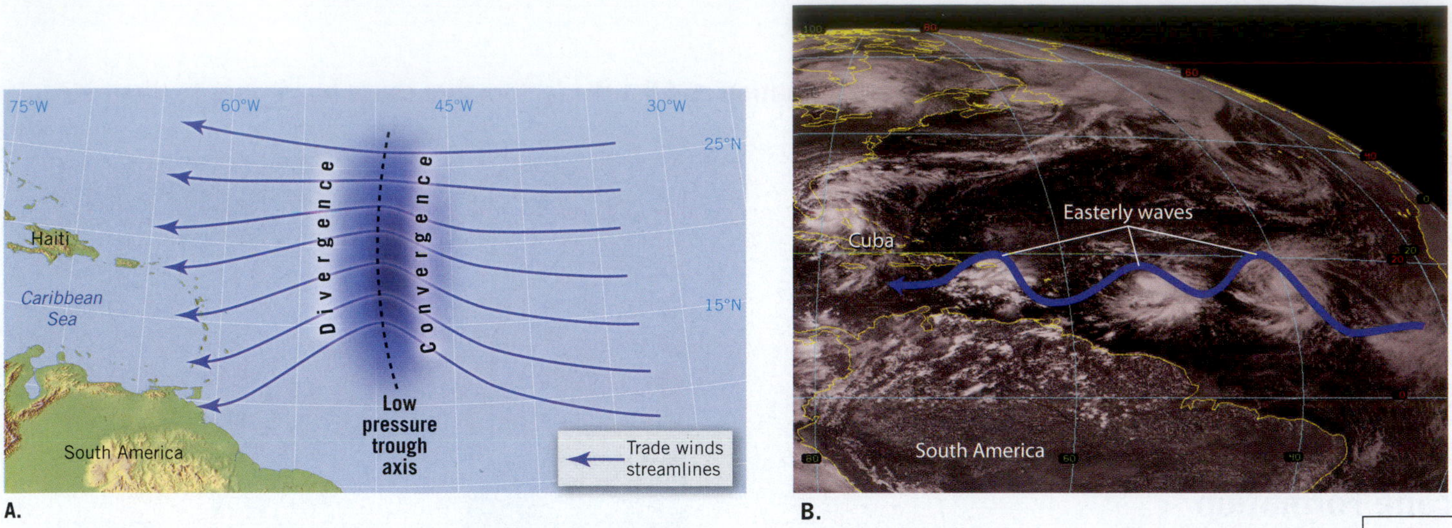


▲ **Figure 11.8 Easterly wave in the subtropical North Atlantic** **A.** Streamlines show low-level airflow. To the east of the wave axis, winds converge as they move slightly poleward. To the west of the axis, flow diverges as it turns toward the equator. Tropical disturbances are associated with the convergent flow of the easterly wave. Easterly waves extend for 2000 to 3000 kilometers (1200 to 1800 miles) and move from east to west with the trade winds at rates between 15 and 35 kilometers (10 and 20 miles) per hour. At this rate, it takes an imbedded tropical disturbance a week or 10 days to move across the North Atlantic. **B.** This satellite image shows several tropical disturbances that developed along a series of easterly waves between Africa and the Caribbean Sea.

Video MM
A Hurricane in the Middle Latitudes
http://goo.gl/r359VH

trade winds, they encounter the cold Canary Current (see Figure 3.8, page 66). If the disturbance survives the trip across the cold stabilizing waters of the current, it is rejuvenated by the heat and moisture of the warmer water of the mid-Atlantic. From this point on, a disturbance may develop into a more intense and organized system; some of these may reach hurricane status.

**Why Tropical Disturbances Dissipate** Even when conditions seem to be right for hurricane formation, many tropical disturbances fail to get stronger. One circumstance that may inhibit further development is a temperature inversion called a **trade wind inversion** that forms in association with the subsidence that occurs in the region influenced by the subtropical high. In the troposphere, a temperature inversion exists when temperatures rise with an increase in altitude rather than fall, which is usually the case. For more about how subsidence can produce an inversion, see the section "Inversions Aloft" in Chapter 13, page 374. A strong inversion diminishes the ability of air to rise and thus inhibits the development of strong thunderstorms. Another factor that works against the strengthening of tropical disturbances is strong upper-level winds. When such winds are present, a strong flow aloft disperses the latent heat released from cloud tops—heat that is essential for continued storm growth and development.

## From Tropical Disturbance to Hurricane

What happens when conditions favor hurricane development? As latent heat is released from the clusters of thunderstorms that make up the tropical disturbance, areas within the disturbance get warmer. As a result, air density decreases and surface pressure drops, creating a region of weak low pressure and cyclonic circulation. As pressure drops at the storm center, the pressure gradient steepens. In response, surface wind speeds increase and bring in additional supplies of moisture to nurture storm growth. The water vapor condenses, releasing latent heat, and the heated air rises. Adiabatic cooling of rising air triggers more condensation and the release of even more latent heat, which causes a further increase in buoyancy. The cycle continues.

Meanwhile, pressure increases at the top of the developing tropical disturbance (see Figure 6.12 and the corresponding discussion of the sea breeze in Chapter 6). This causes air to flow outward (diverge) from the top of the storm. Without this outward flow up top, the inflow at lower levels would soon raise surface pressures and thwart further storm development.

Many tropical disturbances form each year, but more than 90 percent die out without ever organizing into more powerful storms. Recall that tropical cyclones are called hurricanes only when their winds reach 119 kilometers (74 miles) per hour. By international agreement, lesser tropical cyclones are given different names, based on the strength of their winds. When a cyclone's strongest winds do not exceed 63 kilometers (39 miles) per hour, it is called a **tropical depression**. When sustained winds are between 63 and 119 kilometers (39 and 74 miles) per hour, the cyclone is termed a **tropical storm**. It is during this phase that a name is given (Andrew, Katrina, Sandy, and so on). If the tropical storm becomes a hurricane, the name remains the same (**Box 11.2**).

Although only a small fraction of tropical disturbances evolve into tropical depressions, a large percentage of tropical depressions become tropical storms. An even greater percentage of tropical storms strengthen to become hurricanes. Each year between 80 and 100 tropical storms develop around the world. Of them, usually half or more eventually reach hurricane status.

## Hurricane Decay

Hurricanes diminish in intensity when they (1) move over ocean waters that cannot supply warm, moist tropical air; (2) move

## Box 11.2 Naming Tropical Storms and Hurricanes

Tropical storms are named to provide ease of communication between forecasters and the general public regarding forecasts, watches, and warnings. Tropical storms and hurricanes can last a week or longer, and two or more storms can occur in the same region at the same time. Thus, names can reduce the confusion about what storm is being described.

During World War II, tropical storms were informally assigned women's names (perhaps after wives and girlfriends) by U.S. Army Corps and Navy meteorologists who were monitoring storms over the Pacific. From 1950 to 1952, tropical storms in the North Atlantic were identified by the phonetic alphabet—Abel, Baker, Charlie, and so forth. In 1953 the U.S. Weather Bureau (now the National Weather Service) switched to women's names.

The practice of using feminine names continued until 1978, when a list containing both male and female names was adopted for tropical cyclones in the eastern Pacific. In the same year the World Meteorological Organization (WMO) accepted a proposal that both male and female names be adopted for Atlantic hurricanes, beginning with the 1979 season.

The WMO has created six lists of names for tropical storms over ocean areas. The names used for Atlantic, Gulf of Mexico, and Caribbean hurricanes are shown in **Table 11.A**. The lists are ordered alphabetically, and they do not include names that begin with the letters Q, U, X, Y, and Z because of the scarcity of names beginning with those letters. When a tropical depression reaches tropical storm status, it is assigned the next unused name on the list. At the beginning of the next hurricane season, names from the next list are selected, even though many names may not have been used the previous season.

The names for Atlantic storms are used over again at the end of each 6-year cycle unless a hurricane was particularly noteworthy. This is to avoid confusion when storms are discussed in future years. For example, following the hurricane seasons of 2011, 2012, and 2013, Irene, Sandy, and Ingrid were retired. On the lists for 2017, 2018, and 2019, Irma, Sara, and Imelda are found in their place.

**Questions**

1. Why are tropical storms named?
2. What organization is responsible for creating the lists of names?

**Table 11.A** Tropical Storm and Hurricane Names for the Atlantic, Gulf of Mexico, and Caribbean Sea*

| 2014 | 2015 | 2016 | 2017 | 2018 | 2019 |
|---|---|---|---|---|---|
| Arthur | Ana | Alex | Arlene | Alberto | Andrea |
| Bertha | Bill | Bonnie | Bret | Beryl | Barry |
| Cristobal | Claudette | Colin | Cindy | Chris | Chantal |
| Dolly | Danny | Danielle | Don | Debby | Dorian |
| Edouard | Erika | Earl | Emily | Ernesto | Erin |
| Fay | Fred | Fiona | Franklin | Florence | Fernand |
| Gonzalo | Grace | Gaston | Gert | Gordon | Gabrielle |
| Hanna | Henri | Hermine | Harvey | Helene | Humberto |
| Isaias | Ida | Ian | Irma | Isaac | Imelda |
| Josephine | Joaquin | Julia | Jose | Joyce | Jerry |
| Kyle | Kate | Karl | Katia | Kirk | Karen |
| Laura | Larry | Lisa | Lee | Leslie | Lorenzo |
| Marco | Mindy | Matthew | Maria | Michael | Melissa |
| Nana | Nicholas | Nicole | Nate | Nadine | Nestor |
| Omar | Odette | Otto | Ophelia | Oscar | Olga |
| Paulette | Peter | Paula | Philippe | Patty | Pablo |
| Rene | Rose | Richard | Rina | Rafael | Rebekah |
| Sally | Sam | Shary | Sean | Sara | Sebastien |
| Teddy | Teresa | Tobias | Tammy | Tony | Tanya |
| Vicky | Victor | Virginie | Vince | Valerie | Van |
| Wilfred | Wanda | Walter | Whitney | William | Wendy |

*If the entire alphabetical list of names for a given year is exhausted, the naming system moves on to letters of the Greek alphabet (alpha, beta, gamma, and so on). This issue never arose until the record-breaking 2005 hurricane season, when Tropical Storm Alpha, Hurricane Beta, tropical storms Gamma and Delta, Hurricane Epsilon, and Tropical Storm Zeta occurred after Hurricane Wilma.

onto land; or (3) reach a location where the large-scale flow aloft is unfavorable.

Many hurricanes approaching North America from the southeast are turned toward the northeast, away from the continent, by the steering effect of an upper-level trough (see Chapter 9). This change of direction carries the storms toward higher latitudes, where ocean temperatures are cooler and encountering cool, dry polar air masses is more likely. Often the hurricane and a polar front interact, with cold air entering the storm from the west. As a result, the release of latent heat is diminished, upper-level divergence weakens, temperatures in the hurricane's core fall, and surface pressures rise.

**students sometimes ask...**

### When is hurricane season?

Hurricane season occurs at different times in different parts of the world. People in the United States are usually most interested in Atlantic storms. The Atlantic hurricane season officially extends from June through November. More than 97 percent of tropical activity in that region occurs during this 6-month span. The "heart" of the season is August through October (see Figure 11.6), when 87 percent of the minor hurricane (category 1 and 2) days and 96 percent of the major hurricane (category 3, 4, and 5) days occur. Peak activity is in early to mid-September.

Whenever a hurricane moves onto land, it loses its punch rapidly. For example, in Figure 11.3 notice how the isobars show a much weaker pressure gradient on September 6, after Hurricane Fran moved ashore, than on September 5, when it was over the ocean. The most important reason for this rapid demise is the fact that the storm's source of warm, moist air is cut off. When an adequate supply of water vapor does not exist, condensation and the release of latent heat must diminish.

In addition, the increased surface roughness over land results in a rapid reduction in surface wind speeds. This friction causes the winds to move more directly into the center of the low, thus helping to eliminate the large pressure differences.

### ✔ Concept Checks 11.2

1. What is the source of energy that drives a hurricane?
2. During what months do most tropical storms and hurricanes in the Atlantic basin occur? Why?
3. List two factors that inhibit the strengthening of tropical disturbances.
4. Distinguish between *tropical depression, tropical storm,* and *hurricane*.
5. Why does the intensity of a hurricane diminish rapidly when the storm moves onto land?

## 11.3 Hurricane Destruction

**Explain how hurricane intensity is determined and summarize the three broad categories of hurricane destruction.**

Although hurricanes are tropical or subtropical in origin, their destructive effects can be experienced far from where they originate. For example, Hurricane Igor, a storm that formed in the subtropical Atlantic in September 2010 had its greatest impact far to the north, in Newfoundland (**Fig. 11.9**). In 2012, Hurricane Sandy affected the entire eastern seaboard, from Florida to Maine. Damages were estimated at $65 billion, the second costliest hurricane in U.S. history after Katrina in 2005. Destruction was especially great in New Jersey and New York, even though Sandy may no longer have had hurricane status.

The vast majority of hurricane-related deaths and damage are caused by relatively infrequent yet powerful storms. **Table 11.1** lists the deadliest hurricanes to strike the United States between 1900 and 2013. The storm that pounded an unsuspecting Galveston, Texas, in 1900 was not just the deadliest U.S. hurricane ever but the deadliest natural disaster *of any kind* to affect the United States. Of course, the deadliest and most costly storm in recent memory occurred in August 2005, when Hurricane Katrina devastated the Gulf coast of Louisiana, Mississippi, and Alabama. Although hundreds of thousands fled before the storm made landfall, thousands of others were caught by the storm. In addition to the human suffering and tragic loss of life that were left in the wake of Hurricane Katrina, the financial losses caused by the storm are practically incalculable; until then, the $25 billion in damages associated with Hurricane Andrew in 1992 represented the costliest U.S. natural disaster. This figure was exceeded many times over by Katrina's economic impact, which likely exceeded $100 billion.

Although the amount of damage caused by a hurricane depends on several factors, including the size and population density of the area affected and the nearshore ocean-bottom

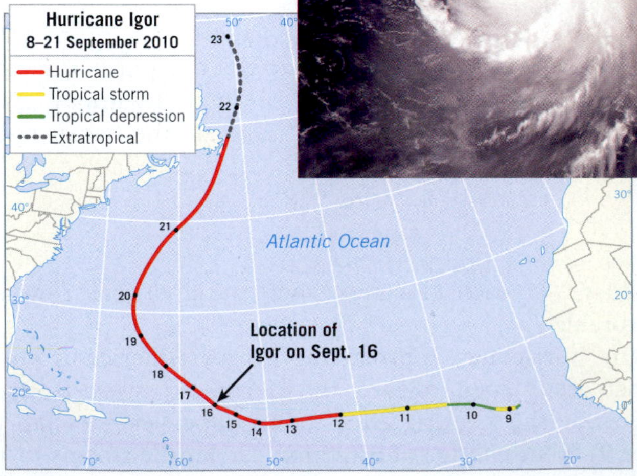


▶ **Figure 11.9 Hurricane Igor** The satellite image shows the storm on September 16, 2010, when it was located at 20° north latitude. Maximum winds were 213 kilometers (132 miles) per hour. As the map shows, shortly after the satellite image was acquired, the storm headed northward. The storm had its greatest impact far from the subtropics, in Newfoundland, where most of the damage resulted from flooding triggered by Igor's heavy rains.

**Video** The 2005 Hurricane Season
http://goo.gl/ZvfcZ

**eye ON THE atmosphere 11.1**

This hurricane occurred in the Atlantic in 2004.

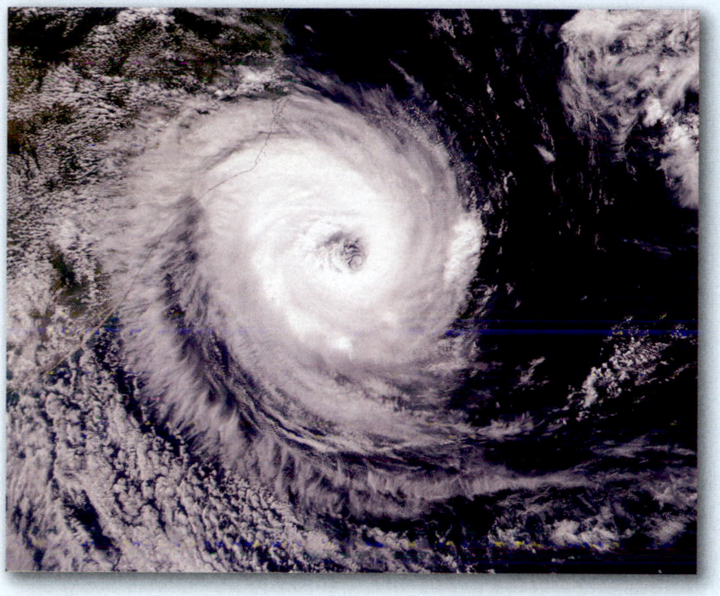

**Questions**

1. Did the storm occur in the North Atlantic or the South Atlantic? How did you figure this out?
2. Did the storm more likely occur in March or September?
3. Are hurricanes in this region common or rare? Explain.

configuration, certainly the most significant factor is the strength of the storm.

## Saffir–Simpson Scale

Based on the study of past storms, the **Saffir–Simpson scale** was established to rank the relative intensities of hurricanes (Table 11.2). Predictions of hurricane severity and damage are usually expressed in terms of this scale. When a tropical storm becomes a hurricane, the National Weather Service assigns it a scale (category) number. Category assignments are based on observed conditions at a particular stage in the life of a hurricane and are viewed as estimates of the amount of damage a storm would cause if it were to make landfall without changing size or strength. As conditions change, the category of a storm is reevaluated so that public safety officials can be kept informed. By using the Saffir–Simpson scale, the disaster potential of a hurricane can be monitored, and appropriate precautions can be planned and implemented.

A rating of 5 on the scale represents the worst storm possible, and a 1 is least severe. Storms that fall into category 5 are rare. Only three storms this powerful are known to have hit the continental United States: Andrew struck Florida in 1992, Camille pounded Mississippi in 1969, and a Labor Day hurricane struck the Florida Keys in 1935.

**Table 11.1** The 10 Deadliest Hurricanes to Strike the U.S. Mainland, 1900–2013

| Rank | Hurricane | Year | Category | Deaths |
|---|---|---|---|---|
| 1. | Texas (Galveston) | 1900 | 4 | 8000* |
| 2. | Southeastern Florida (Lake Okeechobee) | 1928 | 4 | 2500–3000 |
| 3. | Katrina | 2005 | 4 | 1833 |
| 4. | Audrey | 1957 | 4 | At least 416 |
| 5. | Florida Keys | 1935 | 5 | 408 |
| 6. | Florida (Miami)/Mississippi/Alabama/Florida (Pensacola) | 1926 | 4 | 372 |
| 7. | Louisiana (Grande Isle) | 1909 | 4 | 350 |
| 8. | Florida Keys/South Texas | 1919 | 4 | 287 |
| 9. (tie) | Louisiana (New Orleans) | 1915 | 4 | 275 |
| 9. (tie) | Texas (Galveston) | 1915 | 4 | 275 |

Source: National Weather Service/National Hurricane Center.

*This number may actually have been as high as 10,000–12,000.

**students sometimes ask...**

**Are larger hurricanes stronger than smaller hurricanes?**

Actually, there is very little correlation between intensity (either measured by maximum sustained winds or by central pressure) and size (either measured by the radius of gale-force winds or the radius of the outer closed isobar). Hurricane Andrew is a good example of a very intense storm (category 5) that was also relatively small (gale-force winds extended only 150 kilometers [90 miles] from the eye).

Sometimes the intensity of a storm is described using terms such as "major hurricane" or "super typhoon." The National Hurricane Center uses *major hurricane* to describe a storm that reaches maximum sustained 1-minute surface winds of at least 178 kilometers (111 miles) per hour, meaning that it is at least a category 3 storm. The U.S. Joint Typhoon Warning Center uses *super typhoon* for storms in the western Pacific that have sustained winds of at least 210 kilometers (131 miles) per hour, equivalent to at least a category 4 on the Saffir–Simpson

**Table 11.2** Saffir–Simpson Hurricane Scale*

| Scale Number (category) | Central Pressure (millibars) | Winds (km/hr) | Storm Surge (meters) | Damage |
|---|---|---|---|---|
| 1 | ≥980 | 119–153 | 1.2–1.5 | Minimal |
| 2 | 965–979 | 154–177 | 1.6–2.4 | Moderate |
| 3 | 945–964 | 178–209 | 2.5–3.6 | Extensive |
| 4 | 920–944 | 210–250 | 3.7–5.4 | Extreme |
| 5 | <920 | >250 | >5.4 | Catastrophic |

*A more complete version of the Saffir–Simpson hurricane scale can be found in Appendix F, page A-16.

scale. Super Typhoon Haiyan, which struck the Philippines in November 2013, is examined in Box 11.3, later in the chapter.

Damage caused by hurricanes can be divided into three classes: (1) storm surge, (2) wind damage, and (3) inland freshwater flooding.

## Storm Surge

Without question, the most devastating damage in the coastal zone is caused by storm surge. It not only accounts for a large share of coastal property losses but is also responsible for a significant percentage of hurricane-caused deaths. A **storm surge** is a dome of water 65 to 80 kilometers (40 to 50 miles) wide that sweeps across the coast near the point where the eye makes landfall. If all wave activity were smoothed out, the storm surge would be the height of the water above normal tide level (Fig. 11.10). In addition, tremendous wave activity is superimposed on the surge. We can easily imagine the damage that this surge of water could inflict on low-lying coastal areas (Fig. 11.11). Some of the worst surges occur in places like the Gulf of Mexico, where the continental shelf is very shallow and gently sloping. In addition, local features such as bays and rivers can cause surge height to double and the water to increase in speed.

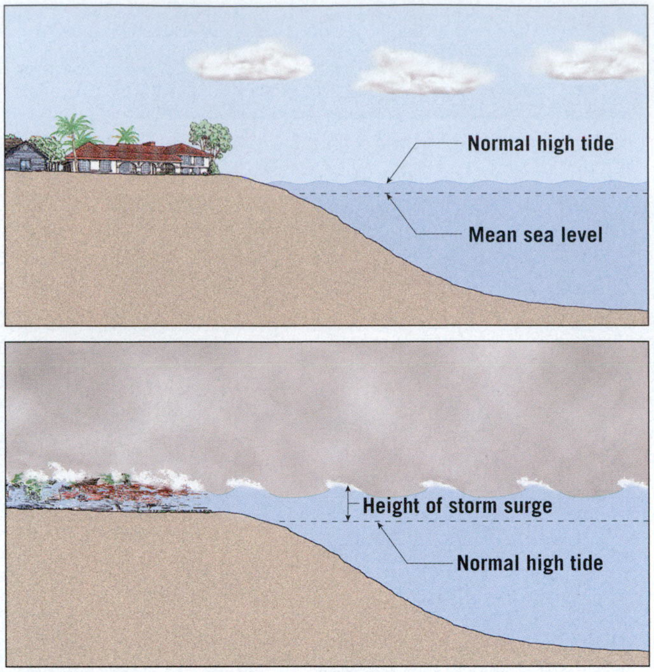


▲ **Figure 11.10 Storm surge** Superimposed upon high tide, a storm surge can devastate a coastal area. The worst storm surges occur in coastal areas where there is a very shallow and gently sloping continental shelf extending from the beach. The Gulf coast is such a place.

In the delta region of Bangladesh, most of the land is less than 2 meters (6.5 feet) above sea level. When a storm surge superimposed on normal high tide inundated that area on November 13, 1970, the official death toll was 200,000; unofficial estimates ran to 500,000. It was one of the worst natural disasters of modern times. In May 1991 a similar event again struck Bangladesh. This time the storm took the lives of at least 143,000 people and devastated coastal towns in its path. As sea level continues to rise in coming decades, low-lying, densely populated coastal areas will become even more vulnerable to the destructive effects of storm surge (Fig. 11.12). There is more discussion about rising sea level in Chapter 14.

A common misconception about the cause of hurricane storm surges is that the very low pressure at the center of the storm acts as a partial vacuum that allows the ocean to rise up in response. However, this effect is relatively insignificant. The most important factor responsible for the development of a storm surge is the piling up of ocean water by strong onshore winds. Gradually the hurricane's winds push water toward the shore, causing sea level to elevate while also churning up violent wave activity.

As a hurricane advances toward the coast in the Northern Hemisphere, storm surge is always most intense on the right side of the eye, where winds are blowing *toward* the shore. In addition, on this side of the storm, the forward movement of the hurricane also contributes to the storm surge. In Figure 11.13, assume that a hurricane with peak winds of 175 kilometers (109 miles) per hour is moving toward the shore at 50 kilometers (31 miles) per hour. In this case, the net wind speed on the right side of the advancing storm is 225 kilometers (140 miles) per hour. On the left side, the hurricane's winds are blowing opposite the direction of storm movement, so the net winds are *away* from the coast, at 125 kilometers (78 miles) per hour. Along the shore facing the left side of the oncoming hurricane, the water level may actually decrease as the storm makes landfall.

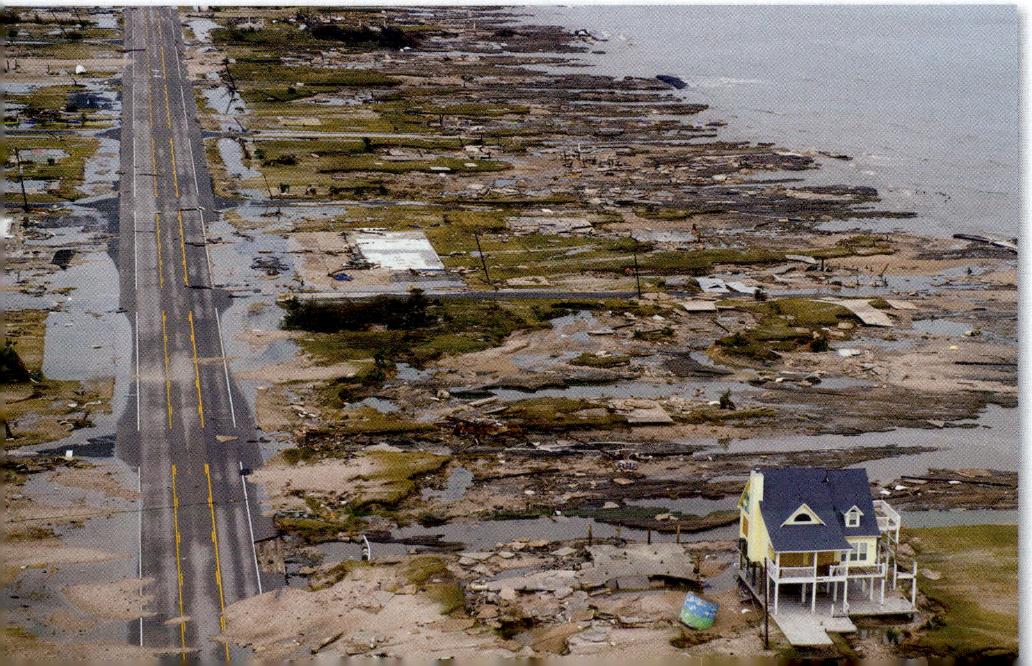

◄ **Figure 11.11 Aftermath of Hurricane Ike** In mid-September 2008, the eye of the storm passed directly over Galveston, Texas. At landfall the storm had sustained winds of 165 kilometers (105 miles) per hour. The extraordinary storm surge caused much of the damage pictured here at Crystal Beach.

# What's Your Forecast?

## More to the Story than the Category by Brian McNoldy, University of Miami, Rosenstiel School of Marine and Atmospheric Science

The approach of a category 4 or 5 hurricane is a frightening prospect, yet hurricanes with lower wind speeds can cause extensive damage when factors such as storm surge are part of the event. Therefore, issuing hurricane warnings involves more than forecasting just the maximum wind speed. Storm surge is the rapid rise of water along a coastline, generated primarily by the onshore winds of a tropical cyclone. To correctly predict this surge, several complex variables must be taken into account, including the intensity, size, forward speed and direction of the storm, the shape of the ocean floor, the topography of the land, and the shape of the coastline. The "storm tide" is a combination of the storm surge and the astronomical tides, so the timing of hurricane landfall relative to normal tides affects the *total inundation.* It is a gross oversimplification to say that stronger storms create a larger storm surge and weaker storms create a smaller storm surge.

**Figure 11.B** shows the five-day track forecast with the "cone of uncertainty" for a real hurricane. When this advisory was issued by the National Hurricane Center, the storm was a fairly average Category 2 hurricane over the Bahamas, but was forecast to weaken slightly, head north, and eventually turn westward into the northeastern United States. This was the time to warn people about the potential for strong winds and significant storm surge.

As you may have guessed, this storm is Hurricane Sandy, the deadliest U.S. storm since Katrina in 2005, and one of the costliest natural disasters in U.S. history (in 2013 dollars).

Sandy formed in the central Caribbean Sea on October 22, 2012 then moved north and briefly strengthened to a Category 3 as it made landfall on the eastern tip of Cuba on October 25. The winds then weakened, but the storm grew in size to become an enormous Category 1 hurricane off the U.S. east coast from October 26 to 29, with maximum sustained winds of 130 kph (80 mph) on the evening of October 29. These winds extended nearly 800 km (500 miles) to the northeast of the storm center when Sandy made landfall, moving more water toward a highly populated urban coastline. Sandy's large size also meant that the hurricane-force winds battered the coast through three 12-hour cycles of high and low tides. A full moon that evening meant that many coastal locations in the path of the storm were expecting a higher-than-usual high tide. The resulting storm tide was the highest ever recorded in the New York metropolitan area. **Figure 11.C** shows the extent of the flooding, which extended beyond the mandatory evacuation zone for a storm of this magnitude and into evacuation zones designated for much more powerful storms.

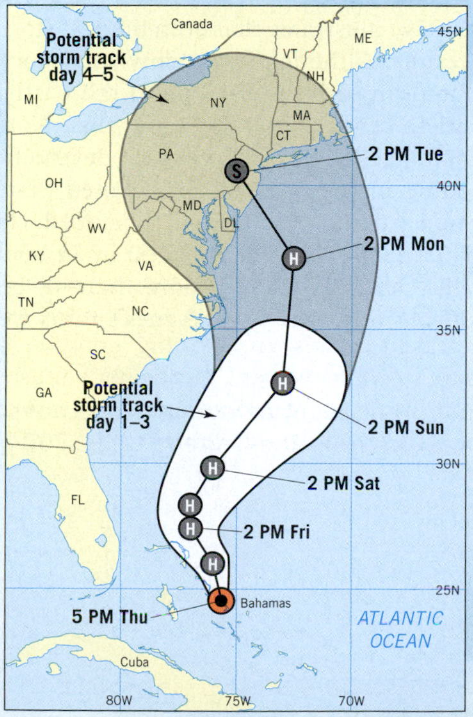


◀ **Figure 11.B. Tropical storm track forecast issued by the National Hurricane Center at 5 P.M. EDT, October 25, 2012.**

▲ **Figure 11.C. Areas flooded by Hurricane Sandy in New York City.** For more details, check the interactive map.

Sandy's Storm Surge Visualized

http://goo.gl/uUSphV

### Questions

1. The track forecast in Figure 11.B was extremely accurate. Based on what you have learned about hurricanes and storm surge in this chapter, where would you expect the strongest winds at landfall? Where would you expect the largest storm surge to occur, and why? Your answer should include sections of states, or significant cities.
2. It was primarily the storm surge, not the strong winds, t so devastating once Sandy made landfall. List the fact made Sandy's storm surge particularly destructive i York City area.

▲ **Figure 11.12 The threat of rising sea level** As sea level rises in response to global warming, low-lying deltas will become even more vulnerable to storm surge. In May 2008 the storm surge from Cyclone Nargis, a category 4 storm, contributed to the devastation of the densely populated Irrawaddy delta in Myanmar (Burma). In the years to come, such places will be even more vulnerable to the effects of tropical cyclones.

## Wind Damage

Destruction caused by wind is perhaps the most obvious of the classes of hurricane damage. Debris such as signs, roofing materials, and small items left outside become dangerous flying missiles in hurricanes. For some structures, the force of the wind is sufficient to cause total ruin. Just read the descriptions of category 3, 4, and 5 storms in Appendix F, page A-16. Mobile homes are particularly vulnerable. High-rise buildings are also susceptible to hurricane-force winds. Upper floors are most vulnerable because wind speeds usually increase with height. Recent research suggests that people should stay below the 10th floor but remain above any floors at risk for flooding. In regions with good building codes, wind damage is usually not as catastrophic as storm-surge damage. However, hurricane-force winds affect a much larger area than storm surge and can cause huge economic losses. For example, in 1992 it was largely the winds associated with Hurricane Andrew that produced more than $25 billion of damage in southern Florida and Louisiana.

▼ **Figure 11.13 An approaching hurricane** Winds associated with a Northern Hemisphere hurricane that is advancing toward the coast. This hypothetical storm, with peak winds of 175 kilometers (109 miles) per hour, is moving toward the coast at 50 kilometers (31 miles) per hour. On the right side of the advancing storm, the 175-kilometer-per-hour winds are in the same direction as the movement of the storm (50 kilometers per hour). Therefore, the *net* wind speed on the right side of the storm is 225 kilometers (140 miles) per hour. On the left side, the hurricane's winds are blowing opposite the direction of storm movement, so the *net* winds of 125 kilometers (78 miles) per hour are away from the coast. Storm surge will be greatest along the part of the coast hit by the right side of the advancing hurricane.

A hurricane may produce tornadoes that contribute to the storm's destructive power. Studies have shown that more than half of the hurricanes that make landfall produce at least one tornado. In 2004 the number of tornadoes associated with tropical storms and hurricanes was extraordinary. Tropical Storm Bonnie and five landfalling hurricanes—Charley, Frances, Gaston, Ivan, and Jeanne—produced nearly 300 tornadoes that affected the southeast and mid-Atlantic states (**Table 11.3**). Hurricane Frances produced the most tornadoes ever reported from one hurricane. The large number of hurricane-generated tornadoes in 2004 helped make this a record-breaking year for tornadoes—surpassing the previous record by more than 300.

## Heavy Rains and Inland Flooding

The torrential rains that accompany most hurricanes represent a third significant threat—flooding. The 2004 hurricane season was very deadly, with more than 3000 fatalities. Nearly all of the deaths occurred in Haiti, as a result of flash floods and mudflows caused by the heavy rains associated with then Tropical Storm Jeanne.

Hurricane Agnes (1972) illustrates that even modest storms can have devastating results. Although it was only a category 1 storm on the Saffir–Simpson scale, it was one of the costliest hurricanes of the twentieth century, creating more than $2 billion in damage and taking 122 lives. The greatest destruction was attributed to flooding in the northeastern United States, especially in Pennsylvania, where record rainfalls occurred. Harrisburg received nearly 32 centimeters (12.5 inches) in 24 hours, and western Schuylkill County measured more than 48 centimeters (19 inches) during the same span. Agnes's rains were

**Table 11.3** Number of Tornadoes Spawned by Hurricanes and Tropical Storms in the United States, 2004

| | |
|---|---|
| Tropical Storm Bonnie | 30 |
| Hurricane Charley | 25 |
| Hurricane Frances | 117 |
| Hurricane Gaston | 1 |
| Hurricane Ivan | 104 |
| Hurricane Jeanne | 16 |

not as devastating elsewhere. Prior to reaching Pennsylvania, the storm caused some flooding in Georgia, but most farmers welcomed the rain because dry conditions had been plaguing them earlier. In fact, the value of the rains to crops in the region far exceeded the losses caused by flooding.

Another well-known example is Hurricane Camille (1969). Although this storm is best known for its exceptional storm surge and the devastation it brought to coastal areas, the greatest number of deaths associated with this storm occurred in the Blue Ridge Mountains of Virginia 2 days after Camille's landfall. Many areas received more than 25 centimeters (10 inches) of rain, and severe flooding took more than 150 lives.

To summarize, extensive damage and loss of life in the coastal zone can result from storm surge, strong winds, and torrential rains. When loss of life occurs, it is commonly caused by storm surge, which can devastate entire barrier islands or zones within a few blocks of the coast. Although wind damage is usually not as catastrophic as storm surge, it affects a much larger area. Where building codes are inadequate, economic losses can be especially severe. Because hurricanes weaken as they move inland, most wind damage occurs within 200 kilometers (125 miles) of the coast. Far from the coast, a weakening storm can produce extensive flooding long after the winds have diminished below hurricane levels. Sometimes the damage from inland flooding exceeds storm-surge destruction.

### ✔ Concept Checks 11.3

1. What is the purpose of the Saffir–Simpson scale?
2. What are the three broad categories of hurricane damage? Provide brief examples of each.

## 11.4 Estimating Hurricane Intensity

**Compare two methods used to determine the intensity of a hurricane and explain why storm intensity may change in the decades to come.**

The Saffir–Simpson hurricane scale appears to be a straightforward tool. However, the accurate observations needed to correctly portray hurricane intensity at the surface are difficult to obtain. Estimating hurricane intensity is difficult because direct surface observations in the eye wall are rarely available. Therefore, winds in this most intense part of the storm have to be estimated. One of the best ways to estimate surface intensity is to adjust the wind speeds measured by reconnaissance aircraft. Other, less accurate techniques make use of data acquired from instruments aboard satellites.

### Dropsondes

Winds aloft are stronger than winds at the surface. Therefore, the adjustment of values determined for winds aloft to values expected at the surface involves *reducing* the measurements made aloft. However, until the late 1990s, determining the proper adjustment factor was problematic because surface observations in the eye wall were too limited to establish a broadly accepted relationship between flight-level and surface winds. In the early 1990s, the reduction factors commonly used ranged from 75 to 80 percent (that is, surface wind speeds were assumed to be between 75 and 80 percent of the speed at 3000 meters [10,000 feet]). Some scientists and engineers even maintained that surface winds were as low as 65 percent of flight-level winds.

Beginning in 1997 a new instrument, called a *Global Positioning System (GPS) dropwindsonde*, came into use. It is often called a **dropsonde** for short. After being released from the aircraft, this package of instruments, slowed by a small parachute, drifts downward through the storm (Fig. 11.14). During the descent, it continuously transmits data on temperature,

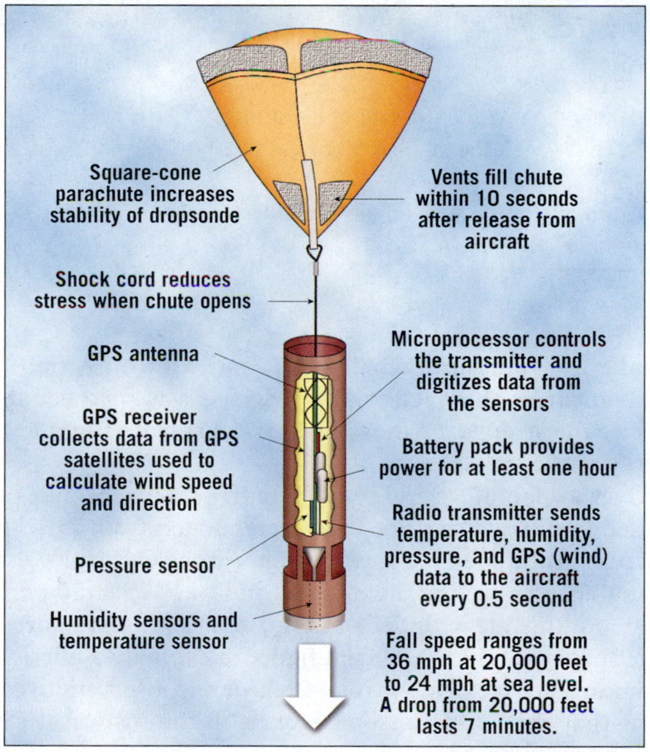
Square-cone parachute increases stability of dropsonde
Vents fill chute within 10 seconds after release from aircraft
Shock cord reduces stress when chute opens
GPS antenna
Microprocessor controls the transmitter and digitizes data from the sensors
GPS receiver collects data from GPS satellites used to calculate wind speed and direction
Battery pack provides power for at least one hour
Radio transmitter sends temperature, humidity, pressure, and GPS (wind) data to the aircraft every 0.5 second
Pressure sensor
Humidity sensors and temperature sensor
Fall speed ranges from 36 mph at 20,000 feet to 24 mph at sea level. A drop from 20,000 feet lasts 7 minutes.

▲ **Figure 11.14 Dropsonde** The Global Positioning System (GPS) dropwindsonde is frequently just called a *dropsonde*. This cylindrical instrument package is roughly 7 centimeters (2.75 inches) in diameter and 40 centimeters (16 inches) long, and it weighs about 0.4 kilogram (0.86 pound). The instrument package is released from an aircraft and falls through the storm via a parachute, making and transmitting measurements of temperature, pressure, winds, and humidity every half-second.

## eye ON THE atmosphere 11.2

This satellite image shows Tropical Cyclone Favio as it came ashore along the coast of Mozambique, Africa, on February 22, 2007. This powerful storm was moving from east to west. Portions of the cyclone had sustained winds of 203 kilometers (126 miles) per hour as it made landfall.

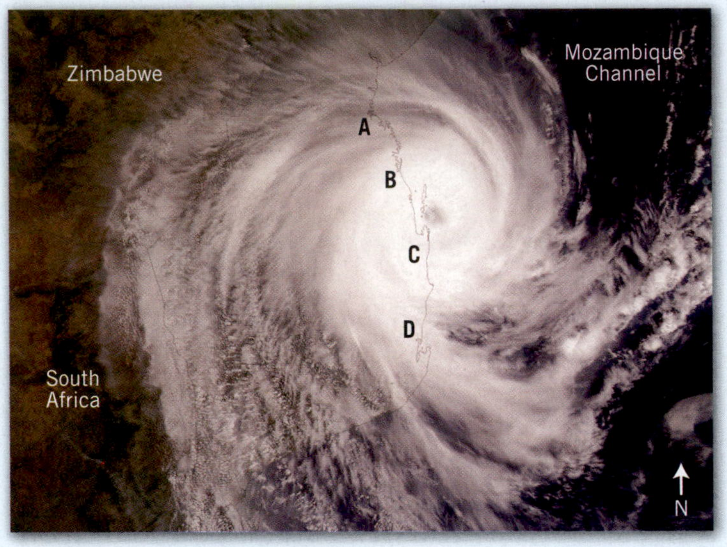


**Questions**

1. Identify the eye and eye wall of the storm.
2. Based on wind speed, classify the storm using the Saffir–Simpson scale.
3. Which one of the lettered sites should experience the strongest storm surge? Explain.

humidity, air pressure, wind speed, and wind direction. The development of this technology provided a way to accurately measure the strongest winds in a hurricane, from flight level all the way to the surface.

Over a span of several years, hundreds of dropsondes were released in hurricanes. The data accumulated from these trials showed that the speed of surface winds in the eye wall averaged about 90 percent of the flight-level winds, not 75 to 80 percent. Based on this new understanding, the National Hurricane Center now uses the 90 percent figure to estimate a hurricane's maximum surface winds from flight-level observations. This means that the winds in some storms in the historical record were underestimated.

For example, in 1992 the surface winds in Hurricane Andrew were estimated to be 233 kilometers (145 miles) per hour. This was 75 to 80 percent of the value measured at 3000 meters (10,000 feet) by the reconnaissance aircraft. When scientists at the National Hurricane Center reevaluated the storm using the 90 percent value, they concluded that the maximum sustained surface winds had been 266 kilometers (165 miles) per hour—33 kilometers (20 miles) per hour faster than the original 1992 estimate. Consequently, in August 2002 the intensity of Hurricane Andrew was officially changed from category 4 to category 5. The upgrade makes Hurricane Andrew only the third category 5 storm on record to strike the continental United States.

## Using Satellite Data

In recent years two methods of using satellite-acquired data to monitor hurricane intensity have been developed. One technique involves using instruments aboard a satellite to estimate wind speeds within a storm. A second method uses satellites to identify areas of extraordinary cloud development, called *hot towers*, in the eye wall of an approaching hurricane.

**Estimating Winds** Instruments called **scatterometers** are a type of radar that can measure the strength of a storm's winds. The image in Figure 11.15 was created using data from the *Oceansat-2* satellite. Scientists applied an experimental technique to determine that Super Typhoon Haiyan's winds peaked at 206 kilometers (128 miles) per hour at the time the measurements were taken. However, the maximum winds were likely stronger because the technique averages data over an area of more than 570 square kilometers (225 square miles), which yields a value lower than the storm's absolute maximum sustained winds. Scientists estimated that maximum wind speeds were about 20 percent higher—about 240 kilometers (150 miles) per hour—at the time the data were acquired.

Scatterometers have a resolution of 25 to 50 kilometers (15 to 30 miles), so they are not capable of resolving a storm's

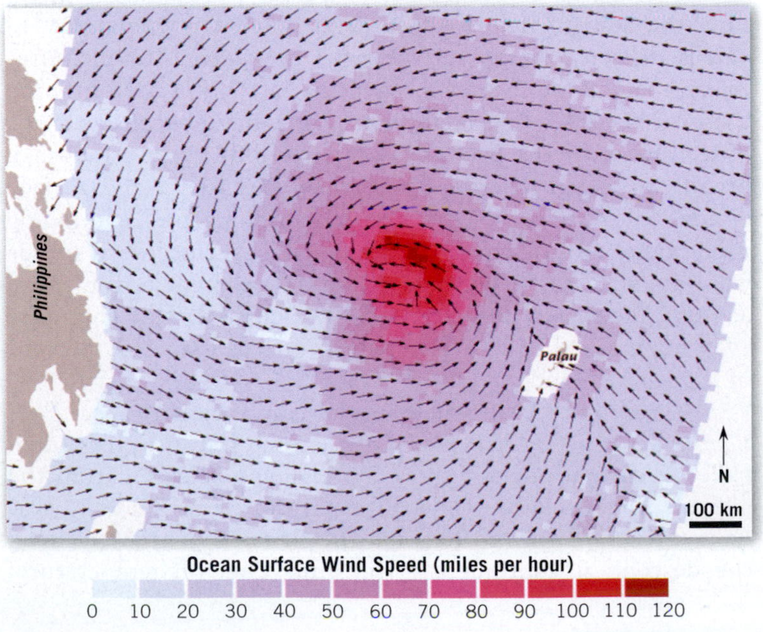


► **Figure 11.15 Measuring Haiyan's winds** The scatterometer aboard the *Oceansat-2* satellite was used to measure the winds of Super Typhoon Haiyan on November 7, 2013. Arrows show wind direction, and colors indicate wind speed, with darker shades indicating stronger winds.

maximum winds. Scatterometers do not directly measure the wind; rather, they work on the principle of wind roughening the ocean's surface. The instrument detects differences in how radiation is scattered by the ocean surface. Then a complex model is used to determine what wind speed would be responsible for that amount of roughness.

### Monitoring Rainfall and Cloud Heights

Figure 11.16 provides a unique satellite perspective. Seeing the pattern of rainfall in different parts of a hurricane is very useful for forecasters because it helps determine the strength of the storm. Scientists have developed a way to process data from the Precipitation Radar (PR) aboard the *Tropical Rainfall Measuring Mission* (*TRMM*) satellite within 3 hours and display it in 3-D. Every time the satellite passes over the same tropical cyclone anywhere in the world, the PR instrument sends data to create a 3-D snapshot of the storm. These images provide information on how heavily the rain is falling in different parts of the storm, such as the eye wall versus the outer rain bands. They also give a 3-D look at cloud heights. Using this technology, NASA scientists discovered that massive cumulonimbus clouds called **hot towers**, sometimes more than 12 kilometers (7 miles) high, develop within an eye wall 6 hours before a hurricane intensifies. This was the case for Hurricane Katrina in Figure 11.16. In February 2014, the *Global Precipitation Mission* (GPM) was launched with more advanced versions of the instruments used by TRMM

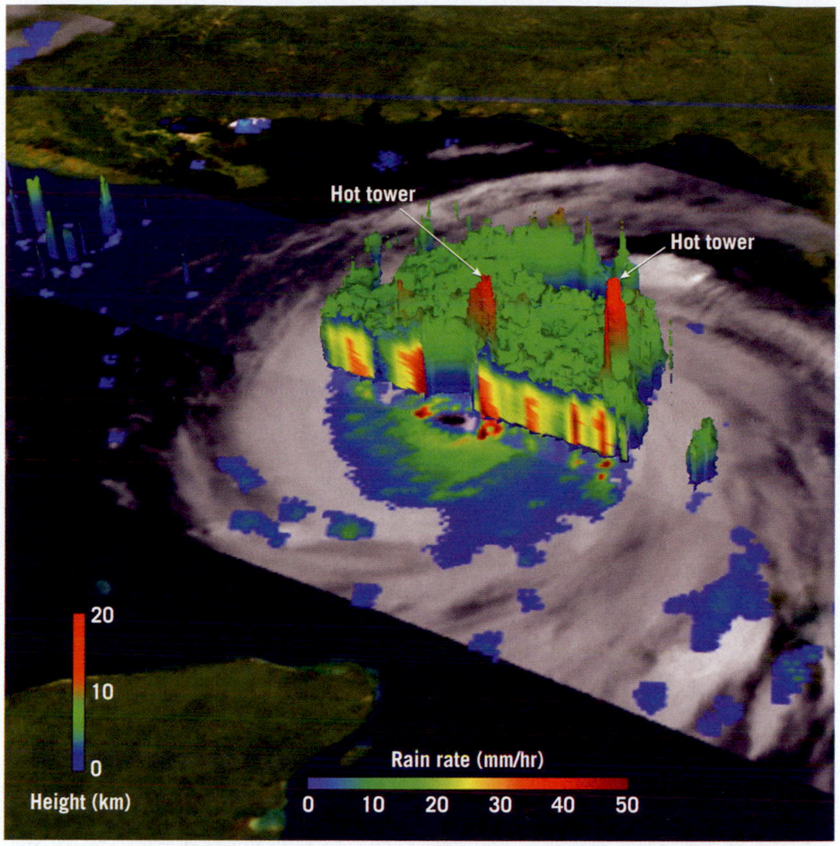


▲ **Figure 11.16 Hot towers** This *Tropical Rainfall Measuring Mission* (*TRMM*) satellite image of Hurricane Katrina was acquired early on August 28, 2005. The cutaway view of the inner portion of the storm shows cloud height on one side and rainfall rates on the other. Two hot towers (in red) are visible: one in an outer rain band and the other in the eye wall. The eye wall tower rises 16 kilometers (10 miles) above the ocean surface and is associated with an area of intense rainfall. Towers this tall near the core often indicate that a storm is intensifying. Katrina grew from a category 3 to a category 4 storm soon after this image was received.

Video MM
Hot Towers and Hurricane Intensification
http://goo.gl/jJmpo

### ✔ Concept Checks 11.4

1. Why is estimating the surface intensity of a hurricane difficult?
2. List and briefly describe three ways of estimating hurricane intensity.

## 11.5 Detecting, Tracking, and Monitoring Hurricanes

**List four tools that provide data used to track hurricanes and develop forecasts.**
**Contrast *hurricane watch* and *hurricane warning*.**

Figure 11.17 shows the paths followed by some notable Atlantic hurricanes. What determines these tracks? The storms can be thought of as being steered by the surrounding environmental flow throughout the depth of the troposphere. The movement of hurricanes has been likened to a leaf being carried along by the currents in a stream, except that for a hurricane, the "stream" has no set boundaries.

In the latitude zone equatorward of about 25° north, tropical storms and hurricanes commonly move to the west, with a slight poleward component. This occurs because of the semipermanent cell of high pressure (called the *Bermuda/Azores High*) that is positioned poleward of the storm (see Figure 7.10B, page 196). On the equatorward side of this high-pressure center, easterly winds prevail, guiding the storms westward. If this high-pressure center is weak in the western Atlantic, the storm often turns northward. On the poleward side of the Bermuda High, westerly winds prevail, steering the storm back toward the east. Often it is difficult to determine whether the storm will curve back out to sea or continue straight ahead and make landfall.

A location only a few hundred kilometers from a hurricane—just a day's striking distance away—may experience clear skies and virtually no wind. Before the age of weather satellites, such a situation made it difficult to warn people of impending storms. The worst natural disaster in U.S. history came as a result of a hurricane that struck an unprepared Galveston, Texas, on September 8, 1900. The strength of the storm, together with the

Video (MM) Improving Hurricane Predictions
http://goo.gl/DLMR8T

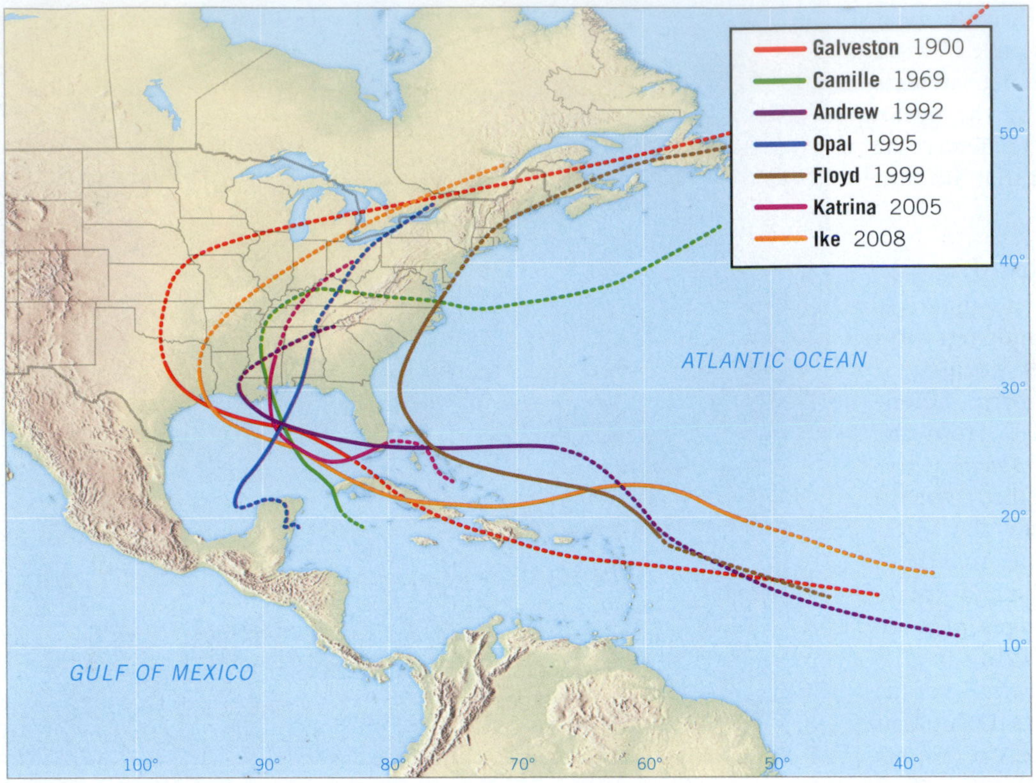

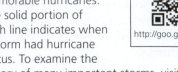

◀ **Figure 11.17 Storm tracks** This map shows a variety of tracks for some memorable hurricanes. The solid portion of each line indicates when a storm had hurricane status. To examine the history of many important storms, visit this interesting interactive site: www.csc.noaa.gov/hurricanes/.

lack of adequate warning, caught the population by surprise and cost the lives of 6000 people in the city and at least 2000 more elsewhere (**Fig. 11.18**).*

In the United States, early warning systems have greatly reduced the number of deaths caused by hurricanes. At the same time, however, property damage rose astronomically. The primary reason for this latter trend is the rapid population growth and accompanying development in coastal areas. Close to 100 million people live within 80 kilometers (50 miles) of a coastline and are thus exposed to the potential destruction associated with a landfalling hurricane.

## The Role of Satellites

Today many different tools provide data that are used to detect and track hurricanes. The information they provide is used to develop forecasts and to issue watches and warnings. The greatest single advancement in tools used to observe tropical cyclones has been the development of meteorological satellites.

Because the tropical and subtropical regions that spawn hurricanes consist of enormous areas of open ocean, conventional observations are limited. The need for meteorological data from these vast regions is now met primarily by satellites. Even before a storm begins to develop cyclonic flow and the spiraling cloud bands so typical of a hurricane, the storm can be detected and monitored by satellites.

The advent of weather satellites has largely solved the problem of detecting tropical storms and has significantly improved monitoring. This was demonstrated in the preceding section. However, we also saw that because satellites are remote sensors,

*For a fascinating account of the Galveston storm, read *Isaac's Storm* by Erik Larson (New York: Crown Publishers, 1999).

▼ **Figure 11.18 Aftermath of the Galveston hurricane of 1900** Entire blocks were swept clean, and mountains of debris accumulated around the few remaining buildings.

A.

B.

◀ **Figure 11.19 Aircraft reconnaissance A.** In the Atlantic basin, most operational hurricane reconnaissance is carried out by the U.S. Air Force Reconnaissance Squadron, based at Keesler AFB, Mississippi. Pilots fly through the hurricane to its center, measuring all basic weather elements as well as providing an accurate location of the eye. They use planes like the one in the background of this image. The National Oceanic and Atmospheric Administration uses small, specially equipped jets (foreground) on research missions to aid scientists in better understanding these storms. **B.** The unstaffed drone aircraft *Global Hawk* is 13.2 meters (44 feet) long and weighs 11,250 kilograms (25,000 pounds). It can climb to heights of 19.5 kilometers (65,000 feet) and remain in the air for up to 30 hours.

Video MM
Interview with a Chase Pilot
http://goo.gl/dNcCfB

it is not unusual for wind-speed estimates to be off by tens of kilometers per hour and for storm-position estimates to have errors. Satellites still cannot precisely determine detailed structural characteristics. A combination of observing systems is necessary to provide the data needed for accurate forecasts and warnings.

## Aircraft Reconnaissance

Aircraft reconnaissance represents a second important source of information about hurricanes. The first experimental flights into hurricanes were made in the 1940s, and the aircraft and the instruments employed have since then become quite sophisticated (**Fig. 11.19A**). When a hurricane is within range, specially instrumented aircraft can fly directly into a threatening storm and accurately measure details of its position and current state of development. Data transmission can be made directly from an aircraft in the midst of a storm to the forecast center, where input from many sources is collected and analyzed.

Measurements from reconnaissance aircraft are limited because they cannot be taken until a hurricane is relatively close to shore. Moreover, measurements are not taken continuously or throughout the storm. Rather, the aircraft provides sample "snapshots" of small parts of the hurricane. Nevertheless, the data collected are critical in analyzing the current characteristics needed to forecast the future behavior of a storm.

A major contribution to hurricane forecasting and warning programs has been an improved understanding of the structure and characteristics of these storms. Although remote sensing from satellites has improved, measurements from reconnaissance aircraft will be required for the foreseeable future to maintain the present level of accuracy for forecasting potentially dangerous tropical storms.

A new form of aircraft reconnaissance is in the developmental stage and makes use of an unstaffed drone aircraft (**Fig. 11.19B**).

The *Hurricane and Severe Storm Sentinel* (*HS3*) is a 5-year program begun by NASA in 2012. This research program is intended to enhance understanding of processes involved in hurricane intensity changes in the Atlantic basin.

## Radar and Data Buoys

Radar is a third basic tool in the observation and study of hurricanes (**Fig. 11.20**). When a hurricane nears the coast, it is monitored by land-based Doppler weather radar (discussed in greater detail in Chapter 10). Doppler radar provides detailed information on hurricane wind fields, rainfall intensity, and storm movement. As a result, local National Weather Service offices are able to provide short-term warnings for floods, tornadoes, and high winds for specific areas. Sophisticated mathematical calculations provide forecasters with important information derived from the radar data, such as estimates of rainfall amounts. A limitation of radar is that it cannot "see" farther than about 320 kilometers (200 miles) from the coast, and hurricane watches and warnings must be issued long before a storm comes into range.

### VORTRAC—A New Hurricane Tracking Technique

Rapidly intensifying storms can catch vulnerable coastal areas by surprise. In 2007 Hurricane Humberto struck near Port Arthur, Texas, after unexpectedly strengthening from a tropical depression to a hurricane in less than 19 hours. In 2004, parts of the southwest coast of Florida were caught off guard when Hurricane Charley's top winds increased from 175 to more than 230 kilometers (110 to 145 miles) per hour in just 6 hours as the storm approached land.

A technique known as VORTRAC (*V*ortex *O*bjective *R*adar *Tr*acking *a*nd *C*irculation) was successfully tested in 2007 and put into use in 2008. It is intended to improve short-term hurricane warnings by capturing sudden intensity changes in potentially dangerous hurricanes during the critical time when storms are nearing land.

Prior to VORTRAC development, the only way to monitor the intensity of a landfalling hurricane was to use aircraft to drop dropsondes into the storm. This can be done only every hour or two, though, so a sudden drop in barometric pressure and the accompanying increase in winds can be difficult to detect in a timely manner.

VORTRAC uses the portion of the Doppler radar network that is along the Gulf and Atlantic coastline from Texas to Maine. Each unit can measure winds blowing toward or away

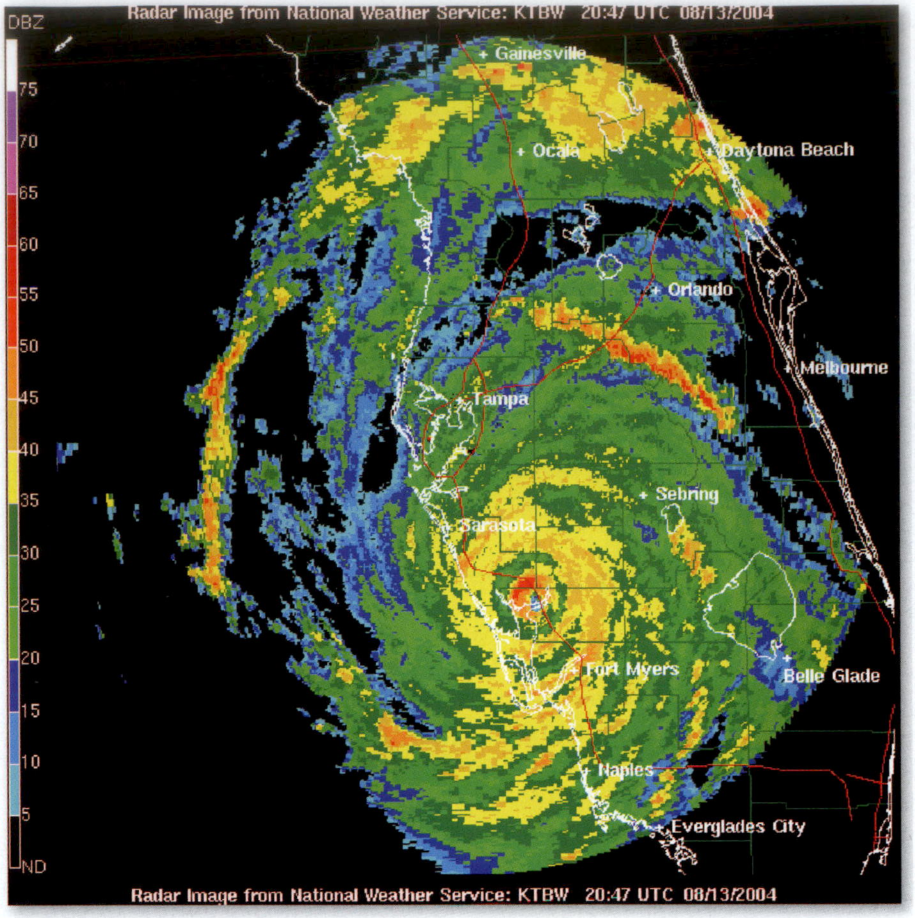


▲ **Figure 11.20 Coastal radar** Doppler radar image of Hurricane Charley over Charlotte Harbor, Florida, just after landfall on August 13, 2004. The range of coastal radar is about 320 kilometers (200 miles).

## Hurricane Watches and Warnings

Using input from the observational tools just described in conjunction with sophisticated computer models, meteorologists attempt to forecast the movements and intensity of a hurricane. The goal is to issue timely watches and warnings.

A **hurricane watch** is an announcement that hurricane conditions are *possible* in a specified coastal area. Because preparedness activities become difficult once winds reach tropical storm force, a hurricane watch is issued 48 hours in advance of the anticipated onset of tropical storm–force winds. By contrast, a **hurricane warning** is issued 36 hours in advance and indicates that hurricane conditions are *expected* somewhere within a specified coastal area. A hurricane warning can remain in effect when dangerously high water or a combination of dangerously high water and exceptionally high waves continue, even though winds may be less than hurricane force.

Two factors are especially important in the watch-and-warning decision process. First, adequate lead time must be provided to protect life and, to a lesser degree, property. Second, forecasters must attempt to keep overwarning at a minimum. This, however, can be a difficult task. Clearly, the decision to issue a warning involves striking a delicate balance between the need to protect the public on the one hand and the desire to minimize the degree of overwarning on the other.

from it, but until VORTRAC was established, no single radar could estimate a hurricane's rotational winds and central pressure. The technique uses a series of mathematical formulas that combine data from a single radar with knowledge of Atlantic hurricane structure in order to map the storm's rotational winds. VORTRAC also infers the barometric pressure in the eye of the hurricane, which is a reliable indicator of its strength (**Fig. 11.21**). Each radar can sample conditions out to a distance of 190 kilometers (120 miles), and forecasters using VORTRAC can update the status of a storm about every 6 minutes.

**Data Buoys** Data buoys provide a fourth method of gathering data for the study of hurricanes (see Figure 12.4, page 327). These remote, floating instrument packages are positioned in fixed locations all along the Gulf and Atlantic coasts of the United States. When you examine the weather maps in Figure 11.3, you can see data buoy information plotted at several offshore stations. Since the early 1970s data provided by these units have become a dependable and routine part of daily weather analysis as well as an important element of the hurricane warning system. The buoys provide the only means of making nearly continuous direct measurements of surface conditions over ocean areas.

▼ **Figure 11.21 VORTRAC** This graph displays the results of a test of the VORTRAC system on Hurricane Humberto, a storm that intensified rapidly as it approached the Texas/Louisiana coast on September 13, 2007. VORTRAC estimates agreed closely with Humberto's actual central pressure, as measured by reconnaissance planes just before landfall. VORTRAC also captured the storm's rapid intensification, indicating that the pressure began falling quickly near the end of a 4-hour span when no aircraft data were available.

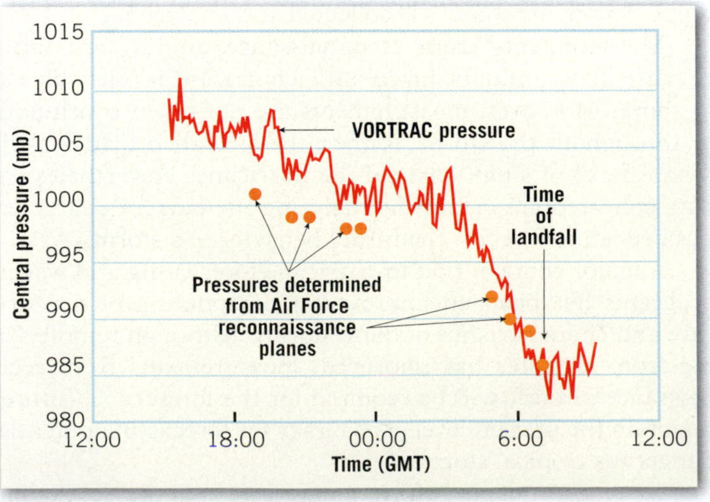

severe& hazardous weather Box 11.1

## Super Typhoon Haiyan

The chapter-opening image shows Super Typhoon Haiyan as it moved across the central Philippines on November 8, 2013. Shortly before the storm struck, the Joint Typhoon Warning Center (JTWC) in Hawaii estimated the winds to be 315 kilometers (195 miles) per hour, making it a category 5 storm on the Saffir–Simpson scale (**Fig. 11.D**). The JTWC is a U.S. Navy and Air Force task force that monitors storms in the Pacific and Indian Oceans. If its estimate was accurate, Haiyan was the strongest storm on record ever to make landfall. Can we be sure about the strength of Haiyan's winds? Scientists cannot be certain because there were few direct wind-speed measurements. Estimates were based on data gathered remotely by satellite. See Section 11.4 for more about using satellite data to determine wind speeds.

Haiyan was the strongest storm on record ever to make landfall.

Whereas Haiyan's sustained winds were estimated to be 315 kilometers (195 miles) per hour, gusts may have reached 370 kilometers (230 miles) per hour. Imagine billions of raindrops and flying debris moving at that speed!

It is a rarity that a storm just happens to reach peak intensity as it is making landfall along a densely populated coast, but that was the case with Haiyan. It was the deadliest Philippine typhoon on record. Officially the death toll reached 6300, but the loss of life was undoubtedly much higher, as thousands remain unaccounted for. In addition, 2 million people lost their homes, and millions more were displaced. Because roads were blocked with debris or completely destroyed, delivery of humanitarian aid to the millions in need was very difficult.

Although winds were extreme, the major cause of loss of life and property was Haiyan's storm surge. At Tacloban, perhaps the hardest-hit city, the wall of water is estimated to have been as high as 7.5 meters (nearly 25 feet). Because much of the city is less than 5 meters (16 feet) above sea level, the result was catastrophic.

In addition to the fierce winds and powerful storm surge, Haiyan produced copious rainfall in the central Philippines. The super typhoon struck just a few days after a tropical storm and a tropical disturbance had passed through the region. Over a 10-day span, the combined rainfall from the three closely-spaced events exceeded 50 centimeters (nearly 20 inches) in many areas, with a peak of more than 68 centimeters (27 inches) in some places (**Fig. 11.E**). Dangerous flooding and mudflows in this mountainous country were widespread.

### Questions

1. Why is there some uncertainty about the strength of Haiyan's winds?
2. Which types of hurricane destruction occurred in the Philippines as a result of Super Typhoon Haiyan?

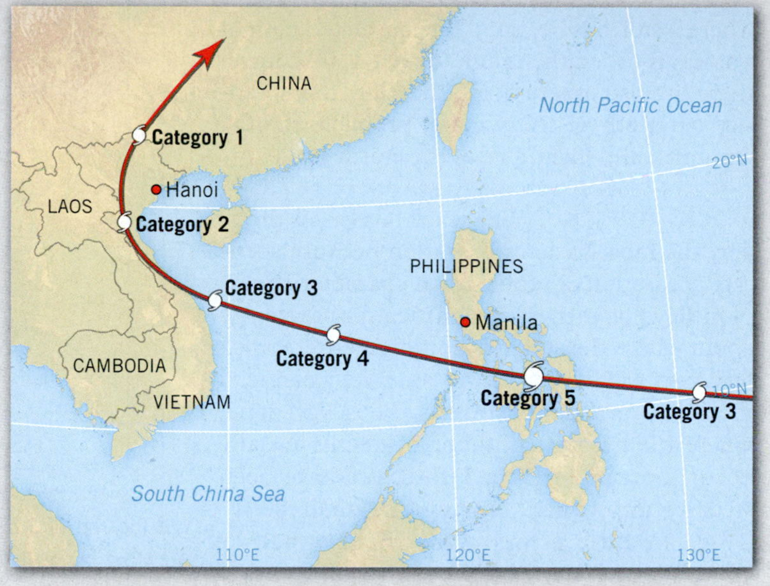


▲ **Figure 11.D Haiyan's path** The map shows the path of this very strong storm. Numbers indicate the intensity of the storm on the Saffir–Simpson scale. After striking the central Philippines, Haiyan made landfall again in Vietnam.

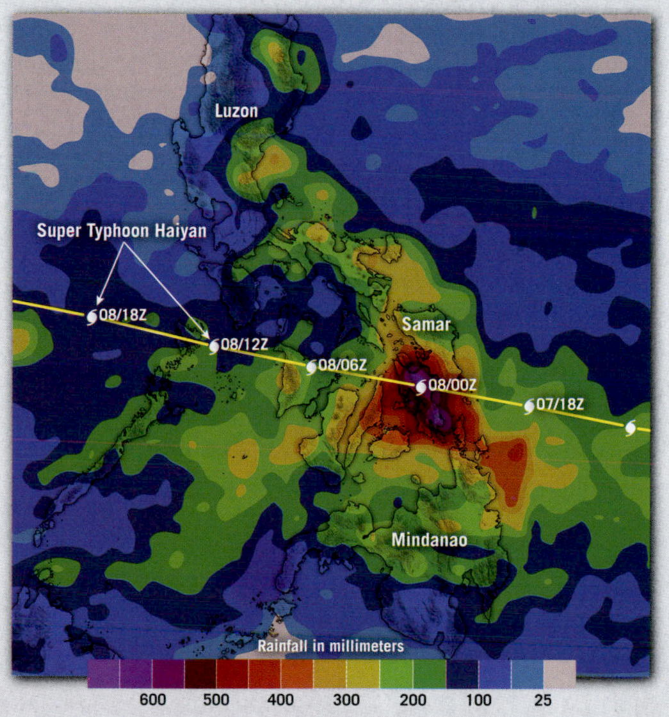


▲ **Figure 11.E Flooding rains** This map shows rainfall totals in millimeters over a 10-day span in November 2013 caused by a tropical disturbance, followed by Tropical Storm Thirty and Super Typhoon Haiyan. Data are from the *Tropical Rainfall Measuring Mission* satellite.

## Hurricane Forecasting

Hurricane forecasts are a basic part of any warning program. Several aspects can be part of such a forecast. We certainly want to know where a storm is headed. The predicted path of a storm is called the **track forecast**. Of course, there is also interest in knowing the intensity (strength of the winds), probable rainfall amounts, and likely size of the storm surge.

**Track Forecasts** The track forecast is probably the most basic information because accurate prediction of other storm characteristics is of little value if there is significant uncertainty about where the storm is going. Accurate track forecasts are important because they can lead to timely evacuations from the surge zone, where the greatest number of deaths usually occur. Fortunately, track forecasts have been steadily improving. During the span 2001–2005, forecast errors were roughly half of what they were in 1990. During the very active 2004 and 2005 Atlantic hurricane seasons, 12- to 72-hour track forecast accuracy was at or near record levels. Consequently, the length of official track forecasts issued by the National Hurricane Center was extended from 3 days to 5 days (Fig. 11.22). Today, 5-day track forecasts are as accurate as the 3-day forecasts of 15 years ago. Much of the progress is due to improved computer models and a dramatic increase in the quantity of satellite data from over the oceans.

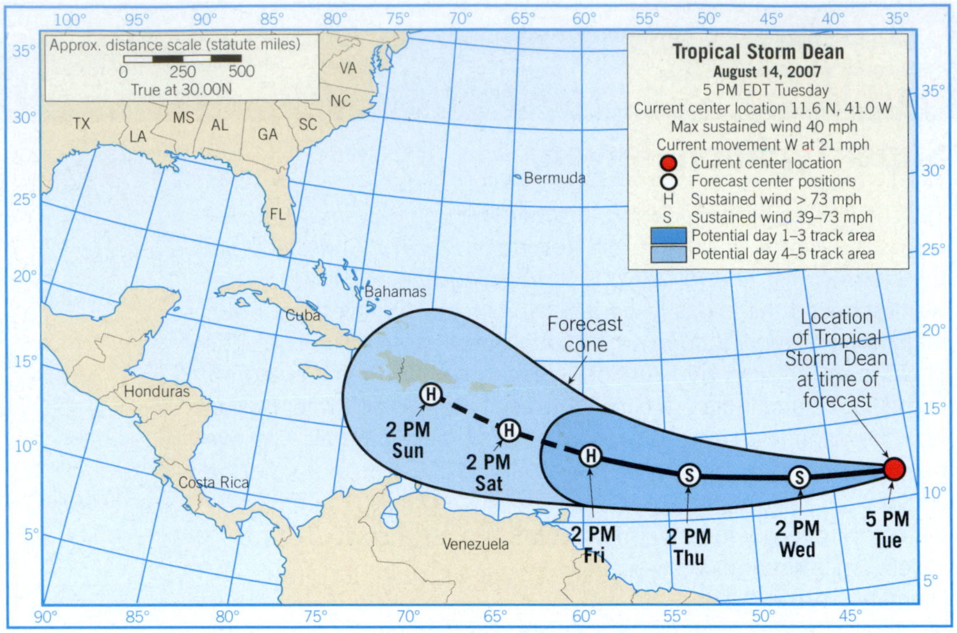


▲ **Figure 11.22 Five-day track forecast for Tropical Storm Dean issued at 5 P.M. Tuesday, August 14, 2007** When a hurricane track forecast is issued by the National Hurricane Center, it is termed a *forecast cone*. The cone represents the probable track of the center of the storm and is formed by enclosing the area swept out by a set of circles along the forecast track (at 12 hours, 24 hours, 36 hours, and so on). The size of each circle gets larger with time. Based on statistics from 2003–2007, the entire track of an Atlantic tropical cyclone can be expected to remain entirely within the cone roughly 60 to 70 percent of the time.

Despite improvements in accuracy, forecast uncertainty still requires that hurricane warnings be issued for relatively large coastal areas. During the span 2000–2005, the average length of coastline under a hurricane warning in the United States was 510 kilometers (316 miles). This represents a significant improvement over the preceding decade, when the average was 730 kilometers (452 miles). Nevertheless, only about one-quarter of an average warning area experiences hurricane conditions.

**Forecasting Other Characteristics** In contrast to the improvements in track forecasts, errors in forecasts of hurricane intensity (wind speeds) have not changed significantly in 30 years. Accurate predictions of rainfall as hurricanes make landfall also remains elusive. However, accurate predictions of impending storm surge are possible when good information regarding the storm's track and surface wind structure are known and reliable data regarding coastal and offshore (underwater) topography are available.

**Video** MM
2013 Atlantic Hurricane Season Outlook
http://goo.gl/FwNuYu

### ✔ Concept Checks 11.5

1. What was the worst natural disaster in U.S. history? Why is such an event unlikely to occur in the United States again?
2. List four tools that provide data used to track hurricanes and develop forecasts.
3. Distinguish between a *hurricane watch* and a *hurricane warning*.
4. What is a track forecast? Why are such forecasts important?

# 11 Concepts in Review Hurricanes

## 11.1 Profile of a Hurricane

▶ Define *hurricane* and describe the basic structure and characteristics of this storm.

**Key Terms:** hurricane, sustained winds, eye wall, rain band, eye

- Hurricanes are intense centers of low pressure that form over warm oceans in the tropics and subtropics. Sustained winds must equal or exceed 119 kilometers (74 miles) per hour.
- Most hurricanes form between latitudes 5° and 30° over all tropical oceans except the South Atlantic and eastern South Pacific. The

North Pacific has the greatest number of storms, averaging 20 per year. In the northwestern Pacific, hurricanes are called *typhoons*, and in the southwestern Pacific and Indian Oceans, they are referred to as *cyclones*.

- A steep pressure gradient generates the rapid, inward-spiraling winds of a hurricane. As the warm, moist air approaches the core of the storm, it turns upward and ascends in a ring of cumulonimbus towers and forms a doughnut-shaped wall called the eye wall.
- At the very center of the storm, called the eye, the air gradually descends, precipitation ceases, and winds subside.

**Q** This image, taken from the International Space Station, shows the inner portion of Hurricane Igor in September 2010. Identify the eye and the eye wall. In which of these zones are winds strongest? In which zone is rainfall intensity greatest?

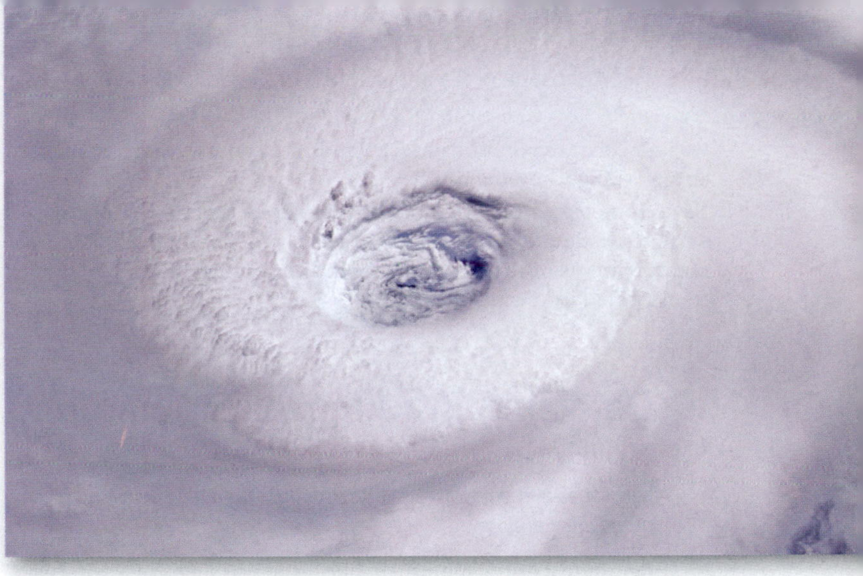

## 11.2 Hurricane Formation and Decay

► Discuss the conditions that promote hurricane formation and list the factors that cause hurricanes to dissipate.

**Key Terms:** tropical disturbance, easterly wave, streamline, trade wind inversion, tropical depression, tropical storm

- A hurricane is a heat engine fueled by the latent heat liberated when huge quantities of water vapor condense. Hurricanes develop most often in late summer, when ocean waters have reached temperatures of 27°C (80°F) or higher and can provide the necessary heat and moisture to the air.
- The initial stage of a tropical storm's life cycle, called a tropical disturbance, is a disorganized array of clouds that exhibits a weak pressure gradient and little or no rotation. Tropical disturbances that produce many of the strongest hurricanes that threaten North America often begin as large undulations or ripples in the trade winds, called easterly waves.
- Each year, only a few tropical disturbances develop into full-fledged hurricanes (storms for which minimum wind speeds of 119 kilometers per hour are required). When a cyclone's strongest winds do not exceed 63 kilometers per hour, it is called a *tropical depression*. When winds are between 63 and 119 kilometers per hour, the cyclone is termed a *tropical storm*.
- Hurricanes diminish in intensity when they (1) move over cool ocean waters that cannot supply warm, moist tropical air; (2) move onto land; or (3) reach a location where large-scale flow aloft is unfavorable.

**Q** This graph shows the distribution of hurricanes in the Atlantic basin between May and December. Why is the occurrence of hurricanes low in early summer?

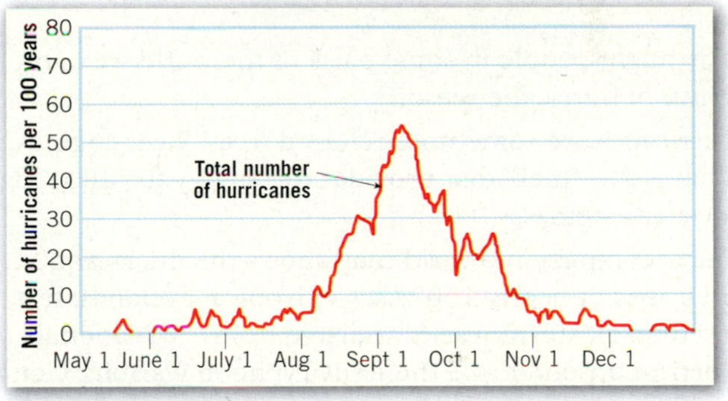


## 11.3 Hurricane Destruction

► Explain how hurricane intensity is determined and summarize the three broad categories of hurricane destruction.

**Key Terms:** Saffir–Simpson scale, storm surge

- Although damages caused by a hurricane depend on several factors, including the size and population density of the area affected and the nearshore ocean-bottom configuration, the most significant factor is the strength of the storm.
- The Saffir–Simpson scale ranks the relative intensities of hurricanes. A 5 on the scale represents the strongest storm possible, and a 1 is the least severe. Category assignments are based on observed conditions at a particular stage in the life of the storm.
- Along the coast, storm surge is responsible for the greatest loss of life and property damage. This flooding wall of water is created by the strong onshore winds associated with certain portions of the storm.
- Unlike storm surge, which is confined to the coast, destruction by wind can cover a broad area. Sometimes wind damage results from tornadoes spawned by the hurricane.
- Torrential rains and flooding are often associated with hurricanes. Long after a storm has come ashore and lost its hurricane-force winds, it may still produce copious rain.

**Q** Which side of an advancing hurricane in the Northern Hemisphere has the strongest winds and highest storm surge—right or left? Explain.

## 11.4 Estimating Hurricane Intensity

► Compare two methods used to determine the intensity of a hurricane and explain why storm intensity may change in the decades to come.

**Key Terms:** dropsonde, scatterometer, hot tower

- Estimating the intensity of an approaching hurricane is difficult because surface observations in the eye wall are seldom available.
- Dropsondes are instrument packages released by reconnaissance airplanes flying high in the storm. During descent, the dropsonde continuously transmits data. Based on abundant dropsonde data,

we know that maximum surface winds are about 90 percent as strong as flight-level winds.

- Scatterometers, a type of radar aboard some ocean-monitoring satellites, estimate surface winds. By applying a complex model, measurements of the roughness of the ocean surface are transformed into wind speeds.
- Instruments aboard the *Tropical Rainfall Measuring Mission* satellite can produce 3-D images that show the development of especially tall rain columns, called hot towers, within a storm. Such towers often signal that a hurricane is intensifying.

## 11.5 Detecting, Tracking, and Monitoring Hurricanes

► List four tools that provide data used to track hurricanes and develop forecasts. Contrast *hurricane watch* and *hurricane warning.*

**Key Terms:** hurricane watch, hurricane warning, track forecast

- North Atlantic hurricanes develop in the trade winds, which generally move these storms from east to west. Early warning systems that help detect and track hurricanes have greatly reduced the number of deaths associated with these violent storms.
- Because the tropical and subtropical regions that spawn hurricanes consist of enormous areas of open oceans, meteorological data from these vast regions are provided primarily by satellites. Other important sources of hurricane information are aircraft reconnaissance, radar, and remote, floating instrument platforms called data buoys.
- Hurricane watches and warnings alert coastal residents to possible or expected hurricane conditions. Two important factors in the watch-and-warning decision process are providing adequate lead time and attempting to keep overwarning to a minimum.
- Track forecasts identify potential areas where hurricane landfall may occur so that evacuations can occur in a timely manner. Track forecasts have improved but still require that hurricane warnings be issued for relatively extensive coastal zones.

## Give it Some Thought

1. Why might people in some parts of the world welcome the arrival of hurricane season?
2. Hurricanes are sometimes referred to as "heat engines." What is the "fuel" that provides the energy for these high-powered engines?
3. The accompanying world map shows the tracks and intensities of nearly 150 years of tropical cyclones. It is based on all storm tracks available from the National Hurricane Center and the Joint Typhoon Warning Center.
   a. What area has experienced the greatest number of category 4 and 5 storms?
   b. Why do hurricanes *not form* in the very heart of the tropics, astride the equator?
   c. Explain the absence of storms in the South Atlantic and eastern South Pacific.

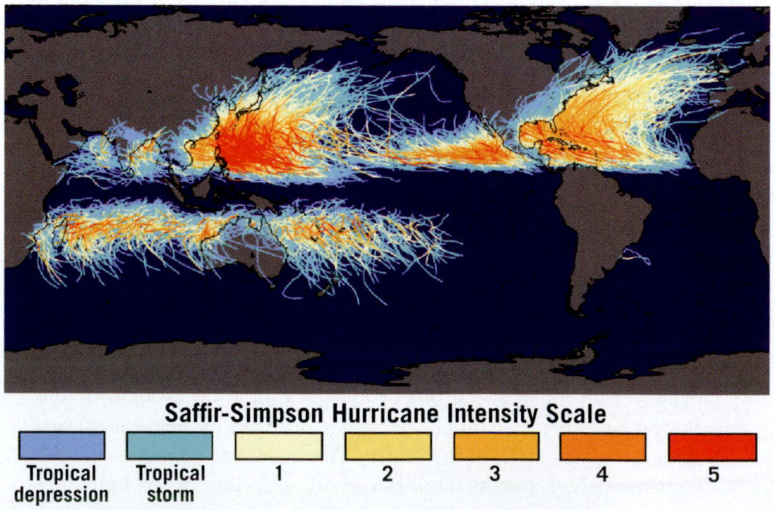


4. Refer to the graph in Figure 11.4B. Explain why wind speeds are greatest when the slope of the pressure curve is steepest.
5. Although observational tools and hurricane forecasts continue to improve, the potential for loss of life due to hurricanes is likely growing. Suggest a reason for this apparent contradiction.
6. Assume that it is late September 2019, and Hurricane Humberto, a category 5 storm, is projected to follow the path shown on the accompanying map. Answer the following questions.

   a. Name the stages of development that Humberto must have gone through to become a hurricane. At what point did it receive its name?
   b. Should the city of Houston expect to experience Humberto's fastest winds and greatest storm surge? Explain why or why not.
   c. What is the greatest threat to life and property if this storm approaches the Dallas–Fort Worth area?

7. Examine the accompanying photo, which shows destruction caused by a strong category 4 hurricane. Which one of the three basic classes of damage was most likely responsible? What is your reasoning?
8. A television meteorologist is able to inform viewers about the intensity of an approaching hurricane. However, the meteorologist can report the intensity of a tornado only *after* it has occurred. Why is this true?

## Problems

Questions 1–5 refer to the weather maps of Hurricane Fran in Figure 11.3.

1. On which of the 2 days were Fran's wind speeds probably highest? How were you able to determine this?
2. **a.** How far did the center of the hurricane move during the 24-hour period represented by these maps?
   **b.** At what rate, in miles per hour, did the storm move during this 24-hour span?
3. The midlatitude cyclone shown in Figure 9.21, page 259 has an east–west diameter of approximately 1200 miles (when the 1008-millibar isobar is used to define the outer boundary of the low). Measure the diameter (north–south) of Hurricane Fran on September 5. Use the 1008-millibar isobar to represent the outer edge of the storm. How does this figure compare to the midlatitude cyclone?
4. Determine the pressure gradient for Hurricane Fran on September 5. Measure from the 1008-millibar isobar at Charleston to the center of the storm. Express your answer in millibars per 100 miles.
5. The weather map in Figure 9.21 shows a well-developed midlatitude cyclone. Calculate the pressure gradient of this storm from the 1008-millibar isobar in western Wyoming to the center of the low. Assume that the pressure at the center of the storm is 986 millibars and the distance is 625 miles. Express your answer in millibars per 100 miles. How does this answer compare to your answer to Problem 4?
6. Hurricane Rita was a major storm that struck the Gulf coast in late September 2005, less than a month after Hurricane Katrina. The accompanying graph shows changes in air pressure and wind speed from the storm's beginning as an unnamed tropical disturbance north of the Dominican Republic on September 18 until its last remnants faded away in Illinois on September 26. Use the graph to answer these questions.
   **a.** Which line represents air pressure, and which line represents wind speed? How did you figure this out?
   **b.** What was the storm's maximum wind speed, in knots? Convert this answer to kilometers per hour by multiplying by 1.85.
   **c.** What was the lowest pressure attained by Hurricane Rita?
   **d.** Using wind speed as your guide, what was the highest category reached on the Saffir–Simpson scale? On what day was this status reached?
   **e.** When landfall occurred, what was the category of Hurricane Rita?

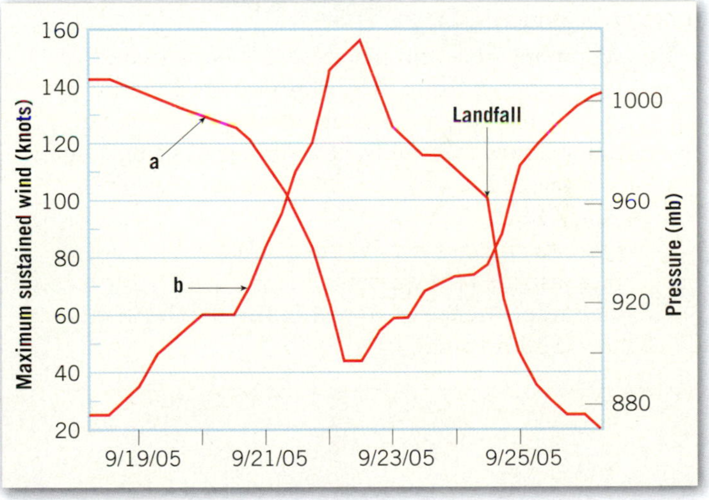

# 12 Weather Analysis and Forecasting

## Focus on Concepts

*Each statement represents the primary learning objective for the corresponding major heading within the chapter. After you complete the chapter, you should be able to:*

**12.1** List and describe the major steps used to generate a weather forecast.

**12.2** Describe the various tools used to acquire weather data.

**12.3** Interpret what a particular pattern in the flow aloft indicates about surface weather.

**12.4** Explain the basis of numerical weather prediction.

**12.5** List and distinguish among the various traditional methods of weather forecasting.

**12.6** Discuss the advantages and disadvantages of infrared imagery and visible imagery generated by weather satellites.

**12.7** Compare and contrast qualitative and quantitative weather forecasts. Differentiate between weather forecasts and 30- and 90-day outlooks.

**12.8** Describe AWIPS and explain how it aids forecasters at local Weather Forecast Offices.

**12.9** Explain why the percentage of accurate forecasts is not always a good measure of forecast skill.

People expect thorough, accurate weather forecasts. The desire for reliable weather predictions ranges from NASA's need to evaluate conditions leading up to a satellite launch to families wondering if the upcoming weekend weather will be suitable for a beach outing. Such diverse industries as airlines and fruit growers depend heavily on accurate weather forecasts. In addition, the designs of buildings, oil platforms, and industrial facilities rely on a sound knowledge of the atmosphere in its most extreme forms, including thunderstorms, tornadoes, and hurricanes. We are no longer satisfied with receiving only short-range predictions but expect accurate long-range forecasts as well.

*Snow covered street in New Haven, Connecticut, February 9, 2013.*

## 12.1 The Weather Business: A Brief Overview

**List and describe the major steps used to generate a weather forecast.**

The first U.S. weather bureau was created in 1870, partially in response to the need for storm warnings on the Great Lakes. Today, the U.S. government agency responsible for gathering and disseminating weather-related information is the **National Weather Service (NWS)**, a branch of the *National Oceanic and Atmospheric Administration* (*NOAA*) (Fig. 12.1). Perhaps the most important services provided by the NWS are forecasts and warnings of hazardous weather, including thunderstorms, floods, hurricanes, tornadoes, winter weather, and extreme heat (Fig. 12.2). According to the *Federal Emergency Management Agency* (*FEMA*), 80 percent of all declared emergencies are weather related. Similarly, the U.S. Department of Transportation reports that more than 6250 vehicular fatalities per year can be attributed to weather. As global populations increase, the economic impact of weather-related phenomena also escalates. As a result, the NWS is under greater pressure to provide more accurate and longer-range forecasts.

### Producing a Weather Forecast

What is a weather forecast? Simply, a **weather forecast** is a scientific estimate of the weather conditions at some future time.

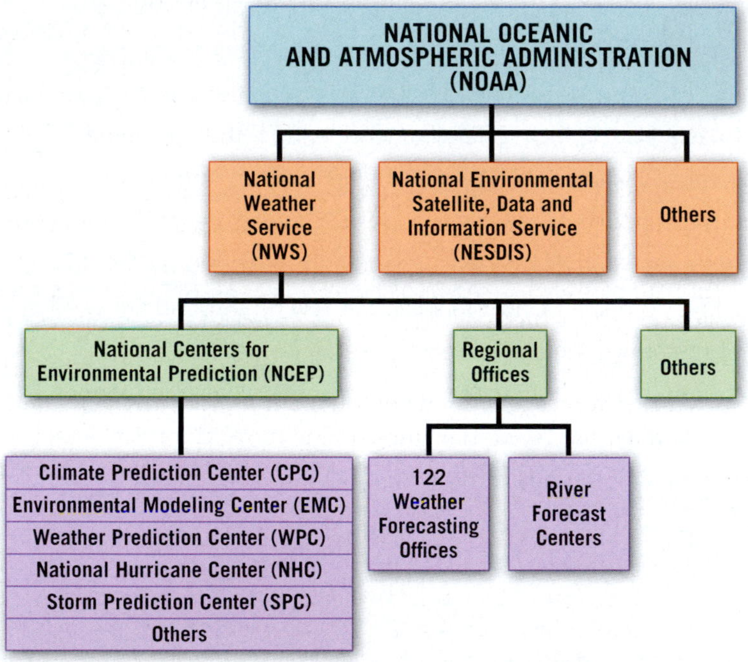


▲ **Figure 12.1 Partial organizational chart for National Oceanic and Atmospheric Administration (NOAA)** This chart shows the relationship between the parent organization (NOAA) and the National Weather Service (NWS) and the NWS branches responsible for collecting and analyzing weather data, as well as producing and disseminating weather forecasts.

Forecasts are usually expressed in terms of the most significant weather variables, including temperature, cloudiness, humidity, precipitation, wind speed, and wind direction. As the following quote illustrates, weather forecasting is a formidable task:

> Imagine a system on a rotating sphere that is 8000 miles wide, consists of different materials, different gases that have different properties (one of the most important of which, water, exists in different concentrations), heated by a nuclear reactor 93 million miles away. Then, just to make life interesting, this sphere is oriented such that, as it revolves around the nuclear reactor, it is heated differently at different times of the year. Then, someone is asked to watch the mixture of gases, a fluid only 20 miles deep, that covers an area of 250 million square miles, and to predict the state of the fluid at one point on the sphere 2 days from now. This is the problem weather forecasters face.*

The first phase of forecasting involves the complicated and detailed process of *collecting* and *assimilating* weather data on a global scale. In the United States, this task is performed by a division of the NWS called the **National Centers for Environmental Prediction (NCEP)**, located in College Park, Maryland. The NCEP is also responsible for using these data to make predictions about the future state of the atmosphere. In Canada the **Canadian Meteorological Centre** performs these tasks.

When billions of pieces of observational data are collected, meteorologists fine-tune the data by correcting as many errors and omissions as possible. This important step ensures the most accurate assessment possible of the current atmospheric conditions, and it also involves smoothing the data to make it compatible with computer models that meteorologists use to make forecasts. Simultaneously, the data are displayed on surface weather maps and upper-level charts that forecasters can easily comprehend. When meteorologists prepare weather maps, they also analyze them and ask questions such as, "Where are the major weather systems located, and how are they changing?" This phase of forecasting is called *weather analysis*.

Next, the NCEP, through its various branches, begins the *predictive*, or *forecast*, *phase*. Modern weather forecasting requires supercomputers programmed to solve basic mathematical equations that describe atmospheric behavior. Using current weather data, these computer models attempt to predict the atmospheric conditions at the end of the forecast period—for example, 12 hours into the future. Experienced forecasters then tweak the computer-generated forecast by using traditional forecasting techniques in hopes of increasing its

*Robert T. Ryan, "The Weather Is Changing . . . or Meteorologists and Broadcasters, the Twain Meet," *Bulletin of the American Meteorological Society* 63, no. 3 (March 1982), 308.

▲ **Figure 12.2 Flooding of Boulder Creek, in downtown Boulder, Colorado, after three days of heavy rain, September 2013.**

accuracy (described in more detail later in the chapter). As part of the forecasting process, the NCEP regularly prepares surface weather maps, upper-level analysis, and various forecast products on national as well as global scales. These materials are supplied through a specially designed communication system to 122 local **Weather Forecast Offices**, where they are used to produce local and regional weather forecasts.

The final phase in the weather business is the dissemination of a wide variety of forecasts. Each Weather Forecast Office issues regional and local forecasts, aviation forecasts, and weather and flood warnings covering its forecast area. Further, all observational data and products (maps, charts, and forecasts) produced by the NCEP are available, mainly via the Internet, to the general public, government agencies responsible for public safety, and private forecasting services such as The Weather Channel, AccuWeather, and Weather Underground.

The demand for highly visual forecasts containing computer-generated graphics has increased proportionately with the use of personal computers and smart phones. The weather animations that appear on most local newscasts are mainly produced by the private sector. The private sector also customizes forecast products to create specialized weather reports tailored for specific audiences. In a farming community, for example, the weather reports might include freeze warnings, while winter forecasts in Denver, Colorado, include the snow conditions at area ski resorts.

Despite the valuable role that the private sector plays in generating and disseminating weather-related information to the public, the NWS is the *official* voice in the United States for issuing warnings during hazardous and life-threatening weather situations. Two major weather centers operated by the NWS serve critical functions in this regard. The Storm Prediction Center (SPC) in Norman, Oklahoma, maintains a constant vigil for severe weather, including thunderstorms and tornadoes, as well as heavy snowfall (see Chapter 10). Hurricane watches and warnings for the Atlantic, Caribbean, Gulf of Mexico, and eastern Pacific are issued by the National Hurricane Center (NHC) located in Miami, Florida (see Chapter 11).

## Who's Who in Weather Forecasting?

The American Meteorological Society defines a **meteorologist** as a person with specialized education, usually a bachelor's degree or higher from a college or university, who uses scientific principles to explain, observe, study, or forecast atmospheric phenomena. The broader term *atmospheric scientist* is sometimes used synonymously with *meteorologist* to describe someone who studies some aspect of Earth's atmosphere. Meteorologists perform various duties such as weather forecasting, atmospheric research, teaching, broadcasting, and providing specialized weather-related products to clients through private-sector companies.

Weather forecasting has always been a primary task performed by meteorologists. Forecasters are employed by a variety of organizations, including the following:

- The NWS, a primary employer of meteorologists in the United States
- The aviation industry, which hires meteorologists to provide pilots with information on weather conditions—like possible turbulence during take-offs, landings, and during flights
- The military, particularly the Air Force and Navy, which uses meteorologists to make weather forecasts for missions around the world
- The National Aeronautical and Space Administration (NASA), which employs forecasters to monitor weather conditions for satellite launches
- Private companies, the fastest-growing area for employing meteorologists, which have various needs for weather forecasts

The duties and responsibilities of meteorologists who work as weather forecasters is one of the main topics of this chapter.

Today, television stations employ more than 2000 meteorologists—both those behind the scenes where forecasts are prepared, as well as broadcast meteorologists who present forecasts and related weather information to viewers. What separates *broadcast meteorologists* from other meteorologists is their ability to describe complex meteorological phenomena to the general public in a way that is easily understood (**Fig. 12.3**). TV and radio stations that do not employ their own meteorologists usually purchase local weather forecasts from private companies. People not trained as meteorologists, often referred to as *weathercasters*, disseminate these forecasts to the public.

The National Center for Atmospheric Research (NCAR) in Boulder, Colorado, employs meteorologists interested in

◀ **Figure 12.3 The duties of a broadcast meteorologist include describing complex meteorological phenomena to the general public in a way that is easily understood.**

research. University and college professors with degrees in meteorology also work in atmospheric research programs, typically funded by government or foundation grants, in addition to their teaching duties. Today more than 100 U.S. and Canadian universities and colleges employ meteorologists or atmospheric scientists.

A related type of atmospheric scientist, called a **climatologist**, studies weather conditions averaged over a period of time. Climatologists often begin their careers as meteorologists, but they can also be trained in a natural science such as physics, oceanography, glaciology, or physical geography. They study both the nature of climates on a local, regional, or global scale and the natural and human activities that impact climate change, topics covered in the next three chapters.

### ✔ Concept Checks 12.1

1. Approximately what percentage of all declared emergencies is weather related?
2. What role does the Storm Prediction Center (SPC) serve?
3. List the major steps involved in providing weather forecasts.
4. How do broadcast meteorologists differ from meteorologists who prepare weather forecasts?

## 12.2 Acquiring Weather Data

**Describe the various tools used to acquire weather data.**

Before weather can be predicted, forecasters must have an accurate picture of current atmospheric conditions. This involves collecting, transmitting, and compiling billions of pieces of observational data. Because the atmosphere is ever-changing, these tasks must be accomplished quickly.

A vast network of weather stations is required to collect the data needed for generating even the shortest-range forecasts. On a global scale, the **World Meteorological Organization (WMO)**, a United Nations agency, is responsible for the international exchange of weather data. This task requires that data from more than 185 participating nations and 6 territories around the globe be collected simultaneously, standardized, and transmitted to the participating countries.

### Surface Observations

Worldwide, about 11,000 observation stations on land, 4000 ships at sea, and 1200 data buoys (**Fig. 12.4**) report atmospheric conditions four times daily, at 0000, 0600, 1200, and 1800 Coordinated Universal Time (UTC), also called Greenwich Mean Time (GMT), which is 6 hours ahead of Central Standard Time (CST). These data are rapidly sent around the globe, using a communications system dedicated to the distribution of weather information.

In the United States, 122 Weather Forecast Offices are responsible for gathering and transmitting local weather information to a central database. In addition to using humans to gather weather data, the NWS operates more than 900 **Automated Surface Observing Systems (ASOS)**. These modern automated systems provide weather observations such as temperature, dew point, wind speed and direction, visibility, and cloud cover and can also detect and determine precipitation amounts (**Fig. 12.5**). In addition, the Federal Aviation Administration (FAA), in cooperation with the NWS, operates automated observation stations at many airports. Automation assists, or in some cases replaces, human observers because it can provide information from remote areas. However, some research has shown that human observers are more reliable at determining certain weather elements, such as cloud cover and sky conditions.

### Observations Aloft

Because weather systems are three-dimensional, upper-air observations are essential for producing reliable forecasts. A

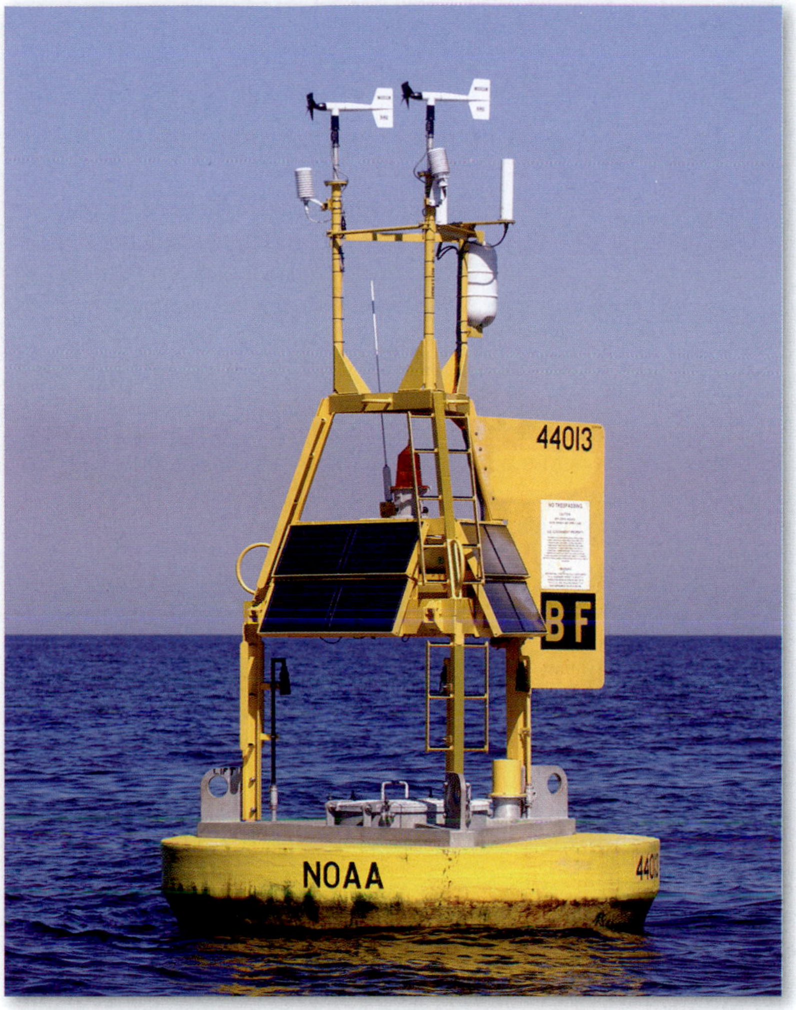


▲ **Figure 12.4 Data buoy used to record atmospheric conditions over a section of the global ocean** The data this buoy collects are transmitted via satellite to a land-based station.

**radiosonde** is a lightweight instrument package carried aloft by a weather balloon that contains sensors to measure temperature, humidity, and pressure. This data is transmitted via radio to the local observation station that launched it, usually a Weather Forecasting Office. Worldwide, about 1300 radiosondes (see Fig. 1.24, page 21t) are usually launched twice daily, so they are aloft about 0000 and 1200 UTC. The vast majority of upper-air observation stations are located in the Northern Hemisphere, with 92 stations operated by the NWS. Ships at sea also launch some radiosondes, and some airplanes drop a similar type of instrument that floats to the ground.

The radiosonde data is referred to as an **atmospheric sounding**, or simply a **sounding**. Meteorologists display the data collected on a special type of thermodynamic diagram called a *Skew-T diagram*—a topic discussed in greater detail later in the chapter.

Radiosondes are a primary source of data used to construct upper-level weather charts. In addition, **rawinsondes**, which are radiosondes tracked by radar, provide data on wind speed and direction at several levels of the atmosphere.

A radiosonde flight typically lasts about 90 minutes, during which it may ascend to heights of more than 35 kilometers (about 22 miles). Because pressure decreases with increases in altitude, the balloon stretches to the size of a two-stall garage and eventually bursts. Then a small parachute opens, and the instrument package slowly descends to Earth. If you find a radiosonde, follow the mailing instructions because the instruments can be reused.

More than 3000 aircraft report on atmospheric conditions—mainly temperature and airflow data—during flight. In addition, satellites equipped with visible and infrared imaging devices have become invaluable instruments for data collection. Satellites equipped with sounding devices provide vertical profiles of temperature and humidity in cloud-free areas and can track the movement of clouds and water vapor.

A number of technical advances have improved our ability to make observations aloft. Special radar units called **wind profilers** can measure wind speed and direction up to 10 kilometers (6 miles) above Earth's surface. These measurements can be taken every 6 minutes, in contrast to the 12-hour interval between balloon launches.

Weather radar is also an important tool for short-range forecasting of severe weather phenomena. **Doppler radar** is extensively used in local forecasting because of its capacity to measure winds and estimate rainfall amounts. Doppler radar also makes it possible to follow the evolution of thunderstorms because the radar can "see" the storm's precipitation structure

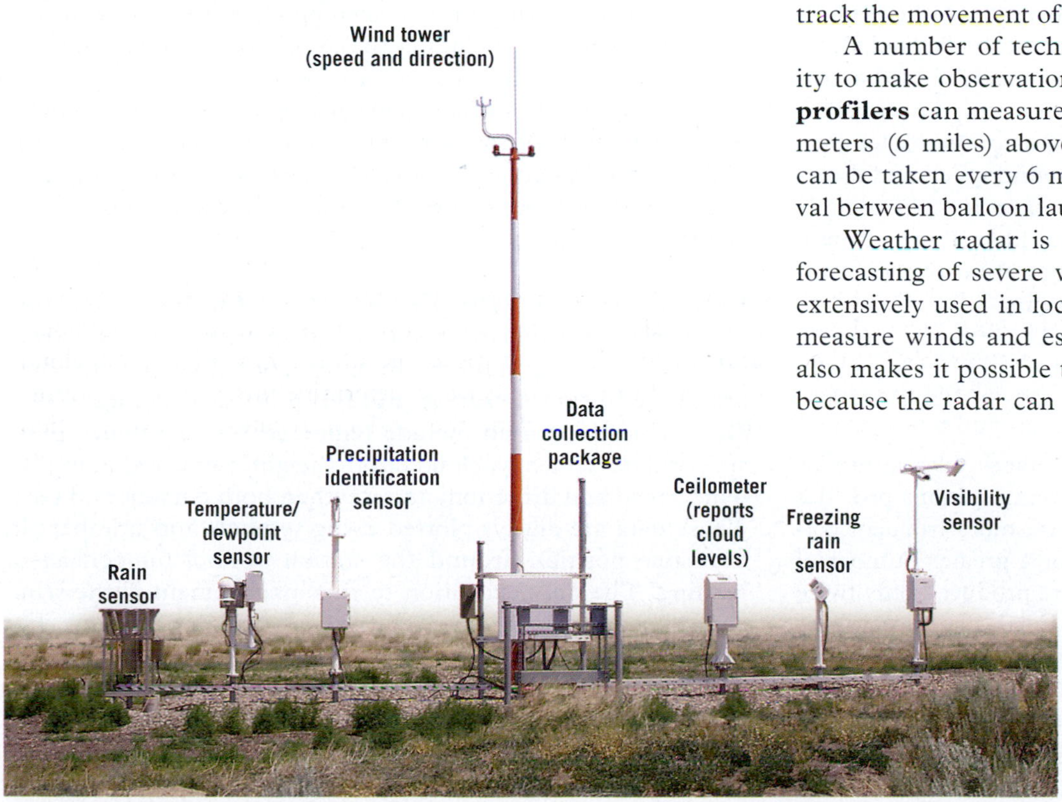


◀ **Figure 12.5 Automated Surface Observing System (ASOS)** Automated observing systems are equipped with a variety of instruments designed to sample the sky for cloud coverage; take temperature and dew-point measurements; determine wind speed and direction; and even detect present weather—such as whether it is raining or snowing.

and track its movement. (For a discussion of Doppler radar applications see Chapter 10, page 293.)

Despite the advances in weather data collection, two difficulties remain. First, observations often have inaccuracies due to instrument and/or transmission error. Second, the network of weather observing stations does not provide consistent coverage around the globe—particularly over the oceans and in remote areas.

### ✔ Concept Checks 12.2

1. What agency is responsible for gathering weather information on a global scale?
2. List the main sources of *surface* weather data.
3. What is the primary source of weather data used to plot upper-level weather charts?

**students sometimes ask...**

### Who was the first weather forecaster?

Benjamin Franklin is often credited with making the first long-term weather predictions in his *Poor Richard's Almanac.* However, these forecasts were based primarily on *folklore* rather than weather data. Nevertheless, Franklin may have been the first to document that storm systems move. In 1743, while living in Philadelphia, rain prevented Franklin from viewing an eclipse. Through later correspondence with his brother, he learned that the eclipse could be seen in Boston, but within a few hours, that city also experienced rainy weather. These observations led Franklin to conclude that the storm that obscured the eclipse in Philadelphia moved up the east coast to Boston.

# 12.3 Weather Maps: Depictions of the Atmosphere

**Interpret what a particular pattern in the flow aloft indicates about surface weather.**

The vast amount of observational data collected by the various national weather services is not only used to predict the weather but also to generate a series of surface weather maps. Because weather systems are three-dimensional, upper-air charts are also drawn for different pressure levels. Forecasters use these tools to observe changes in weather systems over time.

Once these computer-generated weather charts are produced, meteorologists fine-tune them to provide the most accurate description (analysis) of the current weather conditions possible. This task, called **weather analysis**, includes a detailed examination of the atmosphere, specifically the troposphere, where the day-to-day weather occurs. The weather elements examined include variables such as pressure, temperature, dew-point temperature, humidity, wind speed and direction, and cloud cover. The goals of weather analysis include locating developing weather systems, identifying frontal boundaries, and assessing the potential for atmospheric instability that may result in severe weather events.

## Surface Weather Maps

Surface weather maps have several advantages over upper-air charts: The vastly larger number of observing stations provides a more accurate picture of the atmosphere; more frequent collection and assimilation of data allows for a greater number of maps to be generated (upper-level maps are produced only twice daily); and they depict the conditions at Earth's surface, where people live. In addition, fronts are drawn on surface maps but not on upper-air charts. Recall that precipitation patterns are usually associated with fronts, so knowledge of the position and rate and direction of movement of fronts greatly aids forecasters.

A traditional surface map is called a **synoptic weather map** (*synoptic* means "coincident in time") because it displays a synopsis of weather conditions at a given moment. These charts symbolically represent the state of the atmosphere (**Fig. 12.6**). Thus, to the trained eye, a weather map is a snapshot that shows the status of the atmosphere, including data on temperature, humidity, sea-level pressure, and wind. Surface maps are also commonly called **current surface analysis** (or *surface analysis charts*) because they provide a comprehensive look at surface weather conditions, including the locations of fronts and pressure systems—usually on an hourly basis.

More than 200 surface maps and upper-level charts covering several levels of the atmosphere are produced daily by the NWS and its forecast centers. Humans previously completed this tedious task, but computers now analyze and plot the data systematically.

**Constructing a Synoptic Weather Map** Construction of a surface weather map starts with plotting the data from selected observing stations. By international agreement, data are plotted using a **station model**, illustrated in **Figure 12.7**. Data typically plotted include temperature, dew point, pressure and its tendency, cloud cover (height, type, and amount), wind speed and direction, and weather, both current and past. These data are always plotted using symbols and numbers in the same position around the station symbol for consistent reading. The only exception to this arrangement is the wind arrow because it is oriented with the direction of airflow. For example, in Fig. 12.7 the temperature is plotted in the upper-left corner of the sample model, and it will always appear in that location. A more complete weather station model and a key for decoding weather symbols are found in Appendix B.

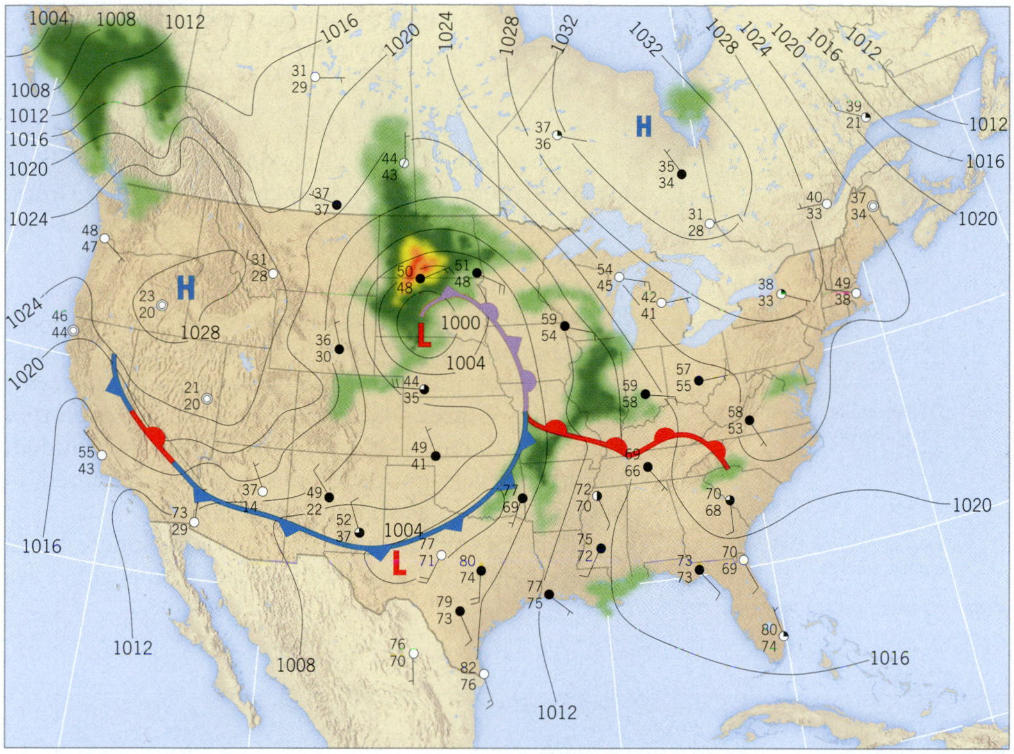


A. Surface weather map.

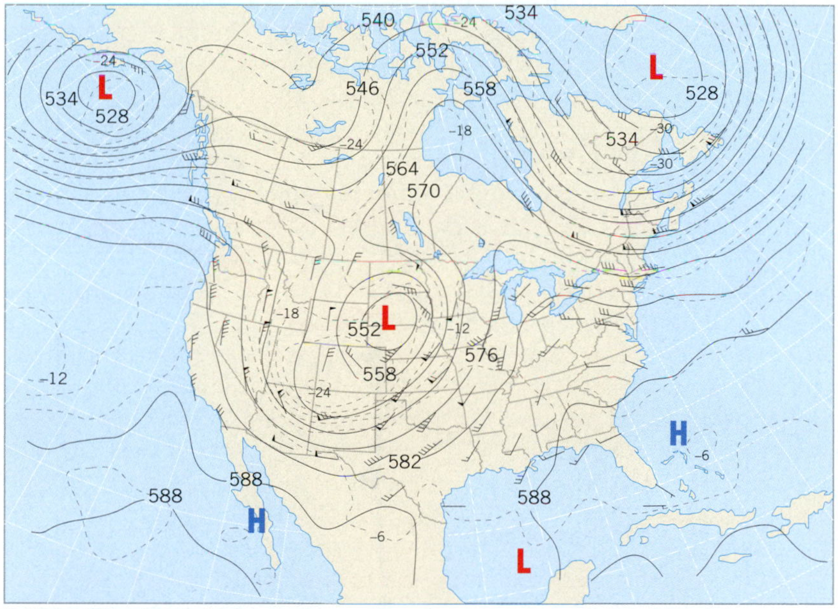


B. 500-millibar level chart.

▲ **Figure 12.6 Simplified synoptic and upper-level weather maps**
**A.** Surface weather map for 7:00 AM Eastern Standard Time, depicting a well-developed middle-latitude cyclone. **B.** A 500-millibar-level map, with height contours in tens of meters, for the same day and time.

Once data have been plotted, isobars and fronts are added to the weather chart (Fig. 12.8). Isobars are usually drawn on surface maps at 4-millibar intervals (1004, 1008, 1012, etc.). The positions of the isobars are estimated as accurately as possible, based on the pressure readings available. Note in Figure 12.8B that the 1012-millibar isobar is about halfway between two stations that report 1010 millibars and 1014 millibars. Frequently, observational errors and other complications require an analyst to smooth the isobars so they conform to the overall picture. Many irregularities in the pressure field are caused by local influences that have little bearing on the larger circulation depicted on the charts. Once the isobars are drawn, centers of high and low pressure are designated.

**Identifying Fronts on a Weather Map** Fronts are boundaries that separate contrasting air masses, so they can often be identified on weather charts by locating zones exhibiting abrupt changes in conditions. Because several elements change across a front, all are examined in order to locate the frontal position accurately. Easily recognized changes that can help identify a front on a surface chart include:

- Marked temperature contrast over a short distance
- Wind shift (in a clockwise direction) over a short distance, by as much as 90°
- Humidity variations that can be detected by examining dew-point temperatures
- Clouds and precipitation patterns that give clues about the positions of fronts

Notice in Figure 12.8B that all the conditions listed are easily detected across the frontal zone. However, not all fronts are as easily defined as the one on our sample map. In some cases, surface contrasts on opposite sides of the front are subdued. When this happens, charts of the upper air, where flow is less complex, become an important tool for detecting fronts.

## Upper-Level Weather Charts

Chapter 9 established the strong connection between cyclonic disturbances at the surface and the wavy flow of the westerlies aloft. The importance of this connection cannot be overstated, particularly as it applies to weather forecasting. In order to understand the development of thunderstorms or the formation and movement of midlatitude cyclones, meteorologists must know what is occurring aloft as well as at the surface.

Upper-air charts are generated twice daily, at 0000 and 1200 UTC. These charts are drawn at 850-, 700-, 500-, 300-, and 200-millibar (mb) levels as height contours (in meters or tens of meters), which are analogous to isobars used on surface maps. Like surface maps, upper-level charts are often referred to as *analysis charts*, or *upper-level analysis*, because they provide an analysis of the upper-level weather conditions.

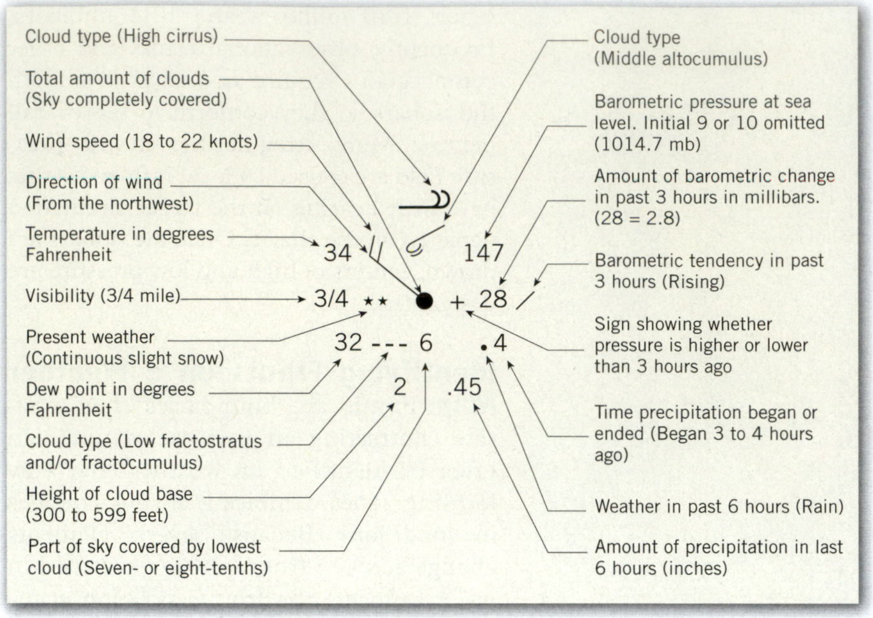


◀ **Figure 12.7 Sample station model showing the placement of commonly plotted data**

Upper-level charts also contain *isotherms* (lines of equal temperature) depicted as dashed lines, and some show humidity as well as wind speed and direction.

**850-Millibar Charts** **Figure 12.9A** is a 850-mb map showing height contours using solid black lines at 30-meter intervals and dashed red isotherms labeled in degrees Celsius. Wind data are plotted using black arrows. If relative humidity is included, humidity levels above 70 percent are indicated in shades of green, as shown in **Figure 12.9B**.

The 850-mb map depicts the atmosphere at an average height of about 1500 meters (1 mile) above sea

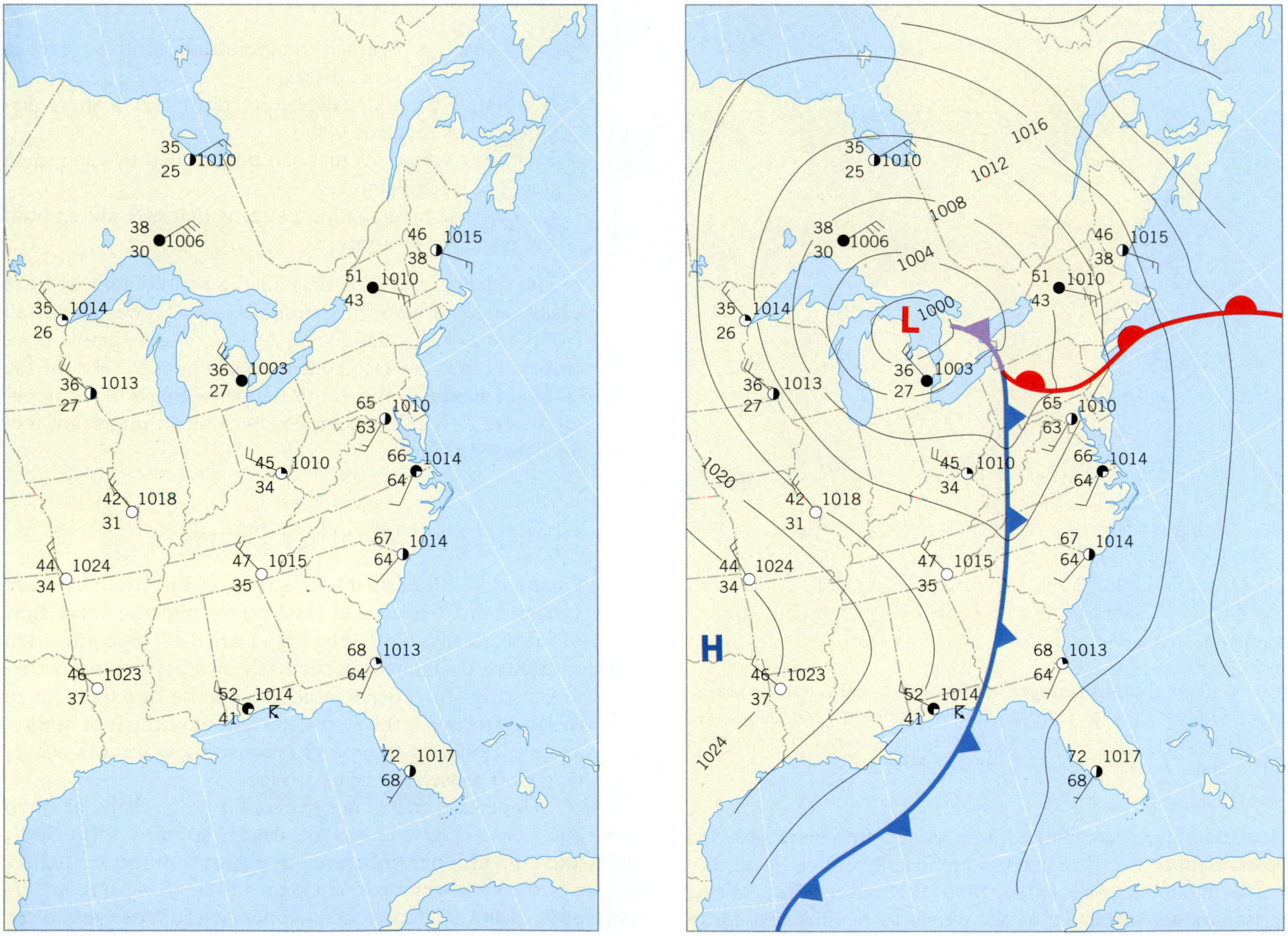


A.

B.

▲ **Figure 12.8 Simplified weather charts** **A.** Stations, with data for temperature, dew point, wind direction, wind speed, sky cover, and barometric pressure plotted. **B.** Same chart showing the isobars and the locations of frontal boundaries.

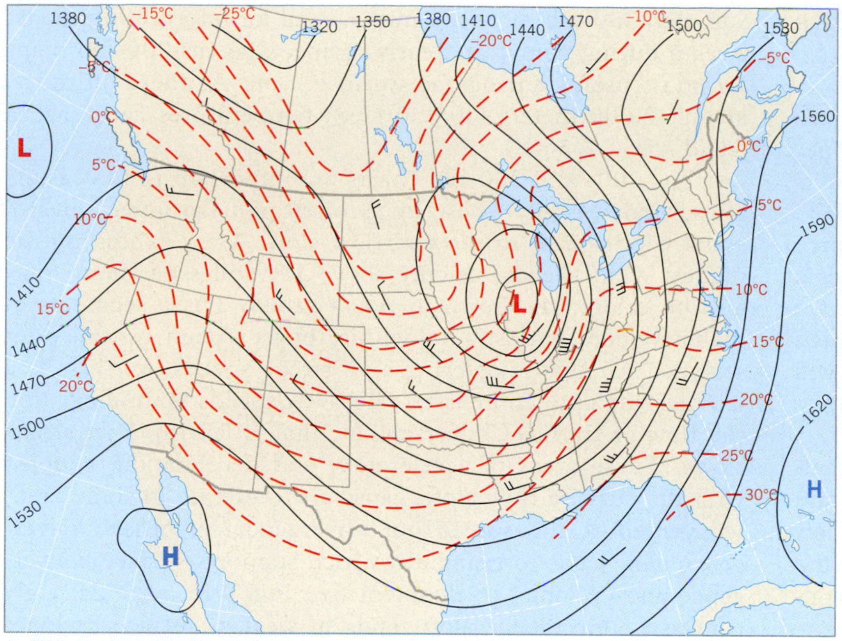


A. 850–mb map showing height contours and isotherms in °C

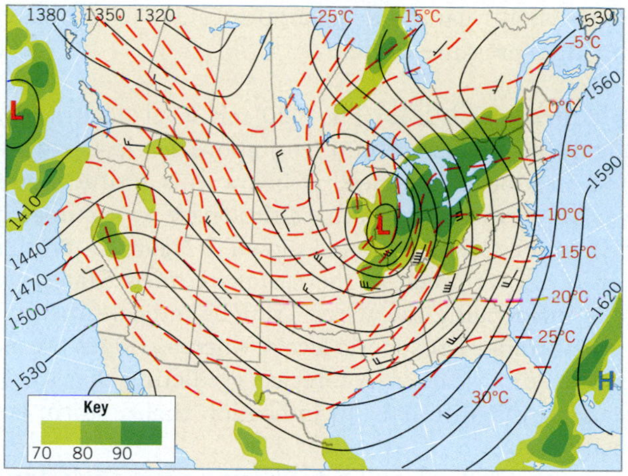


B. Relative humidity at 850–mb level

◄ **Figure 12.9 Typical 850-mb map A.** The solid lines are height contours spaced at 30-meter intervals, and the dashed lines are isotherms in degrees Celsius. **B.** Regions where the relative humidity is greater than 70 percent are depicted in shades of green. **C.** Areas of warm- and cold-air advection are shown with colored arrows.

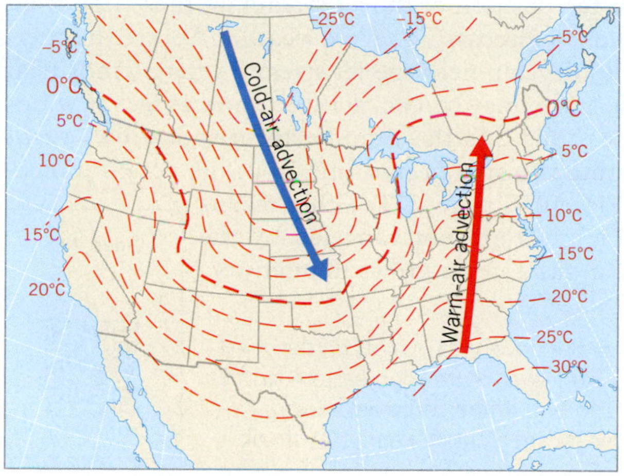


C. Isotherms at 850–mb level

level. In areas near sea level, the height of the 850-mb chart is near the top of *planetary boundary layer*, the layer in which friction and turbulence are common. Within this boundary layer, daily temperature fluctuations are strongly influenced by the warming and cooling of Earth's surface. In high-elevation areas such as Denver, Colorado, the 850-mb level represents surface conditions, and the 850-mb chart is used as a proxy for (in place of) the surface map.

Forecasters regularly examine the 850-mb map to locate areas of possible *cold-air* or *warm-air advection*. Cold-air advection occurs when winds blow horizontally across isotherms from colder to warmer areas. **Figure 12.9C**, a simplified version of Figure 12.9A, shows a blast of cold air moving south from Canada into the Great Plains. As you might expect, cold air advection results in cooler temperatures.

Cold air advection at low levels usually causes air aloft to sink, which enhances atmospheric stability. Recall that stability caused by sinking air is associated with clearing skies, which is expected with the passage of a cold front.

Also notice in Figure 12.9C that warm air is moving from the Gulf of Mexico toward the northeast. This means that the eastern seaboard of the United States will experience rising temperatures over the next day or so. In addition to warm air moving into a cooler region, warm-air advection is generally associated with widespread uplift in the lower troposphere because warm air is less dense (takes up more volume) than the cold air it replaces and causes the air aloft to rise. If humidity in the region of warm-air advection is relatively high, as shown in Figure 12.9B, then lifting could result in cloud formation and possibly precipitation.

Temperatures at the 850-mb level provide other useful information. During winter, for example, forecasters use the 0°C isotherm as the boundary between areas of rain and areas of snow or sleet. Further, because air at the 850-mb level does not experience the daily cycle of temperature changes that occurs at the surface, these maps provide a way to estimate the daily maximum surface temperature. In summer, the maximum surface temperature is usually 15°C (27°F) higher than the temperature at the 850-mb level. In winter, maximum surface temperatures tend to be about 9°C (16°F) warmer than at the 850-mb level, and during the fall and spring, this difference is about 12°C (22°F).

**700-Millibar Charts** The 700-mb flow, which occurs about 3 kilometers (2 miles) above sea level, serves as the steering mechanism for air-mass thunderstorms that tend to occur in the warm summer months. Thus, winds at this level are used to predict the movement of these storms. One rule of thumb with these charts states that when temperatures at the 700-mb level are 14°C (57°F) or higher, thunderstorms will not develop. A warm 700-mb layer acts as a lid that inhibits the upward movement of warm, moist surface air that might otherwise rise to generate towering cumulonimbus clouds. The warm conditions aloft are usually caused by general subsidence associated with a strong high-pressure center.

**500-Millibar Charts** The 500-mb level occurs at approximately 5000 meters (18,000 feet) above sea level—where about half of Earth's atmosphere is below this altitude and half is above. Air temperatures at this altitude are approximately −20°C (−4°F). Notice in the example shown in Figure 12.10A that a large trough occupies much of the eastern United States, whereas a ridge is influencing the West. Troughs indicate the presence of a storm at mid-tropospheric levels, whereas ridges are associated with clear, dry weather. A forecast predicting that a trough will increase in magnitude is an indication that a storm is likely to intensify.

Forecasters find 500-mb charts useful tools for estimating the movement of surface cyclonic storms. These important weather producers tend to travel in the direction of the airflow at the 500-mb level, but at roughly a quarter to half the wind speed. Sometimes an upper-level low (shown as a closed contour in Figure 12.10A) forms within a trough. These lows are associated with counterclockwise rotation and significant vertical lifting, typically resulting in heavy precipitation.

**300- and 200-Millibar Charts** Two of the regularly generated upper-level maps, the 300-mb and 200-mb charts, represent zones near the top of the troposphere. These are the levels at which the details of the *jet stream* can best be observed. Recall from Chapter 7 that the jet stream is a high-velocity river of air that often flows around the globe in the middle latitudes. Because the jet stream is lower in winter and higher in summer, 300-mb maps are most useful for interpreting the jet stream during winter and early spring, whereas 200-mb maps are most useful during the warm season. At these altitudes, about 12,000 meters (40,000 feet), temperatures can reach a frigid −55°C (−56°F).

The 200- and 300-mb charts are important to forecasters for several reasons. These maps often plot **isotachs**, lines of equal wind speed, to show airflow aloft. Areas exhibiting the highest wind speeds, usually above 100 miles per hour, may also be colored as shown in Figure 12.10B. These segments of higher-velocity winds found within the jet stream are called **jet streaks**. Air that is entering a jet streak speeds up—just as a car merging onto an interstate highway speeds up to move into the flow of traffic. Conversely, air that is leaving a jet streak slows. These areas of acceleration and deceleration, coupled with the curving flow aloft, cause air to pile up in some areas (convergence) and spread out (divergence) in others. Divergence aloft leads to rising air, which supports surface convergence and cyclonic development (see Fig. 9.15, page 252). By contrast, convergence aloft tends to weaken surface cyclones and causes them to dissipate.

Locating the position of the jet stream on these upper-air charts also aids in forecasting severe weather. Severe thunderstorms tend to develop near the jet stream, where winds near the top of the storm may be two to three times as fast as winds near the base. This tilts the thunderstorm as it grows vertically, separating the area of updrafts from the area of downdrafts (see Fig. 10.7, page 273). This means that the rising cells are not canceled by the downdrafts, and the storm grows in intensity.

Video MM
Forecasting Upper-Level Winds
http://goo.gl/7goQvk

▼ **Figure 12.10 Comparing 500- and 200-millibar upper-level charts** These charts show the flow pattern aloft on the same time and date but at different heights. **A.** Notice the well-developed trough of low pressure over central and eastern North America on the 500-millibar chart, and a ridge of high pressure positioned over the western portion of the continent. Height contours are spaced at 60-meter intervals. **B.** This 200-millibar chart shows the location of the jet stream (pink) and a jet streak (reddish). Height contours are spaced at 120-meter intervals.

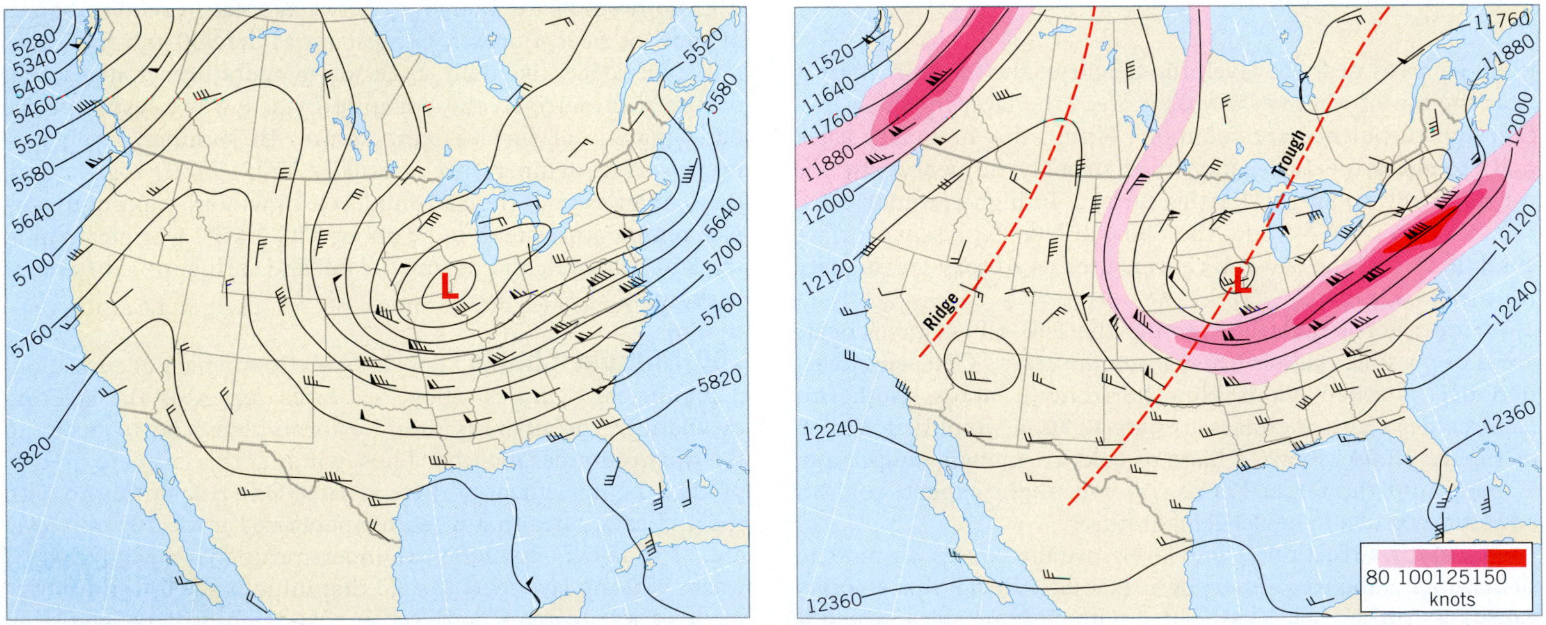


A.

B.

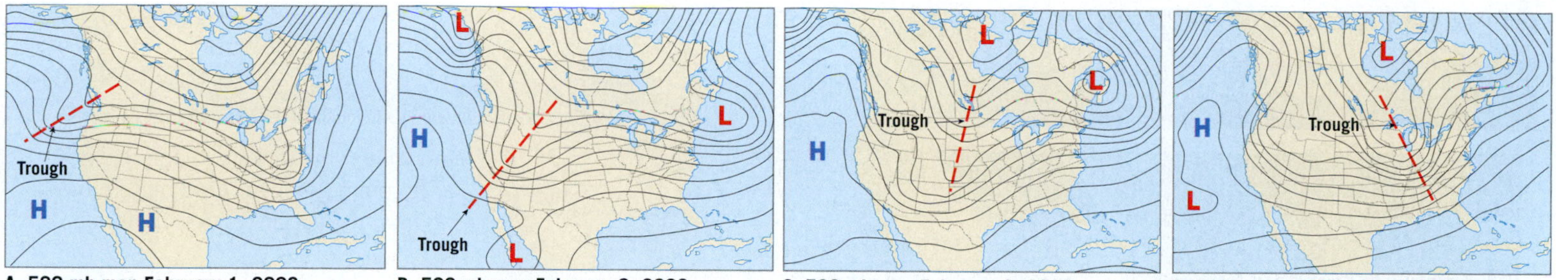


A. 500 mb map February 1, 2003 B. 500 mb map February 2, 2003 C. 500 mb map February 3, 2003 D. 500 mb map February 4, 2003

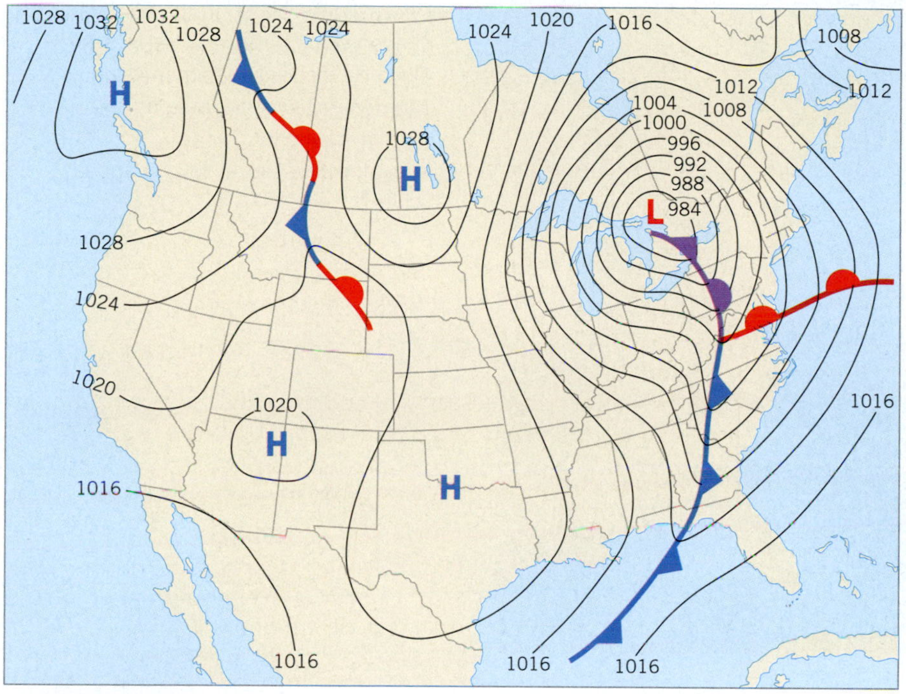


E. Surface map February 4, 2003

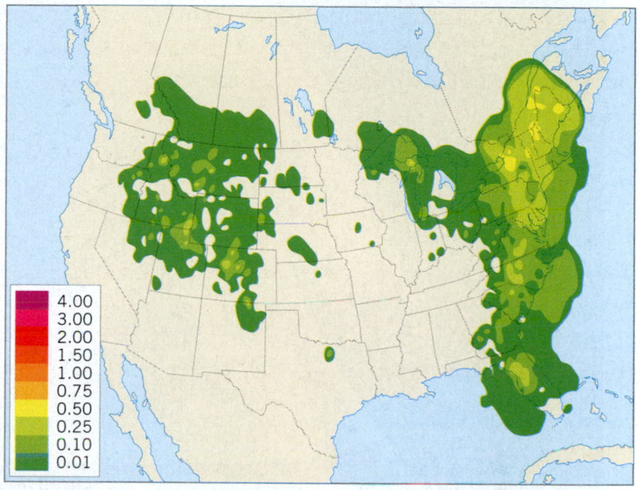


F. Precipitation produced between 7 AM, Feb 4 and 7 AM, Feb 5

▲ **Figure 12.11 The connection between the flow aloft and surface weather** **A–D.** The movement and intensification of an upper-level trough over a 4-day period from February 1 to February 4, 2003. **E.** This surface map shows a strong cyclonic storm that formed earlier in the week and moved, in conjunction with the trough, to the position shown on the February 4 map. **F.** Precipitation map shows precipitation amounts that fell from 7 A.M. February 4 to 7 A.M. February 5, shown in inches.

## The Connection Between Upper-Level Flow and Surface Weather

Thus far, we have looked at how mapping upper-level flow aids in predicting short-lived weather events—for example, how winds aloft steer and support the growth of thunderstorms. Next, we will examine how analyzing long-period changes in the wavy flow of the westerlies impact weather forecasting.

Occasionally, the flow within the westerlies is directed along a relatively straight path from west to east, a pattern referred to as **zonal**. Storms embedded in this zonal flow move quickly across the country, particularly in winter. This situation results in rapidly changing weather, with periods of light to moderate precipitation followed by brief periods of fair weather.

More often, however, the flow aloft consists of long-wave troughs and ridges that have large components of north–south flow, a pattern that meteorologists refer to as **meridional**. Typically, these airflow patterns slowly drift from west to east, but they occasionally stall and may even reverse direction. As this wavy pattern gradually migrates eastward, cyclonic storms embedded in the flow are carried across the country.

Figure 12.11 shows a trough that was centered off the U.S. Pacific coast on February 1, 2003. Over the next 4 days, the trough intensified as it moved eastward into the Ohio Valley. This change in intensity is shown by the height contours, which are more closely packed around the trough on February 4 than on February 1. Embedded in this upper-level trough was a cyclonic system, which grew into the major storm shown on the surface map in Figure 12.11E. By February 5, this disturbance generated considerable precipitation over much of the eastern United States. (Notice that the center of the surface low is located somewhat to the east of the 500-mb trough—as is typical of these weather patterns.)

In general, meridional flow aloft has the potential for generating extreme conditions. During winter, strong troughs tend to spawn large snowstorms, whereas in summer they are associated with severe thunderstorms and tornado outbreaks. By contrast, high-amplitude ridges are associated with record heat in summer and tend to bring mild conditions in winter.

Sometimes these "looped" patterns stall over an area, causing surface patterns to change minimally from one day to the next or, in extreme cases, from one week to the next. Locations in and just east of a stationary or slow-moving trough experience extended rainy or stormy periods, while areas in and just east of a stagnant ridge experience prolonged periods of unseasonably warm, dry weather.

## eye ON THE atmosphere 12.1

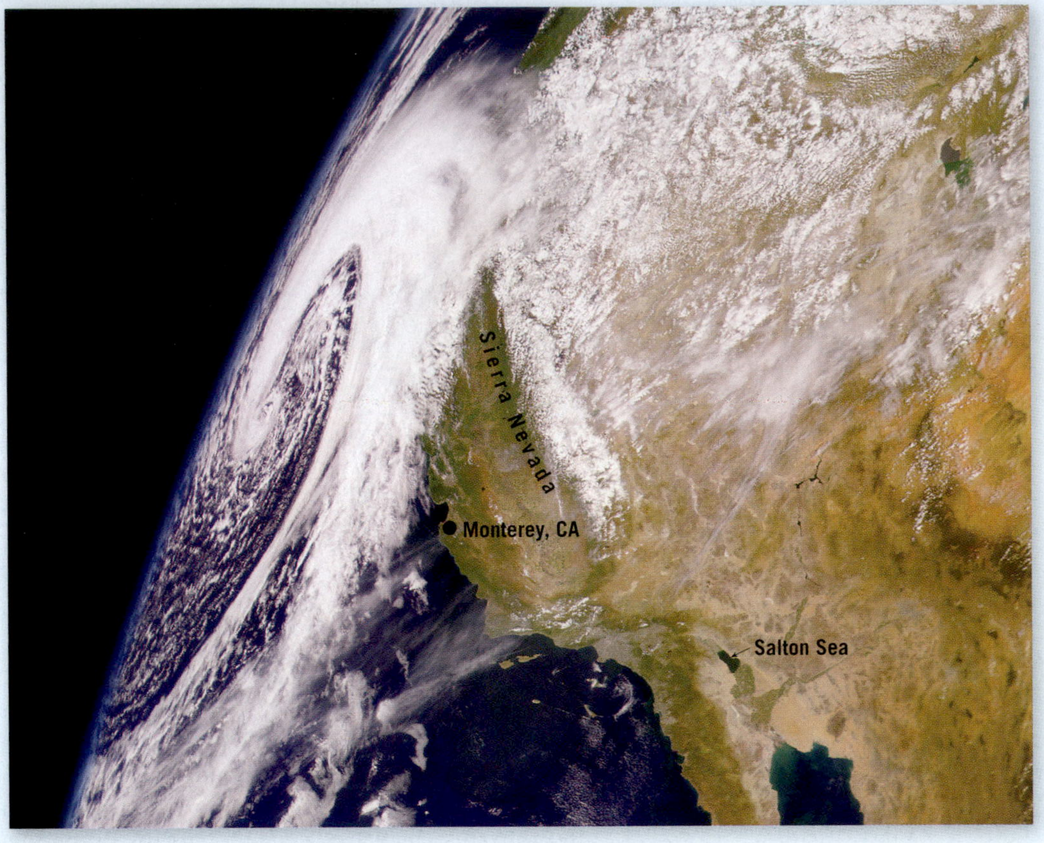


This wintertime satellite image shows a large comma-shaped cloud pattern moving from the Pacific toward California.

**Questions**

1. What name is given to the storm system that resulted in the comma-shaped cloud pattern in the image?
2. Describe the likely weather in the Monterey, California, area as this system passed by.
3. Why does the Sierra Nevada appear white?

Recall from Chapter 7 that the jet stream is strongest (faster wind speeds) in winter and early spring, when the contrast between the warm tropics and the cold polar realm is greatest. As summer approaches, the temperature gradient diminishes, and the westerlies weaken. The jet stream's position also changes seasonally, moving toward the equator in winter. This is why the southern states typically encounter most of their severe weather in winter and spring. (Hurricane season is the exception.) During summer, because of a poleward shift of the jet, the northern states and Canada experience an increase in the number of severe storms and tornadoes. This more northerly storm track also carries most Pacific storms toward Alaska during the warm months, producing a rather long, dry summer season for much of the Pacific coast that lies to the south.

### ✔ Concept Checks 12.3

1. What are two pieces of information that forecasters can glean from 850-mb maps?
2. During winter, if the jet stream is considerably south of your location, will the temperatures where you are likely be warmer or colder than normal?
3. What is a *jet streak*?
4. Which upper-level chart is most effective for observing the polar jet stream in winter?
5. Explain the difference between zonal and meridional airflow.

## 12.4 Modern Weather Forecasting

**Explain the basis of numerical weather prediction.**

Until the late 1950s, all weather maps and charts were plotted manually and served as the primary tools for making weather forecasts. Forecasters used various techniques to extrapolate future conditions from the patterns depicted on the most recent weather charts. One method involved matching current conditions with similar, well-established patterns from the past. From such comparisons, meteorologists predicted how current systems might change in the hours and days to come.

Later, computers were used to plot data and produce surface and upper-air charts. As technologies improved, computers eventually replaced manual creation of weather forecasts. Computers have improved the accuracy and detail of weather forecasts and have lengthened the period for which useful guidance can be given.

## Numerical Weather Prediction

**Numerical weather prediction** is a weather forecast technique that employs mathematical models designed to represent atmospheric processes. (The word *numerical* is somewhat misleading here because all types of weather forecasting are based on some quantitative data and, therefore, could fit in this category.) Numerical weather prediction relies on the fact that the behavior of atmospheric gases is governed by a set of fundamental physical principles or laws (for example, the *ideal gas law* and the *laws of thermodynamics*) that can be expressed as mathematical equations. By solving these equations, a description of the future state of the atmosphere (a forecast) can be established, derived from the current state and expressed in terms of temperature, moisture, cloud cover, and wind. This method is analogous to using a computer to predict the future positions of Mars, using Newton's laws of motion and knowing the planet's current position.

The NWS utilizes a few different numerical models, including a global-scale model called the Global Forecast System (GFS). Because GFS forecast products are freely available on the Internet, they are often used by private forecast services. A standard NWS regional model, which takes into account topographic features such as mountains, is called the North American Mesoscale Model (NAM). Several other organizations also employ excellent global and regional forecast models, including the Canadian Meteorological Centre (CMC), the European Centre for Medium-Range Forecasting (ECMWF), and the United Kingdom Met Office (UKMET), as well as the U.S. military. Most of these forecast products are easily accessed on the Internet, except for UKMET products, which are available to private forecast services at a significant cost. In addition, Japan, Australia, South Korea, China, Russia, and some other countries operate national centers that generate forecasts from their own models. Most of the remaining countries of the world use model output data produced by one of the sources mentioned above to make regional and local forecasts for their respective countries.

Numerical weather prediction uses highly refined mathematical models that attempt to mimic the behavior of the "real" atmosphere. All numerical models employ the same governing equations, but they apply the equations and the parameters in different ways. For example, each model is designed to cover a specific area. Models that are considered "global" in scale usually cover one hemisphere, whereas others are "regional," covering areas such as North America or Europe. Each model has a three-dimensional grid consisting of a large number of points, called *grid points*, separated by a defined distance. A high-resolution model has grid points that are generally separated by less than 5 kilometers, whereas a low-resolution model might have data points located more than 300 kilometers apart. Regardless of area covered or model resolution, all numerical weather prediction models are complex and require considerable computer resources to operate.

**Steps in Numerical Prediction** Numerical weather forecasting begins by assigning current weather measurements (temperature, wind speed, humidity, and pressure) to each of the grid points in the forecast model. This set of values represents the atmospheric conditions at the start of the forecast. Next, a computer simulation solves each equation for as many as 50 levels of the atmosphere—a process referred to as a *computer run*. After literally billions of calculations, a forecast of how these basic weather elements are expected to change over a short time frame (perhaps only 5 to 10 minutes) is generated. When the new values have been calculated, the process begins again, generating a forecast for the next 5 to 10 minutes. These steps are repeated numerous times until the end of the forecast period. Some models produce forecasts on an hourly basis (out to perhaps 12 hours), while other models generate forecasts every 6 hours—with coverage of up to 2 days or longer. Even the simplest models require such a vast number of calculations that they were impractical prior to the development of supercomputers.

Once generated, statistical methods are used to modify machine-generated forecasts by making comparisons of the accuracy of previous forecasts. This approach, known as **Model Output Statistics (MOS**, pronounced "moss"), attempts to correct for errors the model tends to make consistently. For example, certain forecast models may typically predict too much rain, overly strong winds, or temperatures that are too high or too low. MOS also takes into account local conditions that are not built into the forecast models.

MOS-generated forecast products may include expected maximum and minimum temperatures, precipitation probabilities and amounts, chances for thunderstorms, cloudiness, and surface wind speeds and direction. MOS forecasts form the baseline on which NWS and private-sector forecasters attempt to improve.

After the highly refined MOS forecasts are generated, forecasters determine how well they matched the actual weather conditions that developed during the forecast period. This important *postprocessing phase* is necessary for continually adjusting numerical models and improving their predictability.

Using various mathematical models, the NWS produces a variety of forecast charts. Because these machine-generated maps predict atmospheric conditions at some future time, they are often referred to as **prognostic charts**, or simply **progs**. Some forecast charts depict surface weather features such as fronts, high- and low-pressures systems, and expected weather events, as shown in Figure 12.12. Others depict a single weather element such as maximum temperature (see Fig. 12.26), while still others show the airflow pattern aloft at various levels. These forecast products, although indispensable, represent only one step in the complex process of making regional and local forecasts.

**Why Are Numerical Forecasts Imperfect?** Despite the sophistication of numerical models, most still produce forecast errors. Factors affecting their accuracy include the inadequate

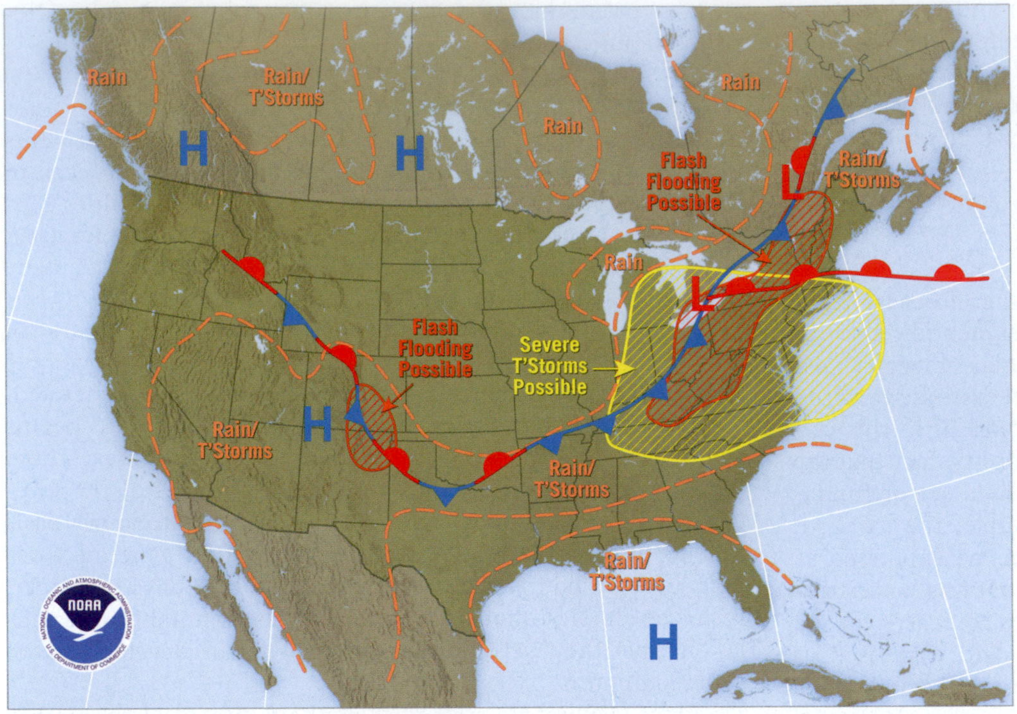


◀ **Figure 12.12 Weather forecast for Sunday, July 27, 2014** This generalized weather forecast was prepared and disseminated by the Weather Prediction Center of the NCEP. It includes areas of high and low pressure and frontal types and locations, and it outlines areas that have potential for precipitation and severe weather.

representation of physical processes operating in the atmosphere, common errors (or omissions) in the initial observations, and the inability of models to depict small weather elements, like tornadoes. The lack of computing power available to the NWS is also often cited as an issue. Forecasters sometimes have to choose between a low-resolution forecast that covers the globe and a high-resolution product that covers a smaller area. In other words, a model that produces high-resolution forecasts has minimal value if its computer run time is too long, rendering the forecasts obsolete.

Although the models are grounded on sound physical laws and capture the major characteristics of the atmosphere, they are designed to simplify the intricacies of a very complex system. The topography in the forecast area is one example of a variable that is not accurately represented in current numerical models. As a result, humans perform the final steps in weather forecasting, using their experience and knowledge of meteorology and also making allowances for known model shortcomings (**Fig. 12.13**).

## Ensemble Forecasting and Uncertainty

One of the most significant challenges for weather forecasters is the chaotic behavior of the atmosphere. Specifically, two very similar weather systems may, over time, develop into two very different weather patterns. One may intensify, becoming a major disturbance, while the other withers and dissipates. To demonstrate this point, Edward Lorenz at the Massachusetts Institute of Technology employed a metaphor known as the *butterfly effect*. Lorenz described a butterfly in the Amazon rain forest, fluttering its wings and setting into motion a subtle breeze that travels and gradually magnifies over time and space. Two weeks later this faint breeze has grown into a tornado over Kansas. Obviously, by stretching the point considerably, Lorenz tried to illustrate that a very small change in initial atmospheric conditions can dramatically affect the resulting weather pattern elsewhere.

To deal with the inherent chaotic behavior of the atmosphere, forecasters rely on a technique known as **ensemble forecasting**. Simply, this method involves producing a number of forecasts using the same computer model but slightly altering the initial conditions, while remaining within an error range of the observational instruments. Essentially, ensemble forecasting attempts to assess how the inevitable errors and omissions in weather measurements might affect forecast accuracy.

One of the most important outcomes of ensemble forecasts is the information they provide about forecast *uncertainty*. For example, a prognostic chart generated using the best available weather data might predict the occurrence of precipitation over a wide area of the southeastern United States within 24 hours. A meteorologist might run the same calculations several times in succession, each time making

Video (MM) Uncertainty in Numerical Models
http://goo.gl/NvqnYP

▼ **Figure 12.13 Meteorologist at the National Hurricane Center, in Miami, Florida** This forecaster is examining satellite imagery of Hurricane Ivan as the eye crosses the Alabama coastline on September 16, 2004. Humans perform the final steps in weather forecasting, using their experience and knowledge of meteorology.

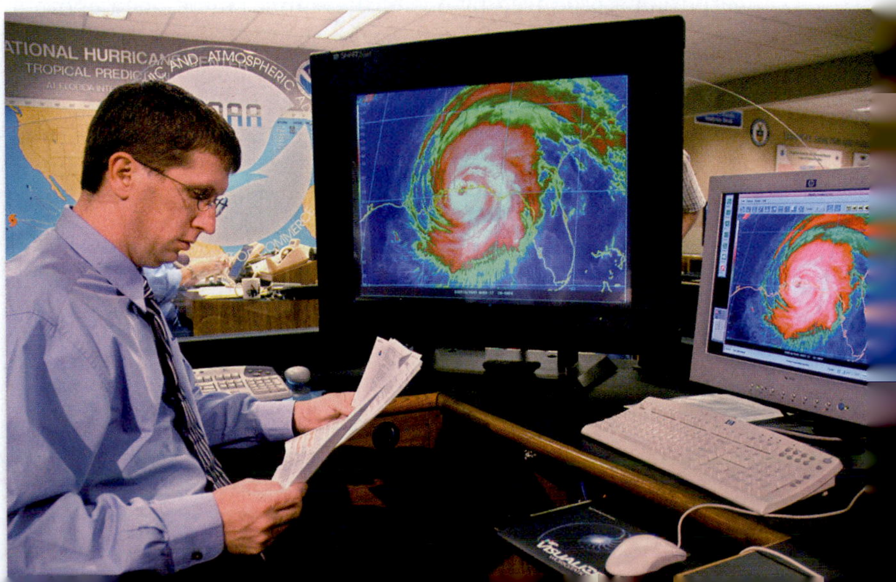

minor adjustments to the initial conditions. If most of these prognostic charts predict a pattern of precipitation in the Southeast, the forecaster will place a high degree of confidence in the forecast. On the other hand, if the progs generated by the ensemble method differ significantly from one run to the next, there will be far less confidence in the forecast.

In summary, numerical weather prediction relies on the fact that the behavior of atmospheric gases is governed by a set of physical principles or laws that can be expressed as mathematical equations. By utilizing data on initial atmospheric conditions, meteorologists solve these equations to predict the future state of the atmosphere. This is a complicated process because Earth's atmosphere is a very complex, dynamic system that can only be roughly approximated using mathematical models. Further, because of the nature of the equations, tiny differences in the data can yield huge differences in outcomes. Nevertheless, these models produce surprisingly good forecasts—much better than those made before the advent of high-speed computers.

### ✔ Concept Checks 12.4

1. Briefly describe the basis of numerical weather prediction.
2. What are prognostic charts?
3. What do computer-generated numerical models try to predict?
4. What additional information does an ensemble forecast provide over a traditional numerical weather prediction?

## 12.5 Other Forecasting Methods

**List and distinguish among the various traditional methods of weather forecasting.**

Although machine-generated prognostic charts form the basis of modern forecasting, other methods are available to meteorologists. These methods, which have stood the test of time, include persistence forecasting, climatological forecasting, the analog method, and trend forecasting.

### Persistence Forecasting

Perhaps the simplest forecasting technique, **persistence forecasting** is based on the tendency of weather to remain unchanged for several hours or even days. If it is raining at a particular location, for example, it is reasonable to assume that it will still be raining in a few hours. Persistence forecasts do not account for possible changes in the intensity or direction of a weather system, nor can they predict the formation or dissipation of storms. Because of these limitations and the rapidity with which weather systems change, persistence forecasts usually diminish in accuracy within 6 to 12 hours, or 1 day at the most.

### Climatological Forecasting

**Climatological forecasting** is another relatively simple way of generating forecasts using climatological data—average weather statistics accumulated over many years. For example, Yuma, Arizona, experiences sunshine approximately 90 percent of its daylight hours; thus, forecasters predicting sunshine every day of the year would be correct about 90 percent of the time. Likewise, forecasters in Portland, Oregon, would be correct about 90 percent of the time by predicting overcast skies in December.

Climatological forecasting is particularly useful when making agricultural business decisions. For example, in the relatively dry north-central portion of Nebraska known as the Sand Hills, center-pivot irrigation made growing corn more feasible. However, farmers were faced with the question of which corn hybrid to plant. A high-yield variety widely used in warmer southeastern Nebraska was the logical choice, but local climatological data showed that because of the cooler temperatures in the Sand Hills, corn planted in late April would not mature until late September, when there is a 50 percent probability of an autumn freeze. Farmers used this important climate information to select a hybrid better suited for the shorter growing season of the Sand Hills.

One interesting use of climatological data is for the prediction of a "white Christmas"—that is, a Christmas with inch or more of snow on the ground. As Figure 12.14 illustrates,

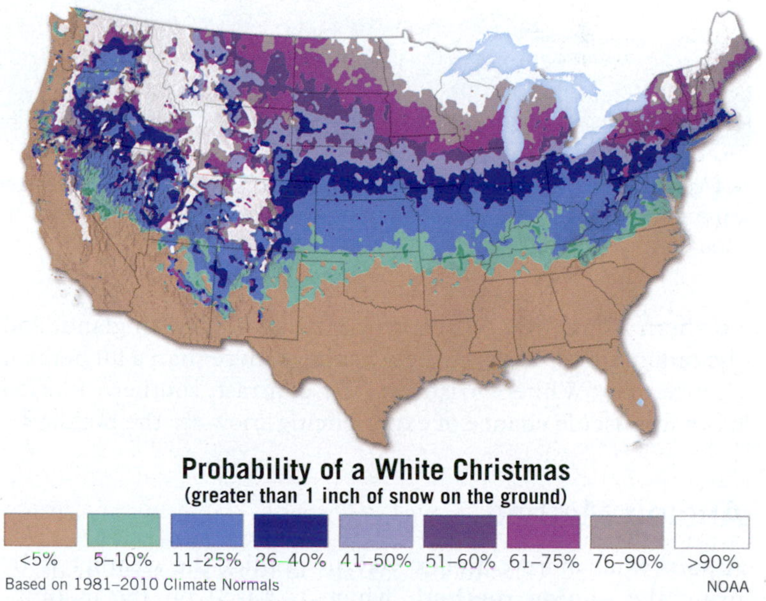


▲ **Figure 12.14 Probability of a "White Christmas"** The probability values are given as a percentage and represent the chance of a Christmas with at least 1 inch of snow on the ground.

▲ **Figure 12.15 Mesoscale phenomena such as this tornado are too small to appear on a prognostic chart** Detection of mesoscale weather events relies heavily on weather radar and geostationary satellites.

northern Minnesota, Wisconsin, Michigan, New England, and the mountainous areas of the West have more than a 90 percent chance of a "White Christmas." By contrast, southern Florida has a minuscule chance of experiencing snow for the holidays.

## Analog Method

A somewhat more complex way to predict the weather is by using the **analog method**, which is based on the assumption that weather repeats itself, at least in a general way. Thus, forecasters attempt to find well-established weather patterns from the past that are similar to a current event. From such a comparison, forecasters predict how the current weather might evolve. The analog method was the backbone of weather forecasting before the advent of computer modeling, and an analog method called *pattern recognition* remains an important tool for improving short-range machine-generated forecasts.

## Trend Forecasting

The **trend forecasting** method involves determining the speed and direction of features such as fronts, clouds, and areas of precipitation. Using this information, forecasters attempt to extrapolate the future position of these weather phenomena. For example, if a line of thunderstorms is moving northeastward at 35 miles per hour, a trend forecast would predict that the storm will reach a community located 70 miles to the northeast in about 2 hours.

Because weather events tend to increase or decrease in speed and intensity or change direction, trend forecasting is most effective over periods of just a few hours. Thus, it works particularly well for forecasting severe weather events that are expected to be short-lived, such as hailstorms, tornadoes, and downbursts, because warnings for such events must be issued quickly and must be site specific.

A type of trend forecasting called **nowcasting** is used to predict very short-lived (usually 6 hours), localized weather events. Nowcasting depends heavily on weather radar and geostationary satellites—important tools for detecting areas of heavy precipitation and clouds that can trigger severe conditions. Using interactive computers capable of integrating data from multiple sources, the primary goal of nowcasting is to issue warnings of severe weather, such as tornadoes, hailstorms, and high winds (Fig. 12.15).

### ✔ Concept Checks 12.5

1. If it is snowing today, what weather might be predicted for tomorrow if the persistence forecasting method is employed?
2. Briefly describe the basis of the analog method of weather forecasting.
3. What term is applied to the very short-range forecasting technique that relies heavily on weather radar and satellites? What type of weather phenomena are typically forecast using this technique?
4. What is the technique that predicts the weather based on average weather statistics accumulated over many years?

# 12.6 Weather Satellites: Tools in Forecasting

**Discuss the advantages and disadvantages of infrared imagery and visible imagery generated by weather satellites.**

Meteorology entered the space age in 1960, when the first weather satellite, *TIROS* 1, was launched. *TIROS* 1, like other weather satellites in this series, was placed into a polar orbit—so that it circles Earth from roughly north to south (Fig. 12.16A). **Polar-orbiting Operational Environmental Satellites (POES)** are low-flying satellites that orbit approximately 850 kilometers (530 miles) above Earth's surface 14 times daily. Earth's rotation causes these satellites to drift about 15° westward, recording a different view with each orbit. The POES system offers the advantage of recording images of the entire planet twice daily and covering a large region in a few hours. Also, because of their relatively low altitude, POES instruments obtain high-resolution images.

Data from POES instruments support weather analysis and forecasting, climate research, global sea-surface temperature measurements, atmospheric soundings of temperature and humidity in cloudless areas, and cloud and precipitation monitoring. The data collected by POES instruments can be assimilated directly into numerical prediction models to aid in producing more reliable short- and medium-range forecasts. The NWS partners with a consortium of European nations so that at least two polar-orbiting satellites are operational at all times—one from the POES series and one European satellite from the MetOp series.

Another class of weather satellites, **Geostationary Operational Environmental Satellites (GOES)**, was first placed in orbit over the equator in 1966 (Fig. 12.16B). As their name implies, GOES instruments remain fixed over a point on Earth because their rate of travel keeps pace with Earth's rate of rotation. In order to remain positioned over a given site, however, the satellite must orbit at a greater distance from Earth's surface (about 35,000 kilometers [22,000 miles]) than polar-orbiting satellites. At these higher altitudes, the GOES images obtained are less detailed than those captured by POES instruments.

The next-generation GOES series, GOES-R, is expected to be launched by 2016. As with the current series, it will be a two-satellite system. The satellites will be positioned at 75°W longitude and 137°W longitude, which will provide enhanced coverage of most of North America and surrounding water bodies.

Other countries, including a consortium of European countries, India, China, Japan, and Russia, have launched geostationary satellites. Collectively, these satellites allow meteorologists to track the movement of large weather systems, like the midlatitude cyclone shown in Figure 12.17, that cannot be adequately followed by weather radar or polar-orbiting satellites. Geostationary satellites easily track swirling cloud masses as they migrate across even the

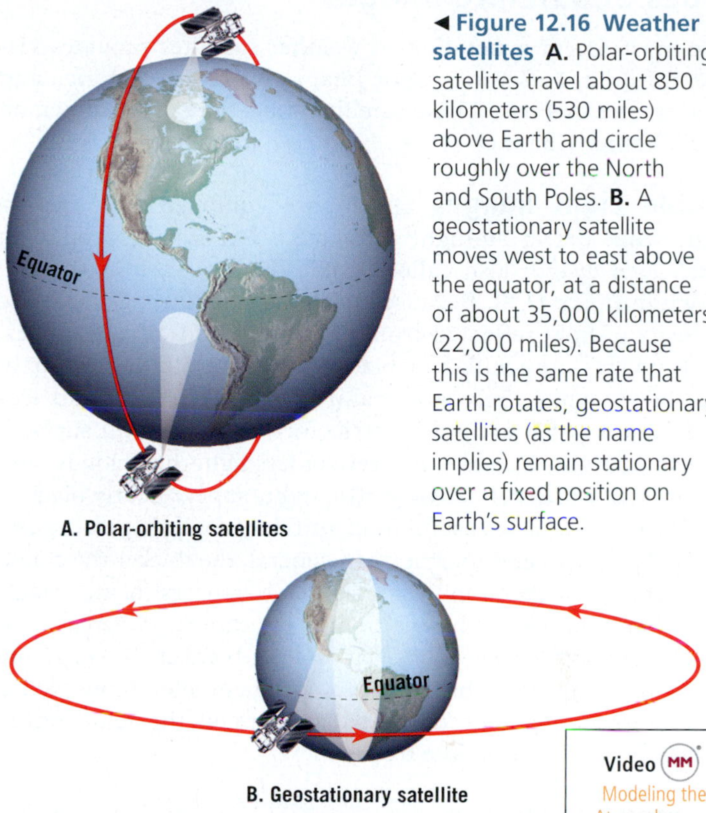


◄ **Figure 12.16 Weather satellites** **A.** Polar-orbiting satellites travel about 850 kilometers (530 miles) above Earth and circle roughly over the North and South Poles. **B.** A geostationary satellite moves west to east above the equator, at a distance of about 35,000 kilometers (22,000 miles). Because this is the same rate that Earth rotates, geostationary satellites (as the name implies) remain stationary over a fixed position on Earth's surface.

Video MM
Modeling the Atmosphere on Your Desktop
http://goo.gl/qUMzma

▼ **Figure 12.17 Weather satellites are invaluable tools for tracking storm systems** This true-color composite image of a mature midlatitude cyclone was produced on June 11, 2011, by an instrument called MODIS (Moderate Resolution Imaging Spectroradiometer).

most remote portions of Earth, and they also play an important role in monitoring the development and movement of tropical storms and hurricanes.

## Types of Satellite Images

The most recent generation of weather satellites provides visible, infrared, and water-vapor images for North America and adjacent ocean areas. These satellite images are often shown on The Weather Channel.

**Visible Light Images** GOES are equipped with instruments that detect sunlight reflected from Earth, producing *visible light images* (also called visible images) such as the one shown in **Figure 12.18**. Because visible light imagery records the intensity of light reflected from cloud tops and other surfaces, these images are similar to black-and-white photos of Earth. The bright-white areas are mainly clouds or snow- and ice-covered areas that are strong reflectors of light. Land surfaces appear dark gray because they reflect less light than clouds, and the oceans, which reflect very little light, appear nearly black.

Visible images are useful in identifying cloud shapes, organizational patterns, and thickness. In general, the thicker the cloud, the higher its albedo, and the brighter it appears in the image. Cumulonimbus clouds have a bumpy appearance, whereas stratus clouds form a continuous flat-looking blanket. **Figure 12.18B** is a visible light image that shows towering cumulonimbus clouds with overshooting tops that have penetrated the temperature inversion that lies above the troposphere.

**Infrared Images** In contrast to visible images, *infrared images* are obtained from radiation emitted (rather than reflected) from objects and, when compared to visible images, are especially useful in determining which clouds are most likely to produce precipitation and/or stormy conditions. Infrared images are produced by computers programmed to display warm objects such as Earth's surface in black and colder objects such as high cloud tops in white. Because high cloud tops are colder than low cloud tops, towering cumulonimbus clouds that usually generate heavy precipitation appear very bright, whereas midlevel clouds appear light gray. Low fair-weather cumulus clouds, stratus clouds, and fog are all relatively warm and therefore appear dark gray or may not be discernable from Earth's surface on infrared images.

Compare Figure 12.18A, a visible image, to **Figure 12.19A**, which is an infrared image of the same region. Both images were acquired on a very hot summer day in 2011 that produced thunderstorms over parts of Minnesota and Wisconsin, as well as in the Southeast. As you can see, the clouds over eastern South Dakota that appear white on the visible image appear gray on the infrared image. Middle clouds at 2000–6000 meters blanket these areas. By contrast, the clouds over eastern Minnesota and western Wisconsin appear bright in both the visible and infrared images. This region has towering cumulonimbus clouds and was experiencing air-mass thunderstorms that brought heavy rain to the area. The areas of isolated storm cells in the southeastern United States are also easily identifiable on the infrared image in Figure 12.19A.

To help meteorologists interpret infrared images, false colorization is sometimes used. The coldest, and thus highest, cloud tops, which typically identify areas of heavy precipitation, are shown in strong colors. Notice the reddish areas over the Minnesota–Wisconsin border in Figure 12.19B. This is the same area that appears as bright white in the infrared image in Figure 12.19A.

**Visible Versus Infrared Imagery** A major advantage of infrared imagery is its ability to detect energy radiated skyward both day and night. As a result, the movement of storms can be monitored 24 hours a day. Visible images, which require reflected sunlight, are not available at night. However, an advantage of visible images is their high resolution—which makes detection of smaller features possible. **Figure 12.20** is a close-up of a portion of the visible image in Figure 12.18. Notice the bands of fair-weather cumulus clouds over eastern Missouri and Iowa

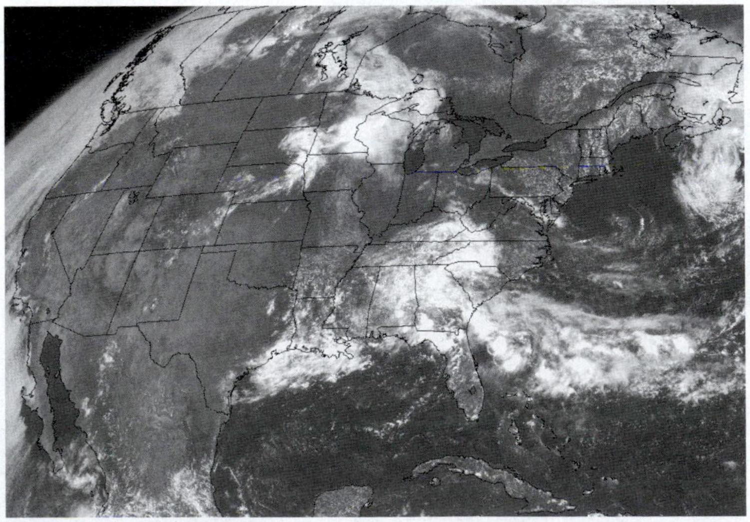

A.

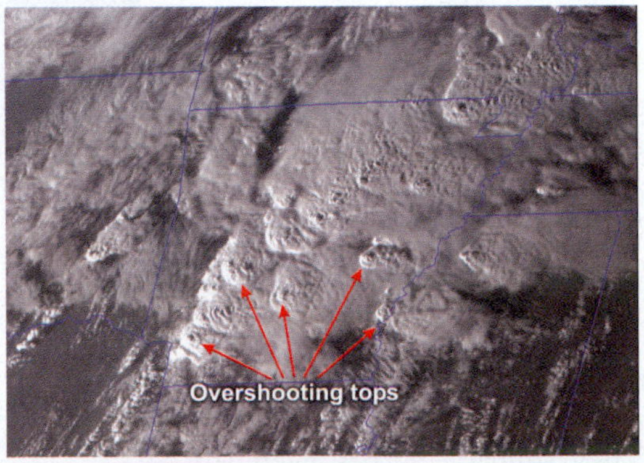


B.

▼ **Figure 12.18 Visible light images A.** Visible light image provided by a GOES of cloud distribution about midday on July 15, 2011. This image records sunlight reflected off various Earth surfaces and appears much like a black-and-white image. **B.** Close-up visible image centered on the state of Iowa shows towering cumulonimbus clouds with overshooting tops.

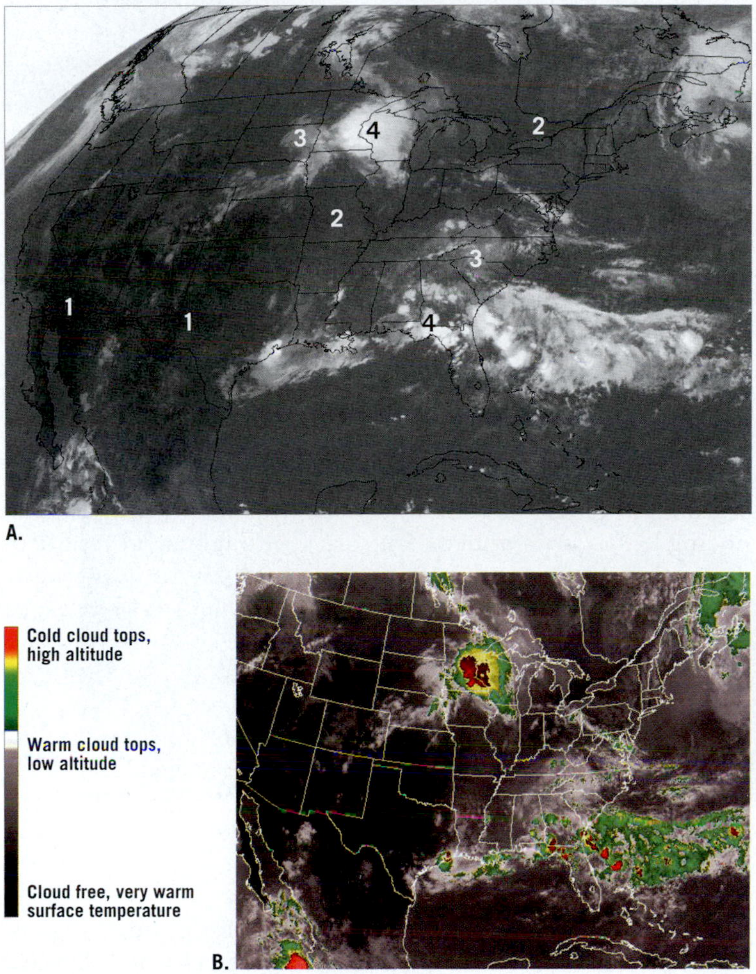


▲ **Figure 12.19 Infrared satellite images** **A.** Infrared image provided by a GOES about midday on July 15, 2011. Numbers 1–4 in this infrared image identify areas that (1) are cloud free; (2) contain low clouds (cumulus) with cloud tops below 2000 meters; (3) are blanketed by cloud tops between 2000 and 6000 meters; (4) feature high, cold cloud tops associated with towering cumulonimbus clouds and thunderstorms, measured by the infrared sensor. **B.** Color-enhanced infrared image for the same time period as part **A.** The dark-reddish color indicates the location of towering cumulonimbus clouds.

and the bumpy tops of the developing cumulonimbus clouds over southeastern Minnesota. Visible images are ideal for showing cloud patterns and storm systems such as midlatitude cyclones and hurricanes.

**Water-Vapor Images** *Water-vapor images* provide yet another way to view Earth. Most of Earth's radiation with a wavelength of 6.7 micrometers is emitted by water vapor. Satellites equipped with detectors for this narrow band of radiation are, in effect, mapping the concentration of water vapor in the atmosphere. Bright-white areas in Figure 12.21 represent regions of high water-vapor concentration, whereas dark areas indicate drier air. Because most fronts occur between air masses having contrasting moisture conditions, water-vapor images are valuable tools for locating frontal boundaries.

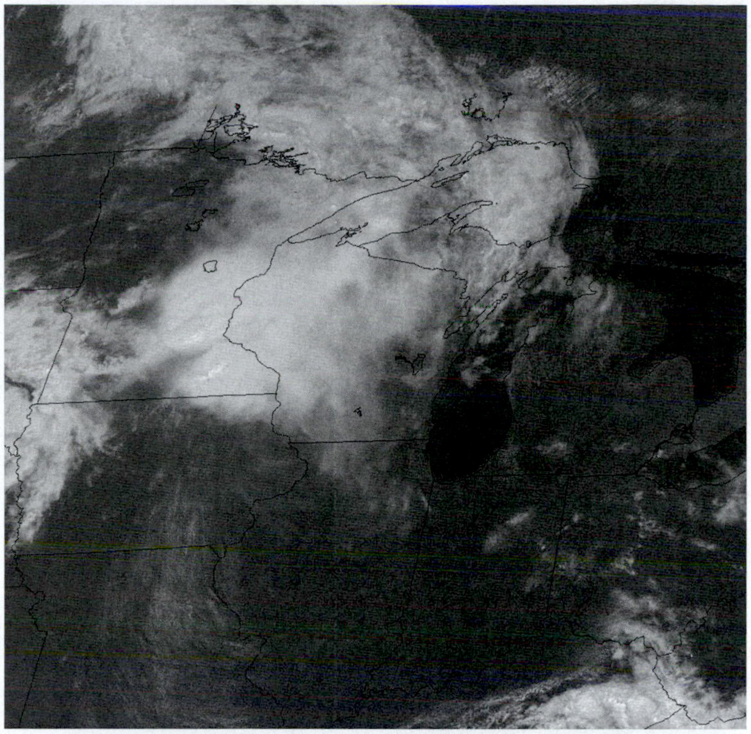

▲ **Figure 12.20 Close-up GOES visible image obtained July 15, 2011** This visible image centers around Wisconsin and was obtained on the same date and time as the images in Figures 12.18 and 12.19. The area of bright clouds over the Minnesota–Wisconsin border were producing heavy precipitation when this image was obtained.

▼ **Figure 12.21 Water-vapor image obtained July 15, 2011** The brighter the white, the greater the atmosphere's water-vapor content. Black areas are driest.

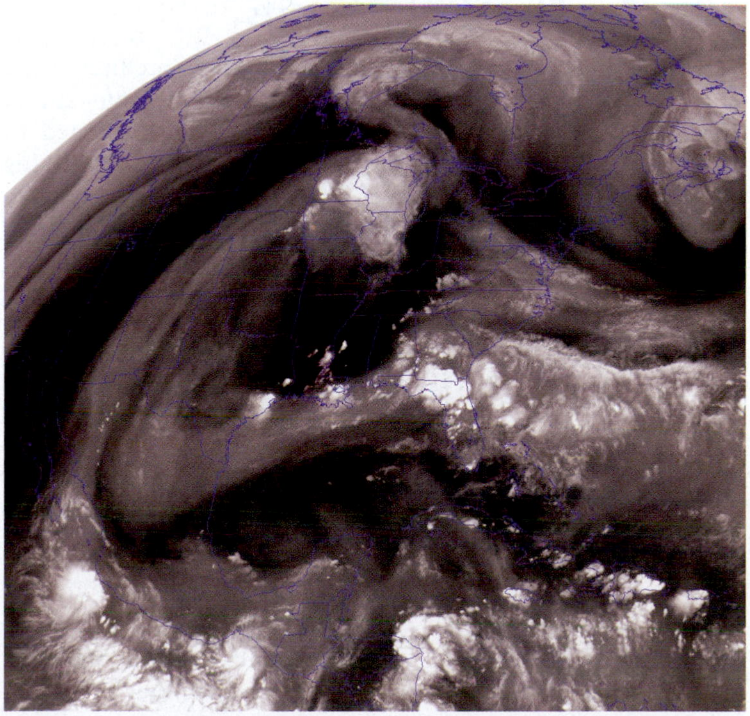

## eye ON THE atmosphere 12.2

These satellite images of the eastern Pacific and a portion of North America were obtained at the same time on July 14, 2001. One of these is an infrared image, and the other is a visible light image.

A.

B.

**Questions**

1. On which image (A or B) are the cloud shapes and patterns more easily recognized?
2. Which one of these images was obtained using infrared imagery? How can you tell?
3. What types of clouds are found in the western Great Plains (just east of the Rockies)?
4. Are the clouds over the eastern Pacific Ocean mainly high clouds or middle- to low-level clouds?

## Other Satellite Measurements

Ingenious developments have made weather satellites more than high-tech TV cameras pointed at Earth from space. Some satellites are equipped with instruments designed to measure temperatures and moisture at various altitudes. This data is an important addition to the upper-air measurements collected by other means. Other satellites allow us to measure rainfall intensity and amounts, both at Earth's surface and above the surface, and to monitor winds in regions where few, if any, conventional instruments are available. Images from the *Tropical Rainfall Measuring Mission* (*TRMM*; see Box 1.1, page 8) and the recently launched *Global Precipitation Measurement Mission* illustrate satellite-based precipitation measurement).

### ✔ Concept Checks 12.6

1. How do satellites help identify clouds that are likely to produce precipitation?
2. What advantage do geostationary satellites have over polar-orbiting satellites? Name one disadvantage.
3. What advantages do infrared images have over visible light images? Name one disadvantage.

## 12.7 Types of Forecasts

**Compare and contrast qualitative and quantitative weather forecasts. Differentiate between weather forecasts and 30- and 90-day outlooks.**

Because many disasters are weather related, probably the most significant professional contribution of a forecaster is issuing a forecast, or warning, of an impending severe weather event. For example, forecasters are not only concerned about the development of a tornado but also its size, location, and intensity. To forecast the path the tornado may take, the characteristics of the parent thunderstorm as well as the nature of the steering winds must be considered.

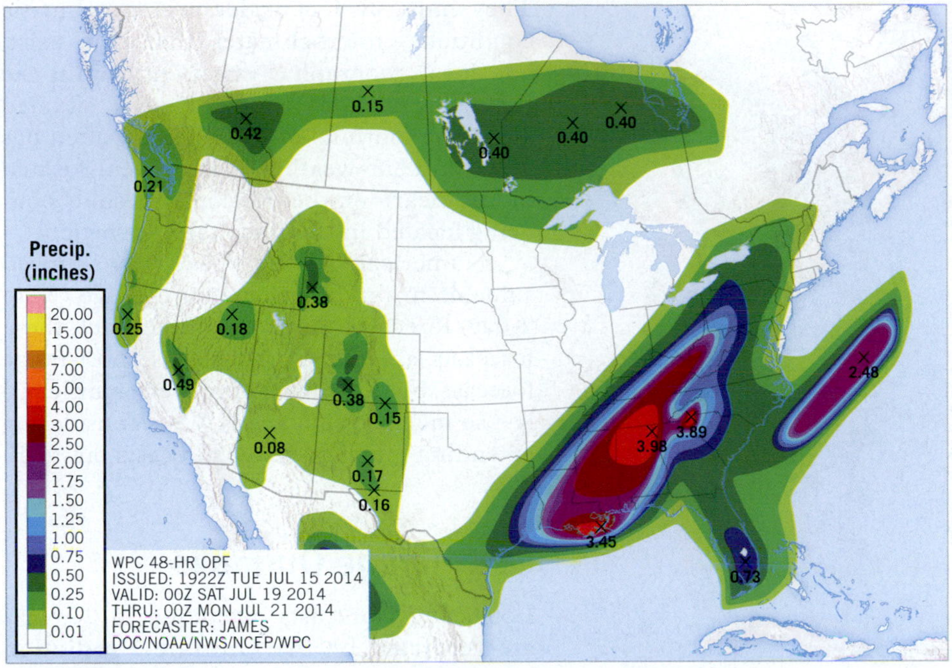


▲ **Figure 12.22 Quantitative Precipitation Forecast (QPF)** This QPF depicts the amount of liquid precipitation expected to fall within a 48-hour period. Precipitation can vary significantly over short distances, especially during thunderstorms, so these forecasts are given as averages. Locations that are expected to receive the largest amount of precipitation (in inches) are marked with the letter *X*.

## Qualitative Versus Quantitative Forecasts

Most forecasts are of one of two types—either qualitative or quantitative.

A **qualitative forecast** describes an aspect of the weather that can be observed but not easily measured, or quantified. Examples include forecast statements like these: "partly cloudy in the morning followed by gradual clearing"; "gusty winds expected in the afternoon"; and "scattered thunderstorms possible after sunset." Qualitative forecasts, although not easy to assess, can be very informative to the public.

**Quantitative forecasts**, by contrast, deal with measurable weather data. An example is a forecast predicting maximum and minimum temperatures. A forecast predicting the amount of precipitation a region will receive over a specified period of time is another example of a quantitative forecast.

Figure 12.22 shows a **Quantitative Precipitation Forecast (QPF)** that depicts the amount of liquid precipitation (in inches) expected to fall over an area in a defined period of time. Because precipitation can vary significantly over short distances, especially during thunderstorms, these forecasts are given as averages for a given area. The forecast shown in Figure 12.22 is for a 48-hour period, with the locations expected to receive the largest amounts of precipitation marked with the letter *X* on the map. Like most other forecasts issued by the NWS, precipitation forecasts can be easily accessed on the Internet by searching for the type of forecast you are interested in acquiring.

The only NWS-issued forecast that is given as a *percentage probability* is precipitation. These forecasts, referred to as **Probability of Precipitation (PoP)**, are often expressed something like ". . . chance of rain 40 percent" or ". . . a 40 percent chance of thunderstorms." What does 40 percent mean? Will it rain 40 percent of the time, or will it rain over 40 percent of the forecast area? Neither. The PoP describes the chance of precipitation occurring in any particular location within the forecast area. PoP is defined as:

$$\text{PoP} = (C \times A) \times 100 \text{ percent}$$

where $C$ is the *confidence* that precipitation will occur *somewhere* in the forecast area, and $A$ equals the *percentage of the area* that will receive measureable precipitation, if it occurs at all.

For example, if the forecaster expresses confidence that there is a 50 percent chance of precipitation in the forecast area and also expects that if rain occurs, it will produce measurable precipitation over 80 percent of the area, the PoP (change of rain) will be 40 percent (PoP = (.5 × .8) × 100 percent = 40 percent). The NWS explains this as a 40 percent chance that precipitation will occur at any given point in the forecast area. Another way to express "40 percent chance of rain" is "on *average*, for all points in the area during the forecast period (usually 12 hours), the chance that rain will occur is 40 percent."

The NWS also produces another type of probability of precipitation forecast for the continental United States and surrounding areas in 6- and 24-hour increments, shown in Figure 12.23A. Notice the area in northeastern Texas and adjacent states with a high probability of receiving precipitation over the next 6 hours. Compare this area to the region that was issued a flash flood warning on the weather map for that same date, in Figure 12.23B. The region under a flood warning corresponds closely to the area on the precipitation map with a high probability of precipitation.

## Short- and Medium-Range Forecasts

What tools forecasters use depends largely on the range of the expected forecast, or how far into the future the forecast extends. The NWS defines weather forecasts based on time, as follows:

- Short-range forecasts extend up to 48 hours (2 days).
- Medium-range forecasts cover periods extending about 3 to 7 days into the future.
- Long-range forecasts are for periods beyond 7 days and have no outer time limit.

As described earlier, *nowcasting*, the shortest forecast period, extends to up to about 6 hours. This includes forecasting weather phenomena such as tornadoes and thunderstorm downbursts that cannot be predicted using numerical forecasting

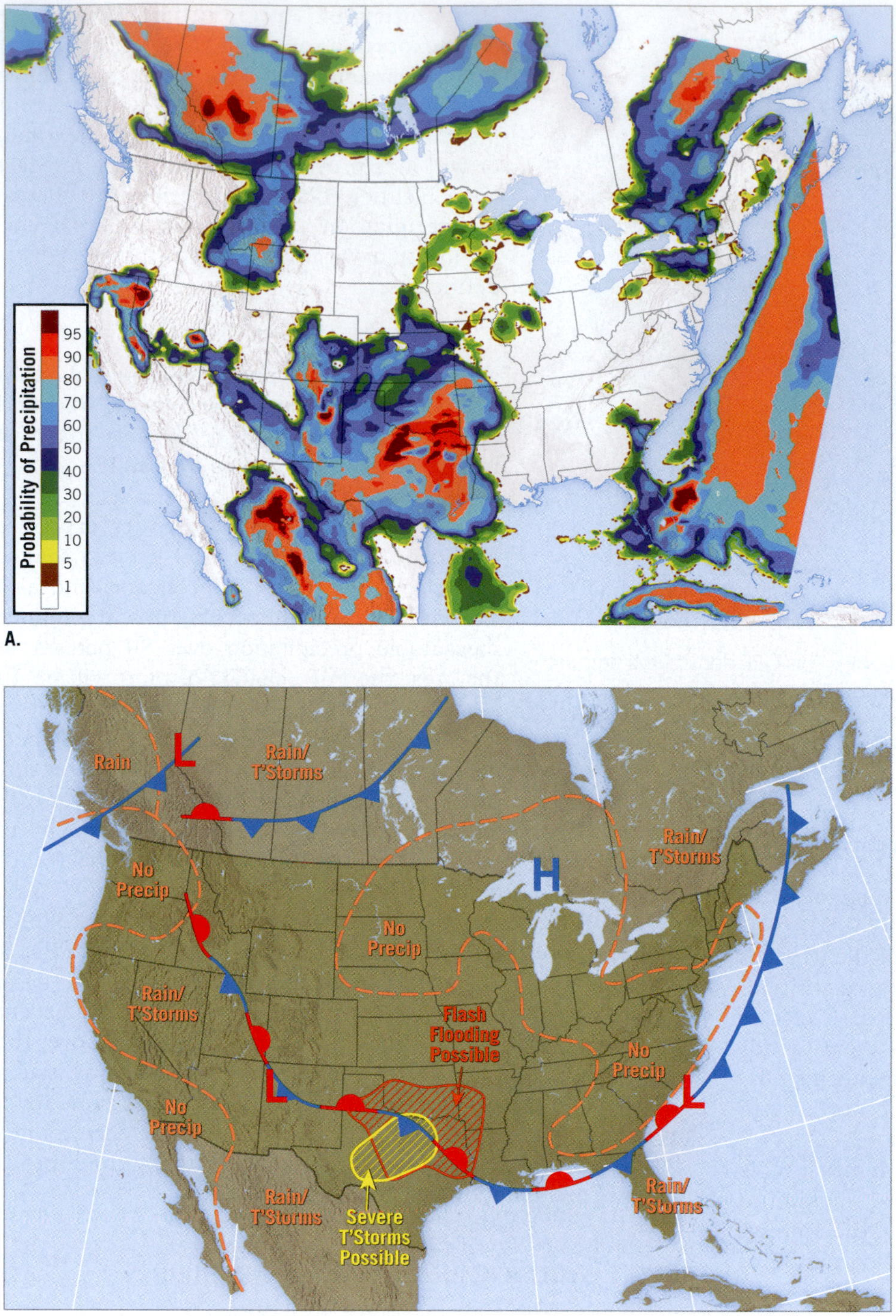


▲ **Figure 12.23 Probability of precipitation forecast map** **A.** Notice the area in northeastern Texas and adjacent states that has a high probability of receiving precipitation over the next 6 hours. Compare this area to the region in B issued for that same date. **B.** The region under a flash flood warning on the weather map corresponds closely to the region on the precipitation map with a high probability of precipitation.

techniques (see Fig. 12.15). To make these short-range forecasts, meteorologists rely heavily on surface weather conditions, weather radar, and satellite images.

Short-range forecasts that extend from the present out to 2 days are based on numerical forecasting techniques. Earlier we discussed how forecasters use their knowledge of typical model behavior and expertise to improve numerical forecasts. They must also consider average climatic conditions, microclimates that may exist, and local geographic variations within the forecast area. For example, Denver, situated along the foothills of the Rockies, often has very different weather than Aspen, located in the heart of the Rockies—despite both being located in the same forecast region.

Numerical models have also proved valuable in producing medium-range (3- to 7-day) forecasts. As you would expect, these forecasts are not as reliable as short-range forecast, but they have improved significantly—so much so that the NWS is considering extending medium-range forecasts out to 10 days.

## Long-Range Outlooks

Long-range forecasts extending to 2 weeks are generated by the **Climate Prediction Center (CPC)** of the NWS. These forecasts depend on a combination of numerical and statistical forecast guidance. The CPC is also responsible for producing monthly and seasonal (3-month) weather maps for temperature and precipitation called 30- and 90-day outlooks. These are not typical weather forecasts; instead, they offer insights into whether it will be drier or wetter and colder or warmer than normal in a particular region of the country.

A series of 13, 90-day outlooks are produced in 1-month increments. The series begins with the outlook for the next 3 months—for example, September, October, and November. Next, a separate 90-day outlook for the period beginning 1 month later (October, November, and December) is constructed. Figure 12.24 shows the temperature and precipitation 90-day outlooks issued for September, October, and November 2011. The temperature outlook for this period calls for warmer-than-usual conditions across much of the southwestern, north-central, and northeastern United States (Fig. 12.24A). The remainder of the country is labeled "equal chances," which indicates that there are no climatic signals for either above- or below-normal conditions during the forecast period. The precipitation outlook shown in Figure 12.24B calls for wetter-than-normal conditions across most of Kansas, Nebraska, and South Dakota, whereas drier-than-normal conditions are expected in a portion of the Southwest from Texas to southeastern Arizona.

Monthly and seasonal outlooks of this type are generated using a variety of criteria. Meteorologists consider each region's climatology—the 30-year average of variables

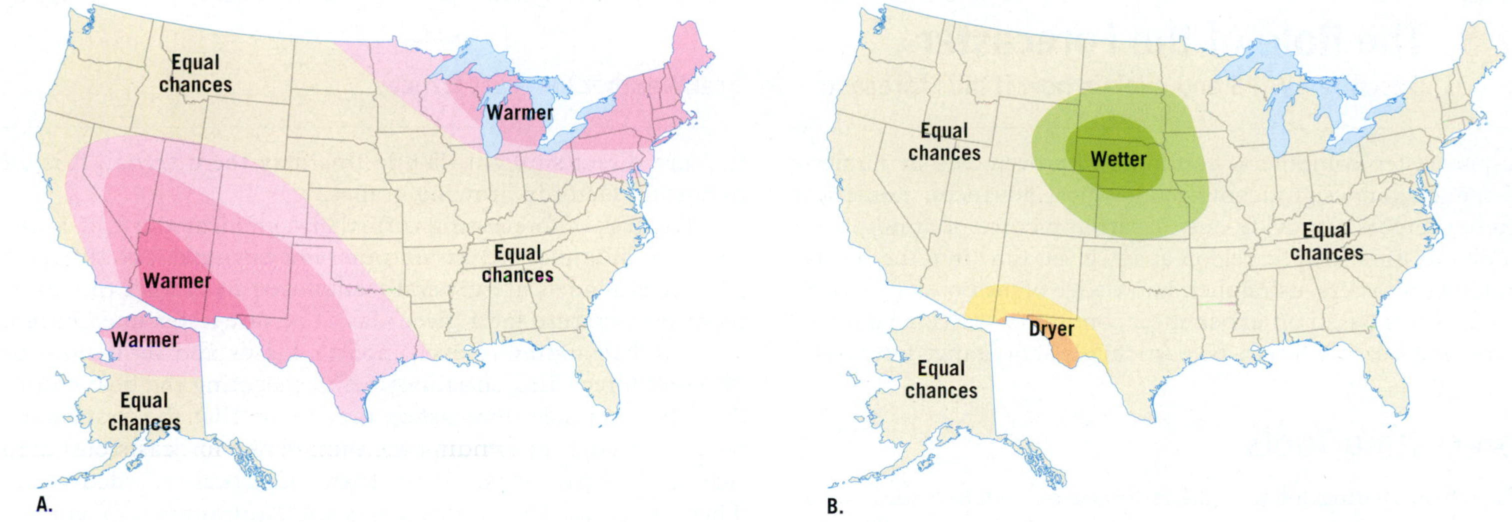


▲ **Figure 12.24 Extended forecast (90-day outlook) maps** These maps depict the **A.** temperature and **B.** precipitation outlooks for September, October, and November 2011. "Equal chances" means no climatic signals for either above- or below-normal conditions.

such as temperature and precipitation. Factors such as snow and ice cover in the winter and persistently dry or wet soils in the summer are considered. Forecasters also take into account current temperature and precipitation patterns. For example, during 2011 and the few preceding years, the U.S. Southwest experienced below-normal precipitation. Based on climatological data, the weather patterns that produce these conditions tend to gradually shift toward the norm rather than change abruptly. Thus, forecasters predicted that this below-normal trend would continue for at least the first several months of 2012.

Recently, sea-surface temperatures in the equatorial Pacific Ocean have been valuable in forecasting temperature and precipitation patterns in various locations around the globe—particularly in the winter season. For more on this relationship, see the section "El Niño, La Niña, and the Southern Oscillation" in Chapter 7.

Despite improved seasonal outlooks, the reliability of extended forecasts has been disappointing. Although some accuracy can be achieved during the late winter and late summer, outlooks prepared for the "transition months" when the weather can fluctuate wildly have shown little reliability.

## Watches and Warnings

The NWS issues *watches* and *warnings* to alert the public about potentially dangerous weather. A **watch** means conditions are right for hazardous weather development, and residents should be aware and ready to act. A **warning** means that dangerous weather has developed and has been located in or moving toward the area. For severe thunderstorms, tornadoes, and flash floods—events that develop quickly—response times are usually very short, so taking immediate shelter is strongly recommended. A hurricane warning means to either evacuate or move to a safe location.

The **Storm Prediction Center (SPC)** in Norman, Oklahoma, has the primary responsibility for forecasting severe weather such as thunderstorms, tornadoes, hail, and lighting. Whenever possible, "outlooks" for areas that might incur severe weather are issued a day prior to the anticipated event. The SPC then works with the local forecast offices of the NWS that are equipped with weather radar to monitor severe weather events as they develop. Before a warning is issued, the NWS wants to be relatively certain a severe weather event has in fact developed. To aid in this process, a volunteer program called SKYWARN was developed in the 1970s. Thousands of volunteers serve as spotters who provide critical information on the location and movement of the severe weather event. Once such an event is reported, the NWS broadcasts a warning via weather radio and contacts governmental authorities responsible for public safety in the affected areas, as well as the private sector.

Tracking and monitoring tropical storms and hurricanes in the Atlantic, Caribbean, and eastern Pacific is the task of the **National Hurricane Center (NHC)** in Miami, Florida. An important forecast challenge for the NHC is population growth along the Gulf and Atlantic coasts, with some areas requiring a 24-hour warning for evacuation. Additional evacuation time is also needed for people located on offshore or barrier islands that have limited access to the mainland.

### ✔ Concept Checks 12.7

1. How is a quantitative forecast different from a qualitative forecast?
2. Explain the difference between a weather watch and a warning.
3. Describe how a tornado warning is different than a warning of a hurricane landfall.
4. What two weather elements are predicted on long-range (monthly) weather charts?

# 12.8 The Role of the Forecaster

**Describe AWIPS and explain how it aids forecasters at local Weather Forecast Offices.**

Despite faster computers, continual improvements in numerical models, and significant technological advances, numerical weather forecasts provide only a partial picture of atmospheric conditions for a specific region at any given time into the future. Human forecasters, using their knowledge of meteorology as well as judgments based on experience, continue to serve a vital role in creating weather forecasts, particularly short-range forecasts.

## Forecasting Tools

The job of issuing local weather forecasts, watches, and warnings to protect life and property rests with the National Weather Service (NWS) and its numerous branches. The NWS provides a local forecaster with the observational data, satellite feeds, numerical model outputs, and the various weather maps discussed earlier in this chapter.

Forecasters at the local Weather Forecast Offices are responsible for modifying numerical predictions by taking into account local conditions and nuances to produce site-specific forecasts. One local condition that forecasters must consider is the existence of microclimates—different climatic zones within a relatively small area. This occurs in Napa Valley, California, a prominent wine-producing district; the southern part of the valley is cooler than the northern part due to its close proximity to the cool waters of San Pablo Bay. On a smaller scale, the south-facing slopes in Napa Valley are warmer than the north-facing slopes because they are exposed to more direct sunlight. While this may seem trivial, it is an important factor in growing grapes.

The task of forecasting is further complicated by the availability of multiple forecast outputs. For example, two different numerical models are generally employed to predict the minimum temperature for a given day. The forecaster must have a basic understanding of each model's biases and limitations in different forecasting situations before selecting the best output data. For example, forecasters may know that the first model tends to predict minimum temperatures for their forecast area that are a degree or two lower than the actual recorded value. Therefore, if the first model predicted a minimum temperature of 45°F and the second model predicted a minimum of 47°F, the forecaster should feel confident using 47°F as the forecast value. In other situations, when no model biases are known for predicting a particular weather element, the forecaster may choose to blend data from both models.

**Advanced Weather Interactive Processing System** To aid in the enormous task of weather forecasting, local weather forecast offices are equipped with an interactive computer system called **Advanced Weather Interactive Processing System (AWIPS)**. AWIPS is an information-processing, display, and communication system that integrates all model and observational data, including satellite and radar data, in one location. An AWIPS workstation consists of three graphical displays and usually one or more personal computers for interacting with text materials (Fig. 12.25). The personal computers are used to access the Internet and e-mail, as well as to look at textual observational data and to communicate with other forecasters. The three AWIPS graphic monitors are connected to a mainframe computer and can be configured into as many as 20 different windows—each displaying different weather information. Forecasters use this system to analyze and integrate the following types of data:

- Numerical weather forecast products
- Doppler radar (the WSR-88D) displays, including dual-polarization radar
- GOES data
- Surface data from an Automated Surface Observing System (ASOS)

◄ **Figure 12.25 Advanced Weather Interactive Processing System (AWIPS)** AWIPS integrates all model and observational data, including satellite and radar data, in one location. An AWIPS workstation consists of three graphical displays and usually one or more personal computers for interacting with text materials.

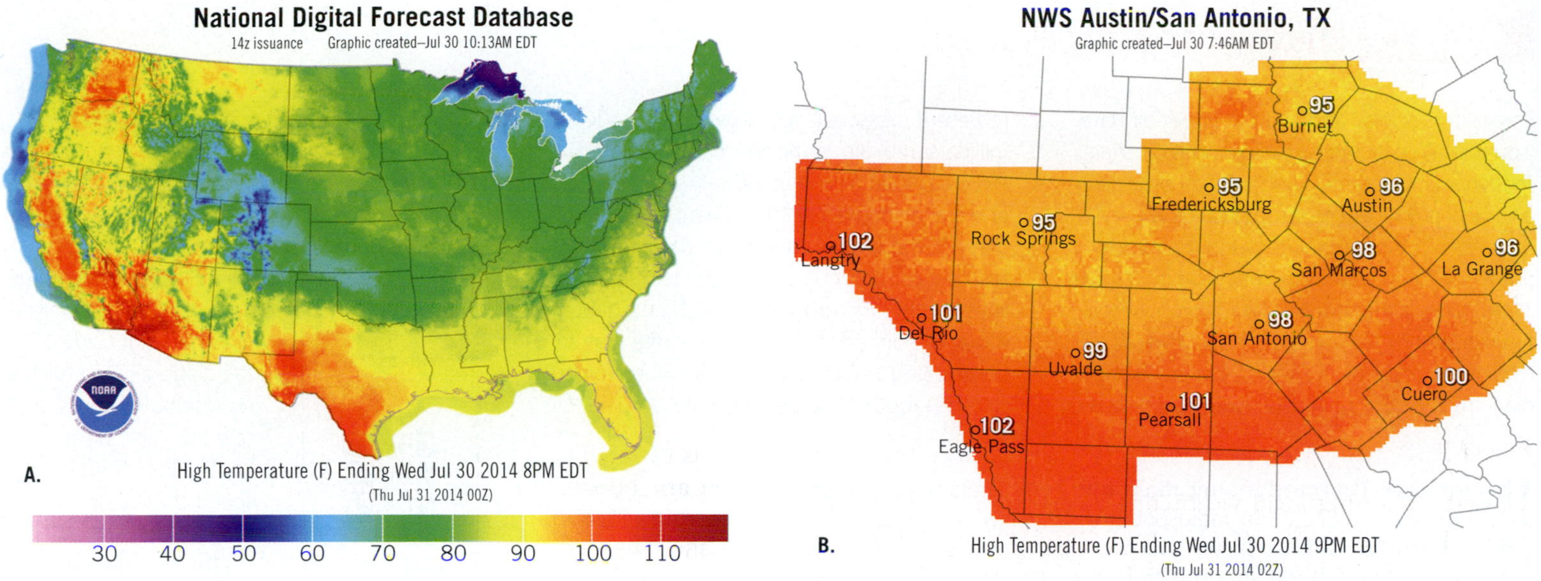


▲ **Figure 12.26 Graphical forecast predicting maximum daily temperatures** **A.** This graphical forecast is for the entire continental United States on July 30, 2014. **B.** The predicted maximum daily temperatures for the Austin/San Antonio area of Texas on that same date.

- Forecast guidance products from the National Hurricane Center and the Storm Protection Center
- Upper-air sounding data from radiosondes launched at nearby sites
- Surface weather maps and upper-level charts.

There are usually multiple AWIPS systems found in each local Weather Forecast Office. A second-generation system, AWIPS II, is currently being implemented to further enhance forecasters' access to weather data.

To use AWIPS, a forecaster imports numerical model prediction data and a variety of observational data into the system—which is equipped with software known as the Interactive Preparation System (IFPS) and the Graphical Forecast Editor (GFE). Then, using tools in the IFPS/GFE, the forecaster manipulates the model guidance for each weather element. Multiple weather elements make up forecast products, including maximum and minimum temperatures, probability of precipitation and amounts, sky cover, and wind speed and direction.

Once forecasters are confident in their modifications to the original guidance, the computer system automatically creates all the forecast products from the graphic display generated using the IFPS/GFE software. In addition to a text forecast, an entire suite of forecast graphics are produced and posted on the Internet. Graphical forecasts for the entire United States, as well as regional and local forecasts for various weather elements, are available by searching online for "NWS graphical forecasts." **Figure 12.26** shows a typical graphic produced for July 30, 2014, showing the predicted maximum daily temperatures for the entire United States and another showing the Austin/San Antonio area of Texas.

These graphic displays are also automatically converted to text, posted on a website, and instantly converted to a computer-generated voice that is broadcast over weather radio. Efficiency in this process—from the time a forecast is generated to the time a watch or warning is disseminated to response personnel and the general public—is critical. AWIPS has significantly aided this process.

**Assessing Potential Atmospheric Instability** In addition to observational data described earlier, forecasters utilize diagrams on which the upper-level data (soundings) collected by radiosondes are plotted. These graphs, called **thermodynamic diagrams**, provide a vertical profile of the temperatures and dew-point temperatures through the atmosphere for a particular time and location. One of the most commonly used thermodynamic diagrams, called a **Skew-T/Log P diagram**, or simply a **Skew-T**, is described in **Box 12.1**. Although Skew-T diagrams look complicated, and using them effectively requires some practice, they are particularly useful when assessing the potential for atmospheric instability—a trigger for severe weather.

Two ways to express the potential for instability are the *lifted index* and the *K-index*. The lifted index is a number generated by hypothetically lifting a parcel of air from Earth's surface to the 500-millibar level, then comparing its temperature to that of the environment aloft. The *lower* the lifted index value, the *more unstable* the atmosphere is in that location (**Fig. 12.27**). This index is useful for predicting severe weather events.

Although similar to the lifted index, the K-index is used mainly to predict air-mass thunderstorms that produce heavy rainfall but that typically are not accompanied by severe weather. The K-index measures thunderstorm potential based on the environmental lapse rate, the moisture content of the lower atmosphere, and the vertical extent of the moist layer. K-indices are also used to determine the potential for flooding; a high value indicates a strong flooding potential.

## Box 12.1 Thermodynamic Diagrams

Forecasters display and analyze upper-air data (soundings) collected by radiosondes on *thermodynamic diagrams*. Such a graph provides a vertical profile of the temperatures and dew-point temperatures through the atmosphere for a particular time and location. **Figure 12.A** shows a thermodynamic diagram similar to that used by the NWS, called a *Skew-T/Log P diagram*, or simply a *Skew-T*. On the Skew-T diagram in Figure 12.A, the horizontal blue lines represent *pressure levels*, labeled every 100 millibars (mb) along the left margin. Slanted (skewed) solid blue lines indicate *temperature* and are labeled every 10°C—hence the name *Skew-T*. The slightly curved thin dashed red lines almost perpendicular to the temperature lines are the *dry adiabats*. Recall from Chapter 4 that the dry adiabatic rate is the rate that an unsaturated parcel of air will cool when forced to rise—about 10°C per 1000 meters. The thin dashed green lines are *wet adiabats*, which depict the rate at which a saturated parcel of air cools as it rises through the atmosphere. In addition, saturation mixing ratios—the amount of water vapor required to saturate air at a given temperature—are shown with thin dashed blue lines. Wind speed and direction are also plotted along the right side of the diagram. One full bar represents 10 knots, and one triangle indicates 50 knots. Note how wind speed changes with height.

Video MM
Forecasting Thunderstorms
http://goo.gl/T26p2r

The sounding shown in Figure 12.A was obtained over Oshkosh, Wisconsin, on April 30, 2012. The two weather elements plotted on this Skew-T diagram are actual air temperatures (solid red line) and dew-point temperatures (solid green line) recorded by a radiosonde carried up. This diagram offers a snapshot of the atmosphere from the surface to the 100-millibar level. Notice that the dew-point temperature line and the temperature line overlap from about the 700-millibar level up to slightly above the 500-millibar level, where they separate. Recall that when air temperature and the dew-point temperature are the same, the air is saturated and water vapor condenses, usually to form clouds. Thus, this sounding indicates the existence of a cloud deck over Oshkosh in a band roughly between the 700- and 440-millibar levels.

**▼ Figure 12.A Thermodynamic diagrams** Forecasters typically display and analyze upper-air data (soundings) collected by radiosondes on a type of thermodynamic diagram called a Skew-T/Log P diagram, or simply a Skew-T. Such a graph provides a vertical profile of air and dew-point temperatures through the atmosphere for a particular time and location.

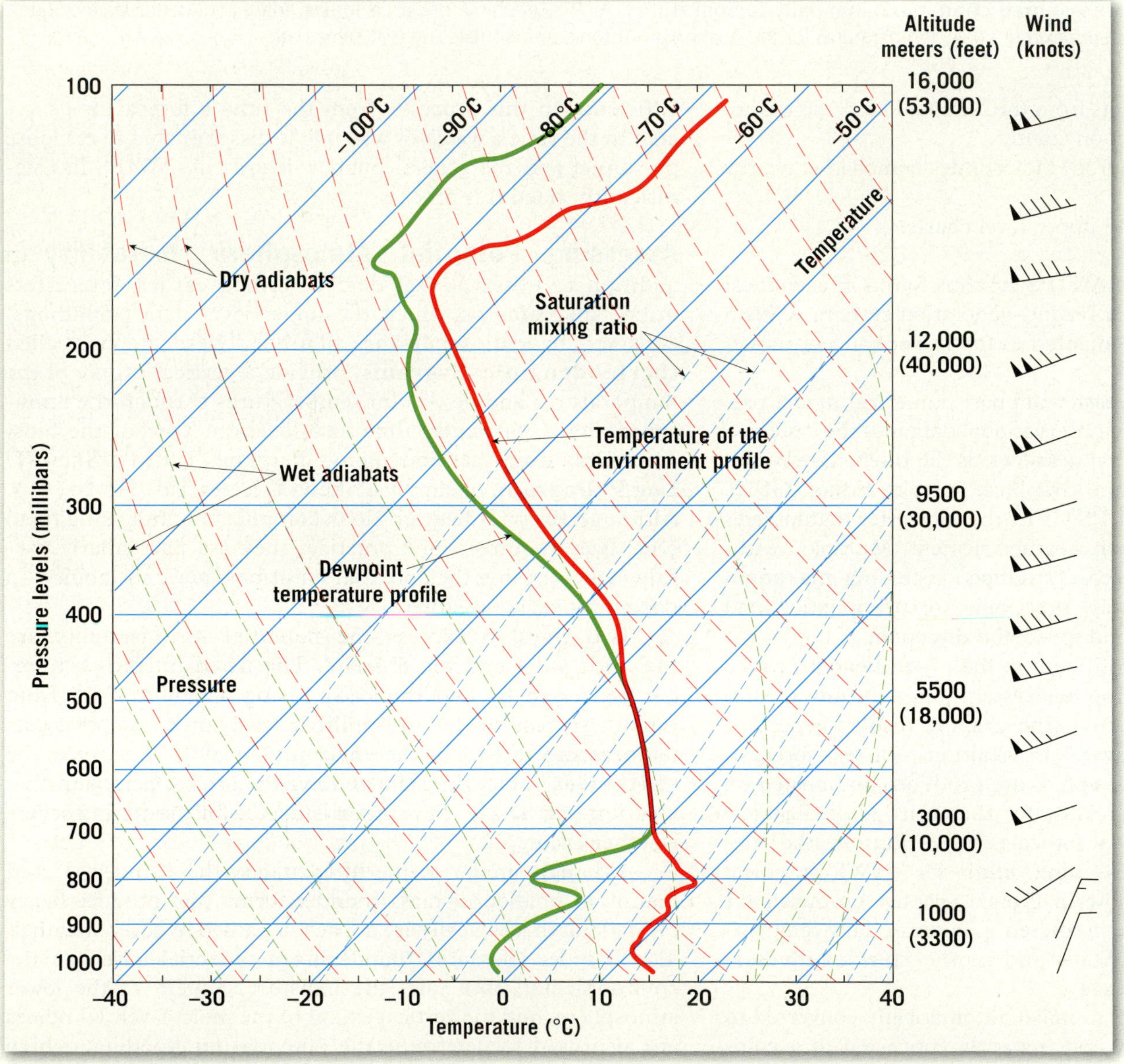

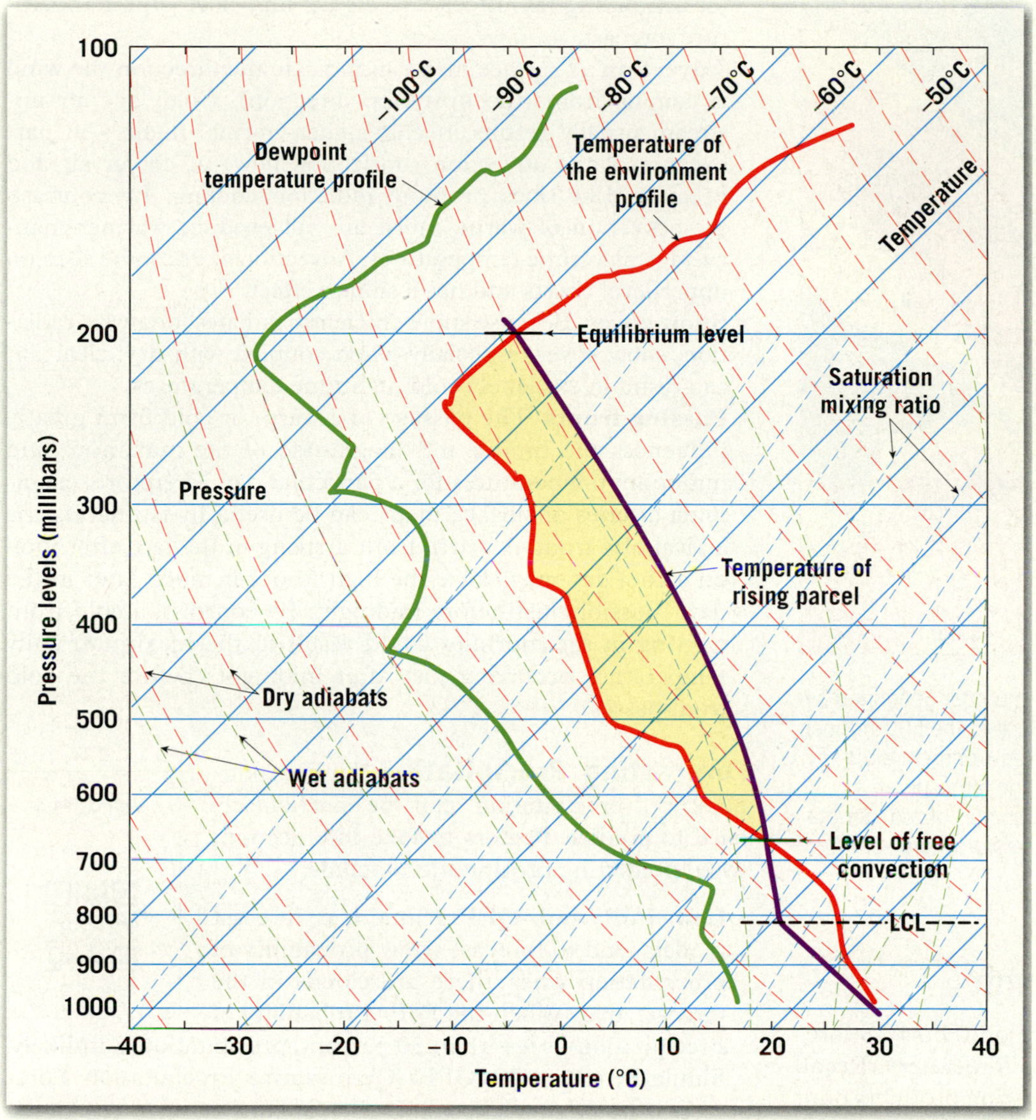


**◀ Figure 12.B Using a Skew-T diagram** Skew-Ts are useful for assessing the potential for instability that can lead to deep atmospheric convection—a condition that may generate towering cumulonimbus clouds and severe weather. The term *deep* means convection that begins near the ground and extends to the top of the troposphere.

Although Skew-T diagrams look complicated, and using them effectively requires some practice, they are particularly useful in assessing potential instability that can lead to deep atmospheric convection—a condition that may generate towering cumulonimbus clouds and severe weather. The term *deep* means convection (mainly upward air movement, but downdrafts are also an important component) that begins near Earth's surface and extends to the top of the troposphere. **Figure 12.B** is Skew-T on which hypothetical dew-point and temperature soundings have been plotted. Using this data, we will assess the stability (or instability) of the atmosphere by locating where a parcel of rising air is warmer than the surrounding atmosphere. We assume that our parcel of air located near Earth's surface has a temperature of 30°C and was forced aloft by the passage of a warm or cold front. The violet line representing the temperature of the rising parcel begins at the surface and continues upward, parallel to the red dashed lines. These two lines are parallel because the rising air is not saturated and therefore cools at the dry adiabatic rate.

At an altitude just below the 800-millibar level, the violet line intersects the lifting condensation level—the height at which condensation and cloud formation begin—marked on the diagram as LCL. Above the LCL, the violet line now runs parallel to the dashed green lines (the lower wet adiabatic rate for saturated air). At about the 650-millibar level, the violet curve intersects and crosses the sounding temperature curve (red line). Therefore, above the 650-millibar level, the rising air is now warmer (less dense) than the temperature of the surrounding air (environment), causing it to become unstable and rise because of its own buoyancy. This level is called the *level of free convection* (LFC). The region of instability and free convection exists until the line representing the temperature of the rising air once again intersects with the sounding temperature (red line) at about the 200-millibar level. At this altitude, the temperature of the rising parcel is again equal to that of its surroundings, causing the parcel to lose its buoyancy. This level is called the *equilibrium level* (*EL*) and is theoretically the height of the thunderstorm. However, deep convection often results in overshooting cloud tops—clouds that rise beyond the equilibrium level.

We can conclude from Figure 12.B that this sounding indicates the potential for deep convection. If this potentially unstable air were forced to rise as it passed over a mountain, or were lifted along a frontal boundary, it could generate severe thunderstorms. This type of information is invaluable to weather forecasters.

## Questions

1. Between roughly what levels (in meters) of the atmosphere did the clouds located over Oshkosh, Wisconsin, on April 30, 2012, occupy? Based on their altitude, what was the cloud type?
2. Notice in Figure 12.A that temperatures generally decrease with height, except in the lower stratosphere. Locate the tropopause on this illustration.
3. Roughly how deep (in meters) is the zone of free convection shown in Figure 12.B?
4. What type of instability is illustrated in Figure 12.B?

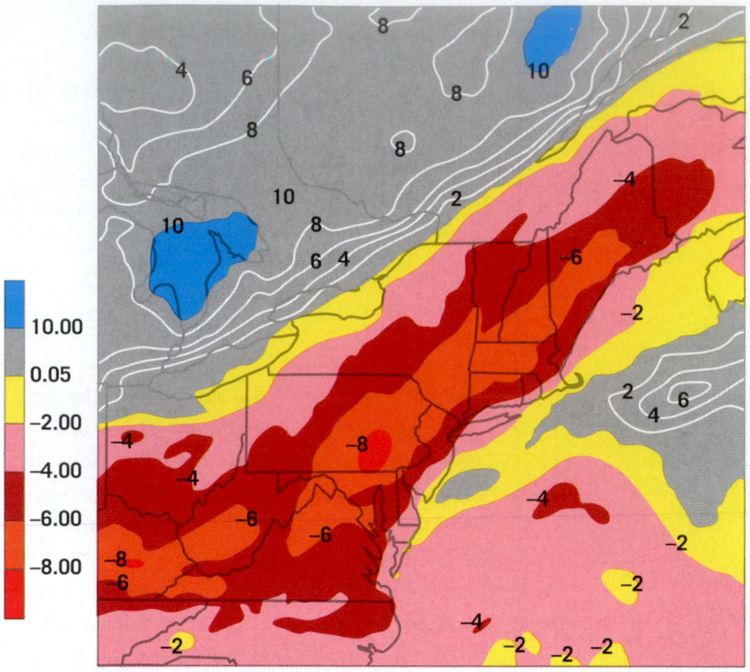


▲ **Figure 12.27 Lifted index** The lifted index is a number generated by hypothetically lifting a parcel of air from Earth's surface to the 500-millibar level, then comparing its temperature to that of the environment aloft. The *lower* the lifted index value, the *more unstable* the atmosphere at that location. Values of −4 or lower (represented by shades of red) denote unstable air with potential to produces severe thunderstorms. By contrast, values above 10 (shown in blue) are very stable.

## Forecasting Using Rules of Thumb

A primary goal of a forecaster is to improve on the MOS forecasts, which are the baseline used by NWS forecasters. Recall that MOS forecasts are highly refined forecasting products built from the data generated from numerical forecast models. One way to accomplish this goal is through the use of simple rules of thumb to aid in forecasting the various weather elements.

**Forecasting Temperature** Weather conditions that are quite simple to predict using model guidance are the maximum and minimum temperatures, which tend to occur in the midafternoon and early in the morning, respectively. If three different models forecast the maximum temperature for tomorrow to be 71°F, 72°F, and 73°F, it is probably safe to forecast an average temperature of 72°F. However, numerical models tend to under- or overestimate temperatures in some situations. Factors to consider when forecasting temperatures are:

- **Extent of cloud cover** Overnight clouds absorb and reradiate longwave terrestrial radiation back to Earth's surface, making the minimum temperature for the day higher than it would be on a cloud-free evening. Daytime clouds reduce surface heating and result in cooler-than-expected daytime high temperatures. Clear skies, on the other hand, allow more solar radiation to reach Earth's surface, producing warmer-than-expected maximum temperatures. Satellite images are useful for estimating the effect that clouds may have on a temperature forecast.
- **Advection** A surface map can indicate the direction the wind is coming from (the upstream direction). Cold, dry air upstream usually results in colder-than-normal nights—in part because of the advection (horizontal flow) of cooler air and because clear skies promote radiation cooling. By contrast, the advection of warm, moist air will produce warmer-than-average nighttime temperatures. Advection aloft can be seen on upper-level charts and has a similar effect.
- **Snow cover** Because snow efficiently radiates longwave radiation, snow cover, especially when coupled with dry, clear air, can result in extremely cold nighttime temperatures.
- **Passing fronts** The passage of a warm or cold front greatly influences the timing and magnitude of the maximum and minimum temperatures for a particular day. Therefore, examining a series of surface maps can be useful in estimating the arrival of a front. A warm front arriving in the late afternoon on a cool day may cause the high to occur much later in the day—possibly just before midnight. By contrast, a cold front arriving in the morning could result in the maximum daily temperature occurring just after midnight (before the cold front passes).

**Forecasting Precipitation** Unlike temperature, precipitation can be particularly hard to predict. Factors to take into account when predicting precipitation include:

Video MM
Forecasting Precipitation
http://goo.gl/DZxFLZ

- **Low PoP and QPF** Numerical forecast models tend to forecast a low probability of precipitation when there are clouds in the forecast area. When the PoP (Probability of Precipitation) is less than 30 percent, precipitation is unlikely. Similarly, when the QPF (Quantitative Precipitation Forecast) is less than a quarter inch, the amount of precipitation predicted for an area is minimal. Stable conditions accompanied by low PoP and QPF readings make precipitation highly unlikely. However, if the air is unstable, as determined by examining thermodynamic diagrams (see Box 12.1), scattered or isolated thunderstorms are more likely.
- **Timing precipitation** Precipitation associated with a cold front usually occurs *with the passage* of the front, whereas precipitation associated with a warm front usually occurs *before its passage*. Weather radar is a tool for determining the rate at which a front, or a line of thunderstorms, is approaching the forecast area and therefore aids in predicting the time that the precipitation will begin and end. Isolated summer thunderstorms are more likely to occur in the afternoon, when solar heating is most intense. Mountain and sea breezes also tend to occur during the warmest part of the day, affecting the timing of precipitation in mountainous and coastal areas.
- **Precipitation quantities** Precipitation amounts tend to be higher with slow-moving or stagnant fronts and with other slow-moving weather systems, such as hurricanes. Frontal precipitation is heavier in the late afternoon or evening, when

solar heating has enhanced the instability of the air forced aloft. Dew-point temperature provides a clue to the water vapor content of the air that will be forced aloft by a front: Very humid air can produce towering cumulonimbus clouds and torrential downpours if it is forced to rise.

In addition to these rules of thumb, a review of Section 3.4 (page 72) on daily temperature changes and Section 4.8 (page 112) on stability offer additional insight into how forecasting skills can be improved.

### ✔ Concept Checks 12.8

1. Describe the role human forecasters play in generating a forecast.
2. Explain AWIPS and briefly describe how it aids forecasters at local Weather Forecasting Offices.
3. What weather phenomenon is the lifting index useful at predicting?
4. Describe how abundant cloud cover affects minimum and maximum temperature forecasts.

## 12.9 Forecast Accuracy

**Explain why the percentage of accurate forecasts is not always a good measure of forecast skill.**

A concept referred to as *forecast skill* is used to measure the effectiveness of a particular forecast method. **Forecast skill** can be described as the improvement that a particular forecast method (or a forecaster) has over a standard forecast technique, such as *climatology*. One example of a climatology forecast would be to predict that the maximum temperature for Peoria, Illinois, on July 28, 2014, would be 85°F, based on historical averages of maximum temperature for that date and location. (*Note:* The actual maximum temperature for that date and location was 76°F.) A perfect forecast result compared to climatology would be given a forecast skill value of 1.0.

When determining forecast skill, the percentage of accurate forecasts cannot be the only consideration. For example, measurable precipitation in Los Angeles is recorded only 11 days each year, on average. This means that, based on climatic data, the chance of rain in Los Angeles is 11 out of 365, or about 3 percent. Knowing this, a forecaster could predict no rain for every day of the year and be correct 97 percent of the time! Although the accuracy is high, the forecast would not indicate skill. For a method to exhibit forecasting skill, it must generate forecasts that are more accurate than those based simply on climatic averages. Using the Los Angeles example, a forecast method must be able to predict rain on at least a few of the days that rain actually occurred in order to demonstrate skill.

The only aspect of a weather forecast that is expressed as a *percentage probability* is rainfall. Statistical data are studied to determine how often precipitation occurred under similar conditions. Although the occurrence of precipitation can be forecast with more than 80 percent accuracy, predictions of the amount, time of occurrence, and duration are not as reliable—particularly when the precipitation occurs in the summer in association with air-mass thunderstorms that cannot be predicted by the current numerical models. A forecast might predict 2 inches of rain for an area, but one town might receive a trace, whereas a neighboring community may be deluged with 4 inches. By contrast, winter precipitation, especially snowfall amounts, can be forecast with much better accuracy. This is because winter precipitation tends occur more slowly and over wider areas that are closely associated with fronts and midlatitude cyclones. Furthermore, forecasts of maximum and minimum temperatures, winds, and pressure distribution tend to have greater accuracy than precipitation forecasts.

How skillful are NWS weather forecasts? In general, very short-range forecasts (0–12 hours) have demonstrated considerable skill, especially for predicting the formation and movement of severe weather such as thunderstorms, hail, and heavy rainfall.

On average, a 5-day forecast using modern techniques is as reliable as a 2-day forecast would have been about 20 years ago. **Figure 12.28** shows that over the past several years the accuracy of 5-day forecasts at predicting airflow patterns aloft has improved significantly. The values closest to 1 on the graph represent higher skill. Figure 12.28 also shows the annual variability of forecast skill. Note that forecast accuracy is much higher in winter than during the midsummer months.

Medium-range forecasts (3–7 days) have also shown significant improvement over the past decade. Yet beyond day 7, the

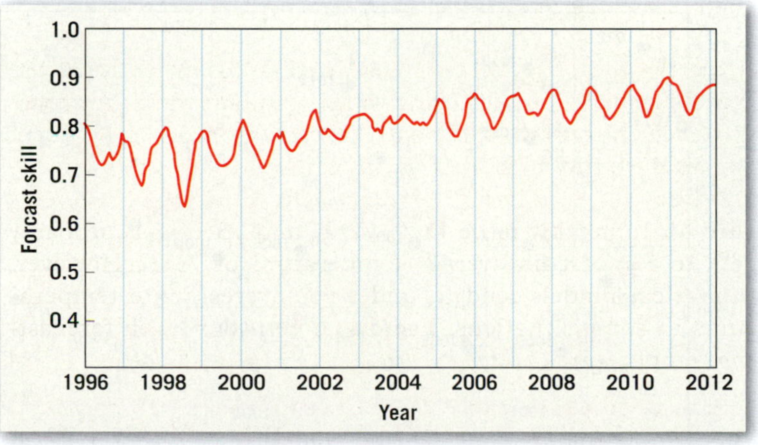


▲ **Figure 12.28 Improvement in forecast skill** The accuracy of 5-day forecasts at predicting airflow patterns aloft has improved significantly. When reading this graph, note that values closest to 1 represent higher skill. Note that forecast accuracy is much higher in winter than during the summer months.

# What's Your Forecast?

## Severe Weather Forecasting

by Adam J. Clark, Research Scientist, Cooperative Institute for Mesoscale Meteorological Studies, University of Oklahoma, and NOAA/National Severe Storms Laboratory

When your favorite television show is interrupted for an announcement that a severe thunderstorm or tornado watch is in effect, where does this information come from? The National Weather Service's Storm Prediction Center (SPC) in Norman, Oklahoma, is responsible for forecasting severe weather over the entire United States up to 8 days in advance. SPC duties include issuing severe thunderstorm and tornado watches, as well as outlooks that predict the likelihood of severe storms on a given day. SPC forecasters rely heavily on numerical weather prediction models for longer-range forecasts and strongly emphasize analysis of current weather conditions to make short-term forecasts. In fact, manual "hand analysis" is still a very important part of forecasting at SPC because it familiarizes forecasters with the current state of the atmosphere and can sometimes help identify subtle but important features that computer-generated analyses miss.

### Observations and Storm-Scale Modeling

Advances in supercomputing technology are revolutionizing all scientific fields. Have you ever wondered how meteorologists use these advances to forecast and model severe weather?

One innovative new tool SPC forecasters use is storm-scale modeling, a type of numerical weather prediction model that uses fine "grids" (analysis of conditions within a small area) to predict individual thunderstorms over regions as large as the continental United States. Although storm-scale models cannot always predict the timing and placement of storms, they are very skillful at predicting the correct storm types (for example, multicell clusters, supercells, or squall lines) on a given day. Storm type is important because each type is closely linked to the type of hazard that can occur. For example, rotating supercells are responsible for most tornadoes and large hail, while most severe wind gusts come from squall lines. In the past, forecasters had to make informed guesses about the expected storm type from models that analyzed conditions on a larger scale. This can be difficult because seemingly similar conditions often produce different storm types. Now storm-scale models can provide more explicit predictions to forecasters, and storm-scale ensembles—running multiple storm-scale models at the same time with slightly different input—can provide information on forecast uncertainty.

### Make a Forecast

On May 11, 2014, observations and model forecasts indicated that the ingredients for severe weather were coming together over the central Great Plains and were expected to initiate storms along a frontal boundary.

Now put yourself in the role of a SPC forecaster.

#### Questions

1. Based on the surface chart valid at 2 P.M. (**Fig. 12.C**), locate the frontal boundary. (Hint: Review Fig. 12.7 and look for a wind shift and a sharp change in air temperature and dew-point temperature on the map.)
2. A forecast made from a series of storm-scale models run simultaneously, valid from 2 to 7 P.M., indicates that almost all model runs predicted strongly rotating supercells (**Fig. 12.D**). What does this say about the forecast uncertainty? Where are the areas of storm rotation relative to the frontal boundary you identified in Figure 12.C?
3. If you were an SPC forecaster and expected the forecast to be correct, what type of severe weather would you expect (for example, high winds, hail, and/or tornadoes)?
4. Use the SPC storm events database to see what happened on this day (www.spc.noaa.gov/exper/archive/events/). Does it match what you expected?

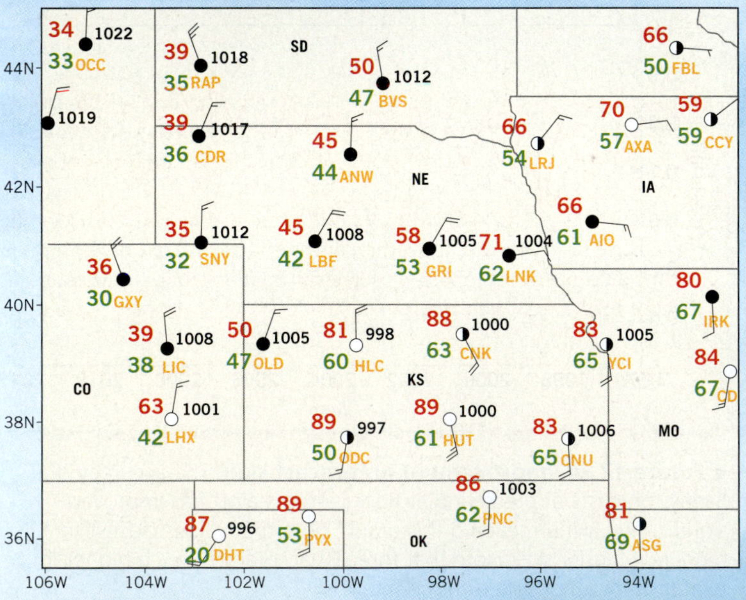


▲ **Figure 12.C. Midwest surface observations for 2 P.M. (LST) May 11, 2014**

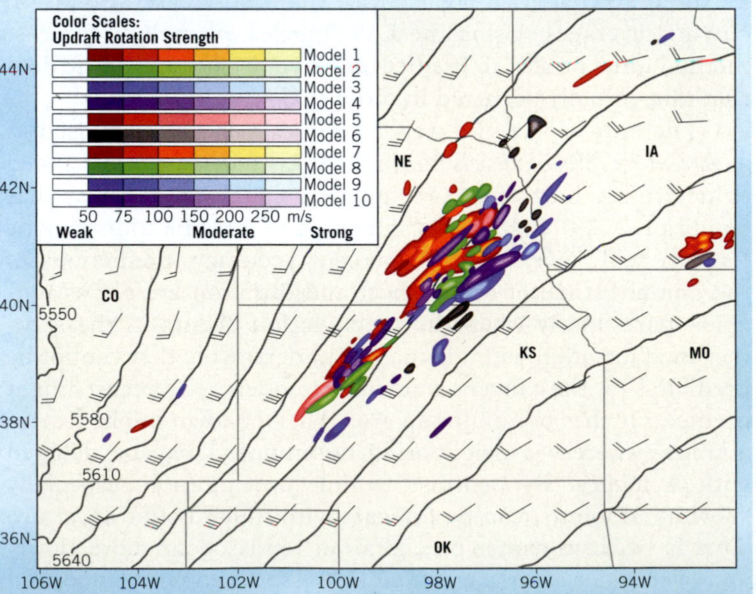


▲ **Figure 12.D. Maximum storm rotation strength over the period 2 to 7 P.M. May 11, 2014, forecast by 10 storm-scale models run at the same time** The different colors indicate different models.

predictability of day-to-day weather, even using these modern methods, proves only slightly more accurate than projections made from climatic data. In fact, it wasn't until 2000 that the NWS extended medium-range forecasts from a maximum of 5 days out to 7 days.

Several factors account for the limited range of modern forecasting techniques. As noted earlier, erroneous data and a network of observing stations that provides only limited coverage of our vast planet are significant hindrances. Not only are large areas of Earth's land–sea surface monitored inadequately, but also global data from the middle and upper troposphere are meager. Moreover, the physical laws of nature are difficult to apply to chaotic systems such as Earth's atmosphere, and current numerical models cannot adequately account for the vast number of atmospheric variables.

In summary, the accuracy of weather forecasts covering short- and medium-range periods has improved steadily over the past few decades, particularly in forecasting the evolution and movement of midlatitude cyclones. Technological developments, improved computer models, and a better understanding of how the atmosphere behaves have added greatly to this success. The ability to accurately predict daily weather beyond day 7, however, remains relatively low.

### ✔ Concept Checks 12.9

1. What is meant by *forecast skill*? Give an example of why the percentage of correct forecasts is not always a good measure of forecast skill.
2. Which weather element is forecast as a percentage probability?
3. List two reasons modern forecasting techniques do not accurately predict the weather beyond 7 days for most regions of the United States.

# 12 Concepts in Review Weather Analysis and Forecasting

## 12.1 The Weather Business: A Brief Overview

▶ List and describe the major steps used to generate a weather forecast.

**Key Terms:** National Weather Service (NWS), weather forecast, National Centers for Environmental Prediction (NCEP), Canadian Meteorological Centre, Weather Forecast Offices, meteorologist, climatologist

- The U.S. government agency responsible for gathering and disseminating weather-related information is the National Weather Service (NWS). Perhaps the most important services provided by the NWS are forecasts and warnings of hazardous weather.
- The process of providing weather forecasts and warnings throughout the United States occurs in three stages. First, observational data are collected and analyzed on a global scale. Second, a variety of techniques are used to establish the future state of the atmosphere—a process called weather forecasting. Finally, forecasts are disseminated to the public, mainly through the private sector.

## 12.2 Acquiring Weather Data

▶ Describe the various tools used to acquire weather data.

**Key Terms:** World Meteorology Organization (WMO), Automated Surface Observing Systems (ASOS), radiosonde, atmospheric sounding (sounding), rawinsonde, wind profiler, Doppler radar

- On a global scale, the World Meteorological Organization (WMO) is responsible for gathering, plotting, and distributing weather data.
- Automated Surface Observing Systems (ASOS) are modern automated systems that provide weather observations such as temperature, dew point, wind speed and direction, visibility, and cloud cover; ASOS also detects and determines precipitation amounts.
- Radiosondes carried aloft by weather balloons contain sensors that measure temperature, humidity, and pressure. Much of the upper-level data used to prepare upper-level weather charts comes from these instrument packages.

## 12.3 Weather Maps: Depictions of the Atmosphere

▶ Interpret what a particular pattern in the flow aloft indicates about surface weather.

**Key Terms:** weather analysis, synoptic weather map, current surface analysis, station model, isotach, jet streak, zonal, meridional

- Assessing current atmospheric conditions, called weather analysis, involves collecting, compiling, and transmitting billions of pieces of observational data.
- The traditional surface map is called a synoptic weather map because it displays a synopsis of the weather conditions at a given moment. To the trained eye, a weather map is a snapshot that shows the status of the atmosphere, including data on temperature, humidity, sea-level pressure, and wind.

- Meteorologists are aware of the strong correlation between cyclonic disturbances at Earth's surface and the daily and seasonal fluctuations in the wavy flow of the westerlies aloft. Upper-air maps are generated twice daily and provide forecasters with a picture of conditions at various levels in the troposphere. These maps are used to predict daily maximum temperature, locate areas where precipitation is likely to occur, and determine whether a thunderstorm will produce a severe weather event.

**Q** The accompanying map shows the isotherms at the 850-mb level with only the 0°C isotherm labeled. Draw an arrow on the map representing the location of warm air advection and another representing cold air advection. Label both arrows appropriately.

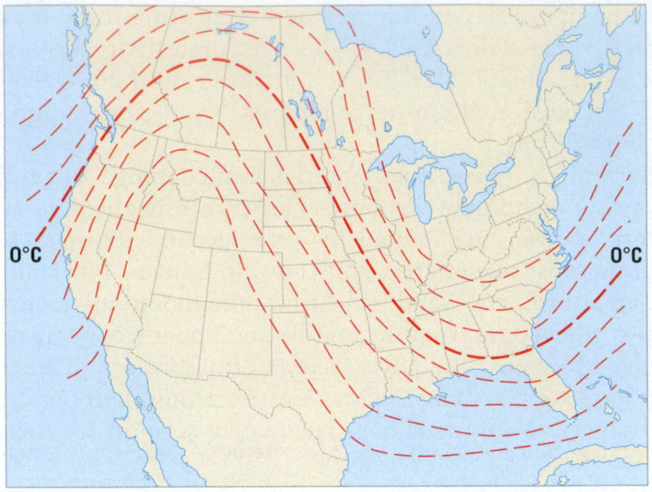


## 12.4 Modern Weather Forecasting ► Explain the basis of numerical weather prediction.

**Key Terms:** numerical weather prediction, Model Output Statistics (MOS), prognostic chart (prog), ensemble forecasting

- Numerical weather prediction, the backbone of modern forecasting, relies on the fact that the behavior of atmospheric gases is governed by a set of physical principles or laws that can be expressed as mathematical equations. If we can solve these equations, we will have a description of the future state (a forecast) of the atmosphere.
- A statistical analysis called Model Output Statistics (MOS) is used to modify and correct these machine-generated forecasts by making comparisons of how accurate previous forecasts have been.
- To address the inherent chaotic behavior of the atmosphere, meteorologists employ ensemble forecasting. This method involves producing a number of forecasts using the same computer model but slightly altering the initial conditions, while remaining within an error range of the observational instruments.

## 12.5 Other Forecasting Methods ► List and distinguish among the various traditional methods of weather forecasting.

**Key Terms:** persistence forecasting, climatological forecasting, analog method, trend forecasting, nowcasting

- Methods used in addition to numerical weather prediction include persistence forecasting, climatological forecasting, the analog method, and trend forecasting.
- A type of trend forecasting called nowcasting is used to forecast very short-lived (usually 6 hours), localized weather events. Nowcasting depends on weather radar and geostationary satellites—important tools for detecting areas of heavy precipitation and clouds that are capable of triggering severe conditions.

**Q** Explain the difference between weather analysis and weather forecasting.

## 12.6 Weather Satellites: Tools in Forecasting ► Discuss the advantages and disadvantages of infrared imagery and visible imagery generated by weather satellites.

**Key Terms:** Polar-orbiting Operational Environmental Satellite (POES), Geostationary Operational Environmental Satellite (GOES)

- Polar-orbiting satellites (POES) and geostationary satellites (GOES) are important tools that allow meteorologists to monitor even the most remote parts of the globe and track the movement of large weather systems.
- These satellites provide visible, infrared, and water-vapor images that are useful in the study of clouds and the distribution of water vapor.

## 12.7 Types of Forecasts ► Compare and contrast qualitative and quantitative weather forecasts. Differentiate between weather forecasts and 30- and 90-day outlooks.

**Key Terms:** qualitative forecast, quantitative forecast, Quantitative Precipitation Forecast (QPF), Probability of Precipitation (PoP), Climate Prediction Center (CPC), watch, warning, Storm Prediction Center (SPC), National Hurricane Center (NHC)

- Qualitative forecasts describe an aspect of the weather that can be observed but not easily measured or quantified, while quantitative forecasts deal with weather data that can be measured, such as maximum and minimum temperatures.

- The only forecast issued by the NWS that is given as a percentage probability is precipitation.
- Long-range weather forecasting relies heavily on statistical averages obtained from past weather events, known as climatic data. Weekly, monthly, and seasonal weather outlooks prepared by the NWS are not weather forecasts in the usual sense but indicate whether a region will experience near-normal precipitation and temperatures.
- A watch means conditions are right for the development of dangerous weather. A warning means that dangerous weather has developed and has been located in or moving toward an area.

**Q** Describe the differences between weather forecasts and 90-day outlooks.

## 12.8 The Role of the Forecaster ▸ Describe AWIPS and explain how it aids forecasters at local Weather Forecast Offices.

**Key Terms:** Advanced Weather Interactive Processing System (AWIPS), thermodynamic diagram, Skew-T/Log P diagram

- The job of issuing local weather forecasts, watches, and warnings rests with the National Weather Service (NWS) and its 122 local Weather Forecast Offices. The responsibility of forecasters at the Weather Forecast Offices is to modify numerical predictions by taking into account local conditions and nuances to produce site-specific forecasts.
- The Advanced Weather Interactive Processing System (AWIPS) is an informational processing, display, and communication system that integrates all model and observational data, including satellite and radar data, in one location.
- The NWS utilizes a thermodynamic diagram called a Skew-T/Log P diagram, or simply a Skew-T, on which temperature and dew-point data from the surface through the top of the troposphere is plotted. These diagrams, along with the lifted index and the K-index, are useful for determining the potential instability of the atmosphere.
- A primary goal of a forecaster is to try to improve on MOS forecasts, the baseline used by NWS forecasters. One way to accomplish this goal is through the use of some rules of thumb.

**Q** In your own words, state a rule of thumb about how cloud cover affects nighttime temperatures.

## 12.9 Forecast Accuracy ▸ Explain why the percentage of accurate forecasts is not always a good measure of forecast skill.

**Key Term:** forecast skill

- Over the past several decades, improvements in observing systems, computer models of physical processes, and assimilation of data into numerical weather prediction systems have steadily improved the ability to predict the evolution of larger-scale weather systems as well as day-to-day variations in temperature, precipitation, and the extent of cloud cover.

## Give it Some Thought

1. The accompanying radar image shows the precipitation pattern (reds and yellow indicate heavy precipitation, and greens indicate light-to-moderate precipitation) associated with a strong midlatitude cyclone. Use the technique called trend forecasting and assume that the cyclone maintains its current strength for the next 24 hours as you complete the following.
   **a.** What state(s) will likely experience thunderstorms associated with a cold front in the next 6 to 12 hours?
   **b.** Will the temperatures in Alabama more likely rise or fall during the next 24 hours? Explain.
   **c.** Will Pennsylvania more likely be warmer or colder in 24 hours than it was when the image was produced? Explain.
   **d.** In which state, New York or Georgia, would cirrus clouds more likely have been overhead when the image was produced?

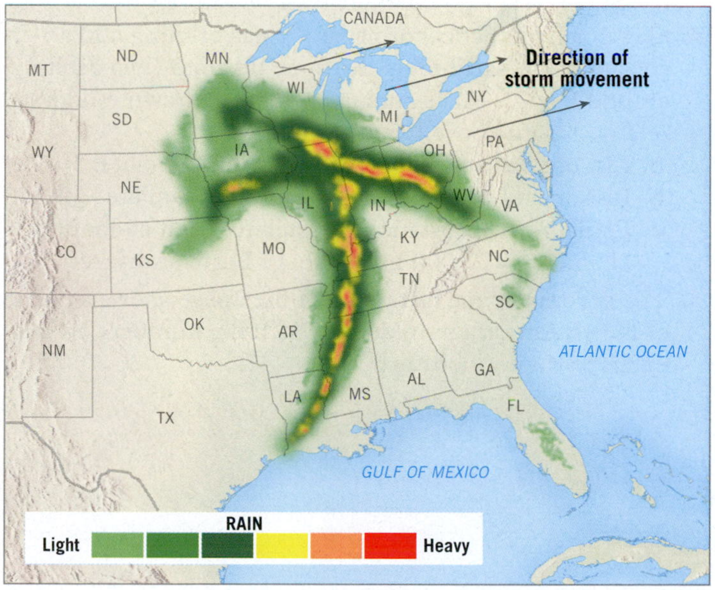

2. The accompanying map is a simplified upper-air chart showing flow aloft and the position of the jet stream. Weather patterns such as this are relatively common over North America. When answering the following questions, assume that long-range weather forecasters had evidence that this flow pattern was going to persist for several months.
   **a.** Would the 90-day outlook for the southeastern United States predict wetter-than-normal, drier-than-normal, or equal chance? Explain your answer.
   **b.** Would the 90-day outlook for the southwestern United States predict wetter-than-normal, drier-than-normal, or equal chance? Explain your answer.

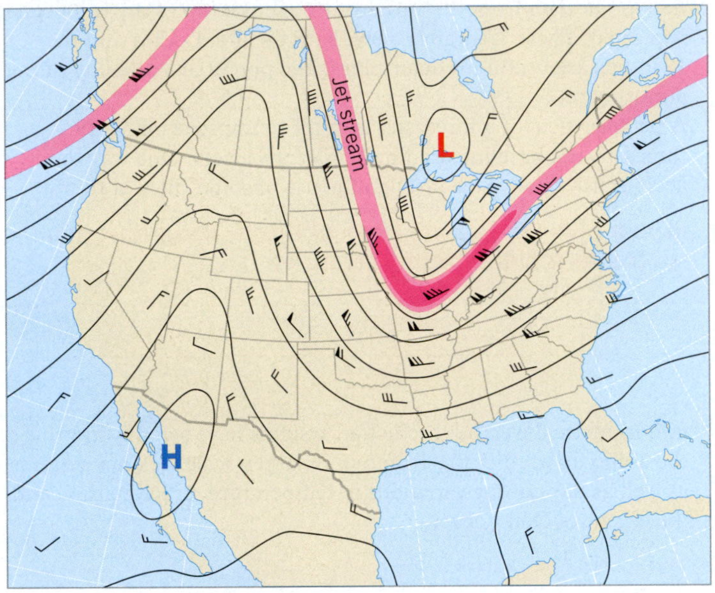


3. Explain the difference between numerical weather prediction and the analog method of weather forecasting.
4. Sketch a hypothetical orbit for a polar-orbiting and a geostationary satellite on the same diagram (two orbits around Earth). Make sure the orbits are drawn to indicate relative height above Earth's surface.
   **a.** List one advantage of a polar-orbiting satellite.
   **b.** List one advantage of a geostationary satellite.
   **c.** Explain why geostationary satellites orbit at about 36,000 kilometers (22,000 miles) above Earth's surface.
5. The accompanying infrared satellite image shows a portion of North America at midday. Match the numbers on the image to descriptions a–d.

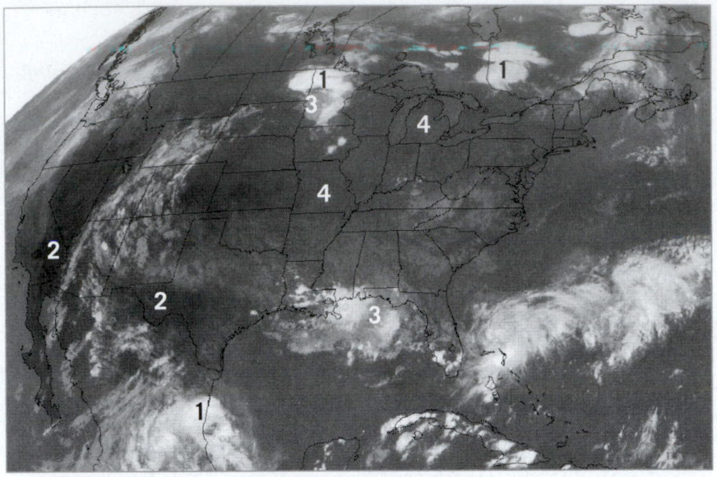


   **a.** These areas are cloud free, and the infrared sensor is measuring very warm surface temperatures.
   **b.** These areas contain low clouds (cumulus), with cloud tops below 2000 meters.
   **c.** These areas are blanketed by middle clouds between 2000 and 6000 meters.
   **d.** These are areas where the infrared sensor is measuring high, cold cloud tops that are likely associated with towering cumulonimbus clouds and thunderstorms.
6. The accompanying maps show the positions of the jet stream on two different midwinter days. On which of these days would the southeastern United States be warmer? Explain your choice.

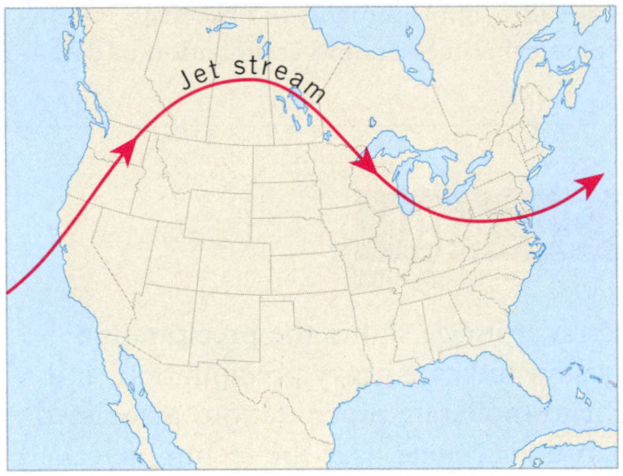


**Day 1**

**Day 2**

## Problems

1. Determine the probability of precipitation (PoP) using the formula below and the following information:

$$\text{PoP} = (C \times A) \times 100 \text{ percent}$$

where $C$ is the *confidence* that precipitation will occur *somewhere* in the forecast area, and $A$ equals the *percent of the area* that will receive measureable precipitation if it occurs.

   **a.** A forecaster expresses confidence that there is an 80 percent chance of precipitation in the forecast area, and it is expected to produce measurable precipitation over 90 percent of the area.

   **b.** A forecaster expresses confidence that there is a 20 percent chance of precipitation in the forecast area, and it is expected to produce measurable precipitation over 10 percent of the area.

2. Many weather reports include a 7-day outlook. Check such a report and jot down the forecast for the last (seventh) day. Then, each day thereafter, write down the forecast for the day in question. Finally, record what actually occurred on that day. Contrast the forecast that was generated 7 days ahead with what actually occurred. Describe in your own words how accurate (or inaccurate) the 7-day forecast was for the day you selected. How accurate was the 5-day forecast for that day? The 2-day forecast?

# 13 Air Pollution

## Focus on Concepts

*Each statement represents the primary learning objective for the corresponding major heading within the chapter. After you complete the chapter, you should be able to:*

**13.1** List several natural sources of air pollution and identify those that are human accentuated.

**13.2** Distinguish between primary and secondary pollutants. List important examples of each and discuss their effects on people and the environment.

**13.3** Summarize trends in air quality since 1980.

**13.4** Describe the influence of wind on air quality. Sketch a graph or diagram showing a temperature inversion and relate it to mixing depth.

**13.5** Discuss the formation of acid precipitation and list some of its effects on the environment.

Air pollution and meteorology are linked in two ways. One concerns the influence that weather conditions have on concentrating, diluting, and dispersing air pollutants. The second connection is the reverse and involves the effects of air pollution on weather and climate. The first of these associations is examined in this chapter. The second and equally important relationship is discussed in Chapter 14 and is the focus of several sections and special-interest boxes.*

---

*See Box 3.3, "How Cities Influence Temperature: The Urban Heat Island," and Box 13.1, "Air Pollution Changing the Climate of Cities." In addition, see the section "Country Breezes" in Chapter 7.

*China is plagued by air quality issues. Major contributors are coal-burning power plants.*

## 13.1 The Air Pollution Threat

**List several natural sources of air pollution and identify those that are human accentuated.**

Air pollution is a continuing threat to our health and welfare. According to a National Research Council report, people living in the most polluted cities in the United States lose an estimated 1.8 to 3.1 years of life due to chronic exposure to particulate matter. In addition, more than 4000 premature deaths occur each year because of the elevated surface ozone concentrations commonly observed in the United States.* Air pollutants also have a negative impact on crop production, costing U.S. agriculture more than $1 billion annually. In other parts of the world, especially developing countries, the negative impact of air pollution on life and agriculture is even more serious.

An average adult requires about 13.5 kilograms (30 pounds) of air each day, compared with about 1.2 kilograms (2.6 pounds) of food and 2 kilograms (4.4 pounds) of water. The cleanliness of air, therefore, should certainly be as important to us as the cleanliness of our food and water.

Air is never perfectly clean. Many natural sources of air pollution have always existed (Fig. 13.1). Ash and gases from volcanic eruptions, salt particles from breaking waves, pollen and spores released by plants, smoke from forest fires and brush fires, and windblown dust are all examples of "natural air pollution" (Fig. 13.2). Ever since people have been on Earth, however, they have added to the frequency and intensity of some of these natural pollutants, especially the last two. For example, the dust storms in Figure 13.3 occurred when strong winds raised dry soil from plowed farm fields.

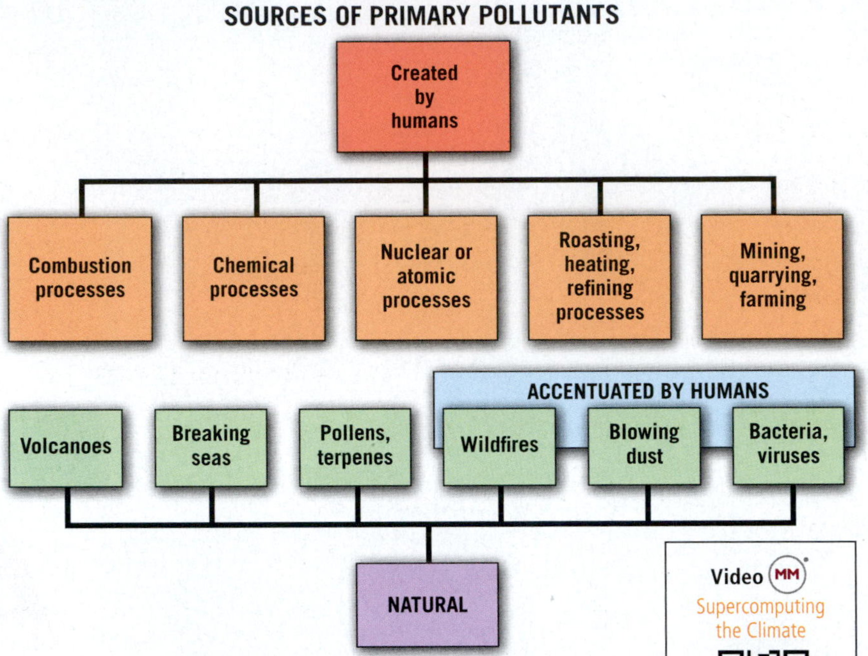


▲ **Figure 13.2 Sources of primary pollutants**

Video (MM) Supercomputing the Climate
http://goo.gl/CxlIQv

*As reported in *Bulletin of the American Meteorological Society*, Vol. 89, No. 6, June 2008. Note that particulate matter is discussed in the section "Primary Pollutants" and that surface ozone, a significant component of urban smog, is discussed in the section "Secondary Pollutants."

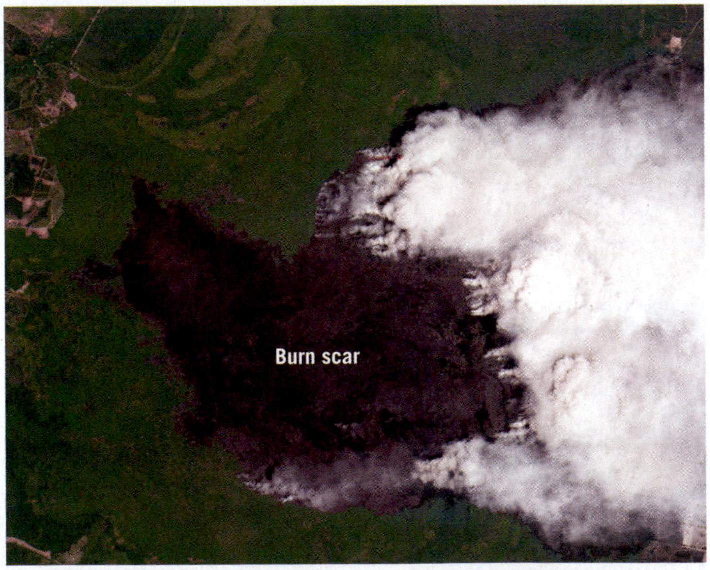


▲ **Figure 13.1 Natural source of air pollution** These plumes of smoke billowing into the sky from a wildfire in southern Georgia are an example of natural air pollution. Lightning started the fire on April 28, 2011, and on May 8, when this satellite image was acquired, nearly 62,000 acres had burned.

With the discovery of fire came an increased number of accidental as well as intentional burnings. Even today, in many parts of the world, fire is used to clear land for agricultural purposes (the so-called slash-and-burn method), filling the air with smoke and reducing visibility. When people clear the land of its natural vegetative cover for any purpose, soil is exposed and blown into the air. Yet when we consider the air in a modern-day industrial city, these human-accentuated forms of pollution, although significant, may seem minor by comparison.

Although some types of air pollution are relatively recent creations, others have been around for centuries. In 1661, when John Evelyn wrote *Fumifugium, or The Inconvenience of Aer and Smoak of London Dissipated, Together with Some Remidies Humbly Proposed*, smoke pollution and foul air obviously plagued Londoners. In his book, Evelyn notes that a traveler, although many miles from London, "sooner smells than sees the city to which he repairs." In fact, London continued to have severe air pollution problems well into the twentieth century. It was only after a devastating smog disaster in 1952 that truly decisive action was taken to clean the air.

London, however, did not have a monopoly on air pollution; many cities were plagued with dirty air with the coming of the Industrial Revolution. In addition to accelerating natural sources, people found many new ways to pollute the air (see

B.

A.

◀ **Figure 13.3 Natural air pollution accentuated by human activities** **A.** This dust storm near Elkhart, Kansas, in May 1937 occurred because the natural vegetative cover that anchored the soil had been removed from a marginal environment so that the land could be farmed. Severe drought made the plowed fields vulnerable to strong winds. It was because of storms like this that portions of the Great Plains were called the *Dust Bowl* in the 1930s. **B.** The summer of 2013 saw Dust Bowl–like conditions in drought-stricken portions of the Great Plains.The event shown here occurred in Lubbock, Texas.

Fig. 13.2) and many new things with which to pollute it. In the mid- to late nineteenth century, the populations of many American and European cities swelled as people sought work in the growing numbers of new foundries and steel mills. As a result, the urban environment became increasingly fouled by the fumes of industry. Charles Dickens vividly describes a late-nineteenth-century factory town in *Hard Times*:

> It was a town of machinery and tall chimneys out of which interminable serpents of smoke trailed themselves forever and ever, and never got uncoiled. It had a black canal in it, and a river that ran purple with ill-smelling dye.

It is clear that poor air quality was not the only environmental pollution that plagued these places! However, it should be noted that the rapid rise in urban air pollution was not necessarily viewed with great alarm. Rather, chimneys belching forth smoke and soot were a symbol of growth and prosperity (**Fig. 13.4**). A well-known lawyer and orator, Robert Ingersoll, is reported to have elicited great cheering and cries of "Good! Good!" from the audience for this statement in an 1880 speech: "I want the sky to be filled with the smoke of American industry and upon that cloud of smoke will rest forever the bow of perpetual promise. That is what I am for." With the rapid growth of the world's population and accelerated industrialization, the quantities of atmospheric pollutants increased drastically.

Since the 1970s, the passage of legislation, development of regulations and standards, and advances in control technology have greatly reduced the frequency and severity of air pollution episodes in the United States and Western Europe. Nevertheless, health authorities are equally concerned about the slow and subtle effects on our lungs and other organs of air pollution levels that are much lower but still present every day, year after year.

**students sometimes ask...**

### What is haze?

Haze is a reduction in visibility caused when light encounters atmospheric particulate matter and gases. Some light is absorbed by the particles and gases, and some is scattered away before it reaches an observer. More pollutants mean more absorption and scattering of light, which limits the distance we can see and can also degrade the color, clarity, and contrast of what we can see. Visibility impairment is one of the most obvious effects of air pollution.

## ✔ Concept Checks 13.1

1. Describe the impact of air pollution on human health.
2. List several examples of natural air pollution. List three that are human accentuated.

▼ **Figure 13.4 Stacks belching smoke and soot** Scenes such as this one of steel mills in western Pennsylvania in the late 1940s were once considered a sign of economic prosperity.

## 13.2 Sources and Types of Air Pollution

**Distinguish between primary and secondary pollutants. List important examples of each and discuss their effects on people and the environment.**

**Air pollutants** are airborne particles and gases that occur in concentrations that endanger the health and well-being of organisms or disrupt the orderly functioning of the environment. Pollutants can be grouped into two categories: primary and secondary. **Primary pollutants** are emitted directly from identifiable sources. They pollute the air immediately upon being emitted. **Secondary pollutants**, in contrast, are produced in the atmosphere when certain chemical reactions take place among primary pollutants. The chemicals that make up smog are important examples. In some cases, the effects of primary pollutants on human health and the environment are less severe than the effects of the secondary pollutants they form.

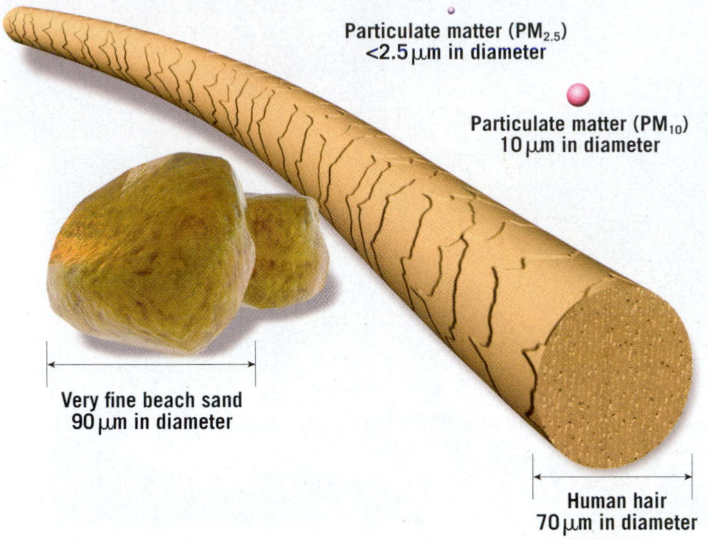


▲ **Figure 13.6 Particulate matter (PM)** This category is a complex mixture of extremely small particles and liquid droplets. The size of particles is directly linked to their potential for causing health problems. Particles 10 micrometers in diameter or smaller frequently pass through the nose and throat and enter the lungs. Once inhaled, these particles can cause serious health effects. $PM_{10}$ stands for "inhalable coarse particles" that are larger than 2.5 micrometers and smaller than 10 micrometers. $PM_{2.5}$ stands for "fine particles" that are 2.5 micrometers and smaller.

### Primary Pollutants

Figure 13.5 depicts the major primary pollutants as percentages (by weight). Sources vary for each pollutant. For example, electricity generation is the most significant source of sulfur dioxide. By contrast, on-road vehicles are the number-one source of carbon monoxide, nitrogen oxides, and volatile organic compounds. Compared to other sources, the tens of millions of cars and trucks on our streets and highways are clearly the greatest contributors. What follows is a brief survey and description of the major primary pollutants.

**Particulate Matter** *Particulate matter* (*PM*) is the general term used for a mixture of solid particles and liquid droplets found in the air. Some particles are large or dark enough to be seen as soot or smoke; others can be detected only with an electron microscope. These particles come in a wide range of sizes: Fine particles are less than 2.5 micrometers in diameter, and coarser-size particles are larger than 2.5 micrometers (Fig. 13.6). These particles originate from many different stationary and mobile sources, as well as from natural sources. Fine particles ($PM_{2.5}$) result from fuel combustion from motor vehicles, power generation, and industrial facilities, as well as from residential fireplaces and woodstoves. Coarse particles ($PM_{10}$) are generally emitted from sources such as vehicles traveling on unpaved roads, materials handling, and crushing and grinding operations, as well as windblown dust. Some particles are emitted directly from their sources, such as smokestacks and cars. In other cases, gases such as sulfur dioxide interact with other compounds in the air to form fine particles.

▼ **Figure 13.5 Emissions estimates of primary pollutants for the United States in 2012** Percentages are calculated on the basis of weight. The total weight in 2012 was 84 million tons.

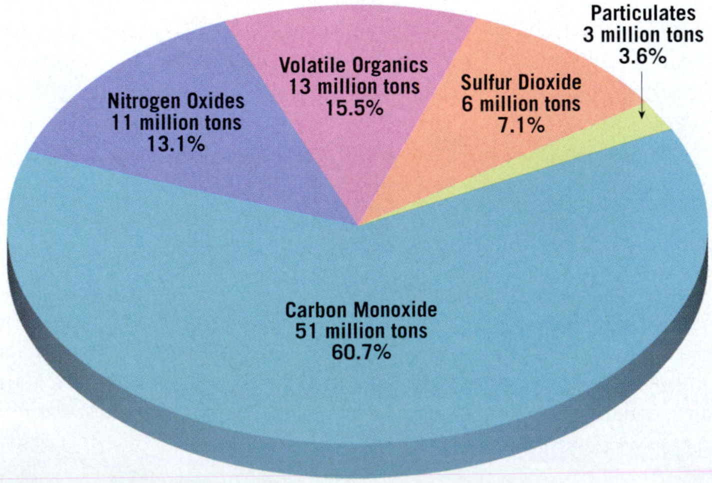


Particulates are frequently the most obvious form of air pollution because they reduce visibility and leave deposits of dirt on the surfaces with which they come in contact. In addition, particulates may carry any or all of the other pollutants dissolved in or absorbed on their surfaces (see Box 13.1).

Originally, total suspended particulate (TSP) was the indicator used to represent this category. It included all particles up to 45 micrometers in diameter. In 1987 the U.S. Environmental Protection Agency (EPA) set new standards related only to particles smaller than 10 micrometers (identified as $PM_{10}$). Then, in 1997, the EPA revised its standards for particulate matter

## Box 13.1 Air Pollution Changing the Climate of Cities

In Box 3.3 (pages 76–77) you saw that air pollution in cities contributes to the heat island by inhibiting the loss of longwave radiation at night. Studies of urban climate have also shown that pollutants may have a "cloud seeding" effect (see Chapter 5) that increases precipitation in and downwind of cities. These influences, however, are not the only ways pollutants influence urban climate.

The blanket of particulates over most large cities significantly reduces the amount of solar radiation reaching the surface. In some cities, the overall reduction in the receipt of solar energy is 15 percent or more, whereas short-wavelength ultraviolet light is decreased by up to 30 percent. This weakening of incoming solar energy is variable. During air pollution episodes, the decrease is much greater than for periods when air quality is good (**Fig. 13.A**). Furthermore, particulates most effectively reduce solar radiation near the ground when the Sun angle is low because the length of the path through the polluted air increases as the Sun angle drops. Thus, for a given quantity of particulate matter, the percentage of solar energy reaching the surface will be reduced the most in high-latitude cities and during the winter.

Compared to surrounding rural areas, the relative humidity in cities is generally 2 to 8 percent lower. One reason is that cities are hotter. Remember from Chapter 4 that as air temperature increases, capacity also rises, and relative humidity drops. A second reason is that less water vapor is supplied to city air by evaporation from the surface. Evaporation is reduced in cities because rainwater rapidly runs off impermeable surfaces, frequently into subsurface storm sewers.

Although relative humidity tends to be lower in cities, the occurrences of clouds and fogs are greater. What is the cause of this apparent paradox? A likely contributing factor is the large quantity of condensation nuclei produced by human activities in urban areas. When hygroscopic (water-seeking) nuclei are plentiful, water vapor readily condenses on them, even when the air is not quite saturated.

▲ **Figure 13.A Air pollution episode in Shanghai, China** It is not difficult to understand why the amount of solar radiation reaching the surface is reduced in cities.

### Questions

1. An air pollution episode at a high-latitude city in winter is likely to reduce the proportion of solar energy reaching the surface more than a similar event in summer. Why?
2. Is the relative humidity in a city more likely to be higher or lower than in a nearby rural area? Explain.

again so that they were based on $PM_{2.5}$. This change was in response to a large amount of research that analyzed the health effects of particulates.

Inhalable particulate matter includes both fine and coarse particles. These particles can accumulate in the respiratory system and are associated with numerous health effects. Exposure to coarse particles is primarily associated with the aggravation of respiratory conditions, such as asthma. Fine particles are most closely associated with health effects such as increased hospital admissions and emergency room visits for heart and lung disease, increased respiratory symptoms and disease, decreased lung function, and even premature death. Sensitive groups that appear to be at greatest risk of such effects include elderly people, children, and individuals with cardiopulmonary disease, such as asthma. In addition to causing health problems, particulate matter is the major cause of reduced visibility in many parts of the United States. Airborne particles can also cause damage to paints and building materials.

**Sulfur Dioxide** *Sulfur dioxide* ($SO_2$) is a colorless and corrosive gas that originates largely from the combustion of sulfur-containing fuels, primarily coal and oil (**Fig. 13.7**). Important sources include power plants, smelters, petroleum refineries, and pulp and paper mills. Once in the air, $SO_2$ is frequently transformed into sulfur trioxide ($SO_3$), which reacts with water vapor or water droplets to form sulfuric acid ($H_2SO_4$). Very tiny particles act as a medium on which the acidic sulfate ion ($SO_4^{2-}$) is carried over long distances in the atmosphere. When it is "washed out" of the air or deposited on surfaces, sulfuric acid contributes to a serious environmental problem known as *acid precipitation*. This issue is the subject of a later section.

High concentrations of $SO_2$ can result in temporary breathing impairment for asthmatic children and adults who are active outdoors. Short-term exposure of asthmatic individuals to elevated $SO_2$ levels while at moderate exertion may result in reduced lung function that may be accompanied by such symptoms as wheezing, chest tightness, or shortness of breath. Other

## eye ON THE atmosphere 13.1

This satellite image from NASA's *Landsat 8* satellite shows a portion of the Kenai Peninsula south of Anchorage, Alaska, on May 20, 2014. Dingy gray smoke from several major wildfires pollutes the air. The well-developed cumulonimbus clouds that are also present owe their origin to the wildfires.

**Questions**

1. According to Figure 13.2, wildfires are a natural source of primary pollutants. This fact implies that the smoke in this image is a "natural pollutant." However, these fires had a human origin. Is this air pollution "natural"? What term or phrase best fits this example?
2. How could the wildfires have been responsible for the development of cumulonimbus clouds?

**students sometimes ask... Burning of coal is considered a significant source of $SO_2$, but do we burn very much coal anymore?**

We do. About 41 percent of the electricity produced in the United States comes from coal combustion (see Fig. 13.7). Coal is also the fuel of choice for many heat-intensive processes, such as producing steel, aluminum, concrete, and wallboard. Many other countries are even more reliant on coal for their energy needs than the United States.

effects associated with longer-term exposures to high concentrations of $SO_2$, in conjunction with high levels of particulate matter, include respiratory illness and aggravation of existing cardiovascular disease.

**Nitrogen Oxides** *Nitrogen oxides* are gases that form during the high-temperature combustion of fuel, when nitrogen in the fuel or the air reacts with oxygen. Motor vehicles and power plants are the primary sources. These gases also form naturally when certain bacteria oxidize nitrogen-containing compounds.

The initial product formed is nitric oxide (NO). When NO oxidizes further in the atmosphere, nitrogen dioxide ($NO_2$) forms. Commonly, the general term $NO_x$ is used to describe these gases. Although $NO_x$ forms naturally, its concentration in cities is 10 to 100 times higher than in rural areas. Nitrogen dioxide has a distinctive reddish-brown color that frequently tints polluted city air and reduces visibility. When concentrations are high, $NO_2$ can also contribute to lung and heart problems. When air is humid, $NO_2$ reacts with water vapor to form nitric acid ($HNO_3$). Like sulfuric acid, this corrosive substance also contributes to the acid rain problem. Moreover, because nitrogen oxides are highly reactive gases, they play an important part in the formation of smog.

**Volatile Organic Compounds** *Volatile organic compounds* (*VOCs*), also called *hydrocarbons*, encompass a wide array of solid, liquid, and gaseous substances that are composed exclusively of hydrogen and carbon. Large quantities occur naturally, with methane ($CH_4$) being the most abundant. Methane, however, does not interact chemically with other substances and has no negative health effects. In cities, the incomplete combustion of gasoline in motor vehicles is the principal source of reactive VOCs. Although some hydrocarbons from other sources are cancer-causing agents, most of the VOCs in city air do not, by themselves, appear to pose significant environmental problems. However, as you will see in a later discussion, when VOCs react with certain other pollutants (especially nitrogen oxides), noxious secondary pollutants result.

**Carbon Monoxide** *Carbon monoxide* (*CO*) is a colorless, odorless, and poisonous gas produced by incomplete burning of carbon in fuels such as coal, oil, and wood. It is the most abundant primary pollutant, with more than three-quarters of U.S. emissions coming from highway vehicles and nonroad equipment.

Although CO is quickly removed from the atmosphere, it can nevertheless be dangerous. Carbon monoxide enters the bloodstream through the lungs and reduces oxygen delivery to the body's organs and tissues. Because it cannot be seen, smelled, or tasted, CO can have an effect on people without their realizing it. In small amounts, it causes drowsiness, slows reflexes, and impairs judgment. If concentrations are sufficiently high, CO

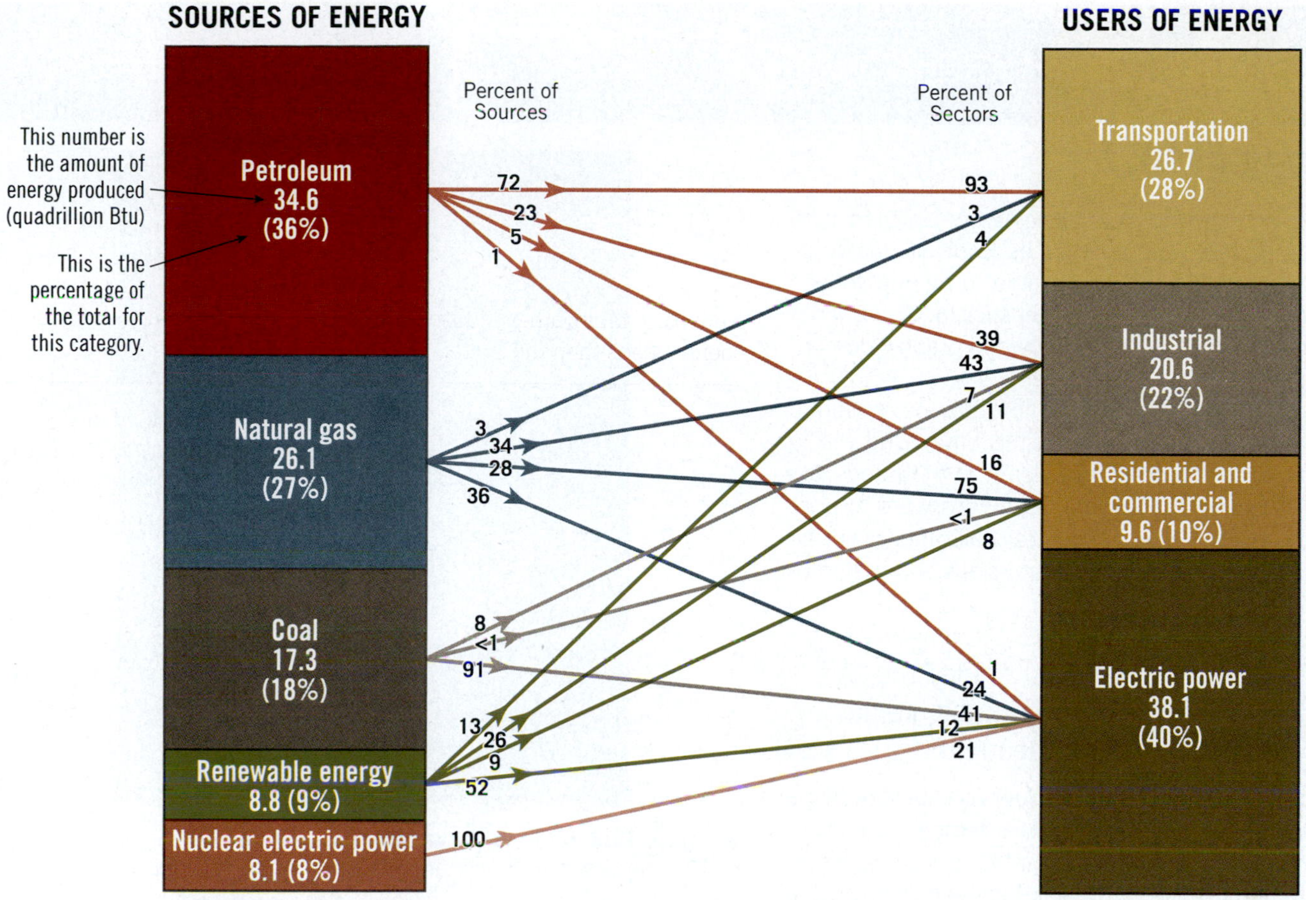


**Reading this double graph:**
The left side indicates what energy sources we use. The right side shows where we use the energy. The lines with numbers that connect the graphs provide more details. Use the top line as an example. It shows that 72% of the petroleum is used by the transportation sector. It also indicates that 93% of the energy used by the transportation sector is petroleum.

◀ **SmartFigure 13.7 U.S. Energy consumption, 2012** Total consumption was 94.9 quadrillion Btu. A quadrillion is 10 raised to the 12th power, or a million million. A major source of air pollutants is the burning of fossil fuels—coal, petroleum, and natural gas.

http://goo.gl/YHkWO

inhalation can be fatal. Carbon monoxide poses a serious health hazard where concentrations can reach high levels, as in poorly ventilated tunnels and underground parking facilities.

The world map in **Figure 13.8** shows average CO concentrations for April 2010. In different parts of the world and in different seasons, the concentrations and sources change significantly. For example, in Africa, seasonal shifts in CO are tied to the widespread agricultural burning that shifts north and south of the equator with the seasons. Fires are also an important source of CO in other regions, such as the Amazon and Southeast Asia. In the United States, Europe, and China, on the other hand, the highest CO concentrations occur in and near urban areas where there are high concentrations of motor vehicles and factories. Wildfires burning over large areas of North America and Russia in some years can also be important sources.

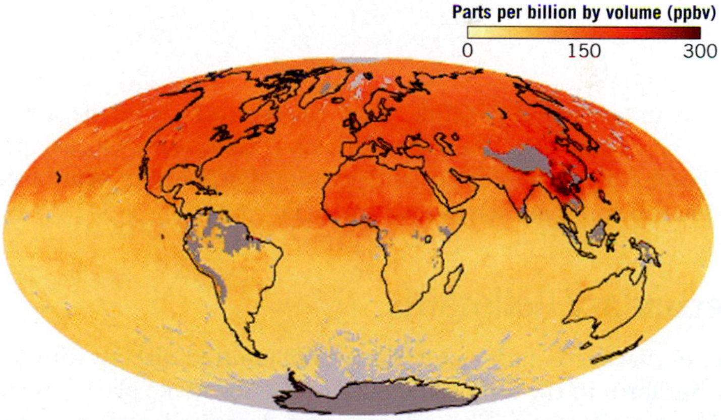


▲ **Figure 13.8 Carbon monoxide** This map shows average concentrations of carbon monoxide (CO) in the troposphere in April 2010 at an altitude of about 3600 meters (12,000 feet). The data were collected by a sensor aboard NASA's *Terra* satellite. Concentrations of CO are expressed in parts per billion by volume (ppbv). Yellow areas have little or no CO. Progressively higher concentrations are shown in orange and red. Places where the sensor did not collect data, perhaps due to clouds, are gray. Satellite observations often show that pollution emitted in one locale can travel great distances and impact air quality far from the original source.

**Video** Global Carbon Monoxide Concentrations
http://goo.gl/AAvWkd

## Lead

*Lead* (Pb) is very dangerous because it accumulates in the blood, bones, and soft tissues. It can impair the functioning of many organs. Even at low doses, lead exposure is associated with damage to the nervous systems of young children.

In the past, automotive sources were the major contributor of lead emissions to the atmosphere because lead was added to gasoline to prevent engine knock. Since the EPA-mandated

severe& hazardous weather Box 13.2

## The Great Smog of 1952

For centuries the fogs of Britain's major cities were polluted with smoke. London was especially notorious for its poor air quality. In the early 1800s London's smoke-laden fog came to be known as a "London particular" (that is, a London characteristic). In 1853 Charles Dickens used the term in *Bleak House* and provided graphic descriptions of London's foul air in several of his novels.

One of London's most infamous air pollution episodes occurred over a 5-day span in December 1952. During this time, acrid yellow smog shrouded the city, bringing premature death to thousands and inconvenience to millions (**Fig. 13.B**). What conditions were responsible for this extraordinary event? As in practically all other major air pollution episodes, it was a combination of emissions and meteorological circumstances.

The weather was unusually cold, and Londoners were burning large quantities of coal to warm their homes. The smoke pouring from these chimneys as well as from the chimneys of London's many factories was not dispersed but rather accumulated in a shallow zone of very stable calm air. The "lid" for this trap was a substantial temperature inversion associated with a high-pressure center that had become established over the southern British Isles. (The link between temperature inversions and air pollution episodes is explored in greater detail in Section 13.4.)

> Acrid yellow smog shrouded the city, bringing premature death to thousands.

▲ **Figure 13.B Midday darkness** London's infamous Great Smog of 1952 persisted for 5 days and was responsible for thousands of deaths.

Fog developed during the day on Friday, December 5. Beneath the well-developed temperature inversion, a very light wind stirred the saturated air to create a fog layer 100 to 200 meters thick. With nightfall came sufficient radiation cooling to produce an even denser fog. Of course, more and more smoke continued to collect in the saturated air. Visibility dropped to just a few meters in many

phaseout of leaded gasoline, lead concentrations in the air of U.S. cities have declined dramatically (**Table 13.1**). Occasional violations of the lead air quality standard still occur near large industrial sources such as lead smelters.

## Secondary Pollutants

Recall that secondary pollutants are not emitted directly into the air but form in the atmosphere when reactions take place among primary pollutants. The sulfuric acid described earlier is one example of a secondary pollutant. After the primary pollutant, sulfur dioxide, is emitted into the atmosphere, it combines with oxygen to produce sulfur trioxide, which then combines with water to create the irritating and corrosive sulfuric acid.

Air pollution in urban and industrial areas is often called **smog**. The term, coined in 1905 by London physician Harold A. Des Veaux, was created by combining the words *smoke* and *fog*. Des Veaux's term was indeed an apt description of London's principal air pollution threat, which was associated with the products of coal burning coupled with periods of high humidity (see **Box 13.2**).

Today, however, smog is used as a synonym for general air pollution and does not necessarily imply the smoke–fog combination. Therefore, when greater clarity is desired, we

students sometimes ask...

### What are toxic air pollutants?

They are chemicals in the air that are known or suspected to cause cancer or other serious health effects, such as reproductive problems or birth defects. These substances are also commonly called "hazardous air pollutants" and "air toxics." The U.S. Environmental Protection Agency regulates 188 toxic air pollutants. Examples include benzene, found in gasoline; perchorethlyene, used in some dry-cleaning facilities; and methylene chloride, used as a solvent and paint stripper.

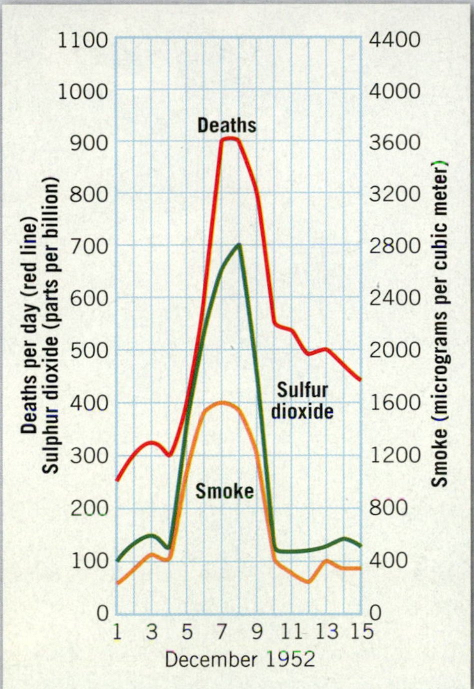


▲ **Figure 13.C Deaths and pollution during the Great Smog of 1952** Smoke and sulfur dioxide were monitored at various sites. The daily averages for 10 of these sites are shown here.

areas. On Saturday, December 6, the weak winter Sun could not "burn away" the fog. The foul yellow mixture of smoke and fog got to be so thick that night that pedestrians could not find their way, even in familiar surroundings. People could not even see their own feet!

Winds did not sweep away the foul air until Tuesday, December 9. In central London the visibility remained below 500 meters continuously for 114 hours and below 50 meters for 48 hours straight. At Heathrow Airport the visibility was less than 10 meters for 48 hours beginning on the morning of December 6.

This infamous episode of December 1952 came to be known as "The Great Smog." Even in a city where "pea soup" fogs were relatively common, this event is legendary and is commonly viewed as the watershed event that gave rise to modern air pollution control in Great Britain as well as elsewhere in Western Europe and North America.

Experts agree that the Great Smog killed about 4000 people that December (**Fig. 13.C**). Furthermore, some researchers believe that an additional 8000 Londoners died in January and February 1953 due to the delayed effects of the smog or to lingering pollution. Other analyses disagree, blaming the excess deaths on influenza. Yet others suggest that many of the deaths may have resulted from an interaction between the smog and the flu. The debate about the effects of London's famous air pollution episode of 1952 is more than just an academic exercise. Although London no longer experiences "great smogs," foul air continues to plague big cities. The World Health Organization estimates that outdoor air pollution is responsible for about 800,000 deaths worldwide each year.

**Questions**

1. What was the source of the pollutants for The Great Smog?
2. Describe a meteorological factor that contributed to the buildup of pollutants.

sometimes find the word "smog" preceded by such modifiers as "London-type," "classical," "Los Angeles–type," or "photochemical." The first two modifiers refer to the original meaning of the word, and the last two to air quality problems created by secondary pollutants.

**Photochemical Reactions** Many reactions that produce secondary pollutants are triggered by strong sunlight and so are called **photochemical reactions.** One common example occurs when nitrogen oxides absorb solar radiation, initiating a chain of complex reactions. When certain volatile organic compounds are present, the result is the formation of a number of undesirable secondary products that are very reactive, irritating, and toxic. Collectively, this noxious mixture of gases and particles is called *photochemical smog.* One of the substances it usually contains is called PAN (peroxyacetyl nitrate), which damages vegetation and irritates the eyes. The *major* component in photochemical smog is ozone. Recall from Chapter 1 that ozone is formed by natural processes in the stratosphere. However, when produced near Earth's surface, ozone is considered a pollutant.

Because the reactions that create ozone are stimulated by strong sunlight, the formation of this pollutant is limited to daylight hours. Peaks occur in the afternoon following a series of hot, sunny, calm days. As you might expect, ozone levels are highest during the warmer summer months. The "ozone season" varies from one part of the country to another. Although May through October is typical, areas in the Sunbelt of the American South and Southwest may experience problems throughout the year. By contrast, northern states have shorter ozone seasons, such as May through September for North Dakota.

The health and environmental effects of ozone are well documented. For example, according to the EPA, health effects attributed to ozone exposure include significant decreases in lung function and increased respiratory symptoms, such as chest pain and cough. Exposure to ozone can make people more susceptible to respiratory infection, result in lung inflammation, and aggravate preexisting respiratory diseases such as

**Table 13.1** Air Quality and Emissions Trends (negative numbers indicate improvements in air quality)

| | Percentage Change in Concentrations | |
|---|---|---|
| | 1990–2012 | 2000–2012 |
| $NO_2$ | –50 | –38 |
| $O_3$ 8-hour | –14 | –9 |
| $SO_2$ | –76 | –65 |
| $PM_{10}$ 24-hour | –39 | –27 |
| $PM_{2.5}$ annual | —* | –33 |
| CO | –75 | –57 |
| Pb | –87 | –52 |
| | **Percentage Change in Emissions** | |
| | **1990–2012** | **2000–2012** |
| $NO_x$ | –52 | –50 |
| VOCs | –45 | –24 |
| $SO_2$ | –76 | –66 |
| $PM_{10}$ | –35 | –10 |
| $PM_{2.5}$ | –57 | –45 |
| CO | –65 | –51 |
| Pb | –80 | –50 |

*Data not available.

asthma. These effects generally occur while individuals are actively exercising, working, or playing outdoors. Children, active outdoors during the summer when ozone levels peak, are most at risk of experiencing such effects. In addition, longer-term exposures to moderate levels of ozone may cause irreversible changes in lung structure, which could lead to premature aging of the lungs.

Ozone also affects vegetation and ecosystems, leading to reduced agricultural crop and commercial forest yields; reduced growth and survivability of tree seedlings; and increased plant susceptibility to disease, pests, and other environmental stresses such as harsh weather. In long-lived species, these effects may become evident only after many years or even decades, thus having the potential for long-term effects on forest ecosystems. Ground-level ozone damage to the foliage of trees and other plants can also decrease the aesthetic value of natural areas.

**Volcanic Smog (Vog)** Although smog is largely a human-induced atmospheric hazard, nature is capable of creating it as well. A prime example occurs in active volcanic areas such as Hawaii. The satellite image in Figure 13.9 shows a dense haze hanging over the Hawaiian Islands. It is a natural phenomenon that has come to be called **vog**, short for *volcanic smog*. In this region vog forms when sulfur dioxide from Kilauea volcano combines with oxygen and water vapor in the presence of strong sunlight. The tiny sulfate particles that make up vog reflect light well, so that vog shows up easily when viewed from space. In 2008, when this event occurred, $SO_2$ concentrations reached unhealthy levels in Hawaii Volcanoes National Park near the summit of Kilauea. In addition to reducing visibility, vog can aggravate respiratory problems such as asthma. Vog is not confined to Hawaii; it can and does occur in other volcanic areas.

▲ **Figure 13.9 Volcanic smog is called vog** The image from NASA's *Aqua* satellite was acquired on December 3, 2008, and shows a dense vog-created haze over the Hawaiian Islands. Vog forms when sulfur dioxide ($SO_2$) from Kilauea volcano on the Big Island of Hawaii combines with oxygen and water vapor in the atmosphere to produce this natural form of smog. Vog is a relatively common occurrence in this region, although it is not usually this thick and widespread.

Video MM
Smog Bloggers
http://goo.gl/s6wVDV

## ✔ Concept Checks 13.2

1. Distinguish between a primary pollutant and a secondary pollutant.
2. List the major primary pollutants. Which one is most abundant? Which two are linked to photochemical smog?
3. What is a photochemical reaction? What is the major component of photochemical smog?
4. What is vog?

## 13.3 Trends in Air Quality

**Summarize trends in air quality since 1980.**

Although Table 13.1 shows that considerable progress has been made in controlling air pollution, the quality of the air we breathe still remains a serious public health problem. Economic activity, population growth, meteorological conditions, and regulatory efforts to control emissions all influence the trends in air pollutant emissions. Up until the 1950s the greatest influences on emissions were related to the economy and population growth. Emissions grew as the economy and population increased. Emissions fell in periods of economic recession. For example, dramatic declines in emissions in the 1930s were due to the Great Depression (Fig. 13.10). Emissions also increase as a result of shifts in the demand for various products. For instance, the tremendous upsurge in demand for gasoline following World War II increased emissions associated with petroleum refining.

In the 1950s U.S. states issued air pollution statutes generally targeted toward smoke and particulate emissions. It was not until the passage of the federal Clean Air Act in 1970 that major strides were made in reducing air pollution. This legislation created the Environmental Protection Agency and charged it with establishing air quality and emissions standards.

### Establishing Standards

The Clean Air Act of 1970 mandated the setting of standards for four of the primary pollutants—particulates, sulfur dioxide, carbon monoxide, and nitrogen oxides—as well as the secondary pollutant ozone. At the time, these five pollutants were recognized as being the most widespread and objectionable. Today, with the addition of lead, they are known as the *criteria pollutants* and are covered by the National Ambient Air Quality Standards (Table 13.2). The primary standard for each pollutant shown in Table 13.2 is based on the highest level that can be tolerated by humans without noticeable ill effects, minus a 10–50 percent margin for safety.

For some of the pollutants, both long-term and short-term levels are set. Short-term levels are designed to protect against acute effects, whereas the long-term standards were established to guard against chronic effects. *Acute* refers to pollutant levels that may be life-threatening within a period of hours or days. *Chronic* pollutant levels cause gradual deterioration of a variety of physiological functions over a span of years. It should be pointed out that standards are estab-

**Table 13.2 | National Ambient Air Quality Standards**

| Pollutant | Standard Value | |
|---|---|---|
| **Carbon monoxide (CO)** | | |
| 8-hour average | 9 ppm* | (10 mg/m³) |
| 1-hour average | 35 ppm | (40 mg/m³)** |
| **Nitrogen dioxide ($NO_2$)** | | |
| Annual arithmetic mean | 0.053 ppm | (100 μg/m³)*** |
| **Ozone ($O_3$)** | | |
| 1-hour average | 0.12 ppm | (235 μg/m³) |
| 8-hour average | 0.08 ppm | (157 μg/m³) |
| **Lead (Pb)** | | |
| Quarterly average | | 0.15 μg/m³ |
| **Particulate < 10 micrometers ($PM_{10}$)** | | |
| 24-hour average | | 150 μg/m³ |
| **Particulate < 2.5 micrometers ($PM_{2.5}$)** | | |
| Annual arithmetic mean | | 15 μg/m³ |
| 24-hour average | | 35 μg/m³ |
| **Sulfur dioxide ($SO_2$)** | | |
| Annual arithmetic mean | 0.03 ppm | (80 μg/m³) |
| 24-hour average | 0.14 ppm | (365 μg/m³) |
| 1-hour average | 75 ppb† | |

*ppm, parts per million.
**mg/m³, milligrams per cubic meter of air. A milligram is one-thousandth of a gram.
***μg/m³, micrograms per cubic meter. A microgram is one-millionth of a gram.
†ppb, parts per billion.
Source: U.S. Environmental Protection Agency, Office of Air Quality Planning and Standards.

▼ **Figure 13.10 Trends in national emissions 1900–2012** Prior to 1970, economic activity and population growth were major factors influencing the emissions of air pollutants. Emissions grew as the economy and population increased, and emissions declined during economic downturns. For example, dramatic declines in emissions in the 1930s were due to the Great Depression. Since 1970, much of the downward trend in emissions has been due to the Clean Air Act.

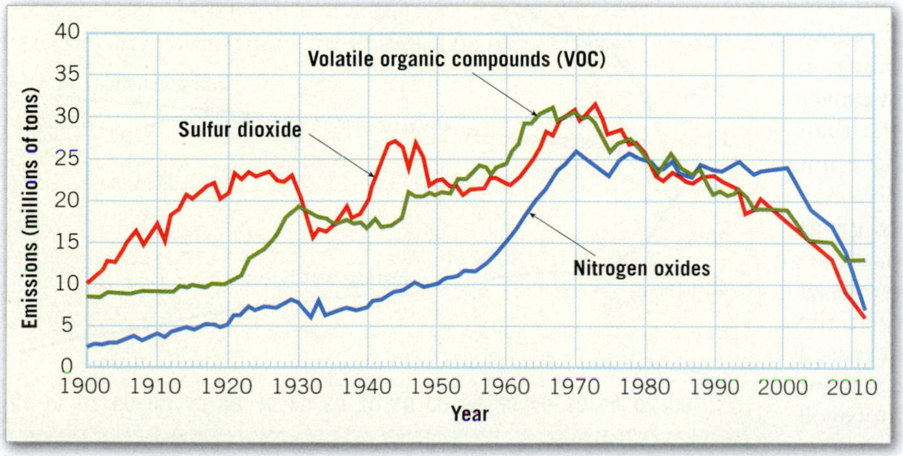

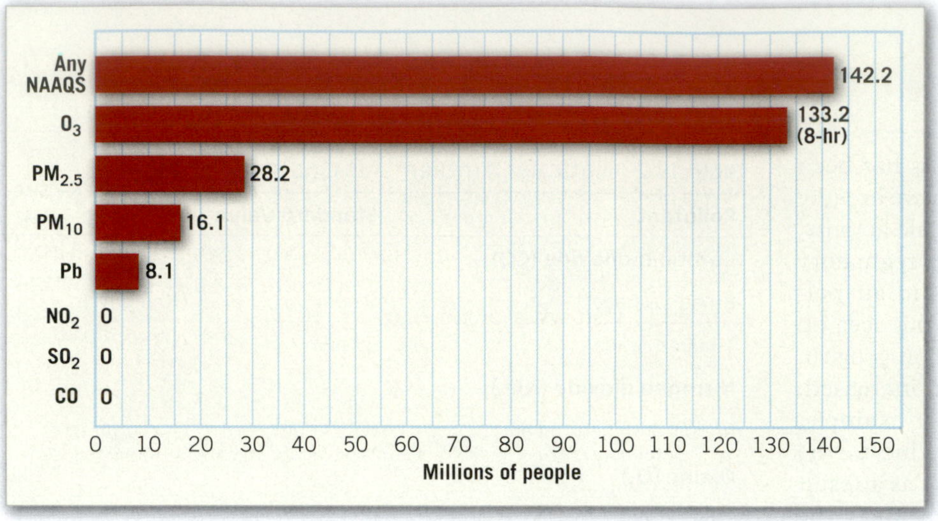


▲ **Figure 13.11 Number of people living in counties with air quality concentrations above the levels of the National Ambient Air Quality Standards (NAAQS) in 2012** For example, 16.1 million people live in counties where $PM_{10}$ concentrations exceed the national standard. Despite substantial progress in reducing emissions, there were still more than 142 million people nationwide who lived in counties with monitored air quality levels above the primary national standards.

lished using human health criteria and not according to their impact on other species or on atmospheric chemistry. Since the original Clean Air Act of 1970 was implemented, it has been amended, and its regulations and standards have been revised periodically.

By 2012 more than 142 million people in the United States resided in counties that did not meet one or more air quality standards (**Fig. 13.11**). It is clear from Figure 13.11 why the EPA describes ozone as our "most pervasive ambient air pollution problem." The number of people living in counties that exceeded the ozone standard is greater than the total number of those living in counties affected by the other six categories of pollutants.

The fact that air quality standards have not yet been met in some places does not indicate a lack of progress. The United States has made significant strides in reducing air pollution. In 2012 emissions of the five major primary pollutants shown in Figure 13.5 totaled about 84 million tons. By contrast, in 1970 when the Clean Air Act first became law, the same five pollutants totaled about 301 million tons. The 2012 total is about 72 percent lower than the 1970 level. As Table 13.1 indicates, downward trends in all pollutants are substantial. This improvement in air quality has been achieved during a time when urban growth has been great. However, methods of control have not been as effective as expected in upgrading urban air quality.

An important reason for the slower-than-expected progress in improving air quality is related to growth. For example, between 1980 and 2012, *on a per-car basis*, emissions of primary pollutants were cut dramatically. However, at the same time this was occurring, the U.S. population increased by 38 percent, and vehicle miles traveled went up by 92 percent (**Fig. 13.12**). In other words, pollution controls have improved air quality, but the positive effects have been partly offset by an increase in the number of vehicles on the road.

## Air Quality Index

The **Air Quality Index (AQI)** is a standardized indicator for reporting daily air quality to the general public. Simply, it is an attempt to answer the question: How clean or polluted is the air today? It provides information on what health effects people might experience within a few hours or days of breathing polluted air. The EPA calculates the AQI for five major pollutants regulated by the Clean Air Act: ground-level ozone, particulate matter, carbon monoxide, sulfur dioxide, and nitrogen dioxide. Ground-level ozone and airborne particulates are the pollutants that pose the greatest risk to human health in the United States.

The AQI scale runs from 0 to 500 (**Fig. 13.13**). The higher the value, the greater the level of air pollution and the greater the health concern. An AQI value of 100 generally corresponds to the national air quality standard for the pollutant. Values below 100 are usually considered satisfactory. When values exceed 100, air quality is considered to be unhealthy—at first for sensi-

▼ **Figure 13.12 Comparison of growth areas and emissions** Between 1980 and 2012 gross domestic product increased 133 percent, vehicle miles traveled increased 92 percent, energy consumption increased 27 percent, and U.S. population increased 38 percent. At the same time, total emissions of the six principal air pollutants decreased about 67 percent.

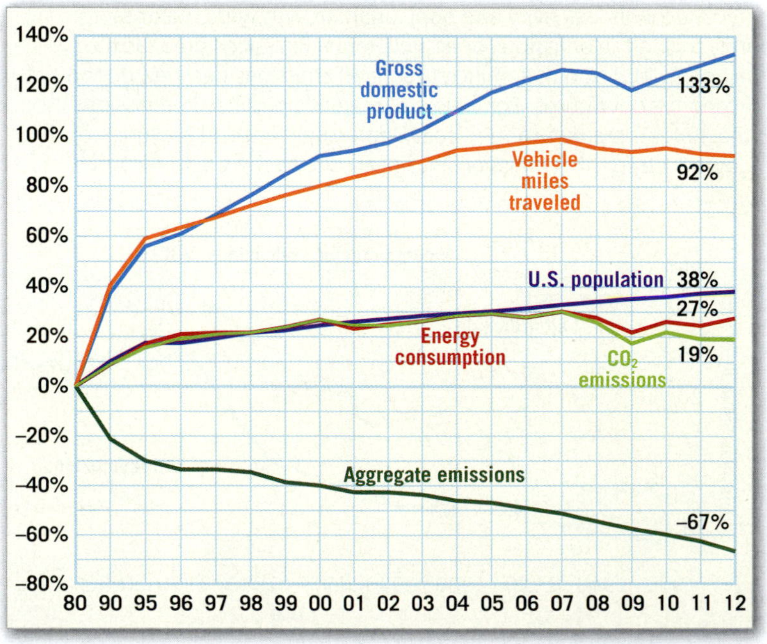

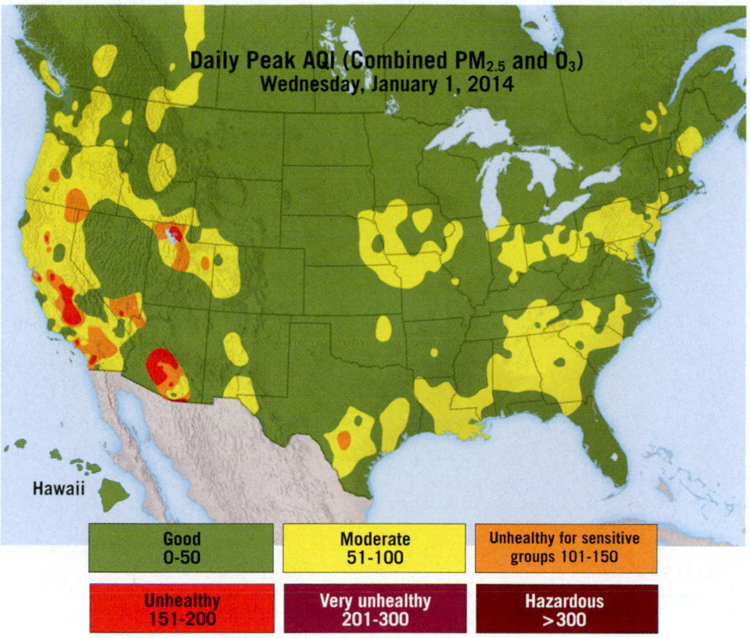


◀ **Figure 13.13 Air Quality Index** The national Air Quality Index (AQI) forecast map for January 1, 2014. Specific colors for each AQI category make interpreting the map relatively easy. To check the current map, go to www.airnow.gov.

tive groups and then, as values increase, for everyone else. Specific colors are assigned to each AQI category to make it easier for people to quickly understand whether air pollution is reaching unhealthy levels in their communities. To access the current AQI, visit www.airnow.gov.

### ✔ Concept Checks 13.3

1. When was the Clean Air Act established? What are its criteria pollutants?
2. Compare emissions of primary pollutants in 1970 with emissions in 2012.
3. What is the Air Quality Index?

## 13.4 Meteorological Factors Affecting Air Pollution

**Describe the influence of wind on air quality. Sketch a graph or diagram showing a temperature inversion and relate it to mixing depth.**

The most obvious factor influencing air pollution is the quantity of contaminants emitted into the atmosphere. Still, experience shows that even when emissions remain relatively steady for extended periods, wide variations in air quality often occur from one day to the next. Indeed, when air pollution episodes occur, they do not generally result from a drastic increase in the output of pollutants; instead, they occur because of changes in certain atmospheric conditions. Box 13.3 provides a good example.

Perhaps you have heard the phrase "The solution to pollution is dilution." To a significant degree, this is true. If the air into which pollution is released is not dispersed, the air will become more toxic. Two of the most important atmospheric conditions affecting the dispersion of pollutants are the strength of the wind and the stability of the air. These factors are critical because they determine how rapidly pollutants are diluted by mixing with the surrounding air after leaving the source.

### Wind as a Factor

The manner in which wind speed influences the concentration of pollutants is shown in Figure 13.14. Assume that a burst of pollution leaves a chimney stack every second. If the wind speed were 10 meters per second (23 miles per hour), the distance between each pollution "cloud" would be 10 meters. If the wind is reduced to 5 meters per second, the distance between "clouds" will be 5 meters. Consequently, because of the direct effect of wind speed, the concentration of pollutants is twice as great with the 5 meters per second wind as with the 10 meters per second wind. It is easy to understand why air pollution problems seldom occur when winds are strong but rather are associated with periods when winds are weak or calm.

A second aspect of wind speed influences air quality: The stronger the wind, the more turbulent the air. Thus, strong winds mix polluted air more rapidly with the surrounding air, thereby causing the pollution to be more dilute.

◀ **Figure 13.14 Effect of wind speed on the dilution of pollutants** The concentration of pollutants increases as wind speed decreases.

**severe& hazardous weather Box 13.3**

# Viewing an Air Pollution Episode from Space

In early October 2010 a high-pressure system settled in over eastern China, and air quality began to deteriorate. By October 9 and 10, China's National Environmental Monitoring Center declared air quality to be *poor* to *hazardous* around Beijing and in 11 eastern provinces. Citizens were advised to take measures to protect themselves. Visibility was reduced to 100 meters (330 feet) in some areas, and news outlets reported that at least 32 people died in traffic accidents caused by the poor visibility. Thousands suffered with asthma and other respiratory difficulties.

Instruments on NASA's *Aqua* and *Terra* satellites captured the natural-color view of this air pollution episode shown in **Figure 13.D**. The milky white and gray covering the right portion of the image is smog, while the brighter white patches are clouds. Two other images from NASA's *Aura* satellite show levels of aerosols (**Fig. 13.E**) and sulfur dioxide (**Fig. 13.F**). The primary source of sulfur dioxide is coal-burning power plants and smelters. Peak concentrations were 6 to 8 times the normal levels for China and 20 times the normal levels for the United States. The Aerosol Index indicates the presence of ultraviolet light–absorbing particles—largely smoke from agricultural burning and industrial processes. Figure 13.E shows that some areas had an index of 3.5. At an index value of 4, aerosols are so dense that you would have difficulty seeing the midday sun.

Thousands suffered with asthma and other respiratory difficulties.

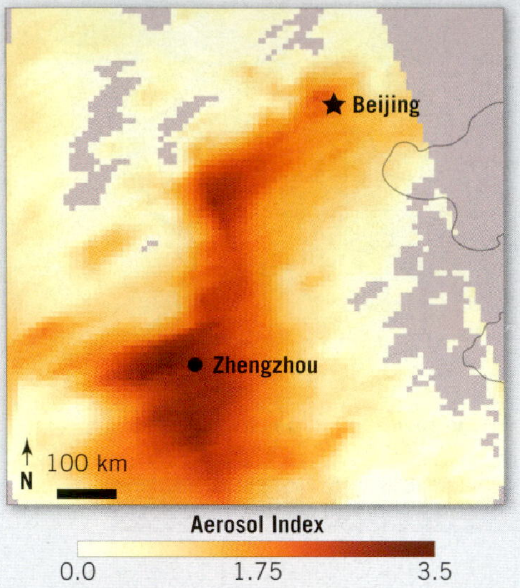


◄ **Figure 13.E Aerosols** This satellite image shows the extremely high levels of aerosols associated with the October 2010 air pollution episode in China. Gray areas lack data.

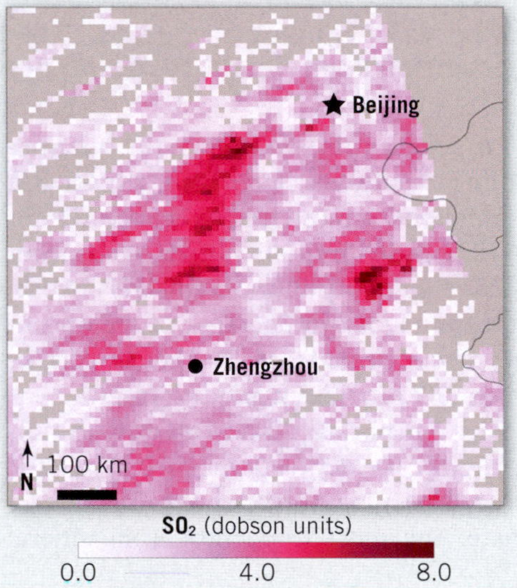


◄ **Figure 13.F Sulfur dioxide ($SO_2$)** Concentrations of this primary pollutant were very high during the October 2010 air pollution episode in China. Gray areas lack data.

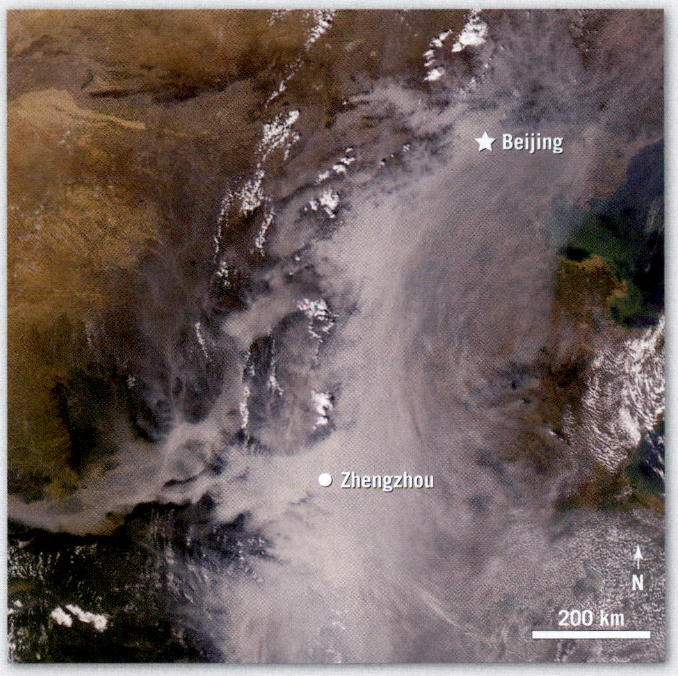


▲ **Figure 13.D Serious air pollution plagues a portion of China** This satellite image from October 8, 2010, captures the extent of the pollution episode.

On October 11, the weather changed. The stagnant air associated with high pressure was replaced when a cold front brought cleansing rain and strong winds that cleared the sky. Clearly, this air pollution episode could not have occurred without emissions from human activities. However, it is also apparent that atmospheric conditions play a key role in causing variations in air quality.

**Questions**

1. In the episode described here, what was the source of the sulfur dioxide pollution?
2. What was the source of most of the aerosols?

▲ **Figure 13.15 Air pollution in downtown Los Angeles** Temperature inversions act as lids that trap pollutants below.

Conversely, when winds are light, there is little turbulence, and the concentration of pollutants remains high.

## The Role of Atmospheric Stability

Whereas wind speed governs the amount of air into which pollutants are initially mixed, atmospheric stability determines the extent to which *vertical* motions will mix the pollution with cleaner air above. The vertical distance between Earth's surface and the height to which convectional movements extend is called the **mixing depth**. The greater the mixing depth, the better the air quality. When the mixing depth is several kilometers, pollutants are mixed through a large volume of cleaner air and dilute rapidly. When the mixing depth is shallow, pollutants are confined to a much smaller volume of air, and concentrations can reach unhealthy levels.

When air is stable, convectional motions are suppressed and mixing depths are small. Conversely, an unstable atmosphere promotes vertical air movements and greater mixing depths. Because heating of Earth's surface by the Sun enhances convectional movements, mixing depths are usually greater during the afternoon hours. For the same reason, mixing depths during the summer months are typically greater than during the winter months.

**Temperature inversions** are situations in which the atmosphere is very stable and mixing depths are significantly restricted. Warm air overlying cooler air acts as a lid and prevents upward movement, leaving the pollutants trapped in a relatively narrow zone near the ground. This effect is dramatically illustrated by the photograph in Figure 13.15. Most of the air pollution episodes cited earlier were linked to the occurrence of temperature inversions that remained in place for many hours or days.

**Surface Temperature Inversions** Solar heating can result in high surface temperatures during the late morning and afternoon that increase the environmental lapse rate and render the lower air unstable (see Chapter 4). During nighttime hours, however, just the opposite situation may occur: Temperature inversions, which result in very stable atmospheric conditions, can develop close to the ground. These surface inversions form because the ground is a more effective radiator than the air above. Therefore, radiation from the ground to a clear night sky causes more rapid cooling at the surface than higher in the atmosphere. Consequently, the coldest air is found next to the ground, yielding a vertical temperature profile resembling the one shown in the upper portion of Figure 13.16. Once the Sun rises, the ground is heated, and the inversion disappears.

Although usually rather shallow, surface inversions may be deep in regions where the land surface is uneven. Because cold air is denser than warm air, the chilled air near the surface gradually drains from the uplands and slopes into adjacent

▼ **Figure 13.16 Surface temperature inversion A.** A generalized temperature profile of a surface inversion. **B.** The temperature profile changes after the Sun has heated the surface.

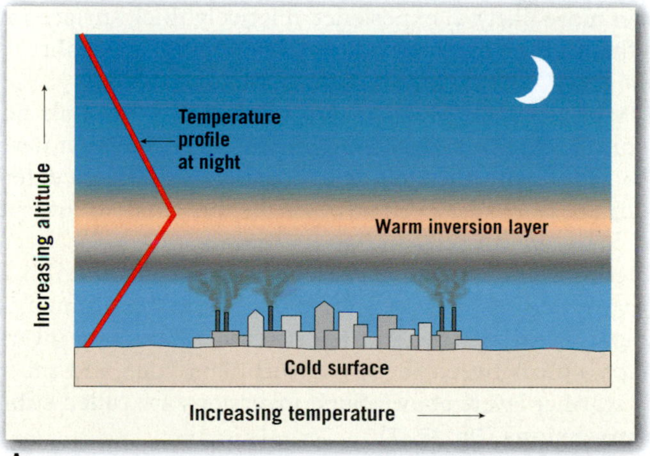


A.

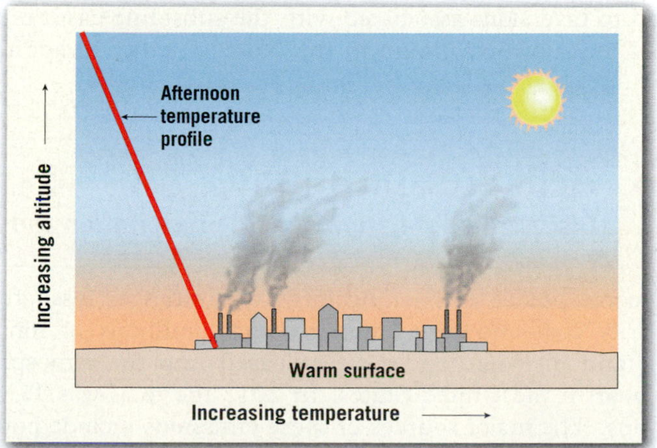


B.

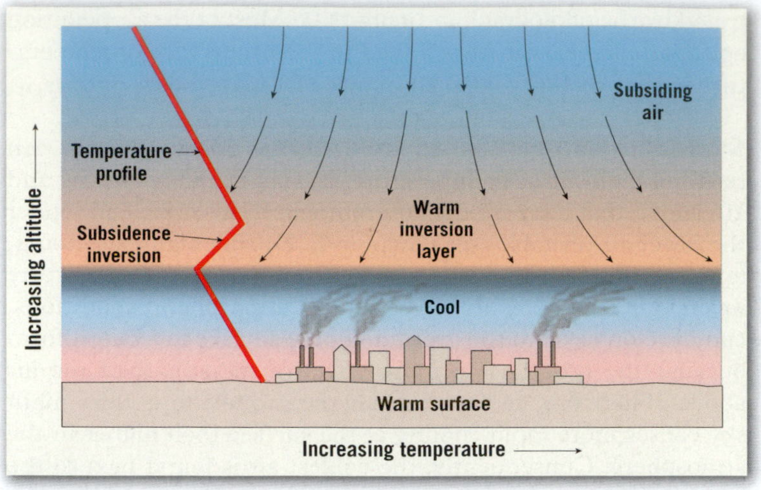


▲ **Figure 13.17 Subsidence inversion** Inversions aloft frequently develop in association with slow-moving centers of high pressure, where the air aloft subsides and warms by compression. The turbulent surface zone does not subside as much. Thus, an inversion often forms between the lower turbulent zone and the subsiding layers above. The red line is a generalized temperature profile of an inversion aloft.

lowlands and valleys. As might be expected, this deeper surface inversion will not dissipate as quickly after sunrise. Thus, although valleys are often preferred sites for manufacturing because they afford easy access to water transportation, they are also more likely to experience relatively thick surface inversions that in turn will have a negative effect on air quality.

**Inversions Aloft** Many extensive and long-lived air pollution episodes are linked to temperature inversions that develop in association with the sinking air that characterizes centers of high air pressure (anticyclones). As the air sinks to lower altitudes, it is compressed, and so its temperature rises. Because turbulence is almost always present near the ground, this lowermost portion of the atmosphere is generally prevented from participating in the general subsidence. Thus, an inversion develops aloft between the lower turbulent zone and the subsiding warmer layers above. Such inversions are called **subsidence inversions** (Fig. 13.17).

The air pollution that plagues Los Angeles is frequently related to inversions associated with the subsiding eastern portion of the subtropical high in the North Pacific. In addition, the adjacent cool waters of the Pacific Ocean and the mountains surrounding the city compound the problem. When winds move cool air from the Pacific into Los Angeles, the warmer air that is pushed aloft creates or strengthens an inversion aloft that acts as an effective lid. Because the surrounding mountains keep the smog from moving farther inland, air pollution is trapped in the basin until a change in weather brings relief. Clearly, the geographic setting of a place can significantly contribute to air quality problems. The Los Angeles area is an excellent example.

In summary, we have seen that when the wind is strong and an unstable environmental lapse rate prevails, the diffusion of pollutants is rapid, and high pollution concentrations will not occur except perhaps near a major source. In contrast, when an inversion exists and winds are light, diffusion is inhibited, and high pollution concentrations are to be expected in areas where there are sources. Air pollution is especially acute in urban areas experiencing frequent and prolonged temperature inversions.

## ✔ Concept Checks 13.4

1. Are most air pollution episodes triggered by a dramatic increase in the output of pollutants? Explain.
2. Describe two ways in which wind influences air quality.
3. What is *mixing depth*? How does it relate to air quality?
4. Contrast the formation of a surface temperature inversion with an inversion aloft.

**students sometimes ask...**

**I've heard that wood-burning fireplaces and stoves can be significant sources of air pollution. Is that actually the case?**

Yes. Wood smoke can build up in areas where it is generated and expose people to high levels of air pollution, especially on cold nights when there is a temperature inversion. Wood smoke contains significant quantities of particulates and much higher levels of hazardous air pollutants, including some cancer-causing chemicals, than smoke from oil- and gas-fired furnaces. Some communities now ban the installation of conventional fireplaces or woodstoves that are not EPA certified.

# 13.5 Acid Precipitation

**Discuss the formation of acid precipitation and list some of its effects on the environment.**

As a consequence of burning large quantities of fossil fuels, primarily coal and petroleum products, millions of tons of sulfur and nitrogen oxides are released into the atmosphere each year in the United States. In 2012 the total was 17 million tons. The major sources of these emissions include power-generating plants, industrial processes such as ore smelting and petroleum refining, and motor vehicles of all kinds. Through a series of complex chemical reactions, some of these pollutants are converted into acids that then fall to Earth's surface as rain or snow. This is referred to as *wet deposition*. By contrast, in areas where the weather is dry, the acid-producing chemicals may become incorporated into dust or smoke and fall to the ground as *dry deposition*. Dry-deposited particles and gases can be washed from the surface by rain, making the runoff more

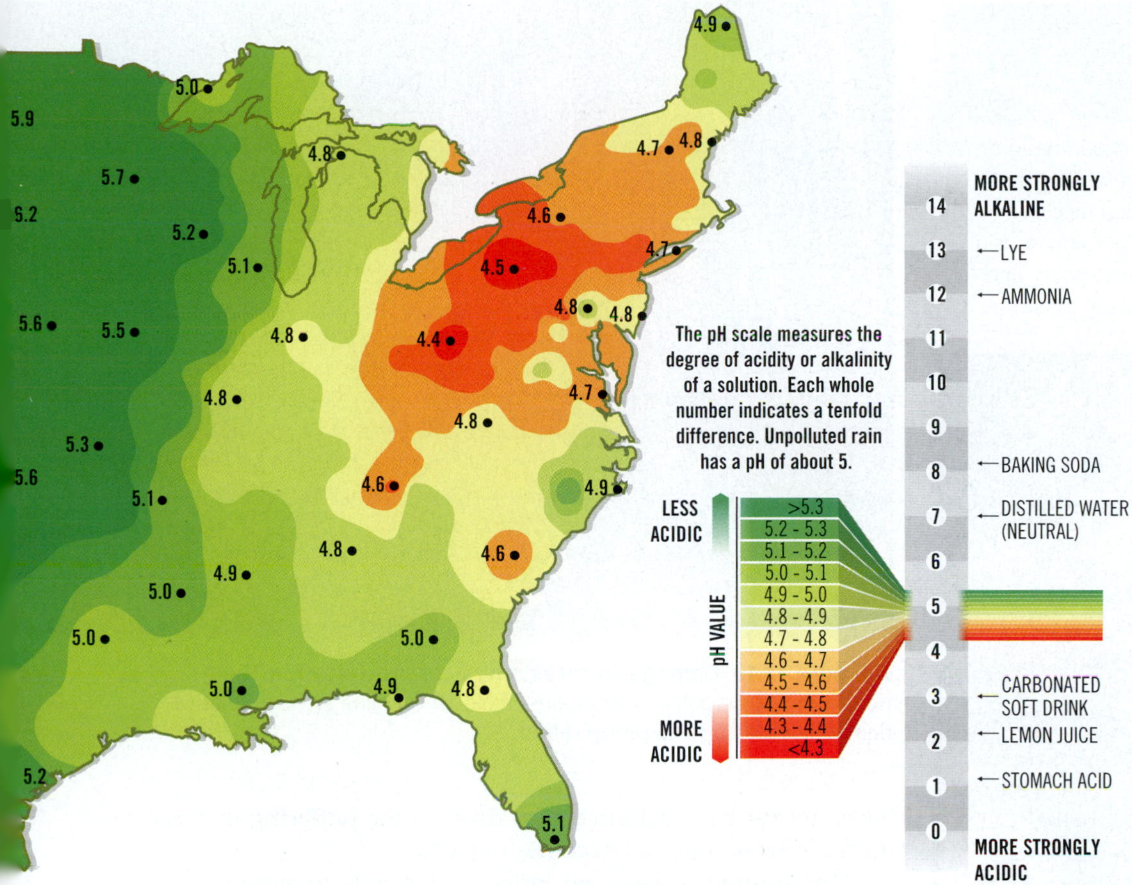


▲ **Figure 13.18 Precipitation pH values for the United States in 2011** In the United States, acid precipitation is most severe in the Northeast.

acidic. About half of the acidity in the atmosphere falls back to the surface as dry deposition.

In 1852 the English chemist Angus Smith coined the term *acid rain* to refer to the effect that industrial emissions had on precipitation in the English Midlands. A century and a half later, this phenomenon is not only the focus of research for many environmental scientists but also a topic with substantial international political importance. Although Smith clearly realized that acid rain causes environmental damage, large-scale effects were not recognized until the middle part of the twentieth century. Eventually, widespread public concern in the late 1970s led to significant government-sponsored studies of the problem. Such research activities continue to examine this unresolved environmental problem.

## Extent and Potency of Acid Precipitation

Rain is naturally somewhat acidic. When carbon dioxide from the atmosphere dissolves in water, it becomes weak carbonic acid. Small amounts of other naturally occurring acids also contribute to the acidity of precipitation. It was once thought that unpolluted rain has a pH of about 5.6 on the pH scale (**Fig. 13.18**). However, studies in uncontaminated remote areas have shown that precipitation usually has a pH closer to 5. Unfortunately, in most areas within several hundred kilometers of large centers of human activity, precipitation has much lower pH values. This rain or snow is called **acid precipitation**.

Widespread acid rain has been known in Northern Europe and eastern North America for some time (Fig. 13.18). Studies have also shown that acid rain occurs in many other regions, including western North America, Japan, China, Russia, and South America. In addition to local pollution sources, a portion of the acidity found in the northeastern United States and eastern Canada originates hundreds of kilometers away, in industrialized regions to the south and southwest. This situation occurs because many pollutants remain in the atmosphere for periods as long as 5 days, during which time the winds may transport them great distances.

One contributing factor is, of all things, a technology that is used to reduce pollution in the immediate vicinity of a source. Taller chimney stacks improve local air quality by releasing pollutants into the stronger and more persistent winds that exist at greater heights. Although such stacks enhance dilution and dispersion, they also promote the long-distance transport of these unwanted emissions. In this way, individual stack plumes with pollution concentrations considered too dilute to be a direct health or environmental threat locally contribute to inter-regional pollution problems. Unfortunately, because atmospheric processes in eastern North America lead to a thorough mixing of pollutants, it is not yet possible to distinguish clearly between the relative impact of distant sources and local sources.

## Effects of Acid Precipitation

Acid rain looks, feels, and tastes just like clean rain. The harm to people from acid rain is not direct. Walking in acid rain, or even swimming in an acid lake, is no more dangerous than walking or swimming in water that has not been affected by acid rain. However, the sulfur dioxide and nitrogen oxide pollutants that cause acid rain do damage human health. These gases interact in the atmosphere to form fine sulfate and nitrate particles that can be transported long distances by winds and inhaled into people's lungs. These particles have been shown to aggravate heart problems and lung disorders such as asthma and bronchitis.

The damaging effects of acid rain on the environment are believed to be considerable in some areas and imminent in others. The best-known effect of acid precipitation is the lowering of pH in thousands of lakes and streams in Scandinavia and eastern North America. Accompanying this have been substantial increases in dissolved aluminum that is leached from the

## eye ON THE atmosphere 13.2

Air pollution creates a translucent veil over China's Sichuan basin on May 6, 2011. The dull gray haze contrasts with bright white clouds and lingering snow on nearby mountains. The pollution results in part from urban and industrial sources in which coal burning is a significant energy source. A week earlier, visibility here was good, and this region was *not* experiencing an air pollution episode. (NASA)

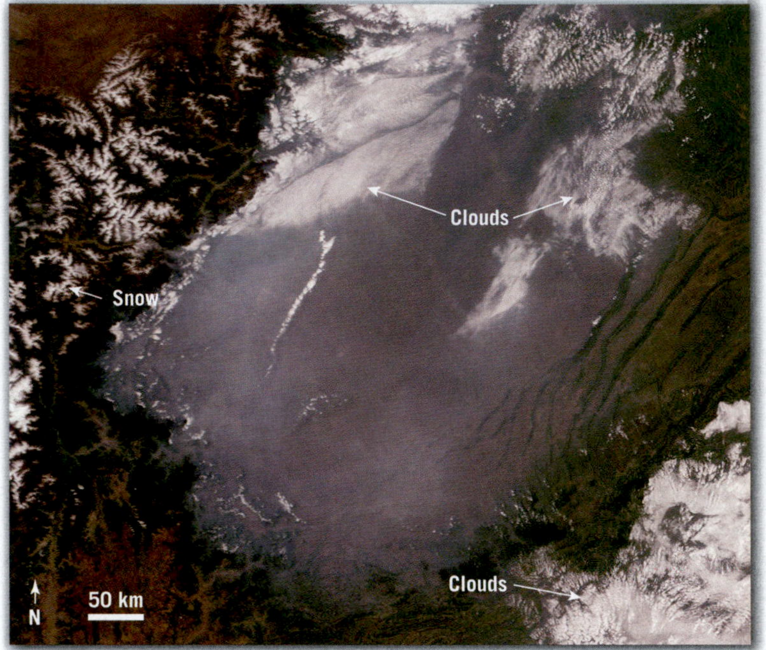


**Questions**

1. What secondary pollutant might you expect to be present in the polluted air on May 6?
2. If we assume that the output of pollutants in this region does not change significantly from week to week, why was the air relatively clear one week and polluted the next?

soil by the acidic water and that, in turn, is toxic to fish. Consequently, some lakes are virtually devoid of fish, whereas others are approaching this condition. Furthermore, ecosystems are characterized by many interactions at many levels of organization, which means that evaluating the effects of acid precipitation on these complex systems is difficult and expensive—and far from complete.

Even within small areas, the effects of acid precipitation can vary significantly from one lake to another. Much of this variation is related to the nature of the soil and rock materials in the area surrounding the lake. Because minerals such as calcite in some rocks and soils can neutralize acid solutions, lakes surrounded by such materials are less likely to become acidic. In contrast, lakes that lack this buffering material can be severely affected. Even so, over a period of time, the pH of lakes that have not yet been acidified may drop as the buffering material in the surrounding soil becomes depleted.

▲ **Figure 13.19 Damage to forests by acid precipitation** These trees in the Appalachian Mountains of North Carolina are examples of damage caused by acid precipitation.

In addition to lakes no longer being able to support fish, research indicates that acid precipitation may also reduce agricultural crop yields and impair the productivity of forests. Acid rain not only harms the foliage but also damages roots and leaches nutrient minerals from the soil (**Fig. 13.19**). Finally, acid precipitation promotes the corrosion of metals and contributes to the destruction of stone structures (**Fig. 13.20**).

In summary, acid precipitation involves the delivery of acidic substances through the atmosphere to Earth's surface. These compounds are introduced into the air as by-products of combustion and industrial activity. The atmosphere is both the avenue by which offending compounds travel from their sources to the sites where they are deposited and the medium in which the combustion products are chemically transformed into acidic substances. In addition to its detrimental impact on aquatic systems, acid precipitation has a number of other harmful effects.

The emission reductions in nitrogen oxides and sulfur dioxide noted earlier in the chapter have not only contributed to improved air quality but have also led to reductions in the acidity of precipitation in many areas (see Table 13.1 and Fig. 13.10). Long-term monitoring of lakes and streams has shown that some acid-sensitive waters have experienced the beginnings of recovery. Nevertheless, although progress has been made, acid precipitation remains a complex and global environmental problem.

## eye ON THE atmosphere 13.3

This tall smokestack is part of a coal-fired electricity-generating plant located in a valley in the rolling hills of West Virginia. Tall stacks are commonly associated with such power plants as well as with many factories. An extensive radiation fog is hugging the ground. (Photo by Michael Collier)

**Questions**

1. Is the time of day shown here more likely early morning or midafternoon? Explain.
2. Sketch a simple graph illustrating the likely vertical temperature profile at the time the photo was taken.
3. What are two reasons that air quality in the local area is better because the stack is present?

▲ **Figure 13.20 Acid rain accelerates chemical weathering** Although we expect rock to gradually decompose, many stone monuments have succumbed prematurely because of accelerated chemical weathering linked to acid precipitation.

Video (MM) Hello Crud http://goo.gl/gDF9pS

### ✔ Concept Checks 13.5

1. Which primary pollutants are associated with the formation of acid precipitation?
2. How much more acidic is a substance with a pH of 4 than a substance with a pH of 6?
3. Based on the map in Figure 13.18, where in the United States is precipitation most acidic?
4. What are some environmental effects of acid precipitation?

# 13 Concepts in Review Air Pollution

## 13.1 The Air Pollution Threat

▸ List several natural sources of air pollution and identify those that are human accentuated.

- Air pollution shortens human life spans and negatively impacts agriculture.
- Air is never perfectly clean. Volcanic ash, salt particles, pollen and spores, smoke, and windblown dust are all examples of "natural air pollution."
- Although some types of air pollution are recent creations, others, such as London's infamous smoke pollution, have been around for centuries.

## 13.2 Sources and Types of Air Pollution ▸

Distinguish between primary and secondary pollutants. List important examples of each and discuss their effects on people and the environment.

**Key Terms:** air pollutant, primary pollutant, secondary pollutant, smog, photochemical reaction, vog

- Air pollutants are airborne particles and gases that occur in concentrations that endanger the health and well-being of organisms or disrupt the orderly functioning of the environment.
- Pollutants can be grouped into two categories: (1) primary pollutants, which are emitted directly from identifiable sources, and (2) secondary pollutants, which are produced in the atmosphere when certain chemical reactions take place among primary pollutants.
- The major primary pollutants are particulate matter (PM), sulfur dioxide, nitrogen oxides, volatile organic compounds (VOCs), carbon monoxide, and lead. Atmospheric sulfuric acid is one example of a secondary pollutant.
- Air pollution in urban and industrial areas is often called smog. Photochemical smog, a noxious mixture of gases and particles, is produced when strong sunlight triggers photochemical reactions in the atmosphere. A major component of photochemical smog is ozone.

**Q** Refer to Figure 13.7. How much of our energy comes from fossil fuels (coal, petroleum, and natural gas)? Which energy user category consumes the most coal?

## 13.3 Trends in Air Quality ▸

Summarize trends in air quality since 1980.

**Key Term:** Air Quality Index (AQI)

- Although considerable progress has been made in controlling air pollution, the quality of the air we breathe remains a serious public health problem.
- Economic activity, population growth, meteorological conditions, and regulatory efforts to control emissions all influence the trends in air pollution.
- The Clean Air Act of 1970 mandated the setting of standards for four of the primary pollutants—particulate matter, sulfur dioxide, carbon monoxide, and nitrogen oxides—as well as the secondary pollutant ozone. In 2012 emissions of the major primary pollutants in the United States were about 72 percent lower than in 1970.

**Q** Have per capita (per person) emissions of air pollutants increased or decreased since 1980? Use Figure 13.12 to support your answer.

## 13.4 Meteorological Factors Affecting Air Pollution ▸

Describe the influence of wind on air quality. Sketch a graph or diagram showing a temperature inversion and relate it to mixing depth.

**Key Terms:** mixing depth, temperature inversion, subsidence inversion

- When air pollution episodes take place, they do not generally result from a drastic increase in the output of pollutants; instead, they occur because of changes in certain atmospheric conditions.
- Two of the most important atmospheric conditions affecting the dispersion of pollutants are (1) the strength of the wind and (2) the stability of the air.
- The direct effect of wind speed is to influence the concentration of pollutants. Strong winds improve air quality by diluting and dispersing pollutants.
- Atmospheric stability determines the extent to which vertical motions will mix the pollution with cleaner air above the surface layer. The vertical distance between Earth's surface and the height to which convectional movements extend is called the mixing depth. Generally, the greater the mixing depth, the better the air quality. A temperature inversion represents a situation in which the atmosphere is very stable and the mixing depth is significantly restricted.
- When an inversion exists and winds are light, diffusion is inhibited and high pollution concentrations are to be expected in areas where pollution sources exist.

**Q** Explain why the air in the top portion of this image is much less polluted than the air near the surface.

## 13.5 Acid Precipitation ▸ Discuss the formation of acid precipitation and list some of its effects on the environment.

**Key Term:** acid precipitation

- In most areas within several hundred kilometers of large centers of human activity, the pH value of precipitation is lower than the usual value found in unpopulated areas.
- Acid precipitation (acidic rain or snow) forms when sulfur and nitrogen oxides produced as by-products of combustion and industrial activity are converted into acids during complex atmospheric reactions.
- The atmosphere is the avenue by which offending compounds travel from sources to the sites where they are deposited, and it is also the medium in which the combustion products are transformed into acidic substances.
- The damaging effects of acid precipitation on the environment include the lowering of pH in thousands of lakes in Scandinavia and eastern North America. Besides producing water that is toxic to fish, acid precipitation has also detrimentally altered complex ecosystems.

## Give it Some Thought

1. In Chapter 1 you learned that we should be concerned about ozone depletion in the atmosphere. But based on what is presented in this chapter, it seems like getting rid of ozone would be a good idea. Can you clarify this apparent contradiction?
2. Table 13.1 shows trends in air quality and emissions. Explain why ozone ($O_3$) appears on the "Percentage change in concentrations" portion of the table but does not appear on the "Percentage change in emissions" portion.
3. The average motor vehicle today emits *much* less pollution than did the vehicles of 30 or 40 years ago. Why have the positive effects of this sharp reduction *not* been as great as we might have expected? Include in your explanation information from one of the graphs in this chapter.
4. Motor vehicles are a major source of air pollutants. Using electric cars, such as the one pictured here, is one way that emissions from this source can be reduced. Although these vehicles emit little or no pollution directly into the air, can they still be connected to the emission of primary pollutants? If so, explain.

5. Assume that you are at an airport in a large urban area. The city is experiencing an air quality alert, and haze reduces visibility. As your plane climbs after takeoff, the air suddenly becomes much cleaner. Provide a likely explanation for the sudden change. You may wish to include a sketch.
6. As the accompanying photo illustrates, air pollution episodes reduce the amount of sunlight reaching Earth's surface. In which of these situations would the reduction likely be greatest (assuming that the level of pollutants is identical in each example)? Explain your choice.
   **a.** High-latitude city in summer
   **b.** High-latitude city in winter
   **c.** Low-latitude city in summer
   **d.** Low-latitude city in winter

7. A glance at a cross-section of a warm or cold front (see Figures 9.2 and 9.3) shows warm air *above* cool or cold air; that is, it shows a temperature inversion. Although temperature inversions are associated with fronts, they have little adverse effect on air quality. Why is this the case?
8. Ozone is sometimes called a "summer pollutant." Why is summertime also "ozone season"?
9. The use of tall smokestacks improves local air quality. However, tall stacks may contribute to pollution problems elsewhere. Explain.

## MasteringMeteorology™

Looking for additional review and test prep materials? Visit the Study Area in *MasteringMeteorology*™ to enhance your understanding of this chapter's content by accessing a variety of resources, including **MapMaster**™ interactive maps, Geoscience Animations, GEODe, *In the News* RSS feeds, flashcards, web links, self-study quizzes, and an eText version of *The Atmosphere*.

# 14 The Changing Climate

## Focus on Concepts

*Each statement represents the primary learning objective for the corresponding major heading within the chapter. After you complete the chapter, you should be able to:*

**14.1** Explain how unraveling past climate changes is related to the climate system and discuss several ways in which such changes are detected.

**14.2** Discuss four hypotheses that relate to natural causes of climate change.

**14.3** Summarize the nature and cause of the atmosphere's changing composition since about 1750. Describe the climate's response.

**14.4** Contrast positive- and negative-feedback mechanisms and provide examples of each.

**14.5** Discuss several likely consequences of global warming.

The focus of this chapter and Chapter 15 is *climate*, the long-term aggregate of weather. Climate is more than just an expression of average atmospheric conditions. In order to accurately portray the character of a place or an area, variations and extremes must also be included. Anyone who has the opportunity to travel around the world will find such an incredible variety of climates that it is hard to believe they could all occur on the same planet.

*Glaciers are sensitive to changes in temperature and precipitation and therefore provide clues about changes in climate. Like most other glaciers in Alaska, Bear Glacier near Seward is steadily retreating back into the mountains.*

# 14.1 The Climate System: A Key to Detecting Climate Change

**Explain how unraveling past climate changes is related to the climate system and discuss several ways in which such changes are detected.**

Climate has a significant impact on people, and we are learning that people also have a strong influence on climate. In fact, today global climate change caused by humans is a major environmental issue. Why is climate change newsworthy? The reason is that research focused on human activities and their impact on the environment has demonstrated that people are inadvertently changing the climate. Unlike changes in the geologic past, which were natural variations, modern climate change is dominated by human influences that are sufficiently large that they exceed the bounds of natural variability. Moreover, these changes are likely to continue for many centuries. The effects of this venture into the unknown with climate could be very disruptive—not only to humans but to many other lifeforms as well. The latter portion of this chapter examines the ways in which humans may be changing global climate.

## The Climate System

Throughout this book you have frequently been reminded that Earth is a complex system consisting of many interacting parts. A change in any one part can produce changes in any or all of the other parts—often in ways that are neither obvious nor immediately apparent. Key to understanding climate change and its causes is the fact that climate is related to all parts of the Earth system. We must recognize that there is a **climate system** that derives its energy from the Sun and includes the atmosphere, hydrosphere, geosphere (solid Earth), biosphere, and cryosphere. The first four were discussed in Chapter 1; the **cryosphere** refers to the portion of Earth's surface where water is in solid form. This includes snow, glaciers, sea ice, freshwater ice, and frozen ground (termed *permafrost*). The climate system *involves the exchanges of energy and moisture that occur among the five spheres*. These exchanges link the atmosphere to the other spheres so that the whole functions as an extremely complex interactive unit. Changes in the climate system do not occur in isolation. Rather, when one part of the climate system changes, the other components react. The major components of the climate system are shown in Figure 14.1.

The climate system provides a framework for the study of climate. The interactions and exchanges among the parts of the climate system create a complex network that links the five spheres. As you will see, because the climate system involves all of Earth's spheres, data from many sources are used to study and decipher climate change.

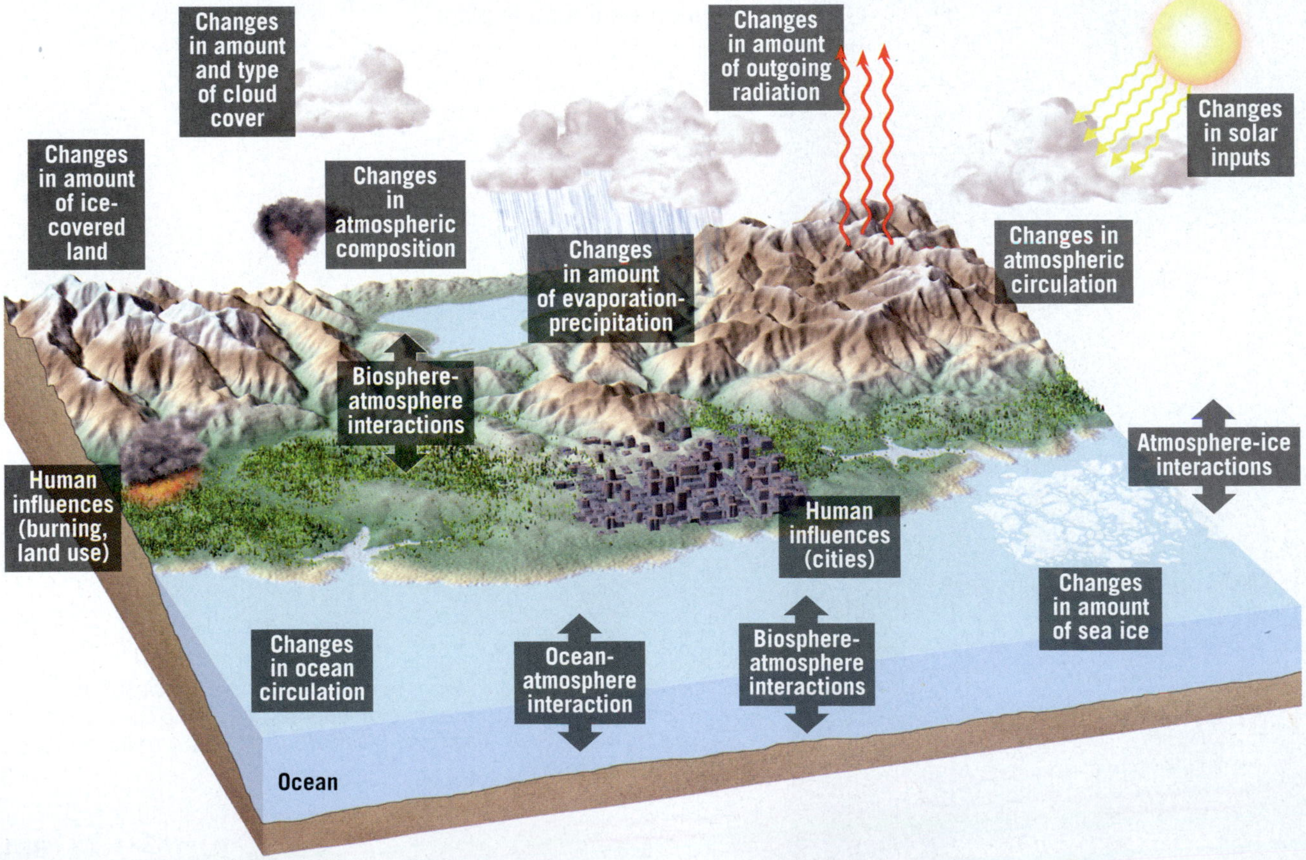


**◄ Figure 14.1 Earth's climate system** Schematic view showing several components of Earth's climate system. Many interactions occur among the various components on a wide range of space and time scales, making the system extremely complex.

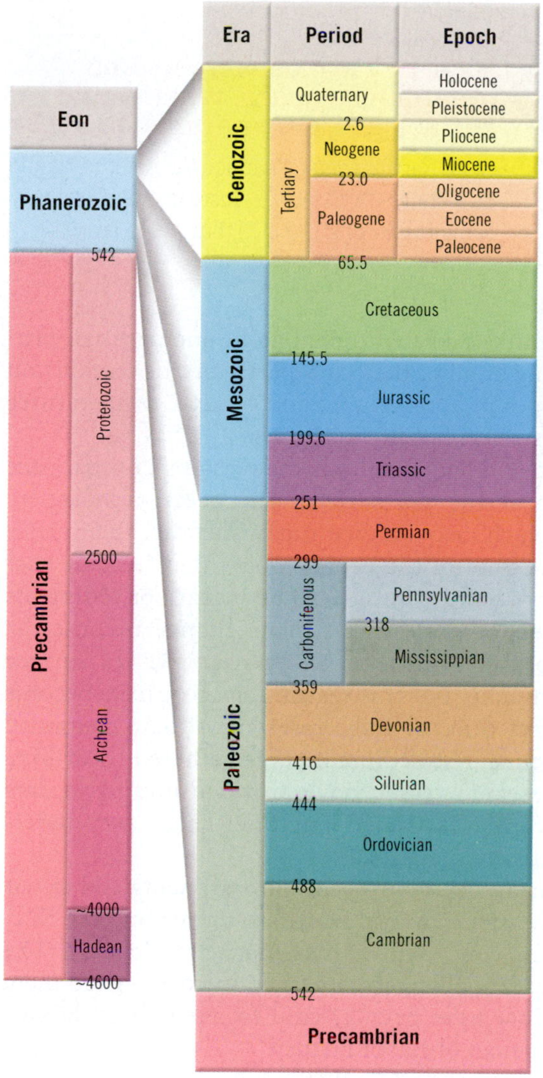


◀ **Figure 14.2 Geologic time scale** The time scale divides the vast 4.6-billion-year history of Earth into eons, eras, periods, and epochs. Numbers refer to millions of years before present. Precambrian time accounts for more than 88 percent of geologic time.

During greenhouse times, there is little, if any, permanent ice at either pole, and relatively warm-temperate climates are found at high latitudes. During icehouse conditions, global climate is cool enough to support extensive ice sheets at one or both poles. Earth's climate has gradually transitioned between these two categories only a few times in the past 542 million years, a span known as the Phanerozoic ("visible life") eon (**Fig. 14.2**). The rocks and deposits of the Phanerozoic eon contain abundant fossils that document major environmental and evolutionary trends. The most recent transition occurred during the Cenozoic era.

The early Cenozoic was a time of greenhouse climates like those the dinosaurs experienced during the preceding Mesozoic era. But by about 34 million years ago, permanent ice sheets were present at the South Pole, ushering in icehouse conditions (**Fig. 14.3**). Climate warmed during the Miocene epoch (about 20 million years ago) as mammal populations reached their greatest diversity. Climate then cooled. In North America, the lush "greenhouse" forests (there were palm trees in Wyoming and banana plants in Oregon) were replaced by open grasslands. Grassland ecosystems are better suited for a cooler, drier "icehouse" climate. By 2 million years ago (the start of the Quaternary epoch), Earth's climate was cold enough to support large ice sheets at both poles. In the Northern Hemisphere, ice advanced nearly as far south as the present-day Ohio River and subsequently retreated to Greenland (**Fig. 14.4**). For the past 800,000 years, this cycle of ice advance and retreat occurred about every 100,000 years. The last major ice sheet advance reached a maximum about 18,000 years ago.

How do we know about these changes? What are the causes? The next sections take a look at how scientists decipher Earth's

▼ **Figure 14.3 Relative climate change during the Cenozoic era** During the past 65 million years, Earth's climate shifted from being a warm "greenhouse" to being a cool "icehouse." Climate is not stable when viewed over long time spans. Earth has experienced several back-and-forth shifts between warm and cold.

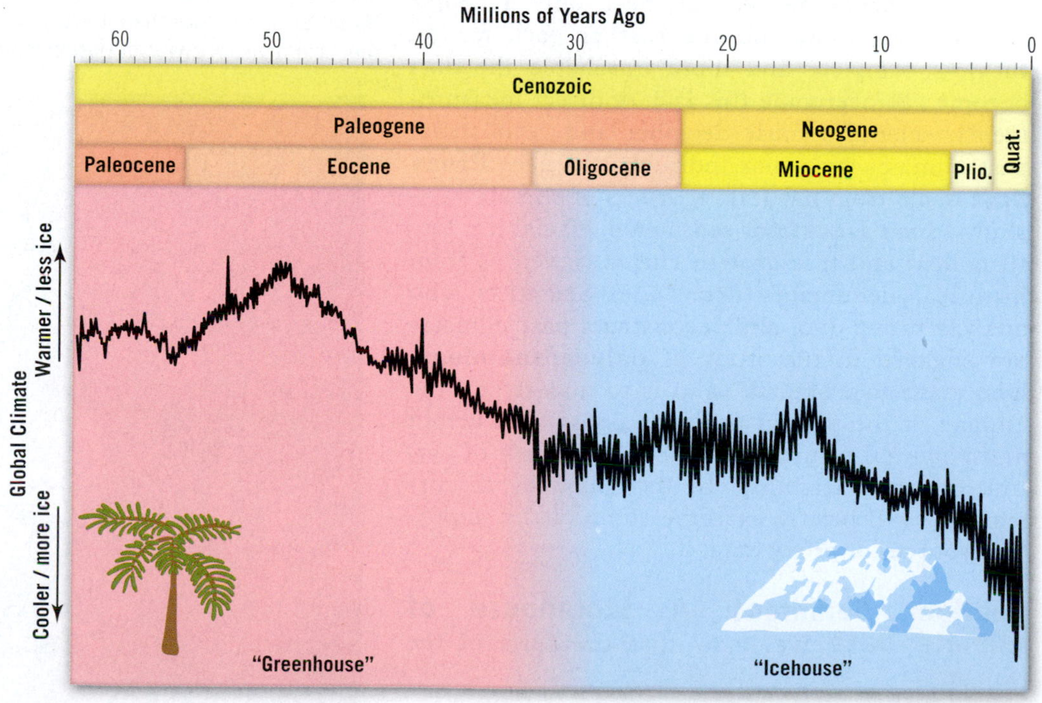


## Detecting Climate Change

Climate not only varies from place to place, as examined in Chapter 15, it is also naturally variable over time. During the great expanse of Earth history, and long before humans were roaming the planet, there were many shifts—from warm to cold and from wet to dry and back again. Practically every place on our planet has experienced wide swings in climate, such as from ice ages to conditions associated with subtropical coal swamps or desert dunes.

**Climates Change** Using fossils and many other geologic clues, scientists have reconstructed Earth's climate going back hundreds of millions of years. Over long time scales (tens to hundreds of millions of years), Earth's climate can be broadly characterized as being a warm "greenhouse" or a cold "icehouse."

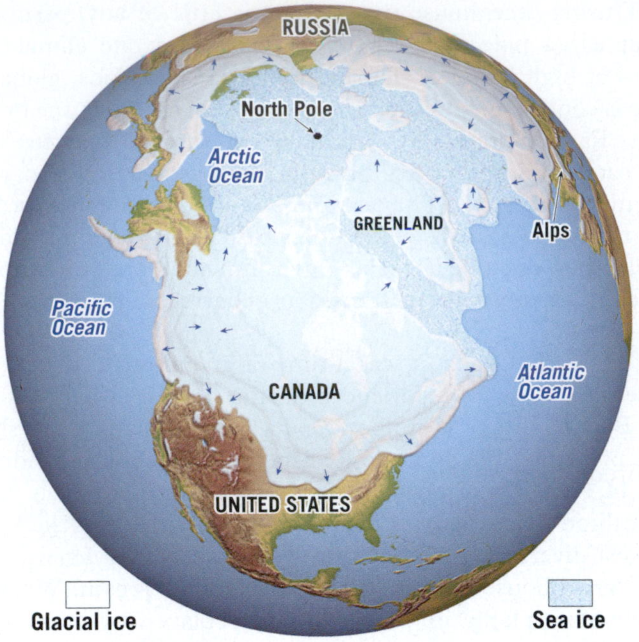


▲ **Figure 14.4 Where was the ice?** This map shows the maximum extent of ice sheets in the Northern Hemisphere during the Quaternary Ice Age.

climate history. We then explore some significant natural causes of climate change.

**Proxy Data** High-technology and precision instrumentation are now available to study the composition and dynamics of the atmosphere. But such tools are recent inventions and therefore have been providing data for only a short time span. To understand fully the behavior of the atmosphere and to anticipate future climate change, we must somehow discover how climate has changed over broad expanses of time.

Instrumental records go back only a couple centuries, at best, and the further back we go, the less complete and more unreliable the data become. To overcome this lack of direct measurements, scientists must decipher and reconstruct past climates by using indirect evidence. **Proxy data** come from natural recorders of climate variability, such as seafloor sediments, glacial ice, fossil pollen, and tree-growth rings, as well as from historical documents (Fig. 14.5). Scientists who analyze proxy data and reconstruct past climates are engaged in the study of **paleoclimatology.** The main goal of such work is to understand the climate of the past in order to assess the current and potential future climate in the context of natural climate variability. In the following discussion, we will briefly examine some of the important sources of proxy data.

**Seafloor Sediment: A Storehouse of Climate Data** We know that the parts of the Earth system are linked so that a change in one part can produce changes in any or all of the other parts. In this section you will see how changes in atmospheric and oceanic temperatures are reflected in the nature of life in the sea.

Most seafloor sediments contain the remains of organisms that once lived near the sea surface (the ocean–atmosphere interface). When such near-surface organisms die, their shells slowly settle to the floor of the ocean, where they become part of the sedimentary record (Fig. 14.6). These seafloor sediments are useful recorders of worldwide climate change because the numbers and types of organisms living near the sea surface change with the climate. For this reason scientists have become increasingly interested in the huge reservoir of data contained in seafloor sediments. The sediment cores gathered by drilling ships and other research vessels have provided invaluable data that have significantly expanded our knowledge and understanding of past climates (Fig. 14.7).

One notable example of how seafloor sediments add to our understanding of climate change relates to unraveling the fluctuating atmospheric conditions of the Quaternary Ice Age. The records of temperature changes contained in cores of sediment from the ocean floor have been critical to our present knowledge of this recent span of Earth history.

**students sometimes ask...**

**If Earth's atmosphere had no greenhouse gases, what would surface-air temperatures be like?**

Cold! Earth's average surface temperature would be a chilly –18°C (–0.4°F) instead of the relatively comfortable 14.5°C (58°F) that it is today.

▼ **Figure 14.5 Ancient bristlecone pines** Some of these trees in California's White Mountains are more than 4000 years old. The study of tree-growth rings is one way that scientists reconstruct past climates.

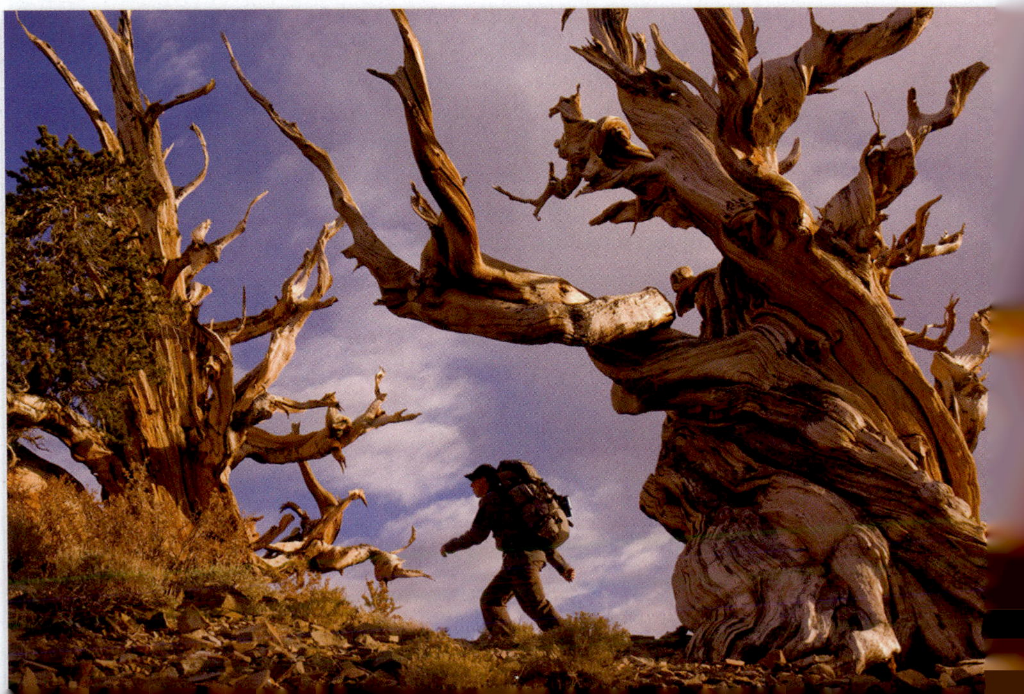

▲ **Figure 14.6 Foraminifera** These single-celled amoeba-like organisms, also called *forams*, are extremely abundant and found throughout the world's oceans. Although the foram record in ocean sediment goes back farther, the remains of these organisms are most commonly used to study climate change during the Cenozoic era. The chemical composition of their hard parts depends on water temperature and the presence or absence of large ice sheets. Because of this relationship, scientists analyze foram shells to estimate ocean temperatures and the existence of ice sheets.

**Oxygen-Isotope Analysis** The isotopes of oxygen in water molecules or in the shells of marine organisms are an important source of proxy data on past climate conditions. **Oxygen-isotope analysis** is based on precise measurement of the ratio between two isotopes of oxygen: $^{16}O$, which is the most common, and the heavier $^{18}O$. A molecule of $H_2O$ can form from either $^{16}O$ or $^{18}O$, but the lighter isotope, $^{16}O$, evaporates more readily from the oceans. Because of this, precipitation (and hence the glacial ice that it may form) is enriched in $^{16}O$. This leaves a greater concentration of the heavier isotope, $^{18}O$, in the ocean water. Thus, during periods when glaciers are extensive, more of the lighter $^{16}O$ is tied up in ice, so the concentration of $^{18}O$ in seawater increases. Conversely, during warmer interglacial periods when the amount of glacial ice decreases dramatically, more $^{16}O$ is returned to the sea, so the proportion of $^{18}O$ relative to $^{16}O$ in ocean water also drops. Now, if we had some ancient recording of the changes of the $^{18}O/^{16}O$ ratio, we could determine when there were glacial periods and therefore when the climate grew cooler.

Fortunately, we do have such a recording. As certain microorganisms secrete their shells of calcium carbonate ($CaCO_3$), the prevailing $^{18}O/^{16}O$ ratio is reflected in the composition of these hard parts. When these organisms die, their hard parts settle to the ocean floor and become part of the sediment layers there. Consequently, periods of glacial activity can be determined from variations in the oxygen-isotope ratio found in shells of certain microorganisms buried in deep-sea sediments.

The $^{18}O/^{16}O$ ratio also varies with temperature. Thus, more $^{18}O$ is evaporated from the oceans when temperatures are high, and less is evaporated when temperatures are low. Therefore, the heavy isotope is more abundant in the precipitation of warm eras and less abundant during colder periods. Using this principle, scientists studying the layers of ice and snow in glaciers have been able to determine past temperature changes.

**Climate Change Recorded in Glacial Ice** Cores of glacial ice are an indispensable source of data for reconstructing past climates (**Fig. 14.8**). Research based on vertical cores taken from the Greenland and Antarctic ice sheets has changed our basic understanding of how the climate system works.

Scientists collect samples with a drilling rig, like a small version of an oil drill. A hollow shaft follows the drill head into the ice, and an ice core is extracted. In this way, cores that sometimes exceed 2000 meters (6500 feet) in length and may represent more than 200,000 years of climate history are acquired for study (**Fig. 14.9A**).

The ice provides a detailed record of changing air temperatures and snowfall. Air bubbles trapped in the ice record variations in atmospheric composition. Changes in the trapped air's carbon dioxide and methane content are linked to

▲ **Figure 14.7 Drilling for seafloor sediment** The *Chikyu* (meaning "Earth" in Japanese) is a state-of-the-art scientific drilling vessel. It can drill as deep as 7000 meters (nearly 23,000 feet) below the seabed in water as deep as 2500 meters (8200 feet). It is part of the Integrated Ocean Drilling Program.

▲ **Figure 14.8 Ice cores contain clues to shifts in climate** This scientist is slicing an ice core from Antarctica for analysis. He is wearing protective clothing and a mask to minimize contamination of the sample. Chemical analyses of ice cores can provide important data about past climates.

Video MM
Climate Change Through Native Alaskan Eyes
http://goo.gl/pzslqe

fluctuating temperatures. The cores also include atmospheric fallout such as windblown dust, volcanic ash, pollen, and modern-day pollution.

Scientists are able to produce a record of past temperatures by means of oxygen-isotope analysis, as described above. A portion of such a record is shown in Figure 14.9B.

**Tree Rings: Archives of Environmental History** If you look at the end of a log, you will see that it is composed of a series of concentric rings. These *tree rings* become larger in diameter outward from the center (Fig. 14.10A). Every year in temperate regions, trees add a layer of new wood under the bark. Characteristics of each tree ring, such as size and density, reflect the environmental conditions (especially climate) that prevailed during the year when the ring formed. Favorable growth conditions produce a wide ring; unfavorable ones produce a narrow ring. Trees growing at the same time in the same region show similar tree-ring patterns.

Because a single growth ring is usually added each year, the age of the tree when it was cut can be determined by counting the rings. If the year of cutting is known, the age of the tree and the year in which each ring formed can be determined by counting back from the outside ring. The dating and study of annual rings in trees is called **dendrochronology.** Scientists are not limited to working with trees that have been cut down. Small, nondestructive core samples can be taken from living trees (Fig. 14.10B).

To make the most effective use of tree rings, extended patterns known as *ring chronologies* are established. They are produced by comparing the patterns of rings among trees in an area. If the same pattern can be identified in two samples, one of which has been dated, the second sample can be dated from the first by matching the ring patterns common to both. Tree-ring chronologies extending back for thousands of years have been established for some regions. To date a timber sample of unknown age, its ring pattern is matched against the reference chronology.

Tree-ring chronologies are unique archives of environmental history and have important applications in such disciplines as climate, geology, ecology, and archaeology. For example, tree rings are used to reconstruct climate variations within a region for spans of thousands of years prior to human historical records. Knowledge of such long-term variations is of great value in making judgments regarding the recent record of climate change.

**Fossil Pollen** Climate is a major factor influencing the distribution of vegetation, so the nature of the plant community occupying an area is a reflection of the climate. Pollen and

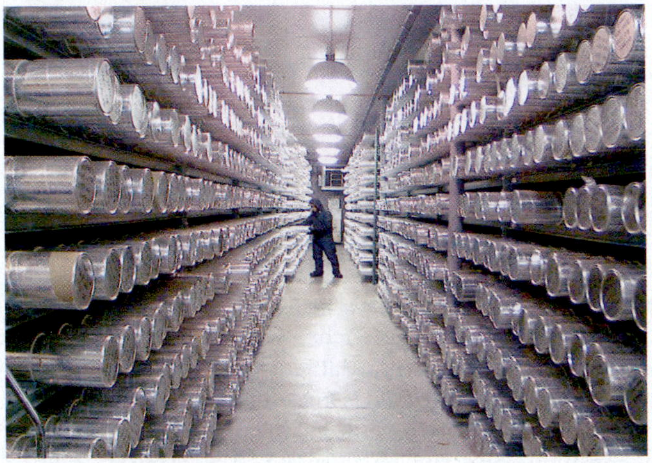

A.

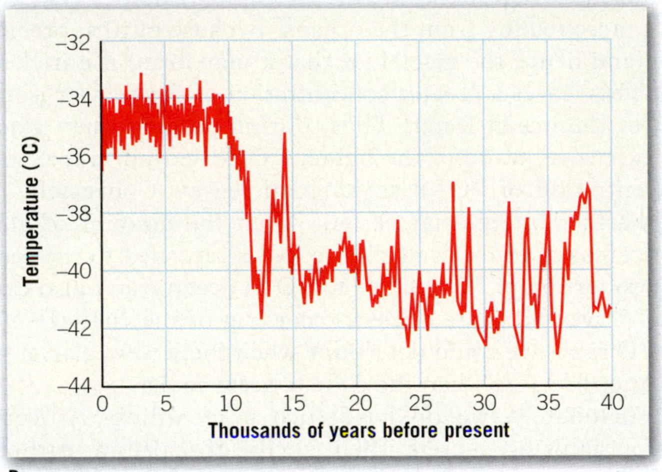


B.

MM
http://goo.gl/J9IGn

▲ **SmartFigure 14.9 Ice cores: Important sources of climate data A.** The National Ice Core Laboratory is a physical plant for storing and studying cores of ice taken from glaciers around the world. These cores represent a long-term record of material deposited from the atmosphere. The lab enables scientists to examine ice cores and preserves the integrity of these samples in a repository for the study of global climate change and past environmental conditions. **B.** This graph, showing temperature variations over the past 40,000 years, is derived from oxygen-isotope analysis of ice cores recovered from the Greenland Ice Sheet.

A.

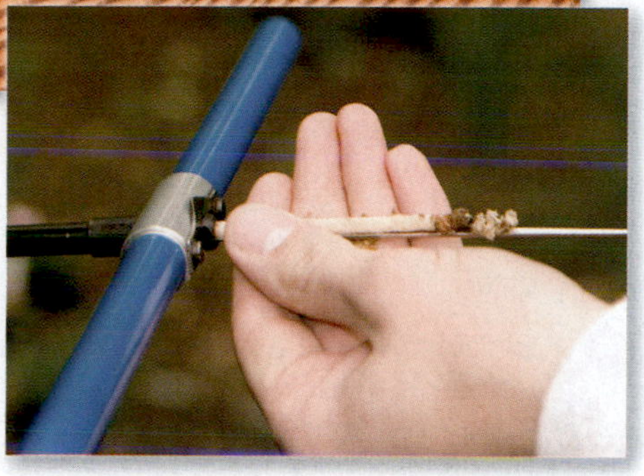

B.

▲ **Figure 14.10 Tree rings A.** Each year a growing tree produces a layer of new cells beneath the bark. If the tree is cut down and the trunk is examined, each year's growth can be seen as a ring. These rings are useful records of past climate because the amount of growth (the thickness of a ring) depends on precipitation and temperature. (Photo by Daniele Taurino/Dreamstime) **B.** Scientists are not limited to working with trees that have been cut down. Small, nondestructive core samples can be taken from living trees.

▲ **Figure 14.11 Pollen** This false-color image from an electron microscope shows an assortment of pollen grains. Note how the size, shape, and surface characteristics differ from one species to another. Analysis of the types and abundance of pollen in lake sediments and peat deposits provides information about how climate has changed over time.

Video MM
20,000 Years of Pine Pollen
http://goo.gl/iCAe1

spores are parts of the life cycles of many plants, and because they have very resistant walls, they are often the most abundant, easily identifiable, and best-preserved plant remains in sediments (**Fig. 14.11**). By analyzing pollen from accurately dated sediments, it is possible to obtain high-resolution records of vegetation changes in an area. Past climates can be reconstructed from such information.

## Corals

Coral reefs consist of colonies of corals, invertebrates that live in warm, shallow waters and form atop the hard material left behind by past corals. Corals build their hard skeletons from calcium carbonate ($CaCO_3$) extracted from seawater. The carbonate contains isotopes of oxygen that can be used to determine the temperature of the water in which the coral grew. The portion of the skeleton that forms in winter has a density that is different from that formed in summer because of variations in growth rates related to temperature and other environmental factors. Thus, corals exhibit seasonal growth bands very much like those observed in trees. The accuracy and reliability of the climate data extracted from corals has been established by comparing recent instrumental records to coral records for the same period. Oxygen-isotope analysis of coral growth rings can also serve as a proxy measurement for precipitation, particularly in areas where large variations in annual rainfall occur.

Think of coral as a *paleothermometer* that enables us to answer important questions about climate variability in the world's oceans. The graph in **Figure 14.12** is a 350-year sea-surface temperature record based on oxygen-isotope analysis of a core extracted from a reef in the Galapagos Islands.

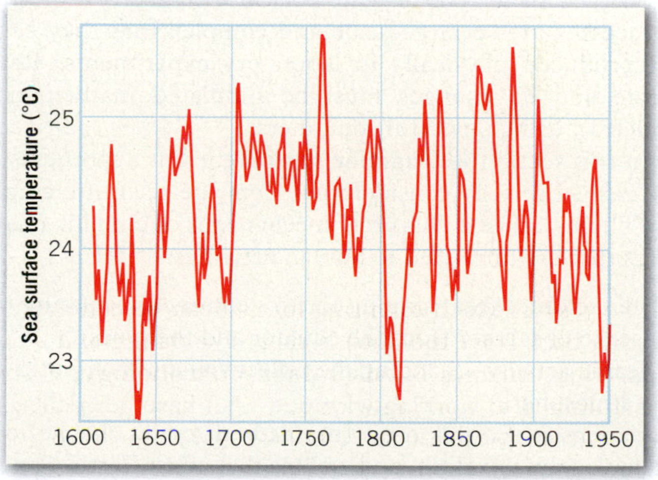


▲ **Figure 14.12 Corals record sea-surface temperatures** Coral colonies thrive in warm, shallow tropical waters. The tiny invertebrates extract calcium carbonate from seawater to build hard parts. They live atop the solid foundation left by past coral. Chemical analysis of the changing composition of coral reefs with depth can provide useful data on past near-surface temperatures. This graph shows a 350-year record of sea-surface temperatures obtained through oxygen-isotope analysis of coral from the Galapagos Islands.

**Historical Data** Historical documents sometimes contain helpful information. Although it may seem that such records should readily lend themselves to climate analysis, that is not the case. Most manuscripts were written for purposes other than climate description. Furthermore, writers understandably neglected periods of relatively stable atmospheric conditions and mention only droughts, severe storms, memorable blizzards, and other extremes. Nevertheless, records of crops, floods, and the migration of people have furnished useful evidence of the possible influences of changing climate (**Fig. 14.13**).

▲ **Figure 14.13 Harvest dates as climate clues** Historical records can sometimes be helpful in the analysis of past climates. The date for the beginning of the grape harvest in the fall is an integrated measure of temperature and precipitation during the growing season. These dates have been recorded for centuries in Europe and provide a useful record of year-to-year climate variations.

Video MM
Climate, Crops, and Bees
http://goo.gl/jNcba

### ✔ Concept Checks 14.1

1. List the five parts of the climate system.
2. What are proxy data, and why are they necessary in the study of climate change?
3. Why are seafloor sediments useful in the study of past climates? Aside from seafloor sediments, list four sources of proxy climate data.
4. Explain how past temperatures are determined by using oxygen-isotope analysis.

## 14.2 Natural Causes of Climate Change

**Discuss four hypotheses that relate to natural causes of climate change.**

A great variety of hypotheses have been proposed to explain climate change. Several have gained wide support, only to lose it and then sometimes to regain it. Some explanations are controversial. This is to be expected because planetary atmospheric processes are so large-scale and complex that they cannot be reproduced physically in laboratory experiments. Rather, climate and its changes must be simulated mathematically (modeled), using powerful computers.

In this section we examine several current hypotheses that have earned serious consideration from the scientific community. These describe "natural" mechanisms of climatic change, causes that are unrelated to human activities:

- Plate tectonics (rearranging Earth's continents, moving them closer to or farther from the equator and the poles)
- Volcanic activity (changing the reflectivity and composition of the atmosphere)
- Variations in Earth's orbit (the natural, cyclic change in our planet's orbit, axial tilt, and wobble that affect Earth–Sun relationships)
- Solar variability (changes in the Sun's radiation output and the effect of sunspots on output)

A later section examines human-induced climate changes, including the effect of rising carbon dioxide levels caused primarily by our burning of fossil fuels.

As you read this section, you will find that more than one hypothesis may explain the same change in climate. In fact, several mechanisms may interact to shift climate. Also, no single hypothesis can explain climate change on all time scales. A proposal that explains variations over millions of years generally cannot explain fluctuations over hundreds of years. If our atmosphere and its changes ever become fully understood, we will probably see that climate change is caused by many of the mechanisms discussed here, plus new ones yet to be proposed.

### Plate Tectonics and Climate Change

Over the past several decades, a revolutionary idea has emerged from the science of geology: **plate tectonics theory.** This theory has gained nearly universal acceptance in the scientific community. It states that the outer portion of Earth is made up of several vast rigid slabs, called *plates,* which move in relation to one another over a weak plastic rock layer below. The plates move incredibly slowly; the average rate at which they move relative to each other is roughly the same rate at which human fingernails grow—perhaps a few centimeters a year.

Most of the largest plates include an entire continent plus a lot of seafloor. Thus, as plates ponderously shift, continents also change position. Not only does this theory allow

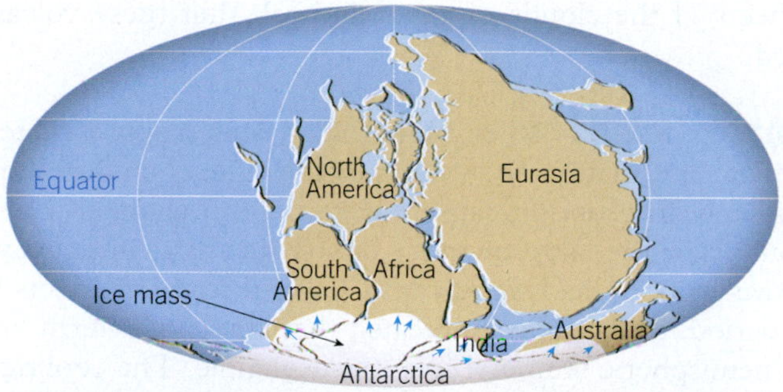


**A. The supercontinent Pangaea showing the area covered by glacial ice near the end of the Paleozoic era.**

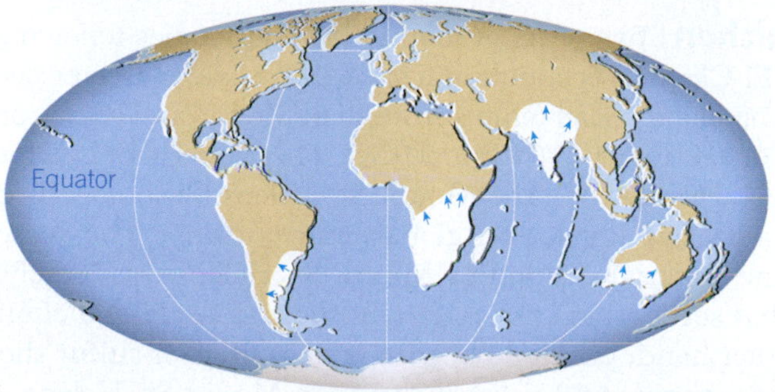


**B. The continents as they appear today. The white areas indicate where evidence of the late Paleozoic ice sheets exists.**

▲ **Figure 14.14 A late Paleozoic ice age** Shifting tectonic plates sometimes move landmasses to high latitudes, where the formation of ice sheets is possible.

geologists to understand and explain many processes and features of Earth's continents and oceans, it also provides climate scientists with a probable explanation for some hitherto unexplainable climate changes.

For example, evidence of extensive glacial activity in portions of present-day Africa, Australia, South America, and India indicate that these regions experienced an ice age about 250 million years ago. This finding puzzled scientists for many years. How could the climate in these presently warm latitudes once have been frigid, like Greenland and Antarctica?

Until the development of the plate tectonics theory, no reasonable explanation existed. Today scientists realize that the areas containing these ancient glacial features were joined as a single "supercontinent" that was located toward the South Pole (**Fig. 14.14A**). Later, as the plates spread apart, portions of the landmass, each moving on a different plate, slowly migrated toward their present locations. Thus, large fragments of glaciated terrain ended up in widely scattered subtropical locations (**Fig. 14.14B**).

We now understand that during the geologic past, plate movements accounted for many other dramatic climate changes, as landmasses shifted in relation to one another and moved to different latitudes. Changes in oceanic circulation must also have occurred, altering the transport of heat and moisture and, hence, the climate as well.

Because the rate of plate movement is so slow, appreciable changes in the positions of the continents occur only over *great* spans of geologic time. Thus, climate changes brought about by plate movements are extremely gradual and happen on a scale of millions of years. As a result, the theory of plate tectonics is not useful for explaining climate variations that occur on shorter time scales, such as tens, hundreds, or thousands of years. Other explanations must be sought to explain these changes.

## Volcanic Activity and Climate Change

The idea that explosive volcanic eruptions might alter Earth's climate was first proposed many years ago. It is still regarded as a plausible explanation for some aspects of climate variability. Explosive eruptions emit huge quantities of gases and fine-grained debris into the atmosphere (**Fig. 14.15**). The greatest eruptions are sufficiently powerful to inject material high into the stratosphere, where it spreads around the globe and remains for many months or even years.

**This satellite image shows the sulfur dioxide ($SO_2$) plume in shades of purple and black. Climate may be affected when large quantities of $SO_2$ are injected into the atmosphere.**

**This image was taken from the International Space Station and shows a plume of volcanic ash streaming southeastward from the volcano.**

► **Figure 14.15 Mount Etna erupting in October 2002** This volcano on the island of Sicily is Europe's largest and most active volcano.

The basic premise is that the suspended volcanic material will filter out a portion of the incoming solar radiation, which in turn will lower temperatures in the troposphere. More than 200 years ago Benjamin Franklin used this idea to argue that material from the eruption of a large Icelandic volcano could have reflected sunlight back to space and therefore might have been responsible for the unusually cold winter of 1783–1784.

Perhaps the most notable cool period linked to a volcanic event is the "year without a summer" that followed the 1815 eruption of Mount Tambora in Indonesia. Tambora's eruption is the largest of modern times. During April 7–12, 1815, this nearly 4000-meter-high (13,000-foot-high) volcano violently expelled more than 100 cubic kilometers (24 cubic miles) of volcanic debris. The impact of these tiny suspended particles is believed to have been widespread in the Northern Hemisphere. From May through September 1816, an unprecedented series of cold spells affected the northeastern United States and adjacent portions of Canada. The region experienced heavy snow in June and frost in July and August. Abnormal cold was also experienced in much of Western Europe.

Three more recent volcanic events have provided considerable data and insight regarding the impact of volcanoes on global temperatures. The eruptions of Washington State's Mount St. Helens in 1980, the Mexican volcano El Chichón in 1982, and Mount Pinatubo in the Philippines in 1991 gave scientists opportunities to study the atmospheric effects of volcanic eruptions with the aid of more sophisticated technology than had been available in the past. Satellite images and remote-sensing instruments allowed scientists to closely monitor the effects of the clouds of gases and ash that these volcanoes emitted.

**Mount St. Helens** When Mount St. Helens erupted, there was immediate speculation about the possible effects on our climate. Could such an eruption cause our climate to change? There is no doubt that the large quantity of volcanic ash emitted by the explosive eruption had significant local and regional effects for a short period. Still, studies indicated that any longer-term lowering of hemispheric temperatures was negligible. The cooling was so slight—probably less than 0.1°C (0.2°F)—that it could not be distinguished from other natural temperature fluctuations.

**El Chichón** Two years of monitoring and studies following the 1982 El Chichón eruption indicated that it had a greater cooling effect on global mean temperature than Mount St. Helens, on the order of 0.3° to 0.5°C (0.5° to 0.9°F). The eruption of El Chichón was *less explosive* than the Mount St. Helens blast, so why did it have a greater impact on global temperatures? The reason is that the material emitted by Mount St. Helens was largely fine ash that settled out in a relatively short time. El Chichón, on the other hand, emitted far greater quantities of sulfur dioxide gas (an estimated 40 times more) than Mount St. Helens. This gas combines with water vapor in the stratosphere to produce a dense cloud of tiny sulfuric acid particles (**Fig. 14.16A**). These aerosols take several years to settle out completely. They lower the troposphere's mean temperature because they reflect solar radiation back to space (**Fig. 14.16B**).

We now understand that volcanic clouds that remain in the stratosphere for a year or more are composed largely of sulfuric acid droplets and not of dust, as was once thought. Thus, the volume of fine debris emitted during an explosive event is not an accurate criterion for predicting the global atmospheric effects of an eruption.

**Mount Pinatubo** The Philippines volcano Mount Pinatubo erupted explosively in June 1991, injecting 25 million to 30 million tons of sulfur dioxide into the stratosphere. The event provided scientists with an opportunity to study the climatic impact of a major explosive volcanic eruption, using NASA's spaceborne Earth Radiation Budget Experiment. During the next year, the haze of tiny aerosols increased albedo and lowered global temperatures by 0.5°C (0.9°F).

**The Impact of Eruptions** It may be true that the impact on global temperature of eruptions like El Chichón and Mount Pinatubo is relatively minor, but many scientists agree that

**A plume of white haze from Anatahan Volcano blankets a portion of the Philippine Sea in April 2005. The haze consisted of tiny droplets of sulfuric acid formed when sulfur dioxide from the volcano combined with water in the atmosphere. The plume is bright and reflects sunlight back to space.**

**A.**

**Net solar radiation at Hawaii's Mauna Loa Observatory relative to 1970 (zero on the graph). The eruptions of El Chichón and Mt. Pinatubo clearly caused temporary drops in solar radiation reaching the surface.**

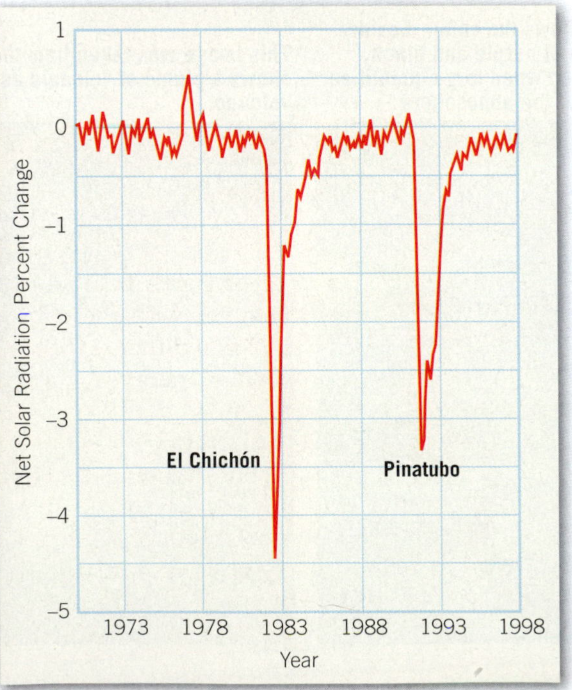


**B.**

◀ **Figure 14.16 Volcanic haze reducing sunlight at Earth's surface** The reflective haze produced by some volcanic eruptions is not volcanic ash but tiny sulfuric acid aerosols.

**students sometimes ask...**

### Could a meteorite colliding with Earth cause the climate to change?

Yes, it is possible. In fact, the most strongly supported hypothesis for the extinction of dinosaurs and many other organisms about 65 million years ago is related to such an event. When a large (about 10 kilometers in diameter) meteorite struck Earth, huge quantities of debris were blasted high into the atmosphere. For months the encircling dust cloud greatly restricted the amount of light reaching Earth's surface. Without sufficient sunlight for photosynthesis, delicate food chains collapsed. When the sunlight returned, more than half of the species on Earth, including the dinosaurs and many marine organisms, had become extinct.

---

the cooling produced could alter the general pattern of atmospheric circulation for a limited period. Such a change, in turn, could influence the weather in some regions. Predicting or even identifying specific regional effects still presents a considerable challenge to atmospheric scientists.

The preceding examples illustrate that the impact on climate of a single volcanic eruption, no matter how great, is relatively small and short-lived. The graph in Figure 14.16B reinforces this point. If volcanism is to have a pronounced impact over an extended period, many great eruptions rich in sulfur dioxide and closely spaced in time must occur. If that happens, the stratosphere will be loaded with enough gases and volcanic dust to seriously diminish the amount of solar radiation reaching the surface. Because no such period of explosive volcanism is known to have occurred in historic times, it is most often mentioned as a possible contributor to prehistoric climatic shifts. Another way in which volcanism may influence climate is described in **Box 14.1**.

## Variations in Earth's Orbit

Geologic evidence indicates that an ice age that began about 3 million years ago was characterized by numerous glacial advances and retreats associated with periods of global cooling and warming. Today scientists understand that the climate oscillations that characterized this ice age are linked to variations in Earth's orbit. This hypothesis was first developed and strongly advocated by Serbian scientist Milutin Milankovitch and is based on the premise that variations in incoming solar radiation are a principal factor controlling Earth's climate.

Milankovitch formulated a comprehensive mathematical model based on the following elements (**Fig. 14.17**):

1. Variations in the shape (**eccentricity**) of Earth's orbit about the Sun
2. Changes in **obliquity**—that is, changes in the angle that the axis makes with the plane of Earth's orbit
3. The wobbling of Earth's axis, called **precession**

Using these factors, Milankovitch calculated variations in the receipt of solar energy and the corresponding surface temperature of Earth back into time, in an attempt to correlate these changes with the climate fluctuations of the ice age. It should be noted that these factors cause little or no variation in the *total* solar energy reaching the ground. Instead, their impact is felt because they change the degree of contrast between the seasons. Somewhat milder winters in the middle to high latitudes mean greater snowfall totals, whereas cooler summers bring a reduction in snowmelt.

Among the studies that added credibility to this hypothesis is one that examined deep-sea sediments.* Through oxygen-

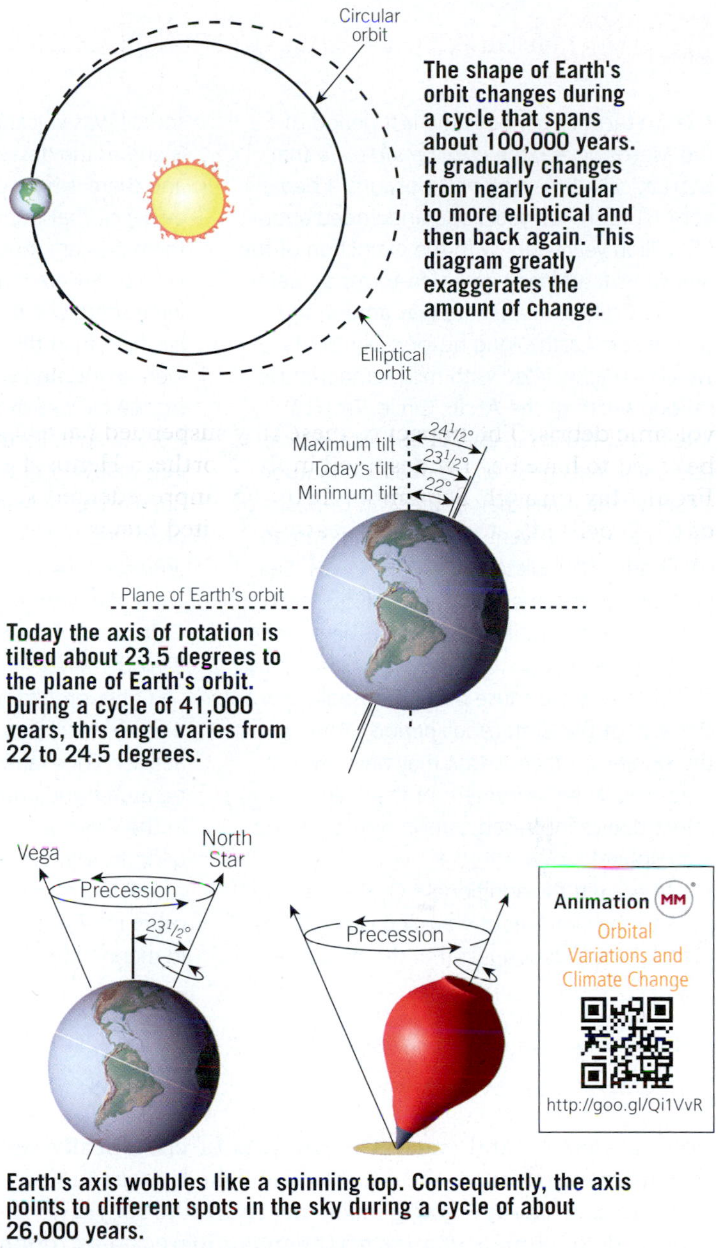


http://goo.gl/e5x6X

▲ **SmartFigure 14.17 Orbital variations** Periodic variations in Earth's orbit are linked to alternating glacial and interglacial conditions during the Quaternary Ice Age.

Circular orbit
The shape of Earth's orbit changes during a cycle that spans about 100,000 years. It gradually changes from nearly circular to more elliptical and then back again. This diagram greatly exaggerates the amount of change.
Elliptical orbit
Maximum tilt
Today's tilt
Minimum tilt
24½°
23½°
22°
Plane of Earth's orbit
Today the axis of rotation is tilted about 23.5 degrees to the plane of Earth's orbit. During a cycle of 41,000 years, this angle varies from 22 to 24.5 degrees.
Vega
North Star
Precession
23½°
Precession
Animation
Orbital Variations and Climate Change
http://goo.gl/Qi1VvR
Earth's axis wobbles like a spinning top. Consequently, the axis points to different spots in the sky during a cycle of about 26,000 years.

*J. D. Hays, J. Imbrie, and N. J. Shackleton, "Variations in the Earth's Orbit: Pacemaker of the Ice Ages," *Science* 194, no. 4270 (1976), 1121–1132. Since the time of this study, subsequent research has supported its basic conclusions.

## Box 14.1 Volcanism and Climate Change in the Geologic Past

The Cretaceous period is the last period of the Mesozoic era, the era of *middle life* that is often called the Age of Dinosaurs. It began about 145 million years ago and ended about 65 million years ago, with the extinction of the dinosaurs (and many other life-forms as well).

The Cretaceous climate was among the warmest in Earth's long history. Dinosaurs, which are associated with mild temperatures, ranged north of the Arctic Circle. Tropical forests existed in Greenland and Antarctica, and coral reefs grew as much as 15° latitude closer to the poles than at present. Deposits of peat that would eventually form widespread coal beds accumulated at high latitudes. Sea level was as much as 200 meters (650 feet) higher than today, indicating that there were no polar ice sheets.

What was the cause of the unusually warm climates of the Cretaceous period? Among the significant factors that may have contributed was an enhancement of the greenhouse effect due to increased carbon dioxide in the atmosphere.

Where did the additional $CO_2$ that contributed to the Cretaceous warming come from? Many geologists suggest that the probable source was volcanic activity. Carbon dioxide is one of the gases emitted during volcanism, and there is now considerable geologic evidence that during the Middle Cretaceous, there was an unusually high rate of volcanic activity. Several huge oceanic lava plateaus were produced on the floor of the western Pacific during this span. These vast features were associated with hot spots, zones where mobile plumes of material rise to the surface from deep in Earth's interior. Massive outpourings of lava over millions of years would have been accompanied by the release of huge quantities of $CO_2$, which in turn would have enhanced the atmospheric greenhouse effect. *Thus, the warmth that characterized the Cretaceous may have had its origins deep in Earth's interior.*

Other probable results of this extraordinarily warm period are linked to volcanic activity. For example, the high global temperatures and enriched atmospheric $CO_2$ in the Cretaceous led to increases in the quantity and types of phytoplankton (tiny, mostly microscopic plants, such as algae) and other life-forms in the ocean. The expansion in marine life is reflected in the widespread chalk deposits associated with the Cretaceous period (**Fig. 14.A**). Chalk consists of the calcium carbonate ($CaCO_3$)–rich hard parts of microscopic marine organisms. Oil and gas originate from the alteration of biological remains (chiefly phytoplankton). Some of the world's most important oil and gas fields are in marine sediments of the Cretaceous, a consequence of the greater abundance of marine life during that warm time.

These links to the extraordinary period of Cretaceous volcanism illustrate the interrelationships among parts of the Earth system. Materials and processes that at first seem to be completely unrelated turn out to be linked. Here you have seen how processes originating deep in Earth's interior are connected directly or indirectly to the atmosphere, the oceans, and the biosphere.

### Questions

1. How was climate during the Cretaceous period different from Earth's present-day climate? Why was it different?
2. What do we see in the present-day landscape that is linked to climate during the Cretaceous period?

isotope analysis and statistical analyses of climatically sensitive microorganisms, the study established a chronology of temperature change going back nearly 500,000 years. This time scale of climate change was then compared to astronomical calculations of eccentricity, obliquity, and precession to determine whether a correlation did indeed exist.

Although the study was involved and mathematically complex, its conclusions were straightforward. The authors found that major variations in climate over the past several hundred thousand years were closely associated with changes in the geometry of Earth's orbit. Cycles of climate change were shown to correspond closely with the periods of obliquity, precession, and orbital eccentricity. More specifically, the authors concluded that changes in Earth's orbital geometry are the fundamental cause of the series of ice ages during the Quaternary period.

The study went on to predict a future trend toward a cooler climate and extensive glaciation in the Northern Hemisphere. But there are two qualifications: (1) The prediction applied only to the *natural* component of climate change and ignored any human influence, and (2) it was a forecast of *long-term trends* because it must be linked to factors that have periods of 20,000 years and longer. Thus, even if the prediction is correct, it contributes little to our understanding of climate changes over briefer periods of tens to hundreds of years because the cycles are too long for this purpose.

If orbital variations explain alternating glacial–interglacial periods, a question immediately arises: Why have glaciers been absent throughout most of Earth's history? Prior to the plate tectonics theory, there was no widely accepted answer. Today we have a plausible answer: because ice sheets can form only on continents, landmasses must exist somewhere in the higher latitudes before an ice age can commence. Thus, it is likely that ice ages have occurred only when Earth's shifting crustal plates have carried the continents from tropical latitudes to more poleward positions.

## Solar Variability and Climate

Among the most persistent hypotheses of climate change have been those based on the idea that the Sun is a variable star and that its output of energy varies through time. The effect of such changes would seem direct and easily understood: Increases in solar output would cause the atmosphere to warm, and reductions would result in cooling. This notion is appealing because it can be used to explain climate change of any length or inten-

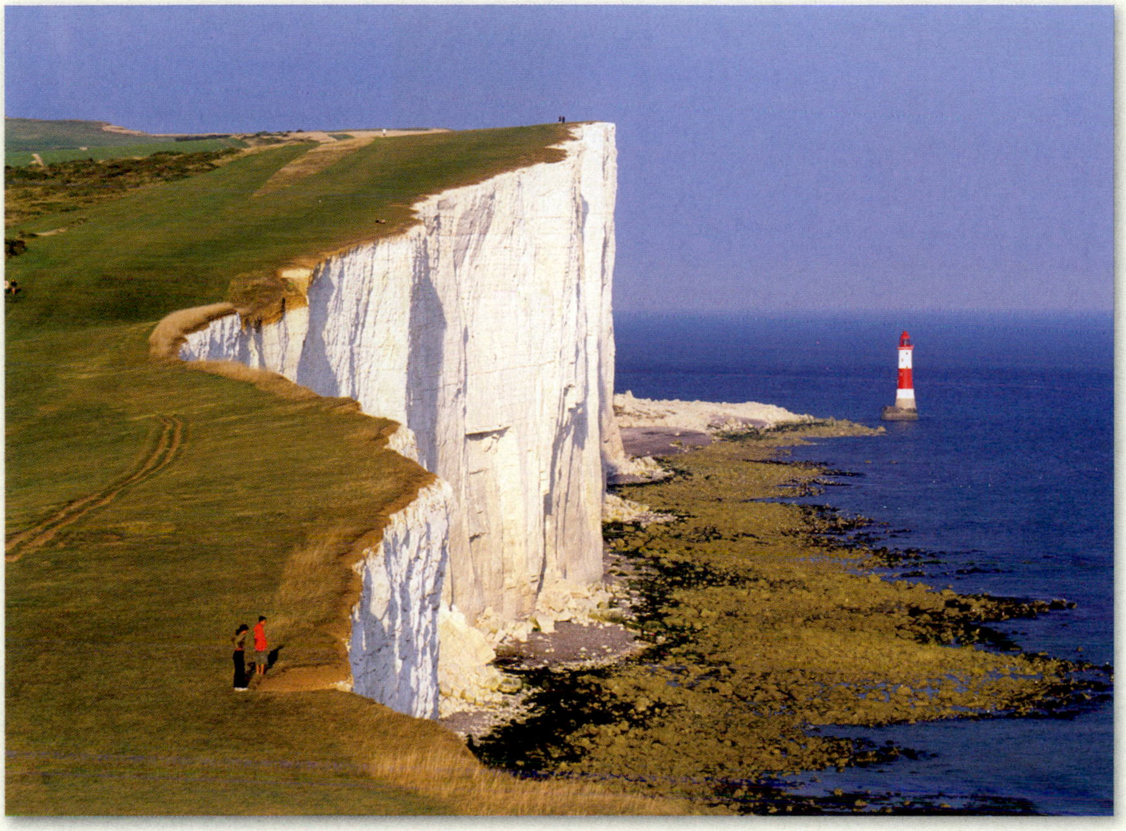

▲ **Figure 14.A White Cliffs of Dover** These famous chalk deposits along the coast of England are composed largely of tiny shells of marine organisms and are associated with the expansion of marine life that occurred during the exceptional warmth of the Cretaceous period. (Photo by Stuart Black/Robert Harding)

sity. However, no major *long-term* variations in the total intensity of solar radiation have yet been measured outside the atmosphere. Such measurements were not even possible until satellite technology became available. Now that we can measure output, we still need many decades of records before we will begin to sense how variable (or invariable) energy from the Sun really is.

Other hypotheses for climate change have related to sunspot cycles. The most conspicuous and best-known features on the surface of the Sun are dark blemishes called **sunspots** (**Fig. 14.18**). Sunspots are huge magnetic storms that extend from the Sun's surface deep into the interior. Moreover, these spots are associated with the Sun's ejection of huge masses

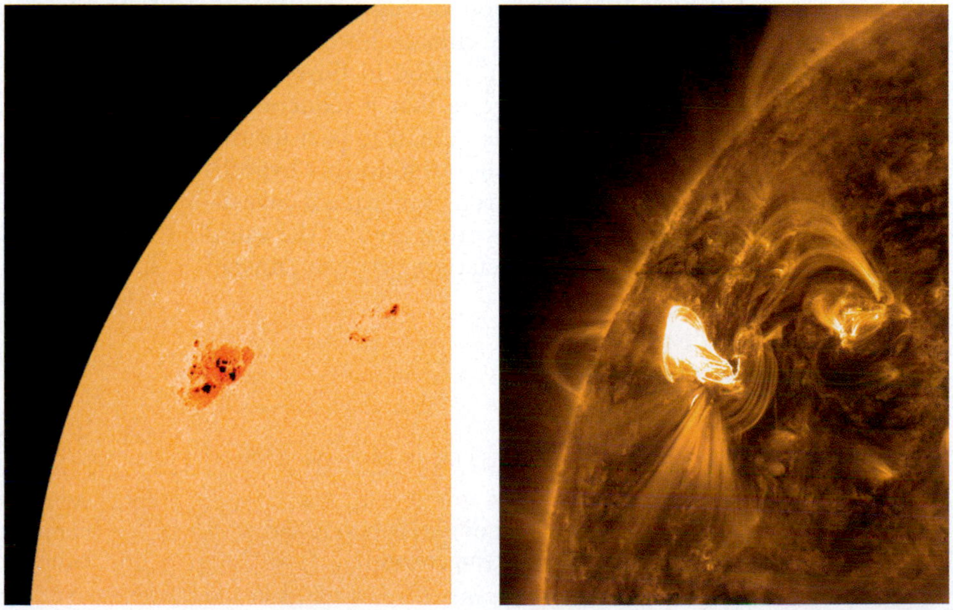
This view shows an approximation of the Sun's surface. The black spots surrounded by deep orange is a sunspot region where magnetic activity is extremely intense.
The instrument that produced this image used ultraviolet, radio, and other parts of the electromagnetic spectrum. Looping lines show solar plasma following magnetic field lines.

► **Figure 14.18 Sunspots** Both images show sunspot activity at the same location on the solar disk at the same time on March 5, 2012, using two different instruments from NASA's Solar Dynamics Observatory.

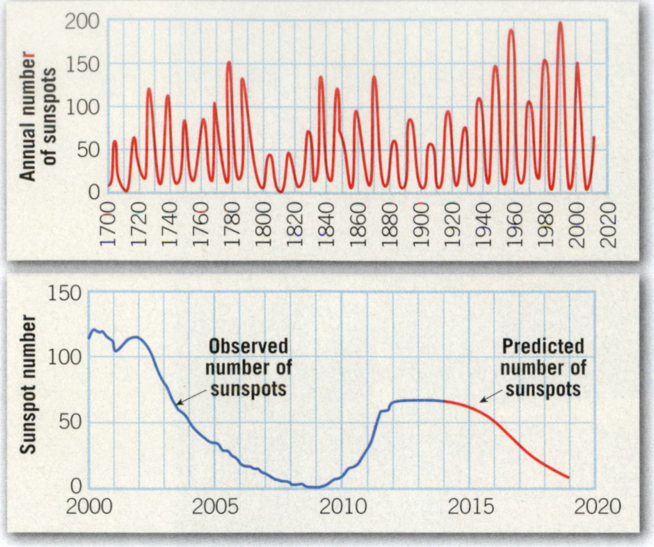


▲ **Figure 14.19 Mean annual sunspot numbers** The number of sunspots reaches a maximum about every 11 years.

of particles that, on reaching Earth's upper atmosphere, interact with gases there to produce auroral displays (see Fig. 1.26, page 23).

Sunspots occur in cycles, with the number of sunspots reaching a maximum about every 11 years (**Fig. 14.19**). During periods of maximum sunspot activity, the Sun emits slightly more energy than during sunspot minimums. Based on measurements from space that began in 1978, the variation during an 11-year cycle is about 0.1 percent. Although sunspots are dark, they are surrounded by brighter areas that apparently offset the effect of the dark spots. It appears that this change in solar output is too small to have any appreciable effect on global temperatures. However, there is a possibility that longer-term variations in solar output may affect climates on Earth.

For example, the span between 1645 and 1715 was a period known as the *Maunder minimum*, during which sunspots were largely absent. This period of missing sunspots closely corresponds with a period in climate history known as the *Little Ice Age*, an especially cold period in Europe. For some scientists, this correlation suggests that a reduction in the Sun's output was likely responsible at least in part for this cold episode. Other scientists seriously question this hypothesis. Their hesitation stems in part from subsequent investigations using different climate records from around the world that failed to find a significant correlation between sunspot activity and climate.

## eye on the atmosphere 14.1

This satellite image shows an extensive plume of ash from an explosive volcanic eruption of Indonesia's Sinabung volcano on January 16, 2014.

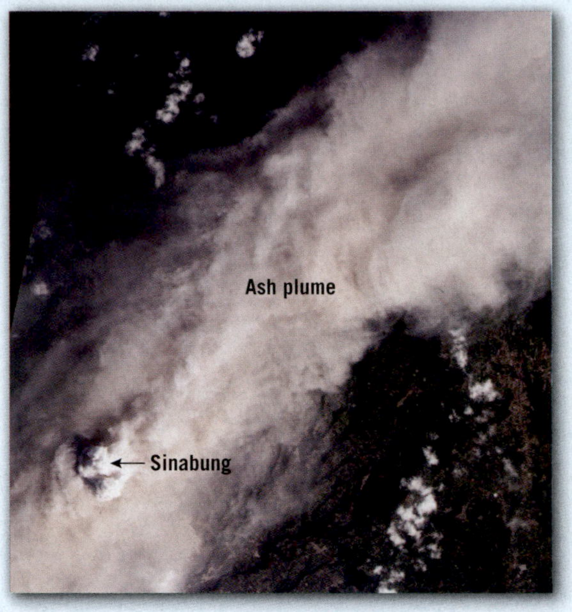


**Questions**

1. How might the volcanic ash from this eruption influence air temperatures?
2. Would this effect likely be long lasting—perhaps for years? Explain.
3. What "invisible" volcanic emission might have a greater effect than the volcanic ash?

**students sometimes ask...**

### Is there a connection between changes in the Sun's brightness and recent evidence of global warming?

Based on recent satellite data, the answer is no. The scientists who carried out this detailed analysis state that the variations in solar brightness measured since 1978, which is as far back as these measurements go, are too small to have contributed appreciably to accelerated global warming over the past 30 years.

## ✔ Concept Checks 14.2

1. How does the theory of plate tectonics help us understand the cause of ice ages?
2. Describe and briefly explain the effect of the El Chichón and Mount Pinatubo eruptions on global temperatures.
3. List and briefly describe the three variations in Earth's orbit that may cause global temperatures to vary.
4. What are sunspots? How does solar output change as sunspot numbers change? Is there a solid connection between sunspot numbers and climate change on Earth?

# 14.3 Human Impact on Global Climate

**Summarize the nature and cause of the atmosphere's changing composition since about 1750. Describe the climate's response.**

So far we have examined potential causes of climate change that are natural. In this section we discuss how humans contribute to global climate change. One impact largely results from the addition of carbon dioxide and other greenhouse gases to the atmosphere. A second impact is related to the addition of human-generated aerosols to the atmosphere.

Human influence on regional and global climate did not just begin with the onset of the modern industrial period. There is evidence that people have been modifying the environment over extensive areas for thousands of years. The use of fire and the overgrazing of marginal lands by domesticated animals have both reduced the abundance and distribution of vegetation. By altering ground cover, humans have modified such important climatological factors as surface albedo, evaporation rates, and surface winds.

▲ **Figure 14.21 Tropical deforestation** Clearing the tropical rain forest is a serious environmental issue. In addition to causing a loss of biodiversity, it is a significant source of carbon dioxide. Fires are frequently used to clear the land. This scene is in Brazil's Amazon basin.

## Rising $CO_2$ Levels

In Chapter 1 you learned that carbon dioxide ($CO_2$) represents only about 0.0400 percent (400 parts per million) of the gases that make up clean, dry air. Nevertheless, it is a very significant component meteorologically. Carbon dioxide is influential because it is transparent to incoming short-wavelength solar radiation, but it is not transparent to some of the longer-wavelength outgoing Earth radiation. A portion of the energy leaving the ground is absorbed by atmospheric $CO_2$. This energy is subsequently re-emitted, part of it back toward the surface, thereby keeping the air near the ground warmer than it would be without $CO_2$. Thus, along with water vapor, carbon dioxide is largely responsible for the *greenhouse effect* of the atmosphere. Carbon dioxide is an important heat absorber, and it follows logically that any change in the air's $CO_2$ content could alter temperatures in the lower atmosphere.

Earth's tremendous industrialization of the past two centuries has been fueled—and still is fueled—by burning fossil fuels: coal, natural gas, and petroleum (**Fig. 14.20**). Combustion of these fuels has added great quantities of carbon dioxide to the atmosphere. The use of coal and other fuels is the most prominent means by which humans add $CO_2$ to the atmosphere, but it is not the only way. The clearing of forests also contributes substantially because $CO_2$ is released as vegetation is burned or decays (**Fig. 14.21**). Deforestation is particularly pronounced in the tropics, where vast tracts are cleared for ranching and agriculture, or subjected to inefficient commercial logging operations. All major tropical forests—including those in South America, Africa, Southeast Asia, and Indonesia—are disappearing. According to United Nations estimates, more than 10 million hectares (25 million acres) of tropical forest were permanently destroyed *each year* during the decades of the 1990s and 2000s.

Some of the excess $CO_2$ is taken up by plants or is dissolved in the ocean, but an estimated 45 percent remains in the atmosphere. **Figure 14.22** is a graphic record of changes in atmospheric $CO_2$ extending back 400,000 years. Over this long span, natural fluctuations varied from about 180 to 300 parts per million (ppm). As a result of human activities, the present $CO_2$ level is about 30 percent higher than its highest level over the past 600,000 years. The rapid increase in atmospheric $CO_2$ since the onset of industrialization is obvious. The annual rate at which atmospheric carbon dioxide concentrations are growing has been increasing over the past several decades.

▼ **Figure 14.20 U.S. energy consumption** The graph shows energy consumption in 2012. The total was 94.9 quadrillion Btu. A quadrillion is 10 raised to the 12th power, or a million million. The burning of fossil fuels represents slightly more than 82 percent of the total.

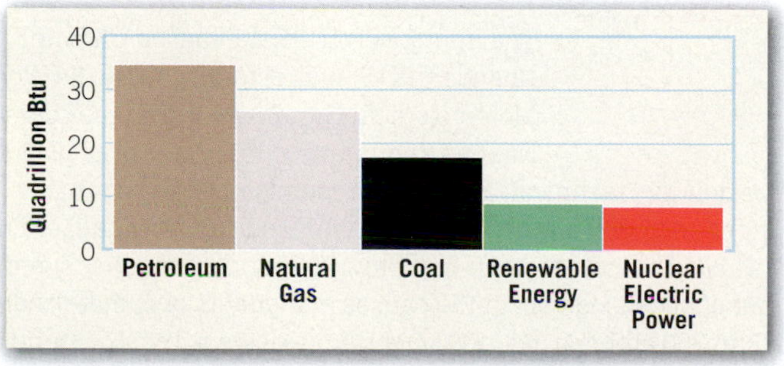

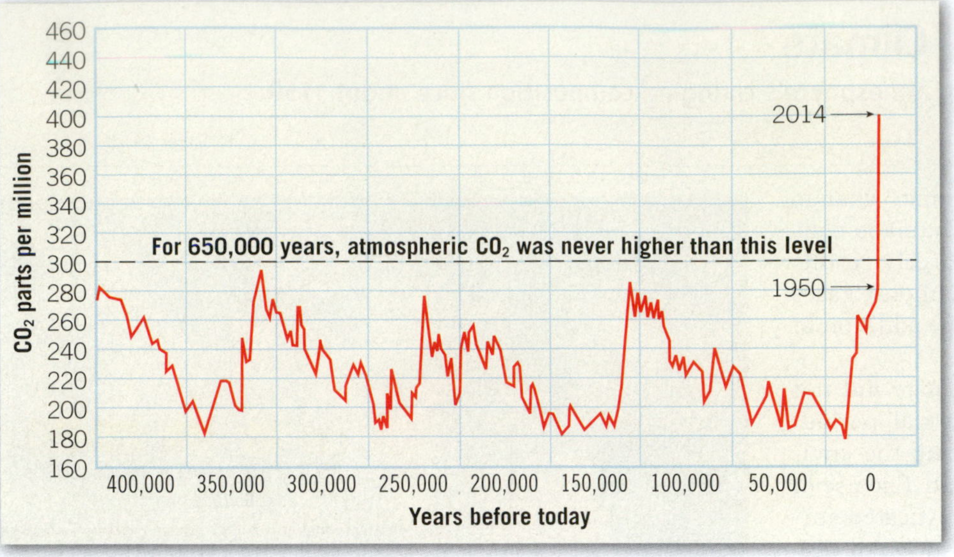


▲ **Figure 14.22 $CO_2$ concentrations over the past 400,000 years** Most of these data come from analyses of air bubbles trapped in ice cores. The record since 1958 comes from direct measurements at Mauna Loa Observatory, Hawaii. The rapid increase in $CO_2$ concentrations since the onset of the Industrial Revolution is obvious. (NOAA)

## The Atmosphere's Response

Given the increase in the atmosphere's $CO_2$ content, have global temperatures actually increased? The answer is yes. According to a 2013 report by the Intergovernmental Panel on Climate Change (IPCC), "Warming of the climate system is unequivocal, as is now evident from observations of increases in global average air and ocean temperatures, widespread melting of snow and ice, and rising global sea level."* Most of the observed increase in global average temperatures since the mid-twentieth century is *extremely likely* due to the observed increase in human-generated greenhouse gas concentrations. As used by the IPCC, *extremely likely* indicates a probability of 95–100 percent. Global warming since the mid-1970s is now about 0.6°C (1°F), and total warming in the past century is about 0.8°C (1.4°F). The upward trend in surface temperatures is shown in **Figure 14.23**. With the exception of 1998, the 10 warmest years in the 134-year record have all occurred since 2000, with 2010 and 2005 ranking as the warmest years on record.

Weather patterns and other natural cycles cause fluctuations in average temperatures from year to year. This is especially true on regional and local levels. For example, while the globe experienced notably warm temperatures in 2013, the continental United States had its 42nd-warmest year. By contrast, 2013 was the hottest year in Australia's recorded history. Regardless of regional differences in any year, increases in greenhouse gas levels are causing a long-term rise in global temperatures. Although each calendar year will not necessarily be warmer than the one before, scientists expect each decade to be warmer than the previous one. An examination of the decade-by-decade temperature trend in **Figure 14.24** bears this out.

*IPCC, "Summary for Policymakers," in *Climate Change 2013: The Physical Science Basis.*

What about the future? Projections for the years ahead depend in part on the quantities of greenhouse gases that are emitted. **Figure 14.25** shows the best estimates of global warming for several different scenarios. The 2013 IPCC report also states that if there is a doubling of the pre-industrial level of carbon dioxide (280 ppm) to 560 ppm, the *likely* temperature increase will be in the range of 2° to 4.5°C (3.5° to 8.1°F). The increase is *very unlikely* (1 to 10 percent probability) to be less than 1.5°C (2.7°F), and values higher than 4.5°C (8.1°F) are possible.

## The Role of Trace Gases

Carbon dioxide is not the only gas contributing to a possible global increase in temperature. In recent years atmospheric scientists have come to realize that human industrial and agricultural activities are causing a buildup of several trace gases that also play

▼ **Figure 14.23 Global temperatures, 1880–2013** With the exception of 1998, 10 of the warmest years in this 134-year temperature record have occurred since 2000.

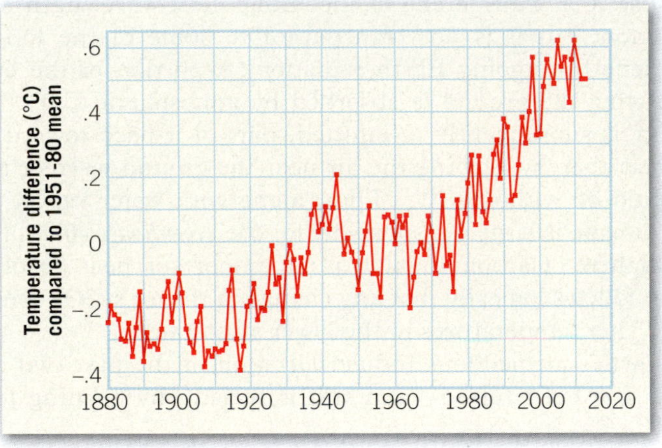


**students sometimes ask...**

### What is the Intergovernmental Panel on Climate Change?

The Intergovernmental Panel on Climate Change (IPCC) was established by the World Meteorological Organization and the United Nations Environment Programme in 1988 to assess scientific, technical, and socioeconomic information that is relevant to an understanding of human-induced climate change. The IPCC is an authoritative group that provides periodic reports regarding the state of knowledge about the causes and effects of climate change. More than 250 authors and 1100 scientific reviewers from 55 countries contributed to *Climate Change 2013: The Physical Science Basis.*

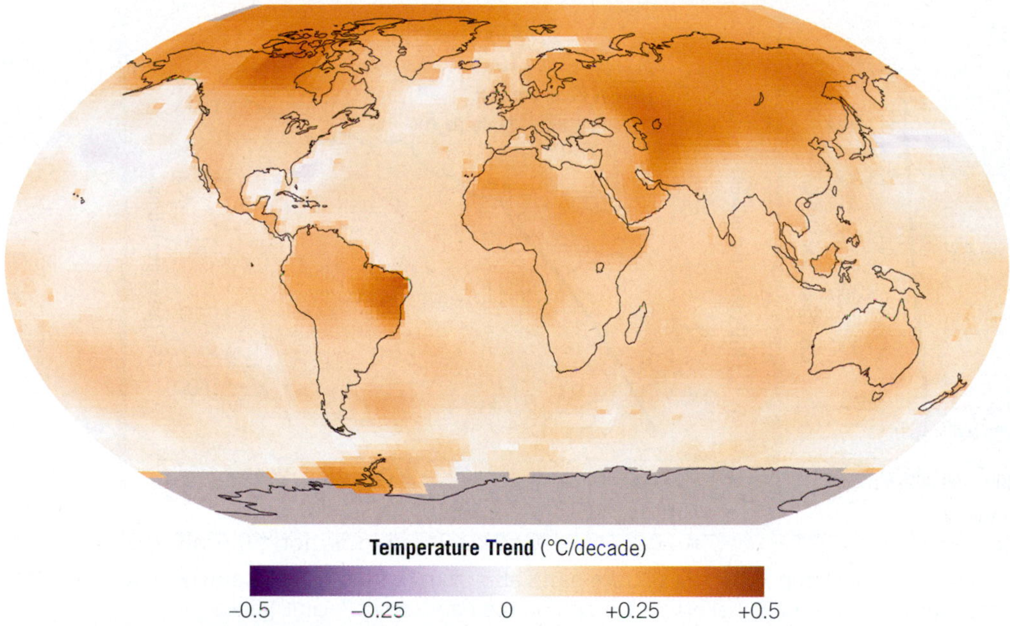


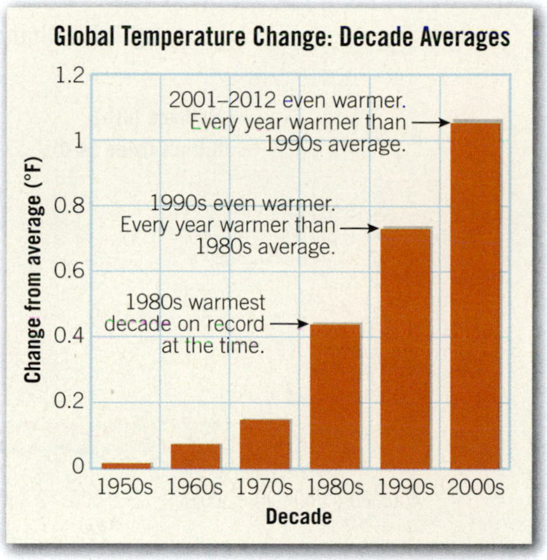


Animation MM Global Warming
http://goo.gl/E1JEYu

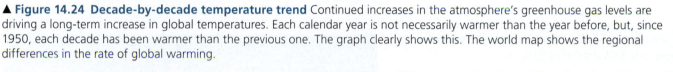

▲ **Figure 14.24 Decade-by-decade temperature trend** Continued increases in the atmosphere's greenhouse gas levels are driving a long-term increase in global temperatures. Each calendar year is not necessarily warmer than the year before, but, since 1950, each decade has been warmer than the previous one. The graph clearly shows this. The world map shows the regional differences in the rate of global warming.

significant roles. The substances are called **trace gases** because their concentrations are so much lower than the concentration of carbon dioxide. The most important trace gases are methane ($CH_4$), nitrous oxide ($N_2O$), and chlorofluorocarbons (CFCs). These gases absorb wavelengths of outgoing radiation from Earth that would otherwise escape into space. Although individually their impact is modest, taken together these trace gases play a significant role in warming the troposphere.

▼ **Figure 14.25 Temperature projections to 2100** The right half of the graph shows projected global warming based on different emissions scenarios. The shaded zone adjacent to each colored line shows the uncertainty range for each scenario. The basis for comparison (0.0 on the vertical axis) is the global average for the period 1980 to 1999. The orange line represents the scenario in which $CO_2$ concentrations were held constant at values for the year 2000.

Video MM Temperature and Agriculture
http://goo.gl/H0y00

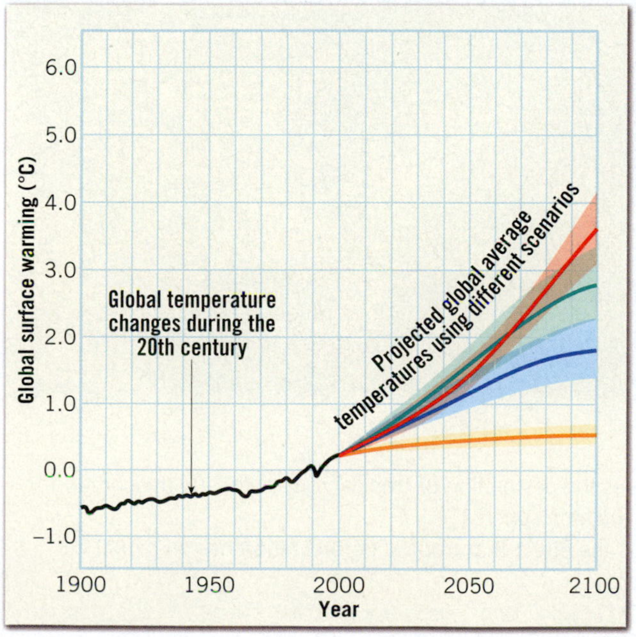


## Methane

Although present in much smaller amounts than $CO_2$, methane's significance is greater than its relatively small concentration would indicate (**Fig. 14.26**). This is because methane is about 20 times more effective than $CO_2$ at absorbing infrared radiation emitted by Earth.

Methane is produced by *anaerobic* bacteria in wet places where oxygen is scarce. (*Anaerobic* means "without air," specifically oxygen.) Such places include swamps, bogs, wetlands, and the guts of termites and grazing animals such as cattle and sheep. Methane is also generated in flooded paddy fields ("artificial swamps") used for growing rice. Mining of coal and drilling for oil and natural gas are other sources because methane is a product of their formation.

The increase in the concentration of methane in the atmosphere has been in step with the growth in human population. This relationship reflects the close link between methane formation and agriculture. As population increases, so do the number of cattle and rice paddies.

## Nitrous Oxide

Sometimes called "laughing gas," nitrous oxide is also building in the atmosphere, although not as rapidly as methane (Fig. 14.26). The increase results primarily from agricultural activity. When farmers use nitrogen fertilizers to boost crop yield, some of the nitrogen enters the air as nitrous oxide. This gas is also produced by high-temperature combustion of fossil fuels. Although the annual release into the atmosphere is small, the lifetime of a nitrous oxide molecule is about 150 years! If nitrogen fertilizer and fossil fuel use grow at

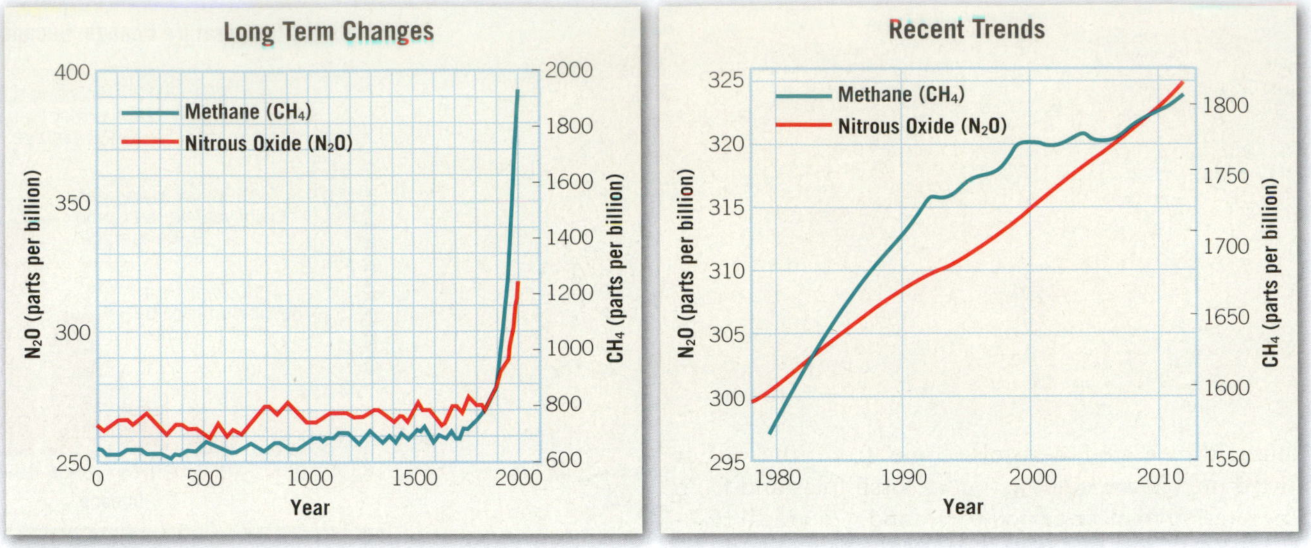


▲ **Figure 14.26 Methane and nitrous oxide** Although $CO_2$ is most important, these trace gases also contribute to global warming. Over the 2000-year span shown here, there were relatively minor fluctuations until the industrial era. The graph on the right shows recent trends.

projected rates, nitrous oxide's contribution to greenhouse warming may approach half that of methane.

**CFCs** Unlike methane and nitrous oxide, chlorofluorocarbons (CFCs) are not naturally present in the atmosphere. CFCs are manufactured chemicals with many uses that have gained notoriety because they are responsible for ozone depletion in the stratosphere. The role of CFCs in global warming is less well known. CFCs are very effective greenhouse gases. They were not developed until the 1920s and were not used in great quantities until the 1950s. Although the Montreal Protocol represents strong corrective action, CFC levels will *not* drop rapidly. (See the section on the Montreal Protocol in Chapter 1.) CFCs remain in the atmosphere for decades, so even if all CFC emissions were to stop immediately, the atmosphere would not be free of them for many years.

**A Combined Effect** Carbon dioxide is clearly the most important single cause for the projected global greenhouse warming. However, it is not the only contributor. When the effects of all human-generated greenhouse gases other than $CO_2$ are added together and projected into the future, their collective impact significantly increases the impact of $CO_2$ alone.

Sophisticated computer models show that the warming of the lower atmosphere caused by $CO_2$ and trace gases will not be the same everywhere. Rather, the temperature response in polar regions could be two to three times greater than the global average. One reason is the fact that the polar troposphere is very stable, which suppresses vertical mixing and thus limits the amount of surface heat that is transferred upward. In addition, an expected reduction in sea ice would contribute to the greater temperature increase. This topic will be explored more fully in the next section.

## eye ON THE atmosphere 14.2

This satellite image from August 2007 shows the effects of tropical deforestation in a portion of the Amazon basin in western Brazil. Intact forest is dark green, whereas cleared areas are tan (bare ground) or light green (crops and pasture). Notice the relatively dense smoke in the left center of the image.

**Questions**

1. How does the destruction of tropical forests change the composition of the atmosphere?
2. Describe the effect that tropical deforestation has on global warming.

## How Aerosols Influence Climate

Increasing the levels of carbon dioxide and other greenhouse gases in the atmosphere is the most direct human influence on global climate. But it is not the only impact. Global climate is also affected by human activities that contribute to the atmosphere's aerosol content. Recall that **aerosols** are tiny, often microscopic, liquid and solid particles suspended in the air. Unlike cloud droplets, aerosols are present even in relatively dry air. Atmospheric aerosols are composed of many different materials, including soil, smoke, sea salt, and sulfuric acid. Natural sources are numerous and include such phenomena as wildfires, dust storms, breaking waves, and volcanoes.

Most human-generated aerosols come from the sulfur dioxide emitted during the combustion of fossil fuels and from burning of vegetation to clear agricultural land. Chemical reactions in the atmosphere convert the sulfur dioxide into sulfate aerosols, the same material that produces acid precipitation (see Chapter 13).

How do aerosols affect climate? Most aerosols act directly by reflecting sunlight back to space and indirectly by making clouds "brighter" reflectors. The second effect relates to the fact that many aerosols (such as those composed of salt or sulfuric acid) attract water and thus are especially effective as cloud condensation nuclei. The large quantity of aerosols produced by human activities (especially industrial emissions) triggers an increase in the number of cloud droplets that form within a cloud. A greater number of small droplets increases the cloud's brightness, causing more sunlight to be reflected back to space.

One category of aerosols, called **black carbon,** is soot generated by combustion processes and fires. Unlike other aerosols, black carbon warms the atmosphere because it is an effective absorber of incoming solar radiation. When deposited on snow and ice, black carbon also reduces surface albedo, thus increasing the amount of light absorbed. Nevertheless, despite the warming effect of black carbon, the overall effect of atmospheric aerosols is to cool Earth.

Studies indicate that the cooling effect of human-generated aerosols offsets a portion of the global warming caused by the growing quantities of greenhouse gases in the atmosphere. The magnitude and extent of the cooling effect of aerosols is uncertain. This uncertainty is a hurdle to advancing our understanding of how humans alter Earth's climate.

It is important to point out some significant differences between global warming by greenhouse gases and aerosol cooling. After being emitted, carbon dioxide and trace gases remain in the atmosphere and influence climate for many decades. By contrast, aerosols released into the troposphere remain there for only a few days or, at most, a few weeks before they are "washed out" by precipitation, limiting their effect on today's climate. Because of their short lifetime in the troposphere, aerosols are distributed unevenly over the globe. As expected, human-generated aerosols are concentrated near the areas that produce them, namely industrialized regions that burn fossil fuels and places where vegetation is burned (**Fig. 14.27**).

▲ **Figure 14.27 Aerosol haze** Human-generated aerosols are concentrated near areas that produce them. Because most aerosols reduce the amount of solar energy available to the climate system, they have a net cooling effect. This satellite image shows a dense blanket of pollution moving away from the coast of China. The plume is about 200 kilometers (125 miles) wide and more than 600 kilometers (375 miles) long.

Video
Global Warming Predictions
http://goo.gl/Z7toq9

## ✔ Concept Checks 14.3

1. Why has the $CO_2$ level of the atmosphere been increasing over the past 200 years?
2. How has the atmosphere responded to the growing $CO_2$ levels? How are temperatures in the lower atmosphere likely to change as $CO_2$ levels continue to increase?
3. Aside from $CO_2$, what trace gases are contributing to global temperature change?
4. List the main sources of human-generated aerosols and describe their net effect on atmospheric temperatures.

# What's Your Forecast?

## Modeling the Impact of Aerosols

by Dr. Don Wuebbles, Department of Atmospheric Sciences, University of Illinois

Atmospheric observations and modeling of aerosols, supported by laboratory studies, indicate that human-generated aerosol emissions play an important role in climate change, as discussed in Section 14.3. These aerosols, commonly released by fossil fuel use and forest burning, also endanger human and ecosystem health, and the international community has called for limiting these emissions. As a result, scientists want to understand how policies to limit aerosol emissions could affect Earth's climate.

To assess the range of possible climate effects of policies to control aerosol pollution, a recent climate modeling study* examined two scenarios that address aerosols (and other "forcings" on climate), modeled over a 50-year period: (1) a "clean" atmosphere where public health measures reduce human-generated emissions of particles by a factor of 10 and (2) a "dirty" atmosphere where newly industrializing nations increase the same aerosol emissions by a factor of 3. Unlike $CO_2$, which is long-lived and therefore well mixed in the atmosphere (essentially the same concentration everywhere), atmospheric aerosols are very short-lived in the atmosphere and are generally removed within a few weeks, so they are not well mixed and tend to have their largest concentrations near where they are emitted.

Graph A in **Figure 14.B** shows surface air temperatures, Graph B shows the average amount of shortwave radiation received at the top of the atmosphere and Graph C shows the precipitation forecast from these models. The black line represents baseline values given current emissions, the red line represents the "clean" case in which aerosols have been removed, and the blue line represents the "dirty" case with increased aerosol emissions. The resulting forecasts were produced by performing multiple runs with a model of Earth's climate system that uses slightly different initial conditions that follow somewhat different pathways. The results provide a strong indication of how well such models represent natural variability in the climate system.

The study shows that the clean case—reducing aerosols by a factor of 10—has far more effect on climate than does adding additional aerosols.

### Questions

1. Based on the graphs from the climate model study in Figure 14.B, describe the relationship between reducing aerosol emissions (the clean case) and the predicted change in global surface temperatures.
2. Does the precipitation response more closely follow the radiation received (Graph A) or the temperature response (Graph B)? Explain?
3. Given the findings presented here, do you think policies to limit aerosol emissions make sense? Why or why not?

---

*M. Dubey, Los Alamos National Laboratory, unpublished results.

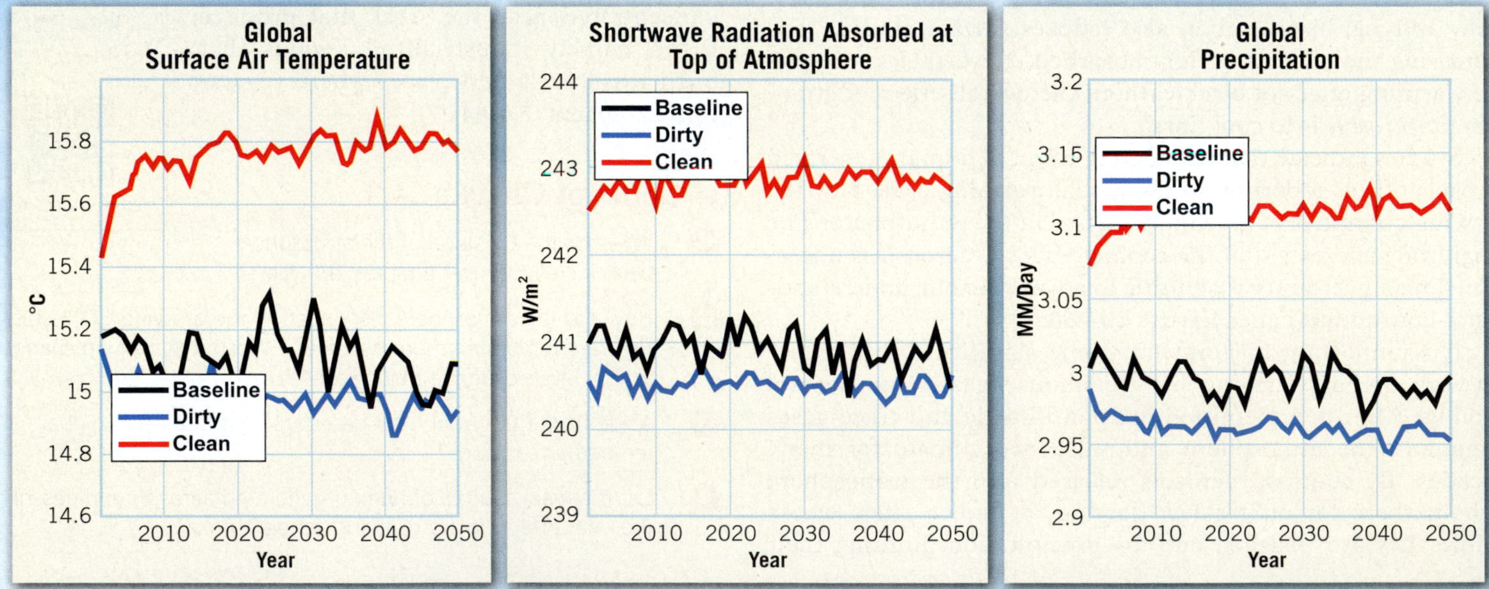


A. B. C.

▲ **Figure 14.B Results of a climate modeling study** **A.** Average annual global surface air temperature **B.** Shortwave radiation at the top of the atmosphere **C.** Average annual global precipitation

# 14.4 Climate-Feedback Mechanisms

**Contrast positive- and negative-feedback mechanisms and provide examples of each.**

Climate is a very complex interactive physical system. Thus, when any component of the climate system is altered, scientists must consider many possible outcomes. These possible outcomes are called **climate-feedback mechanisms.** They complicate climate-modeling efforts and add greater uncertainty to climate predictions.

## Types of Feedback Mechanisms

What climate-feedback mechanisms are related to carbon dioxide and other greenhouse gases? One important mechanism is that warmer surface temperatures increase evaporation rates. This, in turn, increases the water vapor in the atmosphere. Remember that water vapor is an even more powerful absorber of radiation emitted by Earth than is carbon dioxide. Therefore, with more water vapor in the air, the temperature increase caused by carbon dioxide and trace gases is reinforced.

Scientists who model global climate change indicate that the temperature increase at high latitudes may be two to three times greater than the global average. This assumption is based in part on the likelihood that the area covered by sea ice will decrease as surface temperatures rise. Because ice reflects a much larger percentage of incoming solar radiation than does open water, the melting of sea ice replaces a highly reflective surface with a relatively dark surface (Fig. 14.28). The result is a substantial increase in the solar energy absorbed at the surface. This in turn feeds back to the atmosphere and magnifies the initial temperature increase created by higher levels of greenhouse gases.

▼ **Figure 14.28 Sea ice as a feedback mechanism** The image shows the springtime breakup of sea ice near Antarctica. The diagram shows a likely feedback loop. A reduction in sea ice acts as a positive-feedback mechanism because surface albedo decreases, and the amount of energy absorbed at the surface increases.

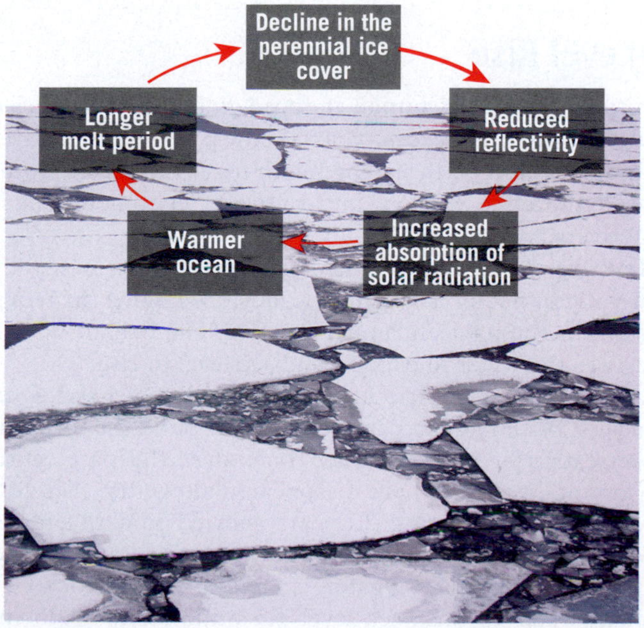


The climate-feedback mechanisms discussed thus far magnify the temperature rise caused by the buildup of greenhouse gases. Because these effects reinforce the initial change, they are called **positive-feedback mechanisms.** However, other effects are classified as **negative-feedback mechanisms** because they produce results that are just the opposite of the initial change and tend to offset it.

One probable result of a global temperature rise would be an accompanying increase in cloud cover due to the higher moisture content of the atmosphere. Most clouds are good reflectors of solar radiation. At the same time, however, they are also good absorbers and emitters of radiation emitted by Earth. Consequently, clouds produce two opposite effects. They are a negative-feedback mechanism because they increase Earth's albedo and thus diminish the amount of solar energy available to heat the atmosphere. On the other hand, clouds act as a positive-feedback mechanism by absorbing and emitting radiation that would otherwise be lost from the troposphere.

Which effect, if either, is stronger? Scientists still are not sure whether clouds will produce a net positive or negative feedback. Although recent studies have not settled the question, they seem to lean toward the idea that clouds do not dampen global warming but rather produce a small positive feedback overall.*

The problem of global warming caused by human-induced changes in atmospheric composition continues to be one of the most studied aspects of climate change. Although no models yet incorporate the full range of potential factors and feedbacks, the scientific consensus is that the increasing levels of atmospheric carbon dioxide and trace gases will lead to a warmer planet with a different distribution of climate regimes.

## Computer Models of Climate: Important yet Imperfect Tools

Earth's climate system is amazingly complex. Comprehensive state-of-the-science climate simulation models are among the basic tools used to develop possible climate-change scenarios. Called *general circulation models* (*GCMs*), they are based on fundamental laws of physics and chemistry and incorporate human and biological interactions. GCMs are used to simulate many variables, including temperature, rainfall, snow cover, soil moisture, winds, clouds, sea ice, and ocean circulation over the entire globe through the seasons and over spans of decades.

*A. E. Dessler, "A Determination of the Cloud Feedback from Climate Variations over the Past Decade," *Science* 330 (December 10, 2010), 1523–1526.

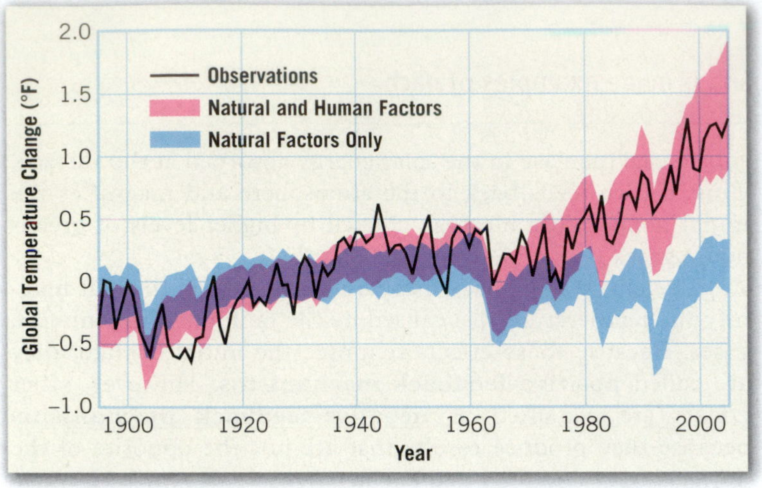


▲ **Figure 14.29 Separating human and natural influences on climate** The blue band shows how global average temperatures would have changed due to natural forces only, as simulated by climate models. The pink band shows model projections of the effects of human and natural forces combined. The black line shows actual observed global average temperatures. As the blue band indicates, without human influences, temperatures over the past century would actually have first warmed and then cooled slightly over recent decades. Bands of color are used to express the range of uncertainty.

In many other fields of study, hypotheses can be tested by direct experimentation in a laboratory or by observations and measurements in the field. However, this is often not possible in the study of climate. Rather, scientists must construct computer models of how our planet's climate system works. If we understand the climate system correctly and construct the model appropriately, then the behavior of the model climate system should mimic the behavior of Earth's climate system (Fig. 14.29).

What factors influence the accuracy of climate models? Clearly, mathematical models are *simplified* versions of the real Earth and cannot capture its full complexity, especially at smaller geographic scales. Moreover, computer models used to simulate future climate change must make many assumptions that significantly influence predictions. They must consider a wide range of possible changes in population, economic growth, fossil fuel consumption, technological development, improvements in energy efficiency, and more.

Despite many obstacles, our ability to use supercomputers to simulate climate continues to improve. Although today's models are far from infallible, they are powerful tools for understanding what Earth's future climate might be like.

### ✔ Concept Checks 14.4

1. Distinguish between positive- and negative-feedback mechanisms.
2. Provide at least one example of each type of feedback mechanism.
3. What factors influence the accuracy of computer models of climate?

## 14.5 Some Consequences of Global Warming

**Discuss several likely consequences of global warming.**

What consequences can be expected as the carbon dioxide content of the atmosphere reaches a level that is twice what it was early in the twentieth century? Because the climate system is complex, predicting the occurrence of specific effects in particular places is speculative. It is not yet possible to pinpoint such changes. Nevertheless, there are plausible scenarios for larger scales of space and time.

As noted, the magnitude of the temperature increase will not be the same everywhere. The temperature rise will probably be smallest in the tropics and increase toward the poles. As for precipitation, the models indicate that some regions will experience significantly more precipitation and runoff. However, others will experience a decrease in runoff due to reduced precipitation or greater evaporation caused by higher temperatures.

Table 14.1 summarizes some of the most likely effects and their possible consequences. The table also provides the IPCC's estimate of the probability of each effect. Levels of confidence for these projections vary from *likely* (67 to 90 percent probability) to *very likely* (90 to 99 percent probability) to *virtually certain* (greater than 99 percent probability).

### Sea-Level Rise

A significant impact of human-induced global warming is a rise in sea level. As this occurs, coastal cities, wetlands, and low-lying islands could be threatened with more frequent flooding, increased shoreline erosion, and saltwater encroachment into coastal rivers and aquifers.

How is a warmer atmosphere related to a rise in sea level? One significant factor is thermal expansion. Higher air temperatures warm the adjacent upper layers of the ocean, which in turn causes the water to expand and sea level to rise.

A second factor contributing to global sea-level rise is melting glaciers. With few exceptions, glaciers around the world have been retreating at unprecedented rates over the past century. Some mountain glaciers have disappeared altogether (Fig. 14.30). A satellite study spanning 20 years showed that the mass of the Greenland and Antarctic ice sheets dropped an average of 475 gigatons per year. (A gigaton is 1 billion metric tons.) That is enough water to raise sea level 1.5 millimeters (0.05 inch) per year. The loss of ice was not steady but was occurring at

**Table 14.1** | Twenty-First-Century Climate Trends

| Projected Changes and Estimated Probability* | Examples of Projected Impacts |
|---|---|
| Higher maximum temperatures; more hot days and heat waves over nearly all land areas (*virtually certain*) | Increased incidence of death and serious illness in older age groups and urban poor<br>Increased heat stress in livestock and wildlife<br>Shift in tourist destinations<br>Increased risk of damage to a number of crops<br>Increased electric cooling demand and reduced energy supply reliability |
| Higher minimum temperatures; fewer cold days, frost days, and cold waves over nearly all land areas (*virtually certain*) | Decreased cold-related human morbidity and mortality<br>Decreased risk of damage to a number of crops and increased risk to others<br>Extended range and activity of some pest and disease vectors<br>Reduced heating energy demand |
| Frequency of heavy precipitation events increasing over most areas (*very likely*) | Increased flood, landslide, avalanche, and mudflow damage<br>Increased soil erosion<br>Increased recharge of some floodplain aquifers due to increased runoff<br>Increased pressure on government and private flood insurance systems and disaster relief |
| Increases in area affected by drought (*likely*) | Decreased crop yields<br>Increased damage to building foundations caused by ground shrinkage<br>Decreased water-resource quantity and quality<br>Increased risk of wildfires |
| Increases in intense tropical cyclone activity (*likely*) | Increased risks to human life, risk of infectious-disease epidemics, and many other risks<br>Increased coastal erosion and damage to coastal buildings and infrastructure<br>Increased damage to coastal ecosystems, such as coral reefs and mangroves |

**Virtually certain* indicates a probability of 99–100 percent, *very likely* indicates a probability of 90–99 percent, and *likely* indicates a probability of 66–90 percent.

*Source:* Based on the Fifth Assessment Report. *Climate Change 2013: The Physical Science Basis, Summary for Policymakers.*

**students sometimes ask...**

### What are scenarios, and why are they used?

A scenario is an example of what might happen under a particular set of assumptions. Using scenarios is a way of examining questions about an uncertain future. For example, future trends in fossil-fuel use and other human activities are uncertain. Therefore, scientists have developed a range of scenarios for how climate may change based on a wide range of possibilities for these variables.

an accelerating rate during the study period. During the same span, mountain glaciers and ice caps lost an average of slightly more than 400 gigatons per year.

Research indicates that sea level has risen about 25 centimeters (9.75 inches) since 1870, with the rate of sea-level rise accelerating in recent years. What about future changes in sea level?

1941

Video MM
Sea Level Rise
http://goo.gl/zRrhgp

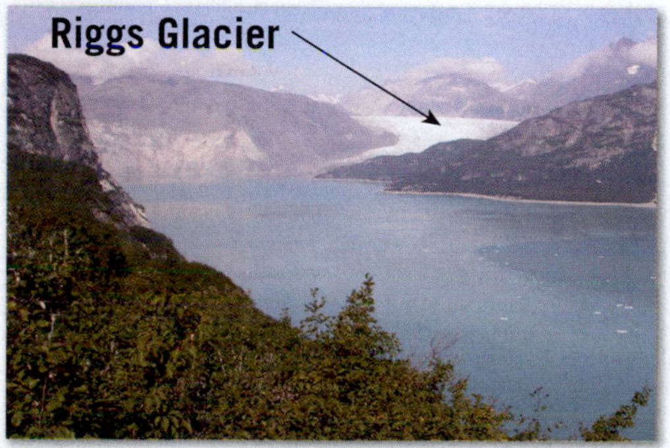

2004

◀ **Figure 14.30 Two images taken 63 years apart from the same vantage point in Alaska's Glacier Bay National Park** Muir Glacier, which is prominent in the 1941 photo, has retreated out of the field of view in the 2004 image. Also, Riggs Glacier (upper right) has thinned and retreated significantly.

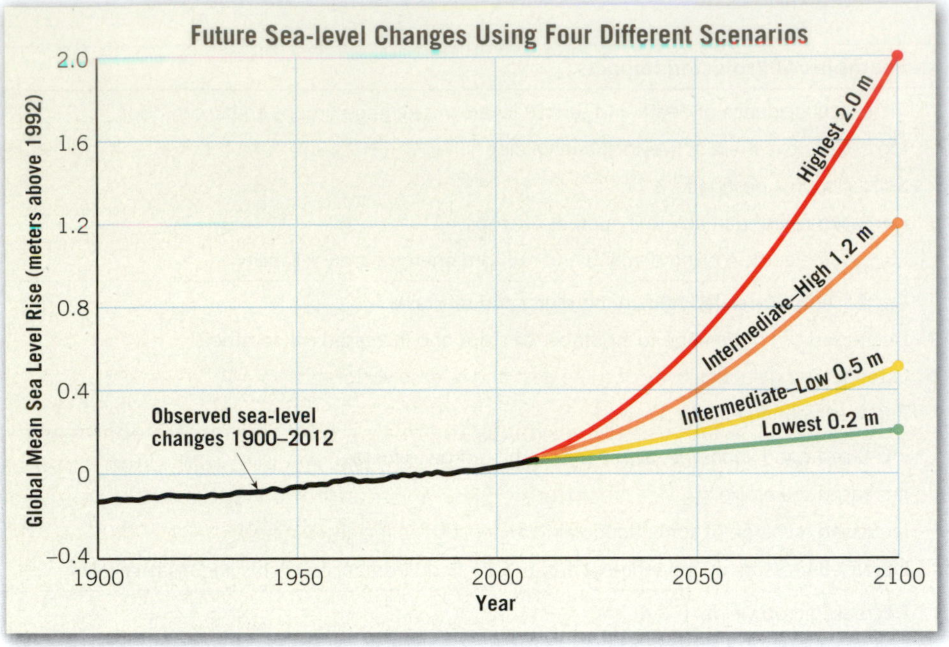


◀ **Figure 14.31 Changing sea level** This graph shows changes in sea level between 1900 and 2012 and projections to 2100 using four different scenarios. Currently the highest and lowest projections are considered to be extremely unlikely. The greatest uncertainty surrounding estimates is the rate and magnitude of ice sheet loss from Greenland and Antarctica. Zero on the graph represents mean sea level in 1992.

Video MM
Retreat of Continental Ice Sheets
http://goo.gl/BqLnkX

As **Figure 14.31** indicates, the estimates of future sea-level rise are uncertain. The four scenarios depicted on the graph represent estimates based on different degrees of ocean warming and ice sheet loss and range from 0.2 meter (8 inches) to 2 meters (6.6 feet). The lowest scenario is an extrapolation of the annual rate of sea-level rise that occurred between 1870 and 2000 (1.7 millimeter/year). However, when the rate of sea-level rise for the period 1993 to 2012 is examined, the annual change is 3.17 millimeters/year. Such data show that there is a reasonable chance that sea level will rise considerably more than the lowest scenario indicates.

Scientists realize that even modest rises in sea level along a *gently* sloping shoreline, such as the Atlantic and Gulf coasts of the United States, will lead to significant erosion and severe permanent inland flooding (**Fig. 14.32**). If this happens, many beaches and wetlands will be eliminated, and coastal civilization will be severely disrupted. Low-lying and densely populated places such as Bangladesh and the small island nation of the Maldives are especially vulnerable. The average elevation in the Maldives is 1.5 meters (less than 5 feet), and its highest point is just 2.4 meters (less than 8 feet) above sea level.

Because rising sea level is a gradual phenomenon, coastal residents may overlook it as an important contributor to shoreline flooding and erosion problems. Rather, the blame may be assigned to other forces, especially storm activity. Although a given storm may be the immediate cause, the magnitude of its destruction may result from the relatively small sea-level rise that allowed the storm's power to cross a much greater land area.

## The Changing Arctic

The effects of global warming are most pronounced in the high latitudes of the Northern Hemisphere. For more than 30 years, the extent and thickness of sea ice have been rapidly declining. In addition, permafrost temperatures (see Tundra Climates in Chapter 15) have been rapidly rising, and the area affected by permafrost has been decreasing. Meanwhile, alpine glaciers and the Greenland ice sheet have been shrinking. Another sign that the Arctic is rapidly warming is related to plant growth (**Fig. 14.33**). A 2013 study showed that vegetation growth at northern latitudes now resembles that which characterized areas 4° to 6° of latitude farther south as recently as 1982. That is a distance of 400 to 700 kilometers (250 to 430 miles). One researcher characterized the finding this way: "It's like Winnipeg, Manitoba, moving to Minneapolis-St. Paul in only 30 years."

**Arctic Sea Ice** Climate models are in general agreement that one of the strongest signals of global warming should be a loss

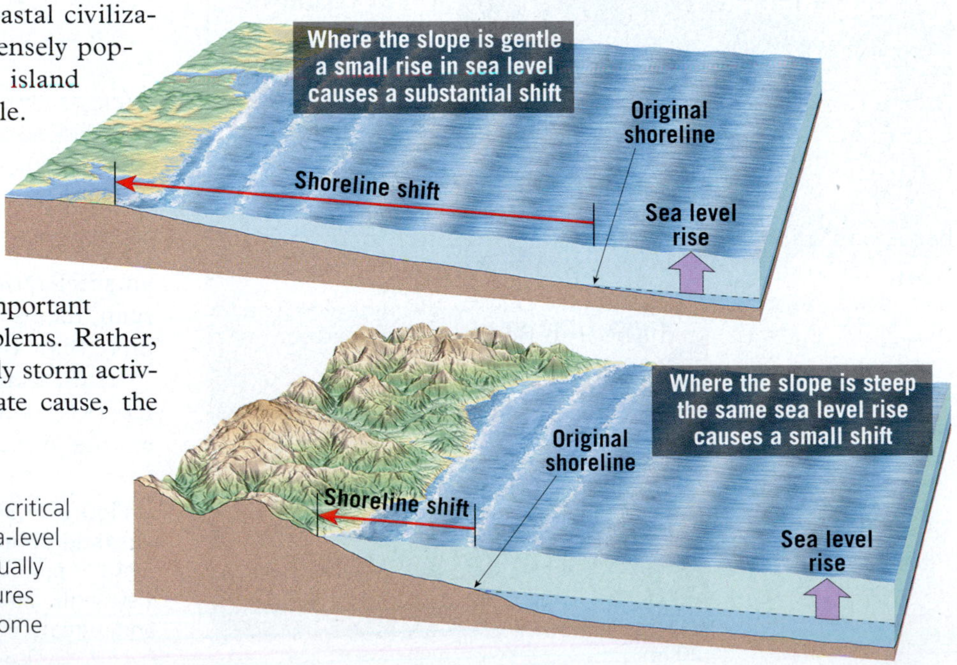


MM
http://goo.gl/DcKrEm

▶ **SmartFigure 14.32 Slope of the shoreline** The slope of the shoreline is critical to determining the degree to which sea-level changes will affect it. As sea level gradually rises, the shoreline retreats, and structures that were once thought to be safe from wave attack become vulnerable.

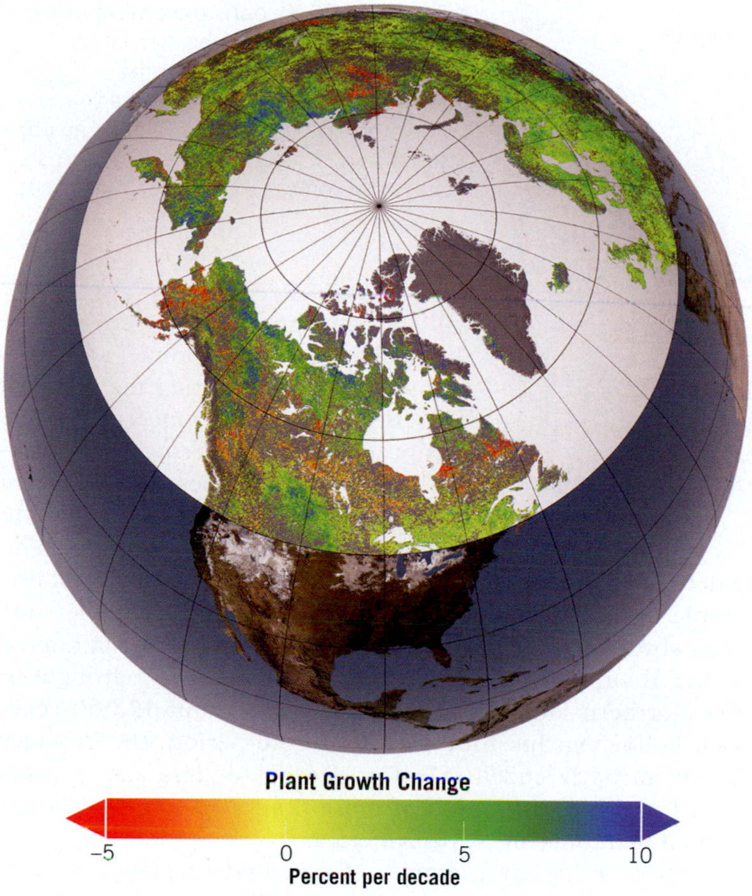


▲ **Figure 14.33 Climate change spurs plant growth beyond 45° north** Of the 26 million square kilometers (10 million square miles) of northern vegetated lands, about 40 percent showed increases in plant growth during the 30-year period ending in 2012 (green and blue on the satellite image).

of sea ice in the Arctic. This is indeed occurring. The map in **Figure 14.34** compares the average sea ice extent for September 2012 to the long-term average for the period 1979–2000. On that date the extent was less than 4 million square kilometers (1.54 square miles)—70,000 square kilometers (27,000 square miles) less than the previous record low, set in September 2007. (September represents the end of the melt period, when the area covered by sea ice is at a minimum.) Not only is the area covered by sea ice declining, but the remaining sea ice has become thinner, making it more vulnerable to further melting.

Models that best match historical trends project that Arctic waters may be virtually ice-free in the late summer by the 2030s. As was noted earlier in this chapter, a reduction in sea ice represents a positive-feedback mechanism that reinforces global warming.

**Permafrost** During the past decade, mounting evidence has indicated that the extent of permafrost in the Northern Hemisphere has decreased, as would be expected under long-term warming conditions. In the Arctic, short summers thaw only the top layer of frozen ground. The permafrost beneath this *active layer* is like the cement bottom of a swimming pool. In summer, water cannot percolate downward, so it saturates the soil above the permafrost and collects on the surface in thousands of lakes. However, as Arctic temperatures climb, the bottom of the "pool" seems to be "cracking." Satellite imagery shows that a significant number of lakes have shrunk or disappeared altogether. As the permafrost thaws, lake water drains deeper into the ground.

Studies in Alaska show that thawing is occurring in interior and southern parts of the state where permafrost temperatures are near the thaw point. As Arctic temperatures continue to rise, some models project that near-surface permafrost may be lost entirely from large parts of Alaska by the end of the century.

Thawing permafrost represents a potentially significant positive-feedback mechanism that may reinforce global warming. When vegetation dies in the Arctic, cold temperatures inhibit its decomposition. As a consequence, over thousands of years, a great deal of organic matter has become stored in the permafrost. When the permafrost thaws, organic matter that may have been frozen for millennia comes out of "cold storage" and decomposes. The result is the release of carbon dioxide and methane—greenhouse gases that contribute to global warming. Thus, like decreasing sea ice, thawing permafrost is a positive-feedback mechanism.

## Increasing Ocean Acidity

The human-induced increase in the amount of carbon dioxide in the atmosphere has some serious implications for ocean chemistry and marine life. Nearly half of the human-generated carbon dioxide ends up dissolved in the oceans. When carbon dioxide ($CO_2$) from the atmosphere dissolves in seawater ($H_2O$), it forms carbonic acid ($H_2CO_3$), which lowers the ocean's pH and changes the balance of certain chemicals found naturally in seawater. (See Fig. 13.18, page 375,

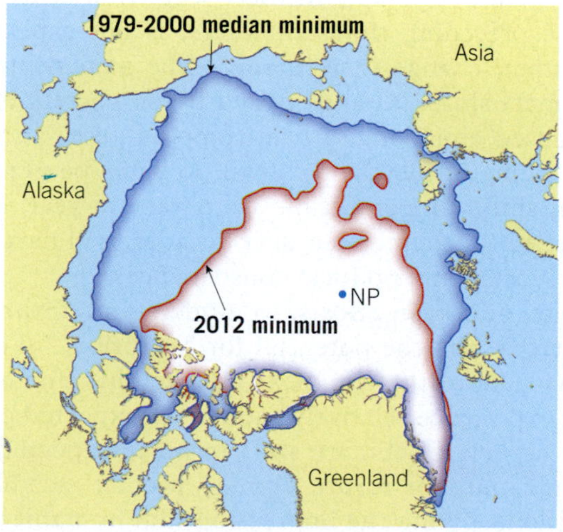


▲ **Figure 14.34 Tracking sea ice changes** Sea ice is frozen seawater. In winter the Arctic Ocean is completely ice covered. In summer, a portion of the ice melts. This map shows the extent of sea ice in early September 2012 compared to the average extent of the period 1979 to 2000. The extent in 2012 was the lowest on record through the year 2013. The sea ice that does not melt in summer is getting thinner.

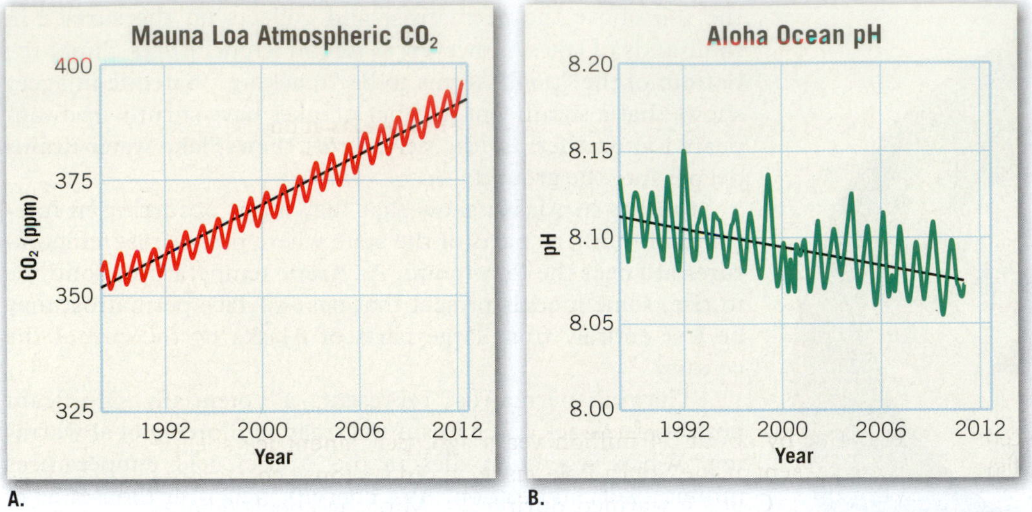


◀ **Figure 14.35 Oceans becoming more acidic** The graphs show the correlation between rising levels of $CO_2$ in the atmosphere as measured at Mauna Loa Observatory (**A**) and falling pH in the nearby ocean (**B**). As $CO_2$ accumulates in the ocean, the water becomes more acidic (pH declines).

to review the pH scale.) In fact, the oceans have already absorbed enough $CO_2$ for surface waters to have experienced a pH decrease of 0.1 since preindustrial times, with an additional pH decrease likely in the future (Fig. 14.35). Moreover, if the current trend in $CO_2$ emissions continues, by the year 2100 the ocean will have experienced a pH decrease of at least 0.3, which represents a change in ocean chemistry that has not occurred for millions of years. This shift toward acidity and the changes in ocean chemistry that result make it more difficult for certain marine organisms to build hard parts out of calcium carbonate. The decline in pH thus threatens a variety of calcite-secreting organisms as diverse as microbes and corals. This concerns marine scientists because of the potential consequences for other sea life that depend on the health and availability of these organisms.

## The Potential for "Surprises"

You have seen that climate in the twenty-first century, unlike during the preceding thousand years, is not expected to be stable. Rather, change is occurring. The amount and rate of future climate shifts depends primarily on current and future human-caused emissions of heat-trapping gases and airborne particles. Many of the changes will probably be gradual environmental shifts, imperceptible from year to year. Nevertheless, the effects, accumulated over decades, will have powerful economic, social, and political consequences.

Despite our best efforts to understand future climate shifts, there is also the potential for "surprises." This simply means that due to the complexity of Earth's climate system, we might experience relatively sudden, unexpected changes or see some aspects of climate shift in an unexpected manner. Many projections of future climate predict steadily changing conditions, giving the impression that humanity will have time to adapt. However, the scientific community has been paying attention to the possibility that at least some changes will be abrupt, perhaps crossing a threshold, or "tipping point," so quickly that there will be little time to react. This is a reasonable concern because abrupt changes occurring over periods as short as decades or even years have been a natural part of the climate system throughout Earth history. The paleoclimate record described earlier in the chapter contains ample evidence of such abrupt changes. For example, one such abrupt change occurred at the end of a time span known as the *Younger Dryas*, a time of abnormal cold and drought in the Northern Hemisphere that occurred about 12,000 years ago. Following this 1000-year-long cold period, the Younger Dryas abruptly ended in a few decades or less and is associated with the extinction of more than 70 percent of large-bodied mammals in North America.

There are many examples of potential surprises, each of which would have large consequences. We simply do not know how far the climate system or other systems it affects can be pushed before they respond in unexpected ways. Even if the chance of any particular surprise happening is small, the chance that at least one such surprise will occur is much greater. In other words, although we may not know which of these events will occur, it is likely that one or more will eventually occur.

The impact on climate of an increase in atmospheric $CO_2$ and trace gases is obscured by some uncertainties. Yet climate scientists continue to improve our understanding of the climate system and the potential impacts and effects of global climate change. Policymakers are confronted with responding to the risks posed by greenhouse gas emissions, knowing that our understanding is imperfect. However, they are also faced with the fact that climate-induced environmental changes cannot be reversed quickly, if at all, due to the lengthy time scales associated with the climate system.

### ✔ Concept Checks 14.5

1. Describe the factors that are causing sea level to rise.
2. Is global warming greater near the equator or near the poles? Explain.
3. Based on Table 14.1, what projected changes relate to something other than temperature?

# 14 Concepts in Review The Changing Climate

## 14.1 The Climate System: A Key to Detecting Climate Change

▶ Explain how unraveling past climate changes is related to the climate system and discuss several ways in which such changes are detected.

**Key Terms:** climate system, cryosphere, proxy data, paleoclimatology, oxygen-isotope analysis, dendrochronology

- Earth's climate system is a complex interchange of energy and moisture that occurs among the atmosphere, hydrosphere, geosphere, biosphere, and cryosphere (ice and snow).
- The geologic record yields multiple kinds of indirect evidence about past climate. These proxy data are the focus of paleoclimatology. Proxy data can come from seafloor sediment, oxygen isotopes, cores of glacial ice, tree rings, coral growth bands, fossil pollen, and even historical documents.
- Trees grow thicker rings in warmer, wetter years and thinner rings in colder, drier years. The pattern of ring thickness can be matched up between trees of overlapping ages for a long-term record of a region's climate.
- Oxygen-isotope analysis is based on the difference between heavier $^{18}O$ and lighter $^{16}O$ and their relative amounts in water molecules ($H_2O$). The $^{18}O/^{16}O$ ratio of ocean water rises during cold times because water containing lighter $^{16}O$ evaporates more readily. During warmer times, water containing $^{18}O$ evaporates, plus melting glacial ice sends some of its $^{16}O$ back to the sea. This makes the marine $^{18}O/^{16}O$ ratio drop. Oxygen isotopes can also be measured in the shells of fossil marine organisms or the water molecules that make up glacial ice. Glacial ice also traps small samples of the atmosphere in air bubbles.

**Q** Why might the sediment core being collected by this oceanographic research vessel be useful to scientists studying climate change?

## 14.2 Natural Causes of Climate Change

▶ Discuss four hypotheses that relate to natural causes of climate change.

**Key Terms:** plate tectonics theory, eccentricity, obliquity, precession, sunspots

- The natural functions of the Earth system produce climate change. The position of lithospheric plates can influence the climate of the continents as well as oceanic circulation. Variations in the shape of Earth's orbit, angle of axial tilt, and orientation of the axis all cause changes in the distribution of solar energy.
- Volcanic aerosols act like a Sun shade, screening out a portion of incoming solar radiation. Volcanic sulfur dioxide emissions that reach the stratosphere are particularly important. Combining with water to form tiny droplets of sulfuric acid, these aerosols can remain aloft for several years.
- Since Earth's climate is fueled by solar energy, variations in the Sun's energy output matter. Solar variability remains unknown, as satellite measuring of its output has been active for only a few decades. Sunspots are dark features on the Sun's surface associated with periods of increased solar energy output. The number of sunspots rises and drops throughout an 11-year cycle. During the peak of the cycle, the Sun puts out about 0.1 percent more energy than during the lowest part of the cycle, but this small cyclical effect is not correlated with the current episode of global warming.

## 14.3 Human Impact on Global Climate

▶ Summarize the nature and cause of the atmosphere's changing composition since about 1750. Describe the climate's response.

**Key Terms:** trace gases, aerosols, black carbon

- Humans have been modifying the environment for thousands of years. By altering ground cover with the use of fire and the overgrazing of land, people have modified such important climatic factors as surface albedo, evaporation rates, and surface winds.
- Human activities produce climate change through the release of carbon dioxide ($CO_2$) and trace gases. Humans release $CO_2$ when they cut down forests or when they burn fossil fuels such as coal, oil, and natural gas. A steady rise in atmospheric $CO_2$ levels has been documented at Mauna Loa, Hawaii, and other locations around the world.
- More than half of the carbon released by humans is absorbed by new plant matter or dissolved in the oceans. About 45 percent remains in the atmosphere, where it can influence climate for decades. Air bubbles trapped in glacial ice reveal that there is currently about 30 percent more $CO_2$ than the atmosphere has contained in the past 650,000 years.
- As a result of the extra heat retention due to added $CO_2$, Earth's atmosphere has warmed by about 0.8°C (1.4°F) in the past 100 years, most of it since the 1970s. Temperatures are projected to increase by another 2° to 4.5°C (3.5° to 8.1°F) in the future.

- Trace gases such as methane, nitrous oxide, and CFCs also play significant roles in increasing global temperature.
- When emitted due to human activities, tiny liquid and solid particles suspended in the air, called aerosols, have an effect on global climate.
- Overall, aerosols reflect a portion of incoming solar radiation back to space and therefore have a cooling effect; yet some aerosols, called black carbon (soot from combustion processes and fires), absorb incoming solar radiation and act to warm the atmosphere. When black carbon is deposited on snow and ice, it reduces surface albedo and increases the amount of light absorbed at the surface.

**Q** Do aerosols spend more or less time in the atmosphere than greenhouse gases such as carbon dioxide? What is the significance of this difference in residence time? Explain.

## 14.4 Climate-Feedback Mechanisms

► Contrast positive- and negative-feedback mechanisms and provide examples of each.

**Key Terms:** climate feedback mechanism, positive-feedback mechanism, negative-feedback mechanism

- A change in one part of the climate system may trigger changes in other parts of the climate system that amplify or diminish the initial effect. These climate-feedback mechanisms are called positive-feedback mechanisms if they reinforce the initial change and negative-feedback mechanisms if they counteract the initial effect.
- The melting of sea ice due to global warming (decreasing albedo and increasing the initial effect of warming) is one example of a positive-feedback mechanism. The production of more clouds (blotting out incoming solar radiation, leading to cooling) is an example of a negative-feedback mechanism.
- Computer models of climate give scientists a tool for testing hypotheses about climate change. Although these models are simpler than the real climate system, they are useful tools for predicting the future climate.

**Q** Changes in precipitation and temperature due to climate change can increase the risk of forest fires. Describe two ways that the event shown in this photo could contribute to global warming.

## 14.5 Some Consequences of Global Warming

► Discuss several likely consequences of global warming.

- In the future, Earth's surface temperature is likely to continue to rise. The temperature increase will likely be greatest in the polar regions and least in the tropics. Some areas will get drier, and other areas will get wetter.
- Sea level is predicted to rise for several reasons, including the melting of glacial ice and thermal expansion. (A given mass of seawater takes up more volume when it is warm than when it is cool.) Low-lying, gently sloped, highly populated coastal areas are most at risk.
- Sea ice cover and thickness in the Arctic have been declining since satellite observations began in 1979.
- Because of the warming of the Arctic, permafrost is melting, releasing $CO_2$ and methane to the atmosphere in a positive-feedback mechanism.
- Because the climate system is complicated, dynamic, and imperfectly understood, it could produce sudden, unexpected changes with little warning.

**Q** This ice breaker is plowing through sea ice in the Arctic Ocean. What spheres of the climate system are represented in this photo? How has the area covered by summer sea ice been changing since 1979? How does this change influence temperatures in the Arctic?

## Give it Some Thought

1. Refer to Figure 14.1, illustrating various components of Earth's climate system. Boxes represent interactions or changes that occur in the climate system. Select three boxes and provide an example of an interaction or change associated with each. Explain how these interactions may influence temperature.
2. Describe one way in which changes in the biosphere can cause changes in the climate system. Next, suggest one way in which the biosphere is affected by changes in some other part of the climate system. Finally, indicate one way in which the biosphere records changes in the climate system.
3. Recent volcanic events, such as the eruptions of El Chichón and Mount Pinatubo, were associated with drops in global temperatures. During the Cretaceous period, volcanic activity was associated with global warming. Explain this apparent paradox.
4. The accompanying photo is a 2005 view of Athabasca Glacier in the Canadian Rockies. A line of boulders in the foreground marks the outer limit of the glacier in 1992. Is the behavior of Athabasca Glacier shown in this image typical of other glaciers around the world? Describe a significant impact of such behavior.

5. It has been suggested that global warming over the past 40 years might have been even greater were it not for the effect of certain types of air pollution. Explain how this could be true.
6. The accompanying graph shows changes in the air's $CO_2$ content at South Pole Station (90° south latitude) and at a similar facility at Barrow, Alaska (71° north latitude). Which line on the graph represents the South Pole, and which represents Barrow, Alaska? Explain how you were able to determine which line is which.

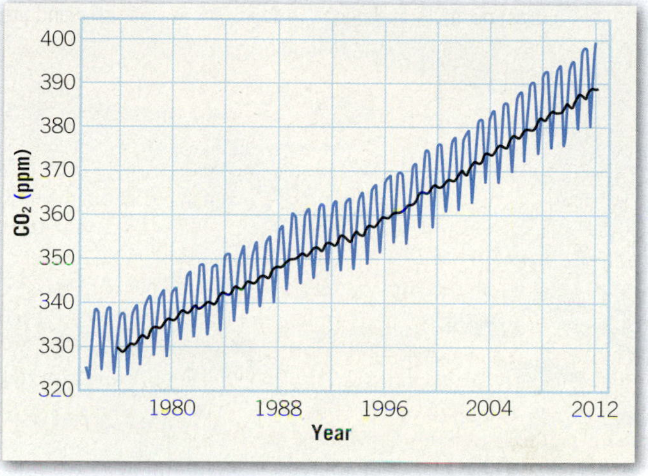


7. During a conversation, an acquaintance indicates that he is skeptical about global warming. When you ask him why he feels that way, he says, "The past couple of years in this area have been among the coolest I can remember." While you assure this person that it is useful to question scientific findings, you suggest to him that his reasoning in this case may be flawed. Use your understanding of the definition of *climate* along with one or more graphs in the chapter to persuade this person to reevaluate his reasoning.

## MasteringMeteorology™

Looking for additional review and test prep materials? Visit the Study Area in *MasteringMeteorology*™ to enhance your understanding of this chapter's content by accessing a variety of resources, including MapMaster™ interactive maps, Geoscience Animations, GEODe, *In the News* RSS feeds, flashcards, web links, self-study quizzes, and an eText version of *The Atmosphere*.

# 15 World Climates

## Focus on Concepts

*Each statement represents the primary learning objective for the corresponding major heading within the chapter. After you complete the chapter, you should be able to:*

**15.1** Explain why classification is a necessary process when studying world climates. Discuss the criteria used in the Köppen system of climate classification.

**15.2** List and briefly discuss the major controls of climate.

**15.3** Compare the two broad categories of humid tropical climates.

**15.4** Contrast low-latitude dry climates and middle-latitude dry climates.

**15.5** Distinguish among the three categories of C climates.

**15.6** Summarize the characteristics of the two categories of D climates.

**15.7** Contrast tundra and ice cap climates.

**15.8** List the characteristics of highland climates.

The varied nature of Earth's surface (oceans, mountains, plains, ice sheets) and the many interactions that occur among atmospheric processes give every location on our planet a distinctive (sometimes unique) climate. However, we cannot describe the climatic character of countless locales; that would require many volumes. Our purpose is to introduce you to the *major climate regions* of the world. We will examine large areas and zoom in on particular places to illustrate the characteristics of these major climate regions. In addition, for regions that are probably unfamiliar to you (the tropical, desert, and polar realms), we briefly describe the natural landscape.

*California's Death Valley is a rain shadow desert. Dry climates cover more land area than any other group.*

# 15.1 Climate Classification

**Explain why classification is a necessary process when studying world climates. Discuss the criteria used in the Köppen system of climate classification.**

In Chapter 1 we mentioned the common misconception that climate is only "the average state of the atmosphere." Although averages are certainly important to climate descriptions, variations and extremes must also be included to accurately portray the character of an area.

Temperature and precipitation are the most important elements in climate descriptions because they have the greatest influence on people and their activities and also have an important impact on the distribution of vegetation and the development of soils. Nevertheless, other factors are also important for a complete climate description. When possible, some of these factors are introduced into our discussion of world climates.

The worldwide distribution of temperature, precipitation, pressure, and wind is, to say the least, complex. Because of how these vary from place to place and time to time, it is unlikely that any two places that are more than a very short distance apart can experience identical weather. The virtually infinite variety of places on Earth makes it apparent that the number of different climates must be extremely large. Having such a diversity of information to investigate is not unique to the study of the atmosphere. It is a problem basic to all science. (Consider astronomy, which deals with billions of stars, and biology, which studies millions of complex organisms.) To cope with such variety, we must devise some means of *classifying* the vast array of data to be studied. Establishing groups of items having common characteristics brings order and manageability to large quantities of information, and it not only aids comprehension and understanding but also facilitates analysis and explanation.

One of the first attempts at climate classification was made by the ancient Greeks, who divided each hemisphere into three zones: *torrid, temperate,* and *frigid* (**Fig. 15.1**). The basis of this simple scheme was Earth–Sun relationships. The boundaries were the four astronomically important parallels of latitude: the Tropic of Cancer (23.5° north), the Tropic of Capricorn (23.5° south), the Arctic Circle (66.5° north), and the Antarctic Circle (66.5° south). Thus, the globe was divided into winterless climates, summerless climates, and an intermediate type that had features of the other two.

Few other attempts to classify climates were made until the beginning of the twentieth century. Since then, many such schemes have been devised. Remember that the classification of climates (or of anything else) is not a natural phenomenon but the product of human ingenuity. The value of any particular classification system is determined largely by its *intended use.* A system designed for one purpose may not work well for another.

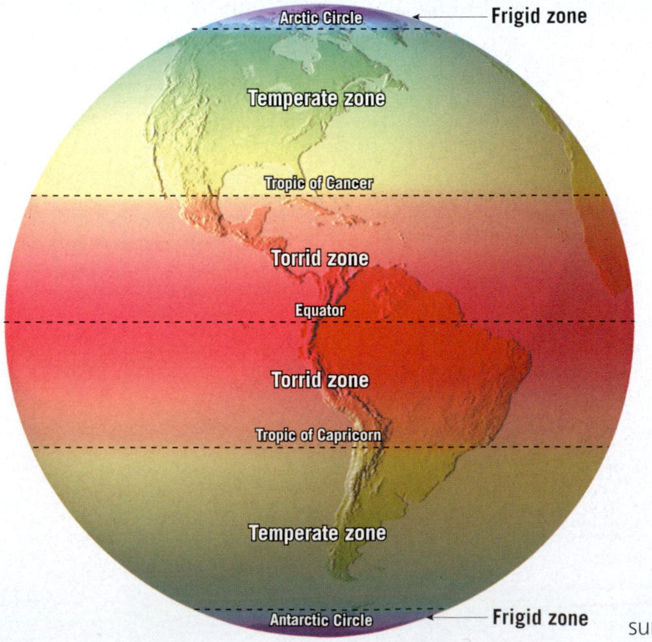


◄ **Figure 15.1 An early climate classification** Among the first attempts at climate classification was one made by the ancient Greeks. They divided each hemisphere into three zones. The winterless *torrid* zone was separated from the summerless *frigid* zone by the *temperate* zone, which had features of the other two.

## The Köppen Classification

In this chapter we use a classification devised by the German climatologist Wladimir Köppen (1846–1940). As a tool for presenting the general world pattern of climates, the **Köppen classification** has been the best-known and most-used system for decades. It is widely accepted for many reasons. For one, it uses easily obtained data: mean monthly and annual values of temperature and precipitation. Furthermore, the criteria are unambiguous, are relatively simple to apply, and divide the world into climate regions in a realistic way.

Köppen believed that the distribution of natural vegetation was the best expression of overall climate. Consequently, the boundaries he chose were largely based on the limits of certain plant associations. He recognized five principal climate groups, each designated by a capital letter:

| | |
|---|---|
| A | ***Humid tropical.*** Winterless climates, with all months having a mean temperature above 18°C (64°F). |
| B | ***Dry.*** Climates where evaporation exceeds precipitation and there is a constant water deficiency. |
| C | ***Humid middle-latitude, mild winters.*** The average temperature of the coldest month is below 18°C (64°F) but above –3°C (27°F). |
| D | ***Humid middle-latitude, severe winters.*** The average temperature of the coldest month is below –3°C (27°F), and the warmest monthly mean exceeds 10°C (50°F). |
| E | ***Polar.*** Summerless climates, where the average temperature of the warmest month is below 10°C (50°F). |

Notice that four of these major groups (A, C, D, and E) are defined on the basis of *temperature.* The fifth, the B group, has *precipitation* as its primary criterion. Each of the five groups is further subdivided by using the criteria and symbols presented in **Figure 15.2**.

| Letter Symbol 1st | 2nd | 3rd | |
|---|---|---|---|
| **A** | | | Average temperature of the coldest month is 18°C or higher. |
| | f | | Every month has 6 cm of precipitation or more. |
| | m | | Short dry season; precipitation in driest month less than 6 cm but equal to or greater than $10 - R/25$ ($R$ is annual rainfall in cm). |
| | w | | Well-defined winter dry season; precipitation in driest month less than $10 - R/25$. |
| | s | | Well-defined summer dry season (rare). |
| **B** | | | Potential evaporation exceeds precipitation. The dry–humid boundary is defined by the following formulas: (Note: $R$ is the average annual precipitation in cm, and T is the average annual temperature in °C.) When R is less than the calculated value the climate is dry.<br>$R < 2T + 28$ when 70% or more of rain falls in warmer 6 months.<br>$R < 2T$ when 70% or more of rain falls in cooler 6 months.<br>$R < 2T + 14$ when neither half year has 70% or more of rain. |
| | S | | Steppe — The BS–BW boundary is 1/2 the dry–humid boundary. |
| | W | | Desert — The BS–BW boundary is 1/2 the dry–humid boundary. |
| | | h | Average annual temperature is 18°C or greater. |
| | | k | Average annual temperature is less than 18°C. |
| **C** | | | Average temperature of the coldest month is under 18°C and above –3°C. |
| | w | | At least 10 times as much precipitation in a summer month as in the driest winter month. |
| | s | | At least three times as much precipitation in a winter month as in the driest summer month; precipitation in driest summer month less than 4 cm. |
| | f | | Criteria for w and s cannot be met. |
| | | a | Warmest month is over 22°C; at least 4 months over 10°C. |
| | | b | No month above 22°C; at least 4 months over 10°C. |
| | | c | One to 3 months above 10°C. |
| **D** | | | Average temperature of coldest month is –3°C or below; average temperature of warmest month is greater than 10°C. |
| | w | | Same as under C. |
| | s | | Same as under C. |
| | f | | Same as under C. |
| | | a | Same as under C. |
| | | b | Same as under C. |
| | | c | Same as under C. |
| | | d | Average temperature of the coldest month is –38°C or below. |
| **E** | | | Average temperature of the warmest month is below 10°C. |
| | T | | Average temperature of the warmest month is greater than 0°C and less than 10°C. |
| | F | | Average temperature of the warmest month is 0°C or below. |

**◀ Figure 15.2 The Köppen system of climate classification** This system uses easily obtained data: mean monthly and annual values for temperature and precipitation. When using this figure to classify climate data, first determine whether the data meet the criteria for the E climates. If the station is not a polar climate, proceed to the criteria for B climates. If the data do not fit into either the E or B groups, check the data against the criteria for A, C, and D climates, in that order.

A strength of the Köppen system is the relative ease with which boundaries are determined. However, these boundaries cannot be viewed as fixed. On the contrary, all climate boundaries shift their positions from one year to the next (Fig. 15.3). The boundaries shown on climate maps are simply average locations based on data collected over many years. Thus, a climate boundary should be regarded as a broad transition zone and not a sharp line.

The world distribution of climates, according to the Köppen classification, is shown in Figure 15.4. We will refer you to this map several times as we examine Earth's climates in the following pages.

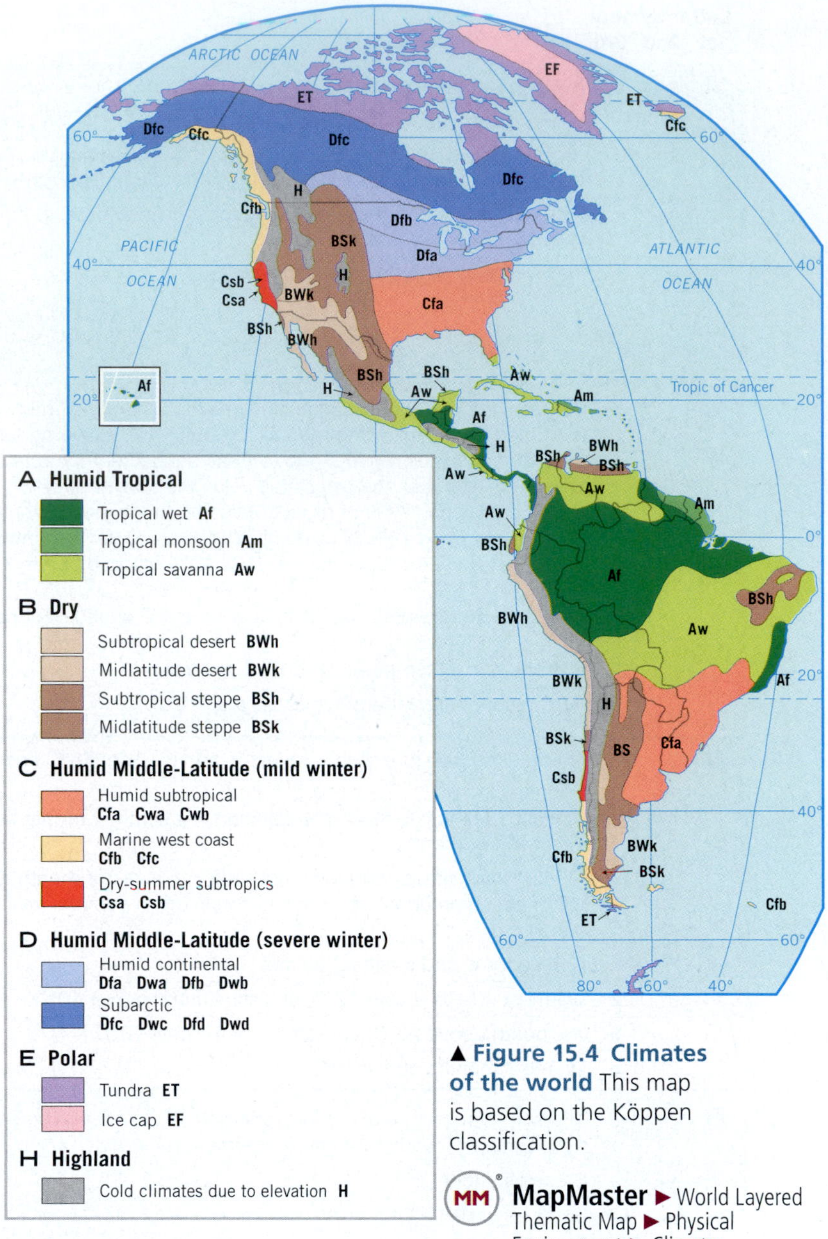


▲ **Figure 15.4 Climates of the world** This map is based on the Köppen classification.

**MapMaster** ▶ World Layered Thematic Map ▶ Physical Environment ▶ Climate

## World Climates—An Overview

After reviewing the controls of climate in Section 15.2, the remainder of this chapter is a tour of world climates. Each discussion will be accompanied by climate diagrams similar to the example in Figure 15.5.

Our tour is organized like this:

- Beginning along the equator, we visit the *wet tropics* (the A climates), studying their temperature and precipitation characteristics, which foster the great tropical rain forests.
- North and south of the wet tropics we look at the *tropical wet and dry* climates (still A climates), including the monsoon areas.
- North and south of the tropical wet and dry areas, we see the *dry* climates (the B climates). Dominating large parts of the subtropics and extending into the interiors of continents in the middle latitudes, deserts and steppes cover nearly one-third of Earth's land surface.
- Moving poleward from the dry subtropical realm, we visit regions exhibiting the *humid subtropical* climate (one of the C climates). These prevail on the eastern sides of continents between 25° and 40° latitude; the southeastern United States is an example.
- On the windward coasts of continents, we next visit *marine west coast* climates (also C climates), such as that of Western Europe.
- We complete our look at C climates with a visit to *dry-summer subtropical* or *Mediterranean* climates, such as those of Italy, Spain, and parts of California.
- In the Northern Hemisphere, where the continents extend into the middle and high latitudes, C climates give way to D climates called *humid continental*. These "breadbasket" areas are hospitable to growing grains and meat that feed much of the world.
- Verging on the polar climates are the *subarctic* climates. These vast zones of coniferous forest in Canada and Siberia are known for their long and bitter winters.
- Around the poles we visit the *polar* climates (the E climates). These are summerless areas of tundra and permafrost or permanent ice sheets.

▼ **Figure 15.3 Conditions change from year to year** Yearly fluctuations in the dry–humid boundary during a 5-year period. The small inset shows the average position of the dry–humid boundary.

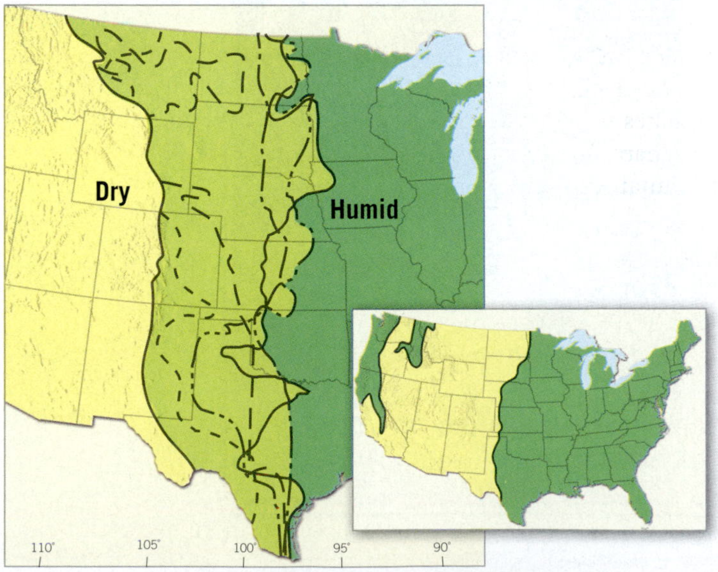

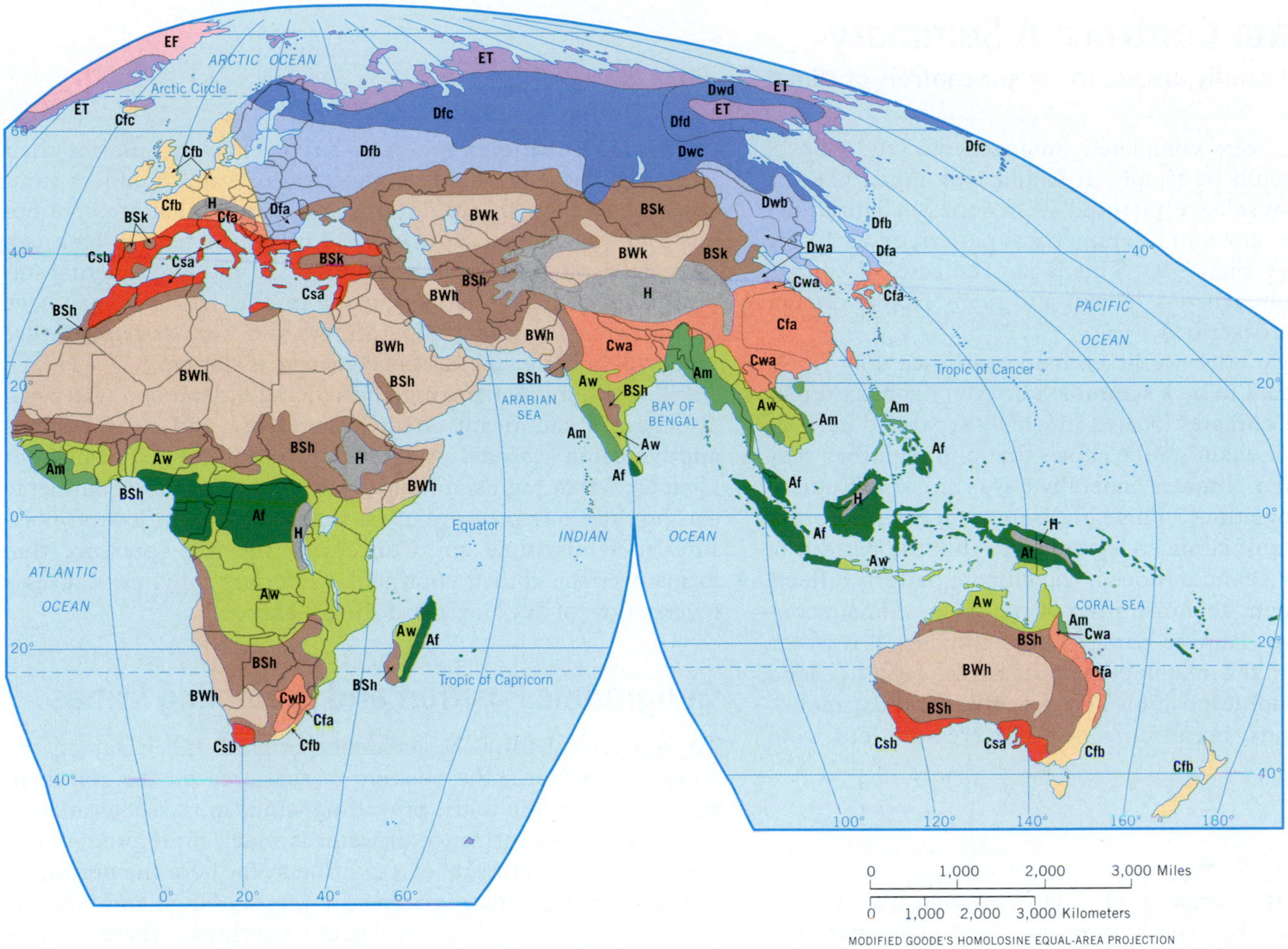


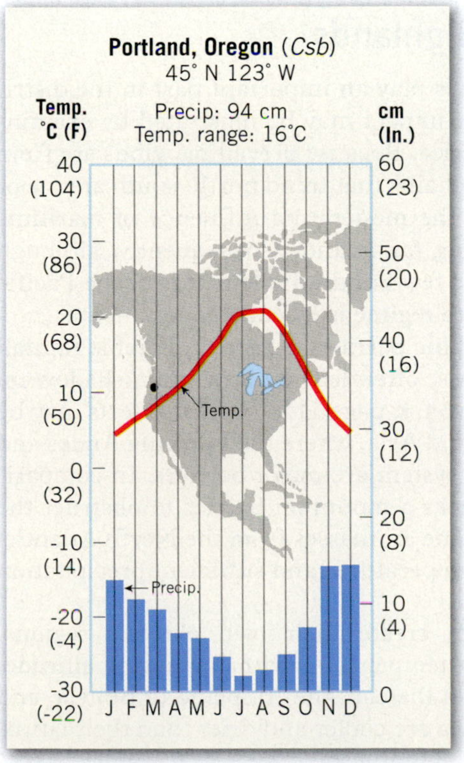

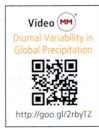

◄ **Figure 15.5 Climate diagram** This diagram for Portland, Oregon, shows important details at a glance. The plot of temperature immediately indicates whether the station is in the Northern Hemisphere or the Southern Hemisphere and whether it is near the equator. The combination line and bar graph provides a visual summary that allows one to readily recognize annual patterns of temperature and precipitation.

**Video** MM
Diurnal Variability in Global Precipitation
http://goo.gl/2rbyTZ

**Video** MM
The Ocean's Green Machines
http://goo.gl/z1MwpP

- Finally, we visit some cool places that are not necessarily near the poles but are chilled by their high elevation. The *highland* climates even occur on mountaintops near the equator and are characteristic of the Rockies, Andes, Himalayas, and other mountain regions.

## ✔ Concept Checks 15.1

1. Why is classification often a necessary task in science?
2. What climate data are needed to classify a climate using the Köppen scheme?
3. Should climate boundaries, such as those shown on the world map in Figure 15.4, be regarded as fixed? Explain.
4. Briefly describe the sequence of climates you would encounter going from equatorial Africa through central Europe and Scandinavia to the North Pole.

## 15.2 Climate Controls: A Summary

**List and briefly discuss the major controls of climate.**

If Earth's surface were completely homogeneous, the map of world climates would be simple. It would look much like the ancient Greeks must have pictured it—a series of latitudinal bands girdling the globe in a symmetrical pattern on each side of the equator (see Fig. 15.1). This is not the case, of course. Earth is not a homogeneous sphere, and many factors disrupt the symmetry just described.

At first glance, the world climate map (see Fig. 15.4) reveals what appears to be a scrambled or even haphazard pattern, with similar climates located in widely separated parts of the world. A closer examination shows that although they may be far apart, similar climates generally have similar latitudinal and continental positions. This consistency suggests an order in the distribution of climate elements and that the pattern of climates is not by chance. Indeed, the climate pattern reflects a regular and dependable operation of the major climate controls. So before we examine Earth's major climates, it is worth reviewing each of the major climate controls: latitude, land and water, geographic position and prevailing winds, mountains and highlands, ocean currents, and pressure and wind systems.

### Latitude

Fluctuations in the amount of solar radiation received at Earth's surface are the single greatest cause of temperature differences. Although variations in such factors as cloud coverage and the amount of dust in the air may be locally influential, recall from Chapter 2 that seasonal changes in Sun angle and length of daylight are the most important factors controlling the global temperature distribution. Because all places situated along the same line of latitude have identical Sun angles and hours of daylight, variations in the receipt of solar energy are largely a function of latitude. Moreover, because the vertical rays of the Sun migrate annually between the Tropic of Cancer and the Tropic of Capricorn, there is a regular latitudinal shifting of temperatures. Temperatures in the tropical realm are consistently high because the vertical rays of the Sun are never far away. As one moves farther poleward, however, greater seasonal fluctuations in the receipt of solar energy are reflected in larger annual temperature ranges.

### Land and Water

The distribution of land and water must be considered second only to latitude in importance as a control of temperature. Recall from Chapter 3 that water has a greater *heat capacity* than rock and soil. Thus, land heats more rapidly and to higher temperatures than water, and it cools more rapidly and to lower temperatures than water. Consequently, variations in air temperatures are much greater over land than over water. This differential heating of Earth's surface has led to climates being divided into two broad classes—marine and continental.

**Marine climates** are considered relatively mild for their latitude because the moderating effect of water produces summers that are warm but not hot and winters that are cool but not cold. In contrast, **continental climates** tend to be much more extreme. Although a marine station and a continental station along the same parallel in the middle latitudes may have similar annual mean temperatures, the annual temperature *range* will be far greater at the continental station.

The differential heating and cooling of land and water can also have a significant effect on pressure and wind systems and hence on seasonal precipitation distribution. Recall from Chapter 7 that high summer temperatures over the continents can produce low-pressure areas that allow the inflow of moisture-laden maritime air. Conversely, the high pressure that forms over the chilled continental interiors in winter causes a reverse flow of dry air toward the oceans.

### Geographic Position and Prevailing Winds

To understand fully the influence of land and water on the climate of an area, the position of that area on the continent and its relationship to the prevailing winds must be considered. The moderating influence of water is much more pronounced along the windward side of a continent, for here the prevailing winds may carry the maritime air masses far inland. On the other hand, places on the lee side of a continent, where the prevailing winds blow from the land toward the ocean, are likely to have a more continental temperature regime.

### Mountains and Highlands

Mountains and highlands play an important part in the distribution of climates. This impact may be illustrated by examining western North America. Because prevailing winds are from the west, the mountain chains that trend north–south are major barriers. They prevent the moderating influence of maritime air masses from reaching far inland. Consequently, although stations may lie within a few hundred kilometers of the Pacific Ocean, their temperature regime is essentially continental.

Also, these topographic barriers trigger orographic rainfall on their windward slopes, often leaving a dry rain shadow on the leeward side (see Box 4.4, page 114). Similar effects may be seen in South America and Asia, where the towering Andes and the massive Himalayan system are major barriers. In comparison, Western Europe lacks a mountain barrier to obstruct the free movement of maritime air masses from the North Atlantic. As a result, moderate temperatures and sufficient precipitation mark the entire region.

Extensive highlands create their own climatic regions. Because of the drop in temperature with increasing altitude, areas such as the Tibetan Plateau, the Altiplano of Bolivia, and the uplands of East Africa are cooler and drier than their latitudinal locations alone would indicate.

## Ocean Currents

The effect of ocean currents on the temperatures of adjacent land areas can be significant. Recall from Chapter 7 that poleward-moving currents, such as the Gulf Stream and Kurashio Currents in the Northern Hemisphere and the Brazil and East Australian Currents in the Southern Hemisphere, cause air temperatures to be warmer than would be expected for their latitudes. This influence is especially pronounced in winter. Conversely, the Canary and California Currents in the Northern Hemisphere and the Peru and Benguela Currents south of the equator reduce the temperatures of bordering coastal zones. In addition, the chilling effects of these cold currents act to stabilize the air masses moving across them. The result is marked aridity and often considerable advection fog.

## Pressure and Wind Systems

The world distribution of precipitation shows a close relationship to the distribution of Earth's major pressure and wind systems. Although the latitudinal distribution of these systems does not generally take the form of simple "belts," we can identify a zonal arrangement of precipitation from the equator to the poles (see Fig. 7.28).

In the realm of the equatorial low, the convergence of warm, moist, and unstable air makes this zone one of heavy rainfall. In the regions dominated by subtropical highs and subsidence, general aridity prevails, creating major deserts. Farther poleward, in the middle-latitude zone dominated by the irregular subpolar low, the influence of the many traveling cyclonic disturbances again increases precipitation. Finally, in polar regions, where temperatures are low and the air can hold only small quantities of moisture, precipitation totals decline.

The seasonal shifting of the pressure and wind belts, which follows the movement of the Sun's vertical rays, significantly affects areas in intermediate positions. Such regions are alternately influenced by two different pressure and wind systems. A station located poleward of the equatorial low yet equatorward of the subtropical high, for example, will experience a summer rainy period as the low pressure migrates poleward and a wintertime dry season as the high moves equatorward. This latitudinal shifting of pressure belts is largely responsible for the seasonality of precipitation in many regions.

### ✔ Concept Checks 15.2

1. List the major climate controls and briefly describe their influence.
2. Which control has the greatest influence on global temperature differences?
3. Contrast continental and marine climates.
4. What is the connection between pressure systems and the world distribution of precipitation?

# 15.3 Humid Tropical (A) Climates

**Compare the two broad categories of humid tropical climates.**

Within the A group of climates, two main types are recognized: wet tropical climates (Af and Am) and tropical wet and dry (Aw). Each climate type has an associated vegetation type.

## The Wet Tropics (Af, Am)

In the wet tropics, constant high temperatures and year-round rainfall combine to produce the most luxuriant vegetation in any climatic realm: the **tropical rain forest** (Fig. 15.6). Unlike the forests that we North Americans are accustomed to, the tropical rain forest is made up of broadleaf trees that remain green throughout the year. In addition, instead of being dominated by a few species, these forests are characterized by many. It is not uncommon for hundreds of different species to inhabit a single square kilometer of the forest. As a consequence, the individuals of a single species are often widely spaced.

Standing on the shaded floor of the forest looking upward, one sees tall, smooth-barked, vine-entangled trees, the trunks branchless in their lower two-thirds, with an almost continuous canopy of foliage above. A closer look reveals a three-level structure. Nearest the ground, perhaps 5 to 15 meters (16 to 50 feet) above, the narrow crowns of rather slender trees are visible. Rising above these relatively short forest species, a more continuous canopy of foliage occupies the height range of 20 to 30 meters (65 to 100 feet). Finally, visible through an occasional opening in the second level, a third level may be seen at the very top of the forest. Here the crowns of the trees tower 40 meters (130 feet) or more above the forest floor.

The environment of the wet tropics just described covers almost 10 percent of Earth's land area (Box 15.1). Figure 15.4 shows that Af and Am climates form a discontinuous belt astride the equator that typically extends 5° to 10° into each hemisphere. The poleward margins are most often marked by diminishing rainfall, but occasionally decreasing temperatures mark the boundary. Because of the general decrease in temperature with height in the troposphere, this climate region is restricted to elevations below 1000 meters (3300 feet). Consequently, the major interruptions near the equator are mostly cooler highland areas.

Also note in Figure 15.4 that the rainy tropics tend to have greater north–south extent along the eastern sides of continents (especially South America) and along some tropical coasts. The greater span on the eastern side of a continent is due primarily to its windward position on the weak western side of the subtropical high, a zone dominated by neutral or unstable air. In other cases, as along the eastern side of Central America, the coast, backed by interior highlands, intercepts the flow of trade winds. Orographic uplift thus greatly enhances the rainfall total.

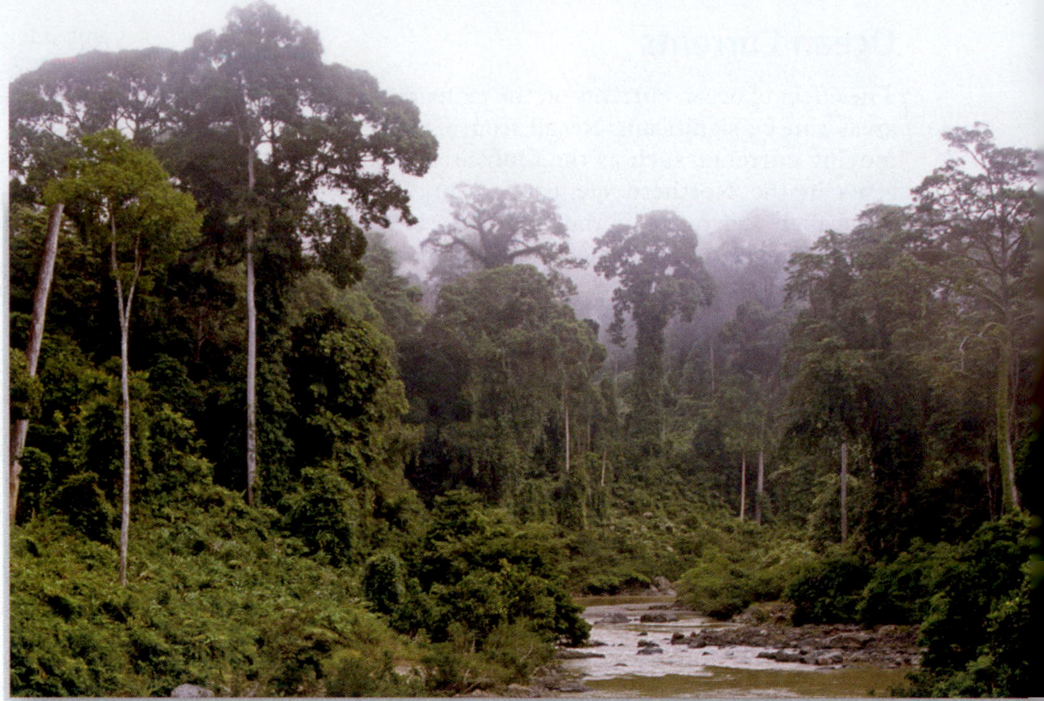

▶ **SmartFigure 15.6 Tropical rain forest** Unexcelled in luxuriance and characterized by hundreds of different species per square kilometer, the tropical rain forest is a broadleaf evergreen forest that dominates the wet tropics. This image shows Borneo's Segama River passing through virgin tropical rain forest.

http://goo.gl/oAGTTr

Data for some representative stations in the wet tropics are shown in **Table 15.1** and **Fig. 15.7A** and **B**. A look at the numbers reveals the most obvious features characterizing the climate in these areas:

1. Temperatures usually average 25°C (77°F) or higher each month. Not only is the annual mean high, but the annual range is very small. (Note the flat temperature curves in the graphs in Fig. 15.7A and B.)
2. Total precipitation for the year is high, often exceeding 200 centimeters (80 inches).
3. Although rainfall is not evenly distributed throughout the year, tropical rain forest stations are generally wet in all months. If a dry season exists, it is very short.

**Temperature** Because places with an Af or Am designation lie near the equator, the reason for their uniform temperature rhythm is clear: Insolation is consistently intense. The Sun's rays are always nearly vertical, and changes in day length are slight throughout the year. Therefore, seasonal temperature variations are minimal. The small difference that exists between the warmest and coolest months often reflects changes in cloud cover rather than in the position of the Sun. In the case of Belém, Brazil, for example, you can see that the highest temperatures occur during the months when rainfall (and hence cloud cover) is lowest.

A striking characteristic of the wet tropics is that *daily* temperature variations greatly exceed *seasonal* differences. Whereas annual temperature ranges in the wet tropics rarely exceed 3°C (6°F), daily temperature ranges are between two and five times greater. Thus, there is a greater variation between day and night than there is seasonally. It is interesting that monthly and daily mean temperatures in the tropics are no greater than those in many U.S. cities during the summer. For example, the highest temperature recorded at Jakarta, Indonesia, over a 78-year period and at Belém, Brazil, over a 20-year period has been only 36.6°C (98°F), compared with extremes of 40.5°C (105°F) at Chicago and 41.1°C (106°F) at New York City.

What makes the wet tropical temperature regime unique is its day-in and day-out, month-in and month-out regularity. Although the thermometer may not indicate abnormal or extreme conditions, the warm temperatures combined with the high humidity and meager winds make apparent temperatures particularly high. The reputation of the wet tropics as being oppressive and monotonous is mostly well deserved.

**Precipitation** The regions dominated by Af and Am climate normally receive from 175 to 250 centimeters (68 to 98 inches)

**Table 15.1** | Data for Wet Tropical Stations

| | J | F | M | A | M | J | J | A | S | O | N | D | YR |
|---|---|---|---|---|---|---|---|---|---|---|---|---|---|
| **Singapore, 1° 21 N; 10 m** | | | | | | | | | | | | | |
| Temp. (°C) | 26.1 | 26.7 | 27.2 | 27.6 | 27.8 | 28.0 | 27.4 | 27.3 | 27.3 | 27.2 | 26.7 | 26.3 | 27.1 |
| Precip. (mm) | 285 | 164 | 154 | 160 | 131 | 177 | 163 | 200 | 122 | 184 | 236 | 306 | 2282 |
| **Belém, Brazil, 1° 18 S; 10 m** | | | | | | | | | | | | | |
| Temp. (°C) | 25.2 | 25.0 | 25.1 | 25.5 | 25.7 | 25.7 | 25.7 | 25.9 | 25.7 | 26.1 | 26.3 | 25.9 | 25.7 |
| Precip. (mm) | 340 | 406 | 437 | 343 | 287 | 175 | 145 | 127 | 119 | 91 | 86 | 175 | 2731 |
| **Douala, Cameroon, 4° N; 13 m** | | | | | | | | | | | | | |
| Temp. (°C) | 27.1 | 27.4 | 27.4 | 27.3 | 26.9 | 26.1 | 24.8 | 24.7 | 25.4 | 25.9 | 26.5 | 27.0 | 26.4 |
| Precip. (mm) | 61 | 88 | 226 | 240 | 353 | 472 | 710 | 726 | 628 | 399 | 146 | 60 | 4109 |

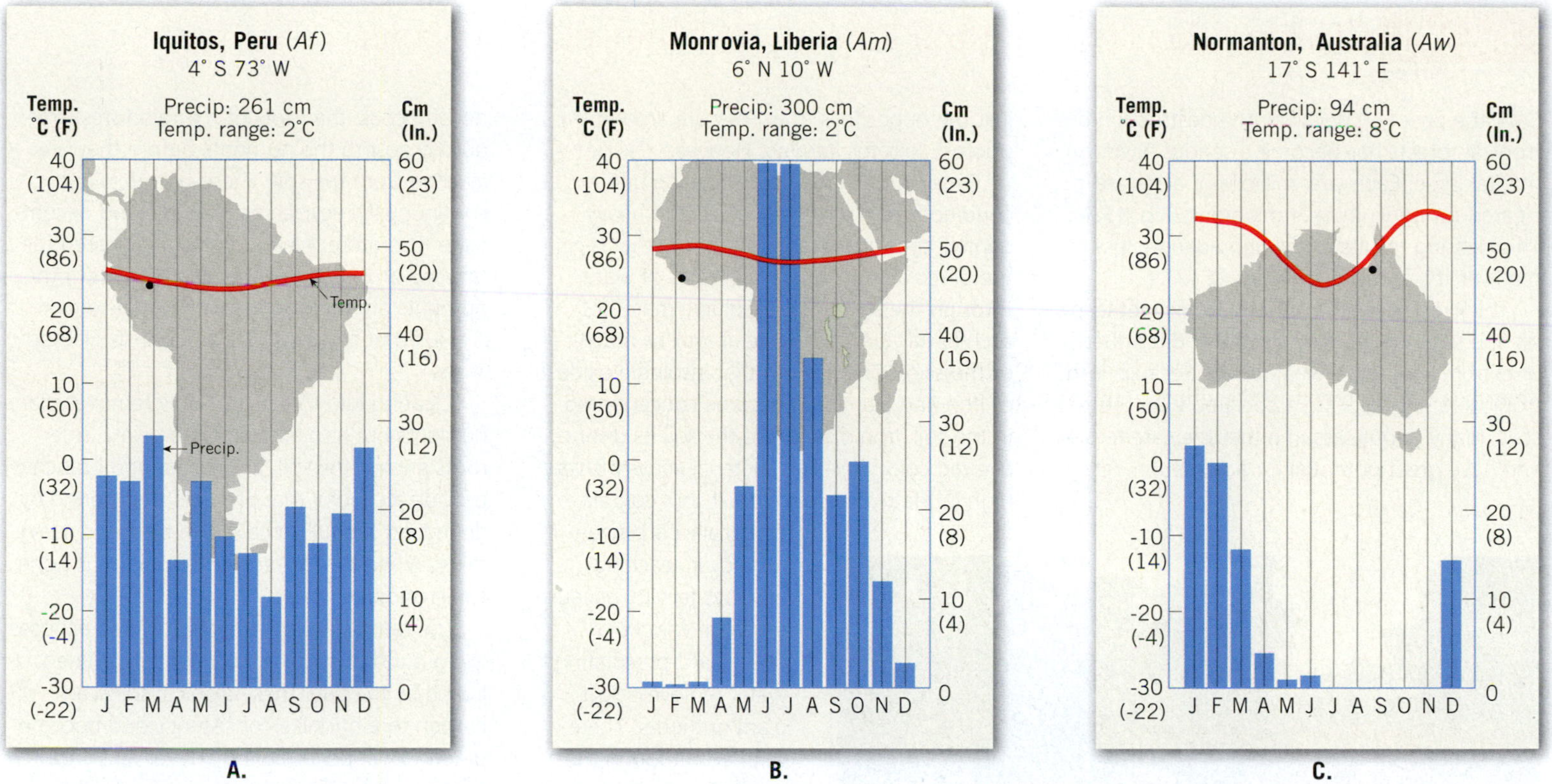


▲ **Figure 15.7 Humid tropical climates** By comparing these three climatic diagrams, the primary differences among the a climates can be seen. **A.** Iquitos, the Af station, is wet throughout the year. **B.** Monrovia, the Am station, has a short, dry season. **C.** As is true for all Aw stations, Normanton has an extended dry season and a higher annual temperature range than the others.

of rain each year. But a glance at the data in Table 15.1 reveals more variability in rainfall than in temperature, both seasonally and from place to place. The rainy nature of the equatorial realm is partly related to the extensive heating of the region and the consequent thermal convection. In addition, this is the zone of the converging trade winds, often referred to as the **intertropical convergence zone,** or simply the **ITCZ.** The thermally induced convection, coupled with convergence, leads to widespread ascent of the warm, humid, unstable air. Conditions near the equator are thus ideal for precipitation.

Rain typically falls on more than half of the days each year. In fact, at some stations, three-quarters of the days experience rain. There is a marked daily regularity to the rainfall at many places. Cumulus clouds begin forming in late morning or early afternoon. The buildup continues until about 3 or 4 P.M., the time when temperatures are highest and thermal convection is at a maximum; then the cumulonimbus towers yield showers. Figure 15.8, showing the hourly distribution of rainfall at Kuala Lumpur, Malaysia, exemplifies this pattern.

Video
Lightning Seasonality
http://goo.gl/Qq8sfA

▼ **Figure 15.8 The distribution of rainfall by time of day at Kuala Lumpur, Malaysia** With its midafternoon maximum, Kuala Lumpur illustrates the typical pattern at many wet tropical stations.

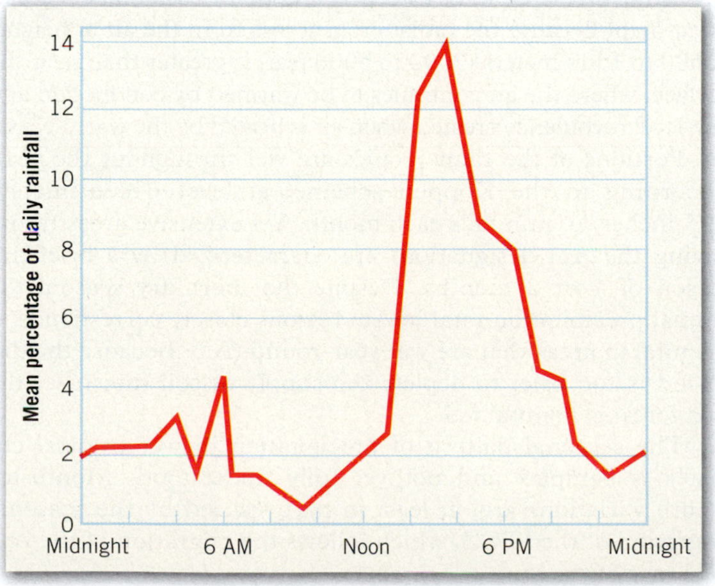


**students sometimes ask...**

### *Isn't* jungle *just another word for* tropical rain forest?

Although both terms refer to vegetation in the wet tropics, they are not the same. The tropical rain forest's high canopy of foliage does not allow much light to penetrate to the ground. As a result, plant foliage is relatively sparse on the dimly lit forest floor. By contrast, anywhere considerable light makes its way to the ground, as along riverbanks or in human-made clearings, an almost impenetrable growth of tangled vines, shrubs, and short trees exists. The familiar term *jungle* is used to describe such areas.

## Box 15.1 Clearing the Tropical Rain Forest—The Impact on Its Soils

Over the past few decades, the destruction of tropical forests has become a serious environmental issue. Each year millions of acres are cleared for agriculture and logging (**Fig. 15.A**). This clearing results in soil degradation, loss of biodiversity, and climate change.

Thick red soils are common in the wet tropics and subtropics. They are the end product of extreme chemical weathering. Because lush tropical rain forests are associated with these soils, many people assume that they are fertile and have great potential for agriculture. Yet just the opposite is true: They are among the poorest soils for farming. How can this be?

Because rain forest soils develop under conditions of high temperature and heavy rainfall, they are severely leached—that is, the percolation of large quantities of water through the soil removes soluble materials such as calcium carbonate as well as much of the silica. The result is that insoluble oxides of iron and aluminum become concentrated in the soil. Iron oxides give the soil its distinctive red color. Because bacterial activity is very high in the tropics, rain forest soils contain practically no humus. Moreover, leaching destroys fertility because the large volume of downward-percolating water removes most plant nutrients. Therefore, even though the vegetation is dense and luxuriant, the soil itself contains few available nutrients.

Most nutrients that support the rain forest are locked up in the trees themselves. As vegetation dies and decomposes, the roots of the rain forest trees quickly absorb the nutrients before they are leached from the soil. These nutrients are continuously recycled as trees die and decompose. Therefore, when forests are cleared for farming or to harvest the timber, most of the nutrients are removed as well. What remains is a soil that contains little to nourish planted crops.

Clearing rain forests not only removes plant nutrients but also accelerates erosion. Tree roots anchor the soil, and leaves and branches provide a canopy that protects the ground by deflecting the full force of the frequent heavy rains. When the protective vegetation is gone, soil erosion increases.

The removal of the vegetation also exposes the ground to strong direct sunlight. When baked by the Sun, these tropical soils can harden to a bricklike consistency and become practically impenetrable by water and crop roots. In only a few years, soils in a freshly cleared area may no longer be cultivable.

▲ **Figure 15.A Tropical deforestation** Clearing the tropical rain forest in Surinam. The thick orange soils are highly leached.

**Questions**

1. Why are tropical rain forest soils red in color?
2. Where are most plant nutrients found in a tropical rain forest?

The cycle is different at many marine stations, with the rainfall maximum occurring at night instead of in the afternoon. The environmental lapse rate is steepest and hence instability is greatest at night because the radiation heat loss from the air at heights of 600 to 1500 meters (2000 to 5000 feet) is greater than near the surface, where the air continues to be warmed by conduction and low-level turbulence created when air is heated by the warm water.

Portions of the rainy tropics are wet throughout the year. According to the Köppen scheme, at least 6 centimeters (2.3 inches) of rain falls each month. Yet extensive areas (those having the Am designation) are characterized by a brief dry season of 1 or 2 months. Despite the short dry season, the annual precipitation total in Am regions closely corresponds to the total in areas that are wet year-round (Af). Because the dry period is too brief to deplete the supply of soil moisture, the rain forest is maintained.

The seasonal pattern of precipitation in wet tropical climates is complex and not yet fully understood. Month-to-month variations are, at least in part, caused by the seasonal migration of the ITCZ, which follows the migration of the vertical rays of the Sun.

MM MapMaster ▶ World Layered Thematic Map ▶ Physical Environment ▶ Climate ▶ Tropical Wet and Dry

## Tropical Wet and Dry (Aw)

Between the rainy tropics and the subtropical deserts lies a transitional climatic region referred to as **tropical wet and dry.** Along its equatorward margin, the dry season is short, and the boundary between Aw and the rainy tropics is difficult to define. Along the poleward side, however, the dry season is prolonged, and conditions merge into those of the semiarid realm.

In the tropical wet and dry climate, the rain forest gives way to the **savanna,** a tropical grassland with scattered deciduous trees (**Fig. 15.9**). In fact, Aw is often called the *savanna climate.* This name may not be appropriate technically because some ecologists doubt that these grasslands are climatically induced. It is believed that woodlands once dominated this zone and that the savanna grasslands developed in response to seasonal burnings by native populations.

### Temperature

The temperature data in **Table 15.2** show only modest differences between the wet tropics and the tropical wet and dry regions. Because of the somewhat higher latitude of most Aw stations, annual mean temperatures are slightly lower.

◄ **Figure 15.9 Tropical savanna grassland** This tropical savanna in Tanzania's Serengeti National Park, with its stunted, drought-resistant trees, was probably strongly influenced by seasonal burnings carried out by native human populations.

In addition, the annual temperature range is a bit greater, varying from 3°C (5°F) to perhaps 10°C (20°F). The daily temperature range, however, still exceeds the annual variation.

Because seasonal fluctuations in humidity and cloudiness are more pronounced in Aw areas, daily temperature ranges vary noticeably during the year. Generally, they are small during the rainy season, when humidity and cloud cover are at a maximum, and large during dry periods, when clear skies and dry air prevail. Furthermore, because of the more persistent summertime cloudiness, many Aw stations experience their warmest temperatures at the end of the dry season, just prior to the summer solstice. Thus, in tropical wet and dry regions of the Northern Hemisphere, March, April, and May are often warmer than June and July.

**Precipitation** Because temperature regimes among the A climates are similar, the primary factor distinguishing the Aw climate from Af and Am is precipitation. Aw stations typically receive 100 to 150 centimeters (40 to 60 inches) of rainfall each year, an amount often appreciably less than in the wet tropics. The most distinctive characteristic of this climate, however, is the *markedly seasonal character of the rainfall*—wet summers followed by dry winters. This is clearly shown by the climate diagram in Figure 15.7C.

The alternating wet and dry periods are due to the latitudinal position of the Aw climate region. It lies between the intertropical convergence zone, with its sultry weather and convective thundershowers, and the stable, subsiding air of the subtropical highs. Following the spring equinox, the ITCZ and the other wind and pressure belts all shift poleward as they migrate with the vertical rays of the Sun (**Fig. 15.10**). With the advance of the ITCZ into a region, the summer rainy season commences, with weather patterns typical of the wet tropics. Later, with the retreat of the ITCZ back toward the equator, the subtropical high advances into the region and brings with it intense aridity.

During the dry season, the landscape grows parched, and nature seems to go dormant as water-stressed trees shed their leaves and the abundant tall grasses turn brown and wither. The dry season's duration depends primarily on distance from

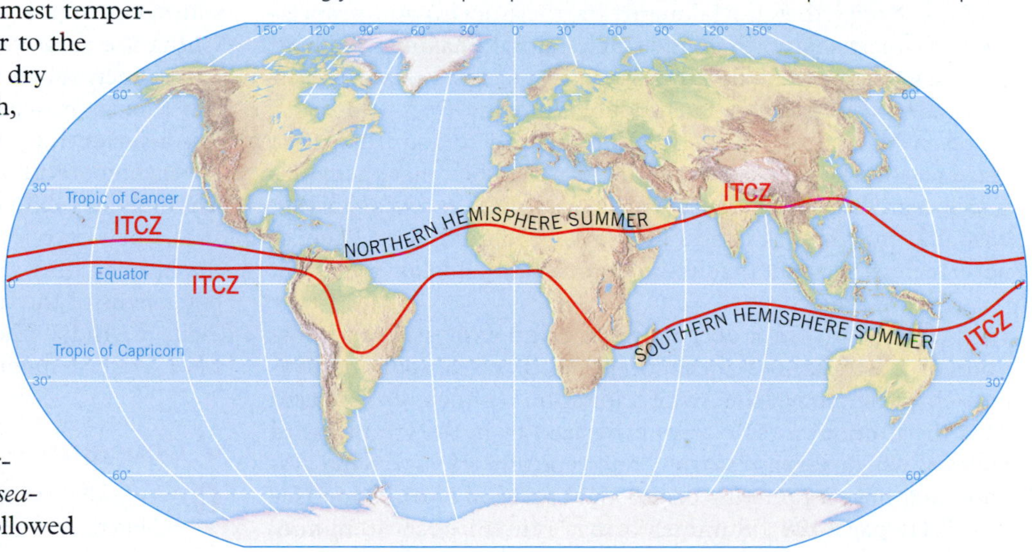


▼ **Figure 15.10 Migrating ITCZ** The seasonal migration of the ITCZ strongly influences precipitation distribution in the tropics and subtropics.

**Table 15.2** | Data for Tropical Wet and Dry Stations

| | J | F | M | A | M | J | J | A | S | O | N | D | YR |
|---|---|---|---|---|---|---|---|---|---|---|---|---|---|
| **Calcutta, India, 22° 32 N; 6 m** | | | | | | | | | | | | | |
| Temp. (°C) | 20.2 | 23.0 | 27.9 | 30.1 | 31.1 | 30.4 | 29.1 | 29.1 | 29.9 | 27.9 | 24.0 | 20.6 | 26.94 |
| Precip. (mm) | 13 | 24 | 27 | 43 | 121 | 259 | 301 | 306 | 290 | 160 | 35 | 3 | 1582 |
| **Cuiaba, Brazil, 15° 30 S; 165 m** | | | | | | | | | | | | | |
| Temp. (°C) | 27.2 | 27.2 | 27.2 | 26.6 | 25.5 | 23.8 | 24.4 | 25.5 | 27.7 | 27.7 | 27.7 | 27.2 | 26.5 |
| Precip. (mm) | 216 | 198 | 232 | 116 | 52 | 13 | 9 | 12 | 37 | 130 | 165 | 195 | 1375 |

**Table 15.3** | Rainfall Regimes and the Movement of the ITCZ in Africa

| | J | F | M | A | M | J | J | A | S | O | N | D |
|---|---|---|---|---|---|---|---|---|---|---|---|---|
| **Malduguri, Nigeria, 11° 51 N** | | | | | | | | | | | | |
| Precip. (mm) | 0 | 0 | 0 | 7.6 | 40.6 | 68.6 | 180.3 | **220.9** | 106.6 | 17.7 | 0 | 0 |
| **Yaounde, Cameroon, 3° 53 N** | | | | | | | | | | | | |
| Precip. (mm) | 22.8 | 66.0 | 147.3 | 170.1 | **195.6** | 152.4 | 73.7 | 78.7 | 213.4 | **294.6** | 116.8 | 22.9 |
| **Kisangani, Dem. Rep. of the Congo, 0° 26 N** | | | | | | | | | | | | |
| Precip. (mm) | 53.3 | 83.8 | **177.8** | 157.5 | 137.2 | 114.3 | 132.0 | 165.1 | 182.9 | **218.4** | 198.1 | 83.8 |
| **Kananga, Dem. Rep. of the Congo, 5° 54 S** | | | | | | | | | | | | |
| Precip. (mm) | 137.2 | 142.2 | **195.6** | 193.0 | 83.8 | 20.3 | 12.7 | 58.4 | 116.8 | 165.1 | **231.1** | 226.0 |
| **Zomba, Malawi, 15° 23 S** | | | | | | | | | | | | |
| Precip. (mm) | 274.3 | **289.6** | 198.1 | 76.2 | 27.9 | 12.7 | 5.1 | 7.6 | 17.8 | 17.8 | 134.6 | **279.4** |
| **Francistown, Botswana, 21° 13 S** | | | | | | | | | | | | |
| Precip. (mm) | **106.7** | 78.7 | 71.1 | 17.8 | 5.1 | 2.5 | 0 | 0 | 0 | 22.9 | 58.4 | 86.4 |

the ITCZ. Typically, the farther an Aw station is from the equator, the shorter the period of ITCZ dominance and the longer the locale will be influenced by the stable subtropical high. Consequently, the higher the latitude, the longer the dry season and the shorter the wet period.

Understanding the movement of the ITCZ is essential to understanding rainfall distribution in the tropics. This is clear from Table 15.3, which shows precipitation data for six African stations. Notice that at Malduguri (farthest north) and Francistown (farthest south), there are single rainfall maxima (red bold print) that occur when the ITCZ reaches its most poleward positions. Between these stations, double maxima (red bold print) represent the passage of the ITCZ on its way to and from these extreme locations. It is important to remember that these statistics represent long-term averages, and that on a year-to-year basis the movement of the ITCZ is far from regular. There is nevertheless no doubt that in the tropics, rainfall follows the Sun.

**The Monsoon** In much of India, Southeast Asia, and portions of Australia, the alternating periods of rainfall and dryness characteristic of the Aw precipitation regime are associated with the **monsoon.** The term is derived from the Arabic word *mausim*, which means "season" and typically refers to wind systems that have a pronounced seasonal reversal of direction (see Fig. 7.11, page 198). Summer brings rainfall because humid, unstable air moves from the oceans toward the land. In winter this reverses, and a dry wind, which originates over the continent, blows toward the sea.

This monsoonal circulation system develops partly in response to the differences in annual temperature variations between continents and oceans. The processes associated with the monsoon are similar to those described in connection with the land and sea breeze (Chapter 7) except that the scales, in both time and space, are much larger.

During spring in the Northern Hemisphere, an irregular area of thermally induced low pressure gradually develops over the interior of southern Asia. It is further strengthened by the poleward advance of the ITCZ (Fig. 15.10). Thus, the *summertime* circulation is from the higher pressure over the ocean toward the lower pressure over the continent. As winter approaches, winds reverse direction as the ITCZ migrates southward, and a deep anticyclone develops over the chilled continent. By *midwinter* dry winds blow from the continent southward to converge on Australia and southern Africa.

**The Cw Variant** Adjacent to the wet and dry tropics in southern Africa, South America, northeastern India, and China are areas that are sometimes designated Cw. Although C has been substituted for A, indicating that these regions are subtropical instead of tropical, the Cw climate is nevertheless a variant of Aw, for the only major difference is somewhat lower temperatures. In Africa and South America, Cw climates are highland extensions of Aw. Because they occupy elevated sites, they have lower temperatures than the adjacent wet and dry tropics. In India and China, Cw areas are middle-latitude extensions of the tropical monsoon realm. In some cases, especially in India, the Cw areas are barely poleward enough to have winter temperatures below those of the A climates.

## ✔ Concept Checks 15.3

1. How does a tropical rain forest differ from a typical middle-latitude forest?
2. Explain each of the following characteristics of the wet tropics:
   - **a.** The climate is restricted to elevations below 1000 meters.
   - **b.** Mean monthly and annual temperatures are high, and the annual temperature range is low.
   - **c.** The climate is rainy throughout the year, or nearly so.
3. What primary factor distinguishes Aw climates from Af and Am climates?
4. What is another name for the tropical wet and dry (Aw) climate?
5. Describe the influence of the ITCZ and the subtropical highs on the annual distribution of precipitation in the Aw climate.

# 15.4 The Dry Climates (B)

**Contrast low-latitude dry climates and middle-latitude dry climates.**

The dry regions of the world cover some 42 million square kilometers (nearly 16.5 million square miles), or about 30 percent, of Earth's land surface. No other climatic group covers such a large land area (**Fig. 15.11**).

The characteristic features of dry climates are their meager yearly rainfall and the unreliability of their precipitation. Generally, the smaller the mean annual rainfall, the greater its variability. As a result, yearly rainfall averages are often misleading.

For example, during one 7-year period, Trujillo, Peru, had an average rainfall of 6.1 centimeters per year (2.4 inches). A closer look reveals that during the first 6 years and 11 months of the period, the station received a scant 3.5 centimeters (1.4 inch) (an annual average of slightly more than 0.5 centimeter). Then during the 12th month of the 7th year, 39 centimeters (15.2 inches) of rain fell, 23 centimeters (9 inches) of it during a 3-day span.

This extreme case illustrates the irregularity of rainfall in most dry regions. Also, there are usually more years when rainfall totals are below the average than years when they are above. As the foregoing example shows, the occasional wet period tends to lift the average. Many of these regions feature drought-tolerant plants adapted to the uncertain precipitation.

## What Is Meant by "Dry"?

It is important to realize that dryness is relative and simply refers to any *water deficiency*. Thus, climatologists define a **dry climate** as one in which the *yearly precipitation is less than the potential water loss by evaporation*. Dryness, then, is not only related to annual rainfall. It is also a function of evaporation, which in turn depends closely on temperature. As temperatures climb, potential evaporation also increases.

For example, 25 centimeters (10 inches) of precipitation may be sufficient to support forests in northern Scandinavia, where evaporation is slight into the cool, humid air and a surplus of water remains in the soil. However, the same amount of rain falling on Nevada or Iran supports only a sparse vegetative cover because evaporation into the hot, dry air is great. Clearly, no specific amount of precipitation can define a dry climate.

To establish a meaningful boundary between dry and humid climates, the Köppen classification uses formulas that involve three variables: (1) average annual precipitation, (2) average annual temperature, and (3) seasonal distribution of precipitation.

The use of average annual precipitation is obvious. Average annual temperature is used because it is an index of evaporation; the amount of rainfall defining the humid–dry boundary is greater where mean annual temperatures are high and less where temperatures are low. The third variable, seasonal distribution of precipitation, is also related to this idea. If rain is concentrated in the warmest months, loss to evaporation is greater than if it is concentrated in the cooler months. Thus, considerable differences exist in the precipitation amounts received at various stations in the B climates.

**Table 15.4** summarizes these differences. For example, if a station with an annual mean of 20°C (68°F) and a summer wet season ("winter dry") receives less than 680 millimeters (26.5 inches) of precipitation per year, it is classified as *dry*. If the rain falls primarily in winter ("summer dry"), however, the station must receive only 400 millimeters (15.6 inches) or more to be

▶ **SmartFigure 15.11**
**Dry climates**
Arid and semiarid climates cover about 30 percent of Earth's land surface. The dry region of the American West is commonly divided into four deserts, two of which extend into Mexico.

http://goo.gl/4K8rJ

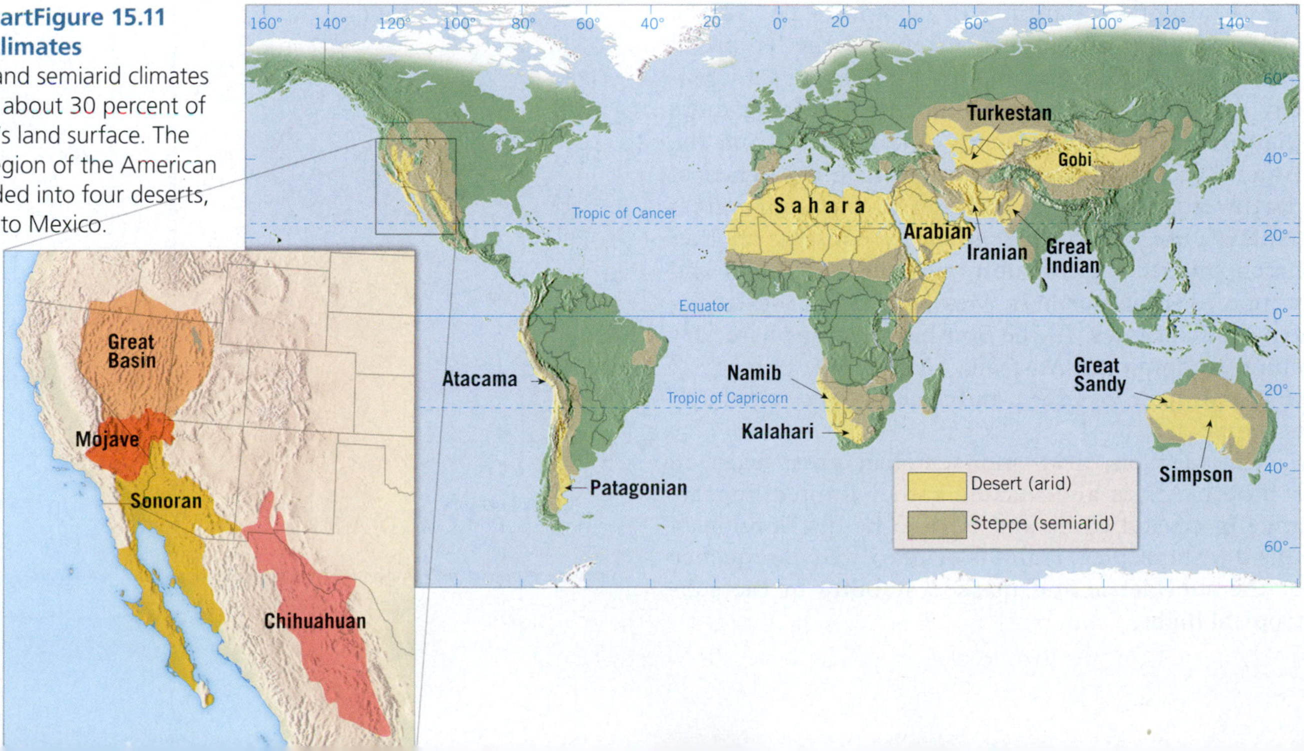


Video
Studying Fires Using Multiple Satellite Sensors
http://goo.gl/qKVJa

**Table 15.4** Average Annual Precipitation at the Dry-Humid Boundary

| Average Annual Precipitation (mm) | | | |
|---|---|---|---|
| **Average Annual Temp. (°C)** | **Summer Dry Season** | **Even Distribution** | **Winter Dry Season** |
| 5 | 100 | 240 | 380 |
| 10 | 200 | 340 | 480 |
| 15 | 300 | 440 | 580 |
| 20 | 400 | 540 | 680 |
| 25 | 500 | 640 | 780 |
| 30 | 600 | 740 | 880 |

considered humid. If the precipitation is more evenly distributed, the figure defining the humid–dry boundary is between the other two.

Within the regions defined by a general water deficiency, there are two climatic types: **arid,** or **desert** (BW), and **semiarid,** or **steppe** (BS). Figure 15.12 presents climate diagrams for both types. Stations A and B are in the subtropics, and C and D are in the middle latitudes. Deserts and steppes have many features in common; their differences are primarily a matter of degree. The semiarid is a marginal and more humid variant of the arid and represents a transition zone that surrounds the desert and separates it from the bordering humid climates. The arid–semiarid boundary is commonly set at one-half the annual precipitation separating dry regions from humid. Thus, if the humid–dry boundary happens to be 40 centimeters, the steppe–desert boundary will be 20 centimeters.

## Subtropical Desert (BWh) and Steppe (BSh)

The heart of low-latitude dry climates lies in the vicinity of the Tropic of Cancer and the Tropic of Capricorn. A glance at Figure 15.11 reveals a virtually unbroken desert environment stretching more than 9300 kilometers (nearly 6000 miles) from the Atlantic coast of North Africa to the dry lands of northwestern India. In addition to this single great expanse, the Northern Hemisphere contains another much smaller area of subtropical desert (BWh) and steppe (BSh) in northern Mexico and the southwestern United States. In the Southern Hemisphere, dry climates dominate Australia. Almost 40 percent of that continent is desert, and much of the remainder is steppe.

In addition, arid and semiarid areas exist in southern Africa and make a rather limited appearance in coastal Chile and Peru. The distribution of this dry subtropical realm is primarily a consequence of the subsidence and marked stability of the subtropical highs.

**students sometimes ask...**

**I heard somewhere that deserts are expanding. Is that actually occurring?**

Yes. The process is called *desertification*, a term that refers to the alteration of land to desertlike conditions as a result of human activities. It commonly takes place on the margins of deserts and is primarily triggered by plowing or overgrazing that removes the sparse natural vegetation in marginal areas. When drought occurs, as it inevitably does in these transition zones, and the vegetative cover falls below the minimum required to hold the soil against erosion, the destruction becomes irreversible. Desertification is occurring in many places but is particularly serious in the region south of the Sahara Desert known as the Sahel.

**Precipitation** Within subtropical deserts, the scant precipitation is both infrequent and erratic. Indeed, no well-defined seasonal precipitation pattern exists in subtropical deserts. The reason is that these areas are too far poleward to be influenced by the ITCZ and too far equatorward to benefit from the frontal and cyclonic precipitation of the middle latitudes. Even during summer, when daytime heating produces a steep environmental lapse rate and considerable convection, clear skies still rule. In this case, subsidence aloft prevents the lower air, with

**▼ Figure 15.12 Climate diagrams for representative arid and semiarid stations** Cairo and Lovelock are classified as deserts; Monterey and Semipalatinsk are steppes. Stations **A** and **B** are in the subtropics, whereas **C** and **D** are middle-latitude sites.

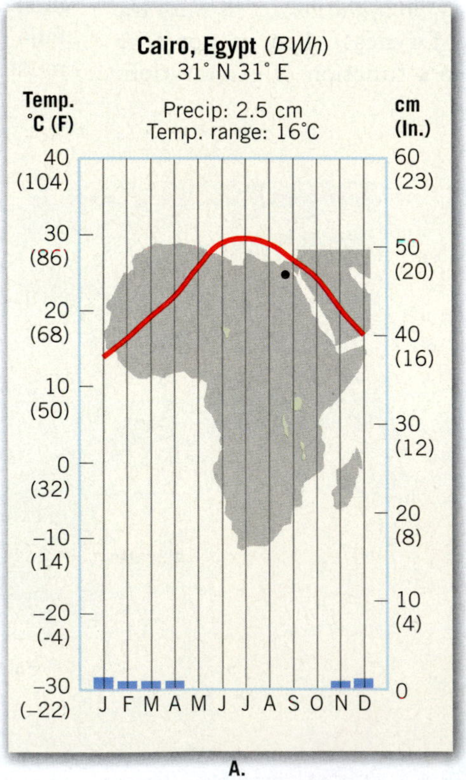

A.

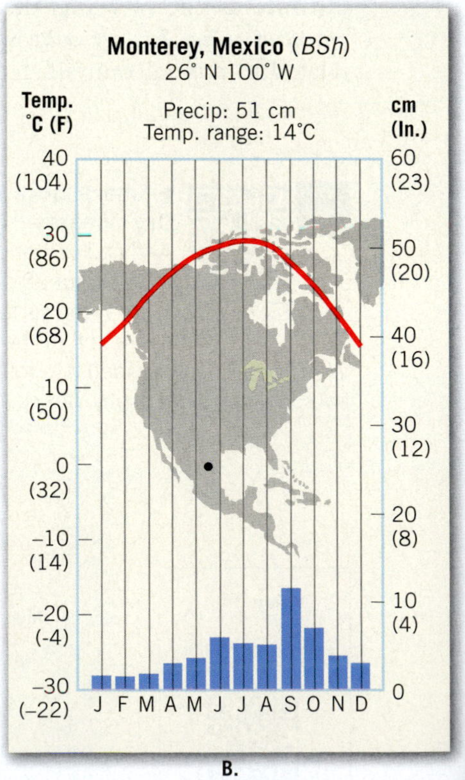

B.

**Table 15.5** | Data for Subtropical Steppe and Desert Stations

| | J | F | M | A | M | J | J | A | S | O | N | D | YR |
|---|---|---|---|---|---|---|---|---|---|---|---|---|---|
| **Marrakech, Morocco, 31° 37 N; 458 m** | | | | | | | | | | | | | |
| Temp. (°C) | 11.5 | 13.4 | 16.1 | 18.6 | 21.3 | 24.8 | 28.7 | 28.7 | 25.4 | 21.2 | 16.5 | 12.5 | 19.9 |
| Precip. (mm) | 28 | 28 | 33 | 30 | 18 | 8 | 3 | 3 | 10 | 20 | 28 | 33 | 242 |
| **Dakar, Senegal, 14° 44 N; 23 m** | | | | | | | | | | | | | |
| Temp. (°C) | 21.1 | 20.4 | 20.9 | 21.7 | 23.0 | 26.0 | 27.3 | 27.3 | 27.5 | 27.5 | 26.0 | 25.2 | 24.49 |
| Precip. (mm) | 0 | 2 | 0 | 0 | 1 | 15 | 88 | 249 | 163 | 49 | 5 | 6 | 578 |
| **Alice Springs, Australia, 23° 38 S; 570 m** | | | | | | | | | | | | | |
| Temp. (°C) | 28.6 | 27.8 | 24.7 | 19.7 | 15.3 | 12.2 | 11.7 | 14.4 | 18.3 | 22.8 | 25.8 | 27.8 | 20.8 |
| Precip. (mm) | 43 | 33 | 28 | 10 | 15 | 13 | 8 | 8 | 8 | 18 | 30 | 38 | 252 |

its modest moisture content, from rising high enough to penetrate the condensation level.

The situation is different in the semiarid transitional belts surrounding the desert. Here a seasonal rainfall pattern becomes better defined and supports more vegetation. As shown by the data for Dakar in **Table 15.5**, stations located on the equatorward side of low-latitude deserts have a brief period of relatively heavy rainfall during the summer, when the ITCZ is farthest poleward. The rainfall regime should look familiar, for it is similar to that found in the adjacent wet and dry tropics, except that the amount is less and the dry period lasts longer.

For steppe areas on the poleward margins of the subtropical deserts, the precipitation regime is reversed. As the data for Marrakech illustrate (Table 15.5), the cool season is the period when nearly all precipitation falls. At this time of year, middle-latitude cyclones often take more equatorward routes and so bring occasional periods of rain.

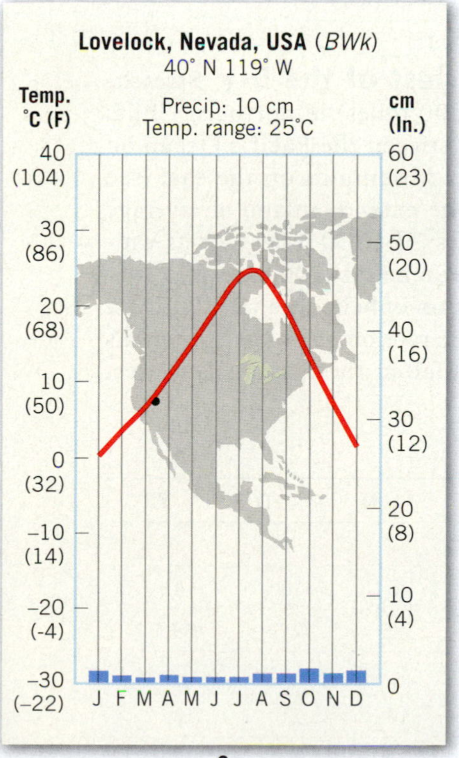


C.

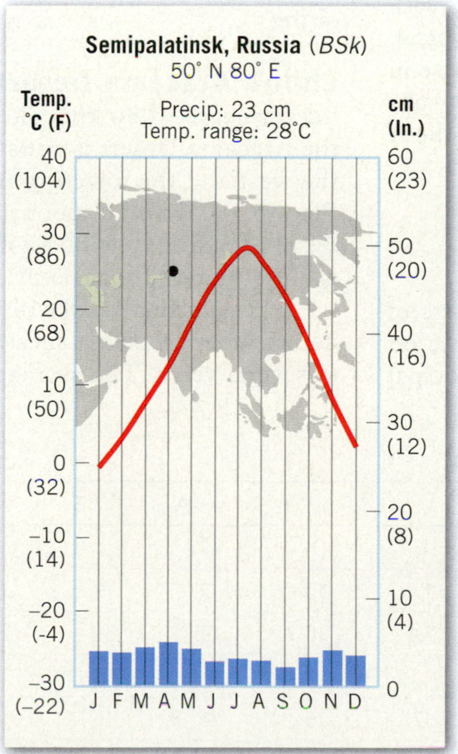


D.

**Temperature** The keys to understanding temperatures in the desert environment are humidity and cloud cover. The cloudless sky and low humidity allow an abundance of solar radiation to reach the ground during the day and permit the rapid exit of terrestrial radiation at night. As would be expected, relative humidities are low throughout the year. Relative humidities between 10 and 30 percent are typical at midday for interior locations.

Desert skies are almost always clear (**Fig. 15.13**). In the Sonoran Desert region of Mexico and the southwestern United States, for example, most stations receive nearly 85 percent of the possible sunshine. Yuma, Arizona, averages 91 percent for the year, with a low of 83 percent in January and a high of 98 percent in June. The Sahara has an average winter cloud cover of about 10 percent, which in summer drops to a mere 3 percent.

A factor contributing to the high ground and air temperatures is that little energy from insolation goes to power evaporation. Thus, almost all the energy goes to heating the surface. In contrast, humid regions are less likely to have such extreme ground and air temperatures, for more energy from solar radiation goes to evaporate water and less remains to heat the ground.

At night, temperatures typically drop rapidly, partly because the water vapor content of the air is fairly low. Ground-surface temperature is also a factor, however. Recall from the discussion of radiation in Chapter 2 that the higher the temperature of a radiating body, the faster that body loses heat. Thus, applied to a desert setting, such environments not only heat up quickly by day but also cool rapidly at night.

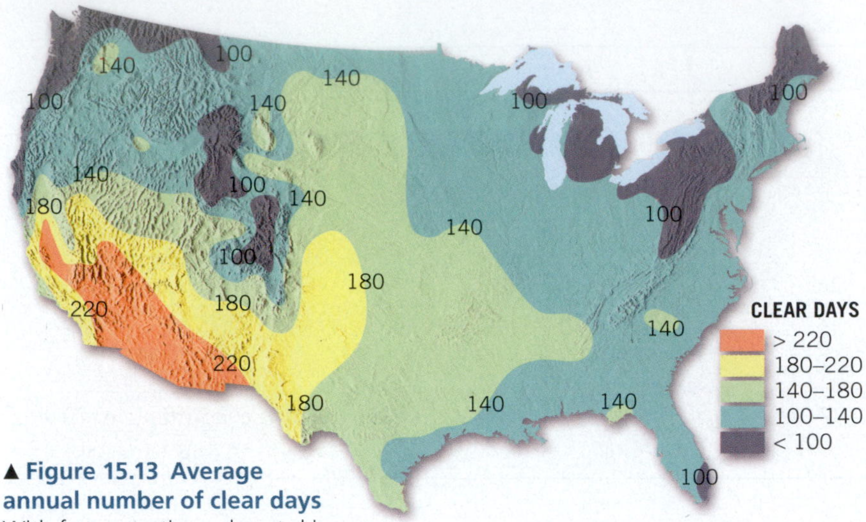


▲ **Figure 15.13 Average annual number of clear days**
With few exceptions, desert skies are typically cloudless and hence receive a very high percentage of the possible sunlight. This is strikingly illustrated in the U.S. Southwest desert.

Consequently, low-latitude deserts in the interiors of continents have the greatest daily temperature ranges on Earth. Daily ranges from 15° to 25°C (27° to 45°F) are common, and they occasionally reach even higher values. The highest daily temperature range ever recorded was at In Salah, Algeria, in the Sahara. On October 13, 1927, this station experienced a 24-hour range of 55.5°C (100°F), from 52.2° to –3.3°C (126° to 26°F).

Because most BWh and BSh areas are poleward of the A climates, annual temperature ranges are the highest among climates in the tropical latitudes. During the low-Sun period, averages are below those in other parts of the tropics, with monthly means of 16° to 24°C (60° to 75°F) being typical. Still, temperatures during the summer are higher than those in the humid tropics. Consequently, annual means at many subtropical desert and steppe stations are similar to those in the A climates.

## West Coast Subtropical Deserts

Where subtropical deserts are found along the west coasts of continents, cold ocean currents have a dramatic influence on the climate, as you will recall from Chapter 7. The principal west coast deserts are the Atacama in South America and the Namib in southern and southwestern Africa. Other areas include portions of the Sonoran Desert in Baja, California, and coastal areas of the Sahara in northwestern Africa.

**A Different Kind of Desert** West coast subtropical deserts deviate considerably from the general image we have of subtropical deserts. The most obvious effect of the cold current is reduced temperatures, as exemplified by the data for Lima, Peru, and Port Nolloth, South Africa (**Table 15.6**). Compared with other stations at similar latitudes, these places have lower annual mean temperatures and subdued annual and daily ranges. Port Nolloth, for example, has an annual mean of only 14°C (57°F) and an annual range of just 4°C (7.2°F). Contrast this with Durban, on the opposite side of South Africa, which has a yearly mean of 20°C (68°F) and an annual range twice that at Port Nolloth.

Although these stations are adjacent to oceans, their yearly rainfall totals are among the lowest in the world. The aridity of these coasts is intensified because the lower air is chilled by the cold offshore waters and hence further stabilized. In addition, the cold currents cause temperatures to often reach the dew point. As a result, these areas are characterized by high relative humidity and much advection fog and dense stratus cloud cover.

The point is that not all subtropical deserts are sunny, hot places with low humidity and cloudless skies. Indeed, the presence of cold currents causes west coast subtropical deserts to be relatively cool, humid places, often shrouded by low clouds or fog.

**Chile's Atacama Desert: Driest of the Dry** Stretching for nearly 1000 kilometers (600 miles) in northern Chile, the Atacama Desert is situated between the Pacific Ocean on the west and the towering Andes Mountains on the east (see Fig. 15.11). This slender arid zone extends inland an average distance of just 50 to 80 kilometers (30 to 50 miles). At its widest, the Atacama spans only 160 kilometers (100 miles).

The Atacama has the distinction of being the world's driest desert. In many places, measurable rain occurs only at intervals of several years. The average rainfall at the Atacama's wettest

**Table 15.6** | Data for West Coast Tropical Desert Stations

| | J | F | M | A | M | J | J | A | S | O | N | D | YR |
|---|---|---|---|---|---|---|---|---|---|---|---|---|---|
| **Port Nolloth, South Africa, 29° 14 S; 7 m** | | | | | | | | | | | | | |
| Temp. (°C) | 15 | 16 | 15 | 14 | 14 | 13 | 12 | 12 | 13 | 13 | 15 | 15 | 14 |
| Precip. (mm) | 2.5 | 2.5 | 5.1 | 5.1 | 10.2 | 7.6 | 10.2 | 7.6 | 5.1 | 2.5 | 2.5 | 2.5 | 63.4 |
| **Lima, Peru, 12° 02 S; 155 m** | | | | | | | | | | | | | |
| Temp. (°C) | 22 | 23 | 23 | 21 | 19 | 17 | 16 | 16 | 16 | 17 | 19 | 21 | 19 |
| Precip. (mm) | 2.5 | T | T | T | 5.1 | 5.1 | 7.6 | 7.6 | 7.6 | 2.5 | 2.5 | T | 40.5 |

A. Orographic lifting leads to precipitation on windward slopes.

B. By the time air reaches the leeward side of the mountains, much of the moisture has been lost, resulting in a *rain shadow desert.*

◄ **Figure 15.14 Rain shadow deserts** Mountains frequently contribute to the aridity of middle-latitude deserts and steppes by creating a rain shadow. **A.** Orographic lifting leads to precipitation on the windward slopes. By the time air reaches the leeward side of the mountains, much of the moisture has been lost. **B.** The Great Basin Desert is a rain shadow desert that covers nearly all of Nevada and portions of adjacent states.

locations is not more than 3 millimeters (0.12 inch) per year. At Arica, a coastal town near Chile's border with Peru, the average annual rainfall is a mere 0.5 millimeter (0.02 inch). Further inland, some stations have *never* recorded rainfall.

Why is this narrow band of land so dry? First, the region is subjected to the dry, subsiding air associated with the semi-permanent cell of high pressure that dominates the eastern South Pacific (see Fig. 7.10, page 196). Second, the cold Peru Current flows northward along the coast and contributes to aridity because it chills and stabilizes the lower atmosphere (see Fig. 7.21, page 204). A third factor that adds to the Atacama's extraordinary aridity is that the Andes Mountains shield the Pacific coast from incursions of humid air from the east.

As is true for other west coast subtropical deserts, the Atacama is a relatively cool place where advection fogs are common. Both phenomena are related to the cold Peru Current. The fogs, called *camanchacas*, form as moist air moves over the cold current and is chilled below its dew point.

## Middle-Latitude Desert (BWk) and Steppe (BSk)

Unlike their low-latitude counterparts, middle-latitude deserts and steppes are not controlled by the subsiding air masses of the subtropical highs. Instead, these dry lands exist principally because of their position in the deep interiors of large landmasses, far removed from the main moisture source—the oceans. In addition, the presence of high mountains across the paths of the prevailing winds further separates these areas from water-bearing maritime air masses (Fig. 15.14). In North America the Coast Ranges, Sierra Nevada, and Cascades are the foremost barriers; in Asia the great Himalayan chain prevents the summertime monsoon flow of moist air off the Indian Ocean from reaching far into the interior.

A glance at Figure 15.11 reveals that middle-latitude desert and steppe climates are most widespread in North America and Eurasia. The Southern Hemisphere lacks extensive land areas in the middle latitudes, so it has a much smaller area of BWk and BSk.

Like subtropical deserts and steppes, the dry regions of the middle latitudes have meager and unreliable precipitation. Unlike the dry lands of the low latitudes, however, these more poleward regions have much lower winter temperatures and hence lower annual means and higher annual ranges of temperature.

The data in Table 15.7 illustrate this point nicely. The data also reveal that rainfall is most abundant during the warm months. Although not all BWk and BSk stations have a summer precipitation maximum, most do because in winter, high pressure and cold temperatures that oppose uplift and precipitation tend to dominate the continents. In summer, however,

**Table 15.7** Data for Middle-Latitude Steppe and Desert Stations

| | J | F | M | A | M | J | J | A | S | O | N | D | YR |
|---|---|---|---|---|---|---|---|---|---|---|---|---|---|
| **Ulan Bator, Mongolia, 47° 55 N; 1311 m** | | | | | | | | | | | | | |
| Temp. (°C) | –26 | –21 | –13 | –1 | 6 | 14 | 16 | 14 | 9 | –1 | –13 | –22 | –3 |
| Precip. (mm) | 1 | 2 | 3 | 5 | 10 | 28 | 76 | 51 | 23 | 7 | 4 | 3 | 213 |
| **Denver, Colorado, 39° 32 N; 1588 m** | | | | | | | | | | | | | |
| Temp. (°C) | 0 | 1 | 4 | 9 | 14 | 20 | 24 | 23 | 18 | 12 | 5 | 2 | 11 |
| Precip. (mm) | 12 | 16 | 27 | 47 | 61 | 32 | 31 | 28 | 23 | 24 | 16 | 10 | 327 |

conditions are somewhat more conducive to cloud formation and precipitation because the anticyclone disappears over the heated continent, and higher surface temperatures and greater mixing ratios prevail.

**students sometimes ask...**

### Aren't deserts mostly covered with sand dunes?

We envision deserts as consisting of mile after mile of drifting sand dunes. It is true that sand accumulates in some deserts, forming striking features. But, perhaps surprisingly, sand accumulations worldwide represent only a small percentage of the total desert area. For example, in the Sahara—the world's largest desert—accumulations of sand cover only *one-tenth* of its area. The sandiest of all deserts is the Arabian, one-third of which consists of sand.

### ✔ Concept Checks 15.4

1. Why is the amount of precipitation defining the dry–humid boundary variable?
2. What is the primary reason for the existence of the dry subtropical realm (BWh and BSh)?
3. Why do ground and air temperatures reach such high values in subtropical deserts?
4. What are the primary factors that cause middle-latitude deserts and steppes?
5. Why are desert and steppe areas uncommon in the middle latitudes of the Southern Hemisphere?

## 15.5 Humid Middle-Latitude Climates with Mild Winters (C)

**Distinguish among the three categories of C climates.**

The Köppen classification recognizes two groups of humid middle-latitude climates. One group has mild winters (the C climates), and the other experiences severe winters (the D climates). The following three sections pertain to the C-type mild winter group. Fig. 15.15 presents climate diagrams of the three types of C climates.

### Humid Subtropical Climate (Cfa)

**Humid subtropical climates** are found on the eastern sides of the continents, in the 25° to 40° latitude range. They dominate the southeastern United States and other similarly situated areas: all of Uruguay and portions of Argentina and southern

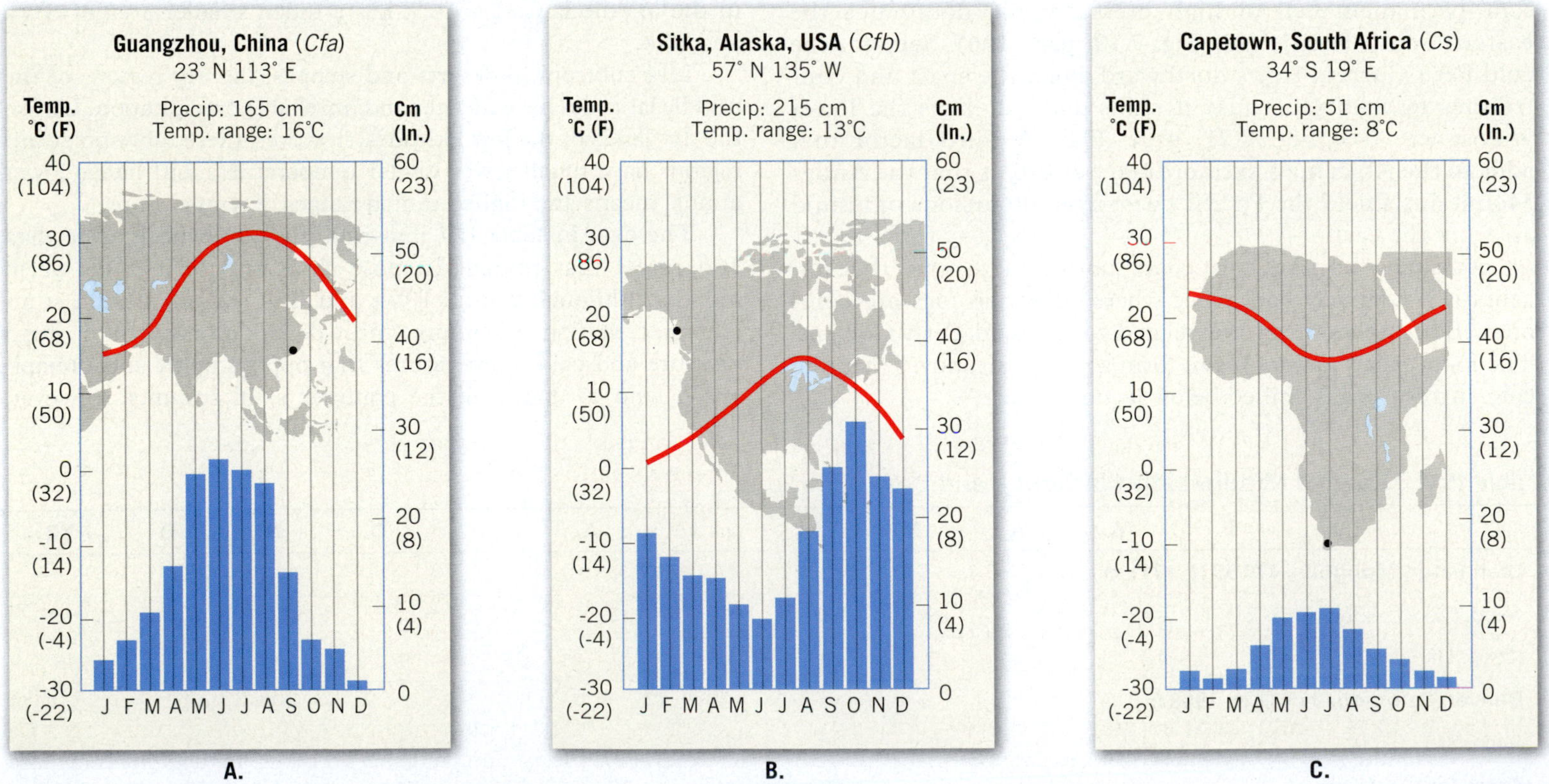


▲ **Figure 15.15 Examples of C climates** **A.** Humid subtropical, **B.** marine west coast, and **C.** dry-summer subtropical.

MapMaster ▶ World Layered Thematic Map ▶ Physical Environment ▶ Climate ▶ Humid Subtropical

Table 15.8 | Data for Humid Subtropical Stations

| | J | F | M | A | M | J | J | A | S | O | N | D | YR |
|---|---|---|---|---|---|---|---|---|---|---|---|---|---|
| **New Orleans, Louisiana, 29° 59 N; 1 m** | | | | | | | | | | | | | |
| Temp. (°C) | 12 | 13 | 16 | 19 | 23 | 26 | 27 | 27 | 25 | 21 | 15 | 13 | 20 |
| Precip. (mm) | 98 | 101 | 136 | 116 | 111 | 113 | 171 | 136 | 128 | 72 | 85 | 104 | 1371 |
| **Buenos Aires, Argentina, 34° 35 S; 27 m** | | | | | | | | | | | | | |
| Temp. (°C) | 24 | 23 | 21 | 17 | 14 | 11 | 10 | 12 | 14 | 16 | 20 | 22 | 17 |
| Precip. (mm) | 104 | 82 | 122 | 90 | 79 | 68 | 61 | 68 | 80 | 100 | 90 | 83 | 1027 |

Brazil in South America, eastern China and southern Japan in Asia, and the east coast of Australia.

In the summer a visitor to the humid subtropics would experience hot, sultry weather of the type expected in the rainy tropics. Daytime temperatures are generally in the lower 30s°C (high 80s°F), but it is not uncommon for the thermometer to reach into the upper 30s°C (90s°F) or even 40°C (more than 100°F) on many afternoons. Because the mixing ratio and relative humidity are high, nights bring little relief. An afternoon or evening thunderstorm is possible, and these areas experience them between 40 and 100 days each year, mostly during summer.

The primary reason for the tropical summer weather in Cfa regions is the dominating influence of maritime tropical air masses. During the summer months, this warm, moist, unstable air moves inland from the western portions of the oceanic subtropical anticyclone. As the maritime tropical (mT) air passes over the heated continent, it becomes increasingly unstable, giving rise to the common convectional showers and thunderstorms.

As summer turns to autumn, the humid subtropics lose their similarity to the rainy tropics. Although winters are mild, frosts are common in higher-latitude Cfa areas and occasionally plague the tropical margins. The winter precipitation is also different in character. Some is in the form of snow, and most is generated along fronts of the frequent middle-latitude cyclones that sweep over these regions. Because the land surface is colder than the maritime air, the air becomes chilled in its lower layers as it moves poleward. Consequently, convectional showers are rare, for the stabilized mT air masses produce clouds and precipitation only when forced to rise.

The data for two humid subtropical stations in Table 15.8 serve to summarize the general characteristics of the Cfa climate. Yearly precipitation usually exceeds 100 centimeters (39 inches), and the rainfall is well distributed throughout the year. Summer normally brings the most precipitation, but there is considerable variation. In the United States, for example, precipitation along the Gulf coast is very evenly distributed. But as one moves poleward or toward the drier western margins, much more falls in summer. Some coastal stations have rainfall maximums in late summer or autumn, when tropical cyclones or their remnants visit the area. In Asia the well-developed monsoon circulation favors a summer precipitation maximum (see the Chinese station in Fig. 15.15A).

Climate data show that summer temperatures are comparable to temperatures in the tropics and that winter values are markedly lower. This is to be expected because the higher-latitude position of the subtropics experiences a wider variation in Sun angle and day length, plus occasional (even frequent) invasions of continental polar (cP) air masses during winter.

## The Marine West Coast Climate (Cfb)

On the western (windward) sides of continents from about 40° to 65° north and south latitude is a climate region dominated by the onshore flow of oceanic air. The prevalence of maritime air masses means mild winters, cool summers, and ample rainfall throughout the year. In North America this **marine west coast climate** (Cfb) extends from near the U.S.–Canada border northward as a narrow belt into southern Alaska (Fig. 15.16). A similar slender strip occurs in South America along the coast of Chile. In both instances, high mountains parallel the coast and prevent the marine climate from penetrating far inland. The largest area of Cfb climate is in Europe, where there is no mountain barrier blocking the movement of cool maritime air from the North Atlantic. Other locations include most of New Zealand as well as tiny slivers of South Africa and Australia.

The data for representative marine west coast stations in Table 15.9 and Figure 15.15B reveal no pronounced dry period, although monthly precipitation drops during the summer. The reduced summer rainfall is due to the poleward migration of the oceanic subtropical highs. Although the areas of marine west coast climate are too far poleward to be dominated by these dry anticyclones, their influence is sufficient to cause a decrease in warm-season rainfall. This is illustrated nicely by two precipitation graphs in Figure 4.D (Box 4.4, page 115)—for Quinault Ranger Station and Rainier Paradise Station.

A comparison of precipitation data for London and Vancouver (Table 15.9) also demonstrates that coastal mountains have a significant influence on yearly rainfall. Vancouver has about two and a half times that of London. In settings like Vancouver's, precipitation totals are higher not only because of orographic uplift but also because the mountains slow the

## eye ON THE atmosphere 15.1

This classic view of Earth from space was taken in December 1968 by an *Apollo 8* astronaut. The image shows the Western Hemisphere. Dense clouds cover much of North America. A large portion of South America is also cloud covered. However, the western margin of South America is cloud free.

**Questions**

1. What is the name of the narrow cloudless desert along the west coast of South America?
2. An ocean current flows just offshore of this desert area. What is the name of the current? Is it warm or cold?
3. How does this desert's close proximity to the Pacific Ocean influence its aridity?
4. In what ways is this desert different from subtropical deserts such as the Sahara and Arabian? Name a desert that has characteristics that are similar to this one.

▼ **Figure 15.16 Marine west coast climate** Fog is common along the rocky Pacific coastline at Olympic National Park, Washington. As the name of this climate implies, the ocean exerts a strong influence, moderating temperatures and providing moisture that supports lush vegetation.

passage of cyclonic storms, allowing them to linger and drop a greater quantity of water.

The ocean is near, so winters are mild and summers are relatively cool. Therefore, a low annual temperature range is characteristic of the marine west coast climate. Because cP air masses generally drift eastward in the zone of the westerlies, periods of severe winter cold are rare. The western edge of North America is especially sheltered from incursions of frigid continental air by the high mountains that intervene between the coast and the source regions for cP air masses. Because no such mountain barrier exists in Europe, cold waves there are somewhat more frequent.

The ocean's control of temperatures can be further demonstrated by a look at temperature gradients (that

MM **MapMaster** ▶ World Layered Thematic Map ▶ Physical Environment ▶ Climate ▶ Marine West Coast

**Table 15.9** | Data for Marine West Coast Stations

| | J | F | M | A | M | J | J | A | S | O | N | D | YR |
|---|---|---|---|---|---|---|---|---|---|---|---|---|---|
| **Vancouver, British Columbia, 49° 11 N; 0 m** | | | | | | | | | | | | | |
| Temp. (°C) | 2 | 4 | 6 | 9 | 13 | 15 | 18 | 17 | 14 | 10 | 6 | 4 | 10 |
| Precip. (mm) | 139 | 121 | 96 | 60 | 48 | 51 | 26 | 36 | 56 | 117 | 142 | 156 | 1048 |
| **London, United Kingdom, 51° 28 N; 5 m** | | | | | | | | | | | | | |
| Temp. (°C) | 4 | 4 | 7 | 9 | 12 | 16 | 18 | 17 | 15 | 11 | 7 | 5 | 10 |
| Precip. (mm) | 54 | 40 | 37 | 38 | 46 | 46 | 56 | 59 | 50 | 57 | 64 | 48 | 595 |

is, changes in temperature per unit distance). Although this climate encompasses a wide latitudinal span, temperatures change much more abruptly when moving inland from the coast than they do in a north–south direction. The transport of heat from the oceans more than offsets the latitudinal variation in the receipt of solar energy. For example, in both January and July, the temperature change from coastal Seattle to inland Spokane, a distance of about 375 kilometers (230 miles), is equal to the variation between Seattle and Juneau, Alaska. Juneau is at about 11° latitude, or roughly 1200 kilometers (750 miles) north of Seattle.

## The Dry-Summer Subtropical (Mediterranean) Climate (Csa, Csb)

The **dry-summer subtropical climate** is typically located along the west sides of continents between latitudes 30° and 45°. Situated between the subtropical steppes on the equatorward side and the marine west coast climate on the poleward side, this climate region is transitional in character. It is the only humid climate that has a pronounced winter rainfall maximum, a feature that reflects its intermediate position (see Fig. 15.15C).

In summer the region is dominated by the stable eastern side of the oceanic subtropical highs. In winter, as the wind and pressure systems follow the Sun equatorward, it is within range of the cyclonic storms of the polar front. Thus, during the course of a year, these areas alternate between being a part of the dry subtropics and being an extension of the humid middle latitudes. Whereas middle-latitude changeability characterizes the winter, subtropical constancy describes the summer.

As is the case for the marine west coast climate, mountain ranges limit the dry-summer subtropics to a relatively narrow coastal zone in both North and South America. Because Australia and southern Africa barely extend to the latitudes where dry-summer climates exist, the development of this climatic type is limited on these continents as well.

Because of the arrangement of the continents and their mountain ranges, inland development occurs only in the Mediterranean basin (Fig. 15.17). Here the zone of subsidence extends far to the east in summer; in winter the sea is a major route of cyclonic disturbances. Because the dry-summer climate is particularly extensive in this region, the name **Mediterranean climate** is often used as a synonym.

MM **MapMaster** ▶ World Layered Thematic Map ▶ Physical Environment ▶ Climate ▶ Dry Summer Subtropical

▼ **Figure 15.17 Italy's Tuscany region** The dry-summer subtropical climate is especially well developed in the Mediterranean region.

**Table 15.10** | Data for Dry-Summer Subtropical Stations

| | J | F | M | A | M | J | J | A | S | O | N | D | YR |
|---|---|---|---|---|---|---|---|---|---|---|---|---|---|
| **San Francisco, California, 37° 37 N; 5 m** | | | | | | | | | | | | | |
| Temp. (°C) | 9 | 11 | 12 | 13 | 15 | 16 | 17 | 17 | 18 | 16 | 13 | 10 | 14 |
| Precip. (mm) | 102 | 88 | 68 | 33 | 12 | 3 | 0 | 1 | 5 | 19 | 40 | 104 | 475 |
| **Sacramento, California, 38° 35 N; 13 m** | | | | | | | | | | | | | |
| Temp. (°C) | 8 | 10 | 12 | 16 | 19 | 22 | 25 | 24 | 23 | 18 | 12 | 9 | 17 |
| Precip. (mm) | 81 | 76 | 60 | 36 | 15 | 3 | 0 | 1 | 5 | 20 | 37 | 82 | 416 |
| **Izmir, Turkey, 38° 26 N; 25 m** | | | | | | | | | | | | | |
| Temp. (°C) | 9 | 9 | 11 | 15 | 20 | 25 | 28 | 27 | 23 | 19 | 14 | 10 | 18 |
| Precip. (mm) | 141 | 100 | 72 | 43 | 39 | 8 | 3 | 3 | 11 | 41 | 93 | 141 | 695 |
| **Santiago, Chile, 33° 27 S; 512 m** | | | | | | | | | | | | | |
| Temp. (°C) | 19 | 19 | 17 | 13 | 11 | 8 | 8 | 9 | 11 | 13 | 16 | 19 | 14 |
| Precip. (mm) | 3 | 3 | 5 | 13 | 64 | 84 | 76 | 56 | 30 | 13 | 8 | 5 | 360 |

**Temperature** Two types of Mediterranean climate are recognized and are based primarily on summertime temperatures. The cool summer type (Csb), as exemplified by San Francisco and Santiago, Chile (Table 15.10), is limited to coastal areas. Here the cooler summer temperatures one expects on a windward coast are further intensified by cold ocean currents.

The data for Izmir, Turkey, and Sacramento illustrate the features of the warm summer type (Csa). At both places winter temperatures are not very different from those in the Csb type. But Sacramento, in California's Central Valley, is removed from the coast, and Izmir is bordered by the warm waters of the Mediterranean; consequently, summer temperatures are noticeably higher at these locations. As a result, annual temperature ranges are also higher in Csa areas.

**students sometimes ask...**

### Do soils differ from one climate to another?

Definitely. Climate is one of the most influential factors in soil formation. Variations in temperature and precipitation determine the type of weathering that occurs and also greatly influence the rate and depth of soil formation. For example, a hot, wet climate may produce a thick layer of chemically weathered soil in the same amount of time that a cold, dry climate produces a thin mantle of mechanically weathered debris. Precipitation also influences the degree to which various materials are leached from the soil, thereby affecting soil fertility. Finally, climate is an important control on the type of plant and animal life present, which in turn influences the nature of the soil that forms.

**Precipitation** Yearly precipitation within the dry-summer subtropics ranges between about 40 and 80 centimeters (16 and 31 inches). In many areas such amounts mean that a station barely escapes being classified as semiarid. As a result, some climatologists refer to the dry-summer climate as *subhumid* instead of humid. This is especially true along the equatorward margins because rainfall totals increase in the poleward direction. Los Angeles, for example, receives 38 centimeters (15 inches) of precipitation annually, whereas San Francisco, 400 kilometers (250 miles) to the north, receives 51 centimeters (20 inches) per year. Still farther north, at Portland, Oregon, the yearly rainfall average is over 90 centimeters (35 inches).

## ✔ Concept Checks 15.5

1. Describe and explain the differences between summertime and wintertime precipitation in the humid subtropics (Cfa).
2. Why is the marine west coast climate (Cfb) represented by only slender strips of land in North and South America but is extensive in Western Europe?
3. How do temperature gradients (north–south versus east–west) reveal the strong oceanic influence along the west coast of North America?
4. What other name is given to the dry-summer subtropical climate?
5. Why are summer temperatures cooler at San Francisco than at Sacramento (Table 15.10)?

# 15.6 Humid Continental Climates with Severe Winters (D)

**Summarize the characteristics of the two categories of D climates.**

The C climates just described have mild winters. By contrast, D climates experience severe winters. In this section, two types of D climates are discussed: the humid continental and the subarctic. Climate diagrams of representative locations are shown in Fig. 15.18. This climate region is dominated by the polar front and thus is a battleground for tropical and polar air masses. No other climate experiences such rapid, nonperiodic changes in the weather. Cold waves, heat waves, droughts, blizzards, and heavy downpours are all yearly events in the humid continental realm. For more about droughts, see Box 15.2.

## Humid Continental Climate (Dfa)

The **humid continental climate** (Dfa), as its name implies, is a land-controlled climate. It is the product of broad continents located in the middle latitudes. Because continentality is a basic feature, this climate does not occur in the Southern Hemisphere, where the middle-latitude zone is dominated by the ocean. Instead, it is confined to central and eastern North America and Eurasia in the latitude range 40° to 50° north.

It may at first seem unusual that a continental climate should extend eastward to the margins of the ocean. However, recall from Chapter 7 that the prevailing atmospheric circulation is from the west, so deep and persistent incursions of maritime air from the east are unlikely.

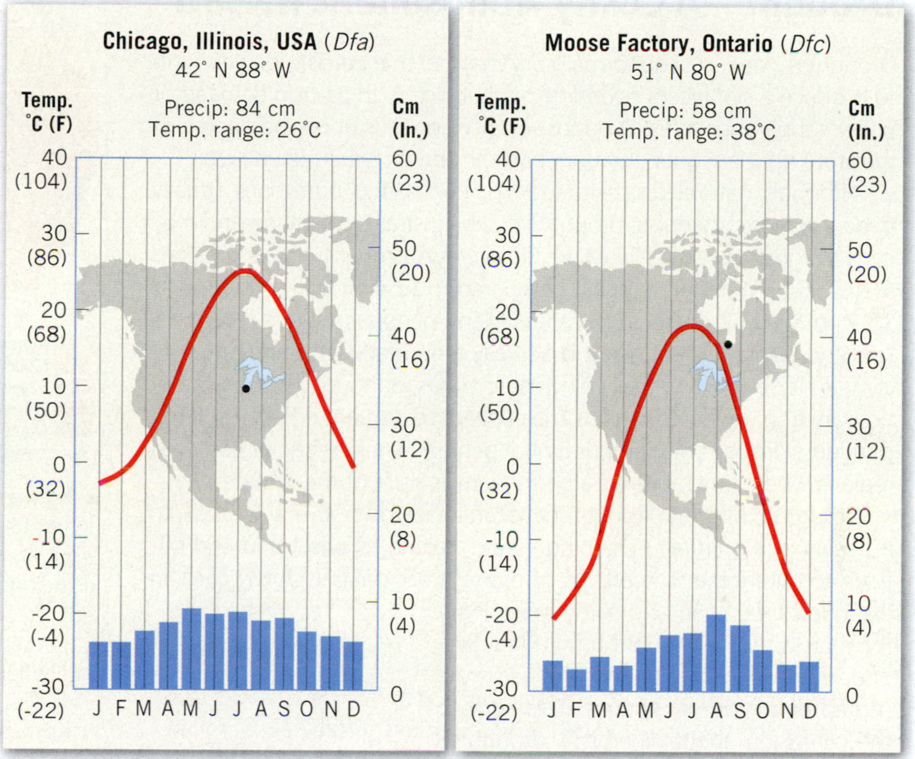


A. B.

▲ **Figure 15.18 Examples of D climates** Climates in this category are associated with the interiors of large landmasses in the mid- to high latitudes of the Northern Hemisphere. Although winters can be harsh in Chicago's humid continental (Dfa) climate, the subarctic environment (Dfc) of Moose Factory is more extreme.

**Temperature** Both winter and summer temperatures in the Dfa climate are relatively severe. Consequently, annual temperature ranges are great. A comparison of the stations in Table 15.11 illustrates this. July means are generally near and often above 20°C (68°F), so summertime temperatures, although lower in the north than in the south, are not markedly

**Table 15.11** Data for Humid Continental Stations

| | J | F | M | A | M | J | J | A | S | O | N | D | YR |
|---|---|---|---|---|---|---|---|---|---|---|---|---|---|
| **Omaha, Nebraska, 41° 18 N; 330 m** | | | | | | | | | | | | | |
| Temp. (°C) | −6 | −4 | 3 | 11 | 17 | 22 | 25 | 24 | 19 | 12 | 4 | −3 | 10 |
| Precip. (mm) | 20 | 23 | 30 | 51 | 76 | 102 | 79 | 81 | 86 | 48 | 33 | 23 | 652 |
| **New York, New York, 40° 47 N; 40 m** | | | | | | | | | | | | | |
| Temp. (°C) | −1 | −1 | 3 | 9 | 15 | 21 | 23 | 22 | 19 | 13 | 7 | 1 | 11 |
| Precip. (mm) | 84 | 84 | 86 | 84 | 86 | 86 | 104 | 109 | 86 | 86 | 86 | 84 | 1065 |
| **Winnipeg, Canada, 49° 54 N; 240 m** | | | | | | | | | | | | | |
| Temp. (°C) | −18 | −16 | −8 | 3 | 11 | 17 | 20 | 19 | 13 | 6 | −5 | −13 | 3 |
| Precip. (mm) | 26 | 21 | 27 | 30 | 50 | 81 | 69 | 70 | 55 | 37 | 29 | 22 | 517 |
| **Harbin, Manchuria, 45° 45 N; 143 m** | | | | | | | | | | | | | |
| Temp. (°C) | −20 | −16 | −6 | 6 | 14 | 20 | 23 | 22 | 14 | 6 | −7 | −17 | 3 |
| Precip. (mm) | 4 | 6 | 17 | 23 | 44 | 92 | 167 | 119 | 52 | 36 | 12 | 5 | 577 |

severe& hazardous weather Box 15.2

## Drought—A Costly Atmospheric Hazard

Drought is a period of abnormally dry weather that persists long enough to produce a significant hydrologic imbalance such as crop damage or water supply shortages. Drought severity depends upon the degree of moisture deficiency, its duration, and the size of the affected area.

Although natural disasters such as floods and hurricanes usually generate more attention, droughts can be just as devastating and may carry a bigger price tag (**Fig. 15.B**). On average, droughts cost the United States \$6 to \$8 billion annually compared to \$2.4 billion for floods and \$1.2 to \$4.8 billion for hurricanes. Many people consider drought a rare and random event, yet it is actually a normal, recurring feature of climate. It occurs in virtually all climate zones, although its characteristics vary from region to region. The concept of drought differs from that of aridity. Drought is a temporary happening, whereas aridity describes regions where low rainfall is a permanent feature of the climate.

Drought is different from other natural hazards in several ways. First, it occurs in a gradual, "creeping" way, making its onset and end difficult to determine. The effects of drought accumulate slowly over an extended time span and sometimes linger for years after the drought has ended. Second, there is no precise and universally accepted definition of drought. This adds to the confusion about whether a drought is actually occurring and, if it is, its severity. Third, drought seldom produces structural damages, so its social and economic effects are less obvious than are damages from other natural disasters.

Drought is a temporary happening, whereas aridity describes regions where low rainfall is a permanent feature of the climate.

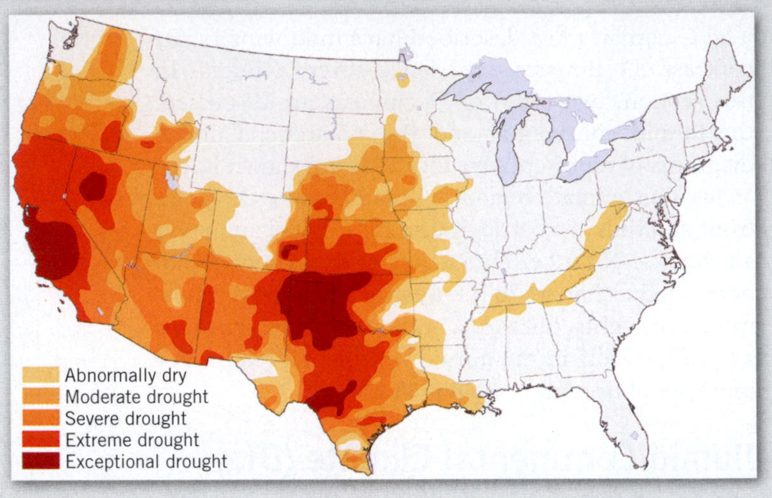


▲ **Figure 15.B Drought-status map for May 6, 2014** On this date nearly 15 percent of the country was experiencing extreme or exceptional drought. Maps such as this are based on a blend of several criteria, including precipitation, soil moisture, water levels in reservoirs, observer reports, and satellite-based maps of vegetation health. To examine current and archived national, regional, and state-by-state drought maps and conditions, go to http://droughtmonitor.unl.edu. (NOAA)

No single definition of drought works in all circumstances. Definitions reflect different approaches to measuring drought: meteorological, agricultural, and hydrologic. *Meteorological drought* deals with the degree of dryness based on the departure of precipitation from normal values and the duration of the dry period. *Agricultural*

different. This is illustrated by the temperature distribution map of the eastern United States in **Figure 15.19**.

The summer map has only a few widely spaced isotherms, indicating a weak summer temperature gradient (Fig. 15.19A). The winter map, however, shows a stronger temperature gradient (Fig. 15.19B). The decrease in midwinter values with increasing latitude is appreciable. Confirming this, Table 15.11 reveals that the temperature change between Omaha and Winnipeg is more than twice as great in winter as in summer.

Because of the steeper winter temperature gradient, shifts in wind direction during the cold season often result in sudden large temperature changes. This is not the case in summer, when temperatures throughout the region are more uniform.

Annual temperature ranges also vary within this climate, generally increasing from south to north and from the coast toward the interior. Comparing the data for Omaha and Winnipeg illustrates the first situation, and comparing New York and Omaha illustrates the second.

**Precipitation** Records from the four stations in Table 15.11 show the general precipitation pattern for Dfa climates. A summer maximum occurs at each station. But it is weakly defined at New York because the east coast is more accessible to maritime air masses throughout the year. For the same reason, New York also has the highest total of the four stations. Harbin, Manchuria, on the other hand, shows the most pronounced summer maximum, followed by a winter dry season. This is characteristic of most eastern Asian stations in the middle latitudes and reflects the powerful control of the monsoon.

The data also reveal that precipitation generally decreases toward the continental interior and from south to north, primarily because of increasing distance from the sources of mT air. Furthermore, the more northerly stations are also influenced for a greater part of the year by drier polar air masses.

Wintertime precipitation is chiefly associated with the passage of fronts connected with traveling middle-latitude cyclones. Part of this precipitation is snow, and the proportion increases with latitude. Although precipitation is often considerably less during the cold season, it is usually more conspicuous than the greater amounts that fall during summer. An obvious reason is that snow remains on the ground, often for extended periods, and rain, of course, does not. Moreover, summer rains are often in the form of relatively short convective showers, whereas winter snows associated with large midlatitude cyclones are more prolonged.

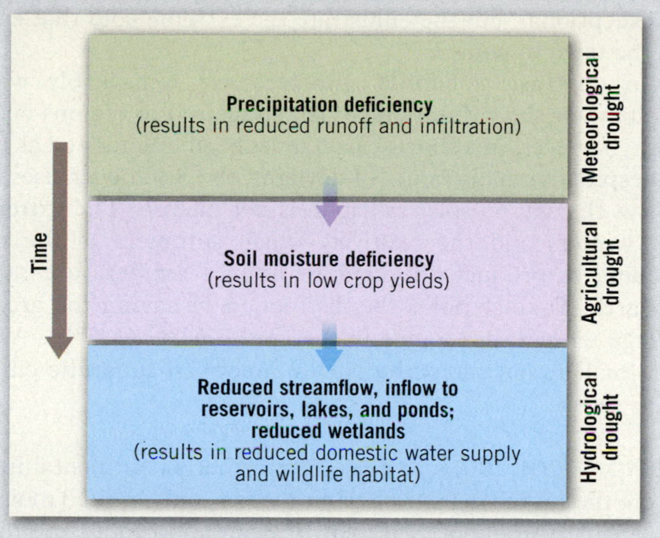


▲ **Figure 15.C Sequence of drought impacts** After the onset of meteorological drought, agriculture is affected first, followed by reductions in streamflow and water levels in lakes, reservoirs, and groundwater. When meteorological drought ends, agricultural drought ends as soil moisture is replenished. It takes considerably longer for hydrologic drought to end.

*drought* is usually linked to a deficit of soil moisture. A plant's need for water depends on prevailing weather conditions, biological characteristics of the particular plant, the plant's stage of growth, and various soil properties. *Hydrologic drought* refers to deficiencies in surface and subsurface water supplies. It is measured as streamflow and as lake, reservoir, and groundwater levels. There is a time lag between the onset of dry conditions and a drop in streamflow or the lowering of lakes, reservoirs, or groundwater levels. So, hydrologic measurements are not the earliest indicators of drought.

There is a sequence of impacts associated with meteorological, agricultural, and hydrologic droughts (**Fig. 15.C**). When meteorological drought begins, the agricultural sector is usually the first to be affected because of its heavy dependence on soil moisture. Soil moisture is rapidly depleted during extended dry periods. If precipitation deficiencies continue, those dependent on surface-water sources and groundwater may be affected.

When precipitation returns to normal, meteorological drought comes to an end. Soil moisture is replenished first, followed by streamflow, reservoirs and lakes, and finally groundwater. Thus, drought impacts may diminish rapidly in the agricultural sector because of its reliance on soil moisture, but they may linger for months or years in other sectors that depend on stored surface or subsurface water supplies. Groundwater users, who are often the last to be affected following the onset of meteorological drought, may also be the last to experience a return to normal water levels. The length of the recovery period depends upon the intensity of the meteorological drought, its duration, and the quantity of precipitation received when the drought ends.

### Questions

1. How is drought different from aridity?
2. What term is used to describe the most serious level of drought (check Fig. 15.B)? In what two states was this level most widespread in May 2014?

▼ **Figure 15.19 Temperatures in the eastern United States A.** During the summer months, north–south temperature variations in the eastern United States are small—that is, the temperature gradient is weak. **B.** In winter, however, north–south temperature contrasts are sharp.

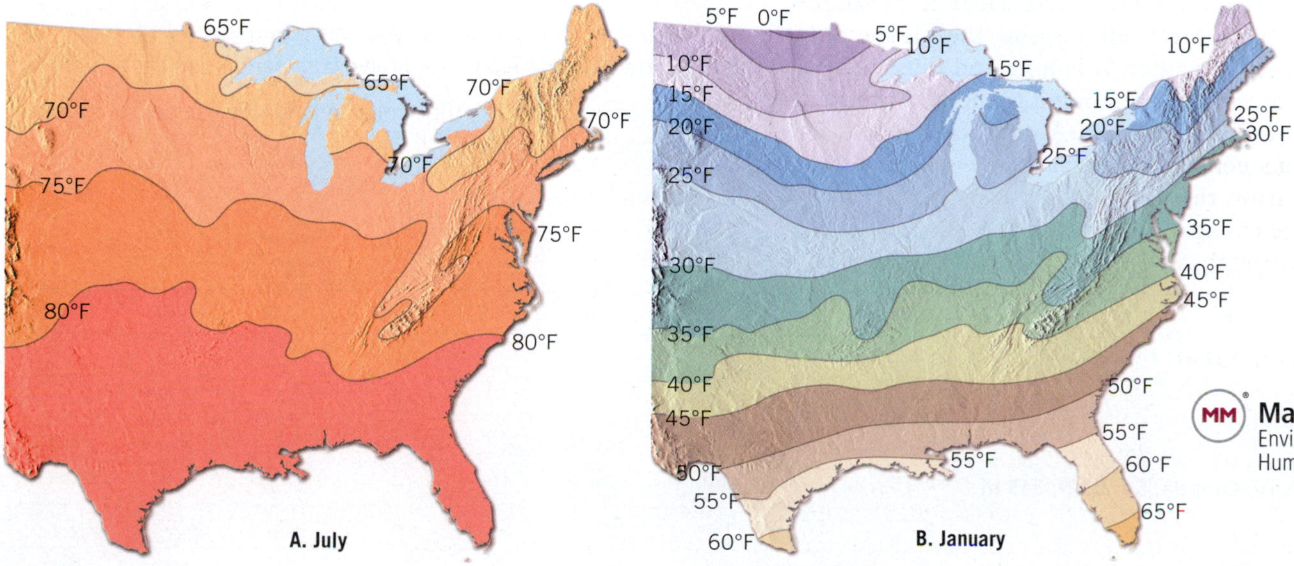


MM MapMaster ▶ Physical Environment ▶ Climate ▶ Humid Contiental

▲ **Figure 15.20 Subarctic climate** The northern coniferous forest, also called the taiga, is associated with subarctic climates. This climate typically experiences the highest annual temperature ranges on Earth.

## The Subarctic Climate (Dfc, Dfd)

North of the humid continental climate and south of the polar tundra is an extensive **subarctic climate.** It covers broad, uninterrupted expanses in North America (western Alaska to Newfoundland) and in Eurasia (Norway to the Pacific coast of Russia). It is often called the **taiga climate,** for it closely corresponds to the northern coniferous forest region of the same name (Fig. 15.20). Although they are scrawny, the spruce, fir, larch, and birch trees in the taiga represent the largest stretch of continuous forest on Earth.

**Temperature** The subarctic is well illustrated by the climate diagram in Figure 15.18B and by the data for Yakutsk, Russia, and Dawson, Yukon Territory, in Table 15.12. Here in the source regions of continental polar (cP) air masses, the outstanding feature is the dominance of winter. It is long and bitterly cold. Winter minimum temperatures are among the lowest recorded outside the ice caps of Greenland and Antarctica. In fact, for many years the world's coldest temperature was attributed to Verkhoyansk in east-central Siberia, where the temperature dropped to −68°C (−90°F) on February 5 and 7, 1892. Over a 23-year period this same station had an average monthly minimum of −62°C (−80°F) during January. These temperatures are exceptional, but they illustrate the extreme cold that envelops the taiga in winter.

In contrast, subarctic summers are remarkably warm, despite their short duration. When compared to regions farther south, however, this short season must be characterized as cool; for despite the many hours of daylight, the Sun never rises very high in the sky, so solar radiation is not intense. The extremely cold winters and the relatively warm summers of the taiga combine to produce the greatest annual temperature ranges on Earth. Yakutsk holds the distinction of having the greatest average temperature range in the world, 63°C (113°F). As the data for Dawson show, the North American subarctic climate is less severe.

**Precipitation** Because these far northerly continental interiors are the source regions of cP air masses, only limited moisture is available throughout the year. Precipitation totals are therefore small, seldom exceeding 50 centimeters (20 inches). By far the greatest precipitation comes as rain from scattered summer convectional showers. Less snow falls than in the humid continental climate to the south, yet there is the illusion of more. The reason is simple: No melting occurs for months at a time, so the entire winter accumulation (up to 1 meter [3 feet]) is visible all at once. Furthermore, during blizzards, high winds swirl the dry, powdery snow into high drifts, giving the false impression that more snow is falling than is actually the case. So although snowfall is not excessive, a visitor to this region could leave with that impression.

### ✔ Concept Checks 15.6

1. Why is the humid continental climate confined to the Northern Hemisphere?
2. Why do coastal stations such as New York City experience primarily continental climatic conditions?
3. Using the four stations in Table 15.11, describe the general pattern of precipitation in humid continental climates.
4. Describe and explain the annual temperature range one should expect in the realm of the taiga.

**Table 15.12** | Data for Subarctic Stations

| | J | F | M | A | M | J | J | A | S | O | N | D | YR |
|---|---|---|---|---|---|---|---|---|---|---|---|---|---|
| **Yakutsk, Russia, 62° 05 N; 103 m** | | | | | | | | | | | | | |
| Temp. (°C) | −43 | −37 | −23 | −7 | 7 | 16 | 20 | 16 | 6 | −8 | −28 | −40 | −10 |
| Precip. (mm) | 7 | 6 | 5 | 7 | 16 | 31 | 43 | 38 | 22 | 16 | 13 | 9 | 213 |
| **Dawson, Yukon Territory, Canada, 64° 03 N; 315 m** | | | | | | | | | | | | | |
| Temp. (°C) | −30 | −24 | −16 | −2 | 8 | 14 | 15 | 12 | 6 | −4 | −17 | −25 | −5 |
| Precip. (mm) | 20 | 20 | 13 | 18 | 23 | 33 | 41 | 41 | 43 | 33 | 33 | 28 | 346 |

# 15.7 The Polar Climates (E)

**Contrast tundra and ice cap climates.**

According to the Köppen classification, **polar climates** are those in which the mean temperature of the warmest month is below 10°C (50°F). Two types are recognized: the tundra climate (ET) and the ice cap climate (EF). Climate diagrams of representative stations are presented in Figure 15.21.

Just as the tropics are defined by their year-round warmth, so the polar realm is known for its enduring cold, with the lowest annual means on the planet. Because polar winters are periods of perpetual night, or nearly so, temperatures are understandably bitter. During the summer, temperatures remain cool despite the long days because the Sun is so low in the sky that its oblique rays produce little warming. In addition, much solar radiation is reflected by the ice and snow or used in melting the snow cover. In either case, energy that could have warmed the land is lost. Although summers are cool, temperatures are still much higher than those experienced during the severe winter months. Consequently, annual temperature ranges are extreme.

Although polar climates are classified as humid, precipitation is generally meager, with many nonmarine stations receiving less than 25 centimeters (10 inches) annually. Evaporation, of course, is also limited. The scanty precipitation is easily understood in view of the temperature characteristics of the region. The amount of water vapor in the air is always small because low mixing ratios must accompany low temperatures. In addition, steep lapse rates are not possible. Usually precipitation is most abundant during the warmer summer months, when the air's moisture content is highest.

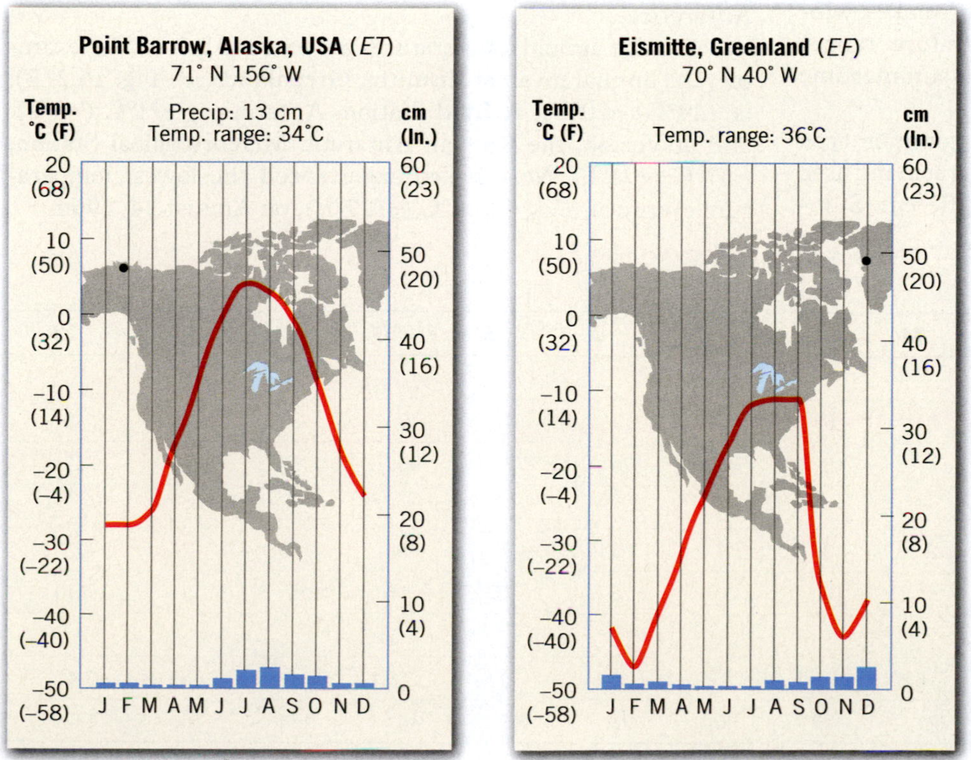


A. B.

## The Tundra Climate (ET)

The **tundra climate** on land is found almost exclusively in the Northern Hemisphere. It occupies the coastal fringes of the Arctic Ocean, many Arctic islands, and the ice-free shores of northern Iceland and southern Greenland. In the Southern Hemisphere no extensive land areas exist in the latitudes where tundra climates prevail. Consequently, except for some small islands in the southern oceans, the ET climate occupies only the southwestern tip of South America and the northern portion of the Palmer Peninsula in Antarctica.

The 10°C (50°F) summer isotherm that marks the equatorward limit of the tundra also marks the poleward limit of tree growth. Thus, the tundra is a treeless region of grasses, sedges, mosses, and lichens. During the long cold season, plant life is dormant, but once the short, cool summer commences, these plants mature and produce seeds very rapidly.

Because summers are cool and short, the frozen soils of the tundra generally thaw to depths of less than 1 meter (3 feet). Large portions of the tundra are characterized by **permafrost,** deep ground that remains frozen year-round (Fig. 15.22).* Strictly speaking, permafrost is defined only on the basis of temperature; that is, it is ground with temperatures that have remained below 0°C (32°F) continuously for 2 years or more. The degree to which ice is present in the ground strongly affects the behavior of the surface material. Knowing how much subsurface ice is present and where it is located is very important in constructing roads, buildings, and other projects in areas underlain by permafrost.

When the activities of people disturb the surface, such as by removing the insulating vegetation mat or by constructing roads and buildings, the delicate thermal balance is disturbed, and the permafrost can thaw. Thawing produces unstable ground that may slide, slump, subside, and undergo severe frost heaving. When a heated structure is built directly on ice-rich permafrost, thawing creates soggy material into which the structure can sink.

The data for Point Barrow, Alaska (Table 15.13 and Fig. 15.21A), on the shores of the frozen Arctic Ocean, illustrate a classic ET station, where continentality prevails. The

*Permafrost is not confined to the tundra. It is also common in subarctic taiga regions.

◄ **Figure 15.21 Examples of E climates** These climatic diagrams represent the two basic types of polar climates. **A.** Point Barrow, Alaska, exhibits a tundra (ET) climate. **B.** Eismitte, Greenland, a station located on a massive ice sheet, is classified as an ice-cap (EF) climate.

MM **MapMaster** ► World Layered Thematic Map ► Physical Environment ► Climate ► Ice Cap

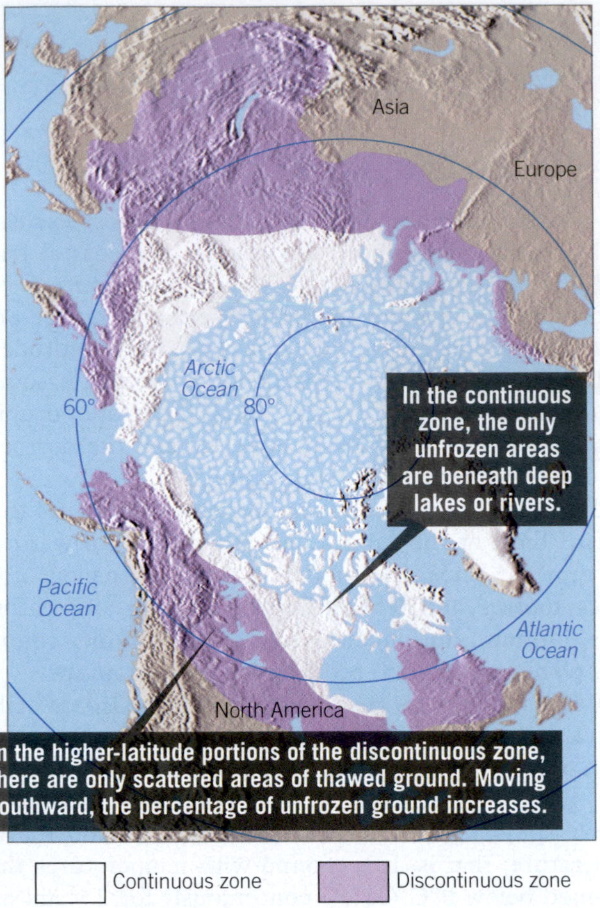


▲ **Figure 15.22 Distribution of permafrost in the Northern Hemisphere** More than 80 percent of Alaska and about 50 percent of Canada are underlain by permafrost.

combination of high latitude and continentality makes winters severe, summers cool, and annual temperature ranges great. Yearly precipitation is small, with a modest summertime maximum.

Although Point Barrow represents the most common type of tundra setting, the data for Angmagssalik, Greenland (see Table 15.13), reveal that some ET stations are different. Summer temperatures at the two stations are equivalent, but winters at Angmagssalik are much warmer, and the annual precipitation is eight times greater than at Point Barrow. The reason is Angmagssalik's location on the southeastern coast of Greenland, where there is considerable marine influence. The warm North Atlantic Drift keeps winter temperatures relatively warm, and maritime polar (mP) air masses supply moisture throughout the year. Because winters are less severe at stations like Angmagssalik, annual temperature ranges are much smaller than at stations like Point Barrow, where continentality is a major control.

Note that tundra climates are not entirely confined to the high latitudes. The summer coolness of this climate is also found at higher elevations as one moves equatorward. Even in the tropics, you can find ET climate characteristics if you go high enough; these are termed *highland climates* and are discussed in detail in the final section of the chapter. When compared with the Arctic tundra, however, winter temperatures in these lower-latitude counterparts become milder and less distinct from summer, as the data for Cruz Loma, Ecuador (see Table 15.13), illustrate.

## The Ice-Cap Climate (EF)

The **ice-cap climate,** designated by Köppen as EF, has no monthly mean above 0°C (32°F). Because the average temperature for all months is below freezing, the growth of vegetation is prohibited, and the landscape is one of permanent ice and snow. This climate of perpetual frost covers a surprisingly large area of more than 15.5 million square kilometers (6 million square miles), or about 9 percent of Earth's land area (**Fig. 15.23**). Aside from scattered occurrences in high mountain areas, it is largely confined to the ice sheets of Greenland and Antarctica.

Average annual temperatures are extremely low. For example, the annual mean at Eismitte, Greenland (see Fig. 15.21B), is −29°C (−20°F); at Byrd Station, Antarctica, −21°C (−6°F); and at Vostok, the Russian Antarctic Meteorological Station, −57°C (−71°F). Vostok also experienced the lowest temperature ever recorded, −88.3°C (−127°F), on August 24, 1960.

**Table 15.13** | Data for Polar Stations

| | J | F | M | A | M | J | J | A | S | O | N | D | YR |
|---|---|---|---|---|---|---|---|---|---|---|---|---|---|
| **Angmagssalik, Greenland, 65° 36 N; 29 m** | | | | | | | | | | | | | |
| Temp. (°C) | −7 | −7 | −6 | −3 | 2 | 6 | 7 | 7 | 4 | 0 | −3 | −5 | 0 |
| Precip. (mm) | 57 | 81 | 57 | 55 | 52 | 45 | 28 | 70 | 72 | 96 | 87 | 75 | 775 |
| **Point Barrow, Alaska, 71° 18 N; 9 m** | | | | | | | | | | | | | |
| Temp. (°C) | −28 | −28 | −26 | −17 | −8 | 0 | 4 | 3 | −1 | −9 | −18 | −24 | −12 |
| Precip. (mm) | 5 | 5 | 3 | 3 | 3 | 10 | 20 | 23 | 15 | 13 | 5 | 5 | 110 |
| **Cruz Loma, Ecuador, 0° 08 S; 3888 m** | | | | | | | | | | | | | |
| Temp. (°C) | 6.1 | 6.6 | 6.6 | 6.6 | 6.6 | 6.1 | 6.1 | 6.1 | 6.1 | 6.1 | 6.6 | 6.6 | 6.4 |
| Precip. (mm) | 198 | 185 | 241 | 236 | 221 | 122 | 36 | 23 | 86 | 147 | 124 | 160 | 1779 |

▲ **Figure 15.23 Ice-cap climate** Greenland and Antarctica are the major examples of this extreme climate. In this image, scientists are conducting research on the Greenland ice sheet.

Video (MM)
Operation IceBridge in Greenland
http://goo.gl/fybvT8

In addition to latitude, the primary reason for such temperatures is the presence of permanent ice. Ice has a very high albedo, reflecting up to 80 percent of the meager sunlight that strikes it. The energy that is not reflected is used largely to melt the ice and so is not available for raising the temperature of air.

Another factor at many EF stations is elevation. Eismitte, at the center of the Greenland ice sheet, is almost 3000 meters above sea level (10,000 feet), and much of Antarctica is even higher. Thus, the permanent ice and high elevations further reduce the already low temperatures of the polar realm.

The intense chilling of the air close to the ice sheet means that strong surface-temperature inversions are common. Near-surface temperatures may be as much as 30°C (54°F) colder than air just a few hundred meters above. Gravity pulls this cold, dense air downslope, often producing strong winds and blizzard conditions. Such air movements, called *katabatic winds* (see Chapter 7), are an important aspect of ice-cap weather at many locations. Where the slope is sufficient, these gravity-induced air movements can be strong enough to flow in a direction that is opposite the pressure gradient.

### ✔ Concept Checks 15.7

1. Although polar regions experience extended periods of sunlight in the summer, temperatures remain cool. Explain.
2. What is the significance of the 10°C (50°F) summer isotherm?
3. Why is the tundra landscape characterized by poorly drained, boggy soils?
4. The tundra climate is not confined solely to high latitudes. Under what circumstances might the ET climate be found in lower-latitude locations?
5. Where are EF climates developed most extensively?

## 15.8 Highland Climates

**List the characteristics of highland climates.**

It is well known that mountain climates are distinctly different from those in adjacent lowlands. Sites with **highland climates** (labeled H in Fig. 15.4) are cooler and usually wetter. The world climate types already discussed consist of large, relatively homogeneous regions. But highland climates are characterized by a great diversity of climatic conditions over small areas. Because large differences occur over short distances, the pattern of climates in mountainous areas is a complex mosaic, too complicated to depict on a world map.

In North America highland climates characterize the Rockies, Sierra Nevada, Cascades, and mountains and interior plateaus of Mexico. In South America the Andes create a continuous band of highland climate that extends for nearly 8000 kilometers (5000 miles). The greatest span of highland climates stretches from western China, across southern Eurasia, to northern Spain, from the Himalayas to the Pyrenees. Highland climates in Africa occur in the Atlas Mountains in the north and in the Ethiopian Highlands in the east.

The best-known climate effect of increased altitude is lower temperatures. Greater precipitation due to orographic lifting is also common at higher elevations. The precipitation map for Nevada in **Figure 15.24** illustrates this nicely. The long, slender zones of highest precipitation coincide with areas of mountainous topography. Despite the fact that mountain stations are colder and often wetter than locations at lower elevations, highland climates are often very similar to those in adjacent lowlands in terms of seasonal temperature cycles and precipitation distribution. **Figure 15.25** illustrates this relationship.

Phoenix, at an elevation of 338 meters (1109 feet), lies in the desert lowlands of southern Arizona. By contrast, Flagstaff is located at an altitude of 2134 meters (about 7000 feet) on the Colorado Plateau in northern Arizona. When summer averages

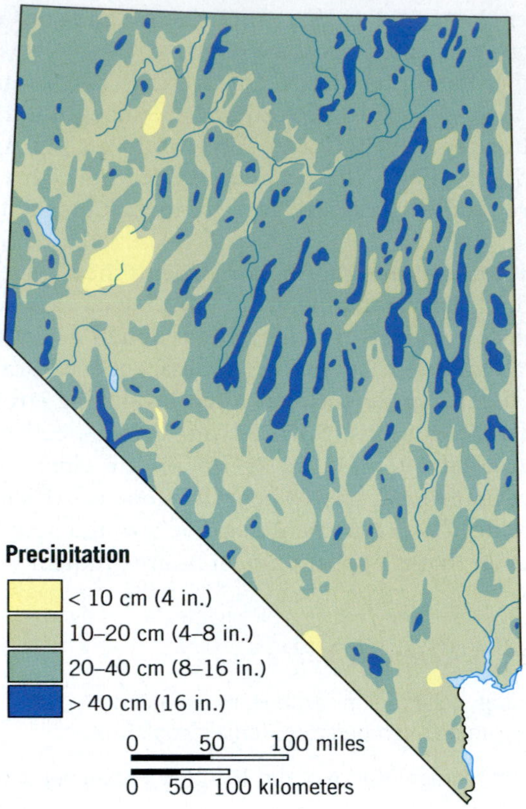


◀ **Figure 15.24 Precipitation map for Nevada** As a part of the Basin and Range region, Nevada is characterized by many small mountain ranges that rise 900 to 1500 meters (3000 to 5000 feet) above the basins that separate them. Where there are mountains, there is more precipitation; where there are basins, there is less precipitation. The relationship is simple and straightforward.

climb to 34°C (93°F) in Phoenix, Flagstaff is experiencing a pleasant 19°C (66°F), a full 15°C (27°F) cooler. Although the temperatures at the two cities are quite different, the annual march of temperature for the two is similar. Both experience minimum and maximum monthly means in the same months. Precipitation data show similar seasonal patterns for both, but the amounts at Flagstaff are higher in every month. In addition, much of Flagstaff's winter precipitation is snow, whereas Phoenix gets only rain.

Because topographic variations are pronounced in mountains, every change in slope with respect to the Sun's rays produces a different *microclimate*. In the Northern Hemisphere, south-facing slopes are warmer and drier because they receive more direct sunlight than do north-facing slopes and deep valleys. Wind direction and speed in mountains can be highly variable and quite different from the movement of air aloft or over adjacent plains. Mountains create various obstacles to winds. Locally, winds may be funneled through valleys or forced over ridges and around mountain peaks. When weather conditions

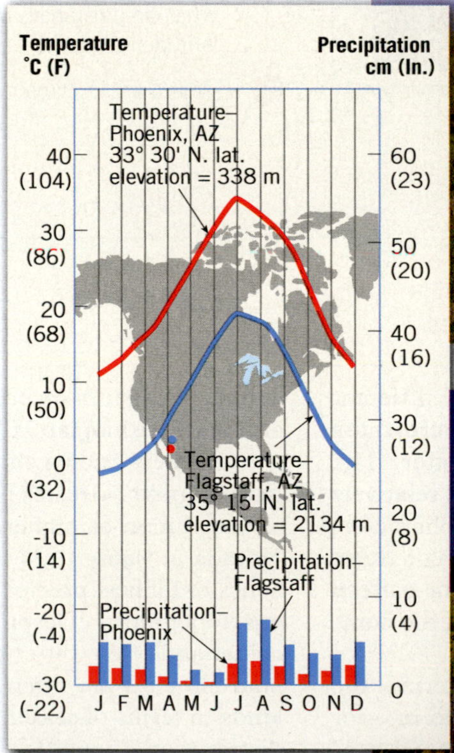

Phoenix, AZ

Flagstaff, AZ

http://goo.gl/zqimSA

▶ **SmartFigure 15.25 Highland climate** The graph includes data for two stations in Arizona and illustrates the general influence of elevation on climate. Flagstaff is cooler and wetter because of its position on the Colorado Plateau, nearly 1800 meters (6000 feet) higher than Phoenix. Only scanty drought-tolerant natural vegetation can survive in the hot, dry climate of southern Arizona, near Phoenix. The natural vegetation associated with the cooler, wetter highlands near Flagstaff, Arizona, is much different from the desert lowlands.

**MapMaster** ▶ World Layered Thematic Map ▶ Physical Environment ▶ Climate ▶ Highland

## eye ON THE atmosphere 15.2

These images show an aerial view and close-up of a portion of the Trans-Alaska Pipeline. This 1300-kilometer- (800-mile-) long structure was constructed in the 1970s to transport oil from Prudhoe Bay on Alaska's North Slope, southward to the ice-free port of Valdez on the Gulf of Alaska. Because temperatures in this region range from cool to frigid, the oil must be heated to flow properly. Pumping stations along the way keep the oil moving through the pipeline.

### Questions

1. The pipeline passes through subarctic (Dfc) and tundra (ET) climate zones. In which climate zone were these photos taken? What clue in the photos helped you figure this out?
2. Notice that in these views, the pipeline is suspended above ground. Based on your knowledge of the climate and landscape of the region, suggest a reason why the pipeline is elevated above ground level.

are fair, mountain and valley breezes are created by the topography itself.

We know that climate strongly influences vegetation, which is the basis for the Köppen system. Thus, where there are vertical differences in climate, we should expect a vertical zonation of vegetation as well. Ascending a mountain can let us view dramatic vegetation changes that otherwise might require a poleward journey of thousands of kilometers. This occurs because altitude duplicates, in some respects, the influence of latitude on temperature and hence on vegetation types. However, we know that other factors—such as slope orientation, exposure, winds, and orographic effects—also play roles in controlling the climate of highlands. Consequently, although the concept of vertical life zones applies on a broad regional scale, the details within an area vary considerably. Some of the most obvious variations result from differences in rainfall or the receipt of solar radiation on opposite sides of a mountain.

To summarize, *variety* and *changeability* best describe highland climates. Because atmospheric conditions fluctuate with altitude and exposure, a nearly limitless variety of local climates occurs in mountainous regions. The climate in a protected valley is very different from that on an exposed peak. Conditions on windward slopes contrast sharply with those on the leeward sides. Slopes facing the Sun are unlike those that lie mainly in the shadows.

## ✔ Concept Checks 15.8

1. The Arizona cities Flagstaff and Phoenix are relatively close to one another yet have contrasting climates. In what ways do they differ, and why?
2. Refer to the precipitation map of Nevada in Figure 15.24. The rainiest areas (>40 centimeters per year) appear as isolated and disconnected zones. Explain this pattern.

# 15 Concepts in Review World Climates

## 15.1 Climate Classification

▶ Explain why classification is a necessary process when studying world climates. Discuss the criteria used in the Köppen system of climate classification.

**Key Term:** Köppen classification

- Climate classification brings order to large quantities of information, which aids comprehension and understanding and facilitates analysis and explanation.
- The most important elements in climate descriptions are temperature and precipitation because they have the greatest influence on people and their activities and also have an important impact on the distribution of vegetation and the development of soils.
- An early attempt at climate classification by the Greeks divided each hemisphere into three zones: torrid, temperate, and frigid. Many climate classifications have been devised, and the value of each is determined by its intended use.
- The Köppen classification, which uses mean monthly and annual values of temperature and precipitation, is a widely used system. The boundaries Köppen chose were largely based on the limits of certain plant associations.
- Five principal climate groups, each with subdivisions, were recognized. Each group is designated by a capital letter. Four of the climate groups (A, C, D, and E) are defined on the basis of temperature characteristics, and the fifth, the B group, has precipitation as its primary criterion.

## 15.2 Climate Controls: A Summary

▶ List and briefly discuss the major controls of climate.

**Key Terms:** marine climate, continental climate

- Order exists in the distribution of climate elements, and the pattern of climates is not by chance. The world's climate pattern reflects a regular and dependable operation of the five major climate controls.
- Latitude relates to seasonal variations in the receipt of solar energy. Temperature differences are largely a function of latitude.
- Land/water influences are an important control. Marine climates are relatively mild, whereas continental climates are typically more extreme.
- Geographic position and prevailing winds affect the degree of land and water influence. The moderating effect of water is more pronounced along the windward side of a continent.
- Mountains and highlands prevent maritime air masses from reaching far inland, trigger orographic precipitation, and, where they are extensive, create their own climate regions.
- Poleward-moving warm ocean currents cause air temperatures to be warmer than otherwise would be expected, whereas equatorward-moving cold currents promote cooler-than-expected temperatures.

## 15.3 Humid Tropical (A) Climates

▶ Compare the two broad categories of humid tropical climates.

**Key Terms:** tropical rain forest, intertropical convergence zone (ITCZ), tropical wet and dry, savanna, monsoon

- Straddling the equator, the wet tropics (Af, Am) exhibit constant high temperatures and year-round rainfall that combine to produce the most luxuriant vegetation in any climatic realm—the tropical rain forest. Temperatures in these regions usually average 25°C (77°F) or more each month.
- Precipitation in Af and Am climates is normally from 175 to 250 centimeters (68 to 98 inches) per year and is more variable than temperature, both seasonally and from place to place. Thermally induced convection coupled with convergence along the intertropical convergence zone (ITCZ) leads to widespread ascent of the warm, humid, unstable air and ideal conditions for cloud formation and precipitation.
- The tropical wet and dry (Aw) climate region is a transitional zone between the rainy tropics and the subtropical steppes. Here, the rain forest gives way to the savanna, a tropical grassland with scattered deciduous trees. Only modest temperature differences exist between the wet tropics and the tropical wet and dry regions. The primary factor that distinguishes the Aw climate from Af and Am is precipitation. In Aw regions the precipitation is typically between 100 and 150 centimeters (40 to 60 inches) per year and exhibits some seasonal character—wet summers followed by dry winters.
- In much of India, Southeast Asia, and portions of Australia, these alternating periods of rainfall and drought are associated with the monsoon, wind systems with a pronounced seasonal reversal of direction. The Cw climate, which is subtropical instead of tropical, is a variant of Aw.

**Q** When does the rainy season occur in a tropical wet and dry climate: winter or summer? Explain.

## 15.4 The Dry Climates (B)

▶ Contrast low-latitude dry climates and middle-latitude dry climates.

**Key Terms:** dry climate, arid (desert), semiarid (steppe)

- Dry regions of the world cover about 30 percent of Earth's land area. Other than meager yearly rainfall, the most characteristic feature of dry climates is unreliable precipitation. Climatologists define a dry climate as one in which the yearly precipitation is less than the potential water loss by evaporation. To define the boundary between dry and humid climates, the Köppen classification uses formulas that involve three variables: average annual precipitation, average annual temperature, and seasonal distribution of precipitation.

- The two climates defined by a general water deficiency are arid, or desert (BW), and semiarid, or steppe (BS). The differences between deserts and steppes are primarily a matter of degree. Semiarid climates are a marginal and more humid variant of arid climates that represent transitional zones that surround deserts and separate them from the bordering humid climates.
- Dominated by subtropical highs, the subtropical desert (BWh) and steppe (BSh) climates lie near the Tropic of Cancer and the Tropic of Capricorn. Within subtropical deserts, the scant precipitation is both infrequent and erratic. In the semiarid transitional belts surrounding the desert, a seasonal rainfall pattern becomes better defined.
- Due to cloudless skies and low humidities, low-latitude deserts in the interiors of continents have the greatest daily temperature ranges on Earth. Where subtropical deserts are found along the west coasts of continents, cold ocean currents produce cool, humid conditions, often shrouded by low clouds or fog.
- Unlike their low-latitude counterparts, middle-latitude deserts (BWk) and steppes (BSk) are not controlled by the subsiding air masses of the subtropical highs. Instead, these climates exist principally because of their position in the deep interiors of large landmasses.

**Q** This photo was taken in Nevada's Great Basin desert, looking west toward the Sierra Nevada range. What factors contribute to the arid conditions in this region?

## 15.5 Humid Middle-Latitude Climates with Mild Winters (C)

▶ Distinguish among the three categories of C climates.

**Key Terms:** humid subtropical climate, marine west coast climate, dry-summer subtropical climate (Mediterranean climate)

- The average temperature of the coldest month in humid middle-latitude climates with mild winters (C climates) is less than 18°C (64°F) but above –3°C (27°F).
- Humid subtropical climates (Cfa) are on the eastern sides of the continents, in the 25° to 40° latitude range. Because of the dominating influence of maritime tropical air masses, summer weather in these regions is hot and sultry, and winters are mild.
- In North America, the marine west coast climate (Cfb) extends from near the U.S.–Canada border northward as a narrow belt into southern Alaska. The prevalence of maritime air masses means mild winters, cool summers, and ample rainfall throughout the year.
- Dry-summer subtropical (Mediterranean) climates (Csa and Csb) are typically found along the west sides of continents, between latitudes 30° and 45°. In summer, the regions are dominated by the stable eastern sides of the oceanic subtropical highs. In winter, as the wind and pressure systems follow the Sun equatorward, they are within range of the cyclonic storms of the polar front.

## 15.6 Humid Continental Climates with Severe Winters (D)

▶ Summarize the characteristics of the two categories of D climates.

**Key Terms:** humid continental climate, subarctic climate, taiga climate

- The average temperature of the coldest month in humid continental climates with severe winters (D climates) is −3°C (27°F) or below, and the average temperature of the warmest month exceeds 10°C (50°F).
- Humid continental climates (Dfa) are land controlled and do not occur in the Southern Hemisphere. They are confined to central and eastern North America and Eurasia, in the latitude range 40° to 50°N. Both winter and summer temperatures in Dfa climates can be characterized as severe, and annual temperature ranges are large. Precipitation is generally greater in summer and generally decreases toward the continental interior and from south to north. Wintertime precipitation is chiefly associated with the passage of fronts connected with traveling midlatitude cyclones.
- Subarctic climates (Dfc and Dfd), often called taiga climates because they correspond to the northern coniferous forests of the same name, are situated north of the humid continental climates and south of the polar tundras. The outstanding feature of subarctic climates is the dominance of winter. Summers in the subarctic, though short, are remarkably warm. The greatest annual temperature ranges on Earth occur here.

## 15.7 The Polar Climates (E)

▶ Contrast tundra and ice cap climates.

**Key Terms:** polar climate, tundra climate, permafrost, ice-cap climate

- Polar climates (ET and EF) are those in which the mean temperature of the warmest month is below 10°C (50°F). Polar climates have the lowest annual mean temperatures on Earth. Although polar climates are classified as humid, precipitation is generally meager, with many nonmarine stations receiving less than 25 centimeters (10 inches) annually.
- Two types of polar climates are recognized. Found almost exclusively in the Northern Hemisphere, the tundra climate (ET), marked by the 10°C (50°F) summer isotherm at its equatorward limit, is a treeless region of grasses, sedges, mosses, and lichens with permanently frozen subsoil, called permafrost. The ice-cap climate (EF) does not have a single monthly mean above 0°C (32°F), prohibiting the growth of vegetation, and the landscape is one of permanent ice and snow.

## 15.8 Highland Climates ▸ List the characteristics of highland climates.

**Key Term:** highland climates

- Highland climates are characterized by a great diversity of climatic conditions over a small area. In North America highland climates occur in the Rockies, Sierra Nevada, Cascades, and mountains and interior plateaus of Mexico. Although the best-known climatic effect of increased altitude is lower temperatures, greater precipitation due to orographic lifting is also common.
- Variety and changeability best describe highland climates. Because atmospheric conditions fluctuate with altitude and exposure to the Sun's rays, a nearly limitless variety of local climates occur in mountainous regions.

**Q** Which of the local winds discussed in Chapter 7 are likely associated with highland climates?

## Give it Some Thought

1. The tropical soils described in Box 15.1 are considered to have low fertility. Nevertheless, these soils support luxuriant rain forest vegetation like that shown in the accompanying photo. Explain how lush plant growth can occur in poor soils.

2. In which of the following climates is annual rainfall likely to be most consistent—that is, which would probably experience the smallest *percentage* change in rainfall from year to year: BSh, Aw, BWh, or Af? In which of these climates is the annual rainfall most variable from year to year? Explain your answers.
3. Albuquerque, New Mexico, receives an average of 20.7 centimeters (8.07 inches) of rainfall annually. Albuquerque is considered a desert when the Köppen classification is applied. The Russian city of Verkhoyansk is located near the Arctic Circle in Siberia. The yearly precipitation at Verkhoyansk averages 15.5 centimeters (6.05 inches), about 5 centimeters (2 inches) less than Albuquerque, yet it is classified as a humid climate. Explain how this can occur.
4. This problem examines two places in North Africa that are adjacent to the Sahara Desert. Both are classified as having a subtropical steppe (BSh) climate. One place is on the southern margin of the Sahara, and the other is north of the Sahara near the Mediterranean Sea.
   a. In which season, summer or winter, does each place receive its maximum precipitation? Explain.
   b. If both stations barely meet the requirements for steppe climates (that is, with only a little more rainfall, both would be considered humid), which station would probably have the lower rainfall total? Explain.
5. Refer to Figure 15.4, which shows climates of the world. Humid continental (Dfb and Dwb) and subarctic (Dfc) climates are usually described as being "land controlled"—that is, they lack marine influence. Nevertheless, these climates are found along the margins of the North Atlantic and North Pacific oceans. Explain why this occurs.
6. In the dry-summer subtropics, precipitation totals increase with an increase in latitude, but in the humid continental climate, the reverse is true. Explain.
7. Although snowfall in the subarctic climate is relatively scant, a wintertime visitor might leave with the impression that snowfall is great. Explain.
8. Refer to the monthly rainfall data (in millimeters) for three cities in Africa. Their locations are shown on the accompanying map. Match the data for each city (A, B, or C) to the correct location (1, 2, or 3) on the map. How were you able to figure this out? *Bonus:* Select a figure in Chapter 7 that would be especially useful in explaining or illustrating why these places have rainfall maximums and minimums when they do.

| | J | F | M | A | M | J | J | A | S | O | N | D |
|---|---|---|---|---|---|---|---|---|---|---|---|---|
| City A | 81 | 102 | 155 | 140 | 133 | 119 | 99 | 109 | 206 | 213 | 196 | 122 |
| City B | 0 | 2 | 0 | 0 | 1 | 15 | 88 | 249 | 163 | 49 | 5 | 6 |
| City C | 236 | 168 | 86 | 46 | 13 | 8 | 0 | 3 | 8 | 38 | 94 | 201 |

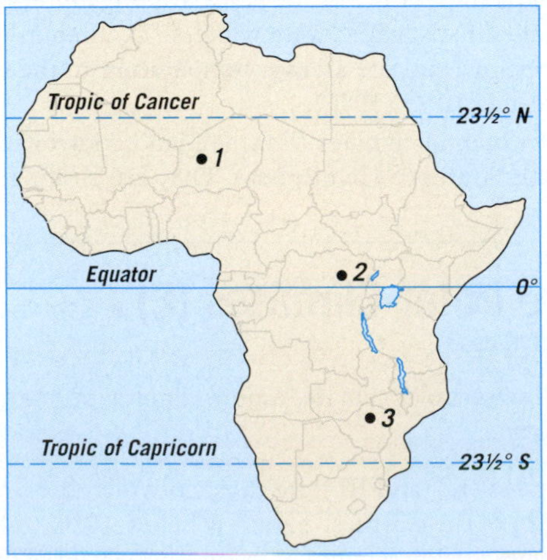

9. Identify three deserts that are visible on this classic view of Earth from space. Briefly explain why these regions are so dry. Identify two other prominent climates in this image. Can you also explain the cause of the band of clouds in the region of the equator?

10. When the rail line in the accompanying photo was built in rural Alaska in the 1940s, the terrain was relatively level. Not long after the railroad was completed, a great deal of subsidence and shifting of the ground occurred, turning the tracks into the "roller coaster" shown here. As a result, the rail line had to be abandoned. Suggest a reason why the ground became unstable and shifted.

## Problems

1. Use Figure 15.2 to determine the appropriate classification for stations A, B, and C in the table below.

| | J | F | M | A | M | J | J | A | S | O | N | D | YR |
|---|---|---|---|---|---|---|---|---|---|---|---|---|---|
| **Station A** | | | | | | | | | | | | | |
| **Temp. (°C)** | −18.7 | −18.1 | −16.7 | −11.7 | −5.0 | 0.6 | 5.3 | 5.8 | 1.4 | −4.2 | −12.3 | −15.8 | −7.5 |
| **Precip. (mm)** | 8 | 8 | 8 | 8 | 15 | 20 | 36 | 43 | 43 | 33 | 13 | 12 | 247 |
| **Station B** | | | | | | | | | | | | | |
| **Temp. (°C)** | 24.6 | 24.9 | 25.0 | 24.9 | 25.0 | 24.2 | 23.7 | 23.8 | 23.9 | 24.2 | 24.2 | 24.7 | 24.4 |
| **Precip. (mm)** | 81 | 102 | 155 | 140 | 133 | 119 | 99 | 109 | 206 | 213 | 196 | 122 | 1675 |
| **Station C** | | | | | | | | | | | | | |
| **Temp. (°C)** | 12.8 | 13.9 | 15.0 | 16.1 | 17.2 | 18.8 | 19.4 | 22.2 | 21.1 | 18.8 | 16.1 | 13.9 | 15.9 |
| **Precip. (mm)** | 53 | 56 | 41 | 20 | 5 | 0 | 0 | 2 | 5 | 13 | 23 | 51 | 269 |

2. Using the maps in Figure 15.19, determine the approximate January and July temperature gradients between the southern tip of mainland Florida and the point where the Minnesota–North Dakota border touches Canada. Assume the distance to be 3100 kilometers (1900 miles). Express your answers in °C per 100 kilometers or °F per 100 miles.

# 16 Optical Phenomena of the Atmosphere

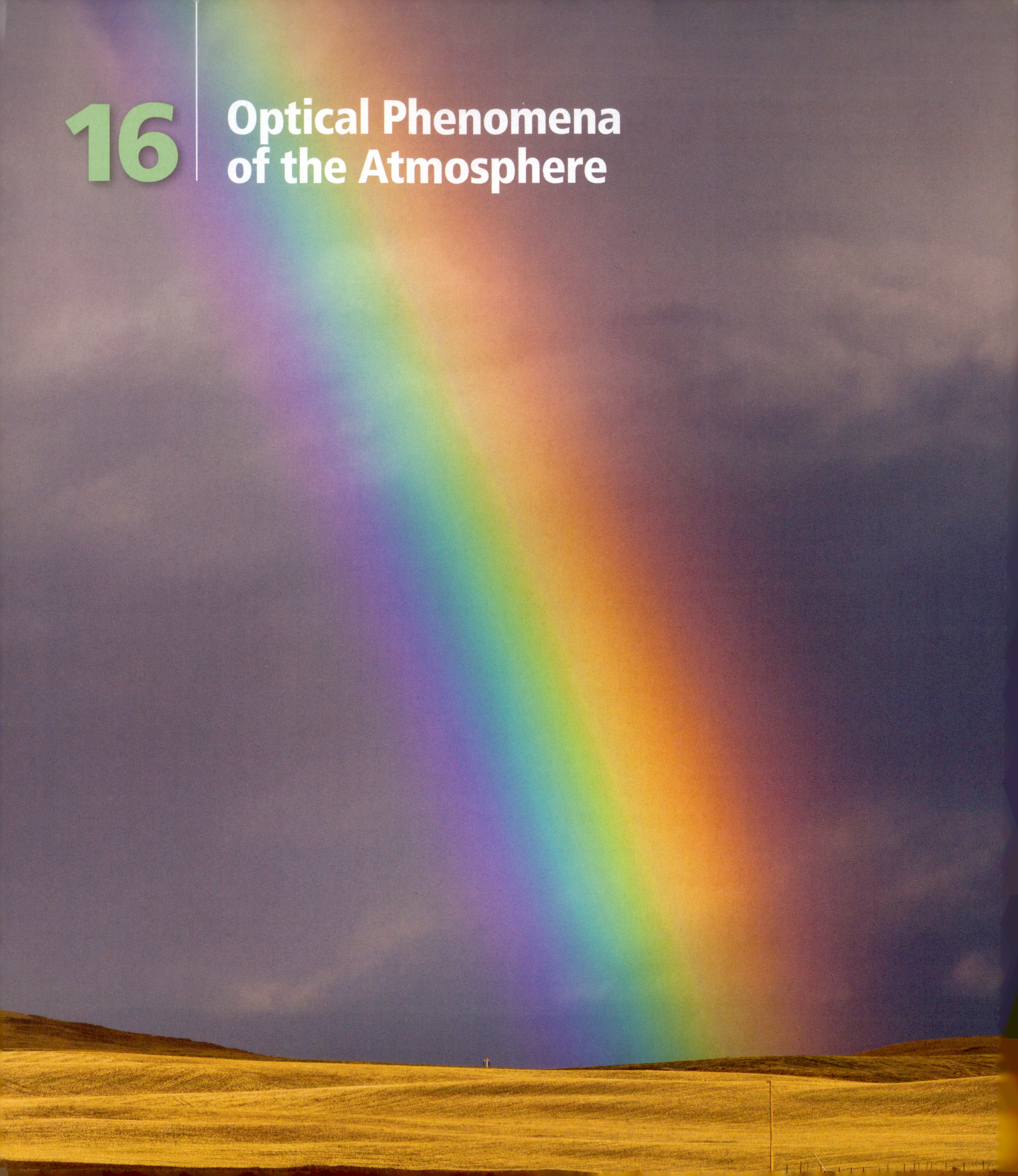

## Focus on Concepts

*Each statement represents the primary learning objective for the corresponding major heading within the chapter. After you complete the chapter, you should be able to:*

**16.1** Explain the principle called the *law of reflection* and discuss how refraction causes white light to separate into the spectrum of colors.

**16.2** Describe how a mirage is created.

**16.3** Sketch a raindrop and show how sunlight travels through it to form a primary rainbow.

**16.4** Distinguish among three different optical phenomena that are produced as a result of the interaction between sunlight and hexagonal ice crystals.

**16.5** Compare and contrast coronas and halos.

One of Earth's most spectacular and intriguing natural phenomena is the rainbow. Its "surprise" appearance and splash of colors make it the subject of both poets and artists, not to mention nearly every casual photographer within reach of a camera. In addition to rainbows, many other stunning optical phenomena occur in our atmosphere. In this chapter, we consider how the most familiar of these displays occur. Knowing when and where to look for these phenomena will allow you to identify each type and, ultimately, witness and appreciate them more frequently.

*Rainbows are among the most common and spectacular of the atmosphere's optical phenomena. This rainbow is near Pincher Creek, Alberta, Canada.*

# 16.1 Interactions of Light and Matter

**Explain the principle called the *law of reflection* and discuss how refraction causes white light to separate into the spectrum of colors.**

The interaction of white (visible) sunlight with our atmosphere creates the numerous optical phenomena we observe in the sky. In Chapter 2 we considered some properties of light and how they contribute to occurrences such as the blue color of the sky and the red color of sunset. This chapter examines other interactions between sunlight and the gases, ice crystals, and water droplets found in the atmosphere that generate still other optical phenomena. We begin with two fundamental ways light and matter interact: reflection and refraction.

## Reflection

Light coming from the Sun toward Earth travels at a uniform speed and in a straight line. However, when light encounters a transparent material, such as a water body, a portion bounces off the surface, and some is transmitted through the material at a slower velocity. The light that bounces back from the surface is *reflected.* Reflected light allows you to see yourself in a mirror. The image that you see in a mirror originates as light that first reflected off you toward the mirror and then was bounced from the silver- or aluminum-coated surface of the mirror back to your eyes. When light is reflected, the rays bounce off the reflecting surface at the same angle at which they met that surface (Fig. 16.1). This principle is called the **law of reflection**. It states that the angle that the incident (incoming) ray makes with a line perpendicular to the reflective surface is equal to the angle that the reflected (outgoing) ray makes to that same line.

On a smooth, highly reflective surface like a mirror, about 90 percent of the parallel incoming rays leave the surface by following outgoing parallel paths. However, when light encounters a slightly irregular surface, the rays are reflected in many directions; this is referred to as *diffuse reflection* (Fig. 16.2). Even something that appears as smooth as a page in this book is sufficiently rough to scatter light in all directions. This type of reflection makes it possible to read the print from almost any angle. Most of the objects in our surroundings are seen by diffuse reflection.

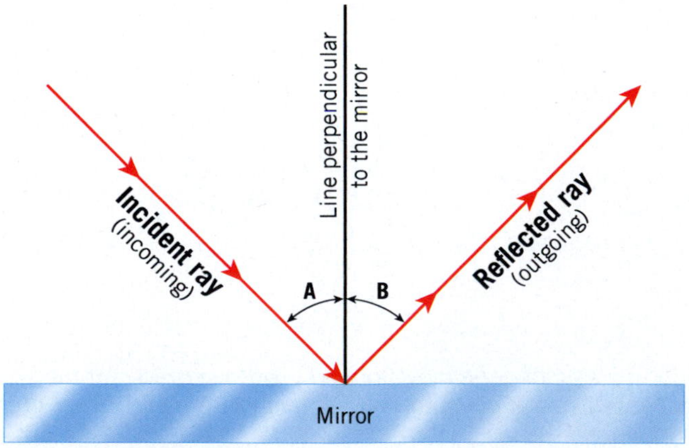


▲ **Figure 16.1 Reflection of light by a mirror** When light rays are reflected from a smooth surface, they bounce off the surface at the same angle at which they met the surface.

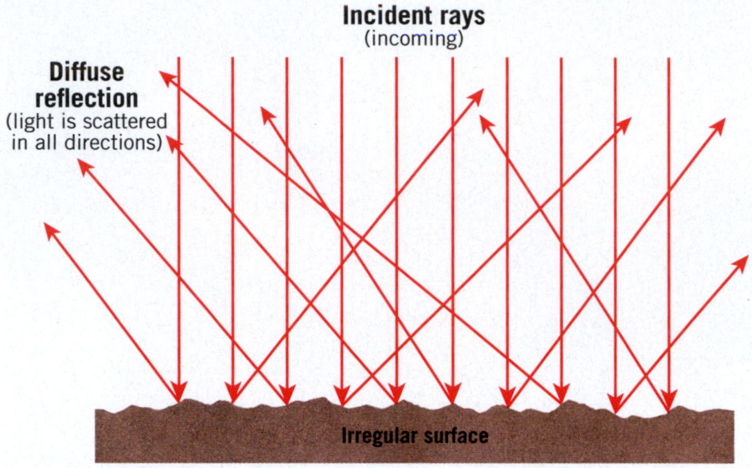


▲ **Figure 16.2 Sunlight striking a rough surface is dispersed (scattered)**

When light is reflected from a very rough surface, the image you see is generally distorted or may appear as multiple images. For example, the reflected image of the Sun when viewed on a wavy ocean surface is not a circular disk but a long, narrow band, as shown in Figure 16.3. What you see is not one image of the Sun but many distorted images of a single light source. This bright band of light is produced because only a portion of the sunlight that strikes each wave is reflected toward the viewer's eyes.

Another type of reflection that is important to our discussion of optical phenomena is called **internal reflection**. Internal reflection occurs when light that travels through a transparent material, such as water, reaches the opposite surface and is reflected back into the same transparent material. You can easily demonstrate this phenomenon by holding a glass of water directly overhead and looking up through the water. You should be able to see clearly through the water because very little internal reflection results when light strikes perpendicular to the surface of a transparent material. While keeping the glass overhead, move it sideways so that you look through the glass at an angle. The underside of the surface of the water takes on the appearance of a silvered mirror. What you are observing here is the internal reflection that occurs when light strikes a water surface at an angle greater than 48° from the vertical (Fig. 16.4). Internal reflection plays an important role in the formation of optical phenomena such as rainbows.

## Refraction

When light strikes a transparent material, the portions that are not reflected are transmitted through the material and

▲ **Figure 16.3 Reflected image of the Sun viewed on a wavy ocean surface** This long, narrow band of sunlight is really multiple distorted images of the Sun reflected from the water surface. Deception Pass Bridge, Washington.

undergo another well-known effect, called *refraction*. **Refraction** is the change of direction of light as it passes obliquely from one transparent medium to another. Refraction occurs because the velocity of light depends on the material that transmits it. In a vacuum, radiation travels at the speed of light (3.0 $\times$ $10^8$ meters per second), but when it travels through air, its speed is slowed slightly. However, in transparent substances such as water, ice, or glass, its speed is considerably slower than in air.*

When light encounters a transparent medium, such as water, at a 90° angle, it travels though that medium in a straight path, as shown in Figure 16.5A. However, when light encounters a transparent medium at an angle of less than 90°, it bends, or refracts (Fig. 16.5B). Refraction can be demonstrated by seeing how a toy car responds when it rolls from smooth flooring onto a carpeted surface. If the toy car meets the carpet at a 90° angle, it slows down but does not change direction. However, if the front wheels of the car hit the carpet at an angle other than 90°, the car not only slows but changes direction, as shown in Figure 16.5C. In this analogy, the directional change occurs because the front wheel that hit the carpet first begins to slow while the other front wheel, which is still on the smooth flooring, continues at the same speed, resulting in a sudden turn of the toy car. Now imagine that the path of the toy car represents the path of light rays, and the smooth flooring and carpet represent air and water, respectively. As light enters the water and is slowed, its path is diverted toward a line extending perpendicularly from the water's surface (Fig. 16.5B). If the light passes from water into air, the bending occurs in the opposite direction—away from the perpendicular.

Refraction can be demonstrated by immersing a pencil half-

▼ **Figure 16.4 Internal reflection** Notice the reflection that appears above the two girls swimming in a pool. Internal reflections are caused by light that is reflected off the interface between the water's surface and the air above.

*The speed of light is a constant; however, as it passes through various substances, light is delayed because of interaction with the electrons in the material. The speed of light as it travels between these intervening electrons is the same as in a vacuum.

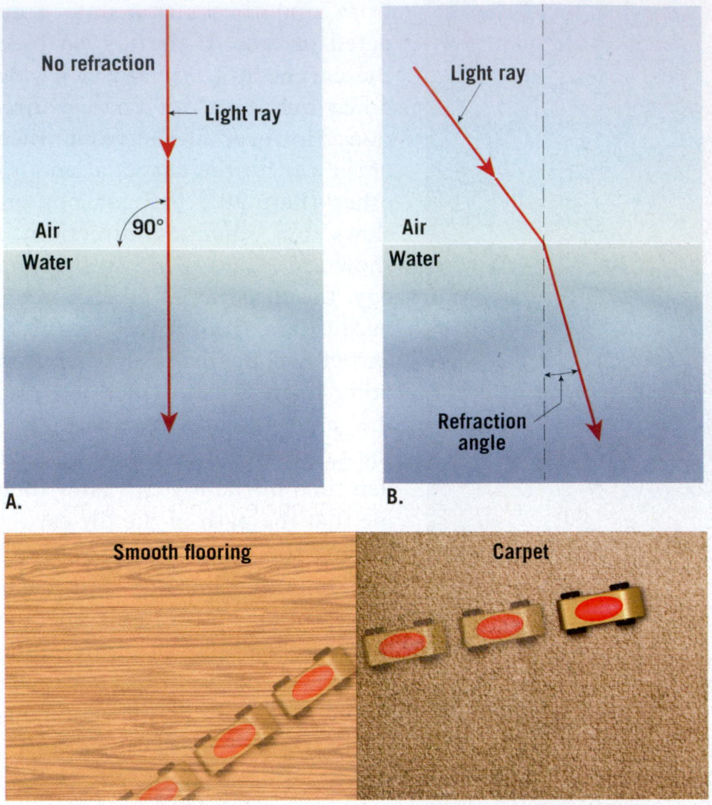


▲ **Figure 16.5 Refraction** **A.** When light rays pass from one transparent material into another at an angle of 90°, the rays travel in a straight path. **B.** Light waves are refracted (bent) when they pass from one transparent material into another at an angle other than 90°. The speed of light is slower in water than in air, which causes light rays to bend toward a line drawn perpendicular to the water surface. **C.** A toy car traveling from a smooth surface to a carpeted floor illustrates the concept of refraction.

way into a glass of water and viewing the pencil obliquely—an angle other than 90° to the surface (**Fig. 16.6A**). How refraction produces the apparent bending of a pencil is illustrated in **Figure 16.6B**. The solid lines in this sketch show the actual path taken by the light, whereas the dashed lines indicate the straight path a human brain perceives. As we look down at the pencil, the point appears to be closer to the surface than it actually is. This occurs because our brain perceives the light as coming along the straight path indicated by the dashed line rather than along the actual bent path. Because all the light coming from the submerged portion of the pencil is bent similarly, this portion of the pencil appears nearer the surface. Therefore, where the pencil enters the water, it appears to be bent upward, toward the surface.

In addition to the abrupt bending of light as it passes obliquely from one transparent substance to another, light also gradually bends as it traverses a material of varying density. As the density of a material changes, the velocity of light changes as well. Within Earth's atmosphere, for example, the density of air usually increases toward Earth's surface. The result of this gradual density change is an equally gradual slowing and bending of light rays. Rays that travel from areas of lower air density to areas of higher air density curve in a direction that has the same orientation as Earth's curvature.

The bending of light caused by refraction is responsible for a number of common optical phenomena. These phenomena result because our brain perceives bent light as if it has traveled to our eyes along a straight path. Try to imagine "looking down" a bent light ray to view a light source located around a corner. If you could see the object, your brain would place that object away from the corner, in "plain sight." On occasion, we see things that are "around the corner." One example is how we view the setting Sun. Several minutes after the Sun has

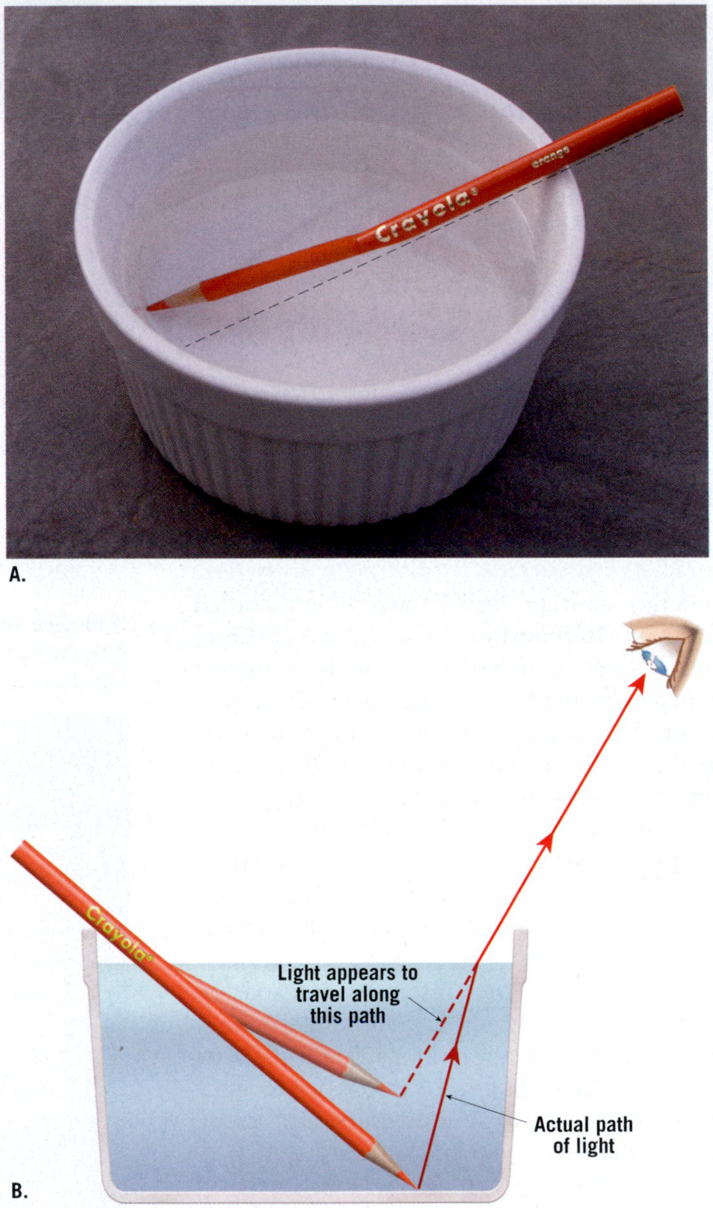


▲ **Figure 16.6 Demonstration of refraction using a pencil**
**A.** Image illustrating refraction of a pencil immersed in water.
**B.** The pencil appears bent because the eye perceives light as if it were traveling along the dashed line rather than along the solid line, which represents the path of the refracted light.

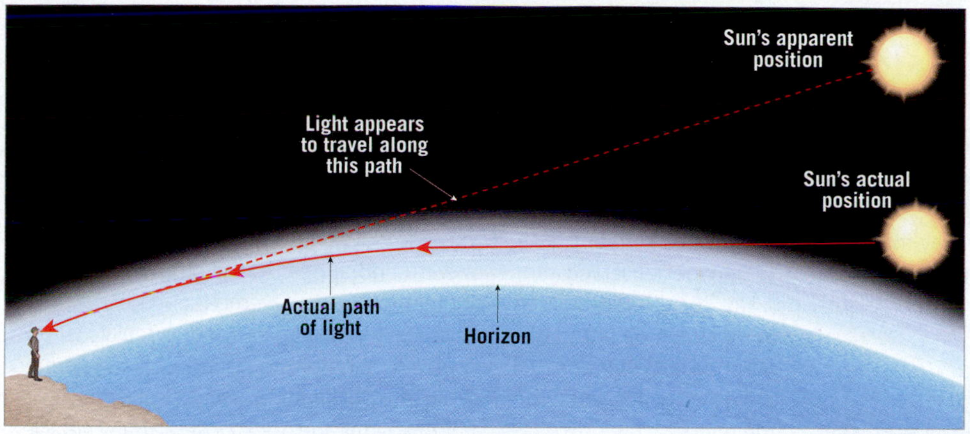


▲ **Figure 16.7 Refraction causes an apparent displacement of the position of the Sun** This displacement is the result of sunlight being refracted as it passes through layers of the atmosphere that have different densities. Because the atmosphere is less dense aloft than it is at the surface, it causes light to bend with the same curvature as Earth's surface.

actually slipped below the horizon, it still appears to us as a full disk. **Figure 16.7** illustrates this situation. It is our *inability to perceive light bending* that causes the position of the Sun to appear above the horizon after it has set.

## ✔ Concept Checks 16.1

1. State the *law of reflection*.
2. Briefly describe the optical phenomenon known as *internal reflection*.
3. What happens to light as it passes obliquely from one transparent medium into another?
4. Provide at least one example of how *refraction* changes the way we perceive an object.

# 16.2 Mirages

**Describe how a mirage is created.**

One of the most interesting optical events common to our atmosphere is the **mirage**. Although this phenomenon is most often associated with desert regions, it can be experienced anywhere (**Box 16.1**). One type of mirage occurs on very hot days, when the air near the ground is much hotter and, therefore, less dense than the air aloft. As noted earlier, a change in the density of air is accompanied by a gradual bending of the light rays. When light from above travels through air that is less dense near the surface, the rays bend in a direction opposite to Earth's curvature. As shown in **Figure 16.8**, this direction of bending causes the light from a distant object to approach the observer from below eye level. Because the brain perceives the light as following a straight path, the object appears below its original position and is often *inverted*, as illustrated with the palm tree in Figure 16.8. The palm tree appears inverted because the rays that originate near the top of the tree are bent more than those originating near the base of the tree.

In the classic desert mirage, a lost and thirsty wanderer encounters an image consisting of an oasis of palm trees and a shimmering water surface on which a reflection of the palms is "seen." Although the trees are real, the water and reflected palms are part of the mirage. Light traveling to the observer through the cooler air above produces the image of the actual trees. The reflected image of the palms is produced from light that traveled downward from the trees and was gradually bent upward as it traveled through the hot (less dense) air near the ground. The image of shimmering water is produced by light that travels downward from the sky and is bent upward. Thus, the water mirage is actually the sky. These desert mirages are called **inferior mirages** because the images appear *below* the true location of the observed object.

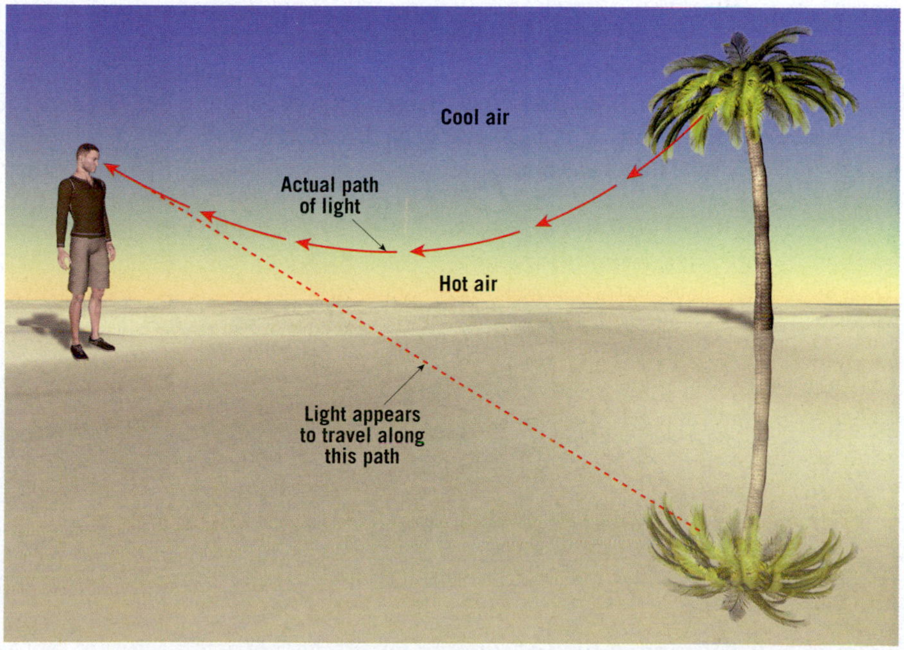


► **Figure 16.8 The classic desert mirage is a type of inferior mirage** In the classic desert mirage, light travels more rapidly in the hot air near Earth's surface than in the cooler air above. Thus, as downward-directed light rays enter the warm air near the surface, they are bent (refracted) upward so that they reach the observer from below eye level, causing objects to appear lower than they actually are.

## Box 16.1 Are Highway Mirages Real?

You have undoubtedly seen a mirage while traveling along a highway on a hot summer afternoon. The most common highway mirages are in the form of "wet areas" that appear on the pavement ahead, only to disappear as you approach (**Fig. 16.A**). Because the "water" always disappears as a person gets closer, many people believe these mirages are only optical illusions. This is not the case. Highway mirages, as well as all other types of mirages, are as real as the images observed in a mirror. As shown in Fig. 16.A, highway mirages can be photographed and therefore are not tricks played by the mind.

▲ **Figure 16.A A classic highway mirage**

What causes the "wet areas" that appear on dry pavement? On hot summer days, the layer of air near Earth's surface is much warmer than the air aloft. Sunlight traveling from a region of colder (more dense) air into the warmer (less dense) air near the surface bends in a direction opposite to Earth's curvature (see Fig. 16.8). As a consequence, light rays that began traveling downward from the sky are refracted upward and appear to the observer to have originated on the pavement ahead. What appears to the traveler as water is really just an inverted image of the sky. This can be verified by careful observation. The next time you view a highway mirage, look closely at any vehicle ahead of you at about the same distance as the "wet" area. Below the vehicle you should be able to see an inverted image of it. Such an image is produced in the same manner as the inverted image of the sky.

### Question

1. Explain the cause of the "wet areas" that sometimes appear in the distance when you are driving on a hot, clear summer day.

Another common type of mirage occurs when the air near the ground is much cooler and, therefore, more dense than the air aloft. Consequently, this effect is observed most frequently in polar regions or over cool ocean surfaces. When the air near the ground is substantially colder than the air above, light bends in the same direction as Earth's curvature. As shown in **Figure 16.9**, this effect allows ships to be seen where ordinarily Earth's curvature would block them from view. This phenomenon, referred to as **looming**, occurs when the refraction of light is significant enough to make the object appear suspended above the horizon. In contrast to a desert mirage, which is an *inferior mirage*, looming is a **superior mirage** because the image is seen *above* its true position.

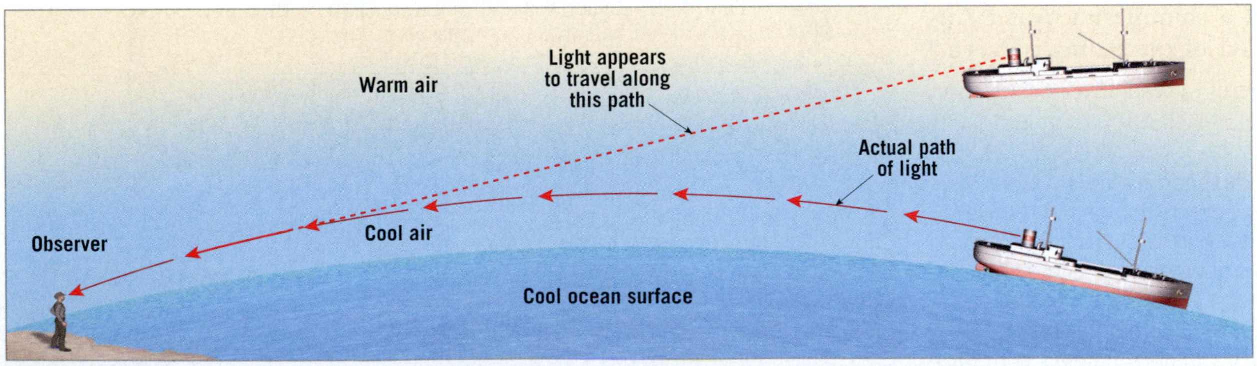


▲ **Figure 16.9 Looming: A type of superior mirage** As light enters a layer of cool air near the surface, it slows and bends downward. This causes objects to be seen above their actual position, hence the term *superior* mirage.

In addition to the easily explained inferior and superior mirages, several more complex mirages have been observed. They occur when the atmosphere develops a temperature profile in which the temperature changes erratically with altitude, causing similar changes in the air's density. Under these conditions, each thermal layer acts like a glass lens. Because each layer bends the light rays somewhat differently, the size and shape of the objects observed through these thermal layers are greatly distorted. You may have observed something similar if you have ever entered a "house of mirrors" at a carnival. While one of the mirrors makes you look taller, others either stretch or compress the image of your body. Mirages are capable of similarly distorting the size and shape of objects.

One type of mirage that changes the apparent size of an object is called **towering**. As the name implies, towering creates a much taller, and often distorted, image of the original object. This optical phenomenon is most frequently observed in coastal areas, where sharp temperature contrasts are common. An interesting type of towering

**eye ON THE atmosphere 16.1**

The accompanying image is an optical phenomenon known as a *green flash* that appears as a green cap above the Sun at sunset or sunrise. Green flashes result from atmospheric changes in density that cause sunlight to bend as it travels from the thinner air in the upper atmosphere into the denser air near Earth's surface, where the observer is located. Green flashes are produced because the light at the blue/green end of the spectrum is bent more than the light in the red/orange range. As a result, the blue rays from the upper limb of the setting Sun remain visible to an observer on Earth after the red rays are obstructed by the curvature of Earth. We should, therefore, expect to see a glimpse of blue/green light at each sunset. Clear conditions and an unobstructed view of the sunset produce the most vivid emerald green flashes. Green flashes are rare, however, because almost all of the blue light and much of the green light is removed from a sunbeam by scattering (see Chapter 2).

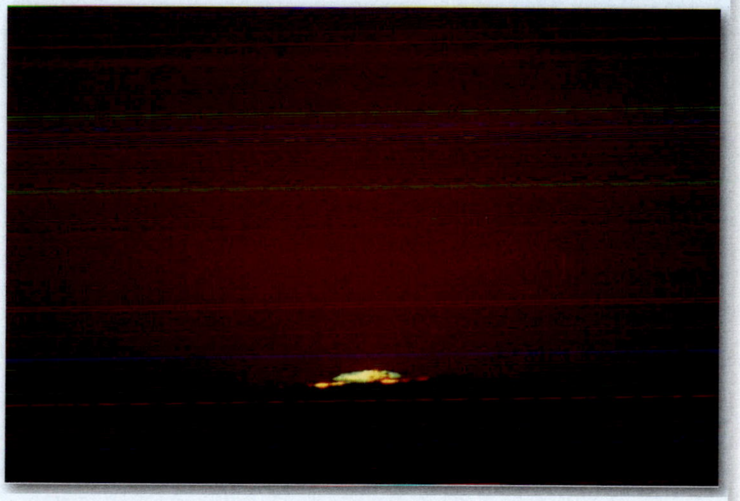

**Questions**

1. What term is used for the bending of light as it travels through a transparent material that exhibits various densities?
2. Green flashes are most commonly photographed over the ocean at sunset. Suggest a reason for this.

is called the *Fata Morgana*, named for the legendary sister of King Arthur, who was credited with the magical power of being able to create towering castles out of thin air (Fig. 16.10). In addition to generating magical castles, the Fata Morgana probably explains the towering mountains that early explorers of the polar regions of the globe observed in the distance but that never materialized.

▲ **Figure 16.10 Fata Morgana** This type of mirage makes objects look taller than they actually are. This particular image is of two icebergs floating on water. Notice that the upper portion of this image (the mirage) is an inverted image of the actual iceberg.

### ✔ Concept Checks 16.2

1. What happens to light if the density of the air through which it travels changes with height?
2. When light travels from warm (less dense) air into a region of colder (denser) air, its path will curve. Does it bend in the same direction as Earth's curvature or in the opposite direction?
3. What is meant by the term *inferior mirage*?
4. How is an inferior mirage different from a superior mirage?

**students sometimes ask...**

**Why is the Moon much larger when it is low on the horizon than when it is overhead?**

This phenomenon is an optical illusion that has nothing to do with the refraction of light. The size of the Moon does not actually change as it rises above the horizon. As meteorologist Craig Bohren stated, "The Moon illusion results from refraction by the mind, mirages from refraction by the atmosphere."*

*Craig F Bohren, *What Light Through Yonder Window Breaks?: More Experiments in Atmospheric Physics* (New York: John Wiley & Sons, Inc), ©1991.

## 16.3 Rainbows

**Sketch a raindrop and show how sunlight travels through it to form a primary rainbow.**

Probably the best-known and most spectacular of all optical phenomena in our atmosphere is the **rainbow** (Fig. 16.11). An observer on the ground sees a rainbow as an arch-shaped array of colors that spans a large segment of the sky. Although the clarity of the colors varies with each rainbow, the observer can usually discern six rather distinct bands of color. The outermost

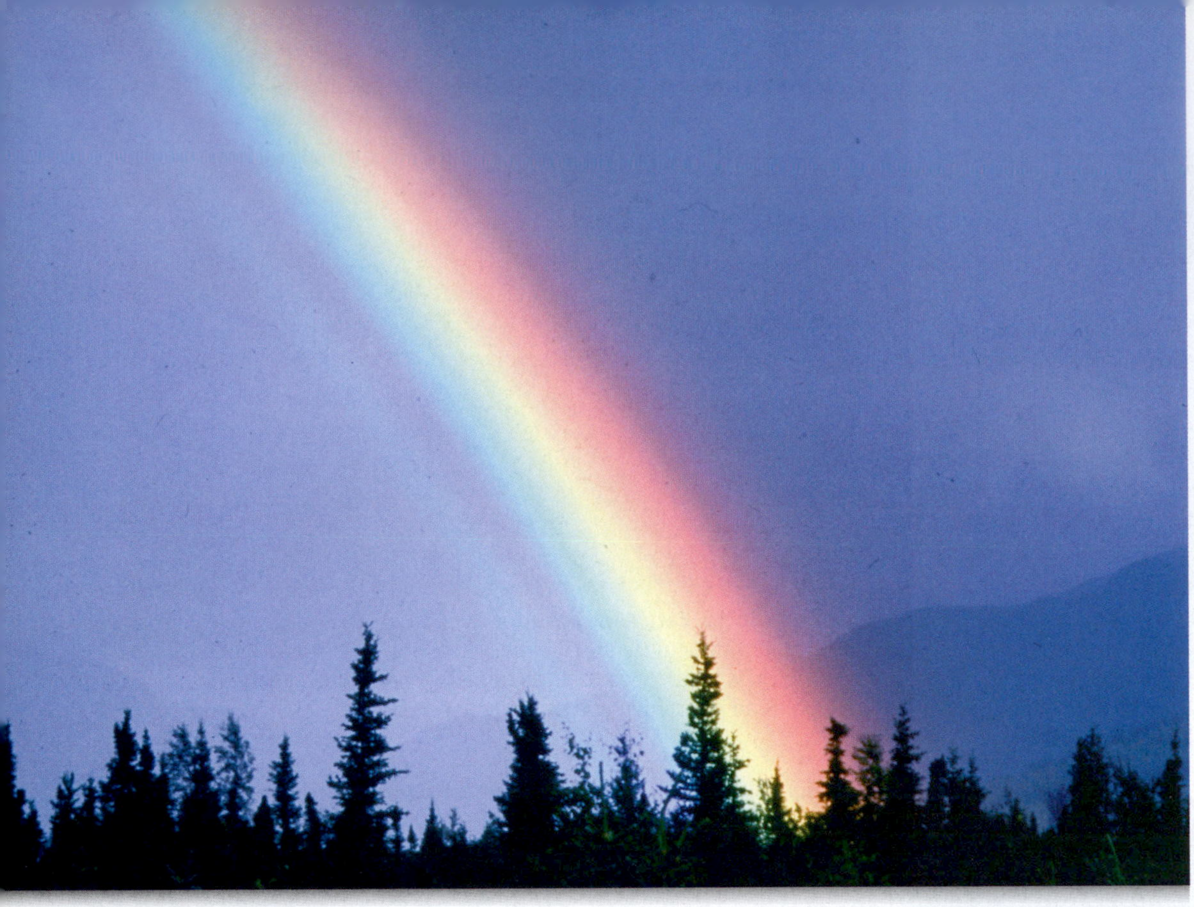

▲ **Figure 16.11 Rainbow** This rainbow appears over spruce trees in the taiga (boreal forest) north of the Alaska Range

band is red and blends gradually to orange, yellow, green, blue, and eventually violet light. These spectacular splashes of color are seen when the Sun is behind the observer and a rain shower is in front. A fine mist of water droplets generated by a waterfall or lawn sprinkler can also generate miniature rainbows.

To understand how raindrops disperse sunlight to generate a rainbow, recall our discussion of refraction. Remember that as light passes obliquely from air to water, its speed slows, which causes it to be refracted (change direction). In addition, each color of light travels at a different velocity in water; consequently, each color will be bent at a slightly different angle. Violet-colored light, which travels at the slowest rate, is refracted the most, whereas red light, which travels most rapidly, is bent the least. Thus, when sunlight (which consists of all colors) enters water, refraction separates it into colors according to their velocity.

Seventeenth-century scientist Sir Isaac Newton used a prism to demonstrate the concept of color separation. Light transmitted through a prism is refracted twice—once as it passes from the air into the glass and again as it leaves the prism and reenters the air. Newton noted that when light is refracted twice, as by a prism, the separation of sunlight into its component colors is quite noticeable (**Fig. 16.12**). We refer to this separation of colors by refraction as **dispersion**.

Rainbows are generated because water droplets act like prisms, dispersing sunlight into the spectrum of colors. Upon entering a droplet, the sunlight is refracted, with violet light bent the most and red the least (**Fig. 16.13A**). Then, when the sunlight reaches the opposite side of the droplet, the rays are internally reflected backward and exit the droplet on the same side they entered. As they leave the droplet, further refraction increases the amount of dispersion and accounts for the nearly complete color separation.

The angle between the incoming (incident) rays and the dispersed colors that constitute the rainbow is 40° for violet and 42° for red light. The other colors—orange, yellow, green, and blue—are dispersed at angles between 40° and 42°. Although each droplet disperses the full spectrum of colors, an observer sees only one color from any single raindrop. For example, if red light from a particular droplet reaches an observer's eye, the violet light from that droplet is visible only to an observer in a different location (Fig. 16.13A). Consequently, each observer sees his or her "own" rainbow, which is generated by a different set of droplets and different rays of light. In effect, a rainbow is produced through the interaction of sunlight with millions of raindrops, each of which acts as a miniature prism.

The curved shape of the rainbow results because the rainbow rays always travel toward the observer at an angle of about 42° from the path of the sunlight (**Fig. 16.13B**). Thus, we experience a 42° semicircle of color across the sky that we identify as the arch shape of the rainbow. If the Sun is higher than 42° above the horizon, an observer

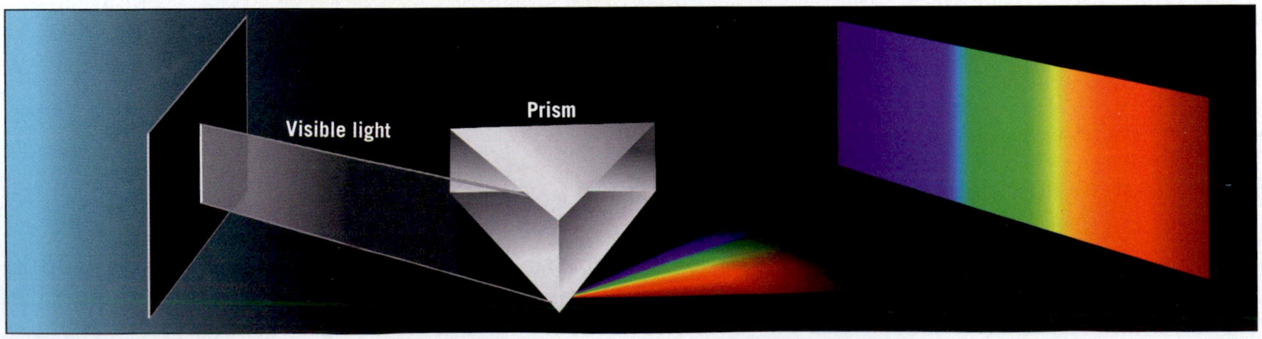


◄ **Figure 16.12 The spectrum of colors is produced when sunlight passes through a prism** Notice that each wavelength (color) of light is refracted (bent) differently.

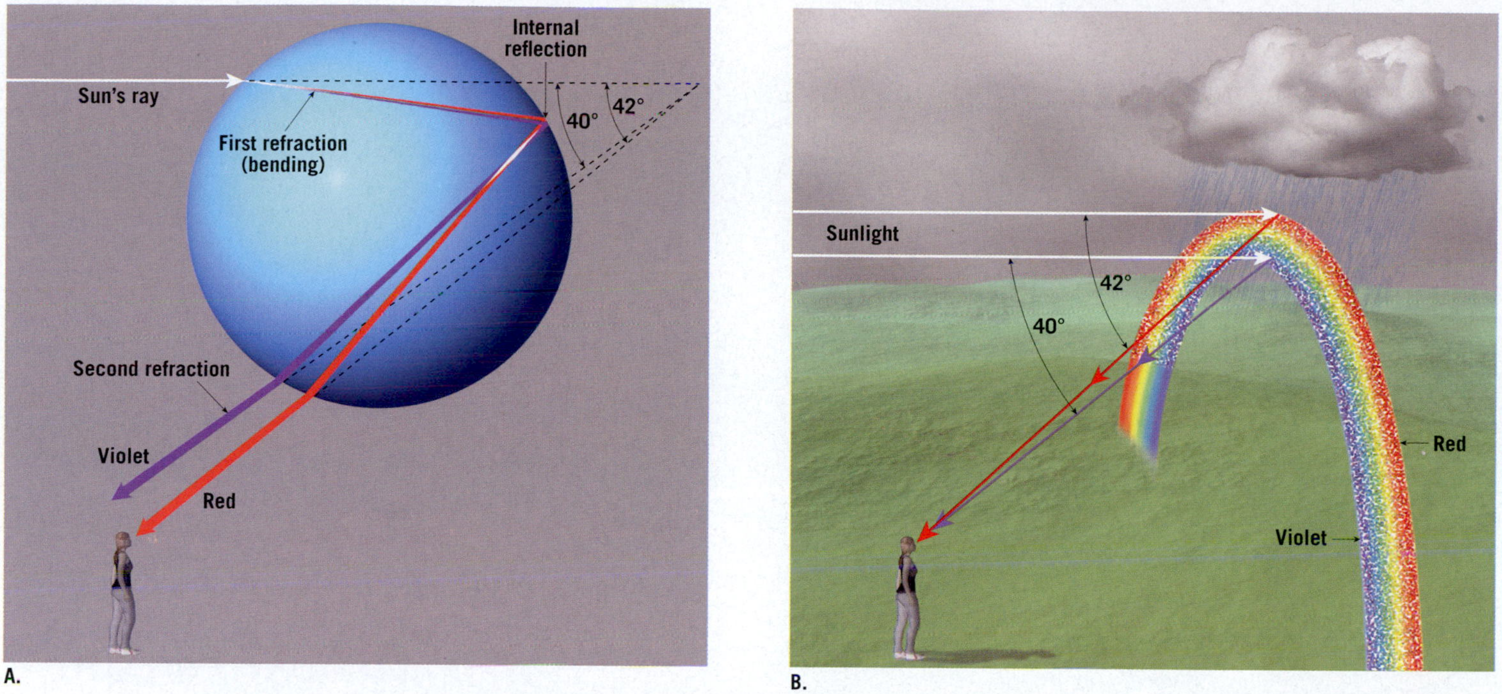


▲ **Figure 16.13 The formation of a primary rainbow** **A.** Color separation results when sunlight is refracted as it enters a raindrop. This light is then internally reflected, and it is refracted again when it leaves the raindrop, resulting in the colors of the rainbow. **B.** It takes millions of raindrops to produce a rainbow. Each observer sees only one color from any one raindrop.

on Earth will not see a rainbow. Under certain conditions, an observer in an airplane will see the rainbow as a full circle.

When a spectacular rainbow is visible, an observer will sometimes see a dimmer secondary rainbow. The secondary bow will be visible about 8° above the primary bow and will suspend a larger arc across the sky.* The secondary bow also has a slightly narrower band of colors than the primary rainbow, and the colors are in reverse order. Red makes up the innermost band of the secondary rainbow and violet the outermost.

The secondary rainbow is generated in much the same way as the primary bow. The main difference is that the dispersed light that constitutes the secondary bow is reflected twice within a raindrop before it exits, as shown in Figure 16.14. The extra internal reflection results in a 50° angle for the dispersion of the color red (about 8° greater than the primary rainbow) and a reverse order of the colors.

The extra reflection also accounts for the fact that the secondary bow is less frequently observed than the primary bow (Fig. 16.15). Each time light strikes the inner surface of the droplet, some of the light is transmitted through the reflecting surface. Because this light is "lost," it does not contribute to the brightness of the rainbow. Although secondary rainbows always form, they are rarely discernible.

Humans use rainbows, like other optical phenomena, as a means of predicting the weather. A well-known weather proverb illustrates this point:

*Rainbow in the morning gives you fair warning,*

*Rainbow in the afternoon, good weather coming soon.*

*For reference, when measuring how far apart objects are in the sky, hold your hand up to the sky. A pinky finger is roughly equal to 1°, three fingers is 5°, and stretched thumb to pinky is about 25°.

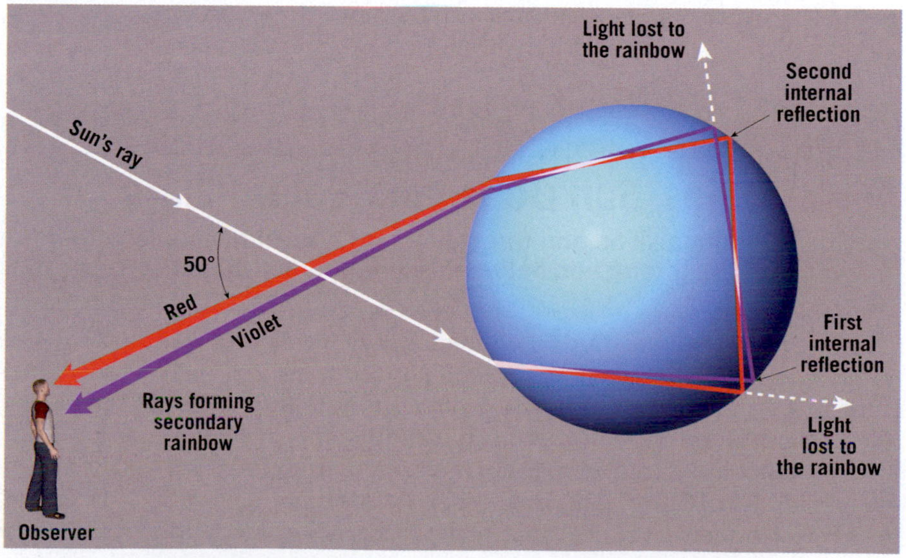


► **Figure 16.14 Formation of a secondary rainbow** A secondary rainbow consists of light rays that experienced two internal reflections rather than one. Notice that the double internal reflection causes the positions of the red and violet rays to be reversed, which accounts for the order of the colors.

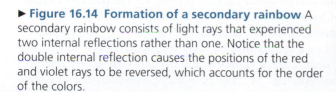

▲ **Figure 16.15 Double rainbow** The secondary rainbow is dimmer than the primary bow, and the order of colors is reversed.

This bit of weather lore relies on the fact that midlatitude weather systems usually move from west to east. Remember that observers must be positioned with their backs to the Sun and facing the rain in order to see a rainbow. When a rainbow is seen in the morning, the Sun is located to the east of the observer, and the clouds and raindrops that are responsible for its formation must therefore be located to the west. Thus, we predict the advance of foul weather when the rainbow is seen in the morning because the rain is located to the west and is traveling toward the observer. In the afternoon, the opposite situation exists: The rain clouds are located to the east of the observer. Therefore, when the rainbow is seen in the afternoon, the rain has already passed. Although this famous proverb has a scientific basis, a small break in the clouds that lets the sunshine through can generate a late-afternoon rainbow. In this situation, a rainbow may certainly be followed shortly by more rainfall.

### ✔ Concept Checks 16.3

1. List the colors of a primary rainbow, in order, from the outer edge to the inner edge.
2. If you were looking for a rainbow in the morning, which direction would you look? Explain.
3. Why is a secondary rainbow less vibrant than a primary rainbow?
4. What term is used to describe the separation of colors that occurs when light is refracted as it passes from one transparent material into another?

## 16.4 Halos, Sun Dogs, and Solar Pillars

**Distinguish among three different optical phenomena that are produced as a result of the interaction between sunlight and hexagonal ice crystals.**

Although halos are a fairly common occurrence, they are rarely seen by casual observers. When noticed, a **halo** usually appears as a narrow whitish ring centered on the Sun or, less often, the Moon (Fig. 16.16). Halos most often appear on days when the sky is covered with a thin layer of cirrus or cirrostratus clouds (see Figure 5.2C, page 127). In addition, halos are seen most often in the morning or late afternoon, when the Sun is near the horizon. Residents of polar regions, where a low Sun and cirrus clouds are common, are frequently treated to views of halos and related phenomena.

The most common halo is called the *22° halo*, named because it has a radius of 22°. Less frequently observed is the larger *46° halo*. Like a rainbow, a halo is produced by dispersion of sunlight. In the case of a halo, however, ice crystals

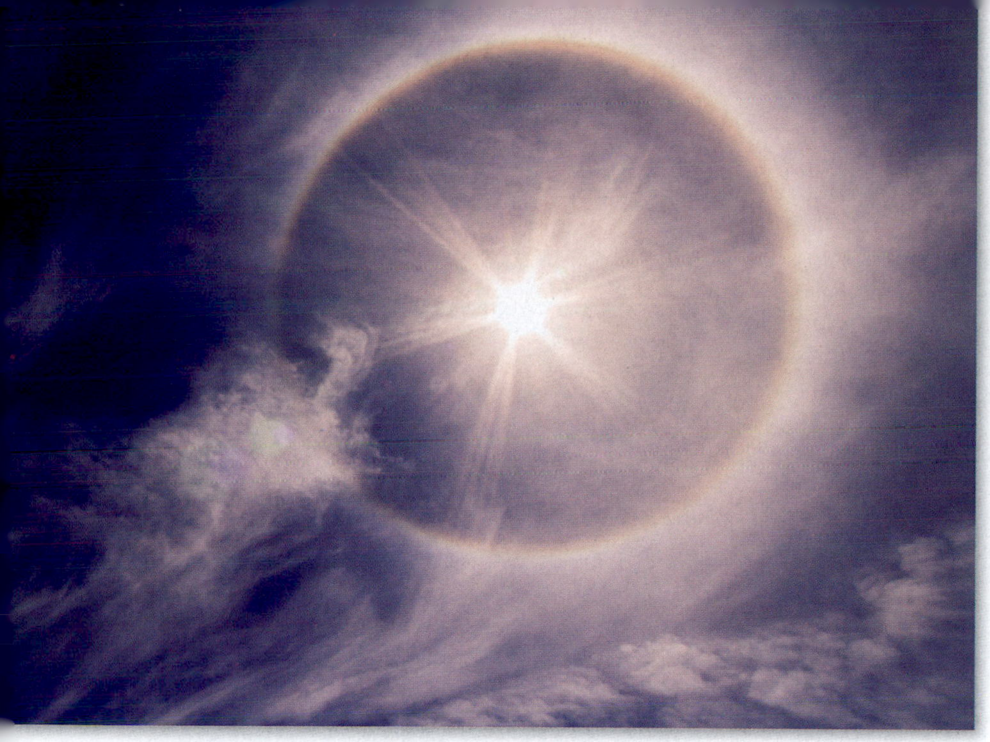

▲ **Figure 16.16 A 22° halo produced by the dispersion of sunlight by ice crystals in cirrostratus clouds**

rather than raindrops refract the light. Thus, the clouds most often associated with halo formation are high clouds. Because cirrus clouds often form as a result of frontal lifting, halos have been accurately described as harbingers of foul weather, as the following weather proverb attests:

*The moon with a circle brings water in her beak.*

Four basic types of hexagonal (six-sided) ice crystals contribute to the formation of halos: plates, columns, capped columns, and bullets (Fig. 16.17). Because the ice crystals responsible for halo formation typically have a random orientation, the

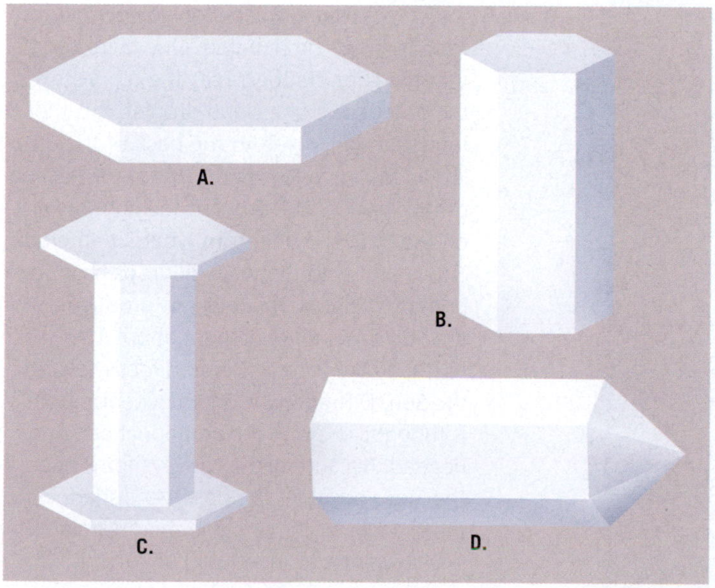


▲ **Figure 16.17 Some common ice crystal shapes found in high clouds** These particular hexagonal shapes contribute to the formation of certain optical phenomena: **A.** plate, **B.** column, **C.** capped column, and **D.** bullet.

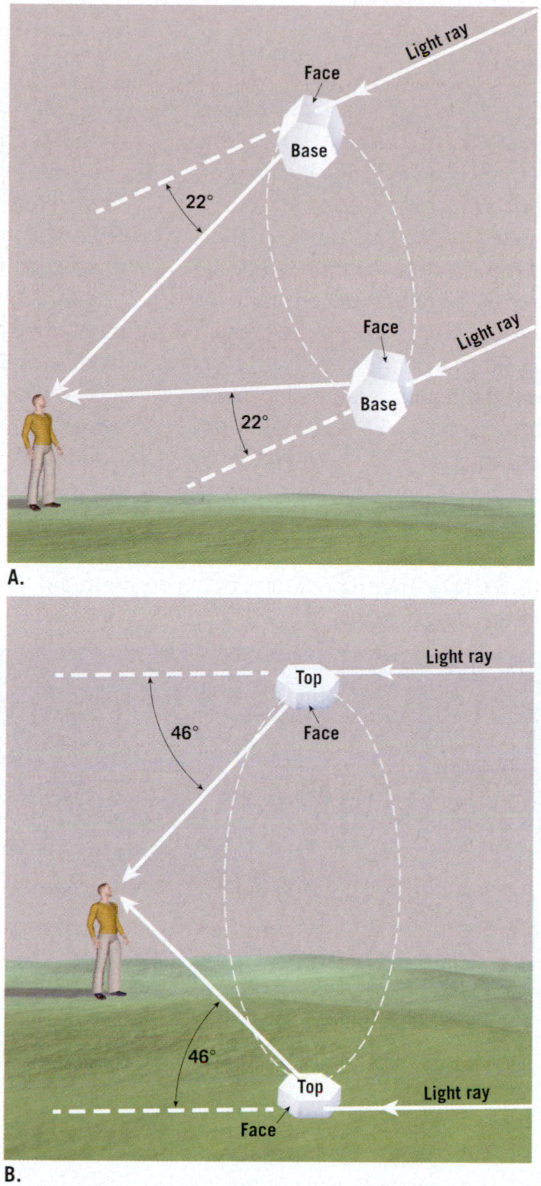


▲ **Figure 16.18 Halos consist of light that is refracted as it passes through ice crystals A.** A 22° halo is produced when the majority of sunlight or moonlight enters a face (rather than the base or top) of an ice crystal and exits one of the opposite faces. **B.** A 46° halo is generated when light enters one of the crystal faces and exits either the top or base of the ice crystal.

diffused light produces a nearly circular halo centered on the illuminating object (Sun or Moon).

The primary difference between 22° and 46° halos is the path that light takes through the ice crystals. The scattered sunlight responsible for the 22° halo enters one face, is refracted, and then exits from a different face, as shown in Figure 16.18A. The angle of separation between the alternating faces of ice crystals is 60°, the same as in a common glass prism. Consequently, ice crystals disperse light in a manner similar to a prism to produce a 22° halo. A 46° halo, by contrast, is formed from light that passes through one face of the crystal, is refracted, and then exits either the top or base of the crystal (Fig. 16.18B). The angle separating these two surfaces is 90°. Light

## Box 16.2 Glories

A *glory* is a spectacle that is rarely witnessed by observers at Earth's surface. However, the next time you are in an airplane and have a window seat, look for the shadow of your aircraft projected on the clouds below. The airplane's shadow is often surrounded by one or more colored rings that constitute a glory (**Fig. 16.B**). Each ring will be colored in a manner similar to a rainbow, with red being the outermost band and violet the innermost. Generally, however, the colors are not as discernible as those of a rainbow. When two or more sets of rings are seen, the inner one will be the brightest and thinnest.

Glories have been observed for centuries and are named for their similar appearance to the iconic halos often seen in religious paintings. In China, this phenomenon is commonly called *Buddha's light*.

Although airplane pilots most commonly see glories, hikers may also see glories when they climb above a cloud or fog layer and have the Sun at their backs. When a hiker's shadow is cast on the cloud or fogbank, the glory will enshroud the observer's head. If two or more persons experience a glory simultaneously, they see only their own shadow surrounded by their own glory.

▼ **Figure 16.B Glory** The colored rings of a glory surround the shadow of an airplane projected on the clouds below.

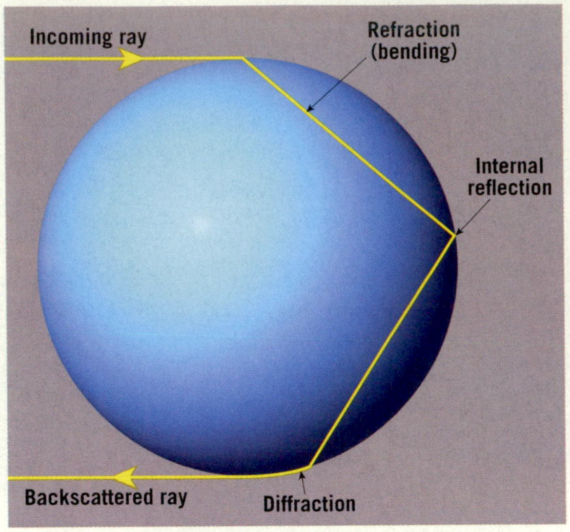


▲ **Figure 16.C Formation of a glory** A glory forms when the Sun's rays strike small, uniform cloud droplets. Although not known with certainty, the glory formation process is thought to involve the refraction, internal reflection, and diffraction of light rays that are eventually backscattered directly toward the Sun.

Glories form by backscattered light in a manner similar to rainbows. However, the cloud droplets that are responsible for glories are much smaller and more uniform in size than the raindrops that produce rainbows.

The scientific explanation of how glories form is still being debated. One hypothesis is that the light that becomes the glory strikes the outer edge of a cloud droplet. It is then refracted and travels to the back of the droplet, as shown in **Figure 16.C**. After reflecting off the back of the droplet, the light ray exits on the opposite side from where it entered. This is similar to the way raindrops produce a primary rainbow. However, to produce a glory, the Sun's rays must be bent about 180°, such that they are backscattered directly toward the Sun. Diffraction is one mechanism that is thought to cause the additional bending needed. Because glories always form opposite the Sun's position, the observer's shadow (or the shadow of the airplane the observer is in) will always be found within a glory.

### Questions

1. Explain the origin of the term *glory*.
2. What material interacts with sunlight to produce a glory?

that passes through ice crystal faces separated by a 90° angle is refracted so it is concentrated 46° from the light source, which accounts for its name. Many other types of halos and partial halos have been observed, all of which owe their formation to the relative abundance of ice crystals of particular shapes and orientations.

Although ice crystals disperse light in the same manner as raindrops (or prisms), halos are generally whitish and do not display the colors of the rainbow. This difference is primarily attributed to the fact that raindrops tend to be uniform in both size and shape, whereas ice crystals vary considerably in size and have imperfect shapes. Although individual ice crystals produce the rainbow of colors in a manner similar to raindrops, the colors tend to overlap and wash each other out. Occasionally, halos display some coloration; in such cases, a bluish ring surrounds a reddish ring.

One of the most spectacular features associated with halos is **sun dogs**. These two bright regions, or "mock suns," as they are often called, can be seen adjacent to a 22° halo and on opposite sides of the Sun (**Fig. 16.19**). Sun dogs form under the same conditions as, and in conjunction with, halos except that their existence depends on the presence of numerous vertically oriented ice crystals. This particular orientation results when elongated ice crystals are slowly descending. Vertically oriented ice crystals cause the Sun's rays to be concentrated (much as with a magnifying glass) in two distinct areas at a distance of about 22° on opposite sides of the Sun.

▲ **Figure 16.19 Sun dogs** Sun dogs form under the same conditions as, and in conjunction with, halos—except that their existence depends on the presence of numerous vertically oriented ice crystals. Vertically oriented ice crystals cause the Sun's rays to be concentrated (much as with a magnifying glass) in two distinct areas at a distance of about 22° on opposite sides of the Sun.

▲ **Figure 16.20 Sun pillars** A Sun pillar is a concentrated shaft of sunlight that is reflected off the bases of platy ice crystals in cold clouds.

Another optical phenomenon caused by falling ice crystals is **sun pillars**. These vertical shafts of light are most often viewed shortly before sunset or shortly after sunrise, when they appear to extend upward from the Sun (**Fig. 16.20**). These bright pillars of light are created when sunlight is reflected toward the observer from the base (bottom) of descending ice crystals that have plate-like shapes oriented like slowly falling leaves. Because direct sunlight is often reddish when the Sun is low in the sky, pillars appear similarly colored. Occasionally, pillars that extend below the Sun can also be viewed.

### ✔ Concept Checks 16.4

1. How are halos and rainbows similar?
2. In what two ways are halos and rainbows different?
3. What is the orientation of the ice crystals that produces a halo? Sun dogs?
4. If a halo is present around the Sun, where would you look to find sun dogs?

## 16.5 Other Optical Phenomena

**Compare and contrast coronas and halos.**

Thus far, we have considered optical phenomena produced when light is reflected and/or refracted (bent) as it travels through a medium such as air, water, or ice crystals. Optical phenomena are also produced when light is bent as it passes around small objects, such as tiny cloud droplets—a process called **diffraction**. Diffraction is described as the interference of any type of wave (here we are concerned with light waves) as it passes near an obstacle comparable in size to its wavelength. Like refraction, diffraction separates white light into the colors of the rainbow.

Two related optical phenomena produced by diffraction are coronas and iridescent clouds. Another optical phenomenon known as a *glory* may involve reflection, refraction, and diffraction (**Box 16.2**).

## eye ON THE atmosphere 16.2

The accompanying image of noctilucent clouds was taken over Bilund, Denmark, on July 15, 2010. *Noctilucent clouds* are thin, wavy clouds that form high in the atmosphere (50–53 miles above Earth) and are observed only for a short time following sunset. Because they are observed only at high latitudes and lie within the mesosphere, they are also known as *polar mesospheric clouds.* Because these usually high clouds are more common today than in the past, some researchers have suggested that they may be a result of global warming.

**Questions**

1. Explain how noctilucent clouds might be illuminated so they are visible to an Earthbound observer after sunset.
2. Do you think noctilucent clouds are composed of water droplets or ice crystals?

▲ **Figure 16.22 Iridescent cloud**

## Coronas

A **corona** most often appears as a bright whitish disk centered on the Moon or Sun. When colors are discernible, the central white disk of the corona is surrounded by one or more concentric rings that exhibit the colors of the rainbow (**Fig. 16.21**). Unique features of coronas include the possible repetition of the corona color sequence and the fact that they are one of the few optical phenomena observed around the Moon more frequently than around the Sun.

A corona is produced when a thin layer of clouds, often altostratus or cirrostratus, veils the illuminating body (Sun or Moon). When the droplets (or sometimes ice crystals) that produce coronas are tiny and uniform in size, diffraction more effectively separates the white light source into the colors of the rainbow. As a result, the coronas generated by these clouds tend to consist of rings that are easily discernable and have the most distinct colors. Coronas produced by larger cloud droplets tend to have muted colors that appear washed out, giving the corona a whitish appearance.

Coronas can be easily distinguished from 22° halos because their color sequence is bluish-white on the inside and reddish on the outside—opposite the halo color sequence. Coronas also form closer to the illuminating body than halos.

A. B.

▲ **Figure 16.21 Coronas A.** Most coronas appear as a bright whitish disk centered on the Moon or Sun. **B.** Coronas that display the colors of the rainbow are rare.

## Iridescent Clouds

**Iridescent clouds** are among the more spectacular and elusive of optical phenomena. Cloud iridescence, most often associated with altostratus, cirrocumulus, or lenticular clouds, appears as areas

**Table 16.1** | Atmospheric Optical Phenomena

| Process | Optical Phenomena | Medium |
|---|---|---|
| Refraction | Mirages | Air |
| | Halos | Ice crystals |
| | Sun dogs | Ice crystals |
| Reflection | Sun pillars | Ice crystals |
| Reflection and refraction | Rainbows | Raindrops |
| Diffraction | Coronas | Cloud droplets |
| | Iridescent clouds | Cloud droplets |
| Scattering | Blue skies | Air |
| | Red sunsets | Air |

of bright colors—violet, pink, and green—generally seen near the edges of a cloud (**Fig. 16.22**). Common examples of iridescence are the spectrum of colors reflected from soap bubbles and from a thin layer of gasoline spilled on a wet surface.

As with the corona shown in Figure 16.21B, the display of colors associated with iridescent clouds is produced by the diffraction of sunlight or moonlight by small, uniform cloud droplets or occasionally small ice crystals. The best time to view iridescent clouds is when the Sun is behind a cloud or just after the Sun has set behind a building or topographic barrier.

**Table 16.1** summarizes the basic optical phenomena described in this chapter, based on what process caused them—reflection and/or refraction or diffraction. The table also includes the optical phenomena produced by scattering, which are described in Chapter 2.

## ✔ Concept Checks 16.5

1. What process is responsible for the colors exhibited by coronas and iridescent clouds?
2. How can coronas be distinguished from 22° halos?

# 16 Concepts in Review Optical Phenomena of the Atmosphere

## 16.1 Interactions of Light and Matter

► Explain the principle called the *law of reflection* and discuss how refraction causes white light to separate into the spectrum of colors.

**Key Terms:** law of reflection, internal reflection, refraction

- The two basic interactions of light with matter are reflection and refraction. The law of reflection states that when light rays are reflected, they bounce off the reflecting surface at the same angle (the angle of reflection) at which they meet that surface (the angle of incidence).
- Refraction is the bending of light due to a change in velocity as it passes obliquely from one transparent medium to another. Furthermore, light will gradually bend as it traverses a layer of air that varies in density.

**Q** What happens to light as it passes obliquely from one transparent medium into another?

## 16.2 Mirages

► Describe how a mirage is created.

**Key Terms:** mirage, inferior mirage, looming, superior mirage, towering

- A mirage is an optical effect of the atmosphere caused by refraction when light passes through air with different densities. The result is that objects appear to be displaced from their true position.
- The classic desert mirage, produced by hot, less-dense air near the surface and cooler air aloft, is called an inferior mirage because it produces an image that appears below its true location.
- When the refraction of light is significant enough to make an object appear suspended above the horizon, it is called a superior mirage. Another type of mirage, known as Fata Morgana, produces large towering images.

**Q** Explain why a mirage disappears when the observer nears the mirage.

## 16.3 Rainbows

► Sketch a raindrop and show how sunlight travels through it to form a primary rainbow.

**Key Terms:** rainbow, dispersion

- Perhaps the most spectacular and most commonly known atmospheric optical phenomenon is the rainbow. When a rainbow forms, raindrops act as prisms and disperse the sunlight into the spectrum of colors in a process called dispersion.
- In a primary rainbow, sunlight is reflected (internal reflection) once within a raindrop. In a dimmer, less frequently observed secondary rainbow, light is reflected twice within a raindrop.

## 16.4 Halos, Sun Dogs, and Solar Pillars

▶ Distinguish among three different optical phenomena that are produced as a result of the interaction between sunlight and hexagonal ice crystals.

**Key Terms:** halo, sun dogs, sun pillar

- A halo is a narrow whitish ring centered on the Sun. Halos occur most often when the sky is covered with a thin layer of cirrostratus clouds. The most common halo is the 22° halo, so named because its radius subtends an angle of 22° from the Sun. Halos are produced by the refraction of sunlight by ice crystals.
- A spectacular effect associated with a halo is sun dogs. These two bright regions, or mock suns, can be seen adjacent to a 22° halo.

## 16.5 Other Optical Phenomena

▶ Compare and contrast coronas and halos.

**Key Terms:** diffraction, corona, iridescent cloud

- Coronas are bright whitish disks centered on the Moon or Sun, and they are the only optical phenomenon more commonly witnessed in association with the Moon than with the Sun.
- Iridescent clouds, like coronas, are formed by diffraction of white light by tiny, uniform cloud droplets, or sometimes ice crystals.

**Q** How can coronas be distinguished from 22° halos?

## Give it Some Thought

1. What particles (cloud droplets, raindrops, or ice crystals) are usually associated with each of these optical phenomena: rainbows, halos, coronas, glories, and sun dogs?
2. The accompanying drawing illustrates how white sunlight is separated into the colors of the rainbow. Match the features numbered 1 through 5 with the following terms: internal reflection, red light, violet light, incident ray, and refracted ray.

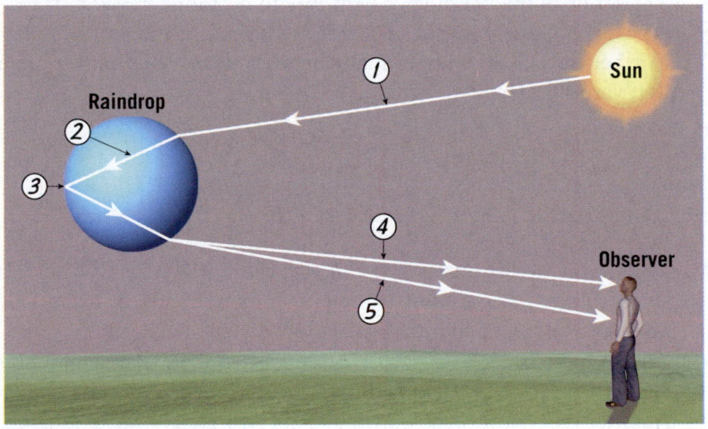


3. If you have ever been close to a large movie theater screen, you may have noticed that the screen is made of small glassy particles oriented at slightly different angles rather than one very smooth surface. Based on what you know about how light is reflected from various surfaces, why do you think this type of screen is used?
4. Explain why the refraction of sunlight causes the length of daylight to be longer than if the Earth had no atmosphere.
5. Explain why a person in Lincoln, Nebraska (41° north latitude), will never see a rainbow at noon on June 21.
6. Name the optical phenomena shown in the accompanying photos.

A.

C.

B.

## MasteringMeteorology™

Looking for additional review and test prep materials? Visit the Study Area in *MasteringMeteorology*™ to enhance your understanding of this chapter's content by accessing a variety of resources, including **MapMaster**™ interactive maps, Geoscience Animations, GEODe, *In the News* RSS feeds, flashcards, web links, self-study quizzes, and an eText version of *The Atmosphere.*

# Appendix A Metric Units

**TABLE A–1** The International System of Units (SI)

**I. Basic units**

| Quantity | Unit | SI symbol |
|---|---|---|
| Length | meter | m |
| Mass | kilogram | kg |
| Time | second | s |
| Electric current | ampere | A |
| Thermodynamic temperature | kelvin | K |
| Amount of substance | mole | mol |
| Luminous intensity | candela | cd |

**II. Prefixes**

| Prefix | Factor by which unit is multiplied | Symbol |
|---|---|---|
| tera | $10^{12}$ | T |
| giga | $10^{9}$ | G |
| mega | $10^{6}$ | M |
| kilo | $10^{3}$ | k |
| hecto | $10^{2}$ | h |
| deka | 10 | da |
| deci | $10^{-1}$ | d |
| centi | $10^{-2}$ | c |
| milli | $10^{-3}$ | m |
| micro | $10^{-6}$ | $\mu$ |
| nano | $10^{-9}$ | n |
| pico | $10^{-12}$ | p |
| femto | $10^{-15}$ | f |
| atto | $10^{-18}$ | a |

**III. Derived units**

| Quantity | Units | Expression |
|---|---|---|
| Area | square meter | $m^2$ |
| Volume | cubic meter | $m^3$ |
| Frequency | hertz (Hz) | $s^{-1}$ |
| Density | kilograms per cubic meter | $kg/m^3$ |
| Velocity | meters per second | m/s |
| Angular velocity | radians per second | rad/s |
| Acceleration | meters per second squared | $m/s^2$ |
| Angular acceleration | radians per second squared | $rad/s^2$ |
| Force | newton (N) | $kg \cdot ms^2$ |
| Pressure | newtons per square meter | $N/m^2$ |
| Work, energy, quantity of heat | joule (J) | $N \cdot m$ |
| Power | watt (W) | J/s |
| Electric charge | coulomb (C) | $A \cdot s$ |
| Voltage, potential difference, electromotive force | volt (V) | W/A |
| Luminance | candelas per square meter | $cd/m^2$ |

**TABLE A–2** Metric–English Conversion

| When you want to convert to | Multiply by: | To find: |
|---|---|---|
| Length | | |
| inches | 2.54 | centimeters |
| centimeters | 0.39 | inches |
| feet | 0.30 | meters |
| meters | 3.28 | feet |
| yards | 0.91 | meters |
| meters | 1.09 | yards |
| miles | 1.61 | kilometers |
| kilometers | 0.62 | miles |
| Area | | |
| square inches | 6.45 | square centimeters |
| square centimeters | 0.15 | square inches |
| square feet | 0.09 | square meters |
| square meters | 10.76 | square feet |
| square miles | 2.59 | square kilometers |
| square kilometers | 0.39 | square miles |
| Volume | | |
| cubic inches | 16.38 | cubic centimeters |
| cubic centimeters | 0.06 | cubic inches |
| cubic feet | 0.028 | cubic meters |
| cubic meters | 35.3 | cubic feet |
| cubic miles | 4.17 | cubic kilometers |
| cubic kilometers | 0.24 | cubic miles |
| liters | 1.06 | quarts |
| liters | 0.26 | gallons |
| gallons | 3.78 | liters |
| Masses and Weights | | |
| ounces | 28.33 | grams |
| grams | 0.035 | ounces |
| pounds | 0.45 | kilograms |
| kilograms | 2.205 | pounds |
| Temperature | | |
| When you want to convert degrees Fahrenheit (°F) to degrees Celsius (°C), subtract 32 degrees and divide by 1.8 (also see Table A–3). | | |
| When you want to convert degrees Celsius (°C) to degrees Fahrenheit (°F), multiply by 1.8 and add 32 degrees (also see Table A–3). | | |
| When you want to convert degrees Celsius (°C) to kelvins (K), delete the degree symbol and add 273. | | |
| When you want to convert kelvins (K) to degrees Celsius (°C), add the degree symbol and subtract 273. | | |

**TABLE A–3** Temperature Conversion Table. (To find either the Celsius or the Fahrenheit equivalent, locate the known temperature in the center column. Then read the desired equivalent value from the appropriate column.)

| °C | | °F | °C | | °F | °C | | °F | °C | | °F |
|---|---|---|---|---|---|---|---|---|---|---|---|
| −40.0 | −40 | −40 | −17.2 | +1 | 33.8 | 5.0 | 41 | 105.8 | 27.2 | 81 | 177.8 |
| −39.4 | −39 | −38.2 | −16.7 | 2 | 35.6 | 5.6 | 42 | 107.6 | 27.8 | 82 | 179.6 |
| −38.9 | −38 | −36.4 | −16.1 | 3 | 37.4 | 6.1 | 43 | 109.4 | 28.3 | 83 | 181.4 |
| −38.3 | −37 | −34.6 | −15.4 | 4 | 39.2 | 6.7 | 44 | 111.2 | 28.9 | 84 | 183.2 |
| −37.8 | −36 | −32.8 | −15.0 | 5 | 41.0 | 7.2 | 45 | 113.0 | 29.4 | 85 | 185.0 |
| −37.2 | −35 | −31.0 | −14.4 | 6 | 42.8 | 7.8 | 46 | 114.8 | 30.0 | 86 | 186.8 |
| −36.7 | −34 | −29.2 | −13.9 | 7 | 44.6 | 8.3 | 47 | 116.6 | 30.6 | 87 | 188.6 |
| −36.1 | −33 | −27.4 | −13.3 | 8 | 46.4 | 8.9 | 48 | 118.4 | 31.1 | 88 | 190.4 |
| −35.6 | −32 | −25.6 | −12.8 | 9 | 48.2 | 9.4 | 49 | 120.2 | 31.7 | 89 | 192.2 |
| −35.0 | −31 | −23.8 | −12.2 | 10 | 50.0 | 10.0 | 50 | 122.0 | 32.2 | 90 | 194.0 |
| −34.4 | −30 | −22.0 | −11.7 | 11 | 51.8 | 10.6 | 51 | 123.8 | 32.8 | 91 | 195.8 |
| −33.9 | −29 | −20.2 | −11.1 | 12 | 53.6 | 11.1 | 52 | 125.6 | 33.3 | 92 | 197.6 |
| −33.3 | −28 | −18.4 | −10.6 | 13 | 55.4 | 11.7 | 53 | 127.4 | 33.9 | 93 | 199.4 |
| −32.8 | −27 | −16.6 | −10.0 | 14 | 57.2 | 12.2 | 54 | 129.2 | 34.4 | 94 | 201.2 |
| −32.2 | −26 | −14.8 | −9.4 | 15 | 59.0 | 12.8 | 55 | 131.0 | 35.0 | 95 | 203.0 |
| −31.7 | −25 | −13.0 | −8.9 | 16 | 60.8 | 13.3 | 56 | 132.8 | 35.6 | 96 | 204.8 |
| −31.1 | −24 | −11.2 | −8.3 | 17 | 62.6 | 13.9 | 57 | 134.6 | 36.1 | 97 | 206.6 |
| −30.6 | −23 | −9.4 | −7.8 | 18 | 64.4 | 14.4 | 58 | 136.4 | 36.7 | 98 | 208.4 |
| −30.0 | −22 | −7.6 | −7.2 | 19 | 66.2 | 15.0 | 59 | 138.2 | 37.2 | 99 | 210.2 |
| −29.4 | −21 | −5.8 | −6.7 | 20 | 68.0 | 15.6 | 60 | 140.0 | 37.8 | 100 | 212.0 |
| −28.9 | −20 | −4.0 | −6.1 | 21 | 69.8 | 16.1 | 61 | 141.8 | 38.3 | 101 | 213.8 |
| −28.3 | −19 | −2.2 | −5.6 | 22 | 71.6 | 16.7 | 62 | 143.6 | 38.9 | 102 | 215.6 |
| −27.8 | −18 | −0.4 | −5.0 | 23 | 73.4 | 17.2 | 63 | 145.4 | 39.4 | 103 | 217.4 |
| −27.2 | −17 | +1.4 | −4.4 | 24 | 75.2 | 17.8 | 64 | 147.2 | 40.0 | 104 | 219.2 |
| −26.7 | −16 | 3.2 | −3.9 | 25 | 77.0 | 18.3 | 65 | 149.0 | 40.6 | 105 | 221.0 |
| −26.1 | −15 | 5.0 | −3.3 | 26 | 78.8 | 18.9 | 66 | 150.8 | 41.1 | 106 | 222.8 |
| −25.6 | −14 | 6.8 | −2.8 | 27 | 80.6 | 19.4 | 67 | 152.6 | 41.7 | 107 | 224.6 |
| −25.0 | −13 | 8.6 | −2.2 | 28 | 82.4 | 20.0 | 68 | 154.4 | 42.2 | 108 | 226.4 |
| −24.4 | −12 | 10.4 | −1.7 | 29 | 84.2 | 20.6 | 69 | 156.2 | 42.8 | 109 | 228.2 |
| −23.9 | −11 | 12.2 | −1.1 | 30 | 86.0 | 21.1 | 70 | 158.0 | 43.3 | 110 | 230.0 |
| −23.3 | −10 | 14.0 | −0.6 | 31 | 87.8 | 21.7 | 71 | 159.8 | 43.9 | 111 | 231.8 |
| −22.8 | −9 | 15.8 | 0.0 | 32 | 89.6 | 22.2 | 72 | 161.6 | 44.4 | 112 | 233.6 |
| −22.2 | −8 | 17.6 | +0.6 | 33 | 91.4 | 22.8 | 73 | 163.4 | 45.0 | 113 | 235.4 |
| −21.7 | −7 | 19.4 | 1.1 | 34 | 93.2 | 23.3 | 74 | 165.2 | 45.6 | 114 | 237.2 |
| −21.1 | −6 | 21.2 | 1.7 | 35 | 95.0 | 23.9 | 75 | 167.0 | 46.1 | 115 | 239.0 |
| −20.6 | −5 | 23.0 | 2.2 | 36 | 96.8 | 24.4 | 76 | 168.8 | 46.7 | 116 | 240.8 |
| −20.0 | −4 | 24.8 | 2.8 | 37 | 98.6 | 25.0 | 77 | 170.6 | 47.2 | 117 | 242.6 |
| −19.4 | −3 | 26.6 | 3.3 | 38 | 100.4 | 25.6 | 78 | 172.4 | 47.8 | 118 | 244.4 |
| −18.9 | −2 | 28.4 | 3.9 | 39 | 102.2 | 26.1 | 79 | 174.2 | 48.3 | 119 | 246.2 |
| −18.3 | −1 | 30.2 | 4.4 | 40 | 104.0 | 26.7 | 80 | 176.0 | 48.9 | 120 | 248.0 |
| −17.8 | 0 | 32.0 | | | | | | | | | |

**TABLE A–4** Wind-Conversion Table (Wind-Speed Units: 1 Mile Per Hour = 0.868391 Knot = 1.609344 km/h = 0.44704 m/s)

| Miles per Hour | Knots | Meters per Second | Kilometers per Hour | Miles per Hour | Knots | Meters per Second | Kilometers per Hour |
|---|---|---|---|---|---|---|---|
| 1 | 0.9 | 0.4 | 1.6 | 51 | 44.3 | 22.8 | 82.1 |
| 2 | 1.7 | 0.9 | 3.2 | 52 | 45.2 | 23.2 | 83.7 |
| 3 | 2.6 | 1.3 | 4.8 | 53 | 46.0 | 23.7 | 85.3 |
| 4 | 3.5 | 1.8 | 6.4 | 54 | 46.9 | 24.1 | 86.9 |
| 5 | 4.3 | 2.2 | 8.0 | 55 | 47.8 | 24.6 | 88.5 |
| 6 | 5.2 | 2.7 | 9.7 | 56 | 48.6 | 25.0 | 90.1 |
| 7 | 6.1 | 3.1 | 11.3 | 57 | 49.5 | 25.5 | 91.7 |
| 8 | 6.9 | 3.6 | 12.9 | 58 | 50.4 | 25.9 | 93.3 |
| 9 | 7.8 | 4.0 | 14.5 | 59 | 51.2 | 26.4 | 95.0 |
| 10 | 8.7 | 4.5 | 16.1 | 60 | 52.1 | 26.8 | 96.6 |
| 11 | 9.6 | 4.9 | 17.7 | 61 | 53.0 | 27.3 | 98.2 |
| 12 | 10.4 | 5.4 | 19.3 | 62 | 53.8 | 27.7 | 99.8 |
| 13 | 11.3 | 5.8 | 20.9 | 63 | 54.7 | 28.2 | 101.4 |
| 14 | 12.2 | 6.3 | 22.5 | 64 | 55.6 | 28.6 | 103.0 |
| 15 | 13.0 | 6.7 | 24.1 | 65 | 56.4 | 29.1 | 104.6 |
| 16 | 13.9 | 7.2 | 25.7 | 66 | 57.3 | 29.5 | 106.2 |
| 17 | 14.8 | 7.6 | 27.4 | 67 | 58.2 | 30.0 | 107.8 |
| 18 | 15.6 | 8.0 | 29.0 | 68 | 59.1 | 30.4 | 109.4 |
| 19 | 16.5 | 8.5 | 30.6 | 69 | 59.9 | 30.8 | 111.0 |
| 20 | 17.4 | 8.9 | 32.2 | 70 | 60.8 | 31.3 | 112.7 |
| 21 | 18.2 | 9.4 | 33.8 | 71 | 61.7 | 31.7 | 114.3 |
| 22 | 19.1 | 9.8 | 35.4 | 72 | 62.5 | 32.2 | 115.9 |
| 23 | 20.0 | 10.3 | 37.0 | 73 | 63.4 | 32.6 | 117.5 |
| 24 | 20.8 | 10.7 | 38.6 | 74 | 64.3 | 33.1 | 119.1 |
| 25 | 21.7 | 11.2 | 40.2 | 75 | 65.1 | 33.5 | 120.7 |
| 26 | 22.6 | 11.6 | 41.8 | 76 | 66.0 | 34.0 | 122.3 |
| 27 | 23.4 | 12.1 | 43.5 | 77 | 66.9 | 34.4 | 123.9 |
| 28 | 24.3 | 12.5 | 45.1 | 78 | 67.7 | 34.9 | 125.5 |
| 29 | 25.2 | 13.0 | 46.7 | 79 | 68.6 | 35.3 | 127.1 |
| 30 | 26.1 | 13.4 | 48.3 | 80 | 69.5 | 35.8 | 128.7 |
| 31 | 26.9 | 13.9 | 49.9 | 81 | 70.3 | 36.2 | 130.4 |
| 32 | 27.8 | 14.3 | 51.5 | 82 | 71.2 | 36.7 | 132.0 |
| 33 | 28.7 | 14.8 | 53.1 | 83 | 72.1 | 37.1 | 133.6 |
| 34 | 29.5 | 15.2 | 54.7 | 84 | 72.9 | 37.6 | 135.2 |
| 35 | 30.4 | 15.6 | 56.3 | 85 | 73.8 | 38.0 | 136.8 |
| 36 | 31.3 | 16.1 | 57.9 | 86 | 74.7 | 38.4 | 138.4 |
| 37 | 32.1 | 16.5 | 59.5 | 87 | 75.5 | 38.9 | 140.0 |
| 38 | 33.0 | 17.0 | 61.2 | 88 | 76.4 | 39.3 | 141.6 |
| 39 | 33.9 | 17.4 | 62.8 | 89 | 77.3 | 39.8 | 143.2 |
| 40 | 34.7 | 17.9 | 64.4 | 90 | 78.2 | 40.2 | 144.8 |
| 41 | 35.6 | 18.3 | 66.0 | 91 | 79.0 | 40.7 | 146.5 |

*(Continued)*

**TABLE A–4** *(Continued)*

| Miles per Hour | Knots | Meters per Second | Kilometers per Hour | Miles per Hour | Knots | Meters per Second | Kilometers per Hour |
|---|---|---|---|---|---|---|---|
| 42 | 36.5 | 18.8 | 67.6 | 92 | 79.9 | 41.1 | 148.1 |
| 43 | 37.3 | 19.2 | 69.2 | 93 | 80.8 | 41.6 | 149.7 |
| 44 | 38.2 | 19.7 | 70.8 | 94 | 81.6 | 42.0 | 151.3 |
| 45 | 39.1 | 20.1 | 72.4 | 95 | 82.5 | 42.5 | 152.9 |
| 46 | 39.9 | 20.6 | 74.0 | 96 | 83.4 | 42.9 | 154.5 |
| 47 | 40.8 | 21.0 | 75.6 | 97 | 84.2 | 43.4 | 156.1 |
| 48 | 41.7 | 21.5 | 77.2 | 98 | 85.1 | 43.8 | 157.7 |
| 49 | 42.6 | 21.9 | 78.9 | 99 | 86.0 | 44.3 | 159.3 |
| 50 | 43.4 | 22.4 | 80.5 | 100 | 86.8 | 44.7 | 160.9 |

# Appendix B Explanation and Decoding of the Daily Weather Map

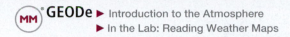

MM® GEODe ▶ Introduction to the Atmosphere
▶ In the Lab: Reading Weather Maps

Weather maps showing the development and movement of weather systems are among the most important tools used by the meterologist. Some maps portray conditions near the surface of Earth and others depict conditions at various heights in the atmosphere. Some cover the entire Northern Hemisphere and others cover only local areas as required for special purposes.

## Principal Surface Weather Map

To prepare the surface map and present the information quickly and pictorially, two actions are necessary: (1) Weather observers and automated observing stations must send data to the offices where the maps are prepared; (2) the information must be quickly transcribed to the maps. In order for the necessary speed and economy of space and transmission time to be realized, codes have been devised for sending the information and for plotting it on the maps.

## Codes and Map Plotting

A great deal of information is contained in a brief coded weather message. If each item were named and described in plain language, a very lengthy message would be required, one confusing to read and difficult to transfer to a map. A code permits the message to be condensed to a few five-figure numeral groups, each figure of which has a meaning, depending on its position in the message. People trained in the use of the code can read the message as easily as plain language (see **Figure B.1**).

The location of the reporting station is printed on the map as a small circle (the station circle). A definite arrangement of the data around the station circle, called the *station model,* is used. When the report is plotted in these fixed positions around the station circle on the weather map, many code figures are transcribed exactly as sent. Entries in the station model that are not made in code figures or actual values found in the message are usually in the form of symbols that graphically represent the element concerned. In some cases, certain of the data may or may not be reported by the observer, depending on local weather conditions. Precipitation and clouds are examples. In such cases, the absence of an entry on the map is interpreted as nonoccurrence or nonobservance of the phenomena. The letter *M* is entered where data are normally observed but not received.

Both the code and the station model are based on international agreements. These standardized numerals and symbols enable a meteorologist of one country to use the weather reports and weather maps of another country even though that person does not understand the language. Weather codes are, in effect, an international language that permits complete interchange and use of worldwide weather reports so essential in present-day activities.

The boundary between two different air masses is called a *front*. Important changes in weather, temperature, wind direction, and clouds often occur with the passage of a front. Half circles or triangular symbols or both are placed on the lines representing fronts to indicate the kind of front. The side on which the symbols are placed indicates the direction of frontal movement. The boundary of relatively cold air of polar origin advancing into an area occupied by warmer air, often of tropical origin, is called a *cold front*. The boundary of relatively warm air advancing into an area occupied by colder air is called a *warm front*. The line along which a cold front has overtaken a warm front at the ground is called an *occluded front*. A boundary between two air masses, which shows at the time of observation little tendency to advance into either the warm or cold areas, is called a *stationary front*. Air-mass boundaries are known as *surface fronts* when they intersect the ground and as *upper-air fronts* when they do not. Surface fronts are drawn in solid black; fronts aloft are drawn in outline only. Front symbols are given in **Table B.1**.

A front that is disappearing or weak and decreasing in intensity is labeled *frontolysis*. A front that is forming is labeled *frontogenesis*. A *squall line* is a line of thunderstorms or squalls usually accompanied by heavy showers and shifting winds (Table B.1).

The paths followed by individual disturbances are called *storm tracks* and are shown by arrows (Table B.1). A symbol (a box containing an X) indicates past positions of a low-pressure center at six-hour intervals. HIGH (H) and LOW (L) indicate the centers of high and low barometric pressure. Solid lines are isobars and connect points of equal sea-level barometric pressure. The spacing and orientation of these lines on weather maps are indications of speed and direction of windflow. In general, wind direction is parallel to these lines with low pressure to the left of an observer looking downwind. Speed is directly proportional to the closeness of the lines (called *pressure gradient*). Isobars are labeled in millibars.

Isotherms are lines connecting points of equal temperature. Two isotherms are frequently drawn on large surface weather maps when applicable. The freezing, or 32°F, isotherm is drawn as a dashed line, and the 0°F isotherm is drawn as a dash–dot line (Table B.1). Areas where precipitation is occurring at the time of observation are shaded.

## Auxiliary Maps

### 500-Millibar Map

Contour lines, isotherms, and wind arrows are shown on the 500-millibar contour level. Solid lines are drawn to show height above sea level and are labeled in feet. Dashed lines are drawn at 5° intervals of temperature and are labeled in degrees Celsius. True wind direction is shown by "arrows" that are plotted as flying with the wind. The wind speed is shown by flags and feathers. Each flag represents 50 knots, each full feather represents 10 knots, and each half feather represents 5 knots.

**FIGURE B.1** Explanation of Station Symbols and Map Entries

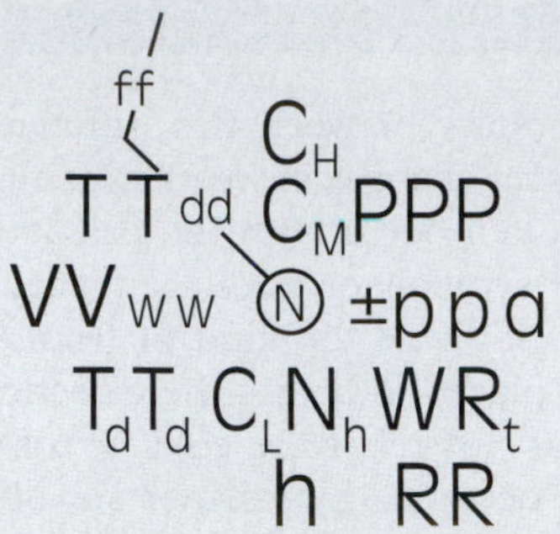

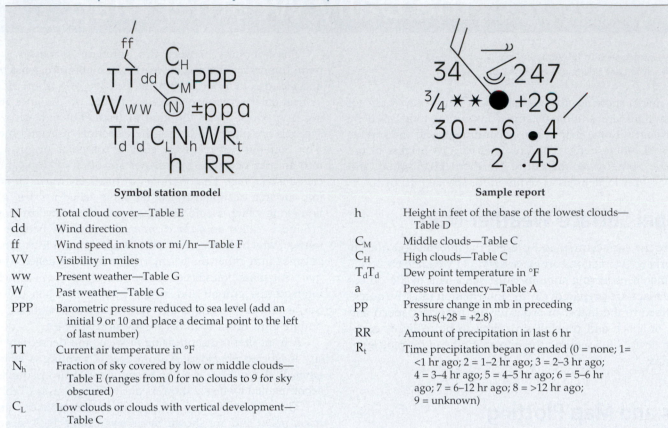

Symbol station model

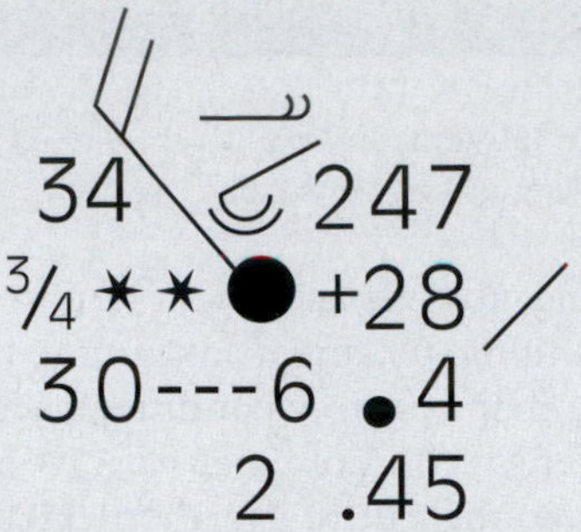


Sample report

| | | | |
|---|---|---|---|
| N | Total cloud cover—Table E | h | Height in feet of the base of the lowest clouds—Table D |
| dd | Wind direction | $C_M$ | Middle clouds—Table C |
| ff | Wind speed in knots or mi/hr—Table F | $C_H$ | High clouds—Table C |
| VV | Visibility in miles | $T_dT_d$ | Dew point temperature in °F |
| ww | Present weather—Table G | a | Pressure tendency—Table A |
| W | Past weather—Table G | pp | Pressure change in mb in preceding 3 hrs(+28 = +2.8) |
| PPP | Barometric pressure reduced to sea level (add an initial 9 or 10 and place a decimal point to the left of last number) | RR | Amount of precipitation in last 6 hr |
| TT | Current air temperature in °F | $R_t$ | Time precipitation began or ended (0 = none; 1= <1 hr ago; 2 = 1–2 hr ago; 3 = 2–3 hr ago; 4 = 3–4 hr ago; 5 = 4–5 hr ago; 6 = 5–6 hr ago; 7 = 6–12 hr ago; 8 = >12 hr ago; 9 = unknown) |
| $N_h$ | Fraction of sky covered by low or middle clouds—Table E (ranges from 0 for no clouds to 9 for sky obscured) | | |
| $C_L$ | Low clouds or clouds with vertical development—Table C | | |

**TABLE B.1** Weather Map Symbols

| Symbol | Explanation |
|---|---|
|  | Cold front (surface) |
|  | Warm front (surface) |
|  | Occluded front (surface) |
|  | Stationary front (surface) |
|  | Dryline |
| -··-··-··-··- | Squall line |
| ——→ ——→ | Path of low-pressure center |
| ⊠ | Location of low pressure at 6-hour intervals |
| — — — — — — | 32° F isotherm |
| —·—·—·—·— | 0° F isotherm |

## Temperature Map (Highest and Lowest)

Temperature data are entered from selected weather stations in the United States. The figure entered above the station dot shows the maximum temperature for the 12-hour period ending 7:00 P.M. EST of the previous day. The figure entered below the station dot shows the minimum temperature during the 12 hours ending at 7:00 A.M. EST. The letter *M* denotes missing data.

## Precipitation Map

Precipitation data are entered from selected weather stations in the United States. When precipitation has occurred at any of these stations in the 24-hour period ending at 7:00 A.M. EST, the total amount, in inches and hundredths, is entered above the station dot. When the figures for total precipitation have been compiled from incomplete data and entered on the map, the amount is underlined. *T* indicates a trace of precipitation (less than 0.01 inch) and the letter *M* denotes missing data. The geographical areas where precipitation has fallen during the 24 hours ending at 7:00 A.M. EST are shaded. Dashed lines show depth of snow on ground in inches as of 7:00 A.M. EST.

**TABLE A** Air Pressure Tendency

| | | |
|---|---|---|
| | Rising, then falling; same as or higher than 3 hr ago | |
| | Rising, then steady; or rising, then rising more slowly | Barometric pressure now higher than 3 hours ago |
| | Rising steadily, or unsteadily | |
| | Falling or steady, then rising; or rising, then rising more rapidly | |
| | Steady; same as 3 hr ago | |
| | Falling, then rising; same as or lower than 3 hr ago | |
| | Falling, then steady; or falling, then falling more slowly | Barometric pressure now lower than 3 hours ago |
| | Falling steadily, or unsteadily | |
| | Steady or rising, then falling; or falling, then falling more rapidly | |

**TABLE B** Cloud Abbreviations

| | |
|---|---|
| St | stratus |
| Fra | fractus |
| Sc | stratocumulus |
| Ns | nimbostratus |
| As | altostratus |
| Ac | altocumulus |
| Ci | cirrus |
| Cs | cirrostratus |
| Cc | cirrocumulus |
| Cu | cumulus |
| Cb | cumulonimbus |

**TABLE C** Cloud Types

| | Low Clouds and Clouds of Vertical Development |
|---|---|
| | Cu of fair weather, little vertical development and seemingly flattened |
| | Cu of considerable development, generally towering, with or without other Cu or Sc, bases all at same level |
| | Cb with tops lacking clear-cut outlines, but distinctly not cirriform or anvil shaped; with or without Cu, Sc, or St |
| | Sc formed by spreading out of Cu; Cu often present also |
| | Sc not formed by spreading out of Cu |
| | St or StFra, but no StFra of bad weather |
| | StFra and/or CuFra of bad weather (scud) |
| | Cu and Sc (not formed by spreading out of Cu) with bases at different levels |
| | Cb having a clearly fibrous (cirriform) top, often anvil shaped, with or without Cu, Sc, St, or scud |

| | Middle Clouds |
|---|---|
| | Thin As (most of cloud layer semitransparent) |
| | Thick As, greater part sufficiently dense to hide Sun (or Moon), or Ns |
| | Thin Ac, mostly semitransparent; cloud elements not changing much and at a single level |
| | Thin Ac in patches; cloud elements continually changing and/or occurring at more than one level |
| | Thin Ac in bands or in a layer gradually spreading over sky and usually thickening as a whole |
| | Ac formed by the spreading out of Cu or Cb |
| | Double-layered Ac, or a thick layer of Ac, not increasing; or Ac with As and/or Ns |
| | Ac in the form of Cu-shaped tufts or Ac with turrets |
| | Ac of a chaotic sky, usually at different levels; patches of dense Ci usually present also |

| | High Clouds |
|---|---|
| | Filaments of Ci, or "mares' tails," scattered and not increasing |
| | Dense Ci in patches or twisted sheaves, usually not increasing, sometimes like remains of Cb; or towers or tufts |
| | Dense Ci, often anvil shaped, derived from or associated with Cb |
| | Ci, often hook shaped, gradually spreading over the sky and usually thickening as a whole |
| | Ci and Cs, often in converging bands, or Cs alone; generally overspreading and growing denser; the continuous layer not reaching 45° altitude |
| | Ci and Cs, often in converging bands, or Cs alone; generally overspreading and growing denser; the continuous layer exceeding 45° altitude |
| | Veil of Cs covering the entire sky |
| | Cs not increasing and not covering entire sky |
| | Cc alone or Cc with some Ci or Cs, but the Cc being the main cirriform cloud |

**TABLE D** Height of Base of Lowest Cloud

| Code | Feet | Meters |
|---|---|---|
| 0 | 0–149 | 0–49 |
| 1 | 150–299 | 50–99 |
| 2 | 300–599 | 100–199 |
| 3 | 600–999 | 200–299 |
| 4 | 1000–1999 | 300–599 |
| 5 | 2000–3499 | 600–999 |
| 6 | 3500–4999 | 1000–1499 |
| 7 | 5000–6499 | 1500–1999 |
| 8 | 6500–7999 | 2000–2499 |
| 9 | 8000 or above or no clouds | 2500 or above or no clouds |

**TABLE E** Cloud Cover

| | |
|---|---|
|  | No clouds |
|  | One-tenth or less |
|  | Two-tenths or three-tenths |
|  | Four-tenths |
|  | Five-tenths |
|  | Six-tenths |
|  | Seven-tenths or eight-tenths |
|  | Nine-tenths or overcast with openings |
|  | Completely overcast (ten-tenths) |
|  | Sky obscured |

**TABLE F** Wind Speed

| | Miles per hour | Knots | Kilometers per hour |
|---|---|---|---|
|  | calm | calm | calm |
|   | 1–2 | 1–2 | 1–3 |
|  | 3–8 | 3–7 | 3–13 |
|  | 9–14 | 8–12 | 14–19 |
| 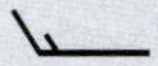 | 15–20 | 8–12 | 14–19 |
|  | 21–25 | 18–22 | 14–19 |
| 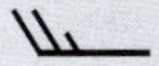 | 26–31 | 23–27 | 41–50 |
| 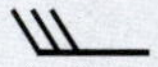 | 32–37 | 28–32 | 51–60 |
|  | 38–43 | 33–37 | 61–69 |
|  | 44–49 | 38–42 | 70–79 |
|  | 50–54 | 43–47 | 80–87 |
|  | 55–60 | 48–52 | 88–96 |
|  | 61–66 | 53–57 | 97–106 |
|  | 67–71 | 58–62 | 107–114 |
|  | 72–77 | 63–67 | 115–124 |
|  | 78–83 | 68–72 | 125–134 |
| 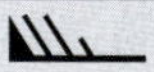 | 84–89 | 73–77 | 135–143 |
|  | 119–123 | 103–107 | 192–198 |

**TABLE G** Weather Conditions

| | | | | |
|---|---|---|---|---|
| Cloud development NOT observed or NOT observable during past hour | Clouds generally dissolving or becoming less developed during past hour | State of sky on the whole unchanged during past hour | Clouds generally forming or developing during past hour | Visibility reduced by smoke |
| Light fog (mist) | Patches of shallow fog at station, NOT deeper than 6 feet on land | More or less continuous shallow fog at station, NOT deeper than 6 feet on land | Lightning visible, no thunder heard | Precipitation within sight, but NOT reaching the ground |
| Drizzle (NOT freezing) or snow grains (NOT falling as showers) during past hour, but NOT at time of observation | Rain (NOT freezing and NOT falling as showers) during past hour, but NOT at time of observation | Snow (NOT falling as showers) during past hour, but NOT at time of observation | Rain and snow or ice pellets (NOT falling as showers) during past hour, but NOT at time of observation | Freezing drizzle or freezing rain (NOT falling as showers) during past hour, but NOT at time of observation |
| Slight or moderate dust storm or sandstorm has decreased during past hour | Slight or moderate dust storm or sandstorm, no appreciable change during past hour | Slight or moderate dust storm or sandstorm has begun or increased during past hour | Severe dust storm or sandstorm, has decreased during past hour | Severe dust storm or sandstorm, no appreciable change during past hour |
| Fog or ice fog at distance at time of observation, but NOT at station during past hour | Fog or ice fog in patches | Fog or ice fog, sky discernible, has become thinner during past hour | Fog or ice fog, sky NOT discernible, has become thinner during past hour | Fog or ice fog, sky discernible, no appreciable change during past hour |
| Intermittent drizzle (NOT freezing), slight at time of observation | Continuous drizzle (NOT freezing), slight at time of observation | Intermittent drizzle (NOT freezing), moderate at time of observation | Continuous drizzle (NOT freezing), moderate at time of observation | Intermittent drizzle (NOT freezing), heavy at time of observation |
| Intermittent rain (NOT freezing), slight at time of observation | Continuous rain (NOT freezing), slight at time of observation | Intermittent rain (NOT freezing), moderate at time of observation | Continuous rain (NOT freezing), moderate at time of observation | Intermittent rain (NOT freezing), heavy at time of observation |
| Intermittent fall of snowflakes, slight at time of observation | Continuous fall of snowflakes, slight at time of observation | Intermittent fall of snowflakes, moderate at time of observation | Continuous fall of snowflakes, moderate at time of observation | Intermittent fall of snowflakes, heavy at time of observation |
| Slight rain shower(s) | Moderate or heavy rain shower(s) | Violent rain shower(s) | Slight shower(s) of rain and snow mixed | Moderate or heavy shower(s) of rain and snow mixed |
| Moderate or heavy shower(s) of hail, with or without rain, or rain and snow mixed, not associated with thunder | Slight rain at time of observation; thunderstorm during past hour, but NOT at time of observation | Moderate or heavy rain at time of observation; thunderstorm during past hour, but NOT at time of observation | Slight snow, or rain and snow mixed, or hail at time of observation; thunderstorm during past hour, but NOT at time of observation | Moderate or heavy snow, or rain and snow mixed, or hail at time of observation; thunderstorm during past hour, but NOT at time of observation |

(Continued)

**TABLE G** Continued

| | | | | |
|---|---|---|---|---|
| Haze | Widespread dust in suspension in the air, NOT raised by wind, at time of observation | Dust or sand raised by wind at time of observation | Well developed dust whirl(s) within past hour | Dust storm or sand-storm within sight of or at station during past hour |
| Precipitation within sight, reaching the ground but distant from station | Precipitation within sight, reaching the ground, near to but NOT at station | Thunderstorm, but no precipitation at the station | Squall(s) within sight during past hour or at time of observation | Funnel cloud(s) within sight of station at time of observation |
| Showers of rain during past hour, but NOT at time of observation | Showers of snow, or of rain and snow, during past hour, but NOT at time of observation | Showers of hail, or of hail and rain, during past hour, but NOT at time of observation | Fog during past hour, but NOT at time of observation | Thunderstorm (with or without precipitation) during past hour, but NOT at time of observation |
| Severe dust storm or sandstorm has begun or increased during past hour | Slight or moderate drifting snow, generally low (less than 6 ft) | Heavy drifting snow, generally low | Slight or moderate blowing snow, generally high (more than 6 ft) | Heavy blowing snow, generally high |
| Fog or ice fog, sky NOT discernible, no appreciable change during past hour | Fog or ice fog, sky discernible, has begun or become thicker during past hour | Fog or ice fog, sky NOT discernible, has begun or become thicker during past hour | Fog depositing rime, sky discernible | Fog depositing rime, sky NOT discernible |
| Continuous drizzle (NOT freezing), heavy at time of observation | Slight freezing drizzle | Moderate or heavy freezing drizzle | Drizzle and rain, slight | Drizzle and rain, moderate or heavy |
| Continuous rain (NOT freezing), heavy at time of observation | Slight freezing rain | Moderate or heavy freezing rain | Rain or drizzle and snow, slight | Rain or drizzle and snow, moderate or heavy |
| Continuous fall of snowflakes, heavy at time of observation | Ice prisms (with or without fog) | Snow grains (with or without fog) | Isolated starlike snow crystals (with or without fog) | Ice pellets or snow pellets |
| Slight snow shower(s) | Moderate or heavy snow shower(s) | Slight shower(s) of snow pellets, or ice pellets with or without rain, or rain and snow mixed | Moderate or heavy shower(s) of snow pellets, or ice pellets, or ice pellets with or without rain or rain and snow mixed | Slight shower(s) of hail, with or without rain or rain and snow mixed, not associated with thunder |
| Slight or moderate thunderstorm without hail, but with rain, and/or snow at time of observation | Slight or moderate thunderstorm, with hail at time of observation | Heavy thundestorm, without hail, but with rain and/or snow at time of observation | Thunderstorm combined with dust storm or sandstorm at time of observation | Heavy thunderstorm with hail at time of observation |

# Appendix C Relative Humidity and Dew-Point Tables

**TABLE C.1** Relative Humidity (Percent)*

| Dry-Bulb (Air) Temperature (°C) | 1 | 2 | 3 | 4 | 5 | 6 | 7 | 8 | 9 | 10 | 11 | 12 | 13 | 14 | 15 | 16 | 17 | 18 | 19 | 20 | 21 | 22 |
|---|---|---|---|---|---|---|---|---|---|---|---|---|---|---|---|---|---|---|---|---|---|---|
| −20 | 28 | | | | | | | | | | | | | | | | | | | | | |
| −18 | 40 | | | | | | | | | | | | | | | | | | | | | |
| −16 | 48 | 0 | | | | | | | | | | | | | | | | | | | | |
| −14 | 55 | 11 | | | | | | | | | | | | | | | | | | | | |
| −12 | 61 | 23 | | | | | | | | | | | | | | | | | | | | |
| −10 | 66 | 33 | 0 | | | | | | | | | | | | | | | | | | | |
| −8 | 71 | 41 | 13 | | | | | | | | | | | | | | | | | | | |
| −6 | 73 | 48 | 20 | 0 | | | | | | | | | | | | | | | | | | |
| −4 | 77 | 54 | 32 | 11 | | | | | | | | | | | | | | | | | | |
| −2 | 79 | 58 | 37 | 20 | 1 | | | | | | | | | | | | | | | | | |
| 0 | 81 | 63 | 45 | 28 | 11 | | | | | | | | | | | | | | | | | |
| 2 | 83 | 67 | 51 | 36 | 20 | 6 | | | | | | | | | | | | | | | | |
| 4 | 85 | 70 | 56 | 42 | 27 | 14 | | | | | | | | | | | | | | | | |
| 6 | 86 | 72 | 59 | 46 | 35 | 22 | 10 | 0 | | | | | | | | | | | | | | |
| 8 | 87 | 74 | 62 | 51 | 39 | 28 | 17 | 6 | | | | | | | | | | | | | | |
| 10 | 88 | 76 | 65 | 54 | 43 | 38 | 24 | 13 | 4 | | | | | | | | | | | | | |
| 12 | 88 | 78 | 67 | 57 | 48 | 38 | 28 | 19 | 10 | 2 | | | | | | | | | | | | |
| 14 | 89 | 79 | 69 | 60 | 50 | 41 | 33 | 25 | 16 | 8 | 1 | | | | | | | | | | | |
| 16 | 90 | 80 | 77 | 62 | 54 | 45 | 37 | 29 | 21 | 14 | 7 | 1 | | | | | | | | | | |
| 18 | 91 | 81 | 72 | 64 | 56 | 48 | 40 | 33 | 26 | 19 | 12 | 6 | 0 | | | | | | | | | |
| 20 | 91 | 82 | 74 | 66 | 58 | 51 | 44 | 36 | 30 | 23 | 17 | 11 | 5 | | | | | | | | | |
| 22 | 92 | 83 | 75 | 68 | 60 | 53 | 46 | 40 | 33 | 27 | 21 | 15 | 10 | 4 | 0 | | | | | | | |
| 24 | 92 | 84 | 76 | 69 | 62 | 55 | 49 | 42 | 36 | 30 | 25 | 20 | 14 | 9 | 4 | 0 | | | | | | |
| 26 | 92 | 85 | 77 | 70 | 64 | 57 | 51 | 45 | 39 | 34 | 28 | 23 | 18 | 13 | 9 | 5 | | | | | | |
| 28 | 93 | 86 | 78 | 71 | 65 | 59 | 53 | 45 | 42 | 36 | 31 | 26 | 21 | 17 | 12 | 8 | 4 | | | | | |
| 30 | 93 | 86 | 79 | 72 | 66 | 61 | 55 | 49 | 44 | 39 | 34 | 29 | 25 | 20 | 16 | 12 | 8 | 4 | | | | |
| 32 | 93 | 86 | 80 | 73 | 68 | 62 | 56 | 51 | 46 | 41 | 36 | 32 | 27 | 22 | 19 | 14 | 11 | 8 | 4 | | | |
| 34 | 93 | 86 | 81 | 74 | 69 | 63 | 58 | 52 | 48 | 43 | 38 | 34 | 30 | 26 | 22 | 18 | 14 | 11 | 8 | 5 | | |
| 36 | 94 | 87 | 81 | 75 | 69 | 64 | 59 | 54 | 50 | 44 | 40 | 36 | 32 | 28 | 24 | 21 | 17 | 13 | 10 | 7 | 4 | |
| 38 | 94 | 87 | 82 | 76 | 70 | 66 | 60 | 55 | 51 | 46 | 42 | 38 | 34 | 30 | 26 | 23 | 20 | 16 | 13 | 10 | 7 | 5 |
| 40 | 94 | 89 | 82 | 76 | 71 | 67 | 61 | 57 | 52 | 48 | 44 | 40 | 36 | 33 | 29 | 25 | 22 | 19 | 16 | 13 | 10 | 7 |

Column heading: Depression of Web-Bulb (Dry-bulb temperature – Wet-bulb temperature = Depression of the wet bulb). Relative Humidity Values.

*To determine the relative humidity, find the air (dry-bulb) temperature on the vertical axis (far left) and the depression of the wet bulb on the horizontal axis (top). Where the two meet, the relative humidity is found. For example, when the dry-bulb temperature is 20°C and a wet-bulb temperature is 14°C, then the depression of the wet bulb is 6°C (20°C−14°C). From **TABLE C.1**, the relative humidity is 51 percent and from **TABLE C.2**, the dew point is 10°C.

**TABLE C.2** Dew-Point Temperature (°C)

| Dry-Bulb (Air) Temperature (°C) | 1 | 2 | 3 | 4 | 5 | 6 | 7 | 8 | 9 | 10 | 11 | 12 | 13 | 14 | 15 | 16 | 17 | 18 | 19 | 20 | 21 | 22 |
|---|---|---|---|---|---|---|---|---|---|---|---|---|---|---|---|---|---|---|---|---|---|---|
| −20 | −33 | | | | | | | | | | | | | | | | | | | | | |
| −18 | −28 | | | | | | | | | | | | | | | | | | | | | |
| −16 | −24 | | | | | | | | | | | | | | | | | | | | | |
| −14 | −21 | −36 | | | | | | | | | | | | | | | | | | | | |
| −12 | −18 | −28 | | | | | | | | | | | | | | | | | | | | |
| −10 | −14 | −22 | | | | | | | | | | | | | | | | | | | | |
| −8 | −12 | −18 | −29 | | | | | | | | | | | | | | | | | | | |
| −6 | −10 | −14 | −22 | | | | | | | | | | | | | | | | | | | |
| −4 | −7 | −12 | −17 | −29 | | | | | | | | | | | | | | | | | | |
| −2 | −5 | −8 | −13 | −20 | | | | | | | | | | | | | | | | | | |
| 0 | −3 | −6 | −9 | −15 | −24 | | | | | | | | | | | | | | | | | |
| 2 | −1 | −3 | −6 | −11 | −17 | | | | | | | | | | | | | | | | | |
| 4 | 1 | −1 | −4 | −7 | −11 | −19 | | | | | | | | | | | | | | | | |
| 6 | 4 | 1 | −1 | −4 | −7 | −13 | −21 | | | | | | | | | | | | | | | |
| 8 | 6 | 3 | 1 | 2 | −5 | −9 | −14 | | | | | | | | | | | | | | | |
| 10 | 8 | 6 | 4 | 1 | −2 | −5 | −9 | −14 | −18 | | | | | | | | | | | | | |
| 12 | 10 | 8 | 6 | 4 | 1 | −2 | −5 | −9 | −16 | | | | | | | | | | | | | |
| 14 | 12 | 11 | 9 | 6 | 4 | 1 | −2 | −5 | −10 | −17 | | | | | | | | | | | | |
| 16 | 14 | 13 | 11 | 9 | 7 | 4 | 1 | −1 | −6 | −10 | −17 | | | | | | | | | | | |
| 18 | 16 | 15 | 13 | 11 | 9 | 7 | 4 | 2 | −2 | 5 | 10 | −19 | | | | | | | | | | |
| 20 | 19 | 17 | 15 | 14 | 12 | 10 | 7 | 4 | 2 | −2 | −5 | −10 | −19 | | | | | | | | | |
| 22 | 21 | 19 | 17 | 16 | 14 | 12 | 10 | 8 | 5 | 3 | −1 | −5 | −10 | −19 | | | | | | | | |
| 24 | 23 | 21 | 20 | 18 | 16 | 14 | 12 | 10 | 8 | 6 | 2 | −1 | −5 | −10 | −18 | | | | | | | |
| 26 | 25 | 23 | 22 | 20 | 18 | 17 | 15 | 13 | 11 | 9 | 6 | 3 | 0 | −4 | −9 | −18 | | | | | | |
| 28 | 27 | 25 | 24 | 22 | 27 | 19 | 17 | 16 | 14 | 11 | 9 | 7 | 4 | 1 | −3 | −9 | 16 | | | | | |
| 30 | 29 | 27 | 26 | 24 | 23 | 21 | 19 | 18 | 16 | 14 | 12 | 10 | 8 | 5 | 1 | −2 | −8 | −15 | | | | |
| 32 | 31 | 29 | 28 | 27 | 25 | 24 | 22 | 21 | 19 | 17 | 15 | 13 | 11 | 8 | 5 | 2 | −2 | −7 | −14 | | | |
| 34 | 33 | 31 | 30 | 29 | 27 | 26 | 24 | 23 | 21 | 20 | 18 | 16 | 14 | 12 | 9 | 6 | 3 | −1 | −5 | −12 | −29 | |
| 36 | 35 | 33 | 32 | 31 | 29 | 28 | 27 | 25 | 24 | 22 | 20 | 19 | 17 | 15 | 13 | 10 | 7 | 4 | 0 | −4 | −10 | |
| 38 | 37 | 35 | 34 | 33 | 32 | 30 | 29 | 28 | 26 | 25 | 23 | 21 | 19 | 17 | 15 | 13 | 11 | 8 | 5 | 1 | −3 | 9 |
| 40 | 39 | 37 | 36 | 35 | 34 | 32 | 31 | 30 | 28 | 27 | 25 | 24 | 22 | 20 | 18 | 16 | 14 | 12 | 9 | 6 | 2 | −2 |

Column heading: Depression of Web-Bulb (Dry-bulb temperature – Wet-bulb temperature = Depression of the wet bulb). Dew-Point Values.

**TABLE C.3** Relative humidity (°F)*

| Dry bulb (°F) | Depression of Web-Bulb (Dry-bulb temperature – Wet-bulb temperature = Depression of the wet bulb) | | | | | | | | | | | | | | | | | | | | |
|---|---|---|---|---|---|---|---|---|---|---|---|---|---|---|---|---|---|---|---|---|---|
| | 1 | 2 | 3 | 4 | 5 | 6 | 7 | 8 | 9 | 10 | 11 | 12 | 13 | 14 | 15 | 16 | 17 | 18 | 19 | 20 | 25 | 30 |
| 0 | 67 | 33 | 1 | | | | | | | | | | | | | | | | | | | |
| 5 | 73 | 46 | 20 | | | | | | | | | | | | | | | | | | | |
| 10 | 78 | 56 | 34 | 13 | | | | | | | | | | | | | | | | | | |
| 15 | 82 | 64 | 46 | 29 | 11 | | | | | | | | | | | | | | | | | |
| 20 | 85 | 70 | 55 | 40 | 26 | 12 | | | | | | | | | | | | | | | | |
| 25 | 87 | 74 | 62 | 49 | 37 | 25 | 13 | 1 | | | | | | | | | | | | | | |
| 30 | 89 | 78 | 67 | 56 | 46 | 36 | 26 | 16 | 6 | | | | | | | | | | | | | |
| 35 | 91 | 81 | 72 | 63 | 54 | 45 | 36 | 27 | 19 | 10 | 2 | | | | | | | | | | | |
| 40 | 92 | 83 | 75 | 68 | 60 | 52 | 45 | 37 | 29 | 22 | 15 | 7 | | | | | | | | | | |
| 45 | 93 | 86 | 78 | 71 | 64 | 57 | 51 | 44 | 38 | 31 | 25 | 18 | 12 | 6 | | | | | | | | |
| 50 | 93 | 87 | 80 | 74 | 67 | 61 | 55 | 49 | 43 | 38 | 32 | 27 | 21 | 16 | 10 | 5 | | | | | | |
| 55 | 94 | 88 | 82 | 76 | 70 | 65 | 59 | 54 | 49 | 43 | 38 | 33 | 28 | 23 | 19 | 11 | 9 | 5 | | | | |
| 60 | 94 | 89 | 83 | 78 | 73 | 68 | 63 | 58 | 53 | 48 | 43 | 39 | 34 | 30 | 26 | 21 | 17 | 13 | 9 | 5 | | |
| 65 | 95 | 90 | 85 | 80 | 75 | 70 | 66 | 61 | 56 | 52 | 48 | 44 | 39 | 35 | 31 | 27 | 24 | 20 | 16 | 12 | | |
| 70 | 95 | 90 | 86 | 81 | 77 | 72 | 68 | 64 | 59 | 55 | 51 | 48 | 44 | 40 | 36 | 33 | 29 | 25 | 22 | 19 | 3 | |
| 75 | 96 | 91 | 86 | 82 | 78 | 74 | 70 | 66 | 62 | 58 | 54 | 51 | 47 | 44 | 40 | 37 | 34 | 30 | 27 | 24 | 9 | |
| 80 | 96 | 91 | 87 | 83 | 79 | 75 | 72 | 68 | 64 | 61 | 57 | 54 | 50 | 47 | 44 | 41 | 38 | 35 | 32 | 29 | 15 | 3 |
| 85 | 96 | 92 | 88 | 84 | 81 | 77 | 73 | 70 | 66 | 63 | 59 | 57 | 53 | 50 | 47 | 44 | 41 | 38 | 36 | 33 | 20 | 8 |
| 90 | 96 | 92 | 89 | 85 | 81 | 78 | 74 | 71 | 68 | 65 | 61 | 58 | 55 | 52 | 49 | 47 | 44 | 41 | 39 | 36 | 24 | 13 |
| 95 | 96 | 93 | 89 | 86 | 82 | 79 | 76 | 73 | 69 | 66 | 63 | 61 | 58 | 55 | 52 | 50 | 47 | 44 | 42 | 39 | 28 | 17 |
| 100 | 96 | 93 | 89 | 86 | 83 | 80 | 77 | 73 | 70 | 68 | 65 | 62 | 59 | 56 | 54 | 51 | 49 | 46 | 44 | 41 | 30 | 21 |
| 105 | 97 | 93 | 90 | 87 | 84 | 81 | 78 | 75 | 72 | 69 | 66 | 64 | 61 | 58 | 56 | 53 | 51 | 49 | 46 | 44 | 34 | 24 |
| 110 | 97 | 93 | 90 | 87 | 84 | 81 | 78 | 75 | 73 | 70 | 67 | 65 | 62 | 60 | 57 | 55 | 52 | 50 | 48 | 46 | 36 | 26 |

*To determine the relative humidity, find the air (dry-bulb) temperature on the vertical axis (far left) and the depression of the wet bulb on the horizontal axis (top). Where the two meet, the relative humidity is found. For example, when the dry-bulb temperature is 70°F and a wet-bulb temperature is 57°F, then the depression of the wet-bulb is 13°F (70°F–57°F). *From Table C.3, the relative humidity is 44 percent and from Table C.4, the dew point is 47°F.*

**TABLE C.4** Dew-Point Temperature (°F)

| Dry Bulb (°F) | Depression of Web-Bulb (Dry-bulb temperature – Wet-bulb temperature = Depression of the wet bulb) | | | | | | | | | | | | | | | | | | | | |
|---|---|---|---|---|---|---|---|---|---|---|---|---|---|---|---|---|---|---|---|---|---|
| | 1 | 2 | 3 | 4 | 5 | 6 | 7 | 8 | 9 | 10 | 11 | 12 | 13 | 14 | 15 | 16 | 17 | 18 | 19 | 20 | 25 | 30 |
| 0 | −7 | −17 | −37 | | | | | | | | | | | | | | | | | | | |
| 5 | −1 | −8 | −20 | | | | | | | | | | | | | | | | | | | |
| 10 | 5 | −1 | −9 | −21 | | | | | | | | | | | | | | | | | | |
| 15 | 11 | 6 | 0 | −8 | −20 | | | | | | | | | | | | | | | | | |
| 20 | 16 | 12 | 8 | 2 | −6 | −17 | | | | | | | | | | | | | | | | |
| 25 | 22 | 18 | 15 | 10 | 6 | −2 | −12 | | | | | | | | | | | | | | | |
| 30 | 27 | 24 | 21 | 17 | 13 | 2 | −7 | −20 | | | | | | | | | | | | | | |
| 35 | 36 | 30 | 27 | 24 | 20 | 16 | 12 | 6 | 1 | | | | | | | | | | | | | |
| 40 | 38 | 35 | 33 | 30 | 27 | 24 | 20 | 16 | 11 | 5 | | | | | | | | | | | | |
| 45 | 43 | 41 | 39 | 36 | 34 | 31 | 28 | 24 | 21 | 16 | 11 | | | | | | | | | | | |
| 50 | 48 | 46 | 44 | 42 | 40 | 37 | 35 | 32 | 29 | 25 | 22 | 17 | | | | | | | | | | |
| 55 | 53 | 51 | 50 | 48 | 46 | 43 | 41 | 39 | 36 | 33 | 30 | 27 | 23 | | | | | | | | | |
| 60 | 58 | 57 | 55 | 53 | 51 | 49 | 47 | 45 | 43 | 40 | 38 | 35 | 32 | 29 | | | | | | | | |
| 65 | 63 | 62 | 60 | 59 | 57 | 55 | 53 | 51 | 49 | 47 | 45 | 42 | 40 | 37 | 34 | | | | | | | |
| 70 | 68 | 67 | 65 | 64 | 62 | 61 | 60 | 57 | 55 | 53 | 51 | 49 | 47 | 45 | 42 | 39 | | | | | | |
| 75 | 74 | 72 | 71 | 70 | 68 | 66 | 65 | 63 | 61 | 59 | 58 | 56 | 54 | 52 | 49 | 47 | 45 | | | | | |
| 80 | 79 | 77 | 76 | 75 | 73 | 72 | 70 | 68 | 67 | 65 | 64 | 62 | 60 | 58 | 56 | 54 | 52 | 50 | | | | |
| 85 | 84 | 82 | 81 | 80 | 78 | 77 | 75 | 74 | 72 | 71 | 69 | 68 | 66 | 64 | 63 | 61 | 59 | 57 | 55 | | | |
| 90 | 89 | 87 | 86 | 85 | 84 | 82 | 81 | 79 | 78 | 76 | 75 | 73 | 72 | 70 | 68 | 67 | 65 | 63 | 62 | 60 | | |
| 95 | 94 | 93 | 91 | 90 | 89 | 87 | 86 | 85 | 83 | 82 | 81 | 79 | 78 | 76 | 75 | 73 | 71 | 70 | 68 | 66 | 56 | |
| 100 | 99 | 98 | 96 | 95 | 94 | 93 | 91 | 90 | 89 | 87 | 86 | 85 | 83 | 82 | 80 | 79 | 77 | 76 | 74 | 73 | 64 | 53 |
| 105 | 104 | 103 | 102 | 100 | 99 | 98 | 97 | 95 | 94 | 93 | 91 | 90 | 89 | 87 | 86 | 85 | 83 | 82 | 80 | 79 | 70 | 61 |
| 110 | 109 | 108 | 107 | 105 | 104 | 103 | 102 | 100 | 99 | 98 | 97 | 95 | 94 | 93 | 91 | 90 | 89 | 87 | 86 | 85 | 77 | 68 |

# Appendix D Laws Relating to Gases

## Kinetic Energy

All moving objects, by virtue of their motion, are capable of doing work. We call this energy of motion, or *kinetic energy.* The kinetic energy of a moving object is equal to one-half its mass (M) multiplied by its velocity (v) squared. Stated mathematically:

$$\text{Kinetic energy} = \frac{1}{2}\text{Mv}^2$$

Therefore, by doubling the velocity of a moving object, the object's kinetic energy will increase four times.

## First Law of Thermodynamics

The first law of thermodynamics is simply the thermal version of the law of conservation of energy, which states that energy cannot be created or destroyed, only transformed from one form to another. Meteorologists use the first law of thermodynamics along with the principles of kinetic energy extensively in analyzing atmospheric phenomena. According to the kinetic theory, the temperature of a gas is proportional to the kinetic energy of the moving molecules. When a gas is heated, its kinetic energy increases because of an increase in molecular motion. Further, when a gas is compressed, the kinetic energy will also be increased and the temperature of the gas will rise. These relationships are expressed in the first law of thermodynamics, as follows: The temperature of a gas may be changed by the addition or subtraction of heat, or by changing the pressure (compression or expansion), or by a combination of both. It is easy to understand how the atmosphere is heated or cooled by the gain or loss of heat. However, when we consider rising and sinking air, the relationships between temperature and pressure become more important. Here an increase in temperature is brought about by performing work on the gas and not by the addition of heat. This phenomenon is called the *adiabatic form* of the first law of thermodynamics.

## Boyle's Law

About 1600, the Englishman Robert Boyle showed that if the temperature is kept constant when the pressure exerted on a gas is increased, the volume decreases. This principle, called *Boyle's law,* states: At a constant temperature, the volume of a given mass of gas varies inversely with the pressure. Stated mathematically:

$$P_1V_1 = P_2V_2$$

The symbols $P_1$ and $V_1$ refer to the original pressure and volume, respectively, and $P_2$ and $V_2$ indicate the new pressure and volume, respectively, after a change occurs. Boyle's law shows that if a given volume of gas is compressed so that the volume is reduced by one-half, the pressure exerted by the gas is doubled. This increase in pressure can be explained by the kinetic theory, which predicts that when the volume of the gas is reduced by one-half, the molecules collide with the walls of the container twice as often. Because density is defined as the mass per unit volume, an increase in pressure results in increased density.

## Charles's Law

The relationships between temperature and volume (hence, density) of a gas were recognized about 1787 by the French scientist Jacques Charles, and were stated formally by J. Gay-Lussac in 1802. *Charles's law* states: At a constant pressure, the volume of a given mass is directly proportional to the absolute temperature. In other words, when a quantity of gas is kept at a constant pressure, an increase in temperature results in an increase in volume and vice versa. Stated mathematically:

$$\frac{V_1}{V_2} = \frac{T_1}{T_2}$$

where $V_1$ and $T_1$ represent the original volume and temperature, respectively, and $V_2$ and $T_2$ represent the final volume and temperature, respectively. This law explains the fact that a gas expands when it is heated. According to the kinetic theory, when heated, particles move more rapidly and therefore collide more often.

## The Ideal Gas Law or Equation of State

In describing the atmosphere, three variable quantities must be considered: pressure, temperature, and density (mass per unit volume). The relationships among these variables can be found by combining in a single statement the laws of Boyle and Charles as follows:

$$PV = RT \quad \text{or} \quad P = \rho RT$$

where $P$ is pressure, $V$ volume, $R$ the constant of proportionality, $T$ absolute temperature, and $\rho$ density. This law, called the *ideal gas law,* states:

1. When the volume is kept constant, the pressure of a gas is directly proportional to its absolute temperature.
2. When the temperature is kept constant, the pressure of a gas is proportional to its density and inversely proportional to its volume.
3. When the pressure is kept constant, the absolute temperature of a gas is proportional to its volume and inversely proportional to its density.

# Appendix E Newton's Laws, Pressure–Gradient Force, and Coriolis Force

## Newton's Laws of Motion

Because air is composed of atoms and molecules, its motion is governed by the same natural laws that apply to all matter. Simply put, when a force is applied to air, it will be displaced from its original position. Depending on the direction from which the force is applied, the air may move horizontally to produce winds, or in some situations vertically to generate convective flow. To better understand the forces that produce global winds, it is helpful to become familiar with Newton's first two laws of motion.

Newton's first law of motion states that an object at rest will remain at rest, and an object in motion will continue moving at a uniform speed and in a straight line unless a force is exerted upon it. In simple terms this law states that objects at rest tend to stay at rest, and objects in motion tend to continue moving at the same rate in the same direction. The tendency of things to resist change in motion (including a change in direction) is known as inertia.

You have experienced Newton's first law if you have ever pushed a stalled auto along flat terrain. To start the automobile moving (accelerating) requires a force sufficient to overcome its inertia (resistance to change). However, once this vehicle is moving, a force equal to that of the frictional force between the tires and the pavement is enough to keep it moving.

Moving objects often deviate from straight paths or come to rest, whereas objects at rest begin to move. The changes in motion we observe in daily life are the result of one or more applied forces.

Newton's second law of motion describes the relationship between the forces that are exerted on objects and the observed accelerations that result. Newton's second law states that the acceleration of an object is directly proportional to the net force acting on that body and inversely proportional to the mass of the body. The first part of Newton's second law means that the acceleration of an object changes as the intensity of the applied force changes.

We define acceleration as the rate of change in velocity. Because velocity describes both the speed and direction of a moving body, the velocity of something can be changed by changing its speed or its direction or both. Further, the term acceleration refers both to decreases and increases in velocity.

For example, we know that when we push down on the gas pedal of an automobile, we experience a positive acceleration (increase in velocity). On the other hand, using the brakes retards acceleration (decreases velocity).

In the atmosphere, three forces are responsible for changing the state of motion of winds. These are the pressure–gradient force, the Coriolis force, and friction. From the preceding discussion, it should be clear that the relative strengths of these forces will determine to a large degree the role of each in establishing the flow of air. Further, these forces can be directed in such a way as to increase the speed of airflow, decrease the speed of airflow, or, in many instances, just change the direction of airflow.

## Pressure–Gradient Force

The magnitude of the pressure–gradient force is a function of the pressure difference between two points and air density. It can be expressed as

$$F_{PG} = \frac{1}{d} \times \frac{\Delta_p}{\Delta_n}$$

where:

$F_{PG}$ = pressure–gradient force per unit mass

$d$ = density of air

$p$ = pressure difference between two points

$n$ = distance between two points

Let us consider an example where the pressure 5 kilometers above Little Rock, Arkansas, is 540 millibars, and at 5 kilometers above St. Louis, Missouri, it is 530 millibars. The distance between the two cities is 450 kilometers, and the air density at 5 kilometers is 0.75 kilogram per cubic meter. In order to use the pressure–gradient equation, we must use compatible units. We must first convert pressure from millibars to pascals, another measure of pressure that has units of (kilograms $\times$ meters$^{-1}$ $\times$ second$^2$).

In our example, the pressure difference above the two cities is 10 millibars, or 1000 pascals (1000 kg/m·s$^{-2}$). Thus, we have:

$$F_{PG} = \frac{1}{0.75} \times \frac{1000}{450{,}000} = 0.0029\,\frac{\text{m}}{\text{s}^2}$$

Newton's second law states that force equals mass times acceleration ($F = m \times a$). In our example, we have considered pressure–gradient force *per unit mass;* therefore, our result is an acceleration ($F/m = a$). Because of the small units shown, pressure–gradient *acceleration* is often expressed as centimeters per second squared. In this example, we have 0.296 cm/s$^2$.

## Coriolis Force as a Function of Wind Speed and Latitude

Figure 6.15 (page 168) shows how wind speed and latitude conspire to affect the Coriolis force. Consider a west wind at four different latitudes (0°, 20°, 40°, and 60°). After several hours, Earth's rotation has changed the orientation of latitude and longitude of all locations except the equator such that the wind appears to be deflected to the right. The degree of deflection for a given wind speed increases with latitude because the orientation of latitude and longitude lines changes more at higher latitudes. The degree of deflection of a given latitude increases with wind speed because greater distances are covered in the period of time considered.

We can show mathematically the importance of latitude and wind speed on Coriolis force:

$$F_{CO} = 2v\,\Omega \sin\phi$$

where:

$F_{CO}$ = Coriolis force *per unit mass of air*

$v$ = wind speed

$\Omega$ = Earth's rate of rotation or angular velocity (which is $7.29 \times 10^{-5}$ radians per second)

$\phi$ = latitude

Note that $\sin\phi$ is a trigonometric function equal to zero for an angle of 0° (equator) and 1 when $\phi = 90°$ (poles).

As an example, the Coriolis force per unit mass that must be considered for a 10-meters-per-second (m/s) wind at 40° is calculated as:

$$F_{CO} = 2\Omega \sin\phi\, v$$

$$F_{CO} = 2\Omega \sin 40° \times 10 \text{ m/s}$$

$$F_{CO} = 2(7.29 \times 10^{-5}\text{s}^{-1})\, 0.64(10 \text{ m/s})$$

$$F_{CO} = 0.00094 \text{ meter per second squared}$$

$$= 0.094 \text{ cm s}^{-2}$$

The result (0.094 $\text{cm/s}^{-2}$) is expressed as an acceleration because we are considering force per unit mass and Force = Mass × Acceleration.

Using this equation, one could calculate the Coriolis force for any latitude or wind speed. Consider **Table E.1**, which shows the Coriolis force per unit mass for three specific wind speeds at various latitudes. All values are expressed in centimeters per second squared ($\text{cm/s}^{-2}$). Because pressure–gradient force and Coriolis force approximately balance under geostrophic conditions, we can see from our table that the pressure–gradient force (per unit mass) of 0.296 $\text{cm/s}^{-2}$ illustrated in the preceding discussion of pressure–gradient force would produce relatively strong winds.

**TABLE E.1** Coriolis Force for Three Wind Speeds at Various Latitudes

| Wind Speed | | Latitude ($\phi$) | | | |
|---|---|---|---|---|---|
| | | 0° | 20° | 40° | 60° |
| (m/s) | (kph) | Coriolis Force ($\text{cm/s}^2$) | | | |
| 5 | 18 | 0 | 0.025 | 0.047 | 0.063 |
| 10 | 36 | 0 | 0.050 | 0.094 | 0.126 |
| 25 | 90 | 0 | 0.125 | 0.235 | 0.316 |

# Appendix F Saffir–Simpson Hurricane Scale

| Scale Number (category) | Central Pressure (millibars) | Wind Speed (kph) | Wind Speed (mph) | Storm Surge (meters) | Storm Surge (feet) | Damage |
|---|---|---|---|---|---|---|
| 1 | ≥980 | 119–153 | 74–95 | 1.2–1.5 | 4–5 | *Minimal.* No real damage to building structures. Damage primarily to unanchored mobile homes, shrubbery, and trees. Also, some coastal-road flooding and minor pier damage. |
| 2 | 965–979 | 154–177 | 96–110 | 1.6–2.4 | 6–8 | *Moderate.* Some roofing material, door, and window damage to buildings. Some trees blown down. Considerable damage to mobile homes. Coastal and low-lying escape routes flood 2 to 4 hours before arrival of the hurricane center. Small craft in unprotected anchorages break moorings. |
| 3 | 945–964 | 178–209 | 111–130 | 2.5–3.6 | 9–12 | *Extensive.* Some structural damage to small residences and utility buildings. Large trees blown down. Mobile homes are destroyed. Flooding near the coast destroys smaller structures with larger structures damaged due to battering by floating debris. Terrain lower than 2 meters above sea level may be flooded inland 13 km or more. Evacuation of low-lying residences within several blocks of the shoreline may be required. |
| 4 | 920–944 | 210–250 | 131–155 | 3.7–5.4 | 13–18 | *Extreme.* Some complete roof structure failures on small residences. Extensive damage to doors and windows. Low-lying escape routes may be cut by rising water 3 to 5 hours before arrival of the hurricane center. Major damage to lower floors of structures near the shore. Terrain lower than 3 meters above sea level may be flooded, requiring massive evacuation of residential areas as far inland as 10 km. |
| 5 | <920 | >250 | >155 | >5.4 | >18 | *Catastrophic.* Complete roof failure on many residences and industrial buildings. Some complete building failures. Severe window and door damage. Low-lying escape routes are cut by rising water 3 to 5 hours before arrival of the hurricane center. Major damage to lower floors of all structures located less than 5 meters above sea level and within 500 meters of the shoreline. Massive evacuation of residential areas on low ground within 8 to 16 km of the shoreline may be required. |

# Appendix G Climate Data

Table G-1 includes data for 51 stations around the world that represent many different climate types. Temperatures are given in degrees Celsius and precipitation in millimeters. Names and locations are given in Table G-2, along with the elevation (in meters) of each station and its Köppen classification. This format was used so that you can use the data in exercises to reinforce your understanding of climate controls and climate classification.

Use Figure 15.2 (page 413) to determine the proper Köppen classification. The flowchart in Figure G-1 will help guide you through the classification process. Once you have classified a station, determine a likely location, based on factors such as mean annual temperature, annual temperature range, total precipitation, and seasonal precipitation distribution. Your location need not be a specific city; it could be a description of the station's setting, such as "middle-latitude continental" or "subtropical with a strong monsoon influence." It would also be a good idea to list the reasons for your selection. You may check your answer by examining the list of stations in Table G-2.

If you simply wish to examine the data for a specific place or data for a specific climate type, consult the list in Table G-2.

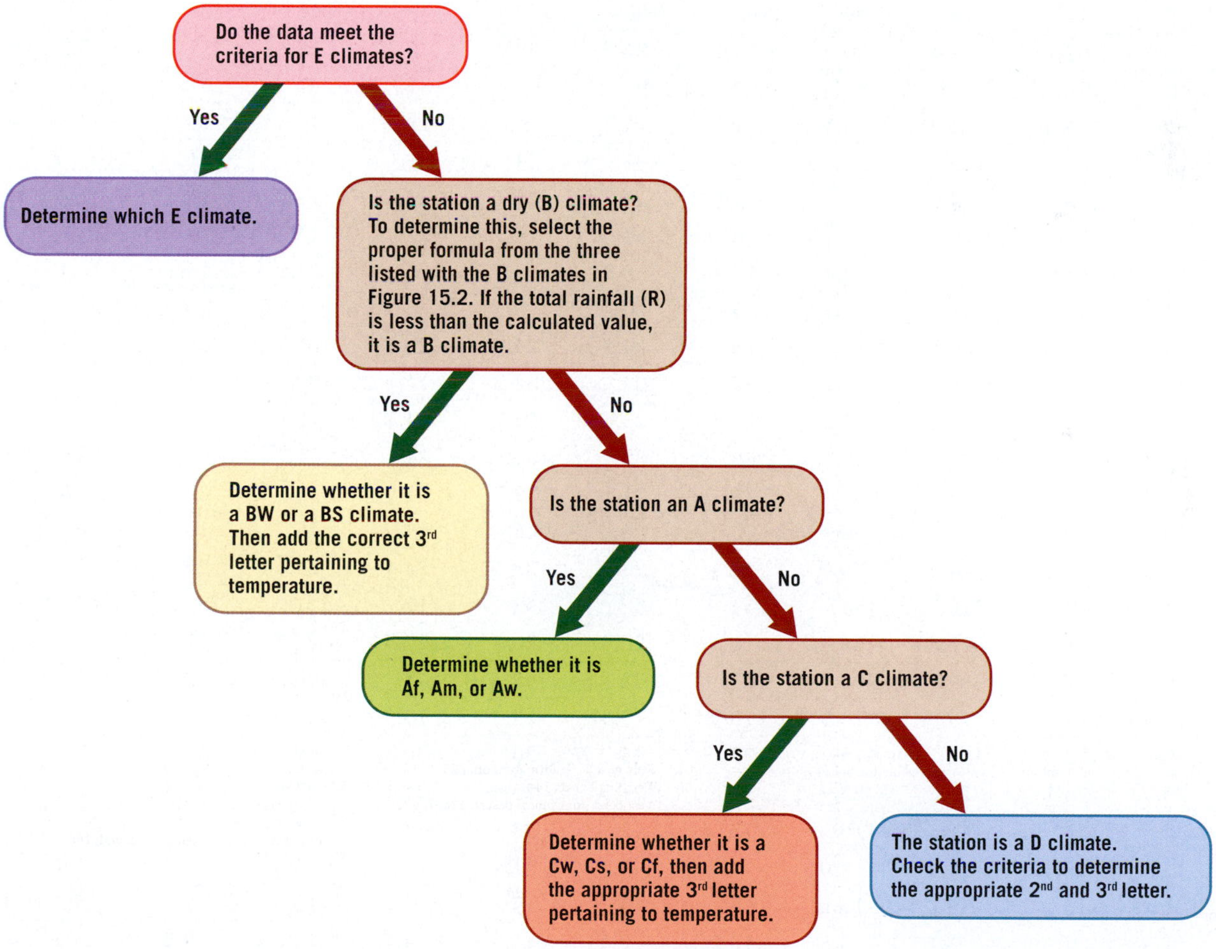


**Figure G–1** Classifying Climates Using Figure 15.2, page 413

**TABLE G–1** Selected Climate Data for the World

| | J | F | M | A | M | J | J | A | S | O | N | D | YR. |
|---|---|---|---|---|---|---|---|---|---|---|---|---|---|
| 1 | 1.7 | 4.4 | 7.9 | 13.2 | 18.4 | 23.8 | 25.8 | 24.8 | 21.4 | 14.7 | 6.7 | 2.8 | 13.8 |
| | 10 | 10 | 13 | 13 | 20 | 15 | 30 | 32 | 23 | 18 | 10 | 13 | 207 |
| 2 | −10.4 | −8.3 | −4.6 | 3.4 | 9.4 | 12.8 | 16.6 | 14.9 | 10.8 | 5.5 | −2.3 | −6.4 | 3.5 |
| | 18 | 25 | 25 | 30 | 51 | 89 | 64 | 71 | 33 | 20 | 18 | 15 | 459 |
| 3 | 10.2 | 10.8 | 13.7 | 17.9 | 22.2 | 25.7 | 26.7 | 26.5 | 24.2 | 19.0 | 13.3 | 10.0 | 18.4 |
| | 66 | 84 | 99 | 74 | 91 | 127 | 196 | 168 | 147 | 71 | 53 | 71 | 1247 |
| 4 | −23.9 | −17.5 | −12.5 | −2.7 | 8.4 | 14.8 | 15.6 | 12.8 | 6.4 | −3.1 | −15.8 | −21.9 | −3.3 |
| | 23 | 13 | 18 | 8 | 15 | 33 | 48 | 53 | 33 | 20 | 18 | 15 | 297 |
| 5 | −17.8 | −15.3 | −9.2 | −4.4 | 4.7 | 10.9 | 16.4 | 14.4 | 10.3 | 3.3 | −3.6 | −12.8 | −0.3 |
| | 58 | 58 | 61 | 48 | 53 | 61 | 81 | 71 | 58 | 61 | 64 | 64 | 738 |
| 6 | −8.2 | −7.1 | −2.4 | 5.4 | 11.2 | 14.7 | 18.9 | 17.4 | 12.7 | 7.4 | −0.4 | −4.6 | 5.4 |
| | 21 | 23 | 29 | 35 | 52 | 74 | 39 | 40 | 35 | 29 | 26 | 21 | 424 |
| 7 | 12.8 | 13.9 | 15.0 | 15.0 | 17.8 | 20.0 | 21.1 | 22.8 | 22.2 | 18.3 | 17.2 | 15.0 | 17.6 |
| | 69 | 74 | 46 | 28 | 3 | 3 | 0 | 0 | 5 | 10 | 28 | 61 | 327 |
| 8 | 18.9 | 20.0 | 21.1 | 22.8 | 25.0 | 26.7 | 27.2 | 27.8 | 27.2 | 25.0 | 21.1 | 20.0 | 23.6 |
| | 51 | 48 | 58 | 99 | 163 | 188 | 172 | 178 | 241 | 208 | 71 | 43 | 1520 |
| 9 | −4.4 | −2.2 | 4.4 | 10.6 | 16.7 | 21.7 | 23.9 | 22.7 | 18.3 | 11.7 | 3.8 | −2.2 | 10.4 |
| | 46 | 51 | 69 | 84 | 99 | 97 | 97 | 81 | 97 | 61 | 61 | 51 | 894 |
| 10 | −5.6 | −4.4 | 0.0 | 6.1 | 11.7 | 16.7 | 20.0 | 18.9 | 15.5 | 10.0 | 3.3 | −2.2 | 7.5 |
| | 112 | 96 | 109 | 94 | 86 | 81 | 74 | 61 | 89 | 81 | 107 | 99 | 1089 |
| 11 | −2.1 | 0.9 | 4.7 | 9.9 | 14.7 | 19.4 | 24.7 | 23.6 | 18.3 | 11.5 | 3.4 | −0.2 | 10.7 |
| | 34 | 30 | 40 | 45 | 36 | 25 | 15 | 22 | 13 | 29 | 33 | 31 | 353 |
| 12 | 12.8 | 13.9 | 15.0 | 16.1 | 17.2 | 18.8 | 19.4 | 22.2 | 21.1 | 18.8 | 16.1 | 13.9 | 15.9 |
| | 53 | 56 | 41 | 20 | 5 | 0 | 0 | 2 | 5 | 13 | 23 | 51 | 269 |
| 13 | −0.1 | 1.8 | 6.2 | 13.0 | 18.7 | 24.2 | 26.4 | 25.4 | 21.1 | 14.9 | 6.7 | 1.6 | 13.3 |
| | 50 | 52 | 78 | 94 | 95 | 109 | 84 | 77 | 70 | 73 | 65 | 50 | 897 |
| 14 | 2.7 | 3.2 | 7.1 | 13.2 | 18.8 | 23.4 | 25.7 | 24.7 | 20.9 | 15.0 | 8.7 | 3.4 | 13.9 |
| | 77 | 63 | 82 | 80 | 105 | 82 | 105 | 124 | 97 | 78 | 72 | 71 | 1036 |
| 15 | 12.8 | 15.0 | 18.9 | 21.1 | 26.1 | 31.1 | 32.7 | 33.9 | 31.1 | 22.2 | 17.7 | 13.9 | 23.0 |
| | 10 | 9 | 6 | 2 | 0 | 0 | 6 | 13 | 10 | 10 | 3 | 8 | 77 |
| 16 | 25.6 | 25.6 | 24.4 | 25.0 | 24.4 | 23.3 | 23.3 | 24.4 | 24.4 | 25.0 | 25.6 | 25.6 | 24.7 |
| | 259 | 249 | 310 | 165 | 254 | 188 | 168 | 117 | 221 | 183 | 213 | 292 | 2619 |
| 17 | 25.9 | 25.8 | 25.8 | 25.9 | 26.4 | 26.6 | 26.9 | 27.5 | 27.9 | 27.7 | 27.3 | 26.7 | 26.7 |
| | 365 | 326 | 383 | 404 | 185 | 132 | 68 | 43 | 96 | 99 | 189 | 143 | 2433 |
| 18 | 13.3 | 13.3 | 13.3 | 13.3 | 13.9 | 13.3 | 13.3 | 13.3 | 13.9 | 13.3 | 13.3 | 13.9 | 13.5 |
| | 99 | 112 | 142 | 175 | 137 | 43 | 20 | 30 | 69 | 112 | 97 | 79 | 1115 |
| 19 | 25.9 | 26.1 | 25.2 | 23.9 | 22.3 | 21.3 | 20.8 | 21.1 | 21.5 | 22.3 | 23.1 | 24.4 | 23.2 |
| | 137 | 137 | 143 | 116 | 73 | 43 | 43 | 43 | 53 | 74 | 97 | 127 | 1086 |
| 20 | 13.8 | 13.5 | 11.4 | 8.0 | 3.7 | 1.2 | 1.4 | 2.9 | 5.5 | 9.2 | 11.4 | 12.9 | 7.9 |
| | 21 | 16 | 18 | 13 | 25 | 15 | 15 | 17 | 12 | 7 | 15 | 18 | 171 |
| 21 | 1.5 | 1.3 | 3.1 | 5.8 | 10.2 | 12.6 | 15.0 | 14.7 | 12.0 | 8.3 | 5.5 | 3.3 | 7.8 |
| | 179 | 139 | 109 | 140 | 83 | 126 | 141 | 167 | 228 | 236 | 207 | 203 | 1958 |
| 22 | −0.5 | 0.2 | 3.9 | 9.0 | 14.3 | 17.7 | 19.4 | 18.8 | 15.0 | 9.6 | 4.7 | 1.2 | 9.5 |
| | 41 | 37 | 30 | 39 | 44 | 60 | 67 | 65 | 45 | 45 | 44 | 39 | 556 |

*(Continued)*

**TABLE G–1** *(Continued)*

| | J | F | M | A | M | J | J | A | S | O | N | D | YR. |
|---|---|---|---|---|---|---|---|---|---|---|---|---|---|
| 23 | 6.1 | 5.8 | 7.8 | 9.2 | 11.6 | 14.4 | 15.6 | 16.0 | 14.7 | 12.0 | 9.0 | 7.0 | 10.8 |
| | 133 | 96 | 83 | 69 | 68 | 56 | 62 | 80 | 87 | 104 | 138 | 150 | 1126 |
| 24 | 10.8 | 11.6 | 13.6 | 15.6 | 17.2 | 20.1 | 22.2 | 22.5 | 21.2 | 18.2 | 14.4 | 11.5 | 16.6 |
| | 111 | 76 | 109 | 54 | 44 | 16 | 3 | 4 | 33 | 62 | 93 | 103 | 708 |
| 25 | −9.9 | −9.5 | −4.2 | 4.7 | 11.9 | 16.8 | 19.0 | 17.1 | 11.2 | 4.5 | −1.9 | −6.8 | 4.4 |
| | 31 | 28 | 33 | 35 | 52 | 67 | 74 | 74 | 58 | 51 | 36 | 36 | 575 |
| 26 | 8.0 | 9.0 | 10.9 | 13.7 | 17.5 | 21.6 | 24.4 | 24.2 | 21.5 | 17.2 | 12.7 | 9.5 | 15.9 |
| | 83 | 73 | 52 | 50 | 48 | 18 | 9 | 18 | 70 | 110 | 113 | 105 | 749 |
| 27 | −9.0 | −9.0 | −6.6 | −4.1 | 0.4 | 3.6 | 5.6 | 5.5 | 3.5 | −0.6 | −4.5 | −7.6 | −1.9 |
| | 202 | 180 | 164 | 166 | 197 | 249 | 302 | 278 | 208 | 183 | 190 | 169 | 2488 |
| 28 | −2.9 | −3.1 | −0.7 | 4.4 | 10.1 | 14.9 | 17.8 | 16.6 | 12.2 | 7.1 | 2.8 | 0.1 | 6.6 |
| | 43 | 30 | 26 | 31 | 34 | 45 | 61 | 76 | 60 | 48 | 53 | 48 | 555 |
| 29 | 12.8 | 13.9 | 17.2 | 18.9 | 22.2 | 23.9 | 25.5 | 26.1 | 25.5 | 23.9 | 18.9 | 15.0 | 20.3 |
| | 66 | 41 | 20 | 5 | 3 | 0 | 0 | 0 | 3 | 18 | 46 | 66 | 268 |
| 30 | 24.6 | 24.9 | 25.0 | 24.9 | 25.0 | 24.2 | 23.7 | 23.8 | 23.9 | 24.2 | 24.2 | 24.7 | 24.4 |
| | 81 | 102 | 155 | 140 | 133 | 119 | 99 | 109 | 206 | 213 | 196 | 122 | 1675 |
| 31 | 21.1 | 20.4 | 20.9 | 21.7 | 23.0 | 26.0 | 27.3 | 27.3 | 27.5 | 27.5 | 26.0 | 25.2 | 24.3 |
| | 0 | 2 | 0 | 0 | 1 | 15 | 88 | 249 | 163 | 49 | 5 | 6 | 578 |
| 32 | 20.4 | 22.7 | 27.0 | 30.6 | 33.8 | 34.2 | 33.6 | 32.7 | 32.6 | 30.5 | 25.5 | 21.3 | 28.7 |
| | 0 | 0 | 0 | 0 | 0 | 2 | 1 | 11 | 2 | 0 | 0 | 0 | 16 |
| 33 | 17.8 | 18.1 | 18.8 | 18.8 | 17.8 | 16.2 | 14.9 | 15.5 | 16.8 | 18.6 | 18.3 | 17.8 | 17.5 |
| | 46 | 51 | 102 | 206 | 160 | 46 | 18 | 25 | 25 | 53 | 109 | 81 | 922 |
| 34 | 20.6 | 20.7 | 19.9 | 19.2 | 16.7 | 13.9 | 13.9 | 16.3 | 19.1 | 21.8 | 21.4 | 20.9 | 18.7 |
| | 236 | 168 | 86 | 46 | 13 | 8 | 0 | 3 | 8 | 38 | 94 | 201 | 901 |
| 35 | 11.7 | 13.3 | 16.7 | 18.6 | 19.2 | 20.0 | 20.3 | 20.5 | 20.5 | 19.1 | 15.9 | 12.9 | 17.4 |
| | 20 | 41 | 179 | 605 | 1705 | 2875 | 2455 | 1827 | 1231 | 447 | 47 | 5 | 11437 |
| 36 | 26.2 | 26.3 | 27.1 | 27.2 | 27.3 | 27.0 | 26.7 | 27.0 | 27.4 | 27.4 | 26.9 | 26.6 | 26.9 |
| | 335 | 241 | 201 | 141 | 116 | 97 | 61 | 50 | 78 | 91 | 151 | 193 | 1755 |
| 37 | −0.8 | 2.6 | 5.3 | 8.5 | 13.1 | 17.0 | 17.2 | 17.3 | 15.3 | 11.5 | 5.7 | 0.3 | 9.4 |
| | 0 | 0 | 1 | 1 | 18 | 72 | 157 | 151 | 68 | 4 | 1 | 0 | 473 |
| 38 | 24.5 | 25.8 | 27.9 | 30.5 | 32.7 | 32.5 | 30.7 | 30.1 | 29.7 | 28.1 | 25.9 | 24.6 | 28.6 |
| | 24 | 7 | 15 | 25 | 52 | 53 | 83 | 124 | 118 | 267 | 308 | 157 | 1233 |
| 39 | −18.7 | −18.1 | −16.7 | −11.7 | −5.0 | 0.6 | 5.3 | 5.8 | 1.4 | −4.2 | −12.3 | −15.8 | −7.5 |
| | 8 | 8 | 8 | 8 | 15 | 20 | 36 | 43 | 43 | 33 | 13 | 12 | 247 |
| 40 | −21.9 | −18.6 | −12.5 | −5.0 | 9.7 | 15.6 | 18.3 | 16.1 | 10.3 | 0.8 | −10.6 | −18.4 | −1.4 |
| | 15 | 8 | 8 | 13 | 30 | 51 | 51 | 51 | 28 | 25 | 18 | 20 | 318 |
| 41 | −4.7 | −1.9 | 4.8 | 13.7 | 20.1 | 24.7 | 26.1 | 24.9 | 19.9 | 12.8 | 3.8 | −2.7 | 11.8 |
| | 4 | 5 | 8 | 17 | 35 | 78 | 243 | 141 | 58 | 16 | 10 | 3 | 623 |
| 42 | 24.3 | 25.2 | 27.2 | 29.8 | 29.5 | 27.8 | 27.6 | 27.1 | 27.6 | 28.3 | 27.7 | 25.0 | 27.3 |
| | 8 | 5 | 6 | 17 | 260 | 524 | 492 | 574 | 398 | 208 | 34 | 3 | 2530 |
| 43 | 25.8 | 26.3 | 27.8 | 28.8 | 28.2 | 27.4 | 27.1 | 27.1 | 26.7 | 26.5 | 26.1 | 25.7 | 27.0 |
| | 6 | 13 | 12 | 65 | 196 | 285 | 242 | 277 | 292 | 259 | 122 | 37 | 1808 |
| 44 | 3.7 | 4.3 | 7.6 | 13.1 | 17.6 | 21.1 | 25.1 | 26.4 | 22.8 | 16.7 | 11.3 | 6.1 | 14.7 |
| | 48 | 73 | 101 | 135 | 131 | 182 | 146 | 147 | 217 | 220 | 101 | 60 | 1563 |

*(Continued)*

**TABLE G–1** *(Continued)*

| | J | F | M | A | M | J | J | A | S | O | N | D | YR. |
|---|---|---|---|---|---|---|---|---|---|---|---|---|---|
| 45 | −15.8 | −13.6 | −4.0 | 8.5 | 17.7 | 21.5 | 23.9 | 21.9 | 16.7 | 6.1 | −6.2 | −13.0 | 5.3 |
| | 8 | 15 | 15 | 33 | 25 | 33 | 16 | 35 | 15 | 47 | 22 | 11 | 276 |
| 46 | −46.8 | −43.1 | −30.2 | −13.5 | 2.7 | 12.9 | 15.7 | 11.4 | 2.7 | −14.3 | −35.7 | −44.5 | −15.2 |
| | 7 | 5 | 5 | 4 | 5 | 25 | 33 | 30 | 13 | 11 | 10 | 7 | 155 |
| 47 | 19.2 | 19.6 | 18.4 | 16.4 | 13.8 | 11.8 | 10.8 | 11.3 | 12.6 | 16.3 | 15.9 | 17.7 | 15.2 |
| | 84 | 104 | 71 | 109 | 122 | 140 | 140 | 109 | 97 | 106 | 81 | 79 | 1242 |
| 48 | 28.2 | 27.9 | 28.3 | 28.2 | 26.8 | 25.4 | 25.1 | 25.8 | 27.7 | 29.1 | 29.2 | 28.7 | 27.6 |
| | 341 | 338 | 274 | 121 | 9 | 1 | 2 | 5 | 17 | 66 | 156 | 233 | 1562 |
| 49 | 21.9 | 21.9 | 21.2 | 18.3 | 15.7 | 13.1 | 12.3 | 13.4 | 15.3 | 17.6 | 19.4 | 21.0 | 17.6 |
| | 104 | 125 | 129 | 101 | 115 | 141 | 94 | 83 | 72 | 80 | 77 | 86 | 1205 |
| 50 | −7.2 | −7.2 | −4.4 | −0.6 | 4.4 | 8.3 | 10.0 | 8.3 | 5.0 | 1.1 | −3.3 | −6.1 | 0.7 |
| | 84 | 66 | 86 | 62 | 89 | 81 | 79 | 94 | 150 | 145 | 117 | 79 | 1132 |
| 51 | −4.4 | −8.9 | −15.5 | −22.8 | −23.9 | −24.4 | −26.1 | −26.1 | −24.4 | −18.8 | −10.0 | −3.9 | −17.4 |
| | 13 | 18 | 10 | 10 | 10 | 8 | 5 | 8 | 10 | 5 | 5 | 8 | 110 |

**TABLE G–2** Locations and Climate Classifications for Table G–1

| Station No. | City | Location | Elevation (m) | Köppen Classification |
|---|---|---|---|---|
| | | **North America** | | |
| 1 | Albuquerque, N.M. | lat. 35°05′N | 1593 | BWk |
| | | long. 106°40′W | | |
| 2 | Calgary, Canada | lat. 51°03′N | 1062 | Dfb |
| | | long. 114°05′W | | |
| 3 | Charleston, S.C. | lat. 32°47′N | 18 | Cfa |
| | | long. 79°56′W | | |
| 4 | Fairbanks, Alaska | lat. 64°50′N | 134 | Dfc |
| | | long. 147°48′W | | |
| 5 | Goose Bay, Canada | lat. 53°19′N | 45 | Dfb |
| | | long. 60°33′W | | |
| 6 | Lethbridge, Canada | lat. 49°40′N | 920 | Dfb |
| | | long. 112°39′W | | |
| 7 | Los Angeles, Calif. | lat. 34°00′N | 29 | BSk |
| | | long. 118°15′W | | |
| 8 | Miami, Fla. | lat. 25°45′N | 2 | Am |
| | | long. 80°11′W | | |
| 9 | Peoria, III. | lat. 40°45′N | 180 | Dfa |
| | | long. 89°35′W | | |
| 10 | Portland, Me. | lat. 43°40′N | 14 | Dfb |
| | | long. 70°16′W | | |
| 11 | Salt Lake City, Utah | lat. 40°46′N | 1288 | BSk |
| | | long. 111°52′W | | |
| 12 | San Diego, Calif. | lat. 32°43′N | 26 | BSk |
| | | long. 117°10′W | | |
| 13 | St. Louis, Mo. | lat. 38°39′N | 172 | Cfa |
| | | long. 90°15′W | | |

*(Continued)*

**TABLE G–2** (*Continued*)

| Station No. | City | Location | Elevation (m) | Köppen Classification |
|---|---|---|---|---|
| 14 | Washington, D.C. | lat. 38°50′N | 20 | Cfa |
| | | long. 77°00′W | | |
| 15 | Yuma, Ariz. | lat. 32°40′N | 62 | BWh |
| | | long. 114°40′W | | |
| | | **South America** | | |
| 16 | Iquitos, Peru | lat. 3°39′S | 115 | Af |
| | | long. 73°18′W | | |
| 17 | Manaus, Brazil | lat. 3°01′S | 60 | Am |
| | | long. 60°00′W | | |
| 18 | Quito, Ecuador | lat. 0°17′S | 2766 | Cfb |
| | | long. 78°32′W | | |
| 19 | Rio de Janeiro, Brazil | lat. 22°50′S | 26 | Aw |
| | | long. 43°20′W | | |
| 20 | Santa Cruz, Argentina | lat. 50°01′S | 111 | BSk |
| | | long. 60°30′W | | |
| | | **Europe** | | |
| 21 | Bergen, Norway | lat. 60°24′N | 44 | Cfb |
| | | long. 5°20′E | | |
| 22 | Berlin, Germany | lat. 52°28′N | 50 | Cfb |
| | | long. 13°26′E | | |
| 23 | Brest, France | lat. 48°24′N | 103 | Cfb |
| | | long. 4°30′W | | |
| 24 | Lisbon, Portugal | lat. 38°43′N | 93 | Csa |
| | | long. 9°05′W | | |
| 25 | Moscow, Russia | lat. 55°45′N | 156 | Dfb |
| | | long. 37°37′E | | |
| 26 | Rome, Italy | lat. 41°52′N | 3 | Csa |
| | | long. 12°37′E | | |
| 27 | Santis, Switzerland | lat. 47°15′N | 2496 | ET |
| | | long. 9°21′E | | |
| 28 | Stockholm, Sweden | lat. 59°21′N | 52 | Dfb |
| | | long. 18°00′E | | |
| | | **Africa** | | |
| 29 | Benghazi, Libya | lat. 32°06′N | 25 | BSh |
| | | long. 20°06′E | | |
| 30 | Coquilhatville, Zaire | lat. 0°01′N | 21 | Af |
| | | long. 18°17′E | | |
| 31 | Dakar, Senegal | lat. 14°40′N | 23 | BSh |
| | | long. 17°28′W | | |
| 32 | Faya, Chad | lat. 18°00′N | 251 | BWh |
| | | long. 21°18′E | | |
| 33 | Nairobi, Kenya | lat. 1°16′S | 1791 | Csb |
| | | long. 36°47′E | | |
| 34 | Harare, Zimbabwe | lat. 17°50′S | 1449 | Cwb |
| | | long. 30°52′E | | |

(*Continued*)

**TABLE G–2** *(Continued)*

| Station No. | City | Location | Elevation (m) | Köppen Classification |
|---|---|---|---|---|
| | | **Asia** | | |
| 35 | Cherrapunji, India | lat. 25°15′N | 1313 | Cwb |
| | | long. 91°44′E | | |
| 36 | Djakarta, Indonesia | lat. 6°11′S | 8 | Am |
| | | long. 106°45′E | | |
| 37 | Lhasa, Tibet | lat. 29°40′N | 3685 | Cwb |
| | | long. 91°07′E | | |
| 38 | Madras, India | lat. 13°00′N | 16 | Aw |
| | | long. 80°11′E | | |
| 39 | Novaya Zemlya, Russia | lat. 72°23′N | 15 | ET |
| | | long. 54°46′E | | |
| 40 | Omsk, Russia | lat. 54°48′N | 85 | Dfb |
| | | long. 73°19′E | | |
| 41 | Beijing, China | lat. 39°57′N | 52 | Dwa |
| | | long. 116°23′E | | |
| 42 | Rangoon, Myanmar | lat. 16°46′N | 23 | Am |
| | | long. 96°10′E | | |
| 43 | Ho Chi Minh City, Viet Nam | lat. 10°49′N | 10 | Aw |
| | | long. 106°40′E | | |
| 44 | Tokyo, Japan | lat. 35°41′N | 6 | Cfa |
| | | long. 139°46′E | | |
| 45 | Urumchi, China | lat. 43°47′N | 912 | Dfa |
| | | long. 87°43′E | | |
| 46 | Verkhoyansk, Russia | lat. 67°33′N | 137 | Dfd |
| | | long. 133°23′E | | |
| | | **Australia and New Zealand** | | |
| 47 | Auckland, New Zealand | lat. 37°43′S | 49 | Csb |
| | | long. 174°53′E | | |
| 48 | Darwin, Australia | lat. 12°26′S | 27 | Aw |
| | | long. 131°00′E | | |
| 49 | Sydney, Australia | lat. 33°52′S | 42 | Cfb |
| | | long. 151° 179E | | |
| | | **Greenland** | | |
| 50 | Ivigtut, Greenland | lat. 61°12′N | 29 | ET |
| | | long. 48° 109W | | |
| | | **Antarctica** | | |
| 51 | McMurdo Station, | lat. 77°53′S | 2 | EF |
| | Antarctica | long. 167°00′E | | |

# Glossary

**Absolute Humidity** The mass of water vapor per volume of air (usually expressed as grams of water vapor per cubic meter of air).

**Absolute Instability** The condition of air that has an environmental lapse rate that is greater than the dry adiabatic rate (1°C per 100 meters).

**Absolute Stability** The condition of air that has an environmental lapse rate that is less than the wet adiabatic rate.

**Absolute Zero** The zero point on the Kelvin temperature scale, representing the temperature at which all molecular motion is presumed to cease.

**Absorptivity** A measure of the amount of radiant energy absorbed by a substance.

**Acid Precipitation** Rain or snow with a pH value that is less than the value for uncontaminated rain.

**Adiabatic Temperature Change** The cooling or warming of air caused when air is allowed to expand or is compressed, not because heat is added or subtracted.

**Advection** Horizontal convective motion, such as wind.

**Advection Fog** Fog formed when warm moist air is blown over a cool surface and chilled below the dew point.

**Aerosols** Tiny solid and liquid particles suspended in the atmosphere.

**Aerovane** A device that resembles a wind vane with a propeller at one end. Used to indicate wind speed and direction.

**Air** A mixture of many discrete gases, of which nitrogen and oxygen are most abundant, in which varying quantities of tiny solid and liquid particles are suspended.

**Air Mass** A large body of air, usually 1600 kilometers or more across, that is characterized by homogeneous physical properties at any given altitude.

**Air-Mass Thunderstorm** A localized thunderstorm that forms in a warm, moist, unstable air mass. This type of storm occurs most frequently in the afternoon in spring and summer.

**Air-Mass Weather** The conditions experienced in an area as an air mass passes over it. Because air masses are large and relatively homogeneous, air-mass weather will be fairly constant and may last for several days.

**Air Pollutants** Airborne particles and gases occurring in concentrations that endanger the health and well-being of organisms or disrupt the orderly functioning of the environment.

**Air Pressure** The force exerted by the weight of a column of air above a given point.

**Air Quality Index (AQI)** A standardized indicator for reporting daily air quality to the general public. It is calculated for five major pollutants regulated by the Clean Air Act.

**Albedo** The reflectivity of a substance, usually expressed as a percentage of the incident radiation reflected.

**Aleutian Low** A large cell of low pressure centered over the Aleutian Islands of the North Pacific during the winter.

**Altimeter** An aneroid barometer calibrated to indicate altitude instead of pressure.

**Altitude (of the Sun)** The angle of the Sun above the horizon.

**Analog Method** A statistical approach to weather forecasting in which current conditions are matched with records of similar past weather events with the idea that the succession of events in the past will be paralleled by current conditions.

**Anemometer** An instrument used to determine wind speed.

**Aneroid Barometer** An instrument for measuring air pressure; it consists of evacuated metal chambers that are very sensitive to variations in air pressure.

**Annual Mean Temperature** An average of the 12 monthly means.

**Annual Temperature Range** The difference between the warmest and coldest monthly means.

**Anticyclone** An area of high atmospheric pressure characterized by diverging and rotating winds and subsiding air aloft.

**Anticyclonic Flow** Winds blow out and flow clockwise about an anticyclone (high) in the Northern Hemisphere, and they blow out and flow counterclockwise about an anticyclone in the Southern Hemisphere.

**Aphelion** The point in the orbit of a planet that is farthest from the Sun.

**Apparent Temperature** The air temperature perceived by a person.

**Arctic (A) Air Mass** A bitterly cold air mass that forms over the frozen Arctic Ocean.

**Arctic Sea Smoke** A dense and often extensive steam fog occurring over high-latitude ocean areas in winter.

**Arid** *See* Desert.

**Atmosphere** The gaseous portion of a planet, the planet's envelope of air; one of the traditional subdivisions of Earth's physical environment.

**Atmospheric Window** Terrestrial radiation between 8 and 11 micrometers in length to which the troposphere is transparent.

**Aurora** A bright and ever-changing display of light caused by solar radiation interacting with the upper atmosphere in the region of the poles. It is called *aurora borealis* in the Northern Hemisphere and *aurora australis* in the Southern Hemisphere.

**Automated Surface Observing System (ASOS)** A widely used, standardized set of automated weather instruments that provide routine surface observations.

**Autumnal Equinox** *See* Equinox.

**Azores High** The name given to a subtropical anticyclone when it is situated over the eastern part of the North Atlantic Ocean.

**Backdoor Cold Front** A cold front moving toward the west or southwest along the Atlantic Seaboard.

**Backing Wind Shift** A wind shift in a counterclockwise direction, such as a shift from east to north.

**Barograph** A recording barometer.

**Barometric Tendency** *See* Pressure Tendency.

**Beaufort Scale** A scale that can be used for estimating wind speed when an anemometer is not available.

**Bergeron Process** A theory that relates the formation of precipitation to supercooled clouds, freezing nuclei, and the different saturation levels of ice and liquid water.

**Bermuda High** The name given to the subtropical high in the North Atlantic during the summer, when it is centered near the island of Bermuda.

**Bimetal Strip** A thermometer consisting of two thin strips of metal welded together, which have widely different coefficients of thermal expansion. When temperature changes, the two metals expand or contract unequally and cause changes in the curvature of the element. Commonly used in thermographs.

**Biosphere** The totality of life-forms on Earth.

**Blizzard** A violent and extremely cold wind, laden with dry snow picked up from the ground.

**Bora** In the region of the eastern shore of the Adriatic Sea, a cold, dry northeasterly wind that blows down from the mountains.

**Buys Ballot's Law** A law which states that with your back to the wind in the Northern Hemisphere, low pressure will be to your left and high pressure to your right. The reverse is true in the Southern Hemisphere.

**Calorie** The amount of heat required to raise the temperature of 1 gram of water 1°C.

**Ceiling** The height ascribed to the lowest layer of clouds or obscuring phenomena when the sky is reported as broken, overcast, or obscured and the clouds are not classified "thin" or "partial." The ceiling is termed *unlimited* when the foregoing conditions are not present.

**Celsius Scale** A temperature scale (at one time called the centigrade scale) devised by Anders Celsius in 1742 and used where the metric system is in use. For water at sea level, 0° is designated the ice point and 100° the steam point.

**Chinook** The name applied to a foehn wind in the Rocky Mountains.

**Circle of Illumination** The line (great circle) separating daylight from darkness on Earth.

**Cirrus** One of three basic cloud forms; also one of the three high cloud types. They are thin, delicate ice-crystal clouds often appearing as veil-like patches or thin, wispy fibers.

**Climate** A description of aggregate weather conditions; the sum of all statistical weather information that helps describe a place or region.

**Climate Change** A study dealing with variations in climate on many different time scales from decades to millions of years, and the possible causes of such variations.

**Climate-Feedback Mechanisms** Several different possible outcomes that may result when one of the atmospheric system's elements is altered.

**Climate System** The exchanges of energy and moisture occurring among the atmosphere, hydrosphere, lithosphere, biosphere, and cryosphere.

**Cloud** A form of condensation best described as a dense concentration of suspended water droplets or tiny ice crystals.

**Cloud Condensation Nuclei** Microscopic particles that serve as surfaces on which water vapor condenses.

**Cloud Seeding** The introduction into clouds of particles (most commonly dry ice or silver iodide) for the purpose of altering the cloud's natural development.

**Clouds of Vertical Development** A cloud that has its base in the low height range and extends upward into the middle or high altitudes.

**Cold Front** The discontinuity at the forward edge of an advancing cold air mass that is displacing warmer air in its path.

**Cold-Type Occluded Front** A front that forms when the air behind the cold front is colder than the air underlying the warm front it is overtaking.

**Cold Wave** A rapid and marked fall in temperature. The National Weather Service applies this term to a fall in temperature in 24 hours equaling or exceeding a specified number of degrees and reaching a specified minimum temperature or lower. These specifications vary for different parts of the country and for different periods of the year.

**Collision–Coalescence Process** A theory of raindrop formation in warm clouds (above 0°C) in which large cloud droplets ("giants") collide and join together with smaller droplets to form a raindrop. Opposite electrical charges may bind the cloud droplets together.

**Condensation** The change of state from a gas to a liquid.

**Conditional Instability** The condition of moist air with an environmental lapse rate between the dry and wet adiabatic rates.

**Conduction** The transfer of heat through matter by molecular activity. Energy is transferred during collisions among molecules.

**Conservation of Angular Momentum, Law of** The product of the velocity of an object around a center of rotation (axis) and the distance of the object from the axis is constant.

**Constant-Pressure Surface** A surface along which the atmospheric pressure is everywhere equal at any given moment.

**Continental (c) Air Mass** An air mass that forms over land; it is normally relatively dry.

**Continental Climate** A climate lacking marine influence and characterized by more extreme temperatures than in marine climates; therefore, it has a relatively high annual temperature range for its latitude.

**Contrail** A cloudlike streamer frequently observed behind aircraft flying in clear, cold, and humid air and caused by the addition to the atmosphere of water vapor from engine exhaust gases.

**Controls of Temperature** Factors that cause variations in temperature from place to place, such as latitude and altitude.

**Convection** The transfer of heat by the movement of a mass or substance. It can take place only in fluids.

**Convection Cell** Circulation that results from the uneven heating of a fluid; the warmer parts of the fluid expand and rise because of their buoyancy, and the cooler parts sink.

**Convergence** The condition that exists when the wind distribution within a given region results in a net horizontal inflow of air into the area. Because convergence at lower levels is associated with an upward movement of air, areas of convergent winds are regions favorable to cloud formation and precipitation.

**Cooling Degree-Day** Each degree of temperature of the daily mean above 65°F. The amount of energy required to maintain a certain temperature in a building is proportional to the cooling degree-days total.

**Coriolis Effect** The deflective effect of Earth's rotation on all free-moving objects, including the atmosphere and oceans. Deflection is to the right in the Northern Hemisphere and to the left in the Southern Hemisphere.

**Corona** A bright, whitish disk centered on the Moon or Sun that results from diffraction when the objects are veiled by a thin cloud layer.

**Country Breeze** A circulation pattern characterized by a light wind blowing into a city from the surrounding countryside. It is best developed on clear and otherwise calm nights when the urban heat island is most pronounced.

**Cryosphere** Collective term for the ice and snow that exist on Earth. One of the *spheres* of the climate system.

**Cumulus** One of three basic cloud forms; also the name of one of the clouds of vertical development. Cumulus are billowy, individual cloud masses that often have flat bases.

**Cumulus Stage** The initial stage in thunderstorm development, in which the growing cumulonimbus is dominated by strong updrafts.

**Cup Anemometer** *See* Anemometer.

**Cyclogenesis** The process that creates or develops a new cyclone; also the process that produces an intensification of a preexisting cyclone.

**Cyclone** An area of low atmospheric pressure characterized by rotating and converging winds and ascending air.

**Cyclonic Flow** Winds blowing in and counterclockwise about a cyclone (low) in the Northern Hemisphere and in and clockwise about a cyclone in the Southern Hemisphere.

**Daily Mean Temperature** The mean temperature for a day that is determined by averaging the hourly readings or, more commonly, by averaging the maximum and minimum temperatures for a day.

**Daily Temperature Range** The difference between the maximum and minimum temperatures for a day.

**Dart Leader** *See* Leader.

**Deposition** The process whereby water vapor changes directly to ice, without going through the liquid state.

**Desert** One of the two types of dry climate—the driest of the dry climates.

**Dew** A form of condensation consisting of small water drops on grass or other objects near the ground that forms when the surface temperature drops below the dew point. Usually associated with radiation cooling on clear, calm nights.

**Dew Point** The temperature to which air has to be cooled in order to reach saturation.

**Diffused Light** Solar energy is scattered and reflected in the atmosphere and reaches Earth's surface in the form of diffuse blue light from the sky.

**Discontinuity** A zone characterized by a comparatively rapid transition of meteorological elements.

**Dispersion** The separation of colors by refraction.

**Dissipating Stage** The final stage of a thunderstorm that is dominated by downdrafts and entrainment leading to the evaporation of the cloud structure.

**Diurnal** Daily, especially pertaining to actions that are completed within 24 hours and that recur every 24 hours.

**Divergence** The condition that exists when the distribution of winds within a given area results in a net horizontal outflow of air from the region. In divergence at lower levels, the resulting deficit is compensated for by a downward movement of air from aloft; hence, areas of divergent winds are unfavorable to cloud formation and precipitation.

**Doldrums** The equatorial belt of calms or light variable winds lying between the two trade wind belts.

**Doppler Radar** A type of radar that has the capacity of detecting motion directly.

**Drizzle** Precipitation from stratus clouds consisting of tiny droplets.

**Dry Adiabatic Rate** The rate of adiabatic cooling or warming in unsaturated air. The rate of temperature change is 1°C per 100 meters.

**Dry Climate** A climate in which yearly precipitation is not as great as the potential loss of water by evaporation.

**Dryline** A narrow zone in the atmosphere along which there is an abrupt change in moisture as when dry continental tropical air converges with humid maritime tropical air. The denser cT air acts to lift the less dense mT air producing clouds and storms.

**Dry-Summer Subtropical Climate** A climate located on the west sides of continents between latitudes 30° and 45°. It is the only humid climate with a strong winter precipitation maximum.

**Dynamic Seeding** A type of cloud seeding that uses massive seeding, a process resulting in an increase of the release of latent heat and causing the cloud to grow larger.

**Easterly Wave** A large migratory wavelike disturbance in the trade winds that sometimes triggers the formation of a hurricane.

**Eccentricity** The variation of an ellipse from a circle.

**Electromagnetic Radiation** *See* Radiation.

**Elements (Atmospheric)** Quantities or properties of the atmosphere that are measured regularly and that are used to express the nature of weather and climate.

**El Niño** The name given to the periodic warming of the ocean that occurs in the central and eastern Pacific. A major El Niño episode can cause extreme weather in many parts of the world. The opposite of *La Niña*.

**Entrainment** The infiltration of surrounding air into a vertically moving air column. For example, the influx of cool, dry air into the downdraft of a cumulonimbus cloud; a process that acts to intensify the downdraft.

**Environmental Lapse Rate** The rate of temperature decrease with height in the troposphere.

**Equatorial Low** A quasi-continuous belt of low pressure lying near the equator and between the subtropical highs.

**Equinox** The point in time when the vertical rays of the Sun are striking the equator. In the Northern Hemisphere, March 20 or 21 is the *vernal*, or *spring, equinox* and September 22 or 23 is the *autumnal equinox*. Lengths of daylight and darkness are equal at all latitudes at equinox.

**Evaporation** The process by which a liquid is transformed into gas.

**Eye** A roughly circular area of relatively light winds and fair weather at the center of a hurricane.

**Eye Wall** The doughnut-shaped area of intensive cumulonimbus development and very strong winds that surrounds the eye of a hurricane.

**Fahrenheit Scale** A temperature scale devised by Gabriel Daniel Fahrenheit in 1714 and used in the English system. For water at sea level, 32° is designated the ice point and 212° the steam point.

**Fall Wind** *See* Katabatic Wind.

**Ferrel Cell** The middle cell in the three-cell global circulation model; named for William Ferrel.

**Fixed Points** Reference points, such as the steam point and the ice point, used in the construction of temperature scales.

**Flash** The total discharge of lightning, which is usually perceived as a single flash of light but which actually consists of several flashes. *See also* Stroke.

**Foehn** A warm, dry wind on the lee side of a mountain range that owes its relatively high temperature largely to adiabatic heating during descent down mountain slopes.

**Fog** A cloud with its base at or very near Earth's surface.

**Forecasting Skill** *See* Skill.

**Freezing** The change of state from liquid to solid.

**Freezing Nuclei** Solid particles that have a crystal form resembling that of ice; they serve as cores for the formation of ice crystals.

**Freezing Rain** *See* Glaze.

**Front** A boundary (discontinuity) separating air masses of different densities, one warmer and often higher in moisture content than the other.

**Frontal Fog** Fog formed when rain evaporates as it falls through a layer of cool air.

**Frontal Wedging** The lifting of air resulting when cool air acts as a barrier over which warmer, lighter air will rise.

**Frontogenesis** The beginning or creation of a front.

**Frontolysis** The destruction and dying of a front.

**Frost** Ice crystals that occur when the temperature falls to 0°C or below. *See also* White Frost.

**Fujita Intensity Scale (F-scale)** A scale developed by T. Theodore Fujita for classifying the severity of a tornado, based on the correlation of wind speed with the degree of destruction.

**Geosphere** The solid Earth, the largest of Earth's four major spheres.

**Geostationary Satellite** A satellite that remains over a fixed point because its rate of travel corresponds to Earth's rate of rotation. Because the satellite must orbit at distances of about 35,000 kilometers, images from this type of satellite are not as detailed as those from polar satellites.

**Geostrophic Wind** A wind, usually above a height of 600 meters, that blows parallel to the isobars.

**Glaze** A coating of ice on objects formed when supercooled rain freezes on contact. A storm that produces glaze is termed an "icing storm."

**Global Circulation** The general circulation of the atmosphere; the average flow of air over the entire globe.

**Glory** A series of rings of colored light, most commonly appearing around the shadow of an airplane that is projected on clouds below.

**Gradient Wind** The curved airflow pattern around a pressure center resulting from a balance among pressure–gradient force, Coriolis force, and centrifugal force.

**Greenhouse Effect** The transmission of shortwave solar radiation by the atmosphere coupled with the selective absorption of longer-wavelength terrestrial radiation, especially by water vapor and carbon dioxide, resulting in warming of the atmosphere.

**Growing Degree-Days** A practical application of temperature data for determining the approximate date when crops will be ready for harvest.

**Gust Front** The boundary separating the cold downdraft from a thunderstorm and the relatively warm, moist surface air. Lifting along this boundary may initiate the development of thunderstorms.

**Hadley Cell** The thermally driven circulation system of equatorial and tropical latitudes consisting of two convection cells, one in each hemisphere. The existence of this circulation system was first proposed by George Hadley in 1735 as an explanation for the trade winds.

**Hail** Precipitation in the form of hard, round pellets or irregular lumps of ice that may have concentric shells formed by the successive freezing of layers of water.

**Halo** A narrow whitish ring of large diameter centered around the Sun. The commonly observed *22° halo* subtends an angle of 22° from the observer.

**Heat** The kinetic energy of random molecular motion.

**Heat Budget** The balance of incoming and outgoing radiation.

**Heating Degree-Day** Each degree of temperature of the daily mean below 65°F is counted as one heating degree-day. The amount of heat required to maintain a certain temperature in a building is proportional to the heating degree-days total.

**Heterosphere** A zone of the atmosphere beyond about 80 kilometers, where the gases are arranged into four roughly spherical shells, each with a distinctive composition.

**High Cloud** A cloud that normally has its base above 6000 meters; the base may be lower in winter and at high-latitude locations.

**Highland Climate** Complex pattern of climate conditions associated with mountains. Highland climates are characterized by large differences that occur over short distances.

**Homosphere** A zone of atmosphere extending from Earth's surface to about 80 kilometers that is uniform in terms of the proportions of its component gases.

**Horse Latitudes** A belt of calms or light variable winds and subsiding air located near the center of the subtropical high.

**Humid Continental Climate** A relatively severe climate characteristic of broad continents in the middle latitudes between approximately 40° and 50° north latitude. This climate is not found in the Southern Hemisphere, where the middle latitudes are dominated by the oceans.

**Humidity** A general term referring to water vapor in the air.

**Humid Subtropical Climate** A climate generally located on the eastern side of a continent and characterized by hot, sultry summers and cool winters.

**Hurricane** A tropical cyclonic storm having minimum winds of 119 kilometers per hour; also known as *typhoon* (western Pacific) and *cyclone* (Indian Ocean).

**Hurricane Warning** A warning issued when sustained winds of 119 kilometers per hour or higher are expected within a specified coastal area in 24 hours or less.

**Hurricane Watch** An announcement aimed at specific coastal areas that a hurricane poses a possible threat, generally within 36 hours.

**Hydrologic Cycle** The continuous movement of water from the oceans to the atmosphere (by evaporation), from the atmosphere to the land (by condensation and precipitation), and from the land back to the sea (via stream flow).

**Hydrophobic Nuclei** Particles that are not efficient condensation nuclei. Small droplets will form on them whenever the relative humidity reaches 100 percent.

**Hydrosphere** The water portion of our planet; one of the traditional subdivisions of Earth's physical environment.

**Hydrostatic Equilibrium** The balance maintained between the force of gravity and the vertical pressure gradient that does not allow air to escape to space.

**Hygrometer** An instrument designed to measure relative humidity.

**Hygroscopic Nuclei** Condensation nuclei having a high affinity for water, such as salt particles.

**Hypothesis** A tentative explanation that is tested to determine whether it is valid.

**Ice Cap Climate** A climate that has no monthly means above freezing and supports no vegetative cover except in a few scattered high mountain areas. This climate, with its perpetual ice and snow, is confined largely to the ice sheets of Greenland and Antarctica.

**Icelandic Low** A large cell of low pressure centered over Iceland and southern Greenland in the North Atlantic during the winter.

**Ice Point** The temperature at which ice melts.

**Ideal Gas Law** The pressure exerted by a gas is proportional to its density and absolute temperature.

**Inclination of the Axis** The tilt of Earth's axis from the perpendicular to the plane of Earth's orbit (plane of the ecliptic). Currently, the inclination is about 23 1/2° away from the perpendicular.

**Inferior Mirage** A mirage in which the image appears below the true location of the object.

**Infrared Radiation** Radiation with a wavelength from 0.7 to 200 micrometers.

**Interface** A common boundary where different parts of a system interact.

**Interference** A phenomenon that occurs when light rays of different frequencies (i.e., colors) meet. Such interference results in the cancellation or subtraction of some frequencies, which is responsible for the colors associated with coronas.

**Internal Reflection** A reflection that occurs when light that is traveling through a transparent material, such as water, reaches the opposite surface and is reflected back into the material. This is an important factor in the formation of optical phenomena such as rainbows.

**Intertropical Convergence Zone (ITCZ)** The zone of general convergence between the Northern and Southern Hemisphere trade winds.

**Ionosphere** A complex atmospheric zone of ionized gases extending between 80 and 400 kilometers, thus coinciding with the lower thermosphere and heterosphere.

**Isobar** A line drawn on a map connecting points of equal barometric pressure, usually corrected to sea level.

**Isohyet** A line connecting places having equal rainfall.

**Isotach** A line connecting places having equal wind speed.

**Isotherm** A line connecting points of equal air temperature.

**ITCZ** *See* Intertropical Convergence Zone.

**Jet Streak** Areas of higher-velocity winds found within the jet stream.

**Jet Stream** Swift geostrophic airstreams in the upper troposphere that meander in relatively narrow belts.

**Jungle** An almost impenetrable growth of tangled vines, shrubs, and short trees characterizing areas where the tropical rain forest has been cleared.

**Katabatic Wind** The flow of cold, dense air downslope under the influence of gravity; the direction of this wind's flow is controlled largely by topography. Also called *fall winds*.

**Kelvin Scale** A temperature scale (also called the *absolute scale*) used primarily for scientific purposes and having intervals equivalent to those on the Celsius scale but beginning at absolute zero.

**Köppen Classification** Devised by Wladimir Köppen, a system for classifying climates that is based on mean monthly and annual values of temperature and precipitation.

**Lake-Effect Snow** Snow showers associated with a *cP* air mass to which moisture and heat are added from below as it traverses a large and relatively warm lake (such as one of the Great Lakes), rendering the air mass humid and unstable.

**Land Breeze** A local wind blowing from the land toward the sea during the night in coastal areas.

**La Niña** An episode of strong trade winds and unusually low sea-surface temperatures in the central and eastern Pacific. The opposite of *El Niño.*

**Lapse Rate** *See* Environmental Lapse Rate; Normal Lapse Rate.

**Latent Heat** The energy absorbed or released during a change of state.

**Latent Heat of Condensation** The energy released when water vapor changes to the liquid state. The amount of energy released is equivalent to the amount absorbed during evaporation.

**Latent Heat of Vaporization** The energy absorbed by water molecules during evaporation. It varies from about 600 calories per gram for water at 0°C to 540 calories per gram at 100°C.

**Leader** The conductive path of ionized air that forms near a cloud base prior to a lightning stroke. The initial conductive path is referred to as a *step leader* because it extends itself earthward in short, nearly invisible bursts. A *dart leader,* which is continuous and less branched than a step leader, precedes each subsequent stroke along the same path.

**Lifting Condensation Level** The height at which rising air that is cooling at the dry adiabatic rate becomes saturated and condensation begins.

**Lightning** A sudden flash of light generated by the flow of electrons between oppositely charged parts of a cumulonimbus cloud or between the cloud and the ground.

**Liquid-in-Glass Thermometer** A device for measuring temperature that consists of a tube with a liquid-filled bulb at one end. The expansion or contraction of the fluid indicates temperature.

**Long-Range Forecasting** Estimating rainfall and temperatures for a period beyond 3 to 5 days, usually for 30-day periods. Such forecasts are not as detailed or reliable as those for shorter periods.

**Longwave Radiation** A reference to radiation emitted by Earth. Wavelengths are roughly 20 times longer than those emitted by the Sun.

**Looming** A mirage that allows objects that are below the horizon to be seen.

**Low Cloud** A cloud that forms below a height of about 2000 meters.

**Macroscale Winds** Phenomena such as cyclones and anticyclones that persist for days or weeks and have a horizontal dimension of hundreds to several thousands of kilometers; also, features of the atmospheric circulation that persist for weeks or months and have horizontal dimensions of up to 10,000 kilometers.

**Marine Climate** A climate dominated by the ocean; because of the moderating effect of water, sites having this climate are considered relatively mild.

**Marine West Coast Climate** A climate found on windward coasts from latitudes 40° to 65° and dominated by maritime air masses. In this climate, winters are mild and summers are cool.

**Maritime (m) Air Mass** An air mass that originates over the ocean. These air masses are relatively humid.

**Mature Stage** The second of the three stages of a thunderstorm. This stage is characterized by violent weather as downdrafts exist side-by-side with updrafts.

**Maximum Thermometer** A thermometer that measures the maximum temperature for a given period in time, usually 24 hours. A constriction in the base of the glass tube allows mercury to rise but prevents it from returning to the bulb until the thermometer is shaken or whirled.

**Mediterranean Climate** A common name applied to the dry-summer subtropical climate.

**Melting** The change of state from solid to liquid.

**Mercury Barometer** A mercury-filled glass tube in which the height of the column of mercury is a measure of air pressure.

**Mesocyclone** A vertical cylinder of cyclonically rotating air (3 to 10 kilometers in diameter) that develops in the updraft of a severe thunderstorm and that often precedes the development of damaging hail or tornadoes.

**Mesopause** The boundary between the mesosphere and the thermosphere.

**Mesoscale Convective Complex (MCC)** A slow-moving roughly circular cluster of interacting thunderstorm cells covering an area of thousands of square kilometers that may persist for 12 hours or more.

**Mesoscale Winds** Small convective cells that exist for minutes or hours, such as thunderstorms, tornadoes, and land and sea breezes. Typical horizontal dimensions range from 1 to 100 kilometers.

**Meteorology** The scientific study of the atmosphere and atmospheric phenomena; the study of weather and climate.

**Microscale Winds** Phenomena such as turbulence, with life spans of less than a few minutes that affect small areas and are strongly influenced by local conditions of temperature and terrain.

**Middle Cloud** A cloud occupying the height range from 2000 to 6000 meters.

**Midlatitude Cyclone** A large low-pressure center with diameter often exceeding 1000 kilometers that moves from west to east and may last from a few days to more than a week and usually has a cold front and a warm front extending from the central area of low pressure.

**Midlatitude Jet Stream** A jet stream that migrates between latitudes 30° and 70°.

**Millibar** The standard unit of pressure measurement used by the National Weather Service. One millibar (mb) equals 100 newtons per square meter.

**Minimum Thermometer** A thermometer that measures the minimum temperature for a given period of time, usually 24 hours. By checking the small dumbbell-shaped index, the minimum temperature can be read.

**Mirage** An optical effect of the atmosphere caused by refraction in which the image of an object appears displaced from its true position.

**Mistral** A cold northwest wind that blows into the western Mediterranean basin from higher elevations to the north.

**Mixing Depth** The height to which convectional movements extend above Earth's surface. The greater the mixing depths, the better the air quality.

**Mixing Ratio** The mass of water vapor in a unit mass of dry air; commonly expressed as grams of water vapor per kilogram of dry air.

**Monsoon** The seasonal reversal of wind direction associated with large continents, especially Asia. In winter, the wind blows from land to sea; in summer, it blows from sea to land.

**Monthly Mean Temperature** The mean temperature for a month that is calculated by averaging the daily means.

**Mountain Breeze** The nightly downslope winds commonly encountered in mountain valleys.

**Multiple-Vortex Tornado** A tornado that contains several smaller intense whirls called suction vortices that orbit the center of the larger tornado circulation.

**National Weather Service (NWS)** The federal agency responsible for gathering and disseminating weather-related information.

**Negative-Feedback Mechanism** As used in climatic change, any effect that is opposite of the initial change and tends to offset it.

**Newton** A unit of force used in physics. One newton is the force necessary to accelerate 1 kilogram of mass 1 meter per second squared.

**Nor'easter** The term used to describe the weather associated with an incursion of *mP* air into the Northeast from the North Atlantic; strong northeast winds, freezing or near-freezing temperatures, and the possibility of precipitation make this an unwelcome weather event.

**Normal Lapse Rate** The average drop in temperature with increasing height in the troposphere; about 6.5°C per kilometer.

**Nowcasting** Short-term weather forecasting techniques that are generally applied to predicting severe weather.

**Numerical Weather Prediction (NWP)** Forecasting the behavior of atmospheric disturbances based upon the solution of the governing fundamental equations of hydrodynamics, subject to the observed initial conditions. Because of the vast number of calculations involved, high-speed computers are always used for NWP.

**Obliquity** The angle between the planes of Earth's equator and orbit.

**Occluded Front** A front formed when a cold front overtakes a warm front.

**Occlusion** The overtaking of one front by another.

**Ocean Current** The mass movement of ocean water that is either wind driven or initiated by temperature and salinity conditions that alter the density of seawater.

**Orographic Lifting** The process in which mountains or highlands act as barriers to the flow of air and force the air to ascend. The air cools adiabatically, and clouds and precipitation may result.

**Outgassing** The release of gases dissolved in molten rock.

**Overrunning** Warm air gliding up a retreating cold air mass.

**Oxygen-Isotope Analysis** A method of deciphering past temperatures based on the precise measurement of the ratio between two isotopes of oxygen, $^{16}O$ and $^{18}O$. Analysis is commonly made of sea-floor sediments and cores from ice sheets.

**Ozone** A molecule of oxygen containing three oxygen atoms.

**Paleoclimatology** The study of ancient climates; the study of climate and climate change prior to the period of instrumental records using proxy data.

**Paleosol** Old, buried soil that may furnish some evidence of the nature of past climates because climate is the most important factor in soil formation.

**Parcel** An imaginary volume of air enclosed in a thin elastic cover. Typically it is considered to be a few hundred cubic meters in volume and is assumed to act independently of the surrounding air.

**Perihelion** The point in the orbit of a planet closest to the Sun.

**Permafrost** The permanent freezing of the subsoil in tundra regions.

**Persistence Forecast** A forecast that assumes that the weather occurring upstream will persist and move on and will affect the areas in its path in much the same way. Persistence forecasts do not account for changes that might occur in the weather system.

**pH Scale** A 0-to-14 scale that is used for expressing the exact degree of acidity or alkalinity of a solution. A pH of 7 signifies a neutral solution. Values below 7 signify an acid solution, and values above 7 signify an alkaline solution.

**Photochemical Reaction** A chemical reaction in the atmosphere triggered by sunlight, often yielding a secondary pollutant.

**Photosynthesis** The production of sugars and starches by plants using air, water, sunlight, and chlorophyll. In the process, atmospheric carbon dioxide is changed to organic matter and oxygen is released.

**Plane of the Ecliptic** The plane of Earth's orbit around the Sun.

**Plate Tectonics Theory** A theory which states that the outer portion of Earth is made up of several individual pieces, called *plates,* which move in relation to one another upon a partially molten zone below. As plates move, so do continents, which explains some climatic changes in the geologic past.

**Polar (P) Air Mass** A cold air mass that forms in a high-latitude source region.

**Polar Climate** A climate in which the mean temperature of the warmest month is below 10°C; a climate that is too cold to support the growth of trees.

**Polar Easterlies** In the global pattern of prevailing winds, winds that blow from the polar high toward the subpolar low. These winds, however, should not be thought of as persistent winds, such as the trade winds.

**Polar Front** The stormy frontal zone separating air masses of polar origin from air masses of tropical origin.

**Polar Front Theory** A theory developed by J. Bjerknes and other Scandinavian meteorologists in which the polar front, separating polar and tropical air masses, gives rise to cyclonic disturbances that intensify and move along the front and pass through a succession of stages.

**Polar High** Anticyclones that are assumed to occupy the inner polar regions and are believed to be thermally induced, at least in part.

**Polar-Orbiting Satellite** Satellites that orbit the poles at rather low altitudes of a few hundred kilometers and require only 100 minutes per orbit.

**Positive-Feedback Mechanism** As used in climatic change, any effect that acts to reinforce an initial change.

**Potential Energy** Energy that exists by virtue of a body's position with respect to gravity.

**Precession** The slow migration of Earth's axis that traces a cone over a period of 26,000 years.

**Precipitation Fog** *See* Frontal Fog.

**Pressure Gradient** The amount of pressure change occurring over a given distance.

**Pressure Tendency** The nature of the change in atmospheric pressure over the past several hours. It can be a useful aid in short-range weather prediction.

**Prevailing Westerlies** The dominant west-to-east motion of the atmosphere that characterizes the regions on the poleward side of the subtropical highs.

**Prevailing Wind** A wind that consistently blows from one direction more than from any other.

**Primary Pollutant** A pollutant emitted directly from an identifiable source.

**Prognostic Chart** A computer-generated forecast showing the expected pressure pattern at a specified future time. Anticipated positions of fronts are also included. They usually represent the graphical output associated with a numerical weather prediction model.

**Proxy Data** Data gathered from natural recorders of climate variability, such as tree rings, ice cores, and ocean-floor sediments.

**Psychrometer** A device consisting of two thermometers (wet bulb and dry bulb) that is rapidly whirled and, with the use of tables, yields the relative humidity and dew point.

**Radiation** The wavelike energy emitted by any substance that possesses heat. This energy travels through space at 300,000 kilometers per second (the speed of light).

**Radiation Fog** Fog resulting from radiation cooling of the ground and adjacent air; primarily a nighttime and early morning phenomenon.

**Radiosonde** A lightweight package of weather instruments fitted with a radio transmitter and carried aloft by a balloon.

**Rainbow** A luminous arc formed by the refraction and reflection of light in drops of water.

**Rain Shadow Desert** A dry area on the lee side of a mountain range.

**Rawinsonde** A radiosonde that is tracked by radio-location devices in order to obtain data on upper-air winds.

**Reflection, Law of** A law that states that the angle of incidence (incoming ray) is equal to the angle of reflection (outgoing ray).

**Refraction** The bending of light as it passes obliquely from one transparent medium to another.

**Relative Humidity** The ratio of the air's water-vapor content to its water-vapor capacity.

**Return Stroke** The electric discharge resulting from the downward (earthward) movement of electrons from successively higher levels along the conductive path of lightning.

**Revolution** The motion of one body about another, as Earth about the Sun.

**Ridge** An elongate region of high atmospheric pressure.

**Rime** A delicate accumulation of ice crystals formed when supercooled fog or cloud droplets freeze on contact with objects.

**Rossby Waves** Upper-air waves in the middle and upper troposphere of the middle latitudes with wavelengths of from 4000 to 6000 kilometers; named for C. G. Rossby, the meteorologist who developed the equations for parameters governing the waves.

**Rotation** The spinning of a body, such as Earth, about its axis.

**Saffir–Simpson Scale** A scale, from 1 to 5, used to rank the relative intensities of hurricanes.

**Santa Ana** The local name given a foehn wind in southern California.

**Saturation** The maximum possible quantity of water vapor that the air can hold at any given temperature and pressure.

**Saturation Vapor Pressure** The vapor pressure, at a given temperature, wherein the water vapor is in equilibrium with a surface of pure water or ice.

**Savanna** A tropical grassland, usually with scattered trees and shrubs.

**Sea Breeze** A local wind that blows from the sea toward the land during the afternoon in coastal areas.

**Secondary Pollutant** A pollutant that is produced in the atmosphere by chemical reactions occurring among primary pollutants.

**Semiarid** *See* Steppe.

**Sensible Heat** The heat we can feel and measure with a thermometer.

**Severe Thunderstorm** A thunderstorm that produces frequent lightning, locally damaging wind, or hail that is 2 centimeters or more in diameter. In the middle latitudes, most thunderstorms form along or ahead of cold fronts.

**Shortwave Radiation** Radiation emitted by the Sun.

**Siberian High** The high-pressure center that forms over the Asian interior in January and produces the dry winter monsoon for much of the continent.

**Skill** An index of the degree of accuracy of a set of forecasts as compared to forecasts based on some standard, such as chance or climatic data.

**Sleet** Frozen or semifrozen rain formed when raindrops pass through a subfreezing layer of air.

**Smog** A word currently used as a synonym for general air pollution. It was originally created by combining the words "smoke" and "fog."

**Snow** Precipitation in the form of white or translucent ice crystals, chiefly in complex branched hexagonal form and often clustered into snowflakes.

**Solstice** The point in time when the vertical rays of the Sun are striking either the Tropic of Cancer (summer solstice in the Northern Hemisphere) or the Tropic of Capricorn (winter solstice in the Northern Hemisphere). Solstice represents the longest or shortest day (length of daylight) of the year.

**Source Region** The area where an air mass acquires its characteristic properties of temperature and moisture.

**Southern Oscillation** The seesaw pattern of atmospheric pressure change that occurs between the eastern and western Pacific. The interaction of this effect and that of El Niño can cause extreme weather events in many parts of the world.

**Specific Heat** The amount of heat needed to raise 1 gram of a substance 1°C at sea-level atmospheric pressure.

**Specific Humidity** The mass of water vapor per unit mass of air, including the water vapor (usually expressed as grams of water vapor per kilogram of air).

**Squall Line** Any nonfrontal line or narrow band of active thunderstorms.

**Stable Air** Air that resists vertical displacement. If it is lifted, adiabatic cooling will cause its temperature to be lower than the surrounding environment; if it is allowed, it will sink to its original position.

**Standard Rain Gauge** A gauge that has a diameter of about 20 centimeters and funnels rain into a cylinder that magnifies precipitation amounts by a factor of 10, allowing for accurate measurement of small amounts.

**Static Seeding** The most commonly used technique of cloud seeding, based on the assumption that cumulus clouds are deficient in freezing nuclei and that the addition of nuclei will spur additional precipitation formation.

**Stationary Front** A situation in which the surface position of a front does not move; the flow on either side of such a boundary is nearly parallel to the position of the front.

**Statistical Methods (Forecasting)** Methods in which tables or graphs are prepared from a long series of observations to show the probability of certain weather events under certain conditions of pressure, temperature, or wind direction.

**Steam Fog** Fog that has the appearance of steam and that is produced by evaporation from a warm water surface into the cool air above.

**Steam Point** The temperature at which water boils.

**Step Leader** *See* Leader.

**Steppe** One of the two types of dry climate; a marginal and more humid variant of the desert that separates it from bordering humid climates. Steppe also refers to the short-grass vegetation associated with this semiarid climate.

**Storm Surge** The abnormal rise of the sea along a shore as a result of strong winds.

**Stratopause** The boundary between the stratosphere and the mesosphere.

**Stratosphere** The zone of the atmosphere above the troposphere characterized at first by isothermal conditions and then a gradual temperature increase. Earth's ozone is concentrated here.

**Stratus** One of three basic cloud forms; also the name of a type of low cloud. Stratus clouds are sheets or layers that cover much or all of the sky.

**Stroke** One of the individual components that make up a flash of lightning. There are usually three to four strokes per flash, roughly 50 milliseconds apart.

**Subarctic Climate** A climate found north of the humid continental climate and south of the polar climate that is characterized by bitterly cold winters and short cool summers. Places within this climatic realm experience the highest annual temperature ranges on Earth.

**Sublimation** The process whereby a solid changes directly to a gas, without going through the liquid state.

**Subpolar Low** Low pressure located at about the latitudes of the Arctic and Antarctic Circles. In the Northern Hemisphere, the low takes the form of individual oceanic cells; in the Southern Hemisphere, there is a deep and continuous trough of low pressure.

**Subsidence** An extensive sinking motion of air, most frequently occurring in anticyclones. The subsiding air is warmed by compression and becomes more stable.

**Subtropical High** Several semipermanent anticyclonic centers characterized by subsidence and divergence located roughly between latitudes 25° and 35°.

**Summer Solstice** *See* Solstice.

**Sun Dogs** Two bright spots of light, sometimes called "mock suns," that sit at a distance of 22° on either side of the Sun.

**Sun Pillar** Shafts of light caused by reflection from ice crystals that extend upward or, less commonly, downward from the Sun when the Sun is near the horizon.

**Sunspot** A dark area on the Sun associated with powerful magnetic storms that extend from the Sun's surface deep into the interior.

**Supercell** A type of thunderstorm that consists of a single, persistent, and very powerful cell (updraft and downdraft) and that often produces severe weather, including hail and tornadoes.

**Supercooled** The condition of water droplets that remain in the liquid state at temperatures well below 0°C.

**Superior Mirage** A mirage in which the image appears above the true position of the object.

**Synoptic Weather Forecasting** A system of forecasting based on careful studies of synoptic weather charts over a period of years; from such studies a set of empirical rules is established to aid the forecaster in estimating the rate and direction of weather-system movements.

**Synoptic Weather Map** A weather map describing the state of the atmosphere over a large area at a given moment.

**Taiga** The northern coniferous forest; also a name applied to the subarctic climate.

**Temperature** A measure of the degree of hotness or coldness of a substance.

**Temperature Gradient** The amount of temperature change per unit of distance.

**Temperature Inversion** A layer in the atmosphere of limited depth where the temperature increases rather than decreases with height.

**Theory** A well-tested and widely accepted view that explains certain observable facts.

**Thermal** An example of convection that involves the upward movements of warm, less dense air. In this manner, heat is transported to greater heights.

**Thermal Low** An area of low atmospheric pressure created by abnormal surface heating.

**Thermistor** An electric thermometer consisting of a conductor whose resistance to the flow of current is temperature dependent; commonly used in radiosondes.

**Thermocouple** An electric thermometer that operates on the principle that differences in temperature between the junction of two unlike metal wires in a circuit will induce a current to flow.

**Thermograph** An instrument that continuously records temperature.

**Thermometer** An instrument for measuring temperature; in meteorology, a thermometer is generally used to measure the temperature of the air.

**Thermosphere** The zone of the atmosphere beyond the mesosphere in which there is a rapid rise in temperature with height.

**Thunder** The sound emitted by rapidly expanding gases along the channel of a lightning discharge.

**Thunderstorm** A storm produced by a cumulonimbus cloud and always accompanied by lightning and thunder. It is of relatively short duration and usually accompanied by strong wind gusts, heavy rain, and sometimes hail.

**Tipping-Bucket Gauge** A recording rain gauge consisting of two compartments ("buckets"), each capable of holding 0.025 centimeter of water. When one compartment fills, it tips, and the other compartment takes its place.

**Tornado** A violently rotating column of air attended by a funnel-shaped or tubular cloud extending downward from a cumulonimbus cloud.

**Tornado Warning** A warning issued when a tornado has actually been sighted in an area or is indicated by radar.

**Tornado Watch** A forecast issued for areas of about 65,000 square kilometers, indicating that conditions are such that tornadoes may develop; a tornado watch is intended to alert people to the possibility of tornadoes.

**Towering** A mirage in which the size of an object is magnified.

**Trace of Precipitation** An amount of precipitation less than 0.025 centimeter.

**Trade Winds** Two belts of winds that blow almost constantly from easterly directions and are located on the equatorward sides of the subtropical highs.

**Transpiration** The release of water vapor to the atmosphere by plants.

**Trend Forecast** A short-range forecasting technique which assumes that the weather occurring upstream will persist and move on to affect the area in its path.

**Tropic of Cancer** The parallel of latitude, 23 1/2° north latitude, marking the northern limit of the Sun's vertical rays.

**Tropic of Capricorn** The parallel of latitude, 23 1/2° south latitude, marking the southern limit of the Sun's vertical rays.

**Tropical (T) Air Mass** A warm-to-hot air mass that forms in the subtropics.

**Tropical Depression** By international agreement, a tropical cyclone with maximum winds that do not exceed 61 kilometers per hour.

**Tropical Disturbance** A term used by the National Weather Service for a cyclonic wind system in the tropics that is in its formative stages.

**Tropical Rain Forest** A luxuriant broadleaf evergreen forest; also the name given the climate associated with this vegetation.

**Tropical Storm** By international agreement, a tropical cyclone with maximum winds between 61 and 115 kilometers per hour.

**Tropical Wet and Dry** A climate that is transitional between the wet tropics and the subtropical steppes.

**Tropopause** The boundary between the troposphere and the stratosphere.

**Troposphere** The lowermost layer of the atmosphere, marked by considerable turbulence and, in general, a decrease in temperature with increasing height.

**Trough** An elongate region of low atmospheric pressure.

**Tundra Climate** A climate found almost exclusively in the Northern Hemisphere and at high altitudes in many mountainous regions. A treeless climatic realm of sedges, grasses, mosses, and lichens dominated by a long, bitterly cold winter.

**Ultraviolet Radiation** Radiation with a wavelength from 0.2 to 0.4 micrometer.

**Unstable Air** Air that does not resist vertical displacement. If it is lifted, its temperature will not cool as rapidly as the surrounding environment, and so it will continue to rise on its own.

**Upslope Fog** Fog created when air moves up a slope and cools adiabatically.

**Upwelling** The process by which deep, cold, nutrient-rich water is brought to the surface, usually by coastal currents that move water away from the coast.

**Urban Heat Island** A city area where temperatures are generally higher than in surrounding rural areas.

**U.S. Standard Atmosphere** The idealized vertical distribution of atmospheric pressure, temperature, and density that represents average conditions in the atmosphere.

**Valley Breeze** The daily upslope winds commonly encountered in a mountain valley.

**Vapor Pressure** The part of the total atmospheric pressure attributable to its water-vapor content.

**Veering Wind Shift** A wind shift in a clockwise direction, such as a shift from east to south.

**Vernal Equinox** *See* Equinox.

**Virga** Wisps or streaks of water or ice particles that fall out of a cloud and evaporate before reaching Earth's surface.

**Visibility** The greatest distance at which prominent objects can be seen and identified by unaided, normal eyes.

**Visible Light** Radiation with a wavelength from 0.4 to 0.7 micrometer.

**Warm Front** The discontinuity at the forward edge of an advancing warm air mass that displaces cooler air in its path.

**Warm-Type Occluded Front** A front that forms when the air behind the cold front is warmer than the air underlying the warm front it is overtaking.

**Water Hemisphere** A term used to refer to the Southern Hemisphere, where the oceans cover 81 percent of the surface (compared to 61 percent in the Northern Hemisphere).

**Wavelength** The horizontal distance separating successive crests or troughs.

**Weather** The state of the atmosphere at any given time.

**Weather Analysis** The stage prior to developing a weather forecast. This stage involves collecting, compiling, and transmitting observational data.

**Weather Forecasting** Predicting the future state of the atmosphere.

**Weather Modification** Deliberate human intervention to influence and improve atmospheric processes.

**Weighting Gauge** A recording precipitation gauge consisting of a cylinder that rests on a spring balance.

**Westerlies** *See* Prevailing Westerlies.

**Wet Adiabatic Rate** The rate of adiabatic temperature change in saturated air. The rate of temperature change is variable, but it is always less than the dry adiabatic rate.

**White Frost** Ice crystals that form on surfaces instead of dew when the dew point is below freezing.

**Wind** Air flowing horizontally with respect to Earth's surface.

**Windchill** A measure of apparent temperature that uses the effects of wind and temperature on the cooling rate of the human body. The windchill chart translates the cooling power of the atmosphere with the wind to a temperature under nearly calm conditions.

**Wind Vane** An instrument used to determine wind direction.

**Winter Solstice** *See* Solstice.

**World Meteorological Organization (WMO)** Established by the United Nations, an organization that consists of more than 130 nations and is responsible for gathering needed observational data and compiling some general prognostic charts.

# Photo, Illustration and Text Credits

**Chapter 1** Chapter Opener 1 pages 2–3 NASA Figure 1.1 page 4 Win McNamee/Getty Images News/ Getty Images Figure 1.2 page 4 Gurinder Osan/AP Images Figure 1.6 page 6 Brian Kersey/UPI/Newscom Text page 9 F. James Rutherford and Andrew Ahlgren, Science for All Americans (New York: Oxford University Press, 1990), p. 7Figure 1.8 page 9 Famartin Text page 9 Louis Pasteur quoted in "Science, History and Social Activism" by Everett Mendelsohn, Garland E. Allen, Roy M. MacLeod, Spring 2001 Text page 9 William Anders Figure 1.9 page 10 NASA Figure 1.10 page 11 Michael Collier Figure 1.12 page 12 (lower right) Michael Collier Figure 1.12 page 12 (lower left) Bernhard Edmaier/Science Source Figure 1.12 page 12 (top right) E.J. Tarbuck Figure 1.13 page 12 (top right) Darryl Leniuk/Age Fotostock Figure 1.13 page 12 (lower left) Age Fotostock/SuperStock Figure 1.14 page 13 Kevork Djansezian/AP Images Figure 1.16 page 14 Terry Donnelly/Alamy Figure 1.B page 15 Greg Vaughn/Alamy Figure 1.C page 15 John Cancalosi/Photolibrary/Getty Images Figure 1.19A page 17 NASA Figure 1.19B page 17 Elwynn/Shutterstock Figure 1.20 page 18 NASA Figure EOA1.1 page 20 Michael Collier Text page 20 Richard Craig, The Edge of Space: Exploring the Upper Atmosphere (New York: Doubleday & Company, Inc., 1968) p. 130Figure 1.22 page 20 David Wall/Alamy Figure 1.24 page 21 David R. Frazier/Danita Delimont Photography/Newscom Figure EOA1.2 page 22 David R. Frazier/Science Source Figure 1.26 page 23 Age Fotostock/Superstock Figure CIR1.1 p. 24 Michael Coller Figure CIR1.3 Michael Collier Figure GIST4 page 26 National Weather Service Figure GIST7 page 27 Rick Wilking/Reuters/Corbis.

**Chapter 2** Chapter Opener 2 pages 28–29 Peter Kunasz/Shutterstock Figure 2.1A page 30 Thomas Barrat/Shutterstock Figure 2.1B page 30 Sony Ho/ Shutterstock Figure 2.7 page 35 Martin Woike/Premium/Age Fotostock Figure EOA2.1 page 37 NASA Text page 42 Based on material prepared by Professor Gong-Yuh Lin, a faculty member in the Department of Geography at California State University, Northridge Figure 2.C page 42 Eddie Gerald/Rough Guides/DK Images Figure 2.19 page 46 NASA Figure 2.20 page 47 Tetra Images/Getty Images Figure EOA2.2 page 47 NASA Figure 2.22A page 50 David Cole/Alamy Figure 2.22B page 50 NASA Figure 2.22C page 50 NASA Figure CIR2.3 page 53 Johner Images RF/ AGE Fotostock Figure GIST7 page 53 NASA Figure GIST11 page 56 USGS.

**Chapter 3** Chapter Opener 3 pages 58–59 Arterra Picture Library/Alamy Figure 3.1A page 60 Andrew A. Nelles/ZUMA Press/Newscom Figure 3.1B page 60 Chuck Eckert/Alamy Figure 3.A page 61 Steve Marcus/Reuters Figure 3.4 page 63 NASA Figure 3.7 page 65 Johns Hopkins University-The Milton S. Eisenhower Library, Special Collections Figure 3.10 page 67 Corbis/Superstock Figure EAO3.1 page 68 Tequilab/Shutterstock Figure 3.16 page 69 Michael Collier Figure EAO3.2 page 71 CoolR/Shutterstock Box 3.2 page 73 Gregory J. Carbone, Professor Carbone is a faulty member in the Department of Geography at the University of South Carolina Figure 3.21A page 74 Daily Telegram, Jed Carlson/AP Images Figure 3.21B page 74 Michael Collier Figure 3.23 page 77 Frank Labua/Pearson Education Figure 3.E page 77 NASA Figure 3.27 page 79 GIPhotoStock/ Science Source Figure 3.28 page 79 Paul Rapson/ Science Source Box 3.4 page 82 IPCC, 2013: Summary for Policymakers. In Climate Change 2013: the Physical Science Basis. Contribution of Working Group 1 to the Fifth Assessment Report of the Intergovernmental Panel on Climate Change. Cambridge and New York: Cambridge Universtiy Press Figure GIST4 page 86 Muzhik/Shutterstock Figure GIST7 page 87 South West Images Scotland/Alamy.

**Chapter 4** Chapter Opener 4 pages 88–89 Christian Beier/Age Fotostock Figure 4.4A (left) page 93 NaturePL/SuperStock Figure 4.4B (right) page 93 Elen_studio/Fotolia Figure EAO4.1 page 96 Michael Collier Figure 4.A (top) page 98 Henryk Sadura/ Fotolio Figure 4.A (bottom) page 98 E.J. Tarbuck Figure 4.12 page 100 Nitr/Fotolia Figure 4.14 (left) page 101 E.J. Tarbuck Figure 4.B page 102 Elena Elisseeva/Shutterstock Figure 4.17 (left) page 105 Dean Pennala/Shutterstock Figure 4.17 (right) page 105 Dennis Tasa Figure 4.B page 106 Age Fotostock/Superstock Figure 4.20 page 107 NASA Figure EOA4.2 page 108 NASA Figure 4.22 page 109 Steve Vidler/Superstock Figure 4.28 page 113 Michael Collier Figure EOA4.3 page 113 Michael Collier Figure 4.30 page 116 Euroluftbild.de/vario Images/ Age Fotostock Figure CIR4.4 (top) page 118 Clynt Garnham Housing/Alamy Figure CIR4.7 (bottom) page 118 Bill Draker/Rolf Nussbaumer Photography/ Alamy Figure GIST2 page 119 Dmitry Kolmakov/ Shutterstock Figure GIST4 page 119 JorgusPhotos/ Fotolia.

**Chapter 5** Chapter Opener pages 122–123 Dennis Frates/Alamy Figure 5.2A page 127 E.J. Tarbuck Figure 5.2B page 127 Jung-Pang Wu/Flickr/Getty Images Figure 5.2C page 127 E.J. Tarbuck Figure 5.3A page 128 E.J. Tarbuck Figure 5.3B page 128 E.J. Tarbuck Figure 5.4 page 128 NOAA Figure 5.5 page 129 E.J. Tarbuck Figure 5.6 page 129 E.J. Tarbuck Figure 5.7 page 129 Doug Millar/Science Source Figure 5.8A page 130 Dennis Tasa Figure EOA5.1 page 130 Daj/Amana Images/Alamy Figure 5.9B page 131 Tim Gainey/Alamy Figure 5.10 page 132 Ed Pritchard/Stone/Getty Images Figure 5.11 page 133 Michael Collier Figure 5.18 page 139 James Steinberg/Science Source Figure 5.19 page 140 Kichigin19/Fotolia Figure 5.21 page 142 Syracuse Newspapers/Dick Blume/The Image Works Figure 5.22B page 142 University Corporation for Atmospheric Research NCAR Figure 5.23 page 142 National Weather Service Figure 5.25 page 143 Marcus Siebert ImageBroker/Age Fotostock Figure 5.A page 144 Kiichiro Sato/AP Images Figure EOA5.3 page 146 H. Raab Figure 5.C page 147 NASA Figure 5.28 page 149 Inga Spence/Alamy Figure 5.29 page 149 AP Images Figure 5.30A page 150 Ryan Anson/ Newscom Figure 5.30B page 150 Zuma Press/ Newscom Figure 5.31 page 151 Dennis Tasa Figure 5.32 page 151 NASA Figure GIST2a-c page 152 E.J. Tarbuck Figure GIST3a-b page 152 E.J. Tarbuck Figure GIST3c NASA Figure GIST11a-b page 154 E.J. Tarbuck Problem 5.5 (top) page 155 Bill Draker/ Rolf Nussabaumer Photography/Alamy Problem 5.6 (bottom) page 155 NOAA.

**Chapter 6** Chapter Opener pages 156–157 Martin Sundberg/UpperCut Images/Alamy Figure 6.3B page 159 Charles E. Winters/Science Source Figure 6.5 page 160 Stuart Aylmer/Alamy Figure EOA6.1 (top) page 172 NASA Figure EOA6.1 page 172 AP Images/Jae C. Hong Figure 6.21 page 173 NASA Figure 6.23A page 175 Lourens Smak/Alamy Figure 6.23B page 175 Vidler Steve/Prisma Bildagentur AG/Alamy Figure EOA6.2 page 175 NASA Figure EOA6.3 page 176 NASA Figure 6.24A page 177 Belfort Instrument Company Figure 6.24B page 177 Lourens Smak/Alamy Figure 6.C page 178 Dazzo/ Radius Images/Getty Images Figure 6.26 page 180 Norbert Eisele-Hein/Imagebroker/Alamy Figure GIST1 page 182 NASA.

**Chapter 7** Chapter Opener pages 184–185 ColsTravel/Alamy Figure 7.1A (left) page 187 Edward J. Tarbuck Figure 7.1B (middle) page 187 Eric Nguyen/Science Source Figure 7.1C (right) page 187 Science Source Figure 7.A page 188 St. Meyers/Science Source Figure 7.4 page 190 Herbert Koeppel/Alamy Figure 7.B page 192 NASA Figure 7.C page 193 Dennis Poroy/AP Photo Figure EOA7.1 page 194 Peter Wey/Shutterstock Figure 7.9 page 195 NOAA Figure 7.12 page 198 Clint Farlinger/Alamy Figure 7.19 page 202 NOAA Figure 7.22 page 204 Jacques Jangoux/Science Source Figure EOA7.2 page 205 NASA Figure 7.26 page 208 Jonathan Wood/Getty Images Figure 7.29 page 210 Science Source Figure EOA7.3 page 211 NASA Figure CIR7.1 page 212 NASA Figure CIR7.9 page 213 Torsten Blackwood/AFP/Getty Images Figure CIR 7.10 page 214 NASA Figure GIST7 page 214 NASA.

**Chapter 8** Chapter Opener pages 216–217 NASA Figure 8.1 page 218 Samo Trebizan/Shutterstock Figure 8.4 page 221 NASA Box 8.2 page 226 Based on Thomas W. Schmidlin and James Kasarik, "A Record Ohio Snowfall During 9-14 November 1996, "Bulletin of the American Meteorological Society, Vol. 80, No. 6, June 1999, p. 1109 Figure 8.B page 226 AP Photo/Tony Dejak Figure EOA8.1 page 227 NASA Figure 8.9 page 227 Michael Dwyer/Alamy Figure EAO8.2 page 228 NASA Figure 8.10 page 229 Rod Planck/Science Source Figure 8.12A page 230 NASA Figure 8.12B page 230 Jebb Harris/Newscom Figure 8.13A page 230 Laurent Baig/Alamy Figure CIR8.4 page 232 Gary Wiepert/Reuters.

**Chapter 9** Chapter Opener pages 234–235 Samuel D. Barricklow/Getty Images Figure 9.4 page 239 Fernando Salazar/AP Photo Figure EOA9.1A-D page 241 E.J. Tarbuck Figure EOA9.1E page 241 Chainfoto 24/Shutterstock Figure EOA9.2 page 246 NASA Figure 9.12 page 248 NASA Figure 9.B page 250 Mike Hollingshead/Science Source Text page 250 Francis Bacon, English philosopher and scientist (1561-1626) Table 9.A page 251 Adapted from the National Weather Service Figure EOA9.3 page 254 NASA Figure 9.C page 256 Jeff Roberson/ AP Photo Figure 9.D page 257 James R Finley/AP Photo Figure 9.19B page 257 Andre Jenny/Alamy Figure 9.21 page 259 NOAA Figure 9.22 page 259 NOAA Figure 9.23 page 260 NOAA Figure 9.24 page 260 Mike McCleary/Bismark Tribune

Figure GIST6 page 264 John Jensenius/National Weather Service.

**Chapter 10** Chapter Opener pages 266–267 AFP/Stringer/Getty Images Figure 10.1A (left) page 268 NASA Figure 10.1B (right) page 268 Iowa State University Figure 10.6 page 271 E.J. Tarbuck Figure 10.8 (bottom) page 273 University Corporation for Atmospheric Research Figure 10.9B page 274 NASA Figure 10.9C page 274 Mike Theiss/National Geographic/Getty Images Figure 10.11 page 275 Mbga9pdf/Wikimedia Box 10.1 page 276 C.C. Elvidge, et al. "U.S. Constructed Area Approaches the Size of Ohio," in EOS Transactions, American Geophysical Union, Vol 85, No. 24 (15 June 2004), 233 Figure 10.A page 276 Andy Cross/Denver Post Getty Images Figure 10.14 page 276 NOAA Figure 10.C page 277 Sue Ogrocki/AP Images Table 10.1 page 278 National Weather Service Figure 10.15 page 278 Tom Ives/AGE Fotostock/SuperStock Figure 10.D page 279 Wendy Schreiber/University Corporation for Atmospheric Research Figure 10.17 page 281 Edward J. Tarbuck Figure 10.18 page 283 Cultura Science/Jason Persoff Stormdoctor/Getty Images Figure 10.19 page 283 Jewel Samad/AFP/Getty Images Figure 10.22D page 285 Gene Rhoden/Weatherpix/Getty Images Figure 10.23 page 285 Paths of Illinois Tornadoes (1916-1969), from Illinnois Tornadoes by John W. Wilson and Stanley A. Changnon, Jr., Circular 103, Illinois State Water Survey Figure 10.24 page 286 JimYoung/Reuters Table 10.2 page 287 National Weather Service, Storm Prediction Center Figure 10.26B page 287 Edeans Figure 10.27A page 288 John Sokich/NOAA Figure 10.27B page 288 Lm Otero/AP Images Figure EOA10.2 page 290 NASA Figure 10.G (left) page 291 NOAA Figure 10.G (middle) page 291 NOAA Figure 10.G (right) page 291 NOAA Figure 10.H (bottom) page 292 NOAA Photo Library, NOAA Central Library; OAR/National Severe Storms Laboratory (NSSL). Photo taken by Harald Richter Figure 10.31B page 293 University Corporation for Atmospheric Research Figure 10.32 page 293 National Geophysical Data Center/NOAA.

**Chapter 11** Chapter Opener pages 298–299 Lance/EOSDIS MODIS Rapid Response Team/NASA Figure 11.1 page 300 Mario Tama/Getty Images News/Getty Images Figure 11.5 page 302 NOAA Figure 11.8B page 304 NOAA Figure 11.9 (right) page 306 NASA Figure EOA11.1 page 307 NASA Table 11.1 page 307 National Weather Service/National Hurricane Center Figure 11.11 page 308 Smiley N. Pool/Newscom Figure 11.12 page 310 Adam Dean/Reuters Figure EOA11.2 page 312 NASA Figure 11.16 page 313 EOS Earth Observing System/NASA Figure 11.18A page 314 AP Images Figure 11.18B page 314 AP Images Figure 11.19A page 315 NOAA Figure 11.19B page 315 NASA Figure 11.20 page 316 NOAA Figure CIR11.1 page 319 NASA Figure GIST3 page 320 NASA Figure GIST7 page 321 Barry Williams/AFP/Getty Images/Newscom.

**Chapter 12** Chapter Opener pages 322–323 Michael Doolittle/Alamy Text page 324 Robert T. Ryan, "The Weather is Changing…or Meterologists and Broadcasters, the Twain Meet," Bulletin of the American Meteorological Society 63, no.3 (March 1982), 308 Figure 12.2 page 325 Dana Romanoff/Getty Images Figure 12.3 page 326 Image Source/Getty Images Figure 12.4 page 327 Michael Dwyer/Alamy Figure 12.5 page 327 Famartin Figure EOA12.1 page 334 NASA Figure 12.13 page 336 Andy Newman/AP Images Figure 12.14 page 337 Data from National Oceanic and Atmospheric Administration Figure 12.15 page 338 A.T.Willett/Alamy Figure 12.17 page 339 NASA Figure 12.18A page 340 NOAA Figure 12.18B page 340 NOAA Figure 12.19 page 341 NOAA Figure 12.20 page 341 NOAA Figure 12.21 page 341 NOAA Figure EOA12.2a,b page 342 NOAA Figure 12.25 page 346 Ryan McGinnis/Alamy Figure 12.C page 352 National Oceanic and Atmospheric Adminstration http://www.spc.noaa.gov/exper/archive/events/ Figure GIST4 page 356 NOAA.

**Chapter13** Chapter Opener pages 358–359 AFP/Getty Images Figure 13.1 page 360 NASA Text page 360 John Evelyn, Fumifugium, or The Inconvenience of Aer and Smoak of London Dissipated, Together with Some Remedies Humbly Proposed, G. Bedel and T. Collins (1661) Figure 13.3A page 361 Library of Congress Prints and Photographs Division [LC-USZ62-47355] Figure 13.3B page 361 Cholder/Shutterstock Text page 361 Robert Ingersoll, "The Works of Robert Ingersoll" New York, C.P. Farrell, 1900 Figure 13.4 page 361 Everett Collection/Superstock Figure 13.A page 363 Doable/Amanaimages RF/Glow Images Figure EOA13.1 page 364 NASA Figure 13.B page 366 Mirrorpix/Newscom Figure 13.9 page 368 MODIS Rapid Response Team/NASA Table 13.2 page 369 U.S. Environmental Protection Agency, Office of Air Quality Planning and Standards Figure 13.D page 372 NASA Figure 13.E page 372 NASA Figure 13.F page 372 NASA Figure 13.15 page 373 Walter Bibikow/Agency Jon Arnold Images/Age Fotostock Figure EOA13.2 page 376 NASA Figure 13.19 page 376 Andre Jenny Stock Connection World/wide/Newscom Figure EOA13.3 page 377 Michael Collier Figure 13.20 page 377 Ryan McGinnis/Alamy Figure GIST4 (left) page 379 David Person/Alamy Figure GIST6 (right) page 379 DC Premiumstock/Alamy.

**Chapter 14** Chapter Opener pages 380–381 Michael Collier Figure 14.5 page 384 Stock Connection Blue/Alamy Figure 14.6 page 385 Biophoto Associates/Science Source Figure 14.7 page 385 Itsuo Inouye/AP Images Figure 14.8 page 386 British Antarctic Survey/Science Source Figure 14.9 page 386 National Ice Core Laboratory Figure 14.10A page 387 Victor Zastol'skiy/Fotolia Figure 14.10B page 387 Gregory K. Scott/Science Source Figure 14.11 page 387 Ami Images/Science Source Figure 14.13 page 388 SGM/Age Fotostock Figure 14.15 page 389 NASA Figure 14.16A page 390 NASA Figure 14.A page 393 Stuart Black/Robert Harding Figure 14.18 page 393 NASA Figure EOA14.1 page 394 NASA Figure 14.21 page 395 Nigel Dickinson/Alamy Figure EOA14.2 page 398 NASA Text page 398 IPCC, "Summary for Policy Makers," in Climate Change 2013: The Physical Science Basis Figure 14.27 page 399 NASA Figure 14.28 page 401 Radius Images/Alamy Table 14.1 page 403 Based on the Fifth Assessment Report Climate Change 2013: The Physical Science Basis, Summary for Policy Makers Figure 14.30 page 403 National Snow and Ice Data Center Figure 14.34 page 405 NASA Figure CIR14.1 page 407 Gary Braasch/ZUMA Press/Newscom Figure CIR14.4 page 408 Michael Collier Figure CIR14.5 page 408 ImageBROKER/Alamy Figure GIST.4 page 409 Hughrocks.

**Chapter 15** Chapter Opener pages 410–411 Michael Collier Figure 15.2A page 413 Earlytwenties/Shutterstock Figure 15.2B page 413 Witold Skrypczak/Alamy Figure 15.2C page 413 Ed Reschke/Photolibrary/Getty Images Figure 15.2D page 413 Martin Sheilds/Alamy Figure 15.2E page 413 J.G. Paren/Science Source Figure 15.6 page 418 Peter Lilja/AGE Fotostock Figure 15.A page 420 Wesley Bocxe/Science Source Figure 15.9 page 421 Kondrachov Vladimir/Shutterstock Figure 15.14A page 427 Dean Pennala/Shutterstock Figure 15.14B page 427 Tasa Graphic Arts Figure EOA15.1 page 430 NASA Figure 15.16 page 430 Richard J. Green/Science Source Figure 15.17 page 431 Stefano Caporali/Tips Images/Age Fotostock Figure 15.20 page 436 Michael P. Gadomski/Science Source Figure 15.23 page 439 Joel Harper Figure 15.25A page 440 Design Pics/Superstock Figure 15.25B page 440 Jim Cole, Photographer/Alamy Figure EOA15.2 page 441 Michael Collier Figure CIR15.4 page 443 Tasa Graphic Arts Figure GIST1 page 444 IK Lee/Alamy Figure GIST9 page 445 NASA Figure GIST10 page 445 U.S. Geological Survey.

**Chapter 16** Chapter Opener pages 446–447 John Sylvester/Glow Images Figure 16.3 page 449 Peter Skinner/Science Source Figure 16.4 page 449 JaySi/Shutterstock Figure 16.6A page 450 E.J. Tarbuck Figure 16.A page 452 Joe Orman Figure 16.10 page 453 Louise Murray/Science Source Figure EOA16.1 page 453 Pekka Parviainen/Science Source Text page 453 Craig F. Bohren, "What Light through Yonder Window Breaks?: More Experiments in Atmospheric Physics," New York: John Wiley & Sons, Inc. © 1991 Figure 16.11 page 454 Michael Giannechini/Photo Researchers Figure 16.15 page 456 Sean Davey/Aurora Open/Glow Images Figure 16.16 page 457 Chainfoto24/Shutterstock Figure 16.B page 458 Laszlo Podor/Alamy Figure 16.19 page 459 Stefan Wackerhagen/imageBroker/Age Fotostock Figure 16.20 page 459 Masaaki Tanaka/Aflo/Glow Images Figure EOA16.2 page 460 NASA Figure 16.21A page 460 E.J. Tarbuck Figure 16.21B page 460 Mike Hollingshead/Science Source Figure 16.22 page 460 Tom Schlatter/NOAA Figure GIST6A page 462 Blickwinkle/Alamy Figure GIST6B page 462 Science Source Figure GIST6C page 462 Ron Dahlquist/Exactostock/Superstock.

**Design elements**: Graphic tab device, PointaDesign/Shutterstock; Severe & Hazardous Weather, David W. Leindecker/Shutterstock; Smart Figure icon, Solarseven/Shutterstock; Students Sometimes Ask, Getty Images; What's Your Forecast, Mexrix/Shutterstock; What's Your Forecast icon, Best Works/Shutterstock.

**Eye on the Atmosphere**: Chapter 1 Neo Edmund/Shutterstock; CG-01 Rudy Umans/Shutterstock; CG-02, Shutterstock; CG-3 Shutterstock; CG-4 Shutterstock; CG-5 Gianni Muratore/Alamy; CG-6 Richard Newstead/Corbis; CG-7 Denise Dethefsen/Alamy; CG-8 NOAA; CG-09 NOAA; CG-10 NOAA, CG-11 Public Domain; CG-12 Jhaz Photography/Shutterstock; CG-13 Brian Cosgrove/DK Images; CG-14 Jim Lee/NOAA; CG-15 Public Domain; CG-16 NOAA.

# Index

# D

# Cloud Guide

**Low Clouds and Clouds of Vertical Development: Cloud Bases 0–2 km (0–6,500 ft)**

JIM LEE/NOAA

**Cumulus** These clouds often have flat bottoms, rounded tops, and a "cellular" structure made up of individual clouds. (The word "cumulus" comes from the Latin word for "heap.") Cumulus clouds tend to grow vertically.

JIM LEE/NOAA

**Cumulus Humilis** Often called fair-weather cumulus, these small white individual masses lack conspicuous vertical development and rarely produce precipitation.

BRIAN COSGROVE/DORLING KINDERSLEY

**Nimbostratus** These low clouds are thick gray layers that contain sufficient water to yield light-to-moderate precipitation.

JIM LEE/NOAA

**Stratocumulus** These are low, layered clouds that have regions of some vertical development. Differences in thickness create varying degrees of darkness when seen from below.

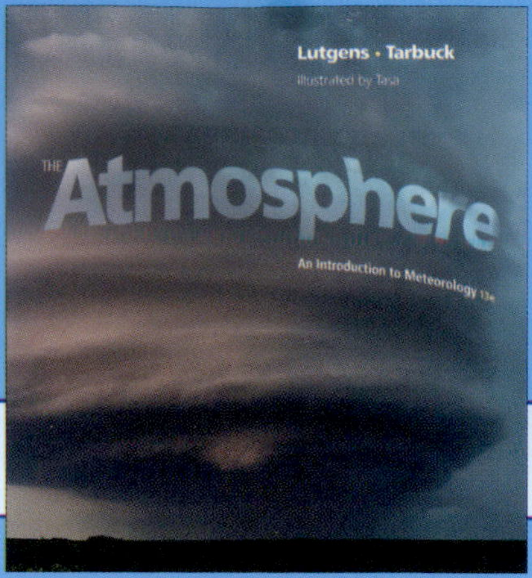

JIM LEE/NOAA

**Cumulus Congestus** These clouds have considerably more vertical development than cumulus humilis. They may produce heavy precipitation, but not the severe weather associated with some cumulonimbus clouds.

SHUTTERSTOCK

**Cumulonimbus** These clouds result from very strong updrafts that may push the cloud tops up to several kilometers into the stratosphere. Their characteristic feature is the anvil, a zone of ice crystals extending outward from the main portion of the cloud.

PUBLIC DOMAIN

**Cumulonimbus with Mammatus** These are dramatic features associated with some cumulonimbus, resulting from strong downdrafts and turbulence along the bases or margins of the clouds.

NOAA

**Cumulonimbus with Wall Cloud** A feature associated with some cumulonimbus clouds. When wall clouds are present, heavy rain, hail, and sometimes tornadoes can be expected.

# Temperature Change

The impact of rising concentrations of atmospheric carbon dioxide and other greenhouse gases will depend heavily on what efforts are made to limit their release by human activities. This map shows a projected outcome of continued heavy reliance on coal and other fossil fuels. Occurring over less than a century, mapped values are fifty to one hundred times faster than anything seen on Earth in the last 65 million years, and will be highly disruptive for both natural ecosystems and human society.

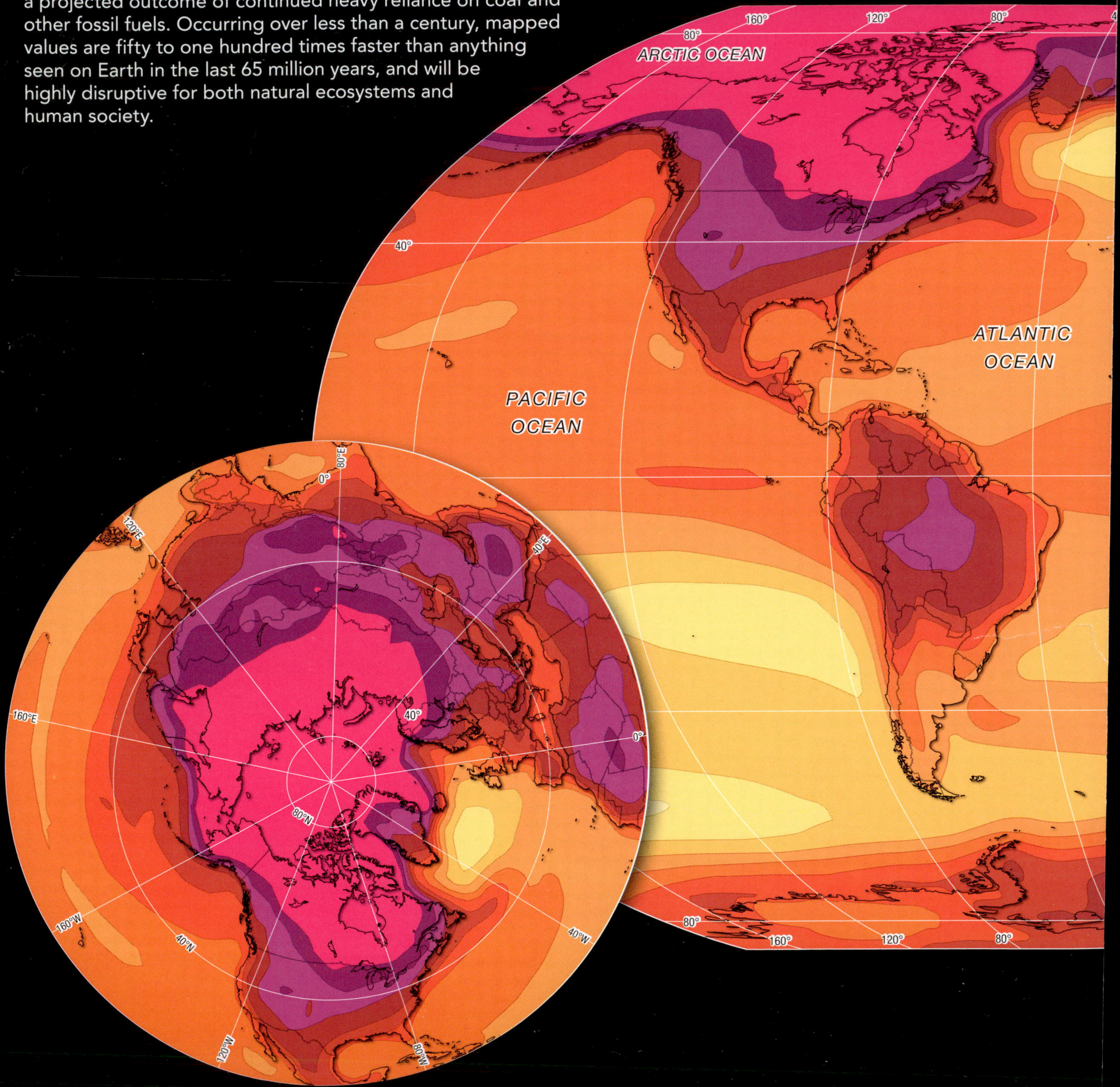

# 2081–2100 (Projected)

THIS MAP REPRESENTS AN AVERAGE BASED ON MORE THAN TWO-DOZEN MODELS OF THE EARTH SYSTEM. NOTICE HOW PROJECTED CHANGES ARE SMALLER OVER OCEANS THAN CONTINENTS, AND ARE VERY LARGE IN THE HIGHER NORTHERN LATITUDES. THE ATMOSPHERIC, OCEANIC, AND LAND SURFACE PROCESSES REPRESENTED IN SUCH MODELS ARE DISCUSSED IN *UNDERSTANDING WEATHER AND CLIMATE,* WITH EXPLANATIONS FOR HOW THESE PROCESSES—ALONG WITH FEEDBACKS AMONG THE VARIOUS EARTH SPHERES—WORK TO PRODUCE THE HIGHLY UNEVEN PATTERN OF HEATING SEEN IN THIS MAP.

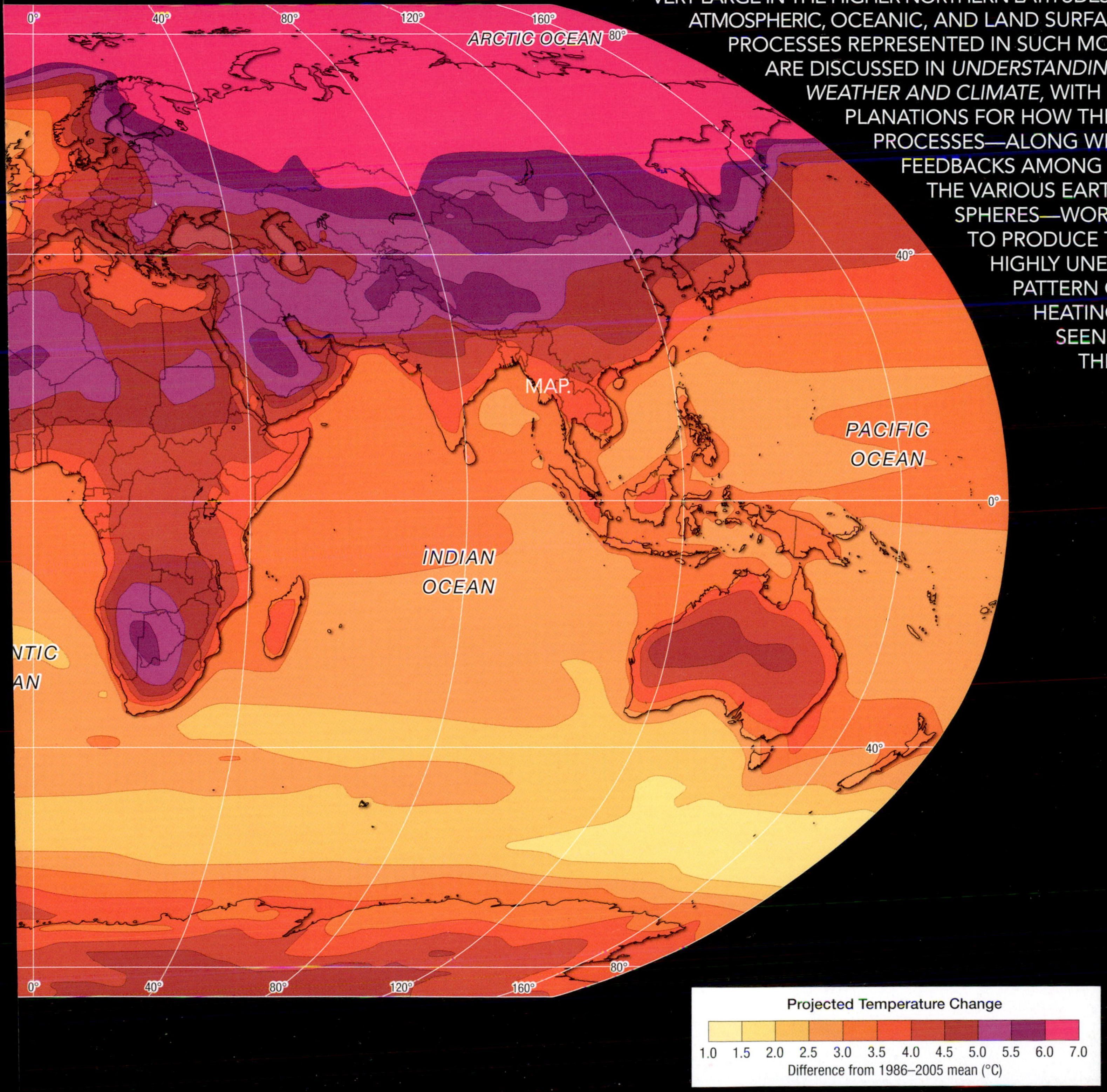


*Source:* Based on data from the IPCC Fifth Assessment Report (2013–2014).